Telephone Switching Systems

For a listing of recent titles in the *Artech House Telecommunications Library*,
turn to the back of this book.

Telephone Switching Systems

Richard A. Thompson

Artech House
Boston • London
www.artechhouse.com

Library of Congress Cataloging-in-Publication Data
Thompson, Richard A., 1942–
Telephone switching systems / Richard A. Thompson.
p. cm. — (Artech House telecommunications library)
Includes bibliographical references and index.
ISBN 1-58053-088-5 (alk. paper)
1. Telephone switching systems, Electronic. I. Title. II. Series.

TK6397 .T52 2000 00-027453
621.387—dc21 CIP

British Library Cataloguing in Publication Data
Thompson, Richard A.
Telephone switching systems. — (Artech House
telecommunications library)
1. Telephone switching systems, Electronic
I. Title
621.3'87

ISBN 1-58053-088-5

Cover design by Igor Valdman

685 Canton Street
Norwood, MA 02062

International Standard Book Number: 1-58053-088-5
Library of Congress Catalog Card Number: 00-027453

10 9 8 7 6 5 4 3 2 1

The book is dedicated to Sandy, my wife of 35 years. She has been my principal source of encouragement and sanity, my partner in every good thing that ever happened to me, and also my colleague in preparing the book.

Table of Contents

Preface

Motivation. The science of telephony and circuit switching is more than 100 years old. It is rich in theory and many people have devoted entire careers to its advancement. But, because the circle of interest used to be limited, this science has not been generally publicized nor taught in universities. Four major changes in the last 20 years have expanded this circle of interest.

- Many corporations, universities, government and military agencies, and other enterprises have developed increasingly specialized telecommunication systems requirements. Once viewed as an overhead of doing business, whose cost had to be controlled, telecommunications is now viewed as a strategic resource. These enterprises must hire telecommunications specialists or use consultants to help them design systems, specify requirements, and deal with suppliers.
- With the deregulation and divestiture of the old Bell System, the many suppliers and manufacturers in the growing telecommunications industry had to and must continue to hire people who understand this science, its ramifications, and its applications.
- Packet switching networks were introduced as an optimal paradigm for computer communications. While packet switching has been better publicized than circuit switching in textbooks and research journals and at conferences, it is because the theory is newer and more work needs to be done. While packet switching is newer than circuit switching, each switching paradigm is optimal for different data types.
- America's national information infrastructure, or "information superhighway," has become, and continues to be, a critical legislative and judicial priority.

These same changes have motivated the initiation of telecommunications programs in many of the nation's major universities. Today, telecommunications, as an academic discipline unto itself, recalls the state of computer science in the early 1970s: programs are not uniform and they appear at various academic levels and within various existing disciplines. Such programs educate the people who will be the telecommunications professionals of the future. These professionals must understand the architectures, the cost, the technology, the industry, the legal and regulatory environment, the user interface, the services, as well as how all these things affect and effect corporate operations.

Professions and academic disciplines can be partitioned vertically and horizontally. *Vertical* disciplines are classical and well defined: sociology, anthropology, psychology, economics, electrical engineering, and computer science, to name a few. *Horizontal* disciplines, and most real professions, cut across the vertical disciplines. A professional social worker must know something about sociology, anthropology, psychology, economics, law, and probably many other things as well. This professional need not be an expert in all these fields, but must understand those topics from all these fields that are relevant to social work.

The study of switching systems, indeed telecommunications itself, is another such horizontal discipline. It cuts across the vertical disciplines of electrical engineering and technology, computer science, information science, cognitive science, communications, business, and law. It is truly an interdisciplinary topic.

Reader. To this end, this book is written for two types of readers:

- Practicing telecommunications professionals, electrical engineers and technologists, and computer and information scientists;
- Graduate and upper-level undergraduate college students in telecommunications, electrical engineering (EE), EE technology, computer science (CS), and information science.

While this book should be readable to those without formal credentials, the reader is expected to possess some core knowledge. More important than proper credentials, the reader needs background (or, at least, determination), a natural curiosity, an engineering perspective, and a level of sophistication equivalent to those who have the proper credentials. Anyone with an EE background will find some of the material in Chapters 2, 3, and 19 to be a little "fluffy." Anyone with a CS background will find my coverage of switching software in Chapters 11, 14, and 18 to be slightly tedious. However, if you don't have either background, you'll appreciate the level I've written to.

It's impossible to write a book that everyone can read. I assume that you already understand how signals move along wires, know a little bit about the asynchronous digital hierarchy (ADH) and synchronous digital hierarchy (SDH) formats, understand how packets are switched, and know something about optical fiber and its concomitant technology. If you know a little bit about these things, you probably have sufficient background for this book. If you find yourself overwhelmed, you may need to read another book before you start this one. Please look at my web page (http://www.tele.pitt.edu/) for some recommendations.

Organization. This is a book about architecture, about designing in the large. So, more than just an ordered collection of material, this book has an architecture that is more than just an outline. Like a good novel, there are many themes and subplots running through it. Look for them: signaling, policy, the design process, and the "partition." One editor I contacted wanted me to package the material into three or four separate books, each for a different market segment, with overlapping chapters and missing chapters. I didn't want to break it up and destroy the themes and subplots that I worked so hard to develop.

One would expect a textbook about a technical subject to be organized along a technical structure — and this one is. However, while the subject is highly technical, the current state of the science of switching is based on much more than just technology and always has been. Switching's own history is extremely relevant to understanding switching's current state. We will spend time on a history lesson and even study, in great detail, three of the significant systems from the past.

The book is organized technically and historically into five parts.

1. *Fundamentals* includes an introduction and overview, and a history and discussion of technological evolution. Two chapters describe the "boundary conditions" on a switching office: the telephone or line side and the network or trunk side. The theory of traffic is also surveyed. Chapters 1 – 5 are fundamental.

2. *Analog Systems* presents the Step-by-Step, Crossbar, and Electronic Switching systems. Although the latter is controlled by a digital computer, it is considered an analog system because it switches analog signals. The theory of interconnecting networks is surveyed. The evolutions of the toll point and of enterprise switching are described, including relevant regulatory issues. Chapters 6 – 11 describe analog systems.

3. *Digital Systems* presents the concepts and architectures of digital switching. After examining private networks and introducing some new concepts, several commercial systems are described. Chapters 12 – 14 describe digital systems.

4. *Broader Issues* surveys the human component as part of a system and some recent concepts and architectures in computer communications. New services and integrated terminals are described, and the human being is presented as a system component. The breakup of the Bell System, also discussed here, had significant architectural impact. The significant architectural differences between packet and circuit switching are described. Chapters 15 – 17 discuss broader issues.

5. *Evolving Systems* discusses three parallel evolutions: (1) the control structure, including the Intelligent Network, (2) evolving access, including wireless telephony and new tethered technologies, and (3) photonics technology, including optical fiber communications and photonic switching. In the last chapter, I describe what I see in the crystal ball. Chapters 18 – 22 describe these evolving systems.

Nontechnical Issues. Few scientific disciplines have been more influenced by as many nontechnical factors as telecommunications switching has been. Thus, a significant portion of the book is devoted to the structure of the industry, regulatory and legal events, human factors, and even the public perception of "Ma Bell." The phone company has been a significant part of Americana: spoofed by entertainers (like Al Capp, Jonathan Winters, and Lily Tomlin), battled by counterculture groups, and the target of politicians seeking the support of the "little guy." Using "blue boxes" and coin telephone fraud, people cheat the phone company with the same malaise with which they cheat insurance companies or the IRS: "It's OK, they won't miss it." Nontechnical issues have had as much architectural impact on switching as the evolving technology has had. The current state of these issues is unbelievable unless their history is understood.

Regulatory history is scattered throughout the text. The systems of their day were as much a function of regulation as they were of technology. Chapter 2 reviews the major regulatory events of the distant past, up to and including the 1956 Consent Decree. Chapter 10 discusses Carterfone, from the late 1960s, and how it brought competition to the customer's premises. Chapter 12 discusses Execunet, from the mid-1970s, and how it brought competition to the long distance network. Chapter 16 discusses (1) Computer Inquiry, and how it separated the Bell System's regulated and unregulated businesses in the early 1980s, (2) the Modified Final Judgment of the 1956 Consent Decree, and how it broke up the Bell System, and (3) the Telecommunications Act of 1996. Chapter 18 describes some of the "enhanced services" issues of the 1970s and 1980s and Chapter 22 discusses regulatory alternatives for our future. All these regulatory issues are presented from the perspective of someone who is both an internal observer and an end user.

Courses. It is important that people learn this material. I hope more universities will offer courses on this material, and I hope the professor who teaches such a course will use this book. For a one-semester course, I recommend using Chapters 3 – 5, 7, 9 – 14, and 16 – 19. The students can skim the other chapters, and the professor will have to go lightly over the software details in Chapters 11, 14, and 18.

I use my book in a two-course sequence at the University of Pittsburgh. I break up the science of telephone switching into hardware and software. In my "Switching Hardware" course, I use Chapters 3 – 10, the hardware parts of Chapter 11, and all of Chapters 12 and 13, the hardware parts of Chapter 14, and all of Chapters 16, 17, 19, and 22. In my "Switching Software" course, I use the software parts of Chapters 11 and 14, and all of Chapter 18. When I teach a course on "photonic communications," I use Chapters 20 and 21 as a supplement to the principal text. I also occasionally teach a course called "Designing Services," and I use Chapter 15 as a supplement to the principal text.

Whatever you do, don't skip Chapter 11. I think the #1 ESS (Electronic Switching System) is as significant to telephone switching as the telephone instrument itself. Even though the 1E is an analog switch and the telephone companies are decommissioning them, you can never fully appreciate switching software unless you force yourself through the 1E's software.

Presentation. One reason that telecommunications switching is difficult pedagogically is that it is generally perceived at many different levels of abstraction: elementary devices, circuits and fundamental modules, interconnection networks (called "fabrics"), architectures of switching systems, architectures of large networks, service provision and terminals, systems engineering and standards, and theory and general philosophy.

Even simple conversations among the practitioners are difficult because participants with different perspectives tend to feel that the others don't appreciate the problem. This textbook attempts to address every level of abstraction, but, obviously, it cannot do this for every topic. The level of abstraction in a particular section should be apparent to the readers, who are asked to be patient while reading a section that may not be presented at the level they prefer.

Don't confuse level of abstraction with level of detail. Any of the eight levels of abstraction above can be skimmed, painstakingly detailed, or intermediately covered. Throughout this book, the level of detail will also vary among the sections. The field has too much excruciating detail for one book to hold. Rather than omit detail entirely and present the material completely as overviews and surveys, scattered details are provided. Special attention to detail is paid to telephone operation, line and trunk circuit operation, ESS software, digital transmission formats, and communications terminals. Frequently, the pedagogy takes the form of an initial, oversimplified description followed by gradual development of greater detail. Conversely, topics like Poisson statistics, digital transmission, computer communications, and photonics are presented as overview surveys.

Relevant theory is surveyed, and not derived. Systems are described by high-level block diagrams and by describing how equipment (and software in modern systems) operates during a call walk-through. While many commercial systems seem similar at the block diagram, details that might differentiate them are taken from the systems manufactured by Western Electric (now Lucent) (1) because of my familiarity with those systems and (2) because information is more readily available. Similarly, the telephone system in the United States is described, with only slight attention paid to other nations.

Cited references and a selected bibliography are provided in a list at the end of each chapter. In 1976 – 1977, celebrating the 100th anniversary of Bell's invention of the telephone, *Computer Magazine, Spectrum,* the *IEEE Proceedings,* and other publications had historical and technical special issues on telephony. Many references are Bell System publications. The Local Switching System Generic Requirements (LSSGR), published by post-divestiture Bellcore, describes the fundamental requirements of local switching systems. In a related issue, within a class environment, the professor is urged to take advantage of the typical hospitality of the local telephone company and arrange for the class to tour a telephone switching office.

Urgency. I've been writing this book for more years than I care to admit. For most of this time, it wasn't a very high priority in my life and I've treated it almost like a hobby. I've always believed that such a book was needed and that it would have a modest market, but it seems there have always been more important things to do. Recent events have changed this.

Major new technologies and approaches are entering the field. Particularly important examples are asynchronous transfer mode (ATM) and Internet protocol (IP), including voice over IP. To utilize and integrate these technologies appropriately, it's important for people in the field to understand circuit switching, and all the advanced services and practices built upon it.

Many people don't understand circuit switching because it is not taught in school and because there is simply not a good book, written like a textbook, with a logical progression and with homework problems. While I can't educate the entire telecommunications industry on the concepts of circuit switching and I can't single-handedly change the course of evolution of the worldwide telecommunications network, I can get this book published, and hope that it helps.

Acknowledgments. Besides Sandy, to whom the book is dedicated, I have to thank my kids, Pam and Jeff, and my parents, relatives, and friends for their tolerance, patience, and encouragement.

Thanks to Amos Joel, under whom I consider myself to be a disciple. Thanks to Chet Day, my first supervisor, and to Ken Stone, my first tutor in telephone switching. Thanks to my dissertation advisor, the late Taylor Booth. Thanks to the former NSF Office of Experimental R&D Incentives for sponsoring a project with Universal Communications Systems of Roanoke and, thereby, encouraging the science of switching as an academic endeavor. Thanks to Virginia Tech, Bill Blackwell, and all those EE 4630 classes. Thanks to the North NJ Section of the IEEE, to John Baka, to the telephone switching class they sponsored in the spring of 1987, and to Marilyn Durner for transcribing those lectures and really starting the text of this book.

Thanks to the "old" Bell Laboratories: to former colleagues from the AIOD, TSPS, and XDS projects, the Getset team, the EYL committee, the photonic switching community, and the DiSCO team — especially to Dave Bergland, Bob Lucky, Nick Maxemchuk, Dave Hagelbarger, Carl Christensen, Doug Bayer, Dewayne Perry, Dave Sincoskie, Jack Horenkamp, Bob Pritchard, Kevin Oye, Rod Alferness, Ivan Kaminow, Andy Chraplyvy, Bill Payne, Scott Hinton, Gaylord Richards, Joe Evankow, Bob Anderson, John Camlet, and Phil Giordano.

Thanks to the leaders of the University of Pittsburgh for their commitment to its telecommunications program, to all those past and future Telcom 2225/2226 classes, to the faculty of the telecommunications program, and especially to Fritz Froehlich and the late Leon Montgomery.

Finally, thanks to Ed Sable, who reviewed the book for Artech House. His constructive criticisms and helpful suggestions have resulted in a better book. Thanks to the staff at Artech House for their assistance during this project, especially Mark, Jessica, Judi, Barbara, and Tina.

1

Introduction

"Write a book. Sing a song. Your creative juices are flowing."
— from a fortune cookie at the Sichuan Restaurant, northeast of Pittsburgh on May 2, 1997

In the beginning, telephone switching systems were created by Bell and Strowger, but not in six days. They evolved, and the forces of "natural selection" were technology, knowledge, regulation, economics, and applications. The ancient history, a discussion of these forces, and a survey of the different systems are presented in the next chapter. Before we can do this, however, we have to define some terms. In the first four sections of this chapter, definitions are given for some elementary concepts: communication(s) (with and without the "s" at the end), telecommunications, system, channel, network, and switching. Then, the last section presents an overview of how a switching system works — by looking at a manual office.

This book is not about telecommunications in general. It is about the networks and systems that have been optimized for point-to-point audio telecommunications — telephony. Because telephone systems are so ubiquitous, anybody who has picked up this book has probably used a telephone. So, you know how it works, but only functionally, and only from the user's end. Section 1.5 begins to make you think about how it works from the other end — by examining the interaction between user and operator, by beginning the discussion of the numbering plan, and by defining a set of generalized required functions.

1.1 TELECOMMUNICATIONS

Telecommunications and system are carefully defined. Switching is described informally here and defined more formally in Section 1.3, after we define channels in Section 1.2.

1.1.1 A Primer

The English verb "to communicate" comes from a Latin word that means "to make common." The noun "communications" (with the "s" at the end), means a "method of communicating." "Communications" is not necessarily a technical term, evidenced by the existence of college courses in communications that prepare English and business majors for the newspaper and advertising industries, respectively. The noun "telecommunications" means a "method of communicating over a distance" and implies technical disciplines and equipment that allow people and machines to communicate, even when separated. While we use the long form "telecommunications" initially, we use the short form "communications" most of the time. The reader who has opened this book expects to read about the "machinery for communicating over a distance."

See Dick "communicate" with Jane. Let Dick have a large chunk of information in his brain and let him want Jane to "have it in common" with him. Dick is unable to produce an external manifestation of this information as a single large chunk in a single instant of time. This is just as well, because Jane would have no way to deal with it, even if Dick could. You might argue that this book, or any document, contains a lot of information in a single large chunk. But, books are written serially and read serially, and they are the communications analog of a primitive store-and-forward system. So, Dick serializes the information and produces an external manifestation as sequential data over some interval of time.

He can use his vocal apparatus to output the information from his brain one sound at a time or he can use his hands to output the information from his brain as one keystroke or written character or smoke puff or drumbeat or flag wave at a time. The generalization is that Dick causes the value of some physical commodity to vary in time within some natural medium that he and Jane have in common. This is not inconvenient to Jane because her ears and eyes, and the mental processes for which they provide input, already restrict her to receiving the information through an audio or visual medium and in this serial form.

See Dick "telecommunicate" with Jane. Let Dick and Jane be so separated that they do not share the natural medium directly. Then the physiologically produced time-varying data cannot get from Dick to Jane directly. Dick and Jane must each have access to a manufactured medium, Dick must arrange for a *channel* to Jane within this intermediate medium, and each must possess an instrument that converts between their own physiologically produced data format and a data format suitable for this medium. Such manufactured media, called networks, are usually made from some electrical technology — electromechanical, electronic, and/or electromagnetic. But, they could even be biologic, fluidic, or photonic.

1.1.2 Switching

In general, "switching" means "turning things on and off." But, the term has different connotations for technologists. To a device physicist, a switching device is one that operates in one of only two states, such as the ON and OFF positions of a light switch or the saturation and cut-off conditions of a transistor. Using these devices, a circuit designer builds switching elements, such as logic gates and flip-flops for Boolean operations and binary memory. Using these elements, a digital designer builds switching circuits by Karnaugh maps, finite-state machines, and other switching design techniques.

However, in the context of telecommunications, "switching" is "connecting and disconnecting the channels in some network" for users, like Dick and Jane. To add to the general confusion, today's telephone switching systems are made from these same devices, elements, and circuits from which computers are made. A formal definition of "switching" is postponed until after "network" and "channel" have been formally defined, later in this section.

1.1.3 Systems

Throughout this book, a "system" is a "collection of interdependent, yet often diverse, components that act together according to a common plan and that serve a common purpose." To a mathematician, however, a system, also called an abstract algebra, is the "combination of a *set,* together with *operators* or *relations* that are closed within the set." The mathematical definition is useful here because it accentuates the two parts of our concept of system. The set corresponds to the collection of components, and the operators represent the procedures by which these components act together.

1.2 CHANNELS

We examine the anatomy of telecommunications. The elementary component is the channel. Any telecommunications system's resource, or infrastructure, is divided up into channels, which are then controlled and allocated by many types of concomitant processes. The *communications divisions* are the resources by which systems are designed, including amplitude, space, time, frequency, and direction. The *communications processes* are the functions that systems perform, including encoding, transmission, bunching, synchronization, duplexing, multiplexing, and switching.

1.2.1 Divisions

Networks and their attached systems provide a global resource that is shared by many sources, destinations, and data. The networks are multidimensional and the principal dimensions are called divisions. These divisions represent the general design space in which communications systems are implemented.

Amplitude division. This is the dimension in which intensity varies. The varied commodity is typically a voltage, but it could be light power or, in a digital system, any commodity that can be turned on and off (represented by an alphabet of <0,1>). In printing this book, the value assigned to each character position comes from the set of typed characters, a finite-valued dimension.

Space division. This is the dimension whose value is defined through spatial separation, internally three dimensional, as with distinct wires or telephones. It is always present because "telecommunications" assumes that users are spatially separated.

Time division. This is the dimension of sequence. It is special because it is also the independent variable in which the data is serialized and of which the data, in the various divisions, is a function.

Frequency division. Also called " spectrum," this dimension is related to amplitude and time through the Fourier transform. Optics brings a new division to communications: wavelength or color. It is related to the frequency division by the speed of the data (light) in the network.

Direction division. This is a two-valued dimension, at least in point-to-point communications. The values in the primer example are Dick-to-Jane and Jane-to-Dick.

Other divisions that occasionally occur in electrical communications systems are polarization, phase, and electromagnetic mode. And, while channels can be defined using these divisions, you'll only see them in a transmission system. I've never seen a switch that interchanges data among channels defined this way.

1.2.2 Processes

The divisions, described above, provide a design space in which the various processes required for communications are implemented. The typical implementations are described in this section. In future systems, these processes could be implemented in alternate divisions.

Transduction. This is the encoding of user-produced data by changing the value of a division, over time. The transduction of Dick's physiologically produced vocal data into a format suitable for the network is real valued and continuous in analog encoding, real-valued pulses at discrete times in sampled encoding, or integer-valued pulses at discrete times in digital encoding. Conversion of the physiologically produced data, regardless of its original form, into a suitable form is part of the transducer. Almost always, the amplitude division is used for transduction. However, in frequency-shift keying (FSK) used in computer modems, binary 0 is encoded as a burst of tone at frequency F_0 and binary 1 by frequency F_1. Optical digital transmission is really *unipolar FSK,* where a 1 is a pulse of light and a 0 is its absence. In a form of sampled encoding, called pulse-width modulation, the real value of a sample is encoded as the width of a pulse and not its height or amplitude.

Transmission. This is the movement of data from a place near Dick to a place near Jane. Transmission naturally involves the space division.

Bunching. This is process of ganging several separate physical channels into one logical channel that fits parallel-formatted data directly. Since channels are precious in most communications systems, parallel transmission is seldom used instead of serial transmission on one channel. Since time is precious in most computer systems, parallel buses are common in computers. Further confusing the issue, the value of the data after transduction may be multidimensional, as a byte, an 8-bit binary word. Such data may be formatted in bit-parallel form, where bunching in the space division is required, or in bit-serial form, yet another use of the time division.

Synchronization. This is necessary because of the delay, and potentially variable delay, involved in transmission. The receiver must adjust its internal timing, at the bit level and word level, to the position in the data format at the transmitter. This procedure is complicated further in multiplexed systems, particularly during start-up and error recovery. Bit-level synchronization in digital transmission is typically performed by electronic circuits that derive clock from the data stream. In a non-return-to-zero (NRZ) binary transmission format, the presence of a voltage denotes a binary 1 and the absence denotes a binary 0. Without bit-level synchronization, the receiver cannot distinguish consecutive 0s or consecutive 1s. In a return-to-zero (RZ) format, each binary 1 is transmitted as a separate pulse, so consecutive 1s could be distinguished. But the receiver still cannot distinguish consecutive 0s without bit-level synchronization. Alternative formats, such as *bipolar* in which a positive pulse is transmitted for a 1 and a negative pulse for a 0, or transmitting the clock in a parallel strobe channel, would not require clock derivation in the receiver.

Duplexing. This is the process by which communications over a channel are two way. The direction division is so obviously involved that this discussion may appear silly. But it is only in photonics or fluidics, where energy packets or material are launched and have momentum, that duplexing and direction are identical. In electronics, except for transmission-line effects, a voltage has no direction. A two-way metallic channel, a pair of wires, holds a net waveform that is the sum of the waveforms produced by the users at each end. The separation of Dick's waveform from this sum at Jane's end of the channel requires equipment that subtracts Jane's waveform from the sum — two-wire to four-wire conversion in a *hybrid.* Sometimes this direction division allows too much crosstalk between the Dick-to-Jane subchannel and the Jane-to-Dick subchannel, and other divisions are used to provide better isolation. Often with speakerphones, transmission is one way only, the channel given to the loudest talker, an amplitude-division precedence scheme controlling direction-division duplexing. In a process called "ping pong," Dick and Jane transmit in rapidly alternating timeslots; a form of time-division duplexing. In Section 4.4, single-frequency (SF) signaling will be described, in which signals are tones with frequency depending on direction — a form of both frequency-division transduction and frequency-division duplexing. A similar scheme, wavelength-division duplexing, has been proposed for two-way photonic transmission over a single fiber.

1.2.3 Multiplexing

In multiplexing, some division provides a sharing mechanism within the network. Each separate and independent slice of this multiplexed division, which could be leased to Dick for the duration of his call to Jane, is still called a channel, but these channels are more logical than physical. No user may exceed the capacity of his assigned piece of the total resource. The space, time, and frequency divisions are typically used for multiplexing. Polarization has also been used successfully, especially in satellite transmission.

Space-division multiplexing. This provides a physically separate path, such as a distinct wire in a cable, to define a separate channel. Since this technique occurs naturally and obviously, the term is seldom seen. The configuration is usually assumed as an unstated default and the exception is distinguished as a bunched, or parallel, channel. The notion of *frequency reuse* in wireless communications is an example of space-division multiplexing. We'll postpone discussing this until Chapter 19.

Frequency-division multiplexing. This is most common in radio and television broadcast. The frequency division is carved into bands that are assigned semi-permanently as channels. Each station's bandwidth may not exceed the bandwidth of the channel. In photonics, wavelength-division multiplexing is similar.

Time-division multiplexing. This is like timesharing in computers. Each user sharing the resource is assigned a turn, or timeslot, usually in a regular cycle, called a frame. In one digital transmission system, called T1, 24 digital-equivalent one-way voice channels are time-multiplexed onto one pair of wires. Each channel consists of a net 64 Kbps, as an 8-bit word 8000 times per second. At 8000 frames per second, each frame must hold one 8-bit word from each user. The data format of a frame is a frame-bit followed by 24 contiguous 8-bit timeslots, for a total of $1 + (24 \times 8) = 193$ bits and a net data rate of $8000 \times 193 = 1.544$ Mbps. While the net resource carries 1.544 Mbps, no user may exceed its allotted 64 Kbps.

Suppose the waveform, illustrated in Figure 1.1, is tapped off a channel inside some network. It could be interpreted as:

- Two consecutive binary 1s with high tolerance in amplitude and pulse width (they could be consecutive bits from a serially formatted word or from separate words in a system of amplitude-division transduction);

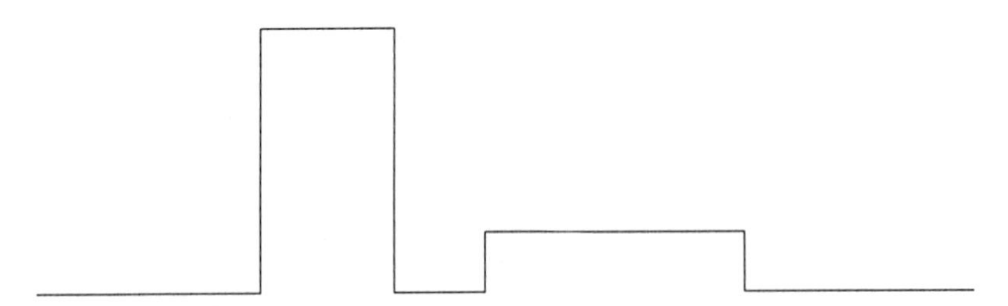

Figure 1.1. A waveform extracted from a communications system.

- A 1 followed by a 0 as above because the amplitude threshold is set between the two pulse heights;
- A 57 followed by a 13 in a system using integer-valued amplitude-division transduction, with a high tolerance in pulse width;
- A 12 followed by a 23 in a system using time-division transduction, like pulse-width modulation,
- A (57,12) followed by a (13,23) in a system using both amplitude-division transduction and time-division transduction simultaneously;
- A real value instead of an integer value in either of the three systems directly above.

In the example, both pulses are implied to be in the information stream from a single user, but all these possibilities pertain as well if the pulses originate from two different users. They appear in the same stream inside the network because of time-division multiplexing. This example has shown amplitude-division transduction, time-division transduction, and time-division multiplexing.

1.3 SWITCHING

Now that we know what a channel is, we can define switching more formally than we did in Section 1.1. In this section, we carefully distinguish between *switching* and *selecting*.

Switching. Switching is the interconnection of several short permanent channels into a single long temporary channel so that the users at the endpoints of the resulting channel may communicate among each other. Since the number of users that are interconnected by a switched channel is usually only two, one on each end, the temporary long channel they require is created by switching together its component links, in series, into a path that reaches between these two end users.

Crosspoints. These are hardware (mechanical, electrical, optical, etc.) devices or software algorithms by which two channels are temporarily connected together to make a single longer channel. Typically, such a switched channel, or circuit, is connected for the duration of the interaction between its users. In the language of computer communications, this interaction is called a session. In the language of telephony, this interaction is called a call. The channel is switched together at the start of the call, and, at the end of the call, the channel is disconnected so that some of the component links might be used by other users for subsequent calls. Note that the channel remains established for its users even if they stop communicating temporarily. This *switching-by-the-call* is assumed through most of this book and, if there is any doubt about this default arrangement, it is specified by the term "circuit switching." Circuit switching, packet switching, and facility switching will be carefully compared and distinguished, but not until Chapter 17.

When an end user transmits information into a switched channel, it goes only to the other user(s) participating in the call. In a switched channel the end users receive *only* that information which is intended for them; they do *not* receive information that is part of other calls. This procedure is in contrast to the operation of broadcast channels or bus-architected local area networks (LANs). Since a radio or television (or a workstation on an Ethernet) receives many channels, the receiver must be tuned (or the workstation must use some addressing procedure) to select a single desired channel from all the received channels. This action is weaker than true switching.

Selecting. Selecting is picking a particular information channel out of many received channels off a common shared super-channel. Selecting is a special case of switching. In a selecting system, all subscribers are permanently connected to the common shared super-channel and each subscriber is physically capable of connecting to any of its internal channels without any arbitration with the system itself. In a true switching system, all subscribers are disconnected from the system unless they originate or terminate a call, in which case they arbitrate with the system in order to be connected, but then only to the participants in this call.

1.3.1 Switching in Three Divisions

Recall that channels are created by dividing a telecommunications resource along the classical divisions — space, time, and frequency. If two channels, both created in the same division, are to be switched together, we'd prefer to perform the switching in the same division in which the channels are defined. The three classical divisions for switching are the same ones typically used for multiplexing: space, frequency, and time.

Space-division switching. This is connecting the spatially separated channels of the space division. The traditional technologies have been electromechanical, electronic, and digital (and optical in research systems). Since the channels are physically separated, the switching typically requires arrays, called fabrics, of switching devices built in one of these technologies.

Frequency-division switching. In its most familiar form, this occurs when a channel is selected on a television or radio. The switching is simplified because there are few transmitted channels, each is simultaneously received by many parties, all channels are broadcast to every receiver in a logical star, and the assignment of channel to transmitting party is dedicated and known. Switching is not centralized, but is distributed and done by each receiving party by channel selection, also found in logical buses. The national telephone network is not analogous because there are many channels, the connections are typically point-to-point, all channels are not broadcast to every receiver (at least in today's technology), and the assignment of channel to transmitter is temporary (at the beginning of the call). So centralized switching is used.

Time-division switching. This allows channels to be switched, as in space, or selected, as in frequency. If every timeslot reaches every receiver, then the receiver can be instructed to retrieve only that data corresponding to its call, analogous to channel selection. However, if the communications system assigns different timeslots to the transmitter and the receiver, then centralized time-division switching, also called timeslot interchange, is required (see Section 13.2). This typically occurs if the network has several stages of time switching or combinations of time switching with switching in other divisions, usually space.

1.3.2 Examples

A small switching system could conceivably be built around frequency-division multiplexing. It could have a coax bus that supports 20 carrier frequencies and that is common to about 100 telephones. When a calling party goes off-hook, some central intelligence assigns it an unused carrier frequency, and when the called party is identified, it is assigned the same channel. All channels are delivered to all users, but only one would be selected. With only 20 carrier frequencies, only 20 simultaneous calls could be supported, probably enough for 100 telephones. For the remainder of the book, selection will be considered a special and separate case of switching.

In discussions of hypothetical time-multiplexed fiber systems for digital television, it is unlikely that every 90-Mbps TV channel can be broadcast down every fiber to every home. Channel selection, as done today in radio or cable systems, will be replaced by centralized switching in logical stars. The user must send a message upstream to a *head end,* in order to select the program channel that should be transmitted back downstream.

In conventional time-division multiplexing, timeslots are assigned regularly. A user would be assigned, for example, timeslot 37 in every frame for the duration of a call. Alternatively, the collection of timeslots could be viewed as a pooled resource available to the users, not at the beginning of a call, but each time a user has a new piece of data to be communicated. This technique, called statistical multiplexing, can be applied to all the divisions, but usually is seen in the time division. Without a regular pattern to the channel assignment, channel switching, usually called circuit switching, is impossible. However, if the destination address is placed in the timeslot by the transmitting party with its data, another form of switching, called packet switching, is possible. Packet switching is rarely discussed in this book because this book is about circuit switching.

1.4 NETWORKS

A channel is a path over which data moves when communication occurs. A network is a system of channels, with emphasis on the connection of its components and not on the components themselves, which are simply treated as ports or endpoints.

1.4.1 Topologies

Simple network topologies are presented, but the reader will find it difficult to map real networks to these textbook structures. Two reasons for this difficulty are that real networks are usually hierarchical and that one must distinguish the physical network from the logical network. These two issues are discussed in later subsections.

Fully connected network (FCN). An FCN with $N = 5$ ports is shown in Figure 1.2(a). The FCN has a dedicated two-way channel for every pair of ports, giving $N(N - 1)/2$ channels. If port A wishes to communicate with port B, A must alert B to expect data and then must transmit the data over the correct channel to B. B must identify this channel and must be ready to receive the data from it. Each port must have a local means to select a channel, analogous to selecting a channel on a television set.

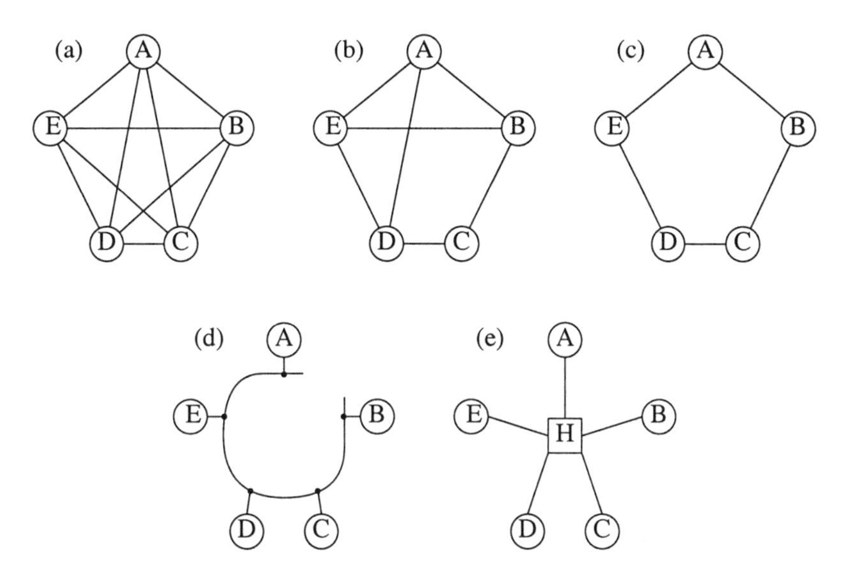

Figure 1.2. Network topologies: (a) FCN, (b) mesh, (c) ring, (d) bus, and (e) star.

Using an FCN as the architecture of a 100,000-port telephone network would require 5 billion channels. Because of the large number of channels required, this architecture is impractical for most networks of realistic size. There are three techniques for making the topology practical.

- Eliminate some or most of the channels and arrange for indirect or tandem transfer of data through intermediate ports — as in a mesh, a ring, or an active star.
- Collapse all the per-port selection functions into one channel from each port to a shared resource, as in a bus or a passive star.
- Architect a hierarchical or sequential network from several layers or stages of simpler networks.

Mesh. This is an arbitrary series/parallel configuration, that includes the FCN as the richest case. Figure 1.2(b) shows the same $N = 5$ ports in a less-than-fully connected network. The ports in a mesh must be smarter than the ports in an FCN because the mesh's ports must route data through the mesh that is not intended for the user they serve directly. In the figure, data from A to C must pass through B or D, enroute.

Ring. Shown in Figure 1.2(c), a ring is the leanest case of a mesh. Every port is connected to the ring by exactly two channels. The ring network is circular, at least logically, if not physically. The routing algorithm inside each port is greatly simplified over the mesh. It is even simpler if the channels in the ring are one way. The main problem with a ring is that, when N is large, data must make many intermediate hops.

Bus. In Figure 1.2(d), each of the $N = 5$ ports has a single short path to a common resource (the bus), that winds around to each port. The ports contend with each other for transmitting on this bus and the transmission must include destination address information. All ports must listen to all transmissions to select those intended only for it. A network that functions like a bus may not physically resemble a bus. The main problem with a bus is that, when N is large, many users compete for a single shared resource.

Star. In Figure 1.2(e), each of the same $N = 5$ ports has a single long path to a central hub. In a passive star, the hub is a common resource, like a large short-circuit, that is functionally and logically equivalent to a bus. One could argue that a network, made with a satellite and N transceiver earth stations, while physically resembling a star, operates more like a bus. The distinction between physical and logical topologies will be discussed later. In an active star, this hub can simultaneously connect any set of pairs of disjoint channels. Because it can switch parallel signals, the active star has more throughput than the passive star or bus or ring. This is proven generally, but mathematically, in Section 17.1. All the selection intelligence, located in the ports in buses and rings, is collapsed to the hub in an active star. For some configurations of N ports, the ring and bus might require a shorter net path length than a star, but an active star is more efficient in its use of intelligence and has much greater throughput than any other topology.

1.4.2 Hierarchical Networks

Real networks, like commercial local area networks and long distance telephone networks, never seem to resemble these simplified textbook cases. One reason is that real networks are usually multilevel or hierarchical.

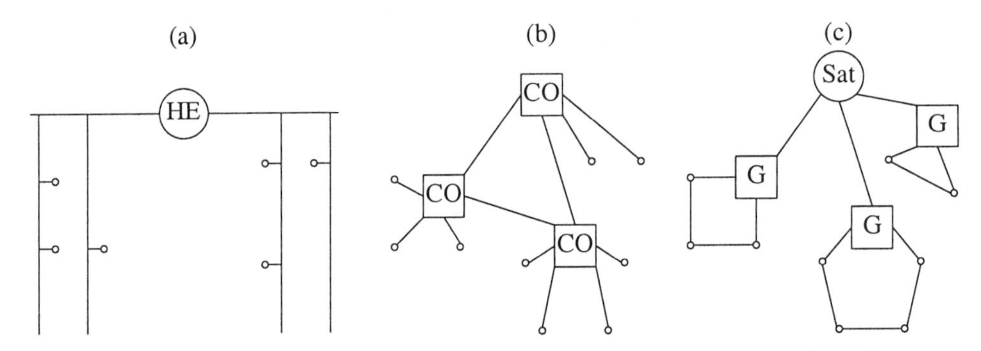

Figure 1.3. Hierarchical networks: (a) CATV, (b) SONET ring, and (c) satellite WAN.

For example, Figure 1.3(a) shows a generic cable TV network. The head end (HE) has a satellite receiver and other forms of community antenna and serves as the hub of a star network of coaxial cables. Each arm (four are shown in the figure) serves a different neighborhood in the service area. A neighborhood is served if a coax passes near the homes, typically by running underground along the streets. A home is served by running a short coax from the house and tapping it into the main coax in the street. Viewed from the head end, the topology is a star. But any of the neighborhood branches is configured like a bus. Actually, each arm of the star has a two-tiered "bus of buses" topology in which each neighborhood bus (Figure 1.3(a) shows four vertically) is connected, in bus fashion, to an arm of the star (Figure 1.3(a) shows two horizontally). The overall topology is a three-level network that could be called a "star of bus of buses." This same topology is typical for modern computer networks: each floor of a building is wired by a single Ethernet bus, the separate floor-wide buses are interconnected by a building-wide Ethernet backbone bus, and all the buildings on a campus are wired together by some kind of metropolitan area network (MAN) — perhaps an ATM physical star (see Section 19.4).

Figure 1.3(b) shows a proposed hierarchical interoffice telephone network. The telephones in a community served by a central office (CO) are connected to that office in a star topology. The figure shows $N = 3$ such switching offices and a few of the telephones they serve. Two proposals for interconnecting switching offices in a contiguous local region called a local access transport area (LATA) are by a ring network or by a star network through a hub. The figure shows the former, and the overall topology is a "ring of stars." The overall topology of the latter would be a "star of stars." The complete national telephone network, once a five-level star, has evolved to a "mesh of stars," as discussed in Chapter 4.

While this example implies that the telephones that are served by a particular switching office are connected to it in a star network. The wires of the *loop plant* are actually deployed physically as a "star of star of stars." Any telephone's physical loop-wire is actually three pairs connected in series, each pair being an arm in one level of the three-level star network. The details are discussed in Section 3.4.

Figure 1.3(c) shows a corporation with $N = 3$ locations around the country. Each location has a LAN, configured as a ring. To intercommunicate among these LANs, each has a special port, called a gateway, that might have a satellite transceiver on it. While some satellite systems may operate like a bus, they have the physical form of a passive star — and the physical topology is a "star of rings."

Frequently the topology of a real network can be understood by breaking it down into hierarchical levels. Usually this is not as simple a task as these examples might imply.

1.4.3 Physical/Logical Networks

Another reason for the difficulty in matching real networks to simple models is that networks may not operate the way they look. One example is a radio or satellite network that has the physical form of a star with the satellite at the hub and the earth stations as the arms. But, if all port transmitters contend for one receiver on the satellite, then the network functions as a passive star. The operation is equivalent to that of a bus, where the network is a simple resource that the ports must share. Two other examples are presented.

One problem with bus networks is that complex procedures are required to resolve the contention for the shared resource among its ports. Contention is worse, and hence communications efficiency is poorer, if the time of transmission or the end-to-end delay in the bus is long. While this is not a problem in a broadcast-only network, like the cable television network in Figure 1.3(a), it is a problem if the ports on the network intercommunicate. While the simplicity of a logical bus network is appealing, efficiency constrains the physical bus to a short length. In Figure 1.1(d), showing a general bus topology, each port is connected by a short wire to a long bus. An alternative is that each port be connected by a long wire to a short bus (or passive star).

It is necessary to distinguish between what a network looks like, physically, and how it works, functionally. Figure 1.4(a) shows a network, in which the bus has been collapsed into a small hub. It has the physical topology of a passive star, but is logically and functionally a bus. Figure 1.4(b) shows the opposite case: a physical bus and logical star. Consider a bus-topology LAN, such as Ethernet. Two terminals and a host

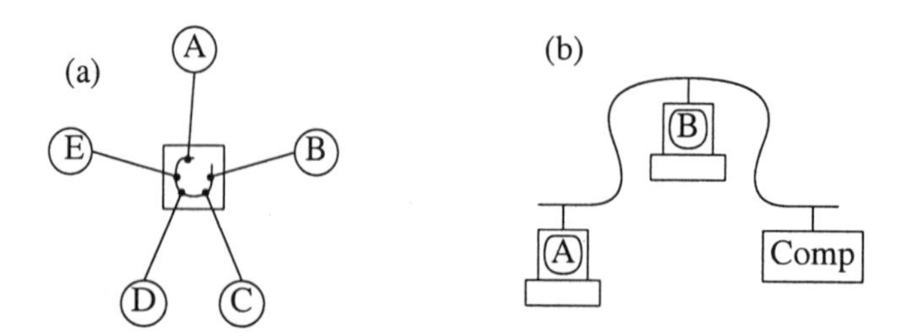

Figure 1.4. Physical/logical networks: (a) star/bus and (b) bus/star.

computer are shown connected to the bus. A program on the computer provides for electronic mail (e-mail). If party A wants to send mail to party B, A types the mail to the e-mail program running on the computer. That program stores the mail for B and sends a message to B's terminal that says: "You have mail." After B requests A's message, the e-mail program delivers it to B's terminal. Logically, A and B and the e-mail program at the hub operate as a passive star. But the star is implemented on a physical bus. Many other configurations are possible. While a logical ring or mesh on a physical bus or physical star doesn't seem useful, a logical bus or logical star on a physical ring or mesh may be.

1.5 POTS — PLAIN OLD TELEPHONE SERVICE

This section describes the user's view of the telephone network and a layman's overview of switching office architectures. It includes more definitions and provides a background from which the remainder of the book may be built.

1.5.1 Residential Telephone Service

The world has about 700 million telephone lines, with distinct directory numbers, and the United States has about half of them. While cellular telephony is very competitive, regular "tethered" telephony is not. In the United States, about 80% of the tethered lines are served by local telephone companies that were once part of the old Bell System and the others by about 1200 independent telephone companies. The term "Bell System" refers to the pre-divestiture era of AT&T, and the term "independent" refers to telephone companies that were never part of pre-divestiture AT&T. Generic terms for telephone company, either a Bell Operating Company (BOC) or an independent, are local exchange carrier (LEC) or telco.

The left side of Figure 1.5 shows a telephone and an extension (alternate telephone with the same directory number) in a residence, each optionally plugged into a modular jack. House wiring connects telephones together and to the protector block (PB). A single line in the figure represents a pair of wires. This protector block is typically found near the fuse box and provides for over-voltage (lightening) protection and manual crossconnect between house wiring and telco wires. Each active pair of telco wires is a telephone line, called a loop, between the pair of house wires serving a telephone (and its extensions) and the switching office, which contains the telco's switching equipment. The architecture of the network of these telephone lines is described in Section 3.4.

Typical modern residential telephone service is flat-rate billed. The monthly charge

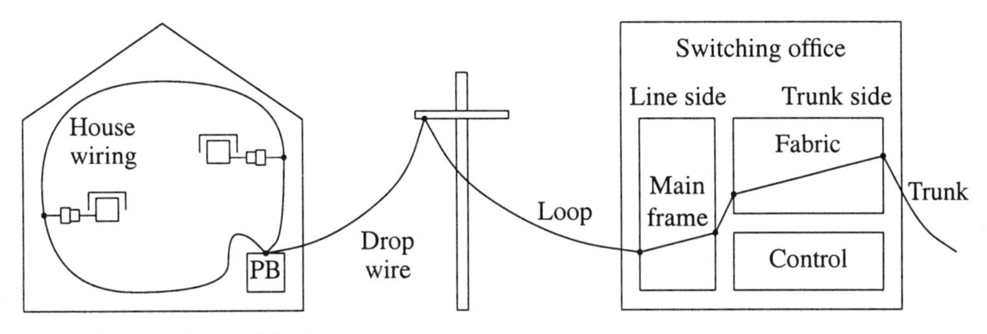

Figure 1.5. Residential telephone service.

covers the lease of the line and the right to make free calls to other parties served by the same office or by certain other offices in a small area. The rate is computed over typical calling patterns and the cost of the common equipment required for the duration of a call. If the user wants to enlarge his flat-rate area, he can subscribe to extended area service. In message-rate billing, each call is individually timed and billed accordingly.

Lately, users of communicating personal computers and home computer terminals place local calls, using dial-up modems, to computers and timeshare services and stay on the line for an unusually long time, free of charge under flat-rate billing. As long as there were not too many, this change from the typical calling pattern was not a serious problem for the telco. However, the increasing use of these instruments and other services, like computerized banking and shopping, is increasingly tying up the common equipment required for the duration of a call in a way inconsistent with the computation of the flat-rate charge. The telcos will have to do something in the future: either encourage data services to move off the traditional telephone network or return to message-rate billing.

Nonstandard telephones lines, like those serving coin telephones or premises switching equipment (Chapter 10), have always required special class of service inside the central office. Now a richer variety of nonstandard telephones, such as credit card telephones, and special services, such as high-speed data, are available. Furthermore, the features available from the standard telephone have increased enormously. The user may subscribe to features like abbreviated dialing, incoming call waiting, and call forwarding.

1.5.2 Operators

Human telephone operators, also called attendants, have played a long and varied role in the history of telephony.

Local operators run a manual central office, a function made almost entirely extinct by automatic exchanges. In regions served by such offices, these attendants are reached whenever a subscriber goes off-hook. These manual exchanges, run by local operators, are described a little later in this section. For manual assistance on a local call, the local operator (who may not be physically local) is usually reached by dialing 0.

These attendants operate from manual switchboards in premises switching equipment, a function made largely extinct by the automatic private branch exchange (PBX). They are reached whenever a subscriber goes off-hook or an incoming call enters a PBX. The PBX and direct-in dialing are described in Chapter 10.

Toll operators assist on emergency, collect, third party, and other kinds of operator-assisted calls. Today, in the United States, toll operators are reached when a subscriber dials 00 or 0+. Toll offices and the Operator Service Position System are described in Chapters 4, 5, 9, 12, and 14.

Information operators provide directory assistance. Intra-area attendants are reached when a subscriber dials 411. Attendants for other area codes are reached when a subscriber dials that area code followed by 555-1212. Years ago, these attendants sat in a room full of telephone directories. Now, these attendants sit at computer terminals and access directories on data bases.

1.5.3 The Numbering Plan

Every user, or subscriber, of a telephone line has a directory number so the line may be reached by automatic equipment. The first three characters, now all digits, of a directory number identify the telephone office into which the extension is connected and is called the office code. The last four digits correspond to a subscriber's line in the serving office. So, there are a maximum of 10,000 lines in an office or exchange. Even in the late 1960s some locations required operator-assistance on long distance calls, and the numbering plan had directory numbers with alphanumeric office codes, like PEnnsylvania 6-5000. Some directory numbers had fewer than, or more than, seven digits.

With the automation of long distance calling, called direct distance dialing (DDD), came a massive change in the national numbering plan. North America was partitioned into number plan areas (NPA) and each area was assigned an area code. The "homey" directory number format was changed to a uniform seven-digit form, a change that was unpopular at the time. Before 1+ dialing became universal for long distance calls, office codes had the format, NNX, where N is a digit between 2 – 9 and X is any digit 0 – 9. This provides 640 office codes in an NPA — including codes like 911, 555, and 936 which are reserved for calling emergencies, information, time, and weather. Furthermore, with Centrex services (Section 10.5), many corporations received their own office code, distinct in the NPA.

Before 1+ dialing for long distance calls, area codes had the format, NKX, where N and X are as above and K is a 0 – 1. That no office code has a 0 or 1 as the second digit distinguishes them from area codes. This format allows for 160 codes, but some of them are used for special services, like 800, 900, 911, and 411. The familiar assignment of area codes to NPAs is found in the front section of any telephone book. Since the original assignment of area codes, some states have grown and NPAs have divided. Just as subscribers resisted office code changes, area code changes are also emotionally charged. One highly publicized uproar occurred when subscribers in Brooklyn, NY, lost the 212 area code. Quite recently, we used up the last one of these codes and the numbering plan had to change again.

Numbers beginning with 1 or 0 are special. A leading 1, where it was required, has always been a prefix indicating that the following digits are for a toll call. The 1 prefix has not always been required, depending on the type of central office. But recently, the 1 prefix has become universal and it indicates, not only a toll call, but a toll call that does not require operator assistance. Now that the leading 1 is required for a toll call, office codes and area codes can all have the format NXX format.

Formerly, dialing 0 (not as a prefix), placed a call to the operator. More recently, with the Traffic Service Position System (TSPS) (Sections 5.4.1 and 9.3) and automatic overseas dialing, 0 has been changed to be a prefix on a regular telephone number that indicates a request for operator assistance — for collect, credit card, or person-to-person calling — on the call whose digits follow the prefix. Dialing 00, or just 0 with a long timeout, calls the operator directly and 01 is a prefix for more specialized calls, such as specifying a nondefault long distance carrier (Section 12.6) or automatic overseas dialing (with a 011 prefix). A subscriber-dialed overseas credit-card call over a nondefault carrier requires that the subscriber dial many digits.

In this era of divestiture, the long distance providers, the telcos, and other independent companies will introduce competing operator services. Another change will be required so that the subscriber may specify the desired operator. It is important to note that the numbering plan is constantly changing and is not always universal. In today's unregulated era, there is not even agreement on who has control over the numbering plan. But, fortunately, computer program control brings to telephone switching offices a greater ease in implementing these changes.

1.5.4 A Generalized Switching Office

A switching office is introduced by defining its boundary conditions, giving a general block diagram, and briefly describing a manual office. A switching office can also be called a central office, a wire center, or an exchange. One office provides communications service to 10,000 telephone lines, all lines with the same office code. An office is distinguished from a building: a typical building houses more than one office and the equipment for several offices in a building is typically integrated together.

An office is two sided, and the sides are called the line side and the trunk side. A line, or loop, is a path between an office and a telephone while a trunk is a path between two offices. The details of the line side are covered in Chapter 3 and of the trunk side, including the national network architecture, in Chapter 4. Some subsystems, internal to the office, may also be two sided. The sidedness may be defined differently, such as originate/terminate or transmit/receive, but the two sides of the overall office are the line side and the trunk side.

Let subscriber A, whose directory number is (201) 464-1234, place a call to subscriber B, whose directory number is (201) 464-3456. A is called the calling party and B the called party. The switching office determines from A's dialed digits, which do not include the 201, that the call is intraoffice and local (not billed). The connection is made entirely within the 464 office between A and B. However, if A dials C at 665-5678, then the system arranges for a previously idle interoffice trunk between the 464 and 665 offices, for a connection in the 464 office between A and the 464 end of the trunk, and for a connection in the 665 office between C and the 665 end of the trunk. If the 464 office and the 665 office share the same building, this trunk remains inside this building. If the 464 office and the 665 office are in separate buildings, the trunk must interconnect them. If A dials D on (212) 322-7890, then the system determines a route along a sequence of offices in the national network, arranges for previously idle trunks between the offices in the route, and establishes appropriate connections in those offices. The purpose of this book is to describe the details of how these connections are made.

The right side of Figure 1.5 shows a generalized block diagram of a switching office, with the line side on the left and the trunk side on the right. The lines terminate in the office on a manual crossconnect, called the main distribution frame (MDF or main frame). This simplifies the administration of lines with line appearances on the fabric.

The fabric, or connection network, provides generally arbitrary connections between any pair of lines or trunks. The term "fabric" is recent and not completely accepted, but is used throughout this book to avoid ambiguity with the term "network," used to describe the interconnection of offices, a higher level of connectivity than that among lines and trunks within an office. Control is shown in the figure as a centralized block, but it can be completely distributed within the fabric or implemented as a lumped subsystem, depending on the style of the equipment. Another variation is the modular architecture, in which an entire office is an interconnection of many identical modules, each having the architecture shown in the figure. (See Section 2.4.)

1.5.5 Manual Office

We relate a manual office to Figure 1.5: the manual switchboards, and the wiring of lines and trunks to them, represent the fabric, and the human operators embody the control. When a subscriber lifts the receiver on his telephone, a lamp is lit near a jack on at least one switchboard. The attendant inserts a plug on one end of a patch cord into A's jack, connects the headset, and says: "Number please." A then gives B's directory number or name to the operator, who inserts the plug on the other end of the same patch cord into B's jack and throws a switch that causes B's telephone to ring. This procedure is further complicated if A's line is multipled to jacks on more than one switchboard or if B's jack does not appear on the same switchboard as A's jack. When B answers, ringing stops, the operator disconnects from the call, and can serve another one. When A hangs up, the operator unplugs both jacks on the cord.

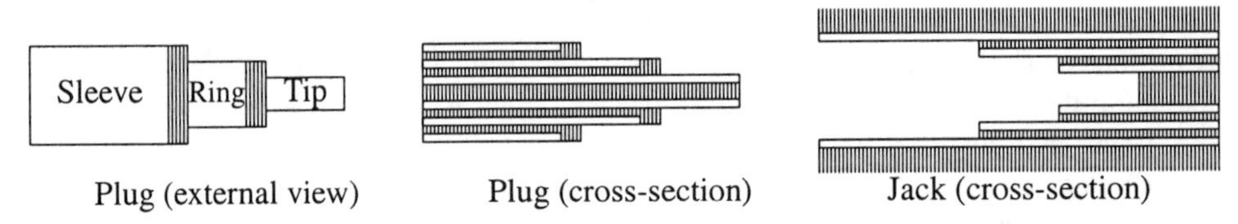

Figure 1.6. Plug and jack on a patch cord.

A plug and jack configuration is shown in Figure 1.6. A plug is shown on the left and its cross-section in the center. The cross-section of the jack is shown on the right. In Figure 1.6, the clear areas represent conducting metal and the hatched areas represent insulation. The conducting areas labeled Tip, Ring, and Sleeve on the plug are coaxial cylinders. When the plug is inserted in the jack, each cylinder in the plug makes electrical contact with a corresponding coaxial cylinder in the jack. That this contact is spring loaded is not shown in Figure 1.6.

The cord contains three wires that are connected to the tip, ring, and sleeve on the plug on each end of the cord. The electrical waveform carrying the subscribers' data is transmitted over the two wires connected to the tip and ring on each end. The wire connected to the sleeve on each end carries a DC signal to light the lamp indicating the state of the call. This terminology of "tip, ring, and sleeve" has outlived the manual switchboard and persists today as the name of the three wires switched through any fabric. The wire terminals inside a telephone are labeled T and R, for connecting the tip and ring, the names still used for the two wires in the line between the telephone and the switching office. The convention is that the red wire is the ring and the green wire is the tip (See Chapter 3).

1.5.6 Office Functions

Eight functions of a switching office are described by following a call through a manual office in more detail. The equivalent implementation of each function in an automatic office is inserted parenthetically.

If A wants to call B, A lifts the handset on the telephone and, by turning a crank and generating a small AC signal (or automatically), causes a lamp to be lit. The operator, at the other end of the line sees the lamp and knows that this particular line wants service. The function of watching a line to see if it requests service is *attending*. Attending is done per-line, over every idle line in the office. In automatic offices, attending is expensive and has been implemented in a variety of ways in the different switching systems.

Then, the operator says, "Number please" to A (or automatic equipment returns a dial tone to A). A gives B's directory number to the operator, or A dials that number. The operator (or the automatic switching office) must perform the *receiving* function.

Now the operator (or the automatic office) must determine whether the number is valid, the location of the called party's jack (or fabric appearance), and whether the connection is intraboard, intraoffice, unbilled, or toll. Then the fixed procedure (or program) is followed for completing the call. This function is *control.* The remaining functions are part of this procedure.

For an interboard call, the operator must check whether the called party is busy or idle. If B is busy, the operator reports that to A (or A hears busy tone) and a connection is not completed. In modern offices, A and/or B can subscribe to new services that allow A to wait until B hangs up or that direct a special tone to B indicating an incoming call. This basic function is *busy testing*. For interboard or interoffice calls, busy testing is indirect: A may be connected to the board or office that serves B but would not be connected through it to B's line.

If B is idle, the operator causes B's telephone to ring, the *alerting* function. In a manual office and in some of the automatic offices, A is connected to B's line and hears the same signal that rings B's telephone. In other automatic offices, A hears a different signal, called an audible ring tone, while B's telephone is ringing. Again, for interboard or interoffice calls, alerting is indirect.

When B answers the telephone, ringing (and audible ring tone back to A) stops, the operator drops off the call, and the connection is complete. The call is *connected.* Depending on the system or the type of call, other forms of connecting provided for accepting A's dialed digits, ringing B's telephone, giving A audible ring tone, or connecting A to another board or to a trunk to another office.

If A's call could not be completed at one switchboard, the operator must talk to another operator at the board or office serving B. After carrying out this *transmitting* function, the operator at the other end can complete the call. Automatic systems must perform a similar function.

Finally, the operator must monitor the active call to determine, at least, when the call is over, the *supervision* function. Supervision also occurred when the operator determined when B answered a ringing telephone. Supervision is similar to attending, both involve watching or monitoring, but supervision is done per-active-call and alerting is done per-idle-line. This distinction is important enough to separate the functions. Supervision is described in general in Chapter 3.

SELECTED BIBLIOGRAPHY

Becker, F. K., "Telephone Systems." Chapter 7 in *Telecommunications: An Interdisciplinary Text,* (L. Lewin, ed.), Dedham, MA: Artech House, 1984.

Bellamy, J., *Digital Telephony,* Second Edition, New York, NY: Wiley, 1991.

Blalock, W. M., "North American Numbering Plan Administrator's Proposal on the Future of Numbering in World Zone 1," *Bellcore Letter,* Letter No. IL-92/01-013, Jan. 6, 1992.

Brooks, J., *TELEPHONE — The First Hundred Years,* New York, NY: Harper & Row, 1976.

Clarke, A. C., et al., *The Telephone's First Century — and Beyond,* New York, NY: Thomas Y. Crowell, 1977.

Freeman, R. L., *Telecommunication System Engineering,* Third Edition, New York, NY: Wiley, 1996.

Joel, A. E., Jr., "Classification and Unification of Switching System Functions," *Proc. Int. Switching Symposium,* 1972, pp. 446 – 453.

Joel, A. E., Jr., "The Past 100 Years in Telecommunications Switching," *IEEE Communications Magazine,* (issue on 100 Years of Communications Progress), Vol. 22, No. 5, May 1984.

Joel, A. E., Jr., "What Is Telecommunications Circuit Switching?" *Proceedings of the IEEE,* (issue on Telecommunications Circuit Switching), Vol. 65, No. 9, Sept. 1977, pp. 1237 – 1253.

Lindstrom, A. H., and R. Karpinski, "Telecom History 101," *Telephony Magazine,* (90th anniversary issue), Vol. 220, No. 24, Jun. 17, 1991.

McDonald, J. C. (ed.), *Fundamentals of Digital Switching,* Second Edition, New York, NY: Plenum, 1990.

Noll, A. M., *Introduction to Telephones and Telephone Systems,* Norwood, MA: Artech House, 1986.

Pearce, J. G., *Telecommunication Switching,* New York, NY: Plenum Press, 1981.

Pierce, J. R., *Signals: The Telephone and Beyond,* San Francisco, CA: W. H. Freeman and Company, 1981.

Pierce, J. R., "Telephony — A Personal View," *IEEE Communications Magazine,* (issue on 100 Years of Communications Progress), Vol. 22, No. 5, May 1984.

Schindler, G. E. (ed.), *A History of Engineering and Science in the Bell System,* Murray Hill, NJ: Bell Telephone Laboratories, 1982.

Switching Systems, New York, NY: American Telephone & Telegraph, 1961.

Talley, D., *Basic Electronic Switching for Telephone Systems,* Rochelle Park, NJ: Hayden Book Company, 1975.

Talley, D., *Basic Telephone Switching Systems,* Rochelle Park, NJ: Hayden Book Company, 1969.

Tracey, L. V., "30 Years: A Brief History of the Communications Industry," *Telecommunications,* Vol. 31, No. 6, Jun. 1997.

2

Background

"Tomorrow, and tomorrow, and tomorrow creep in this petty pace until the last syllable of recorded time; and all our yesterdays light fools the way to dusty death. . . . It is a tale told by an idiot; full of sound and fury, signifying nothing." — from *Macbeth* by William Shakespeare

Today's U.S. Public Switched Telephone Network (PSTN) was not created "ab ovo" in six days by Alexander Graham Bell. Far from it. It evolved and is a classic illustration of how the fittest survive. The history or genealogy, the story of this evolution, is an essential part of the story of the system. The architecture of modern switching systems is dependent on regulation and industry image, the evolution of technology and its economics, and the increasing knowledge of architectural design. These three issues are developed, initially, in this chapter. The first section describes what I mean by telecommunications' ancient history. Section 2.2 describes the telecom industry and Sections 2.3 through 2.5 respectively present the evolutions of relevant technology, architecture, and some of the relevant systems.

2.1 ANCIENT HISTORY

If you take courses in Western civilization, you'll find that historians divide history into two parts: ancient and modern. The boundary between the two parts is usually the Roman Empire. The Pax Romana was an era of several hundred years during which things were stable (at least, superficially so). The history of telecommunications also had a Pax Romana, which coincided with the Pax Romana of the history of 20th century America — it was the 1950s, the "Happy Days." As we proceed, we'll see much correlation between the history of telecommunications and the history of the United States and the world. This section presents the ancient history of telecommunications, the significant events during the period of time that begins with Morse and Bell and that ends with the 1956 Consent Decree.

This ancient history of telecommunications is rich and interesting, filled with intrigue and unusual characters. But, like the settling of the West and like George Washington, many of the events and characters in this history have become legendary and exaggerated with the passing of time. Some stories are not true and some are only close to the truth. I find myself torn between my duty to present the facts or the perception. I have opted for the latter, partly because it's more interesting, but mostly because it is important that the reader understand how this industry is perceived. I believe that many events of the modern era were partly prejudiced by the "Ma Bell" image, and that was only partly deserved. Discussions of subsequent events, technologies, and regulations are distributed throughout the rest of the book.

2.1.1 Morse

Samuel Morse invented the telegraph in 1832, and tapped the first message: "What hath God wrought?" Many telegraph companies formed in the United States, but by 1866 many independent telegraph companies had been acquired by one company, Western Union (WU). The telephone business would have a similar history of many independent companies for 30 years, with one emerging dominant. Furthermore, WU would be a significant player in the history of the telephone industry. Although we tend to think of the transatlantic cable as a modern achievement, one was completed in 1858 to carry telegraph signals.

2.1.2 Bell

On March 7, 1876, Alexander Graham Bell filed a patent for a device he called the telephone. Bell's principal invention was the earpiece, the predecessor of today's audio speaker. Elisha Gray, under contract with Western Union, lost the rights to a similar invention by only three hours. Thomas Edison, also under contract with Western Union, previously had been granted the patent on the mouthpiece part of the telephone. Only three days after filing the patent, Bell and his assistant, Thomas Watson, demonstrated a working telephone when Bell supposedly spilled battery acid on himself and yelled over the prototype telephone: "Mr. Watson come here. I want you."

Bell had been a teacher of the deaf and he intended his work to aid the hearing-impaired. Bell married one of his students, Mabel Hubbard, and some suggest that Mabel was the original "Ma Bell." Her father, Gardiner Hubbard, became Bell's principal backer, the president of the first telephone company, and the business acumen behind it.

Bell and Hubbard offered to sell the patent rights for the telephone to Western Union for $100,000. Figure 2.1 shows a letter, probably a fake, supposedly written by a group commissioned by WU to advise them on this purchase. Despite having supported the work of Gray and Edison, Western Union expressed little outward interest in the telephone. Whether their reason was genuine or stemmed from a personal grudge between Hubbard and Ordend, the president of WU, Western Union declined to buy Bell's patent; in hindsight, one of the largest business blunders of all time.

May of 1877 saw the first telephone exchange, in Boston, operated by a burglar alarm company. This company had run wires to six client banks, in a star network, and contracted with them that if any of their alarms tripped, somebody at the hub would run across the street to alert the police. Because the banks disabled the alarms during the day, the owner suggested placing Bell's telephones on these same wires during the day and operating a primitive switchboard at the hub so the banks could intercommunicate easily.

In July of 1877 the Bell Telephone Company formed, with Gardiner Hubbard as president. The Bell Telephone Company manufactured telephones and sold the instruments and the right to use them. One client company was the New England Telephone Company, formed in 1878, and it opened the first commercial telephone

November 15, 1876

Chauncey M. Depew
President, Western Union Telegraph Co.
New York City

Dear Mr. Depew:

This committee was formed at your request to consider the purchase of U.S. Patent 174,465 by the Western Union Company. Mr. Gardiner G. Hubbard and Mr. A. G. Bell, the inventor, have demonstrated their device, which they call the "telephone," for us, and discussed their plans for its use.

The "telephone" purports to transmit the speaking voice over telegraph wires. We found that the voice is very weak and indistinct, and grows even weaker when long wires are used between the sender and receiver. Technically, we do not see that this device will ever be capable of sending recognizable speech over a distance of several miles.

Messrs. Hubbard and Bell want to install one of their "telephone" devices in virtually every home and business establishment in the city. This idea is idiotic on the face of it. Furthermore, why would any person want to use this ungainly and impractical device when he can send a messenger to local telegraph office and have a clear written message sent to any large city in the United States?

The electricians of our own company have developed all the significant improvements in the telegraph art to date, and we see no reason why a group of outsiders, with extravagant and impractical ideas, should be entertained, when they have not the slightest idea of the true practical problems involved. Mr. G. G. Hubbard's fanciful predictions, while they sound very rosy, are based upon wild-eyed imagination and a lack of understanding of the technical and economic facts of the situation, and a posture of ignoring the obvious technical limitations of his device, which is hardly more than a toy, or a laboratory curiosity. Mr. A. G. Bell, the inventor, is a teacher of the hard of hearing, and his "telephone" may be of some value for his work, but it has too many shortcomings to be seriously considered as a means of communication.

In view of these facts, we feel that Mr. G. G. Hubbard's request for $100,000 for the sale of this patent is utterly unreasonable, since the device is inherently of no value to us. We do not recommend the purchase.

Figure 2.1. The legendary Chauncey Depew letter.

exchange in New Haven, connecting 21 telephones. An operator listened to all phone calls to determine the end of a call.

In 1880, the New England Telephone Company and the Bell Telephone Company merged to become the American Bell Telephone Company. This name reappeared a century later as the original name of AT&T's subsidiary, created because of the Computer

Inquiry II ruling (Section 16.2). American Bell introduced commercial long distance service between Boston and Providence. Since this service predated the invention of the amplifier, calls were greatly attenuated.

The Western Electric Company (WECo) had been an electrical equipment manufacturer for, and under corporate control of, Western Union. In 1882, American Bell secretly bought controlling interest in Western Electric. This maneuver, which might be seen as predatory today, gave American Bell needed additional manufacturing capability and dealt a crippling blow to Western Union, from which it never really recovered. A legal battle between American Bell and Western Union was settled by the U.S. government. Through a protracted negotiation, arbitrated by the federal government, the young telephone company was granted monopoly power for telephone and Western Union for telegraph. Almost a century later, Western Union would try, unsuccessfully, to use the spirit of this agreement to claim monopoly right to data communications and to restrict the Bell System from that arena.

By 1885, there were over 300 licensed telephone companies, and the telephone was only nine years old. The American Bell Telephone Company formed a subsidiary, called the American Telephone and Telegraph Company (AT&T) to provide its long distance service. Despite the settlement with Western Union, the word "telegraph" was deliberately included in the company's name.

2.1.3 Strowger

1889 brings another legend. Before automatic exchanges and Yellow Pages, subscribers completed many calls by conversation, frequently friendly, with the operator, often known. For example, someone making arrangements for a funeral might ask the operator for connection to "an undertaker." Almond Strowger was an undertaker in Kansas City and he believed that a competitor's wife, a telephone operator, was unfairly steering business to her husband.

He was inspired to invent an automatic way for people to make telephone calls, without having the human element involved. This system, built from a device called the Strowger switch, is called Step-by-Step. This clever system, described in Chapter 6, not only has been economically viable for almost a century, especially in small exchanges, but it had two properties that are regaining importance today: it was cellular and self-routing. In many ways, this invention was as big a breakthrough as the telephone.

Like so many inventors of this industrial age, Strowger went into business. The first automatic telephone exchange, a Step-by-Step office, was installed in 1892 by the Strowger Automatic Telephone Exchange Company. That company, now named the Automatic Electric Company, is the manufacturing branch of the General Telephone and Electronics Corporation. Gee! No, GTE.

Until the Strowger patent expired in 1906, the Bell System was 100% manual and didn't operate any automatic exchanges. By waiting until 1906, AT&T avoided paying patent royalties to Strowger, but their first Step offices were installed by Strowger's company.

2.1.4 The Bell System

In 1893, 17 years after 1876, the Bell patent expired and many independent telephone companies arose. The only leverage American Bell had over these independents was that its subsidiary, AT&T, the nation's only long distance carrier, refused to connect to any of them. The independents complained to the federal government and the controversy lasted 20 years. In 1900 the parent company, American Bell Telephone, adopted the name of its subsidiary and became the American Telephone & Telegraph Co.

Theodore Vail became the president of AT&T in 1908. By 1911, AT&T was organized into the Bell System, the corporate structure that remained almost unchanged until divestiture in 1984 (Section 16.3). The organization featured many state and regional subsidiaries responsible for local service, the Long Lines division for long distance service, the Western Electric subsidiary for manufacturing, and a General Departments staff, all reporting to a small headquarters organization. Vail is credited for the desire to provide universal service [1], equal telephone service to every home in the United States. He positioned Ma Bell as kindly and protective of its subscribers, but cunning and ruthless to its competitors.

AT&T bought out Western Union in 1911. Urged on by the independents and acting through new antitrust legislation, the U.S. Department of Justice (DoJ) reached an agreement with AT&T in 1913. In exchange for being allowed to retain Western Electric, AT&T relinquished Western Union and agreed to connect the independents. This agreement with the federal government was the beginning of an 80-year continuous negotiation between the DoJ and the Bell System, mainly over AT&T's ownership of WECo.

In 1913, AT&T introduced repeaters in its long distance trunks. They were made from vacuum tubes, only recently invented by DeForest in 1907. The specific significance is that much greater distances could be spanned and existing spans were less attenuated. The general significance is the beginning of the Bell System's association with the leading edge of technology. In 1923, AT&T created Bell Telephone Laboratories as its subsidiary responsible for research and development.

Another agreement with the Justice Department occurred in the 1930s. This time, to be allowed to retain Western Electric, the Bell System agreed to withdraw from all "non-telephone" businesses. One was the Westrex Division of WECo, which made the sound tracks in many moving pictures. The principle behind this agreement was a fear that the Bell System could use its vast resources to underprice competitors in such unregulated businesses, a theme that would recur. In 1934, the Federal Communications Commission (FCC) was established in the federal government's legislative branch to oversee the U.S. communications industry by allocating radio spectrum and regulating interstate telephony. Unfortunately for the Bell System, the creation of the FCC did not diminish the DoJ's interest. From this time forward, the Bell System would have to deal with two government entities that, as typical bureaucracies, had different goals and even seemed to compete with one another.

2.1.5 The Consent Decree of 1956

In 1948, once again the DoJ tried to force the Bell System to divest Western Electric. One of the DoJ's complaints was that Bell Lab's engineers were paid directly by AT&T from revenues accrued from U.S. telephone subscribers through their telephone bill. The DoJ argued that, since these engineers performed R&D for WECo, they should be paid directly by WECo from revenues accrued from the sale of the equipment designed by these engineers. AT&T argued that many Bell Labs' engineers and scientists performed fundamental research that supported the nation's high-tech economy and provided systems engineering functions that benefited the non-Bell telephone companies as much as those inside the Bell System. AT&T maintained that, since Bell Labs was a "national resource," paying Bell Labs' employees through the telephone bill was similar to paying federal labs' employees through U.S. tax revenue.

AT&T and the DoJ contested this issue, and many others, in an eight-year federal case before finally arriving at an agreement, called the Consent Decree of 1956. Formally, this was a temporary compromise, which the two parties agreed would be tried out for an unspecified period of time and that either party could reopen in the future. It was reopened in 1984 (Section 16.3).

AT&T agreed that those engineers in Bell Labs who developed new products for Western Electric would be paid by WECo. But, AT&T was allowed to use revenues from its rate payers, called R&SE revenue, to pay those Bell Labs' engineers and scientists who performed research and systems engineering. However, since Bell Labs was now being viewed as a national resource, the fruits of all its employees should be national property. Specifically, AT&T agreed not to collect royalties on any patents awarded to Bell Labs personnel, whether supported by R&SE revenue or not, and retroactive to the date of the start of the trial in 1948. AT&T couldn't have known it at the time, but this retroaction included a 1949 patent awarded to three Bell Labs' employees: Brattain, Bardeen, and Shockley. Thus, in agreeing to the 1956 Consent Decree, AT&T collected no royalties for the invention of the *transistor*.

The 1956 Consent Decree continued the policy of separating price and cost. All prices would continue to be set by regulators and the price for some equipment or service might have little relationship to its cost. Prices were regularly adjusted so that local calling with dial telephones, residential and coin, Vail's definition of universal service, was priced well below actual cost. Regulated prices for business telephone service, long distance calling, and even Touch Tone calling were allowed to be well above actual cost so that AT&T and the other telcos could ensure a reasonable rate of return to their investors. While this pricing situation may have been stable (as long as the system remained closed), AT&T was always nervous about being vulnerable. This pricing process never made good business sense, but it was encouraged by people who believed it was the only way that government could ensure the universality of telephone service. Alternative ways of doing this are discussed in Section 22.4.

I have made the point several times already, and will emphasize this point throughout the book: the telecom industry is a microcosm of its times. To illustrate, this era in the history of telecommunications is compared to the greater context of global human 20th century history. In the United States, the 1950s were the "Happy Days," a time of innocence and prosperity. I wish everyone could have had a childhood like I had, like Richie Cunningham or Ricky Nelson, in a traditional family in a traditional middle-class neighborhood. In the United States in the 1950s, the Golden Rule was alive and well, Ike was president, Elvis was king, and the telephone company was kindly old Ma Bell.

But, things are not always as they seem. Many barriers were broken between the late 1940s and the early 1960s: by Jackie Robinson and Edmund Hillary, by Roger Bannister and Roger Maris, by Chuck Yeager and Yuri Gagarin. While we thought these things were barriers at the time, they really weren't barriers at all, or they shouldn't have been. This suggested that we weren't as limited as we thought we were and that maybe much more change was waiting ahead of us. If the world was really so peaceful, why was I having air-raid drills in elementary school and why were we airlifting supplies into

Berlin? If the United States was so innocent, why was President Eisenhower sending U.S. troops into Little Rock and why did he warn the nation about its own military-industrial complex? Then, Ike had a heart attack, the Russians launched Sputnik, and the Happy Days were gone — or had they only been an illusion? Finally, to remove any doubt that the bubble had burst, the United States entered into war in Vietnam.

The telecom industry was similarly serene, on the outside. But, under the surface, direct distance dialing would radically change telephony, programming technology would radically change system control, and the transistor would radically change everything electronic. People will argue that during this period, with the telecommunications industry dominated by the Bell System, the country and its individual citizens were not well served. While a free market is almost always the best form of economy, Ma Bell built a telephone network for the United States that was far superior to that of any other nation, and Americans were far better served by the Bell System in this period than the citizens of any other country were by their telephone network or than Americans have been in any other period in this country. Ma Bell was kindly and protective of her customers and her employees, but she was cunning and ruthless towards those who got on her turf, especially to those who wanted to "skim the cream" that subsidized universal service. The separation of price and cost was an accident waiting to happen. Finally, the 1968 Carterfone case would so radically change the rules that the stability of the 1956 Consent Decree would have to end.

We have covered the ancient history, up to the 1950s Pax Romana, and will continue the history of regulation in later chapters. Carterfone and Execunet are described in Sections 10.6 and 12.6, respectively. These regulatory decisions allowed competition in customer-premises equipment and in long distance service, respectively. Since they didn't resolve the separation between price and cost, the entire regulatory structure became unstable and divestiture became inevitable. Chapter 16 is devoted to the 1981 Computer Inquiry II ruling and to the 1984 Modified Final Judgment (MFJ) of the 1956 Consent Decree, the official breakup of the Bell System. Enhanced services are described in Chapters 15 and 18, the Telecommunications Act of 1996 is discussed in Section 16.6, and future regulation is philosophized in Section 22.4.

2.2 THE TELECOM INDUSTRY

The worldwide telecommunications industry is a complex trillion-dollar business. While some of the businesses are extremely nationalistic, many of the companies in this industry are themselves large and global. In part this industry has gradually evolved, and in part it experienced sudden and dramatic changes, as in the early 1980s by the breakup of the Bell System, described in Chapter 16.

This industry is categorized into six components: carriers, telecom equipment manufacturers, regulatory bodies, service providers, computer networking equipment manufacturers, and the users. The carriers are introduced here and elaborated throughout the book. The telecom equipment manufacturers are described in this section. The regulatory bodies in the United States, and the different worldwide regulatory structures, are described in this section. Since third-party services and computer networks are relatively recent phenomena, they are described later, in Chapters 15 and 17, respectively. Users are discussed here and elaborated in Chapter 15.

2.2.1 Carriers

The carriers provide the channels used in communications. A carrier may:

- Be a corporation or a government agency;
- Serve a specific region, or a set of disjoint regions, or any region it chooses to serve;
- Serve its region exclusively or in competition with other carriers;
- Be an independent enterprise, a subsidiary of some larger company, or a large conglomerate or holding company;
- Own its transmission infrastructure, the physical lines and the equipment on each end, in which case it provides service over some lines and leases some lines to other carriers or may only provide service over lines that it leases from another carrier;
- Offer direct access to end users or it may be accessible only through other carriers;
- Offer channels over different types of media — twisted pair, coaxial cable, optical fiber, radio, or satellite;
- Offer circuit switching, packet switching, or channel selection, but usually only one of these is offered by any one carrier.

Local Exchange Carriers. The telephone company, as perceived from a typical residence, is formally called a local exchange carrier (LEC). By definition, LECs offer direct access to end users. Traditionally, the LECs offered circuit-switched telephone service over two-wire twisted-pair telephone lines that they owned and operated. But, they also own a lot of fiber and wireless channels, they lease facility-switched channels, and they are starting to provide video service and packet-switched data. Since today's LEC industry is changing rapidly, first we describe the industry as it was when it was stable — in the mid-1980s, right after divestititure. Then we discuss the recent changes.

In the mid-1980s, the United States had more than 1500 separate telephone companies. These corporations were, and still are, regulated by several state and federal agencies. An LEC was considered a natural monopoly: a company that is granted a charter, called a "right of eminent domain," to serve a specific geographical region, free of competition. These LECs are described in three categories: small independent companies, large multiregion non-Bell companies, and the Bell Operating Companies that were part of AT&T.

Some LECs were, and still are, relatively small companies that provided telephone service within a specific contiguous region. These companies were independent of the Bell System (or its successors) and were not affiliated with any other larger corporations. Rochester Telephone was the largest of these independents, serving about 1 million lines in greater Rochester NY. Other independents that served more than 100,000 users were: Lincoln Tel. & Tel. (Nebraska), Commonwealth Telephone of Pennsylvania, Florida Telephone, and Winter Park Telephone (Florida). There were also many mid-sized companies, like North Pittsburgh Telephone, which still serves about 60,000 lines in a suburban region north of Pittsburgh. And there were hundreds of very tiny companies, like West Buxton Telephone Company, which serves only several hundred lines in the region around Bonny Eagle Pond in Maine. Many of these enterprises, especially the very small ones, were mom and pop companies. In the mid-1980s, more than 1500 of these independent companies served only about 4% of the telephone lines in the United States.

Other independent non-Bell LECs were, and still are, large corporations that provide telephone service to millions of lines in disjoint regions around the United States. While each of these companies is independent of the (old) Bell System, each is a small version of the (old) Bell System that evolved similarly to how the Bell System did. Many of their individual regions had been comprised of independent companies that were bought out by the parent company. The largest of these companies, GTE, is so large that it serves more lines than some of the Baby Bells. Other large independents included (and, in some cases, still include): United Telecom, ConTel, Central Telephone, Mid-Continent Telephone, Century Telephone, and Allied Telephone (AllTel). In the mid-1980s, these multiregion independents served about 13% of the telephone lines in the continental United States. Combined, all these unaffiliated and multiregional independents served only about 17% of the telephone lines in the continental United States. Since they owned about 50% of the switching offices in the United States, they obviously served small regions and their offices were small.

So, in the mid-1980s, about 83% of Americans received telephone service from one of the more than 20 Bell Operating Companies (BOCs) that used to be affiliated with the old Bell System. New York Tel, Southern New England Tel (Connecticut), New Jersey Bell, Diamond State Tel (Delaware), the Chesapeake & Potomac companies (of Maryland, Washington, DC, Virginia, and West Virginia), Bell of Pa, Ohio Bell, Cincinnati Bell, Michigan Bell, Indiana Bell, Illinois Bell, Wisconsin Bell, Nevada Bell, and Pacific Tel (California) each provided telephone service within its respective state. New England Tel, Southern Bell, South Central Bell, Southwest Bell, Northwest Bell, and Mountain Bell each provided telephone service over some multistate region of the United States.

For most of this century, AT&T owned these LECs and the term BOC attests to their membership in the old Bell System. In 1984, AT&T divested its local telephone business and surrendered ownership of its BOCs but didn't release them as independents. Seven corporations were created in 1984 to take AT&T's place as the owners of these various BOCs. Each of these Baby Bells was given jurisdiction over a distinct geographical region of the United States, not to be a LEC in its region, but to be the holding company that owned the BOCs in its region. These seven Regional Bell Operating Companies (RBOCs) were: Ameritech, Bell Atlantic, Bell South, Nynex, Pacific Telesys (PacTel), Southwest Bell, (SBC), and US West. Two of AT&T's original BOCs, SNETCo and Cincinnati Bell, had not been 100% owned by AT&T and were not assigned to an RBOC. They became independent in 1984. The details of divestiture are given in Section 16.3.

So, the typical state in the United States was a patchwork quilt of geographical regions, each the serving area of some different telephone company. In every state, some regions were assigned to the state's BOC, some regions to some of the multiregional independents, and some regions to some small unaffiliated independents. While none of the contiguous 48 states lacked a BOC, none of these states was 100% served by its BOC. Ohio was the only state that ever had two BOCs — Cincinnati Bell serving greater Cincinnati and Ohio Bell serving much of the rest of the state.

These traditional LECs provided traditional twisted-pair telephone lines to subscribers within their respective chartered regions, they provided originating and terminating telephone service, and they provided the talking connection for local calls and local access for long distance calling. But in recent years, emerging technology, new regulations, marketing, and business mergers have changed the local exchange industry.

- Most of the RBOCs have eliminated the BOCs for everything except regulatory dealings with the respective states. For example, in Pennsylvania, while Bell of Pa still exists, customers perceive that they deal directly with Bell Atlantic.
- Many independents have been acquired by larger companies. SBC acquired PacTel and SNETCo and, as this book is going to press, SBC is trying to acquire Ameritech. Bell Atlantic merged with Nynex and, as this book is going to press, Bell Atlantic is trying to acquire GTE.
- The competition encouraged by the Telecom Act of 1996 (Section 16.6) has eroded (only slightly, so far) the LECs' monopoly status in a region. Even AT&T is trying to get back into local telephony.
- New technologies, like Hybrid Fiber/Coax and the many kinds of wireless (Chapter 19), have changed the nature of telephone access and significantly eroded the LECs' traditional monopoly business.

Cellular telephony has grown very rapidly and has become a significant direct competition to the LECs. While some LECs may also be cellular carriers, it is typically through a subsidiary and almost always in competition with other cellular carriers. While much cellular telephone usage is new business, the LECs have lost calls to the cellular carriers. We'll discuss cordless, mobile, cellular, and the emerging low earth orbit satellite (LEOS) systems in Section 19.3.

Competitive access providers (CAPs), also called bypass carriers (for reasons explained in Chapter 16), also provide local telephone access over their own facilities. CAPs typically lease private lines in competition with the LEC so business clients in dense areas can have PBX trunks direct to their long distance carrier without going through the LEC. But some CAPs also provide competitive access to the LECs' switching offices. We'll discuss all this further in Chapter 16.

Some community antenna television (CATV) carriers recently have begun to offer telephone service to their subscribers. Some cable bandwidth, normally allocated to television signals, is split up into many voice channels. Subscribers plug their telephones into jacks on modified set-top boxes. Some CATV carriers provide dial tone by their own switching office and some simply provide competitive access to a LEC switching office.

Because they are required to provide universal service, the LECs must complete calls for dial-up modem connections between computers and between PCs and the Internet. Revenue to the LEC is far below cost for such calls and the long connection times affect their regular telephone service. We'll explain this impact in Chapter 16. Adding insult to injury, the Internet, with its government subsidy and LEC-subsidized access price, provides Internet telephony in competition to the long distance carriers and the LECs themselves. We'll describe this in Section 19.5.

While the LECs are encountering competition in their primary business, they are also getting into other businesses. Most LECs are looking at carrying television signals in competition with the CATV carriers. LECs also offer ISDN and other digital services for better connectivity between computer equipment.

Typically, the subscribers that the LECs lose to competitors are the subscribers that make a lot of calls and are located in densely populated areas where they are easily served. This erosion of the customer base that subsidizes POTS for others will have to have an eventual effect on universal service.

Interexchange Carriers. Until the 1970s, long distance telephone calls in the United States were connected over one nationwide long distance monopoly network that was owned and operated by the Bell System. Even subscribers served locally by an independent telco used AT&T's Long Lines network for long distance calls. In the 1970s, competing parallel national networks emerged, initially called other common carriers (OCCs). Section 12.6 describes this evolution and the birth of MCI and Sprint. In 1984, long distance was defined as inter-LATA (a LATA is defined in Section 16.5), and the companies that provide this service came to be called interexchange carriers (IXCs). Toll, long distance, interexchange, and inter-LATA calling will all be carefully defined and distinguished in Chapters 9 and 16. The RBOCs were specifically precluded from the IXCs' business, and AT&T was specifically precluded from the local telephone business. But this too has changed recently.

So, in the United States now, IXCs are corporations that provide long distance telephony competitively in any region they choose to serve. End users typically do not connect directly to an IXC, but they are connected to the user's preselected IXC by the user's LEC when the user dials a long distance call. Some IXCs (like MCI) are independent companies, some (like AT&T) were affiliated with parent companies, and some (like Sprint) have been bought and sold to several owners. While the larger ones own their own transmission facilities, primarily fiber and digital microwave, they didn't always, and many IXCs lease lines from one another. While the IXCs primarily offer circuit-switched telephony, they also lease out facility-switched channels and they are starting to offer packet-switched wide area networks (WANs).

The IXC industry has also had many recent mergers and acquisitions. While many of the older IXCs got started by leasing facilities from the more established carriers, a surprisingly large number of relatively new companies have jump-started into the industry by fibering the entire country first. Despite all the new competitors, market share hasn't changed much since the mid-1980s. AT&T's Communications division has typically handled about 50 – 55% of the interexchange business in the United States. While AT&T's market share has fallen from 100% in the last 25 years, the business has changed so much and interexchange calling has grown so much that AT&T is still doing very well in this business. However, AT&T's profit from this business has fallen off because: (1) when AT&T had a long distance monopoly, price was deliberately regulated well above cost to help subsidize POTS, and (2) even though the interexchange carriers try to avoid competing on cost (they compete on gimmicks), many astute subscribers and most enterprises do select on cost.

The market has also been relatively stable with MCI at 25 – 30%, Sprint at 10 – 15%, and hundreds of others sharing the remaining 5%. The FCC and state regulatory commissions still monitor the IXC business, with AT&T being more carefully regulated than the others. The RBOC's entry into this market should be significant, especially for calls completely within an RBOC region or for targeted interregion calls (like New Jersey to New York City).

The global equivalent is similar to the U.S. model in providing network links; however, there is no true global IXC. Motorola might have provided the first one with its Iridium project, described in Section 19.3, but Iridium has not been a success. There are many transoceanic cables and satellite channels, and almost as many arrangements for owning them. But there is no international body to adjudicate nationalistic concerns and to look out for the consumers' best interests.

Consider placing an international call from Los Angeles to Budapest. The originating party in Los Angeles dials the 011 international dialing prefix, followed by the country code for Hungary, the Hungarian area code for Budapest, and the local telephone number. These numbers are signaled to a switching office in this user's LEC (assume it's PacTel). Software in this switching office determines that the call must be routed to this subscriber's default IXC (assume it's AT&T). Software in AT&T's IXC Network determines that the call must be routed to AT&T's transatlantic gateway switching office in Baltimore. Note that, while this call passes through the US West, Bell South, and Bell Atlantic regions, it does not use their facilities and they get no revenue; only PacTel and AT&T receive revenue, on the U.S. side. The corresponding process does not apply on the European side. Since the call is routed through Spain, France, Italy, and Austria, every one of these countries collects a significant toll. Europe needs the equivalent of an IXC, but the nations in Europe resist surrendering the revenues from such calls in order to lower the overall end-to-end price to the end users outside their countries.

Competition is now the norm in the United States' IXC business and it will probably become the norm worldwide. While Internet telephony has some strong advocates, many technical issues (described in Section 19.5) remain to be resolved and many of the price advantages are based on an artificial economy. Still, Internet telephony is becoming significant enough that some LECs and IXCs are having to decide whether to provide it or even whether to totally switch over to it. Low earth orbit satellite (LEOS) systems, described in Section 19.3, could be major competitors to traditional terrestrial wire-line carriers. We'll discuss the future global network in Chapter 22.

Nonvoice Carriers. Television signals are transmitted over three different types of networks.

- Local broadcasters transmit FM radio signals from large antennas. These signals are received by rooftop antennas and rabbit ears, and channels are selected at the television set or VCR. This process is problematic if the transmitting antenna is low or has low power or if the subscriber is separated by distance or hills. But the subscriber doesn't have to pay any carrier. Nationwide television networks transmit national programming to their local affiliates, over leased lines or satellite feed, so these affiliates can broadcast the signal locally.
- CATV does not stand for cable television; it stands for community antenna television. The cable company has large receiving antennas, for local broadcast signals and for satellite signals, at their head end (the architectural equivalent of a switching office). These received signals are remodulated and encrypted and transmitted over a local distribution network, typically coaxial cable, in a chartered

region. Cable carriers are granted monopoly status so that the community is not spoiled by multiple competing cable networks. The subscriber pays the cable carrier for access and needs a set-top box for reception, decryption, and channel selection. The subscriber gets improved reception of broadcast signals and access to program providers, like ESPN and HBO, which are not broadcast locally.

- In the third alternative, the subscriber can have his own satellite dish and directly receive the signals that are down-linked to the local broadcasters and to the CATV carriers. The traditional analog downlink systems, with the large dish in the back yard, are being rapidly replaced by the digital systems with the smaller window-mounted dish. Since the signals are encrypted, this party is supposed to pay a monthly fee for the right to decrypt these signals. However, this subscriber needs one of the methods above to receive local programming, since this content is not broadcast by satellite.

Television communications use the PSTN only rarely because the PSTN is optimized for voice communications, not television. A television signal requires about 1000 times more bandwidth than voice, the connection duration is typically longer than a telephone call, and television is typically broadcast while voice is typically point-to-point. While a few television signals do use transmission facilities that are leased from the LECs and IXCs, these signals would never be switched through telephone switching systems. While television communications could use the PSTN, apparently it has been more economical to construct separate networks, optimized for television communications, than to try to force television communications over a less than optimal network. We can see an early example of a general principle: while radically different types of communication might be efficiently transported over a common infrastructure, it seems to be inefficient to switch radically different types of communication through a common network. This is a very important point and we'll come back to it in Chapters 17 and 19.

Since computer communication is as different from voice as television is, the same concept has traditionally applied also to computer networks. We used to observe this principle in private internal networks, where computer communication used a LAN and voice used a PBX. More recently, however, some enterprises have moved their voice traffic onto their LANs. We also observe this principle in large networks, public and private, where computer communication uses a WAN, like the Internet, and voice uses the PSTN. Similarly, more recently, we're seeing some voice over traditional WANs, like frame relay or the Internet. While computer communication and voice have often been transported over a common infrastructure, they have traditionally been switched separately. The Internet, which has become the world's public wide area carrier for computer communication, leases facilities from the PSTN carriers, but uses packet switching instead of the PSTN's circuit switching. There are few public local area carriers for computer communication anywhere in the world. But all this may change — proposals for voice over ATM or IP are discussed in Chapter 19.

Public computer communication not only shares PSTN facilities, but it is also switched through the PSTN. In Chapter 17 we'll try to convince you that it is more efficient for voice and computer communication to be switched separately. While we don't always do what is globally optimal, in this particular case, we do switch voice and computer communication through the same local access network, but not because it is efficient. We do this, in spite of its inefficiency, because today's regulations force the LECs to provide access for dial-up modem calls, and at no charge if the call is local. Computer communications users are not going to voluntarily migrate to an alternate local access, even if it's more efficient, if they have to pay some charge they're not currently paying. The LECs are being forced by regulation to absorb this inefficiency and in this sense, voice subsidizes computer communications worldwide. During the time that computer communication was insignificant compared to voice, this was not a big financial burden. But, this is changing and — somehow, someday — the regulators worldwide will have to deal with it.

2.2.2 Telecom Equipment Manufacturers

We discuss here the industry of the companies that manufacture equipment for the voice network and its infrastructure. Four types of telephone communications equipment are manufactured: (1) central office switching equipment, (2) transmission equipment for the infrastructure of physical channels, (3) premises switching equipment, and (4) telephones.

Central office switching equipment is extremely complex, and its design, especially the software design in computer-controlled systems, requires a large up-front commitment of resources that only a large corporation can afford. While transmission equipment is less complex, competition forces these systems to lag the state of the art so closely that only corporations large enough to afford research can compete.

Before divestiture, AT&T's BOCs and its Long-Lines division purchased their switching and transmission equipment almost exclusively from Western Electric, the Bell System's manufacturing subsidiary. GT&E and IT&T had similar organizations, but on a smaller scale: their carriers purchased equipment almost exclusively from wholly owned manufacturing affiliates. GT&E had enough LEC regions that it also operated a small IXC, and GT&E's manufacturing branch was Strowger's Automatic Electric Company. IT&T operated an international IXC and manufactured equipment for sale to international LECs and U.S. independents. Canada, Germany, Japan, and several other nations had arrangements that were similar to the Bell System in the United States or, at least, operated so nationalistically that they were effectively similar. For example, Bell Canada purchased its equipment almost exclusively from Northern Telecom, Deutsches Bundespost purchased its equipment almost exclusively from Siemens, and Nippon Telephone & Telegraph purchased its equipment almost exclusively from Nippon Electric Co. The LECs in other nations and the U.S. independents purchased from the open market: in the United States mostly from GT&E, IT&T, and some unaffiliated manufacturers, like Stromberg-Carlson.

Currently, there are only a few corporations worldwide capable of making central office switching equipment and transmission equipment. While this business was nationally parochial once, with a limited market for their product and extremely high R&D costs, these corporations now compete worldwide. This market is fiercely competitive, with failures and mergers announced regularly, and more expected.

- When divestiture separated the BOCs from Western Electric in the 1980s (Section 16.3), the BOCs began purchasing from the open market. Western Electric (then renamed AT&T Technologies and now separate from AT&T as Lucent Technologies) sold more on the open market.

- GT&E dropped the & from their name, sold their switch manufacturing business to what is now Lucent, bought Sprint, and then sold it again. As this book is going to press, GTE is possibly being taken over by Bell Atlantic.
- IT&T dropped the & from their name and sold off their international IXCs and their manufacturing business (to Alcatel). ITT owns hotel chains and insurance companies, but reentered telecommunications by acquiring Western Union.
- Siemens bought Stromberg-Carlson.

The major companies that manufacture switching systems for the worldwide market are: Siemens, Ericsson (in Sweden), Alcatel, NEC, Northern Telecom (now called Nortel), and Lucent. Sales to third-world nations, especially wealthy ones, are important and this makes standardization important. These same companies also compete, along with several others, in the manufacture of transmission equipment.

Most of the corporations that make central office switching equipment also manufacture PBXs. Because this equipment is simpler and easier to develop, there are many other manufacturers as well, especially in the market with small systems. In the United States, this business was opened to non-Bell manufacturers by the FCC's Carterfone decision (Section 10.6). After 30 years, it is now well shaken out, but still highly competitive.

In the United States, the manufacture of telephones and, especially the way telephones are marketed, also changed dramatically over this same 30-year interval. Once an instrument leased from the telco, the telephone has become a consumer product, marketed like a toaster or electronic calculator. The telephone instrument has become more complicated with the addition of new features like recorded answering, cordless connection between base and handset, and mobile/cellular service. All these instruments are typically sold in consumer electronics retail stores.

2.2.3 Regulatory Structures and Agencies

In many countries, the telephone company is a part of a Post, Telephone, and Telegraph (PTT) branch of the federal government. While this arrangement might seem equivalent to the American structure of government-chartered natural monopolies, these PTTs have typically not provided an equivalent quality of service. While the old Bell System was almost the same as a government agency, Ma Bell provided better telecommunications service than any PTT ever provided — perhaps because AT&T's stockholders demanded a return on their investment. Many countries are privatizing their PTT telephone service, and, in accordance with recent world trade agreements, they are also opening their LECs and IXCs to competition.

In the United States, the telephone industry is regulated by different agencies in several branches of multiple levels of government.

- Each state has a Public Utility Commission (PUC) that regulates the electric power industry, the telephone industry, and other monopolies within its state's boundary. The PUCs tend to act in the best interests of consumers, with a heavy bias toward their short-term best interests. When electric or telephone rates are allowed to rise, the citizens tend to blame the governor and the legislators who appointed the commissioners, and they want to be reelected.

- The Federal Communications Commission is an agency in the legislative branch of the United States' federal government. The FCC regulates radio, particularly spectrum allocation, and the telephone industry, particularly interstate, and other telecom issues, like the emerging digital television standards. The FCC, for example, still regulates long distance rates, especially AT&T's.
- While most of the rules for the telecom industry come from the FCC, many rules have also come from legislation passed in state legislatures and the U.S. Congress and from rulings resulting from lawsuits tried in many different levels of the court system.
- Since some of the most significant rules affecting the Bell System came from cases in federal court where the litigant was the U.S. Department of Justice, the DoJ has effectively been one of the most significant regulators of the telecom industry.

Finally, the story of government's effect on the telecom industry is not complete without mentioning the Rural Electrification Agency (REA). While the REA publishes standards for telephone systems, it is not really a regulator. As an agency within the U.S. Department of Agriculture, the REA provided much of the investment for the infrastructure that brought electric power and POTS to rural Americans. Many small independent LECs were funded by the REA.

2.2.4 Users

Users are discussed in the context of the system, including their input/output behavior, in Chapter 15. Here we discuss users in the context of the telecom industry. The users are typically viewed in two camps: residential and enterprise. Most users want high-quality service, reliability, accountability, customizable unbundled options, and cost-based prices. This industry does not, and never has, come close to this expectation. We'll discuss this in Chapter 22.

Since most enterprises have a telecom department, enterprise users get accountability and advocacy. Enterprise users typically believe that the current competitive environment is better than the previous Bell-dominated environment. Residential users have no real advocates. Residential users typically see lower quality and higher prices and they wonder why we ever busted up Ma Bell.

2.3 EVOLUTION OF TECHNOLOGY

The give-and-take between switching system architectures and technological evolution is worth observing and understanding. The earliest and simplest technology, with which the reader is assumed least familiar, is reviewed in the greatest depth.

2.3.1 Electrometallic Technology

The most primitive electrical technology is that of metallic conductors, contacts, switches, and electromechanical relays. Electric current is carried by metal conductors — typically copper wires covered by a nonconducting insulator. If the conducting path is broken, either accidentally, or deliberately by using a switch, then the current ceases to flow. Current can also be diverted from its intended destination by a short circuit. Ends of wires, mounting posts, device leads, and the fine conducting paths on printed wiring boards are semipermanently connected by spring tension, twisting, wrapping, or soldering. Temporary connections are made through electrical contacts. Contacts are prone to corrosion and dust and can draw an arc, which can pit the surface, while opening and closing under high voltage. Less reactive metals, like gold, are typically used.

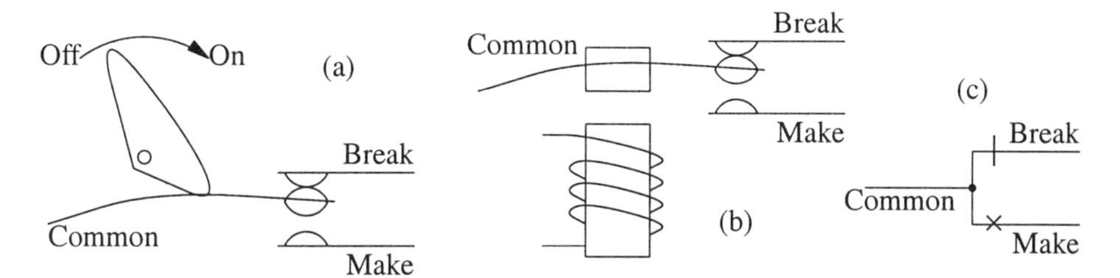

Figure 2.2. Switches: (a) manual, (b) relay, and (c) schematic convention.

A metallic mechanical switch has two functional parts: (1) a mechanical means for changing the switch's state and latching it in the ON or OFF position, and (2) an electrical contact which makes or breaks the conducting path through the switch. In an N-pole switch, conduction is transferred from the common lead to one of N other leads. Figure 2.2(a) shows a double-pole switch, also called a make/break switch or a transfer switch. In a multiple-throw switch, the same mechanism controls more than one path, making some and breaking others.

A relay, shown in Figure 2.2(b), is a switch that is operated electromagnetically. The figure shows a multicontact wire-spring relay, equivalent to a multiple-throw switch. It shows the electromagnet, the bar that mechanically moves the springs and different types of contacts. Contacts not only make, break, or transfer with the operation of the relay, but also the timing may be different to allow designing around race conditions. The nomenclature is: early-make, make, late-make, early-break, break, late-break, make-before-break, and break-before-make. If a copper cylinder is placed around the coil of the electromagnet, the magnetic field will be slow to collapse. While this does not significantly delay the operation of the relay, it does delay its release after being operated. The relay is said to be slow release.

The Bell System's convention [2] for switch and relay contacts, used throughout this book, is shown in Figure 2.2(c). The single line crossing the line of the wire represents a normally closed contact, a contact that opens or breaks when the switch or relay is operated. The × in the line of the wire represents a normally open contact, a contact that closes or makes when the switch or relay is operated. The relay's electromagnetic coil is represented by a rectangle with the name of the relay printed inside. Multiple contacts on the same switch or relay are each identically labeled with the name of this switch or relay.

Logical OR is implemented by parallel contacts and logical AND by series contacts [3 – 6]. For example, Figure 2.3 shows a circuit with four relays — A, B, C, and L. The L-relay is operated when current flows through its windings, when A is operated, and either B is operated or C is not. The figure on the left is a wiring diagram with the contacts shown close to the relays that operate them. The figure in the center is a schematic diagram showing only the relevant contacts. The Boolean expression for operating the L-relay is $L = A\ (B + C')$. Besides implementing digital logic, a relay can be latched, implementing a 1-bit memory, if its electromagnet is kept in operation by one of its own contacts. While relay logic circuits implement general logic and memory, they

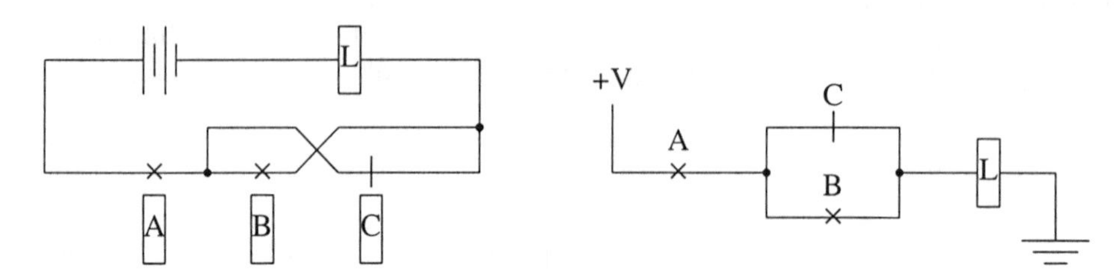

Figure 2.3. A typical logic circuit in electromechanical technology.

are especially suited for functions like priority selection chains, commutating, and switching fabrics.

2.3.2 Switching Fabrics

Spatial analog switching fabrics are built from large-motion switches, from small-motion switches, from wire-spring relays, and from semiconductor devices. They are typically implemented in coordinate or grid configurations.

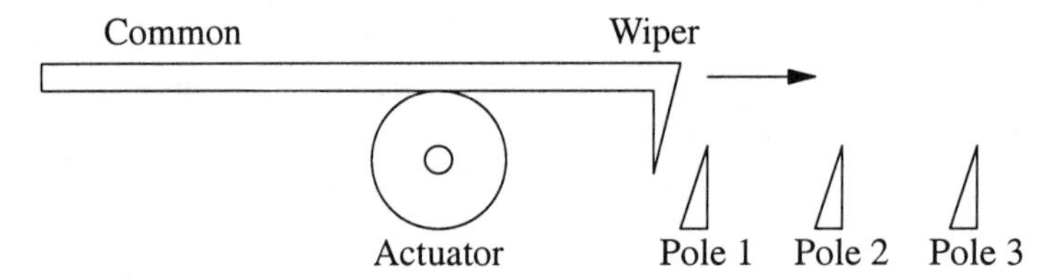

Figure 2.4. A traveling switch.

A *large-motion switch,* also called a traveling switch, is shown in Figure 2.4. As the name implies, it is characterized by large motions. It is an N-pole relay in which the common contact is mechanically carried by a wiper across the separate stationary poles. To connect the common to the jth output, the wiper must be physically moved j places. Examples are the Strowger switch used in the Step-by-Step system (Section 2.5.1 and Chapter 6) and the switch used in the fabric of the Panel system (Section 2.5.2).

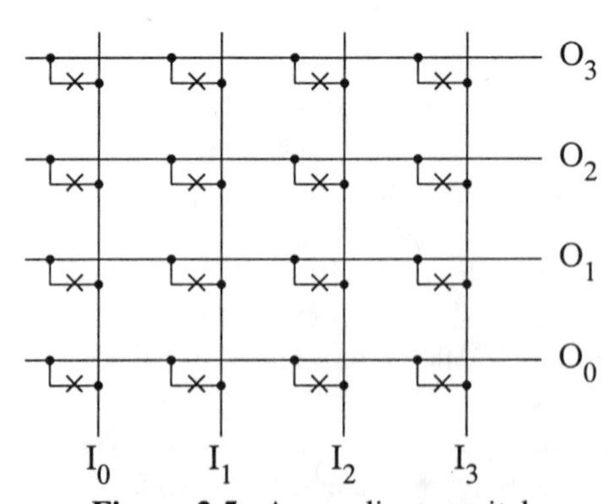

Figure 2.5. A coordinate switch.

A *coordinate switch,* shown in Figure 2.5, is characterized by small-scale motion, or no motion at all. It is described logically as two parallel planes, one with M parallel vertical conductors and the other with N parallel horizontal conductors, and MN contacts, one bridging each crossover. These contacts are called crosspoints. The coordinate switch is M-by-N, equivalent to M independent N-pole switches, with corresponding poles multipled together. To connect the ith input to the jth output, the contact at coordinate $[i,j]$ must be closed. The contacts are electromechanical or electronic switching devices. The Crossbar switch used in the fabric of Crossbar systems (Section 2.5.3 and Chapter 8) is logically a coordinate switch, but with a mechanical implementation. The fabrics of most Electronic systems (Section 2.5.4 and Chapter 11) are made from true coordinate switches.

2.3.3 Electronics Technology

The vacuum tube had significant impact in communications switching systems as a linear device used for amplifying signals. While the tube could have been used in nonlinear operation as a binary device in logic circuits, like gates or flip-flops, such usage was rare. Vacuum tubes could have been used in analog gates and then functioned as switching elements, but no such implementation is known.

The hysteretic magnetic core had great impact in the computer industry as a memory element and, as such, indirectly impacted the switching industry. Direct applications include its use as switch-hook sensing in electronic systems (Section 11.3) and in implementing hardware for number translation (Section 9.2).

The transistor replaced the vacuum tube in most applications for amplifying signals. Of greater significance, as in the computer industry, was its use in nonlinear operation as a binary device in logic circuits, like gates and flip-flops. Semiconductor devices, like the PNPN switch, the field-effect transistor, or even the bipolar transistor, have seen surprisingly limited application as switching elements in coordinate fabrics. The major advantage of metallic switches over semiconductor switches in this application is the former's ability to support the high-power signal used to ring the telephone.

The greatest impact of semiconductor transistor technology, as in the computer industry, is from integration. The evolution of the scale of integration, leading to today's custom VLSI, is responsible for cost reductions and digital implementations of previously difficult engineering problems. That modern communications systems and computing systems are made from the same silicon raw material tempted communications manufacturers, like Western Electric (when it was part of the Bell System), to enter the computer business and computer manufacturers, like IBM, to enter the communications business. The increase in computer communications further emphasized that these two industries may be evolving into a single large one. This predicted confluence partially fueled the regulatory events of the 1980s in the United States.

2.3.4 Digital Electronics

Figure 2.6 shows block diagram representations for two elementary devices needed for the communications switching of digital signals. On the left is the digital shutter, which acts like a simple switch contact. The input data x to the digital shutter is a binary-valued signal, and the digital shutter is controlled by a one-bit code c. When $c = 0$, the input is disconnected from the output, and when $c = 1$, the input is connected to the output. Thus, the binary-valued output y is a logical-0 when $c = 0$ and duplicates the logical value of x when $c = 1$. Output y is a Boolean function of c and x and has four logical cases: $y = 0$ when $cx = 00$ or $cx = 01$ because $c = 0$ causes $y = 0$ regardless of the value of x; $y = 0$

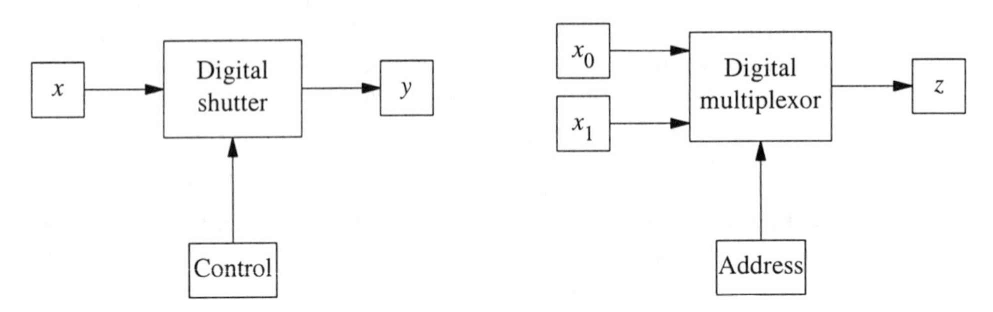

Figure 2.6. Two basic digital switching elements.

when $cx = 10$ because $c = 1$ causes $y = x = 0$; $y = 1$ when $cx = 11$ because $c = 1$ causes $y = x = 1$.

The functional behavior of the digital shutter is identical to that of logical AND. The digital shutter is implemented in digital electronics by a single two-input AND gate. Since the inputs to a two-input AND gate are equivalent, x and c can be connected to either gate input. In Figure 2.6, the x and c signals can be interchanged. While this may seem like much ado about a simple concept, the philosophical implications are enormous. The basic fundamental building block for digital communications and the basic fundamental building block for digital computing are the same device.

The digital multiplexor (mux), on the right of Figure 2.6, is an extension of the digital shutter to multiple inputs. The mux has N different digital input signals: $x_0, \cdots, x_{N-1}$. The mux is controlled by a K-bit address, $a_{K-1} \cdots a_0$, where $N = 2^K$. If the K-bit address equals the binary code of n, where $0 \le n \le N-1$, then mux output z is connected to input x_n. That is, z has the same binary value that x_n has. A two-input digital multiplexor is controlled by a 1-bit code a_0. An address of $a_0 = 0$ causes output $z = x_0$ and an address of $a_0 = 1$ causes output $z = x_1$. Its implementation is discussed first as the logical OR of N shuttered inputs and second as a general design of a digital circuit.

The implementation of a two-input digital mux requires a digital shutter on each input. The digital shutter on x_1 is controlled by a_0 directly. Its output $y_1 = 0$ when $a_0 = 0$ and $y_1 = x_1$ when $a_0 = 1$. The digital shutter on x_0 is controlled by the logical complement of a_0. This shutter's output $y_0 = x_0$ when $a_0 = 0$ and $y_0 = 0$ when $a_0 = 1$. The digital multiplexor's output z is simply the logical OR of the two shutter outputs y_0 and y_1. When $a_0 = 0$, the mux output $z = y_0 + y_1 = x_0 + 0 = x_0$. When $a_0 = 1$, the mux output $z = y_0 + y_1 = 0 + x_1 = x_1$.

An alternative procedure yields the same implementation. The canonic sum-of-products expression of any Boolean function implies a canonic or-of-ands implementation in digital electronics. The implementation follows from the one-points of the Boolean function. The two-input digital mux has three binary inputs — a_0, x_1, and x_0. The output z equals logical 1 in four of the resulting eight cases. Connecting z to x_0 when x_0 equals logical 1 (the value of x_1 is irrelevant) is covered by $a_0x_1x_0 = 001$ *or* 011. Connecting z to x_1 when x_1 equals logical 1 (the value of x_0 is irrelevant) is covered by $a_0x_1x_0 = 110$ *or* 111. The minimal or-of-ands implementation requires two two-input AND gates that feed one two-input OR gate. An AND gate with inputs $\bar{a}_0 x_0$ covers the 001 and 011 cases. An AND gate with inputs $a_0 x_1$ covers the 110 and 111 cases. Since a digital shutter and a two-input AND gate are identical, the two implementations are identical.

A general *N*-bit digital multiplexor has one canonic form that includes a separate address decoder and has a second canonic form in which the decoding is integrated. A digital decoder is a device with *K* input wires and *N* output wires, where $N = 2^K$. A *K*-bit address is applied to the input and exactly one of the *N* outputs has the value of logical 1; a different output has a 1 for each of the *N* combinations of the *K*-bit input. An *N*-bit mux, with separate decoding, has a two-input AND gate (digital shutter) on each data input wire. One gate input to the *n*th AND gate comes from x_n and the other comes from the *n*-output of the decoder. An *N*-bit mux, with integrated decoding, has a (*K*+1)-input AND gate on each data input wire. One gate input to the *n*th AND gate comes from x_n and the remaining *K* inputs come from the appropriate combination of each of the address wires or their complements so that these *K* inputs are all 1s when the address is the binary code for *n*.

A spatial digital fabric, which switches digital signals and is controlled by digital signals, has a two-input AND gate for every coordinate. The *i*th input to the fabric is wired to one of the gate inputs in each of the *N* gates in the *i*th column of the fabric. The other gate input to each gate is for the digital control signal for the [*i,j*]th crosspoint. Each of the outputs of the *M* gates in the *j*th row is connected to different input of the *j*th *M*-input OR gate. The output of this OR gate is the *j*th output of the fabric. If control lead [*i,j*] = 1, then output *j* has the digital value of input *i*. Each row of the digital fabric is seen to be an *N*-input digital multiplexor.

In both the analog and digital coordinate fabrics, a given input wire may be multipled to many crosspoint contacts or to one input of many different gates. If contacts [*i,j*] and [*i,k*] in the analog fabric are operated, or if digital control leads [*i,j*] and [*i,k*] both = 1 in a digital fabric, then input *i* is fanned out to outputs *j* and *k*. This works in electrical technology as long as analog loss is within acceptable limits. This works in digital technology as long as the fan-out of input *i* is within the specifications of the logic family. However, analog electrical technology and digital electronic technology are not equally amenable to the corresponding multipling of the output wires. In the analog electrical fabric, output wire *j* may be multipled to many crosspoint contacts. If the contacts at coordinates [*i,j*] and [*h,j*] are operated in the analog fabric, then output *j* is the analog sum of analog inputs *i* and *h*.

Some digital logic families allow a limited number of gate outputs to be multipled together. Usually logical 0 prevails and multipling *n* gate outputs together is equivalent to connecting them to an *n*-input AND gate. If digital control leads [*i,j*] and [*h,j*] are both = 1 in a digital fabric, then output *j* is the logical AND of inputs *h* and *i*. While the analog output may be intelligible as a conference call, the digital output is not. However, many digital logic families don't allow the multipling of gate outputs, especially for large numbers of gates. In these families, the OR gate is necessary for each fabric output. If the logic family provides a gate, modified for tri-state operation, then the gate outputs can be multipled and each fabric output resembles one wire of a bus. Even with tri-state gates, only one gate is allowed to be operational at a time and summing still does not occur as in analog technology.

2.3.5 Photonics Technology

A new technology, photonics, is now emerging. The conventional wisdom regarding this new technology is that it offers a means for improving performance of conventional architectures. In particular, high data rates are possible on point-to-point links. But, we believe that this technology has much greater significance.

> *Photonics will do to telecommunications what the transistor did to computing: it will lower the cost, raise the performance, make the availability almost universal, and dramatically affect the everyday life of the entire human race.*

We spend so much effort in this book on the relationship between technology and architecture because we believe that photonics technology will eventually cause us to re-architect switching systems.

Consider the history of the architecture of building walls. When bricks were made of dried mud, the weight of the wall would crumble the bricks at the bottom and walls were necessarily short. Adding straw to the mud, kiln-fired clay, cinder block, concrete, and reinforced concrete were a succession of new technologies that enabled a stronger brick and allowed walls to be higher. Then, along came steel. An architect with a small mind conceives of a brick made of steel and figures out how high a wall can be built. An architect with an open mind thinks: "Maybe I should think about a different way to build a wall?"

2.3.6 Software Technology

In hardware systems, built from electromechanical or digital electronic technologies, the system's input/output response and intelligence are determined by how the various devices are wired together. This changed fundamentally in the mid-1960s with the introduction of stored program control as the means of implementing system intelligence. While software systems are still built with digital electronics technology, the components are either digital computers or computer peripherals that are directly controlled by digital computers.

Not only has switching software closely followed the state of the art of software science, many of the advancements in software science have resulted from research and development of switching software and by the efforts of scientists and engineers employed by various telecommunications companies. The Bell System had been making digital computers, for internal use as controllers of switching equipment, for 15 years before it was allowed to market them commercially. Early switching software was written in assembly language and scheduling of program segments was controlled directly by the program designers. The earliest switching software was implemented before the invention of the modern operating system. More modern switching software is written in a high-level language and for an operating system that supports multiprocessing. The most modern systems use many processors instead of one centralized processor and, so, the software environment is partitioned into multiple machines, with communicating processes on communicating processors.

Three advantages that stored program control has over wired logic control are:

- Complex features are technically and economically feasible;
- Changes and new features are implemented by rewriting a program instead of rewiring a system;
- Such changes are implemented in *N* offices by distributing *N* copies of the new program instead of rewiring the *N* offices.

The complexity of switching software surprises even the most experienced of program designers and always has. As more and more new features have been added, this complexity has multiplied because of feature interaction in the software. The development of modern switching software typically requires hundreds and even thousands of programmers. The problem of crowd control further compounds the already complex task.

Many of the large telecommunications manufacturers are very interested in software development environments, and experts believe that the company that best manages software development is the one that wins in the marketplace. Software development has been described as a monster with which all manufacturers must live. The company that finds the silver bullet will slay the monster and capture the market because of lower cost and rapid delivery of new features.

2.3.7 Reliability

Electromechanical devices wear out after a certain number of operations. One effect is that friction takes its toll on moving parts. Large motion switches wear out sooner than small motion switches. Another effect is that, if two open contacts have different voltages, as the contacts approach each other a spark jumps before the contacts physically touch. This spark pits and corrodes the surface of the contact a tiny bit more with each operation of the switch or relay. Electronics technology, with no moving parts nor spark gaps, is inherently more reliable.

Certain faults will affect only one call and may exist in a system for a long time until discovered by routine maintenance procedures or by a pattern of customer complaints. Other faults could make an entire switching office inoperable. In a well-designed system, these are extremely infrequent. Such faults are random and independent over multiple offices. While this may seem so obvious as to be silly, we emphasize that if two switching offices were built from identical technology and at the same time, the occurrence of a random fault in one office does not suggest that the same fault will exist in the other.

Impact by Software. Another advantage to stored program control is that the software can include programs that automatically test the equipment, quarantine faulty equipment, and aid in its repair. By knowing the failure rates of components and by appropriately scheduling the fault testing programs, the reliability of a switching system can be statistically controlled. Typical specification for switching systems is one hour of down time every 40 years.

But stored program control has brought with it an extremely subtle and insidious side effect [7]. Software can be faulty. Instead of wearing out, the faults are design bugs that the designers made and that were not detected in the testing and verification process during the manufacture of the software. While design bugs occur in hardware also, they are usually detected during development and are seldom encountered in the field.

Because software is so much more complex than hardware, testing is computationally more difficult and is less likely to find all the bugs. Usually these unfound bugs lurk in the seldom-traveled paths in the program. We frequently find software design bugs in the field. The insidiousness occurs when a program is released. It is distributed to all switching offices made from the same equipment model. If a particular set of circumstances causes this seldom-traveled path to be traveled, the system crashes. If this set of circumstances is observed nationally, then every system in the nation crashes. Software faults are neither random nor independent.

Impact by Photonics. As we'll discuss in detail in Chapters 20 and 21, one of optical fiber's most useful characteristics is its extremely high transmission capacity. Optical fibers can be highly multiplexed so that one fiber can carry the same number of telephone channels that used to require many copper or radio links. Since its introduction in the 1980s, optical fiber has caused a reduction in the number of physical channels in the PSTN's national network. While this does effect a beneficial cost savings, it has the added effect that each of these fibers becomes more strategic. As the PSTN's infrastructure has evolved over the last 20 years, it has fewer alternate routes and is more vulnerable to a cut fiber. In the last 10 years, the United States has experienced major outages that wouldn't have been so disastrous when transmission was more dispersed.

2.4 EVOLUTION OF ARCHITECTURE

Driven by the march of technology, changing regulatory action, improved understanding of theory, changing applications, and market forces, the architectures of communications systems have evolved over the years in many ways. The two areas of evolution discussed here are the system's overall structure and the control of the system's switching fabric. System structures have evolved in their granularity and in the implementation of their intelligence, and fabric control has evolved in its placement and its method.

The reader is warned that this section is general and philosophical. But the concepts presented here are important for understanding the differences among the various systems and the architectural principles that have been learned over the years.

2.4.1 Switching System Granularity

This interesting generalization of system architectures appears to have a cyclic evolutionary pattern. Systems have evolved through a sequence of four stages and appear to have evolved back to the first stage again. Continued evolution to the second stage appears on the horizon.

A *modular* architecture has an iteration of large, semi-identical, semi-independent subsystems that are loosely connected through a simple fabric. Each module is a small version, or microcosm, of the whole system. A manual office, described in Section 1.5, is modular. The operator and switchboard represent a module and the multipling of lines and trunks to several switchboards represents the simple fabric. The simplest case, an office with one switchboard and no multipling, is common. More recently, some modern digital offices also have a modular architecture, in which several semi-independent switching modules are connected together through a common center-stage subsystem. While the manual office features distributed modules in a bus network and the modern offices feature hierarchical modules in a star network, both are modular.

A *cellular* architecture has an iteration of small, semi-identical, mutually dependent elements that are tightly connected through an underlying fabric. It is more finely grained than a modular architecture. An individual cell, like an individual cell in the human body, is not a microcosm of the whole and cannot exist alone. A Step-by-Step office, described in Chapter 6, is built from a complex interconnection of Strowger switches. In the future, VLSI technology and wafer-scale integration may lead to systolic arrays, or cellular automata. There may be a future application in switching systems, particularly with optics (Chapter 20).

A *lumped* architecture has a collection of different subsystems with specialized functions connected together and carefully orchestrated. It is the traditional system (Section 1.1), like the collection of organs in the digestive tract of the human body, each with a specialized function but all working together.

A *centralized* architecture has a collection of many simple subsystems connected, in a star network, to a complex common node. It is the complement of a modular architecture, where the node is simple and the end points are complex. Examples from the human body are the circulatory system centered around the heart, the nervous system centered around the brain, and the hand with fingers centered around the palm. This architecture was effective in the early computer era, when processors and memory were expensive. One computer would reside at the hub of a system, surrounded by "dumb" peripherals.

As in the discussion of network architectures in Section 1.4, a given system may have different architectures of granularity at different levels or layers of its structure. For example, a modern digital switching office may have a modular architecture in which each switching module is a fully functional microcosm of the whole. But, each switching module may have a centralized architecture, with dumb peripherals controlled by a common switch processor, within each module. In addition, while a centralized architecture has many dumb peripherals all controlled by a central processor, the collection of peripherals has a lumped structure. At the highest level of architecture, all switching offices have a lumped structure when viewed as a collection of subsystems that provide main frame crossconnect, power, heating and cooling, transmission facilities on the loop and trunk sides, billing equipment, administrative access, and fabric/control. All the examples in the preceding paragraphs pertain to the fabric/control subsystem in a switching office.

2.4.2 Implementation of System Intelligence

Granularity describes how intelligence is distributed among the components of a system. In this section, we describe the technology by which that intelligence is implemented.

There have been three significant implementations of intelligence in switching systems:

- *Human,* where an operator provides the control, as in manual systems;
- *Wired logic,* where control sequence and data formats are generated according to how relays and wires are connected, as in electromechanical systems;
- *Stored program,* where control sequence and data formats are generated through computer programs and stored binary information, as in computer-controlled systems.

2.4.3 Fabric Control Placement

Control of the switching fabric can reside directly with the subscriber or indirectly with some intermediate operator or equipment. Indirect control can be exerted from a register or some other form of centralized common equipment.

Direct control. The switching fabric can be controlled directly by the subscriber. Switches are operated directly by dial pulses generated by the subscriber's telephone, avoiding intermediate storage or processing of the dialed digits. Philosophically, in a direct-controlled cellular system, each cell contains enough intelligence to interpret a dialed digit and establish a connection to a next cell, which interprets the next digit. This intelligence in a cell is only used during dialing and is idle for most of the duration of a call. Modern self-routing fabrics (Section 7.5), used in some prototype packet systems, are reminiscent of direct-controlled fabrics. The fabric control is in a packet and is established per-packet instead of from the subscriber and established per-call, but the control concept is similar.

Register control. In communications switching terminology, a register is a piece of equipment, or subsystem, that provides intermediate storage for dialed digits. In register control, after receiving all the digits from the subscriber, the register proceeds to interpret and process these digits, and then it controls the fabric. The manual office is representative, where the operator performs the function of the register. Evolution from direct control to register control involves extracting the intelligence from the cells in a direct-controlled cellular system and collapsing the cellular system into a fabric subsystem. The extracted intelligence appears in lumped registers that are scheduled to operate on a call only during dialing and to drop off the call after they have established the desired connection through the fabric.

Common control. This philosophy of extracting intelligence evolves one step further. In a register-controlled system, the intelligence for interpreting and processing the dialed digits and controlling the fabric is resident in the registers. It is idle during the collection of the digits and is only used after dialing is completed. In a common-control system, this intelligence is extracted from the registers, leaving them only enough intelligence to collect digits and determine the end of dialing (nontrivial), and deposited in common equipment, typically a processor in a system with a centralized architecture.

2.4.4 Fabric Control Methods

The previous section described three places to find the intelligence for controlling the fabric. This section describes two methods by which that intelligence controls the fabric.

In *progressive* control, the connection through the fabric is established one layer, or stage, at a time. This method is obviously associated with direct control, where each dialed digit establishes a connection through a corresponding stage of the fabric. But even registers or other common equipment could exert their control over the fabric similarly, as in the register-controlled Panel system (Section 2.5.2). Self-routing fabrics are progressive controlled.

In *end-marked* systems, a lumped subsystem or processor-resident program determines the set of switch closures in the fabric required to connect a call. If progressive control is serial, then end-marked control is parallel. The computation is based on knowing the two ports, or endpoints, on the fabric that must be connected and on the other connections currently established in the fabric. All these switches are then closed simultaneously, in parallel. A manual office is representative.

2.4.5 Cost

The cost of an air conditioner, automobile, or switching office has two fundamental components: first cost and continuing cost. First cost is usually dominated by the cost of purchasing the equipment. While we may purchase an air conditioner with a single lump payment, we typically purchase an automobile with a loan that is paid off in monthly installments. Similarly, the purchase price of telecommunications equipment can be translated into an equivalent annual cost using an annuity formula based on the expected lifetime, regulated depreciation, predicted inflation, and the company's rate for stock dividends. However, the purchase price is typically viewed as a first cost. Other, less relevant, first costs include staff overhead, sales tax, space allocation, shipping, and installation. Continuing costs for telecommunications equipment are typically partitioned into three categories: operation, administration, and maintenance (OA&M). Operating expenses include electric power, heat and air conditioning, and space allocation. Administrative expenses include assignment and utilization bookkeeping, installing new subscribers, allocating and reallocating resources, and special procedures (like tracing calls for the police). Maintenance includes preventative exercising, handling complaints and trouble reports, fault testing, diagnosis, and repair.

During the past 100 years, equipment costs have fallen and people costs have risen. Both effects mean that the net cost of equipment has gradually changed from being dominated by the first cost of purchase to where OA&M has become very significant. This subsection examines the relationship between first cost and architectural style. The total cost picture also includes the cost of OA&M.

Before proceeding, we must distinguish unit cost and incremental cost. Consider building a new home. Let the land cost $100,000 and suppose the general contractor charges about $10,000 for each room. Consider a graph of total cost versus the number of rooms. It would be a straight line, beginning at $100,000 for a null house and proceeding to the upper right with a slope of $10,000 per room. From this graph, a 10-room house would cost $200,000. The incremental cost is the localized slope of the graph. So, the incremental cost of a 10-room house is $10,000 per room, and a constant in our example. The unit cost is the slope of the line between a given point on the graph and the graph's origin. So, the unit cost of the 10-room house is $20,000 per room, but changes dramatically from $110,000 per room for a one-room house to $15,000 per room for a 20-room house.

Figure 2.7 shows three graphs for first cost of equipment versus the number of lines. The graphs represent systems with cellular, centralized, and modular architectures. In a cellular architecture, the system cost is related almost solely to the number of lines. While the graph in Figure 2.7 shows a linear relationship, a truer representation would be some weighted sum of linear and logarithmic components. We would expect the incremental cost of a large cellular system to be less than the incremental cost of a small one. But except for a small cost of common equipment, incremental cost and unit cost are virtually identical in cellular architectures.

In a centralized architecture, even a null system with no lines requires the relatively large expense of the common equipment. Then, from this starting point, total cost increases as the number of lines increases. Since system intelligence is distributed throughout the cells of a cellular system and is concentrated in the common equipment of a centralized architecture, we expect the incremental cost of a centralized system to be

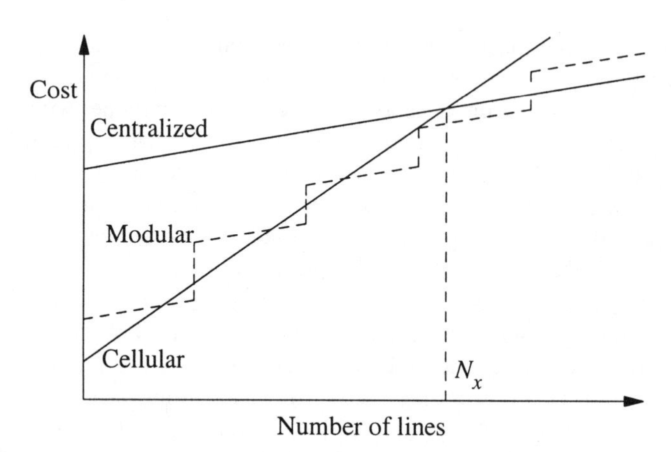

Figure 2.7. Cost of equipment versus office size.

less than the incremental cost of a cellular system. At some number of lines, N_x, a cellular and centralized system cost the same. If the number of lines $n < N_x$, the centralized system costs more than a cellular system because the cost of the common equipment is shared by relatively few lines; the unit cost is high. If the number of lines $n > N_x$, the centralized system costs less than the cellular system because the costs of the common equipment are shared by relatively many lines; the unit cost is low.

A modular architecture exhibits the characteristics of the cellular architecture on the scale of modules and the characteristics of the centralized architecture on the scale of lines within each module. A modular system with one module and no lines has less common equipment than a null centralized system because a null module is a smaller piece of equipment than the large common equipment for a centralized system. But, such a modular system has more common equipment than a cellular system because of the presence of the null module. With one module, cost increases as the number of lines increase at about the same incremental cost as a centralized system until the number of lines fills the capacity of a module. Then a second module must be added, causing a large step increase in cost. The cost characteristic for a modular system resembles a staircase where each riser is the cost of another module and each step has an incremental slope. Frequently, the first riser may be larger than the others because the module interconnection network is purchased when the second module is purchased. The graph for a modular architecture can cross over the graph for a cellular architecture several times. It can also cross over the graph for a centralized architecture several times. In general, cellular architectures are cheapest in small sizes, centralized architectures are cheapest in large sizes, and modular architectures are cheapest in between. The cost characteristic of a lumped architecture is more complex to analyze, but typically falls in between the characteristics of the centralized and modular systems.

Deferring cost usually has great benefit. Consider a community of 2000 telephone lines that is expected to grow rapidly — doubling every five years for 15 years. If the telco that serves this community decides to purchase new switching equipment, it would not equip for 16,000-line capacity in the first year. It would equip for 2500 lines in the first year and plan for growth, purchasing additional equipment as it is needed. When the system has grown to 16,000 lines, after 15 years, it would be difficult to identify its original first cost. Complicating this issue, equipment from one manufacturer may be easier to grow than equipment from another. So deciding on a brand of equipment from competing manufacturers would be a complex problem.

2.4.6 The Partition

Related to grow-ability, is change-ability. In past times, the telephone system was well defined and changes, however minor, required years to implement. In the current environment, the telephone subscriber is so bombarded by so much change that confusion reigns. This is very exciting and is probably good for the consumer in the long run, but change has cost. Some architectures are inherently less expensive to change than others and the equipment of some manufacturers is inherently less expensive to change than that of other manufacturers. The main motivation behind stored-program control is that change is thought to be easier to implement than in wired-logic systems.

Signaling is the communication of information about a call, distinguished from the information carried by the call. In the initial systems, the call itself, the control of the call, and the call's signaling were all tightly integrated. But, as we study the evolution of technology, architectures, and systems, we'll see that the call itself, its control, and its signaling have become gradually more separated. It is important to observe this gradual separation as we proceed through the book. I call this evolving separation the partition.

2.5 EVOLUTION OF SYSTEMS

All switching systems were not created in the first seven days after Bell's invention of the telephone. Over 100 years, with advancing technology, increased understanding of the theory, and a changing market, new systems appeared. Some displaced older ones and others failed to do that because the properties of the older ones had been underestimated; a true Darwinian evolution based on natural selection and survival of the fittest. This section summarizes these systems, places them historically, and characterizes them with respect to the technologies and architectural properties described in the previous sections.

TABLE 2.1. Characteristics of switching systems

<table>
<tr><th>System</th><th>Man</th><th>Step</th><th>Panel</th><th>#1XB</th><th>#5XB</th><th>Electr</th><th>Digital</th></tr>
<tr><td>Granularity</td><td>modul</td><td>cellu</td><td colspan="3">lumped</td><td>central</td><td>modular</td></tr>
<tr><td>Intelligence</td><td>human</td><td colspan="4">wired</td><td colspan="2">programmed</td></tr>
<tr><td>Fabric</td><td>cords</td><td colspan="2">traveling</td><td colspan="2">small-motion</td><td>relay</td><td>electr</td></tr>
<tr><td>Control loc</td><td>regstr</td><td>direct</td><td colspan="2">register</td><td colspan="3">common</td></tr>
<tr><td>Cont meth</td><td>endmk</td><td colspan="3">progressive</td><td colspan="3">end-marked</td></tr>
</table>

Table 2.1 tabulates manual and automatic systems against their characteristics. The fabric technologies tabulated in the center row were described in Section 2.3. The architectural characteristics tabulated in the other four rows were described in Section 2.4. While the systems in these columns are briefly described in this section, some are described in detail in later chapters.

2.5.1 Step-by-Step

The Step-by-Step system is the first automatic telephone system. It was invented in 1889 by Almond Strowger. The first installation was in 1892 by the Strowger Automatic Electric Company. The first Bell system installation was not until 1919, but by Strowger Automatic Electric. The first Step-by-Step office that Western Electric manufactured and installed was in 1921.

Step-by-Step is a cellular system with wired logic. Each cell is the traveling electromechanical switch invented by Strowger. The Strowger switch and Step-by-Step's architecture are described in detail in Chapter 6. The mechanisms in the Strowger switches that establish network connections are driven directly by the subscriber at the dial telephone, one dialed digit at a time; in direct, progressive control.

Initially intended to replace manual offices, Step-by-Step has been a survivor, despite many "Stamp Out Step" campaigns by switch designers and telcos. Even as late as 1965, more than half the telephones in the United States were served by Step-by-Step offices and new Step-by-Step offices were still being installed. By the 1980s, new installation ceased in the United States, and the percentage of telephones served dropped to about 25%. But, about 40% of the individual offices in the United States were Step-by-Step. Step-by-Step offices are typically small, mostly owned by small independent telcos. While still quite popular internationally, Step is practically extinct in the United States. Step-by-Step is the classical cellular office, economical in small sizes.

Step-by-Step offices are relatively expensive to maintain and difficult to provision for new services (even Touch Tone is difficult to provide), but firmly entrenched as existing plant and still economical in small offices. Partly because Step-by-Step is still common (globally), but mostly because its design is rich in useful concepts, Step is described in detail in Chapter 6. The architectural concept has reappeared in packet switching systems in what are now called self-routing networks, as described in Section 7.5.

2.5.2 Panel

The Panel system was the next automatic system. Panel was invented by engineers at AT&T and its first installation was in 1930. The economics of granularity-versus-size, shown in Figure 2.7, was understood then, and Panel was specifically designed for application in large offices.

The panel switch has a rectangular array of contacts on a flat panel, mounted perpendicular to the floor, and a set of wipers carried by long vertical bars that extend below the panel. Switch outputs are multipled across a row of contacts and each switch input is wired to a wiper. Input i connects to output j by causing the ith bar to move up to its jth position so that its wiper touches the contact there. A bar moves when a clutch pushes it against a continually rotating roller, analogous to the operation of early electric typewriters, — truly a traveling switch. Panel offices had the reliability of such mechanical devices; rollers and clutches wore out and required frequent maintenance. The technology was appropriate for the number of operations that telephone systems typically performed at the time that Panel was designed, but proved inadequate for the number of operations that Panel offices would be required to perform. The rapid growth of telephone service outgrew this clutch-and-roller technology.

Panel's contribution was in being the first lumped granularity, indirect-controlled telephone system. Immediately after going off-hook, the subscriber is connected to a register through the same fabric that connects the subscriber during calls. The subscriber dials the complete number of the called party into this register, which then interprets these digits and causes the connection through the fabric. This interpretation addresses the rigidity of Step-by-Step. The register exerts its control over the fabric by starting a bar in motion and counting the contacts as they are touched by the wiper, a method called revertive pulsing. Panel, like Step-by-Step, has wired-logic and its control is progressive, even though it is indirect via a register. If another office, even a non-Panel office, must complete a call into a Panel office, then that first office must perform revertive pulsing. This requirement was an unfortunate mistake because revertive pulsing came to be required in all later generations of switching equipment.

A Panel office requires a large overhead of common equipment like registers, and mechanisms like motors and rollers, but has a small cost of per-line equipment. Consistent with the cost curve of Figure 2.7, Panel was economical (compared to Step-by-Step) for large exchanges, like the famous Pennsylvania 6 office in New York City. By 1965, the few remaining Panel systems served only about 10% of U.S. telephones. By the early 1980s, Panel was extinct. Those Panel offices that were not replaced by Crossbar were replaced by Electronic Switching Systems. The Bigelow exchange in Newark, NJ was the last Panel office, going out of service in the late 1970s. A single Panel switch is displayed in the Smithsonian Institute. Panel has some good general concepts, like lumped granularity and register control, but it has some bad ones, like requiring revertive pulsing in all neighboring offices, and it was implemented in a poor technology. This book has no detailed discussion of Panel because it is extinct and because Panel is architecturally similar to Crossbar, which is discussed in detail.

2.5.3 Crossbar

There were three significant species of Crossbar offices in the United States, called Number k Crossbar (XBk), for k = 1, 4, and 5. All were designed at Bell Laboratories. Number 1 Crossbar was first installed in 1938 and had modest success as an alternative for Panel in new large offices. Number 2 Crossbar was designed, but never manufactured. The next Crossbar system was designed as a toll office, also called a class-4 office (Chapters 4 and 9). It would be the first automatic toll equipment, intended to replace the then current generation of manual equipment, the 4A Toll Switchboard. With all these 4s associated with toll switching, the numeral 3 was deliberately skipped and this system was called Number 4 Crossbar. Number 5 Crossbar was first installed in 1948 and was highly successful and popular as a mid-sized local office, just in time for the growth of the suburbs in the United States.

The Crossbar switch was invented in Sweden, and Crossbar systems were installed in Europe before they were in the United States. It is a 10-by-20 coordinate switch, made in electromechanical technology, but it is physically small. Most important, the small motions of the switch's mechanisms allow many operations with high reliability. The Crossbar switch and Number 5's architecture are described in detail in Chapter 8.

Number 1 Crossbar has a similar architecture to Panel, even to the detail of retaining revertive pulsing. The major difference is that the fabric is built from a small-motion switch, bringing a corresponding increased reliability over Panel. Thus XB1 has lumped granularity, wired-logic intelligence, and progressive register control, all like Panel. About halfway through the period when many large offices were being installed, Number 1 Crossbar replaced Panel as the system of choice.

An immediately noticeable distinction of the architecture of Number 4 Crossbar is that the registers are not connected through the system's fabric. This results from a problem unique to Step-by-Step offices where the digits provided to the toll office are a continuation of the subscribers dialing the telephone. The register must be connected in the interdigit interval, and system requirements do not allow enough time for a connection through a general fabric. Number interpretation is more complicated, and changes more frequently, in a toll office than in a local office. Thus, the designers chose to extract some of the intelligence that performs this interpretation from the individual registers and into more common equipment that each register can access when necessary. This first step in architectural evolution toward common control simplified the equipment in each register, of which there are many in an office, and concentrated the equipment that is most frequently changed. Only several hundred XB4 offices were installed simply because there were not many toll offices in the United States. Most of these systems were subsequently replaced by their electronic counterpart, ESS Number 4 (4, again).

This extraction of intelligence from the registers evolved further in the design of Number 5 Crossbar. Here there are two different types of common equipment, both called markers. One type assists a register between the time that a subscriber requests service and receives dial tone, and the other type assists the register in recognizing the end of dialing and in interpreting the dialed digits. XB5 has common control, but still with lumped granularity, along with the properties of wired-logic intelligence and end-marked control of its fabric. With the nationwide trend of people moving to suburbs in the United States during the 1950s, the telcos went through a period when mid-sized offices were installed in these suburbs; and Number 5 Crossbar was the system of choice. Step-by-Step remained the system of choice in small, predominantly rural, offices.

After 1965, XB1 was no longer installed in new offices, and, like Panel, XB1 was extinct by the 1980s, having been replaced by electronic systems. By 1965, Crossbar offices served 40% of U.S. telephones — 15% by XB1 and 25% by XB5. By 1990, this figure dropped to about 15%, almost all XB5. Now, even Number 5 Crossbar is also virtually extinct in the United States and there are no new installations of any species of Crossbar office. I thought one Crossbar system should be detailed in this book and, since Number 5 was most innovative and popular, it is described in Chapter 8.

2.5.4 Electronic

Computer-controlled telephone systems were born in the 1950s in the Research Area at Bell Laboratories. A prototype system, called ESSEX, was beta-tested in 1960 in Morris, IL. However, the progression from a research prototype to a production system was much more difficult than expected. The preliminary development of the system required breakthroughs in processor design, programming languages, compiling, real-time scheduling, and other efforts that have since become whole branches of the discipline now known as computer science. The first Electronic Switching System, or 1ESS, was installed in Succasunna, NJ in 1965. The conversion of the Pennsylvania 6 office in New York City followed shortly afterwards. Later similar designs include ESS #2 and #3 and similar products from other vendors, but none was as successful as the original #1 and its successor, #1A.

The primary motivation for a computer-controlled switching system was the observation that change was expensive and the belief that change would be simplified if control were implemented in software. Rather than simply replace the markers in Crossbar offices with computers, which was suggested, a new system architecture evolved in which the concept of centralizing intelligence was taken to its extreme. Like Number 5 Crossbar, the fabric control of 1ESS is common and end-marked and the fabric is a metallic-contact switch that connects analog signals. But the fabric is co-ordinate, not small-motion, and the system is architected as a central computer surrounded by simple peripherals that are incapable of independent operation. The equipment is wired together, but control intelligence is extracted from the wiring and implemented in software in the computer. As computers evolved, the detailed design of these systems has changed over 20 years. 1ESS evolved to a more modern version, the 1A ESS. But system architectures have not changed much, except that some intelligence, implemented in microprocessors, has migrated into the peripherals.

Computer-controlled systems were designed for the large and mid-sized office, replacing existing Panel offices and as an alternative to Crossbar in a new office. By 1985, these successful and popular systems, predominantly the 1A ESS, were serving about 50% of U.S. telephones. The general concepts of process control computers and 1ESS are described in detail in Chapter 11.

2.5.5 Digital

The first Digital Switching System was Western Electric's 4ESS. It was designed as a toll switch and was first installed in 1972 as an alternative to XB4 in new installations and later as a replacement for existing XB4 systems. As the telecommunications industry became open and competitive, other vendors, notably Northern Telecom with its DMS10, predicted a market for digital switching in local offices before AT&T did. The first DMS10 was installed in the late 1970s and Western Electric's first digital local switch, the 5ESS, was first installed in the early 1980s.

The revolution of low-cost digital electronics has affected switching system architectures by more than just lowering the cost of the computers that control them. Digital switching systems are characterized by digitization of speech signals and fabrics that connect low-power digital signals, through networks of logic gates and electronic cross-points in coordinate switches. Architecturally, these systems are still electronic and computer controlled, except that an evolution away from complete central control and toward a modular architecture has been coincident.

With large computers at the hub and other common equipment at the hub and in the modules, digital systems are designed for the mid-sized and large office. However, remote modules, slaved off main systems in distant offices, are cost-effective in small-office applications. Most of the electromechanical systems still surviving in the 1990s have been replaced, and usually by a digital system. Now, digital systems serve more than 50% of U.S. telephones today, and this figure is growing rapidly as 1Es and 1As are replaced. The 4ESS and 5ESS are described in Chapter 14.

2.5.6 Popularity

The popularity of the various local switching systems is indicated in Table 2.2. The table shows the percentage of telephones in the United States that were served by each type of system. Three years — 1965, 1980, and 1995 — are tabulated. The figures in the 1995 column aren't even estimated — they're educated guesses.

The deployment strategy of the ESS over the 1965 – 1980 interval is apparent from the figures. ESS obviously replaced all Panel and Number 1 Crossbar offices and many of the nation's larger Step offices. While the percentage of lines served by Number 5

TABLE 2.2. Popularity of local switching systems

Office type	1965	1980	1995
Step-by-Step	53%	25%	10%
Panel	7%	—	—
Crossbar 1	15%	—	—
Crossbar 5	25%	25%	—
Electronic	—	50%	40%
Digital	—	—	50%

Crossbar remained the same over this interval, the number of lines served increased significantly because the total number of lines in the nation increased. In the years since 1980, the percentage of lines served by digital offices has increased markedly. Digital offices have replaced almost all Crossbar offices, some Step offices, and even many older ESS offices.

EXERCISES

2.1 These questions [8] illustrate how much telephony effects American culture.

- What do Pat Nixon, Aristotle Onassis, and Dustin Hoffman have in common?
- Who answers these telephone numbers: (a) (212) PEnnsylvania 6-5000 (Glenn Miller, 1943)? (b) (202) 456-1414?
- Name the popular songs that feature these telephone lines: (a) "Hours of time on the telephone line to talk about things to come . . ." (b) "A telephone that rings, But who's to answer?" (c) "No phone, no pool, no pets, . . ." (d) "But she'll just hear that phone keep ringing off the wall, that's all." (e) "When I call you up, your line's engaged . . ."
- What fictional characters are associated with the following: (a) "One ringy dingy," (b) The shoe phone, (c) Public telephone booths, and (d) "Phone home"?
- Identify these classic movies with telephone-related titles: (a) Adapted from a John O'Hara novel, Liz Taylor portrayed a call girl and won her first Oscar, 1960. (b) A bedridden Barbara Stanwyck overhears the plotting of her own murder by husband Burt Lancaster, 1948. (c) Alfred Hitchcock suspense thriller, starring Grace Kelly and Ray Milland, featured a giant phone as a foreground prop and a conveniently placed sewing basket, 1954. (d) Chicago reporter, played by James Stewart, helps a scrubwoman prove her son's innocence in the death of a policeman, 1948.

2.2 The next six exercises are intended for readers who are familiar with digital design. Subsequent chapters of this book don't depend on being able to do these exercises. Consider the series-parallel relay circuit in Figure 2.8 [3, Problem 4.1(a)]. By inspecting the figure, write the Boolean expression for electrical

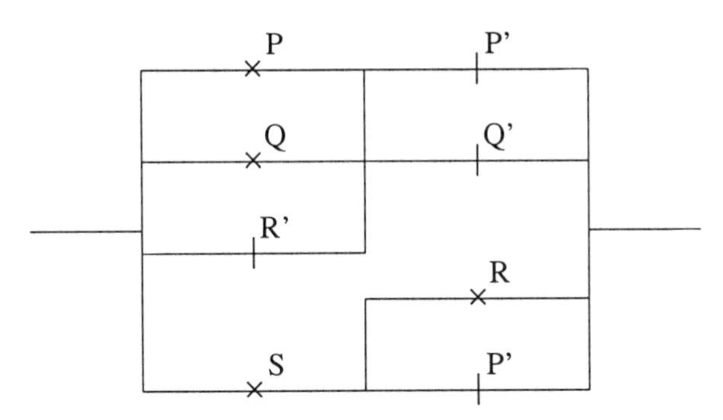

Figure 2.8. Series-parallel relay contact circuit [3].

conduction, left to right. Simplify the Boolean expression. Sketch an equivalent series-parallel relay contact circuit that uses as few contacts as possible. Sketch the equivalent implementation with nand gates.

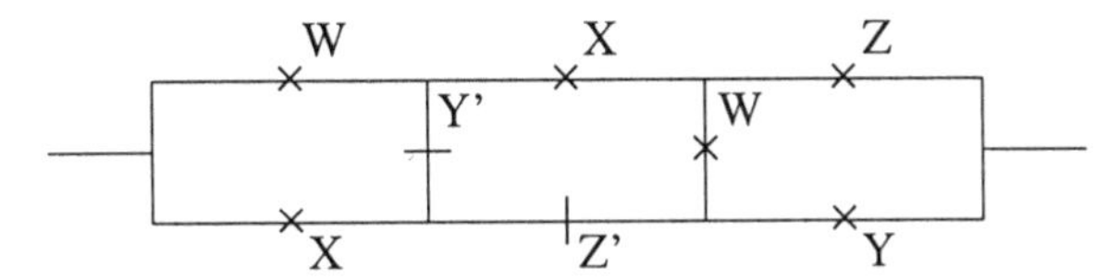

Figure 2.9. Subtly complicated relay contact problem [3].

2.3 Consider the subtly complicated relay circuit in Figure 2.9 [3, Problem 4.1(c)]. Write the truth table for this circuit. Write the Boolean expression for electrical conduction, left to right. Sketch an equivalent series-parallel circuit that uses as few contacts as possible. Sketch the equivalent implementation with nand gates.

2.4 Show the gate implementation for a four-input digital multiplexor with a separate decoder (also show the gate implementation of the decoder). Show the gate implementation for a four-input digital multiplexor with integrated decoding.

2.5 The internal components of a K-stage shift register include: K clocked D-type flip-flops and K two-input digital multiplexors. Consider the kth stage, $0 \leq k \leq K-1$. Each flip-flop has a data input wire d_k, a data output wire q_k, and a clock lead c_k. Each mux has x_{0k} and x_{1k} input wires, a single wire a_k for its 1-bit address, and an output wire z_k. The external wires are a K-bit parallel input word $P_0, \ldots, P_{K-1}$, a 1-bit serial input wire I, a 1-bit control wire SP that determines serial (if $SP = 0$) or parallel (if $SP = 1$) operation, a clock input C, and a K-bit parallel output word $Q_0, \ldots, Q_{K-1}$. Show the internal block diagram for a four-stage shift register. Hint: all the c_k are tied together internally and all the a_k are tied together internally.

2.6 A conventional ripple-carry adder is an example of a cellular digital circuit. Consider a different example, a binary comparator. The circuit compares the magnitudes of two N-bit binary numbers. The numbers are $A = a_{N-1} \cdots a_0$ and $B = b_{N-1} \cdots b_0$. The ith cell examines the ith bit in each number, gets data from the (i–1)th cell, and passes data on to the (i+1)th cell. The ith cell has three outputs for the (i+1)th cell: g_i indicates that A > B up to bit i, e_i indicates that A = B up to bit i, and l_i indicates that A < B up to bit i. Design a single cell and show how the total circuit is connected.

2.7 Repeat the previous exercise, except let each cell examine two consecutive bits from each binary number.

2.8 In Figure 2.7, let the equation for the "cellular" line be $1M + 200n$, for the "modular" line be $2M + 333Km + 67n$, and for the "centralized" line be $4M + 33n$. All equations are in dollars, n is the number of lines, and m is the number of modules, at 5000 lines per module. (a) For which range of "Number of lines" is the cellular architecture least expensive? (b) For which range of "Number of lines" is the modular architecture least expensive? (c) For which range of "Number of lines" is the centralized architecture least expensive? (d) A new office is required for an area that currently serves about 8000 lines, but whose population is expected to triple during the lifetime of the office. Which architecture would you recommend? (e) A new office is required for an area that currently serves about 18,000 lines, but whose population is expected to triple during the lifetime of the office. Which architecture would you recommend?

2.9 Based on Table 2.2 and assuming that the number of telephones remained constant (uniformly across all system types) from 1965 to 1980, what percent of 1965 Step subscribers changed to Electronic by 1980? What percent of 1965 Panel subscribers changed to Electronic by 1980? What percent of 1965 XB1 subscribers changed to Electronic by 1980? What percent of 1965 XB5 subscribers changed to Electronic by 1980? Why wasn't the oldest system replaced the most by Electronic systems? Based on Table 2.2 and assuming that the number of telephones remained constant (uniformly across all system types) from 1980 to 1995, what percent of 1980 Step subscribers changed to Digital by 1995? What percent of 1980 XB5 subscribers changed to Digital by 1995? What percent of 1980 Electronic subscribers changed to Digital by 1995?

2.10 Repeat the previous exercise, except assume that the total number of telephone lines increased by 50% from 1965 to 1980 and again from 1980 to 1995. In both time periods, assume that all the growth occurred in non-Step serving areas and that all new lines were assigned to the new system introduced in the period.

REFERENCES

[1] Dziatkiewicz, M., "Universal Service — A Taxing Issue," *Wireless Business & Technology,* Vol. 5, No. 1, Jan. 1999.

[2] Rey, R. F. (ed.), *Engineering and Operations in the Bell System,* Second Edition, Murray Hill, NJ: AT&T Bell Laboratories, 1977.

[3] Caldwell, S. H., *Switching Circuits and Logical Design,* New York, NY: John Wiley & Sons, 1963.

[4] Keister, W., A. E. Ritchie, and S. H. Washburn, *The Design of Switching Circuits,* New York, NY: D. Van Nostrand Co., 1951.

[5] Shannon, C. E., "A Symbolic Analysis of Relay and Switching Circuits," *Transactions of the AIEE,* Vol. 57, 1938, pp. 713 – 723.

[6] Washburn, S. H., "An Application of Boolean Algebra to the Design of Electronic Switching Circuits," *AIEE Transactions, Part 1, Communication and Electronics,* Vol. 72, Sept. 1953, pp. 380 – 388.

[7] Musa, J. D., "Software Reliability," *IEEE Spectrum Magazine,* (issue on Supercomputers: A Special Report), Vol. 26, No. 2, Feb. 1989.

[8] *Games Magazine,* Feb. 1985.

SELECTED BIBLIOGRAPHY

Andrews, F. T., "The Heritage of Telegraphy," *IEEE Communications Magazine,* (issue on Telecom at 150), Vol. 27, No. 8, Aug. 1989.

Becker, F. K., "Telephone Systems," Chapter 7 in *Telecommunications: An Interdisciplinary Text,* (L. Lewin, ed.), Dedham, MA: Artech House, 1984.

Bellamy, J., *Digital Telephony,* Second Edition, New York, NY: Wiley, 1991.

Brooks, J., *Telephone — The First Hundred Years,* New York, NY: Harper & Row, 1976.

Clarke, A. C., et al., *The Telephone's First Century — and Beyond,* New York, NY: Thomas Y. Crowell, 1977.

"Common Heritage," *AT&T Focus Magazine,* Apr. 25, 1989.

Freeman, R. L., *Telecommunication System Engineering,* Third Edition, New York, NY: Wiley, 1996.

Henck, F. W., and B. Strassburg, *A Slippery Slope — The Long Road to the Breakup of AT&T,* Westport, CT: Greenwood Press, 1988.

Hills, M. T., *Telecommunications Switching Principles,* Cambridge, MA: MIT Press, 1979.

Howard, P. K., *The Death of Common Sense — How Law Is Suffocating America,* New York, NY: Warner Books, 1994.

Joel, A. E., Jr., "Classification and Unification of Switching System Functions," *Proc. Int. Switching Symposium,* 1972, pp. 446 – 453.

Joel, A. E., Jr., *Electronic Switching: Central Office Switching Systems of the World,* New York, NY: IEEE Press, 1976.

Joel, A. E., Jr., "Public Digital Switching Systems," *IEEE Communications Magazine,* (issue on Digital Switching), Vol. 21, No. 3, May 1983.

Joel, A. E., Jr., "The Past 100 Years in Telecommunications Switching," *IEEE Communications Magazine,* (issue on 100 Years of Communications Progress), Vol. 22, No. 5, May 1984.

Joel, A. E., Jr., "What Is Telecommunications Circuit Switching?" *Proceedings of the IEEE,* (issue on Telecommunications Circuit Switching), Vol. 65, No. 9, Sept. 1977, pp. 1237 – 1253.

Kellagher, J., "U.S. Central Office Switch Market," *Telecommunications Magazine,* (issue on Switching Inside and Outside the Network), Vol. 21, No. 2, Feb. 1987.

Lindstrom, A. H., and R. Karpinski, "Telecom History 101," *Telephony Magazine,* (90th anniversary issue), Vol. 220, No. 24, Jun. 17, 1991.

Lucas, J., "What Business Are the RBOCs In?" *TeleStrategies Insight Magazine,* Sept. 1992.

McDonald, J. C. (ed.), *Fundamentals of Digital Switching,* Second Edition, New York, NY: Plenum, 1990.

Noll, A. M., *Introduction to Telephones and Telephone Systems,* Norwood, MA: Artech House, 1986.

Pearce, J. G., *Telecommunication Switching,* New York, NY: Plenum Press, 1981.

Pierce, J. R., *Signals: The Telephone and Beyond,* San Francisco, CA: W. H. Freeman and Company, 1981.

Pierce, J. R., "Telephony — A Personal View," *IEEE Communications Magazine,* (issue on 100 Years of Communications Progress), Vol. 22, No. 5, May 1984.

Schindler, G. E., Jr. (ed.), *A History of Engineering and Science in the Bell System,* Murray Hill, NJ: Bell Telephone Laboratories, 1982.

Switching Systems, New York, NY: American Telephone & Telegraph, 1961.

Talley, D., *Basic Electronic Switching for Telephone Systems,* Rochelle Park, NJ: Hayden Book Company, 1975.

Talley, D., *Basic Telephone Switching Systems,* New York, NY: Hayden Book Company, 1969.

Telecommunications Transmission Engineering, Volume 1 — Principles, New York, NY: American Telephone & Telegraph, 1977.

Telecommunications Transmission Engineering, Volume 3 — Networks and Services, New York, NY: American Telephone & Telegraph, 1977.

Telephony (90th Anniversary Issue), Vol. 220, No. 24, Jun. 17, 1991.

Temin, P., with L. Galambos, *The Fall of the Bell System,* New York, NY: Cambridge University Press, 1987.

Tracey, L. V., "30 Years: A Brief History of the Communications Industry," *Telecommunications,* Vol. 31, No. 6, Jun. 1997.

Transmission Systems for Communications, New York, NY: Bell Telephone Laboratories, 1964.

Weisman, D., "The Impact of Silicon Integration on the Communications Industry," *Telecommunications,* Vol. 32, No. 3, Mar. 1998.

Weiss, M. B. H., and P. Bernt, *The Regulation and Deregulation of US Telecommunications,* Textbook under contract with Lawrence Erlbaum and Associates, Mahwah, NJ, C. Sterling, (ed.).

3

The Line Side

"A telephone that rings and no one answers." — from the song "These Foolish Things Remind Me of You"

This chapter is the first of two in which the boundary conditions on a local switching system are presented. This chapter describes the line side of a local office, the side that interfaces to the subscribers. The next chapter, Chapter 4, describes the trunk side, the side that interfaces to the network. The reader is assumed to have some familiarity with the fundamentals of analog transmission.

The telephone is described in considerable detail in the first two sections, with the internal electronics presented in Section 3.1 and subscriber signaling in Section 3.2. The network of wires between the telephone and the switching office is described in considerable detail in the next two sections, with typical house wiring presented in Section 3.3 and the telco's loop plant in Section 3.4. Per-line office functions, described in Section 3.5, include a typical line circuit, dialed digit reception, and supervision. Non-POTS lines, described in Section 3.6, include coin telephones, customer premises and remote switching, and loop carrier systems.

3.1 THE TELEPHONE

The telephone is described here in detail, beginning with the physical design and then the electrical design. Other data equipment, like televisions and computer terminals, are described, less thoroughly, in Chapter 15. The schematic diagram for the telephone is developed gradually, but not completely, beginning with the audio parts and successively adding central battery, the hook-switch, side-tone, and finally some miscellaneous parts. The development continues into the next two sections, where ringing and signaling, respectively, are added.

3.1.1 Physical Design

The classic telephone instrument has two physical parts, the base and the handset. Somewhere on the base is a cradle or switch-hook, in which the handset rests when the telephone is not in use. This cradle has a spring-loaded plunger, connected to an internal switch, called the hook-switch. When the handset is on-hook, the switch-hook is down and the hook-switch is in its normally open position. When the handset is off-hook, the switch-hook is up and the hook-switch closes. A cord with internal wires electrically connects the base to the handset. In cordless telephones, this connection is a low-power radio link. Some telephones have a switch-hook button on the handset.

In most modern telephones the speaker, or earpiece, and the microphone, or mouthpiece, are on the handset and they physically conform to the human head. On some early telephones, the mouthpiece was on the base unit — which was shaped like, and called, a candlestick — and the handset consisted only of a small speaker held to the subscriber's ear. With SpeakerPhone, however, the earpiece (speaker) is found in the base unit or in some separate adjunct equipment, but not in the handset. A signaling device, either a 10-position dial or a numeric keypad, is located on either the base or the handset, depending on the telephone model. Also depending on the model, this signaling device is optionally lighted.

The telephone is connected externally by a cord containing at least two wires, but usually four and frequently more — 50 in the Key Telephone described in Section 10.1. Telephones used to be connected directly to house wiring by the telco's installer, but now the familiar modular jack is used.

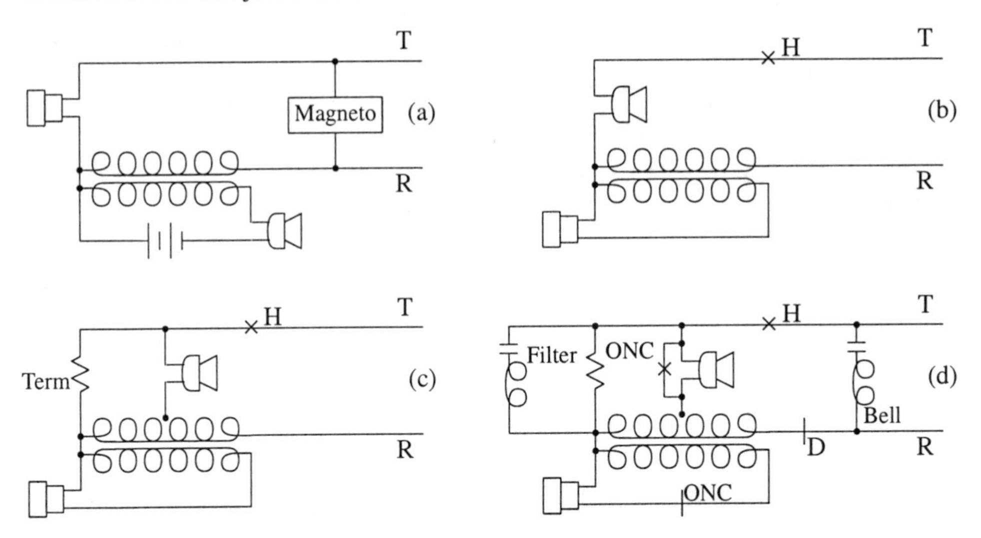

Figure 3.1. Telephone schematic: adding detail from (a) to (d).

3.1.2 Audio Circuit

The development of the schematic diagram for the telephone begins with an over-simplified circuit drawing in Figure 3.1(a), showing the earpiece, mouthpiece, local battery, and magneto. The earpiece, like most audio speakers, works like Bell's original invention. It has a flexible cone of paper with a piece of iron at the point and anchored at the base, and a gap between the piece of iron and an electromagnet. Varying electric current flowing through the electromagnet varies the magnetic field, moving the cone and generating audible air motion.

The mouthpiece, like any carbon microphone, works like Edison's original invention. Granules of carbon are placed between two parallel plates, and the total resistance of the sandwich varies with any pressure that compresses the plates together and packs the carbon granules closer together. The mouthpiece translates acoustic waves in the air, caused by the subscriber's speech, into changes in electric current. A typical AC audio signal has amplitude less than 1 volt, but an exceptionally loud sound could appear as great as ± 2 volts.

3.1.3 Battery

Operation of the mouthpiece in the telephone requires DC, called talking battery. If the battery is inside the telephone, as seen in Figure 3.1(a), then the DC must be isolated from the line by AC coupling the audio signal, by a capacitor or transformer, as shown. A crank-operated generator across the line gives the subscriber a means to light a lamp for alerting the attendant.

Local power operation is shown in Figure 3.1(a). In central power operation, shown in Figure 3.1(b), talking battery is provided at the switching office, to the right of, but not shown on, the figure. Several of the issues in powering the residential telephone are listed below.

- AC-to-DC conversion from house power means that the telephone would cease operation during power failures.
- A dry-cell battery will not last long and must be replaced, either by the subscriber or by telco personnel.
- A wet-cell battery is long lasting, but is potentially dangerous.
- A rechargeable battery would work, but they are only recently economical and are appearing in modern electronic multifunction telephones.
- Local power requires the generator and crank for alerting.
- With central power, a DC monitor at the switching office provides a simple means for alerting and supervision.
- But, with central power, the voltage at the telephone depends on its distance from the source.
- Central power allows for providing more reliable service.

Since voltage uniformity is the only advantage to local power, central power is almost universal in the United States. DC power is provided to the telephone over the same two wires that carry the audio signal, and AC coupling occurs in the switching office and not in the telephone, as indicated in Figure 3.1(b). A significant part of any switching office is the DC power plant, consisting of AC-to-DC converters, many batteries for regulation and backup during short power outages, and frequently a gasoline- or diesel-powered generator for backup during long power outages. The local switching office is typically one of the largest consumers of electric power in any community because it powers not only the switching equipment, but also all the telephones connected to it.

Nominal talking battery is 48V DC at the switching office. The polarity is negative with respect to earth ground. A measurement at the telephone, when it's on-hook, is slightly smaller because of current leakage in the wires between the telephone and the office. When the telephone is off-hook and current flows through it, the voltage across the telephone is much smaller, typically as small as 12V, because of resistance in these wires. Some care must be exercised around opened telephones or exposed telephone wires. 48V DC is high enough to be respected by everyone, and to threaten children or those who may be sensitive to electrical shock. The shock can be especially uncomfortable if fingers are wet or if you are in the practice of stripping insulation with your teeth. Subscribers are warned not to use telephones in the bath or pool. The warning is compounded if there is a possibility that the switching office might ring the telephone (Section 3.2.1). Ring signal is more powerful than talking battery and the combination gives an electrical shock greater than that of household electricity.

3.1.4 Switch-Hook

As stated above, the on-hook/off-hook status of the handset operates a plunger and the plunger operates the hook-switch. This normally open switch is in series between the central battery and the current conducting parts of the telephone, as seen in Figure 3.1(b). When the telephone is idle, the handset is on-hook, the hook-switch is open, and no current flows down the wires from the battery in the switching office. When telephone service is desired, the subscriber takes the handset off-hook, the hook-switch closes, and current flows from the battery through the telephone. In the switching office, the state of the hook-switch is easily determined by any type of current detecting device.

In the real world, leakage current flows in the wires even when the telephone is on-hook and the resistance of the wires causes telephones far from the office to conduct less current then telephones close to the office. The typical threshold for off-hook is about 25 mA, and the typical DC resistance of the telephone is about 500 ohms.

3.1.5 Side-Tone

Consider the telephone, with schematic shown in Figure 3.1(b), connected to another like it over a long, lossy transmission line. The subscriber of the telephone, listening at the earpiece, hears his/her own voice at a certain level of loudness because the signal from the mouthpiece flows directly through the earpiece. But, this subscriber hears the party at the other end of the line at a lower level because the line has attenuated the other party's signal. This phenomenon is especially annoying to the subscriber if he/she is in a room with operating machinery, like a dishwasher, or other form of background noise. This process of the subscriber hearing his/her own voice is called side-tone and it is desirable that the intensity of the side-tone match the intensity of the other party's signal.

In Figure 3.1(c), an impedance-matching termination network is added and one lead to the mouthpiece has been moved to the center-tap of one side of the transformer. The signal coming down the line from the other party couples through the transformer to the earpiece and the signal from the mouthpiece still proceeds down the line to the other party. But, any signal from the mouthpiece to the earpiece must couple through the transformer at the center-tap. If the mouthpiece is perfectly center-tapped on its winding, the signal cancels through the transformer and nothing will be heard.

But, since subjective testing has shown that typical telephone subscribers prefer some side-tone, the winding is tapped slightly off-center and the imbalance allows a small amount of the subscriber's own voice and background noise to couple to the earpiece. A problem, left as an exercise, occurs in extension telephones that is similar to the 100% side-tone in Figure 3.1(b).

3.1.6 Odds and Ends

Telephones that are physically far from the switching office will receive a weaker audio signal than telephones that are close. This effect is partially compensated by using a varistor, a device whose resistance varies with the current through it, because these same telephones are separated from the talking battery by a long wire and receive less DC. Carbon-filament light bulbs have a similar characteristic as a varistor and were used in trunk equipment in switching offices before the semiconductor varistor was invented.

The complete schematic, shown in Figure 3.1(d), has been described except for the bell, circuitry associated with dialing, the D and off-normal contacts, and the filter. These are discussed in the next two sections. In most telephones, the discrete electrical elements in the schematic are not visible because they are physically packaged as a unit. The bell and the hook-switch are wired to this unit, and two wires, typically green and red, connect to the telephone's external jack. These two wires bring talking battery, ring signal, and the two-way audio signal from the switching office to the telephone's internal components. In telephones with lamps, two additional internal wires, typically yellow and black, carry low-voltage AC.

3.2 SUBSCRIBER SIGNALING

The bell is added to the development of the schematic diagram of the telephone. Ringing signal is described and party lines are discussed. The description of the telephone is completed by discussing signaling. Signaling is the transmission of information about a call, such as hook-switch status (supervision), called or calling party identity, and billing or routing information. After signaling a change in supervision status by lifting the telephone's handset off the cradle, the calling party must signal the identity of the called party to the office at the start of a call, typically by transmitting the called party's directory number. Two classical means by which a subscriber signals this called number are by dial pulse and by Touch Tone. While Touch Tone is a registered trademark of AT&T, this term is much more commonly used than the generic term dual-tone multi-frequency (DTMF).

3.2.1 Ringing

The bell is shown in the schematic diagram of the telephone in Figure 3.1(d). Since the telephone must ring when on-hook, the bell is connected across the line on the office side of the hook-switch. Since DC conduction indicates an off-hook telephone, the bell is AC-coupled across the line by a series capacitor. The bell's winding is such a high inductance that it has little effect on the audio signal.

With the telephone on-hook, the bell contributes most of the AC impedance seen down the line from the switching office. During the regulatory period when telephone instruments were unregulated and sold commercially, but subscriber's were charged extra for extension telephones (Section 10.6), it was common practice for the telcos to check each subscriber's line impedance periodically to verify the number of telephones. It was also common practice for subscribers to thwart this check by disconnecting the bell in their unreported "illegal" extension telephones.

The large acoustic power that allows the subscriber to hear the telephone ring, requires a large electrical power delivered to the bell. Compared to the other signals encountered, ring signal is a high-voltage low-frequency AC signal. The nominal 86V

RMS signal is 120V peak-to-peak, almost the magnitude of household electricity. Superimposed on the –48 battery, ring signal swings between about +70V and –170V. The effective voltage applied to the telephone wires during ringing, what someone feels if touching the wires, is 86 + 48 = 134V, slightly larger than household electricity.

At 20 Hz, the signal itself is inaudible. The ring signal is typically sinusoidal, generated in the switching office by rotating machinery, an electrical generator. Some small systems generate a square-wave signal. Typically the ring signal is switched to a line that must be rung through the fabric, requiring that the crosspoints and the fabric wiring be suitable for such high-power signals. Low-power crosspoints and fabric wiring are possible if ring signal is switched by a special relay associated with each line and not through the fabric.

The signal is modulated over a 6-second period — on for 2 seconds and off for 4 seconds. The ring generator has three phases, with one-third of the telephones in an office rung off each phase. The reader has probably had the experience of placing a call and having the called party answer before any ringing was heard (by the calling party). This only happens in switching offices in which the audible ring tone heard by the calling party is a separate signal from the ring signal to the called party. The ringing of the called party's telephone may be on a different phase from the audible ring tone heard by the calling party.

3.2.2 Dialing

The subscriber dials a digit by placing his finger in the appropriate one of 10 holes in the dial, turning the dial to the stop, and releasing the dial, allowing it to return. If the subscriber dials digit *N*, then *N* binary pulses are transmitted down the line to the office during this return. Note the careful distinction between "digit" and "pulse," there being *N* pulses in the digit *N*. The timing of the dial pulses is controlled by a spring-and-governor mechanism in the dial.

Two switches in the telephone, shown in Figure 3.1(d), are associated with the dial. A ratchet mechanism in the dial causes the dial contact to open and close for every dial-pulse. The off-normal contact (onc) disconnects the earpiece from the circuit and another contact shorts out the mouthpiece while the dial is being turned and released.

While the dial is returning, its mechanisms cause the dial contact to break and make in a period of about .1 second, about 10 pulses per second. The duty cycle is about 50%. By its position in the schematic of Figure 3.1(d), the dial contact is seen to act identically as the hook-switch. The DC from talking battery is interrupted with every break. Even though the operation of the dial contact is a square waveform, the LC circuit at the left of Figure 3.1(d) acts as a low-pass filter and rounds the current waveform. Crosstalk in wires and electromagnetic interference (as in a television set) are reduced during dialing.

Equipment in the switching office counts the breaks in the line to determine the dialed digit, but also times each break. A long break interval is interpreted as the subscriber hanging up. Since the dial contact is closed while the subscriber turns the dial forward, the office equipment sees a make interval of this duration between digits. The make interval between the last pulse of one digit and the first pulse of the next digit is much longer than the make interval between consecutive pulses of the same digit. The interpulse and interdigit timing is not extremely critical, and tolerances depend on the type of switching office. Timing is generally forgiving enough that the curious subscriber may simulate dialing by manually operating the switch-hook rapidly.

3.2.3 Dual-Tone Multiple Frequency

DTMF pulsing is another type of subscriber signaling, more commonly known by AT&T's registered trade mark Touch Tone. A telephone equipped for DTMF has a numeric keypad instead of a dial. The keypad typically has 12 keys — the 10 digits, plus * and #. Some telephones have keypads with only the 10 digit keys and some have keypads with 16 keys. When a key is pressed, an audible musical chord, consisting of two frequencies (hence, dual-tone), is placed on the tip and ring. Each key produces a different pair of tones.

TABLE 3.1. Dual-tone multifrequency

697	1	2	3
770	4	5	6
852	7	8	9
941	*	0	#
	1209	1336	1477

The frequencies are partitioned as four in the low band and three in the high band. The chord produced by each key contains one frequency from each band. Table 3.1 shows how the seven different frequencies are assigned to each row and to each column of keys in the rectangular DTMF keypad. While the conventional 12 digits require only seven tones, 1633 Hz is also available. The pair of tones produced by a given key is found by the frequencies corresponding to the key's row and column in Table 3.1. For example, pressing 1 produces a chord with a 697-Hz tone and a 1209-Hz tone, simultaneously. The seven tones are unusual frequencies and were carefully selected to avoid accidental detection in human speech, harmonics, or sum or difference tones.

A DTMF receiver is connected to the tip and ring in the switching office while the subscriber is assumed to be signaling. It has a narrow-band filtered tone detector for each of the seven tones. A digit is received when enough audio energy is detected within the bandpass of exactly two tone detectors.

The tone generators in a DTMF telephone are wired so that, if two keys in the same row or column are pressed simultaneously, only the single tone common to the two keys is produced. For example, pressing 1 and 4 simultaneously produces only the 1209-Hz tone. And, for example, pressing 1 and 5 simultaneously produces no tones.

Some of the older DTMF telephones were polarization dependent. If installed with the tip and ring reversed, the entire telephone still works, except for the DTMF pad. In many switching offices, a battery-reversal on tip and ring is a common signal that the called party has answered and billing may proceed. This action intentionally disables the DTMF pad in these older telephones. However, many new services, like bank-by-phone and some non-AT&T long distance services, require callers to continue using the DTMF pad after calling the listed service point. Such services work because the subscriber's telephone places the DTMF tones on the same tip and ring as speech and the switching office has disconnected its tone detectors, assuming subscriber signaling is done. To allow such services, newer DTMF telephones are insensitive to polarization. Operation of such services for subscribers with dial telephones is much more difficult because most switching offices do not repeat dial pulses, except during the interval when the subscriber is expected to be dialing. Without repeating, the service point would receive an attenuated and damped voltage signal with no true physical breaks in DC current.

3.3 HOUSE TELEPHONE WIRING

The network of wires between the telephone and the switching office is described in this section and the next one. This section describes the interior wiring in a typical residence. The next section describes the architecture of the wiring between the residence and the switching office.

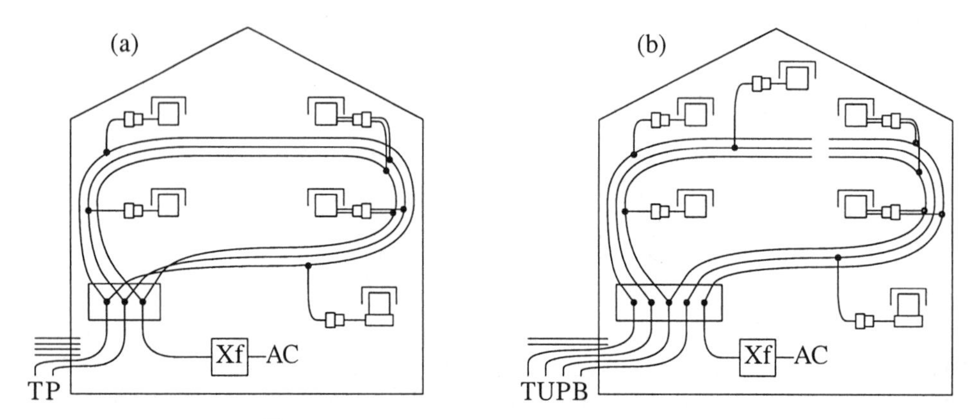

Figure 3.2. The wiring in my previous home.

House wiring is illustrated by a real example, a single-family, up-scale two-story residence, built in 1976. While the details typically would be different for older and newer houses and for multiple-family dwellings, the concepts are similar. Wiring in a business or office is much more complex and is described in Section 10.2. This residence and its house wiring are illustrated in Figure 3.2. I lived in this house during a time in my life when my household was more dynamic than it is now; and I personally reconfigured the house wiring as my needs changed. Figure 3.2(a) illustrates this house and its telephone wiring when my family had only two active telephone lines.

A fuse-like device protects the house wiring, the subscriber's equipment, and the house itself from over-voltage on the line, like a lightening hit or accidental contact with a power line. When this device is mounted outside the home, as in most newer homes, it is called the telephone network interface (TNI). When this device is found inside the home, as in most older homes, it is called the protector block. The TNI serves as the electrical connector block between the house wiring, which I own, and the drop wire, the part of the telco's loop plant that enters my home. This pair is called the drop wire even if it enters the home through an underground conduit. Since the drop wire enters this particular home through the same conduit as the electrical utility's power wires, the protector block is found near the fuse box on a plywood panel in my basement, shown in the lower left of the figure. Meeting at this protector block, shown in the lower left of the house outline in Figure 3.2(a), are:

- The telco's drop wires, six pairs in a gelatin-packed, metal-sheathed cable, shown entering from the lower left of Figure 3.2(a);
- My house wiring, two soft-sheathed cables with three pairs in each;

- A third soft-sheathed cable, with one pair, that runs to a small transformer plugged into a nearby AC receptacle, shown in the lower center of the figure.

In the figure, a pair of wires is represented by one line. At the protector block, there appear to be two cables for the house wiring. But these are two ends of the same cable, snaking inside the walls through my house and accessible in every room through a hole in the wall covered by a plastic cover plate. By wiring the home in a physical ring, every room may be accessed even if the wire breaks (once). The color and pattern on their insulator identify the wires and pairs. For example, in the cable carrying my house wiring, the blue wire with the white stripe and the white wire with the blue stripe are a pair. Again, in the figure one line represents a pair of wires.

When I was a child, I lived in a home that was built in 1952 in a "lower scale" neighborhood. The drop wire and the house wiring had only one pair. The house wiring was a star, not a loop, and there were no modular jacks. The house wiring was a single pair that ran directly to the single black dial-pulse telephone. But my parents, children of the Great Depression, never considered owning an extension telephone or paying extra for Touch Tone. Their entire generation of Americans was most unlikely to subscribe for a second line.

The home I live in now has six pairs in a drop wire that is shared by two residences; and the interior wiring has four pairs. Oddly, the interior wiring is not a loop. This house was built in 1985 and second lines were becoming quite common, especially for computer modems. But the significant change, typical of most newer homes in the United States, is that the protector block's function is now on the exterior of the building. A small plastic box, the TNI, is accessible by telco linemen without requiring entry into the home.

At the time of Figure 3.2(a), I subscribed for two separate lines with different directory numbers from the telephone company. The principal line (labeled P in the figure) had the number listed in the public telephone directory and I had extensions in the living room (on the left in the figure), kitchen (on the right), and master bedroom (on the upper right). The so-called "teen phone" (labeled T in the figure) was an unlisted second line that was billed with the principal line (Section 9.1) and I had extensions in a bedroom (on the upper left in the figure) and in a basement den (on the lower right). The telephones in the master bedroom and the kitchen had lighted dials.

The two separate lines were accommodated by two of the six drop-wire pairs; the remaining four pairs were unterminated and unused. In the figure, a single line represents a pair of wires. Each active drop-wire pair terminated on the protector block and connected, from there, to both ends of a house-wire pair. Both ends of the third house-wire pair were tied to the wires that run to the transformer. A telephone is installed to a particular line by electrically connecting the green and red wires from the telephone to the appropriate pair anywhere along the house wire. An extension is installed by connecting another telephone to the same house-wire pair.

A typical installation is illustrated by a lighted wall-mounted telephone in the kitchen (on the right of the figure). It connects directly to the house wiring by mounting, mechanically and electrically, onto a modular jack on a cover plate in the kitchen. In the wall behind the cover plate, the green and red wires from the modular jack, represented by the upper single line of the two shown in the figure, are tied to the pair carrying the principal line. The yellow and black wires from the jack, represented by the lower single line of the two shown in the figure, are tied to the pair carrying the transformed AC. Tabletop telephones are installed similarly except that the cover plate has a modular jack and the installation is completed by a cord with a modular plug on each end. In the wall behind the cover plate, appropriate wires from the modular jack are tied to appropriate house-wire pairs. If the telephone is located far from the cover plate, a cord may be run from the cover plate along the baseboard to a small box containing a modular jack.

My house wiring was installed, while my house was being built, by telco personnel, appropriately called installers. In the pre-divestiture era, even for simply relocating a telephone within the home, the installer ran baseboard cables and made connections in the wall behind the cover plate. In the post-divestiture era, this is now the home owner's responsibility, although telco installers or commercial electricians can be hired.

At one time my home had the more complex configuration shown in Figure 3.2(b). Besides a principal line (labeled P on the figure), my home had two teen lines (labeled T and U), and a business Centrex line (labeled B) that I used with a modem and a computer terminal. The telco had enough drop wires to bring these four lines into my home, but this residence did not have enough house-wire pairs. My children's bedrooms were both on the left side of the house, and my den and both lighted telephones were on the right side. So I cut the wires at the appropriate place, shown on the upper right of Figure 3.2(b), effectively separating the house wiring serving the left and right sides of the house. At the protector block, I connected the three pairs, now serving only the left side of the house, to the drop wires carrying the principal line and the two teen lines. I connected the three pairs, now serving only the right side of the house, to the pair carrying transformed AC and to the drop wires carrying the principal line and the business line. I rewired the connections behind many of the cover plates and carefully recorded what I had done.

Perhaps I have provided too much detail on something as mundane as house wiring, but it is an area of personal sensitivity. I have the type of extended family that, when I received my B.S. in EE, assumed I could repair their television sets and wire their do-it-yourself family rooms for electricity. I'm sure the reader is familiar with such expectations from others and the embarrassment and difficulties in explaining the difference between academic and practical knowledge. The best way to avoid this is to learn how to do these jobs. These same dear people, hearing that I am so expert in telephone switching that I have written a book on the subject, naturally expect that I can install and repair their telephones. I can and I do and I'm glad to, but we know this activity is independent of being an academic expert in switching. If you, the reader, allow your relatives to know that you are reading this book, maybe the extra detail in this section, and in the previous sections, will prove useful.

3.4 LOCAL LOOP

The wires between the telephone and the switching office are described in this section and the previous one. The previous section described the wiring inside a residence. This section describes the wired local access network between a residence and the switching office, known collectively as the loop plant. The first subsection describes the dual nature of this local network; how the wire pairs act simultaneously as a loop in a DC electrical circuit and as a transmission line for AC electrical signals. Then, after presenting some measured electrical characteristics of the wire pairs in the local loop, the network's physical architecture is described. Finally, the last subsection describes the main frame, where the wire pairs terminate in the switching office.

3.4.1 Telephone Lines

This subsection presents the physical and electrical characteristics of the traditional telephone line — including resistance, characteristic impedance, bandwidth, and high-frequency impairments. The traditional analog telephone line has a dual nature: it acts simultaneously as a loop in a DC electrical circuit and as a transmission line for AC electrical signals. The electrical characteristics are presented in this dual context.

Physically, analog POTS uses a pair of twisted wires between the telephone and the switching office. A single wire, with a common ground reference, would be too noisy and lossy for decent transmission, especially in lines that can be several miles long. Untwisted wire pairs would also be too noisy. While wire gauges from #22 through #28 can be found, typically, #26-gauge wire is used; #28-gauge wire is a little too lossy, #22- and #24-gauge wire is a little too bulky and expensive. Typically, buildings and the "last 100 meters" outside use cables that contain 1 – 6 twisted pairs inside. Some enterprises commonly use 25-pair cables to each office. The external cables, like you see on poles, contain hundreds of twisted pairs and the thick cables, found in the office's cable vault, contain thousands of pairs.

DC loop. The traditional telephone line is a simple series electrical circuit with a 48V battery and a current detector at the office end, a single-pole switch and a resistor inside the telephone, and the wire resistance in each of the line's two wires. When the telephone is on-hook, its internal switch is open and no current flows in the loop. When the telephone is off-hook, its internal switch is closed and current flows around the loop. Thus, DC current in the loop provides the signaling by which the current detector in the switching office is aware of the on-hook/off-hook state of the telephone. This DC loop signaling provides three functions described in Section 1.5 — the attending function signals a new request, answer supervision signals off-hook during ringing, and call supervision signals hangup at the end of a call. Subscriber signaling by dial pulsing also interrupts the DC, but for intervals that are too short for the current detector to see. Telephone operation and dial pulsing are described later.

Section 1.5 mentioned that the two wires are called tip and ring after the convention for how they were connected to the traditional plug. Inside the telephone and in the wiring behind wall jacks, the convention is that the insulating jackets on the loop wires are colored red and green (sometimes the black and yellow wires are used). It is also conventional that the 48V battery's positive terminal be connected to local ground; most automobile batteries are also negative with respect to ground. Summarizing these conventions, the tip wire is green and is usually connected to ground and the ring wire is red and is usually connected to -48V.

At #26-gauge, copper wire has about 40 ohms of electrical resistance per 1000 ft. By convention, the current detector in the switching office requires a minimum of 25 mA to signal that a telephone is off-hook. To draw a minimum current of 25 mA from a 48V battery, the maximum net resistance in the loop is 48V/25mA = 1920 Ω. Since the typical DC resistance of a telephone is about 500 ohms, only 1420 Ω is left for the two wires in the loop. At #26-gauge, the maximum loop length is:

$$\frac{710\ \Omega}{40\ \Omega/\text{kft}} = 17,750\ \text{ft}$$

This conventional maximum unrepeatered loop length in the United States, given by the term "18 kilo-feet" (an odd combination of the Metric and English systems), is about 5.4 km (a little more than 3.3 miles).

Operation of the twisted pair as a DC loop is so fundamental to network operation that the wires and other equipment between the telephones in a community and switching office that serves this community are collectively called the loop plant.

AC transmission line. The twisted pair traditionally carries audio-frequency signals, typically in both directions. Besides the electrical analog of speech, which flows from the mouthpiece in the speaker's telephone to the earpiece in the listener's telephone, this line also carries AC signaling for DTMF tones from telephone to switching office (the signal by which an office makes a telephone ring), and various call progress tones that the switching office signals back to the user at his telephone.

This line is subject to attenuation, interference, crosstalk, and reflections like any other transmission line. A copper twisted-pair transmission line has a characteristic impedance of about 180 ohms. To reduce line reflection, which sounds like an echo, the telephone's AC impedance must match the line's characteristic impedance. The bandpass of a typical analog POTS line is in the 8 – 12 kHz range, much higher than the bandwidth of the signal we receive today (I'll explain why when we talk about digital switching).

High-frequency impairments. High-frequency impairments commonly found in many telephone pairs are bridged taps and load coils. Both are electrical effects that were deliberately placed in the loop plant to alter the loop's electrical performance. But, a loop that is used for digital signaling, such as for digital loop carrier multiplexing or for ISDN, must be free of bridged taps and load coils.

When a subscriber wishes to terminate his telephone service, the telephone number is disabled in the switching office, but telco linemen may not disconnect all the segments of the subscriber's loop. When a different subscriber's loop is subsequently connected, one of these segments might be used. But, the lineman may not bother to disconnect this segment from the loop in which it was previously a part. The disconnection may not actually occur until other segments of that original loop are used to still other loops. While this practice saves time on every disconnect order, we find that many loops branch into unterminated segments at the pedestal, the distribution point, and the main frame. These extraneous branches are called bridged taps. Bridged taps have little significant practical effect on analog voice telephone signals, but at high frequencies they can cause reflections because of transmission line impedance mismatch.

Load coils are inductors that have been installed into a wire path to adjust the spectrum of its bandpass. Deliberately adding the proper amount of inductance can decrease the midband loss because of resonance with the wire's capacitance. But, load coils usually increase the loss away from the resonant frequency; thus sacrificing bandwidth and exacerbating frequency distortion at very high frequencies. This effect is illustrated by actual measurements in the next section.

Load coils, which were installed many years ago in long loops to enable audio transmission without requiring amplifiers, have turned out to be problematic. While shoddy record-keeping has inhibited their removal, most telcos have removed them because they are especially detrimental to digital signaling, as with ISDN, for example.

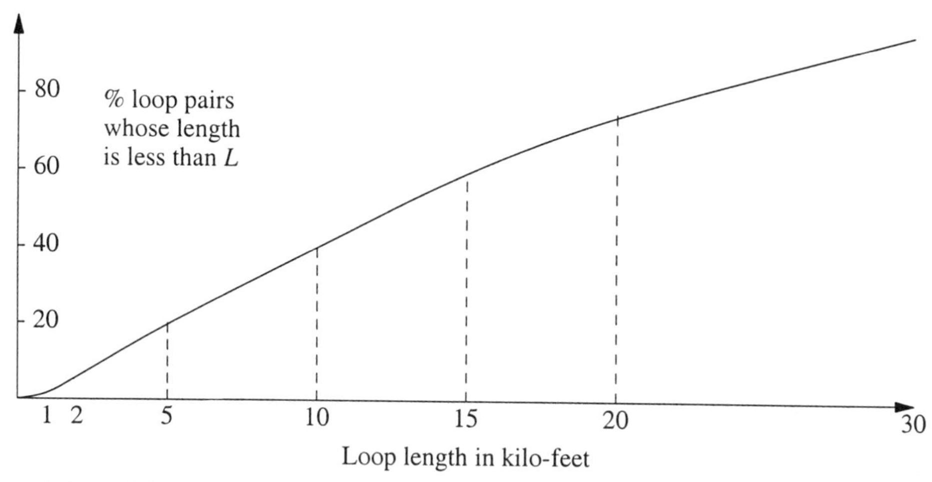

Figure 3.3. Distribution of Bell System's 1980 loop lengths. © 1984 AT&T [1].

3.4.2 Measurements

Figure 3.3 summarizes the Bell System's 1980 loop census [1]. The y axis is the percentage of loop pairs whose total length (the sum of all three segments) is less than the value L on the x axis. While there is no reason to expect loop length to be uniformly distributed, the distribution turns out to be surprisingly uniform, especially between 2000 and 15,000 ft. For example, only 5% of all loops are shorter than 2000 ft and 20% of all loops are shorter than 5 kft. The median loop length is around 12 kft and the mean loop length is about 13.4 kft, with a standard deviation of about 1200 ft. In 1980, 25% of all loops were longer than 20 kft and typically had load coils or amplifiers. Loops in the United States are typically longer than in most other countries.

Loops in the United States are almost all twisted pair. While some pairs use aluminum wire, most use copper. Typical wire gauges are 19, 22, 24, and 26. The typical number of pairs in a loop cable range from the common 6-pair drop wire to extremely thick cables, some as large as 3600-pair, found in the feeder plant.

Attenuation of a wire pair, plotted in Figure 3.4, is seen to be strongly dependent on frequency. The x axis in Figure 3.4 is broken, showing two different scales. Attenuation is plotted from 0 – 4 kHz at the left and from 10 – 60 kHz at the right.

The attenuation of an unloaded loop is seen to rise smoothly with frequency [2]. A 3-mile pair attenuates a pure 1-kHz signal by 6 dB, a pure 3-kHz signal by 9 dB, and a pure 20-kHz signal by 18 dB. The percentages of the respective transmitted signal that reach the other end of the pair are: 25%, 12.5%, and 1.6%.

Adding a load coil to the pair is seen to dramatically flatten the attenuation up to 3 kHz, but the inductor practically blocks any signal above 4 kHz. You can see from Figure 3.4 why load coils were so desirable for voice transmission, but are so detrimental for ISDN at 144 Kbps (Section 13.4), ADSL at 1 MHz (Section 19.1), or even calling party

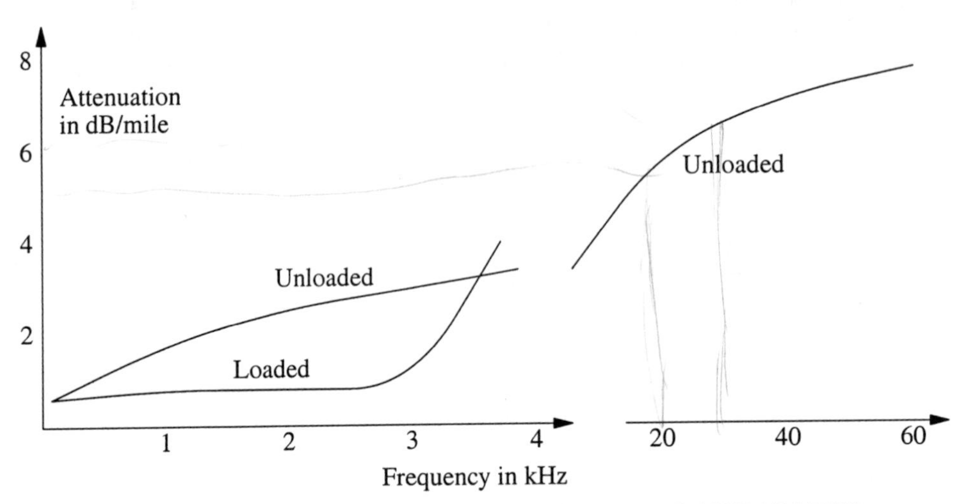

Figure 3.4. Typical loop attenuation versus frequency. © 1977 AT&T [2].

ID at 12 kHz (Section 18.1). While the numerical values in this figure and the next one represent a 22-gauge cable pair, a similar shape would be expected for other gauges. Higher gauge pairs would have higher attenuation and higher delay.

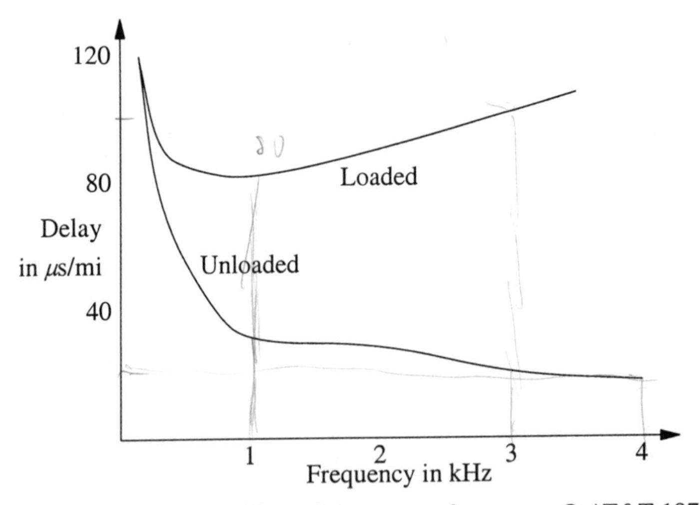

Figure 3.5. Typical loop delay versus frequency. © AT&T 1977 [2].

Delay in an unloaded loop is seen in Figure 3.5 to fall smoothly with frequency [2]. Recalling that the speed of light in air is very close to one nanosecond per foot, Figure 3.5 shows that a pure 4-kHz signal travels at 20 microseconds per mile, about one-fourth the speed of light. A pure 1-kHz signal is twice as slow and a pure 200-Hz signal is four times slower. Load coils are seen to significantly slow a signal.

Noise has also been carefully measured on real trunks [1]. While absolute signal power is typically measured in decibels relative to 1 milliwatt (dBm), absolute noise power is typically measured in decibels for random noise relative to 1 picowatt (dBrn).

So, 1 dBrn = –90 dBm. Since not all noise is equally annoying to humans during analog voice transmission, a subjective C-message weighting is typically applied over the noise spectrum to measure the effective amount of noise in the voice range. Such weighted noise power is given in decibels for C-message random noise relative to 1 picowatt (dBrnC). In a census [1] taken in 1980 over sampled loops shorter than 20 kft, 50% of all tested trunks had measured noise less than 92 dBrn and 61 dBrnC at the telephone, and 90% had measured noise less than 106 dBrn and 80 dBrnC. When measured at the central office, the noise level is typically about 20 dB smaller.

It was observed [1] that most of the total noise is caused by induction from electric power lines at 60 Hz and its harmonics. C-message weighted noise is dominated by the 9th harmonic at 540 Hz. When measured lines were grouped by urban, suburban, and rural, urban lines were typically 5 dB less noisy and rural lines were typically 5 dB more noisy than the Bell System average — probably because urban lines are typically shorter and rural lines are typically longer than the average. Noise was seen to vary significantly during the day, measuring about 5 dB greater than the daily mean during working hours.

But, reader beware. Suppose we wish to deliver a pure 60-kHz signal, at a signal-to-noise ratio of 20 dB, to all telephone subscribers whose loops are shorter than 20 kft. If we assume a maximum noise of 110 dBrn (which equals +20 dBm) at the receiver, we would deduce that the signal intensity at the receiver needs to be +40 dBm. So, at 8 dB/mile of attenuation over 4 miles, we might deduce that we need a transmit power of about +70 dBm. This analysis is perfectly correct, but only if one subscriber is served in all the pairs in a cable. Transmitting at +70 dBm will induce crosstalk, at 60 kHz, in all the other pairs in the cable and significantly raise the noise in all the other receivers (see discussion of ADSL in Section 19.1).

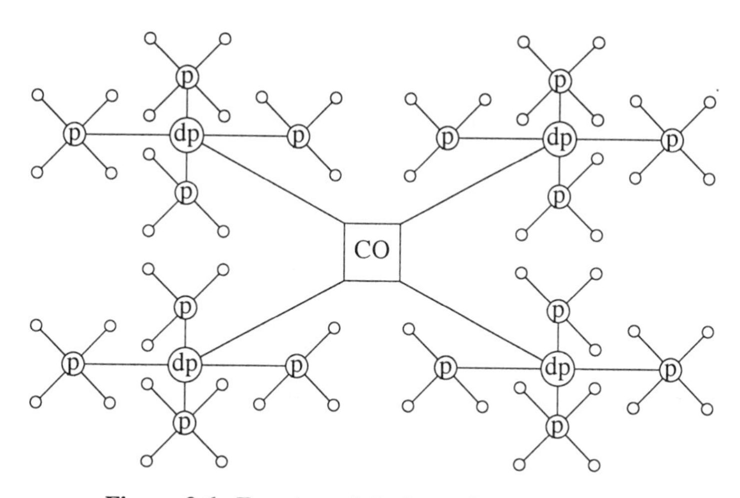

Figure 3.6. Top view of the loop plant architecture.

3.4.3 Topology

Figure 3.6 illustrates the top view of the three-level star architecture of the telco's loop plant. Figure 3.6 shows a CO in the center of its serving area. Feeder cables branch out from this CO in four directions, each terminating in a distribution point (dp in the figure). Distribution cables branch out from each distribution point in four directions, each terminating in a pedestal (p in the figure). Drop wires branch out from each pedestal, in three or four directions, each terminating in a residence (small unlabeled circle in the figure).

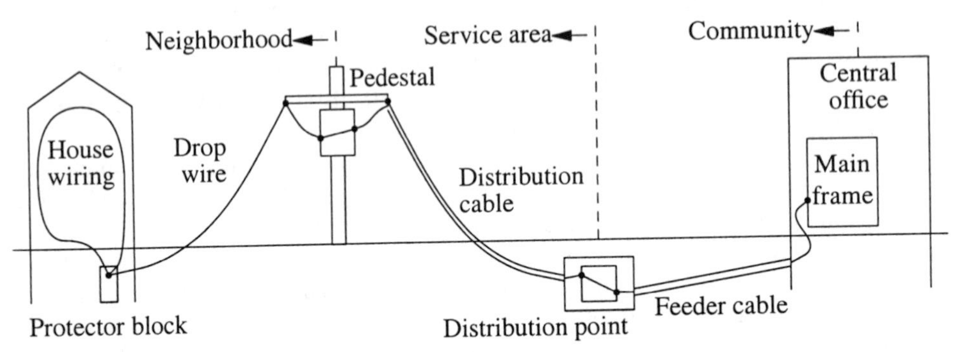

Figure 3.7. Side view of the loop plant architecture.

Figure 3.7 illustrates the side view of the three-level star architecture of the telco's loop plant. Figure 3.7 shows the CO at the extreme right. Only one distribution point is shown, and the single feeder cable that connects it to the CO. Only one pedestal is shown, and the single distribution cable that connects it to the distribution point. Only one residence is shown, and the single drop wire that connects it to the pedestal.

Examine both figures as this description proceeds. The primary arms of the star, collectively called the feeder plant, run from the switching office at the hub of the network to secondary nodes, called distribution points. Figure 3.7 shows only one feeder cable, running to one distribution point. A switching office has many feeder cables, each carrying thousands of pairs, or their digital equivalent channels, between it and many distribution points. Each feeder cable "feeds" a different part of the community, called a service area. Each distribution point is a large crossconnect field, typically underground in a vault and accessible through a manhole. Each service area consists of a distribution point, at its hub, and a star-of-stars subnetwork.

The secondary arms of the star, collectively called the distribution network, run from these distribution points to tertiary nodes, called pedestals. Figure 3.7 shows only one distribution cable, running to one pedestal. Each distribution point has many distribution cables, each carrying hundreds of pairs, or their electronic equivalents, between it and the pedestal that serves a neighborhood. Each pedestal, at the hub of a small star network, is a small crossconnect field, typically located on the ground, but shown in Figure 3.7 on a utility pole.

The tertiary arms of the loop network are the drop wires. Figure 3.7 shows only one drop wire, running from a pedestal to one protector block. The six pairs described in the previous section, called drop wires even though they are underground, run from my house to one of these pedestals.

The architecture of the loop plant has evolved as a compromise between two large expenses to the telco: the investment in physical plant and equipment and the labor costs of operation, administration, and maintenance (OA&M). The six pairs of drop wires serving my home do not run all the way to the switching office. This would be wasteful of physical plant, since four are unused. However, if I should order another teen line, the telco would not run a new pair from my home all the way to the switching office. This would be extremely costly for telco personnel, called linemen, to dig up streets and lay new conduit and cable.

The number of pairs in a feeder cable optimizes this tradeoff within its service area. The number of pairs in a distribution cable optimizes this tradeoff within the neighborhood it serves. The number of pairs in a drop wire optimizes this tradeoff within a home. The crossconnect panels in the distribution points and in the pedestals provide the necessary flexibility. Installing a new line in my home is a six-step procedure, proceeding left-to-right across Figure 3.7.

1. A telephone must be connected inside the home by a cable with modular plugs to a modular jack.
2. The modular jack must be tied behind the cover plate to a spare pair in the house wiring.
3. This house-wiring pair must be tied at the protector block, or TNI, to the house-end of a spare pair in the drop wire.
4. The network-end of this drop-wire pair must be tied inside the pedestal to the subscriber-end of a spare pair in the distribution cable.
5. The office-end of this distribution pair must be tied at the distribution point to the subscriber-end of a spare pair in the feeder cable.
6. The office-end of this feeder pair must be tied, via the main frame, to the fabric inside the switching office.

These six steps merely provide the physical connection. The four unused drop wires into my home are unterminated at my protector block and are also unterminated in the pedestal.

The telco tries to leave enough spare pairs to allow simple growth but not so many that the investment is wasted. The loop plant represents a large telco investment in right-of-way, conduit, poles, vaults, cable, and other equipment. The telco is typically aware of planning and zoning board activities in its serving communities and requires long-term notification about new construction for housing developments and apartment complexes.

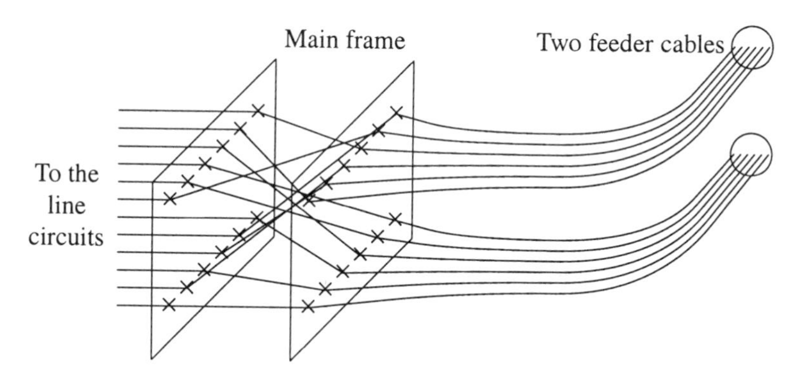

Figure 3.8. Main distribution frame.

3.4.4 Main Distribution Frame

The feeder cables typically enter the switching office building in the basement. They are housed in large-diameter watertight sheathes under positive air pressure by an air compressor in the office. If a feeder cable acquires a small hole, it will bubble rather than have water leak in, and air flow from the compressor indicates a problem. All the pairs in all the feeder cables are fanned out and terminated on one side of a huge crossconnect field, called the main distribution frame (MDF) or main frame for short. The main frame, illustrated in Figure 3.8, is 14 ft high, complete with sliding ladders, and runs the length of the central office building. Each appearance is fused to protect switching equipment from over-voltage. Thus, the pairs to the loop plant and eventually to the subscriber's telephones terminate on the line side of this MDF. Wiring inside the office consists of many pairs that terminate, at one end, on the office side of the main frame and, at the other end, on line circuits and one or more appearances on the switching fabric.

In progressive-controlled systems, like Step-by-Step, the directory number determines the subscriber's appearance on the terminating side of the fabric (Chapter 6). The pair on the line side of the MDF connects through the loop plant to a given subscriber. This pair must be crossconnected at the MDF to the pair on the office side of the MDF, which runs to the appearance on the terminating side of the fabric corresponding to the subscriber's directory number. This same pair on the line side must be crossconnected at the MDF to another pair on the office side, which runs to an appearance on the originating side of the fabric. This originating appearance, including per-line equipment called a line circuit (next section), is arbitrary and is chosen to balance load (Chapter 5). In end-marked fabrics, like Crossbar Number 5 and electronic systems, a subscriber needs only one fabric appearance, called his equipment location. This appearance is arbitrary and is selected when the subscriber's pair is crossconnected at the main frame. These end-marked systems require a means for the equipment to translate both ways between directory number and equipment location for terminating calls and for billing.

At this point the reader may appreciate the labor-intensive effort required by the telco to install a new line and may understand why the telcos charge so much and appeal often for commissions to raise that charge. Besides the many house and loop connections described earlier, telco personnel, called framemen, must crossconnect the subscriber's pair at the main frame. Depending on the technology of the office, translation equipment is wired or translation data is programmed and class of service (Section 3.5.5) is wired or programmed. Records of the directory numbers, loop plant, main frame and other office wiring, and translations and class of service must be maintained meticulously.

3.5 PER-LINE FUNCTIONS

Most equipment in a switching office is pooled and temporarily assigned and connected to a subscriber's line for the duration of dialing or for the duration of a call. However, per-line equipment is dedicated to a subscriber's line, even when the line is not in use on a call. Per-line equipment discussed so far in the home and loop plant includes the telephone and any extensions, the various segments of the pair, and assorted connectors, jacks, and fuses. Per-line equipment inside the switching office includes the line circuit and equipment for providing busy indication and class of service. While per-line equipment is typically simple and inexpensive, the net cost is high because there is so much of it.

This section also describes the circuit that receives dialed digits inside the CO and several circuits that provide for call supervision.

3.5.1 Line Circuit Description

Line circuits provide talking battery, busy indication, and attend a line, responding to off-hook for call origination and answer supervision. The line circuit described in this section is taken from a Step office. A line circuit in a crossbar office is slightly simpler. Line circuits in electronic and digital offices are described in Chapters 11 and 14.

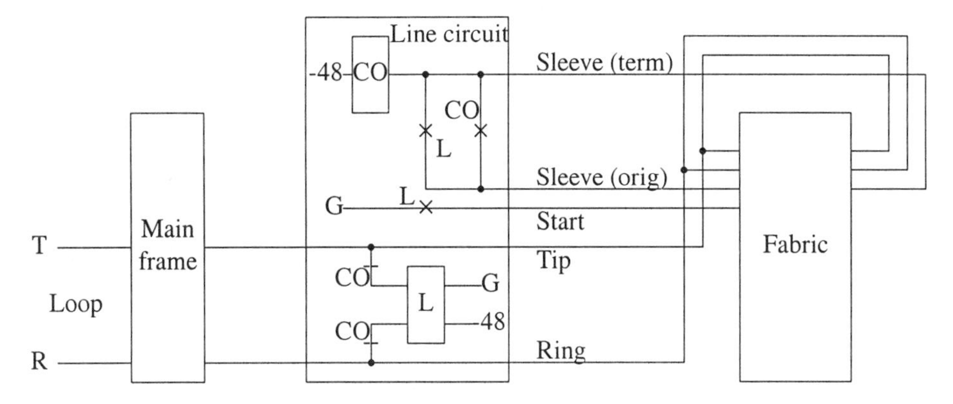

Figure 3.9. Line circuit.

Figure 3.9 shows the tip and ring of one loop pair, crossconnected at the main frame, its dedicated line circuit, and the originating and terminating side of the fabric. A third wire, called the sleeve (Figure 1.6), interconnects the line circuit with the two sides of the fabric. A fourth wire, the start, interconnects the line circuit only with the originating side. Tip, ring, and sleeve are all switched through the fabric from the calling party to the called party. The line's tip and ring carry voice both ways between both parties. The sleeve wire does not extend to the subscriber — it is introduced inside the office to control the fabric.

The line circuit has two relays, the line relay and the cut-off relay. The function of the line relay is to track the state of the hook-switch in the telephone. The function of the cut-off relay is to disconnect the line relay from the loop when the latter should not be attending. With the telephone on-hook, both relays are in their normally released state. The originating and terminating sides of the fabric present open circuits on all wires. Talking battery is applied to the tip and ring through the resistance of the line relay's dual coils. The sleeve and the start present an open state to the originating side of the fabric, indicating that the line is quiescent. The sleeve presents battery through the cut-off relay's coil at the terminating side, indicating that the line is idle. The line relay attends the telephone.

3.5.2 Line Circuit at Origination

Let the subscriber, whose loop and line circuit are shown in Figure 3.9, decide to originate a call. When he raises the handset, the hook-switch closes, completing the DC circuit in the loop. Current flows from ground, through the upper coil of the line relay, down the tip, through the resistance of the telephone, back up the ring, through the lower coil of the line relay, to the -48V battery. The line relay operates, echoing the state of the telephone's hook-switch and closing the two normally open contacts of the line relay. One contact causes the line circuit to present a ground on the start wire to the originating side of the fabric. The other contact connects the sleeve appearance on the originating side to the coil of the cut-off relay and the sleeve appearance on the terminating side. As will be shown in Section 6.3, ground on the start is the signal for the originating side to connect to the line.

- The originating side assumes supervision control of the line by attaching a relay, analogous to the line relay, across tip and ring.
- Dial tone is placed on the line, indicating that the office is ready for the subscriber to begin dialing.
- The originating side presents a ground on the sleeve, indicating to the terminating side that this line is busy and signaling the line circuit that it should drop off the line.

The ground state presented at the originating appearance of the sleeve completes the DC path through that second contact of the line relay, operating the cut-off relay. A normally open cut-off relay contact, paralleling that second line relay contact, closes. Through this contact, the ground state presented at the originating appearance of the sleeve is presented to the terminating side. This grounded sleeve marks the line busy at the terminating side and holds the cut-off relay operated for the duration of the call. Two normally closed cut-off relay contacts open, removing the line relay from the loop. The line relay releases, removing the ground state from the start.

The system is now ready for the subscriber to begin dialing. We continue from this point in Section 3.5.6.

3.5.3 Line Circuit at Answer

Let a call be placed to the subscriber, whose loop and line circuit are shown in Figure 3.9. Before the tip, ring, and sleeve carrying the call through the fabric are connected to this called party, the state of this subscriber's sleeve is tested at the terminating side.

If this subscriber's telephone is on-hook, the line circuit presents battery on the sleeve to the terminating side. This condition at this moment indicates to the fabric that the called party is idle. Among the three wires carrying the call through the fabric, the tip and ring are not connected yet at the terminating side to the tip and ring of the called party. Instead, the terminating side of the fabric switches ring signal and an answer supervision relay, a relay circuit similar to the line relay, onto the called party's tip and ring. The called party's telephone rings. Audible ring tone is switched onto the tip and ring carrying the call and is heard through the fabric by the calling party.

At the same time, the sleeve in the three wires carrying the call through the fabric, held in a ground state at the calling party's appearance on the originating side of the fabric, is connected to the sleeve of the called party at the terminating side. Current flows from this ground at the originating side of the call, down the sleeve through the fabric, to the sleeve at the terminating side of the called party's line circuit, through the coil of the cut-off relay, to battery. The cut-off relay of the previously idle called party operates, disconnecting the called party's line relay.

If the called party does not answer, the calling party abandons the call by hanging up. The open on the loop is detected at calling party's appearance on the originating side of the fabric and the ground is removed from the sleeve. This releases all the contacts in the fabric and releases the cut-off relays in the line circuits of both the calling and called parties. The call is dropped. If the called party answers, the act is not interpreted as a request to originate a call because the line relay does not stimulate the called party's appearance on the originating side. Instead, that answer supervision relay, at the called party's appearance on the terminating side of the fabric, operates. This causes the fabric to remove the ring signal and audible ring tone and to connect the tip and ring of the wires carrying the call through the fabric to the called party's tip and ring at their appearance on the terminating side.

3.5.4 Busy Indication

Reconsider a call placed to the subscriber, whose loop and line circuit are shown in Figure 3.9. If this called party's telephone is off-hook, the sleeve at his appearance on the terminating side of the fabric is in a ground state.

If this subscriber is busy because he had originated a call, Section 3.5.2 showed that the ground state was presented to this sleeve appearance by the subscriber's appearance on the originating side of the fabric through a cut-off relay contact in the line circuit. If the called party is busy because he had answered a call, Section 3.5.3 showed that the ground state was presented to this sleeve appearance by the calling party's appearance on the originating side of the fabric through the call connection through the fabric.

This grounded sleeve condition at this moment indicates to the fabric that the called party is busy. The terminating side of the fabric switches busy tone onto the tip and ring carrying the call and it is heard through the fabric by the calling party. If the called party hangs up after this, but before the calling party abandons the call, there is no means by which the call can be completed.

3.5.5 Class of Service

Other per-line equipment in the office provides for class of service. COS is information about a line that aids the office to process a call. Examples include: whether the subscriber has a dial-pulse or Touch-Tone telephone, coin telephone or not, party-line ringing procedure, whether the line connects to a telephone or to remote switching equipment, restrictions on originating or terminating, message-rate or flat-rate billing, and if the bill is past due.

Each type of switching equipment implements COS differently. The subject is discussed in the chapters corresponding to the types.

3.5.6 Dialing In

Continuing from Section 3.5.2, the subscriber has gone off-hook, the line circuit has responded to the off-hook and then been cut off, supervision and talking battery have been applied from "deeper" within the CO, dial tone has been applied, and now the subscriber begins to dial his telephone.

The schematic in the upper part of Figure 3.10 is a simplified drawing of an electro-

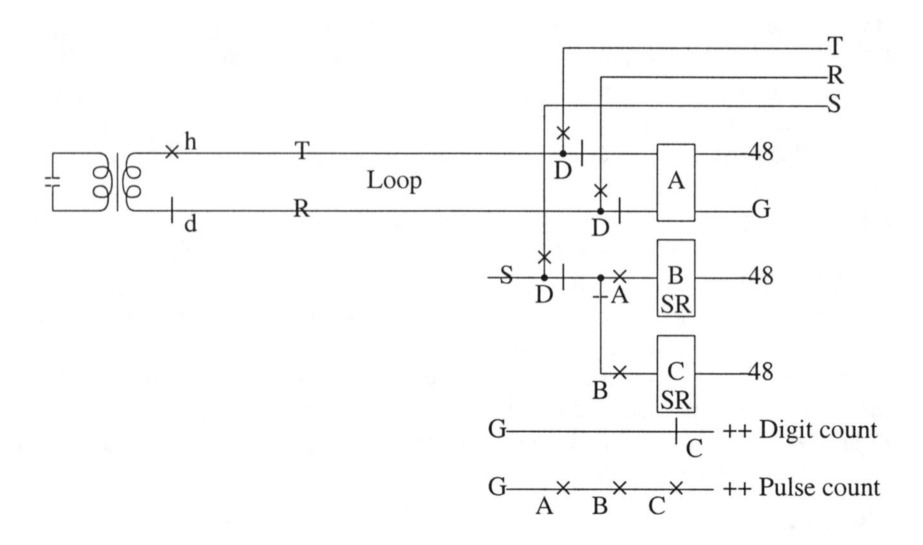

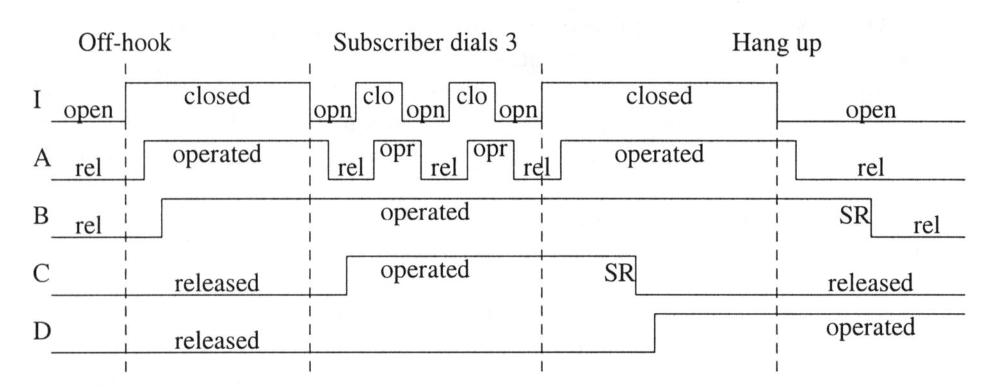

Figure 3.10. Digit receiver circuit and pulse waveforms.

mechanical circuit that receives dial pulses. The upper left of Figure 3.10 shows the dial contact in the telephone and the tip and ring runs from the telephone, across the figure, to this circuit. The A, B, and C relays are shown in the upper right of Figure 3.10, with several make and break contacts from each relay. Recall that a directory number is a sequence of dialed digits and that a dialed digit is a sequence of dial pulses. So, the circuit must count the pulses in each digit and perform timing to determine digit boundaries and the end of dialing.

In Step-by-Step, many circuits like the one in Figure 3.10 are distributed throughout the fabric. A different circuit receives each dialed digit. One such circuit, located inside the fabric close to the originating side, receives all the pulses of the first dialed digit. After receiving and interpreting that digit, the circuit and its associated switch connect the call one stage deeper into the fabric to another such circuit that acts on the next digit. This continues until the path has been directed through the fabric to the terminating side and the called party. So a seven-digit call uses seven of these circuits (not strictly true, as seen in Section 6.4), and a one-digit call, a 0 for the operator or a 1 to connect the path to an outgoing trunk, uses only one such circuit. Each pulse in a dialed digit moves a mechanism in the electromechanical switch associated with the digit receiving circuit. The circuit must provide a contact operation per dial pulse to actuate the switch. Second, it must signal the end of the digit, so the next stage can be ready for the next digit. Third, it must distinguish an on-hook and, if received, must cause the partial connection through the fabric to be dropped.

In Crossbar and other register- and common-controlled systems, the circuit of Figure 3.10 is part of a dial pulse receiver (DPR). There are many DPRs in the office, all wired to the trunk side of the fabric. After a subscriber, wired to the line side of the fabric, goes off-hook, he is connected through the entire fabric to an idle DPR. This DPR returns dial tone and receives every pulse in every dialed digit from this subscriber. The pulses in each digit are counted and the digits are stored in the DPR until dialing is complete. As part of this DPR, the circuit in Figure 3.10 must provide a contact operation per dial pulse to advance the pulse counter in the DPR. Second, it must signal the end of the digit so the DPR can store the present pulse count and reset the pulse counter for the next digit. Third, it must distinguish an on-hook and, if received, must cause the call to be dropped. The circuit shown in Figure 3.10 is similar in Step-by-Step and in Crossbar, except that it is replicated per-switch in the Step-by-Step fabric and per-DPR in Crossbar.

The circuit of Figure 3.10 must distinguish three timing intervals: the time between pulses, the time between digits, and a timeout indicating abandonment. The A relay corresponds to the L relay in the line circuit. It provides talking battery to the telephone through its windings and tracks the state of the hook-switch or dial contact in the telephone. One contact in the A relay provides a switch closure for every received dial pulse. The B and C relays provide for the other timing requirements. These relays, altered for slow-release, remain operated during the interval of one dial pulse. The D relay, not shown in Figure 3.10, acts like the cut-off relay in the line circuit. In Step-by-Step, this D relay operates during the interval between the digit just received and the next digit. Make contacts of this D relay disconnect the corresponding A relay and transfer tip, ring, and sleeve to the next stage in the fabric. A corresponding action does not occur in Crossbar.

3.5.7 Dialed Digit Reception

The lower part of Figure 3.10 shows five waveforms for the interval around the received dial pulses for a single dialed digit 3. The top waveform indicates the current flow from tip to ring. The lower four waveforms indicate the state of the A, B, C, and D relays, respectively, where a high level represents an operated relay or closed make contact. When tip and ring are connected to this A relay, either in the interval preceding this digit in Step-by-Step or at the beginning of dialing in Crossbar, the A relay provides talking battery to the telephone and A operates. Since the subscriber hasn't started dialing this digit yet, the dial contact in the telephone is still in the make position. A make contact of the A relay causes the B relay to operate, but the C relay is still released.

Dial pulsing begins after the subscriber releases the dial. The dial contact in the telephone opens and closes with a period of 0.1 second. Current in the loop flows and stops with each dial pulse and the A relay releases and operates correspondingly. The simultaneous operation of the A, B, and C relays provides the external signal that drives the switching mechanism in Step-by-Step or that advances the pulse counter in the DPR in Crossbar. Thus, the A relay tracks the telephone's dial contact. If the A relay were supervising the call, the switching office would disconnect the call with the first dial pulse received. But the slow-release B relay, and not the A relay, provides supervision and the 50-ms interval of a single dial pulse is insufficient time for B to release. So, B remains operated during the dial pulses, as seen in the third waveform at the bottom of Figure 3.10. Suppose the subscriber hangs up in the middle of dialing. Since the dial contact and the hook-switch have a similar effect on loop current, this on-hook resembles an extended dial pulse and the slow-release B relay would release after about 120 ms. A contact, not shown in Figure 3.10, of the B relay provides the external signal that causes the switching office to drop the call. Thus, the B relay tracks the telephone's hook-switch.

In the schematic, since the C relay's winding is wired through one of B's make contacts and one of A's break contacts, C operates when B is operated and A is released. This combination occurs with the arrival of the first pulse in the sequence for this digit, as seen in the fourth waveform in Figure 3.10. Since C is also slow-release, the 50-ms interval between dial pulses is insufficient time for C to release. So C, like B, remains operated during the dial pulses. At the end of the last pulse in the sequence for this digit (the third pulse in the upper waveform of Figure 3.10), the dial contact in the telephone remains operated and loop current flows while the subscriber cranks the dial for the next digit. After 120 ms into that interval, the C relay releases, as seen in the fourth waveform in Figure 3.10. This can only occur at the end of the pulse sequence representing the digit. Thus, the C relay tracks the telephone's off-normal contact. A contact, shown in Figure 3.10, of the C relay provides the external signal that signifies the end of each digit. In Step-by-Step, this signal causes the operation of the D relay (not shown in Figure 3.10), and the progression of the call one stage deeper into the fabric. In Crossbar, the signal causes the DPR to save the contents of the pulse counter and to advance the digit counter.

3.5.8 Supervision

After a call is set up, talking battery must still be applied to both parties throughout the call. If applied through the *supervision relay*, functioning like a line relay or A relay, the state of the party's hook-switch may be tracked. This state provides the switching office a means to ascertain call completion through contacts of the supervision relay, not shown in the figures below. Besides determining the time to drop the switch train through the fabric, this signal also provides the time for end of billing on those calls that are billed. The supervision relay is slow-release so subscribers' accidents with the switch-hook or dial don't cause a call to be dropped.

Figure 3.11 illustrates three different configurations of the supervision relay, depending on whether supervision control is assigned to both parties in (a), only the calling party in (b), or to either party in (c). *Both-party supervision,* illustrated in Figure 3.11(a), has the simplest supervision circuit. Since current flows from talking battery through the supervision relay to both parties, S cannot release until both parties hang up. The major drawback to both-party supervision and to *called-party supervision* is that these arrangements give the called party, the party not billed, control over the end of billing and the disconnect. If, for whatever reason, the called party does not hang up at the end of a billed call, the calling party receives a large inexplicable telephone bill. Furthermore, with calls through more than one office, the talking battery would be greatly attenuated if it were provided from one central point on the call, say the office of the

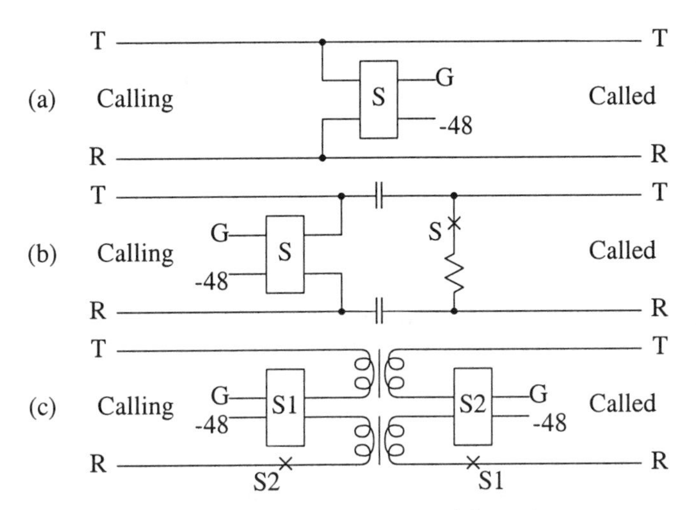

Figure 3.11. Alternative supervision schemes.

calling party. While the long DC path implied by Figure 3.11(a) would work in an intraoffice call, it would not work in an interoffice call. The talking connection, especially for multioffice calls, is AC coupled. Any DC signaling, like dial pulsing and state-of-supervision, must be detected on the calling side of the AC coupling capacitor or transformer and then repeated on the called side (Section 4.4).

A supervision circuit, like the one in Figure 3.11(b), gives the calling party control over the end of billing and provides talking battery from a party's home office on interoffice calls. *Calling-party supervision* is provided by decoupling the DC through a capacitor, in Figure 3.11(b), or a transformer. Such a circuit appears somewhere in the fabric along the switch train. If the called party hangs up, the supervision relay remains operated and the call is not dropped. If the calling party hangs up, the supervision relay releases and the call is dropped. Calling-party supervision is the most common form of supervision, especially in the old electromechanical offices. But it is not without its problems, as exemplified by yet another unsubstantiated legend.

Legend has it that there were two butchers in the same neighborhood in Brooklyn, back in the days of telephoned orders and delivery boys on bicycles. Butcher A discovered that he could linger at home, waiting until his competitor B opened his store, and then could call B's store from his home, placing a crank order or feigning a wrong number. Leaving his home telephone off-hook, A opened his store and B's telephone would be busy all day long. We have all had the experience of hanging up and then lifting the receiver to place another call, only to find the original call still connected. We probably have not noticed that this event occurs only when we are the called party on the original call.

The ideal is *either-party supervision,* provided by a circuit like that in Figure 3.11(c). When either party hangs up, the corresponding supervision relay releases and the office may drop the call and stop billing. Figure 3.11(c) shows transformer coupling, but capacitive coupling could be used. Either-party supervision requires a more expensive supervision circuit and requires monitoring twice as many relays.

I close this section and chapter by sharing a fantasy. I would like to have a switch on my telephone that provides me called-party supervision on demand. I would be willing to busy my telephone line for an hour or two if it would penalize the people and computers that place junk telephone calls. As a public service, I could lay the receiver down, neither hanging up nor listening, while they are billed and their lines and equipment are tied up. Of course, the people who place what many of us consider to be junk calls, and who think it is their right to do so, don't share this fantasy.

3.6 NON-POTS LINES

Five types of non-POTS lines are briefly described: unswitched private lines, loops to coin telephones, lines to customer-premises switching equipment, lines to the telco's remote switching equipment, and loop carrier. All but the unswitched private lines appear on the fabric in the switching office. All are distinguished by class of service.

3.6.1 Unswitched Private Lines

Suppose two subscribers in the same community wish to be interconnected by a permanent wire pair, not necessarily for voice communication. Rather than dig up the streets and run this line themselves, they may each lease a wire pair to the central office and arrange with the telco for the lines to be connected at the main frame. Neither line is crossconnected through the main frame to the fabric. The typical application is to connect alarms in banks, stores, and private homes to the police department or a private alarm agency.

If the two subscribers are served by separate central offices, they may lease a trunk between the two COs and arrange for connection to it through the main frame in each CO. Neither the leased line nor the leased trunk appear on the fabric in either CO. Leased lines and trunks are discussed at length in Chapter 12.

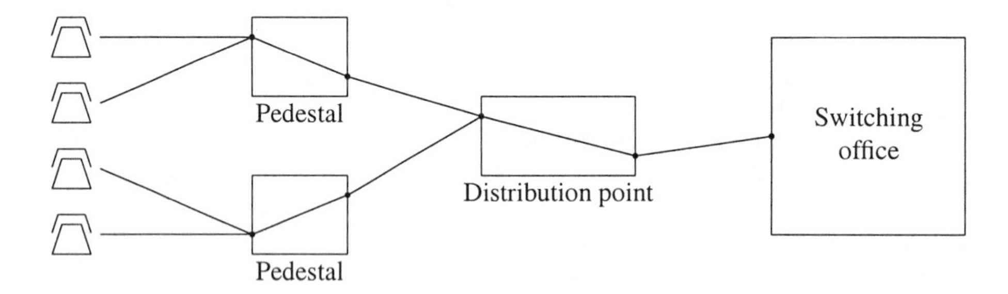

Figure 3.12. Four-party line.

3.6.2 Party Lines

A *party line* is a collection of telephones, with separate directory numbers and in separate residences, but sharing one line to the switching office. Figure 3.12 shows a four-party line, with four subscribers sharing a common tip/ring loop pair. In this example, the two upper subscribers are neighbors, served by a common pedestal, and the two lower subscribers are neighbors, served by a different common pedestal. Each pair of neighbors shares a common pair in the distribution cable between the distribution point and their pedestal. All four subscribers share a common pair in the feeder cable between the distribution point and the CO. Party lines can share different segments of the loop, including a common drop wire.

Party lines are no longer common, mainly because the telcos no longer encourage them by offering rates significantly lower than private line service. The two-party line was most common, but four- and eight-party lines were offered. With only one party allowed to use the line at a time, a subscriber determines whether the line is in use by going off-hook and listening. The greatest technical challenge is ringing an individual telephone, called selective ringing, instead of ringing all the parties on the line.

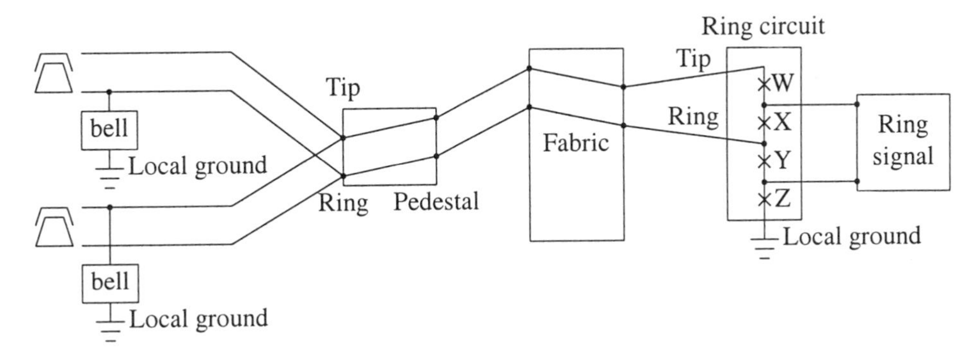

Figure 3.13. Ringing a two-party line.

The typical party line had two parties and a simple method for selective ringing. In Figure 3.1(d), the bell is wired between the tip and ring wires. As illustrated at the left in Figure 3.13, in a two-party line, the bells in each party's telephones are wired differently. The telco's installer connects the bells in all the tip party's telephones (lower left in Figure 3.13) between the tip wire and local ground (a water pipe). The installer connects the bells in all the ring party's telephones (upper left in Figure 3.13) between the ring wire and local ground. The ring circuit (at the right in Figure 3.13) in the switching office connects the ring signal to the called party's tip and ring in one of three ways:

- Across tip and ring for single-party lines (by operating contacts W and Y in the ring circuit of Figure 3.13);
- Between the tip wire and building ground to ring the tip party of a two-party line (by operating contacts W and Z in the figure);
- Between the ring wire and building ground to ring the ring party (by operating contacts X and Z in the figure).

Other less common means for selective ringing use the polarity of the talking battery or use filters to distinguish between a 20-Hz ringer and a 30-Hz ringer. Typically, an eight-party line used *coded ringing*. The ring generator can produce short bursts of ring signal besides the regular 2-second burst and each telephone on the party line is rung with a different code of long and short bursts. All parties are rung, but the intended protocol is that the party that answers is the one whose code is rung.

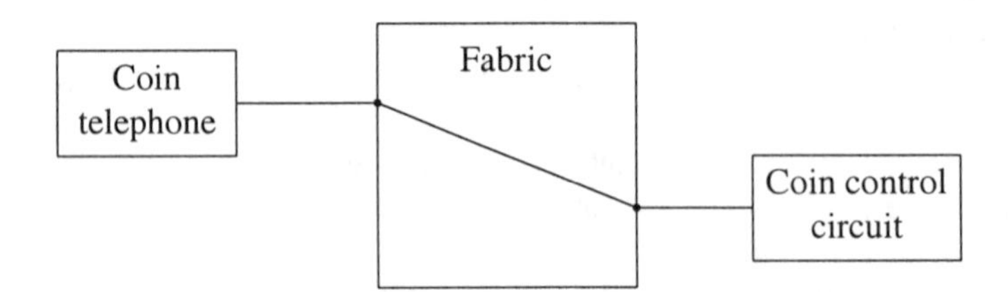

Figure 3.14. A coin telephone connected to a coin control circuit.

3.6.3 Coin Telephones

Coin-operated public telephones are controlled much differently from POTS telephones. During dialing, a POTS telephone must signal its hook-switch and dialing (or toning) state with a circuit in the switching office that is specially designed for that purpose and is attached to the call only for the duration of dialing. A coin telephone has a similar need, except that coin status must also be signaled. So, at the time of off-hook, a different, more complex, coin control circuit is connected (as shown in Figure 3.14) instead of the simple circuit used for POTS digit reception. In Step-by-Step offices, loops to coin telephones are physically segregated on the originating side of the fabric and typically, also on the terminating side, by assigning coin telephones directory numbers of 99xx. Digit receiving circuits in this part of the fabric are made more complex. In common-control offices, loops to coin telephones are distinguished by class of service and connected through the fabric to different circuits.

The problems that the telcos face with telephone fraud are epitomized in coin-operated public telephones. Many of us remember, as children, sticking a pin through the cord between handset and base and touching the pin to the base in an effort to simulate the DC signal associated with coin insertion. Or, perhaps, we remember calling Mom from the movies and, if we made the conversation brief enough, the coin telephone would return our dime. These tactics no longer work. The telcos shielded the cord, so you can't stick a pin through it. Then, the telco changed to audio signaling, bongs and gongs as the coins go in, instead of DC signaling. But I've heard of people who can whistle into the coin telephone and simulate the gong associated with coin insertion. There seems to be an ongoing battle between people who try to cheat the telco and the telco's designers of coin telephones and coin control circuits. We will describe long distance fraud in Sections 4.4.2 and 8.5.4. Two complicated business issues facing the telephone industry are introduced here: the incompatibility of deregulation with pricing that is not based on cost and the cost of changing procedures.

Cheating, stealing, and vandalism are unfortunately common in public telephone service. Considering the cost of lost revenue and repairs, the cost to the telcos of providing coin telephone service is greater than the income derived, even at the outrageous rate of 25 cents for the first 3 minutes. This fact is accentuated in neighborhoods where the provision of coin telephones is probably most important. The coin telephone tariff is a simple example of the problems that arise when regulated prices are not based on cost, usually because of social engineering by the regulating bodies. The generality will be discussed in Chapter 16. Illustrating the entire problem with de-regulation of telephony, if a telco is required to lose money in one branch of its business, for whatever honorable reason, this telco must compensate by artificially overcharging in another branch. This works in a closed system, and it did until competition was allowed.

For years, the operating procedure with coin telephones was that, first, the subscriber deposits the coin and then dial tone signals that dialing may begin. Another unsubstantiated legend has it that someone once had an automobile accident late at night on a highway in Illinois. This person, injured in the accident, limped down the highway to a coin-operated public telephone, only to find that he did not have any coins in his pocket. This person's litigation against the telco led to what is simply called "dial tone first." Now the procedure in most coin telephones is that, first, the subscriber goes off-hook and receives dial tone and, then, must insert coins for calls that are not operator-assisted. No coins are required to call the operator. This simple change in procedure was expensive to the telcos because coin control circuits all over the United States had to be physically changed. In wired-logic telephone systems, the changes were designed for each different type of coin control circuit, instructions were written and distributed, and craftsmen made the changes. In computer-controlled telephone systems, however, the change in procedure was implemented as a simple program change and the updated program was distributed.

3.6.4 Private Branch eXchange Trunks

Businesses with a need for many telephones have a number of options available to them, as described in Chapter 10. One of these options, the PBX with equipment on the customer premises, presents an unusual interface to the line side of a central office. A PBX is switching equipment, typically on the customer's premises and typically much smaller than a central office. The telephones in a business terminate on the line side of the PBX. PBX trunks connect the PBX to the central office on which it is said to "home." The PBX end of a PBX trunk terminates on the trunk side of the PBX, and the CO end terminates on the line side of the central office. In many ways a PBX appears to a central office as a central office appears to a toll office.

When a PBX subscriber goes off-hook, the PBX returns dial tone and accepts the dialed digits. If the call is for another subscriber in the same PBX, the PBX completes the call and rings the called party. If the calling party wishes to call someone outside the PBX, he first dials a 9 (typically). This signals the PBX to complete a connection to a trunk. The circuit on the PBX end of the trunk closes the loop to the CO. Since the CO end of the trunk terminates on the line side of the CO, this closure looks like any other off-hook and the CO returns dial tone. The calling party heard one dial tone from the PBX and dialed 9 and now hears a second dial tone, this one from the CO. The calling party dials the number of the called party and the CO receives these digits and acts on them.

This PBX resides in some company or other kind of enterprise. This enterprise's telephone number is assigned at the CO to this block of trunks. Whenever a calling party outside the PBX calls the enterprise's telephone number, the CO connects the call to any one of these trunks that is idle and connects ringing. In a simple PBX, the ringing alerts an attendant at the enterprise. The attendant answers and gives the company's name and tries to determine which telephone within the PBX the calling party wishes to reach. If the PBX has the *direct in-dialing* service, the calling party could directly dial the telephone number of the desired party inside the PBX. The CO translates all PBX telephone numbers into the same block of trunks on its line side and connects the called party to an idle trunk. Responding to a special class of service, the CO dials the last four numbers of the called party's telephone number down the trunk into the PBX. The PBX, expecting these digits, completes the call. We provide more detail in Section 10.4.

3.6.5 Remote Switching

The telco may provision remote switching equipment that operates in conjunction with a CO like a PBX does. While such equipment is generically called a remote switch unit, there are three architectural variations depending on the call-handling capability of the remote equipment.

A *remote concentrator* (RC) is relatively simple equipment that merely optimizes the number of channels that are needed in the loop plant extended to some community that is very remote from a CO. The RC takes advantage of the fact that most telephone subscribers are idle most of the time. N loops could serve M subscribers, where $N < M$, with a small probability that some subscriber might be temporarily denied service. When a subscriber goes off-hook, the RC connects the line from the given telephone to some idle loop to the CO. The RC does not return dial tone and does not receive the subscriber's dialed digits — they are transmitted on to the CO. Incoming calls to subscribers served by an RC are handled like direct in-dialed calls to a PBX. A call between two neighbors served by the same RC is completed through the CO and ties up two separate lines in the loop plant between the CO and the RC.

A *community dial office* (CDO) is more like a PBX in that it processes the subscribers' dialed digits and can complete an intra-CDO call without using lines to the CO. A variation on the CDO occurs in stored-program switching offices that have a modular architecture.

A *remote switching module* (RSM) is architecturally similar to any other module in a modular switching system, such as a 5ESS (Section 14.2). But instead of physically residing in the CO, it is located in a remote community. It receives the subscribers' dialed digits, just like a CDO, but must communicate with the office's central control, like any other module must, in order to interpret these digits. Another application of RSMs is that a telco might install one in an office containing old equipment. If a subscriber desired some new service that the old equipment might not be able to provide, he might be moved to the RSM and would actually be served by another CO.

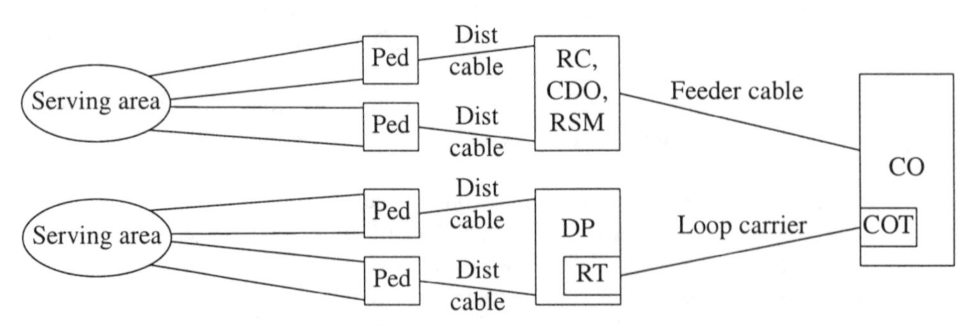

Figure 3.15. Remote switching and loop carrier.

Figure 3.15 illustrates the concepts from this section and the next one. The upper half of Figure 3.15 shows a remote switch unit (RSU) in one of the distribution points in some CO's loop plant. Subscribers in the area served by the two upper pedestals in Figure 3.15 are concentrated to a few feeder pairs by the RSU. If the RSU is a CDO or

RSM, these subscribers can call among each other without their physical connection ever reaching the CO. If the RSU is an RSM, the feeder cable is a special cable that also carries signaling and interprocessor messages (Section 14.2).

3.6.6 Loop Carrier

A particular subscriber's loop may not be a physical copper twisted pair all the way from the telephone to the serving CO. While the telephone end of any loop is a copper twisted pair, intermediate segments, particularly in the feeder plant, may be equivalent channels in some digital or fiber-optic carrier system.

The lower half of Figure 3.15 shows a loop carrier system between the CO and the lower distribution point. This loop carrier system might use some of the pairs in the feeder cable or it might use a separately installed optical fiber that parallels the feeder cable. The multiplexing equipment on the CO end of the loop carrier system, called the central office terminal (COT), resides inside the CO. If the CO's switching equipment is a digital switch, it probably has a direct interface for the loop carrier, eliminating the need for the COT. The multiplexing equipment on the subscriber end of the loop carrier system, called the remote terminal (RT), resides somewhere in the loop plant. Figure 3.15 shows the typical installation, with the RT in the vault housing a distribution point.

At level 1 of the North American Digital Signaling Hierarchy, 24 digital channels are multiplexed onto a 1.5-Mbps stream. (In Europe, 32 channels are multiplexed up to 2 Mbps.) This 1.5-Mbps rate can be used on a copper twisted pair, as long as it is free of bridge taps and load coils and a digital repeater is installed every mile. Since digital transmission is unidirectional, two pairs are needed between the two terminals of the DS1 digital carrier equipment. Typically a third pair is dedicated as a spare.

Suppose one DS1 terminal is placed in the CO, the COT. Then the CO side terminates on 24 appearances on the main frame. The loop side uses three twisted pairs in some feeder cable. Let the RT be placed in the distribution vault at the other end of the feeder cable. Then, the subscriber side of the RT terminates on 24 appearances on the feeder side of the patch panel in the distribution point. Then 24 wires from distribution cables would be assigned to these digital equivalent channels. The 24 subscriber loops use three physical pairs in the feeder cable instead of 24. The *pair gain,* the ratio of equivalent channels to physical lines, is eight.

Loop carrier is used to avoid digging up the streets when more channels are required in some existing cable route. It is common in the feeder plant and is even appearing in the distribution plant (the remote terminal would be in the pedestal). Remote terminals are frequently associated with remote concentrators and digital RSMs typically connect to their host systems via digital carrier. Digital switching equipment in the CO might directly interface to the digital carrier, eliminating the need for the CO terminal.

EXERCISES

3.1 In tone ringing, an audio signal is applied to the loop at the CO instead of a high voltage AC signal. A speaker replaces the bell in the telephone. Show a simple change to the telephone schematic that allows the earpiece to be used also as this ringing speaker. What problem does this modification pose for extension telephones?

3.2 One problem with conventional telephones as extensions is that two parties in the same location on separate telephones hear each other much louder than either hears a remote party, similar to the side-tone problem in a conventional telephone. Design a two-person telephone set.

3.3 Two extensions are connected to the same loop. With both off-hook, can either party dial into the CO? Can either party Touch Tone into the CO? Explain why this is different.

3.4 An undesired nonlinearity in the loop could cause intermodulation of Touch Tone signals. This could result in sum or difference frequencies. These sums or differences could cause trouble if any were close to any of the existing single Touch Tone frequencies. Which digits are most troublesome? Similarly, if the waveform of the local AC power signal varied slightly from being sinusoidal, then harmonics would arise and they could be induced into a telephone line. Which single Touch Tone frequency is closest to an integer multiple of 60 Hz?

3.5 We could extend the maximum unrepeatered loop length above 18 kft by using a lower wire gauge in the loop's twisted pair. Based only on drawing 25 mA at an off-hook telephone: (a) How long may a loop be if the wire's resistance is 30 ohms per 1000 feet? (b) What is the maximum resistance per 1000 ft in each of the two wires of the loop if we wish to extend the loop to 25 kft?

3.6 Consider signal attenuation, but only as a "voltage divider" among the telephone's resistance and the resistance of the two loop wires — ignore impedance, reflection, and other AC effects. Suppose a signal is transmitted from a CO to a telephone over a #26-gauge 18-kft twisted pair. What percentage of the transmitted signal appears at the telephone (drops across the telephone's resistor)? If two extension telephones are both off-hook, what percentage of the transmitted signal appears at each telephone?

3.7 A 3-kHz signal is transmitted over an 18-kft unloaded loop. From Figure 3.4, how many decibels of attenuation does it suffer? What percentage of the transmitted signal power reaches the telephone? Comparing this answer to the previous exercise, what percentage of the total attenuation is due to DC effects and what percentage is due to AC effects?

3.8 If maximum allowable attenuation is 15 dB, how far can you transmit a 20-kHz signal over a typical unloaded loop? A 30-kHz signal?

3.9 From Figure 3.3, approximately what percentage of loops in the United States in 1973 exceeded 18 kft? Assuming a Gaussian distribution, with a mean and standard deviation given in Section 3.4.2, what answer do you get? Is the Gaussian assumption accurate for predicting loop length?

3.10 Consider the dispersion of a square-wave pulse over a typical unloaded loop. If a 1-kHz square wave is transmitted over a 3-mile loop, what is the net delay of the (1-kHz) fundamental frequency? Of the (3-kHz) third harmonic? Which Fourier component arrives first? How wide is a 50% duty-cycle (0.5-ms) pulse at the received end?

3.11 Repeat the previous exercise for a loaded loop.

3.12 Repeat the two previous exercises for a 500-Hz square wave signal.

3.13 Considering only the dispersion between the first and third harmonics, what is the maximum loop length for carrying a 1-kHz square-wave signal before intersymbol interference is detected beween the first and third harmonics? Do this exercise for loaded and unloaded loops.

3.14 Suppose a CO has 20 feeder cables, each distribution point has 20 distribution cables, and each pedestal has 40 drop pairs. (a) Assuming 50% of the drop wires are unused spares, how many pairs are needed in each distribution cable, in each feeder cable, and at the CO's main frame? (b) Assuming 10% overcapacity in the distribution and feeder plant, how many pairs are needed in each distribution cable, in each feeder cable, and at the CO's main frame?

3.15 Let a DT relay be included in the pulse repeating circuit of Figure 3.9. Show the connection of tip and ring to a dial tone source through a contact of this DT relay. Show a circuit with contacts of the A, B, and C relays that operates and releases this DT relay.

3.16 Show a circuit for called-party supervision using transformer coupling. Show a circuit for either-party supervision using capacitive coupling.

REFERENCES

[1] Batorsky, D. V., and M. E. Burke, "1980 Bell System Noise Survey in the Loop Plant," *AT&T Bell Laboratories Technical Journal,* May/Jun., 1984. Copyright © 1984 AT&T. All rights reserved. Portions reprinted with permission.

[2] *Telecommunications Transmission Engineering, Volume 2 — Facilities,* New York: American Telephone & Telegraph Company, 1977. Copyright ©, 1977, AT&T, used by permission.

SELECTED BIBLIOGRAPHY

Bellamy, J., *Digital Telephony,* Second Edition, New York, NY: Wiley, 1991.

Fihe, J. I., and G. E. Friend, *Understanding Telephone Electronics,* Dallas, TX: Texas Instruments, 1983.

Freeman, R. L., *Telecommunication System Engineering,* Third Edition, New York, NY: Wiley, 1996.

Freeman, R. L., *Telecommunication Transmission Handbook,* Third Edition, New York, NY: Wiley, 1991.

Howard, P. K., *The Death of Common Sense — How Law Is Suffocating America,* New York, NY: Warner Books, 1994.

Manhire, L. M., "Physical and Transmission Characteristics of Customer Loop Plant," *The Bell System Technical Journal,* Jan. 1978.

McDonald, J. C. (ed.), *Fundamentals of Digital Switching,* Second Edition, New York, NY: Plenum Press, 1990.

Noll, A. M., *Introduction to Telephones and Telephone Systems,* Norwood, MA: Artech House, 1986.

Rey, R. F. (ed.), *Engineering and Operations in the Bell System,* Second Edition, Murray Hill, NJ: AT&T Bell Laboratories, 1977.

Talley, D., *Basic Telephone Switching Systems,* New York, NY: Hayden Book Company, 1969.

Telecommunications Transmission Engineering, Volume 1 — Principles, New York: American Telephone & Telegraph Company, 1977.

Transmission Systems for Communications, Murray Hill, NJ: Bell Telephone Laboratories, 1964.

4

The Trunk Side

"Gosh, Toto. I don't think we're in Kansas any more." — said by Dorothy in the *Wizard of Oz*, by Frank Baum.

This chapter is the second of two in which the boundary conditions on a local switching system are presented. The previous chapter described the line side of a switching office — a line, or loop, is a communications channel between the switching office and the subscriber. This chapter describes the trunk side of a local office — a trunk is a communications channel between two switching offices. The line side interfaces to the subscribers that home on the office, and the trunk side interfaces to the interexchange network.

This chapter proceeds in top-down fashion. In the first section, at the highest architectural level, the classical hierarchical structure of the national network is described. Sections 4.2 and 4.3 distinguish *trunks* from *trunk circuits* and describe various types of trunks. At the lowest architectural level, the last two sections describe a classical trunk circuit and two common types of interoffice signaling.

4.1 HIERARCHICAL INTEREXCHANGE NETWORK

Any switching network is a collection of nodes, where the switching occurs, and links, which carry data between the nodes. Since links require physical paths, or infrastructure, data can move only between nodes that are physically interconnected. Network links are extracted from this infrastructure.

While most networks allow more than one switched route between any pair of end points, most networks have a preferred routing procedure. AT&T's classical long distance network had a hierarchical structure with five tiers of nodes. While this architecture has been replaced, only recently, the classical architecture is described in this section. The new architecture is described in Section 16.5.

In Chapter 1, the U.S. national telephone network was described as a geographical layout of area codes and also as a geographical layout of the domains of various telecommunications corporations. This chapter takes the perspective of the logical organization, or architecture, of the interexchange network — the network of trunks that interconnect the nation's switching offices.

Figures 3.6 and 3.7 illustrated the top view and the side view, respectively, of the telephone loop plant. The two figures below illustrate the top view and the side view, respectively, of AT&T's classic long distance network.

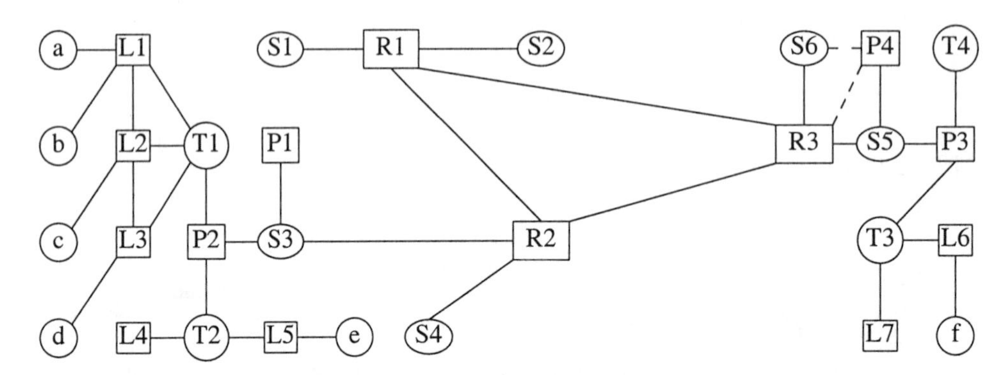

Figure 4.1. Geographical layout of a hypothetical long distance network.

Figure 4.1 represents a geographical view, looking down on a hypothetical network that represents a simplification of AT&T's classic long distance network. The lines represent the connectivity of the infrastructure, the groups of trunks in the network. This hypothetical network is intended to be a microcosm of the U.S. national network, demonstrating its salient features without showing the entire structure. The actual structure, with many thousands of nodes, is too large.

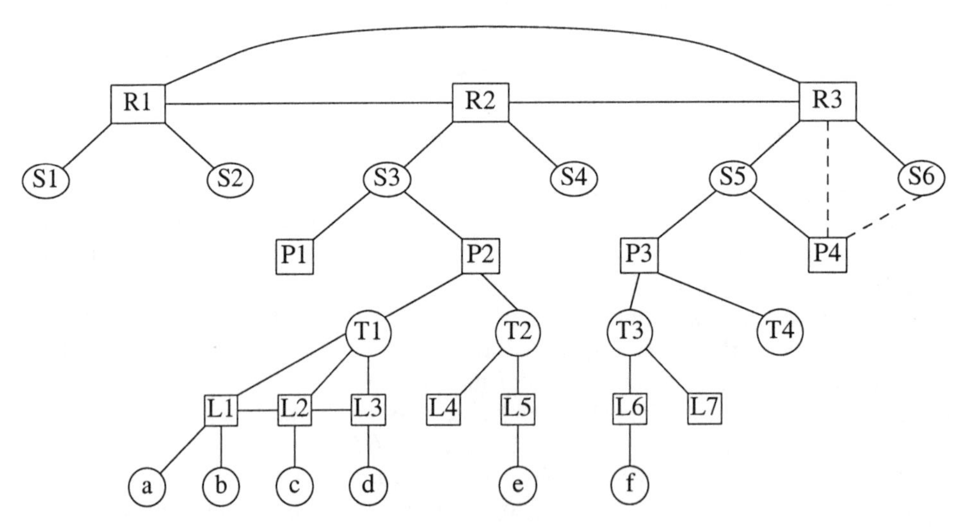

Figure 4.2. Hierarchical view of the same network.

Figure 4.2 represents a hierarchical view of the same hypothetical network. It shows the same set of offices, with the same connectivity, but in a diagram that better shows the hierarchical architecture. The reader is urged to examine both figures while reading the description that follows.

The boxes labeled L1 – L7 represent local switching exchanges. The offices in the levels above L1 – L7, and all the trunks that interconnect them, form the interexchange network. The offices in the levels between L1 – L7 and T1 – T4, and all the trunks that interconnect them, form the *intratoll network.* The offices in the levels above T1 – T4, and all the trunks that interconnect them, form the *intertoll network.*

Figures 4.1 and 4.2 are redrawn in each of the following subsections. The boxes are shown in all subsequent figures, but the only lines shown on the subsequent figures correspond to the route of the call described in the corresponding subsection.

The discussion in this section proceeds in a bottom-up fashion. The first subsection discusses the connection of a call between two subscribers in the same community. This call uses only a small part of this hypothetical microcosm network. The next two subsections discuss the connection of a call between two subscribers in adjoining communities. This call requires little more of this microcosm network. The next-to-last subsection discusses the connection of a call between two subscribers on different coasts of the United States. This call requires much of this microcosm network. The last subsection describes the real network, the generalization of this microcosm.

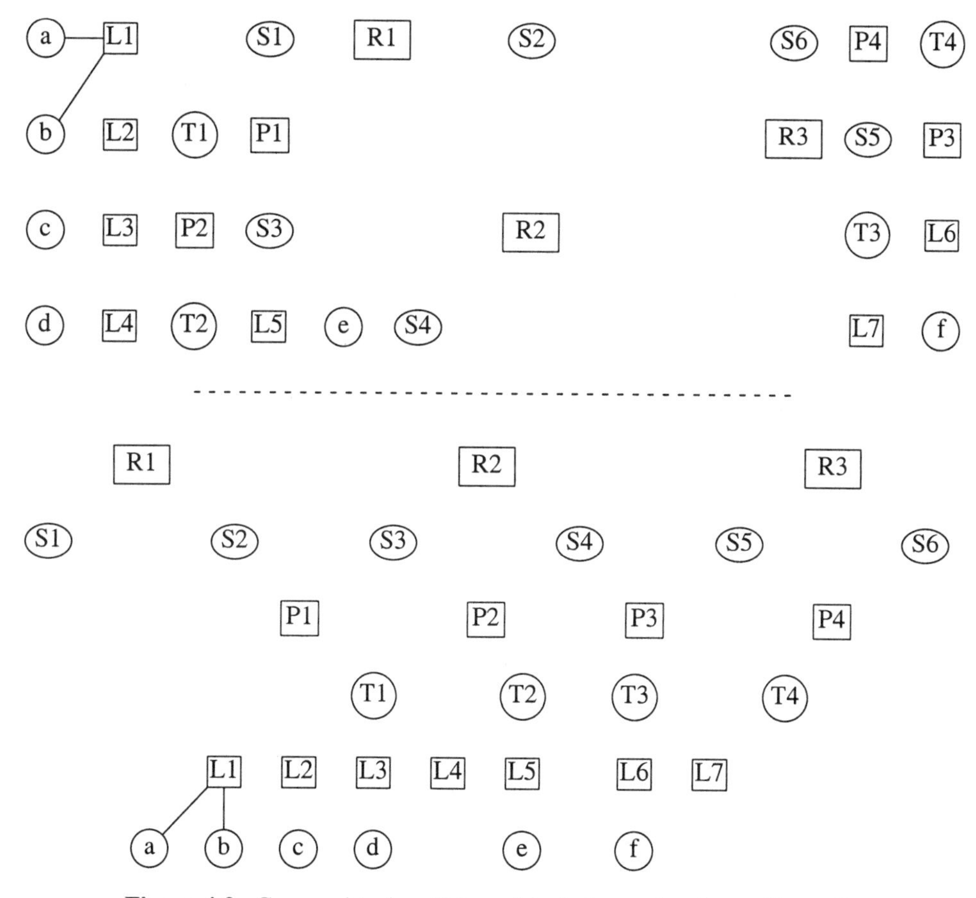

Figure 4.3. Geographical and hierarchical views of an intraoffice call.

4.1.1 A Local Call

In Figures 4.1 and 4.2, the small circles with lower-case letters a – f represent six telephone subscribers in this microcosm network. A residential subscriber's telephone is connected by the local loop to the local switching center, or central office, that serves the community.

A call between two subscribers, a and b each homing on the same CO, is called a local intraoffice call. In Figure 4.3, the call is set up by, and switched through, the local switching center, L1, that serves both subscribers. The call proceeds from a to L1 to b. The principal topic of this book is the architectures of these central offices, when we finally get past this introductory material. In the geographical view in the upper half of Figure 4.3, since a and b are geographically close, living in the same community, the connection of a call between them is confined to this community. In the hierarchical view in the lower half of Figure 4.3, since a and b home on the same CO, L1, the connection of a call between them is confined to L1 and below in the hierarchy of the microcosm network. Unless a has message-rate billing, the call is not billed.

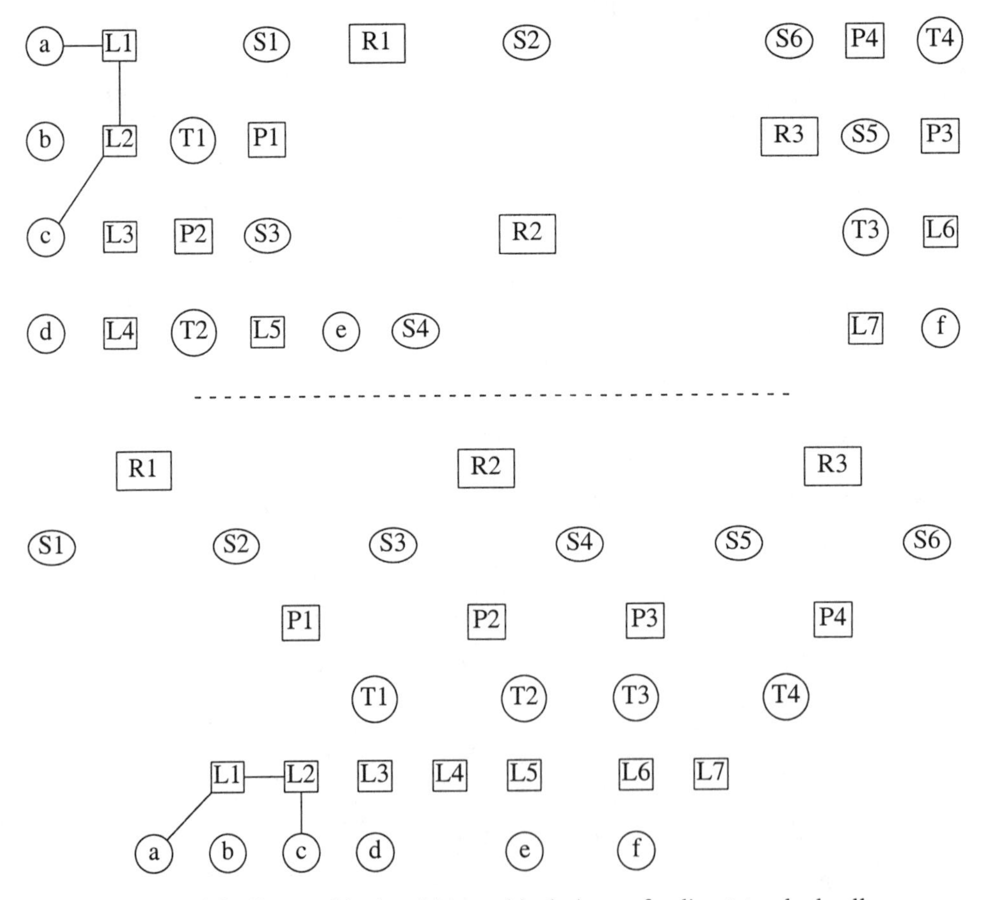

Figure 4.4. Geographical and hierarchical views of a direct-trunked call.

A call between two subscribers, a and c, each homing on a different CO, is called an interoffice call. If the two offices, L1 and L2 in Figure 4.1, are geographically close, then a and c are in each other's local calling area and the call is still called a local call. The switching offices involved in a local interoffice call may or may not be directly connected. In Figure 4.4, the two offices, L1 and L2, are directly interconnected — a group of trunks is provided between L1 and L2. This call is set up by, and switched through, both central offices, L1 and L2. The call proceeds from a to L1 to L2 to c. In the geographical view in the upper half of Figure 4.4, since a and c are geographically close, living in adjoining communities, the connection of a call between them is confined to these communities. In the hierarchical view in the lower half of Figure 4.4, although a and c home on different switching offices, L1 and L2, these offices are directly connected and the call is confined to L1 and L2 and below, in the hierarchy of the microcosm network. Unless a has message-rate billing, the call is not billed.

In Chapter 1, switching centers were justified by noting the impracticality of interconnecting all 350 million telephones in the United States with a unique channel for every pair of telephones. Similarly, it would be impractical to interconnect the 40,000 local switching centers in the United States with a unique multichannel route for every pair of central offices, a fully connected network. There are assumed to be enough calls between nearby central offices to justify the moderate expense of interconnecting pairs of such local centers directly by a short multipair cable, supporting a large set of trunks. There are probably not enough calls between widely separated central offices to justify the great expense of interconnecting pairs of such local centers directly by a long multipair cable, supporting a small set of trunks. Thus, a complex, hierarchical network is required, even to interconnect central offices.

4.1.2 A Tandem-Switched Call

Consider an interoffice call between two subscribers, a and d, each homing on a different CO in Figures 4.1 and 4.2. Suppose the two offices L1 and L3 are geographically close, but not directly connected. This call is set up by, and switched through, L1 and L3 and some intermediate office, either T1 or L2. If this intermediate office is T1, a *toll center* above the two central offices in the hierarchical view in Figure 4.2, the call is said to be toll switched. If this intermediate office is L2, a local center equivalent to the two central offices in the hierarchical view in Figure 4.2, the call is said to be tandem switched.

First, consider the tandem switched call. Consider planning the trunk facilities among the three offices: L1, L2, and L3. Let there be enough traffic among them to justify direct trunking between each pair. As indicated in the geographical view in the upper half of Figure 4.5, suppose L2 is located physically between L1 and L3. The trunks between L1 and L3 would probably use the same facilities — such as right-of-way, conduit, and transmission equipment — as the trunks between L1 and L2 and between L2 and L3. A direct connection between L1 and L3 would probably run through the basement of L2. From a in L1, a call to c in L2 would use the part of these facilities between L1 and L2, and a call to d in L3 would use the rest of these same facilities. This interoffice call between a and d would be local and direct-trunked, just like the call between a and c that was described in the previous subsection. Similarly, from c in L2, a call to d in L3 would use the part of the facilities between L2 and L3, and a call from a in L1 to d in L3 would use the rest of these same facilities.

Tandem switching in L2 allows that the facilities between L1 and L2 need not be

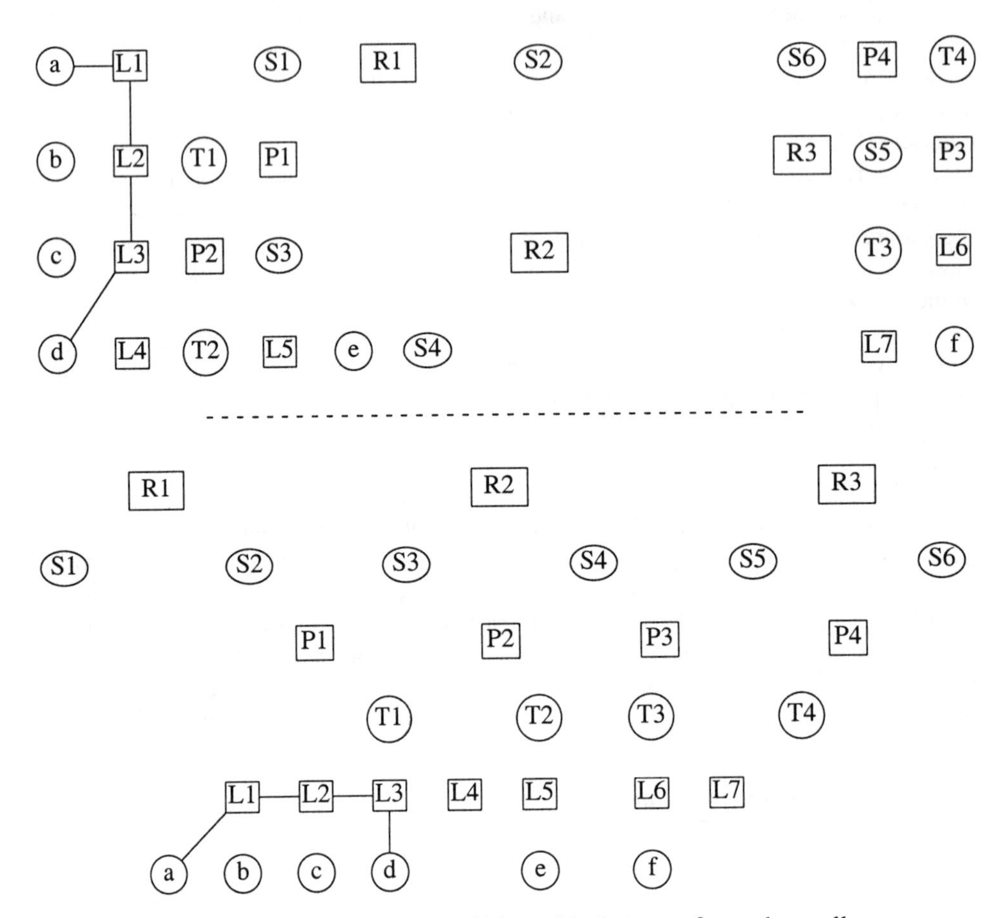

Figure 4.5. Geographical and hierarchical views of a tandem call.

segregated into L1-L2 and L1-L3 groups and the facilities between L2 and L3 need not be segregated into L1-L3 and L2-L3 groups. Section 5.3.1 shows that such combined trunk groups are more efficient than segregated groups. Thus, tandem switching in L2 may save facilities between L1 and L2 and between L2 and L3 over a corresponding segregated trunking arrangement. If L2 provides a tandem function, the interoffice call between a and d is local and tandem switched, different from a direct-trunked call.

A tandem-switched call between a and d proceeds from a to L1 to L2 to L3 to d. In the geographical view in the upper half of Figure 4.5, while a and d live in nearby communities, the connection of a call between them is not confined to these communities, but extends to the community where L2 is located and to intermediate communities through which the inter-CO trunks pass. In the hierarchical view in the lower half of Figure 4.5, a and d home on different switching offices, L1 and L2, but they are tandem switched and the call is confined to L1, L2, L3, and below. A tandem office is typically not a distinct office, but is a local switching center with the additional feature that it provides for the through-switching of its incoming trunks to its outgoing trunks. The tandem function may be integrated with the local switching function or it may be a logical and/or physical partition of the central office. Such a tandem switch is part of a mesh architecture by which hierarchically equivalent local telephone offices are interconnected.

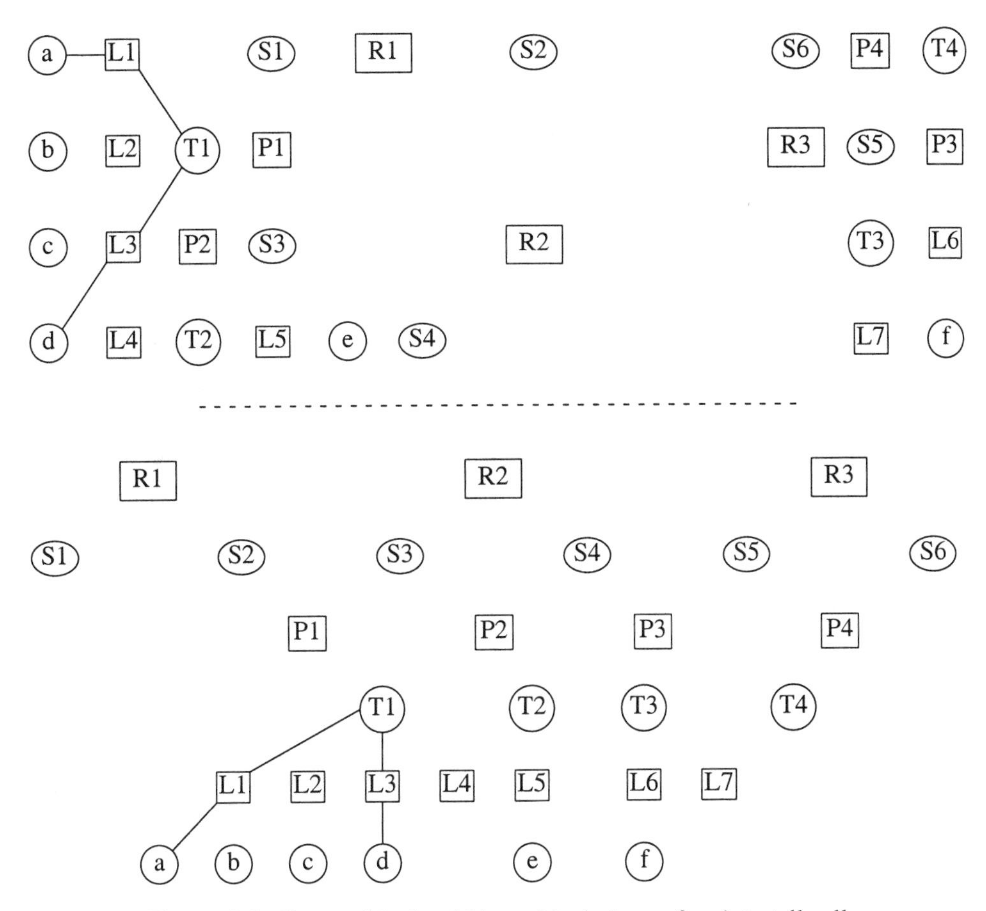

Figure 4.6. Geographical and hierarchical views of an intratoll call.

4.1.3 An Intratoll Call

Unlike a tandem office, a toll office is not a local switching center, nor part of one. Telephones do not home directly on a toll office, or any of the exchanges above it in the hierarchy in Figure 4.2. The function of these offices is to provide for the through-switching of its incoming trunks to its outgoing trunks. Furthermore, as its name implies, the toll center typically provides for computing charges for nonlocal, billed calls. A toll center is at the hub of a star architecture by which nearby local telephone offices are interconnected. Every local switching center homes on at least one toll center in a star network. If the tandem function also exists in this network, changing it from a star to a mesh, it coexists with this toll function, as indicated by L2 and T1 in Figures 4.1 and 4.2.

Consider a toll-switched, interoffice call between two subscribers, a and d, each homing on a different CO in Figure 4.6. Suppose the two offices L1 and L3 are geographically close, but not directly connected. The previous section considered a tandem-switched connection of this call, and this subsection considers a toll-switched connection. This call is set up by, and switched through, L1 and L3 and the intermediate office T1. In the hierarchical view in the lower half of Figure 4.6, T1 is the lowest common hub on which a and d home. The call proceeds from a to L1 to T1 to L3 to d. In the geographical view in the upper half of Figure 4.6, while a and d live in nearby communities, the connection of a call between them is not confined to these communities, but extends to the community where T1 is located and to intermediate communities through which the intratoll trunks pass. In the hierarchical view in the lower half of Figure 4.6, a and d home on different switching offices, L1 and L3, but these offices home on the same toll office, T1, and the call is confined to T1 and below.

If d resides in a's local calling area, the call is still local, not billed, even though it may be switched through the toll center. If d resides outside a's local calling area, the call is not local, must be billed, and would almost always be switched through the toll center. Such a call is called a toll call, as opposed to a local call. The term "long distance" is common, but it is not well defined. A call between a and d, in nearby communities in the same toll serving area, cannot be called long distance, but it may be a toll call. A call from coast to coast in the United States is certainly a long distance call, but it is not clear what the boundary is. The process for computing charges for toll calls is described in Section 9.1.

4.1.4 An Intertoll Call

A call between two subscribers, a and e in Figures 4.1 and 4.2, used to be called an intertoll call. In the post-divestiture United States, this kind of call is now called inter-LATA (we'll describe LATAs later). Not only do a and e home on different local switching centers, L1 and L5, but these offices home on different toll centers, T1 and T2. These toll centers home on a common office, P2, above T1 and T2 in the hierarchical view in the lower half of Figure 4.7. In this hierarchical view, P2 is the lowest common hub on which a and e home. In Figure 4.7, the call is set up by, and switched through, the local switching centers L1 and L5, the toll centers T1 and T2, and this primary center P2. The call proceeds from a to L1 to T1 to P2 to T2 to L5 to e.

In the geographical view in the upper half of Figure 4.7, since a and e are geographically separated, living in distant communities, but probably in the same state, the connection of a call between them is not confined to these communities, but extends to the communities where T1, T2, and P2 are located and to intermediate communities through which the intratoll and intertoll trunks pass. The call extends into the intertoll network. In the hierarchical view in the lower half of Figure 4.7, since a and e eventually home on the same primary center, P2, the connection of a call between them is confined to P2 and below in the hierarchy of the microcosm network.

A call between two subscribers, a and f in Figures 4.1 and 4.2, is also an intertoll call. Unlike the previous call between a and e, this call is a long distance call. It assumed that a and f do not home on any common center in the hierarchy of the microcosm. The regional centers, R1, R2, and R3, are hierarchically equivalent and are interconnected in a mesh network. If they were interconnected in a star network, through a common national center at the hub, then this hypothetical center would be the lowest common center in the hierarchy at which a and f home. In Figure 4.8, the call is set up by, and switched through, the local switching centers L1 and L6, the toll centers T1 and T3, the primary

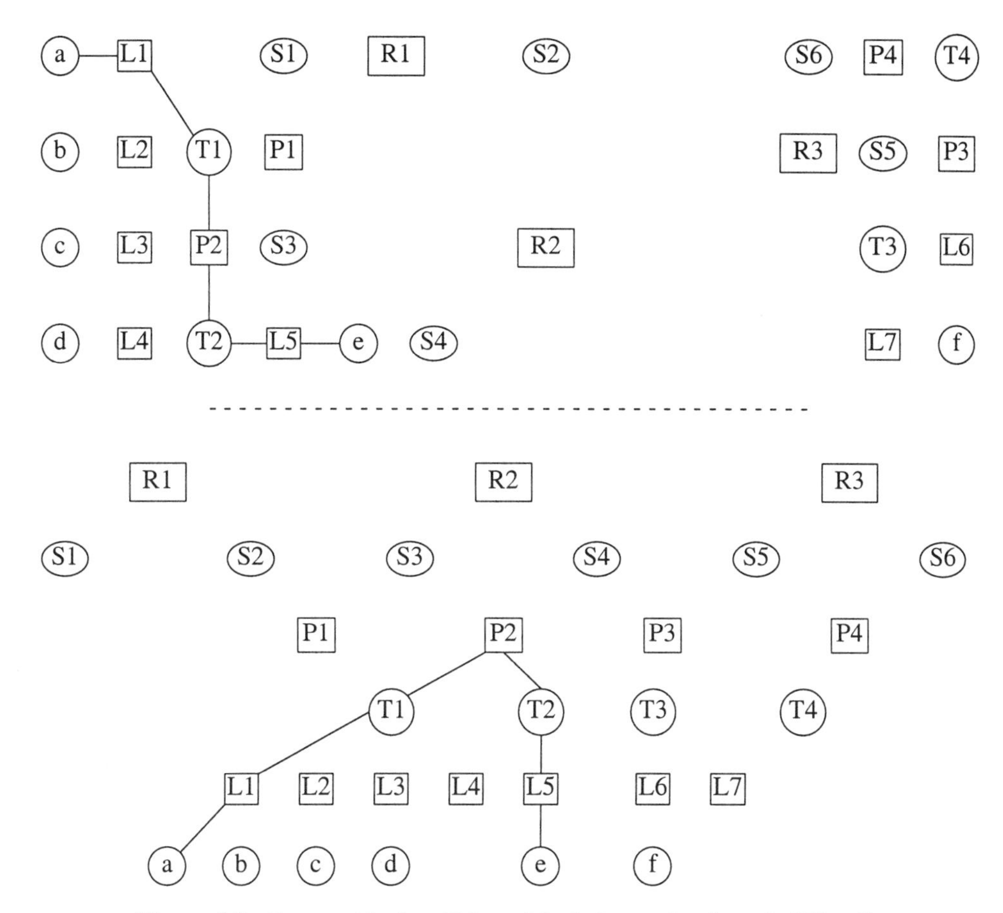

Figure 4.7. Geographical and hierarchical views of an inter-LATA call.

centers P2 and P3, the sectional centers S3 and S5, and the regional centers R2 and R3. The call proceeds from a to L1 to T1 to P2 to S3 to R2 to R3 to S5 to P3 to T3 to L6 to f.

In the geographical view in the upper half of Figure 4.8, since a and f are geographically separated, perhaps living on opposite coasts of the United States, the connection of a call between them is not confined to their communities, but extends to the communities where T1, T3, P2, P3, S3, S5, R2, and P3 are located and to intermediate communities across the United States through which the intratoll and intertoll trunks pass. In the hierarchical view in the lower half of Figure 4.8, since a and f never home on a common center, the connection of a call between them is not confined to any one subtree in the hierarchy of the microcosm network.

The network of Figures 4.1 and 4.2 generalizes to the U.S. intertoll network as it was, before long distance competition, when the only intertoll network was the one that was designed, owned, and controlled by the old Bell System. This traditional architecture of the U.S. telephone network is described in the next subsection. The U.S. intertoll network was architected as a mesh of four-level stars, with a switching office at each node in the star's hierarchy.

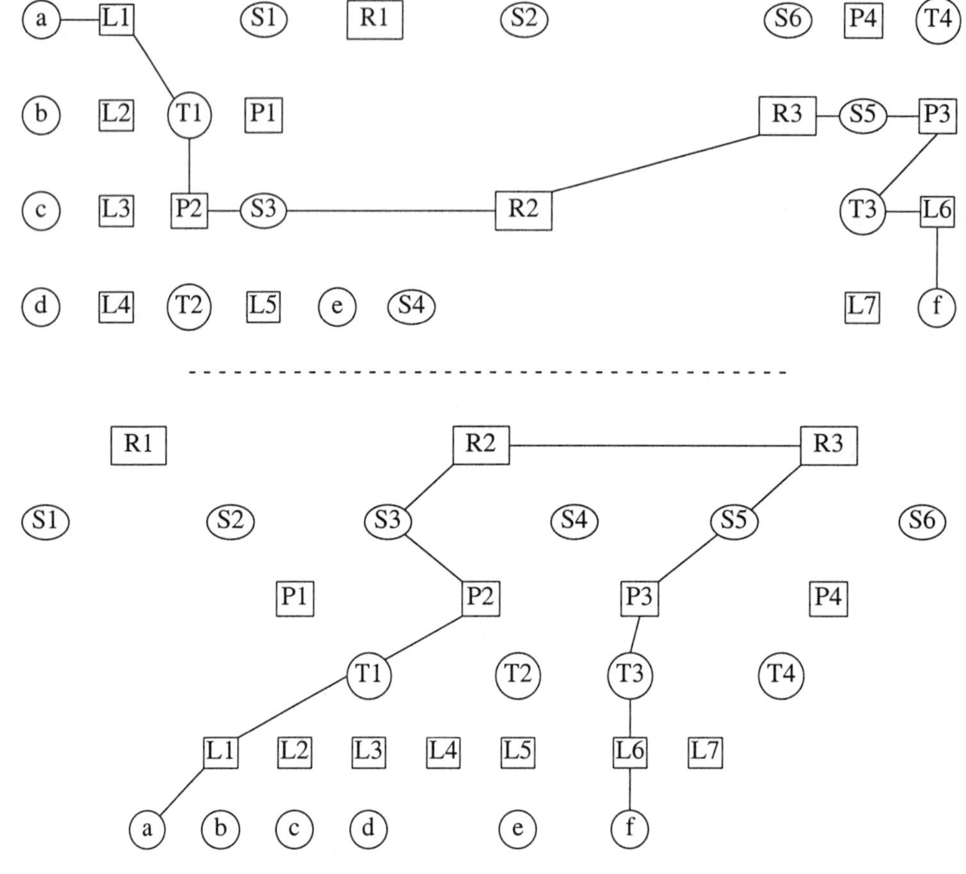

Figure 4.8. Geographical and hierarchical views of a long distance call.

4.1.5 The Traditional Toll Hierarchy

The 10 switching centers at the top of this hierarchy, called class 1 offices or regional centers, were geographically distributed over the United States and were interconnected in a mesh network. R1, R2, and R3 represent them in Figures 4.1 and 4.2. Each regional center is the hub of a star connection to about four class 2 offices, or sectional centers. In Figures 4.1 and 4.2 they are represented by S1 – S6. Each sectional center is the hub of a star connection to about five close class 3 offices, or primary centers. P1 – P4 represent them in Figures 4.1 and 4.2. Each primary center is the hub of a star connection to about nine nearby class 4 offices, or toll centers. T1 – T4 represent them in Figures 4.1 and 4.2. Each toll center is the hub of a star connection to about 30 adjoining class 5 offices, or local centers. L1 – L7 represent them in Figures 4.1 and 4.2. These class 5 offices are the central offices that connect, in yet another star, to the subscribers.

Table 4.1 tabulates real numbers for each level in this hierarchy. The "Bell" column shows the number of offices of each class in the Bell System in 1981. The "Non-Bell" column shows the number of offices of each class that were owned and operated by independent telcos. For example, in 1981 the United States had 174 primary centers, nine

TABLE 4.1. Toll hierarchy and quantities (1981)

Hierarchy	Name	Bell	Non-Bell
Class 1	Regional center	10	—
Class 2	Sectional center	35	—
Class 3	Primary center	165	9
Class 4	Toll center	1,040	476
Class 5	Local center	20,846	23,507

of which were not under Bell System control. Notice especially that, in 1981, more than half (53%) of the 44,353 class 5 offices in the United States were not part of the Bell System. In Section 2.2, we mentioned that the independents serve only about 20% of the U.S. telephones. Thus, these independent COs tend to be much smaller on the average than those in the Bell System.

This structure, a mesh of five-level stars, is not strictly followed. Consider P4 in Figures 4.1 and 4.2. While P4 normally homes on S5, indicated by the solid line, P4 has an alternate path to S6, indicated by a dashed line. Such alternate connections are common and serve two purposes. They provide fault tolerance, so failure of any center does not isolate all subscribers that home on class 5 offices hierarchically below the failed center. These alternate connections also provide alternate routing for calls if the preferred routes are all busy.

While P4 normally homes on S5, indicated by the solid line, P4 has another alternate path to R3, indicated by another dashed line. If the calling rate is high and the distance is short between two offices, direct connections can skip stages in the strict hierarchy. Particularly if a class 1, class 2, and class 3 office are all housed in the same building, a skipped connection between the class 3 office and the class 1 office is common. Finally, tandem switching, shown as L2 in Figures 4.1 and 4.2, also disrupts the pure star structure of the national network.

As competition appeared in the long distance business in the 1980s (Section 12.6), the national network gradually evolved into a set of parallel networks, each network owned and operated by a different corporation. Further confusing this structure, many of the links used by the non-Bell networks were leased from the Bell System. With divestiture, the Bell System's Long Lines network became the AT&T Communications network. In a recent architectural change, the traditional hierarchy of toll offices has been flattened. In a network architecture [1] called dynamic nonhierarchical routing (DNHR), the intertoll network has become a mesh and network control is distributed, instead of hierarchical. In this new DNHR architecture, the numerical classification of class 1 – 5, as tabulated in Table 4.1, becomes meaningless. However, just as the tip and ring naming convention for wires has long outlived the manual switchboard, the term "class 5 office," as a naming convention for a local central office, has outlived this traditional toll hierarchy.

4.2 TRUNKS

While a *line* is a channel between a switching office and a subscriber, a *trunk* is a channel between two switching offices. A trunk is the fundamental element of the interoffice network, described in the previous section. Like lines, trunks can be switched or leased.

Like leased lines, leased trunks are discussed in Chapter 12. Here we limit the discussion to switched trunks, trunks that appear on the fabric of a switching office. In this section, we distinguish trunk channels, trunk circuits, trunk groups, and trunk carriers. We also distinguish one-way from two-way trunks and two-wire from four-wire trunks.

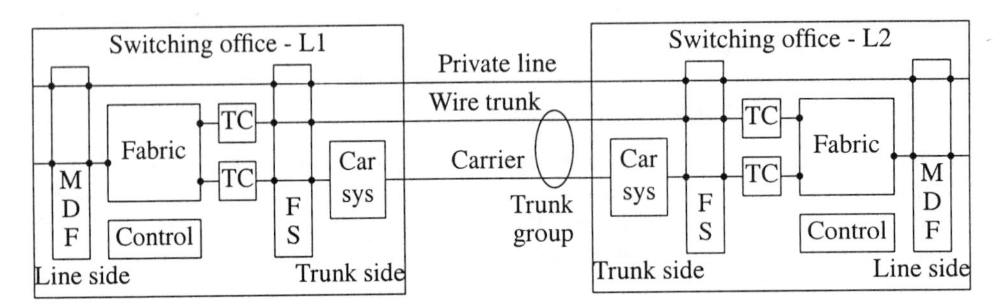

Figure 4.9. Terminology.

Figure 4.9 shows typical interoffice trunking and its confusing terminology. A *trunk channel*, usually simply called a trunk, is the communications path between two switching offices. Like a line, a trunk is a pair of wires or its equivalent in some technology like analog or digital electronics or photonics. Similar to a line circuit, a *trunk circuit* is the electrical interface between the trunk and the switching office. While a line has one line circuit, on the end of the line in the switching office on which the line homes, a trunk has at least two trunk circuits, one in the switching office on each end.

A set of trunks between the same pair of switching offices is called a *trunk group*. If the trunk group between office L1 and office L2 contains N trunks, then the telephone company can support no more than N simultaneous telephone calls between subscribers that home on L1 and subscribers that home on L2. Over the years, a large variety of *carrier systems* have been invented and a large number have been deployed in the national network. Using various means of modulating and multiplexing, these systems provide various numbers of channels, or equivalent trunks, over various kinds of transmission facility.

This section is about trunk channels. Two subsections describe one-way versus two-way trunks and two-wire versus four-wire trunks. Two subsections describe the provision of multiple trunks over a single facility, as phantom trunks and as channels in carrier systems. The next two sections describe trunk circuits and then inter-office signaling.

4.2.1 One-Way and Two-Way Trunks

Figure 4.10 has two diagrams of two local exchanges, L1 and L2, with direct trunking. In the upper diagram, direct trunks between L1 and L2 use two independent groups of one-way trunks, represented by one trunk in each group. In the lower diagram, direct trunks between L1 and L2 use one group of two-way trunks, represented by one trunk. Consider a call that proceeds from left to right across each diagram in Figure 4.10. If the *sidedness* of the fabrics in L1 and L2 is determined by call setup direction, then any trunk appears on the terminating side of the fabric in the office of an originating party and on the originating side of the fabric in the office of a terminating party.

The calling party, wired to the originating side of the fabric in L1, is connected through this fabric to a trunk circuit that is wired to the terminating side of this fabric.

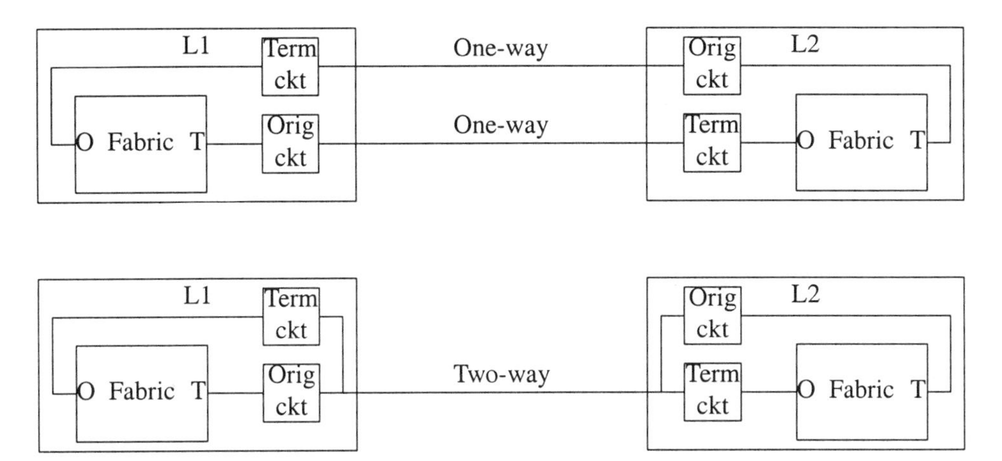

Figure 4.10. One-way and two-way trunks.

The called party, wired to the terminating side of the fabric in L2, is connected through that fabric to a trunk circuit that is wired to the originating side of this fabric. These two trunk circuits must be wired to the same trunk, the lower one in the upper diagram and the only one shown in the lower diagram. Since the call originates in L1 and terminates in L2, the trunk circuits used on this call are an originating circuit in L1 and a terminating circuit in L2. Confusion arises because originating circuits appear on the terminating side of a fabric and terminating circuits appear on the originating side of a fabric.

A trunk is not called one-way because speech or data signals can only flow in one direction over it. This distinction occurs in the two separate pairs of a four-wire trunk. The upper diagram of Figure 4.10 distinguishes two one-way trunks connecting two switching offices. Each can carry only calls that originate in one office and terminate in the other, respectively. The *call setup flow* is one-way, not the data flow. A one-way trunk has its originating circuit wired to the terminating side of the fabric in the office where a call originates and its terminating circuit wired to the originating side of the fabric in the office where a call terminates. Such a trunk allows people in this first office to place calls to people in this latter office. Calls from people in the latter to people in the first also use one-way trunks, between the same pair of offices, but wired and equipped the opposite way. When a one-way trunk is in use, there are no wasted equipment nor idle fabric appearances. But, as we will discuss in Section 5.3.1, since the trunk groups are less efficient, more one-way trunks are required than would be required in a two-way trunking arrangement.

The lower diagram of Figure 4.10 shows that a two-way trunk has an originating circuit and a terminating circuit on each end. It is wired to both the originating and terminating sides of the fabrics in each office it connects. These circuits are frequently integrated into a single dual-function trunk circuit that handles both originating and terminating traffic. Two-way trunks are prone to an occasional problem, called glare, that is not found in one-way trunks. Occasionally, the CO on each end of an idle trunk may try to place a call to the other CO at practically the same time. Each CO could seize its end of the same two-way trunk for originating its call to the other office. After this glaring trunk times out, both subscribers would have to redial.

While two-way trunks are more versatile and more expensive, during any given call half of the equipment is idle and half of the fabric appearances are idle. Section 5.3.1 will show that, because they are more versatile than one-way trunks, two-way trunks are more efficient. Let calling statistics between two switching offices L1 and L2 indicate the need for N_1 one-way trunks in a group from L1 to L2 and N_2 one-way trunks in a group from L2 to L1. Equivalent service is provided by a single group of $N_3 < N_1 + N_2$ two-way trunks. If the cost of cable is far greater than the cost of trunk circuits, then saving a few total trunks may be more economical than saving half the equipment and fabric appearances with every trunk.

4.2.2 Two-Wire and Four-Wire Trunks

The difference between the "wire-ness" and "way-ness" of trunks is confusing. A two-wire trunk provides a single pair and allows data flow in both directions over this pair. A four-wire trunk provides two pairs and restricts data flow to a different direction in each pair. Two-wire and four-wire trunks can be either one-way or two-way trunks. If the trunks in Figure 4.10 represent two-wire trunks, each line in the figure represents a pair of wires. If the trunks in Figure 4.10 represent four-wire trunks, each line in the figure represents two pairs of wires.

In the handset, containing the earpiece and the mouthpiece of the telephone instrument, communications is four-wire. The cable between the handset and the base of the telephone contains four wires, one pair connects the earpiece to the circuit shown in Figure 3.1(d), and the other pair connects to the mouthpiece to this circuit. In the base, the four-wire communications in the handset is converted by a transformer circuit to the two-wire communications in the house wiring, loop, and central office. A similar circuit, called a hybrid transformer, is used at any interface that requires a two-wire to four-wire conversion, such as in a trunk circuit that interfaces a four-wire trunk to a typical two-wire switching office.

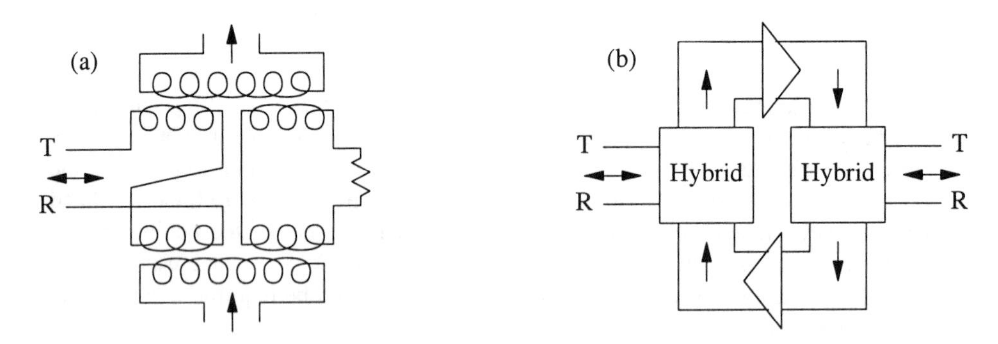

Figure 4.11. (a) Hybrid transformer and (b) two-wire analog repeater.

Figure 4.11(a) shows a hybrid, with two-wire communications on the left and four-wire communications at the top and bottom. Four-wire communications becomes important, especially on the trunk side of a switching office, when spans are so long that signals must be amplified. Unfortunately, nature doesn't easily give us bidirectional amplifiers. *Negative-admittance* amplifiers have been used to amplify signals that travel in both directions simultaneously. But the most common means of amplifying a bidirectional signal is to use conventional one-direction amplifiers in a four-wire configuration. A two-wire analog repeater, shown in Figure 4.11(b), consists of two hybrids and two conventional amplifiers in opposite directions.

A two-wire trunk requiring N two-wire amplifiers has $2N$ hybrids. A four-wire trunk has twice as much wire, but only two hybrids, one on each end. It is far easier to provide amplification in four-wire trunks than in two-wire trunks. This simplicity often pays for the cost of the added wire.

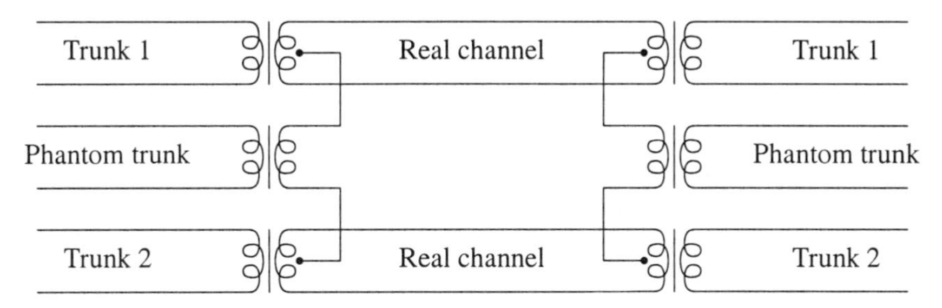

Figure 4.12. Phantom trunk.

4.2.3 Phantom Trunks

Figure 4.12 shows an arrangement that provides three equivalent trunks over just two physical channels. The virtual trunk used by the center trunk circuits is called a phantom trunk. The phantom is connected to center-taps on the transformers in the two physical trunks. The upper pair in Figure 4.12 acts in parallel as the tip of the phantom trunk. The lower pair in Figure 4.12 acts in parallel as the ring of the phantom trunk. As long as the center-taps are perfect, there is no crosstalk between the phantom and the two physical trunks.

Phantom trunks are AC-coupled by nature, making DC signaling difficult. Furthermore, the increased use of trunk carrier systems, particularly digital carrier, makes phantom trunk arrangements old fashioned.

4.2.4 Carrier Systems

Carrier systems provide some means for multiplexing many channels over a single physical transmission medium. Two generalized forms of transmission medium are:

- Guided-wave media, such as twisted pair, coax, and optical fiber;
- Directional radio, such as between microwave radio towers or between earth stations and satellites.

If the guided-wave medium is an existing two-wire path, a pair, the installation of an N-channel carrier system on such a pair expands the pair from one channel to N channels, commonly called pair gain. The first guided-wave carrier systems were installed in the 1920s — these evolving systems were named alphabetically (A carrier, B carrier, etc.) These early systems were analog, using amplitude modulation and frequency multiplexing. J carrier was the last of the analog AM carriers to use conventional twisted pairs, placing 12 channels on two pairs. L carrier used coaxial cable and evolved in a series: L1 – L5. L5 carrier had 13,200 analog voice-grade channels on one coax. At 4 kHz of analog bandwidth plus 1 kHz of guard band between channels, L5 carrier required $13,200 \times 5$ kHz = 66 MHz of total bandwidth. K carrier, N carrier, and O carrier also used twisted pairs but they had to be special dedicated wires, not ordinary telephone grade wire.

Channels on analog carriers were organized hierarchically, into groups and super-groups. N individual channels, at 5 kHz each, were frequency multiplexed into a group. Then, M group channels, at $5N$ kHz each, were themselves frequency multiplexed into a super-group. This structure simplified the add/drop of entire channel groups at hops along the carrier and was the predecessor for the DS1/DS2/... structure of the asynchronous digital hierarchy and the OC1/OC3/... structure of the synchronous digital hierarchy.

Little analog carrier remains in the United States, having been replaced by digital carrier. After analog-to-digital conversion, by sampling an analog signal 8000 times a second and pulse-code modulating at 7 or 8 bits per sample, the digital equivalent of an analog voice channel, either 56 or 64 Kbps, is called a DS0 or bearer channel. The digital carrier hierarchy, now called the asynchronous digital hierarchy (ADH), multiplexes 24 DS0 channels up to DS1 at 1.544 Mbps, four DS1 channels up to DS2 at 6.3 Mbps, seven DS2 channels up to DS3 at 45 Mbps, and six DS3 channels up to DS4 at 272 Mbps. While DS1 and DS3 transmission is still very common in the United States, the other layers of the old ADH have almost vanished, replaced by the synchronous digital hierarchy (SDH). In SDH previously called Sonet, OC1 at 52 Mbps (OC stands for optical carrier) is approximately equivalent to one DS3. Then, the hierarchy multiplexes: three OC1 channels up to OC3 at 155 Mbps, four OC3 channels up to OC12 at 625 Mbps, four OC12 channels up to OC48 at 2.5 Gbps, and four OC48 channels up to OC192 at 10 Gbps.

With guided-wave carrier systems, the physical medium (twisted pairs, coax, or fiber) must be specifically installed. This can be very expensive, especially if it requires obtaining new right-of-way. Directional radio (microwave) has been the other general form of analog transmission medium. Early analog microwave carrier used frequency modulation (FM), but was quickly replaced by microwave carrier, using some form of digital modulation like digital phase modulation (also called phase shift keying) and digital quadrature amplitude modulation (QAM). Two extremely popular digital microwave carrier systems, TD2 and TD3, have the familiar large repeaters on towers, typically seen on mountain tops. They allow 25-mile spacing between the transmitters and receivers, and a route requires neither digging trenches nor stringing poles. The microwave bands, allocated by the FCC, are between 2 – 18 GHz. There are typically about 1,500 channels in one of those carriers.

An alternate implementation of microwave carrier is by satellite, typically in geosynchronous orbit. Satellite orbits are discussed in Section 19.3. While satellite communications has been used in telephony, the great distance of geosynchronous orbits makes round-trip speed-of-light delay so great that users don't like it for telephony. Satellites in geosynchronous orbits are very useful for data communications and especially for broadcast television. Telephony via satellites in low earth orbits is also described in Section 19.3.

4.3 TRUNK CIRCUITS

The channels on the network side of a switching office interface with other offices and with the national network. These channels are called trunks. Just as every subscriber line on the line side has a dedicated circuit, called the line circuit, every trunk on the trunk side also has a dedicated circuit, called the trunk circuit. While the line circuit monitors the state of the subscriber's hook-switch, the trunk circuit has several additional functions. Another difference is that pairs in the trunk side are typically long enough that repeaters and amplifiers are needed.

Lines and trunks support two kinds of communications: the data *in* the call and the data *about* the call. Subscribers' speech or data is carried, in one direction over one pair in a four-wire trunk and in both directions over a two-wire trunk or a subscriber line, during the time in which the call is set up. The data about the call helps the system set up the call, bill the call, and tear it down when it's done. This data about the call is called signaling.

4.3.1 Signaling

Signaling is the transmission of data that sets up, controls, or describes a call. Signaling is discussed further in Section 18.1. Two kinds of information must be signaled between the offices on each end of a call.

- Call setup signaling includes the called party's directory number and, sometimes, the calling party's number (for billing). The called number must be signaled from the calling party's central office to the called party's central office. In Step-by-Step, the dial pulses from the calling party's telephone proceed over the trunk to the other office, and there they direct the call to the called party. In a Crossbar office, where the calling party has dialed the entire called number into the office, that number is later signaled over the trunk to the other office. In either case, long trunks may require pulse repeating.
- Supervisory signaling communicates the on/off-hook status of remote telephones. Also signaled between the two central offices, this information informs the offices to start and stop billing for the call and to set up or drop the switch train that completes the call through the fabrics in each office.

Signaling has evolved through four generations: inband signaling, out-of-band digital signaling, common channel signaling, and networked signaling. This chapter describes the early generations. Signaling that uses the same channel as the voice is called inband signaling. Signaling that uses a different channel from the voice is called out-of-band signaling. Inband and out-of-band signaling have two forms each: DC and AC.

- In inband DC signaling, call setup information is communicated by loop pulsing (similar to dial pulsing) and supervisory information by battery reversals. In the remaining subsections of this section, the traditional trunk circuit is gradually developed, including circuits for such inband DC signaling.
- Inband AC signaling is described in Section 4.4. Call setup information is communicated by multifrequency pulsing (MF), which is similar to Touch Tone. Supervisory information is signaled by single-frequency signaling (SF), also called E&M signaling.

- Out-of-band signaling, including common channel signaling and networked signaling, is usually digital. The computers that control respective offices send digital messages to each other over a data channel between the two offices. These methods are described in later chapters. The most modern method, networked signaling, is rapidly becoming universal, and is described in Section 18.1.
- While progress tones, like dial tone and audible ring tone, should be included as part of signaling, they typically are not included but are always implemented using the inband AC signaling method.

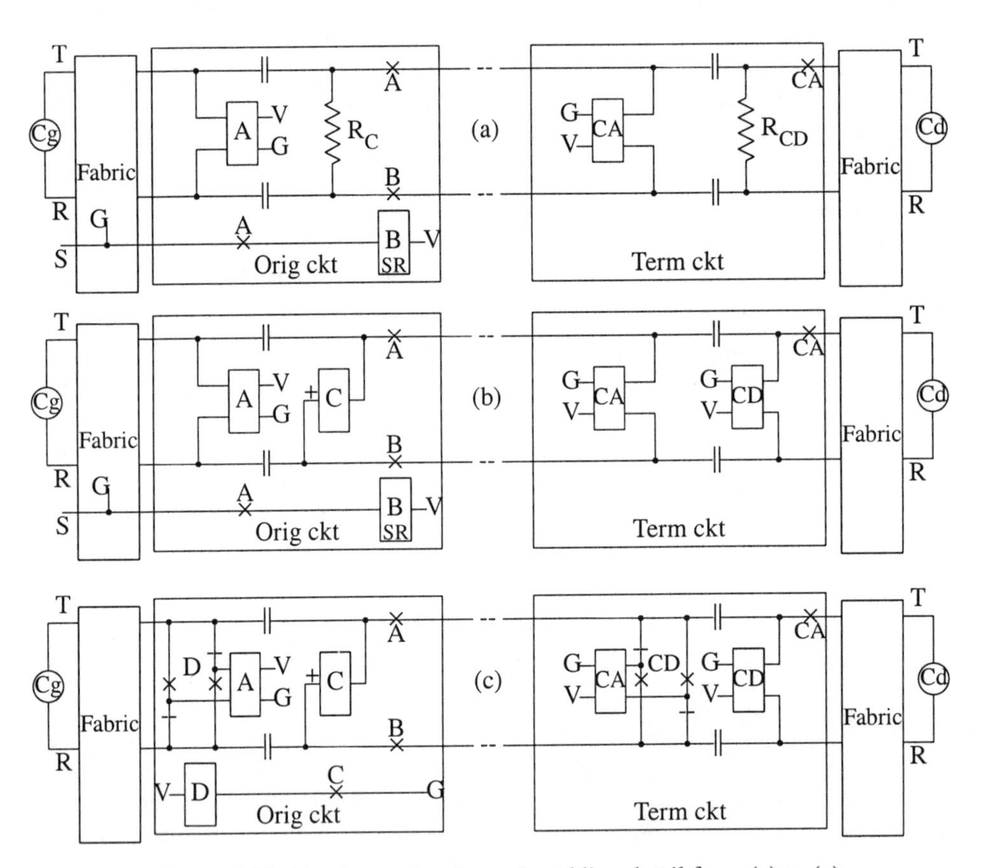

Figure 4.13. Trunk circuit schematic: adding detail from (a) to (c).

4.3.2 Trunk Circuits for DC Signaling

In DC signaling, both supervision and number transmission are accomplished by repeating the DC status of a line around the AC coupling between the line's segments. Figure 4.13 shows three stages of the gradual development of a trunk circuit. Trunk circuits are complex and this discussion, while appearing lengthy, is still incomplete. The *originating trunk circuit* is seen to the left of center in Figure 4.13(a), (b), and (c). On the left edge of this circuit the tip, ring, and sleeve are wired to the right side of the fabric (which could be the fabric's terminating side) in the calling party's office. The calling party's tip and ring are shown on the left edge of this fabric (which could be its originating side). On the right edge of this originating circuit, the tip and ring are wired to the wire pair of the one-way two-wire trunk at the center of the figures. The *terminating trunk circuit* is seen to the right of center in Figure 4.13(a), (b), and (c). On the left edge of this circuit the tip and ring are wired to the wire pair of the one-way two-wire trunk. On the right edge of this circuit the tip, ring, and sleeve are wired to the left side of the fabric (which could be this fabric's originating side) in the called party's office. The called party's tip and ring are shown on the right edge of this fabric (which could be its terminating side).

The originating circuit in Figure 4.13(a) shows DC isolation and AC coupling via two capacitors, a terminating resistor R_C, a conventional A relay, a slow-release B relay, and several make contacts on each relay. The terminating circuit in Figure 4.13(a) shows DC isolation and AC coupling via two capacitors, a terminating resistor R_{CD}, the CA relay, and one make contact. Between the calling and called parties, the pair is partitioned by coupling capacitors into three DC loops: one between the calling party and the A relay, one between R_C and the CA relay, and one between R_{CD} and the called party.

When the calling party's office determines that the party intends to call somebody in another office, the calling party is connected through the fabric to the originating circuit of a trunk that connects to that other office. Assume the calling party has a dial phone and that the dialed digits are transmitted directly to the called party's office. When the calling party is connected to the trunk circuit, talking battery and supervision are provided by the A relay.

As in the discussion in Section 3.5, the DC connection of the loop to the off-hook telephone causes A to operate and that causes the slow-release B relay to operate. The contacts of the A and B relays on the trunk side of this originating circuit simulate the dial contact and hook-switch, respectively, of the telephone. The loop in the trunk closes and current through R_C operates the CA relay in the terminating circuit on the other end. A contact of the CA relay closes and, resembling a telephone off-hook, current through R_{CD} signals a new request in the called party's office. That office responds to this request as it would to a telephone off-hook, except that it does not return dial tone.

4.3.3 Loop Pulsing

Inband DC signaling is the most primitive and the earliest form and has two components. Loop pulsing is used to transmit call setup information (called party's directory number) and reverse-battery signaling is used to communicate supervisory information.

In a Step-by-Step office, the called party's directory number is transmitted directly to the other office as the calling party dials that number. In a Crossbar or other common-control office, this information is outpulsed indirectly after having been stored in a register or in the common control itself. The details will be provided in the chapters that describe the corresponding systems.

Figure 4.13(a) shows that the dialing telephone, or *outpulser*, is connected through the fabric to the A relay in the originating trunk circuit. Since that relay releases and operates with every dial pulse, the contact of the A relay in the trunk loop repeats each dial pulse. Since the CA relay in the terminating trunk circuit releases and operates with every dial pulse, the contact of the CA relay on the office side repeats each dial pulse. This produces the dial pulsing that drives the fabric in the other office.

4.3.4 Reverse-Battery Signaling

The second component of inband DC signaling is reverse-battery signaling, used to communicate supervisory information (the on/off-hook status of a single telephone). It is called reverse-battery signaling because the two states of a telephone hook-switch map to the two polarities of a 48V battery across the tip and ring of the trunk.

The trunk circuits of Figure 4.13(a) are further developed in Figure 4.13(b). The load resistor R_{CD} in the office side of the terminating circuit in the called party's office is really the coils of the CD relay. The load resistor R_C in the trunk side of the originating circuit in the calling party's office is really the coils of the polarized C relay. Figure 4.13(b) shows that this C relay only operates if current flows through its windings from ring to tip. The CD relay provides the talking battery to the called party through that party's office. If that telephone is on-hook, while ringing, CD is not operated. Furthermore, in this state, the polarity of the DC on the trunk loop, provided through the windings of the CA relay, is the opposite from the polarity that operates the C relay.

The trunk circuits are further developed in Figure 4.13(c) as far as we will go. The CA relay on the trunk side of the terminating circuit in the called party's office is connected to the trunk loop through two transfer contacts of the CD relay. The A relay on the office side of the originating circuit in the calling party's office is connected to the fabric through two transfer contacts of the C relay.

When the called party answers, the CD relay operates, reversing the connection of the CA relay to the tip and ring in the trunk. This new polarity causes the C relay to operate, simulating the called party's off-hook to the calling party's office. Operation of the C relay starts billing and other procedures, depending on the type of office and type of call. Operation of the C relay reverses the connection of the A relay to the tip and ring in the calling party's fabric. Depending on the type of office, polarity may reverse all the way back to the calling party's telephone. Battery reversal to Touch Tone telephones was discussed in Section 3.5.

At the end of the call, if the calling party hangs up first, the A relay releases, the slow-release B relay eventually releases, and the CA relay releases, simulating an originating telephone hangup. If the called party hangs up first, the CD relay releases, the trunk loop reverts to its original polarity, and the C relay releases, simulating a terminating telephone hangup. In either case, billing is stopped and each office drops the connection through its fabric between its respective party and the trunk circuit on its end of the trunk.

4.4 INTEROFFICE SIGNALING

Inband DC signaling was described in the previous section. This section describes inband AC signaling. Inband AC signaling followed DC signaling and has two components. Multifrequency pulsing (MF), similar to Touch Tone, is used to transmit call setup information and single-frequency signaling (SF), also called E&M signaling, is used to communicate supervisory information. This section describes the SF or E&M signaling first and then MF pulsing later.

4.4.1 SF or E&M Signaling

In SF signaling, the supervisory or on/off-hook status of a telephone at the far end of a trunk is communicated by the absence or presence, respectively, of a tone on the trunk. The tone has a single frequency (hence, the name), but a different frequency at either end of the trunk. The tones are inaudible because they are transmitted on both wires in the pair with respect to ground, typically by center-tapping a transformer.

(a)

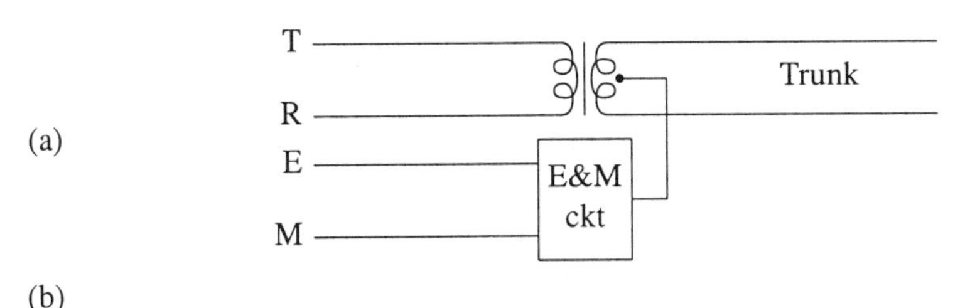

(b)

Originating office				Terminating office			
Calling	E	M	Tone	Tone	E	M	Called
Idle	Open	Ground	None	None	Open	Ground	Idle
Calling	Open	Battery	2400	None	Ground	Ground	Ringing
Talking	Ground	Battery	2400	2600	Ground	Battery	Answer
Hangup	Ground	Ground	None	2600	Open	Battery	Talking
Talking	Open	Battery	2400	None	Ground	Ground	Hangup

Figure 4.14. Single-frequency signaling: (a) circuit and (b) sequences.

Figure 4.14(a) shows a typical arrangement for SF signaling. The trunk circuit has two additional wires interfacing to its office. Using the *receive* wire, or E-lead, the trunk circuit informs its office that the telephone in the office at the other end of the trunk is off-hook. Using the *transmit* wire, or M-lead, the office instructs the trunk circuit to send a tone down the trunk informing the other end that the telephone in this office is off-hook. These two wires could have been named R and T, for "receive" and "transmit," but these names would have been confused with the universal T and R names for tip and ring. History is unclear whether the wires were named E and M for "rEceive and transMit" or for "Ear and Mouth." The single frequencies are 2400 Hz on one end and 2600 Hz on the other end. Each trunk circuit has filters so it only hears the opposite tone from the one it produces. With no metallic return path for these AC tones, they are communicated with respect to the local ground at each of the two central offices.

Figure 4.14(b) shows a typical signaling sequence. If the calling party is idle and the called party is idle, neither tone is present on the trunk. When the trunk is seized, the originating office puts battery on the M wire, causing the trunk circuit to put 2400 Hz on the pair. The trunk circuit at the other end detects this tone. It applies ground to its E wire, signaling its office of a new call request. After the call is set up and the called party answers, its office puts battery on its M wire, and the trunk circuit in the called party's office puts 2600 Hz on the pair. The trunk circuit in the calling party's office detects this tone. It applies ground to its E wire, signaling this office that the called party has answered. Both tones are present on the trunk during the call. When either party hangs up, the respective office grounds the M wire, removing the tone and informing the office at the other end of the hangup. When both parties hang up, both tones are gone.

4.4.2 Multifrequency Pulsing

Another kind of inband AC signaling is multifrequency pulsing, used to transmit call setup information. In Crossbar, or other common control systems, the calling party dials the called party's entire directory number into a register or other common equipment. If the office interprets that the call goes to another office, the calling party is disconnected from the register and connected through the fabric to the appropriate trunk circuit. Then, the called party's directory number is transmitted to the other office. This may be DC loop-pulsed over the trunk, as in Section 4.3.3, or it may be MF-pulsed over the trunk. A circuit, called an outpulser or outgoing sender, is temporarily connected to the trunk circuit. The register's contents, and possibly other information, are transmitted by a sequence of chords, similar to Touch Tone dialing.

TABLE 4.2. Multifrequency signaling

900	1				
1100	2	3			
1300	4	5	6		
1500	7	8	9	0	
1700	ST3	ST2	KP	ST1	ST0
	700	900	1100	1300	1500

Format = KP N N X X X X X ST#

Table 4.2 shows the tones and the signaling format. MF pulsing is similar to Touch Tone, but predates it. Each character is transmitted as two tones of the six available: 700, 900, 1100, 1300, 1500, and 1700 Hz. Unlike Touch Tone, all pairs are used, 15 combinations: 10 digits, a start character (KP), and four stop characters (ST0 – ST3). The format is: the start character, the called party's directory number (NNX-XXXX) with optional area code, optionally the calling party's directory number, optionally some class of service or billing information, and one of four stop characters. The choice of stop character is a simple means of signaling information on how to process the call. A simple call terminates with ST0, but a call requiring operator intervention at the other office, for example, uses a different stop character.

Many readers will recall hearing a rapid sequence of tones immediately after dialing. You may recall that there were at least nine tones, sometimes twice as many, although they were almost too rapid to count, and that the first and last tones had higher pitch than those between them. You probably did not make the association of the presence of nine tones with the placement of nonbilled calls to other offices, of the presence of many more tones with the placement of long distance or other billed calls (calling party number is required by remote billing equipment), or of the absence of tones with intraoffice calls. Those tones were the characters of MF pulsing. And now that I've reminded you of this, you may be thinking that you haven't heard these tones lately.

In the classic fabric configuration during outpulsing, the calling party and the outpulser are simultaneously connected to the trunk circuit. While the characters were MF-pulsed to the office at the other end of the trunk, they were audible to the calling party. But, people learned to build "blue boxes." If you could hear the MF tones, you could generate them yourself through the telephone mouthpiece. If you outpulsed the right sequence at the right time, overloading the signals outpulsed by the local office, you could have your call billed to some innocent party. At first, blue boxes were a mere annoyance, but the schematics were distributed to, among others, many college EE students. While a tribute to American ingenuity and underground distribution, the telephone company had to do something.

A nationwide effort in the 1970s, thousands of telephone offices were changed so that the calling party was not connected to his trunk circuit until after outpulsing was completed. Chapter 11 describes how it is easier and cheaper to make such changes to computer-controlled offices than to wired-logic offices. While MF pulsing is still used on some trunks, because of blue boxes, we no longer hear those once familiar tones. Networked signaling, described in Section 18.1, has largely replaced MF pulsing and permanently solved the blue box problem.

EXERCISES

4.1 Use the following approximate numbers in this exercise: 200 million residential telephone subscribers in the United States, 80% served by Bell telcos and 20% served by independents. Let there be 40,000 COs in the United States, 50% owned by former Bell telcos, and 50% owned by independents. The seven RBOCs own an average of three Bell telcos each. Suppose there are 1200 independents in the United States. (a) How many subscribers are served by an average CO? By the average Bell CO? By the average independent's CO? (b) How many subscribers are served by the average RBOC? By the average Bell telco? By the average independent telco? (c) How many COs are owned by the average RBOC? By the average Bell telco? By the average independent telco?

4.2 Suppose the seven RBOCs merged again and created a long distance network analogous to AT&T's classic five-level network. (a) Assuming this network has seven regional centers, instead of the 10 indicated in Table 4.1, show these regional centers on a sketch of the continental United States, each in a city near the geographical middle of its RBOC region. (b) On your map, interconnect these seven centers by interregional links so that (1) Ameritech's regional center has links to all other centers except BellSouth, (2) SBC's regional center has links to all other centers except the two in the northeast, and (3) the other five regional centers each has links to three neighbors. (c) Partition each RBOC region into four sections, one section in each diagonal corner, and show a sectional center in a city near the geographical middle of its respective section. (d) On your map, interconnect each sectional center by a link to the regional center in its RBOC region and also to a nearby regional center in a neighboring RBOC region. (e) How many routes, passing through two or fewer regional centers, are there from the sectional center that serves Pittsburgh to the sectional center that serves San Diego?

4.3 Let two COs, m miles apart, require direct trunking. Let a copper pair cost \$500/mile. Let a one-way loop-signaling trunk circuit cost \$200 at each end and a two-way loop-signaling trunk circuit cost \$500 at each end. Let traffic require 25 one-way trunks in each direction or 44 two-way trunks (Section 5.3.1). Over what range of m are one-way or two-way trunks used?

4.4 Continuing, let a phantom transformer cost $210 at each end (each pair handles three trunks on two physical pairs) and let E&M signaling circuits cost $1350 at each end of a trunk. Over what range of m are phantom E&M trunks more economical than loop-signaling trunks?

4.5 Still continuing, let unamplified analog transmission be unacceptable for $m > 15$ miles in two-wire and $m > 20$ miles in four-wire. Assume that $m < 30$ miles. Let in-office hybrids cost $100 at each end of a pair, unidirectional amplifiers cost $2000, and bidirectional analog repeaters cost $14,000 (including hybrids and amplifiers). Over what range of m are two-wire or four-wire pairs used?

REFERENCE

[1] Ash, G. R., "Dynamic Network Evolution, with Examples from AT&T's Evolving Dynamic Network," *IEEE Communications Magazine,* Vol. 33, No. 7, Jul. 1995.

SELECTED BIBLIOGRAPHY

Bellamy, J., *Digital Telephony,* Second Edition, New York: Wiley, 1991.

Brooks, J., *Telephone — The First Hundred Years,* New York, NY: Harper & Row, 1976.

Clarke, A. C., et al., *The Telephone's First Century — and Beyond,* New York, NY: Thomas Y. Crowell, 1977.

Freeman, R. L., *Telecommunication System Engineering,* Third Edition, New York, NY: Wiley, 1996.

Freeman, R. L., *Telecommunication Transmission Handbook,* Third Edition, New York, NY: Wiley, 1991.

McDonald, J. C. (ed.), *Fundamentals of Digital Switching,* Second Edition, New York, NY: Plenum Press, 1990.

Noll, A. M., *Introduction to Telephones and Telephone Systems,* Norwood, MA: Artech House, 1986.

Rey, R. F. (ed.), *Engineering and Operations in the Bell System,* Second Edition, Murray Hill, NJ: AT&T Bell Laboratories, 1977.

Schindler, G. E. (ed.), *A History of Engineering and Science in the Bell System,* Murray Hill, NJ: Bell Telephone Laboratories, 1982.

Talley, D., *Basic Telephone Switching Systems,* Rochelle Park, NJ: Hayden Book Company, 1969.

Telecommunications Transmission Engineering, Volume 2 — Facilities, New York: American Telephone & Telegraph Company, 1977.

Telecommunications Transmission Engineering, Volume 3 — Networks and Services, New York: American Telephone & Telegraph Company, 1977.

Transmission Systems for Communications, Murray Hill, NJ: Bell Telephone Laboratories, 1964.

5

Traffic Theory

> *"We should regard the present state of the universe as the effect of its anterior state and as the cause of the one which is to follow. Given, for one instant, an intelligence which could comprehend all the forces by which nature is animated and the respective situation of the beings who compose it — an intelligence sufficiently vast to submit these data to analysis — it would embrace in the same formula the movements of the greatest bodies of the universe and those of the lightest atom. For it, nothing would be uncertain and the future, as the past, would be present to its eyes. . . . The curve described by a simple molecule of air or vapor is regulated in a manner just as certain as the planetary orbits; the only difference between them is that which comes from our ignorance. Probability is relative, in part to this ignorance, in part to our knowledge."* — Pierre Simon, Marquis de Laplace

This chapter is a survey or summary of a theory by which an adequate number of servers is provided for a certain level of user demand for service. The theory is related to the operations research used by industrial engineers in inventory control and factory layout and the queuing theory used by network engineers in designing buffers for packet-switching systems. This survey is restricted to the simple case where the server system provides enough servers so that, while a user might be blocked when requesting service, it would be within some acceptably small probability.

The pedagogy is to provide the reader a working knowledge and develop some intuition through informal explanation and examples. In the first section, the parameters — traffic, servers, and blocking — are defined, discussed, and quantified. The statistics of these parameters are presented in the second section. Economy of scale is explained in Section 5.3, and extended examples are given in the last section.

5.1 SERVERS

Any system providing service to a pool of users is characterized by three parameters:

- Capacity, serving ability, or number of servers provided in the system;

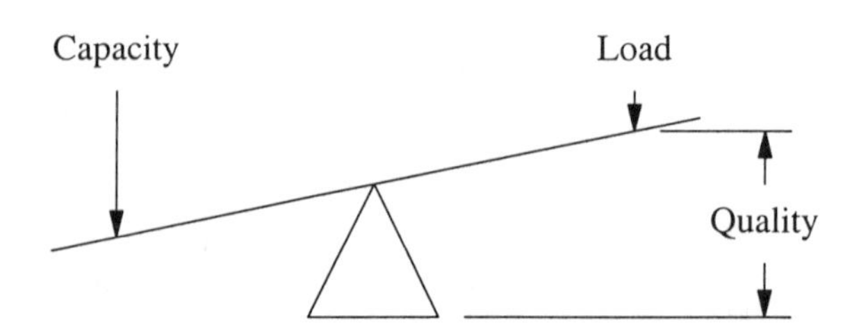

Figure 5.1. Capacity, load, and quality.

- Load, usage, or demand placed on the system by the users;
- Quality, performance, or grade of service provided by the system.

Figure 5.1 intends to provide intuition into how these three parameters trade off against each other.

5.1.1 Analogies and Applications

Consider an airplane. Capacity is the horsepower of the engine, load is the weight of the cargo, and service is the altitude and speed attainable. Consider a heating system or cooling system. Capacity is the BTU rating of the furnace or air conditioner, load is the outside temperature, and service is the inside temperature. While these examples illustrate the tradeoff among the three parameters, neither corresponds well to a network or other switching application.

Consider a complex sewer system. Capacity is the diameter of the pipes and load is the varying instantaneous depth of the fluid. Service is the probability that the fluid flows freely, without backing up if the instantaneous depth momentarily exceeds the size of the pipes. This example is more analogous with a network because of flow and conservation. The flow of waves through the sewer pipes corresponds to the movement of calls through a network. Just as fluid is input only at specified sources and output at specified drains, and flows with conservation, calls through a network enter from the calling parties, terminate on various servers or called parties, and move through a network with conservation.

Other examples are even more analogous with networks because capacity is provided by an integer number of servers. One measures the capacities of these various server systems by counting workstations in a manufacturing assembly line, spaces in a parking lot, airport check-in clerks, lanes in a highway, tellers in a bank, barbers in a shop, check-out clerks in a supermarket, or channels in a trunk group.

5.1.2 Capacity

Capacity is a measure of a system's ability to provide service — typically quantified as the number of servers. Capacity is the parameter that is typically designed into a server system. It is frequently the dependent variable, whose value is determined by estimating the load and specifying the quality. Under constant load, one may change the grade of service by causing a corresponding change in the capacity. To maintain a constant grade of service under changing load, capacity must change correspondingly.

Capacity is the size of a server system, often measured as the number of servers, and is the chief contributor to the cost of providing service. There is a delicate balance between cost and quality that can be sensitive to the market. Paying users may avoid one server system or use a competitor's system if quality is not acceptable or if cost is too high. In telephony, one measures capacity by counting dial-pulse receivers in a central office, operators in a toll office, internal links in a fabric, or channels in a trunk group.

5.1.3 Load

Load is a measure of the amount of service demanded of a system — typically quantified as the amount of call traffic, the instantaneous or average number of simultaneous calls. It is the demand for simultaneous potential service, assuming the server system has enough capacity to provide it. Load can be measured as service demanded simultaneously by the users, but under the condition that the system's quality is perfect. Load can be measured as the number of servers that are simultaneously busy, but an infinite number of servers must be assumed. The reality that a server system only provides a limited number of servers does not affect load, only quality. Load must be known or estimated before the server system is designed. To maintain a constant grade of service, if capacity is increased, then load may be increased. If it is required to raise the grade of service under constant capacity, then load must somehow be lowered either by reducing the number of users, their frequency of use, or their time spent being served. In a telephone switching environment, load is illustrated by the instantaneous number of users simultaneously dialing, operators assisting on calls, busy internal links in a fabric, or interoffice calls in progress.

Load is a random variable whose value is the number of simultaneous users that are potentially served at some instant of time. Load is formally called traffic intensity and is described statistically by an assumed probability density function. Traffic intensity is measured in Erlangs, named after an early pioneer of this theory. A traffic intensity of k Erlangs on a server system means that only k simultaneous users are potentially served.

Three components contribute to traffic intensity: the number of users, the average rate with which each user requests service, and the duration, or holding time, of such a service request. Many of the highways in suburban regions are congested because they were designed years ago for a certain level of automobile traffic. With changing demographics, more people live near these roads, the typical person takes more drives per day, and drives further (or, more exactly, spends more time on the road). Since the server system cannot distinguish the effects of the number of users from the rate with which the average user requests service, these components are lumped into a single component. The effect of 600 users, each making one request per hour, is identical to that of 10 users, each making one request per minute. The net rate with which new requests reach the server system is called the arrival rate.

In telephony, the load on a trunk group, for example, depends on the average rate of arrival of new calls and on the average length of such calls. If the arrival rate doubles, but the holding time halves, the traffic intensity is unchanged. Arrival rate could be measured in calls per minute and holding time could be measured in minutes per call. In the product of these two, their dimensions cancel. Though traffic intensity is measured in Erlangs, it is a dimensionless quantity. The terminology CCS, representing 100 call-seconds, is the product of arrival rate in calls per hour times holding time in seconds per call divided by 100. One Erlang = 36 CCS. This terminology is old-fashioned.

5.1.4 Quality

Quality is a measure of goodness of the service provision — typically quantified as the probability that a service request will be blocked. Quality is usually a requirement, given by the specifiers of the server system. It is usually overspecified, initially, and then negotiated down when the cost of a server becomes known. If capacity is held constant, increased loading causes decreased grade of service. Under constant load, increased capacity causes increased grade of service.

Consider server systems where the users queue for servers, such as getting in line to reach an airline baggage check-in clerk, bank teller, barber, supermarket checkout clerk, or the counter in a so-called "fast food" establishment. Quality could be measured as the inverse of the average delay spent in a queue or as the probability that the time in the queue is under some annoyance threshold, d. In a telephone switching environment, quality is usually a special case of this latter measure, with $d = 0$. If a server is not available immediately, a request for service is said to be blocked, and the frequency with which this happens is called the probability of blocking, P_b. In a telephone switching environment, quality is almost always measured by $1 - P_b$. But three clarifications are called for:

- A formal quantitative definition of blocking;
- A generally acceptable threshold for P_b;
- When the measurements should be made.

By convention and practice, a server system is said to be blocked when all the servers are in use, even if there is not an outstanding request that cannot be met. So the probability of blocking is formally defined as the probability that the current request uses the last server. The traditional standard in the telephone industry is $P_b = 0.01$. In applications where servers are especially expensive, like operators or long distance trunks, a value of 2 – 5% may be used. In applications where multiple servers are needed, like links in a fabric to obtain an end-to-end connection, a value as low as 0.001 may be needed, per server, to obtain $P_b = 0.01$, overall.

5.1.5 Peak Busy Hour

One could compute average hourly rainfall by dividing annual rainfall by the number of hours in a year. If the capacity of a sewer system is designed around this figure, the system may be blocked during most of April, is likely to block during any heavy rainfall, and will undoubtedly block during a hurricane. However, a sewer system with the capacity for the so-called 100-year rain is probably overdesigned and too expensive. A compromise would be to provide enough capacity for the average heaviest rain of the year. A similar issue occurs in telephony.

Computing traffic intensity as the load over a normal 24-hour period would greatly underestimate the load when it was important. But computing it as the average load between 2:00 and 3:00 P.M. on Mother's Day would overestimate it. The convention is to compute traffic intensity as the load during the busiest hour of a normal day, called the peak busy hour. A system with enough capacity to handle this load will be idle at 2:00 A.M. and may experience some blocking on Mother's Day, but it will provide acceptable quality most of the time. One day of extraordinarily high traffic intensity was November 22, 1963, the day President Kennedy was assassinated.

In a typical residential switching office, peak busy hour occurs between 9:30 and 10:30 A.M. and again between 3:00 and 4:00 P.M. Stories about the housewife getting her husband off to work and her kids off to school, cleaning up after breakfast, and then talking on the phone for an hour may be sexist and offensive to some, but some coincident effect is seen at the telephone company. The late afternoon peak is similarly attributed to kids calling their friends after getting home from school. However, similar peaks are seen in the typical business switching office and in the long distance network. The reason for this pattern is not so easily explained. The long distance network also experiences a large peak between 8:00 and 9:00 P.M., and a short burst of activity between 11:00 and 11:30 P.M., when the rates change. These peaks are felt in the local exchanges that provide access to the long distance network.

5.1.6 Disposition of Blocked Calls

In general, the way in which blocked call requests are disposed will affect the performance of a server system. Three assumptions have evolved in the classical development of the theory:

- If a call request is blocked, the call immediately vanishes from consideration. Whether the user gives up or the call is handled by alternate means, the server system does not get a later request for the same call. In other words, if a call, whose holding time would have been T seconds, is blocked, then its revised holding time is 0 seconds.
- If a call request is blocked, the user queues or waits until a server becomes idle. The delay in the queue does not effect the holding time of the call. In other words, if a call, whose holding time would have been T seconds, is blocked, then after the call is eventually served, its revised holding time is still T seconds.
- If a call request is blocked, again the user queues until a server becomes idle. But the holding time of the call is shortened by the delay in the queue. In other words, if a call, whose holding time would have been T seconds, is blocked and delayed by U seconds, then after the call is eventually served, its revised holding time is $T - U$ seconds.

In a telephone environment, the differences among these three assumptions are minor because they are most noticeable at high P_b, which is avoided anyway. While the first two assumptions may be relevant for certain cases, they respectively overestimate and underestimate P_b in general. Thus, while the third assumption may seem to be the silliest of the three, a P_b calculation based on this assumption lies between the calculations based on the other two. Empirically, the third assumption has worked well over the years and is the usual convention in the telephone industry.

If blocking occurs in a network or fabric on a call that cannot be alternately routed, the calling party is usually signaled with audible reorder tone, also called "fast busy." Sometimes, if a switching office becomes extremely busy, users could encounter a long delay between lifting the receiver and hearing a dial tone. Under extreme overloads, an office may cycle through 30-second intervals in which it ignores different groups of users in an attempt to reduce effective load.

5.2 BLOCKING

The Poisson and exponential probability laws are discussed in the first three subsections. Derivation of statistical formulae is beyond the intent of this book. The interested reader is referred to any elementary text on probability theory. While unfamiliar readers may become lost in the language of probability theory that is contained in this section, they are urged to read through to glean some background. Section 5.2.4 is the core of this chapter, and, while total understanding is not necessary, a working knowledge is important for the rest of this book.

5.2.1 Poisson Arrivals

Events are assumed to arrive during some interval of time, and we need a probabilistic way to describe this process. Intuitively, we need a function where the probability that there would be an arrival, and the expected number of arrivals, both increase with the length of the time interval. Furthermore, we expect that large numbers of arrivals would be unlikely and, if the mean is nontrivial, then small numbers of arrivals would also be unlikely. The Poisson probability law is used classically in the statistics of arrivals. This law describes the arrivals of some randomly repeated event in some fixed interval of time. The events can be quasi-instantaneous, such as the number of hits per millisecond on a Geiger counter or the number of raindrops per second falling on a puddle. When the Poisson probability law is used in the context of service requests, only the leading edge of the request, its arrival, is modeled. The duration, or holding time, is modeled separately.

Let N be a random variable, an algebraic symbol that assumes a random value. Let the values that N assumes be integers. Then a probability density function, $p(N = k)$, gives the probability that the random variable, N, has the numeric value of integer, k. The probability density function of the Poisson law is:

$$p(N = k) = \frac{a^k}{k!}\, e^{-a}$$

Here, N is the number of arrivals in some interval of time, a random variable. The expression $p(N = k)$ is the probability that exactly k requests arrive in this interval. The parameter "a" scales the equation to the size of the interval and also turns out to be the mean value of the random variable, N, as shown below. Two similar computations with this formula are: verifying that the sum over all k of $p(N = k)$ is 1 and calculating the mean.

$$\sum_{k=0}^{\infty} p(N = k) = (1 + a + \frac{a^2}{2} + \frac{a^3}{6} + \cdots)\, e^{-a} = e^{+a}\, e^{-a} = 1.0$$

$$\bar{N} = \sum_{k=0}^{\infty} k\; p(N = k) = (0 + a + a^2 + \frac{a^3}{2} + \frac{a^4}{6} + \cdots)\, e^{-a} = a$$

The Poisson probability density function, with $a = 3$, is drawn as the solid curve resembling a bar chart in Figure 5.2. Since the function only has value for integer arguments, a more accurate representation would be as weighted impulses in Figure 5.2. For this curve, the x axis is the number of arrivals, k, and the y axis is the probability of k arrivals in some interval. A probability distribution function, $P(N \leq k)$, gives the probability that the random variable N has a numerical value that is less than or equal to integer k. The probability distribution function is the integral of the probability density

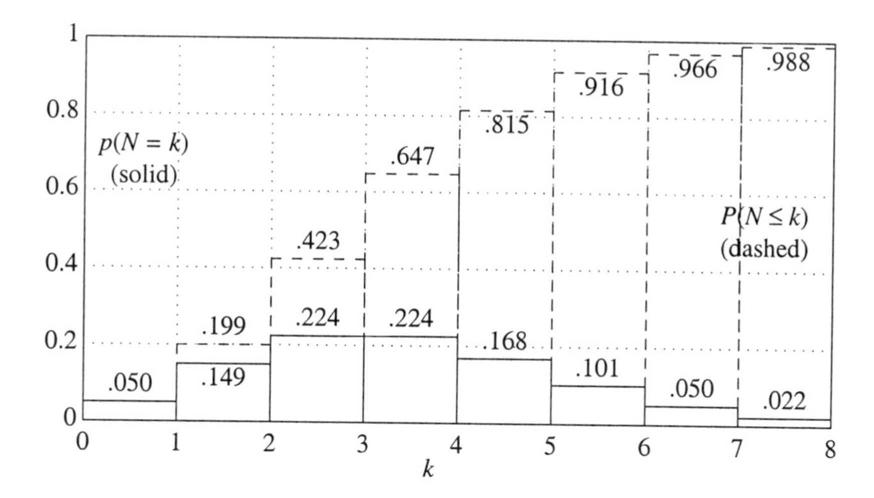

Figure 5.2. Poisson density and distribution functions for $a = 3$.

function. The Poisson probability distribution function, again with $a = 3$, is drawn as the dashed curve in Figure 5.2. For this curve the y axis is the probability of k or fewer requests in this interval.

5.2.2 Exponential Holding Times

Two classical assumptions are made about *holding times* in the telephone switching environment. The simplest assumption is that holding time is constant. This may seem unrealistic, but it is true enough for servers like audible tone generators and even for Touch Tone receivers. Compared to the variation in the lengths of telephone calls, there is little variation in how long people listen to a busy tone or even in how long it takes people to Touch Tone a telephone number. A constant holding time would be a poor assumption for servers like trunks or links in a network or fabric.

The more realistic assumption is that the holding time, t, is a random variable that is weighted heavily for durations that are smaller than the mean value and weighted lightly for durations that are increasingly greater than the mean value. Telephone engineers have classically assumed that holding time obeys an exponential probability law. If T is a random variable with real values, t, then the probability density function of the exponential law is:

$$p(t) = \frac{1}{h} e^{-t/h}$$

The value of a probability density function over a real-valued random variable does not have the physical significance as over an integer-valued random variable. The probability that random variable T has the real value t, exactly, is infinitesimally small. The parameter "h" scales the equation and, again, is the mean value of the random variable. Integrating over all positive real numbers,

$$\int_0^\infty p(t)\, dt = 1 \quad \text{and} \quad \bar{T} = \int_0^\infty t\, p(t)\, dt = h$$

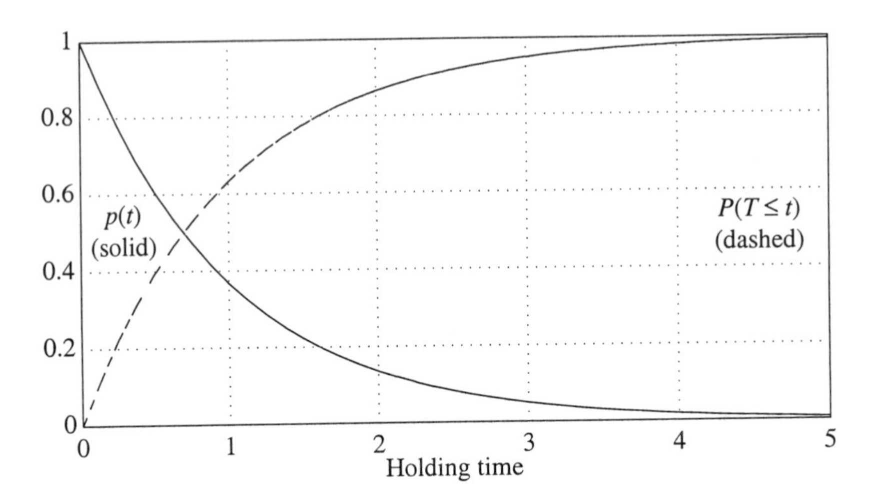

Figure 5.3. Exponential probability density and distribution functions.

The exponential probability density function, with $h = 1$, is drawn as the solid curve in Figure 5.3. The value of a probability distribution function, $P(T \leq t)$, over a real-valued random variable does have physical significance. $P(T \leq t)$ is the probability that the random variable T has a numerical value that is less than or equal to real number, t. The exponential probability distribution function, again with $h = 1$, is drawn as the dashed curve in Figure 5.3. As a model for holding times of network links and trunks, the exponential law is slightly unrealistic around $t = 0$, but it is classically used.

It can be shown that the exponential and Poisson laws are reciprocal. If the arrivals of some event obey the Poisson law, then the time between successive arrivals obeys the exponential law.

5.2.3 Traffic Intensity

Consider overlapping events with random starting times and random durations. Consider a random variable, I, defined as the number of simultaneously active events. Traffic intensity, discussed in Section 5.1 and measured in Erlangs, is such a random variable. Let the events start (arrive) under a Poisson law and let each event's duration obey an exponential law (or let them be constant). Under these assumptions, it has been shown that I, the number of simultaneously active events, obeys a Poisson law. Thus, not only are arrivals Poisson distributed, but traffic intensity, the number of simultaneously active events, is also Poisson distributed.

Figure 5.2 also illustrates the probability density function, $p(I = k)$, and probability distribution function, $P(I \leq k)$, of traffic intensity I. The parameter, $a = 3$ Erlangs, is the average traffic intensity, the average number of potentially busy servers or of users simultaneously receiving potential service. The probability that the load is zero, that no user wants service, that no server is busy, is $e^{-3} = 0.050$. The probability that one server is busy is 0.149 and that one or fewer is busy is $0.050 + 0.149 = 0.199$. The probability that two servers are busy is 0.224 and that two or fewer are busy is $0.199 + 0.224 = 0.423$. The probability that three servers are busy is also 0.224 and that three or fewer are busy is $0.423 + 0.224 = 0.647$. This value, three, is the mean value of the random variable. The

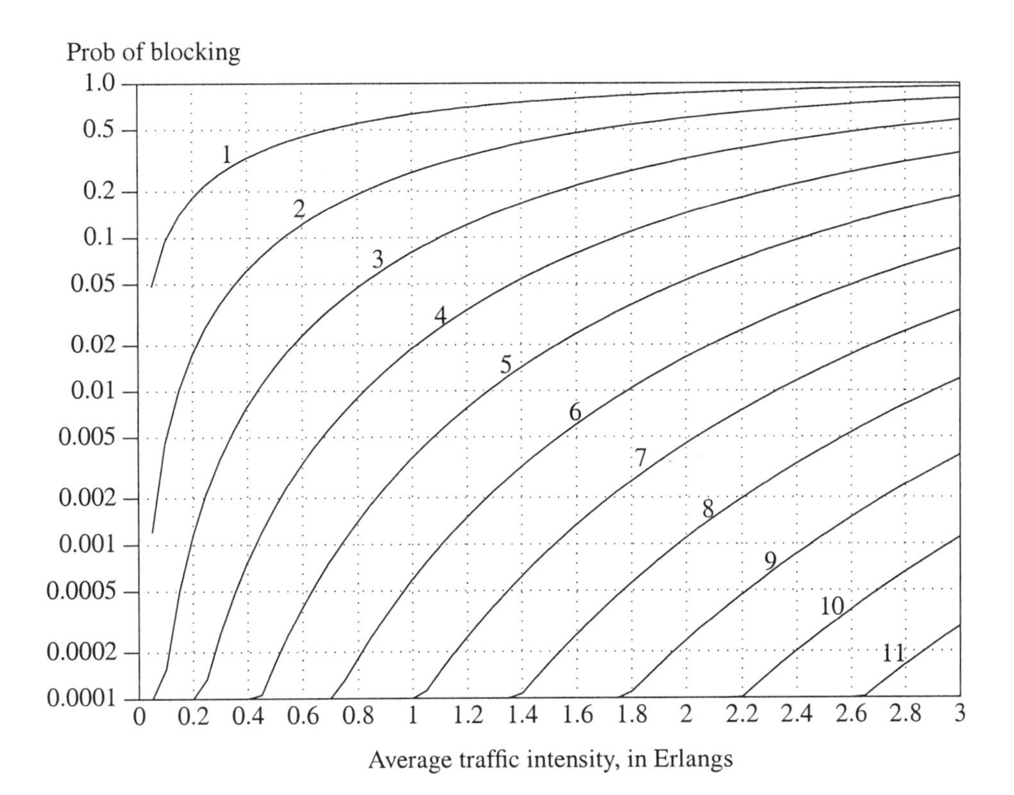

Figure 5.4. Poisson curves for TI ≤ 3.

probability that four servers are busy is 0.168 and that four or fewer are busy is 0.647 + 0.168 = 0.815. The probability that five servers are busy is 0.101 and that five or fewer are busy is 0.815 + 0.101 = 0.916.

Recall that the convention for probability of blocking includes using the last server, so the next user cannot be served. If a system has k servers, the formal probability of blocking is the probability that there is a demand for k or more servers,

$$P_b = P(I \geq k) = 1 - P(I \leq k - 1)$$

If there are only k servers, the probability that more than k are in use is, of course, 0. However, the probability of demand for more than k servers is greater than 0, even if the demand cannot be met. So, referring to Figure 5.2 with $a = 3$, if a server system has six servers, then its probability of blocking is $P_b = 1 - P(I \leq 5) = 1 - 0.916 = 0.084$. In practice, in the real world, nobody makes computations like this. It is all done graphically.

5.2.4 Charts

Consider a two-dimensional space, where the x axis is average traffic intensity and the y axis is probability of blocking. In this space, a curve is plotted that shows probability of blocking as a function of average traffic intensity, for a given number of servers. A separate curve is required for every value of k, the number of servers. A family of curves is typically plotted on one graph.

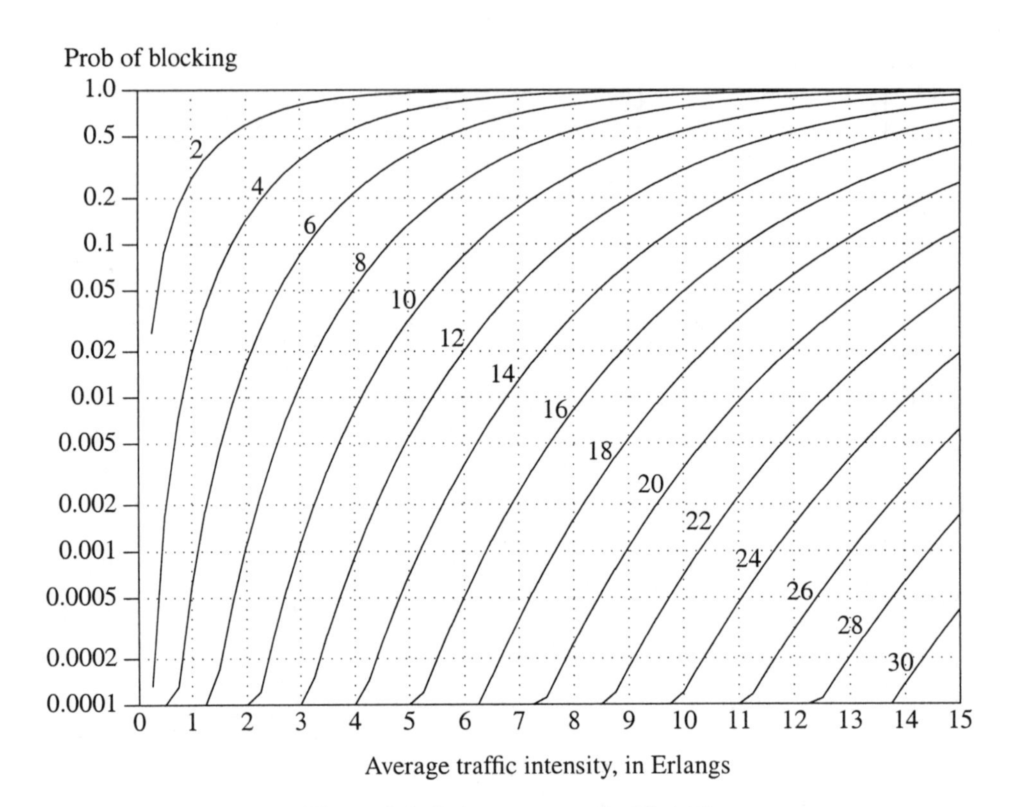

Figure 5.5. Poisson curves for TI ≤ 15.

The Poisson-distributed traffic intensity statistics are plotted in Figures 5.4, 5.5, and 5.6. Each graph shows probability of blocking over a different range of traffic intensity, with separate curves for different values of number of servers. In Figure 5.4, the traffic intensity ranges from 0 to 3 Erlangs, and the number of servers ranges from 1 to 11. In Figure 5.5, the traffic intensity ranges from 0 to 15 Erlangs, and the indicated number of servers ranges from 2 to 30 by twos. In Figure 5.6, the traffic intensity ranges from 0 to 75 Erlangs, and the indicated number of servers ranges from 10 to 100 by tens. One typically uses the chart that best matches the scale of the problem.

The right edge of Figure 5.4 corresponds to an average traffic intensity of 3 Erlangs, that three users are potentially served on the average. The curve corresponding to six servers crosses that vertical line at a horizontal value somewhere between 0.05 and 0.1. While it is difficult to read even one significant figure off the chart, $P_b = 0.084$ was computed in the discussion with Figure 5.2. The reader is encouraged to find this point and several others.

- On Figure 5.4, 1 Erlang with four servers has $P_b = 0.02$.
- On Figure 5.5, 9 Erlangs with 20 servers has $P_b = 0.001$.
- On Figure 5.5, 12.5 Erlangs with 23 servers (interpolating a missing curve) has $P_b = 0.005$.

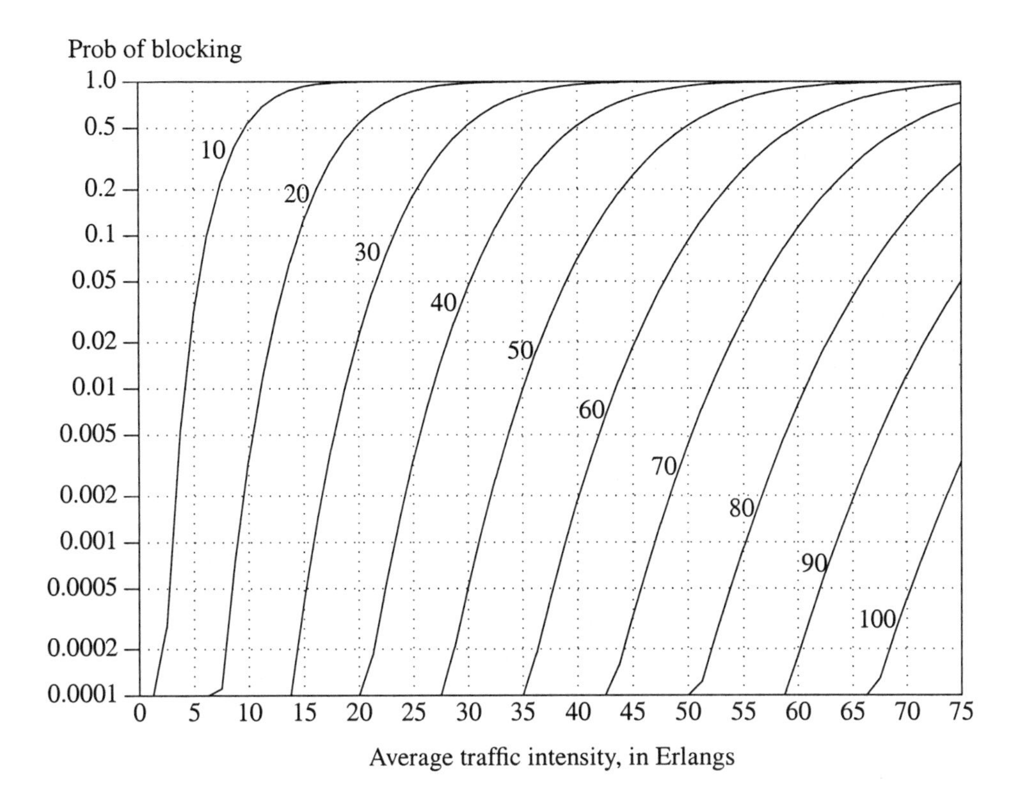

Figure 5.6. Poisson curves for TI ≤ 75.

- On Figure 5.6, 65 servers at $P_b = 0.02$ can handle 50 Erlangs.
- On Figure 5.4, to provide $P_b = 0.02$ over 3 Erlangs, between seven and eight servers would be required. If $P_b = 0.03$ is not acceptable, then seven servers are insufficient and eight must be provided, giving $P_b = 0.012$.

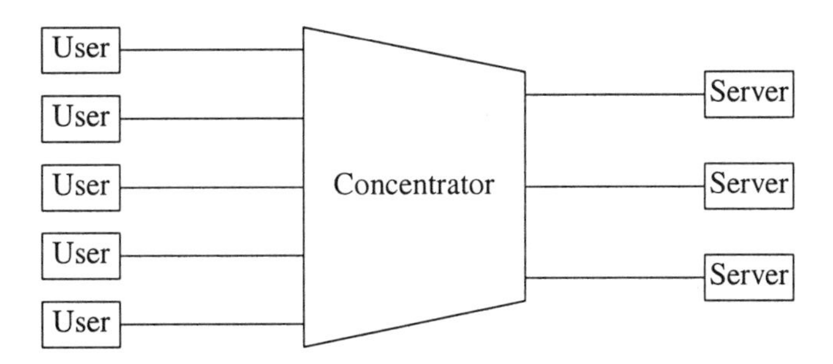

Figure 5.7. Concentrating five users to three servers.

5.2.5 Example of Network Concentration

Consider a switching fabric, or interconnection network, in which M users are connected to ports at the left and K servers are connected to ports at the right. Figure 5.7 illustrates where $M = 5$ users at the left compete for $K = 3$ servers at the right. If $K < M$, this fabric is said to concentrate (Section 7.2.3) the M users to the K servers. No more than K of the M users may be simultaneously served. Assume that the fabric has "full access," that each user can be connected through this fabric to any of the K servers, as long as the server is idle. The server might be a trunk, dial-pulse receiver, intermediate link through the fabric, or any similar resource required for a telephone call.

Let 1000 subscribers be connected to the originating side of such a fabric. Let the average subscriber originate one call every 2 hours during the peak busy hour. Let the average call duration (or, more precisely, the average holding time of the server) be 3 minutes. The average traffic intensity per subscriber is:

$$\bar{I}_{\text{sub}} = \frac{1}{2 \times 60} \frac{\text{calls}}{\text{minute}} \times 3 \frac{\text{minutes}}{\text{call}} = 0.025 \text{ Erlangs}$$

The total average traffic intensity is $\bar{I}_{\text{tot}} = 1000 \times 0.025 = 25$ Erlangs. From Figure 5.6, 25 Erlangs at $P_b = 0.005$ requires 40 servers. Thus, the 1000 subscribers may be concentrated down to only 40 links, and the chance that a call is blocked is only 1 in 200 during the peak busy hour. This example is continued in Section 5.3.4, where the assumption of full access is removed. Blocking in fabrics is generally complex and is briefly illustrated in Section 7.4.

5.3 ECONOMY OF SCALE

This extremely important issue is introduced by an example comparing one-way trunks and two-way trunks. Occupancy and access size are discussed and the network concentration example, from the previous section, is continued to illustrate these concepts. Extended examples are given in the next section of this chapter to further illustrate economy of scale.

5.3.1 One-Way and Two-Way Trunks

Exercise 4.3 called for comparing the cost of two groups of 25 one-way trunks versus the cost of one group of 44 two-way trunks. The justification for these quantities is illustrated. Let there be 15 Erlangs of average traffic intensity for calls that originate in office A and terminate in office B. Similarly, let there be 15 Erlangs of average traffic intensity for calls that originate in office B and terminate in office A.

In general, the number of servers in a group is computed from the net traffic from all the users that can access the group. If the calls between A and B are served by two independent groups of one-way trunks, the distinct trunk groups serve separate sets of users. The total number of one-way trunks is found by first computing the number of trunks in each group and then adding the group sizes together. From Figure 5.5, 15 Erlangs at $P_b = 0.01$ requires 25 trunks in each group. The total number of one-way trunks is 50.

However, if these calls are served by one group of two-way trunks, the traffic from A to B is *pooled* with the traffic from B to A. The single trunk group serves calls from A to B and from B to A. The total number of two-way trunks is found by first adding the traffic for the two groups together and then computing the number of trunks for this total traffic. The total traffic is 30 Erlangs. From Figure 5.6, 30 Erlangs at $P_b = 0.01$ requires that the total number of two-way trunks is 44.

This decrease in the number of trunks is not without added cost. Recall that a one-way trunk has one trunk circuit on each end and that a two-way trunk has two trunk circuits on each end. If A and B are far apart, the cost of trunk circuits is less significant than the cost of the channels and two-way trunks are more economical, even though their circuits are more expensive. If A and B are close together, the cost of trunk circuits is a more significant part of the total cost and one-way trunks are more economical, even though more are required.

Going from one-way trunks to two-way trunks illustrates a general effect, called "economy of scale." This decrease in the number of trunks is possible by expanding the access to a group of servers or pooling groups of servers that were once separated. Economy of scale is effected by combining two specialized groups into one more general group, resulting in requiring fewer total servers.

5.3.2 Occupancy

Another way of stating the generalization above is that few large groups are more efficient than many small groups. This efficiency or utilization quantifies economy of scale — and is called occupancy. Occupancy could be viewed as a fourth parameter of server systems, but it is a simple function of two of the other parameters, proportional to load and inversely proportional to capacity. However, if capacity and load increase simultaneously, in a way that holds occupancy constant, as the system size increases, the grade of services improves, and dramatically (Exercise 5.17). Conversely, if capacity and load increase simultaneously, and grade of service is held constant, as system size increases, occupancy improves dramatically. Occupancy is the probability that the average server is in use and it is measured as the average traffic intensity per server.

In Figure 5.4, four servers at $P_b = 0.01$ can handle about 0.8 Erlangs of average traffic intensity. The corresponding occupancy is 0.8/4 = 0.2 Erlangs per server and is represented on Figure 5.8 by the point labeled 4. Figure 5.8 shows a locus of points taken from Figures 5.4, 5.5, and 5.6. The point labeled k ($k = 2, 3, 4, \ldots, 50$) on Figure 5.8 represents the occupancy of k servers at $P_b = 0.01$. The mathematics of the curve that passes through these points is beyond this text. This locus asymptotically approaches occupancy = 1.0 for a large server groups. Quantifying economy of scale is apparent.

5.3.3 Access Size

Suppose a single time-multiplexed bus could be distributed to every telephone user in the United States. If the typical user loads the system at 0.05 Erlangs, how many timeslots are required? At 0.05 Erlangs each, 100 million users cause 5M Erlangs, total. While the charts provided in this text don't go that high, Figure 5.8 suggests that 6M channels would provide $P_b > 0.01$. At 64 Kbps per digital voice channel, the total rate on some kind of highly multiplexed bus must be 6M × 64K = 384 Gbps. Conventional fiber-optic transmission systems are far from this figure, but conventional single-mode optical fiber just may have this much bandwidth (if we ever learn how to use it), and architectures for photonic systems have been proposed to take advantage of such a large throughput (see Chapters 20 and 21). While this immense data rate might be supported some day, loss, delay, and economic penalties make such a scheme impractical anyway.

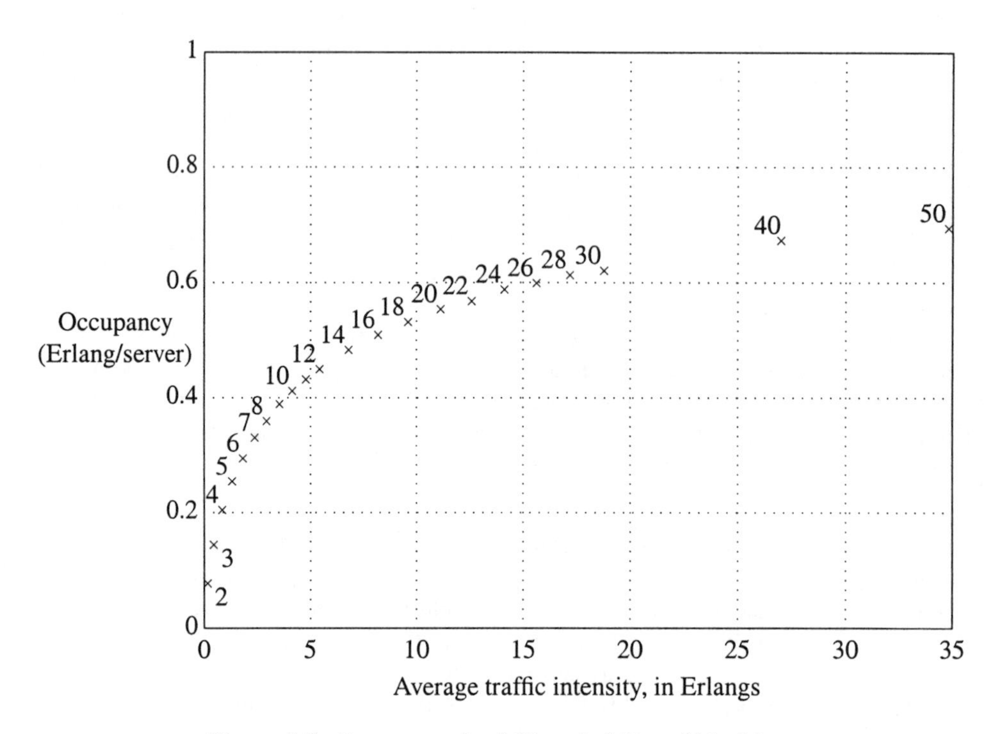

Figure 5.8. Occupancy for 0.01 probability of blocking.

Because such full access is impractical, the national telephone network has the complex structure that is the subject of this book. This total network requires many times more than 6M channels because each subscriber cannot directly access all the channels.

5.3.4 Limited Network Access Example

It was shown in Section 5.2.5 that 1000 subscribers could be served by 40 servers through a nonblocking, concentrating fabric with full access. Suppose such a 1000 × 40 full-access fabric cannot be built. Let the largest single fabric be limited to 100 ports. The 1000 subscribers must be physically partitioned into 10 distinct groups, with each group connected to the left side of a different fabric. Each group of subscribers generates 25/10 = 2.5 Erlangs of average traffic intensity. From Figure 5.4, 2.5 Erlangs at $P_b = 0.005$ requires eight servers. Each of the 10 independent fabrics must be 100 by 8 and, the total number of servers is 8 × 10 = 80, instead of 40.

If the full-access fabric were possible to build, it might be impractical, even though economy of scale reduces the required number of servers. Consider a simple fabric that connects the users to the servers and let one relay contact be required to connect each user to each server. A full-access fabric requires 1000 × 40 = 40,000 contacts. A fabric that connects one group of 100 servers to its eight servers requires 100 × 8 = 800 contacts. The total fabric, implementing a partial access system, requires 10 × 800 = 8000 contacts. In general, full access requires fewer servers than limited access, but is more expensive to implement. This example is continued throughout Section 7.2.

5.4 EXAMPLES

5.4.1 Traffic Service Position System

Let an area served by some sectional center (previous chapter) contain 15 toll offices. Let each toll office have its own group of toll operators. Let the average traffic intensity to each group of toll operators be 4 Erlangs during the peak busy hour. From Figure 5.5, $P_b = 0.02$ shows that nine operators would be needed in each toll center at these times. Since this represents a limited access system, we must add servers, not traffic, to find the total capacity. During the peak busy hour, nine toll operators in each of 15 toll offices, or 135 total, are needed in this sectional area. Each operator would be occupied at only 44%.

A system called Traffic Service Position System (TSPS), described in Section 9.3.2, has two features relevant to this chapter [1]. The first significant feature of TSPS is that it pools the toll operators inside some large geographic area. Instead of being associated with particular toll offices, the toll operators are part of a section-wide pool to which all the toll offices have access. This full access is extremely expensive to implement (Section 9.3.2). However, with the full access provided by TSPS, the number of toll operators is found by adding traffic and not servers. The total average traffic intensity is $4 \times 15 = 60$ Erlangs in the entire sectional area. From Figure 5.6, $P_b = 0.02$ shows that 76 servers are required. Furthermore, the toll operators are now occupied at 79%.

The second feature of TSPS is that it shortens the amount of dialogue between the caller and the toll operator. For example, consider the procedure for placing a long distance collect call. Before TSPS, the operator had to ask for the called party's telephone number, when placing the call, and for the calling party's name, to ask later if the called party will accept the charges. With TSPS, the operator must still ask for the calling party's name, but this calling party had previously dialed the called party's telephone number. This procedure helps automate the placing of the call and reduces the holding time on the operator. If the holding time on a toll operator is halved, then the total average traffic intensity in this sectional area is halved to 30 Erlangs. From Figure 5.6, 30 Erlangs at $P_b = 0.02$ requires 42 servers, reduced from 76. Occupancy reduces slightly to 71%.

The economics are illustrated with numbers that are only made up. Let the number of attendants required during an entire day equal twice the number required in the shift that includes the peak busy hour. Let the loaded annual salary (take-home pay plus cost of benefits) be \$60,000/year. Without TSPS, toll operators in this sectional area cost 135 $\times$ 2 $\times$ \$60K = \$16.2M per year. With TSPS, they cost 42 $\times$ 2 $\times$ \$60K = \$5.04M per year. Presumably, the \$11 million a year savings more than covers the amortized first cost plus annual operating cost of a TSPS. While some of the net savings is from automation that reduces holding time, much of the savings is from reducing the required number of operators through full access and economy of scale.

5.4.2 Toll Trunking

Consider N central offices all served by the same toll center. Two values of N are considered, $N = 4$ and $N = 10$. In addition, two intratoll network architectures are considered, a simple star with no direct inter-CO trunking and a star supplemented with direct inter-CO trunking between every pair of offices. Thus, there are four cases, for each value of N and each architecture. Let all the inter-CO channels be two-way trunks.

Let there be 36 Erlangs of average external traffic intensity in each direction at each of the N switching offices. Let half be intratoll and half be intertoll. Let the intratoll traffic in each direction at each switching office be split evenly over the other $N - 1$ offices. The number of trunk channels will be computed below for all four cases. If one assumes that the cost of a trunk is dominated by the cost of trunk circuits and testing, and that physical channels are inexpensive, then this number of trunks is proportional to the total cost of the trunking. This assumption gets closer to the truth every year, as multiplexing lowers the long-haul cost and the cost of OA&M becomes more significant.

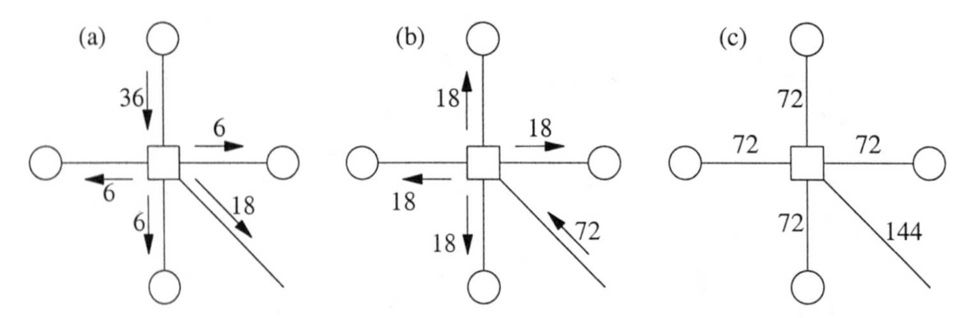

Figure 5.9. Four COs with hubbed trunking.

Four COs with Hubbed Trunking. In the first case under consideration, $N = 4$ central offices are connected by a star network to a common toll office at the hub. Since there is no direct inter-CO trunking, all inter-CO calls are placed through this hub. Every intra-toll call requires two trunks, one from the originating CO to the toll office and one from the toll office to the terminating CO. The average traffic intensity of each group of two-way trunks is illustrated on Figure 5.9(c). These numbers are explained by two derivations, one that begins with Figure 5.9(a) and one that begins with Figure 5.9(b).

Figure 5.9(a) shows the simple star architecture, where the numbers and arrows represent the average inter-CO traffic intensity that only originates in this toll area and only from the CO at the north. All this outgoing 36 Erlangs flows to the toll office, and from there, 18 flows outward to the intertoll network and the other 18 splits evenly to the other three switching offices. An illustration of incoming traffic that terminates in this same CO would be identical to Figure 5.9(a) except that the arrows would be reversed. Similar illustrations could be done for the other three offices. Each line on Figure 5.9(a) corresponds to a group of two-way trunks, and the total traffic in each group is found by adding the numbers associated with this line from each of the eight illustrations represented by Figure 5.9(a). The resultant is Figure 5.9(c), where the numbers represent the total two-way traffic flow.

Figure 5.9(b) shows the disposition of the average intertoll traffic intensity that is only incoming to this toll area. All this incoming 72 Erlangs flows to the toll office from the national network and from there splits evenly to the four switching offices. An illustration of outgoing intertoll traffic would be identical to Figure 5.9(b) except that the arrows would be reversed. The total intertoll traffic in each trunk group is found by adding the numbers associated with the corresponding line from each of the two illustrations represented by Figure 5.9(b). After 36 Erlangs of intratoll traffic is added to each CO-to-hub trunk group, the resultant again is Figure 5.9(c).

Thus, there are 72 Erlangs of average traffic intensity in each group of two-way trunks between a CO and the toll office. This results from 18 Erlangs in each of four categories: originating intratoll, originating intertoll, terminating intratoll, and terminating intertoll. The traffic in each category can be added because the trunks are two-way and all traffic flows through the hub. From Figure 5.6, 72 Erlangs at $P_b = 0.01$ requires 93 servers. With 93 two-way trunks in the group between the hub and each of the four serving offices, the total number of intratoll trunks in this toll area is $4 \times 93 = 372$ trunks.

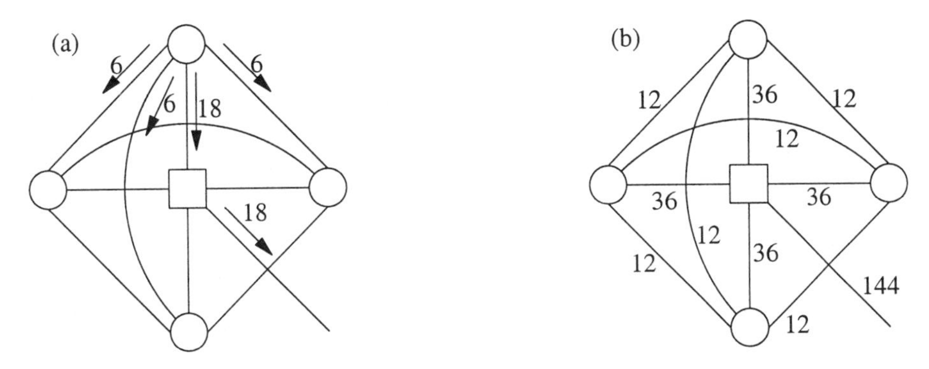

Figure 5.10. Four COs with direct trunking.

Four COs with Direct Trunking. In the second case under consideration, let direct trunking be added between every pair of central offices within this toll area. There are $N \times (N-1)/2 = 6$ groups of direct two-way trunks: interconnecting N with E, N with S, N with W, E with S, E with W, and S with W. Let an intratoll call between a given pair of offices use one trunk from the group that interconnects this pair, instead of two trunks from the groups that interconnect this pair with the hub. At each office, half the traffic goes to the hub and on to the national network. For this 36 Erlangs each way, Figure 5.6 shows we need 52 servers. The other half of the traffic, 36 Erlangs, is distributed from each office over the three other offices. The average traffic intensity in each group of two-way trunks is illustrated on Figure 5.10(b). These numbers are derived, beginning with Figure 5.10(a).

Figure 5.10(a) shows the original star architecture, plus the six additional groups of direct two-way trunks. The numbers and arrows represent the average inter-CO traffic intensity that only originates in this toll area and only from the CO at the north. Of this outgoing 36 Erlangs, 18 flows to the intertoll network via the toll office at the hub and the other 18 splits evenly to the other three switching offices over the direct trunks to those offices. An illustration of incoming traffic that terminates in this same CO would be identical to Figure 5.10(a) except that the arrows would be reversed. Similar illustrations could be done for the other three offices. Each line on Figure 5.10(a) corresponds to a group of two-way trunks, and the total traffic in each group is found by adding the numbers associated with this line from each of the eight illustrations represented by Figure 5.10(a). The resultant is Figure 5.10(c), where the numbers represent the total two-way traffic flow.

Thus, each of the six groups of direct two-way trunks, which interconnects the pairs of central offices, carries 12 Erlangs. From Figure 5.5, 12 Erlangs at $P_b = 0.01$ requires

22 servers. Each of the four groups of two-way trunks, that interconnects the central office with the toll office at the hub, carries 36 Erlangs. From Figure 5.6, 36 Erlangs at $P_b = 0.01$ requires 52 servers. Thus, there are $6 \times 22 = 132$ total inter-CO trunks and $4 \times 52 = 208$ total CO-to-toll trunks, for a grand total of $132 + 208 = 340$ two-way trunks in the entire toll area.

In comparing these 340 total trunks to the 372 total trunks in the first case, two effects work against each other. In the hubbed architecture of the first case, an intratoll inter-CO call uses two trunks and in the direct-trunking architecture of the second case, such a call only uses one trunk. Thus, the first case is expected to have more total trunks than the second; and it does. But, if 50% of the calls use an extra trunk, intuition might suggest that the first case has 50% more trunks than the second. The difference is not this large because this first effect is mitigated by the second. The first case exhibits greater economy of scale than the second case because the second case has more groups of more specialized servers than the first. For $N = 4$, this second effect is weaker than the first effect. In the next subsection, with $N = 10$, this second effect will be stronger than the first.

Ten Switching Offices. In the third case under consideration, there is only hubbed trunking, but among $N = 10$ central offices. As in the first case, the 72 Erlangs of average total traffic intensity, between each CO and the toll office at the hub, requires 93 servers. With 93 servers in 10 groups, instead of 4, there are 930 total two-way trunks. Again, two trunks are required for every intratoll call.

In the fourth case under consideration, direct trunking is added to the previous case. There are $N \times (N - 1)/2 = 45$ groups of direct two-way trunks. In the second case illustrated in Figure 5.10(a), the 18 Erlangs of intratoll traffic originating from the north CO splits into 6 Erlangs to each of the other three central offices. In this fourth case, with $N = 10$, this same 18 Erlangs splits into 2 Erlangs to each of the other nine offices. By reasoning similar to that of the second case, the total configuration, corresponding to Figure 5.10(b), has 4 Erlangs on each of the 45 groups of direct two-way trunks and 36 Erlangs on each of the 10 groups of CO-to-toll two-way trunks.

From Figure 5.5, 4 Erlangs at $P_b = 0.01$ requires 10 servers in each group of direct two-way trunks. As in the second case, the 36 Erlangs of average total traffic intensity, between each CO and the toll office at the hub, requires 52 servers. Thus, there are $45 \times 10 = 450$ total inter-CO trunks and $10 \times 52 = 520$ total CO-to-toll trunks, for a grand total of $450 + 520 = 970$ two-way trunks in the entire toll area. This fourth case requires more total trunks than the third case. $N = 10$ is large enough that the effect of economy of scale overcomes the effect that an extra trunk is used on half the calls.

EXERCISES

5.1 Consider a highway that is used primarily as an automobile artery for daily commuters. Which characteristics of the highway correspond to capacity, load, and quality?

5.2 Assume the average telephone user is busy on one 3-minute call during the peak busy hour. (a) What is the average traffic intensity per user? (b) What is the average net traffic intensity for a 10K-line office?

5.3 Suppose a switching system's operating cost is 50% call setup and 50% usage. Call setup cost is proportional to arrival rate and usage cost is proportional to traffic intensity. (a) What happens to total cost if arrival rate doubles, but holding time halves? (b) What happens to total cost if arrival rate halves, but holding time doubles?

5.4 Consider Figure 5.5 and the curve for six servers. Consider two similar curves: one for the case where blocked calls vanish and one for the case where all blocked calls retry and hold as long as they would have if they hadn't been blocked. All three curves approximately coincide for low probability of blocking. But, for high probability of blocking, one curve would diverge above the curve in Figure 5.5 and one would diverge below it. Which is which?

5.5 From Figure 5.2, what is the probability (a) that $N = 3$ or 4? (b) That $N = 4$ or 5? (c) That N is an odd number?

5.6 Draw a graph with two curves, similar to those in Figure 5.2, for the Poisson density and distribution functions when $a = 4$.

5.7 From Figure 5.3, what is the probability (a) that T is between 1 and 3? (b) That $T < 3$ in an exponential density whose mean is 2?

5.8 Consider a pool of Touch Tone receivers, with TI = 2.4 Erlangs. (a) What is the probability of blocking if the pool has seven servers? (b) How many servers are needed for $P_b < 0.005$?

5.9 Consider a trunk group, where the probability of blocking must be less than 0.02. (a) How much traffic intensity can be supported if the group has 18 trunks? (b) How many trunks are needed to support 13 Erlangs of traffic intensity?

5.10 Consider an optical fiber whose channels are time- and wavelength-multiplexed. Let there be four wavelengths and 24 timeslots. (a) How much traffic can be supported for $P_b < 0.005$? (b) What is the probability of blocking for a traffic intensity of 70 Erlangs?

5.11 A trunk group with 12 trunks supports TI = 4 Erlangs. (a) What is the P_b? (b) How many servers are needed for $2 \times$ the P_b in (a)? (c) How many servers are needed to support twice as much traffic as in (a)? (d) How much traffic can be supported by twice as many servers as in (a)?

5.12 A bank measures customer statistics during its busiest hour every day. (a) If the bank averages 90 customers during this peak hour, what is the mean arrival rate? (b) If the average teller transaction is 2 minutes, what is the mean traffic intensity? (c) In order to stay even with this rate of service requests, what is the minimum number of tellers the bank should employ during this peak hour? (d) If the bank employs one more than this minimum number, what is the probability that an arriving customer is blocked? Note that, because of the odd definition of blocking, this blocked customer may still be served.

5.13 Consider a pedestal that serves a neighborhood of 30 homes. Assume that the homes average 1.5 lines each. (a) If the pedestal is served by a 60-pair cable, what is the probability that a new subscription request will be blocked? (b) As more customers buy home PCs and subscribe to second lines, suppose line density rises to 1.6 lines per home. What then, is the probability that a new subscription request will be blocked?

5.14 A parking lot with 80 places is measured to block with $P_b = 0.008$. (a) What traffic intensity does the lot handle? A law requires that enough places be dedicated for the handicapped to provide the same grade of service as the original lot. If one person out of 600 is handicapped, (b) what is the traffic load on this lot from handicapped persons? (c) How many places are required? Let every two handicapped places take the room of three regular places. After losing this number of places, (d) what is the new grade of service on the general population? (e) What is the change over the previous grade of service? (f) What is the occupancy of the original lot and of the new handicapped places? (g) What is the probability that all handicapped places are empty?

5.15 In Figure 5.5, for each curve representing N servers, find the point corresponding to $\text{TI} = 0.5 \times N$ Erlangs. Connecting these points yields a curve of constant occupancy — in this case, an occupancy of 50%. It should be a straight line and should illustrate that, at constant occupancy, larger systems provide better service.

5.16 A server system is required to meet three specifications: (1) number of servers is 15 or smaller, (2) occupancy is 0.5 or greater, (3) probability of blocking is 0.02 or less. On a photocopy of Figure 5.5, indicate the area of operation.

5.17 From Figure 5.7, for $P_b = 0.01$, (a) how much traffic is supported by a system with 16 servers and what is the occupancy? (b) How much traffic is supported by a system with half as many servers and what is the occupancy? (c) How much traffic is supported by a system with twice as many servers, and what is the occupancy?

5.18 The network concentration example of Section 5.3.4 is continued. Let 9 of the 10 separate concentrating fabrics be constructed using eight links, as calculated. (a) How many Erlangs of average traffic intensity can these eight links handle? (b) How many subscribers can be connected to the left side of each concentrating fabric? (Hint: more than 100.) (c) What are the dimensions of each of these nine concentrating fabrics? (d) How many of the 1000 subscribers are left over? (e) How many servers are needed to handle those left over? (Hint: fewer than eight.) (f) What are the dimensions of this 10th concentrating fabric? (g) How many total servers does this system of 10 fabrics need?

REFERENCE

[1] Jaeger, R. J., and A. E. Joel, Jr., "TSPS No. 1 Systems Organization and Objectives," *The Bell System Technical Journal,* Vol. 49, 1970, p. 2417.

SELECTED BIBLIOGRAPHY

Bellamy, J., *Digital Telephony,* Second Edition, New York, NY: Wiley, 1991.

Brooks, J., *Telephone — The First Hundred Years,* New York, NY: Harper & Row, 1976.

Clarke, A. C., et al., *The Telephone's First Century — and Beyond,* New York, NY: Thomas Y. Crowell, 1977.

Cooper, R. B., *Intro to Queueing Theory,* New York, NY: North Holland, 1981.

Duffy, F., and R. A. Mercer, "A Study of Network Performance and Customer Behaviour During Direct Distance Dialing Call Attempts in the U.S.A.," *The Bell System Technical Journal,* Vol. 57, No. 1, 1978.

Freeman, R. L., *Telecommunication System Engineering,* Third Edition, New York, NY: Wiley, 1996.

Kolmogorov, A. N., *The Foundations of Probability,* Second Edition, New York, NY: Chelsea, 1956.

McDonald, J. C. (ed.), *Fundamentals of Digital Switching,* Second Edition, New York, NY: Plenum Press, 1990.

Mina, R. R., *Introduction to Teletraffic Engineering,* Chicago, IL: Telephony Publishing Corporation, 1974.

Molina, E. C., "Application of the Theory of Probability to Telephone Trunking Problems," *The Bell System Technical Journal,* No. 6, 1927, pp. 461 – 494.

Noll, A. M., *Introduction to Telephones and Telephone Systems,* Norwood, MA: Artech House, 1986.

Rey, R. F. (ed.), *Engineering and Operations in the Bell System,* Second Edition, Murray Hill, NJ: AT&T Bell Laboratories, 1977.

Schindler, G. E., Jr. (ed.), *A History of Engineering and Science in the Bell System,* Murray Hill, NJ: Bell Telephone Laboratories, 1982.

Switching Systems, New York, NY: American Telephone & Telegraph, 1961.

Syski, R., *Introduction to Congestion Theory in Telephone Systems,* Edinburgh, Scotland: Oliver and Boyd, 1960.

Talley, D., *Basic Telephone Switching Systems,* Rochelle Park, NJ: Hayden Book Company, 1969.

6

Step-by-Step

> *"One small step for man; one giant leap for mankind."* — Neil Armstrong, when he stepped off the LEM *Eagle* onto the surface of the moon

The Step-by-Step system is the oldest architecture of the automatic exchanges and was an extremely successful commercial system. But, because of Step's economy in small offices (Section 2.4.5), it is still found in several countries and it is only fairly recently that Step has come close to extinction inside the United States. While Step's ancient architecture inherently lacks the power of common control, Step offices have been modernized for Touch Tone and even computerized to provide some new services and modern signaling. Since Step was summarized in Section 2.5.1, the reader who is in a hurry could skip this chapter entirely. But I urge you to skim it, at least, because Step has historical value, an interesting "cellular" architecture (as defined in Section 2.4.1), and a novel originating-to-terminating sidedness. Even more relevant, two concepts from Step could be applied to our most modern systems.

- The process by which a call is progressively set up in the fabric, one stage for each dialed digit, carries over to modern self-routing packet-switched networks, described in Section 7.5.
- Procedures that distribute wear uniformly, like Step's *slipped multipling* wiring pattern, are useful models for modern distributed wireless LANs, where the architectural objective is to distribute power consumption uniformly [1, 2].

The overall architecture is discussed in Section 6.1 and the Strowger switch is described in Section 6.2. Three different configurations of a Strowger switch provide the implementations for three kinds of stages in the Step-by-Step fabric. These stages are the line finder, selector, and connector and they are described in Sections 6.3, 6.4, and 6.6, respectively. Digit absorbing in selectors is discussed in Section 6.5. An example of a small Step-by-Step office is given in Section 6.7 and a discussion of Step's pros and cons is found in Section 6.8. Section 6.8 begins the discussion of what I call the "partition," a discussion which continues through the closing sections of Chapters 8, 11, 14, and 18.

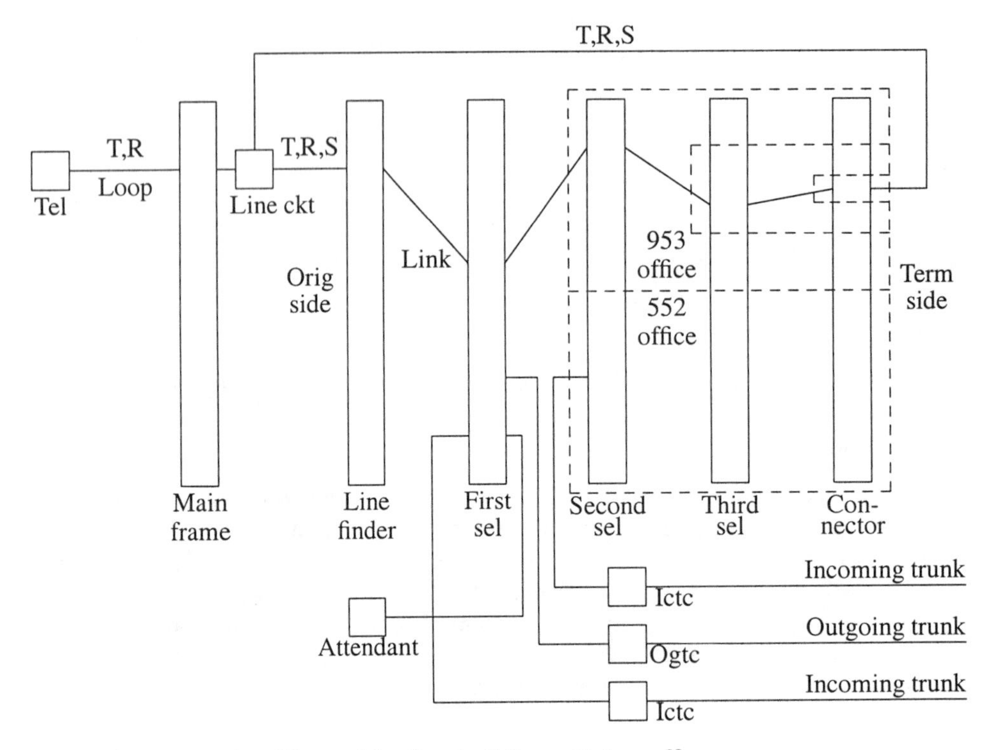

Figure 6.1. Step building with two offices.

6.1 ARCHITECTURE

Step's architecture is described in this section, with details provided in later sections. A block diagram is shown in Figure 6.1 and subsections in this section describe a connection of a telephone to the system, the fabric, operation during call setup, and trunking into and out of a Step-by-Step office. Figure 6.1 illustrates the architecture of a Step-by-Step building containing two offices. Step-by-Step has a two-sided fabric, where the two sides are the *originating side* and the *terminating side.* The direction of placing a call is from left to right across Figure 6.1.

6.1.1 Telephone Connection

A telephone that can both originate and terminate calls must be connected to both sides of the fabric. Once a call connection is established through the fabric, with the calling party on the originating side and the called party on the terminating side, speech goes both ways over the talking path.

The tip and ring wires connect a subscriber's telephone with a line circuit (Section 3.5) in the office via the loop plant and a cross-connection at the main frame (Section 3.4). One line circuit is dedicated to each line served by the office. A third wire, the sleeve, joins the tip and ring at the line circuit, and these three wires connect to the originating and terminating sides of the fabric (Figure 3.4).

The tip, ring, and sleeve of each subscriber's originating appearance are wired from its line circuit to the input side of the *line finder* stage, as shown on the left of Figure 6.1. The location on the fabric of this originating appearance is independent of the subscriber's directory number. The tip, ring, and sleeve of each subscriber's terminating appearance are wired from its line circuit to the output side of the *connector* stage, as shown on the right of Figure 6.1. The location on the fabric of this terminating appearance corresponds directly to the subscriber's directory number.

6.1.2 Fabric

The first stage is called the line finder stage, shown on the left in Figure 6.1. At this stage, the subscribers' originating appearances are concentrated down (Section 5.2.5) to a smaller number of *first selector links,* wired between the output side of this stage and the input side of the *first selector* stage. Each active call uses a first selector link, for the duration of the call. Thus, the number of subscribers that are active simultaneously, on calls they originated, cannot exceed the number of first selector links in the office.

The intermediate stages of the fabric are called *selector* stages. The usual three are shown in the center of Figure 6.1. The line finder stage, first selector links, and first selector stage are common to a Step-by-Step building. The output side of the first selector stage is wired, by second selector links, to the input side of the second selector stage. The subfabric that includes the second and third selector stages and the connector stage is segmented into the offices in the building. Two offices are indicated in Figure 6.1, a 552- office and a 953- office.

The output side of the second selector stage is wired, by third selector links, to the input side of the third selector stage. The subfabric that includes the third selector stage and connector stage is segmented into thousands-groups within the offices in the building. One such segment is indicated within the 953- office in Figure 6.1. The rightmost stage of the fabric is called the connector stage. The output side of the third selector stage is wired, by connector links, to the input side of the connector stage. The connector stage is segmented into hundreds-groups within the offices in the building. One such segment is indicated within the 953- office in Figure 6.1.

6.1.3 Call Setup

Immediately after going off-hook, the calling party is connected, through the line finder stage to a first selector link, to the first selector stage of the fabric. This first selector stage returns dial tone back over this switch path to the subscriber, who then dials over this same path into this first selector stage. Immediately after the first selector stage receives the first dialed digit, dial tone is disabled.

The first selector stage extends the switch path to one of its outputs, depending on the digit or digits dialed by the subscriber. If the subscriber dials 0, the switch path extends to an output that is wired to an attendant, shown as the lowest output off the first selector stage in Figure 6.1. The procedure after the subscriber dials 1 is discussed in the next subsection. If the subscriber dials 2 – 9, the first selector stage waits for two more digits (usually), accepting the entire office code of the called party's directory number. If this office code represents a switching office in this building, the first selector stage extends the switch path to one of its outputs, wired by a second selector link to the second selector stage in this building. The switch path extends to that particular segment of this second selector stage that corresponds to the office code dialed by the subscriber. This is indicated on Figure 6.1 by the highest indicated output of the first selector stage, wired to the 953- segment of the second selector stage. At this point in the call setup, the switch path has reached the called party's office.

The next digit dialed by the subscriber proceeds over this switch path, on through the first selector stage, into this second selector stage. The switch path then extends to an output of the second selector stage, wired by a third selector link, to that segment of the third selector stage corresponding to the office code and to the thousands digit that was just dialed by the subscriber. The next digit dialed by the subscriber proceeds over this switch path, on through the first and second selector stages, into this third selector stage. The switch path then extends to an output of the third selector stage, wired by a connector link, to that segment of the connector stage corresponding to the office code and thousands digit and to the hundreds digit that was just dialed by the subscriber. The next two digits dialed by the subscriber proceed over this switch path, on through the first, second, and third selector stages, into this connector stage. The switch path then extends to that output of the connector stage corresponding to the called party's directory number, the called party's terminating appearance on the fabric.

There may be many variations on this theme. In a small exchange, with less than 1000 subscribers, directory numbers would only require six digits and the fabric would not have the equivalent of the second selector stage, for switching on the thousands digit. In a PBX (a Step-by-Step PBX, called the 701, was a popular system), subscribers do not dial an office code for an intraoffice call and the fabric would not have the equivalent of the first selector stage for switching on the office code. But the first stage of selectors would trunk out to the local switching center on a 9 (Section 10.4.2).

6.1.4 Trunking

Recall from Section 1.5 that the *sidedness* of a switching building is specified as the line side and the trunk side. However, the sidedness of the fabric in a Step-by-Step building is specified as the originating side and the terminating side. This distinction is carefully illustrated in Figure 6.1. Consider an outgoing call to another office, first by direct trunking and then by toll switching.

Return to that step in the call setup where the first selector stage awaits the subscriber's first dialed digits. Let the subscriber dial the three-digit office code, into the first selector stage, of a switching office that is located in another building, but one that has direct trunks to this building. The first selector stage extends the switch path to an output that is wired to the originating trunk circuit of an outgoing trunk that connects to that remote office. This call is represented in Figure 6.1 by the middle output off the first selector stage, connected to the middle trunk on the trunk side of the building. In the office at the other end of this trunk, the trunk interfaces to a terminating trunk circuit that is wired to the appropriate segment of the second selector stage. The subscriber's next digits are repeated over this trunk and advance the switch path through this distant office.

If there are no direct trunks to the other office, the call must be toll-switched and the subscriber's first dialed digit must be a 1. Then, the switch path extends through the first selector stage to an originating trunk circuit that interfaces to an interoffice trunk to the toll office that serves the calling party's building. This trunk is also represented in Figure 6.1 by the middle output off the first selector stage. The subscriber's next digits are repeated over this trunk and delivered directly to this toll office.

Similarly, there are two kinds of incoming call from another office, by direct trunking or by toll switching. Incoming trunks from an office with direct trunking are represented by the upper trunk on the trunk side of Figure 6.1. A terminating trunk circuit is typically connected to the second selector stage because the calling party would have dialed the first three digits into his home office to reach this trunk, and only the last four digits would come over the trunk. A terminating trunk circuit, for an incoming trunk from a toll office, is connected to the first selector stage, represented by the lowest incoming trunk in Figure 6.1. Since the called party's entire directory number is expected, the first selector is needed to switch on the office code.

Unlike lines, incoming trunks are not concentrated through a stage of line finders. One reason is that they are assumed to be busy anyway. A two-way local interoffice trunk has its originating circuit connected to the output of the first selector and its terminating circuit connected to the input of the second selector. A two-way intratoll trunk has its originating circuit also connected to the output of the first selector, but its terminating circuit is connected to the input of the first selector.

At the position in the fabric where the *switch train* reaches a toll trunk, the identity of the calling party is unknown. If the calling party places a toll call, which must be billed, the equipment in the toll office that computes charges must, somehow, identify the calling party. This information can be obtained by operator intercept at the toll office or by automatic number identification (ANI) equipment in the Step-by-Step office. In the latter case, it is outpulsed over the same trunk to the toll office. ANI and the billing process are described in Section 9.2.3.

6.2 STROWGER SWITCH

The generic Strowger switch is described. General operation is explained and three distinct operating modes are discussed. These three modes are shown to correspond to the application of the switch as a line finder, selector, or connector.

6.2.1 Description

The cover of a Strowger switch is the size and shape of a rural mailbox, but mounted vertically. A long rod, which protrudes vertically out of the bottom of the switch's cover, carries three to six wipers. When the switch is positioned on a frame, the wipers are in proximity to a semi-cylindrical bank of contacts that are part of the frame. Removing the cover reveals a complicated structure of electromechanical actuators, electromechanical relays, wires, switch contacts, and this vertical rod.

The upper half of the switch's internal structure holds six relays labeled A – F. With only minor physical changes, the same Strowger switch can be used as a line finder, selector, or connector. The function of these relays depends on whether the switch acts as a line finder, selector, or connector. For example, when the switch acts as a selector or connector, relays A – D perform the supervision and dial repeating functions described in Section 3.5.7.

The bottom half of the switch's internal structure holds three actuators that control the vertical and rotary position of the rod. Each operation of the *vertical actuator* moves the rod upward by one additional vertical notch on a ratchet attached to the rod. Ten vertical actuations, as might be caused by 10 dial pulses, move the rod upward about 1.25 inches. Each operation of the *rotary actuator* moves the rod clockwise by one additional longitudinal notch on a different ratchet attached to the rod. Ten rotary actuations, as might be caused by 10 dial pulses, move the rod from a 10 o'clock position to a 2 o'clock position. A third actuator disengages the other two actuators from their respective ratchets. When this *release actuator* operates, the spring at the top of the rod, rotates the rod counter-clockwise back to its 10 o'clock position. Simultaneously, gravity drops the entire rod down to the *rest* position, below the first vertical position on the ratchet. Switch operation has a characteristic "clackety-clack" sound, and switch release, when this release actuator operates, has a characteristic loud "clunk." A building containing thousands of Strowger switches is physically large and extremely noisy.

Any Strowger switch has several contact stacks that are optional, depending on its specific application. For example, one contact stack, controlled by a bushing mounted at the top of the rod, operates when the rod is brought to the rest position by the release actuator and releases when the vertical actuator begins to move the rod upward. These contacts indicate the idle/busy status of the switch.

Two (or three, if the switch is a line finder) bushings, about 1.5 inches apart, are mounted on the bottom of the rod where it protrudes outside the cover. A pair of electrical wipers is mounted on each bushing. Two flexible wires, for a tip and a ring, connect to the wipers on the lower bushing. These wipers brush against contacts, which are mounted on posts that are arranged in a semi-cylindrical bank, with 10 vertical rows and 10 radial positions, similar to the structure and operation of the rotary traveling switch in Figure 2.3(a). Each post has a contact on the top and the bottom, for separate electrical conduction with the two wipers. A third flexible wire, for a sleeve, connects to a wiper on the upper bushing. These wipers brush against contacts in a separate semi-cylindrical bank, physically above the bank that provides contacts for the tip and ring. These tip, ring, and sleeve wipers travel together, vertically and radially, as the rod is moved by the actuators. The two banks of contacts that these wipers brush against are physically part of the frame on which the Strowger switches are mounted.

6.2.2 Equipment Frames

Inside a switching office, each switch's rod and bushings are exposed below the switch's cover, but the actuators and A – F relays are covered. When covered, a Strowger switch has the size and appearance of a rural mailbox, except it is mounted vertically, and the cover is typically pale gray. A Strowger switch is mounted on the front side of an equipment frame. Each switch's contact banks are part of this frame and all the inter-switch wiring appears on the back of the frame. Switches are mounted side-by-side on a frame and, depending on its height, a frame may have several levels of switches.

These frames stand on the central office floor, physically adjacent in rows, with aisles between the rows for personnel access. All the interframe wiring is carried in troughs above the frames. The line finder stage, selector stages, and the connector stage of Figure 6.1 are made from similar frames, each made using similar switches. The floor plan of a typical building of Step-by-Step switching offices is similar to the diagram in Figure 6.1. With row after row of high frames, the feeling of being inside such an office is like being inside a tall library room filled with large gray books. The noise level, however, bears no such resemblance to that in a library.

A Step-by-Step building, containing one Step office that serves 4000 lines, needs about 1000 Strowger switches. Mounting the switches 6 inches apart on a three-level frame requires 167 linear feet of frames. Allowing 6 feet between rows of frames, this office requires over 1000 square feet, just for the fabric. A 4000-line main frame, frames for 4000 line circuits and several hundred trunk circuits, tone and ring generators, and power equipment all would require at least that much area again.

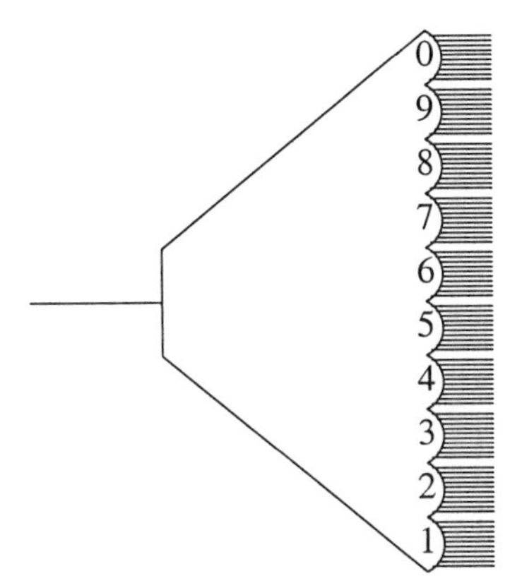

Figure 6.2. Schematic symbol for a Strowger switch.

6.2.3 Operation and Modes

The vertical and rotary actuators and their ratchet mechanisms are made to move the rod to one of 100 positions, arranged as 10 vertical by 10 rotary positions. Mounted on this rod, the *ganged* wipers, one each for the tip, ring, and sleeve, are directed to one of 100 contact positions, triply ganged in two semi-cylindrical banks. A Strowger switch brings one three-wire common to one of 100 triple-contacts, electrically a triple-throw, 100-pole switch. Figure 6.2 shows a schematic symbol for a Strowger switch. The single line at the left represents a three-wire "common." The 10 bumps on the right represent the 10 vertical levels, and the 10 outputs on each bump represent each level's 10 rotary positions. Each of the 100 outputs on the schematic represents three wires.

Once directed to a position, the ratchet mechanisms hold it for the duration of a call. Removal of the electrical ground from the sleeve connection through the fabric is the signal to the release actuator in each switch used for a given call. This action, called "dropping the switch train," disconnects the call by breaking the electrical connection of the tip, ring, and sleeve through the fabric and normalizes all the switches used for the connection of the call. These normalized switches are made available for another call. An intraoffice call through a typical Step office uses five Strowger switches, one in each stage of the fabric in Figure 6.1. With calling party supervision, when the calling party hangs up, these five switches release almost simultaneously with a loud "clunk."

Both the vertical and rotary actuators are operated in one of two modes, called select and hunt. In *select mode,* the respective actuator is directed by dial pulses from a telephone. Since each dial pulse is an open circuit followed by a closed circuit, the receiving actuator is operated and released by each pulse. Each pulse directs the respective pawl one position along the respective ratchet. If the calling party dials a 7 into a switch whose vertical/rotary actuator is in select mode, then this actuator moves the rod by seven vertical/rotary positions.

In *hunt mode,* the respective actuator is directed by the electrical state of the contacts that are touched by the wipers moved by this actuator. The wipers sweep over all contacts in the vertical/rotary direction until one with a specified electrical condition is found, typically an ungrounded sleeve contact. Using hunt mode, for example, an Nth selector searches over all its outputs to find a link to an idle $(N + 1)$th selector. Since actuator current must be pulsed for each position that the rod is moved, an actuator in hunt mode requires additional mechanisms that an actuator in select mode does not need. One mechanism operates contacts that temporarily break the current to this actuator's coil and another indicates the final position of the rod's motion.

The operating mode of a Strowger switch is described by the pair [V,R], where the switch's vertical actuator operates in mode V and its rotary actuator operates in mode R. Only three of the four possible pairs of *switch modes* are used: [hunt, hunt], [select, hunt], and [select, select], corresponding respectively to the switch being used as a line finder, selector, or connector.

6.3 LINE FINDERS

The originating side of the fabric in a Step-by-Step office is called the line finder stage. The line finder stage is a collection of line finder groups and each group is a collection of individual line finder switches. An individual line finder is a Strowger switch, configured to operate in [hunt, hunt] mode. The purpose of the line finder stage is to provide concentration. Without it, an office would require a dedicated first selector for each subscriber's line. Since lines are usually idle, such dedicated equipment would be wasteful.

A line finder group is described and its operation is discussed, with two over-simplifications. The simplifying assumptions are lifted and class of service is described.

6.3.1 Line Finder Group

Figure 6.3 shows more detail than Figure 6.1 of a small part of the originating side of a Step-by-Step fabric. Two line circuits are shown at left. The line finder stage, shown at right, is partitioned into line finder groups. Only one line finder group is shown, and it is shown to have only two line finders. The line finders face in the opposite direction from the selectors and connectors in the Step office. The common of each line finder is the tip, ring, and sleeve of a link to the common of a dedicated first selector. Line finders and first selectors are wired as a pair, connected common-to-common. While a line finder has 10 vertical positions, only four are depicted on the schematic in Figure 6.3 of each line finder. While a line finder connects to 100 lines (shown below to be 200), the connection to only two is depicted. Besides containing a set of line finder switches, a line finder group also has 10 start circuits, with only one shown in Figure 6.3, and one allotter circuit, as shown. The set of subscriber lines served by a line finder group is also considered to be part of the group.

As a pedagogical tool, temporarily assume that a line finder provides access to the originating appearance of 100 lines. It will be shown that a line finder serves 200 lines. Let all the lines in the building be arbitrarily partitioned into groups containing 100 (200) lines. Each group of lines is part of a different group of line finders. The tip, ring, and sleeve of each subscriber line are connected to one three-wire pole on each line finder in the line finder group that serves this group of lines. The start wires from all the line circuits in a group are connected to the lone allotter circuit in the line finder group. This allotter circuit has a separate control wire to each line finder.

Let the 100 (200) lines in each group be arbitrarily partitioned into start groups, each containing 10 (20) lines. A start circuit serves each different start group. The two lines in Figure 6.3 are not only in the same line finder group, they are also in the same start

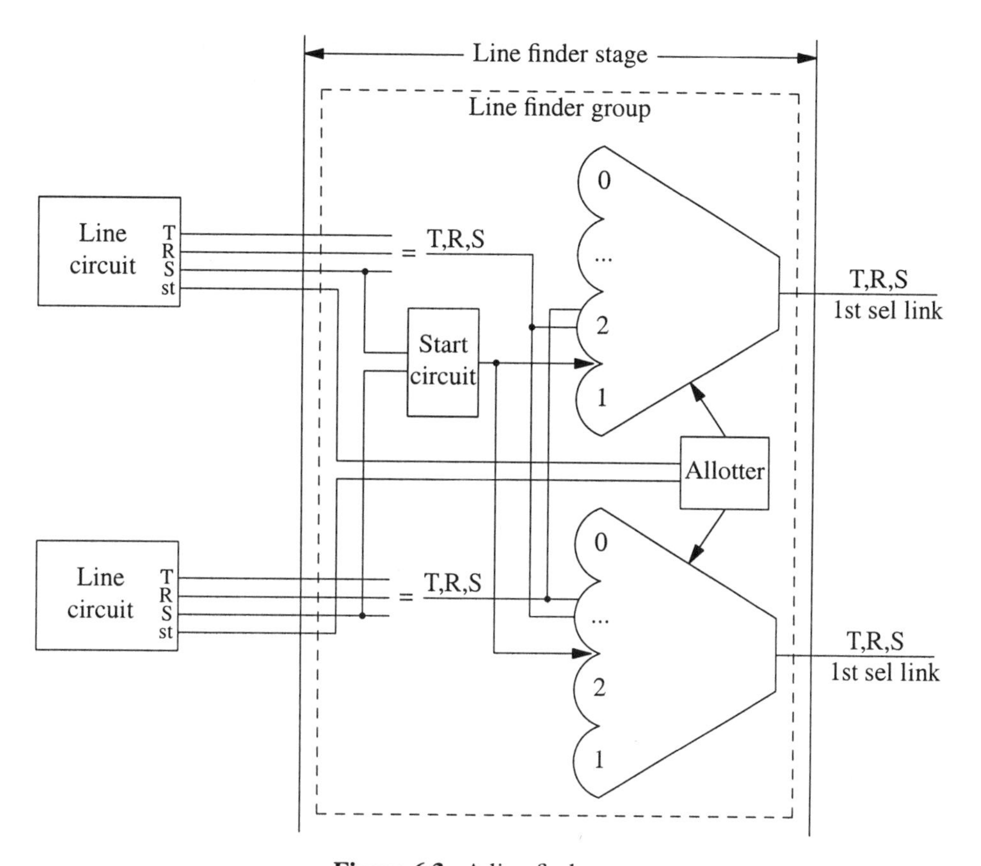

Figure 6.3. A line finder group.

group. When the lines in a start group are connected to the poles on a line finder, they are always connected to a common row of poles, and in the same order. Each start group's common row is a different numbered row on every line finder. The two lines in Figure 6.3 are served by the indicated start circuit and they are connected to row 2 on the upper line finder and row "..." on the lower one. The upper line is connected before the lower one in the rotary direction in both rows.

When a Strowger switch is configured as a line finder, a *vertical commutator* is mounted on the rod and it brushes along a bank of 10 vertically mounted contacts, one contact corresponding to each row of poles. The output from each start circuit is wired to that vertical contact on each line finder that corresponds to the row of poles to which this start circuit's input lines are connected. In Figure 6.3, the start circuit's output is wired to the vertical contact corresponding to row 2 on the upper line finder and to the vertical contact corresponding to row "..." on the lower one.

6.3.2 Simplified Operation

Referring back to Section 3.5.2, when a telephone goes off-hook, the line relay in its line circuit operates, placing battery on the sleeve wire and ground on the start wire. All other sleeve wires in this group of lines are grounded if the line is busy and served, or open if the line is idle. Battery occurs only on the sleeve wire of a line that has just gone off-hook but hasn't been "found" yet. The start circuit, serving the start group of which this line is a member, responds to the battery on one of its input sleeve wires. It places ground on that vertical contact on each line finder that corresponds to the row of poles to which its start group's input lines are connected.

The allotter circuit, serving the group of which this line is a member, responds to the ground on one of its input start wires. It selects an idle line finder in the group and signals it, over the control wire, to begin operation. In hunt mode, the vertical actuator on this selected idle line finder raises the rod until the vertical commutator brushes the vertical contact that was grounded by the start circuit. The tip, ring, and sleeve wires of the line that just went off-hook are connected to one of 10 poles in this row. After reaching the grounded vertical contact, the vertical actuator stops and the rotary actuator begins. Also in hunt mode, this rotary actuator rotates the rod until the sleeve wiper brushes some sleeve contact that has battery on it.

Several events occur at the instant this off-hook line is found. The rotary actuator stops, with its tip, ring, and sleeve wipers electrically brushing the tip, ring, and sleeve contacts of the desired line in the semi-cylindrical bank of poles. A relay connects ground to the sleeve wire, operating the cut-off relay in the line circuit and indicating that the line is busy to the terminating side of the fabric. The tip, ring, and sleeve are connected by a link to that first selector that is directly wired to this line finder. Supervision passes to this first selector's A relay and another relay operates to connect dial tone across tip and ring.

6.3.3 Access to 200

A real line finder serves 200 lines, not 100 as described so far. Since each line has a tip, ring, and sleeve, some form of triple-contact pole is required in the switch bank. To serve 100 lines, a Strowger switch requires three wipers and 300 electrical contacts. Section 6.2 described three wipers mounted on two bushings, with two wipers mounted on the lower bushing and one mounted on the upper. In rotary motion, the double wipers straddle the corresponding physical posts, brushing against electrical contacts mounted on the top and bottom of a post. So, the 300 corresponding contacts are configured as a lower bank of 100 posts with double contacts and an upper bank of 100 posts with single contacts. To expand a Strowger switch to serve 200 lines, it requires six wipers on three bushings and 600 electrical contacts. The 600 contacts are configured as three banks of 100 posts, each with double contacts.

A line finder's contacts are configured as three 10-by-10 semi-cylindrical banks of dual contacts. This arrangement provides six electrical connections, enough for the tip, ring, and sleeve of two telephone lines, for each of its 100 mechanical positions. The wipers on the upper bushing brush the tip and ring of one line, the wipers on the lower bushing brush the tip and ring of the other line, and the wipers on the middle bushing brush the sleeves of both lines. With 20 lines on a row, instead of 10, each start circuit serves 20 lines. A line finder finds the line requesting service by first locating the vertical position that had its vertical contact grounded by this start circuit, and then locating the rotary position at which some sleeve contact has the battery condition. Whether the upper or lower sleeve at this position has battery determines whether the tip and ring brushed by the wipers on the upper or lower bushing are connected, with this sleeve, to the switch's common wires.

The reason for serving 200 lines, instead of 100, is to expand the *access* of lines to line finders in this stage of concentration (Section 5.2.5 and Section 7.2). It was seen in Section 5.3 that larger access provides greater efficiency. For example, let the typical line be busy an average of 1.8 minutes out of every hour on calls this line originates. Calls that terminate on this line do not tie up a line finder. Then, each subscriber produces 0.03 Erlangs of traffic intensity at the line finder stage in the Step office. If there were 100 subscribers in a line finder group, the group must handle 3 Erlangs total. By Figure 5.4, eight line finders would serve this group of 100 lines with a P_b = 0.012 and 37.5% occupancy. Instead, with 200 subscribers in a line finder group, the group must handle 6 Erlangs total. By Figure 5.5, 13 line finders would serve this group of 200 lines with a P_b = 0.009 and 46% occupancy. By this example, placing 200 lines in a line finder group instead of 100, saves three line finders for every 200 subscribers for roughly the same P_b, a savings of 60 Strowger switches in a 4000-line Step office. While an individual line finder provides 200:1 concentration, the group described above provides for 200:13 ≈ 15:1 concentration.

6.3.4 Allotting

Since one line finder, and its link to a first selector, is required for each originated call, the number of line finders in a line finder group is determined by the intensity of originating traffic from the group served. The 200 lines are concentrated "down" to this number, typically 10 – 15.

When a subscriber lifts his handset to originate a call, the line circuit for this line places a ground signal on the start wire. The allotter responds to this ground signal and selects a line finder in the line finder group to find the line that has just requested service. The allotter is a circuit of relay contacts, wired in a scheme called a preference chain (Exercise 6.3). The allotter examines all the line finders in its group that are idle and selects the one in which this line is wired on the lowest row. This selection criterion reduces the time spent in vertical hunt and distributes wear over all the line finders in a group.

Even with the access at 200 lines, the quality of service provided by a line finder group is sensitive to variations in traffic intensity. Consider subscribers, like brokers, who originate more than the average number of telephone calls or who connect via modems to computers and have calls with a long holding time. When such a subscriber is added to a line finder group, or when existing group members suddenly change their calling patterns, it can be noticeable to the other subscribers in the group. If all the line finders in a group are busy, a subscriber may go off-hook and not get a dial tone.

For example, consider a group served by 13 line finders, based on 0.03 Erlangs per subscriber. Let five of these subscribers be off-hook for hours on modem-connected calls. The remaining 195 subscribers are served by the eight remaining line finders. At 0.03 Erlangs each, the net traffic intensity is 0.03 × 195 = 5.85 Erlangs and, from Figure 5.5, P_b = 0.25, highly unacceptable. The telco must either add more line finders to this group or move some originating appearances to another group. Moving line appearances on the originating side of a Step fabric is not a problem, except for the labor cost involved and possible problems with preserving class of service (next section).

6.3.5 Class of Service

Some lines have certain conditions, limitations, or privileges that require special handling. Examples are coin telephones, PBX trunks, restricted calling, subscription to Touch Tone, wide area telephone service (WATS), speed calling, and other services, or even indication that the bill is unpaid. Indication for special handling is called class of service. One way to identify such conditions in a Step-by-Step office is for all lines with the same originating class of service to be wired to a common line finder group. One line finder group might serve all the coin telephones in an office, another group might serve all the PBX trunks, and Touch Tone subscribers might also be segregated into separate line finder groups from dial-pulse subscribers.

Another way to identify such conditions is to associate an n-bit binary code to each line. A coding method, sometimes used in Step-by-Step offices, is for each line finder to have a fourth bushing and wiper pair on its rod, corresponding to a fourth bank of posts with two contacts per post. The contacts are binary-coded with open or ground. If the two lines corresponding to each [horizontal, rotary] position have the same class of service, then both corresponding contacts in the fourth bank are used for both lines together to encode four distinct class of service conditions. If the two lines corresponding to each [horizontal, rotary] position have different class of service, then only one corresponding contact in the fourth bank is used for each line to encode two distinct class of service conditions.

6.4 SELECTORS

In the Step-by-Step fabric, the *selector subfabric* is a collection of selector stages, where the *N*th selector stage is a collection of *selector groups* and each group is a collection of selector switches. An individual selector is a Strowger switch operated in [select, hunt] mode. The purpose of the selector subfabric is to route the *switch train* as the calling party dials the directory number of the called party.

The selector subfabric is described and its operation is discussed, but with two over-simplifications. The operation of a selector stage, group, and switch are described. Digit disposition, network blocking, and interstage wiring are discussed.

6.4.1 Selector Subfabric

Assume that the selector and connector stages from Figure 6.1 are redrawn in Figure 6.4 in greater detail. While Figure 6.1 correctly shows three selector stages and one connector stage, Figure 6.4 incorrectly shows seven selector stages because two pedagogical simplifications are made. The *N*th stage is assumed to be responsible for the *N*th digit of directory number dialed by the calling party. Each selector stage receives and absorbs one dialed digit in the number and extends the switch path to the next stage in the fabric. The handling of the first three digits by the first selector stage is described in Section 6.5, and the handling of the last two digits by the connector stage is described in Section 6.6.

Under these two simplifications, if a seven-digit directory number is assumed, seven selector stages are needed in this hypothetical selector subfabric. The first three selector stages would be common to a building, because the office code of the called party's directory number is not known until the switch path reaches this point in the fabric. The last four selector stages would be segmented into the separate offices within a building.

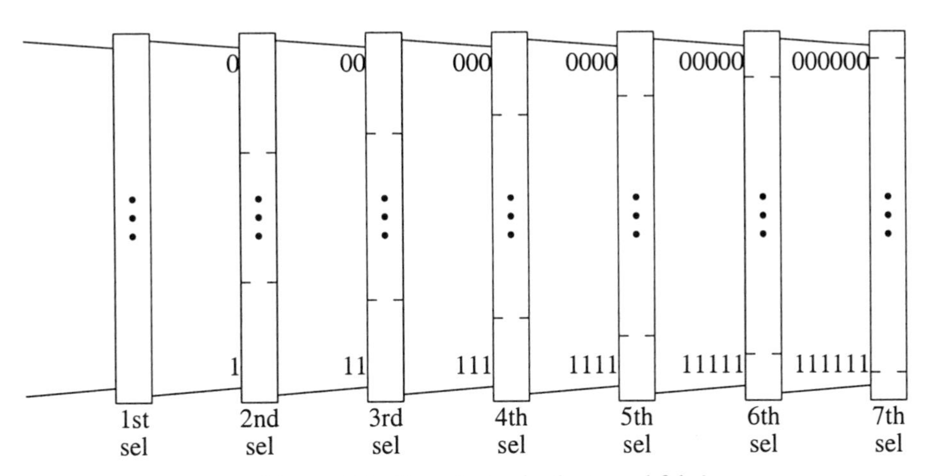

Figure 6.4. Hypothetical selector subfabric.

The outputs of the seventh selector stage are the subscriber's terminating appearances on the fabric. Each subscriber's tip, ring, and sleeve are connected, between his dedicated line circuit and the terminating side of the fabric, as shown on the right of Figure 6.1. This termination must be to a unique output of the unique segment of the final selector stage, which corresponds to the subscriber's directory number.

The first selector stage in Figure 6.4 has only one selector group, and this stage/group serves all calls through the fabric. The first dialed digit steers the switch train to the appropriate group in the second selector stage. Typically, digits 1 and 0 are used for toll and operator calls, respectively, and do not steer the switch train into the second selector stage. The second selector stage in Figure 6.4 is partitioned into 10 selector groups. These groups are identified by 1, 2, ..., 9, 0 in the figure, corresponding to the first dialed digit; but some (at least 1 and 0) are not equipped. Each group serves all calls through the fabric in which the called party has the corresponding same first digit in its directory number. The second dialed digit steers the switch train to the appropriate group in the third selector stage, and so forth. The seventh selector stage in Figure 6.4 is partitioned into one million selector groups. These groups are identified by six-digit numbers, corresponding to the first six dialed digits; but most groups are not equipped. Each group serves all calls through the fabric in which the called party has the corresponding same first six digits in its directory number. The seventh dialed digit steers the switch train to the called party.

For the moment, ignore the fact that calls, in which the leading digit is 0 or 1, trunk out of the fabric from intermediate stages. If we assume, instead, that all calls proceed through all the selectors in the fabric, then the Nth selector stage in Figure 6.4 is partitioned into 10^{N-1} selector groups. These groups are identified by $(N - 1)$-digit numbers, corresponding to the first $N - 1$ dialed digits; but many may not be equipped. Each selector group in the Nth selector stage serves all calls through the fabric in which the called party has the corresponding same first $N - 1$ digits in its directory number. The next dialed digit steers the switch train to the appropriate group in the $(N + 1)$th selector stage.

Consider the selector group reached by the dialed number $d_1 d_2 \cdots d_N d_{N+1}$ in this

$(N+1)$th selector stage. The separate selectors in this group are reached from different rotary positions on the d_{N+1}th row of any selector from the same group in the Nth selector stage. This group in the Nth selector stage was reached by the dialed number $d_1 d_2 \cdots d_N$. Because as many as 10 selectors in this Nth selector group are wired to the same $(N+1)$th selector group, hunting for an individual idle selector is necessary.

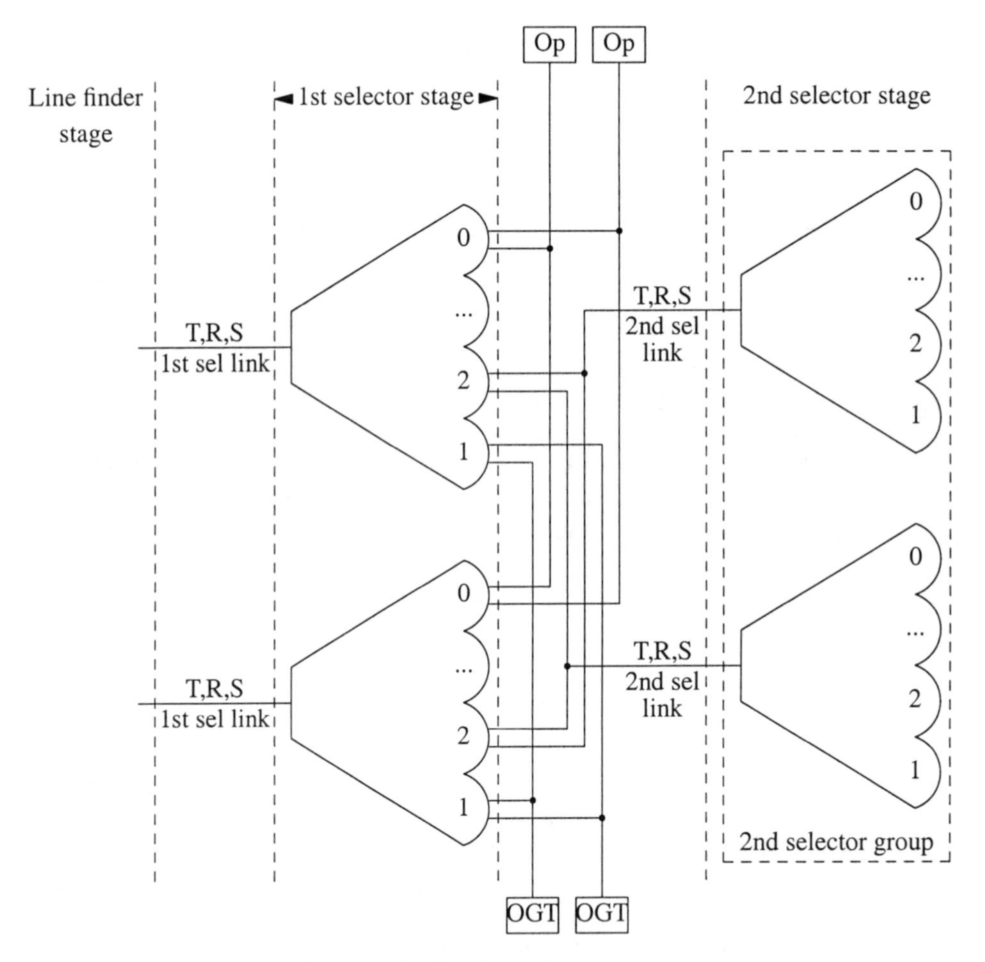

Figure 6.5. The first selector stage.

6.4.2 Simplified Operation

Figure 6.5 shows a Step-by-Step building in more detail, including the line finder stage to the left, the first selector stage in the center, the second selector stage to the right, and several operators and outgoing trunk circuits.

A first selector is paired with every line finder in the office, input-to-input with the switch in each [line finder, first selector] pair facing in the opposite direction. The tip, ring, and sleeve wires connecting a switch-pair together are called the first selector link. One first selector link and switch-pair are required for the duration of every call, assigned at the time of origination. The first selector returns talking battery and dial tone to the calling party, and accepts the first dialed digit (it will be shown in Section 6.5 to accept the first three).

Every stage of selectors contains many selector switches, all operating in [select, hunt] mode. A digit dialed into a selector directly moves its shaft vertically, one position for each pulse in the digit. At the end of the digit, indicated by the released C relay, vertical movement ceases and rotary hunt begins. The selector then searches radially for an open circuit on a sleeve contact, indicating a path out to an idle selector in the appropriate group in the next stage of the fabric.

The top row of the bank of contacts in each first selector, reached by dialing 0, is connected to 10 or fewer local operators. The bottom row, reached by dialing 1, is connected to 10 or fewer outgoing trunks to the toll office. These trunks carry calls to those offices to which this office does not have direct trunking, including long distance calls, typically outside the flat-rate billing area. Other trunks, to nearby offices with direct trunking, are connected to selectors deeper in the fabric (not a true statement) because they are reached by dialing the three-digit office code into this office. Dialing instructions, provided in the telephone book, inform the subscriber whether the 1-prefix is required. Intermediate rows on the bank of contacts in each first selector, corresponding to a first digit of 2 – 9, are connected to groups of second selectors according to the dial plan in the office.

6.4.3 Digit Disposition and Network Blocking

Each dialed digit is handled in this oversimplified selector subfabric in one of five ways:

- The digit advances the switch train one stage deeper in the fabric or to an outgoing trunk and the subscriber continues dialing;
- The incomplete call is blocked inside the fabric for lack of enough trunks or selectors in the next selector group and the subscriber hears path-busy tone;
- The digit is inconsistent with the office dial plan and the subscriber hears reorder tone;
- The digit completes the call to an idle called party and the subscriber hears ring-back tone;
- The digit completes the call to a busy called party and the subscriber hears busy tone.

Strowger switches, equipped for hunt mode in the rotary direction, have a tab on the shaft that operates a contact stack when the shaft reaches the 2 o'clock position. If the shaft hunts completely across a row without finding an idle path to the next selector group, a contact operated by this tab connects path-busy tone to the tip and ring.

Selector positions reached by invalid digit sequences have open-circuit sleeve connections so they appear idle and stop hunt mode. The corresponding tip and ring are wired directly to reorder tone. Completion of dialing occurs in the connector stage and is described in Section 6.6.

6.4.4 Multipling in Selector Groups

If the paths to the operators were wired to the 0-row of every first selector in the same order, a pattern called uniform multipling, the operators would be nonuniformly busy. The operator wired to the first position would be the first one hunted by all traffic and would be the busiest, and so forth. Similarly, mechanical equipment, like trunks and next selectors, would wear unevenly. So the same row of every selector switch in the same group, while wired to the same set of equipment, is deliberately wired to that equipment in a different order, in a pattern called slipped multipling.

In Figure 6.5, consider the two indicated second selector links between the two selectors in the first selector stage and the two selectors in the 2-group of the second selector stage. The upper first selector hunts over these second selectors so that it would choose the lower one first, if it were idle. The lower first selector hunts over these second selectors so that it would choose the upper one first, if it were idle.

Suppose traffic requires more than 10 operators in the building, or more than 10 trunks in a trunk group, or more than 10 switches in a group of second selectors. Since a given row of a given selector can only reach 10 servers, a similar multipling issue arises. Such server groups, and the selector groups that reach them, could be wired as subgroups of 10 each; but this would violate the full-access principle. Instead, one first selector is wired to servers 1 through 10, the second to servers 2 through 11, the third to servers 3 through 12, and so forth. This pattern distributes traffic better than subgrouping would.

For example, let a first selector stage have 324 selector switches. If there are nine outgoing trunks to the toll office, the 1-level of each first selector is wired to all nine trunks, with the last rotary position permanently marked busy. Thirty-six selectors would be wired to hunt in the order 1, 2, 3, ..., 9; 36 would be wired to hunt in the order 2, 3, ..., 9, 1; 36 to hunt in the order 3, ..., 9, 1, 2; and so forth. If, instead, there are 12 outgoing trunks to the toll office, the 1-level of each first selector is wired to only 10 of the 12 trunks. Twenty-seven selectors would be wired to hunt in the order 1, 2, 3, 4, ..., 10; 27 would be wired to hunt in the order 2, 3, 4, ..., 10, 11; 27 to hunt in the order 3, 4, ..., 10, 11, 12; 27 in the order 4, ..., 10, 11, 12, 1; and so forth.

6.5 DIGIT ABSORBING

An optimization mentioned above is that the first three dialed digits, the office code, typically do not require three stages of selectors. In most Step-by-Step offices, the first selector alone handles the entire office code and the remaining selector stages (and connectors) handle the remaining digits, typically four. This optimization is best explained by example.

TABLE 6.1. Three hypothetical step communities

Community	Office code(s)
Blacksburg	552, 953
Christiansburg	382
Radford	639, 741

Consider buildings in three communities: Blacksburg, Christiansburg, and Radford. Let each be served by Step-by-Step buildings with mutual direct trunking. Let Blacksburg and Radford each be large enough to need two offices and let the office codes be as tabulated in Table 6.1. For the reader familiar with these three communities in southwestern Virginia, the office codes have been slightly altered to make the example work out better. Furthermore, Steppers do not really serve these three communities.

If the first digit dialed by any subscriber is a 9, the only valid second digit is 5 and the only valid third digit is 3. If the second and third digits used selector stages, its switches would be sparsely wired and only serve as dial plan filters. The motivation for optimizing out the second and third selectors is seen.

6.5.1 Counter-Based Digit Absorbing

Assume there is a digit counter in each first selector that allows it to distinguish the first three digits. Each first selector can act on any digit in one of three ways:

- Absorbing (A) means the selector does nothing except increment the digit counter.
- Blocking (B) means the selector connects reorder tone because this digit in this position is inconsistent with the dial plan.
- Connecting (C) means the corresponding row on the selector's output is wired to more equipment.

The dial plan in Blacksburg requires that all first selectors connect (C) to toll trunks and operators if the first digit dialed is 1 and 0, respectively. Wires to originating trunk circuits and operator switchboards are physically connected (C) to the rotary positions on the 1 and 0 levels of these Strowger switches. These first selectors must absorb (A) first digits of 3, 5, 6, 7, and 9 and second digits of 3, 4, 5, and 8 as assumed parts of valid office codes. These first selectors must connect (C) to each of the direct-trunked offices off the third digits of 1, 2, 3, and 9. All other digits would be blocked (B).

TABLE 6.2. Treatment of three digits of office code

Digit	1st	2nd	3rd	1st	2nd	3rd
1	C	B	C	C	B	B
2	B	B	C,C	B	B	B
3	A	A	C	A	A	C
4	B	A	B	B	C	B
5	A	A	B	A	A	B
6	A	B	B	A	B	B
7	A	B	B	A	B	B
8	B	A	B	B	C	B
9	A	B	C	A	B	C
0	C	B	B	C	B	B

Summarizing, each first selector in the Blacksburg building would be controlled according to the leftmost columns of the table in Table 6.2. However, the same selector cannot connect (C) to the 552- and 382- offices off the same third digit, neither can it physically connect (C) to toll trunks and to the 741- office off a dialed 1. While there is a logical difference, they require the same rotary positions on the same levels of the same physical switches. One solution is for selectors to connect (C) to the 741- office and to the 382- office off the second digit of their office codes. Those offices, at the other end, must absorb (A) their respective third digit in their second selector stage.

There can be at most one C in any row of the table. The rightmost columns of the table in Table 6.2 has C in the second column of row 4 and B in the third column of row 1 because this office trunks out to the 741- office on the 4 instead of the 1. The table has C in the second column of row 8 and would have B in the third column of row 2 because this office trunks out to the 382- office on the 8 instead of the 2. The entry in the third column of row 2 is C and not B because this office trunks out to the 552- office on the 2. Similar tricks that can be used are to connect (C) on the first digit or to change the dial code for accessing toll trunks.

Notice that the subscriber who dials 359 is trunked out to the 639- office because a first selector cannot distinguish those two sequences of digits; this is felt to be acceptable. Digressing, the reader should appreciate the complexity of implementing a change in the dial plan whenever a new office code is added; every first selector in a building must be physically rewired.

6.5.2 Counter-Free Digit Absorbing

If the counter is removed or not used, the first selector cannot distinguish first, second, and third digits. The following three rules translate between counter-based arrangements and counter-free arrangements. If the counter-based arrangement has a B in all three positions, then the counter-free treatment is B. If the counter-based treatment has A and B in all three positions, then the counter-free treatment is A. If the counter-based treatment has a C in any (one) position and the other two are B, then the counter-free treatment is C. A conflict occurs if the counter-based arrangement has a C in any (one) position and either of the other two is A. Applying these rules to the rightmost columns in the table of Table 6.2, this conflict occurs in rows 3 and 9.

TABLE 6.3. Counter-free arrangement of the office codes

Digit	Action
1	C to toll
2	C to 552
3	A
4	C to 74(1)
5	A
6	C to 6(39)
7	A
8	C to 38
9	C to 9(53)
0	C to 0

This conflict is resolved if the Blacksburg building trunks to the 639- office on the leading 6 and to its own 953- office on the leading 9. Each of those offices must absorb the last two digits of its respective office code. The arrangement of office codes by a counter-free first selector is shown in Table 6.3. If the 639- office has less than 8000 subscribers, its second selectors can absorb the 3 and 9, counter-free, if these numbers are avoided as thousands digits for directory numbers. If the 953- office has less than 8000 subscribers, its second selectors can absorb the 5 and 3, counter-free, if these numbers are avoided as thousands digits for directory numbers.

Without counters in the first selectors, the following dial sequences all select the 552-office: 552, 972, 97972 and just 2. This is viewed as acceptable and the 2 shortcut is even a benefit. Such a shortcut is usually made public within the community. While a subscriber with directory number 552-1097 can be dialed up with 2-1097 in a local call, a subscriber making a nonlocal call to this number cannot use the shortcut.

6.5.3 Implementation

The first selector stage in a Step-by-Step office is more easily implemented if office codes have been chosen carefully so all out-trunking is on the third digit and all digit absorbing is counter-free. When out-trunking is not on the third digit, the target second selectors must perform digit absorbing, preferably counter-free. Counter-based digit absorbing, in first or second selectors is more difficult and expensive.

A Strowger switch may have an auxiliary shaft, attached to the main shaft between the vertical ratchet and the rotary return spring, behind the main shaft and projecting above it. A bushing at the top of this auxiliary shaft has 10 flexible metal tabs on each side. These tabs may be bent inward with pliers. The *N*th tab on one side, if not bent back, operates switch contacts when the shaft is in the *N*th vertical position.

In counter-based digit absorbing, one side is used as a 1-bit binary code for the first digit and the other side is used as a 1-bit binary code for the second digit. In counter-free digit absorbing, both sides are used as a 2-bit binary code to encode the A, B, and C states. Every time the dialing plan changes, these tabs must be changed on all selectors equipped for digit absorbing.

6.6 CONNECTORS

The terminating side of the fabric in a Step-by-Step office is called the connector stage. The connector stage is a collection of connector groups and each group is a collection of individual connector switches. An individual connector is a Strowger switch configured to operate in [select, select] mode. The purpose of the connector stage is to provide *expansion* from few active paths to many terminating appearances.

Use of a connector stage instead of the hypothetical sixth and seventh selector stages in Figure 6.5 is discussed. A connector group is described and its operation is discussed, when the called party is both idle and busy.

6.6.1 The Truth about the Selector Subfabric

According to the description of the hypothetical selector subfabric, shown in Figure 6.4, the sixth and seventh stages accept the last two digits of the called party's directory number, as dialed by the calling party. Also, according to this description, a selector accepts its appropriate digit during its vertical motion, in select mode, and then finds an idle next selector during its rotary motion, in hunt mode.

If a seventh selector really worked this way, there would be nothing for it to hunt over during rotary motion, because the seventh, and last, digit would have uniquely identified the called party during vertical motion. So, the seventh selector stage is eliminated and the rotary motion of all the sixth selectors is changed to accept the seventh digit in select mode. This merger of the hypothetical sixth and seventh selectors, into one stage whose switches operate in [select, select] mode, is called the connector stage. The outputs of the connectors are the subscriber's terminating appearances.

Consider the hypothetical selector subfabric, shown in Figure 6.4. Section 6.5 showed that those stages, identified in Figure 6.4 as the second and third selector stages, really don't exist because the first selector stage handles the first three digits. It has just been shown that the stage, identified in Figure 6.4 as the seventh selector stage, really doesn't exist because the sixth selector stage handles the last two digits. Renaming and renumbering the selector stages, the hypothetical subfabric, shown in Figure 6.4, becomes the subfabric, shown in Figure 6.1.

- The first, second, and third selector stages in Figure 6.4 become the first selector stage in Figure 6.1 and it handles the first three digits.
- The fourth selector stage in Figure 6.4 becomes the second selector stage in Figure 6.1 and it handles the fourth digit.
- The fifth selector stage in Figure 6.4 becomes the third selector stage in Figure 6.1 and it handles the fifth digit.
- The sixth and seventh selector stages in Figure 6.4 become the connector stage in Figure 6.1 and it handles the last two digits.

In a small office, serving fewer than 1000 subscribers, the subscribers could be assigned a six-digit directory number. If the office is a Stepper, it would not have a third selector stage. In a PBX, subscribers do not dial an office code for intraoffice calls. If the PBX is a Step-by-Step PBX, like the 701, it would not have the equivalent of the first selector stage.

6.6.2 Connector Groups

The outputs of the connectors are the subscriber's terminating appearances. Just as each subscriber has originating appearances on several line finders, every line finder in one line finder group, each subscriber has terminating appearances on several connectors, every connector in one connector group. While originating appearances are arbitrarily assigned to a line finder group, terminating appearances are assigned to a connector group according to the subscriber's directory number.

Consider the subscriber with directory number 876-5432 and let the office with office code 876 be a Stepper. One connector group in the connector stage in that 876- office is reached by dialing 876-54. The tip, ring, and sleeve from the terminating side of this subscriber's line circuit are wired to each Strowger switch in this connector group. The connection is to the second post in the third row of the contact banks on each connector in this group. There is no odd multipling; all connectors in a group are wired identically.

The number of connectors in a connector group is determined by the traffic intensity of the 100 subscribers served by the group; the 100 people in the office whose directory number has the same five-digit prefix (876-54). Copying the example from Section 6.3.3, let the typical line be busy an average of 1.8 minutes out of every hour on calls this line terminates. Calls that originate from this line do not tie up a connector. Then, each subscriber is responsible for 0.03 Erlangs of traffic intensity at the connector stage in the Step office. With 100 subscribers, maximum, in a connector group, a full group must handle 3 Erlangs total. By Figure 5.4, eight connectors would serve this group of 100 lines with a $P_b = 0.012$ and 37.5% occupancy. While an individual connector provides 1:100 expansion, the group described above provides for 8:100 ≈ 1:13 expansion.

The quality of service provided by a connector group is sensitive to variations in traffic intensity, especially with the access at only 100 lines. Consider subscribers, like pizza parlors, who terminate many more than the average number of telephone calls or host computers that terminate calls with long holding times from callers with terminals and modems. When such a subscriber is added to a connector group, or when existing group members suddenly change their calling patterns, it can be noticeable to the other subscribers in the group. If all the connectors in a group are busy, a potential call will be blocked and the calling party will hear path-busy tone. The called party doesn't know, directly, that service has been affected. Since most people don't distinguish the tones for path-busy and called-party busy, the people that call a subscriber in an underserved connector group, accuse him of talking on the telephone too much.

For example, consider a connector group served by eight connectors, based on 0.03 Erlangs per subscriber. Let only two of the subscribers be the called party for hours on modem-connected calls. The remaining 98 subscribers are served by the six remaining connectors. At 0.03 Erlangs each, the net traffic intensity is $0.03 \times 98 = 2.94$ Erlangs and, from Figure 5.4, $P_b = 0.08$, a little too high. The telco must either add more connectors to this group or move some terminating appearances to another group. Moving line appearances on the terminating side of a Step fabric, however, is a large problem, because each moved party must agree to change his directory number.

6.6.3 Operation of the Selector Subfabric

Consider that some subscriber has lifted the handset of his telephone with the intention of calling the party with directory number 876-5432. Let both parties be served by the same Step-by-Step building, not necessarily the same office. This calling party's line is found by the operation described in Section 6.3.2 and is connected to a first selector by the operation described in Section 6.4.2. This calling party hears a dial tone.

This calling party dials 8, stepping this first selector up eight levels. A tab at the top of its shaft forces this first digit to be absorbed, the switch normalizes, and dial tone is removed. This calling party dials 7, stepping this first selector up seven levels. Another tab at the top of its shaft forces this second digit to be absorbed and the switch normalizes again. This calling party dials 6, stepping this first selector up six levels. A different tab at the top of its shaft forces this third digit to be connected and the switch begins rotary motion in hunt mode. Rotary motion stops on the first post in this sixth row that has an open-circuit condition on its sleeve. Tip, ring, and sleeve are connected to a second selector link and onto a previously idle Strowger switch in the second selector stage. The switch is in that group that corresponds to the 876- dial prefix; the left edge of the 876- office's fabric in this building's subfabric. The D relay operates in this first selector and the A relay operates in this second selector — advancing supervision.

This calling party dials 5 and this fourth digit steps this second selector up five levels, where the switch begins rotary motion in hunt mode. Rotary motion stops on the first post in this fifth row that has an open-circuit condition on its sleeve. Tip, ring, and sleeve are connected to a third selector link and onto a previously idle Strowger switch in the third selector stage. The switch is in that group that corresponds to the 876-5 dial prefix. The D relay operates in this second selector and the A relay operates in this third selector — advancing supervision. This calling party dials 4 and this fifth digit steps this third selector up four levels, where the switch begins rotary motion in hunt mode. Rotary motion stops on the first post in this fourth row that has an open-circuit condition on its sleeve. Tip, ring, and sleeve are connected to a connector link and onto a previously idle Strowger switch in the connector stage. The switch is in that group that corresponds to the 876-54 dial prefix. The D relay operates in this third selector and the A relay operates in this connector — advancing supervision.

6.6.4 Connector Operation

The example continues. This calling party dials 3 and this sixth digit steps this connector up three levels, where the switch waits for the next digit. This calling party dials 2 and this seventh digit rotates this connector by two positions.

Let the called party be idle; the case where the called party is busy is covered in the next paragraph. The line circuit has caused an open-circuit condition on the sleeve wired to this position. The sleeve of the switch train is connected to the sleeve of the called party. The ground condition on the sleeve of the switch train operates the cut-off relay in the called party's line circuit. The connector retains supervisory control of the call, connects ring signal directly to the called party's tip and ring and, after attenuation, to the tip and ring of the switch train for the calling party. When the called party answers, the ringing signal and the attenuator are disconnected, and the tip and ring of the switch train are connected directly to the tip and ring of the called party.

Reconsider the call to the subscriber with directory number 876-5432. Let this called party be busy when the calling party's seventh digit, a 2, rotates the chosen connector by two positions along its third level. The called party's line circuit has caused a ground condition on the sleeve wired to this position. Then, this connector connects the busy tone to the tip and ring of the switch train which is heard by the calling party. Even if the called party hangs up immediately afterwards, the call will not be completed.

Sometimes a connector can be made to revert to hunt mode over one row or even part of one row. A hunt group is a set of consecutive posts on a rotary that are hunted over by a connector. The application is discussed in Section 10.3.1.

6.7 EXAMPLE

Consider a Step system that serves a small community with 500+ people.

6.7.1 Initial Deployment

This system has a small main frame and 500+ line circuits. Assume that each user is busy 6% of the time, on the average, during peak busy hour. If originating and terminating traffic is balanced, then each user originates 0.03 Erlangs of traffic intensity and each user terminates 0.03 Erlangs of traffic intensity. So, the fabric must handle 500 $\times$ 0.03 = 15 Erlangs of traffic load on its originating side and another 15 Erlangs of traffic load on its terminating side.

Since a line finder group serves at most 200 originating parties, three line finder groups are required to handle any number of originating parties between 401 – 600. If we assume that originating traffic divides uniformly over the three groups, then each group must handle 15/3 = 5 Erlangs. Using P_b = 0.01, Figure 5.5 shows that 5 Erlangs requires 11 servers. Thus, each line finder group needs 11 switches and the line finder stage has 33 switches total. Since every line finder switch is wired directly to its own dedicated first selector switch, the first selector stage also has 33 switches.

6.7.2 Trunking

Let the office code of this office be 654. Let the office code of two nearby local offices, to which this office has direct trunks, be 873 and 952. Let the toll prefix be 1 and the operator code be 0. Thus, all 33 first selectors are wired to absorb digits 5 – 9. Links to the second selectors are wired to the 4-level on all the first selectors. Trunks to the 873- office are wired to the 3-level on all the first selectors. Trunks to the 952- office are wired to the 2-level on all the first selectors. Trunks to the toll office are wired to the 1-level on all the first selectors. Links to the operators are wired to the 0-level on all the first selectors. In this example, we'll ignore operator traffic.

Let's assume that originating traffic is split 60/40 between intraoffice and interoffice calls, respectively. Let's further assume that interoffice calls are split 50/25/25 among calls to the toll office and calls to each of the two nearby direct-trunked offices, 873 and 952. Thus, this office's originating 15 Erlangs are split up as: 9 Erlangs (intraoffice) to the second selector stage, 3 Erlangs to the trunk group connecting to the toll office, 1.5 Erlangs to the trunk group connecting to the 873- office, and 1.5 Erlangs to the trunk group connecting to the 952- office. Let's assume that the toll office and the two local offices are close enough that one-way trunks are economically justified. Then, using $P_b = 0.02$ for trunks, Figure 5.4 shows that the outgoing toll trunk group requires eight trunks and the outgoing trunk groups to offices A and B require five trunks each.

Thus, on all 33 first selectors the first eight rotary positions on the 1-level are connected to originating trunk circuits on outgoing toll trunks, the first five rotary positions on the 2-level are connected to originating trunk circuits on outgoing trunks to the 952- office, the first five rotary positions on the 3-level are connected to originating trunk circuits on outgoing trunks to the 873- office, the higher positions on these 1-, 2-, and 3-levels are connected to the network busy tone, and all 10 rotary positions on the 4-level are connected to second selector links (more on this below).

The wiring to the trunk groups is in slipped multipling. For example, if the five trunks to the 952- office are called a, b, c, d, and e, on six of the 33 first selectors the first five rotary positions on the 2-level are wired to a, b, c, d, and e, respectively; on six of the 33 first selectors the first five rotary positions on the 2-level are wired to b, c, d, e, and a, respectively; on seven of the 33 first selectors the first five rotary positions on the 2-level are wired to c, d, e, a, and b, respectively; on seven of the 33 first selectors the first five rotary positions on the 2-level are wired to d, e, a, b, and c, respectively; and on seven of the 33 first selectors the first five rotary positions on the 2-level are wired to e, a, b, c, and d, respectively.

6.7.3 Fabric

Back to the output of the first selectors, we've seen that this office's originating subscribers generate 9 Erlangs of intraoffice traffic. This traffic flows over second selector links to a pool of second selectors that handles intraoffice calls. If we had full-access connectivity between the first and second selectors, Figure 5.5 shows that 17 second selectors would be needed. But we don't have full access because this pool of servers is reached only from the 4-level of the first selectors (after the originating subscriber dials 654); so each first selector can access only 10 servers. Economy of scale suggests that the number of intraoffice second selectors is ≥ 17.

If we segregated the 33 first selectors into two distinct groups and allocated a separate pool of second selectors to each of these two distinct groups, then each second selector group would have to handle 4.5 Erlangs and would require 11 second selectors per group (10 of them would cause $P_b = 0.02$). Thus, 22 second selectors in a 2-group segregated-access arrangement will handle the traffic; but this still doesn't satisfy the first selectors physical limitation of 10 servers per row. Twenty second selectors in a 2-group segregated-access arrangement will almost handle the traffic and still satisfies the first selectors physical limitation. If we segregated the 33 first selectors into three distinct groups and allocated a separate pool of second selectors to each of these three distinct groups, then each second selector group would have to handle 3 Erlangs and would require eight second selectors per group. Thus, 24 second selectors in a 3-group segregated-access arrangement will handle the traffic and still satisfy the first selectors' physical limitation.

Since a slipped-multipling arrangement performs worse than a full-access arrangement, but better than a totally segregated-access arrangement, intuition suggests that the number of servers should be between 17 and 24. Intuition and experience suggest that using slipped multipling to distribute intraoffice traffic, the required number of servers is roughly in the middle of this range. While 20 servers were insufficient for segregated access, it is probably a sufficient number for a slipped multiple. Since three groups of eight each can handle 3 Erlangs each, the number of intraoffice second selectors is ≤ 24. Another way of looking at this is to observe that if 20 servers in two segregated groups block at 0.02, then 20 servers using slipped multipling ought to block at a lower probability. So, we guess that ≈ 20 switches should handle this load.

So, first selectors 1 and 21 are wired from their 4-level to second selectors 1 – 10; first selectors 2 and 22 are wired from their 4-level to second selectors 2 – 11, ...; first selectors 11 and 31 are wired from their 4-level to second selectors 11 – 20; first selectors 12 and 32 are wired from their 4-level to second selectors 12 – 20 and then 1; first selectors 13 and 33 are wired from their 4-level to second selectors 13 – 20 and then 1 – 2; first selector 14 is wired from its 4-level to second selectors 14 – 20 and then 1 – 3, ...; and first selector 20 is wired from its 4-level to second selectors 20 and then 1 – 19.

Let's assume that interoffice traffic is balanced, with equal originating and terminating traffic in each destination. The incoming toll trunk group also has eight trunks and the incoming trunk groups from the 873- and 952- offices also have five trunks each. If we a assume that this office's 654 office code is absorbed at the originating office, then all incoming trunks must connect to second selectors. While we might consider an arrangement similar to line finders to provide concentration between the trunks and the second selectors, trunks can be assumed to be very highly occupied (if they are not, we would need fewer of them). Thus, each of the 18 incoming trunks is connected to its terminating trunk circuit and then to a dedicated second selector.

This office has 38 second selectors — 20 in the pool that is slip-multipled to the 4-level of the 33 first selectors for intraoffice calls and 18 that are each directly wired to an incoming trunk. This office does not have a third selector stage, and the people in the community have a six-digit telephone number.

Since a connector group handles up to 100 terminating appearances and since this office has slightly more than 500 telephones, the office has six connector groups. The 500+ telephone numbers are between 654-111 and 654-600 (0 is 10). There is allocation for 600 numbers. If the traffic to the 500+ terminating parties is distributed uniformly over the six connector groups, then each group must handle:

$$\frac{15 \text{ Erlangs of total traffic}}{6 \text{ groups}} = 2.5 \text{ Erlangs/group}$$

and each group requires eight switches. There are 48 connectors in the office.

This building/office requires 33 + 33 + 38 + 48 = 152 Strowger switches.

6.7.4 Growth and Change

Let this community grow gradually. With any growth, the capacity of the power and tone plants must be increased, the main frame must be expanded, line circuits must be added, and more trunks might be needed. In addition, as the number of subscribers approaches 600, switches must be added gradually to the various groups in the various stages of the fabric. In addition to this, when size passes 600, 700, and so on, new line finder and connector groups must be added, and some terminations on the fabric may be rearranged. In addition to this, when size passes 1000, a third selector stage must be added and seven-digit dialing must be introduced to the community.

While a Strowger switch is not expensive, installing and rewiring switches and otherwise changing a Step-by-Step office is labor intensive. The level of skill and training of the framemen that do this work is extremely high. Compounding this, Step is highly susceptible to external effects. Changes in the dial plan, growth in neighboring communities, and even changes in traffic patterns in neighboring offices cause major changes in a Step office. The addition of a new office code to some building causes a major rewiring of all the first selectors in all the old offices in that building and in any Step offices in other buildings with direct trunking to this building.

6.8 DISCUSSION

Issues presented are cost versus size, growth, OA&M, call tracing, and Touch Tone provision. General pros and cons of Step-by-Step systems are discussed.

6.8.1 Cost versus Size

Step is the epitome of the cellular architecture, discussed in Section 2.4.5. The cost-per-line curve is that of the cellular system in Figure 2.7.

- In small offices, Step-by-Step is less expensive than common control or modular systems. Step has little common equipment, equipment required in a switching office regardless of its size, compared to the other systems that will be described.
- As a cellular system, Step grows in small increments by the addition of individual Strowger switches. Thus the cost-per-line curve is smoother than that of a common control or modular system.
- The average slope of this cost-per-line curve, the incremental cost per subscriber, is higher than that of the common control or modular systems.
- In large offices, Step-by-Step is more expensive than common control or modular systems. In a large common control or modular office, the cost of the common equipment is spread over enough subscribers that the cost per line is less than that of Step.

- The cost-per-line curve of Step crosses over that of a typical common control or modular system. Over the years, this crossover point has moved to the left, toward smaller office sizes.

6.8.2 OA&M

Operation, Administration, and Maintenance (OA&M) is the cost of running a switching office. Most of this cost is labor personnel, the highly skilled framemen and switchmen. OA&M has many cost components.

- New subscribers, and any additional switches they need in line finder and connector groups, must be added. Subscriber appearances on the fabric must be moved because of changes in class of service or traffic balancing.
- Digit absorbing must be reconfigured and previously unused first selector rows must be equipped in response to changes in the dial plan.
- Detection and location of points of high blockage in the fabric are difficult. Interpretation of subscriber complaints requires experience and thorough knowledge of the office. Switches must be added to switch groups that are suspected to be undersized.
- Detection and location of faults in the fabric are also difficult. Again, patterns must be derived from subscriber complaints. Since Step has no built-in automatic testing procedures, the fabric must be manually tested periodically to locate faulty switches and broken links. These faults must be repaired manually.
- Repairs and changes are time-consuming and must be performed on-line; the office cannot be brought down. Class of service is temporarily changed and faulty switches are simulated busy by holding contacts operated with toothpicks. Permanent changes are performed later.

All the varieties of switching office have an average annual OA&M expense. But cost of OA&M is higher in Step than in more modern systems. More OA&M is required because the architecture is more sensitive to change. The OA&M is more labor intensive because electromechanical equipment is more difficult to keep working, requires significant preventive maintenance, and the changes and fixes are performed manually with soldering irons, wire strippers, and plyers.

Step requires a large amount of floor space. While this has a direct expense, it also contributes to increased time in simply walking from one point in the fabric to the next. Furthermore, all this labor takes place in an extremely noisy environment.

On the positive side, Step-by-Step is simple to understand. Training the craftsmen for Step-by-Step is easier than for ESS. Craftsmen are not specialized to pieces of equipment because so much of the equipment is made from the same piece part.

6.8.3 Limitations

The Step architecture is further limited because (1) calls are set up through the fabric directly as the number is dialed, and (2) there is no common place where the called number resides so it might be read, altered, interpreted, or translated.

Obviously, the fabric has many different paths that provide a switched connection between a given pair of calling and called parties. In Step, if the switch-train is blocked at some point in the fabric, it cannot back up and try an alternate path. In a fabric with end-marked control, the controller investigates all alternate paths until one is found that completes the desired connection.

Call tracing is difficult without common control. When the police request a call trace for a given party, the framemen in a Step office are prepared for rapid response. When this party is called, these framemen inspect the connectors in the called party's connector group, looking for the one in use on this call. Then they inspect the third selectors that are wired to this connector, and so forth. They run around the office with the wiring diagram of the fabric until they have worked their way backwards through the fabric to the line circuit of the calling party. In the typical kidnapping story on television, the called party must keep the bad guy on the line so the telephone company has time to trace the call. This drama is passé, because it only relates to electromechanical offices.

Let a subscriber served by a Stepper dial 1 or the office code of a distant local office that is direct trunked. Unlike dialing out of a PBX (Section 10.4.4), the current numbering plan does not require this subscriber to pause or to wait for dial tone from this receiving office before continuing to dial. This receiving office detects a seizure on this trunk and must be prepared to receive dialed digits immediately. If this receiving office is also a Stepper, the trunk is directly wired to a selector switch, another reason that such trunks are not wired to the line finder stage. If this receiving office has common control, a dial pulse receiver must be connected to the trunk. But a typical connection through the fabric takes longer to complete than the interdigit interval. So a special arrangement is necessary, described in Sections 8.4.5 and 9.2.5. When the transmitting office has common control, the receiving office can signal its readiness to receive the digits and this special arrangement is unnecessary.

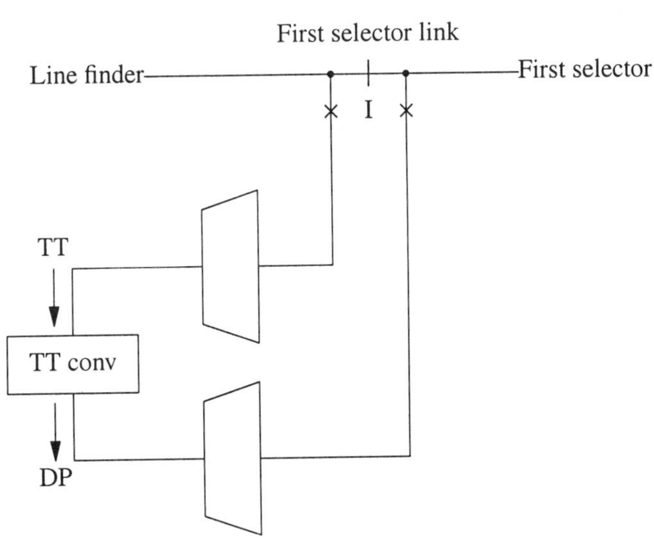

Figure 6.6. Touch Tone interception.

6.8.4 New Services

Touch Tone digits cannot pulse a Strowger switch. Step offices are equipped for Touch Tone by changing certain first selector links so that additional equipment may be added, in series, during dialing. This change includes inserting an *interception circuit,* two make contacts and one break contact of an interception relay, in the paths of the tip and ring of the link. The additional equipment includes a Touch Tone converter (TTC), that accepts Touch Tone digits and converts them into dial-pulse digits, and matched concentrators. See Figure 6.6, where a line represents a tip/ring twisted pair.

Immediately after going off-hook, a Touch Tone subscriber's line is found by a line finder and connected to a first selector link. The I relay is operated and an idle Touch Tone converter is assigned. The matched concentrators connect the line finder side of the link to the Touch Tone input of this TTC and the first selector side of this same link to the dial pulse output of this same TTC. The Touch Tone converter returns dial tone to this subscriber. As Touch Tone digits are input to the TTC, they are recorded and dial-pulsed into the first selector. Since the Touch Tone digits may be input faster than the corresponding dial pulses can be output, the TTC stores the intermediate digits. After the TTC has output the last dial-pulse digit, several seconds after the subscriber is done dialing, the I relay is released. Then, the link is cut through, and the TTC is available for another call.

Every first selector link that any Touch Tone subscriber might reach must be changed for Touch Tone interception. By segregating all Touch Tone subscribers into common line finder groups, the number of links that must be changed is reduced.

Step modernization was proposed many times. The first selector link would be intercepted, as in Touch Tone provisioning, except that the Touch Tone digits received from the subscriber would be examined by a processor before they were dial-pulsed into the Step fabric. The processor could reside in the Step office or one remote processor might serve several offices via a data link between it and each TT-DP converter. Adding intelligence to Step in this way enables the kinds of services that were introduced with ESS and also enables a primitive form of alternate routing. In abbreviated dialing, the subscriber stores, say, 10 full directory numbers in a shared memory and then uses, say, #8 to signify a request to call the eighth of those numbers. Abbreviated dialing and other similar services are obviously impossible in a regular Step office and require this kind of interception. In addition, the processor could initiate automatic redial in case the fabric blocked the initial call attempt.

6.8.5 The Partition

Step has no physical partition between switching hardware and control because the signaling path is inseparable from the data path and the control subsystem is inseparable from the connection subsystem. But a logical partition exists; it is a temporal partition.

The temporal duration of a call is partitioned into setup time and connect time. During call setup, the integrated system is in control mode and expects signaling only; you can't speak and expect anything to happen or expect anyone to hear you. During connect time, the integrated system is in talking mode and expects conversation only; you can't dial your telephone and expect anything to happen. You can't talk during dialing and you can't dial during talking.

This partition requires three triggers (triggers will be discussed again in Chapter 18) to effect the transitions from: idle mode to call setup mode to talking mode and back to idle mode. These triggers are supervisory signals. When the calling party raises the handset off his nonringing telephone, he triggers the system from idle mode to call setup mode. When the called party answers his ringing telephone, he triggers the system from call setup mode to talking mode. When the calling party hangs up (assuming calling-party supervision) he triggers the system from talking mode back to idle mode.

One of the biggest economic problems with Step (or with any distributed-control cellular architecture) is that the control intelligence remains associated with the call not just during call setup, but for the entire duration of the call. And, in spite of the control intelligence being present, it can only be used when the mode is right.

EXERCISES

6.1 On a phototcopy of the block diagram of a Step office in Figure 6.1, add the connection to a two-way toll trunk (see Figure 4.5) and a two-way interoffice direct trunk.

6.2 Consider an electronic integrated circuit, called a selector, that has one three-wire input and 10 three-wire outputs [3 – 5]. A selector accepts one bit-serial BCD digit (after conversion from Touch Tone) over two of the input wires and then connects its input pair to the output pair that corresponds to the digit. The third wire in all ports is used for supervision and busy indication. Subsequent signals, including other BCD digits, pass through this connection. Consider another electronic integrated circuit, called a hunter, also with one three-wire input and 10 three-wire outputs. When a hunter is connected at its input, after testing the third input wire, it connects the input pair to an idle output pair in some priority order. Subsequent signals, including other BCD digits, pass through this connection. For each of the three applications of the Strowger switch, sketch its equivalent implementation using these electronic circuits.

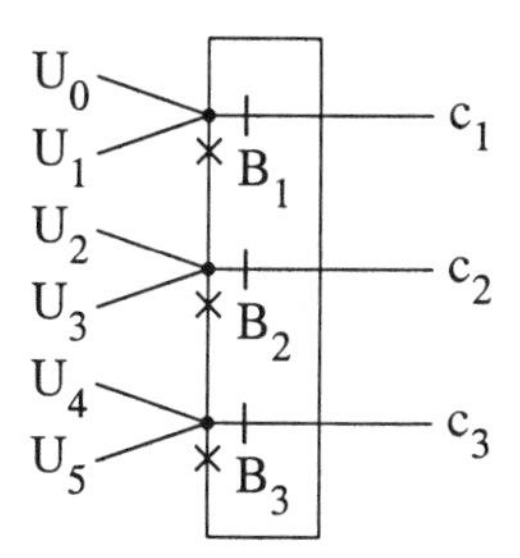

Figure 6.7. Preference chain.

6.3 Consider the relay contact circuit for an allotter preference chain shown in Figure 6.7. In the figure, the rectangle is not a box, but is a wire from the make contact of B_3 to the common of B_1. Since this is not a series/parallel circuit, it is difficult to analyze and should be appreciated as a clever design. Let U_0 – U_5 be the start wires from six line circuits. Let B_1 – B_3 be the rest position transfer contacts, indicating the busy/idle status of three line finders. Let c_1 – c_3 be the outputs of this allotter, the control wires to these same three line finders. For each line

(0 – 5) specify the first, second, and third preference, as determined by this circuit, for the three line finders.

6.4 Suppose a Step office has 20 first selectors and 20 local operators. Sketch two different wiring patterns. (a) Ten of the first selectors connect to one group of 10 operators, and the other 10 first selectors connect to the other 10. (b) The 20 operators are ordered and slipped multipling is used so that each first selector begins hunting with a different operator.

6.5 Describe how the following classes of service would be implemented in a Step PBX that homes on a Step CO. (a) Internal calls only, no dial-9 calls. (b) Internal and external local calls only, no long distance. (c) Originate only, no terminating calls. (d) Terminate only, no originating calls.

6.6 In the extended example of Section 6.7, if all the one-way trunks were replaced by two-way trunks, how many trunks would be needed in each trunk group? How does this change the number of switches in the office?

6.7 Suppose a Step office has 150 Touch Tone subscribers, which are segregated into one line finder group. (a) If these subscribers average 1.5 originations during the PBH and they average 4 minutes on each call, what is the TI on the LFG? (b) How many line finders are required in the LFG for $P_b < 0.01$? (c) If these subscribers average 10 seconds to dial a telephone number, what is the TI on the group of TT-DP converters? (d) How many TT-DP converters are required? (e) What are the dimensions of the two concentrators in Figure 6.6?

6.8 Completely lay out the fabric of a base-3 Step-by-Step central office that serves 27 subscribers. Use the following assumptions.

- Assume you live in a base-3, or ternary, country, where the dial telephones have three holes in the dials instead of 10. Subscribers can only dial 1, 2, or 0.
- In this country, Strowger switches are manufactured, not as 10 × 10, but as 3 × 3, with nine contacts on the back of a switch. Thus, there are nine subscribers per line finder group or connector group. There is no doubled access by tripled banks in line finders.
- Each subscriber in this office has a 4-digit directory number, 2-xxx (each x = 0, 1, or 2), and originates an average of one 3-minute call every 4 hours.
- All interoffice calls are toll switched and the toll access code is 1. The operator access code is 0. Forty-five percent of all originating traffic is intraoffice, 45% is outgoing to toll, and 10% is to the operator. There is one incoming toll trunk, carrying 0.076 Erlangs, that connects directly to a second selector.
- Use $P_b = 0.01$ throughout, except for 0.1 at the operators.
- Show all switches, operators, start circuits, allotters, and trunk circuits as blocks, and all the wiring that interconnects them. Use the symbol of Figure 6.2 for a Strowger switch, except with three bumps instead of 10.

Use slipped multipling, but let the tip, ring, and sleeve leads be represented by one line.

- Show two typical subscribers, connected to line circuits (as blocks) and to the originating and terminating sides of the fabric. Identify their directory numbers.

Hints: (a) first, find the number of line finders in a line finder group and the number of connectors in a connector group and then the number of groups; (b) next, find the number of outgoing trunks and operators; (c) you should need to use only five connections at the back of each first selector; (d) you should need to use 21 Strowger switches in the total office.

REFERENCES

[1] Bambos, N., "Toward Power-Sensitive Network Architectures in Wireless Communications," *IEEE Personal Communications,* Vol. 5, No. 3, Jun. 1998.

[2] Woesner, H., et al., "Power-Saving Mechanisms in Emerging Standards for Wireless LANs," *IEEE Personal Communications,* Vol. 5, No. 3, Jun. 1998, pp. 40 – 48.

[3] Thompson, R. A., "A Chronology of the Design and Development of an Electronic Telephone System," *Report to Office of Experimental R&D Incentives of the National Science Foundation,* Nov. 1973.

[4] Thompson, R. A., "Experiment Definition: R&D Incentives of an Electronic Telephone System," *Final Report to the Office of Experimental R&D Incentives of the National Science Foundation,* May 1974.

[5] Thompson, R. A., "The Management of Industrial R&D at the University," *IEEE SouthEastCon,* 1975.

SELECTED BIBLIOGRAPHY

Brooks, J., *Telephone — The First Hundred Years,* New York, NY: Harper & Row, 1976.

Clarke, A. C., et al., *The Telephone's First Century — and Beyond,* New York, NY: Thomas Y. Crowell, 1977.

Joel, A. E., Jr., *Electronic Switching: Central Office Switching Systems of the World,* New York, NY: IEEE Press, 1976.

Miller, K. B., *Telephone Theory and Practice, Vol. II,* New York, NY: McGraw Hill, 1933.

Noll, A. M., *Introduction to Telephones and Telephone Systems,* Norwood, MA: Artech House, 1986.

Pearce, J. G., *Telecommunication Switching,* New York, NY: Plenum Press, 1981.

Rey, R. F. (ed.), *Engineering and Operations in the Bell System,* Second Edition, Murray Hill, NJ: AT&T Bell Laboratories, 1977.

Schindler, G. E. (ed.), *A History of Engineering and Science in the Bell System,* Murray Hill, NJ: Bell Telephone Laboratories, 1982.

Talley, D., *Basic Telephone Switching Systems,* New York, NY: Hayden Book Company, 1969.

7

Interconnection Fabrics

"The head bone's connected to the neck bone, the neck bone's connected to the shoulder bone, ..." — from the song *Dry Bones*

A switching office's *interconnection network,* or fabric, is that subsystem through which the calling and called parties' lines are physically or logically connected together. In Step, the fabric is integrated with its control, but in most systems the fabric is a distinct physical subsystem that is controlled by another distinct physical subsystem. In a manual office, for example, the fabric is the patch-panel with its double-ended cords and the separate controller is the operator. This chapter is a survey of interconnection fabrics. Various fabric architectures are discussed at the block-diagram level of detail. Concrete examples will appear in the descriptions of the Crossbar, Electronic, and Digital systems in the following chapters.

Several concepts, briefly described in Chapter 2, are defined in the first section of this chapter. In the second section, multistage fabrics are developed in a progression from a single-stage to a four-stage architecture. Many digressions are introduced during this progression. In Section 7.3, the hierarchy of blocking in fabrics is developed and several classical nonblocking fabric architectures are explained. Blocking in fabrics is discussed in Section 7.4, and Section 7.5 describes self-routed packet switching through a *Banyan network.*

7.1 FUNDAMENTALS

Preliminary discussions include technologies for space-division fabrics, various definitions and terminology, and *sidedness.*

7.1.1 Technology

Figure 7.1 shows a simple 2×2 fabric, as it would be implemented in three different technologies. The leftmost implementation is electrometallic, using wires and single-pole single-throw relay contacts. The center implementation is digital electronic, using AND and OR gates. The rightmost implementation is photonic, using passive splitters, light modulators (shutters), and passive combiners.

The technology must provide a means by which each input to the fabric reaches

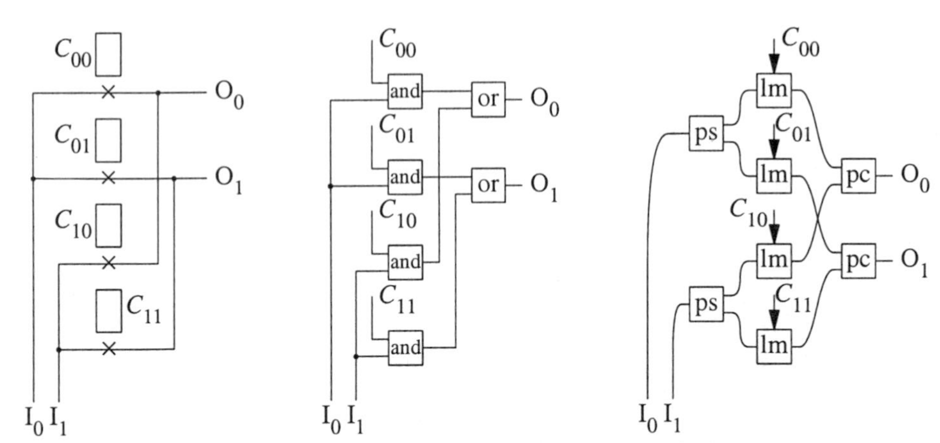

Figure 7.1. Implementations in three technologies.

many crosspoints. In a large-motion traveling switch, as in Figure 2.4, some actuator physically carries an input wire to an electrical post associated with a crosspoint. In electrometallic and electronic technologies, an input wire is simply fanned out. In photonics and transmission-line technologies, fan-out requires a device for impedance matching and/or power splitting, not just a simple electrical contact. The technology must also provide a means by which each output from the fabric is reached from many crosspoints. In electrometallic and some digital electronic technologies, an input wire is simply "bused" or "OR-tied." Other digital electronic technologies don't allow OR tying and require an OR gate that has a separate fan-in from each crosspoint. In photonics and transmission-line technologies, fan-in also requires a device (actually the same device as is used for fan-out).

In each implementation in Figure 7.1 the four crosspoints are controlled by four signals: C_{00} connects I_0 to O_0, C_{01} connects I_0 to O_1, C_{10} connects I_1 to O_0, and C_{11} connects I_1 to O_1. In the electrometallic technology, the control signal operates a relay or otherwise closes a metallic contact or solid-state switch. In the digital electronic technology, the control signal enables an AND gate so that the binary state of its output matches the binary state of its other input. In the photonic technology, the control signal opens a mechanical or solid-state optical shutter.

7.1.2 Terminology and Definitions

The terms "connect" and "connection" are used to indicate the completion of a path or channel through a fabric, subfabric, or switch. These terms are not used to indicate the location of wires within a fabric, subfabric, or switch. Thus, a wire between points A and B does not connect A and B together, in our use of the word. The two points were connected when the wire was installed during the manufacture of the system. Although a little awkward, I'll use "wire" as a verb in such a context and say that A and B are wired together. The term "wire" must be loosely interpreted because the actual physical channel could be an optical fiber or a string between two tin cans.

While generalized fabrics for multicasting are beginning to appear in the literature, very little theory has been developed for fabrics that support narrowcast (one-to-several) or broadcast (one-to-many) circuit-switched connections; this area is quite open. The discussion in this chapter, indeed the entire book, is confined to fabrics that support point-to-point communications. Such fabrics provide a connection between a single port on the left side to a single port on the right side. Telephone features, like add-on and conferencing, are not handled directly in the fabrics but by external circuits with multiple fabric appearances.

The cost of a fabric is complicated to measure. While size, packaging, power, and complexity of control all contribute to a fabric's cost, a simple first-order measure of cost is to count the total number of crosspoint contacts. This count is made on the various fabric architectures herein, but the reader is warned that it is only a part of total cost.

Fabrics will be distinguished by their number of stages, but this apparently simple concept is difficult to define rigorously. A stage of a fabric is a physical or logical subfabric through which any end-to-end path must be connected. If a stage is a simple matrix switch, or its equivalent, then a stage corresponds to the operation of a single crosspoint, and a path through a k-stage fabric requires the operation of k crosspoints. The Step-by-Step fabric, described in the previous chapter, has five stages: line finders, first selectors, second selectors, third selectors, and connectors. While each stage has many Strowger switches, a given call connection uses only one contact in only one switch in each stage. Confusing the issue, fabrics are frequently viewed hierarchically and a stage in one discussion may actually be a subfabric that is itself comprised of several stages.

In a full-access fabric, every input can be connected to every output. Usually a total fabric has full access, but its component subfabrics, modules, and switches may not. In Step, any line on the originating side has access to any line on the terminating side. But, in the first selector stage, for example, every first selector link has access to a maximum of 100 second selector links, typically less than full access. Section 5.3 showed that limited access affects performance and efficiency of server groups. It will also affect the architecture of interconnection fabrics. This theme will recur throughout this chapter.

Consider two ports, one on each side of a fabric, that are both idle and that must be interconnected through the fabric. If other existing calls have so congested the fabric that no idle path through the fabric is available for this call, then the fabric is said to block. While fabrics can be designed to be nonblocking, such fabrics are expensive. Probability of blocking, P_b, is an important fabric parameter.

In general, a fabric provides many different connection paths between any pair of its endpoints. Providing many alternate paths improves reliability and lowers P_b in blocking fabrics. While the fabric in Step-by-Step has many alternate paths between any given pair of endpoints, *direct control* of the fabric makes automatic retrial impossible.

Fabric control was discussed in Section 2.4. Three kinds of control placement are *direct, register,* and *common control.* Two methods of connection setup are *progressive* and *end-marked.* In this chapter, we discuss the general complexity of the control algorithm, but not its implementation.

Switching must take place in every division in which channels are defined (space, time, frequency, wavelength, polarization, etc.). While this chapter concentrates on the space division, many of the concepts can be extended to the other divisions. Digital time-division switching and circuit switching in both space and time are described in Chapter 13. Further generalization is presented in Chapter 21, where we describe wavelength-division switching and optical switching in space, time, and wavelength.

7.1.3 Sidedness

In Section 1.5, the two sides of a switching office were given as the line side and the trunk side. However, the sidedness of the fabric inside a switching office is not necessarily defined the same way. Four alternates are given.

- Lines/Trunks. Wires to telephones and user devices appear on one side of the fabric and wires to other switching offices appear on the other side. This is the same sidedness as the overall switching office. The fabrics in Number 5 Crossbar and 1ESS use this sidedness for interoffice calls.
- Originating/Terminating. Wires to telephones and trunks that can place calls appear on one side of the fabric and wires to telephones and trunks that can receive calls appear on the other side. Any telephone or trunk that can originate and terminate calls must have an appearance on both sides. The direction from inputs to outputs through the fabric coincides with the direction of placing a call through the fabric. The fabric in Step-by-Step uses this sidedness.
- Mouth/Ear. Wires to mouthpieces and data transmitters appear on one side of the fabric and wires to earpieces and data receivers appear on the other side. Any telephone or trunk that talks and listens must have an appearance on both sides. The direction from inputs to outputs through the fabric coincides with the direction of the flow of information through the fabric. These fabrics are half-duplex and any full-duplex call needs two separate connections through the fabric, one over which A transmits to B and one over which B transmits to A. Digital fabrics use this sidedness because data cannot flow backwards through a digital gate.
- External/Internal. Wires to telephones, trunks, and other external devices appear on one side of the fabric and wires to internal supervision circuits appear on the other side. These fabrics are called "folded." The fabrics in Number 5 Crossbar and 1ESS use this sidedness for intraoffice calls.

7.2 MULTISTAGE FABRICS

The number of stages in a fabric is called its "depth." Fabrics are developed in this section from one-stage fabrics to four-stage fabrics. Many digressions will appear in this progression. One-stage fabrics digress into definitions of concentration, distribution, and expansion and into a discussion of multipling. Two-stage fabrics digress into folded fabrics, limited access, and the perfect shuffle pattern. A generalization of three-stage fabrics is postponed to the next section. Four-stage fabrics digress into modularity and wiring plan. Two different architectures of a 1000×1000 fabric were described in Chapter 5; fabrics of this size will be used as examples throughout this section.

This section shows how a fabric's number of stages relates to five other properties of the fabric:

- Total number of contacts;
- Access;

- Number of alternate paths;
- Probability of blocking;
- Control complexity.

These five characteristics will be revisited as we progress through fabrics with increasing numbers of stages.

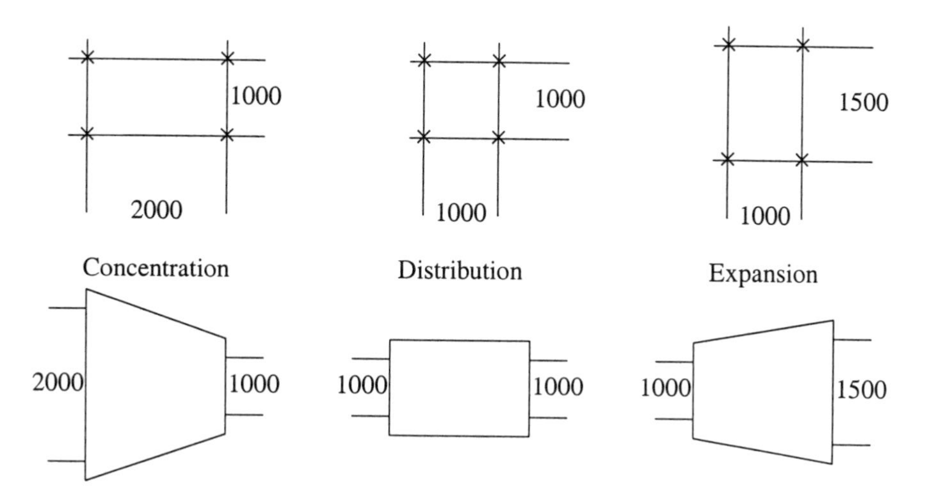

Figure 7.2. Physical and schematic representations.

7.2.1 One-Stage Matrix

Figure 7.2 generalizes Figure 7.1 in several ways. The upper diagrams give a physical generalization, and the lower diagrams give a schematic generalization. In Figure 7.1, regardless of the technology, the inputs were shown as vertical lines entering the fabric from below, and the outputs were shown as horizontal lines exiting the fabric from the right. The three upper diagrams in Figure 7.2 follow this same rule, where the fabrics have 2000, 1000, and 1000 inputs, respectively, and 1000, 1000, and 1500 outputs, respectively. The lower diagrams show schematic representations of these same three fabrics, where the inputs are shown on the left and outputs on the right. The shapes of the three fabrics are drawn as trapezoids, where the length of the left and right edges corresponds to the number of ports on that edge.

Consider a one-stage fabric, configured as a rectangular matrix. Let it have M input lines, each wired to a vertical conductor in the matrix, and N output lines, each wired to a horizontal conductor. The fabric has MN crosspoint contacts, one at each two-dimensional coordinate in the matrix. Thus a 1000×1000 matrix requires one million crosspoints. Frequently, $N = M$ and the ith user or endpoint is wired to both the ith input and the ith output. If these users are never connected to themselves, the one-stage fabric requires $N^2 - N$ crosspoints because the crosspoints on the main diagonal are unnecessary.

To connect the ith input, $1 \le i \le M$, to the oth output, $1 \le o \le N$, the crosspoint contact at coordinate $[i,o]$ must be closed. A convention dilemma is seen. It seems natural to assign the ordering of the two-dimensional name of a crosspoint as [input,output]. But, the naming convention for points in a matrix is usually [row,column] and the convention in fabrics is usually to assign inputs to columns and outputs to rows. These two conventions together produce the unnatural assignment [output,input] to crosspoints. Since the assignment of inputs to columns and outputs to rows is not universal, we will use the [input,output] convention to name crosspoints.

If input i is idle, every contact (N of them) in the ith column is open. Even when i is busy, $N-1$ contacts remain idle. An analogous statement applies to the outputs. The occupancy of the crosspoints is low, especially if N and M are large, and the one-stage architecture is seen to be inefficient. However, since crosspoint $[i,o]$ is dedicated to connecting input i to output o and has no other function, this connection can always be made. As long as the endpoints are idle or available, one-stage fabrics do not block. Since crosspoint $[i,o]$ is the only means of connecting input i to output o, one-stage fabrics have no alternate paths. If a contact malfunctions, the corresponding connection cannot be made. This inherent limitation of one-stage fabrics can be compensated by building a total fabric from redundant parallel subfabrics and multipling each input and output to each subfabric.

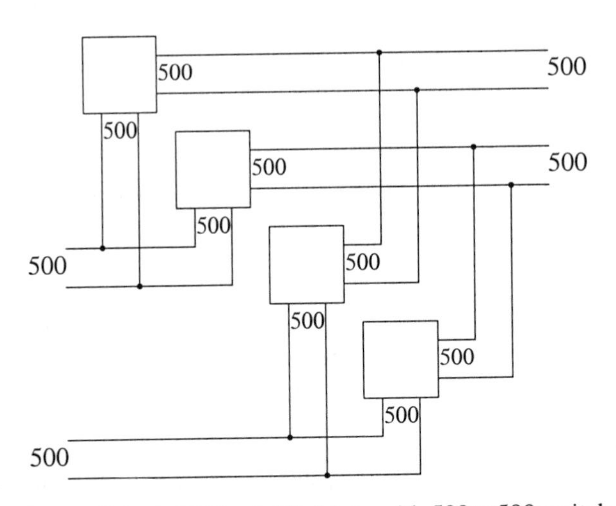

Figure 7.3. A 1000×1000 fabric with 500×500 switches.

7.2.2 Multipling

Before leaving one-stage fabrics, we study the effect of switches with limited access. A 1000×1000 one-stage fabric is a 1000×1000 matrix with one million crosspoints. However, suppose a physical 1000×1000 switch is not possible. Figure 7.3 shows an equivalent fabric in which each physical switch is a 500×500 module, or subfabric. Each input to the fabric is multipled to an input port on two different switches. Each output from the fabric is multipled from an output port on two different switches. The inputs and outputs are partitioned into two groups of 500 each. The overall fabric requires $2 \times 2 = 4$ of these physical switches, where switch $[j,p]$ is used to connect an input from input group j to an output from output group p, where $j = 0$ or 1 and $p = 0$ or 1.

Generalizing, let $M = r \times m$ and $N = s \times n$. Then, if an $M \times N$ fabric is built using physical $m \times n$ switches, the fabric requires $r \times s$ of them. Each input to the fabric is multipled to an input port on r different switches and each output from the fabric is multipled from an output port on s different switches. The overall fabric control is slightly more complex than in equivalent fabric built as a single full-access switch. Let input $[i,j]$ be the ith input in the jth input group, where $1 \leq i \leq m$ and $1 \leq j \leq r$. Let output $[o,p]$ be the oth output in the pth output group, where $1 \leq o \leq n$ and $1 \leq p \leq s$. Then input $[i,j]$ is connected to output $[o,p]$ by closing crosspoint contact $[i,o]$ in switch $[j,p]$.

Since each limited-access switch has $m \times n$ crosspoints and there are $r \times s$ switches, a one-stage fabric built with limited-access switches has $(rs) \times (mn)$ total crosspoints. The equivalent fabric built as a single full-access switch has $(rm) \times (sn)$ total crosspoints, the same number. The two implementations have the same number of contacts, the same number of paths through the fabric (namely, 1), and the same probability of blocking (namely, 0.0). In one-stage fabrics, the only effects from limited access are to require more physical switches and to increase the control complexity slightly. However, in fabrics with more than one stage, the use of switches with limited access has a significant effect on the overall architecture of the fabric.

7.2.3 Concentration, Distribution, and Expansion

A switching fabric with M inputs and N outputs performs *concentration, distribution,* or *expansion,* depending on the inequality relation between M and N. The three diagrams across the bottom of Figure 7.2 are schematic representations of these three types of fabrics. In later subsections, we will discuss multistage fabrics, in which each stage is a subfabric. These three classifications are especially important in describing these subfabrics or individual switches in larger fabrics.

Distribution, if $M = N$. Such a fabric distributes its M inputs over its N outputs. In fabrics with originating/terminating or mouth/ear sidedness, in which users are normally wired to both sides of the fabric, such a fabric distributes its users among themselves. In large multistage fabrics, stages of distribution compensate for the limited access of the switches. The schematic representation of a distribution fabric is shown in the lower center diagram of Figure 7.2.

Concentration, if $M > N$. Such a fabric concentrates its M inputs down to a smaller number of outputs. In large multistage fabrics, a stage of concentration funnels low-traffic lines down to few links with high occupancy. The links are *servers,* their number is traffic engineered, and the user may be occasionally blocked (Sections 5.2.5 and 5.3.4). In the Step-by-Step fabric, the line finder stage concentrates the users' originating appearances down to a few first selector links (Section 6.3). The schematic representation of a concentration fabric is shown in the lower left diagram of Figure 7.2.

Expansion, if $N > M$. Such a fabric expands its M inputs out to a larger number of outputs. In large multistage fabrics, a stage of expansion dissipates high-traffic links over many lines with low occupancy. In the Step-by-Step fabric, the connector stage expands the few connector links out to the users' terminating appearances (Section 6.6). The schematic representation of an expansion fabric is shown in the lower right diagram of Figure 7.2.

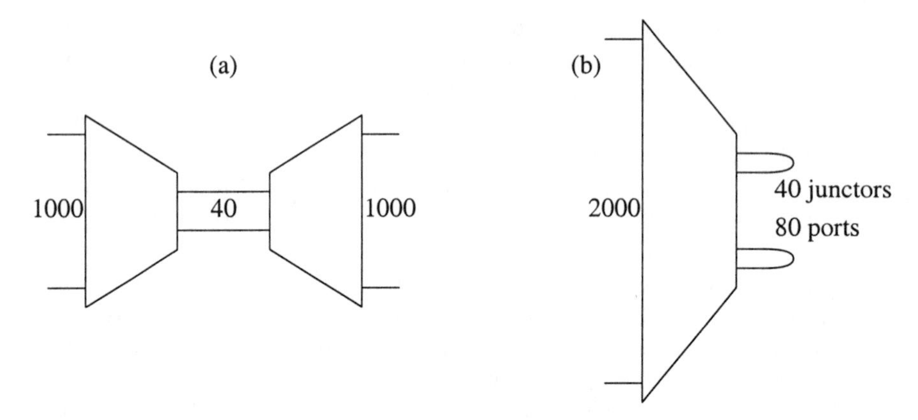

Figure 7.4. (a) A Two-stage fabric and (b) corresponding folded fabric.

7.2.4 Two-Stage Fabrics

Figure 7.4(a) illustrates a two-stage full-access 1000×1000 fabric. The first stage concentrates the 1000 inputs down to 40 junctors and the second stage expands the 40 junctors out to the 1000 outputs. The first stage and the calculation of the number of junctors were described in Section 5.2.5. If each stage is built as a single full-access matrix switch, each such switch requires $1000 \times 40 =$ 40,000 crosspoints. The total fabric has 80,000 crosspoints, only 8% of the number of crosspoints required in the corresponding single-stage fabric. In general, a two-stage concentrator/expander fabric, with M inputs, N outputs, and G junctors, requires $(M + N) \times G$ crosspoints.

The huge cost savings that concentrator/expander fabrics have over one-stage fabrics has a corresponding performance penalty: the fabric blocks. But the only blocking occurs in the concentration stage, where the N users compete for the G junctors, $G < M$. The number of junctors is easily calculated from the traffic assumptions and desired grade of service. Compensating for a finite probability of blocking, however, the two-stage fabric has many alternate paths.

In the fabric of Figure 7.4(a), any input can reach any output using any junctor. A fabric with G junctors has G alternate paths. Frequently, *fault tolerance* is a more important requirement of the fabric than the small chance that the fabric might block, especially when P_b can be designed to any specification. Thus, the two-stage fabric is generally preferred, especially in large-access switches where economy of scale permits G to be significantly smaller than M or N. A two-stage fabric, with $M = N = G$, has many *alternate paths* and $P_b = 0.0$, but it has twice as many crosspoints as a single-stage fabric, at the same cost as a fabric built from two parallel one-stage subfabrics.

Connecting the ith input, $1 \le i \le M$, to the oth output, $1 \le o \le N$, requires one junctor and two crosspoints. First, the fabric controller assigns any idle junctor to this call; let the gth one be used, $1 \le g \le G$. Then, the fabric controller operates one crosspoint in each stage: at coordinate $[i,g]$ in the concentrator and at coordinate $[g,o]$ in the expander.

7.2.5 Folded Fabrics

The two-stage fabric of Figure 7.4(a) is built from two 1000×40 switches, where the first stage connects 1000 inputs to one end of 40 junctors and the second stage connects 1000 outputs to the other end of the same 40 junctors. Instead, let it be built from one 2000×80 switch that connects 2000 inputs and outputs to 80 junctor-ends. Figure 7.4(b) illustrates this conversion to external/internal sidedness. Instead of bridging between two adjacent stages of a fabric, these junctors are folded back so that both ends appear on the same edge of the same stage of the fabric. This style of architecture is called a folded fabric and such folded-back junctors are called "hair pins" in the vernacular of fabric designers.

In a folded fabric, the junctors serve as a two-port circuit by which two lines may be connected together for a call. This circuit can also provide for the supervision of the call. So, a folded fabric may be viewed as a one-stage fabric that concentrates lines down to two-port supervision circuits. Line A is connected to line B in a folded fabric by finding an idle supervision circuit (junctor), connecting A to one end, and connecting B to its other end. The fabrics in Number 5 Crossbar (Section 8.3) and in 1 ESS (Section 11.3) are folded for intraoffice calls, but have lines/trunks sidedness for interoffice calls.

With respect to probability of blocking, fault tolerance, and control complexity, the two-stage fabric of Figure 7.4(a) is equivalent to the folded fabric of Figure 7.4(b). But, where the two-stage fabric of Figure 7.4(a) has 80,000 crosspoints, the folded fabric of Figure 7.4(b) has $2000 \times 80 = 160{,}000$. This number is misleading because Figure 7.4(b) is too general. If the fabric of Figure 7.4(a) has originating/terminating sidedness, each line appears on each side of the fabric. In the equivalent folded fabric, each line would have one appearance (on the left side), the total fabric would be 1000×80, and would have 80,000 crosspoints.

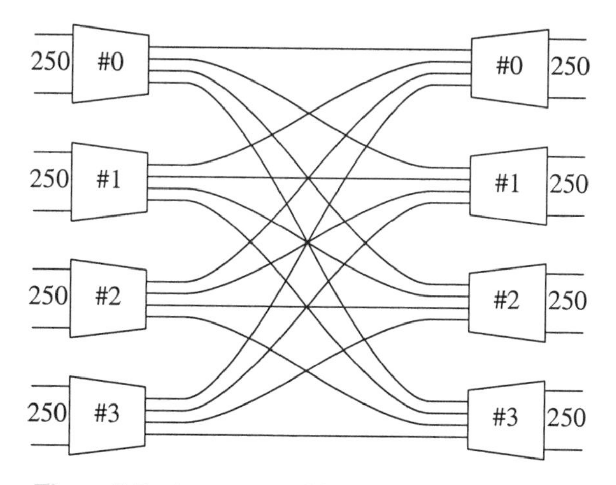

Figure 7.5. A two-stage fabric with limited access.

7.2.6 Limited Access

Figure 7.5 shows a two-stage concentrator/expander fabric built from limited-access switches. With 250 ports on one side of a switch, each stage requires four switches to provide access to the 1000 ports on each side of the fabric.

Assume each subscriber generates 0.025 Erlangs of traffic intensity. The total traffic intensity through one limited-access switch in the first stage is 250 × 0.025 = 6.25 Erlangs. If all traffic had the same destination and $P_b = 0.005$ were required, then Figure 5.5 shows that 14 servers (junctors) would be required. But, since the switches in the second stage have limited access, the traffic has four different destinations and the junctors must be partitioned into four server groups. Each line in the center of Figure 7.5 represents a group of junctors. If the generated traffic is split evenly over the four limited-access switches in the second stage, each junctor group carries 6.25/4 = 1.56 Erlangs. Figure 5.4 shows that six servers (junctors) are required in each group. There are 4 × 6 = 24 junctors at the inner edge of each switch in Figure 7.5, for a total of 4 × 24 = 96 junctors between the two stages. In Figure 7.5, the total count of crosspoints is 250 × 24 × 8 = 48,000. By comparison, in Figure 7.4(a), the equivalent fabric using full-access switches requires only 40 junctors, but has 80,000 crosspoints.

If the access is limited to 100 ports on one side of a switch, each stage requires 10 switches to reach 1000 lines. The per-switch traffic intensity is 100 × 0.025 = 2.5 Erlangs. The junctors at each switch are partitioned into 10 groups, each carrying 2.5/10 = 0.25 Erlangs. Figure 5.4 shows that three junctors provide $P_b = 0.005$ in each group. There are 10 × 3 = 30 junctors at the inner edge of each switch and 10 × 30 = 300 total junctors between the two stages. The total count of crosspoints is 100 × 30 × 20 = 60,000; less than the 80,000 required for the equivalent full-access fabric of Figure 7.4(a), but more than the 48,000 required for the equivalent 250-access fabric of Figure 7.5.

As the switch access decreases, the number of crosspoints would decrease proportionately, except that the reduced economy of scale increases the number of junctors. At large values of switch access, the second effect is smaller than the first and the crosspoint count in the total fabric decreases with decreasing access. At small values of switch access, the second effect is larger than the first and the crosspoint count in the total fabric increases with decreasing access.

Since any junctor can be used for any call in the fabric of Figure 7.4(a), the number of alternate paths is 40. In the fabric of Figure 7.5, a call between A and B must use one of the six junctors between the switch in the first stage on which A appears and the switch in the second stage on which B appears. Thus, there are only six alternate paths in Figure 7.5. If switch access is reduced to 100, there are only three alternate paths. In general, consider connecting a call from port i in switch j on the left side to port o in switch p on the right side. Of all the links between switch j on the left and switch p on the right, suppose the gth link is idle and is selected by the control algorithm. The left end of this link terminates on port $[g,p]$ on the inner edge of switch j and on port $[g,j]$ on the inner edge of switch p. Two crosspoints contacts are closed: the one at the intersection of i and $[g,p]$ in switch j and the one at the intersection of $[g,j]$ and o in switch p. The probability of blocking and the control complexity are both subtly dependent on the access size of the switches.

7.2.7 Perfect Shuffle

Successive stages are usually interconnected in a wiring pattern called the "perfect shuffle." The analogy to shuffling a deck of playing cards provides an orderly way of perceiving the interconnection pattern and of labeling the interconnecting wires.

Consider a generalized deck of playing cards with $s \times k$ cards. Create s stacks in order by taking k cards at a time from the original deck. Create a reordered deck by taking the top card from each stack in the same order, placing it at the bottom of the new deck, and repeating until all the stacks are empty. In an ideal shuffle of a conventional deck of playing cards, $s = 2$ and $k = 26$. The original ordering of the generalized deck is [1,1], [1,2], ..., [1,k], [2,1], [2,2], ..., [2,k], ..., [s,1], [s,2], ..., [s,k]. The final ordering of the deck is [1,1], [2,1], ..., [s,1], [1,2], [2,2], ..., [s,2], ..., [1,k], [2,k], ..., [s,k]. The perfect shuffle is also seen to be analogous to *matrix transposition.*

Consider the 16 links in the center of Figure 7.5. They begin on the interior edge of the four switches in the left stage and terminate on the interior edge of the four switches in the right stage. The line, labeled [s,k], that begins on the kth level on the interior edge of the sth switch in the left stage, ends on the sth level on the interior edge of the kth switch in the right stage. Reading the 16 line labels from top to bottom along the interior edge of the four switches in the left stage gives a sequence identical to the ordering of the original deck with $s = k = 4$. Reading the 16 line labels from top to bottom along the interior edge of the four switches in the right stage gives a sequence identical to the reordered deck. This interstage wiring pattern is called the perfect-shuffle pattern and will appear repeatedly throughout this chapter and the rest of the book.

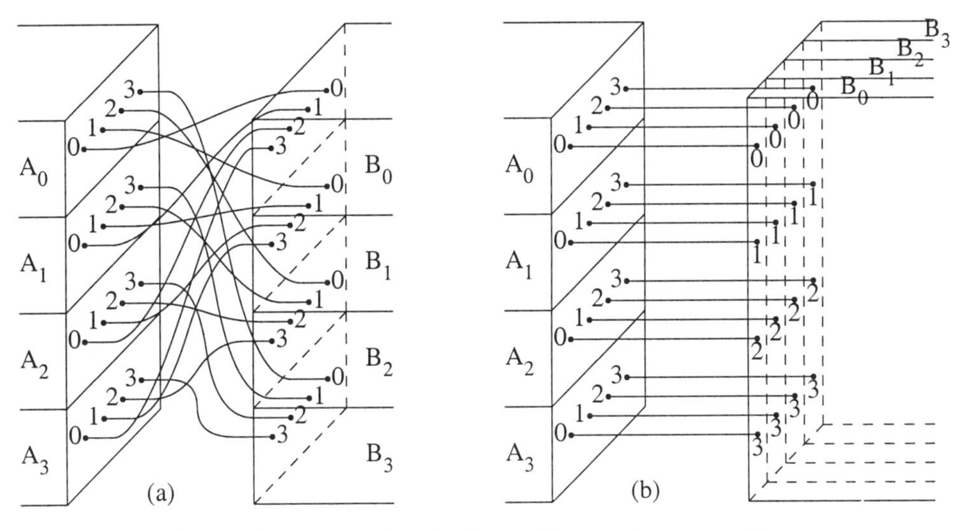

Figure 7.6. (a) Perfect shuffle and (b) matrix transposition.

Figure 7.5 is redrawn in three dimensions in Figure 7.6(a). The four switches in each stage are stacked with switch 0 on top and switch 3 on the bottom. In the left stack the four ports on each switch are labeled 0 – 3 from front to back, but in the right stack the four ports on each switch are labeled 0 – 3 from back to front. The perfect-shuffle pattern is a tangled web of spaghetti between the two stacks. If the rightmost stack is rotated 90 degrees so that the top comes out of the page, then all the wires line up as shown in Figure 7.6(b).

7.2.8 Three-Stage Fabrics

A two-stage concentrator/expander full-access fabric was shown in Figure 7.4(a) and described in Section 7.2.4. It is an improvement over the one-stage matrix fabric because (1) it uses fewer crosspoints with higher occupancy and (2) it provides alternate paths for fault tolerant operation. Its minor drawbacks are that (1) it may occasionally block and (2) the control is slightly more complex. Adding a third stage to the center of the fabric, providing distribution between the concentrator and the expander in Figure 7.4(a), provides no benefit.

However, when switch access is limited, the two-stage concentrator/expander fabric breaks vertically, as shown in Figure 7.5 and described in Section 7.2.6. The junctors that exit each concentrator are partitioned into limited-access groups, each group serving a different expander. Economy of scale increases the total number of junctors. Adding a stage of distribution to the center of this fabric, even using limited-access switches, has the effect of combining the junctor groups and negating this need for more junctors because of so many groups of junctors with limited access.

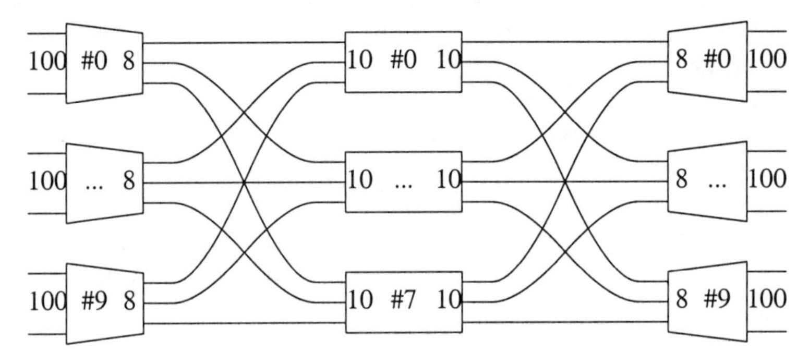

Figure 7.7. A fabric with three stages.

Figure 7.7 shows a three-stage concentrator/distributor/expander fabric built from limited-access switches. With 100 ports on the outer edge of each concentrator and expander, these stages require 10 switches to provide access to the 1000 ports on each side of the fabric. Let the junctors between the concentration stage and the distribution stage be called a-links and the junctors between the distribution stage and the expansion stage be called b-links. Because of the distribution stage, any a-link may reach any switch in the expansion stage, and any b-link may be reached from any switch in the concentration stage. Thus, the a-links and b-links are not partitioned.

Assume each subscriber generates 0.025 Erlangs of traffic intensity. The total traffic intensity through one limited-access switch in the first stage is $100 \times 0.025 = 2.5$ Erlangs. Figure 5.4 shows that eight servers (junctors) provide $P_b = 0.005$, at the concentration stage and the expansion stage. Blocking probability in the overall fabric may be slightly higher (Section 7.4). There are $10 \times 8 = 80$ a-links and 80 b-links. In the previous subsection, we saw that a two-stage fabric needs 300 servers if switch access is limited to 100. Adding the distribution stage to the middle of such a fabric greatly reduces the total number of junctors. Since a full-access two-stage fabric (Figure 7.4(a)) only needed 40 servers, using limited-access switches requires more junctors, even with a center stage of distribution.

The distribution stage is shown in Figure 7.7 with eight limited-access switches, each with size 10×10. A distribution stage containing 10 8×8 switches or 5 16×16 switches would also terminate the 80 a-links and 80 b-links but would not support the perfect-shuffle pattern at the a-links and at the b-links. The total count of crosspoints is

$100 \times 8 \times 20 + 10 \times 10 \times 8 = 16{,}800$; fewer than the count for any two-stage fabric (and far fewer than the count for a one-stage fabric). Although this three-stage fabric has more switching modules, they are smaller because there are fewer junctors.

In the full-access two-stage fabric of Figure 7.4(a), any of the 40 junctors can complete any call, and so, the number of alternate paths is 40. In the limited-access two-stage fabric, with 100×30 concentrator and expander switches, a call between any port on concentrator j and any port on expander p can use any of the three junctors that are wired between switches j and p. In a three-stage fabric with a perfect-shuffle pattern, consider a call between any port on concentrator j and any port on expander p. If distribution switch g is arbitrarily selected, there is only one a-link between switches j and g and only one b-link between switches g and p. The selected path requires these two particular junctors and a connection in switch j between the left port and the a-link, a connection in switch g between the a-link and the b-link, and a connection in switch p between the b-link and the right port. But, there are eight such paths, one through each of the eight distribution switches. A fabric with G distribution switches in its center stage has G alternate paths for any call.

In two-stage fabrics, probability of blocking is a simple single-server calculation using the charts in Figures 5.4 – 5.6 from Section 5.2. A more theoretical discussion of three-stage fabrics, in Section 7.3.2, will show that fabrics can be designed to be nonblocking, rearrangeably nonblocking, and blocking depending on some simple relationships among the numbers of inputs and outputs on individual switches. In three-stage fabrics that may block, the P_b calculation is complicated because each of the G alternate paths requires the availability of two specific links. Probability of blocking in multistage fabrics is discussed in Section 7.4. In three-stage fabrics, as in two-stage fabrics, the large number of alternate routes also implies a high degree of fault tolerance. The failure of any crosspoint does not preclude any connection, it simply degrades the blocking performance, if the path selection algorithm can avoid the faulty crosspoint.

The fabric of Figure 7.7 has $10 \times 10 \times 8 = 800$ more crosspoints than the corresponding two-stage fabric made with 100×8 switches, an increase from 16,000 to 16,800 crosspoints, or only 5%. In general, a symmetric three-stage fabric with r $n \times m$ concentrators, r $m \times n$ expanders, and m $r \times r$ distributors, has $mr(2n + r)$ total crosspoints. The corresponding symmetric two-stage fabric has $2nmr$ total crosspoints.

Let the three stages be labeled X, Y, and Z. Input $[i,j]$ is connected to level i on switch X_j; $0 \le i \le 99$ and $0 \le j \le 9$. Output $[o,p]$ is connected to level o on switch Z_p; $0 \le o \le 99$ and $0 \le p \le 9$. To interconnect input $[i,j]$ to output $[o,p]$, select path g, $0 \le g \le 7$, one through each switch in the middle stage Y. Let Y_g be a middle switch in which the junctor between X_j and Y_g is idle and the junctor between Y_g and Z_p is also idle; $0 \le g \le 7$. If no such Y_g satisfies both requirements, the call is blocked. It may be possible to move existing calls to alternate links in order to free up available links for this new call. If path g is available, the control algorithm must close crosspoint contacts $[i,g]$ in switch X_j, $[j,p]$ in switch Y_g, and $[g,o]$ in switch Z_p.

7.2.9 Four-Stage Fabrics

There would be no advantage gained if another stage of distribution were added to the three-stage fabric of Figure 7.7. However, Figure 7.8 shows a four-stage fabric with yet another new twist. Consider building the 1000×1000 fabric with switches whose access

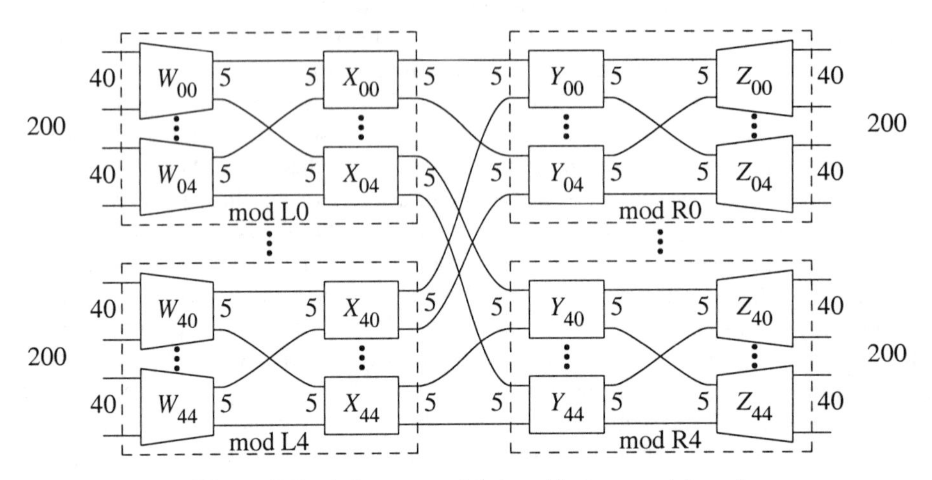

Figure 7.8. A four-stage fabric with a poor wiring plan.

is limited to 40 ports. At 0.025 Erlangs per port, the total traffic intensity in a switch is $0.025 \times 40 = 1.0$ Erlangs. At $P_b = 0.005$, Figure 5.4 shows that five servers are required. Thus, each switch in the concentration stage at the left and in the expansion stage at the right has 40 ports for users on its outer edge and five ports for links on its interior edge. To serve 1000 users on each edge of the fabric, 25 such switches are needed in the left and right stages. If the a-links, between the two leftmost stages, and the c-links, between the two rightmost stages, are wired in a perfect-shuffle pattern, then each of the two middle stages would require five switches, where each switch is 25×25.

Instead, consider a different wiring pattern in which the perfect shuffle at the a-links and c-links is partitioned. Consider, first, partitioning each of the four stages of the fabric into five segments. Each segment in the concentration stage and in the expansion stage contains five switches each, where each switch is 40×5. Let the a-links, between the two leftmost stages of the fabric, and the c-links, between the two rightmost stages, be confined to their respective segments. Let a subfabric module be comprised of one segment of an outer stage, the corresponding segment from the next inner stage, and the perfect shuffle pattern that wires these two segments together. Then, the perfect shuffle pattern within a module requires that the interior stage contain five switches in the segment, where each switch is 5×5. The entire fabric consists of five modules on the left and five modules on the right. The five modules on any one side are all identical and modules on the right are the mirror image of modules on the left. The manufacturing advantage, and corresponding cost savings, in constructing intermediate modules is significant.

The modules are interwired at the b-links, between the two interior stages. This is the only place in the fabric where full access occurs. Let the four stages be called W, X, Y, and Z and let X_{kj} be switch j in module k in stage X. In the wiring pattern shown in Figure 7.8, the b-link that is wired at its left end to interior port i on switch X_{kj} in module k is wired at its right end to interior port k on switch Y_{ji} in module j. Other interwiring patterns are possible and may even be better; see Exercise 7.4. This example has been carefully selected so that there are the same numbers of modules in the left and right sides of the fabric, switches in the left and right sides of each module, and ports on the interior edges of all switches; namely, five. Thus, at least numerically, the b-link wiring pattern is totally flexible. In general, the dimensions in the fabric may restrict the wiring rules.

Let port [*i,j,k*] on the left side of the fabric be wired to exterior port *i* on switch W_{kj} in left module *k*, and let port [*o,p,q*] on the right side of the fabric be wired to exterior port *o* on switch Z_{qp} in right module *q*. A connection between [*i,j,k*] and [*o,p,q*] has five alternate paths, depending on the choice of b-link between module *k* on the left and module *q* on the right. In this particular b-link wiring plan, the left ends of these five b-links terminate on the interior edge of switch X_{kq} and their right ends terminate on interior port *k* on switch Y_{qg}; for $0 \le g \le 4$. Select b-link number *g*. Four crosspoints must be operated: crosspoint [*i,q*] in switch W_{kj}, crosspoint [*j,g*] in switch X_{kq}, crosspoint [*k,p*] in switch Y_{qg}, and crosspoint [*g,o*] in switch Z_{qp}. The reader who has carefully followed this has undoubtedly noticed a remarkable increase in the complexity of the control algorithm.

This wiring plan, however, has an extremely subtle and extremely serious drawback. While each alternate path uses a different *Y*-switch in module *q*, they all use the same *X*-switch in module *k*, namely X_{kq}. In fact, all five alternate paths use the same a-link, the only a-link between switch W_{kj} and switch X_{kq}. A connection between any port on switch W_{kj} and any port on module *q* must use this same a-link. If the example connection is established, none of the other 39 ports on W_{kj} can be connected to any of the other 199 ports on module *q* because the required a-link is in use. This fabric blocking is not so much caused by concentration or by server allocation as by a poor wiring plan.

The total number of contacts in this fabric is $40 \times 5 \times 50 + 5 \times 5 \times 50 = 12{,}250$, a further reduction from any of the three stage fabrics with this same number of ports. The total number of junctors is $5 \times 5 \times 5 = 125$ between each successive stage, for 375 altogether. We see that there is economical advantage in making a fabric from a large number of small switches. Control complexity is much higher than in fabrics with fewer stages. The number of alternate paths is determined by the segmentation. Probability of blocking is strongly determined by wiring plan.

7.3 NONBLOCKING FABRICS

Fabrics are categorized by a hierarchy of "blockingness" that has four levels. Two important general structures, the Clos and Benes architectures, are discussed. The Clos architecture is extended to operate simultaneously in the space and time divisions in Section 13.3. Both the Clos and Benes structures are revisited when we discuss photonic switching in Chapters 20 and 21.

7.3.1 Blocking Hierarchy

A fabric architecture may, in the interest of cost, be so "lean" that calls may occasionally block or it may be so "rich" that calls never block. The fabric architecture may be just rich enough that "nonblockingness" is effected by moving calls, that are already connected in the fabric, to alternate paths in order to free up a path for some new call. Or, the fabric might be so rich that any new call can always be connected without rearranging already existing connections. In this latter case, the fabric architecture may be just rich enough that nonblockingness is effected as long as paths are selected by some algorithm. Or, the fabric might be so rich that any new call can always be connected, regardless of how paths are selected. This discussion leads to the classical hierarchy of fabric blockingness.

- *Wide-sense nonblocking.* A fabric is nonblocking in the wide sense if there is a path between any two idle ports for any existing configuration of connections in the fabric and no matter how paths were selected for these existing connections.
- *Strict-sense nonblocking.* A fabric is nonblocking in the strict sense if the existence of such a path depends on using some given path selection algorithm.
- *Rearrangeably nonblocking.* A fabric is rearrangeably nonblocking if an arbitrary idle party on the left can always connect to an arbitrary idle party on the right. However, it may be necessary to move existing calls to alternate paths.
- *Blocking.* A fabric is blocking if a path cannot be guaranteed.

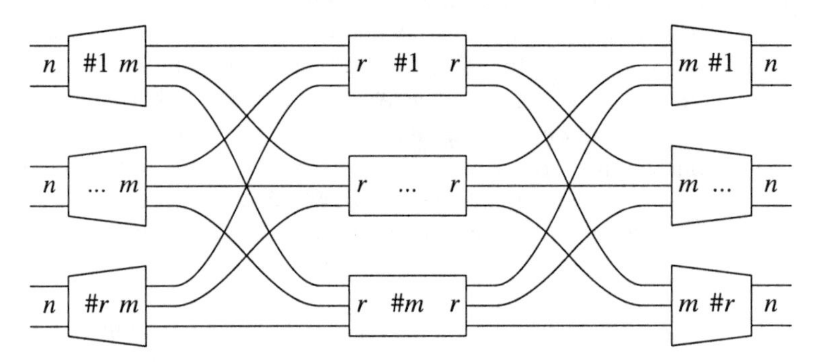

Figure 7.9. Generalized Clos architecture.

7.3.2 The Clos Architecture

The Clos (pronounced "Close," with the hard s, as in "That was close.") architecture is a generalized three-stage fabric with a perfect shuffle interwiring the stages [1]. Figure 7.9 is a generalization of Figure 7.7, but with the concentration and expansion reversed in the left and right stages. Assume the fabric is symmetric, with equal numbers of input and output ports. In Figure 7.9, the fabric has r switches in the left and right stages and m switches in the center stage. Each switch in the left and right stages has n outer ports and m inner ports. So the total number of ports on each edge of the overall fabric is $N = r \times n$. Each switch in the center stage is $r \times r$. There are m paths between any input and any output, one path through each switch in the center stage. With $2r$ switches that are each $n \times m$ and m switches that are each $r \times r$, the total number of contacts is $mr(2n + r)$. Two classical theorems have been proven about Clos fabrics [2, 3].

- If $m \geq 2n - 1$, then a Clos fabric is wide-sense nonblocking. Each switch in the left and right stages must be $n \times (2n - 1)$, at least. In each of these switches, the number of internal ports is almost twice the number of external ports; a very rich fabric. But, fortunately, the condition is not a necessary condition; it only guarantees that the fabric will be nonblocking, and in the wide sense. There are three-stage nonblocking fabrics that don't meet this inequality.

- If $m \geq n$, then a Clos fabric is rearrangeably nonblocking. Each switch in the left and right stages must be $n \times n$ square switches, at least. This condition is a necessary condition, because if there were fewer than n internal ports on any outer switch, then this switch is a concentrator and a port on the outer edge could be blocked from reaching any of the m ports on the inner edge.

Historically, this theory has had very little practical application. Wide-sense nonblocking Clos fabrics are too rich and inefficient to be used in practical systems. In large systems, the low occupancy is too expensive, and in small systems three-stage fabrics are not necessary. Rearrangeably nonblocking Clos fabrics are potentially practical, but the complexity of path selection has been prohibitive. Virtually nonblocking fabrics, described in Section 7.4.3, are much less expensive and practically as good.

7.3.3 Rearrangement

Rearrangeably nonblocking fabrics generally have higher occupancy and are more economical than fabrics that are nonblocking in the wide sense or in the strict sense. These advantages have two corresponding penalties: physically moving a call to another path through a fabric can cause a glitch in the data carried, and rearrangement greatly complicates the algorithm that controls the fabric.

Consider moving a call from one path to another in a space-only fabric. Three temporal procedures are *break/make, make/break,* and *simultaneous.* In the break/make procedure, the initial path is disconnected first and then the call is reconnected through the alternate path. Depending on the fabric's technology, the interval with no connection can last between several nanoseconds and several milliseconds. In the make/break procedure, the alternate path is connected before the initial path is disconnected. Instead of an interval with no connection, this procedure causes an interval with two parallel connections.

Depending on the technology, notably in the new photonic technologies, such parallel paths cause transmission mismatches similar to those caused by bridged taps in the loop plant or interference in coherent light. Furthermore, this procedure may not always work. For example, the addition of call A to a fabric might require the rearrangement of calls B and C. If the alternate path for call B uses a crosspoint currently used by call C and the alternate path for call C uses a crosspoint currently used by call B, then neither call B nor call C can be moved first. In either procedure, if the data transmission is digital, between several bits and several million bits can be lost, depending on the data rate. If the data transmission is analog voice, a "click" might be heard.

The solution, of course, is instantaneous simultaneous rearrangement of all calls that must be moved to alternate paths. This is, of course, physically impossible in a space-only fabric. But in a digital system, if the device speed of the crosspoints is small compared to the bit interval, then simultaneous rearrangement is virtually instantaneous. If the data transmission on all lines is RZ and synchronous, then operation could be enhanced by precisely timing the rearrangement to occur between two successive bits or frames or during some intentionally introduced *guard bands* in the transmission format.

The control algorithm for a fabric or subfabric is responsible for selecting a path for a new call given the configuration of existing connections. If the algorithm is controlling a rearrangeably nonblocking fabric, the bookkeeping is further complicated because some calls must be disconnected and reconnected before new calls are connected. The added bookkeeping complexity is not the hard part of controlling a rearrangeably nonblocking fabric. It is difficult enough to design an algorithm that knows which path to select. It is extremely difficult to design an algorithm that knows which calls to move first. An algorithm is described in Section 7.3.6 in which all existing calls are moved every time a new call is added. This is a largely open area of research and seems particularly amenable to artificial intelligence theory.

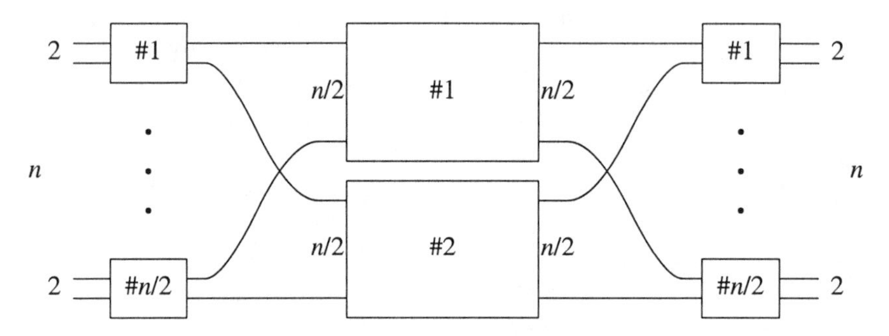

Figure 7.10. A Benes fabric.

7.3.4 The Benes Architecture

The Benes (pronounced "Be'nesh") architecture is a special case of the Clos architecture [4 – 7]. It has $m = n = 2$, $r = n/2$, a recursive construction algorithm, and is made uniformly from many identical switching elements. Because $m = n$, Benes fabrics are rearrangeably nonblocking.

The fundamental element is a nonblocking 2×2 point-to-point switch. This switch, called a beta element, has only two configurations [8]. In the BAR state, left port 0 is connected to right port 0 and left port 1 is connected to right port 1. In the CROSS state, left port 0 is connected to right port 1 and left port 1 is connected to right port 0. The beta element's state is controlled by one binary bit, with value logical 0 for the BAR state and logical 1 for the CROSS state.

The basis of the recursive construction is a 2×2 fabric. The 2×2 Benes fabric has a single stage and this stage has a single beta element. The fabric's two inputs are wired to the beta element's two left ports and the fabric's two outputs are wired to the beta element's two right ports.

The inductive step of the recursive construction shows the construction of an $N \times N$ Benes fabric out of two $N/2 \times N/2$ Benes fabrics and additional beta elements. A three-stage Clos architecture is implemented with $n = m = 2$, $N = r \times n$, and $r = N/2$. Thus, the center stage has $m = 2$ switches that each have dimension $r \times r = N/2 \times N/2$ and each outer stage has $r = N/2$ switches that each have dimension $n \times m = 2 \times 2$. The stages are interwired in a perfect-shuffle pattern.

Consider first a 4×4 Benes fabric. It has $N/2 = 2$ beta elements in its left stage and two beta elements in its right stage. It has two subfabrics in its center stage, where each is 2×2. Thus, each subfabric in the center stage is a single beta element. The 4×4 Benes fabric has three stages, with two beta elements in each stage. The stages are interwired by a perfect-shuffle pattern. The four fabric inputs are wired to the two ports on the left edges of the two switches in the left stage. The four fabric outputs are wired to the two ports on the right edges of the two switches in the right stage.

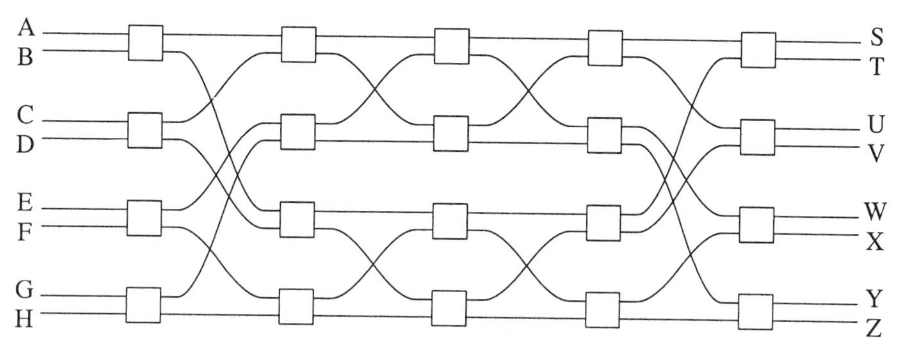

Figure 7.11. 8 × 8 Benes fabric.

Figure 7.11 shows an 8 × 8 Benes fabric. Column 1 holds the left stage containing four beta elements and column 5 holds the right stage containing four beta elements. Columns 2 – 4 hold the center stage containing two 4 × 4 Benes subfabrics. Each 4 × 4 subfabric was described in the previous paragraph. A 16 × 16 Benes fabric has two of the fabrics in Figure 7.11 in its center stage and has eight beta elements in each of its left and right stages.

In general, an $N \times N$ Benes fabric is a rectangular array of 2 × 2 switches The array has $N/2$ rows of switches and $2 \log_2 N - 1$ columns of switches. N is a power of 2 and the log is to the base 2. There are a total of $N \times \log_2 N - N/2$ beta elements.

I'm not aware of a single practical system ever deployed commercially that was architected around a Benes fabric. This is due partly to the complexity of the control algorithm (next subsection) and partly to the technologies that have been used to construct interconnection fabrics. There is no natural 2 × 2 switching device that is fabricated from electromechanical, solid-state, or digital logic technologies. So the Benes architecture has been an intellectual curiosity. However, photonic technology provides a natural 2 × 2 switching device, and the Benes architecture is receiving considerable attention by the photonic switching research community.

7.3.5 Control of Benes Fabrics

A classical algorithm that controls Benes fabrics rearranges all existing connections with each added new call and has quite high computational complexity [9 – 12]. Since the left and right stages of the Benes architecture each consist of $N/2$ 2 × 2 switches, every switch in these stages terminates two ports on the fabric. Let a port's *dual* be the other port on the outer edge of the fabric with which it shares a switch.

The center stage of the Benes fabric contains two subfabrics to which the 2 × 2 switches in the outer stages are interwired by perfect-shuffle pattern. Observe that the links on the interior edge of any switch in an exterior stage are wired to opposite interior subfabrics. Let a port and its dual both be actively connected, each on its own separate call. If one of the ports is connected through one of the interior subfabrics, then its dual must be connected through the opposite interior subfabric. This observation suggests a control algorithm for Benes fabrics.

Assume that the fabric is fully active; that every left port must be connected to some right port. If a port is idle, then arbitrarily assign it to an idle port on the opposite edge of the fabric. Start with an arbitrary port, i_0, on an arbitrary side of the fabric, the left. Now, i_0 must be connected to some port on the right side of the fabric, to $C(i_0) = o_0$. Complete this call by a path through an arbitrary interior subfabric, the upper one. Let $D(o_0) = o_1$ be the dual of the port to which i_0 is connected.

Now, o_1 must be connected to some port on the left side of the fabric, to $C(o_1) = i_1$. Complete this call by a path that must pass through the lower interior subfabric, the opposite interior subfabric from the one used by the connection between i_0 and o_0. Let $D(i_1) = i_2$ be the dual of the port to which o_1 is connected. Now, i_2 must be connected to some port on the right side of the fabric, to $C(i_2) = o_2$. Complete this call by a path that must pass through the upper interior subfabric, the opposite interior subfabric from the one used by the connection between i_1 and o_1. And so forth, until all the connections are made. Whenever a new call must be added, it's possible that the entire configuration might be rearranged.

As described so far, the algorithm merely assigns a connection to the upper or lower interior subfabric. So, the algorithm would have to be repeated within each interior subfabric, at every layer of nesting. The time complexity of this algorithm has been shown to have order $T = KN \log_2 N$. Thus, the time complexity is between linear and quadratic as a function of N. This could be a problem in large fabrics. So, for example, quadrupling the number of ports on each edge of a Benes fabric causes an eight-times increase in the time to run the control algorithm. Further, since the entire algorithm must be run for each new call, this more complex algorithm must be run four times more frequently. Thus, total required computer time is increased 32 times. An eight-times increase in the number of ports causes a $24 \times 8 = 192$-times increase in real-time demand. A 16-times increase in ports causes a $64 \times 16 = 1024$-times increase in real time demand, and so forth.

While this classical algorithm could require that all calls be moved as a result of a newly added call, even a theoretically optimal algorithm could have a large number of rearrangements. The worst-case number of calls that might have to be moved to accommodate a newly added call has been shown to equal $N/2 - 1$ in a Benes fabric with N ports on each edge. While this original algorithm has been improved and alternate algorithms have been proposed, the development of algorithms for controlling rearrangeable fabrics is a largely open research area.

7.4 BLOCKING FABRICS

Before we begin a discussion of blocking within a fabric, we must distinguish two kinds of fabric blocking:

- *Access blocking.* This kind of blocking occurs because a network's ports are concentrated at the fabric's exterior stages of switching. While ports may block at the concentrators, the interior of the fabric may be nonblocking for those calls that are not blocked at the outer stages. Whether such a network is blocking or nonblocking with external concentration is moot.
- *Path blocking.* In this kind of blocking, calls can be blocked at any stage in a fabric, not just in the concentrators at the fabric's edges.

However we choose to define fabric blocking, nonblocking fabrics are very rich and have traditionally been too expensive to implement in real systems. Some of the most modern fabric architectures, the *permutation fabrics* used for self-routing packet switching (described in detail in the next subsection), have extremely poor blocking probabilities. So, while real systems currently have fabrics that do block, the economics of switching systems (Section 18.2.4) is changing so much that the "fabric component" has become a practically insignificant part of a system's total price. We may see nonblocking fabrics — or, at least, fabrics that exhibit access blocking only — in future systems.

Accurate mathematics for computing the probability of blocking, especially path blocking, in a multistage fabric is very complex. Even a cursory discussion is a very advanced topic. An appreciation for the complexity is provided in this section. In the first subsection, three simplifying assumptions are described that, when applied, make the calculation tractable, at least. However, these assumptions are not completely realistic. The second subsection summarizes a more accurate, but much more complicated, approach to finding probability of blocking. This section closes with a discussion of fabrics that, while not nonblocking, are called "virtually nonblocking."

7.4.1 An Overly Simplified Approach

Three assumptions will be made to simplify the computation of the probability of blocking in a multistage fabric. While not completely ridiculous, these assumptions can be unrealistic.

- All paths are equilikely. Let there be G different paths through a fabric to interconnect some port on the left with some port on the right. The path selection algorithm is assumed to be equally likely to select any of the paths. Then, the probability that a given path is selected is $P_p = 1/G$. However, these algorithms are not designed to locate all paths and then randomly select one. Instead, they proceed serially through some ordered list of paths and stop when they find the first one that works. The actual probability of selecting a given path depends on where this path resides in the list.

- The various stages have independent blocking statistics. A path through a multistage fabric requires one link between every adjacent stage of switching. In a fabric with L stages, any path requires $L - 1$ links. It is assumed that the probabilities of blocking for a link at stage X and for a link at stage Y are independent. However, suppose the calls through a fabric are distributed such that the left sides of the calls are concentrated in switch X_2 in the left stage of a three-stage fabric, and the right sides of the calls are concentrated in switch Z_{14} in the right stage. Now, consider a new call whose left port is also on switch X_2, but whose right port is not on switch Z_{14}. Obviously, fewer a-links will be available for this call than b-links. The statistics at various stages are not independent.

- Links in the same stage have independent blocking statistics. The availability of all the links in the same stage of switching are all equal and independent of one another. However, obviously, the probabilities that a-link g is busy and that h is busy are neither equal nor mutually independent.

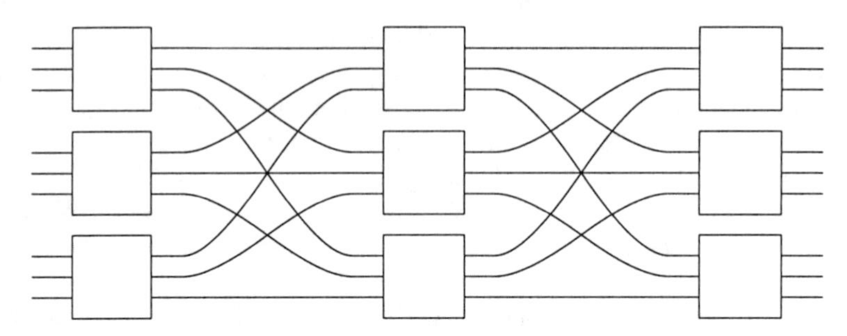

Figure 7.12. Illustrating probability of blocking in a fabric.

The effect of these three assumptions on a computation is illustrated on a small three-stage fabric. Figure 7.12 shows a 9×9 three-stage fabric. It has three switches in each stage and all switches are 3×3. The stages are interwired in a perfect-shuffle pattern. Since $n = m$ on the switches in the outer stages, the fabric is rearrangeably nonblocking. Thus, P_b computed below is actually the probability that no path is immediately available or the probability that at least one existing call will have to be moved. A general formula for probability of fabric blocking is derived for this extremely simple fabric.

$$\begin{aligned} P_b(\text{fabric}) &= P_b(\text{path1}) \times P_b(\text{path2}) \times P_b(\text{path3}) \\ &= P_b(\text{path})^3 \\ &= [1 - P_c(\text{path})]^3 \\ &= [1 - P_c(a - \text{link}) \times P_c(b - \text{link})]^3 \\ &= [1 - P_c(\text{link})^2]^3 \\ &= [1 - [1 - P_b(\text{link})]^2]^3 \end{aligned}$$

Since there are three paths between any pair of ports, the probability of fabric blocking is the probability of being blocked on all three paths. By the first assumption, this equals the cube of the probability of being blocked on any one path. But, the probability of blocking on a path plus the probability of connection on the same path must equal 1. Since a given path requires a specific a-link and a specific b-link, the probability of connection on a path equals the product of the probabilities of connection on each link. By the second assumption, this product is simply the square of the probability of connection on either link. By the third assumption, all links in any stage are equivalent. But, the probability of connection on a link plus the probability of blocking on the same path must equal 1.

Assume that each left port originates 0.333 Erlangs of traffic intensity. With three left ports on the left edge of each switch in the left stage, each switch generates 1 Erlang. The probability of being blocked at an a-link is the probability of blocking with 1 Erlang and three servers, and $P_b(\text{link}) = 0.08$ from Figure 5.4. So,

$$\begin{aligned} P_c(\text{link}) &= 1 - 0.08 = 0.92 \\ P_c(\text{path}) &= [P_c(\text{link})]^2 = 0.846 \\ P_b(\text{path}) &= 1 - 0.846 = 0.154 \\ P_b(\text{fabric}) &= P_b(\text{path})^3 = 0.036 \end{aligned}$$

The three assumptions may be unrealistic, but only slightly so. Furthermore, it has been argued that the assumptions somehow cancel each other's effect. The resulting figures typically are not all that unreasonable. The reader can imagine trying to derive a more accurate formula for probability of fabric blocking, even on a "toy" example such as this one. The formula above, derived on the basis of the three assumptions, is easily extended to fabrics with more paths and more stages.

Finally, it should be observed that the analysis of packet-switched networks is based on equivalent assumptions. Most of the analytical methods assume that successive packets in a network are statistically independent — in time and in destination. But, obviously, if I send you a packet, then: (1) I am more likely than average to send you another one, (2) you are more likely than average to send me a packet, and (3) we are each more likely to send our subsequent packets sooner than we would on average. Furthermore, if packet traffic is heavier than average in one part of a packet-switched network, like the Internet, then it is likely to be heavier than average in other parts of this network. None of these packet dependencies is typically considered when packet-switched networks are analyzed. The assumption that packets are statistically independent is analogous to the assumption that fabric links are independent — and the assumption is at least equally unrealistic.

7.4.2 Virtually Nonblocking Fabrics

While nonblocking fabrics may be expensive, especially in large sizes, a fabric could contain so many links that the probability of blocking is extremely low. A fabric is called virtually nonblocking if its probability of blocking is so low that a block is unlikely to occur within the lifetime of the product or, even, its users.

Consider a fabric that terminates 1000 subscribers and let each subscriber originate one 3-minute call every 2 hours, on the average. The traffic intensity is 3/120 = 0.025 Erlangs per user and 25 Erlangs for the entire fabric. But, considering originations only, since each user originates 0.5 calls per hour, the fabric handles 500 originations per hour, or $500 \times 24 \times 365.24 = 4.4$ million originations per year.

Consider a fabric that is not rich enough to be nonblocking, but is rich enough in equivalent servers that the probability of blocking is extremely small. Since the average number of blocked calls in some interval T is given by $\bar{N}_b(T)$ = (the total number of calls during T) $\times$ (P_b of each call), setting $\bar{N}_b$ to one blocked call per year, we find that $P_b = 1/4.4\text{M} = 2.3 \times 10^{-7}$ Thus, if the fabric is rich enough that its probability of blocking is 2.3×10^{-7}, then it blocks one call per year on the average. While this fabric is not nonblocking, it is certainly nonblocking for all practical purposes and we call it virtually nonblocking. Many commercial systems, that are advertised as nonblocking, are actually only virtually nonblocking.

In describing highly reliable equipment, the probability of failure is typically so low that the figure is unintuitive and unappreciated. The statistics are typically presented in terms of *mean time between failures* instead of *probability of failure.* Similarly, with virtually nonblocking fabrics, the probability of blocking is so low that the statistics can be presented in terms of mean time between blocks instead. For this example, the mean time between blocked calls, $\bar{T}_b$, is 1 year.

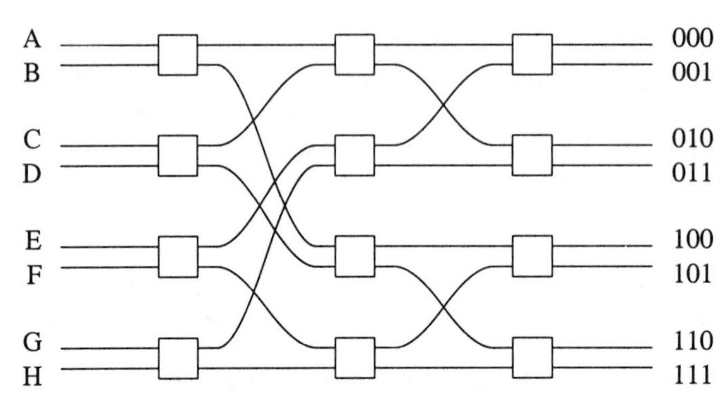

Figure 7.13. The Banyan network.

7.5 SELF-ROUTED PACKET FABRICS

In self-routed packet-switched networks, a packet is steered to its proper output port by the hardware of some specially architected switching fabric [13 – 17]. This section describes one class of switching fabrics, the Banyan, which can be used this way.

7.5.1 Banyan Networks

Banyan networks have the fewest stages of switching, while still providing full input-to-output connectivity, of any switching fabric architecture. While Banyans are much more relevant to packet switching than to circuit switching, this section introduces Banyans in a circuit-switched context. Then, after discussing their name and presenting a mathematical description, a Banyan's ability to perform self-directed fast packet switching is described in the fourth and fifth subsections.

If a Banyan network has N ports on the left side and N ports on the right side, and the switching elements are all $p \times p$ switches, then the number of stages, K, is given by:

$$K = \log_p N^+$$

where N^+ is the smallest power of 2 that is $\geq N$. Figure 7.13 shows an 8×8 Banyan network, built using 2×2 switches. It has three stages of switching. The K-stage Banyan is constructed from two $(K - 1)$-stage Banyans, plus a common first stage, similar to the recursive construction of the Benes network, described in Section 7.3.4. So, all Banyan networks have a congruent wiring plan — with perfect-shuffle wiring between the first two stages of each nested interior subnetwork.

Since the Banyan fabric provides only one path from a given left port to a given right port, the Banyan is not a good architecture for circuit switching. But it's even worse than this. As a circuit switch, the Banyan is a blocking fabric — and it doesn't even have concentration in the first stage. Worse, the Banyan network blocks in every stage because of its wiring pattern. For example, in Figure 7.13, if left port A needs a connection to right port 000, the only path is the one the proceeds straight across the top of the fabric. The upper switch in each stage must be put in its BAR state. Having made this connection, now try to connect left port B to right port 001. In fact, left port B can't be connected to right ports 001 – 011 and left ports C – D can't be connected to right port 001 either.

The 8 × 8 Benes network (Figure 7.11) has $2K - 1 = 5$ stages and the 8 × 8 Banyan network in Figure 7.13 has $K = 3$ stages. Comparing the structures of the two architectures, the first K stages or the last K stages of any Benes network is a Banyan network. In the class of switching networks built from 2 × 2 switches, the K stages of the $N \times N$ Banyan network is the minimum for full access (so every input can reach any output), but the $2K - 1$ stages of the Benes network is the minimum to be rearrangeably nonblocking. It is the additional stages in the Benes Network that provide the additional network paths and improved blocking characteristic.

7.5.2 The Name

Common to tropical climates worldwide, banyan trees are unusual because they grow secondary trunks from their branches back down to the ground. If you turn Figure 7.13 sideways and compare it to a banyan tree, you can see how this network architecture got its name.

Unfortunately, this term "Banyan" has also been used to name a general family of network architectures. The family has three members — called the baseline, the omega, and the banyan. All three members of the Banyan family have the same number of stages and number of switches, only their interconnection wiring is slightly different. For example, the left and right half of a Benes network is each a species of the Banyan genus, one side is a baseline network and one side is the banyan network. Any species with K stages can be easily transformed into another species with K stages by simply moving switches around logically.

7.5.3 Permutations

Let the elements of a set be listed in some order. If a set has N elements, then any list of these elements must have N positions. For example, two of the six different orderings of $S = \{a, b, c\}$ are $L_1 = < abc >$ and $L_2 = < bac >$. A *permutation* is a reordering of a list [18,19]. The algebraic expression $L_1 \times P = L_2$ signifies that permutation P reorders L_1 into L_2.

Permutations are conveniently represented with the notation of *permutation cycles.* A permutation is a set of cycles where: (1) a cycle is a closed subset of the positions in a list, (2) each element in a cycle is permuted to the element that follows it, and (3) the last element is permuted to the first (hence, a cycle). For example, the cycle $(ijk...mn)$ means that the element in position i in the original list moves to position j in the new list, the element in position j in the original list moves to position k in the new list,..., the element in position m in the original list moves to position n in the new list, and the element in position n in the original list moves to position i in the new list. The cycle (i) means that the element in position i in the original list stays in position i in the new list. Every list position appears in exactly one cycle in the set of cycles that represents a permutation.

Permuting $L_1 = < abc >$ into $L_2 = < bac >$ requires that the first element moves to the second position, the second element moves to the first position, and the third element stays in its original position. The set of permutation cycles that represents this permutation is $\{(12)(3)\}$. The permutation $\{(1)(2)...(N)\}$ is the *identity permutation,* under which every element stays in its original position.

As the first of two examples using Figure 7.13, consider the left column of beta switches. The ports on their left edges have been labeled <ABCDEFG> from top to bottom. Suppose all four switches in the left column are put in the CROSS state. Let the labels of the ports on the right edges of these four switches be determined by the label of the port on the left edge to which each right port is connected through its respective switch. Thus, this list of right port labels is a permutation of the list of left port labels. The permutation performed by this first column of switches, if all four are in the CROSS state, is represented by {(12)(34)(56)(78)}. Then, reading the labels of these right ports from top to bottom gives <BADCFEHG>. Not only does a column of switches permute the port labels, but the link wiring does also. For example, the perfect-shuffle wiring in the a-links between the first two stages causes a permutation represented by {(1)(253)(467)(8)}.

As a second example using Figure 7.13, if all four switches in the left column are in the BAR state, then the permutation performed by this column is the identity and the eight interior ports are labeled <ABCDEFG>. Now, consider the eight ports on the left edges of the four switches in the middle column. If they are labeled according to the label of the port input to which they are attached, then the port labels from top to bottom are <ACEGBDFH>. Note that permutation performed by columns of switches is variable, depending on the states of the switches. Permutations performed by link wiring are fixed. The b-links are described by {(1)(23)(4)(5)(67)(8)}. Consider configuring the fabric of Figure 7.13 such that all 12 switches are in the BAR state. Then the permutation from fabric inputs to fabric outputs is represented by {(1)(25)(3)(47)(6)(8)} and the fabric inputs are reordered at the fabric outputs as <AECGBFDH>.

7.5.4 Packet Switching in a Banyan Network

Now, consider packet switching, instead of circuit switching, in a Banyan. A packet is walked through the three-stage Banyan network of Figure 7.13. Let the switches operate on individual bits analogous to how a Strowger switch operates on dialed digits. Recall the operation of Step-by-Step during call setup: a Strowger switch absorbs the first digit dialed into it, uses that digit to select an output, and then it lets all subsequent digits pass through to this output. In the 2×2 switches of Figure 7.13, the switch uses the first bit received from either input port to select one of the two output ports. This first bit is not transmitted through to this selected output, but all subsequent bits are. On Figure 7.13's right side, number the eight output ports 0 – 7 top-down. Consider a packet entering from the left side into any of the eight input ports — arbitrarily select input port 3, the lower port on switch 1 in the leftmost column. Suppose the first 3 bits in the packet's header are 101.

This first switch absorbs this leading 1 and uses it to select its own output 1, its lower output port. The remaining bits in the header, 01, and all subsequent bits in the packet pass through this switch, from its lower input port to this lower output port. From the network's wiring diagram, these bits arrive at the upper input port on switch 2 in the center column. This second switch absorbs the leading 0 and uses it to select its own output 0, its upper output port. The remaining bits in the header, 1, and all subsequent bits in the packet pass through this switch, from its upper input port to this upper output port. From the network's wiring diagram, these bits arrive at the upper input port on switch 2 in the rightmost column. This third switch absorbs the leading 1 and uses it to select its own output 1, its lower output port. While no bits remain in the header, all subsequent bits in the packet pass through this switch, from its upper input port to its lower output port.

So, the packet, with its header stripped off, arrives at the right edge of the overall network, at the network's output port 5. This example generalizes to Banyans of any size, to packets that enter on any input port, and to packets with any binary code in their header.

> *Let a packet enter an* N × N *Banyan network on any input port. If this packet's header contains the* K*-bit binary code for* n, *the packet exits the network on output port* n.

This Banyan fabric is just one of many architectures for self-routing packet switches.

7.5.5 Collision Control

Simultaneous packets, that enter the Banyan on different ports, will "collide" if they have the same destination port. Furthermore, since the Banyan fabric is not nonblocking, packets with different destinations might still collide inside the fabric. Two paradigms for collision control within the Banyan fabric are *queuing* and *deflection*. Queuing internal to the Banyan fabric is described first and deflection routing is discussed briefly later.

Queues Inside the Banyan. Earlier in this section, we walked a packet through a Banyan. That packet entered the Banyan on port 3 and exited on port 5. Suppose a second packet entered the fabric during the time that the first packet, described above, was transmitting through. Consider several different cases.

- Let this second packet have 011 in its header and let it enter the fabric on port 7. The two packets would direct themselves through the fabric simultaneously, without interfering with each other, and each would arrive at its respective output port.
- Let this second packet have 011 in its header and let it enter the fabric on port 2. Even though both packets arrive at switch 1 in the leftmost column (over different input ports), all the switches in the fabric could be designed to steer packets simultaneously, as long as they go to different output ports. The two packets would direct themselves through the fabric, simultaneously, and each would arrive at its respective output port.
- Let this second packet enter the fabric on input port 7, as in the first example, but let this packet have 101 in its header. Now, since the two packets in the fabric are both trying to reach output port 5, they will collide there. In this case, some kind of "gatekeeper" on the input side could examine the incoming packets and control the collision, somehow.
- Let this second packet enter the fabric on input port 7, as in the first and third examples, but let its header read 100. Now, there is no apparent collision — at least, not at the fabric's outputs. However, both packets will direct themselves to switch 2 and, once there, they collide at this switch's output because both packets must transmit to this switch's upper output port. In this case, the gatekeeper approach gets very sophisticated.

Packets can be queued at the fabric's input ports or internally, inside any of the switches [20]. Internal queuing requires buffers and queue control within every switch in the fabric, but input queuing requires an overall control that can determine that a collision will occur inside the fabric. Neither implementation is very good, especially regarding how they scale.

Deflection Inside the Banyan. Rather than providing for queues and buffers internal to the Banyan fabrics, congestion is more easily controlled in Banyans by using deflection. With collision control by queuing, if two packets must be steered to the same switch output port, only one is steered there and the other is queued. With collision control by deflection, if two packets must be steered to the same output port of the same switch, only one is steered there and the other steered to the other port (the wrong port). The fabric can be augmented to correct the misdirection, or the larger network, in which the Banyan is just one node, can be relied on to steer the packet correctly.

Several researchers have proposed improving the blocking performance of circuit-switched fabrics by cascading more stages. For example, while the three-stage 4×4 Benes fabric is rearrangeably nonblocking, adding a similar fourth stage makes the resulting four-stage fabric completely nonblocking. In one proposal, a similar technique improves the performance of any single self-directed $N \times N$ packet fabric, by connecting k similar $N \times N$ packet fabrics in cascade.

The jth destination node has k ports, wired to the jth output port on each Banyan. Since $j < N/2$, the extra ports at the output of each Banyan are connected to the input ports of the next successive Banyan. If a packet is deflected in one of the Banyans, it is routed to the next Banyan, which tries to steer it to its proper output, and so forth through each successive stage of Banyans until the packet is successfully delivered or the fabric runs out of stages. By making k big enough, any desired level of overall packet blocking can be achieved. The really nice feature of this architecture is that the owner can pay for any desired level of performance.

EXERCISES

7.1 Redraw the fabric in Figure 7.5 as a folded fabric. Let each limited-access switch be 250×32. Show how the right edges of the switches are interconnected by junctors.

7.2 A two-stage 1000×1000 fabric was discussed in Section 7.2.6. Three implementations were described, depending on the access limitations of the modules in each stage: (1) if each stage has one module that reaches 1000 ports, the resulting fabric has 80K crosspoints; (2) if each stage has four modules that each reach 250 ports, the resulting fabric has 48K crosspoints; and (3) if each stage has 10 modules that each reach 100 ports, the resulting fabric has 60K crosspoints. Consider fabrics in which the number of modules in each stage is r, for $r = 2, 3, 5, 6, 7, 8, 9$. The number of ports that are wired to the external edge of each module is $1000/r$. Use $P_b = 0.005$ to find the number of links in each link-group. (a) Compute the total number of crosspoints for these seven fabrics. (b) Construct a graph that plots the total number of crosspoints against r, the number of modules with uniformly limited access, for $1 \leq r \leq 10$. (c) What is the optimum number of modules?

7.3 Repeat the previous exercise for three-stage fabrics. Let the number of center-stage modules equal the number of modules in each edge-stage. Compute the number of links per link-group based on $P_b = 0.002$.

7.4 There are six regular patterns for wiring the b-links in the four-stage modular fabric of Figure 7.8. Consider the b-link whose left side terminates on level i of switch X_{kj} in left module k. Its right side can terminate on: (a) level i of switch Y_{kj} in right module k, (b) level i of switch Y_{jk} in right module j, (c) level j of switch Y_{ki} in right module k, (d) level j of switch Y_{ik} in right module i, (e) level k of switch Y_{ji} in right module j, or (f) level k of switch Y_{ij} in right module i. Pattern (e) is described in Section 7.2.9 and is seen to have congestion at the a-links. Patterns (a) and (c) have limited access because left module k has access only to right module k. Describe any problems with patterns (b), (d), and (f). Hint: draw a simple fabric for each case, with two modules per stage and two switches per stage, where each switch is 2×2. Is there a best pattern?

7.5 For patterns (b), (d), and (f) in the previous problem, tabulate the rules for switch operation for connecting left port $[i,j,k]$ to right port $[o,p,q]$.

7.6 Suppose each outer switch in Figure 7.8 terminates 56 ports on its exterior edge instead of 40. How many a-links or c-links terminate on the interior edges of each W switch or Z switch? If there are six W switches and Z switches in a module, how many modules are there? You should end up requiring that the X and Y stages are configured as six switches in a module, where each switch is 6×6. For the wiring of the b-links, between the X and Y switches, consider the six ports on the interior edge of each X and Y switch as three pairs of ports. Let the pair of b-links whose left ends are wired to ports $2i$ and $2i + 1$ on X_{kj} have their right ends wired to ports $2k$ and $2k + 1$ on Y_{ij}, where $0 \le i \le 2$, $0 \le j \le 5$, and $0 \le k \le 2$. How does this pattern compare to the patterns in Exercise 7.4?

7.7 A small three-stage switching fabric has three switches on each edge. Each switch has three ports on its exterior edge and five ports on its interior edge. (a) Add appropriately sized switches to the center stage to complete the fabric following the Clos architecture. (b) Show all interconnection wiring. (c) How many possible different paths exist between a left port and a right port? (d) What phrase describes this fabric's blocking characteristic?

7.8 Draw a Clos fabric that has $n = 2$, $m = 3$, and $r = 4$. What phrase describes its blocking characteristic?

7.9 Draw a Clos fabric that has $n = 4$, $m = 6$, and $r = 2$. What phrase describes its blocking characteristic?

7.10 Consider four alternate implementations of a 12×12 Clos fabric: (1) $n = 2$ and $r = 6$, (2) $n = 3$ and $r = 4$, (3) $n = 4$ and $r = 3$, and (4) $n = 6$ and $r = 2$. For wide-sense nonblocking, what is the minimum value for m in each case? Using this minimum value of m, calculate the total number of crosspoints for each case. Which is cheapest?

7.11 Repeat the preceding problem, but for a 12×12 Clos fabric that is rearrangeably nonblocking.

7.12 Suppose a new switching device is naturally 3×3. Generalize the Benes architecture to 3×3 switches and sketch a 27×27 switching fabric.

7.13 This exercise illustrates that Benes networks are not better than rearrangeably nonblocking. (a) Sketch a 4×4 Benes network. Label the left ports top to bottom as A, B, C, and D. Label the right ports top to bottom as W, X, Y, and Z. (b) With all ports idle, suppose A wants to talk to W. How many paths are available for this call? (c) Arbitrarily configure the three upper switches for this call. The problem gives the same result if you had selected the lower path. Show the configured switches and the A-to-W call on your sketch. (d) With this A-to-W call still connected, suppose D wants to talk to Z. How many paths are available for this call? (e) Unlike the first call, here the choice of path is not arbitrary. If you choose the lower path, which subsequent calls would be blocked? (f) If you choose the upper path, which subsequent calls would be blocked? Show the configured switches, the A-to-W call, and the D-to-Z call on your sketch. (g) With these two calls still connected, suppose C wants to talk to Y. How many paths are available for this call? (h) Show the configured switches and the three calls on your sketch. (i) Don't do it, but can you connect the remaining idle ports to each other? (j) Suppose D is finished talking to Z and now wants to talk to X. Disconnect the D-to-Z call. Show the configured switches, the A-to-W call, and the C-to-Y call on your sketch. (k) How many paths are available for the D-to-X call? (l) Rearrange either existing call. (m) How many paths are available for the D-to-X call? Show the configured switches and the three calls on your sketch.

7.14 This exercise illustrates how fabrics are enriched by making them *thicker.* (a) Sketch a 4×4 Benes network. Label the left ports top to bottom as A, B, C, and D. Label the right ports top to bottom as W, X, Y, and Z. (b) Add a third 2×2 switch to the center stage and convert all the edge switches to 2×3. (c) What are the values for m, n, and r for this Clos network? (d) According to the Clos inequalities, what is the blocking category for this 4×4 fabric? We'll test to see if this fabric is strict-sense or wide-sense nonblocking. (e) With all ports idle, suppose A wants to talk to W. How many paths are available for this call? (f) Arbitrarily configure the three upper switches for this call. The problem gives the same result if you had selected any other path. Show the configured switches and the A-to-W call on your sketch. (g) With this A-to-W call still connected, suppose D wants to talk to Z. How many paths are available for this call? (h) Which path leaves the most switches in unspecified states? (i) In a strict-sense nonblocking, the choice of path would be arbitrary. We test this by selecting either of the non-optimal paths. Arbitrarily connect this D-to-Z call through the middle switch in the center stage. Show the configured switches, the A-to-W call, and the D-to-Z call on your sketch. (j) With these two calls still connected, suppose C wants to talk to Y. How many paths are available for this call? Which path leaves the most switches in unspecified states? (k) Instead of C-to-Y, consider C-to-X as a third call. How many paths are available for this call? (l) Again, we test for strict-sense by selecting the worst case. So, connect this C-to-

X call. Show the configured switches and the three calls on your sketch. (m) Don't do it, but can you connect the remaining idle ports to each other? (n) How many paths would be available for this call? (o) Suppose D is finished talking to Z and now B wants to talk to Z. Disconnect the D-to-Z call. Show the configured switches, the A-to-W call, and the C-to-X call on your sketch. (p) How many paths are available for the B-to-Z call? (q) Pick the "worst" one and show the configured switches and the three calls on your sketch. (r) Don't do it, but can you connect the remaining idle ports to each other? (s) How many paths would be available for this call? (t) We haven't proven it, but do you think the fabric might be strict-sense or wide-sense nonblocking?

7.15 This exercise illustrates how fabrics are enriched by making them *deeper.* We'll add an extra stage to a Benes network. Then, we'll test if the resulting fabric is better than rearrangeably nonblocking and, if it is, whether it's strict-sense or wide-sense nonblocking. (a) Sketch a 4×4 Benes network. Label the left ports top to bottom as A, B, C, and D. Label the right ports top to bottom as W, X, Y, and Z. (b) Break the wiring between the second and third stages and insert a fourth column of 2×2 switches there. Wire this added column of switches, using the same perfect-shuffle wiring pattern as between the other stages. (c) With all ports idle, suppose A wants to talk to W. How many paths are available for this call? (d) Arbitrarily configure the four upper switches for this call. The problem gives the same result if you had selected any other path. Show the configured switches and the A-to-W call on your sketch. (e) With this A-to-W call still connected, suppose D wants to talk to Z. How many paths are available for this call? (f) If we were to select the path that goes through the four lower switches, would any subsequent calls be blocked? (g) Is this fabric wide-sense nonblocking? (h) Arbitrarily connect this D-to-Z call through the upper switch in the second stage and the lower switch in the third stage. The problem gives the same result if you had selected the opposite path. Show the configured switches, the A-to-W call, and the D-to-Z call on your sketch. (i) With these two calls still connected, suppose C wants to talk to Y. How many paths are available for this call? (j) Connect this C-to-Y call. Show the configured switches and the three calls on your sketch. (k) Don't do it, but can you connect the remaining idle ports to each other? (l) How many paths would be available for this B-to-X connection? (m) Suppose D is finished talking to Z and now wants to talk to X. Disconnect the D-to-Z call. Show the configured switches, the A-to-W call, and the C-to-Y call on your sketch. (n) How many paths are available for the D-to-X call? (o) Show the configured switches and the three calls on your sketch. (p) Don't do it, but can you connect the remaining idle ports to each other? (q) How many paths would be available for this B-to-Z connection? (r) We haven't proven it, but do you think the fabric might be strict-sense nonblocking?

7.16 Using the same network size and call statistics from Section 7.4.2, (a) what is the rate of calls blocked if $P_b = 2 \times 10^{-9}$? (b) What blocking probability results in blocking one call per month?

7.17 Give the permutation cycles for the interstage wiring between each successive pair of stages in an 8×8 Benes network (Figure 7.11).

7.18 Sketch a network that has K 8×8 Banyans (Figure 7.13) in cascade, with a bank of eight circuits inserted between each successive Banyan. Circuit C_{mk} is inserted between output m of Banyan k and input m of Banyan $k + 1$, where $0 \leq m \leq 7$ and $1 \leq k \leq K$. Besides being wired to the mth input of the next Banyan, each C_{mk} between each stage (for all k) is also wired to output device m. Each circuit accepts a packet from the preceding Banyan and examines the address in the header (assume it is not lost in the Banyan). If the address is m, the circuit delivers the packet to its output device (m). But, if the address is not m (because the packet was deflected in the preceding Banyan), the circuit delivers the packet to input m on the next Banyan. Suppose the packets at inputs 0 – 7 have destination addresses 0,1,2,6,0,6,0,2, respectively. (a) Walk these eight packets through this network, assuming that packets with the smallest addresses have the highest priority. (b) Which packets are permanently deflected if $K = 1$? (c) If K = 2, 3, 4? (d) With three packets destined for output 0, we obviously need at least K = 3 Banyans. Is it always true that the maximum number of packets destined for any one output equals the maximum number of Banyans that would be used? (e) What is the value of K so there is never any blocking?

REFERENCES

[1] Clos, C., "A Study of Non-Blocking Switching Networks," *The Bell System Technical Journal,* No. 32, 1953, pp. 402 – 424.

[2] Duguid, A. M., "Structural Properties of Switching Networks," *Brown University Progress Report,* BTL-7, 1959.

[3] Paull, M. C., "Reswitching of Connection Networks," *The Bell System Technical Journal,* No. 41, 1962, pp. 833 – 855.

[4] Benes, V. E., *Mathematical Theory of Connecting Networks and Telephone Traffic,* New York, NY: Academic Press, 1965.

[5] Benes, V. E., "On Rearrangeable Three-Stage Connecting Networks," *The Bell System Technical Journal,* No. 41, 1962, pp. 1481 – 1492.

[6] Benes, V. E., "Optimal Rearrangeable Multistage Connecting Networks," *The Bell System Technical Journal,* Vol. 43, No. 4, Part 2, Jul. 1964, pp. 1641 – 1656.

[7] Waksman, A., "A Permutation Network," *Journal of the ACM,* Vol. 15, No. 1, Jan. 1968, pp. 159 – 163.

[8] Joel, A. E., Jr., "On Permutation Switching Networks," *The Bell System Technical Journal,* Vol. 47, No. 5, Jun. 1968, pp. 813 – 822.

[9] Chiang, L. S., "Faster Benes Network Algorithm," Master's Thesis, University of Pittsburgh, 1993.

[10] Hwang, F., "Control Algorithms for Rearrangeable Clos Networks," *IEEE Transactions on Computers,* Vol. COM-31, No. 8, Aug. 1983.

[11] Lee, K. Y., "A New Benes Network Control Algorithm," *IEEE Transactions on Computers,* Vol. C-36, No. 6., Jun. 1987, pp. 768 – 772.

[12] Opferman, D. C., and N. T. Tsao-Wu, "On a Class of Rearrangeable Switching Networks: Part 1: Control Algorithm," *The Bell System Technical Journal*, Vol. 50, No. 5, May/Jun. 1971, pp. 1579 – 1600.

[13] Batcher, K. E., "Sorting Networks and Their Applications," *Proc. AFIPS Joint Computer Conference*, Vol. 32, Spring 1968, pp. 307 – 314.

[14] Kim, H. S., and A. Leon-Garcia, "Non-Blocking Property of Reverse Banyan Networks," *IEEE Transactions on Communications*, Vol. 40, No. 3, Mar. 1992, pp. 472 – 476.

[15] Nassimi, D., and S. Sahni, "A Self-Routing Benes Network and Parallel Permutation Algorithms," *IEEE Transactions on Computers*, Vol. C-30, No. 5, May 1981, pp. 332 – 340.

[16] Nassimi, D., and S. Sahni, "Parallel Algorithms to Set Up the Benes Permutation Networks," *IEEE Transactions on Computers*, Vol. C-31, No. 2, Feb. 1982, pp. 148 – 154.

[17] Raghavendra, C. S., and R. V. Bopppana, "On Self-Routing in Benes and Shuffle-Exchange Networks," *IEEE Transactions on Computers*, Vol. 40, No. 9., Sept. 1991, pp. 1057 – 1064.

[18] Benes, V. E., "Permutation Groups, Complexes, and Rearrangeable Connecting Networks," *The Bell System Technical Journal*, No. 43, 1964, pp. 1619 – 1640.

[19] Stone, H. S., *Discrete Mathematical Structures and Their Applications*, Chicago, IL: Science Research Associates, 1973.

[20] Karol, M. J., M. G. Hluchji, and S. P. Morgan, "Input versus Output Queueing on a Space-Division Packet Switch," *IEEE Transacions on Communications*, Vol. COM-35, No. 12, Dec. 1987, pp. 1347 – 1356.

SELECTED BIBLIOGRAPHY

Bellamy, J., *Digital Telephony*, Second Edition, New York, NY: Wiley, 1991.

Broomell, G., and J. R. Heath, "Classification Categories and Historical Development of Circuit Switching Topologies," *Computing Surveys*, Vol. 15, No. 2, Jun. 1983, pp. 95 – 133.

Hui, J. Y., *Switching and Traffic Theory for Integrated Broadband Networks*, Boston, MA: Kluwer, 1990.

Kim, H. S., and A. Leon-Garcia, "Non-Blocking Property of Reverse Banyan Networks," *IEEE Transactions on Communications*, Vol. 40, No. 3, Mar. 1992, pp. 472 – 476.

Kruskal, C. P., and M. Snir, "The Performance of Multistage Interconnection Networks for Multiprocessors," *IEEE Transactions on Computers*, Vol. 32, No. 12, Dec. 1983.

Lea, C. T., "Multi-Log *N* Networks and Their Applications in High-Speed Electronic and Photonic Switching Systems," *IEEE Transactions on Communications*, Vol. 38, No. 10, Oct. 1990, pp. 1740 – 1749.

Lee, C. Y., "Analysis of Switching Networks," *The Bell System Technical Journal*, No. 34, 1955, pp. 1287 – 1315.

Lee, T. T., "Non-Blocking Copy Networks for Multicast Packet Switching," *IEEE Journal on Selected Areas in Communications*, Vol. 6, No. 9, Dec. 1988, pp. 1455 – 1467.

Lee, T. T., M. S. Goodman, and E. Arthur, "Broadband Optical Multicast Switch," *XIII Int. Switching Symposium*, Stockholm, Sweden, 1990.

Marcus, M. J., "The Theory of Connecting Networks and Their Complexity: A Review," *Proceedings of the IEEE*, (issue on Telecommunications Circuit Switching), Vol. 65, No. 9, Sept. 1977, pp. 1263 – 1270.

McDonald, J. C. (ed.), *Fundamentals of Digital Switching*, Second Edition, New York, NY: Plenum Press, 1990.

Oie, Y., et al., "Survey of Switching Techniques in High-Speed Networks and Their Performance," *Technical Report No. 89-37*, Dept of Information and Computer Science, University of California, Irvine.

Padmanabhan, K., and D. H. Laurie, "A Class of Redundant Path Multistage Interconnection Networks," *IEEE Transactions on Computers*, Vol. 32, No. 12, Dec. 1983.

Richards, G. W., "Theoretical Aspects of Multi-Stage Networks for Broadband Networks," *Tutorial Presented at Infocom '93,* San Francisco, CA, Mar. 28 – 29, 1993.

Schindler, G. E. (ed.), *A History of Engineering and Science in the Bell System,* Murray Hill, NJ: Bell Telephone Laboratories, 1982.

Shyy, D. J., and C. T. Lea, "Log(N, m, p) strictly Nonblocking Networks," *IEEE Transactions on Communications,* Vol. 39, No. 10, Oct. 1991, pp. 1502 – 1510.

Switching Systems, New York, NY: American Telephone & Telegraph, 1961.

Turner, J. S., "Design of a Broadcast Packet Switching Network," *IEEE Transactions on Communications,* Vol. 36, No. 6, Jun. 1988, pp. 734 – 743.

8

Crossbar

"From each according to his ability; to each according to his need."
— from the *Communist Manifesto,* by Karl Marx

This chapter expands the discussion of Crossbar systems from Section 2.5.3 and describes the species called Number 5 Crossbar, the most popular Crossbar system in the United States. This electromechanical system, simply called Number 5, was specifically designed to be a residential suburban CO. By 1964, Number 5 Crossbar offices served 25% of all U.S. subscribers. In the two decades since then, Crossbar systems were gradually replaced by Electronic and Digital systems until their penetration decreased to about 15% of tethered lines by the mid-1980s. This replacement accelerated in the 1990s, and Crossbar systems, like Step, are practically extinct in the United States. Also like Step, Crossbar systems are still found in other countries.

Since Crossbar systems were summarized in Section 2.5.3, the reader who is in a hurry could skip this chapter entirely. But, like Chapter 6, I urge you to skim it, at least. Number 5's details are included here for several reasons.

- As a significant system in the United States, Number 5 has historical value.
- Number 5 has some clever design details, particularly in the fourth stage of the fabric and in how connectors enable signaling among the various subsystems.
- Number 5's architecture represents a *lumped* system in which various pieces of equipment have specific and specialized roles in the total system. The process by which the control of a call is progressively handed off from one specialized piece of equipment to another one carries over to how calls are handled by various specialized software modules in modern switching software. Since the software walk-throughs in Chapters 11 and 14 are difficult, they will be easier to follow if you "practice" on the hardware walk-throughs here.
- This chapter continues the discussion of the *partition.* In the evolution from Step to Number 1 Crossbar (and Panel), intelligence was extracted out of the fabric and concentrated in devices, called senders, that stored and interpreted the subscribers' dialed digits. In the evolution from Number 1 to Number 5, intelligence is further extracted and concentrated into primitive electromechanical computers, called

markers. The device that stores the subscribers' digits, called an originating register in Number 5, does not have the intelligence to interpret these digits. Each OR must request a marker when it is time for digits to be interpreted. While Number 5 is a slight migration toward a *centralized* architecture, we consider Number 5 to have a *lumped* architecture because the marker does not participate in virtually every operation.

The coverage of Number 5 in this chapter is less detailed than that of Step in Chapter 6. Section 8.1 describes the overall architecture of a Number 5 Crossbar office, and Section 8.2 describes the fundamental element, the Crossbar switch. Number 5's fabric is described in Section 8.3. Section 8.4 presents the procedure and interworking of the subsystems during the walk-through of an intraoffice call, an outgoing interoffice call, and an incoming interoffice call. Number 5 Crossbar is discussed generally in Section 8.5.

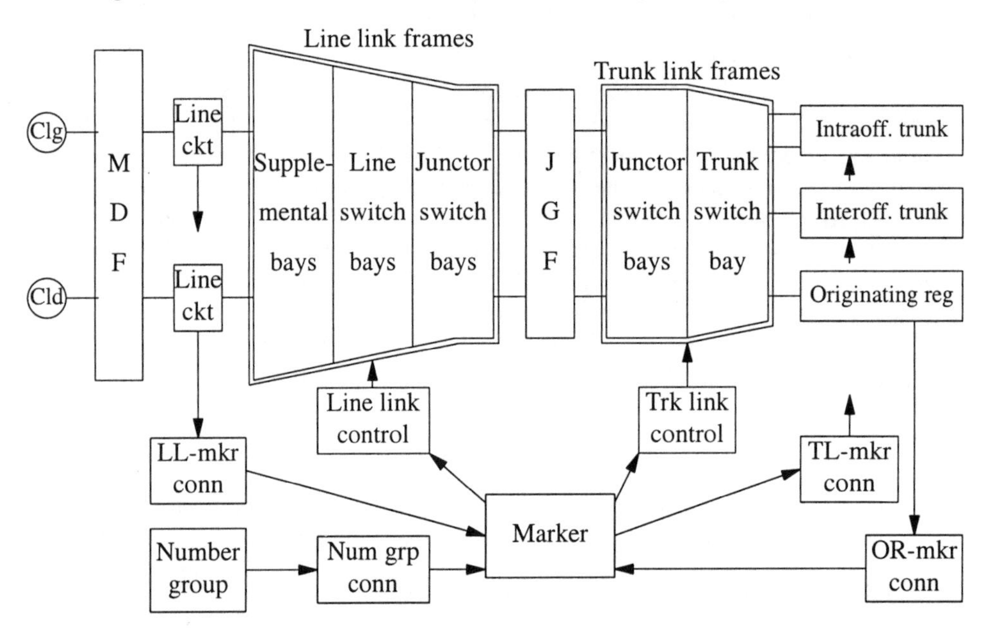

Figure 8.1. Number 5 Crossbar block diagram.

8.1 SYSTEM ARCHITECTURE

Figure 8.1 shows an incomplete block diagram of a Number 5 Crossbar switching office. More blocks are added in Section 8.4, when interoffice calls are discussed, and in the next chapter, when billing is discussed. The line side of the office is on the left and the trunk side is on the right. Lines are cross-connected at the main distribution frame (MDF) and connected to line circuits that are similar to, but simpler than, those in Step. Since the sidedness of the fabric is lines/trunks, the line circuits interface to only one port on the fabric; on the exterior edge of the line side of the fabric.

Number 5's fabric is shown across the upper center of Figure 8.1 as line link frames and trunk link frames, cross-connected through the junctor grid frame (JGF). An office has one complete line link frame for each 300 – 500 lines and one complete trunk link frame for each 160 trunks and service circuits. Each such frame in the fabric consists of 2 – 3 bays, and each bay contains 10 of the 20 × 10 electromechanical small-motion Crossbar switches from which the system derives its name.

The system contains some special-purpose one-stage fabrics, called links or connectors, through which various subsystems communicate with each other. For example, after the *j*th originating register has received all of the subscriber's dialed digits, these digits are transmitted to the *i*th marker through the multiple contacts of the [*i,j*]th crosspoint in the originating register – marker connector. These special-purpose links and connectors are simple one-stage fabrics that are built from these same 20 × 10 Crossbar switches. Generally, links have three-wire crosspoints for analog transmission over tip, ring, and sleeve and connectors have multicontact crosspoints for parallel data transmission. Note that the term "link" is used two ways: as a path segment between two successive stages in the general fabric and as a special-purpose one-stage fabric.

Other circuits shown in Figures 8.1, 8.7, and 8.8 are identified as registers, senders, trunk circuits, markers, or controllers. These are electrical/electromechanical circuits, made predominantly from electromechanical relays. Originating registers hold the originating subscriber's dialed digits. Senders hold digits for outpulsing to another office. Circuits like registers and senders are small enough that four to eight of them fit in one bay of equipment, and the number of these circuits in the office is traffic-engineered. Intraoffice trunk circuits are simple supervision circuits (Section 3.4.8) for intraoffice calls. Interoffice trunk circuits are the one-way and two-way trunk circuits that interface to interoffice trunks. Since trunk circuits are used for most of the duration of a call, there are many more trunk circuits than there are registers or senders.

There are two kinds of markers, called dial-tone markers and completing markers. They provide call processing during two different phases of setting up a call. A mid-sized Crossbar office may have two to three dial-tone markers and five to six completing markers, each occupying several bays of relays. The line link controller and trunk link controller are relay circuits that translate coded network requests from the markers into signals to the appropriate select and hold relays in the fabric. The number group is a bay of electromagnetic equipment used for translating the called party's directory number into the location of its port on the fabric.

The appearance of the interior of a Crossbar office is dominated by two kinds of equipment bays: bays full of relays and bays full of Crossbar switches. While not nearly as loud as a Step office, a Crossbar office is still quite noisy.

8.2 CROSSBAR SWITCH

A Crossbar switch is about 2 feet long and 6 inches tall. Six bars, running horizontally across the front of the switch, are the "cross bars" by which the switch is named. They are mounted about 1.5 inches in front of the body of the switch. Like axles, the bars are held in position by the frame at the left and right ends of the switch. While they are free to rotate to three different positions, spring tension holds them in a rotary *home* position. A pair of *wings* is attached to one end of each crossbar, one above the bar and one below the bar. An electromagnet is mounted behind, and perpendicular to, each wing. If the upper electromagnet is activated, it attracts the upper wing, rotating the crossbar 10 – 15 degrees so the top of the bar rotates slightly backwards. If the lower electromagnet is activated, it attracts the lower wing, rotating the crossbar 10 – 15 degrees so that the bottom of the bar rotates slightly backwards. These 12 electromagnets are called select magnets because they select which bar is rotated and in which direction.

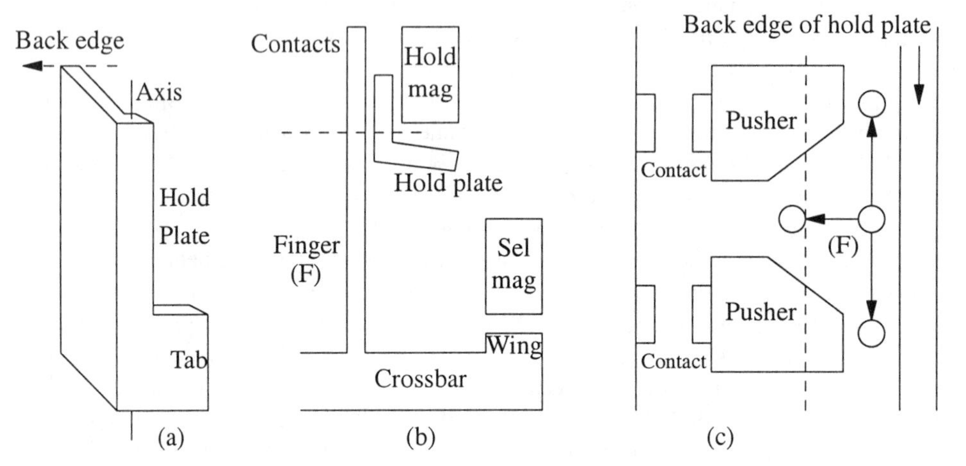

Figure 8.2. Crosspoint mechanics: (a) hold plate, (b) finger, and (c) contacts.

Twenty more electromagnets are mounted perpendicular to the front of the switch, across the bottom of the switch. Each of these hold magnets has an associated hold plate, shown isolated in schematic in Figure 8.2(a). The hold plate is L-shaped both in its front face and in its horizontal cross-section. It is free to rotate about an axis that runs vertically through its front edge. The tab at the lower right of the front face of each hold plate lies in front of its associated hold magnet and is spring-loaded away from it by a fraction of an inch. When a hold magnet is activated, the tab on its hold plate is attracted into the page in Figure 8.2(a), the entire hold plate rotates 10 – 15 degrees about its axis, shifting its back edge to the left a fraction of an inch.

Two crosspoints are associated with the intersection of a crossbar and a hold plate, one crosspoint for each rotation of the crossbar. In the two-dimensional array of crosspoints, the abscissa of a crosspoint is determined by a hold magnet and its ordinate by a select magnet. One of the six crossbars is dedicated for testing, a function that will not be discussed in this chapter. Ignoring this crossbar, the rest of the total switch is a 20×10 array of crosspoints. The contacts associated with a given crosspoint are closed by the mechanics in the body of the switch that responds to the rotation of a crossbar (by operating one of its select magnets) and to the lateral shift of the back edge of a hold plate (by operating its hold magnet). A simplified explanation of this mechanism follows.

Twenty flexible fingers protrude into the switch's mechanism from the back of each crossbar; one finger for each hold plate. Figure 8.2(b) shows a horizontal cross-section, looking down from the top. It shows one crossbar and one of its fingers extending back into the body of the switch. A cross-section along the dashed line in Figure 8.2(b) and a close-up view of the front of the switch body behind the crossbars are presented in schematic in Figure 8.2(c). Figure 8.2(c) shows the cross-section of this finger in the center, the back edge of a hold plate to its right, and the two crosspoint contacts to its left. The left edge of Figure 8.2(c) is stationary and, while the two *pushers* can move horizontally to the left, they are spring-loaded to the position shown, with the contacts open. The rotation of the crossbar moves the finger a fraction of an inch into the vertical gap between the hold plate and the pushers. In Figure 8.2(c), the two unlabeled circles, above and below the cross-section of the finger, represent the finger's positions when the respective select magnets are operated.

Consider three cases.

- If neither select magnet is operated, the crossbar is not rotated and the finger is in the center position indicated in Figure 8.2(c). If the hold magnet is operated, the back edge of the hold plate moves to the left in front of the two pushers (dashed vertical line). It pushes the flexible finger into the horizontal gap between the two pushers (third unlabeled circle). The pushers don't move and the contacts don't close.
- If the lower select magnet is operated, the crossbar rotates so that its upper edge comes out of the page and the finger moves to the upper position indicated in Figure 8.2(c). If the hold magnet is operated, the back edge of the hold plate moves to the left in front of the two pushers (dashed vertical line). It pushes the flexible finger into the upper pusher, which shifts to the left and the upper contacts close. The lower pusher doesn't move and the lower contacts don't close.
- If the upper select magnet is operated, the crossbar rotates so that its lower edge comes out of the page and the finger moves to the lower position indicated in Figure 8.2(c). If the hold magnet is operated, the back edge of the hold plate moves to the left in front of the two pushers (dashed vertical line). It pushes the flexible finger into the lower pusher, which shifts to the left and the lower contacts close. The upper pusher doesn't move and the upper contacts don't close.

Ignoring the test crossbar, a Crossbar switch is an array of 20×10 crosspoints. While Figure 8.2(c) implies that each crosspoint has a single make contact, a pusher can close several parallel contacts: typically three (for tip, ring, and sleeve) in most of the switches in the fabric and in special-purpose links, or six in the switches in the trunk switch bays of the fabric and in special-purpose connectors.

The operation of a select or hold magnet has a characteristic "clunk." The contacts associated with a crosspoint remain closed as long as the hold magnet is operated. The fingers are flexible enough and have enough friction between the back edge of the hold plate and the pushers that a finger is held in place after the select magnet is released, or even after the complementary select magnet is operated. A closed contact opens only when the hold magnet releases. The release of a hold magnet has an associated "ping," as the finger springs back to its rest position.

The contacts are interconnected in the back of the switch. Typically, one wire of a switch input line is connected to one side of a corresponding contact in the 10 crosspoints associated with one hold magnet, and one wire of a switch output line is connected to the other side of a corresponding contact in the 20 crosspoints associated with one select magnet. This *banjo wiring* runs vertically and horizontally and interconnects the wiring posts across the back of a switch. A Crossbar switch, with three contacts per crosspoint, wired this way functions as a 20×10 switch for three-wire lines. If the horizontal banjo wiring is cut down the center of the switch, a single mechanical Crossbar switch functions as two separate 10×10 switches.

For those Crossbar switches in the fabric, a hold magnet is kept operated by the ground on the sleeve that it connects. When a subscriber hangs up to end a call, the line circuit removes the ground from the sleeve, and the Crossbar switches that completed the call automatically drop the switch train.

8.3 FABRIC ARCHITECTURE

The Number 5 Crossbar fabric, shown in the upper center of Figure 8.1, is described. The first three subsections describe the architecture of the overall fabric: its logical and physical structure, its modularization, and a minimum fabric. The last two subsections describe the line link frames and trunk link frames, respectively, in greater detail. These are the two elementary switching modules from which the total fabric is built. The architecture of Number 5's switching fabric is complicated because its logical, physical, and hierarchical descriptions all seem to be different.

8.3.1 Logical and Physical Structure

Logically, the fabric has four stages of switching, with lines on one side and trunks on the other. The leftmost and rightmost logical stages are both concentrators and the two interior stages are distributors.

Physically, the fabric has five stages of equipment, roughly analogous to the four logical stages of switching. Each physical stage consists of many bays of of equipment, where each bay contains 10 Crossbar switches. All the bays look essentially alike. Figure 8.1 shows a physical stage of supplemental bays, a physical stage of line switch bays, and a physical stage of junctor switch bays, all on the line side of the fabric. And it shows a physical stage of junctor switch bays and a physical stage of trunk switch bays, both on the trunk side. Note that two of the physical stages have the same name and must be distinguished by specifying their side of the fabric.

The leftmost logical stage of switching, a concentration stage, is implemented physically by the supplemental bays, the line switch bays, and the left half of the junctor switch bays on the line side. The second logical stage of switching, a distribution stage, is implemented by the right half of the junctor switch bays on the line side. The third logical stage, also a distribution stage, is implemented by the entirety of the junctor switch bays on the trunk side. The rightmost logical stage of switching, another concentration stage, is implemented by the trunk switch bays.

The fabric wiring between the first two logical stages, called a-links in the general four-stage fabric of Section 7.2.9, are called line links in the fabric of Number 5 Crossbar. These line links interconnect the left and right halves of the junctor switch bays on the line side. The links between the two interior stages, called b-links in Section 7.2.9, are called junctors in Number 5. They interconnect the junctor switch bays on the line side with the junctor switch bays on the trunk side. The physical interconnection uses a large wiring frame, called the junctor grid frame (labeled "JGF" in Figure 8.1). The links between the two rightmost stages, called c-links in Section 7.2.9, are called trunk links in Number 5. These trunk links interconnect the trunk switch bays and the junctor switch bays on the trunk side.

Summarizing up to this point in the discussion, every bay contains 10 physical Crossbar switches. Each physical stage of the fabric contains many bays. The entire fabric contains five physical stages that function as four logical stages.

8.3.2 Switching Modules

In Section 7.2.9, a switching module was shown to be a subfabric that contains parts of more than one logical stage of switching. The fabric of Number 5 is modularized and the switching modules are called frames. A switching module on the line side is called a line link frame (LLF), and a switching module on the trunk side is called a trunk link frame (TLF).

The line side of the fabric has the same number of line switch bays and junctor switch bays. For the moment, assume that the number of supplemental bays is also equal to this value. Each LLF contains one supplemental bay, one line switch bay, one junctor switch bay on the line side, and all the interconnecting wires, including the line links. A fabric with R_l bays in each of the physical stages on the line side is organized as R_l LLFs. This is not strictly true because, as will be discussed in Section 8.3.4, a supplemental bay is optional in an LLF. The line links in a given LLF interconnect the switches between the two logical stages within the LLF. None of the LLFs are interconnected.

The trunk side of the fabric has the same number of trunk switch bays and junctor switch bays. Each TLF contains one trunk switch bay, one junctor switch bay on the trunk side, and all the interconnecting wires, including the trunk links. A fabric with R_t bays in each of the physical stages on the trunk side is organized as R_t TLFs. The trunk links in a given TLF interconnect the switches between the two logical stages within the TLF. None of the TLFs are interconnected.

The R_l LLFs are wired to the R_t TLFs by the junctors. When a frame is added to the fabric, the junctor wiring must be changed on the floor of the switching office. A cross-connection frame, called the junctor grid frame (JGF), facilitates this activity.

8.3.3 A Minimum Fabric

One fully equipped LLF terminates up to 500 lines on its exterior edge and 100 junctors on its interior edge. If an LLF is not equipped with an optional supplemental bay, then it terminates up to 300 lines, but still 100 junctors. In either case, an LLF concentrates its lines down to 100 junctors in two logical stages of switching. Each TLF terminates up to 160 trunks and service circuits on its exterior edge and 200 junctors on its interior edge. A TLF slightly concentrates its 200 junctors down to its trunks and service circuits in two logical stages of switching.

A minimum fabric has one TLF, which terminates up to 160 trunks and service circuits. But this TLF also terminates 200 junctors on its interior edge. Since one LLF terminates 100 junctors on its interior edge, this minimum fabric contains up to two LLFs. The 100 junctors from each LLF and the 200 junctors to the TLF are wired together at the JGF. This minimum fabric is shown in Figure 8.3.

Suppose the fabric is doubled to four line link frames, $LLF_0 - LLF_3$, and two trunk link frames, $TLF_0 - TLF_1$. The JGF would need considerable rewiring. Half the junctors (50) from each of the original LLFs remain wired to half the junctors (100) to the original TLF. The other 50 junctors from each original LLF would have to be rewired at the JGF to 100 junctors to the added TLF. The other 100 junctors to the original TLF would have to be rewired at the JGF to 50 junctors from each of the added LLFs. The remaining 50 junctors from each of the added LLFs would be wired at the JGF to the remaining 100 junctors to the added TLF. Thus, the 100 junctors from each LLF would be split between the two TLFs, 50 junctors from each LLF wired to each TLF. Conversely, the 200

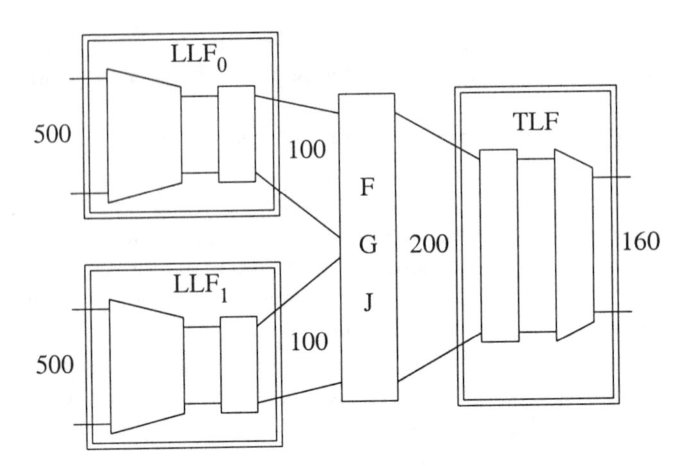

Figure 8.3. Minimum fabric.

junctors to each TLF would be split among the four LLFs, 50 junctors wired to each at the JGF.

An office has one LLF for each 300 – 500 lines and one TLF for each 160 trunks and service circuits. In general, the fabric in any Number 5 office has about twice as many line link frames as it has trunk link frames.

8.3.4 Line Link Frame

The sidedness of the Crossbar fabric is lines/trunks. Each subscriber has one appearance on the Crossbar fabric, on the exterior edge of some line link frame. This one appearance is used for originating and terminating calls. A line link frame is a subfabric with two logical stages of switching. It concentrates up to 500 lines down to 100 junctors, which are wired to the JGF. The first logical stage concentrates the 500 lines to 100 internal links, called line links, and the second logical stage distributes these 100 line links to the 100 junctors.

Any LLF contains two or three physical bays of equipment: one (line side) junctor switch bay, one line switch bay, and one optional supplemental bay. Physically, each bay contains 10 Crossbar switches, where each switch is a physical/mechanical 20×10 array of crosspoints. The LLF, shown in Figure 8.4, has 10 separate Crossbar switches in a junctor switch bay, 10 separate Crossbar switches in a line switch bay, and 10 separate Crossbar switches in an optional supplemental bay. A two-bay LLF, without a supplemental bay, terminates a maximum of 300 lines. A three-bay LLF, with a supplemental bay, terminates a maximum of 500 lines. Figure 8.4 shows one three-bay LLF.

A line switch bay and a supplemental bay each contain 10 switches that are each electrically 20×10, consistent with the physical/mechanical structure of the switches. Figure 8.4 shows only one of the 10 Crossbar switches in each bay, as a dashed rectangle. This *j*th Crossbar switch in the line switch bay terminates 20 lines on its 20 verticals. The *j*th Crossbar switch in the supplemental bay terminates another 20 lines on its verticals. In Figure 8.4, each vertical and each horizontal represents three wires: tip, ring, and sleeve. Each $\times$ on Figure 8.4 represents three electrical contacts.

The back of any Crossbar switch contains 30 *banjo wires* because each of the 10

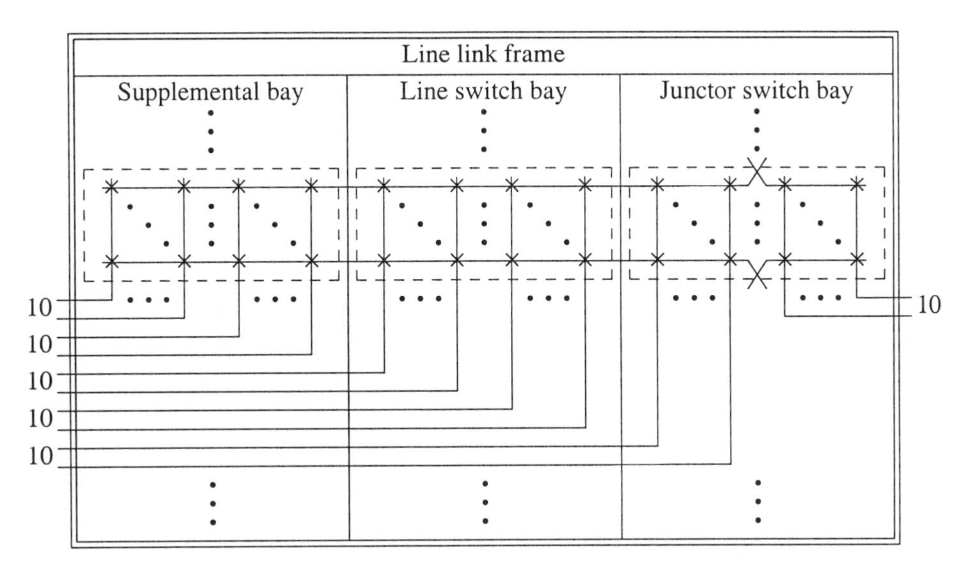

Figure 8.4. Line link frame.

horizontals has a tip, ring, and sleeve. The 30 banjo wires on the *j*th Crossbar switch in the supplemental bay are directly wired to the 30 banjo wires on the *j*th Crossbar switch in the line switch bay and to the 30 banjo wires on the *j*th left *half-switch* in the junctor switch bay. The select magnets on these three Crossbar switches are operated together. Thus, these two-and-a-half Crossbar switches are logically equivalent to a single 50×10 concentration switch. This extension of the banjo wiring is called a "build-out." A line that terminates on any of the 50 verticals on these two-and-a-half switches can be connected to any of the 10 common horizontals by operating the appropriate select and hold magnets.

The 10 physical Crossbar switches in a junctor switch bay (on the line side) are equivalent to 20 switches that are each 10×10, electrically. In the junctor switch bay, the banjo wiring is cut between the 10th and 11th verticals on each of the 10 switches. So, each Crossbar switch consists of two 10×10 arrays of crosspoints that share a common mechanism for control. As will be seen in Section 8.3.5, the trunk switch bays are wired in an even stranger arrangement. The indicated Crossbar switch in the junctor switch bay of Figure 8.4 is split in this manner. The left half-switch terminates 10 lines on its verticals. The right half-switch terminates 10 junctors on its verticals.

Each LLF is a two-stage subfabric, where the first stage concentrates 500 lines down to 100 line links in 10 limited-access switches. This stage of any LLF contains 10 separate 50×10 concentrators. The first logical stage of switching in the overall fabric contains a collection of separate 50×10 concentrators. These concentrators are organized as 10 per LLF.

The second logical stage of switching in each LLF distributes the 100 line links over 100 junctors also in 10 limited-access switches. This stage of any LLF contains 10 separate 10×10 distributors. Each of these switches is implemented in the right half-switch of the 10 Crossbar switches in the junctor switch bay. Each LLF terminates 100 junctors on the 10 verticals of each of these 10 switches. Figure 8.4 shows 10 junctors terminating on one switch. The 10 horizontals from each 50×10 concentrator are wired to the 10 horizontals in each 10×10 distributor in a perfect-shuffle pattern. These 100 wires are called line links, after which the frame is named. These line links fill the gap where the banjo wiring was cut between the 10th and 11th verticals in all 10 Crossbar switches in the junctor switch bay. Figure 8.4 shows two line link terminations on each side of this gap on one Crossbar switch.

Each 50×10 logical switch in the first stage of a LLF concentrates 50 lines down to 10 line links. Figure 5.5 shows that 10 servers handle 4 Erlangs of traffic intensity with $P_b = 0.008$. Distributing 4 Erlangs over 50 lines suggests that this concentration is acceptable if each line produces 4/50 = 0.08 Erlangs on the average. Since each line has one fabric appearance for originating and terminating calls, this 0.08 Erlangs must represent the sum of per-line originating and terminating traffic intensities; the total off-hook traffic load. This is a reasonably conservative figure for conventional telephony at a conventional residential line.

However, heavy users (like businesses during the workday), can be 50 – 100% busier than 0.08 Erlangs per line. So, the degree of concentration offered by a 50×10 switch is suspect, at least. At 0.12 Erlangs each, 50 users produce 6 Erlangs of total load. Figure 5.5 shows that 10 servers on 6 Erlangs yields $P_b = 0.08$, unacceptably high.

From Figure 8.4, if the supplemental bay is missing, a built-out concentrator in the first logical stage occupies the *j*th Crossbar switch in the line switch bay and the *j*th left half-switch in the junctor switch bay. The logical switch has 30 verticals, instead of 50. Even at 0.15 Erlangs per line, 30 users produce a total load of 4.5 Erlangs and 10 servers provide $P_b = 0.018$, a barely acceptable figure. So, the supplemental bay is optional. Conventional users are segregated onto LLFs that are equipped with a supplemental bay. Heavy users are segregated onto LLFs that are not equipped with a supplemental bay.

8.3.5 Trunk Link Frame

Each conventional trunk and each service circuit has one appearance on the Crossbar fabric, on the external edge of some trunk link frame. Intraoffice calls are completed through an intraoffice trunk (IOT) that has two appearances on the trunk side of the fabric. Each party in an intraoffice call is connected, through the fabric, to a separate appearance of the same IOT.

A TLF is a subfabric with two logical stages of switching. It concentrates 200 junctors down to 160 trunk appearances. The first logical stage distributes 200 junctors to 200 internal links, called trunk links, and the second logical stage concentrates these 200 trunk links down to the 160 trunk appearances. Any TLF contains two physical bays of equipment: one (trunk side) junctor switch bay and one trunk switch bay. Physically, each bay contains 10 Crossbar switches, where each switch is a physical/mechanical 20×10 array of crosspoints.

Figure 8.5 shows only one of the 10 Crossbar switches in the junctor switch bay of a TLF. The physical/mechanical 20×10 switch is represented as a dashed rectangle. In Figure 8.5 each vertical and each horizontal represents three wires: tip, ring, and sleeve. Each × on Figure 8.5 represents three electrical contacts. As in the junctor switch bay in the LLF, the 10 physical Crossbar switches in a junctor switch bay in a TLF are equivalent to 20 switches that are each 10×10, electrically. The banjo wiring is cut between the 10th and 11th verticals on each of the 10 switches. So, each Crossbar switch consists of two 10×10 arrays of crosspoints that share a common mechanism for control.

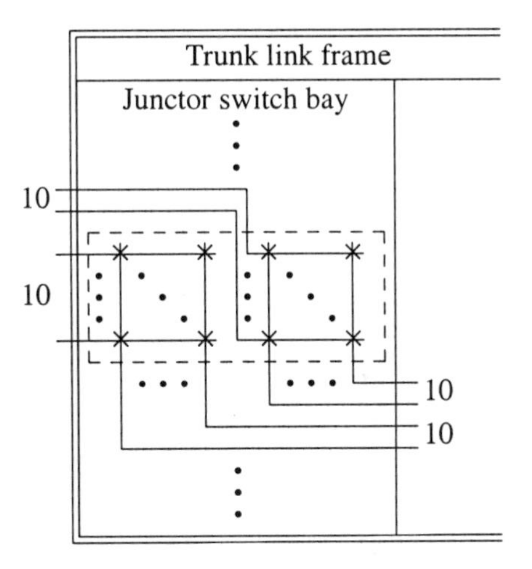

Figure 8.5. Trunk link frame.

Each of the 20 10×10 switches in a junctor switch bay terminates 10 junctors on its 10 horizontals and terminates 10 trunk links on its 10 verticals. In each of these 20 switches, 10 junctors are distributed over 10 trunk links. Figure 8.5 shows 20 junctor terminations on the left and 20 trunk link terminations on the right. The junctor switch bays from all the TLFs in the office, when taken together, constitute the fabric's third logical stage of switching.

Each TLF is a two-stage subfabric, where this first stage distributes 200 junctors over 200 trunk links in 20 limited-access 10×10 switches. The second logical and physical stage in a TLF is called a trunk switch bay. All the trunk switch bays, taken together, constitute the fourth logical stage of switching in the total fabric.

Figure 8.6 shows a single trunk switch in a trunk switch bay. In every switch in every trunk switch bay in every TLF, the banjo wiring has six wires per horizontal instead of the three wires per horizontal that is used in all the other switches in the fabric. On the previous figures each line represented three wires, but in Figure 8.6 each line represents a single wire. Twenty 3-wire trunk links are connected to the twenty 6-wire horizontals, in a slightly bizarre way. Sixteen 3-wire trunks or service circuits are connected to the upper eight 6-wire verticals (2 – 9) in a very bizarre way.

On horizontal levels 0 and 1, the posts in the back of the switch are not banjo wired. Instead of connecting a trunk link to a given vertical, the trunk link is connected to horizontals 0 and 1 where they cross the given vertical. On horizontal 0 the three-wire trunk link is connected to only the left three wires of the six-wire vertical and on horizontal 1 to only the right three wires. So, the three-wire trunk link at V1 in Figure 8.6 is connected to the six-wire vertical on the right side of Figure 8.6; to its left three wires by operating crosspoint [V1,0] and to its right three wires by operating crosspoint [V1,1]. The lowest crossbar controls this choice.

Each of the other horizontals is banjo wired. Two three-wire trunks or service circuits terminate on each six-wire horizontal. For example, in Figure 8.6, trunks T14 and T15 terminate on horizontal 9. If crosspoint [V1,9] is operated, then T14 is connected to the left three wires in V1's vertical and T15 is connected to the right three

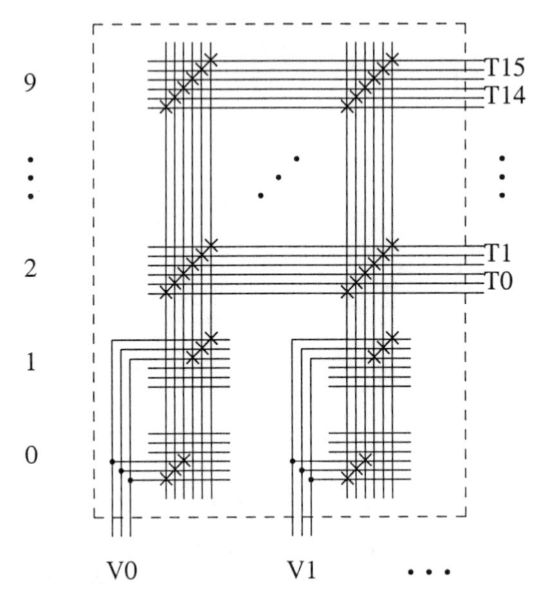

Figure 8.6. Configuration of a trunk switch.

wires in V1's vertical. To connect the trunk link at V1 to trunk T14 it is necessary to operate two crosspoints simultaneously: [V1,0] and [V1,9]. The connection of a trunk link to an output port requires the operation of two crosspoints, both in the same column. Two select magnets are operated and then the common hold magnet. While a Crossbar switch in a trunk switch bay is physically a six-wire switch with 20 inputs and 10 outputs, it is logically a three-wire switch with 20 inputs and 16 outputs.

While the various bays in the fabric of a Number Five may appear similar, they may be distinguished by close inspection. One examines the horizontal banjo wiring on the back of the 10 crossbar switches in the bay. If the banjo wiring goes straight across all 20 verticals, then the switch is part of a line switch bay or a supplemental bay in an LLF. If the banjo wiring is split between the 10th and 11th verticals, then the switch is part of a junctor switch bay in either an LLF or a TLF. If the two lowest verticals have no banjo wiring, and/or the switch has six contacts per crosspoint instead of three, then the switch is part of a trunk switch bay on a TLF.

8.4 CALL WALK-THROUGHS

The operation and interaction of Number 5's various subsystems are illustrated by call walk-throughs. The sequence of operations, between the time that a calling party goes off-hook and the time that the dialed number is translated, is presented in the first two subsections. Two different outcomes of this translation are described in the third and fourth subsections: if the called party is served by the same office as the calling party or if the called party is served by a different office. The final subsection describes the sequence of operations for an incoming call from a calling party served by a distant office to a called party served by this office.

8.4.1 Getting Dial Tone

The walk-throughs in these next three subsections refer to equipment illustrated in Figure 8.1. Every subscriber line is connected to a dedicated line circuit, which is similar to the line circuit in Step except that tip, ring, and sleeve terminate only on the line side of the fabric. An idle line is wired to talking battery through the windings of its line relay. When a calling party goes off-hook, this line relay operates, causing a signal to alert the line link marker connector to establish a connection between the line circuit and an idle dial tone marker.

A marker is a primitive wired-logic computer made with electromechanical relays. There are two kinds of markers: called dial tone markers and completing markers, each with a specialized function. A Number 5 Crossbar office has several of each kind of marker, their numbers determined by the peak busy traffic load in the office. The specialized function of a dial tone marker is to establish a connection through the fabric between any subscriber that has just gone off hook and an idle originating register. The marker gets its name because the originating register returns dial tone to the subscriber. So, the dial tone marker connects the calling party to dial tone.

Per-line information, called class of service, that is wired into each line circuit is read by the selected dial tone marker through the multiple contacts of the operated crosspoint in the line link marker connector. The address of this crosspoint identifies, for the dial tone marker, the location of the calling line's port on the line side of the fabric. So, the dial tone marker knows the "fabric address" of the calling party by [vertical, switch, bay] and whether the calling party uses dial pulse or Touch Tone, or is a specialized line like a PBX trunk or coin telephone.

The dial tone marker causes the trunk link marker connector to establish a connection between this marker and a service circuit, called an originating register. The selected originating register must be idle and must be of the appropriate type for the calling party's class of service. The dial tone marker transmits the calling party's fabric address and class of service to this originating register through the multicontact crosspoint in the trunk link marker connector. While this originating register has no use for this information, it is stored in this originating register because it will be needed later in the processing of the call.

The originating register has a tip, ring, and sleeve termination on the trunk side of the fabric. From the address of the operated crosspoint in the trunk link marker connector, the dial tone marker derives the fabric address of the originating register's termination on the trunk side. Addressing the calling party and the selected originating register by their respective [vertical, switch, bay], the dial tone marker determines an idle path through the fabric between the calling party and the selected originating register. The dial tone marker transmits instructions to the line link controller for the LLF on which the calling party's line terminates and to the trunk link controller for the TLF on which the originating register is terminated. These controllers operate the requested crosspoints in their respective frames and the calling party is connected through the fabric to the selected originating register.

The connection through the fabric is three-wire (for tip, ring, and sleeve). This connection, called the switch train, requires the continuous operation of four hold magnets in the fabric, one in each logical stage of switching. The ground side of each of these four hold magnets is wired to the sleeve wire of the switch train itself. Thus, the switch train remains connected as long as the sleeve path is grounded. While the connection of a path through the fabric requires a specific set of commands from a marker to the link controllers, the disconnection of this path occurs immediately when the sleeve is opened.

The originating register provides supervision and talking battery to the tip and ring through the windings of a supervisory relay. In the originating register, the sleeve is grounded through a contact of this supervisory relay. The grounded sleeve holds up the switch train and also operates the line circuit's cut-off relay, causing the line relay to disconnect. In an originating register, in its normal state, the office's dial tone source is connected to the tip and ring through the operated contacts of a dial tone relay. So, when the connection between calling party and originating register is made, calling party supervision transfers from the line circuit to the originating register, the originating register grounds the sleeve to busy the calling party and to hold up the switch train, and the calling party hears dial tone.

At this time, the dial tone marker's job is done. Its connector crosspoints are opened and it is marked "idle." It is available to process another new off-hook. The dial tone marker's *holding time* is the short interval between the time when the calling party goes off-hook and when he hears dial tone. So, relatively few dial tone markers can handle an entire office, even if the office has many subscribers and they are very busy.

8.4.2 Number Translation

The calling subscriber dials or "tones" the directory number of the desired called party into the originating register. Upon receiving the first dial pulse or the first Touch Tone digit, the originating register releases its dial tone relay and the calling party no longer hears dial tone. A dial pulse originating register receives and counts the calling party's dial pulses, performs interdigit timing, and stores all the dialed digits. A Touch Tone originating register receives and stores all the "toned" digits.

Somehow, the originating register must know when dialing is done. Determining the total number of expected digits, based on the first one or more received digits, is called pretranslation. Number 5 was designed, and many systems were installed, before Touch Tone and before direct distance dialing. So, in the initial design, every originating register in every Number 5 office had the simple pretranslation: expect exactly seven dial-pulse digits unless the first dialed digit is a 0. The 1 prefix was not used for toll calls and long distance calls were completed manually by dialing 0 and giving the long distance number to the operator verbally.

Direct distance dialing (DDD) was the new numbering plan that introduced the area code and allowed direct-dialed long distance calling. Number 5's pretranslation algorithm was redesigned for new installations. At extremely great expense, every originating register in every existing Number 5 office had to be rewired *in situ* for the new pretranslation algorithm. The DDD pretranslation algorithm was as follows: expect one digit if the first digit is 0, expect 10 digits if the second digit is 0 or 1, and expect seven digits otherwise. This works because all area codes had three digits, where the middle digit is 0 or 1. This pretranslation no longer works because the North American numbering plan changed again during the mid-1990s.

The originating register has an internal timer that times out if the first pulse or digit is not received within 30 seconds after the initial connection. This timer is not restarted after each digit. So, after at least one digit is received, the originating register remains connected until it thinks that dialing is done. If the calling party recognized a misdial during dialing, he would hang up and start again. If the calling party hangs up, the supervisory relay in the originating register releases and opens the sleeve. The switch train drops (the four hold magnets release), and the calling party's line circuit normalizes. This is called abandoning the call.

While an originating register can pretranslate a dialed number, it does not have enough intelligence to use the called party's directory number to determine how to complete the call. This latter function, called translation, is performed by the other kind of marker, the completing marker. After receiving the last digit, the originating register alerts the originating register marker connector to establish a connection between this originating register and an idle completing marker. Through the operated multicontact crosspoint in the originating register marker connector, the originating register transmits the called party's directory number to this selected completing marker. It also transmits the calling party's fabric address and class of service, which had been stored in the originating register by the dial tone marker.

The completing marker examines this directory number that has just been dialed and determines which type of call has been requested. It could be an intraoffice call, a direct-trunked interoffice call, a toll-trunked interoffice call, a call to a toll or information operator, or some kind of special call. This call walk-through continues in the next subsection as if the called party's directory number contains the office code of this office. It continues from this point in Section 8.4.4 as if the called party is determined to be served by another switching office.

8.4.3 Intraoffice Call

The completing marker knows its own office code. Assume for this subsection, that the three-digit office code in the dialed number is the code for this office. Then, the called party is served by the same office as the calling party and the call is an intraoffice call. After making this determination, the completing marker can complete the call (hence, its name). Connecting the calling and called parties through the fabric requires knowledge of the fabric addresses of the two parties. While the completing marker knows the fabric address of the calling party, it knows only the directory number of the called party, not its fabric address.

The *number group* is a subsystem in every Number 5 office that translates from directory number to fabric address for the 10,000 subscribers served by the office. The number group is a large manually wired cross-connect bay. Inside a Number 5 office, the installation of a new subscriber requires not only a manual cross-connect at the main distribution frame, but also a manual cross-connect at the number group. The completing marker alerts the number group connector to establish a two-way connection between this completing marker and the number group. The completing marker transmits the last four digits of the called party's directory number to the number group through a multicontact crosspoint in the number group connector. The number group transmits the [vertical, switch, bay] fabric address of the called party back to the completing marker.

If the called party is idle, then the completing marker completes the intraoffice call between the calling and called parties. The completing marker locates an idle intraoffice trunk, a supervisory "hairpin" with two separate appearances on the trunk side of the fabric. Using the fabric addresses of the calling party and the IOT's calling party port, the completing marker finds an interconnecting path through the fabric. Using the fabric addresses of the called party and the IOT's called party port, the completing marker finds an interconnecting path through the fabric.

The completing marker transmits instructions to the line link controller for the LLF on which the calling party terminates, to the line link controller for the LLF on which the called party terminates, and to the trunk link controller for the TLF on which the IOT terminates. These controllers operate the indicated crosspoints to establish the fabric connection between the calling party and the IOT's calling party port, and between the called party and the IOT's called party port. In the fabric connection between the calling party and the originating register, the sleeve path is opened. This action releases the four hold magnets that held up the switch train for this connection. The originating register is disconnected from the calling party and is marked "idle."

If the called party had been busy, the completing marker would not have used an IOT for the call. Instead, it would have completed a connection through the fabric between the calling party on the line side and a trunk-side service circuit that provides busy tone. This service circuit would ground the sleeve to hold up the switch train and would supervise the calling party while he listens to busy tone. When the calling party hangs up, the supervisory relay in the service circuit releases and opens the sleeve. The switch train drops (the four hold magnets release) and the calling party's line circuit normalizes. The call is finished.

Since the completing marker has done its job for now, its connector crosspoints are opened and it is marked "idle." Supervision and talking battery are provided to both parties through windings of supervisory relays on their respective ports on the IOT. The tip wires at the two ports and the ring wires at the two ports are wired together through relay contacts that are open. Thus, the calling and called parties are not connected to each other. In its normal state, an IOT provides ring signal to its called-party port through a normally closed contact of its ring relay and provides audible ringing tone to its calling-party port through another normally closed contact of its ring relay.

While the calling party in a Step office directly hears the attenuated actual signal that rings the called party, the calling party in a Number 5 office, and most other electromechanical, electronic, and digital offices, hears a tone that simulates this sound. Since the ring signal is on for 2 seconds and off for 4 seconds, there are three possible phases. Any switching office has a single *tone and ring* subsystem that produces a ring signal and the variety of audible tones. Ring signal and audible ring tone are produced in three phases, and each phase is wired to approximately one-third of the IOTs in the office. In a given IOT, the calling party port is not necessarily wired to the same phase of audible ring tone as the phase of ring signal to which the called party port is wired. This would explain the phenomenon where, occasionally, the called party answers during the first ring while the calling party never heard a "ring." This same phenomenon can occur in Electronic and Digital offices, but never in Step offices.

If the called party does not answer, the calling party would hang up. The supervisory relay in the IOT's calling party port would release and open both sleeves. Both switch trains would drop and both line circuits would normalize. The call is finished.

If and when the called party answers the call, the supervisory relay operates in the IOT's called-party port. This action provides supervision and talking battery to the called party and operates the IOT's ring relay. Ring signal is disconnected from the called-party port and the audible ringing tone is disconnected from the calling-party port. The contacts that had opened each port's tip wires from each other and each port's ring wires from each other are now closed. The calling and called parties are now interconnected and may talk to each other.

When either party hangs up to finish the call, the supervisory relay releases in the respective port of the IOT. The release of either supervisory relay opens both sleeves. Both switch trains drop and both line circuits normalize. The call is finished.

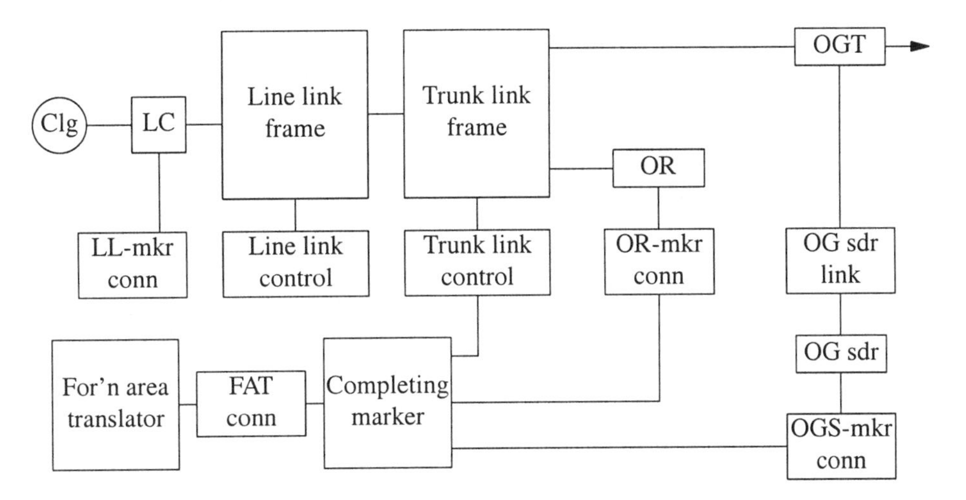

Figure 8.7. Interoffice outgoing call.

8.4.4 Outgoing Interoffice Call

As stated at the start of the previous section, the completing marker knows the office code of its own office. When the completing marker receives the called party's directory number from the originating register, it checks the office code. If the office code is different from that of this office, then the called party is not served by this office. Then, this is an *outgoing interoffice* call. Figure 8.7 shows an abbreviated architecture of a Number 5 office. Some subsystems that are shown in Figure 8.1 are excluded from Figure 8.7. But some additional subsystems are shown in Figure 8.7 that were not included in Figure 8.1. These additional subsystems are used for outgoing interoffice calls.

In the intraoffice call, the number group translated the last four digits of the called party's directory number into the line-side fabric address of the called party. In the outgoing interoffice call, the foreign area translator translates the first three digits of the directory number into the trunk-side fabric address of an appropriate outgoing interoffice trunk circuit (or the originating circuit of a two-way interoffice trunk). If the office code is that of a switching office with which this office has direct trunking, then the fabric address returned by the foreign area translator is that of a trunk circuit from the group of outgoing trunks (or that of the originating circuit of a trunk in the group of two-way trunks) between these two offices. For any other office code (back when there was only one long distance company), the fabric address returned by the foreign area translator would have been that of a trunk circuit from the group of trunks between this office and the toll office. In the modern era, this translation is greatly complicated by the distinction between the LEC's toll office and the IXC's point of presence and by "equal access" (next chapter) to more than one long distance company.

The completing marker alerts the foreign area translator connector to establish a two-way connection between this completing marker and the foreign area translator. The completing marker transmits the first three digits of the directory number dialed by the calling party to the foreign area translator through the multiple contacts of the operated crosspoint in the foreign area translator connector. The foreign area translator transmits the corresponding [vertical, switch, bay] fabric address back to the completing marker. Based on the fabric addresses of the calling party and the selected outgoing interoffice trunk circuit, the completing marker determines an idle path through the fabric that will interconnect them. The completing marker transmits connection instructions to the line link controller for the LLF on which the calling party terminates and to the trunk link controller for the TLF on which the selected outgoing interoffice trunk circuit terminates.

The sleeve is opened in the path through the fabric between the calling party and the originating register, dropping this switch train. So, the calling party is disconnected from the originating register and connected instead to the selected outgoing interoffice trunk circuit. This trunk provides supervision and talking battery to the calling party and grounds the sleeve to hold up the new switch train. When the outgoing interoffice trunk circuit is connected, its supervisory relay operates because the DC loop is closed through the calling party's hook switch. In a DC signaling trunk, a contact of this relay simulates a hook switch on the trunk side of this trunk circuit, and the switching office at the other end of this trunk sees an origination request. In an E&M signaling trunk (Section 4.4.1), the trunk circuit requests an origination in the office at the other end of the trunk by the appropriate single frequency signal.

Outgoing interoffice calls are more complicated because, to complete the call, the office at the other end of the trunk must know the called party's directory number. This information is currently stored in the completing marker in this office. In addition, if the call is a toll call requiring itemized billing, then the billing equipment at the toll office must know the identity of the calling party. While this information is also stored in this same completing marker, the calling party is identified there by his fabric address. The billing equipment requires the calling party's directory number. If the call is a toll call, the calling party's fabric address is translated to his directory number; the opposite translation from that of the number group. Discussion of this translation is postponed to Section 9.2.4. Depending on the type of call, the necessary information is transmitted to the office at the other end ot this trunk by a service circuit called an outgoing sender.

The completing marker alerts the outgoing sender marker connector to establish a connection between this completing marker and an idle outgoing sender. The completing marker transmits the called party's directory number and the identity of the outgoing interoffice trunk to the outgoing sender through the multiple contacts of the operated crosspoint in the outgoing sender marker connector. If the call is a toll call, the completing marker also transmits the calling party's directory number and class of service. All this information is stored in this outgoing sender. Since the completing marker is finished, its connector crosspoints are opened and it is marked "idle."

The outgoing sender alerts the outgoing sender link to establish a connection between this outgoing sender and the identified outgoing interoffice trunk. The connection through this outgoing sender link is a simple two-wire tip and ring connection and not a multicontact connection. In the outgoing interoffice trunk circuit, the tip and ring from the outgoing sender link is wired directly to the tip and ring of the call. The outgoing sender transmits the called party's directory number, by multifrequency signaling, through the outgoing sender link and the outgoing interoffice trunk circuit to the office at the other end of this trunk. If the call is a toll call, the outgoing sender also transmits, by multifrequency pulsing, the calling party's directory number and class of service to the office at the other end of this trunk.

Since the calling party is connected to this outgoing interoffice trunk through the fabric, this multifrequency pulsing is (once was) audible, as a rapid sequence of chords. As discussed in Section 8.5, this is no longer true. Since the outgoing sender is finished, its link crosspoints are opened and it is marked "idle." The outgoing interoffice trunk continues to provide supervision and talking battery for the calling party. The user hears audible ring tone or busy tone, as appropriate, over this trunk from the switching office at its other end. If the calling party hangs up, the supervisory relay in the trunk releases and the office at its other end sees an open circuit. If the called party answers, answer signal is transmitted from the other end as a battery reversal or E&M signal.

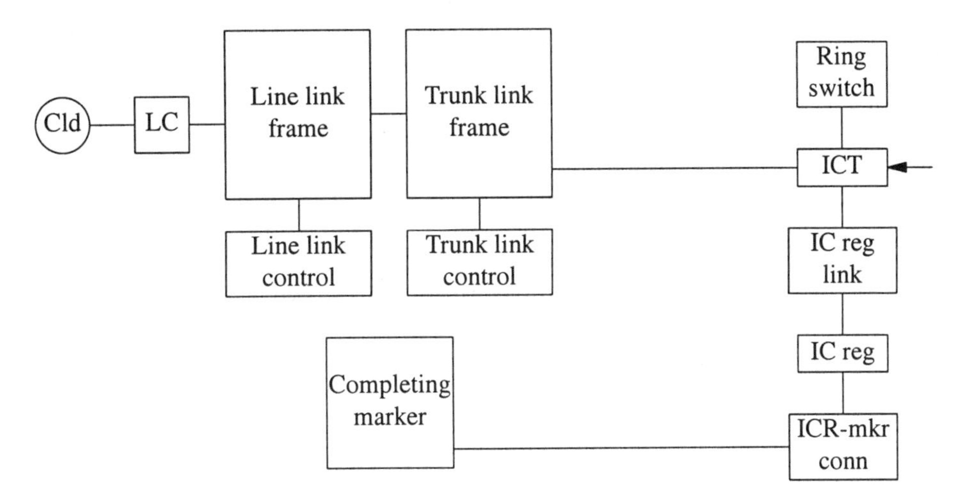

Figure 8.8. Incoming interoffice call.

8.4.5 Incoming Interoffice Call

The previous section described an outgoing interoffice call. The switching office at the other end of the outgoing interoffice trunk perceives this same call as an incoming interoffice call. It is a call for a subscriber served by this office that was originated by a subscriber served by another office. Figure 8.8 shows another abbreviated architecture of a Number 5 office. Some subsystems that are shown in Figure 8.1 or 8.7 are excluded from Figure 8.8. But some additional subsystems are shown in Figure 8.8 that were not included in Figures 8.1 nor 8.7. These additional subsystems are used for incoming interoffice calls.

An incoming interoffice trunk circuit (or the terminating circuit of a two-way interoffice trunk) recognizes a call termination request by the loop closure (DC signaling) or by the appropriate single frequency change (E&M signaling). The incoming interoffice trunk circuit responds by alerting the incoming register link to establish a connection to an idle incoming register. The connection is two-wire, connecting the tip and the ring.

The function of the incoming register on this call is similar to the function of the originating register on a call that originates from the line side, except that the incoming register receives multifrequency tones instead of Touch Tones or dial pulses, the incoming register provides neither dial tone nor supervision, and the signaling connection is through the one-stage incoming register link instead of through the office's fabric. The incoming register stores the fabric address of the incoming interoffice trunk circuit to which it has just been connected. The incoming register receives and stores the called party's directory number, transmitted by multifrequency signaling from the originating office. In the case of a toll call, it also receives and stores the calling party's directory number and class of service. The call walk-through of a toll call is described in Chapter 9. Pretranslation is not necessary because one of the multifrequency stop (ST) digits is sent to indicate the end of the transmission.

After receiving all the digits from the originating office, the incoming register alerts the incoming register marker connector to establish a connection between this incoming register and an idle completing marker. The incoming register transmits the trunk-side fabric address of the incoming interoffice trunk circuit and the called party's directory number to the completing marker through the multiple contacts of the crosspoint in the incoming register marker connector. The incoming register is finished. It releases all link and connector connections and is marked "idle."

The completing marker completes this incoming interoffice call like it did the intraoffice call in Section 8.4.3. It uses the number group to translate the called party's directory number into a line-side fabric address. It determines the busy/idle status of the called party. Based on this line-side fabric address of the called party and the trunk-side fabric address of the incoming interoffice trunk circuit, it determines an idle path through the fabric that interconnects them. It transmits connection orders to the appropriate line link controller and trunk link controller. Then, this completing marker opens all connector connections and is marked "idle."

The incoming interoffice trunk circuit's function on this incoming interoffice call is similar to the intraoffice trunk circuit's function on the intraoffice call in Section 8.4.3. It provides supervision and talking battery to the called party and grounds the sleeve wire that busies the called party's line circuit and holds up the switch train. It also supervises the incoming trunk and signals when the called party answers and hangs up. Through some contacts of its ring relay, it applies ring signal to the fabric-side tip and ring to ring the called party's telephone and it applies audible ringing tone to the trunk-side tip and ring for the calling party.

8.5 DISCUSSION

The call walk-throughs in Step and Crossbar have fundamental differences. The cellular and progressive properties of Step give its call walk-through a transitional, almost evolutionary, nature. While the details of the Strowger switch and the Step fabric are not trivial, the high-level procedure for establishing a phone call is easy to understand. The lumped nature of the Crossbar architecture gives its call walk-through an abruptness with many handoffs. Different subsystems are involved in their specialized roles during appropriate times of the establishment of a phone call. The call walk-through in Crossbar is more complicated than in Step because of the interaction of many different parts with different functions.

8.5.1 OA&M

Related to this fundamental difference in the call walk-throughs, the isolation and location of faults is very different between Step and Crossbar. Location of faults in Step depends on which digit in the telephone number was being dialed at the time that the fault is detected or on patterns of telephone numbers that might have the same digit in the same position in the number. Location of faults in Crossbar depends on the particular function that fails. The craftsmen that maintain a Step office, or that implement changes in one, must understand the overall operation of the fabric. The craftsmen that maintain a Crossbar office, or that implement changes in one, must understand the function and operation of the subsystems. There are several significant maintenance differences between Step and Crossbar.

The moving parts of a Crossbar switch have less mass and move a smaller distance than the moving parts of a Strowger switch. This small-motion electromechanical technology provides more switch operations and higher reliability over the life of the switch. But, because of its small structures, the state of a Crossbar switch is more difficult to observe by physical inspection than that of a Strowger switch or Panel switch. The Crossbar switch is designed with a sixth Crossbar and is actually a 20×12 switch. The two additional horizontals are testing appearances and are not wired to links or junctors in the usable fabric. This additional mechanism and additional horizontals allow for automated online testing and greatly simplify manual testing. Step has no such facility, and manual testing requires simulating a call placement and physical inspection of the switches by walking through the office.

The marker performs online fabric testing during call setup. If a test fails, the marker not only knows that it fails, but it also knows the fabric appearances and selected path of the failed call. If a trunk or service circuit detects some fault immediately after being connected, it can communicate with the marker before the marker drops off the call. Because of this common control, relevant testing information is available, at least temporarily. Crossbar offices have a subsystem, called a trouble reporter, and markers transmit this information when faults are suspected. The trouble reporter prints a trouble ticket and drops it in a bin and, depending on the severity of the fault, may ring an alarm. The switchmen process the reported faults. This kind of facility is impossible in Step because there is no central repository of information about a call. This facility is more advanced in a computer-controlled office because the relevant information about a call can be stored in the computer's memory for the duration of the call. In Crossbar, once the call is connected, the marker drops off the call and it forgets this relevant information so that it is free to handle a subsequent call. Furthermore, with computer control, we will see that more significant testing can be performed and that greater automation is implemented.

While the maintenance of an office is expensive in manpower, so are administrative tasks, like installing new subscribers and load balancing by relocating subscriber line appearances. In Step and Crossbar, installing a new subscriber requires a cross-connect at the main distribution frame that wires the subscriber's line to a line circuit and to an arbitrary port on the line side of the fabric. In Step, a new subscriber's line must also be wired to the terminating side of the fabric, and to the specific port that corresponds to the subscriber's directory number. While the subscriber has only one fabric appearance in Crossbar, installing a new subscriber also requires manual wiring in the number group so that a marker may translate a called party's directory number into his address on the line side of the fabric. Manual wiring is also required, in both Step and Crossbar, in cross-connect translation frames for translating line-side address into directory number for billing. These frames are described in the next chapter. Load balancing, by changing subscriber fabric appearances, requires similar manual wiring. In Step, of course, load balancing on the terminating side is impossible without changing the subscriber's directory number.

8.5.2 The Fabric

Even though the Step fabric has many different paths between given calling and called parties, the system is only capable of attempting to use one of them. If a certain selector is unable to locate an idle link to a next selector, the control architecture precludes backing up to a previous selector and going around the blockage. In Crossbar, the marker is capable of examining many candidate paths to find one that is completely unblocked. Without this alternate routing capability, the Crossbar fabric would need to be richer to provide the same probability of blocking.

The Crossbar fabric is said to be extremely efficient. I've never seen any data supporting this position, but it is believed by people with considerable experience. My own intuition suggests that the fabric would be better if there were a slight expansion in the trunk switch bay instead of a slight concentration there. Since the same fabric resources are used to connect lines to many different groups of trunks and service circuits, these resources are pooled, but the trunks and service circuits are specialized. Economy of scale suggests (to me, anyway) that fewer links are needed than ports for the same grade of service.

8.5.3 Incoming Signaling

When a Crossbar office receives signaling from one of its own subscribers, it makes a connection through its fabric between this subscriber, on the line side, and an originating register, which has an appearance on the trunk side. It would seem to follow, then, that when the office receives signaling from a distant office, that it would make a fabric connection between the incoming trunk, on the trunk side, and an incoming register, which would have an appearance on the line side. But, instead of using the general-purpose fabric to establish the connection for receiving incoming signaling, the incoming register link is used as a special-purpose fabric for this connection, and incoming registers terminate on this small special fabric instead. The reason is that the time required establishing a connection through the general fabric is much greater than the time required establishing a connection through a special link. The former takes longer because of all the handoffs and the involvement of a marker. The time difference is typically less than 1 second, but that turns out to be quite significant.

The delay in making the fabric connection does not affect subscriber signaling because telephone users have learned to wait for dial tone before starting their signaling. A long connection delay would not affect incoming signaling from another Crossbar office because the incoming register could signal the outpulser, with a *wink* for example, that it was ready to receive. The problem arises when the distant office is a Stepper, because the incoming signaling is transmitted directly by the calling party in that other office. Consider the subscriber in a Step office who places a call to a subscriber in a nearby Crossbar office. The subscriber simply dials a seven-digit directory number. The first three digits, the office code, are dialed into the first selector stage in the Step office, and the last four digits are dialed into an incoming register in the remote Crossbar office. Between the time of the last pulse of the third digit and the time of the first pulse of the fourth digit, the Step office must seize an outgoing trunk, the Crossbar office must respond, and an incoming register must be connected. The subscriber does not wait for this to happen; he keeps on dialing. Strict conformance with the Number 5 architecture would have required that the Step subscriber wait for a second dial tone, from the incoming register in the Crossbar office, before dialing the fourth digit. It would have been too confusing for the subscribers to wait for a second dial tone on some directory numbers and not on others.

A typical Number 5 office has three different incoming register links, depending on the signaling format. Through one dedicated incoming register link, incoming trunks from Step offices and dial-pulse incoming trunks from other Crossbar offices are connected to a group of incoming registers that receive dial pulse signaling. Through another dedicated incoming register link, multifrequency incoming trunks from other Crossbar offices are connected to a group of incoming registers that receive multifrequency signaling. Through yet another dedicated incoming register link, incoming trunks from Panel and Number 1 Crossbar offices are connected to a group of incoming registers that use revertive pulse signaling.

8.5.4 Blue Boxes

For an outgoing interoffice call, the outgoing sender outpulses the called party's directory number and possibly other information about the call, including the calling party's identification. The typical format is multifrequency signaling and the path is over a connection through the outgoing sender link to the tip and ring of the outgoing trunk. This used to occur while the calling party was connected to the outgoing trunk through the fabric. For many years, subscribers served by Number 5 offices became accustomed to hearing a rapid sequence of tones as the multifrequency chords were outpulsed.

But if the subscriber can hear these tones, then the clever and fraudulent subscriber can interfere with them. Illegal devices, known as Blue Boxes, were designed and circulated through an underground network. The legend is that the original invention was by a 13-year-old kid who built one in his basement. The Blue Box was connected like an extension telephone and it produced MF signaling that overpowered that of the outgoing sender. Calls could be caused to be fraudulently billed to the wrong party, or billing equipment could be tricked into billing a long distance call as a simple toll call. Blue Box fraud was stopped by modifying all the Number 5 offices in the United States so that the subscriber was not completely connected to the outgoing trunk during outpulsing. Then, when outpulsing is finished, the connection between the subscriber and the outgoing trunk is completed. You don't recall hearing the MF pulsing tones lately, do you?

MF pulsing was also audible in ESS offices, which were also susceptible to Blue Box fraud. The modification to ESS was a software change that was much easier to implement than the wiring changes required in every Number 5 office.

8.5.5 Modernization

Modernization of Number 5 Crossbar was debated in the late 1950s. Markers could have been replaced by modern electronic computers with stored program control. This action would have prolonged the longevity of Number 5, but probably would have restricted the market for the Number 1 ESS. A "computerization" of Number 5 did occur, but much later.

8.5.6 The Partition

In Crossbar, as in Step, the signaling path is still inseparable from the data path. But unlike Step, notice how Crossbar's control subsystem is separate from its connection subsystem. The separation is still a little blurred because the connection subsystem is used to establish control connections during signaling between the user and the control subsystem.

Crossbar's control subsystem, besides being separate from its connection subsystem, separates itself into functional parts. Careful study of Panel and Number 1 Crossbar would reveal an evolution as equipment became gradually more specialized, finally culminating in the Number 5 Crossbar architecture. The underlying philosophy is to require the use of the minimum amount of equipment during each temporal phase of the call. As opposed to how Step ties up all of call control during the entire call, the Crossbar architects tried to eliminate tying up specialized equipment during phases of a call when it is not needed on the call. Now, in Crossbar, the user can't talk during dialing or dial during talking, not only because the appropriate equipment isn't responding, but because the appropriate equipment isn't even connected to the user. Crossbar's increased specialization of equipment will be seen again in subsequent telephone systems (Chapters 11 and 14), but in their software architectures.

EXERCISES

Consider a "base-four" Crossbar office. Let the Crossbar switches be 4 × 4 and let a bay contain four such switches. Consider only three-wire switches; ignore the clever use of multiwire switches in the trunk switches in the trunk switch bay in the trunk link frame of real Crossbar offices. The five exercises below examine two alternative fabrics for a 96-line Crossbar office fabricated in this base-four technology.

8.1 A line link frame contains at least two such bays, called a line switch bay and a junctor switch bay, interconnected as a two-stage subfabric. Subscriber lines are wired to verticals on the line switches in the line switch bay, and junctors are wired to verticals on the junctor switches in the junctor switch bay. In each line link frame, the 16 line links are connected in a perfect-shuffle pattern from the four horizontals on each of the four line switches in the line switch bay to the four horizontals on each of the four junctor switches in the junctor switch bay. The line switch bay may be built out an arbitrary number of times, by banjo wiring any number of supplemental bays to the line switch bay. With no build-out, each line switch distributes four lines over four line links. With a one-bay build-out, each logical line switch concentrates eight lines down to four line links. With a two-bay build-out, each logical line switch concentrates 12 lines down to four line links, and so forth. If each line in the office averages 0.07 Erlangs in the peak busy hour and $P_b = 0.015$ is required, specify the build-out configuration for each line link frame. Show how an office with two such built-out line link frames would terminate 96 subscriber lines.

8.2 A trunk link frame contains exactly two bays, called a junctor switch bay and a trunk switch bay. Consider, first, that these bays are interconnected as a one-stage subfabric instead of a two-stage subfabric. In each trunk link frame, let the 16 trunk links be banjo wired from the four horizontals on each of the four junctor switches in the junctor switch bay to the four horizontals on each of the four trunk switches in the trunk switch bay. In other words, the junctor switch bay is a build-out of the trunk switch bay. Each built-out junctor/trunk switch concentrates eight junctor terminations, on its verticals, down to four trunk ports, on its horizontals. Since each such trunk link frame terminates 32 junctors, while each line link frame terminates 16 junctors, then one trunk link frame serves two line link frames. Show the fabric block diagram and junctor wiring for the 96-line office using the line link frame from Exercise 8.1.

8.3 Now, allocate trunks and service circuits to the 16 ports on the trunk side of the fabric from Exercise 8.2. Assume that traffic on the trunk side is distributed as follows: 10% to originating registers, 50% to intraoffice trunk circuits, 20% to a group of direct-switched interoffice trunks, and 20% to a group of interoffice toll trunks. The limited number of ports on the trunk side is exacerbated by the fact that each intraoffice trunk requires two ports. Ignoring all other types of trunks and service circuits, allocate an appropriate number of originating registers, intraoffice trunks, and interoffice trunks for the two separate trunk groups. Try to minimize the worst-case P_b over these four separate server groups. Are these acceptable blocking probabilities?

8.4 As an alternative architecture to Exercise 8.2 above, consider instead that the bays in a trunk link frame are interconnected as a two-stage subfabric. In each trunk link frame, let the 16 trunk links be connected in a perfect-shuffle pattern from the four horizontals on each of the four junctor switches in the junctor switch bay with the four horizontals on each of the four trunk switches in the trunk switch bay. Each junctor switch distributes four junctor terminations, on its verticals, to four trunk links, on its horizontals. Each trunk switch distributes four trunk links, on its horizontals, to four trunk ports, on its verticals. Since each such trunk link frame terminates 16 junctors, *N* trunk link frames serve *N* line link frames. Show the fabric block diagram and junctor wiring for the 96-line office using the line link frames from Exercise 8.1.

8.5 Now, allocate trunks and service circuits to the 32 ports on the trunk side of the fabric from Exercise 8.4. Assume that traffic on the trunk side is distributed as in Exercise 8.3. Ignoring all other types of trunks and service circuits, allocate an appropriate number of originating registers, intraoffice trunks, and interoffice trunks for the two separate trunk groups. Try to minimize the worst-case P_b over the four separate server groups. Are these acceptable blocking probabilities?

SELECTED BIBLIOGRAPHY

Aro. E., "Stored Program Control-Assisted Electromechanical Switching — An Overview," *Proceedings of the IEEE,* (issue on Telecommunications Circuit Switching), Vol. 65, No. 9, Sept. 1977, pp. 1313 – 1323.

Brooks, J., *Telephone — The First Hundred Years,* New York, NY: Harper & Row, 1976.

Clarke, A. C., et al., *The Telephone's First Century — and Beyond,* New York, NY: Thomas Y. Crowell, 1977.

Joel, A. E., Jr., *Electronic Switching: Central Office Switching Systems of the World,* New York, NY: IEEE Press, 1976.

Noll, A. M., *Introduction to Telephones and Telephone Systems,* Norwood, MA: Artech House, 1986.

Pearce, J. G., *Telecommunication Switching,* New York, NY: Plenum Press, 1981.

Rey, R. F. (ed.), *Engineering and Operations in the Bell System,* Second Edition, Murray Hill, NJ: AT&T Bell Laboratories, 1977.

Schindler, G. E., Jr. (ed.), *A History of Engineering and Science in the Bell System,* Murray Hill, NJ: Bell Telephone Laboratories, 1982.

Talley, D., *Basic Telephone Switching Systems,* New York, NY: Hayden Book Company, 1969.

9

Evolution of the Toll Point

> *"No man is an island, entire of itself. Any man's death diminishes me, because I am involved in mankind. And, therefore, never send to know for whom the bell tolls. It tolls for thee."* — from *Devotions,* by John Donne

The toll point served for many years as the location of the toll area's operator pool, the collection place for billing information, the second tier in the switching hierarchy, and the entry into the long distance network. This chapter opens with a discussion of billing and then presents the architecture of the toll point as it was, and had remained relatively unchanged, for the first half of the twentieth century. Two "master plans" followed: *automation* and *moderization.*

A grand plan for automating the toll point began in the early 1950s. This plan included the addition of area codes, direct dialing of long distance telephone calls, automated billing procedures, modifications to existing local switching systems, and an automatic switching system (Number 4 Crossbar) designed specifically for the toll point. North Americans knew this plan as direct distance dialing. By the time all the nation's toll points were automated, in the mid-1960s, it was time to modernize them and another grand plan was started.

Modernization included a minor change in the national number plan, operator automation, updating the technology of the automatic billing equipment, and a computer-controlled digital switching system (Number 4 ESS) for the toll point. By the time all the nation's toll points were modernized, in the early 1980s, divestiture was upon us and the toll point disintegrated.

Of all the levels in the national switching hierarchy, the toll point was most affected by divestiture. After divestiture, the toll point and its four-part functionality were split between the LECs and AT&T's long distance business. Jumping ahead in time, the last section of this chapter describes the remnants of the toll point that we have today.

Like Chapters 6 and 8, this is another of those chapters that you could skip if you're in a hurry. Again I urge you to at least skim this one, and to carefully read at least the last section. This chapter begins a subplot, which winds throughout the book and which tells a very important lesson in architecture:

If you're going to colocate unrelated functions,
make sure you do it so they're separable later.

9.1 BILLING AND THE 1940s TOLL POINT

The telephone bill evolved gradually over the first 70 years of the Bell System. Since the toll point has always been important to the billing process, its function and architecture also evolved gradually. In this section, the telephone bill and the process for collecting billing information, as they were in the late 1940s, are described first. Then, the function and architecture of the toll point are described, as they were at this same time. The late 1940s are selected because the entire system was stable and the grand plan for *toll automation* had not yet begun.

We must carefully distinguish two words that are often used interchangeably: "cost" and "price." The cost of a product is the total expense required to produce and sell it. The price of a product is its market value, the amount that customers pay when they buy the product. The difference is "profit," and in a free economy if a product's price is less than its cost, the vendor doesn't stay in business very long.

9.1.1 Government Regulation of a Monopoly

In the early years of the telephone industry, telephone service moved from luxury status to being a necessity, especially in commerce, but even in managing a residence. Like water, natural gas, and electric service, telephone service was a *natural monopoly.* In a natural monopoly, the customer realizes greater economy with a single supplier in a serving area because the delivery system is not replicated and the investment in the single delivery system is shared by everyone in the area being served. Such natural monopolies must be regulated by governments.

The typical regulating body is filled by political appointees, and the mission of regulating price frequently gets confused with other political and social agendas. It is difficult for any utility to be granted a rate increase, even if it is justifiable and in the public's long-term best interest, during an election year when the governor is running for another term. It is difficult for a utility to be granted a rate increase for some service, such as coin telephone charges, that affects the poor more than the wealthy, even if it is justified by cost. Even though coin telephone charges have increased from 10¢ to 25¢ or 35¢ in the last 40 years, the cost (including the cost incurred by vandalism and theft) of providing coin telephones in downscale neighborhoods is probably still not covered.

For many years, Ma Bell and the rest of the telephone industry operated under a pricing system that contained subsidies and cross-payments and in which the price of a service was not always related to its cost. While POTS was offered universally at a low price, the telephone companies were attractive to investors because they could charge extra for luxury services like extension telephones and Touch Tone and for nonuniversal services like enterprise switching and long distance. While most residential users liked this system, most enterprises didn't. When various regulatory agencies mandated deregulation and competition in the profitable parts of the pricing system, the subsidies were threatened. But the precedent was set for two things: (1) some kinds of services were somehow deemed deserving to be subsidized by other kinds of services, and (2) since the amount of a subsidy never appears as a line item on any budget, it can never be perfectly identified, quantified, or targeted. Encouraging competition in the arenas that subsidized the favored service meant that not only did the overall pricing system have to change, but it was likely that the entire industry would change — and we're still struggling through this change. This theme was introduced in Section 2.1.5, continues here, and flows through several of the chapters ahead.

By the late 1940s all states had Public Utility Commissions (PUCs), and the federal government had its Federal Communications Commission (FCC). The negotiation between the Bell System and each of the states was such an important part of the telephone business that many of the operating companies, the divisions of the Bell System that provided local telephone service, were organized along state boundaries. Frequently, the various commissions would concede that a telephone company was justified in requiring additional revenue. But instead of raising the rate on some basic universal service that might already be priced under cost, the added revenue would be allowed by raising the rate on a less sensitive service (like long distance) that affects businesses and the more affluent. While the telephone companies were always nervous about pricing some services under cost and subsidizing them by charging higher rates on other services, the total rate structure worked as long as the monopoly was intact and the pricing system was closed.

9.1.2 Cost Components of Telephone Service

Residential telephone service has five cost components. These components have never been individually itemized on the telephone bill, and the price of telephone service has never reflected the actual costs of these components nor has it even been based on this structure.

Access and accessibility. This cost includes the installation of a new telephone, the dedicated channels and per-line circuits, and planned overcapacity. Customers should expect to pay for the labor required to be physically and logically connected at the time that new telephone service is requested. While we do pay for this, the price has historically been far below the actual cost. Dedicated channels in the loop plant and per-line circuits in the switching office are allocated whether the customer is on-hook or off. Customers should expect to "rent" these channels and circuits as part of a flat-rate monthly charge, independent of call usage. Single-party lines are more expensive than multiparty lines because the cost of channels is shared in the latter. While customers who live far from the switching office require a longer channel than those who live close, it is not fair to charge for this difference, and an average price is charged to everyone. While customers do not pay for overcapacity, it does not seem unreasonable to rent the unused pairs in the drop cable at least, but we do not. Even people with no telephone service might be expected to be accountable for the accessibility of the pedestal in the neighborhood. Every residence in a community pays for a new sewer system through municipal taxes and then pays separate connection and usage charges. But the telephone company has never had the right to bill noncustomers. While there is a significant cost for placing a call to a called party who is either busy or does not answer, the telephone companies historically have never been allowed to charge to recover this cost.

Usage-sensitive costs. These costs of basic service have been charged in three different ways. At a coin telephone, the user pays for each call at the time that the call is placed. However, such a charge-per-use must be enough to cover not only the dedicated costs but also the sizeable cost of theft, vandalism, and repair. In message-rate billing, customers are charged in a monthly bill for each local call that they originate. In flat-rate billing, customers pay a fixed price every month for unlimited calling within a certain local calling area. In the last several years the telephone companies have experienced difficulty with flat-rate billing. A large part of this cost is the investment required to deploy equipment. The equipment deemed necessary is determined by long-standing statistical patterns of usage. When an ever increasing number of customers with computer terminals and dial-up modems place uncharged local calls to computers and Internet service providers, and stay on the line for hours, the need to deploy more equipment requires a price increase that is unfairly borne by all subscribers.

On-premises equipment. For basic service, this cost includes the cost of the telephone instrument. For many years the telephone itself belonged to the telephone company and its rental was included in the monthly charge for residential service. Similarly, switching equipment on the business customer's premises belonged to the telephone company and it was rented as part of subscribing to Centrex (next chapter). When the telcos lost this revenue, as a result of the Carterfone decision (Section 10.6), the entire monopoly pricing structure began to crumble.

Supplemental services. Optional features and non-POTS services have costs for usage and equipment that include rental of telephone extensions and other special equipment, Touch Tone service, customer switching equipment, and optional features like call waiting, call forwarding, abbreviated dialing, calling party identification, personal telephone number, and many more.

Touch Tone service is a classic case where price is inconsistent with cost. Mainly because of the greatly decreased holding time on digit receiving circuits, it is generally believed that Touch Tone subscribers place a smaller cost burden on the telephone company than dial-pulse subscribers do. Yet, Touch Tone subscribers have always paid a premium rate. In part, the Bell System wished to spread the manufacture of Touch Tone telephones over several years, but the principal reason for the premium charge was that people who could afford to buy a new telephone were asked to subsidize POTS for people who could not (or would not). The Touch Tone premium is a monthly charge that is independent of call usage. Whether this basis was consistent with actual cost was a silly argument because there was no added cost.

The price of most of the new optional features has both a monthly subscription component for "right-to-use" and a usage-sensitive charge per use. Except for abbreviated dialing, where the customer stores dialing codes in some computer memory in the switching office, most of the new features have little cost associated with their provision or access. While call processing is slightly complicated when one of these new features is used, the complication is by no means proportional to the premium charged. In a truly unregulated competitive market, these new features would be quite inexpensive. But then, in such a market, POTS would be very much more expensive. This pricing structure has caused many of these new features to be implemented in the customer premises equipment (CPE) where, in many cases, it costs more but is priced lower than providing a corresponding service in the network.

Supplemental calls. Additional costs for non-POTS calls include charges for long distance and, more recently, charges for operator-assisted calls. This cost component is separated from the rest of the telephone bill and is "detail-billed," with each call identified. One of the principal functions of the toll point has been to collect information for the detail-billing of toll calls. The entire network architecture would be quite different if toll calls had not been detail-billed.

Great economy can be realized if the peak busy hour can be flattened. Electric power utilities want usage-sensitive pricing to encourage more uniform use of generating facilities. In the telecommunications industry, migrating traffic from the peak busy hour to less busy times of day can reduce the required number of trunks, channels, service circuits, network links, and other servers. From the opposite perspective, if a given capital investment is required for equipment that provides an acceptable grade of service during the peak busy hour, then increasing the occupancy of this equipment during off hours, even at reduced rates, leads to increased revenue. Rates based on time of day or day of week have influenced subscriber calling habits significantly.

9.1.3 The Billing Process

The billing process has two stages. First, the relevant per-call information is recorded on-line in some switching office. Then, after this information is retrieved from the switching office where it is collected, the bills are prepared at a data processing center.

Message rate recording was automated early. A message register is a small visual counter, like an automobile odometer, that was attached to the line circuit of every message-rate subscriber. The reverse battery signal, indicating called party answer, incremented the counter. These message registers were physically aggregated. Periodically, the entire group of message registers was photographed and the photograph was sent to the data processing center where the bills were prepared.

Flat-rate billing allows the subscriber to place an unlimited number of calls without charge to parties within some local dialing area. The charge is based on typical calling patterns.

Wide Area Telephone Service (WATS) allows subscribers to change the boundaries of the dialing area within which their calls are not detail-billed. In "out-WATS," the subscriber pays an increased monthly flat-rate charge for the right to originate calls, free of a usage charge, to parties who reside in a larger calling area. These boundaries are not arbitrary, but conform to increasingly larger calling areas, including entire states or potentially the entire nation.

In "in-WATS," the subscriber agrees to pay the charges for received calls from originating parties within some calling area. This service is more difficult to implement because the originating party's billing equipment must recognize that a call should be billed as a collect call. In America's current numbering plan, in-WATS subscribers are assigned directory numbers with 800 as the equivalent area code (888 was recently added to supplement 800). Many businesses subscribe to in-WATS over the entire nation and advertise their 800 number as "toll free." In-WATS has become extremely popular and much more complicated in the current era of multiple telephone companies and the Intelligent Network (Chapter 18).

9.1.4 The Toll Point in the 1940s

Figure 9.1 shows the architecture of the toll point as it was in the late 1940s. The class-4 offices were all manual, staffed by toll operators. The latest technology was called the 4A Toll Switchboard, a minor variation of the fourth generation of the manual office specifically designed for the toll point. For some reason, the number 4 has been associated with the toll point throughout its history.

Consider the subscriber who wished to place a long distance call. If this subscriber's local class-5 switching office was manual, then either: (1) the local operator completed a long distance call directly, or (2) the local operator connected the subscriber to a toll

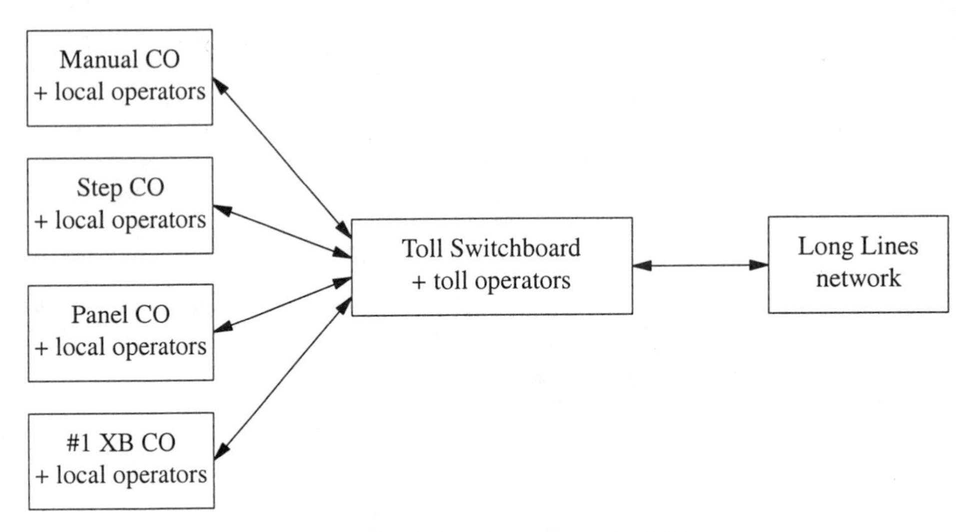

Figure 9.1. Toll architecture in the 1940s.

trunk and the toll operator then completed the call after conversing with the subscriber. If the subscriber's local office was a Stepper, then this subscriber dialed 0 or 1, the first selector connected a toll trunk, and the toll operator completed the call after conversing with the subscriber. If this subscriber's local office was a common-control office like Panel or some version of Crossbar, then this subscriber dialed 211 (typically) into a register, the number was pretranslated as complete, the office connected the subscriber to a toll trunk, and the toll operator completed the call after conversing with the subscriber.

In the initial numbering plan, subscribers dialed 0 to reach an operator for information, long distance calls, collect calls, or other assisted calls. The local operator, housed in the local central office building, performed all these functions. Later, in some locations, operators were specialized: the information and long distance functions were performed by separate pools of operators, both housed at the toll point, and the other functions were still performed by the local operator. In this arrangement, the subscriber would have to distinguish the toll and the local operator by dialing 1 or 0, respectively, or further distinguish the information and long distance operator by dialing 411 or 211, respectively. It was apparent almost immediately that great savings were possible by pooling the local operators and performing all the operator functions from the toll point. In an arrangement without local operators, the subscriber did not have to distinguish the local operator from the toll operator but might still use 0, 411, and 211 to distinguish assistance, information, and long distance.

Consider the subscriber who places a long distance call in the area served by some manual toll office. After the subscriber is connected to a toll trunk by the local switching office, the trunk seizure causes a lamp to light on a switchboard at the toll point. One of the operators there inserts a plug on one end of a cord into the jack associated with this lamp and answers the call with "Operator" or "Number, please." The subscriber then tells this operator the seven-digit directory number and the city and state of the desired called party and identifies his own directory number. This toll operator then inserts the plug on the other end of the cord into a jack corresponding to an idle trunk to the appropriate office: another local office for an intratoll call or a class-3 office for an intertoll call. The operator writes a *call attempt note* on a notepad to the effect that a call from a given directory number to another given directory number is using the trunk number corresponding to the jack just used.

For an intertoll call, this trunk seizure causes a lamp to light on a switchboard at the class-3 office. One of the operators there answers the seizure, converses with the original operator, learns the desired destination of the call, and repeats this procedure. The call proceeds up and down the toll hierarchy and across the country until it reaches the toll office of the called party. When the operator in this office seizes a trunk to the called party's end office, this final operator dials the seven-digit directory number into this end office (assuming that it's an automatic office). The originating subscriber hears audible ring tone from the called party's end office.

At this point in the call, all the operators disconnect and disengage from the call except the original toll operator. This operator watches for answer supervision. If the called party answers, then the calling party's toll operator writes a *call answer note* indicating trunk number and time of answer. When either party hangs up, this same operator writes a *call completion note* indicating trunk number and time of hang-up. All the operators in the entire switch train unplug their respective cords.

At least once a day, all the note papers were collected from all the toll operators and brought to a centralized data processing facility. Toll calls were tracked manually by the trunk number. Consider the sequence of these call notes on a given notepad. A call attempt note for a certain trunk number would be followed by call notes for other trunk numbers. But, the next call note with this same trunk number is either another call attempt note or is a call answer note. In the former case, the original attempt was not answered and no billing entry is made. In the latter case, the original attempt was answered and the three consecutive call notes with the same trunk number provide enough information to determine whom to bill and by how much. In this manner, the bills were prepared manually and mailed out. Each toll call was detail-billed on the monthly bill based on the three entries on the operator notepads.

9.2 AUTOMATION AND THE 1960s TOLL POINT

By the early 1950s it was apparent that automation was necessary in two areas of telephony: long distance calling and billing. Since these areas are related, they were automated together under one of the largest projects ever attempted by any industry in the private sector. A very complex master plan was developed and this plan was carefully orchestrated. The entire Bell System and all the independents were mobilized. Billions of dollars were invested — fortunately, this era of Eisenhower and the Cleavers (a fictional family from a 1950s television show) was a time of relative prosperity. The subscribers knew it as "direct distance dialing."

9.2.1 Direct Distance Dialing

DDD required a major change in the national numbering plan, a corresponding subscriber education program, complete overhaul and automation of the billing process, new equipment and major modifications to each different type of local switching office, the installation of these changes in every switching office in the country, and the development and installation of automatic switching equipment at the toll point.

One aspect of the new numbering plan was the *enumeration* of the office code. Like post offices, telephone offices were named, sometimes after the community that they served. For example, a resident in a certain neighborhood in Chicago, Illinois might have had the directory number: Kildare-1234. While this convention is sufficient in a manual switching office, it requires some form of abbreviation with automatic switching and dial telephones. The letters Q and Z are dropped from the useable alphabet and the remaining 24 letters are assigned to the numbers 2 – 9 on the telephone dial, three letters per number. If seven-digit dialing is used, Kildare-1234 is dialed as KIL-1234. While it is numerically equivalent to 545-1234, KIL and not 545 is spoken and written in the telephone book. If the Kildare exchange grew to more than 10,000 lines, then the original office code would be changed to KI 5 and a new office code, perhaps KI 6, would be added.

This convention caused some problems as local offices were automating and the number of lines increased. For example, the KI(ldare) 6 and LIN(den) office codes both have abbreviations that are numerically equivalent. So, office codes had to be selected carefully and many subscribers experienced changes in directory number. This assignment of letters to the dial restricted the total set of allowable office codes in a state to 640 because every office code had to be numerically equivalent to a unique NNX code, where N is 2 – 9 and X is 0 – 9. But, not all of these 640 codes were useable — for example, try to think of an office name whose first two letters are numerically equivalent to 95 (W, X, or Y followed by J, K, or L). Even if all 640 codes could be used, it was clear that some more populous states were going to run out. So, the office code name was dropped and KIldare 5-1234 was officially changed to 545-1234. Many people didn't like this and we heard comments like: "Why don't they get rid of my name, too, and just give me a number?" and "What would you expect from the phone company?" While this enumeration of office codes could have allowed the pool of valid office codes to expand to NXX, a different approach was used.

The nation was partitioned into number plan areas (NPAs), each with its own area code. These codes were taken from the pool of NKX numbers, where K is 0 or 1. Populous states were assigned several NPAs, and the entire pool of NNX office codes was available in each NPA. The real cleverness of this numbering plan was that 10-digit dialing was not required for calls within the same NPA and pretranslation was still relatively simple:

> *If the second digit in a telephone number is a 0 or 1, then the first three digits must be an area code and the subscriber will dial a total of 10 digits. If the second digit is 2 – 9, then the first three digits must be an office code and the subscriber will dial a total of seven digits.*

Subscribers served by Steppers still had to dial a leading 1 on toll calls in order to be connected to a toll trunk. While some thought was given to requiring this leading 1 uniformly across the entire nation, the idea was dismissed because it was thought that Step would soon be extinct. In hindsight, this was a bad choice: 40 years later, Step was still not extinct, and 20 years after that, another change in the numbering plan would universally require the leading 1 universally anyway.

9.2.2 Automatic Message Accounting

Automatic message accounting (AMA) equipment was added to many switching offices. AMA equipment can be located in the local switching office (LAMA) or can serve a region centrally from the Toll Office (CAMA). It consists of a tape recorder and hardware that interfaces this recorder to the trunks and common equipment in the local or toll office. Tape recorder technology of this era used punched paper tape.

The AMA recorder automates the notepaper activity that was performed by the toll operator. Every attempted toll call causes one entry on the AMA tape. This *call attempt entry* associates the called party's directory number, the calling party's directory number, and the identification of the trunk over which the call is placed. If the called party does not answer, there is no subsequent second or third entry with this trunk ID. If the called party does answer, the *call answer entry* records the trunk ID and the time of this answer to the next 10th of a minute. The *call disconnect entry* records the trunk ID and the time of the first hang-up to the previous 10th of a minute.

The punched paper tape is removed from the AMA recorder in the early morning hours and is delivered to the data processing center. At these new data processing centers, the manual preparation of telephone bills was converted over to computers, which were just becoming commercially available. These new computers read the AMA tapes and automatically processed the telephone bills.

This AMA process was not as simple as it may appear initially because of the operating procedures in the various types of switching equipment. In Number 5 Crossbar, while the calling party's identity and class of service are recorded in the originating register and would be available for the call attempt entry in the event of a toll call, the subscriber is identified by fabric address and not directory number. The situation is worse in Panel and Number 1 Crossbar because the calling party information is not stored and the calling party is anonymous when the dialed digits are translated and known to be a toll call. In Step, at the point in the fabric where the subscriber has indicated that a toll call is in progress, this subscriber has become anonymous by the hunting process in the line finder stage.

In all these systems, without a major change, the calling party's directory number is not known. The changes that needed to be made in Step, Panel, and Number 1 Crossbar offices was quite fundamental. The changes in a Step office are described in the next subsection. The changes in a Number 5 Crossbar office are simpler and are described in Section 9.2.4.

9.2.3 Automatic Number Identification in Step

In Step, once the switch train progresses past the line finder stage, the calling party is totally anonymous. Not even a fabric address is known at this point in the switch train, where some first selector is hunting for an idle toll trunk. This was never a problem as long as billing was flat-rate or even message-rate. But the requirement for automatic detailed billing of toll calls "broke" the architecture. The inventors and architects of Step never anticipated that automatic equipment would be required to make the call attempt entry, including the calling party's directory number, at the beginning of every toll call.

While the fundamental architecture of Step was not changed, an adjunct, called automatic number identification (ANI), was designed and manufactured. ANI equipment had to be added to every Step office in the country, and a similar subsystem was added to

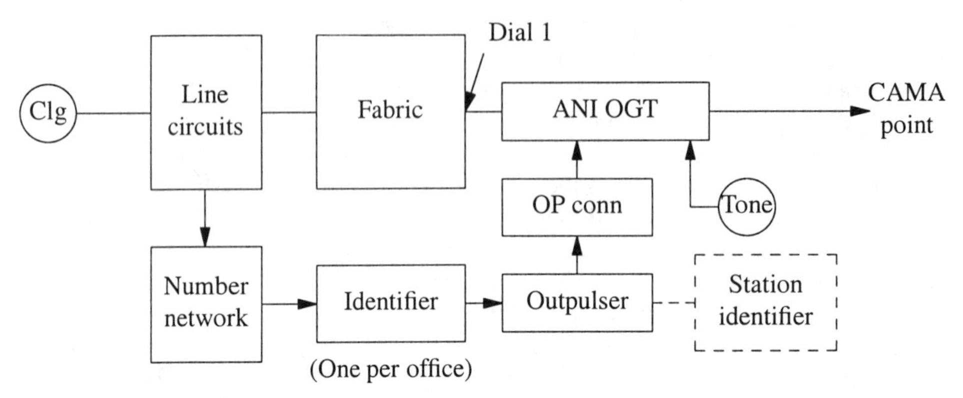

Figure 9.2. Step, Panel, or Number 1 Crossbar equipped with ANI.

Panel and Number 1 Crossbar offices. ANI's function is to recover the identity of the calling party, by directory number, and to transmit this information, over the selected toll trunk, to the CAMA equipment at the toll point. ANI is described here and CAMA is described in the next subsection.

The line circuits and the Step-by-Step fabric are shown across the top of Figure 9.2. All outgoing toll trunks were modified for ANI/CAMA operation. This relatively simple change provided access for dial-pulse or multifrequency signaling over the trunk to the CAMA/toll point. In Step, the calling party signals the called party's directory number to the CAMA/toll point directly from his telephone. The ANI outpulser signals the calling party's directory number over this same trunk.

The ANI equipment is shown across the bottom of Figure 9.2. A number network is associated with each different 10K-line office of 10,000 directory numbers. Each number network's 10,000 input wires are attached to the sleeve wires in the 10,000 line circuits in its office. The number network is described below. Each Step building (a building may house several offices) has one ANI identifier. The ANI identifier gets its input from the offices' number networks and delivers its output to the ANI outpulsers. The station identifier was added later and is discussed in the next chapter.

Consider a subscriber, served by a Stepper, who places a toll call in this ANI/CAMA environment. The subscriber dials a 1 prefix and the first selector handling the call hunts for an idle outgoing ANI/CAMA toll trunk. This subscriber continues dialing the directory number, possibly including an area code, of the intended called party. These digits proceed through the switch train in the Step fabric, out over the toll trunk (after being repeated in the originating circuit), and into the Number 4 Crossbar office at the toll point. When dialing is completed, the toll office "winks" to signal the local Step office to transmit the calling party's directory number. A wink is a momentary battery reversal on the trunk (a prolonged battery reversal signals called party answer).

The originating circuit of the ANI/CAMA trunk recognizes the wink and connects to an idle ANI outpulser in the same way that an outgoing trunk in Number 5 Crossbar connects to an idle outpulser. The ANI outpulser bids for the ANI identifier and waits for its signal of readiness. When the ANI identifier is ready to serve this call, it signals the ANI/CAMA OGT to operate a relay that places a 5800-Hz tone on the sleeve wire on the fabric side of the trunk circuit. Since the entire building has only one ANI identifier, only one sleeve connection through the entire fabric carries this tone. This tone is present on the sleeve wire in only one line circuit on the originating side of the fabric and is detected on only one of the 10,000 inputs on only one of the number networks.

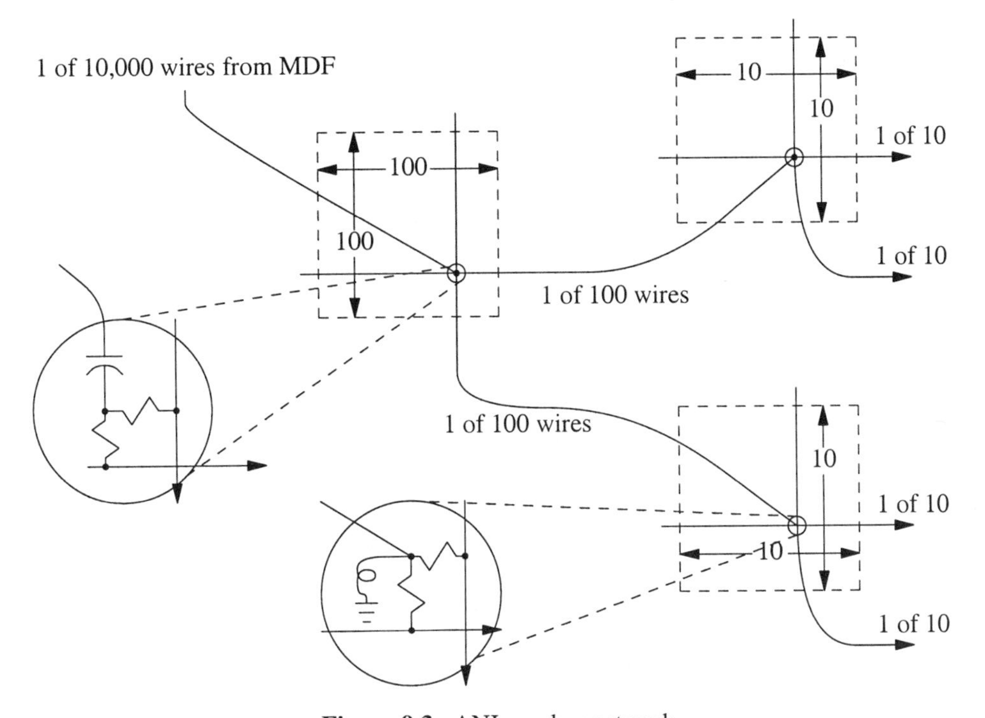

Figure 9.3. ANI number network.

An ANI number network for one office of 10,000 lines is shown in Figure 9.3. Ten thousand resistor-capacitor (RC) circuits are arranged in a square array of 100 × 100. The wire from the capacitor in each RC circuit is connected to the sleeve wire in the appropriate line circuit. It is this connection that effectively assigns a line circuit to a directory number. Establishing this connection is added to the procedure of installing any new subscriber or changing a subscriber's directory number. The RC circuit at coordinate $[i,j]$ has two resistors, one from the common point to the ith horizontal wire and another from the common point to the jth vertical wire.

Two hundred resistor-inductor (RL) circuits are arranged in two square arrays of 10 × 10 each. The capacitor and inductor values are selected for series resonance at 5800 Hz. In one of the 10 × 10 arrays of RL circuits, the wire from the common of each RL circuit is connected to one of the 100 horizontal wires in the 100 × 100 array of RC circuits. The other 10 × 10 array of RL circuits is similarly wired to the 100 vertical wires in the 100 × 100 array of RC circuits. In each 10 × 10 array, the RL circuit at coordinate $[i,j]$ has two resistors, one from the common point to the ith horizontal wire and another from the common point to the jth vertical wire.

Consider a toll call from the subscriber with directory number 756-1234. The 5800-Hz tone proceeds from the ANI/CAMA OGT to the sleeve wire on the fabric side of this trunk, along the sleeve connection through the fabric to its originating side, to the sleeve wire in the calling party's line circuit, and along one of the 10,000 wires to the number network that corresponds to the 756- office. This particular wire had been connected to the capacitor in the RC circuit at coordinate [14,23] in the 100 × 100 array. At this RC circuit, the tone signal divides and proceeds through each resistor. The tone is present only on wire 14 of the 100 horizontal wires and only on wire 23 of the 100 vertical wires in the array at the left of Figure 9.3.

This 14th horizontal wire is connected to the common point in the RL circuit at coordinate [1,4] in the 10 × 10 array in the upper right of Figure 9.3. At this RL circuit, the tone signal divides and proceeds through each resistor. The tone is present only on wire 1 of the 10 horizontal wires and only on wire 4 of the 10 vertical wires in the array at the upper right of Figure 9.3. Associated with these 10 horizontal wires, a set of 10 sense amps and a binary encoder indicates that the thousands digit of the calling party's directory number is 1. Associated with these 10 vertical wires, another set of 10 sense amps and a binary encoder indicates that the units digit of the calling party's directory number is 4.

The 23rd vertical wire from the array at left is connected to the common point in the RL circuit at coordinate [2,3] in the 10 × 10 array in the lower right of Figure 9.3. At this RL circuit, the tone signal divides and proceeds through each resistor. The tone is present only on wire 2 of the 10 horizontal wires and only on wire 3 of the 10 vertical wires in the array at the lower right of Figure 9.3. Associated with these 10 horizontal wires, a third set of 10 sense amps and a binary encoder indicates that the hundreds digit of the calling party's directory number is 2. Associated with these 10 vertical wires, a fourth set of 10 sense amps and a binary encoder indicates that the tens digit of the calling party's directory number is 3.

By considering which number network produced the result, the ANI identifier assigns 756 as the office code. The "1234" is appended, the seven-digit directory number is forwarded to the ANI outpulser, and the ANI identifier disconnects and proceeds to serve another call. The ANI outpulser transmits the seven-digit directory number, by dial pulses or by multifrequency signaling, to the CAMA/toll point over the ANI/CAMA trunk. The ANI outpulser disconnects and proceeds to serve another call. The processing at the CAMA/toll point follows the same procedure as CAMA with Number 5 Crossbar, which is described next.

9.2.4 AMA with Number 5 Crossbar

Number 5 Crossbar is the first of the popular switching systems (and the only one in the 1950s) in which the calling party is not anonymous at the time that the system must make this first AMA entry. The only minor inconvenience is that the calling party is known by the fabric address and not by the directory number. So the only modification to the architecture is that an extra translation is required in the processing of toll calls.

With Step, Panel, and Number 1 Crossbar local offices, the AMA equipment was located at the toll point in an arrangement called central automatic message accounting (CAMA). The local office was modified to transmit the calling party's directory number to this CAMA/toll point over the same trunk that would be used for the call. A Number 5 Crossbar office could also be modified to operate in conjunction with a CAMA/toll point. However, the option was also available with Number 5 to locate the AMA equipment at the local office in an arrangement called local automatic message accounting (LAMA). Both arrangements are described in this subsection.

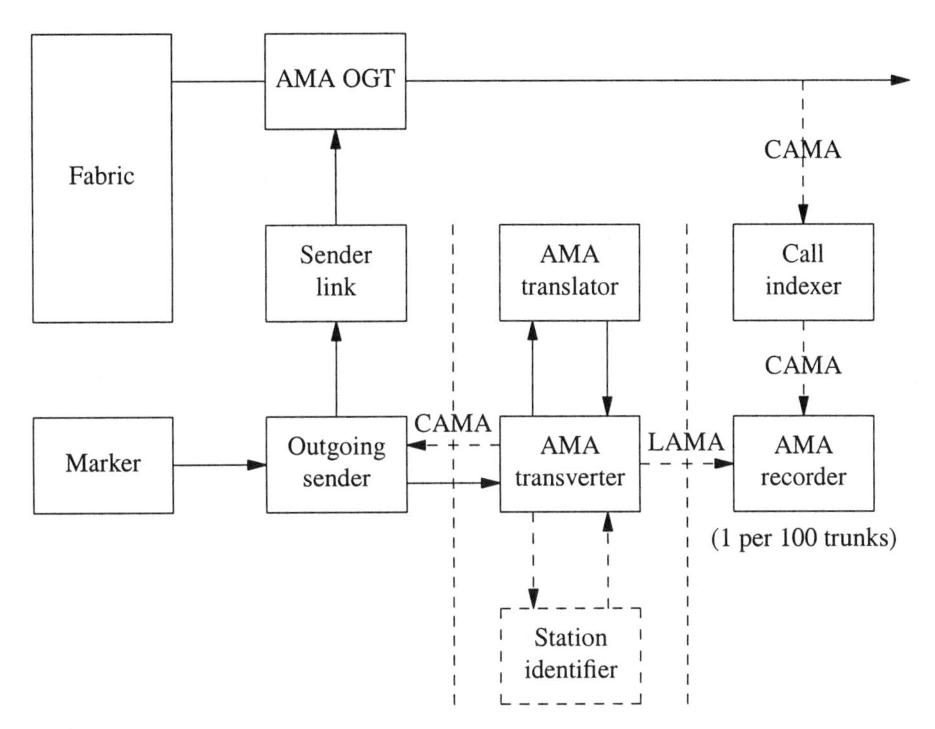

Figure 9.4. LAMA and CAMA with Number 5.

Figure 9.4 shows a partial block diagram of a Number 5 Crossbar office, modified for AMA. The two vertical dashed lines in Figure 9.4 separate the original Number 5 equipment on the left, the adjunct equipment for AMA in the center, and the AMA equipment itself on the right. In a LAMA arrangement, the AMA equipment on the right of Figure 9.4 is located in the building with the local office. In a CAMA arrangement, this same AMA equipment is located at the CAMA/toll point instead. Only one minor variation distinguishes LAMA and CAMA operation in walking a toll call through a Number 5 Crossbar office. The adjunct equipment consists of a transverter and an AMA translator. The transverter is a primitive message passing subsystem. The station identifier, shown below the transverter in Figure 9.4, was added later and is described in the next chapter.

The AMA translator had 40 large magnetic cores, organized in four groups of 10 each. Its operation is reminiscent of magnetic core computer memory, a new technology at the time. In this AMA translator, line equipment number is decoded into a pulse of current on one of 10,000 wires. Each wire is strung serially through one core in each of the four groups depending on the value of each digit in the four-digit directory number. This pulse of current on one wire induces voltages across four of the 40 cores and these voltages are encoded into directory number. The wiring of the AMA translator for the correct translation was added to the tasks involved in installing a new subscriber.

Consider a subscriber served by a Number 5 office modified for AMA, who originates a toll call. After dialing the second digit into the originating register (OR, not shown in Figure 9.4). This OR performs its new pretranslation algorithm and determines how many digits to expect. This OR, like every OR in every Number 5 office in the country, has also been modified to accept up to 10 digits. After receiving all of the expected number of dialed digits, this OR passes call information to an idle completing marker (CM), and is disconnected. Via the foreign area translator (also not shown), this CM identifies a trunk group. An idle AMA outgoing trunk is selected from this trunk group and the subscriber is connected to it through the fabric. In the CAMA arrangement, all such trunks proceed to the CAMA/toll point. In LAMA, the trunk can proceed directly to another local office.

Then, this CM passes call information to an idle outgoing sender and then it drops off the call. This outgoing sender connects to the selected AMA OGT through the sender link and MF pulses the called party's directory number. Simultaneously, it passes all the call information to the transverter, which passes only the calling party's line equipment number to the AMA translator. The calling party's directory number is returned from the AMA translator back to the transverter. At this point in the call walk-through, the procedure diverges depending on whether the AMA arrangement is local or central.

In a LAMA arrangement, the transverter passes the called and calling party's directory numbers and the trunk identity to the local AMA recorder. The call attempt entry is recorded for this call. In a CAMA arrangement, the transverter passes the calling party's directory number back to the outgoing sender. This outgoing sender appends the calling party's directory number to the message it is still in the process of MF pulsing to the CAMA/toll point. The switching equipment there receives called and calling party's directory numbers over a trunk it can identify and has all the call information needed to pass to the central AMA recorder. The call attempt entry is recorded for this call. During the 1950s and early 1960s, when MF pulsing was audible, the alert Number 5 subscriber could determine whether the arrangement was LAMA or CAMA by the duration of the MF pulsing on toll calls.

The call indexer is part of the AMA equipment and it interfaces the AMA recorder to the AMA trunks. The call indexer appears on the trunk side of a LAMA CO and on the CO side of a CAMA/toll point. The call indexer monitors all the trunks for supervisory signaling. It creates the call answer and call disconnect entries, including trunk ID and time to the 10th of a minute, for the AMA recorder. The tapes are removed from all AMA recorders in the early morning hours and delivered to the centralized data processing facility. Since the time-of-day clock in the AMA recorder advances every 10th of a minute (or 6 seconds), with extremely short toll calls it is possible that the call answer and call disconnect entries might have the same time stamp. So, occasionally, very short calls are not billed.

9.2.5 Number 4 Crossbar

A version of Crossbar switching equipment, specifically designed for the toll point, was called Number 4 Crossbar. Since the toll point functions as the gateway to the national network for the central offices that home on it, the sidedness of a toll office is the *central office side* and the *network side*. Intertoll trunks, which connect a class 4 toll office with the class 3 office on which it homes and with other neighboring class 4 offices, interface on the network side. Two-way and one-way toll trunks, which connect a toll office with the COs that home on it, interface on the CO side. If a toll office doubles as a CAMA point, as many did, these toll trunks on the CO side are the other end of the Step ANI trunks and the Number 5 CAMA trunks described earlier in this section.

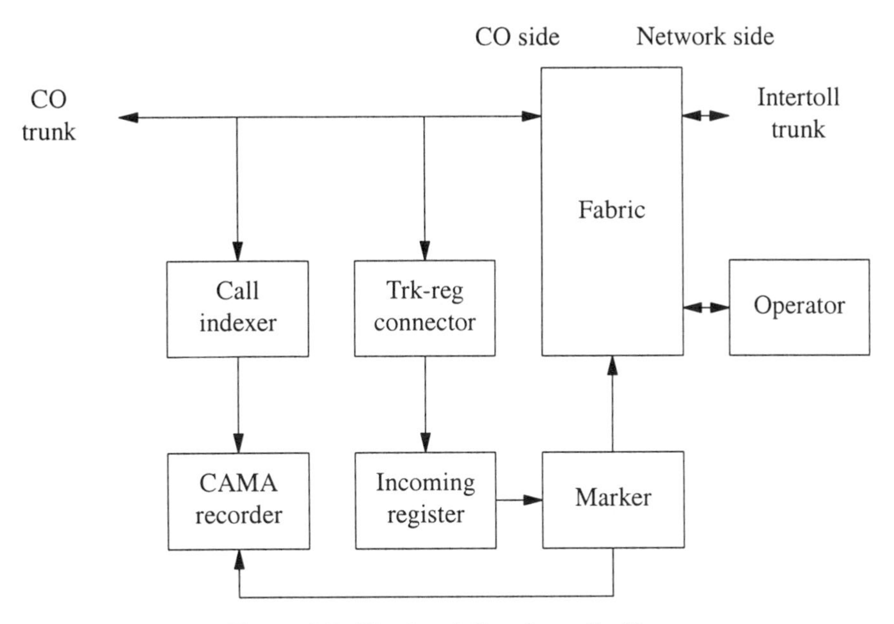

Figure 9.5. Number 4 Crossbar toll office.

The general architecture of Number 4 Crossbar is illustrated in Figure 9.5. The fabric was a typical Crossbar fabric, not unlike Number 5's fabric described in the previous chapter. The markers were equivalent in function and design to the completing markers in Number 5. While the markers passed call information to the CAMA recorder for the call attempt entry, the call answer and call disconnect entries were created through the call indexer, which monitored the CO trunks for supervisory signaling. Incoming registers, each including an MF receiver, appeared on special connectors and not on the main fabric because of timing requirements with trunks from Step offices, as discussed in Section 8.4.5. Toll operators appeared on the network side of the fabric, and a fabric connection was required to connect one to a CO trunk.

Consider a conventional direct-dialed long distance call. If the subscriber is served by a local Step office, he dials 1, a three-digit area code (if the called party is another NPA), and a seven-digit directory number. The leading 1 causes the first selector to connect to an idle ANI/CAMA outgoing trunk. In the Number 4 Crossbar office at the toll point, the trunk-register connector detects this trunk seizure and it connects an incoming register to the trunk. The remaining 10 or seven digits of the desired called party are dialed by this subscriber directly into this incoming register. When pretranslation indicates that dialing is complete, the incoming register causes the trunk's terminating circuit to wink for the calling party's identification. The ANI equipment in the local Step office responds and the calling party's directory number is dial- or MF-pulsed over this same trunk into this same incoming register. Now, the incoming register bids for an idle marker, passes to it all the call information (including trunk ID), and drops off the call. The marker passes the call attempt entry to the CAMA recorder and completes the call.

If this subscriber is served by a local Number 5 Crossbar office instead, he dials the three-digit area code (if the called party is in another NPA) and the seven-digit directory number, without the leading 1. The completing marker in this Number 5 office connects this subscriber and an outgoing sender to an idle toll trunk. This trunk seizure causes the trunk register connector in the Number 4 Crossbar toll office to connect an idle incoming register to the other end of this toll trunk. If the Number 5 local office is part of a CAMA arrangement, the outgoing sender MF-pulses the calling and called party's directory numbers and the call proceeds as above. If the Number 5 local office has LAMA, this incoming register receives only the called party's directory number and the call proceeds as above except that the call attempt entry is not made because it was made on the LAMA recorder in the local office.

If the subscriber served by a Step office dials 0, he is connected to an idle local operator by the first selector, or, if a toll area's local operators are pooled at the toll point, he is connected to one of a special group of toll trunks. Seizures on these trunks cause the marker in the Number 4 office to connect the trunk through the fabric to an operator. If the subscriber served by a Number 5 local office dials 0 or 211, the outgoing sender MF-pulses a code into the incoming register indicating the need for an operator, and a separate trunk group is not necessary. The MF-pulsing format was described in Section 4.4.2. Calls from hotels or other PBX's that originate over a PBX trunk into a Number 5 office are marked by class of service so that the outgoing sender sends the code for operator intervention for calling party identification (more on this in the next chapter). Similarly, the ANI outpulser or outgoing sender MF-pulses the code for operator intervention in the event of a failure in the ANI equipment or AMA translator.

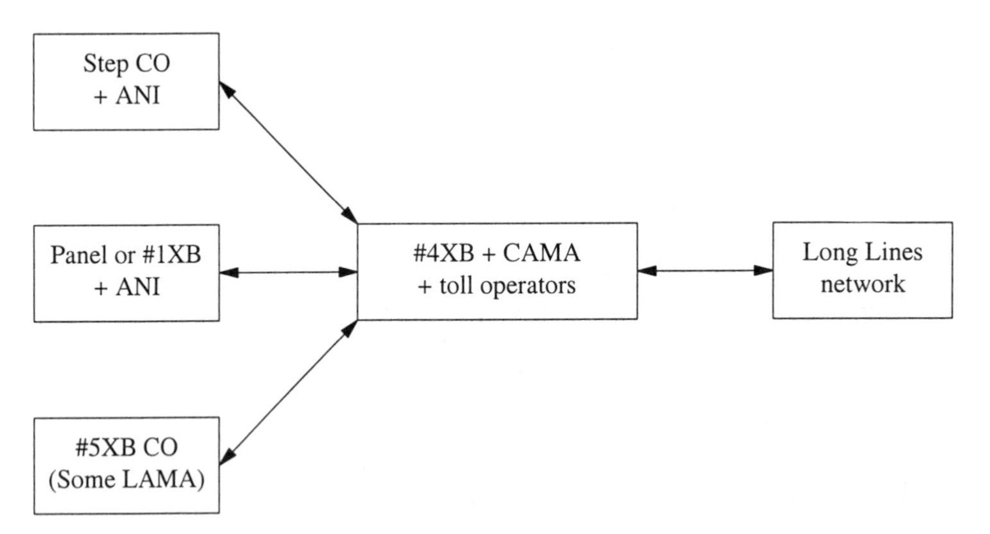

Figure 9.6. Toll architecture in the 1960s.

9.2.6 The Toll Point in the 1960s

By the early 1960s, the nation's toll points had all evolved to the common general architecture shown in Figure 9.6. All the manual toll offices had been automated, and all with Number 4 Crossbar. Most of the nation's Step, Panel, and Number 1 Crossbar offices had been augmented with ANI equipment, and the toll points that served these offices doubled as CAMA Points. All of the nation's Number 5 Crossbar offices were equipped with AMA translators and had either LAMA or CAMA arrangements. Direct distance dialing was practically universal. The local operator was almost extinct: almost every operator function was performed by operators, pooled at the toll point, who served an entire toll area. The billing process was automated in modern data processing centers.

The evolution to this common architecture was very natural. Nobody really noticed, at that time, that four related, but completely separable, functions were performed through this same architecture. The toll point served as a:

- Hub among neighboring central offices for intratoll calls;
- Gateway to the (only) national network for intertoll calls;
- Operator pool for greater economy of scale over a toll area's operators;
- Central point for AMA recording.

Since the architecture of Figure 9.6 was relatively optimal for all four of these functions, it is only with hindsight that their separate nature is observed. While this architecture was stretched in the era of toll modernization, it was not until divestiture in 1984, when these functions were split between separate companies, that this common architecture breaks. Today, these four functions above are each performed by separate equipment.

9.3 MODERNIZATION AND THE 1980s TOLL POINT

It took 15 years to implement toll automation across the entire United States. DDD was so successful that the cost of long distance calls fell dramatically. Even though the long distance service was subsidizing local service, most long distance prices dropped so much, especially in comparison to other costs, that subscribers made many more calls than they used to. Even with DDD and with pooling local operators at the toll point, subscriber demand on operators remained high because of the general increase in traffic. The architecture was barely able to handle the huge increase in long distance traffic.

9.3.1 Four Technologies

Four technologies evolved between 1950 and 1965, any one of which was significant enough to suggest a modernization. So, by the time all the toll points were finally automated, it was time to start over again to modernize them.

Magnetic recording on ferrite-coated plastic tape made large technological advances in this era. Reel-to-reel analog tape recorders became popular in the consumer audio market. Digital magnetic tape became the common means of mass memory in computer systems, replacing punched paper tape because of higher bit density and much higher reliability. Punched paper tape was plagued with problems because of paper chaff and moving mechanical parts. A new digital magnetic tape AMA recorder was designed for new and growth applications, but considerable thought was given to a national program of replacing all the old punched paper tape AMA recorders.

Digital electronics began with discrete-component resistor-transistor logic (RTL) and evolved to diode-transistor logic (DTL) and then to transistor-transistor logic (TTL). The integrated circuit was invented and the cost implications were very obvious. It was clear that electromechanical equipment was dated and any new installation using equipment made from Crossbar switches and relay logic was probably foolish.

Program control through digital electronic computers had obvious advantages over wired-logic control through markers (see Chapter 11). The #1 Electronic Switching System (ESS) was under development, and clearly the arguments that suggested a computer-controlled local office would apply at least as well to the toll point. Think of all the money that would have been saved in the DDD campaign if all those local and toll switching systems had been programmable.

Digital transmission had cost and reliability advantages over conventional and multiplexed analog transmission. The T1 Carrier System, in which 24 analog voice channels are analog-to-digital converted and multiplexed up to 1.544 Mbps, was introduced and was extremely popular for *pair gain*. With the huge growth in long distance calling, pair gain was needed most in the trunk plant. With so much digital carrier in the trunk plant, perhaps a digital fabric at the toll point might provide some kind of synergy or overall cost savings.

Did it make sense to respond to each of these technologies separately, or did it make more sense to institute another huge national program and respond to all four technologies simultaneously? The toll point was modernized in two stages. First, the Traffic Service Position System (TSPS) was introduced, and then Number 4 Crossbar was replaced by #4 ESS, a computer-controlled system with a digital fabric.

9.3.2 Traffic Service Position System

Pooling local operators into the toll point enabled a huge cost savings through economy of scale. But while the occupancy of toll operators was much higher than that of local operators, it was still quite low. Perhaps, pooling operators even higher up in the network hierarchy could effect even more cost savings. Even more significant than the economy of scale, installing a system which automates the operator's activity on an assisted call should reduce the operator's holding time and permit even fewer operators in the area served by a sectional center. Some artificial numbers were presented in Section 5.4.1 to give a sense of how effective this pooling and reduced holding time would be. But would it really be economical to switch all operator-assisted telephone calls all the way up the network hierarchy to the class-3 sectional center?

An alternate architecture [1] suggested placing an *intercept circuit* in every toll trunk in a class-3 region. This intercept circuit is similar to the circuit described in Section 6.8.4 that intercepts the first selector link, between a line finder and its mated first selector, to provide Touch Tone service in a Step office (Figure 6.6). Then, all these intercept circuits would feed one system that was logically between each CO and its toll point. The physical arrangement was quite complicated and expensive.

Figure 9.7 illustrates how one TSPS can serve every toll trunk in a class-3 region. This region has only four buildings, shown in dashed boxes in Figure 9.7. The building in the lower right holds a central office and a toll office. This toll office serves the CO in the same building and the CO in the building shown in the upper right. The building in the center of Figure 9.7 holds a CO, a toll office, a regional office, and the TSPS. This toll office serves the CO in this same building and the CO in the building in the lower left of Figure 9.7. This regional office serves the two toll offices. While intrabuilding trunks are relatively inexpensive, appropriate conduit and transmission facilities are provisioned to handle the interbuilding trunks.

Consider installing TSPS into this environment. The four boxes labeled "I" represent intercept circuits, one associated with each of four representative toll trunks.

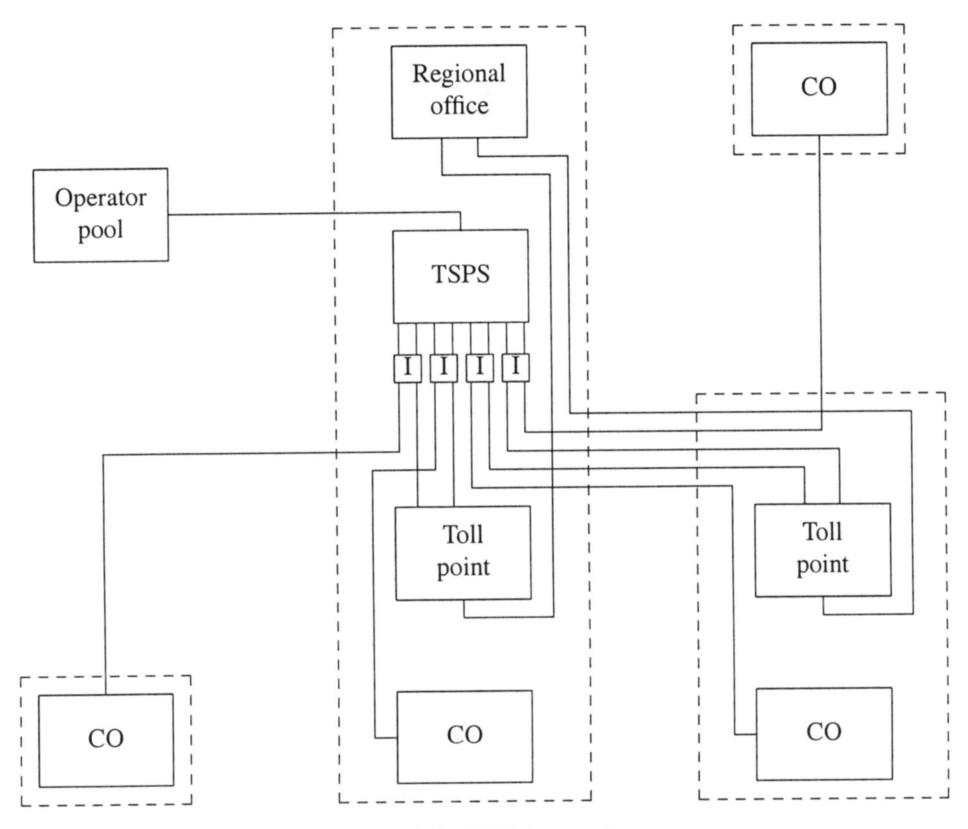

Figure 9.7. TSPS in a region.

The leftmost intercept circuit intercepts a toll trunk between the CO in the lower left and the toll office in the center. The next-leftmost intercept circuit intercepts a toll trunk between the CO and the toll office that both reside in the building in the center. The rightmost intercept circuit intercepts a toll trunk between the CO in the upper right and the toll office in the lower right. The next-rightmost intercept circuit intercepts a toll trunk between the CO and the toll office that both reside in the building in the lower right.

The conduit between the building in the center and the building in the lower right contains five representative channels. Only the upper channel, representing a trunk between the toll office at right and the regional office, was required before the TSPS was installed. This is typical of the manner in which channels had to be squandered in order to implement this architecture. The saving grace is that a new conduit was typically not needed and the additional channels were typically acquired through pair gain on existing physical cables.

It wasn't clear that the cost of the TSPS and all those additional channels would cover the savings effected by the economy of scale of higher-order pooling of operators. Perhaps the TSPS should be more complex than serving as a simple *operator concentrator?* By the late 1960s, the electronic program-controlled Electronic Switching System had been developed and successfully installed. TSPS would be built in this same technology and with a similar architecture. While additional complexity will increase the

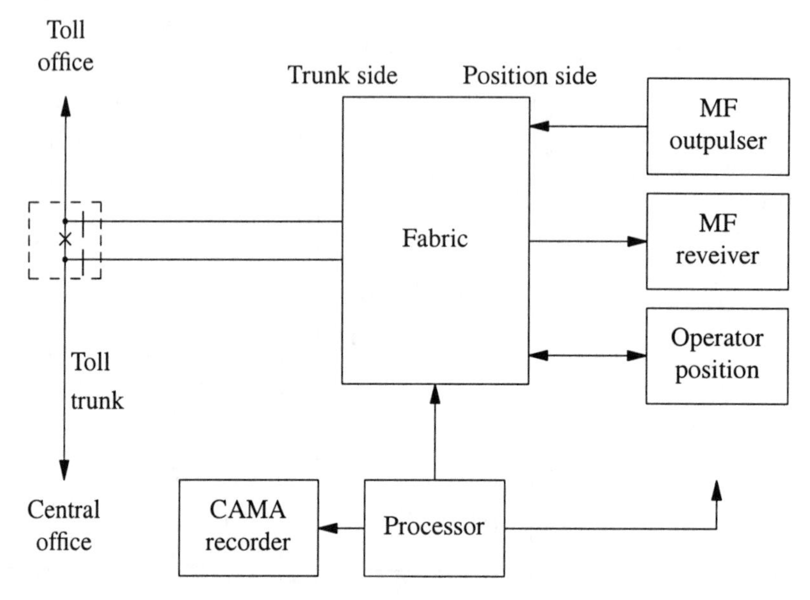

Figure 9.8. Traffic service position system.

system's cost, it will allow greater functionality, which could lead to lower overall cost in at least three ways.

- Instead of intercepting the trunk on command from the toll point, when its marker determines that an operator is required, the trunk could be intercepted during outpulsing. This way the computer program control in the TSPS processes the call instead of the marker in the toll point. Modernization of the toll point can be postponed, thus retaining the investment and depreciation schedule of the Number 4 Crossbar equipment.
- If the TSPS is going to know the call information, it can be an alternate CAMA point. The new magnetic tape AMA recorders can be gradually installed at the TSPS, and the old paper tape AMA recorders can be gradually decommissioned at the toll points.
- Program control and knowledge of the call information suggest the possibility of assisting the operator with the call, thus reducing the holding time on the operators. This was discussed in Section 5.4.1. New operator consoles were designed that displayed calling and called numbers before the operator even got on the call.

The national numbering plan was modified for 0+ calling, in which 0 became a dial prefix instead of a complete call. Subscribers wishing to place a collect call, for example, dialed the DDD directory number with a 0 prefix.

Figure 9.8 shows the general architecture of a TSPS. TSPS hardware and software is similar to that of the #1 ESS. Since the #1 ESS is not described until Chapter 11, this description of TSPS is kept at a very general level. The sidedness of a TSPS has trunks on one side and operator positions and service circuits on the other side. The representative trunk in Figure 9.8 connects the CO at the lower left to the toll point at the upper left. The dashed box contains the intercept circuit. Every trunk has two appearances on the TSPS fabric, one appearance for the CO segment of the trunk and one appearance for the toll segment of the trunk. MF receivers, on the position side of the fabric, receive the MF signaling that the MF outpulser in the CO transmits over the trunk (thinking it's going to the toll point). MF outpulsers, on the position side of the fabric, transmit the MF signaling that was intercepted and possibly modified to the MF receivers in the toll point.

The regional pool of operators also appears on the position side of the TSPS fabric. The new operator console has panels on which the TSPS processor displays calling and called directory numbers. The operator controls the intercept circuit and may connect to either segment of the trunk, or both. Most important, the consoles may be remote from the building that houses the TSPS. While the operators are logically pooled at the TSPS for an entire class-3 region, typically they are not physically pooled. Thus, operators can have a working environment in an office building that would be more pleasant and safer than the typical neighborhood where the telephone company's buildings are found. This feature makes it easier to hire and retain operators. The consoles are typically remoted into several office buildings that are scattered around the serving region.

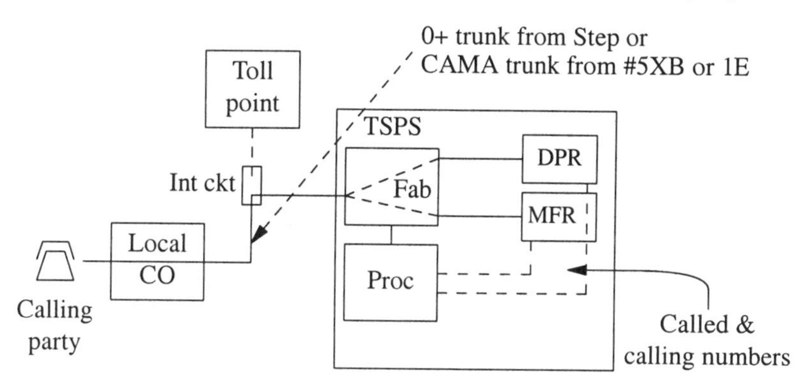

Figure 9.9. TSPS intercepts the interoffice signaling to the toll point.

Suppose the subscriber, at left in Figure 9.9, places a collect call to some party in another NPA. Under the new numbering plan, the subscriber dials 0, followed immediately by the area code and directory number of the called party. If the calling party's local office is Step-by-Step (Chapter 6), the leading 0 causes the first selector to connect the subscriber to an ANI toll trunk in a 0+ trunk group. Since the trunk's intercept circuit is in the *intercepted state,* the seizure is detected by the TSPS and not the toll point, and the trunk's CO leg is connected through the TSPS fabric to a dial pulse receiver. The remaining 10 digits dialed by the subscriber are received by the TSPS and not by the toll office (yet). At the end of dialing, the DPR winks, and the ANI outpulser MF pulses the calling party's directory number over this same trunk into a multifrequency receiver in the TSPS.

If the calling party's local office is Number 5 Crossbar (Chapter 8), the subscriber dials his 11-digit number into an originating register. The completing marker causes the subscriber and an outgoing sender to be connected to a CAMA trunk. Since the trunk's intercept circuit is in the intercepted state, the seizure is detected by the TSPS and the CO leg is connected through the TSPS fabric to an MF receiver. The outgoing sender MF-pulses the 11 digits (including the 0) that the subscriber dialed and the calling party's directory number toward the toll office, but they are intercepted by the TSPS. If the calling party's local office is a #1 ESS, the call processing and signaling are similar.

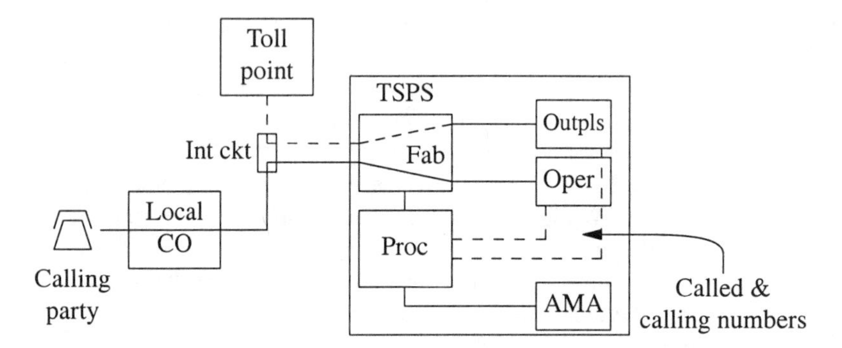

Figure 9.10. TSPS operator talks to the calling party.

The program in the TSPS processor interprets the digits, disconnects the DP or MF receiver, locates an idle operator, displays the calling and called party's directory numbers on this operator's console, and connects this operator to the CO segment of the trunk. Before TSPS, the operator at the toll point started a long conversation with "Number, please." But with TSPS assistance, since the number is known, the conversation between the operator at the TSPS and subscriber is brief. Over the connection shown in Figure 9.10, the operator says, "How may I help you?" The subscriber replies, "Collect, please." The operator asks, "Whom shall I say is calling?" The subscriber replies, "Joe Jones." The operator says, "Thank you, Mr. Jones."

Then the operator signals the processor to complete the *assisted call* through the toll point. TSPS connects an MF outpulser to the toll segment of this trunk (the dashed connection in Figure 9.10) and causes the called party's directory number to be transmitted. The call proceeds at the toll point as if these digits had come directly from the CO. The TSPS processor makes a *collect call attempt* entry on the CAMA recorder.

After outpulsing, the trunk's toll segment is disconnected from the TSPS outpulser and is connected to the operator. The operator hears audible ring tone from the distant office (at right in Figure 9.11) as the called party is rung. The calling party hears audible ring tone from the TSPS. Depending on how busy or how suspicious the operator is, the operator can verify in a directory that the indicated calling party's directory number does, in fact, belong to someone named Jones.

When the called party answers, that party and the operator converse about a willingness to accept the charges. If called party agrees, the operator signals the processor and the *call answer* entry is made on the CAMA recorder. The operator is disconnected from the trunk and is free to process another call, and the intercept circuit is placed in the *cut-through* state. The TSPS is removed from the trunk except that its call indexer responds when a supervision change indicates that the *call disconnect* entry is needed on the CAMA recorder. The TSPS operator intercepts non-AIOD toll calls from a PBX and hotel/motel calls in a similar manner.

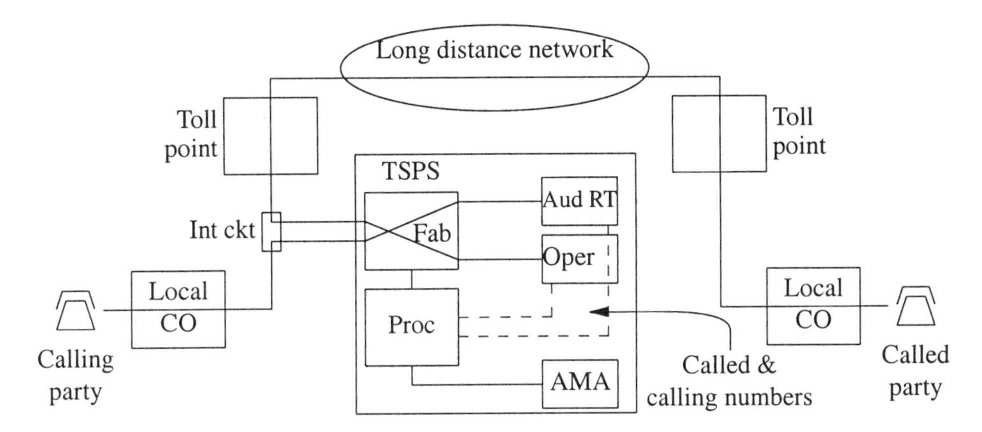

Figure 9.11. TSPS operator talks to the called party.

9.3.3 Effect on the Numbering Plan

America's national number plan was changed by TSPS. The change had to be introduced regionally, as TSPS was gradually deployed and only where it was deployed. Until TSPS, 0 had always caused a one-digit call to the operator. Beginning with TSPS, 0 became a prefix. Now, a call to the operator requires dialing 00, or a single 0 followed by a time-out. Operator-assisted calls, such as collect calls, are dialed with a 0 prefix, called 0+ dialing. Long distance calls that did not require operator assistance required a 1 prefix in TSPS regions, but 1+ dialing was still not universal.

As 0 became a prefix, instead of a one-digit call, some new dial prefixes, beginning with 01, became available. So, 011 was adopted as the prefix for international dialing. Only in the past couple of years, has 1+ dialing become a universal part of America's national number plan [2]. Now a leading 1 is the prefix for any call to another NPA, as long as operator intervention is not requested. The 10-digit toll call is in the past.

A call within your NPA requires seven digits, where the first digit can't be 0 or 1.
A call to another NPA requires 11 digits, where the first digit must be a 0 or 1.

9.3.4 Other Attendants

DDD changed the dialing code for information. Before DDD, the subscriber dialed 411 for a local information operator, who needed access to a national set of telephone books. After DDD, the subscriber needed to call the information operator in the NPA of the desired called party and the code changed to 1+NKX+555+1212, where NKX is the area code of the destination NPA.

A system, similar to TSPS, was designed and installed nationally for increased automation and distribution of information operators. Like TSPS, this capability is now built into the software of modern digital switching systems. The same principles for economy of scale and desirable work location apply to information operators as apply to toll operators. These same principles have been also been applied to the design of 911 systems.

Service observing is a practice in which specialized attendants listen to the early part of random users' conversations in order to determine the quality of randomly sampled connections. While service observing attendants were carefully selected and only the first 30 seconds of any call were sampled, service observing has always been a controversial practice. A system, similar to TSPS, was also designed and installed nationally for increased automation and distribution of service observing operators.

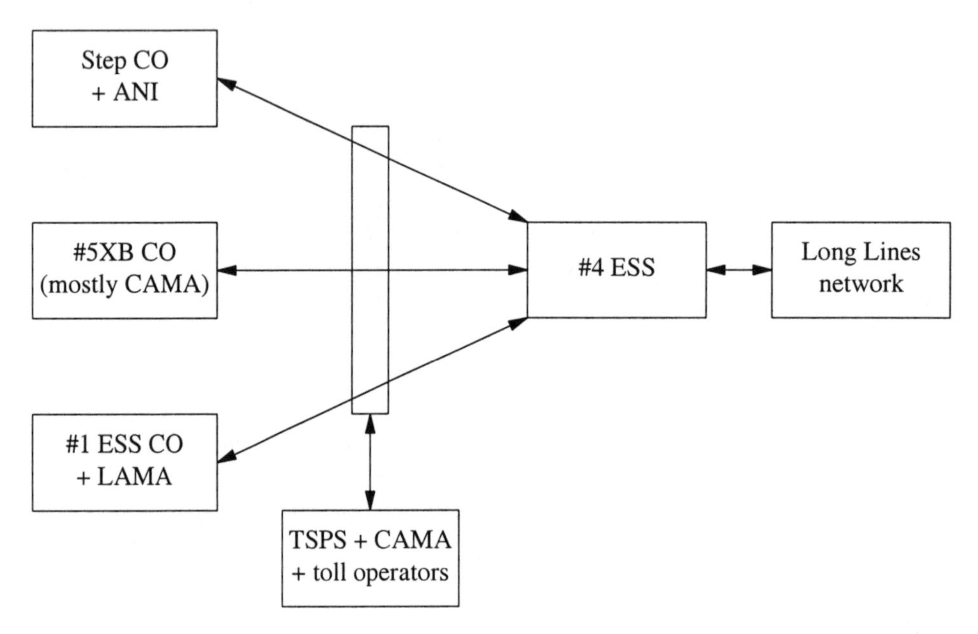

Figure 9.12. Toll architecture in the 1980s.

9.3.5 The Toll Point in the 1980s

Figure 9.12 shows the general architecture of the toll point after the toll modernization campaign was completed, by the early 1980s. Be careful because this figure is a little ambiguous: TSPS and the 4E wouldn't necessarily coexist in the same toll area. Since 4E deployment was typically expensive and typically accompanied a massive digitization of the concommitant transmission plant, the 4E was gradually deployed in a nationwide program. But since the operating companies were eager for the cost savings of modernization, TSPS would be installed in the interim. TSPS would be used in a given region until all the region's Number 4 Crossbar toll points were replaced by the 4E. Since the 4E had magnetic-tape CAMA and operator assistance software [3, 4], the TSPS could be decommissioned after all the region's toll points were modernized to 4Es — but it frequently was not. Once TSPS and/or the 4E were universal, the entire nation was using 0+ calling for operator assistance.

Eventually, the Number 4 Crossbar equipment in over 1000 class-4 offices was replaced by the digital electronic #4 ESS. A description of the 4E is postponed until Section 14.1. By this time, Panel is extinct and Number 1 Crossbar is almost extinct. Three significant local office technologies remain. Step offices are abundant and all are equipped with ANI. Number 5 Crossbar is equally popular, most of them using the CAMA arrangement with TSPS as the CAMA point. The #1 ESS and its technological successor, the #1A ESS, are most popular for most of these offices using the LAMA arrangement.

The four functions, once concentrated at the toll point, were now split between the TSPS and the toll point. The TSPS serves (1) as the access to the operator pool and (2) as the CAMA point; the toll office serves (3) as the CO hub and (4) as the gateway to the national network. While the four functions were partitioned into two locations, divestiture would require that the four functions be partitioned between two companies. As described in the next section, these splits were unfortunately inconsistent.

9.4 DISINTEGRATION

Divestiture will be described in great detail in Chapter 16. The important point here is that in 1984 the Bell System was broken into eight different companies. Seven of these are the Regional Holding Companies that own the Bell Operating Companies which provide local telephone service. The eighth is the new AT&T, a long distance company, among other things. In dividing the hierarchical national network among these companies, it was clear that the class-5 offices should go to the respective local telephone company and that class-1 through class-3 offices should go to AT&T. But who should get the class-4 offices and the TSPSs?

As the local area's gateway to the long distance network, a class-4 office should have gone to AT&T. But, as the intratoll hub among an area's local exchanges, a class-4 office should have gone to the local Bell Operating Company. As the regional pool for long distance operators, a TSPS should have gone to AT&T. But, as a CAMA point, where billing information is collected for an entire class-3 region, a TSPS should have gone to the local Bell Operating Company. The class-4 offices and the TSPSs were divided up in ways that were inconsistent from one region to the next.

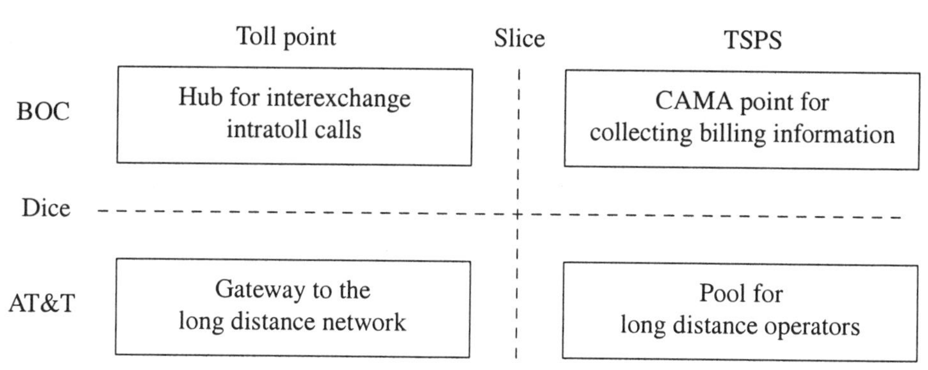

Figure 9.13. Disintegration of the four toll functions.

Figure 9.13 illustrates the disposition of those four functions that seemed so logically to belong together and that were integrated and colocated into the toll point so many years before. With the advent of some new technology, the desire to introduce it into the network gradually, and the rapidly declining ratio of the cost of personnel to the cost of transmission, TSPS was introduced in the early 1970s as a temporary fix. The four functions were "sliced" apart, with two going to TSPS and two remaining at the toll point. Then, with divestiture in 1984, the four functions were "diced" even further apart, with two going to AT&T and two going to the newly independent BOCs.

Those class-4 offices that went to AT&T after divestiture became the *gateway,* called the point of presence (POP), into AT&T's network for what used to be a toll region. If AT&T didn't get the toll point in some region, then a POP was added — and not necessarily another 4E. AT&T and most of the other IXCs now typically use a digital class-5 switching office for their POP. Besides having additional programs to handle the gateway function, a system like a 5E already comes with software to handle subscribers and comes with with Centrex software, neither of which were available in the 4E. Athough they lack the local connectivity, the IXCs could provide competitive local switching, especially to bypassed (Section 16.5.4) Centrex and PBX customers. Eventually, AT&T flattened its classic five-level hierarchy into an architecture called dynamic nonhierarchical routing, in which routing is done in the most efficient way without adhering to a hierarchy [5]. DNHR is described in Section 16.5.2.

A class-4 office that went to a BOC after divestiture became this BOC's *interexchange hub* for what used to be a toll region. If the BOC didn't get the toll point in some region, then they had to add one — not necessarily a 4E and not necessarily in a physical star. Most of the BOCs also now typically use a digital class-5 switching office for their hub. This switch is now typically called a tandem office, a name (Section 4.1.2) that confuses people familiar with the predivestiture architecture.

But if a BOC lost the physical toll point in divestiture, it had to also redeploy its interexchange network, not just replace the toll point. Many of the BOCs, and other LECs, architected their interexchange networks around a Sonet ring. If the IXC gateways are on this Sonet ring with the LEC's class-5 offices, then the LEC doesn't need a hub switch. But typically the IXC gateways are not in the LEC's Sonet ring and the LECs deploy a tandem office for IXC connections and for some interexchange switching. Since this tandem office is also on the LEC's Sonet ring, the interexchange network is a physical ring, but a logical star. Consistent with the history of the toll point, by the time it's all straightened out, it will be time to reconfigure it again, this time because of the Intelligent Network, described in Chapter 18.

There is a lesson in Figure 9.13 for all architects everywhere. Functionality that seems sensible to colocate, perhaps should not be fully integrated, because the technological and economic forces that colocated them might change. And if the functions were too integrated together, they may be difficult (and expensive) to separate. We'll see overintegration again with switching software. We'll see it overintegrated in Chapter 11, we'll see a little bit of separation in Chapter 14, and we'll see it break into little pieces in Chapter 18.

EXERCISES

9.1 Consider a 10,000-line switching office in a residential serving area. Suppose the telco spends $400K per year on routine system operation and software upgrades and another $400K per year to pay the mortgage on its total investment in this office and its peripheral loop plant and trunk plant. Suppose the cost of new installations in this serving area exceeds the revenue the telco is allowed to charge by $50K per year. Supppose, because of vandalism and underutilization, the telco loses another $50K per year on required coin telephones in the area. Suppose the telco spends another $50K per year expanding its facilities so it can handle those Internet calls that provide no added revenue. Suppose the telco collects from this area $150K per year in fees on leased lines, another $150K per year in long distance access fees, and $50K per year for each of (1) Touch Tone fees, (2) billed interexchange calls, and (3) charges for enhanced services (calling party ID, call forwarding, etc.). (a) Assuming the telco pays a 50% tax on its profits, how much does·this telco need to charge each of its 10,000 subscribers every month in order to return 8% of revenue to its investors? (b) Discuss how close this fee is to the actual cost of providing flat-rate POTS.

9.2 Two kinds of electromechanical switching offices must be converted from central billing to local billing. (a) What changes are necessary in a Step office to convert it from ANI CAMA to LAMA? (b) What changes are necessary in a Number 5 Crossbar office to convert it from CAMA to LAMA?

9.3 Incoming registers had to be placed on the Number 4 Crossbars CO side to be compatible with subscriber signaling through Step offices. The NTSC standard format for color television had to be made compatible with black-and-white television sets. Discuss the similarities between these two designs.

9.4 The time-of-day clock in AMA recorders increments every 6 seconds. For an extremely brief telephone call, it is possible that the call answer entry and the call disconnect entry could have the same time stamp. Such a call would not be billed, even though it may have been completed. If a call lasts more than 6 seconds, the two entries would have different time stamps and would be billed. Sketch a graph that shows probability of billing on the y axis and call duration on the x axis.

REFERENCES

[1] Jaeger, R. J., and A. E. Joel, Jr., "TSPS No. 1 Systems Organization and Objectives," *The Bell System Technical Journal,* Vol. 49, 1970, p. 2417.

[2] Blalock, W. M., "North American Numbering Plan Administrator's Proposal on the Future of Numbering in World Zone 1," *Bellcore Letter,* Letter No. IL-92/01-013, Jan. 6, 1992.

[3] Basso, R. J., et al., "OSPS System Architecture," *AT&T Technical Journal,* Vol. 68, No. 6, Nov./Dec. 1989, pp. 9 – 24.

[4] Kaskey, B., "A New Operator Services Feature for the 5ESS Switch," *AT&T Technology Magazine,* (issue on Electronic Messaging), Vol. 4, No. 2, 1989.

[5] Hurley, B. R., C. J. R. Seidl, and W. F. Sewell, "A Survey of Dynamic Routing Methods for Circuit-Switched Traffic," *IEEE Communications Magazine,* (issue on Network Planning), Vol. 25, No. 9, Sept. 1987.

SELECTED BIBLIOGRAPHY

Bellamy, J., *Digital Telephony,* Second Edition, New York, NY: Wiley, 1991.

Brooks, J., *Telephone — The First Hundred Years,* New York, NY: Harper & Row, 1976.

Clarke, A. C., et al., *The Telephone's First Century — and Beyond,* New York, NY: Thomas Y. Crowell, 1977.

Jacobsen, C. R., and R. L. Simms, "Toll Switching in the Bell System," *Proceedings of the International Switching Symposium,* 1976, pp. 132 – 134.

Noll, A. M., *Introduction to Telephones and Telephone Systems,* Norwood, MA: Artech House, 1986.

Rey, R. F. (ed.), *Engineering and Operations in the Bell System,* Second Edition, Murray Hill, NJ: AT&T Bell Laboratories, 1977.

Schindler, G. E., Jr. (ed.), *A History of Engineering and Science in the Bell System,* Murray Hill, NJ: Bell Telephone Laboratories, 1982.

Talley, D., *Basic Telephone Switching Systems,* Rochelle Park, NJ: Hayden Book Company, 1969.

"Telecom Billing '92," *Conference on Developing Billing Systems for Tomorrow's Advanced Services,* Washington, DC, Jul. 23 – 24, 1992.

Telecommunications Transmission Engineering, Volume 3 — Networks and Services, New York, NY: American Telephone & Telegraph, 1977.

10

Enterprise Switching

"I'd rather fight than switch." — from a 1960s cigarette commercial

Chapter 3 described the line side of a switching office, and Section 3.7.3 described PBX trunks. This chapter continues this discussion. Switching is provided, physically and logically, at the central office for the typical residential subscriber. This provision may be inadequate and undesirable for subscribers in businesses and other large organizations because of the high degree of intraorganization calling and the high demand for special calling features. Thus, switching is frequently provided physically, or at least logically, closer to these subscribers than the central office. While this expanded service is typically inappropriate for residential subscribers, it is often used by businesses, schools, hotels, hospitals, government agencies, and other organizations in which mutually interactive people are housed in close proximity. We use the term "enterprise" as a generic term for such organizations. This chapter describes the switching services provided for an enterprise.

We described the residence line side in Chapter 3. In an enterprise environment, two typical differences are the *key telephone* and the *loop architecture.* The first two sections briefly describe the key telephone and the typical wiring plan on a campus of multistory buildings. The next two sections cover a progression of functionality, from the hunt group to the key system to the private branch exchange. Section 10.5 describes *Centrex* and the implementation of two important customer switching services: direct inward dialing (DID) and automatic identified outward dialing (AIOD). Section 10.6 covers an important regulatory event, the Carterfone decision, and its ramifications. Like the previous chapter, the coverage begins early in the history of switching and then goes forward almost to the present. In the last section, we review our progress through the timeline.

10.1 MULTILINE KEY TELEPHONE

Key telephones, shown schematically in Figure 10.1, are seen commonly in the enterprise environment. On the exterior, a key telephone resembles the traditional telephone except for six clear buttons, typically located in line along the lower front part of the base unit. Each button has a lamp underneath it. The cable, at the back of the base unit, has many pairs, typically 25 (many of which are not used). In a common business office wiring plan, $N \leq 6$ pairs are wired to each of N different telephones. At each key telephone, the

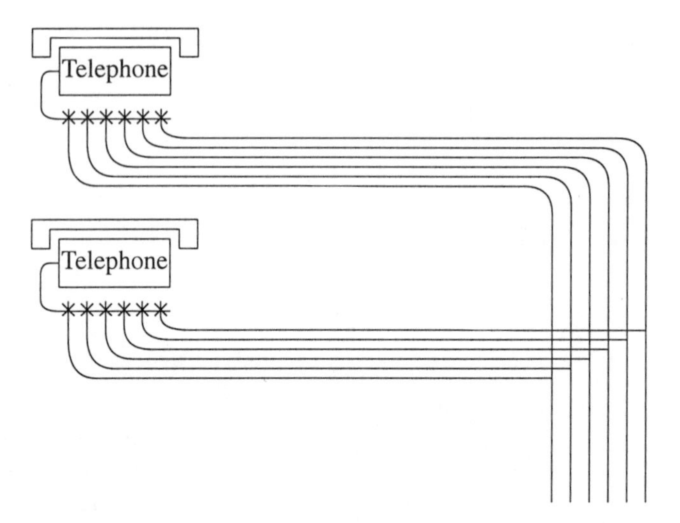

Figure 10.1. Schematic for a multiline key telephone set.

user knows the state of any of the *N* pairs by the lamp that corresponds to this pair, and can access any of the *N* pairs by depressing the button that corresponds to this pair.

The group of users, called a pickup group, that share a set of *N* lines are typically related by belonging to the same department or residing in the same office. While each telephone is the primary instrument for a different individual in this pickup group, each individual can access any of the *N* lines. A subscriber at a key telephone may initiate a call on any of the lines by first depressing the unlit button for the desired line, and then going off-hook. A subscriber at a key telephone may add onto an existing call, as an extension, by first depressing the corresponding lit button, and then going off-hook. A subscriber at a key telephone may answer an incoming call on any of the lines, by first depressing the corresponding blinking button, and then going off-hook.

Typically, all six buttons are not used for line selection. Frequently, the leftmost button is clear red and is wired internally to provide the hold function. When the user presses the hold button, a 1-kOhm resistor is connected, inside the key telephone, across the tip and ring of the currently selected line. This temporarily terminates the call and maintains supervision, while the user is free to access a different line, either to answer it or to place another call. When the user reselects the line that had been on hold, the resistor is disconnected and the user is reconnected.

Many key telephone features are available by custom wiring the key telephone. These features are requested via the buttons on the key telephone and are implemented directly internal to the key telephone.

- Line selection is the default feature for any button.
- The currently selected line can be put on hold.
- An external device, like a speakerphone or a modem, can be connected to the selected line. Key telephones were commonly used with the 212 Data Set.

- The key telephone's user is connected to an external speaker so the user can page someone.
- An internal conference bridge can be connected to the currently selected line.
- In the "buzz" feature, one of the buttons on one telephone is wired directly to an audible buzzer in another telephone. This way, for example, a manager can buzz for his secretary or a secretary can buzz the boss that some incoming call, a call the secretary answered because the boss didn't, really does require immediate attention.

While key telephones are still very common, many offices now have analog and digital multiline telephones that are functionally equivalent but have many more features, many of which might no longer be implemented internally inside the telephone. For example, hold can be provided inside the switching system, instead of internally in a key telephone. User A somehow signals to switching software that the call should be suspended, that user B should be put on hold, and that user A should be given dial tone or should be freed from this call to answer another incoming call.

To accommodate key telephones, the typical office was wired with the familiar thick gray 25-pair cable. Since modern digital telephones don't use all 25 pairs, and even many key telephones never used them all either, unused pairs in office wiring are available for other applications, like a 10-base-T Ethernet connection between a PC and an Ethernet LAN.

10.2 ENTERPRISE WIRING

Chapters 3 and 4 described the boundaries of a local switching office: the line side in Chapter 3 and the trunk side in Chapter 4. Enterprise switching equipment, similarly, has a line side and a trunk side. The trunk side of an enterprise switch terminates the set of channels that are wired to the line side of the local switching office on which the enterprise homes. What seems to be a trunk to an enterprise switch looks like a line to the local class-5 switching office. These channels may be connected to a logical network, defined within the PSTN, called an enterprise network, which is described in Section 12.4. The line side of an enterprise's switching system terminates the building or campus wires that are wired to office telephones. This enterprise wiring has a typical architecture that is similar to the three-tiered star of the typical residential loop plant.

Figure 10.2 represents one building on some enterprise's campus. This building has three floors and a basement. The enterprise's switching system, assumed to be a PBX, (defined later in this chapter) is assumed to be located in another building on campus. This PBX is the enterprise's analogy to a telco's central office. Just like a telco's central office, the campus' PBX is typically the hub of a three-level hierarchical star physical network. The per-building wiring frames, typically found in the basement of each building on the campus, are the enterprise's analogies to the distribution points in a telco's loop plant. The per-floor wiring closets are the enterprise's analogies to the pedestals in a telco's loop plant.

Large multipair cables, passing through the cross-campus conduit, connect the PBX's main frame to the wiring frame in each building. These cables are the enterprise's analogy to the feeder cables in a telco's loop plant. One such feeder cable carrying six

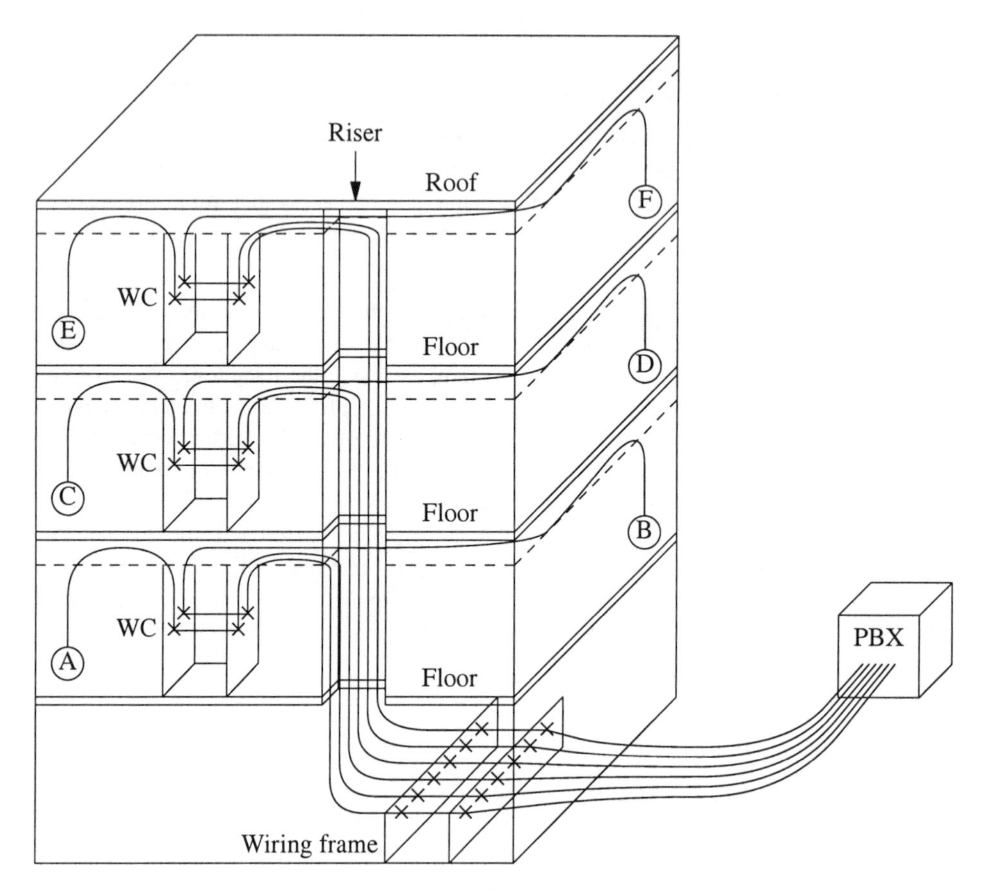

Figure 10.2. Building and campus wiring.

pairs, each pair represented by a single line, is illustrated at the lower right in Figure 10.2. Smaller multipair cables, passing up the vertical risers in each building, connect the wiring frame in the basement of each building to the wiring closets on each floor of the building. These cables are the enterprise's analogy to the distribution cables in a telco's loop plant. Three such distribution cables, each carrying two pairs, are illustrated in the building riser at the center of the figure. Still smaller multipair cables typically pass through the space above the drop-ceiling on each floor, to connect the wiring closet on the floor to the telephones on this same floor. These cables are the enterprise's analogy to the drop wire cables in a telco's loop plant. Six such drop wires are illustrated in the figure, two on each floor, connecting to telephones A through F.

In Figure 10.2, trace the wiring from B's telephone back to the PBX. A cable proceeds from the telephone on B's desk to a jack in the wall in B's office (not shown in the figure). A conventional tip/ring telephone uses a conventional two-pair cable, as in the residence, typically with a modular plug on each end, connected to modular jacks in the telephone and in the wall. A key telephone set uses a thick 25-pair cable, typically built into the telephone and "connectorized" near the wall. A digital set uses a three- or four-pair cable, also with modular plugs and jacks on each end.

In Figure 10.2, B's office is located at the back right of the first floor. From the modular jack on the wall in B's office, B's pair goes up behind the wall to the space above the drop-ceiling, through a wiring trough across this space, to the top of the first-floor wiring closet, and down inside this wiring closet to a termination block. B's pair is jumpered in this first-floor wiring closet to one of the pairs in the riser cable that terminates on another block in this same closet. B's pair now proceeds inside one of the riser cables down the building's riser shaft, typically near the elevator shaft, down to the basement and over to a termination block at the building's wiring frame. B's pair is jumpered at this wiring frame to one of the pairs in the campus feeder cable, which also terminates on a block at this wiring frame. B's pair now proceeds inside one of the campus feeder cables through a conduit across the campus to the main frame at the PBX. B's pair is jumpered at this main frame to another pair that leads to a PBX line card.

10.3 HUNT GROUPS AND KEY SYSTEMS

In the simplest customer functionality, (POTS), one black dial telephone is wired to an analog tip/ring loop. The addition of extension telephones provides a little more functionality and is the upper limit in functionality for the typical residential subscriber. Three additional levels of functionality, available to the enterprise subscriber, are the *hunt group,* the *key telephone system,* and the *private branch exchange* (PBX).

As I've stated several times so far, if this book is about any one thing, it's about architecture. This section and the next one illustrate an important architectural principle:

The services that a system might offer are often enabled or constrained by the system's architecture.

This section and the next one present a succession of increasing architectural complexity for enterprise switching systems and describe the corresponding increasing functionality.

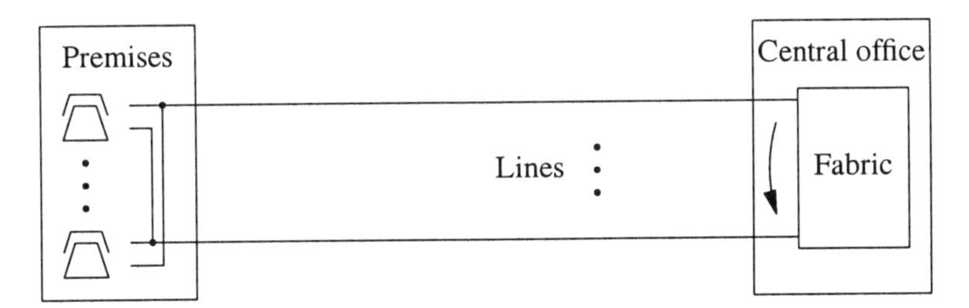

Figure 10.3. Hunt group.

10.3.1 Hunt Group

A hunt group, shown in Figure 10.3, consists of 10 (typically) or fewer telephone lines with a special arrangement on each end. While each line has its own directory number, only one of these numbers is listed under the enterprise's entry in the telephone directory. On the subscriber's end, a key telephone is associated with each line. While N telephones require N lines, each key telephone can access six (at most) of the N lines. Each key telephone is the primary instrument on one line but acts like an extension telephone for another five (or fewer).

In the central office, the N lines are grouped like a trunk group, but on calls into the enterprise the lines are selected in priority order — not in the order of most idle nor uniformly, as in a slipped multipling arrangement. The arrangement for hunt groups in Step was discussed in Section 6.6. On the terminating side of the Step fabric, all the lines in a hunt group have consecutive directory numbers and they all appear on the same row (and in the same sequence) on each connector in the appropriate connector group. These connectors are modified to revert to hunt mode after selecting to the first position, corresponding to the listed directory number of the enterprise. Hunt groups are still commonly called rotaries, after their implementation in Step. Hunt groups are implemented in marker- and program-controlled switching systems by a special class of service on terminating calls. On terminating calls, lines in a hunt group are selected by the marker or processor using the same procedure that this marker or processor uses for selecting a trunk from a trunk group. On originating calls, any system simply processes calls individually.

Wm. F. Vesley Co. was an automobile parts distributor in Garwood, NJ. Inside the store was a warehouse full of auto parts and a long counter with four clerk positions. The enterprise had a five-line hunt group with the directory listing 789-0123. Each clerk position had a set of parts catalogs and a key telephone. These four lines were identified inside the store by their last digit: 3, 4, 5, and 6. The fifth key telephone, identified as 7 inside the store, was in the bookkeeper's office. It acted as an overflow line for the four clerks, and it was the preferred line when anyone made an outgoing call. Anyone outside the enterprise, who needed to speak directly to the bookkeeper, simply dialed 789-0127. There was no hunting if this line was busy.

Suppose a local auto mechanic needed a fuel pump for an 1989 Ford Escort. He would dial 789-0123. The switching office would hunt over the group of lines and connect the line from the service station to the first idle line to the store. If lines 3 and 4 happened to be busy at the time, this call would get line 5. At the store, the common bell rang and the lamp for line 5 flashed on all five key telephones. One of the four clerks, typically the one who had been idle the longest, selected the key for line 5 on his key telephone and answered this incoming call. After conversing with the mechanic, this clerk identified the part number from the catalog, found the part in the warehouse, wrote out a purchase order, and placed the part and the PO on the counter. The next available driver delivered the order.

The hunt group has the simplest functionality among the three arrangements discussed in this section and the next. It is not possible for anybody, inside or outside the enterprise, to call a specific individual at the enterprise — neither automatically nor even through a manual switchboard. It is extremely inconvenient to dedicate one person inside the enterprise to the job of answering all incoming calls and forwarding calls to the requested party. An equivalent function can be implemented in a hunt group by using an auxiliary paging system. In a hunt group, a simple feature like call transfer, was implemented at the Vesley Company by yelling across the store, "Bill, pick up on line 6."

10.3.2 Key System

Consider the different ways for handling an incoming call to a person who is not at his desk or is busy on his telephone.

- *Call intercept* is the telephone service in which someone else answers all your incoming calls, typically to screen them.
- *Call pickup* is the telephone service in which someone else answers your incoming calls when you're busy or not at your desk.
- *Recorded answer,* also called voice mail, is an alternative to pickup and can be provided by a per-desk machine or by a common service built into the switching system. In an enterprise the drawbacks to call answer are that calls are not screened nor deflected to someone else and nobody will try to find you if the call is important.
- *Call forwarding* is another related telephone service, but it is inconvenient to initiate every time you leave your desk and more inconvenient to reset each time you return.

The simplest, but most inconvenient, form of call pickup is typically used in a large office that houses several desks. Everyone in this office has his own key telephone, mutually wired to each of the other telephones in the pickup group. A slightly more convenient form of call pickup requires additional equipment, called a key system, or the equivalent software in the CO or PBX (next section).

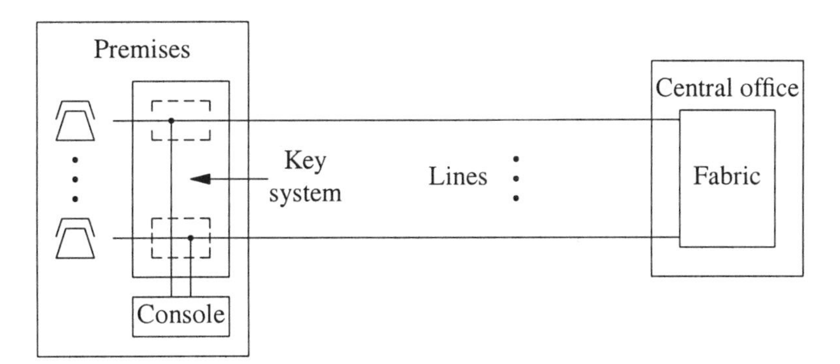

Figure 10.4. Key system.

Key telephone service is provided for regular subscribers (not a hunt group) of a central office or a PBX, as shown in Figure 10.4. Such subscribers are grouped and served by a common operator, called the attendant. Additional equipment on the enterprise's premises includes a key system and a multiline attendant console, sometimes called a call director. Within the key system, the line between the switching system and each subscriber is intercepted by a circuit similar to that used for Touch Tone in Step (Section 6.8.4) and for TSPS service on toll trunks (Section 9.3.2). This intercept circuit allows the attendant access to the subscriber's end or the switch's end of each line.

Each line in a key system is programmed for one of two procedures: *pickup* or *intercept.*

- With pickup, incoming calls ring through to the given line. The key system automatically switches the call, after a predetermined number of rings, and connects the attendant to the switch end of the line.

- With intercept, every incoming call to the given line rings at the attendant's console first and not at the called telephone. This way the attendant can screen incoming calls for this party, typically an executive in the enterprise.

Functionally equivalent service is available as a software feature in any modern program-controlled central office or PBX, without requiring the physical key system equipment. In some advanced services (Section 14.3.7), subscribers store a *pickup chain* in the processor's database, a sequence of directory numbers to which an incoming call is transferred after a predetermined number of rings.

A key system provides a useful service for incoming calls. But, it cannot provide concentration of telephones onto lines and two subscribers cannot be interconnected within the key system. This next level of functionality requires Centrex or a PBX.

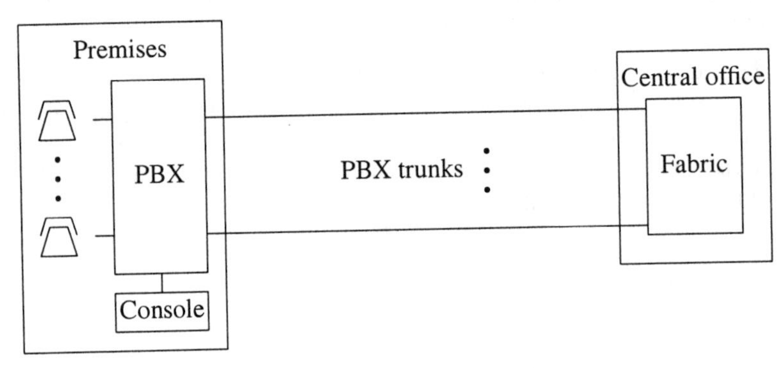

Figure 10.5. Private branch exchange.

10.4 THE PBX

A private branch exchange is switching equipment that resides on the enterprise's premises, as shown in Figure 10.5. It provides call processing, connects internal calls, and concentrates all the telephones in the enterprise down to a reduced number of lines (called PBX trunks) to the central office. The PBX's line side terminates all the telephones in the enterprise. The PBX's trunk side terminates the PBX trunks that connect the PBX to other PBXs or to the central office on which it homes.

This section begins with the primordial PBX, the manual switchboard, and briefly presents other PBX technologies. Then, the typical PBX numbering plan and second dial tone are described. The difficulties with outgoing toll calls and with incoming calls are discussed.

10.4.1 PBX Equipment Evolution

The technology and architecture of PBXs have a history that is similar to those of central offices. The primordial PBX was manual, similar to the manual exchange described in Section 1.5. But unlike class-5 switching offices, which are no longer manual, many small enterprises still have manual switchboards. Manual enterprise switching was so common that many people still use the term "switchboard" as a synonym for PBX, even a modern automatic one.

Inside an enterprise with a manual switchboard, the calling party went off-hook and one of the switchboard attendants answered. If calling someone inside the enterprise, the calling party identified the desired called party by their name and the attendant completed the call, as described in Section 1.5. If the enterprise were small enough and the switchboard were located centrally enough, the attendant might know that the desired called party isn't in his office and is in someone else's office, and the attendant might ask the calling party if he wants to call his desired party at the other location. This kind of service vanished with automation and, only recently, have we begun to think of automatic ways to bring it back.

Calling someone outside the enterprise had two implementations. (1) The calling party provides the telephone number. Here, the attendant selects an idle outside line (which is wired to the local switching office like a subscriber line), dials the requested number into this switching office over this line, and then connects the line to the calling party at the switchboard. (2) The calling party asks for an outside line. Here, the attendant connects the calling party to an idle outside line and the calling party dials the number himself. Incoming callers dial the PBX's main number into the local switching office. This office assigns the call to an idle outside line, which are all configured in a hunt group.

Some people, particularly in Europe, use the term "PBX" as a synonym for a manual switchboard and use "PABX" (for private automatic branch exchange) to specify the modern implementation. In the early years, automation of PBXs lagged automation of central offices by many years probably because operators were required for many years anyway to handle incoming calls. In the modern era, new technology is typically introduced into PBXs before it is introduced into central office switching equipment (Section 10.6.3 and Section 18.2.3). PBXs tend to be quite small — a 1000-line system, while small for a central office, would be a large PBX.

Consistent with the cost/size discussion in Section 2.4.5, Step-by-Step technology was very relevant to the PBX market. An extremely popular Step-by-Step PBX was called the 701. Size and flexibility were very important issues as well, and Crossbar systems, scaled-down versions of Number 5, were also popular. The 756 and 757 were very popular Crossbar PBXs. In the post-Carterfone era (Section 10.6), when the PBX industry became competitive, many of the early non-Bell PBXs were Crossbar systems, even into the late 1970s. Nippon Electric (NEC), for example, made a very competitive and popular Crossbar system during this period.

Electronics technology and computer control first appeared in a centralized architecture as the #1 ESS (next chapter), designed for large switching offices. Western Electric built a scaled-down version, the #2 ESS, for the small central office market and a variation on the 2E, called the ESS-101, for the large PBX market. In the 1970s, when the PBX market became very competitive, many companies offered analog PBXs that were controlled by a second generation minicomputer, and which were targeted for small and intermediate-sized offices.

One very popular switch with an interesting architecture was Western Electric's Dimension PBX. Various aspects of Dimension are described throughout the book. Dimension's switching fabric is based on pulse amplitude modulation, like conventional pulse code modulation, but without the digital coding. AT&T's attempt to market Dimension as a digital switch is discussed in Section 13.1.3.

While PBXs had always interfaced to conventional 500 sets (POTS telephones) and

2500 sets (key telephones), many manufacturers in this era introduced fancy telephones with customized features and proprietary interfaces. WECo introduced its proprietary Multi-button Electronic Telephone, the "Met Set," to work with its Dimension PBX. All the manufacturers marketed these fancy telephones as a way to provide new services to business users and to, presumably, increase business productivity. But they were also interested in creating such a large embedded base of expensive proprietary telephones inside an enterprise so the enterprise could never afford to switch to another PBX vendor. This market strategy is still being used, and enterprise telecom managers would be wise to insist on standard interfaces, like ISDN, for new telephones.

Digital PBXs were introduced in the late 1970s. Western Electric's System 75 PBX, now part of Lucent's Definity family, is described in detail in Section 14.3. Now, 20 years after the industry introduced digital switching, PBX technology hasn't changed very much. And because it hasn't, the entire PBX function may be subsumed into an enterprise's computer communications network. PBX manufacturers don't seem determined to prevent this from happening.

Before the competitive era, AT&T (via Western Electric) was the dominant PBX vendor. After competition was allowed, AT&T's market share dropped steadily to the 25 – 35% range. Other companies — like Rolm, Northern Telecom, Tai, and many more — competed successfully with AT&T using products similar to the Western Electric products that are described here. Again, as in the discussion of central office systems, Western Electric products are described in this text mainly because their descriptions are most readily available.

10.4.2 PBX Numbering Plan

An important feature of an automatic PBX is that the subscribers may call each other using very short internal directory numbers. The PBX itself processes such an intra-PBX call, and not the central office. An enterprise with fewer than 70 lines could easily adopt a numbering plan in which all the lines have two-digit directory numbers. Typically no PBX directory numbers begin with 0 nor 9 because 0 is used to call the attendant and 9 (sometimes, also 8) is used to connect to an outside line.

A typical two-floor motel has three-digit room numbers and assigns directory numbers to match the room numbers: 1xx and 2xx. Guests at the motel dial 0 to call the attendant at the front desk, dial 9 for an outside line (to the LEC) used only for local calls, and dial 8 for an outside line (to some IXC) for long distance calls.

10.4.3 PBX Trunks

Like a class-5 local switching office, a PBX has a line side and a trunk side. The channels that terminate as trunks on a PBX terminate as lines at the switching office on which the PBX homes. While you won't see this terminology, a PBX could be thought of as a class-6 switching office.

Just as trunks (on a class-5 office) can use DC or E&M supervision to indicate the seizure on one end and request for connection on the other, PBX trunks also have two classic varieties of supervision. A *loop start* trunk uses classical DC supervision. To indicate a new outgoing call, in the trunk circuit on the PBX side, a relay contact on the CO side of the AC coupler is operated by PBX control. This contact causes DC current to flow in the trunk's loop, which signals the switching office exactly like a telephone does. A *ground start* trunk also uses DC supervision, but with a voltage signal instead of a current signal. To indicate a new outgoing call, in the trunk circuit on the PBX side, a relay contact on the CO side of the AC coupler is operated by PBX control. But in this type of circuit, the contact connects one side of the pair to local ground, a signal that is detected on the CO side.

10.4.4 Second Dial Tone

For an outgoing call, a PBX subscriber originates a call to a party who is not a subscriber in this same PBX. When the outgoing caller lifts the handset, he hears dial tone from the PBX, and, in the typical PBX numbering plan, he dials 9 into the PBX. The PBX connects his line to an idle PBX trunk, wired to the line side of the central office, and simulates off-hook on this trunk either by loop start or by ground start. The central office responds, as it would to an off-hook signal on any line, by returning dial tone. The PBX subscriber has been instructed to wait for this second dial tone before dialing the called party's directory number into the central office.

In most modern PBXs, the entire directory number is initially dialed into the PBX. Since many enterprises have private voice networks and subscribe to alternate forms of WATS, "WATS boxes" and other network route optimization features may alter the called number before the PBX outpulses it to the CO. Even though the PBX subscriber may no longer dial directly into the CO, PBX subscribers are so accustomed to hearing second dial tone after the 9 (or the 8) that the PBX provides it, even though it is completely unnecessary.

Many enterprises have multiple locations, which they wish to interconnect by a private network. An enterprise may lease private lines and trunks from a carrier to provide direct connections between its PBXs at the various locations. These channels are called tie lines, and tie-line access must be built into the PBX's number plan. If the PBXs provide tandem switching of the enterprise's tie lines, then the enterprise can provide an enterprise-wide network [1 – 5]. With appropriate number translation in each PBX, an enterprise might even have an enterprise-wide numbering plan that is independent of location. With appropriate intelligence in the PBX controller's software, an enterprise might automatically switch outside long distance over its enterprise network to the PBX that is closest to the called party before using the public network to complete the call. Such enterprise networks are described in Section 12.4.

10.4.5 Outgoing Toll Calls

Suppose this outgoing caller places a toll call. Unless some precaution is taken, the ANI or AMA equipment in the CO can identify only that the call originated over a specific PBX trunk. So, the call is billed to the enterprise's directory number. At month's end, when the enterprise gets the bill, the enterprise cannot verify that the call was placed nor can it identify which employee placed the call. Since those in management at most enterprises don't like this, some precaution is taken. Unidentified billing is especially unacceptable in a hotel or hospital where the charges for the toll call must not only be identified to an internal extension, they must be known at the end of the call (rather than at the end of the month) so they can be charged to the room before the customer checks out.

In one common precaution, all the PBX trunks are assigned a class of service at the central office such that the toll operator intercepts any toll calls placed over one of these trunks. The toll operator requests calling party identification or room number (if the call is from a hotel) and causes it to be entered on the AMA recorder. As a special feature for hotels, billing information may be transmitted back over the toll trunk from the toll point to the PBX at the hotel when the call is finished. If the enterprise suspects that PBX subscribers, such as hotel customers, may be giving false identifications, in an alternate precaution, the PBX attendant provides the intercept because the room number can be determined directly.

In the second common precaution, the PBX trunks are partitioned into two separate groups. The PBX trunk group used for local unbilled outgoing calls is accessed inside the PBX by dialing 9 and is assigned a "toll restricted" class of service at the CO. The PBX trunk group used for outgoing toll calls is accessed within the PBX only through the attendant and is assigned a regular class of service at the CO. The outgoing caller who places a toll call dials 0 and requests that the attendant complete the call. The attendant records the subscriber's identification. The third precaution, the most common and most modern one, uses the automatic identified outward dialing feature, one of the two Centrex features described in the next section.

10.4.6 Incoming Calls

An enterprise has a listed directory number so that people outside the PBX may call the enterprise; viewed by the PBX as an incoming call. The central office connects such a calling party to an idle PBX trunk, on the CO's line side, and rings this trunk. An attendant at the PBX answers the incoming call, converses with the calling party, and then "extends" the call to the appropriate party within the PBX. Even if the subscriber placing an incoming call knows the directory number (or room number in a hotel or hospital) of the called party, if the PBX does not have direct inward dialing, then the incoming caller must use the enterprise's main number and request a connection through the PBX attendant. Direct inward dialing is the other Centrex feature described in the next section.

The PBX subscriber who wishes call intercept may use the PBX attendant for the intercept service, or a group of PBX subscribers might use a key system. Key systems are used as PBX adjuncts in the same way that they are used as central office adjuncts (previous section).

10.5 CENTREX

"Centrex" is another term that often is used carelessly as a synonym for "PBX" — but we will be more precise [6,7]. Enterprise switching service has two alternative implementations: by using a PBX on the enterprise's premises or by providing the service as an extension of POTS directly from the central office. The latter is called Centrex. Two calling features, direct inward dialing (DID) and automatic identified outward dialing (AIOD), are very important to many enterprises. Years ago these two features were only available in the Centrex implementation of enterprise switching. When these features came to be provided also in a PBX implementation, the term "Centrex" came to be used for either implementation, as long as the two desirable features were provided. Now these features are practically universal in both the PBX and CO implementations and the term "Centrex" (or its marketing cousin, "ESSX") is used once again to identify the CO implementation.

10.5.1 CO-Centrex

Until the late 1960s, business customers did not own their PBXs. They contracted for customer switching service from the telephone company and its implementation was left to the telco. The choice was based on economics and features. If the business' location is extremely far from the geographical location of the switching office, the cost to connect each telephone in the business directly to the CO would be very high. The savings from concentrating PBX lines down to PBX trunks could more than offset the cost of placing switching equipment on the customer's premises. Customer switching service would be implemented using a PBX.

However, if the business' location is extremely close to the geographical location of the switching office, the cost to connect each line directly to the CO would be very low. The savings of concentrating lines down to PBX trunks would be small compared to the cost of PBX equipment on the customer's premises. Customer switching service could be implemented directly from the CO as a virtual PBX. Customer switching service would be implemented using Centrex.

A group of central office telephone lines can be isolated by physical segregation in a Step-by-Step central office or by class of service in a marker- or computer-controlled CO. Any of these lines can be allowed to use a customized PBX numbering plan for calls to parties within the same group. They can dial 9, and receive second dial tone, to originate outside calls, even to called parties served by this same CO but outside this isolated group of lines. But the economics of PBX cost versus concentration savings was not the only factor in choosing between the PBX and Centrex implementations. Until the late 1960s, DID and AIOD were available only with Centrex.

The only difference between a Centrex line and a regular residence line is the numbering plan that the subscriber uses. Like residence lines, Centrex lines are assigned directory numbers that are unique in the NPA, and they have fabric addresses that are known to, and accessible by, the central office. DID occurs naturally in Centrex just as it occurs naturally with residence telephone service. A calling party outside a business can call a Centrex line inside the business by simply dialing the desired called party's unique directory number. This directory number is normally signaled to the CO where it is translated into a fabric address, in this CO, and the connection is made; just like calling any residence line. With incoming calls in the PBX implementation, the directory number was not typically unique in the NPA. Furthermore, even if it were, the called number is known at the CO, where it was signaled, and not at the PBX, where another connection is needed.

AIOD also occurs naturally in Centrex, just as it occurs naturally with residence telephone service. A Centrex subscriber who originates a toll call is identified by the CO's ANI or AMA translator exactly as a residence subscriber is identified. While the bill is sent to the business, the calling party is identified on each detailed entry. With outgoing calls in the PBX implementation, the CO can identify only that a toll call came over a certain PBX trunk and cannot identify the line that is connected to this trunk within the PBX.

After the DDD/toll automation program was started, a related system-wide program was instituted so that DID and AIOD could be available from the PBX implementation. By offering the same features from either implementation of customer switching service, the choice of implementation could be a simple economic decision. To emphasize this, the term "Centrex" was expanded to include any implementation of customer switching service that provided DID and AIOD.

Classical Centrex came to be called CO-Centrex because customer switching service with DID and AIOD is provided naturally from the central office; there is no PBX on the customer's premises. The term was redundant because "Centrex" identified the CO-resident implementation of enterprise switching. The classical PBX implementation came to be called CU-Centrex if the PBX, or customer unit, was modified to provide DID and AIOD. This term was an oxymoron because the CU-resident implementation of enterprise switching service was not previously called Centrex.

10.5.2 Direct Inward Dialing with CU-Centrex

To provide direct inward dialing into a true PBX, it is necessary, but not sufficient, that each PBX subscriber's line have a unique directory number in the NPA. Allocating a unique directory number to every PBX subscriber would have strained the pre-DDD national numbering plan even more than the growth of telephone service was already straining it. So DID in CU-Centrex was practically impossible until the DDD numbering plan was in place, with its NNX office codes and NPA area codes. In addition, DID in a true PBX requires that this directory number be signaled to the PBX on incoming calls.

This particular change to CO procedure is a rare case where providing the change from a Step-by-Step CO was easier than from any other kind of CO. Consider a Step CO, with office code 234, that provides CU-Centrex to a 70-line PBX serving a business whose listed directory number is 234-5601. While the 70 lines could be assigned the directory numbers 234-5610 through 234-5679, the numbers 234-5680 through 234-5699 are also lost to the NPA. Numbers 234-5601 through 234-5600 would be a hunt group to the PBX attendants. The PBX would complete intra-PBX calls on the basis of the last two digits of the full directory number: 10 through 79.

In the Step office, up to 10 incoming (to the PBX) trunks could appear on the six-level of every Strowger switch in the 234-5xxx third selector group. These PBX trunks take the place of the links that would normally (for residence lines) connect to the 234-56xx connector group. This connector group is missing from the CO and is essentially replaced by the PBX. Since the Step fabric is progressive, instead of dialing these last two digits over a link into some connector in the CO, the calling party dials these last two digits over a PBX trunk into the PBX. The two-digit signaling for DID is accomplished naturally.

Providing DID from a nonprogressive central office requires a modified procedure; many completing markers were rewired, as they were with DDD. The problem is that the entire directory number of the desired called party is signaled into an originating register in the CO (or equivalent hardware or software). Then the completing marker (or equivalent dialing interpretation program) determines that some PBX needs part of this number. Crossbar offices had a procedure for outpulsing on the trunk side but not on the line side, where the PBX trunks terminate.

For DID from a non-Step CO, a special PBX trunk, called a DID trunk, performs dial-pulse signaling from its originating circuit on the CO end. In addition, in Number 5 Crossbar, the completing marker and the line side of the Crossbar office are modified. Recognizing a special class of service on call termination, the completing marker passes the last four digits of the called party's directory number to an outpulser on the line side of the CO. This outpulser seizes the selected DID trunk, alerting the PBX, and it outpulses these four digits over this trunk to the PBX in the same way that an outpulser on the trunk side outpulses over an outgoing trunk to the toll point or to another CO. A corresponding procedure is written into the call processing programs in any computer-controlled CO. The PBX senses the seizure on the incoming DID trunk, connects a digit receiver, and receives the signaling in the same manner in which it came from a Step office.

10.5.3 Automatic Identified Outward Dialing with CU-Centrex

AIOD was much more difficult to implement for CU-Centrex than DID was. The worst situation for CU-Centrex DID occurred only in Crossbar offices and required the modification of all the completing markers in an office, the replacement of PBX trunk originating circuits in the CO with DID originating circuits, and the addition of outpulsers and a connecting link to the CO's line side. The corresponding situation for CU-Centrex AIOD occurred in every kind of CO and required adding (1) an ANI subsystem to the PBX, (2) a new subsystem to the CO, and (3) a data link that interconnected them.

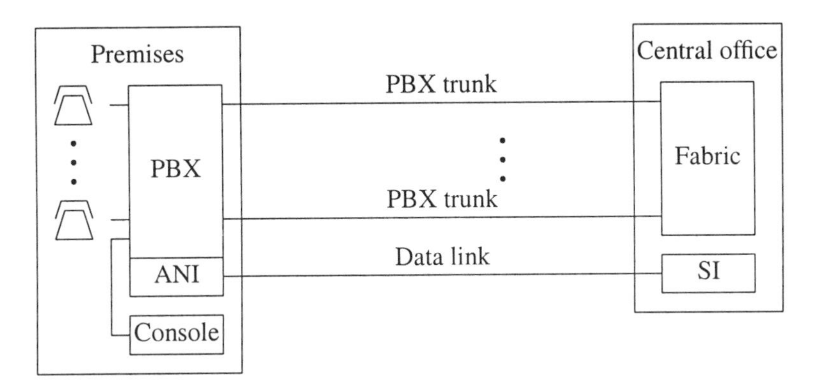

Figure 10.6. PBX environment modified for AIOD.

The new subsystem in the CO is called the station identifier (SI). In a Step-by-Step office, this SI is an adjunct to the ANI outpulsers (see Figure 9.2). In a Crossbar office, this SI is an adjunct to the AMA transverter; (see Figure 9.4). Figure 10.6 shows how the PBX environment from Figure 10.5 is modified with an ANI subsystem at the PBX, an SI at the CO, and a data link between them.

A separate subsystem that provides automatic number identification was designed and manufactured for every different kind of PBX. An ANI subsystem was installed in every PBX for which CU-Centrex service was provided. ANI in the 701 Step-by-Step PBX was a minor variation on the ANI subsystem for Step-by-Step central offices. ANI in the 756 and 757 Crossbar PBXs was a minor variation on the AMA translator in Number 5 Crossbar. ANI became a standard feature in the later 800 family of PBXs and in all subsequent commercial PBXs. In a computer-controlled PBX, ANI is built into the call processing programs, and the translation from equipment number to directory number is part of the computer's memory. The principal difference between PBX ANI and CO ANI is that the PBX ANI signals over a dedicated data link from the PBX to the CO instead of signaling over the trunk itself. Signaling for AIOD was the first application of *common channel signaling* (discussed in Section 18.1), although this term was not used at the time.

ANI is invoked in the PBX whenever a PBX trunk is seized at the PBX end because some PBX subscriber dials 9. The ANI subsystem identifies the PBX directory number of the calling PBX subscriber that has just been connected to the given PBX trunk. This is done by a tone over the sleeve connection and a number network in a 701 Step PBX or by translation equipment similar to an AMA translator in a Crossbar PBX or by a table lookup in the computer memory of a processor-controlled PBX. A digital message is transmitted over the data link from the PBX ANI subsystem to the SI in the CO. This data link transmission occurs while, or even slightly before, the called party's number is signaled over the PBX trunk. The message contains <trunk number, station number> for the PBX trunk that was just seized and the telephone number of the PBX subscriber that seized it. The transmission occurs for every trunk seizure, whether or not the calling party actually places a toll call (in a Step PBX, the PBX has no way of knowing).

The station identifier was shown in Figures 9.2 and 9.4 as an adjunct to ANI in Step or to LAMA and CAMA in Number 5 Crossbar, respectively. Discussion of the SI was postponed to this point in the book. Its three components are a data link receiver, an associative memory, and an interface to the ANI outpulsers in a Step office or to the AMA transverter in a Number 5 Crossbar office. When the SI receives a <trunk number, station number> message from a PBX ANI subsystem, it finds the word in the associative memory that currently holds this trunk number and it writes the new message in this word. By this process, the identity of the PBX subscriber who just seized the trunk replaces the identity of the PBX subscriber who used it previously. When the SI receives a message through its interface from the CO, this message consists only of a trunk number. The SI finds the word in the associative memory that holds this trunk number, reads out the station number, and transmits this station number through this same interface back to the CO.

Besides adding the SI to the office, the ANI outpulsers or the AMA transverter were slightly modified. They had to test the calling party identity that was returned by the number networks or the AMA translator, respectively. If this identification is recognized as a PBX trunk, they must perform the additional step of passing this trunk number to the SI and receiving the station number that is returned.

If a PBX subscriber dials 9, the PBX connects him to an idle PBX trunk and the called party's directory number is signaled over this trunk into the central office. Meanwhile, the PBX ANI subsystem transmits the <trunk number, station number> message to the SI. This message is stored in the SI before the central office has received all the digits of the called party's number over the PBX trunk. If this call from the PBX is a toll call, the CO begins to follow the procedure that it normally follows for a toll call from a residence subscriber.

In a Step CO, an ANI outpulser reads the identification of the calling party from the number network. In a Number 5 Crossbar CO, the AMA transverter reads the identification of the calling party from the AMA translator. In either type of CO, this calling party identification is tested to see if it is a PBX trunk number. If it is not, the calling party identification is signaled to the CAMA point, following the usual procedure from Section 9.2.3. If the calling party's identification (read by the ANI outpulser from a step number network or by the AMA transverter from a Number 5 AMA translator) is a PBX trunk number, it would be marked as an "indirect" identification of the calling party. Before signaling to the CAMA point, this Step ANI outpulser or Number 5 AMA transverter signals this trunk number to the SI. The SI completes the identification by finding the number of the PBX extension that is currently using the PBX trunk whose number it just received. The SI signals this station number back to the Step ANI outpulser or Number 5 AMA transverter. Instead of transmitting the trunk number to the CAMA point, the Step ANI outpulser or Number 5 AMA transverter transmits this station number, the PBX directory number of the calling party, to the CAMA point as the calling party identification.

10.5.4 A Modern PBX

In a modern PBX, the user inside the enterprise dials the called party's entire telephone number into the PBX controller's software. PBXs still return a second dial tone after a leading 8 or 9, not because the dialer must be sure that some other switch is ready to receive the rest of the dialed digits, but only because PBX users expect to hear a second dial tone. Since modern switching software always knows the identity of the calling party, ANI is anachronistic in a modern PBX. But AIOD still requires cooperation between PBX control and CO control.

One way to provide AIOD is for PBX software to log the association of calling party to assigned trunk. Then, when the telco sends the monthly bill, if billing is itemized to trunk number, software at the enterprise can correlate the bill to the calling party. Another way to provide AIOD in a modern PBX still uses a data channel between PBX control software and program in the CO's control software that performs the SI frame's equivalent function.

Frequently, an enterprise that needs several PBX trunks would lease an entire T1 between its PBX and the CO's line side. This AIOD data link would be one of the channels in the T1. Inband and out-of-band signaling are described in Section 18.1. AIOD was the first instance of a third type of signaling, called common channel signaling, in which information about a call is not transmitted over the same channel (the bearer channel) that carries the call, but is transmitted over a separate channel that is common to a group of bearer channels. The four types of signaling are discussed in Section 18.1.

10.6 CUSTOMER PREMISES COMPETITION

The Bell System existed, and the rest of the telecom industry coexisted, under monopoly regulation in the United States for many years. While the specifics were reregulated several times (Section 2.1), culminating in the 1956 Consent Decree, the underlying concept of subsidizing universal POTS remained. While the Bell System was big and slow to change, the United States had the best telecommunications in the world and Bell Laboratories was producing some of the best research in the world. Above all, it was stable, perhaps too stable for some. Competition was sought by many, and, while competition is probably a good thing in general, competition in this regulatory environment brought instability.

Competition began in the customer premises market, including PBXs and the telephone instrument itself. In this section we discuss how this happened and how this created an unstable environment [8 – 11]. The arm's-length subsidiaries of the early 1980s, divestiture in 1984, and the Telecom Bill of 1996 all would be the inevitable result of the events in the late 1960s.

10.6.1 The Monopoly Environment

This subsection discusses two characteristics of a monopolistic telecommunications environment: the responsibility for end-to-end integrity and the typical pricing compromise between universal service and adequate investor return.

First we discuss end-to-end integrity. Before competition, the Bell System manufactured every piece of equipment in its network from end to end, and guaranteed system integrity from end to end. These two concepts were, and still are, linked. If you make the equipment and maintain it yourself, if you build your own network out of your own equipment and manage this network yourself, then you are responsible. If something doesn't work, it's your fault; you can't blame somebody else, and you have to fix it.

While Ma Bell was slow and big and easily disliked, she was reliable and she took care of her customers (as long as they didn't object to her prices). As corny as it may sound, "quality" and "service" were more than just marketing techniques or buzz words — they were part of the corporate spirit. If you worked for Ma Bell, you felt like you were serving your country — I worked there then, and I know. But now it's gone — not just broken into pieces; this maternalism is gone, forever, from all the pieces. We will talk about how this happened in pieces (appropriately) throughout the remainder of the book. Here we describe how it all started.

Before we begin the discussion of subsidization, the second characteristic of a monopolistic telecommunications environment, we must clearly understand the subsidization of universal service (POTS) and we must clearly understand the difference between *cost* and *price.*

- The cost of some commodity or service is the expense incurred by the vendor who provides it.
- The price of some commodity or service is the amount of money that the user pays to purchase it.

The price of any kind of telecommunications service was specified by the layers of regulatory agencies and was not based on the actual cost of providing the service. While it seems like this was a bizarre way to run a business, it is justified. If A lives across the street from the telephone office and B lives 4 miles away from the telephone office, clearly A's service costs less than B's. Is it fair to make B pay a higher price than A pays, just because the telephone company decided to build its telephone office closer to A's house? Since most of us think not, we have a price structure that is based on the average cost. The telephone company's lost revenue in providing service to B is compensated by its increased revenue in providing service to A. A subsidizes B's service. However, most of us are not so socialistic that we're willing to subsidize this standard price for telephone service to C, who chooses to live in the wilderness, hundreds of miles from his nearest neighbor.

While the telephone may have been a luxury item in its early days, it quickly became a necessity. Because of its importance in commerce, safety, and society, one could argue that every American has the right to have basic telephone service at a reasonable price. America's regulators decided that basic telephone service, called universal service or POTS, would be priced below its cost. Nonbasic services would be priced well above their cost so the telephone companies would make enough money to compensate for the revenue lost in providing POTS. In hindsight, this probably should not have been done — at least, not to the degree that it was done. We'll discuss alternatives in Section 22.4.

So, universal POTS — local calls over a single residential line from a black dial telephone — was subsidized by non-POTS services. Coin telephone service was also priced below cost, typically under half of its cost. Non-POTS services — like long distance, enterprise switching, leased lines, designer telephones, extensions, and Touch Tone — were priced artificially high.

While AT&T and all the other telcos were monopolies, they were also publicly owned corporations. These companies acquired capital, the revenue needed to invest in new equipment and channels, by selling stocks and bonds. Their shareholders expected a reasonable rate of return on their investment. AT&T stock, in particular, was always regarded as a safe investment that would provide a handsome and predictable dividend.

If the cost of running the telephone business rose a little or if revenue declined a little, then the telcos would either have to reduce their dividends or petition the various agencies for some means to increase prices. Typically, the telco would ask to raise the price of POTS, to bring its price more in line with its cost. And typically, the regulators would decline this and agree to increase the price of something else. It was ironic that whenever the phone company tried to raise the price of POTS, some regulator would defend "poor old Grandma," who wouldn't be able to afford it. And whenever the regulators tried to lower the telcos' dividends, the phone companies would defend "poor old Grandma," the archetype stockholder. After many years of this, prices for most non-POTS services were much higher than their costs. It would be easy to compete in these non-POTS markets.

10.6.2 The Carterfone Decision

The pricing structure for telephone service was unstable, and this landmark regulatory decision tipped it over. I will argue that the telecom industry we have today is a natural result of this decision. We have to describe the Hush-a-Phone case, what a Carterfone was, and the strategies employed by the lawyers on both sides.

Hush-a-Phone. In Section 3.1.5, in the discussion of side-tone, we noted that background noise (as from a household appliance) is annoying on the telephone, especially on extensions. Hush-a-Phone was a passive cup-shaped device that was attached physically to the telephone's mouthpiece. Acoustically, it could block some of the ambient background noise, while slightly amplifying the speaker's voice. It did not connect to the telephone's electrical circuits — it attached physically and only affected the signal acoustically. Hush-a-Phone was retailed, but not by the Bell System. It was installed easily, but by the user and not by a telco employee. Hush-a-Phone became popular in business offices that had many people on the telephone simultaneously and in other loud environments, like machine shops and factory floors.

Since the telephone was Bell System property, AT&T took the position that only the Bell System should be allowed to manufacture, retail, and install any device that alters its properties. While Ma Bell claimed to be concerned about customer complaints that could be Hush-a-Phone's fault, she also saw an opportunity to price another thing well above its cost and she was deeply concerned about establishing a precedent for future attachments to her property. AT&T's public concerns about quality and system integrity and its attempt to block Hush-a-Phone retailing received little public sympathy. AT&T was viewed as being petty and was accused of bullying the small company that manufactured the Hush-a-Phone. A civil case was appealed all the way to the U.S. Supreme Court. AT&T lost. Hush-a-Phone retailing was approved and allowed. Ma Bell saw the first tiny crack in her regulatory stability.

Carterfone. Telephony and radio have always had an interesting relationship, with great potential synergy that is only now fruitful. We'll discuss the progression from cordless to mobile to cellular in Section 19.2. But, until the late 1960s, mobile telephones and ship-to-shore systems were essentially ham radios, with no direct access to the Public Switched Telephone Network (PSTN). When a mobile subscriber called a regular telephone subscriber, he radioed the telephone company's mobile operator, gave the called number verbally (a manual system), and the operator placed the call and patched the radio channel through at a switchboard.

The Carter Electronics Company developed an automated system, called the Carterfone, to give mobile subscribers automatic direct access to the PSTN. The Carterfone's hub was much like a PBX. Carter, the company's founder and president, wanted to manufacture, retail, and install Carterfone systems through his own company, and AT&T adamantly opposed it. All the Hush-a-Phone arguments and concerns applied here, but Carterfone was much more threatening to AT&T. Carter wanted his equipment, which was not manufactured by the Bell System, to be given a direct electrical connection to Bell System wires. AT&T refused to connect to Carterfone systems, and Carter's attorneys brought the issue to the Federal Communications Commission.

During the FCC proceedings, and in the publicity surrounding it, two profound issues were raised. Each of these was more important than Carterfone itself: (1) Was AT&T analogous to an electric power utility or to a bus company? (2) Should Ma Bell protect (or was she overprotecting) her technical integrity?

The Telephone Utility. Carter's side presented a very compelling analogy. They took the position that the telephone company was just another utility — like the gas company, the water company, or the electric company. Such utilities were allowed — in fact, encouraged — to be monopolies because it is too inefficient for more than one of each kind of utility to provide service in the same geographical region. Such a "natural monopoly" should only apply to the access infrastructure of pipes or wires and to the service delivered over them. It should not apply to the appliance in the user's home.

While the gas company provides natural gas service to my home, I purchase my water heater and my kitchen range from a retailer. While the water company provides water service to my home, I purchase my faucets, toilets, and bath tubs from a retailer. While the electric company provides electric power to my home, I purchase lamps and toasters from a retailer. So, while the telephone company should provide telephone service to my home or business, I should be able to purchase my telephone or my PBX from a retailer. The Carterfone system was positioned as just another appliance that ought to be allowed to be connected to the "telephone utility." AT&T provided two counterarguments.

- The telephone business is more complicated and its equipment is more delicate than the gas, water, or electric businesses. They couldn't allow just anybody to design, manufacture, install, and maintain equipment that connects to the Bell System because they might damage its sensitive equipment. Furthermore, if something were to fail, it would be difficult to determine whether the fault is in the appliance or in the utility. AT&T's position here was weakened because they had used it so absurdly in the Hush-a-Phone case.

- This utility analogy opens AT&T's telephone and PBX businesses to competition. While Ma Bell publicly claimed that she wanted to offer competitive prices, she was not allowed to do this because of regulated POTS subsidization. So, if these businesses were deregulated, AT&T would lose a source of POTS subsidy. If the FCC allows this, it will raise the price of Grandma's telephone service.

Cream Skimming. AT&T's side produced a different, but equally compelling, analogy. Suppose your grandfather wanted to operate a public bus route. He determined that, between point A and point B, there was enough demand that he could run buses every 15 minutes between 7 A.M. and 9 A.M. and between 4 P.M. and 6 P.M. on Monday through Friday. He calculated that he could make a reasonable profit if he charged 5¢ per ride. He petitioned some local government agency for a license.

Now, suppose this agency had been eager to provide bus service for their community anyway, but they wanted it 24 hours a day, 7 days a week. They told Grandpa that they would grant him the license, but only if he operated around the clock. They all agreed that, at 10¢ a ride, Grandpa's buses would make enough money during the rush hours that they would compensate for money lost during the other times. They agreed that if the bus business isn't doing well, or is doing too well, they would adjust the fares. So, they did it.

Over the years, Grandpa and then Dad ran a successful business. Fares increased slowly, mostly because of inflation. The service was excellent — never an accident and always on time — for 90 years. The owners were active in the community and everybody was happy. Then one day a hippie with a VW van (Ma Bell viewed Carter with the same disdain that the 1960s establishment viewed the hippies.) decided to offer competing bus service, but only during the rush hour, and proposed to charge only half the fare that our bus company charges. The competitor petitions for a license. Is this fair?

The FCC's Ruling. In 1968 the FCC ruled in favor of Carter and ordered the Bell System to connect to Carterfone systems. While this was bad enough for AT&T, the FCC went further. The FCC adopted the telephone utility analogy and dismissed the bus company analogy. Their decision had five far-reaching ramifications.

- It allowed competition in the arena of anything that connected to telephone wires inside the customers premises. Customer premises equipment (CPE) included telephones and PBXs.
- It destroyed forever the concept that one company could guarantee end-to-end integrity of the network.
- It eliminated one of the major sources of revenue to the telcos for subsidizing POTS and it threatened to undermine the entire tariff structure.
- It established a precedent for competition in other arenas — like long distance (see Section 12.6).
- It made divestiture inevitable.

It's not clear why the FCC's ruling went so far beyond the limits of the particular case in front of them. One could guess that Carter's lawyers succeeded in putting more at stake in an attempt to worry AT&T into settling out of court. It is even more unclear why AT&T's lawyers didn't just settle this thing with Carter before it ever got to the FCC. AT&T could have given Carter an exclusive exception, since he wasn't really competing with them directly. AT&T's lawyers should have been able to read the public mood on this, should have foreseen the FCC's decision and known its long-range ramifications, and — most of all — should have avoided allowing this decision to be made by a group of lawyers who clearly didn't understand the technology or the industry.

Discussion. The Carterfone decision was another microcosm of its time. In the late 1960s, the United States was involved in a very unpopular war in Viet Nam and the public mood was against government, big business, and the military/industrial complex. The FCC and the federal courts were filled with commissioners and judges who were less likely to rule in favor of big business than those appointed during other presidential administrations. And no business was bigger than Ma Bell. AT&T tried to warn the country of the ramifications, but the public was unmoved. In the United States, in 1968, nobody felt sorry for AT&T. The public was on Carter's side and they were pleased when Carter's David slew AT&T's Goliath. AT&T warned of rising telephone rates because of this cream skimming. The public sentiment was to accuse AT&T of being a poor loser and of using this as an excuse to raise rates. One editorial accused Ma Bell of being "in menopause." But AT&T had been right, and the next 15 years saw significant erosion of the revenues that were used to subsidize POTS. That nobody competed with the Bell System in the POTS market (the "skim milk"), but so many tried to compete in so many other markets (the "cream") shows that AT&T had been right. But the deed was done, the precedent was set, and nobody outside the Bell System seemed to care.

But, people inside the Bell System during the 1970s and 1980s wondered what happened and they looked for someone to blame. Their villain was not Judge Green nor Baxter, not Carter nor MCI's McGowan (Section 12.6.2), not the FCC nor the Supreme Court. Their villain was AT&T's own lawyers and upper management in the mid-1960s, who decided that AT&T should not settle with Carter. Their decision forced the case to the FCC, with a result that they should have been able to predict. The Carterfone decision by the FCC in 1968:

- Set a precedent for subsequent regulations (discussed in Sections 12.6 and 16.6);
- Began unraveling the POTS subsidization, which is still not completed in the United States.

10.6.3 Customer Premises Equipment

Beginning in 1968, any telephone appliance located on the customer's premises could be manufactured anywhere, purchased retail, and installed by anybody. The principal areas of competition were the telephone instrument and the PBX. But, there were certain conditions. While Ma Bell lost the war in the Carterfone decision, she did win one battle. The FCC agreed that the electrical integrity of the Public Switched Telephone Network had to be protected.

The technical issues that were raised included interpulse timing and other temporal specifications, adherence to frequency tolerance (Touch Tone, MF, E&M, etc.), proper impedance and resistance (line termination, ringing, off-hook), end-to-end loss and delay (which can cause echo), and protection from damaging voltages. But, all lines are "fused" with over-voltage protection at the main frame anyway, and much telco equipment (especially in the small independents) didn't meet their own specs anyway. Many people argued that the telephone companies were using these technical issues as an excuse to be as uncooperative as possible. Initially, this protection of integrity was implemented differently in the telephone market and the PBX market.

Telephone Registration. The price of the telephone instrument was separated from the price of telephone service. Subscribers were given the choice of: (1) returning the phone company's telephone to the telco, (2) purchasing this telephone from the phone company, or (3) leasing it from the phone company. For those who chose to lease, the price for basic telephone service was decreased, and the rental charge was separate. If you elected to purchase, you paid a one-time price and your telephone was yours, and it was yours to maintain. While the rental price was very high, the phone companies would repair (for free) only the telephones that they owned. So, most Grandmas leased their telephones, and the price of universal POTS increased for many.

Manufacturers who wished to sell the telephone instrument in America's consumer marketplace had to submit designs and prototypes to an independent laboratory where they were certified. Once certified, or registered, the telephone was tagged with a registration number. The telephone subscriber who purchased and installed any telephone, Bell or non-Bell, was supposed to call the phone company and report the new telephone and its registration number — and report whether this new telephone replaced an existing telephone or was being installed as another extension telephone.

Since extension telephones were viewed as luxuries, they had always been overpriced (to subsidize POTS). Initially, immediately after the Carterfone decision, the phone companies still charged subscribers for extension service. Since the phone company typically did not lease out the telephone instrument that was installed at the extension, did not add another line into the residence, nor did anything to the existing line — and since the public didn't appreciate the phone company's problem with subsidizing universal POTS — most people saw no reason to have to pay the phone company for an extension telephone. So, if a subscriber wanted to add an extension telephone, he could buy the telephone retail, install it at home, and it would work (of course) immediately — whether or not he reported the new extension. If the subscriber reported the new telephone, his monthly bill was increased. If he did not report it, his bill was not increased. People learned quickly, and despite sternly worded statements from the phone company in the monthly bill, many extension telephones were not reported.

A similar thing happened with Touch Tone service. If all of your home telephones were uniformly dial or tone, then your class of service caused you to be connected to the right kind of digit receiver when you went off-hook. But since most of us had some dial phones and some push-button phones, if any phones had Touch Tone, you were supposed to tell the phone company, so they could give you a Touch Tone class of service (and charge you more every month because Touch Tone is not POTS). The phone companies didn't tell us this, but Touch Tone receivers could also detect dial pulses. And, in fact, in many central offices (typically not in Step), dial and tone reception was done by the same service circuit, because of the operational simplicity and economy of scale.

However, suppose you had dial class of service, bought a new Touch Tone telephone and installed it, and didn't report it. Your new phone would typically still work, but the

Touch Tone charge would appear on next month's phone bill and you would hereafter have Touch Tone class of service (and would pay for it). In other words, if the central office received Touch Tone signals from a line with a dial COS, then they caught you. The true causality was not, typically, that class of service determined the detection of signaling. The true causality was that signal detection determined COS. And the true reason for this particular COS was to determine whether you could be charged the incremental Touch Tone luxury tax. Why were the phone companies so sneaky? The regulatory bodies wanted POTS to be subsidized, but they were destroying the sources of the subsidy and public opinion was not sympathetic.

So many extension telephones were never reported that the phone companies initiated a routine testing procedure. Recall from Section 3.2.1 that when a telephone is on-hook, the only impedance seen from the line is the coil for the bell. *N* telephones on the same tip and ring look like *N* bell coils in parallel. So, from the length of your loop and the number of telephones you reported, the net inductance seen from the main frame could be accurately predicted and was accurately measured. Threatening letters were mailed to people whose measured loop inductance was inconsistent with the number of telephones they reported. Often, the telephone company just raised the monthly bill without saying anything, like they did with Touch Tone. People learned quickly that the test counted the number of telephone bells in the house, not the number of telephones. Most of us learned how to open up an unreported telephone and disconnect its ringer.

Gradually the tariff changed and the registration procedure evolved into today's standardization procedure. After several years, the extra charge for extension telephones was dropped and subscribers didn't have to report new telephones. But the phone companies still charge a premium for Touch Tone.

The biggest complaint with this arrangement occurred when something broke. If you couldn't get dial tone, for example, and you called the phone company to complain about it, you had to convince them that you had already verified that the problem was not with your telephone and was not in the house wiring — unless you still leased your telephone from your phone company. You had to (and still do) disconnect your telephone and try it at a neighbor's house and at the protector block. If your telephone works at a neighbor's, then your telephone is OK. If your telephone works at the protector block, then your loop is OK. While this logic is simple for you and me, it was a significant problem for many subscribers, especially those with old telephones, without modular jacks, that were hard-wired to the house wiring (especially wall-mounted telephones). It's still frustrating now, but we've become accustomed to it. It was very frustrating in 1968 and it hurt Ma Bell's image even more.

The Interface. The Carterfone decision led also to competition in the PBX market. Instead of purchasing monthly Centrex service from the phone company (and the phone company selected its implementation), Centrex was viewed differently, depending on its implementation. While CO-Centrex didn't change with the Carterfone decision, the price of CU-Centrex was broken into two components: the monthly rental of the PBX and its telephones, and the monthly price of interconnecting the PBX to the CO. Because DID and AIOD no longer required CO-Centrex and because the Dimension PBX was so popular and successful, many Centrex customers had the CU implementation. During the time between the Carterfone decision and the activation of follow-up regulation in the FCC and PUCs, an enterprise had a three-way choice for Centrex: (1) it could pay monthly for CO-Centrex service, (2) it could pay monthly for CU-Centrex service (including Dimension rental), or (3) it could purchase a retail PBX and pay monthly for DID trunks that the phone company had to provide. Price (not cost) heavily favored the third option.

In the residence market, the boundary between the phone companies' property and the subscribers' property was established easily at the protector block. However, "protecting the network's integrity" took on a different, and much more complex, implementation in the enterprise market. Initially, AT&T insisted on placing an active interface between any non-Bell PBX and the PBX trunks to the CO. The regulatory bodies agreed, and the interface had to be installed, by the phone company, on the premises with the PBX.

The interface was so large (probably by design) that it was often a larger box than the PBX to which it interfaced. It was designed to operate off locally derived electrical power, and since it had no option for battery backup in case of power outage, when there was a power failure, the interface opened the PBX trunks. So, even if the PBX was battery-backed, nobody inside the enterprise would be able to place "dial-9" calls during a power outage, unless the enterprise had provided a DC-to-AC converter for its interface. While DID trunks were available for Bell and non-Bell PBXs, the interface did not provide the common channels for AIOD signaling or for time-and-charges signaling back to hotels. So a hotel, or any other enterprise that wanted IOD, had to use operator intercept. Of course, the enterprise could have AIOD if they purchased or leased a Bell PBX. Finally, the phone companies were notoriously slow to install the interface.

Ma Bell was accused of using the interface to deliberately impede the progress of its rivals in the PBX market. She was accused of similar impeding with the equal access issue that MCI would face in another couple of years (Section 12.6.5). While protection of network integrity was never dismissed, its implementation gradually evolved. After several years, the interface became obsolete and was replaced by a procedure involving registration, like what was used with telephones. Then eventually this too evolved to the standardization procedure still in use today. This is still, by no means, a solved problem.

The Interconnect Industry. Before the Carterfone decision, a PBX was telephone company equipment that resided on the enterprise's premises. The CO/CU implementation of Centrex was based on the economics of whether the concentration of lines-to-trunks justified the extra equipment on the premises. When this CO/CU-Centrex decision became complicated by the offering of DID and AIOD services, PBXs and serving COs were modified so that these services were available in both implementations, and the CO/CU decision was again based strictly on equipment optimization.

After the Carterfone decision, not only did the telephone instrument become an openly competitive consumer electronic commodity, but the PBX market was also opened to free competition. Because customer switching service had always been deliberately overpriced to subsidize POTS, this market looked very lucrative. This industry was called the "interconnect industry." Many companies rushed into the market by manufacturing, retailing, installing, and maintaining PBXs for any kind of enterprise. Because the Bell System's participation in this market remained regulated for many years after the Carterfone decision, the competition was biased against Bell System PBXs. Ma Bell compensated for this by using the interface as an impediment.

Initially, AT&T was not allowed to sell its Dimension PBX, only lease it. Because depreciation rules in United States tax laws made it favorable for a business to buy and own a PBX (and depreciate it), the so-called "interconnect companies" had a very clear advantage, and they thrived. Because there was little expertise in the United States (outside the Bell System), at least initially, international companies with existing products were the first to enter the market. Companies like Nippon Electric (NEC) and Northern Telecom (now calling themselves Nortel) did very well. Many new companies started up to import, install, and maintain PBXs. Other new companies started up to design and manufacture new PBXs — like Rolm and TAI. They competed among each other on price, reputation, and features.

The phone companies' penetration of the PBX market fell from 100% before 1968 to around 20% (typically, leased Dimensions) in only a few years. When AT&T was allowed to sell PBXs, they recovered a little, with Dimensions and with a newer PBX line. So, after the initial frenzy, there were many more companies than the market could support. A period of shakeout, consolidation, and takeovers followed. For example, Rolm was taken over by IBM when IBM decided to enter the "confluence war" with AT&T (Section 16.2.3). When IBM lost this war (AT&T also lost — it was a war nobody won), Rolm was sold to Siemens, a German conglomerate that makes appliances, heavy electrical equipment, and telephone switching equipment.

Western Electric's traditional culture of high-quality manufacturing and methodical business procedures made it difficult for them to compete on the basis of price or on features. AT&T became the high-cost, high-quality vendor in the PBX market. With modern processor-control, new PBX features are implemented by designing and provisioning new programs. This task is much simpler and faster in a small company than in a large one and in the small call processing software environment of a PBX than in large central office switching software. Thus, while Western Electric's PBXs tended to lag the market a little in the timing of introducing new features, Centrex really lagged (and still does to this day). This lag is much of the motivation behind the migration to the Intelligent Network, discussed in Chapter 18. With its Definity (Section 14.3) and Merlin lines, AT&T (now Lucent) mounted a gradual comeback, but controls only about one-third of the U.S. PBX market share.

Earlier, this chapter stated that PBXs tended to lag behind central offices in technology and in features. This changed dramatically in this competitive market and PBXs typically led CO switches in new features. Digital PBXs appeared years before digital central offices. Increased functionality and lower price are, of course, the reason to let the market be competitive. In this case, we got both — but we also got increased instability and the inevitability of more regulatory changes.

EXERCISES

10.1 Consider a business establishment with a six-line hunt group. Why must the listed directory number end with a digit 1 – 5 if the serving switching office is a Step-by-Step? Equivalently, if the listed directory number ends in 0, how do we know that the serving office is not a Stepper?

10.2 Sketch a block diagram for CO-Centrex system embedded in the Step-by-Step central office with office code 345. Let the business' listed directory number be

345-2001. Let the PBX have nine attendants, numbered 345-2001 through 345-2009. Let the PBX have 690 lines, numbered 345-2011 through 345-2699. Let the PBX have the equivalent of 10 outgoing PBX trunks, reached by dialing 9 within the virtual PBX. Show how typical lines appear on a segregated line finder group and on a segregated connector group. Show the subfabric for the virtual PBX. Show how the originating side of the 10 equivalent outgoing trunks appears on the first selector outputs in this dedicated subfabric and how the terminating side of these 10 outgoing trunks appears in the general Step office. Show how the attendants would be organized as a hunt group.

10.3 Sketch a block diagram for the Step-by-Step central office in Exercise 10.2, except let it provide CU-Centrex to the same business. Let the PBX have 10 outgoing PBX trunks, accessed by dialing 9 within the PBX, and 10 incoming PBX trunks. Show how these 20 trunks terminate in the Step CO.

10.4 Suppose a large Centrex enterprise has 15 separate internal departments, each with its own computer center. Trying to encourage employees to work from home, the company sets up a modem pool and a hunt group for each computer center. (a) If each computer center has 5 Erlangs of peak busy hour load and allows peak busy hour $P_B = 0.05$, how many total hunt-group lines are required? (b) If the enterprise merges its modem pools behind a PBX, as a concentrator, how many total hunt-group lines are required?

10.5 Some enterprise's telecom manager wishes to avoid paying the LEC's extra charge for DID trunks but wants to provide DID for external callers with Touch Tone telephones. He provides a second group of incoming trunks and his PBX provides second dial tone for incoming calls on these trunks. (a) Why is this scheme restricted to Touch Tone subscribers and problematic for callers with dial telephones? (b) Describe the enterprise's listing in the telephone directory. (c) What might happen if incoming calls are allowed to place dial-9 calls?

10.6 Some enterprise's telecom manager is asked to provide three categories of outside calling for the enterprise's staff. Some extensions have unrestricted access for outside calls, both flat-rate local calls and long distance calls. Other extensions are allowed access only for outside local calls, and a third category of extensions is not allowed to place any kind of outside call. The telecom manager arranges for two separate groups of PBX trunks. One group is accessed within the PBX by dialing 9 and another group is accessed by dialing 8. Describe how class of service at the PBX and at the CO could be arranged to implement the three categories of outside calling. There are two different arrangements.

REFERENCES

[1] Hester, S., "Managing the Combined Virtual Network," *Data Communications,* Vol. 8, No. 12, Dec. 1979, pp. 41 – 53.

[2] Gould, E. P., and C. D. Pack, "Communications Network Planning in the Evolving Information Age," *IEEE Communications Magazine,* (issue on Network Planning), Vol. 25, No. 9, Sept. 1987.

[3] Langford, G., "Planning the Multi-National Network," *Telecommunications Magazine,* Vol. 23, No. 11, Nov. 1989.

[4] "Connecting the Global Enterprise," *Telecommunications Magazine,* complete issue, Vol. 23, No. 11, Nov. 1989.

[5] Kirvan, P., *The Best of Paul Kirvan on Communications Management,* Nokomis, FL: Nelson Publishing, 1995.

[6] Landry, R., "Centrex — A New Concept of PBX Services," *Bell Telephone Magazine,* Autumn 1961, p. 10.

[7] Shea, P. D., "Centrex Service — A New Design for Customer Group Telephone Service in the Modern Business Community," *Transactions of the AIEE,* Vol. 80, Pt. 1, 1961, p. 474.

[8] Temin, P., with L. Galambos, *The Fall of the Bell System,* New York, NY: Cambridge University Press, 1987.

[9] Henck, F. W., and B. Strassburg, *A Slippery Slope — The Long Road to the Breakup of AT&T,* Westport, CT: Greenwood Press, 1988.

[10] Dziatkiewicz, M., "Universal Service — A Taxing Issue," *Wireless Business & Technology,* Vol. 5, No. 1, Jan. 1999.

[11] Weiss, M. B. H., and P. Bernt, *The Regulation and Deregulation of US Telecommunications,* Textbook under contract with Lawrence Erlbaum and Associates, Mahwah, NJ,

SELECTED BIBLIOGRAPHY

Bahr, W. J., "Alternatives for PBX and Key System Integration," *Telecommunications Magazine,* (issue on Switching Inside and Outside the Network), Vol. 21, No. 2, Feb. 1987.

Brooks, J., *Telephone — The First Hundred Years,* New York, NY: Harper & Row, 1976.

Clarke, A. C., et al., *The Telephone's First Century — and Beyond,* New York, NY: Thomas Y. Crowell, 1977.

"Computer Telephony: Say Goodbye to Key Phone Systems," *Communications News,* Vol. 35, No. 2, Feb. 1998.

Junker, S. L., and W. E. Noller, "Digital Private Branch Exchanges," *IEEE Communications Magazine,* (issue on Digital Switching), Vol. 21, No. 3, May 1983.

McDonald, J. C. (ed.), *Fundamentals of Digital Switching,* Second Edition, New York, NY: Plenum Press, 1990.

Noll, A. M., *Introduction to Telephones and Telephone Systems,* Norwood, MA: Artech House, 1986.

Rey, R. F. (ed.), *Engineering and Operations in the Bell System,* Second Edition, Murray Hill, NJ: AT&T Bell Laboratories, 1977.

Schindler, G. E., Jr. (ed.), *A History of Engineering and Science in the Bell System,* Murray Hill, NJ: Bell Telephone Laboratories, 1982.

Telecommunications Transmission Engineering, Volume 3 — Networks and Services, New York, NY: American Telephone & Telegraph, 1977.

Tracey, L. V., "30 Years: A Brief History of the Communications Industry," *Telecommunications,* Vol. 31, No. 6, Jun. 1997.

11

Program Control — #1 ESS

> *"I think there's a world market for maybe five computers."* — Thomas J. Watson, chairman of IBM in the 1950s

This is one of the biggest chapters in the book because, I think, the #1 ESS is probably the most important and significant switching system since the Step-by-Step system 80 years earlier. Since the #1 ESS, telephone switching has completely changed from a hardware endeavor into a software endeavor. So, if you don't understand switching software, you know less than half about modern switching. And since switching software is most transparent in the #1 ESS, this chapter is one of the most important chapters in the book.

We begin with an introduction to electronic switching, followed by a brief discussion of modern computer control of any general process. Then, the #1 ESS' hardware and software architectures, respectively, are described in the two sections after these. In Section 11.5, a call is walked through this hardware and software, simultaneously. A section on the #1 ESS' maintenance philosophy is followed by a general discussion.

Section 11.4, describing the software architecture of the #1 ESS, is one of the most important parts of this book. You cannot claim to understand switching in the modern era without appreciating the complexity of switching software. This complexity has two components: *feature interaction* and the behavior of *execution time.* Feature interaction will not be very apparent from a discussion of #1 ESS software, but the behavior of execution time will be. Feature interaction is an extremely important topic that will be discussed in depth in Section 18.2.2.

In the newest systems, like Lucent's #5 ESS, Nortel's DMS-10, and Siemens EWSD, the behavior of their execution time is hidden behind their operating systems. Hiding this behavior simplifies the task of developing programs for these systems, a benefit for programmers, but it clouds a true appreciation of how these systems work. Conversely, in a system with a monolithic program architecture, like the #1 ESS, the behavior of its execution time is painfully apparent. While this makes a more complex environment in which programming professionals must design software, it makes the best pedagogical instrument for explaining the behavior of a processor's execution time.

11.1 INTRODUCTION

The first commercial processor-controlled telephone office was called the Electronic Switching System (ESS). As later systems were developed, they were called #2, #3, and so forth. So, this first system came to be called the #1 ESS. Several years after its original development, the #1 ESS was redesigned using more modern technology, and this newer model was called the #1A ESS. Now, even the abbreviated names of these two models are further abbreviated, to the 1E and the 1A, respectively. The 1E is the older, original model. The 1A is the newer, more modern model. Both have a centralized architecture, with nonintelligent peripherals surrounding a processor.

Scientists and engineers in the research area at Bell Laboratories started a project in the mid-1950s to investigate the feasibility of controlling a switching office by a digital computer. A research prototype was even alpha tested at Morris, Illinois, and product development started in the late 1950s. Crossbar modernization, by replacing markers with digital computers, was considered, but an entire new system was developed instead, using several new technologies. A new switching device, the *ferreed switch,* was used in the fabric. A new current-sensing device, the *ferrod,* was used in supervisory circuits. A new fabrication technology, the *ferrite sheet,* provided a highly integrated (for that time) magnetic core random-access memory (RAM). Another new magnetic technology, the *piggyback twister,* provided a new kind of computer memory that we now call electrically alterable read-only memory (EAROM).

While computers of this era were primitive by today's standards, software and programming techniques were "Neanderthal." ESS programs were written in a very powerful assembly language, using a lot of indirection, macros, and subroutines. A high-level language was not used because Algol, Cobol, and Fortran (version I) were only themselves just being invented. ESS has no operating system, per se, because the first modern operating system was itself only under development at IBM (and would appear in their 7090 computer).

ESS development took much longer than expected, partly because of problems with these new technologies, but mostly because the programming effort was greatly underestimated. Similar underestimation occurred in two other contemporary large programming projects — the operating system for the 7090 and the airline reservation system that was the predecessor of American Airlines' Sabre system. Thirty years later, we're still not very good at estimating large programming efforts — you can imagine how poor they were at estimating these first ones. The first ESS office was installed in Succasunna, New Jersey in 1965 — about a year before it was really perfected.

The #2 and #3 ESS were briefly discussed in Section 2.5.4. The #4 ESS, or 4E, is a digital system that was developed specifically for use in toll offices. The 4E was discussed briefly in Section 9.3.5 and will be described in more detail in Section 14.1. The #5 ESS, or 5E, is a digital system that was developed mainly as a local central office. The 5E, featuring a *modular* architecture, is described in Section 14.2.

Is ESS an analog system or a digital system? Some of the marketeers in the late 1970s and early 1980s tried to argue that the 1E/1A and the Dimension PBX were digital because they were controlled by digital computers. But the marketplace didn't accept this definition. These systems are analog because the information they switch is in analog form. A digital switch converts analog information (like conventional voice) to digital form (using pulse code modulation, for example) before the system can switch it.

An analog switch converts digital information (like computer data) to analog form (using a modem, for example) before the system can switch it. But, one must take care not to confuse analog with electromechanical. While the 1E/1A is analog like its electromechanical predecessors, it is computer-controlled like its digital successors.

11.2 COMPUTERIZED PROCESS CONTROL

ESS is described in the context of modern process control. We present the general architecture for the computer control of a generalized process and discuss how a process' inputs and outputs are observed by the controlling computer and how this computer manipulates the process' control points. This section is a logical development, not an historical development. ESS was not developed as a special case of this general theory. To the contrary, many of the concepts of ESS were extended later to the computer control of other processes. ESS was a pioneering effort from which the theory of modern process control has been a generalization.

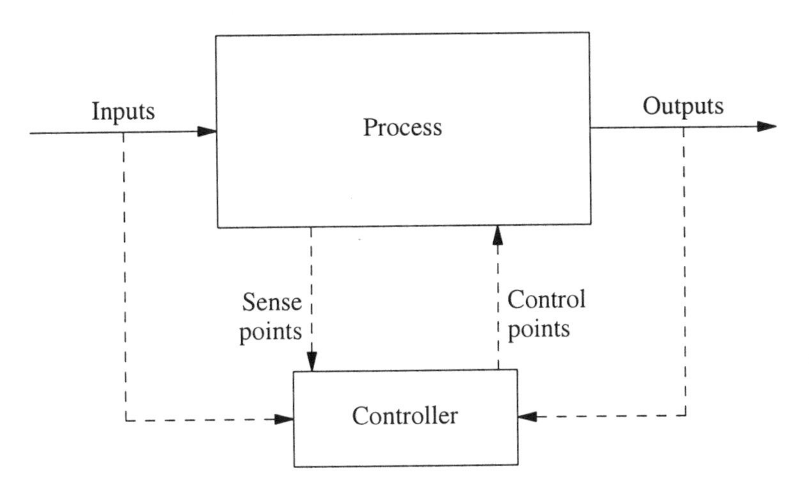

Figure 11.1. Generalized architecture for process control.

11.2.1 General Architecture

Figure 11.1 shows a high-level block diagram of a process control system. The process (an oil refinery or automobile engine or telephone switching office) is at the top of the figure. The controller, a digital computer, is at the bottom of the figure. Every process has inputs, outputs, internal sensing points, and control points. From the perspective of the process, there is a distinction between its inputs and its outputs. But from the perspective of the controller, the process' inputs and outputs, and its internal points, are just quantities to be sensed.

The values of these inputs, outputs, and sense points are data for the controller. The controller uses this input data, and data about the "state" of the process, which it has stored internally, and performs a computation according to its internal program. This computation produces the data that updates the process state, stored internally, and the data that the controller outputs to the process. This controller output data regulates the process by manipulating the process' control points. Consider three examples.

- Inputs to an oil refinery are crude oil, catalysts, and energy. Outputs from an oil refinery are gasoline, kerosene, diesel fuel, other petro-chemicals, and waste products. Input, output, and internal fluid flows are sensed by the controller, as are the temperature and gas pressure in various chambers and the content of waste gases. Control points in an oil refinery are fluid flow valves and temperature regulators.
- Inputs to an automobile engine are gasoline, air, coolant, and oil. Outputs from an automobile engine are its exhaust gas and the torque that turns the automobile's wheels. A variety of transducers measure various temperatures, pressures, fluid levels, speeds, and even the content of the exhaust gas. These transducers provide input to the controller. Control points in an automobile engine regulate the air/gasoline mixture and the timing of the spark advance.
- Inputs to a telephone switching office are originated calls. Lines and trunks are the conduits over which calls flow, analogous to the pipes in an oil refinery and the fuel line and exhaust pipe in an automobile engine. Outputs from a telephone switching office are terminated calls. Supervisory sensors provide call status information to the controller, and similar devices provide data about dialed or in-pulsed digits. Control points in a telephone switching office are switches and relays in the fabric and in the various line, trunk, and service circuits.

11.2.2 Controller I/O

A sensor is a transducer that converts some measured quantity in the process into digital data that the controller can read as input. A control point is a transducer that converts digital data that the controller produces as output into some setting or movement that manipulates, regulates, or otherwise exerts some control over the process. From the controller's perspective, the sensors are simple computer input peripheral devices and the control points are simple computer output peripheral devices.

Two classes of i/o peripherals are *analog* and *binary.* An analog quantity has a real-number value that changes in continuous time. A controller input, like the gas pressure in a chemical reactor, and a controller output, like the spark advance setting for an automobile engine, are examples of analog i/o. But digital computers are incompatible with real numbers and with continuous time. The real-number value at the process and the digital value at the controller are made compatible by including analog-to-digital converters in the peripheral subsystem. Continuous time in the process and discrete time in the controller are made compatible by sampling i/o often enough that the changes in the discrete values of consecutive samples are relatively smooth.

A binary quantity has a two-state ON/OFF value that also changes in continuous time. A controller input, like the state of a limit switch or a telephone supervisory sensor, and a controller output, like the signal to open an overflow valve or to operate a relay in a switching fabric, are examples of binary i/o. While the two-state value of a binary quantity is easily compatible with some individual bit in some digital word, the controller's incompatibility with continuous time is extremely important, even with binary quantities. One can imagine the serious consequences if some temperature or pressure alarm sensor in a nuclear power plant is examined by its controller on a schedule of once every 3 hours.

Two temporal mechanisms have evolved that govern the time when a computer reads its input data. Computer input can be *event-driven* or *polled.* An event, like the operation of a limit switch or the moment that an analog quantity exceeds some threshold, can interrupt the normal operation of a computer. This normal operation, where programs are well behaved and wait their turn for execution, is called base level. Some base-level program, which was executing at the time of the interrupt, is suspended so another program can respond to the event that caused the interrupt. Depending on the urgency of the event, the interrupting program might request that some base-level program eventually process the input or it might process the input itself, immediately.

In event-driven input, the data takes the initiative and demands the attention of the computer. In polled input, the computer takes the initiative and the data is not read until the computer gets around to it. Two forms of polling are *deferred polling* and *timer-driven polling.* In deferred polling, the program that reads the input data is just another one of those well-behaved programs that waits its turn to execute in base level. In timer-driven polling, an external clock periodically interrupts the computer and causes the execution of the program that reads the input data.

A disadvantage of deferred polling is that an urgent input might wait a long time to be read, especially if the computer gets very busy and many base-level programs are scheduled for execution. This same property can be an advantage because, when the computer is very busy, it might be wise to delay finding more work to do. Interrupts, whether event-driven or timer-driven, are characterized by the overhead of gracefully suspending the executing program so that its execution can resume when the interrupting program is done. In addition, interrupts are typically prioritized so that very urgent events can interrupt other interrupt-level programs, while less urgent events may only interrupt base-level programs. A process with thousands of inputs would never be designed so that each one could cause an event-driven interrupt.

In ESS, fault detection causes event-driven interrupts, new off-hooks on subscriber lines are read under deferred polling, and other supervisory and signaling data are read under timer-driven polling.

11.2.3 Real-Time Programming

The actual passage of real time is relevant to whatever process is being controlled. Cooking time in chemical reactors, spark delay in automobile engines, and interpulse timing in telephone offices are all examples of events in real time in their respective processes. However, the computer controlling any such process is completely asynchronous from the external world's calendar and clock. A process controller's concept of time is internal, local, and relative to the variations and fluctuations of its internal clock circuit. Doubling the instruction rate in the microprocessor that controls an automobile engine should not cause the engine to idle twice as fast. Real-time programming is the design of programs and program environments in which external real time is a relevant parameter.

Though time in the controller is asynchronous from time in the process it controls, it would still be simple to design and program a process controller that monitors a single sensor and regulates a single control point. Since the typical process control system has many sensors and many control points but only one controller, this lack of synchronization between process and controller is exacerbated by a need for apparent simultaneity in the control of all the subprocesses. Typically, many separate programs share computer time in a time scale that is fine enough so that the various subprocesses are controlled with imperceptible delay. It would not be acceptable for the controller of an automobile engine to allow 5 minutes adjusting spark phase, and then 5 minutes adjusting gasoline/air mixture, and then 5 minutes displaying information on the driver's console, and so forth, through perhaps 20 different tasks. It would not be technically possible to cycle through these same 20 tasks allowing 5 nanoseconds on each task. The actual time scale lies between these two extremes, and the actual schedule of these tasks is typically much more complicated than simply taking turns.

The potential scenario has the controller executing a given program for several milliseconds, then stopping it and executing another one for several milliseconds, and so forth, through some predetermined schedule. However, it's far more complicated than this. Some programs have internal real-time breaks in which, having produced some output to a control point, the program would not continue until the new output has had a chance to alter the process. Such programs are not timed and stopped — they stop themselves. In addition, successive intervals of program executions may actually be the same program executing on behalf of a succession of clients or different subprocesses. The schedule is seldom predetermined. Typically, various programs request one another.

In this context, the definition of "program" is elusive. We think of a program as a sequence of computer instructions that performs some function. But a program may have many discontinuous segments, each separated by a real-time break. These segments may not follow one another in a single-threaded predetermined sequence. Many programmers may design a program, and other programmers, who weren't even born when the original program was first designed, may make subsequent changes to it. The same program may execute for many different clients, where these clients can be other programs or subprocesses or users of the process being controlled.

Time sharing has two classical implementations: *multiprocessing* and *multitasking.* In multiprocessing, if two clients require the execution of the same program, two separate copies, called processes, of the program are created and each process is assigned its own client. In multitasking, if two clients require the execution of the same program, this single copy of the program executes distinctly for each different client. The 1E uses multitasking, and the 5E (Section 14.2) uses multiprocessing. The Definity PBX, described in Section 14.3, uses a compromise.

Since multitasking requires only one copy of each program, computer memory is conserved. However, a program in a multitasking environment must be carefully designed. In particular, memory must be carefully allocated to the specific client, and data cannot be stored arbitrarily in scratchpad memory. Such programs are called re-entrant, because they are reentered with new clients, even if they are currently active with old clients.

The bizarre characteristics of real-time process control are summarized.

- I/O can be event-driven or polled, and polled i/o can be deferred or timer-driven.
- The controller must acquire the process' clock and the time sharing must be scaled finely enough that multiple sequential programs appear virtually simultaneous.
- Because the various programs take real-time breaks, the various continuous segments must be scheduled to run when the breaks are over.

- Because the various programs are not mutually independent, they must be able to request one another's execution.
- Data must be carefully maintained so that a program with many clients doesn't alter the wrong client's data, and client data must be available to more than one program.
- The time-sharing schedule is very complicated.

11.3 HARDWARE ARCHITECTURE

ESS hardware is described. After presenting an overall block diagram, some details of the fabric and the peripheral unit subsystem are given. System configurations are illustrated for dial pulsing, intraoffice ringing, outpulsing, and incoming calls.

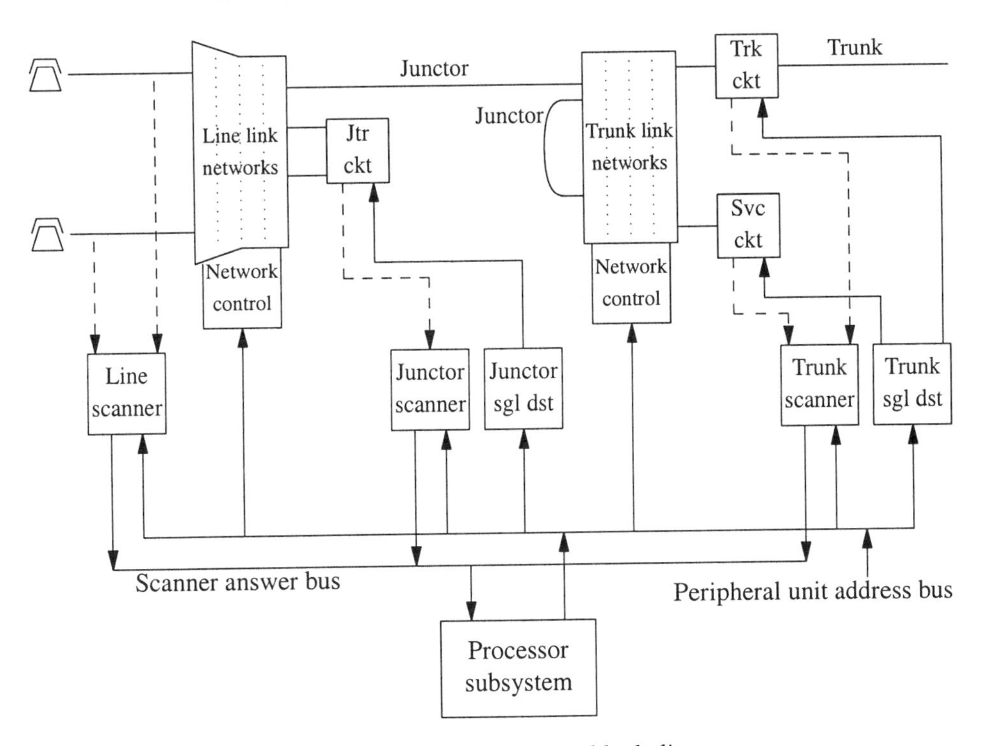

Figure 11.2. #1 ESS hardware block diagram.

11.3.1 Overall Block Diagram

Figure 11.2 shows a block diagram of an ESS switching office. While many switching-illiterate people might think that Figure 11.2 is the block diagram of an ESS office, it is merely a block diagram of ESS' hardware, and it represents much less than half the story of ESS.

The top third of Figure 11.2 shows a block diagram of a conventional looking telephone office. The office has a line side and a trunk side and its fabric has this same sidedness. The loop plant terminates on the line side through a main distribution frame, not shown in the figure. Trunk circuits and a variety of service circuits — including dial pulse receivers, Touch Tone receivers, ring generators, audible ring tone circuits, and MF outpulsers and receivers — all terminate on the trunk side.

Not apparent from Figure 11.2 is the fact that the various subsystems in this top third of the figure are all much "dumber" than their counterparts in a Crossbar office. An ESS line circuit, for example, consists only of a simple current detector, called a ferrod, and a cut-off relay so the ferrod can be removed from its line when supervision is transferred to a junctor or trunk circuit. These line ferrods are aggregated into the line scanner in Figure 11.2. An ESS Touch Tone receiver cannot store the digits it receives — it simply reads them for the processor and they are stored in the processor's memory. Not only is an ESS dial pulse receiver not able to store digits, it cannot even count the pulses in a digit — it simply detects pulses for the processor and they are counted by a program in the processor, and then the digits are stored in the processor's memory.

Three important subsystems from Figure 11.2 are described in the next three subsections. ESS' processor subsystem is represented at the bottom of Figure 11.2. The hardware of this subsystem is described in Section 11.3.2. The software is described later in Section 11.4. Figure 11.2 shows a collection of circuits between the processor subsystem at the bottom and the telephone switching process at the top. These various scanners, signal distributors, pulse distributors (not shown), and network controllers collectively comprise the peripheral unit subsystem. These peripheral units present a digital i/o interface to the processor and an analog electrical or electromechanical interface to the telephone switching equipment. This peripheral subsystem is described in Section 11.3.3. The blocks labeled "Line link networks" and "Trunk link networks" represent ESS' eight-stage fabric. This fabric is described in Section 11.3.4.

11.3.2 Processor Subsystem

The ESS processor subsystem is shown at the extreme bottom of Figure 11.2. This subsystem contains two or four processors, each with two classes of memory units. The basic processor is duplicated for reliability and may, optionally, be augmented by a front-end i/o processor, which is also duplicated. The processor's architecture is only a little different from other conventional general-purpose computers of its era. It has more internal hardware registers than are customarily found in such computers and these registers are all general-purpose — there is no specialized accumulator. Some unusual and clever computer programming commands were designed into the processor's instruction set (more on this in the next section). The original 1E processor had a 36-bit data bus and instruction word, and its internal clock ran at 1 MHz. Most of us have personal computers that are much more powerful, but it was truly state-of-the-art for its time. The 1A processor has a much faster instruction rate.

The computer memory for this processor subsystem is partitioned into the *call store* and the *program store.* Line busy/idle status bits, the fabric map, dialed digits, and other temporary data are stored in the call store. It is volatile memory from which the processor may read and write. The 1E uses ferrite-sheet technology in its magnetic call store units. The later 1A uses more modern RAM in its electronic call stores.

Computer programs, fixed translation tables, and class of service information are stored in the program store. It is semipermanent memory from which the processor may only read. While this memory may not be written by the processor, it may be written manually using the memory unit writer. This way, new releases of the program and

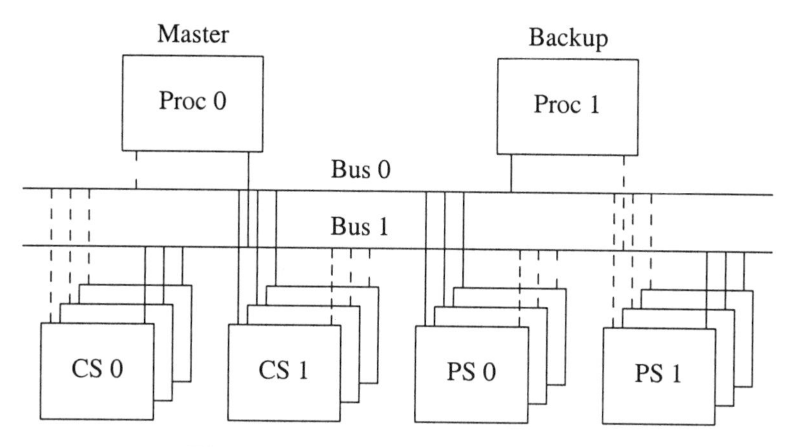

Figure 11.3. ESS processor subsystem.

updated translation information may be installed. The 1E uses piggyback twister technology in its magnetic program store units. The 1A uses more modern EAROM in its electronic program stores.

The call store and the program store are each comprised of many separate memory units with a fixed number of memory words per unit. A common memory address bus and memory data bus tie all the memory units to the processor. Not only is every memory unit duplicated for reliability, but the entire memory bus is also duplicated. Figure 11.3 shows a processor pair, a duplicated call store subsystem, and a duplicated program store subsystem, all interconnected by a duplicated memory bus.

A *master* configuration of the processor subsystem might use processor 0, memory bus 1, call store 0, and program store 1 — indicated by the solid lines connected to processor 0 in Figure 11.3. The *backup* configuration would use processor 1, memory bus 0, call store 1, and program store 0 — indicated by the dashed lines connected to processor 1 in Figure 11.3. The two configurations operate independently, but in parallel and simultaneously. Comparators make sure that both configurations always perform identical operations. If the processor subsystem has the optional front-end processor, called the signal processor, then this additional system has a similar duplicated configuration.

The processor subsystem also includes a master control center (MCC), not shown in Figure 11.3, which provides a human interface between the processor and the craftsmen who operate the 1E system. The MCC contains an alarm control and display panel, a teletype (remember those?), a trunk and line test panel, the LAMA recorders, and the memory unit writer. The alarm control and display panel has the lamps and bells by which the processor informs the craftsman of system status and some special buttons and switches by which the craftsman can exert control over the system. The teletype allows for more general and verbose i/o. The trunk and line test panel has appearances on both sides of the fabric and allows the craftsman to connect test equipment through the fabric to suspicious lines and trunks. The LAMA recorders were discussed in Section 9.2.2.

11.3.3 Peripheral Unit Subsystem

The peripheral unit (PU) subsystem acts as the interface between the digital ESS processor and the analog telephone switching process that it controls. From the processor's perspective, this PU subsystem is a collection of digital i/o peripherals and an intricate structure of digital buses and wires. From the perspective of the switching process, this PU subsystem is a collection of current-detecting ferrods and control signals that operate and release relays and that set and reset flip/flops. This subsystem is complicated — partly to enable the necessary fault tolerance and partly because it resembled the IBM 7090's i/o channel subsystem, which also seemed to be more complicated than was necessary.

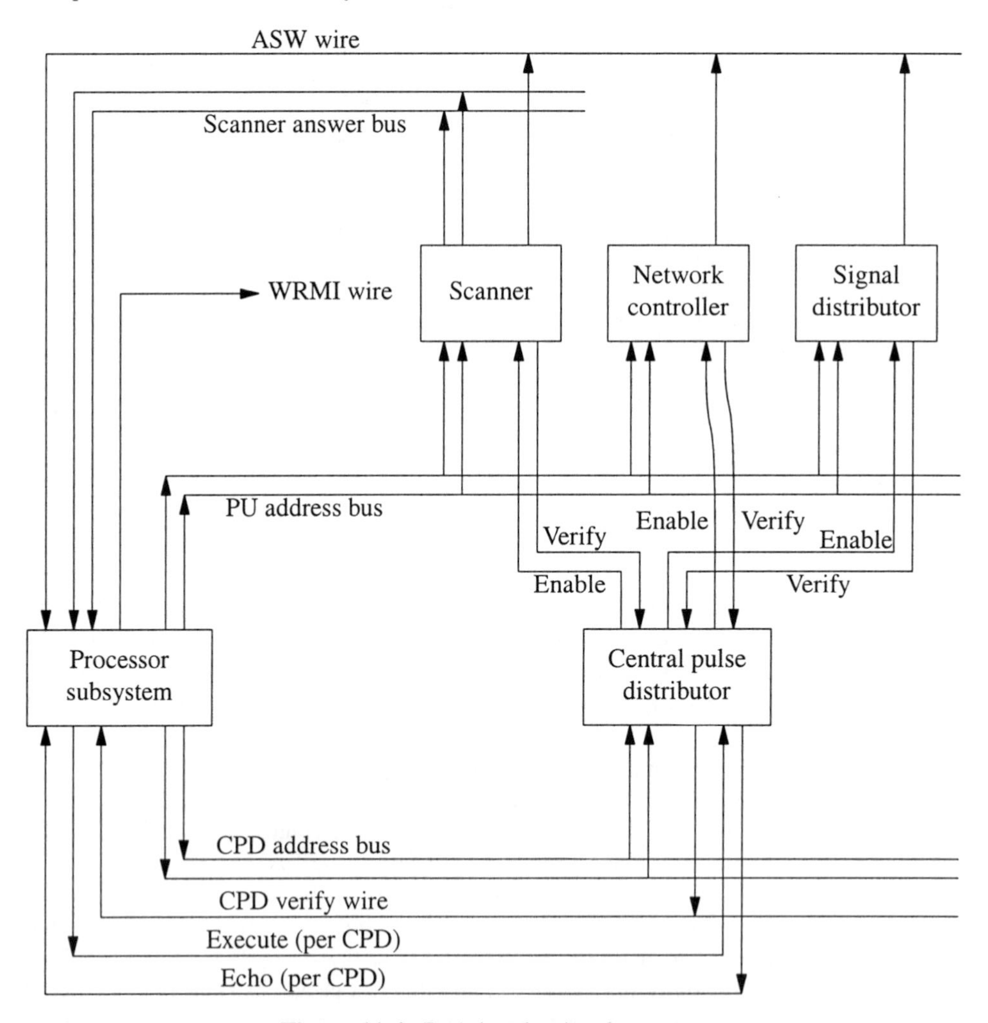

Figure 11.4. Peripheral unit subsystem.

Figure 11.4 shows an expanded view of this PU subsystem. While the figure only shows one central pulse distributor, one scanner, one network controller, and one signal distributor, there are actually many of each of these and the ones shown are only representative. The buses in Figure 11.4 are also duplicated (not shown) in an arrangement similar to the processor/memory subsystem. This PU subsystem is described here in much greater detail than that given to the rest of the ESS hardware. While the PU subsystem really is that much more complex, the extra steps and partial results to be described here characterize the design philosophy that enables the maintenance programs to isolate faults to specific circuits.

Scanners. A ferrod is an electromagnet that has four coils around a soft iron core. Two of the coils are the tip and ring of some line, trunk, or circuit in a balanced, or common mode, configuration. The other two coils carry signals for *strobe* and *sense.* If supervision current flows through tip and ring, the soft core is saturated; and a small strobe pulse produces an insignificant sense voltage. If supervision current does not flow through tip and ring, the soft core is not saturated; and a small strobe pulse produces a significant sense voltage. If the strobe coils to 16 ferrods are wired serially, then these 16 ferrods can be sensed, in parallel, by one strobe pulse. With several milliseconds of delay between applying the strobe and reading the digitized response, ferrods must be sensed infrequently.

Because ferrods are used for supervision, they are a logical part of line circuits, junctor circuits, trunk circuits, and service circuits. Ferrods are physically aggregated into units called scanners, each containing 1024 ferrods, organized in a matrix with 64 rows and 16 columns. A scanner has a 6-bit input address, an enable input, and a 16-bit output. The input address steers the enable signal so it strobes one of the 64 rows. The parallel sense voltages from the 16 ferrods in this row are amplified, digitized, and placed on the 16-bit output bus.

The PU subsystem contains one line scanner for every 1024 lines, one junctor scanner for every 512 intra-LLN junctors (either-party supervision requires two ferrods per junctor), and one trunk scanner for every 1024 scan points in trunk circuits and service circuits. Like junctor circuits, most trunk circuits have two supervisory scan points. While most service circuits have only a single supervisory scan point, Touch Tone and multifrequency receivers (TTR and MFR, respectively) have an additional scan point for each frequency detector (7 in a TTR and 15 in an MFR).

SDs, NCs, and CPDs. Typical of most process control environments, the ESS has two kinds of binary control points to which the processor must deliver an output: to devices with electronic speeds or to electromechanical devices. Electronic devices can respond to the short pulses that are produced directly by the processor's i/o peripherals, but electromechanical devices require more sustained signals for their operation. In the ESS, a central pulse distributor (CPD) provides output pulses for electronic devices in the process, and a signal distributor (SD) provides signals to electromechanical devices. The distinction between CPDs and SDs is a little confusing because a CPD is used to provide the enable pulse to an SD.

Line circuits, junctor circuits, trunk circuits, and service circuits are all built with electromechanical relays that are physically part of their respective circuit. The control logic that operates and releases these relays, while also logically part of its respective circuits, is all physically aggregated into units called signal distributors. Similarly, the control logic that operates and releases the relays in the various LLNs and TLNs is also physically aggregated into units called network controllers (NCs). Each SD and NC contains 1024 logic elements, each capable of controlling one relay, organized in a matrix with 32 rows and 32 columns. An SD or NC has a 10-bit input address and an enable input. The input address steers the enable signal so it operates or releases one of its SD's or NC's 1024 relays. Since it can take 20 milliseconds to operate or release an electromechanical relay, assignments to SDs and NCs must be spread out temporally.

The PU subsystem contains: one line signal distributor for every 1024 relays in the system's line circuits, one line link network controller for every LLN containing 1024 relays, one junctor signal distributor for every 1024 relays in the various junctor circuits, one trunk link network controller for every TLN containing 1024 relays, and one trunk signal distributor for every 1024 relays in the various trunk circuits and service circuits. Different circuits have different numbers of relays.

Many of the line circuits, junctor circuits, trunk circuits, and service circuits are also built with flip/flops that are physically part of their respective circuit. The control logic that sets and resets these flip/flops, or that produces other electronic pulses, while also logically part of its respective circuits, is all physically aggregated into units called central pulse distributors. Each CPD contains 1024 logic elements, each capable of producing an electronic pulse that can set or reset one flip/flop or perform some other function in an electronic circuit, organized in a matrix with 32 rows and 32 columns. A CPD has a 10-bit input address and an execute input. The input address steers the execute signal so it produces one of the 1024 electronic pulses associated with the particular CPD. Since flip/flops can be operated at electronic speeds, assignments to CPDs can be executed rapidly.

The PU subsystem contains one central pulse distributor for every 1024 electronic control points in the system's periphery. Besides setting and resetting peripheral flip/flops, other electronic control points that receive a pulse from a CPD are the enable inputs to each scanner, network controller, and signal distributor.

Verification. Each scanner, signal distributor, and network controller produces two signals to verify that it thinks it is operating correctly. Each such PU produces a *verify* output that signals that it has received an enable signal. The verify output is wired back to the only CPD that ever enables this particular PU. In addition, each scanner, signal distributor, and network controller also produces an all-seems-well (ASW) signal after it thinks it has scanned the requested row or operated the requested relay. The ASW output from each PU is wired to a common wire that is input to the processor.

Each central pulse distributor also produces two signals to verify that it thinks it is operating correctly. Each such CPD produces an *echo* output that signals that it has received an execute signal. The echo output from each CPD is wired back to the processor. In addition, each central pulse distributor also produces a verify signal after it receives a verify signal from any PU. The verify output from each CPD is wired to a common wire that is input to the processor.

Scanners, network controllers, and signal distributors are susceptible to noise on their enable inputs. Because all these peripheral units read their addresses from the same peripheral unit address bus (PUAB), a noise spike on an enable would cause a PU to take some action that was based on whatever data happened to be on the PUAB. To help safeguard against this, the processor produces a "we really mean it (WRMI)" signal, that is wired to every PU. A PU takes action only when its enable pulse coincides with the WRMI signal. PUs are still susceptible to noisy enables during the WRMI signal.

While individual ferrods, relay control elements, and electronic pulse elements are not duplicated, the core part of all peripheral units is duplicated. Like the processor subsystem's memory bus in Figure 11.3, the three buses in Figure 11.4 (the CPD address bus, the peripheral unit address bus, and the scanner answer bus) are all duplicated. Unlike Figure 11.3, this duplication is not shown in Figure 11.4.

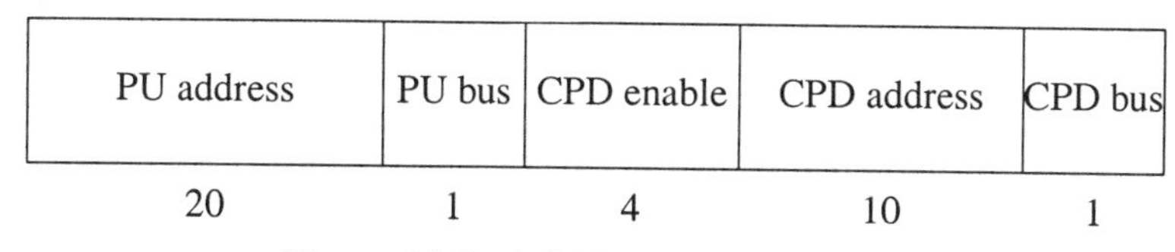

Figure 11.5. A 36-bit peripheral order.

Peripheral Order. Figure 11.5 shows the format of a peripheral order. For each peripheral unit, the 20-bit PU address is more than enough to specify which of its 64 rows of ferrods should be scanned or which of its 1024 individual relays should be operated or released. For each CPD, the 10-bit CPD address specifies which one of its 1024 outputs should produce a pulse. A 4-bit CPD enable is decoded into the CPD execute pulse that determines which CPD should respond to the CPD address bus. So, a system has a maximum of 16 CPDs, or 16K pulse points. Two additional bits in a peripheral order specify which of the duplicated PU address buses and CPD address buses should be used.

Consider a peripheral order walk-through where some program requests the status of some particular row of 16 scan points. This program invokes a subroutine to assemble the 36-bit peripheral order. A peripheral order buffer execution program executes the peripheral order by the following sequence of events.

1. The processor's CPD bus choice latch is set according to the CPD bus bit in the order. All CPDs read this latch (not shown in Figure 11.4) to determine which CPD address bus to read from.

2. The processor's PU bus choice latch is set according to the PU bus bit in the order. All PUs read this latch (not shown in Figure 11.4) to determine which PU address bus to read from.

3. The 10-bit CPD address from the order is placed on the specified CPD address bus. While all CPDs read this bus, only the selected one responds to it.

4. The 20-bit PU address from the order is placed on the specified PU address bus. While all PUs read this bus, only the selected one responds to it.

5. The 4-bit CPD enable from the order is decoded and the processor pulses one of its 16 CPD execute wires. This pulse identifies which CPD should read the CPD address bus. This CPD uses this pulse as a strobe to read the CPD address off the specified CPD address bus.

6. If the address that this selected CPD reads from the CPD address bus is interpreted as valid, this CPD returns a pulse on its echo wire. The processor expects to receive a pulse on that one out of 16 echo wires that corresponds to the execute wire on which it transmitted a pulse.

7. This selected CPD decodes the CPD address that it just read off the CPD address bus. It transmits a pulse on the corresponding output wire. This particular wire would be connected to the enable input on the particular scanner that contains the desired row of ferrods. This pulse identifies which PU, a scanner in this case, should read the specified PU address bus.

8. The processor transmits a pulse signifying "we really mean it" over the WRMI wire that is common to all critical electronic signal points. The simultaneous arrival of this WRMI pulse and the enable pulse from the CPD is used by this PU as a strobe to read the PU address off the PU address bus.

9. If the address that this selected scanner reads from the PU address bus is interpreted as valid, this scanner returns a pulse on its verify wire. The CPD expects to receive a pulse on that one verify wire that corresponds to the enable wire on which it transmitted a pulse.

10. When this CPD receives the pulse over the appropriate verify wire, it transmits a pulse onto the CPD verify wire that is common to all CPDs. The processor expects to receive this pulse.

11. This selected scanner decodes the PU address that it just read off the PU address bus. This identifies a particular row of 16 ferrods, and their states are sensed in parallel, producing a 16-bit scan word. This scanner transmits this 16-bit word onto the scanner answer bus.

12. If the scanner's internal operation had proceeded normally, this selected scanner transmits an ASW pulse over the ASW wire that is common to all PUs. The processor uses this pulse on the ASW wire to strobe the 16-bit scan word off the scanner answer bus.

11.3.4 Fabric

In this subsection, the ESS fabric is described generally. Several representative connections are illustrated by call walk-throughs in the next subsection. You should be impressed by how simple, obvious, and unenlightening this all is. That's because ESS hardware is simple, obvious, and unenlightening. It's worth going through it, however, because this background will make it just a little easier when we walk a call through ESS' software. When we finally put the hardware and software together, you will find that the walk-through is not simple nor obvious. You will, hopefully, find it to be enlightening.

The switching element in the ESS fabric is the ferreed switch, a miniature make-contact, packaged in a small evacuated bottle. Ferreeds are assembled into arrays with 32 rows and 32 columns, each array having its own network controller. Every connection through the fabric is two-wire, for tip and ring — the ESS fabric has no sleeve wire.

The overall fabric has eight stages of switching, where each stage contains many of these arrays. Like the four-stage fabrics that were discussed in Sections 7.2.9 and 8.3, these arrays are assembled into subfabrics, called networks. Each network is a four-stage subfabric, where each stage is built from these 32×32 arrays of ferreeds. There are two kinds of networks: a line link network (LLN) and a trunk link network (TLN). Each LLN has two stages of concentration followed by two stages of distribution. Each TLN has four stages of distribution.

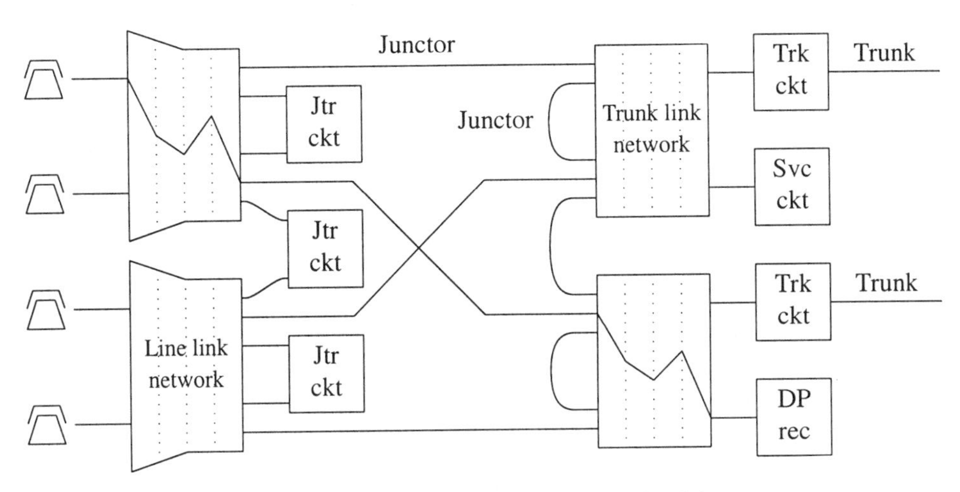

Figure 11.6. Fabric connection for dial pulsing.

The overall fabric consists of two half-fabrics: the line-side half-fabric is a collection of LLNs and the trunk-side half-fabric is a collection of TLNs. Figure 11.6 magnifies the detail of the ESS fabric. The figure shows two line link networks in the line-side superstage, two trunk link networks in the trunk-side superstage, and some representative junctors. A typical ESS office has more than two LLNs and more than two TLNs.

The LLNs and TLNs are all interconnected by junctors: some connect an LLN to a TLN and some connect two LLNs or two TLNs together. Every connection from the line side to the trunk side of the fabric uses eight stages of switching and a junctor that interconnects the LLN on which the line appears with the TLN on which the trunk appears. Four of the stages of switching, in this LLN, connect the given line to the line-side of the selected junctor. Four of the stages of switching, in this TLN, connect the trunk side of this junctor to the given trunk. Figure 11.6 also shows a representative connection, between the telephone in the upper left and the dial pulse receiver in the lower right.

While the overall fabric acts as a two-sided eight-stage fabric for connecting lines to trunks, each half-fabric acts as a folded four-stage fabric for its respective port-to-port connections. Every connection between two lines in the same office uses just four stages in the line-side half-fabric and a junctor that interconnects the LLN on which line 1 appears with the LLN on which line 2 appears. Every connection between two trunks in the same office uses just four stages in the trunk-side half-fabric and a junctor that interconnects the TLN on which trunk 1 appears with the TLN on which trunk 2 appears.

The junctors that interconnect two LLNs are more complex than the other kinds of junctors. These junctors include a junctor circuit that provides talking battery and supervision to both lines that are connected to it for an intraoffice call. The other kinds of junctors don't need junctor circuits because talking battery and supervision are provided by the trunk or service circuit on the trunk side of the fabric.

Since the dense packaging of the ferreed switches in the networks makes switch replacement difficult and expensive, their life is extended by ensuring that they are always switched "dry." Consider a series electrical circuit, or loop, with a DC battery, a load resistor, and a mechanical switch contact all wired in series. If the contact is open, no current flows, but as the mechanical contact gradually closes, a spark jumps across the contacts just before they finally meet. This spark damages the face of the contacts, and after many thousand repetitions, the contacts have such high resistance and such a highly pitted surface that they must be replaced. This inevitable problem is resolved economically in the ESS by placing an additional contact, one that is easily accessed, in the switch train.

All trunk circuits, service circuits, and junctor circuits provide talking battery through the contacts of a cut-through relay, in the circuit between the battery and the fabric appearance. Because these cut-through contacts are open when the fabric connection is made, the contacts in the fabric are "dry" when they close (talking battery has not yet been applied to the loop). After operating the fabric relays, the cut-through contact is closed, completing the talking connection. Since these cut-through contacts are the only contacts that arc, they are made more robust and more accessible for easier replacement.

11.3.5 Connections

Assume that the subscriber at the upper left of Figure 11.6 goes off-hook and that the ESS processor subsequently scans this line's current detector, a ferrod wired to this subscriber's line circuit but physically located in one of the line scanners. Furthermore, assume that some program senses this change in line status and that another program later determines, from a class of service data table in the processor's program store, that this line corresponds to a dial pulse subscriber. This program would then find an idle dial pulse receiver (DPR) and an idle path through the fabric between this DPR on the trunk side of the fabric and this line's appearance on the line side.

Dialing. This fabric connection is shown in Figure 11.6 as the solid line through the four stages of the upper line link network and the solid line through the four stages of the lower trunk link network. This fabric path uses one of the junctors that interconnects these two subfabric units. To make this connection, the ESS processor sends a peripheral order that cuts off the line scanner ferrod corresponding to this subscriber's line circuit, four orders to the network controller for the upper left LLN, four orders to the network controller for the lower right TLN, and last (and after a slight delay) an order to a trunk signal distributor that cuts through the DPR. By establishing this connection, talking battery and supervision are provided by the DPR and a trunk scanner ferrod that is wired to this DPR but is physically located in a trunk scanner. Furthermore, when this line and this DPR are initially connected, dial tone would be audible through a relay contact in the dial pulse receiver.

The ESS processor must now scan the trunk scanner ferrod associated with this DPR often enough (every 10 msec) so that programs can sense the makes and breaks of individual dial pulses. After the first dial pulse is detected by the DPR's trunk scanner ferrod and sensed by a program in the processor, this program would open the dial tone contact in the DPR and the subscriber would no longer hear the dial tone. Dial pulses are detected, counted, and stored by programs in the ESS processor.

Intraoffice call. Suppose this calling party dials the directory number of another subscriber served by this same ESS. A program in the ESS processor interprets the dialed digits and processes the call. First, this program translates the dialed directory

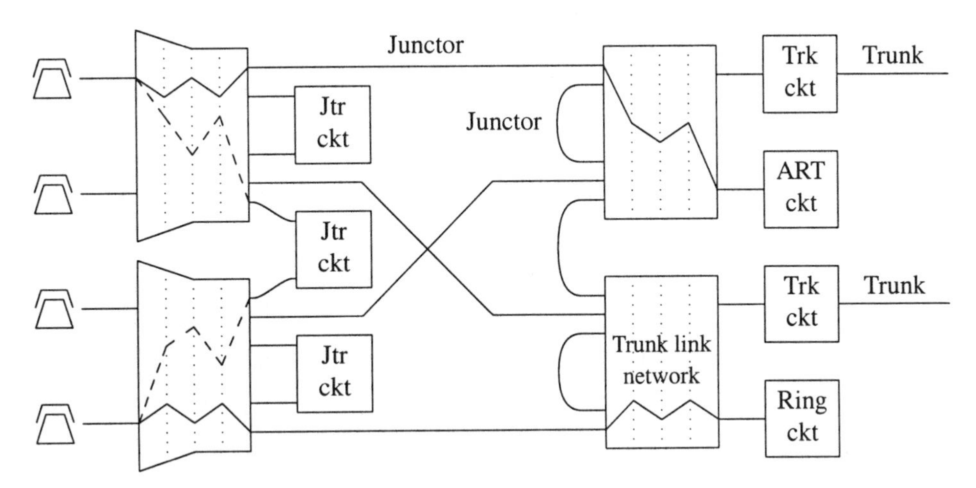

Figure 11.7. Fabric connection for ringing an intraoffice call.

number into a line-side address on some LLN and finds a talking path between the calling and called parties through the fabric. To find this talking path, this call processing program must locate an idle intra-LLN junctor from the junctor group that connects the calling party's LLN with the called party's LLN, compute a path through one LLN between the calling party and one end of this junctor circuit, and compute a path through the other LLN between the called party and the other end of this junctor circuit. This fabric path, from calling party to junctor circuit to called party, is shown as a dashed line in Figure 11.7. It is marked "busy" in the processor's fabric map, but the connection is not actually established yet. There would be no point in ringing the called party if no talking connection through the fabric could be found later.

Next, this call processing program locates an idle ring-generator circuit (RGC), computes a fabric path between this RGC and the called party, locates an idle audible ring tone circuit (ARTC), and computes a fabric path between this ARTC and the calling party. The program sends out the peripheral orders to cause these two fabric paths to be connected and to cause the previous dial pulsing path to be disconnected. These two connections are shown as the solid lines in Figure 11.7. In this configuration, talking battery, supervision, and ring signal are provided through the fabric to the called party by the selected RGC, and talking battery, supervision, and audible ring tone are provided through the fabric to the calling party by the selected ARTC. The supervisory trunk scanner ferrods associated with this RGC and with this ARTC are periodically scanned (every 25 msec) by the ESS processor.

Interoffice calls. Instead of an intraoffice call, suppose this calling party dials the directory number of a called party who is served by another switching system. After dialing is completed, the call processing programs cause the calling party to be disconnected from the DPR, but then process this call differently. Instead of reserving an intraoffice connection through an intra-LLN junctor, the call processing programs reserve a connection through the fabric between the calling party and an appropriate outgoing trunk circuit, as shown by the dashed path in Figure 11.8. Instead of establishing the ringing connection illustrated previously, the call processing programs establish a

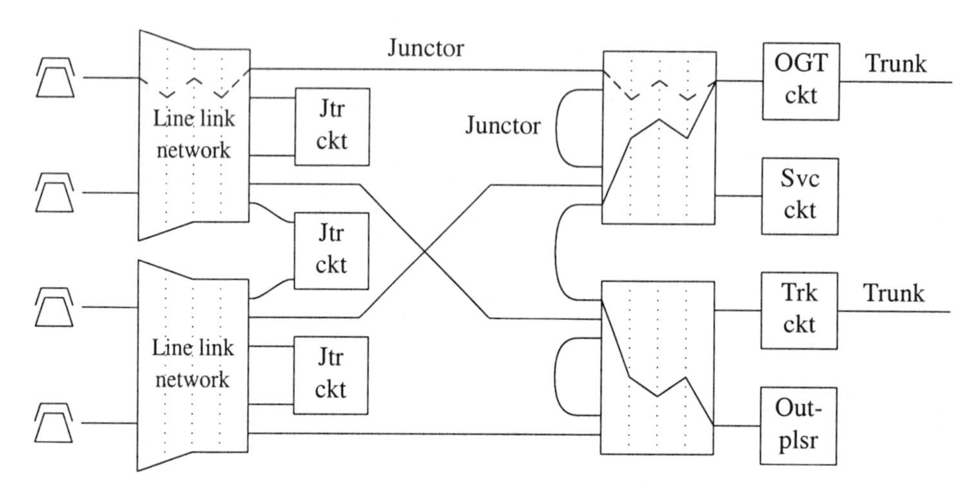

Figure 11.8. Fabric connection for outpulsing.

connection using an intra-TLN junctor between an MF outpulser and this selected outgoing trunk circuit, as shown by the solid path in Figure 11.8. After the called party's directory number is transmitted, and the calling party's number is in a CAMA environment, the call processing programs cause the solid path to be disconnected and the dashed path to be connected. The calling subscriber hears audible ring tone over the trunk from the distant office.

If the called party answers, the RGC's trunk scanner ferrod detects the presence of DC. The call processing program senses this change in supervision the next time the ESS processor scans the ferrod. The call processing program sends the peripheral orders that cause both the dashed talking path to be connected and the two solid ringing paths to be disconnected. Talking battery and supervision are now provided by the selected junctor circuit, and its supervisory junctor scanner ferrods (one on each leg) are periodically scanned by the ESS processor looking for call completion. When an on-hook is detected and sensed, or if the calling party hangs up before the called party ever answers the call, the program causes all established and reserved paths to be disconnected.

The call processing and the fabric connections are still different if this ESS office terminates an interoffice call instead of originating one. If a distant office seizes an incoming trunk, the trunk scanner ferrod associated with this trunk circuit, located in one of the trunk scanners, detects DC. After it is scanned and sensed, a call processing program causes a fabric connection, using an intra-TLN junctor, between this trunk circuit and an MF receiver, as shown in Figure 11.9. After MF pulsing is completely received, the digits are interpreted and this connection is released. A path is reserved between the incoming trunk circuit and the called party, but two other paths are connected first. An RGC is connected to the called party and an ARTC is connected to the incoming trunk circuit. When the called party answers, the supervision change is detected by the RGC's trunk scanner ferrod and the stimulated call processing program releases the ringing connection and establishes the previously reserved talking connection.

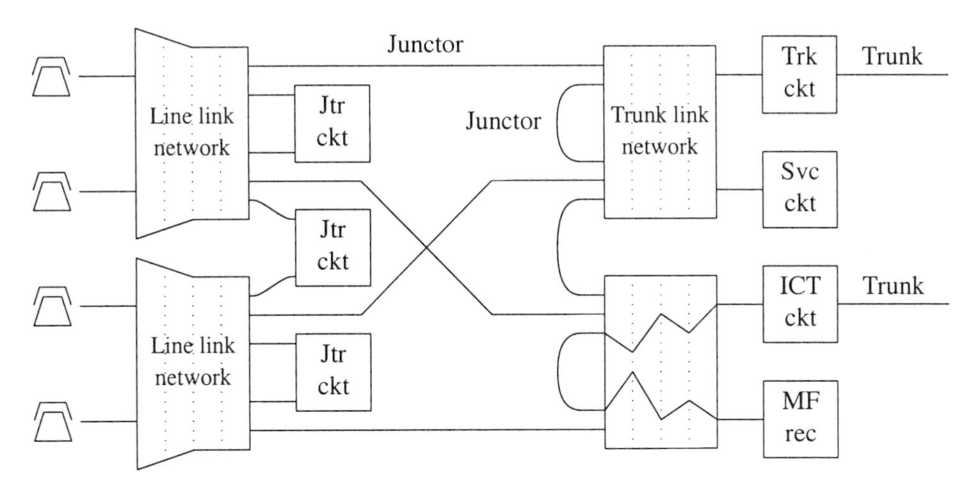

Figure 11.9. Fabric connection for inpulsing.

11.4 SOFTWARE ARCHITECTURE

The architecture of the ESS software is described. This long section summarizes many of the basic concepts that characterize modern real-time operating systems. This description is top-down: the overall structure is described at the most general level in the first subsection and then greater and greater detail is added in subsequent subsections.

11.4.1 Allocating Execution Time

At the most general level of operation, execution time in the ESS processor is distributed among classes of programs responsible for:

- Timer-driven i/o;
- Call processing;
- Maintenance of the system.

These classes are left imprecisely defined in this preliminary discussion. The assignment of programs to these three classes will be refined as this section proceeds.

Allocating the processor's execution time to these three classes of programs is not constant. If a component should fail, a flurry of maintenance activity would be expected. If a UFO should appear in the sky, a flurry of call processing would be expected: "Hey, Joe, did you see that?" The variation in the allocation of execution time over these three classes of programs has both a random nature and a logical nature.

Consider an ESS office with a size and traffic intensity such that its subscribers initiate telephone calls at the rate of 600 off-hooks per minute. Since idle subscriber lines are polled every 0.1 seconds (nominally), this rate corresponds to an average of one new off-hook per polling interval. If these new off-hooks were distributed in time so that they actually occurred regularly at a rate of one every 0.1 seconds (statistically, extremely unlikely), the allocation of execution time among these three classes of programs in a 1-minute interval would be smooth, regular, and uniform. If these new off-hooks were distributed in time so that all 600 occurred in the same 0.1-second interval and none followed for the remainder of a minute (statistically, even more unlikely), the allocation of execution time among these three classes of programs in this 1-minute interval would be highly irregular. Such random variations are ignored in the following discussion; only the smoothed average behavior is considered.

In this ESS office, suppose the average traffic intensity changed such that only 60 new off-hooks occurred in each 1-minute interval (a regular rate is still assumed, but now it is one every second). Then, the allocation of execution time among the three classes of programs in this 1-minute interval would still be regular, but it would be different from when the off-hook rate was 10 times higher. The allocation of execution time over different classes of programs is highly correlated to traffic intensity. This relationship is both difficult and important to understand.

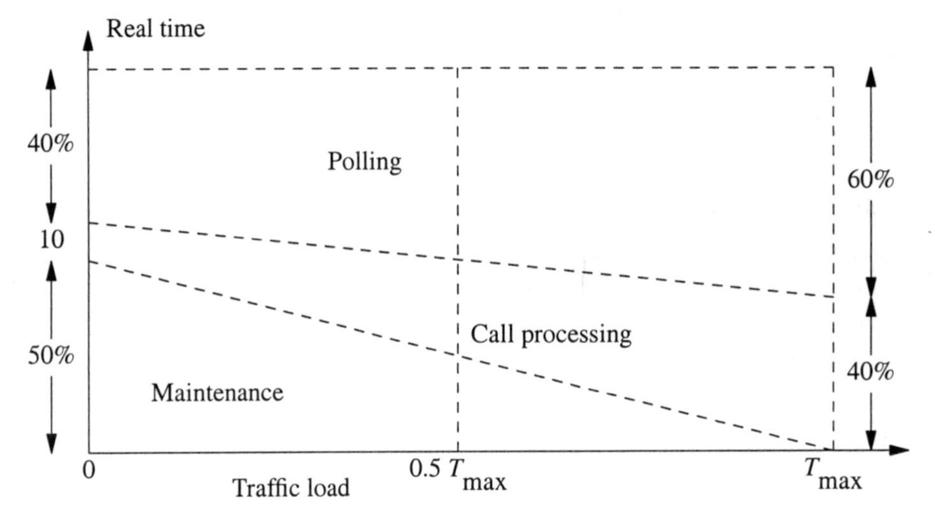

Figure 11.10. Allocating execution time for different traffic load.

Figure 11.10 shows how the processor's execution time is allocated to the programs in these three classes as a function of smoothed traffic load on the system. The y axis represents processor execution time, and the x axis represents the system's traffic load. The left side of Figure 11.10 represents a system at 3 A.M. with virtually 0 Erlangs of call traffic. The right side represents a system in its peak busy hour whose processor is barely keeping up with its call traffic. The numbers in Figure 11.10 are fictitious and any correlation to actual numbers on actual systems is purely coincidental. While Figure 11.10 suggests that the assignment of execution time is a linear function of traffic intensity, there is no reason to expect linearity (the relationship is probably not linear).

Polling programs are scheduled for execution at a constant frequency. But a program that looks for work, and finds some, requires a longer execution time than the same program that looks for work, and finds none. Assume that polling programs, as a class, consume 40% of the processor's execution time just in the overhead of looking for work — scanning sensors for new off-hooks and changes in supervision and reading internal buffers for timer-driven outputs to control points. On the left side of Figure 11.10, representing 0 traffic load, since no work would ever be found, the total execution time allocated to polling is only the time spent looking for work. Assume that polling programs, on the average, consume 50% more execution time for each execution when the traffic load is T_{max} than when it is 0. So, the execution time allocated to polling programs increases to 60% on the right side of Figure 11.10.

Consuming 40 – 60% of execution time just looking for work might seem to be inefficient. But, the best measure of efficiency is the amount of execution time consumed, per task found. Let n be the average number of tasks found by the polling programs in a typical 5-millisecond interval. When traffic is light, polling programs consume 2 milliseconds (40%) of a 5-millisecond interval and might find only $n = 1$ task — for 2 milliseconds per task found. When traffic is heavy, polling programs consume 3 milliseconds (60%) of a 5-millisecond interval but might find $n = 60$ tasks — for 0.05 milliseconds per task found. If i/o were interrupt-driven and if every i/o interrupt consumed 0.1 milliseconds of processing overhead, then one task consumes 0.1 milliseconds and 60 tasks consume 6 milliseconds. Thus, interrupt-driven i/o would be 20 times more efficient than timer-driven i/o at the low traffic load, but is half as efficient at the high traffic load. Since the high-traffic condition governs the architecture, efficiency is only significant when the system is busy. Execution time can be squandered when traffic load is light because the processor doesn't have much to do anyway.

While some call processing programs are also routinely and regularly scheduled, most are scheduled for execution only when changes in calls are sensed. Assume that routine, regularly scheduled call processing programs consume 10% of execution time and that the stimulated call processing programs, and the incremental execution time of routine programs, consume an additional 30% of execution time at T_{max}. On the left side of Figure 11.10, representing 0 traffic load, since no calls would be processed, the total execution time allocated for call processing is only 10% of execution time consumed by routine programs. On the right side of Figure 11.10, representing a traffic load of T_{max}, call processing requires 40% of execution time.

Notice that at a traffic load of T_{max} the combination of polling and call processing consumes 100% of execution time. Since no additional execution time is available, T_{max} is defined as the system's maximum traffic load. If traffic intensity should ever exceed T_{max}, the ESS processor would not be able to keep up with the amount of work that it would be required to do. Work can be queued so that short-term bursts of high traffic are effectively smoothed so that they don't overload the processor. A very long interval of high traffic, however, could overflow the queues and then the entire software system crashes. The decline in performance at high loads is neither gradual nor graceful.

While not all maintenance is deferrable, assume for this discussion that it is, at least in the average and over the short term. So, maintenance programs are assigned a low priority and are executed whenever the processor has any spare execution time that is not needed for polling nor for call processing. On the left side of Figure 11.10, with 0 traffic load, 50% of execution time is available for maintenance. On the right side, with maximum traffic load, no execution time is available for maintenance.

Consider a traffic load at half the maximum value. Interpolating between the two sides of Figure 11.10, polling consumes 50% of execution time, call processing requires 25% of execution time, and 25% of execution time is left over for maintenance.

While Figure 11.10 is a useful visualization of the allocation of execution time and the dependence of this allocation on traffic load, it does not illustrate the temporal behavior. An alternate visualization is presented in the next subsection which shows time as a variable and traffic load as a parameter.

11.4.2 Temporal Behavior

In Section 5.1, we saw that net traffic intensity depends on two statistical parameters: arrival rate of new off-hooks and holding time (call duration). Similarly, here, some program's net execution time depends on the traffic load through two separate but mutually dependent mechanisms: how often a program executes and how long it runs when it does execute. First, programs are scheduled for execution at a frequency that depends on traffic. Second, the net time that a program consumes in each execution changes with traffic. These mechanisms are different in the three classes of programs.

- Call processing and maintenance programs are executed at base level in a prioritized sequence that is discussed later. Polling programs are timer-driven by a clock that interrupts base level every 5 milliseconds.
- Whenever any polling program or call processing program runs, it executes all the work that is queued up for it. So, the duration of a particular execution of some program depends on the amount of work queued up for it. This queue length depends on traffic load, but also on how much time has passed since the last time this program executed.
- Maintenance programs, on the other hand, are broken into 10-millisecond segments and are assigned a low priority in base level. When maintenance programs are finally allowed to run, only one segment of one program is executed and then call processing programs are run again.

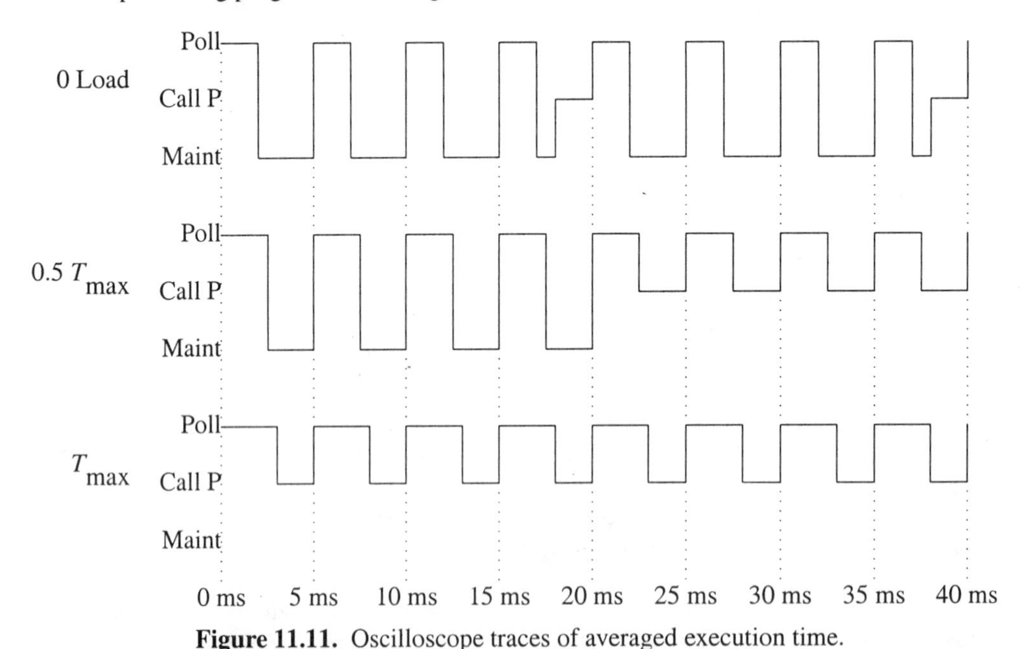

Figure 11.11. Oscilloscope traces of averaged execution time.

Figure 11.11 is an alternate visualization of how the processor's execution time is allocated among these three classes of programs. The figure shows how elapsed time distributes among the three classes, and it shows this for three different traffic load conditions. Suppose the processor produced an electronic signal with a value of 0V whenever maintenance programs execute, 0.3V whenever call processing programs execute, and 0.6V whenever polling programs execute. Figure 11.11 shows a 40-millisecond oscilloscope trace for this signal with the oscilloscope's vertical deflection set at 1V per inch and its horizontal deflection set at 10 milliseconds per inch. Only the averaged conditions are illustrated — a true display of such a hypothetical signal would be irregular. For example, the falling edges at the end of each polling interval would occur randomly.

The lower waveform in Figure 11.11 represents a typical 40-millisecond interval under maximum traffic load. Since polling programs execute every 5 milliseconds and they consume 60% of execution time, they use 3 milliseconds out of every 5-millisecond interval. The rest of execution time, 2 milliseconds out of every 5-millisecond interval, is spent in base level. Since virtually 0% of execution time is left over for maintenance programs, virtually all of base-level execution time is spent processing calls. On the average, every 5-millisecond interval is identical: 3 milliseconds for polling and 2 milliseconds for call processing. Maintenance programs virtually never run.

The middle waveform in Figure 11.11 represents a typical 40-millisecond interval under half maximum traffic load. Since polling programs still execute every 5 milliseconds, but now they consume 50% of execution time, they use 2.5 milliseconds out of every 5-millisecond interval. The rest of execution time, 2.5 milliseconds out of every 5-millisecond interval, is spent in base level. Since call processing programs require 25% of execution time, they use 10 milliseconds out of the 40-millisecond interval. Since 25% of execution time is left over for maintenance programs, they use 10 milliseconds (one complete program segment) out of the 40-millisecond interval. In the first 5-millisecond interval, some maintenance program segment begins execution and retains control of base level for its entire duration, 10 milliseconds of accumulated base-level execution time (20 milliseconds of elapsed time), until the end of the fourth segment. When maintenance finally relinquishes control of base level, call processing requires 10 milliseconds of accumulated base-level execution time to handle the work that has queued up over the 40 milliseconds since call processing last ran. In the middle waveform of Figure 11.11, maintenance uses the first four 5-millisecond intervals and call processing uses the last four. The mean time between successive executions of a maintenance program segment is 40 milliseconds.

The upper waveform in Figure 11.11 represents a typical 40-millisecond interval under 0 traffic load. Since polling programs still execute every 5 milliseconds, but now they consume 40% of execution time, they use 2 milliseconds out of every 5-millisecond interval. The rest of execution time, 3 milliseconds out of every 5-millisecond interval, is spent in base level. Since call processing programs require 10% of execution time, they require 4 milliseconds out of the 40-millisecond interval. Since 50% of execution time is left over for maintenance programs, they use 20 milliseconds (two complete program segments) out of the 40-millisecond interval. Consider the first half of the 40-millisecond interval. Some maintenance program segment begins execution in the first 5-millisecond interval and retains control of base level for its entire duration, 10 milliseconds of accumulated base-level execution time, until part way through the fourth interval. When maintenance finally relinquishes control of base level, call processing requires 2 milliseconds of accumulated base-level execution time to handle the work that has queued up over the 20 milliseconds since call processing last ran. Call processing consumes the last 2 milliseconds of the fourth 5-millisecond interval. The second half of the 40-millisecond interval is identical. The mean time between successive executions of a maintenance program segment is 20 milliseconds.

11.4.3 Interrupts

Interrupt-driven programs are prioritized differently than programs that run at base level. Interrupt programs are prioritized by their *order of criticality* to overall system operation. Interrupt priorities are called levels and are assigned the letters A through H and J (letter I is skipped). An event that triggers a program assigned to E level, for example, can interrupt any base-level program or any program assigned to interrupt levels F, G, H, or J; but cannot interrupt programs assigned to levels A, B, C, or D.

Base-level programs are not critical — the overall system will still operate if base-level programs are suspended. While timer-driven polling programs are critical to call processing, their suspension would not crash the overall system. While some polling programs run at interrupt levels G and H, most are assigned to J level. In the waveforms of Figure 11.11, call processing and maintenance programs run at base level and polling programs run as shown because a 5-millisecond clock causes a J-level interrupt. Fault-detection interrupts are more critical because certain faults could crash the overall system. Fault-detection programs are assigned to interrupt levels B through F. At A level, the interrupt caused by the manual override "emergency action" button has priority over all programs. Interrupt levels A through G are discussed in Section 11.6.

A clock circuit, external to the processor, causes a processor interrupt every 5 milliseconds, usually at J level. While the programs that execute during such an interrupt consume an average of 2 – 3 milliseconds (see Figure 11.11), occasionally an interrupt could execute longer than 5 milliseconds. So, these 5-millisecond interrupts occur at J level or H level, depending on whether the processor is currently executing programs at base level or J level, respectively. Polling programs are partitioned into high and low priorities, where high-priority programs are most critical and have very short execution times and low-priority programs are less critical or have relatively long execution times. While the total elapsed time to execute all high-priority programs will not exceed 5 milliseconds, the total elapsed time to execute all high- and low-priority programs could exceed this figure occasionally.

When a J-level interrupt occurs, the current address of the base-level program and the contents of the processor's hardware registers are saved in memory. Then, all active high-priority programs are executed. Then, all active low-priority programs are executed. Then, the processor's hardware registers are restored to the values they had before the interrupt, and the base-level program that was interrupted continues at the address that was stored. When an H-level interrupt occurs, the current address of the J-level program and the contents of the processor's hardware registers are saved in memory. Then, all active high-priority programs are executed. Then, the processor's hardware registers are restored to the values they had before the interrupt, and the J-level program that was interrupted continues at the address that was stored. This system of J- and H-level interrupts ensures that at least the high-priority programs are executed regularly every 5 milliseconds.

Up to this point, most of the discussion about J-level i/o has been about polling and scanning, the "i" of "i/o." While it might seem that the "o" of "i/o" could be performed asynchronously in base level, delivering outputs synchronously from programs at J level provides both increased performance and greater simplicity. This is seen by considering a typical output peripheral, like one of the network controllers.

An output peripheral unit handles one peripheral order at a time and it requires a certain fixed execution time to do it. If many virtually simultaneous programs (or repeated executions of the same program) issue peripheral orders asynchronously to the same peripheral unit, the orders would need to be queued in the unit and would wait their turn for execution. Even a given program, acting on behalf of a given client, must issue four distinct peripheral orders to the same network controller to establish a connection between the client's line appearance and a junctor. Time of completion of any one order would not be known by the program that requests the order unless a system of completion messages were provided.

Instead, peripheral orders are effectively queued in the processor. The various base-level programs write their output peripheral orders into *peripheral order buffers* (POBs) and then they suspend themselves until all the orders have been executed. These POBs are unloaded synchronously at J level. Since execution time in the peripheral units is never greater than 25 milliseconds, POB execution is scheduled every 25 milliseconds, or every fifth J-level interrupt. Peripheral orders are delivered to the peripheral units so that no unit receives more than one order in each 25-millisecond interval. Completion messages are not necessary.

Besides greatly underestimating the software development effort for ESS, its system engineers also greatly overestimated the execution-time capacity of the ESS processor. These two important issues are, of course, intimately related — if a program turns out to be larger than expected, then its execution time will probably be longer than expected and it will probably consume more execution time than expected. While some early predictions had the ESS processor controlling a switching office of several hundred thousand lines, initially, the 1E processor barely supported several thousand lines.

One way to boost the real-time capacity of a system is the technological solution of building a faster processor. A processor whose instruction cycle time is five times faster can support greater than five times more traffic. The real-time capacity increases more than the instruction cycle time because fixed program overhead consumes a smaller percentage of execution time in a faster processor. So, because the 1A has a faster, more modern processor than the 1E, 1A offices are typically larger than 1E offices.

Another way to boost the real-time capacity is to use multiple cooperating processors. The simplest implementation is to augment the main processor with a single additional special-purpose processor, called the signal processor, that handles the i/o. ESS' signal processor would be called a "front-end" in today's vernacular. The SP handles most of the J-level work and the main processor handles most of the base-level work. The third ESS ever installed was a system with a signal processor that replaced the Panel system in the famous PEnnsylvania 6 switching office in New York City. Using an SP as a front-end machine is standard in the 4E (Section 14.1).

With the commercial availability of microprocessor chips in the late 1970s, service circuits and other peripherals were made more intelligent, but this intelligence was still programmable. The collection of microprocessors in a switching office, like a 5E, can be thought of as a large distributed front-end processor. Just like any ESS with a signal processor, the main-frame processor in a 5E requires a smaller percentage of its execution time for i/o than does a 1E (without a front-end).

11.4.4 Deferability

Deferability is the delay that can be tolerated between the time when a program is requested to run and the time when the program actually begins to run. Base-level programs are prioritized by *order of deferability*, instead of by order of criticality. Base-level priorities are assigned the letters I and A through E. Because the letters A through E are used to name two different things, base-level priorities and interrupt levels, the reader must be careful — a better naming convention would have been helpful. The least deferrable base-level programs are assigned to I, next least to A, next least to B, and so forth. Maintenance programs that perform diagnosis, exercising, and auditing are most deferrable and are assigned to priority E. In the waveforms of Figure 11.11, call processing programs run at base-level priorities I and A through D and maintenance programs run at base-level priority E. Base-level priorities A through E run in the repeating sequence:

A B A C A B A D A B A C A B A E A B A C A B A D A B A C A B . . .

First, all A-priority programs execute all the work they have. Then, all B-priority programs execute all the work they have. Then, instead of proceeding to C-priority programs next, the A-priority programs run again. Then, the schedule proceeds to C priority, another A priority, then back to B, another A, and so forth. In priority E, instead of executing all work, as in priority levels A through D, only one 10-millisecond program segment is allowed to run.

A program, called main, keeps track of this ABACABA... scheduling of base-level priority intervals. This ABACABA... schedule pertains only to base level and only to priorities A through E in base level. The entire sequence of priority intervals is interrupted every 5 milliseconds for J-level polling. The highest priority of base-level work, called interject, is scheduled outside this ABACABA... sequence.

New work for an A-priority program in the second A interval in the sequence above, that wasn't scheduled for this program during the first A interval, could have several sources. It could have been scheduled by another A-priority program that ran later in the first A-priority interval or earlier in the second A interval or by some B-priority program that just ran or, most likely, by some polling program that ran during some intermediate J-level interrupt. Programs at A priority have the opportunity to execute during every other priority interval. B-priority programs have this opportunity during every fourth priority interval (except at the repetition boundary); C-priority programs every eighth; D-priority programs every sixteenth; and E-priority programs every thirty-second. While the order of the priority intervals is described by the ABACABA... sequence, the time scale for this sequence depends on traffic load.

Consider a software system under gradually increasing traffic load. Since a call processing program is more likely to have work in heavy traffic than in light, its average execution time increases. Because J-level interrupts gradually consume more execution time, the elapsed time spent in any one call processing program increases even more. As all call processing programs execute longer, the interval between consecutive executions of the same program increases. This longer interval causes even more work to queue up for each program. The elapsed time to complete one ABACABA... cycle is called E-to-E time. It is a strong function of traffic load, perhaps a quadratic. Figure 11.11 shows that E-to-E time can be as small as 20 milliseconds, but an artificial upper limit also exists.

A program, which runs at E priority, resets an external "sanity timer" every time E priority occurs. If this timer reaches 1.4 seconds without being reset, the timeout causes a B-level fault-detection interrupt (not to be confused with a B-priority base-level program), and the overall system is declared "insane." Thus, the maximum E-to-E time is 1.4 seconds. Since D priority is scheduled twice as often as E priority, the maximum D-to-D time is 0.7 seconds. Continuing, the maximum C-to-C time is 0.35 seconds, the maximum B-to-B time is 0.175 seconds, and the maximum A-to-A time is 0.0875 seconds. Typical average deferment times are much less than the maximum times.

If some program is specified to have a maximum tolerable deferment of 0.2 seconds between the time its execution is scheduled and the time it actually executes, it could be assigned to B priority of base level. However, if some program is determined to have a maximum tolerable deferment of 0.05 seconds between its scheduled and execution times, then assignment to A priority won't even be good enough. Such a program would be assigned to interject, I priority. Interject programs are given the opportunity to execute after the completion of each individual program segment in each of the A through E priorities in base level. Because no base-level program segment consumes more than 10 milliseconds of base-level execution time, the maximum interject deferment is the 20 milliseconds (approximately) of elapsed time required to complete a 10-millisecond program segment (see Figure 11.11). If some program is determined to have a maximum tolerable deferment of 0.01 seconds between its scheduled and actual execution times, then assignment to interject won't even be good enough and it must be run as a J-level interrupt (it would be executed in every other 5-millisecond J-level interrupt).

For example, active dial pulse receivers are scanned nominally every 10 milliseconds. At 10 pulses per second and a 50% duty cycle, each pulse is in break for 50 milliseconds and in make for 50 milliseconds. So, the signal's state changes approximately every five times that the DPR is scanned. However, if the program that scans DPRs is deferred 50 milliseconds past its scheduled time for execution, an entire pulse might be missed and a wrong number would result. A 10- or 20-millisecond deferment is tolerable and DPR scan must be performed at J level.

On the other hand, idle lines are scanned (the attending function) nominally every 100 milliseconds. So, the longest nominal delay between the time that a subscriber lifts his receiver and the starting time of the program that determines this event is 100 milliseconds, the shortest delay is 0, and the average nominal delay is 50 milliseconds. This is an acceptable delay in a system under normal operation. In fact, the line scan program could even be deferred by as much as several hundred milliseconds and the subscriber would not perceive an unusual delay in hearing a dial tone. But, what happens if the starting time of a regularly scheduled program is deferred past the time of its next regular start?

Some programs, like timetable administrators (Section 11.4.8) must eventually execute for each time they are scheduled to execute. With other programs, like DPR scan or line scan, if a scheduled execution is deferred past the time of its next regularly scheduled execution, one of the executions is simply canceled. In fact, in the special case of line scan, such a cancellation is beneficial and line scan is deliberately assigned an exceptionally high deferment tolerance.

Load shedding is the procedure of intentionally ignoring new work when a system is already overloaded handling the work it currently knows about. In Crossbar offices, when traffic load on markers gets too high, the battery is temporarily disconnected from all the line relays in an entire frame for intervals as long as 10 – 20 seconds. Since new off-hooks would not be detected during such an interval, dial tone can be delayed that long and many subscribers will simply hang up and try again later.

Load shedding is implemented in ESS by simply assigning line scan to what seems to be an exceptionally low priority in base level and then allowing the normal scheduling to run its course. While line scan execution is scheduled every 100 milliseconds in ESS, it is assigned to D priority of base level. When the ESS processor is very busy, D-to-D time can get as great as 700 milliseconds. Not only could line scan be deferred for 700 milliseconds, but six intermediate scheduled executions would be shed. Furthermore, because line scan would find more new requests when it finally does execute, its overhead is better distributed and it is more efficient. The only penalty is a slightly annoying delay for dial tone when the system is very busy.

A similar intentional deferment occurs with maintenance programs by assigning them to E priority of base level. The only difference between line scan deferment and maintenance deferment occurs in a lightly loaded system. Because line scan is requested for execution every 100 milliseconds, it would never execute more frequently than this — even if the system had nothing better to do. Line scan can only execute less frequently than this; when the system is heavily loaded. Because there are hundreds of maintenance program segments and all of them are queued for execution, some maintenance work is always available. When the system is lightly loaded, maintenance programs naturally fill the available execution time.

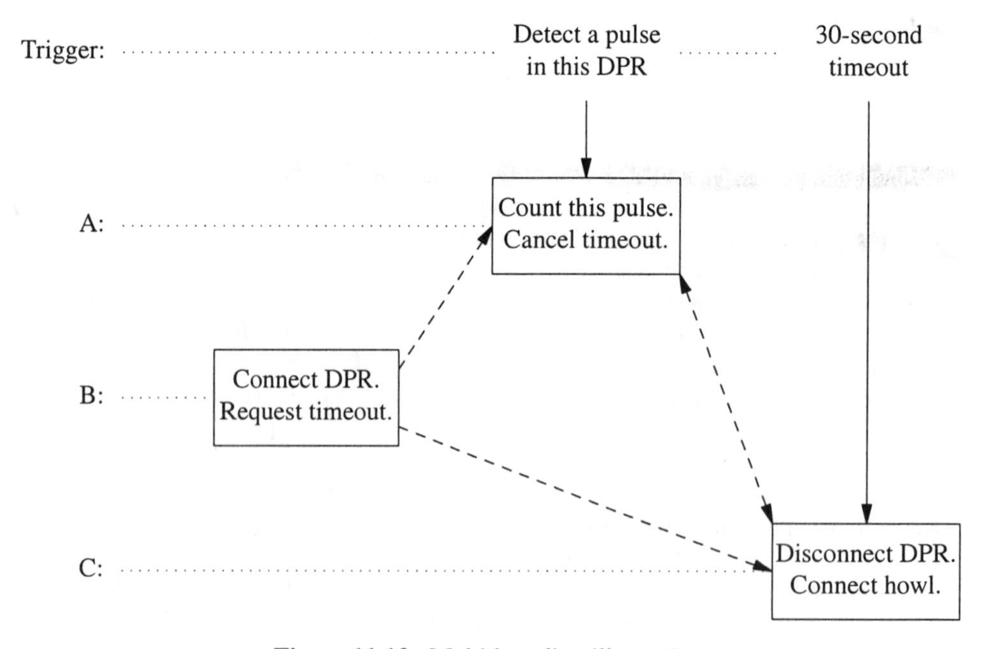

Figure 11.12. Multithreading illustration.

11.4.5 Multithreading

Multithreaded programming is illustrated by an example. Assume that a subscriber has gone off-hook, the event has been scanned, the line ferrod is cut off, the subscriber is identified as a dial pulse customer, and a network connection has just been requested between this subscriber's line appearance and an idle DPR. There can be no further activity on this call until the peripheral order buffer (POB) execution program, at J level, executes all the peripheral orders to actually effect the requested network connection. Even if the subscriber hangs up, the processor won't know it until after the DPR is connected to the line and the supervisory ferrod in the DPR is scanned.

The dialing connection — lines (DCL) program, which has processed this call to this point, suspends itself and some other program has the opportunity to run (or this same program runs on behalf of another subscriber). While DCL is not finished with this call, its initial execution-time segment is finished. A subsequent segment will run later in response to an external event, some action by the subscriber. One such action that causes DCL to be reactivated, on behalf of this subscriber, is when the subscriber's first dial pulse is detected by the trunk scan program. What happens if this first dial pulse never arrives?

In the event of an accidental off-hook or a short circuit in the loop plant that simulates an off-hook (called a permanent signal), it is not wise to allow the line to tie up a DPR for days, or even hours. Normal call processing requires a 30-second timeout. If 30 seconds pass without a dial pulse being detected, the DPR is disconnected and made available for another call and the subscriber's line appearance is connected to an audible permanent signal tone (also called howl) circuit. DCL had to take care of this contingency in its first segment before it suspended. So, this first segment of DCL arranged for the future activation of two different subsequent segments, two "threads" of DCL.

Figure 11.12 shows these three segments in time and by base-level priority. The initial program segment is shown at the left, running at B priority of base level. The two alternate threads, whose future activations are arranged by this initial segment, are shown at the center and at the right. The alternate thread that counts dial pulses is shown in the center of Figure 11.12, running at A priority of base level. The alternate thread that disconnects the dialing connection and replaces it with a howl connection is shown at the right, running at C priority of base level. The dial pulse detection thread activates when a dial pulse is detected in the DPR used for this call. The permanent signal thread activates 30 seconds after the suspension of the program segment that connected the DPR.

However, these two threads are even more tangled than this because they perform mutually exclusive actions on this call. Both threads cannot be allowed to execute — only one of them. If a dial pulse is detected before the 30-second timeout, then dial pulse counting executes and, among its functions, it must undo the arrangements that the initial segment made for the activation of the permanent signal thread. If 30 seconds elapse before a dial pulse is ever detected, then permanent signal connection executes and, among its functions, it must undo the arrangements that the initial segment made for the activation of dial pulse counting. Both dial pulse counting and permanent signal connection typically wait for their respective triggers on more than one call. So, the undoing of arrangements must be made with respect to the specific call. These various arrangements will be described later.

11.4.6 Task Dispensers

The H-level interrupt, both J-level interrupt priorities, and the six base-level priorities each has its own task dispenser (TD). Call processing programs that are assigned to the I and A through D priorities, and maintenance programs that are assigned to the E priority, are clients of the task dispenser in their assigned priority. Whenever a priority level is executed, the corresponding task dispenser distributes all the scheduled work to all of its clients (except in E). Different task dispensing mechanisms are described in Section 11.4.8.

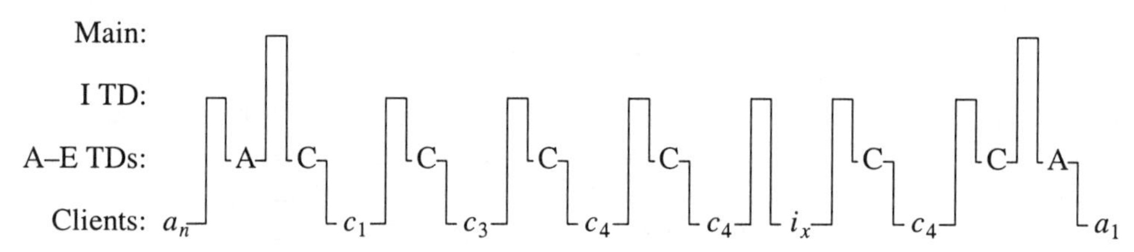

Figure 11.13. Typical base level program sequence.

Figure 11.13 shows a typical sequence of program executions. It shows an entire typical C-priority interval, including the end of the previous interval and the beginning of the next interval — both of which must be A-priority intervals. Suppose call processing program c_1, a C-priority client, is required to execute in every C-priority interval. Suppose no new work is scheduled for c_2, one new job is scheduled for c_3, and three new jobs are scheduled for c_4. Suppose that new work appears for interject client i_x during the second execution of c_4.

Consider the left side of Figure 11.13. After the last A client concludes and interject work is sought, the A-priority task dispenser (ATD) finds no more work for any of its clients, so it returns to main. Main determines that C priority is next in the ABACABA... schedule, so the C-priority task dispenser (CTD) executes. CTD sees that its client c_1 must execute, which it does, and then interject work is sought. Figure 11.13 illustrates that the interject task dispenser is executed at the conclusion of every client program. CTD sees that c_2 has no work, but that c_3 does. So, CTD gives this new job to c_3, it executes, and then interject work is sought. Then, CTD gives a new job to c_4, it executes, and then interject work is sought. Then, CTD gives a second new job to c_4, it executes, and then interject work is sought. This time interject work is found, for i_x, which executes and returns to the interject task dispenser. No other interject work is found. Then, CTD gives a third new job to c_4, it executes, and then interject work is sought. Then, CTD finds no more work for any of its clients, so it returns to main. Main determines that A priority is next in the ABACABA... schedule, so ATD executes. ATD sees that its client a_1 must execute, which it does; and Figure 11.13 ends.

11.4.7 Memory Management

A large amount of ESS processor memory is partitioned into many special-purpose blocks of consecutive words. Such memory blocks are buffers, hoppers, tables, registers, cells, and lists. Each variety of blocks has many special-purpose types and each type can be an individual unit or can have many individual units. The various properties of buffers, hoppers, tables, registers, cells, and lists are shown in Table 11.1 and expanded upon in the following paragraphs.

TABLE 11.1. Tabulated properties of memory blocks

Type	Long	Struc	Dens	Type	Dur	Contents
Buffer	fixed	simple	part	many	temp	data
Hopper	fixed	simple	part	uniq	temp	data, ptrs
Table	fixed	simple	alloc	uniq	perm	data, adds
Regist	fixed	format	alloc	many	temp	data, ptrs
Cell	fixed	format	full	uniq	temp	pointers
List	vari	simple	full	uniq	temp	regs, bufs

Blocks. *Buffers* are fixed-length blocks of consecutive memory that hold consecutive simple one-word entries. The entries are data that is stored temporarily, typically to be passed from one program to another. A buffer must be long enough for its worst-case use, but it is not necessarily filled each time it is used. Each type of buffer has many units. One popular type is the peripheral order buffer (POB). Any base-level program that initiates relay operations and fabric connections does so by acquiring an idle POB, loading it consecutively with peripheral orders, and queuing it for the POB execution program that runs at J level. Any active base-level program segment that issues peripheral orders needs one POB for each active call. So, there are many POBs in the processor's memory and their number is engineered by traffic considerations.

Hoppers are also fixed-length blocks of consecutive memory that hold consecutive simple one-word entries. But in a hopper, each entry is some form of data that triggers the execution of some specific program. A hopper must also be long enough for its worst-case use, and it is not necessarily filled each time it is used. Unlike buffers, each type of hopper is a single unit with its own special purpose. An example is the line execution results hopper (LERH). Line scan is a program that scans subscriber lines and loads the address of each new off-hook into consecutive entries in the LERH. Each subscriber address in the LERH is processed later by a separate execution of the DCL program. Many programs are similarly hopper-driven, but each such program is driven by its own unique hopper. Hopper administration is described in more detail below.

Tables are also fixed-length blocks of consecutive memory that hold consecutive simple one-word entries. While the data in buffers and hoppers is temporary and is loaded in the order of occurrence, the data in a table is semipermanent and its location in the table is some relevant address. Unlike buffers and hoppers, tables are always filled. Like hoppers, but unlike buffers, each type of table is a single unit with its own special purpose. One example is the line translation table, where the entry addressed by a subscriber's scanner address contains his directory number and class of service. Another example is a transfer table, where the ith entry is the start address of the ith program in some ordered set of specific programs, such as the ith client of some task dispenser.

Registers are also fixed-length blocks of consecutive memory but, instead of holding consecutive simple one-word entries, registers temporarily hold a variety of special-purpose data in a strict format. Like buffers, each type of register has many units. As one example, every active call in an ESS has a temporarily associated originating register (OR), a name right out of Number 5 Crossbar. An OR is acquired for each new origination by the DCL program and it remains with the call until disconnect. The programs that process the call load the OR with the subscriber's fabric address, directory number, class of service, dialed digits, current network connection, and various pointers.

The size of the full-access pool of ORs is engineered by traffic considerations. As another example, every service circuit in an ESS has a permanently associated circuit register. When a call processing program needs a service circuit for a call, it acquires an idle circuit register and the corresponding hardware is simultaneously acquired.

Cells are also fixed-length blocks of consecutive memory. Like hoppers and tables, but unlike buffers and registers, each type of cell is a single unit with its own special purpose. Using a strict format, like a register, a cell is temporary storage for the various pointers used in administering hoppers and lists. Every distinct hopper and list has its own *head cell.* The process by which head cells administer hoppers and linked lists is described below.

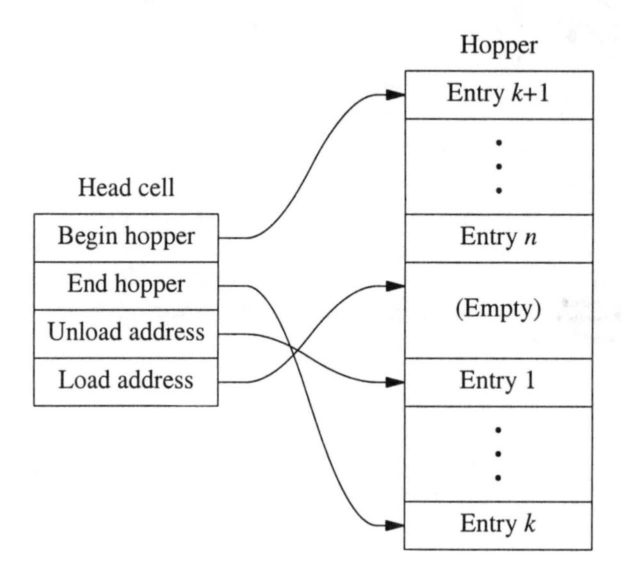

Figure 11.14. Hopper administration.

Lists are quite different from the previous memory structures. A list has variable-length memory and is not a block of memory, but is a mechanism for logically interconnecting blocks of memory like buffers and registers. Like hoppers, tables, and cells, each type of list is a single unit with its own special purpose. For example, the OR idle linked list holds all originating registers that are not currently active. Any program that needs an OR simply removes the oldest one from this list. The administration of one-way and two-way linked lists is described below.

Adminstration Programs. Hoppers are more complex to administer than might be thought because the structure is logically circular. A mechanism is needed that logically connects the end of the hopper around to its beginning. Figure 11.14 shows a head cell and the hopper that it administers. The head cell holds four addresses, all pointing into the hopper: begin hopper points to the beginning of the hopper's memory block, end hopper points to the end of the hopper's memory block, unload address points to the oldest entry in the hopper, and load address points to the word after the newest entry in the hopper.

The program that loads the hopper must store its new data in the load address, unless the hopper is full. This program must also be careful with an empty hopper and it must update the load address by considering the hopper's wrap-around. The algorithm is illustrated by the pseudo-code below.

```
        "Algorithm to Load a Hopper"

        "Check if hopper is full"
If Hopper-empty flag is OFF AND Load Address = Unload Address
        Abort

        "Check if hopper is empty"
If Hopper-empty flag is ON
        Set Load Address = Begin Hopper
        Set Unload Address = Begin Hopper
        Set Hopper-empty flag to OFF

Write new data into the word to which the Load Address points

        "Update the Load Address"
If Load Address equals End Hopper
        Set Load Address = Begin Hopper
Else
        Increment Load Address by 1
```

This program, which adds new data to the hopper, places it into the word to which the load address points. It then increments the load address so it points to the next location in the hopper. The program that takes data out of the hopper reads from the word to which the unload address points. It then increments the unload address so it points to the next location in the hopper.

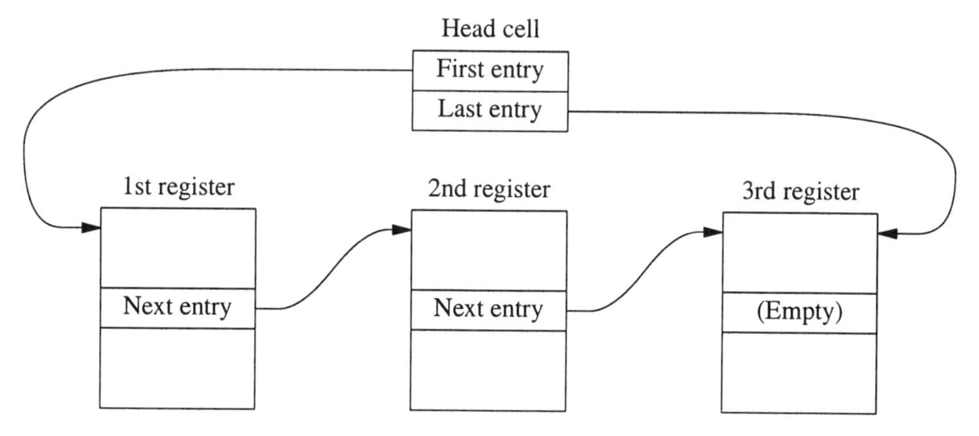

Figure 11.15. One-way linked list with three entries.

If the hopper is full, the load address and the unload address point to the same word. In this case the hopper-loading program should abort. The hopper-unloading program should abort if the hopper is empty. The hopper-unloading program knows it has read the last entry in the hopper when the incremented unload address equals the load address. It sets a flag, typically load address = 0. Both programs test this flag as a special case.

While a hopper is a way to queue a fixed number of one-word data entries, a linked list is a way to queue any number of registers. Figure 11.15 shows the structure of a one-way linked list. The list's head cell contains pointers to the register that has been on the list for the longest time and to the register that was placed on the list most recently. Every register on the list contains one pointer word, which points to the next register on the list. Unlike a hopper, a linked list can never be full. The program that loads a new register onto the list appends this new register to the end of the list. The program that unloads a queued register off the list takes the register from the front of the list, the register that has been on the list for the longest time. Both programs must take special care with an empty list. Both programs follow, illustrated in pseudo-code.

```
          "Algorithm to Load a Linked List"

          "Check if list is empty"
If List-empty flag is ON
          "If so, the New register is made First on list"
          Store New register address in Head cell + 0
          Set List-empty flag to OFF
Else
          Read Last register address from Head cell + 1
          "Let P = position in register of list-pointer word"
          Store New register address in Last register + P

          "In either case:"
Store New register address in Head cell + 1
```

Typically, the list-empty flag does not require a special bit. Since low memory addresses typically hold bootstrap loading programs, it is safe to assume that memory address 0000 is never the first word of a legitimate register. So, last register address = 0000, in the contents of head cell + 1, could be used as an ON list-empty flag, to indicate that the list is empty and requires special handling. This flag is set by the program that unloads the list whenever the unloaded register matches the first register. This flag is cleared by the program that loads the list as part of its special handling of an empty list. Besides setting the flag when it empties the list, the program that unloads the list also checks this flag so it can abort if the list is already empty.

Unfortunately, list administration is more complicated than has been suggested so far. Besides loading a list and unloading a list, a third activity is to delete a register from the list. There are many situations in which a register has been loaded onto a list, but then some event occurs that cancels the normal unloading of this register. Some program, stimulated by such an event, must delete this register from the linked list and mend the list so the registers below it are still linked to the registers above it.

Such a program is given the address of the register to be deleted. The pointer word in this register contains the address of the register immediately below it on the list. The program must search the entire list to find the register immediately above it on the list. Then, the pointer word in the register above the deleted register is changed from the

```
        "Algorithm to Unload a Linked List"

        "Check if list is empty"
If List-empty flag is ON
        Abort

Read First register address from Head cell + 0

        "Check if end of list"
Read Newest register address from Head cell + 1
If First register address = Last register address
        Set List-empty flag to ON
Else
        "Update First entry on list"
        "Let P = position of List.pointer.word in register"
        Read Next register address from First register + P
        Store Next register address in Head cell + 0

        "In either case:"
Return First register address
```

address of the deleted register to the address of the register below it on the list. Through all of this the program must make special cases out of the first and last registers on the list. The pseudo-code for such a program is left as an exercise. If deleting a register from the list is not a rare event, typically a two-way list will be used instead of a one-way list.

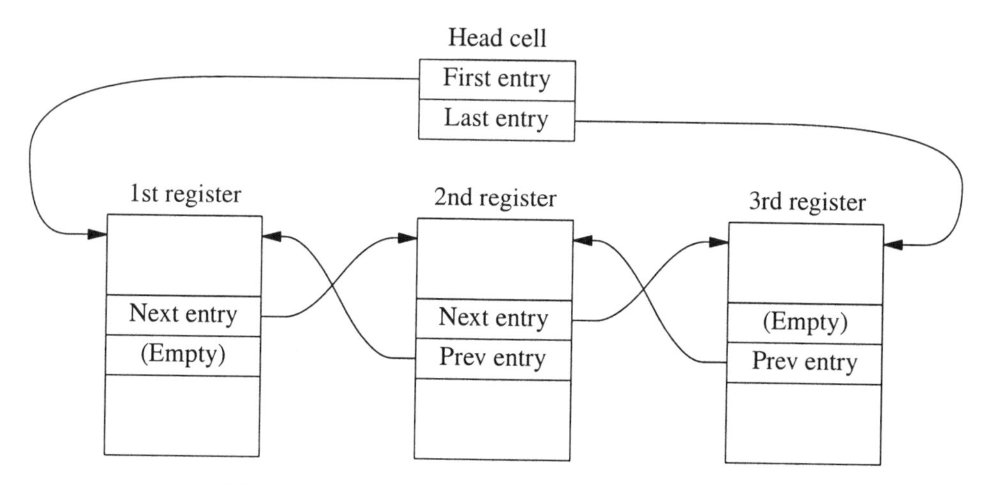

Figure 11.16. Two-way linked list with three entries.

Figure 11.16 illustrates a two-way linked list. Each register on a two-way list requires two pointer words, one that points to the next register on the list and one that points to the prior register on the list. While loading and unloading are made a bit more complicated in a two-way list, deleting a register from the list is greatly simplified over a one-way list because the program need not search the list to find the register that points to the register to be deleted.

11.4.8 Administration of Program Control

Most programs are clients of task dispensers and, as such, fall into two categories. One category of task dispenser clients is those reentrant programs that are executed as needed, performing their respective function on a different task with each distinct execution. The task dispenser knows the mechanism by which the individual tasks are scheduled for a given reentrant client, typically by a hopper or a linked list. For example, one of the clients of the B-priority task dispenser is DCL and it is driven by the LERH, a hopper that is loaded by line scan. The B-priority task dispenser tends to its various clients. When it gets to DCL, it loops on the LERH. It reads the next word from the LERH and passes it to DCL. When DCL suspends, it returns to its task dispenser which gets the next word. The loop ends when the LERH is empty.

The second category of task dispenser clients is those programs that are not re-entrant and data-driven, but execute once when scheduled. They are scheduled regularly, randomly, or periodically. One regularly scheduled client is sanity timer reset, which executes one time each time the E-priority task dispenser executes. A randomly scheduled client is line scan, which is executed once by the D-priority task dispenser, but only if it had been previously scheduled for execution. A periodically scheduled client is POB execution, a client of the task dispenser that administers low-priority J-level interrupt tasks and which executes once every fifth J-level interrupt. Since they are neither data-driven nor reentrant, this second category of client programs is not administered by its task dispensers through hoppers or linked lists. The administration of regularly scheduled clients is trivial. The randomly scheduled and periodically scheduled clients are administered by *ordered bits.*

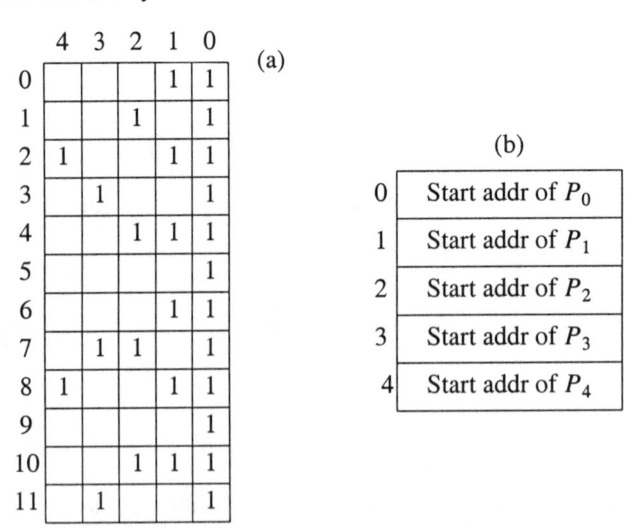

Figure 11.17. (a) Ordered bits buffer and (b) transfer table.

Client administration by ordered bits is illustrated in Figure 11.17. The two pieces of the administrative mechanism are the ordered bits buffer (OBB), shown at left, and its corresponding transfer table (XT), shown at right. Each task dispenser has two of these two-piece mechanisms; one for its randomly scheduled clients and one for its periodically scheduled clients. The clients in each category of each task dispenser are arbitrarily ordered. The *i*th such client in each category is assigned to the *i*th column in its task dispenser's OBB for its subcategory and to the *i*th word in its corresponding XT. The task dispenser determines whether to execute its *i*th client or not by the 1/0 logical value of the *i*th bit in the OBB. This task dispenser reads the starting address of this *i*th client from the *i*th word of the XT.

First, consider the randomly scheduled clients. While Figure 11.17 shows that an OBB has *n* words, this is true only for periodically scheduled clients. The OBB for randomly scheduled clients has only one word and it would be located in random-access (writeable) memory. If the *i*th bit in this word is 0 when the task dispenser runs, then the *i*th client would not be executed this time. Any program that wishes to activate the *i*th client, must set the *i*th bit in this OBB to 1. If the *i*th bit in this word is 1 when the task dispenser runs, then the *i*th client is executed this time and the *i*th bit is reset afterward.

11.4.9 1E Programming

This particular function is selected here to illustrate the details of writing programs in the ESS assembly language. Because ESS program development predates practical operating systems, the programming environment was primitive, but transparent. Similarly, in the early 1960s, the only practical high-level languages were Cobol and Fortran (now called Fortran I) and compilers were very inefficient. While programs for more modern switching systems have been written in high-level languages like C and CHILL (a special-purpose language designed for switching systems), the #1 ESS was programmed in its own assembly language. This language was quite complicated, with several fields per instruction and many macros and subroutines.

The general format of instructions in the assembly language is:

```
LADD:  OPC(*)        OADD,INX,OBJ
```

where OPC is the mnemonic code for the machine operation, followed by * if the addressing is indirect, LADD is the mnemonic address of the local memory location, OADD is the mnemonic address of the object memory location, INX identifies any index register that might be used, and OBJ identifies any object register that might be used. Consider four consecutive programming instructions that begin at mnemonic address LOOP in the program store:

```
LBEG:  DZRMO         DOBB,,Z
       JMPN          LOUT,,Z
       JMPS*         DOXT,Z,K
       JUMP          LBEG
LOUT:
```

The DZRMO operation (detect and zero the rightmost one) reads the word at memory location DOBB, the D-priority OBB, in the call store. The rightmost bit that has value 1 is changed to 0, its binary-coded bit position is placed in register Z, and memory location DOBB is written with the new value. If the word at DOBB has no 1s, then -1 is loaded in the Z register by the DZRMO operation. The JMPN operation is a conditional jump, where the condition is a negative quantity in the Z register. The four instructions are part of a loop that terminates when all the 1s in DOBB have been processed. The JMPS* operation is an indirect jump to subroutine. It reads from memory location DOXT, the D-priority ordered bits transfer table, indexed by the contents of register Z. This word is the memory address that program control jumps to, as if it were the starting address of a subroutine. The address of the next memory word is placed in register K and is used by the subroutine as a return address. The last instruction in the subroutine is:

```
JUMP        0,K
```

For example, let DOBB be the second word (# 1) in the buffer at the left of Figure 11.17 and let DOXT be the transfer table at the right. The first time through the OBB loop, the data at DOBB is 00101, as shown in Figure 11.17. After executing the DZRMO instruction, DOBB contains 00100 and register Z contains the binary code for 0, because the rightmost 1 has been 0ed and this rightmost 1 was found in the 0-bit position. The net address for the JMPS* instruction is DOXT + 0 and program P_0 is executed as a subroutine. The next time through this OBB loop, the data at DOBB is changed to 00000 and register Z contains the binary code for 2, because the rightmost 1 has been 0ed again and this rightmost 1 was found in the 2-bit position this time. Now, the net address for the JMPS* operation is DOXT + 2 and program P_2 is executed as a subroutine. Because there are no more 1s in DOBB, this OBB loop would end the next time through.

Next, consider the periodically scheduled clients. The OBB for these clients does have n words, as shown in Figure 11.17, and they are located in the program store. Each task dispenser maintains its own ordered bits buffer index (OBBI), an N-counter that counts from 0 to $N - 1$ and cycles around. The value of N is the least common multiple of the periods of all the periodically scheduled clients. For example, in Figure 11.17, let program P_0 be required every time the task dispenser runs, P_1 be required every second time, P_2 every third time, P_3 every fourth time, and P_4 every sixth time. Because the least common multiple of these periods is 12, the OBB has $N = 12$ words.

The N words in the OBB are permanently written with the cyclic pattern of client activations. In the OBB of Figure 11.17, because column 0 has a 1 in every row, program P_0 executes every time the task dispenser runs. Because column 1 has a 1 in every second row, P_1 executes every second time the task dispenser runs, when the OBBI has even-numbered values. Because column 2 has a 1 in every third row, P_2 executes every third time, when the OBBI = 1, 4, 7, or 10. Because column 3 has a 1 in every fourth row, P_3 executes every fourth time, when the OBBI = 3, 7, or 11. Because column 4 has a 1 in every sixth row, P_4 executes every sixth time, when the OBBI = 2 or 8.

Each time the task dispenser runs, it increments its OBBI and uses its new value, j, $0 \leq j \leq N - 1$, to index the OBB. The ith client executes (or not) during this jth phase depending on the 1/0 value of the ith bit of the jth word. For example, consider three consecutive phases of the task dispenser whose OBB is shown in Figure 11.17, specifically phases 10, 11, and 0. When $j = 10$, the task dispenser executes P_0, P_1, and P_2. When $j = 11$, the task dispenser executes P_0 and P_3. When $j = 0$, the task dispenser executes P_0 and P_1.

The programming instructions are slightly different for the periodic OBB. Before the OBB loop is entered, the instructions

```
        READ     DOBI,,J
        READ     DOBB,J,I
        WRITE    TEMP,,I
```

read the value of the *N* counter into the J register and copy the data from DOBB + contents of J into temporary memory. The *j*th OBB is stored in TEMP before entering the OBB loop so that the permanent copy is not be altered by the DZRMO operation. Inside the OBB loop the corresponding four instructions are:

```
LBEG:   DZRMO    TEMP,,Z
        JMPN     LOUT,,Z
        JMPS*    DOXT,Z,K
        JUMP     LBEG
LOUT:
```

11.4.10 Timing

Frequently a program or segment is triggered by the timeout of a real-time clock. For example, in the multithreading illustration of Figure 11.12, the segment that disconnects the DPR and connects permanent signal tone is triggered by the timeout of a 30-second clock that begins timing when the DPR is first connected. Such programs or segments are clients of a timeout task dispenser (TOTD). A TOTD is just a special case of a task dispenser that administers a linked list and is, itself, a random client of some task dispenser that administers an ordered bits buffer. Thus, while these TOTDs are important enough to justify a detailed explanation, this explanation also serves as an example of the operation of several of the task-dispensing mechanisms described above.

Each timeout task requires a register, typically an OR. Four memory words are required for special use in any register that is undergoing a timeout: for the forward pointer and backward pointer of a linked list (timing linked lists are usually two-way), for the starting address of the program that is triggered by the timeout, and a decrementing tick counter. Each software clock with a different period requires a dedicated linked list and a dedicated TOTD that administers it.

A timeout of $T = m \times t$ seconds is initiated by setting the tick counter in a register to m and placing the register on the linked list that corresponds to period t. Then, every t seconds when the software clock ticks, the t-second TOTD loops through its linked list and subtracts 1 from the tick counter in each linked register. When the decremented tick counter equals 0, the timeout is accomplished. The TOTD would then remove the register from this linked list and would execute a subroutine jump to the address that was previously stored in the register. After this client program is finished, it returns to the TOTD, which continues processing its linked list.

For example, suppose the normal processing of some call requires that 8 seconds elapse between the end of one program segment and the start of a second one. Suppose the program designer decides to use a 200-millisecond software clock to accomplish the timing. This first program segment places the starting address of this second segment in the address word of the OR that corresponds to the particular call, places the integer 40 in this OR's tick counter, places this OR on the end of the 200-millisecond timing linked list, and then it returns to its task dispenser.

Every 200 milliseconds, the 200-millisecond TOTD processes every OR on its linked list. Each time it reaches this OR, it decrements the tick counter. The 40th time it processes this OR occurs after slightly more than 8 seconds of real time has passed since this OR was first placed on this linked list. In this 40th processing of the OR the tick counter decrements to 0, the OR is removed from the 200-millisecond timing linked list, and control is passed to the second segment.

This 200-millisecond TOTD (we'll call it AClock) is itself a random client of the A-priority ordered bits buffer task dispenser. As a program that runs in A priority of base level, it may be deferred by as much as 87.5 milliseconds in a barely sane system. So, a nominal 8-second timeout has an actual elapsed time between 8.0 seconds and 8.0875 seconds. Such a timeout is typically specified as 8.04 ± 0.04 seconds. Note that the deferments of 40 consecutive executions of AClock do not accumulate in determining the net actual timeout. Only the deferment of the 40th execution is relevant.

Besides this 200-millisecond software clock period, the #1 ESS maintains software clocks every 100 milliseconds and every 500 milliseconds. The corresponding TOTDs run under different base-level priorities, making a choice of deferments available by which the program designer can match any specified timeout tolerance. The 100-millisecond TOTD (we'll call it IClock) runs in interject priority, where its maximum deferment is 20 milliseconds. The 500-millisecond TOTD (we'll call it CClock) runs in C priority, where its maximum deferment is 350 milliseconds. If the 8-second timeout had used IClock instead of AClock, the tick counter would have been set to 80 and the timeout tolerance would have been 8.01 ± 0.01 seconds. Since the OR would be processed 80 times instead of 40, the timing administrative overhead would be approximately double that of AClock. If the 8-second timeout had used CClock instead of AClock, the tick counter would have been set to 16 and the timeout tolerance would have been 8.17 ± 0.17 seconds. Since the OR would be processed 16 times instead of 40, the timing administrative overhead would be approximately 40% that of AClock.

The overall structure of these three software clocks begins with the 5-millisecond hardware clock that triggers J-level interrupts. A program we'll call JClock is a periodic client of the low-priority J-level ordered bits buffer task dispenser. JClock runs every 20th interrupt, or every 100 milliseconds. Corresponding to this tick of the 100-millisecond software clock, JClock sets the trigger for IClock by setting the appropriate bit in the interject-priority random-client ordered bits buffer.

IClock does not actually run until the next time that the interject task dispenser runs. IClock has three functions: it acts as the 100-millisecond TOTD, it determines whether or not to request AClock and CClock, and it triggers line scan by setting the appropriate bit in the D-priority random-client ordered bits buffer. IClock maintains a 10-counter that is incremented cyclically each time IClock runs. If IClock's internal counter is even, IClock sets the trigger for AClock by setting the appropriate bit in the A-priority random-client ordered bits buffer. AClock, the 200-millisecond TOTD, does not actually run until the next time that main invokes the A-priority task dispenser. If IClock's internal counter equals 0 or 5, IClock sets the trigger for CClock by setting the appropriate bit in the C-priority random-client ordered bits buffer. CClock, the 500-millisecond TOTD, does not actually run until the next time that main invokes the C-priority task dispenser.

11.5 CALL WALK-THROUGH

The program architecture is partially illustrated; we can't possibly present the whole thing. Then a call is walked through the entire system, hardware and software.

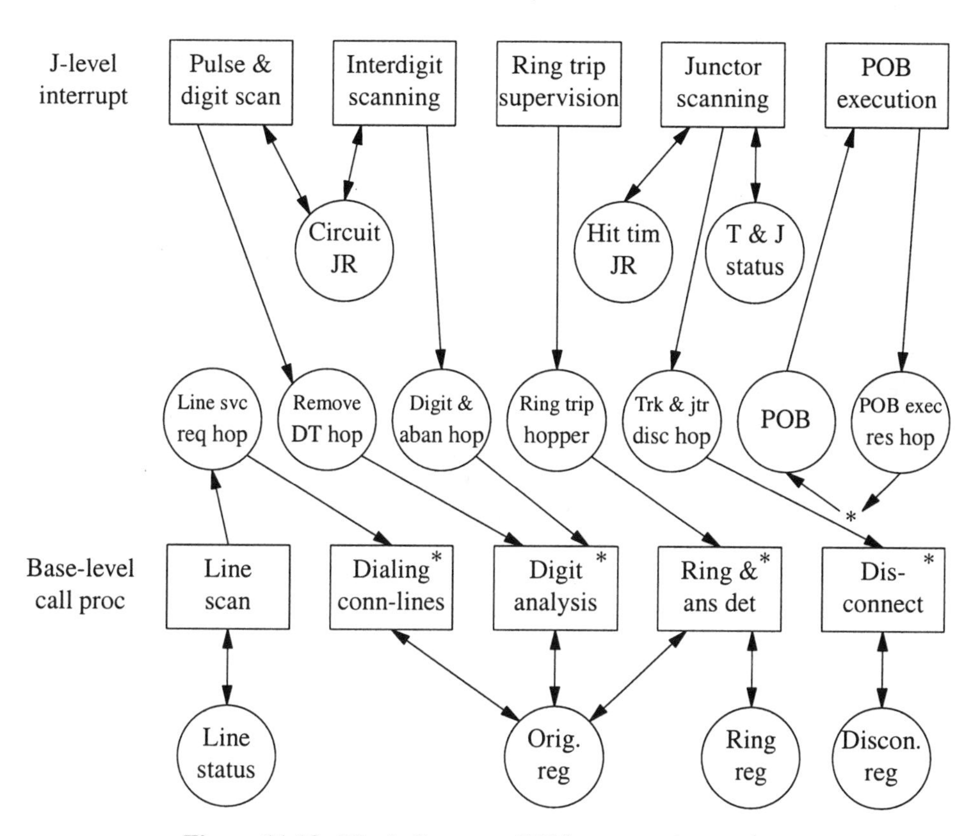

Figure 11.18. Block diagram of ESS program interactions.

11.5.1 Program Interaction

Figure 11.18 shows how some of the ESS call processing programs are organized and how they interact with each other. Programs are shown as boxes, while circles are used to indicate buffers, hoppers, tables, and registers. The boxes in the second row represent five of the many programs that run in the various priorities of base level. In general, all of these programs have many real-time segments, and the various segments of a program don't necessarily run at the same base-level priority. The boxes in the fifth (top) row represent five of the many programs that run in the two priorities of J level. Most of these programs are periodic clients that run in periods that are multiples of 5 milliseconds. For example, pulse and digit scan runs every 10 milliseconds and POB execution runs every 25 milliseconds.

The circles in the first (bottom) row represent some of the tables and registers that are used by the base-level call processing programs in the second row. As discussed above, one originating register is associated with every active call in the office for its duration. ORs are used by many of the call processing programs in base level. Other registers, like ring registers and disconnect registers, are temporarily assigned to calls during certain phases of the call. The line status table holds the *last look* at the scan points for every line in the office. Line scan scans every line nominally every 100 milliseconds and compares each scan point's *new look* against this last look. If line scan determines a new off-hook, it updates the line status table and places the identity of the originating party's scan point in the line service request hopper.

The circles in the third (middle) row represent some of the hoppers and buffers that act as interfaces and triggers between base-level programs and J-level programs. The line service request hopper is loaded by line scan and serves as the task hopper for DCL. The remove dial tone hopper is loaded by pulse and digit scan in J level whenever the first dial pulse or Touch Tone digit is detected. This remove dial tone hopper is the task hopper for the segment of digit analysis that issues the peripheral order that opens the dial tone relay in the service circuit used for the call. Other hoppers shown in Figure 11.18 similarly queue tasks that are requested by J-level programs and are implemented by base-level programs.

Because peripheral orders typically are not executed in base level, most base-level programs simply write their peripheral orders into peripheral order buffers and these POBs are unloaded, and the orders are actually executed, by the POB execution program that runs in J level. If the base-level program that loaded a POB needs to know that all its peripheral orders have been executed, this program loads a special order, with a return address, at the end of the list of orders in the POB. If the POB execution program sees this special order, it loads the accompanying return address onto the POB execution results hopper. A base-level task dispenser unloads the PERH, effectively causing a return to the base-level program that had originally loaded the POB.

The circles in the fourth row represent some of the tables and registers that are used by some J-level programs. Circuit junior registers are associated with calls during the time that their corresponding service circuit is connected. The trunk and junctor status tables serve the same function for trunks and junctors that the line status table serves for lines. Notice that trunks and junctors are scanned in J level and that lines are scanned in D priority of base level. It will be useful for the reader to refer to Figure 11.18 throughout the partial call walk-through in rest of this section.

A call is walked through ESS' hardware and software only as far as the detection of the calling subscriber's off-hook, the connection through the fabric to a dial pulse receiver, and the subscriber's entry of dial pulse signaling. While calls were walked through Step and Crossbar from off-hook to called party answer, the reader is probably already so saturated with software detail that too much more won't be read anyway.

11.5.2 Scanning the Line

The subscriber lifts the receiver of his telephone, which closes the hook switch and completes the DC path for talking battery. Central office battery is connected to this line through the balanced windings of a ferrod in one of the line scanners, and the flow of DC saturates the soft magnet in this ferrod. However, even though this line's scan point is actuated, the processor is not aware of the action until it scans the ferrod.

The software activity that eventually leads to scanning this ferrod begins in one of the J-level interrupts. Every 20 J-level interrupts, or 100 milliseconds, the program we called JClock executes as a periodic client of the low-priority J-level ordered bits buffer task dispenser. The execution of JClock that is relevant to our call is the first one that occurs after the scan point actuates. Thus, as much as 100 milliseconds of real time may pass between the time of the off-hook and the time that JClock executes. JClock's only function is to request the execution of the program we called IClock, by setting the appropriate bit in the random-client ordered bits buffer that is administered by the interject-priority task dispenser. When JClock completes, it returns to JTD, the J-level task dispenser.

When the program that is currently executing in base level finally completes, it returns to its task dispenser. Before finding any more tasks for its own clients, this task dispenser allows the interject-priority task dispenser (ITD) to look for tasks. ITD executes IClock during this interval, but as much as 20 milliseconds of real time may pass between JClock's request for IClock and IClock's actual execution. IClock administers the 100-millisecond timing linked list, increments its internal 10-counter and determines whether to request the executions of the programs we called AClock and CClock, and requests the execution of line scan by setting the appropriate bit in the random-client ordered bits buffer that is administered by the D-priority task dispenser (DTD). If line scan had not had a chance to run since the previous time that IClock set this bit, then this bit would already be set. When IClock completes, it returns to ITD.

Following its ABACABA... schedule, main eventually executes DTD. If IClock's request for line scan had just missed a D-priority interval, as much as 700 milliseconds of real time could pass between this request and the next D-priority interval. While this much latency is possible, it is unlikely, even in a busy system. While line scan is nominally scheduled every 100 milliseconds, if the office is so busy that D-to-D time approaches 700 milliseconds, then line scan would actually execute every 700 milliseconds. The line scan bit in the D-priority ordered bits buffer would be set seven times for every time that line scan actually executed. DTD causes line scan to execute.

Inside a nested loop, line scan issues an output command to the *r*th row of the *u*th line scanner. The CPD address identifies the peripheral unit as this *u*th line scanner and the PU address identifies this *r*th row. The states of the 16 ferrods in this *r*th row of this *u*th line scanner are read in parallel onto the 16-bit scanner answer bus and into one of the processor's internal registers. Using a bit-wise EXOR instruction, this 16-bit new look is compared with the corresponding last look, read from the line status table (LST). This new look replaces the last look in the LST. If the *b*th bit of the new look is 1 and the *b*th bit of the last look was 0, then a new off-hook has been found. A DZRMO instruction on the EXOR result identifies the binary code for bit-position *b* of this new request. The scanner number *u* and row number *r* and bit position *b* are concatenated into a word "u.r.b," called the line equipment number (LEN). This same binary word, if partitioned into subwords in a different way, represents the frame number of a line link frame and the internal switch number and horizontal level number for this line's line-side address on the fabric. This LEN is loaded onto the line service request hopper (LSRH). If the line service request hopper is full, then line scan abruptly terminates.

This subprogram lies inside a loop on the 64 rows in the *u*th line scan unit, and this inner loop is nested inside an outer loop on all the line scanners in the office. In a typical office, line scan consumes about 10 milliseconds of accumulated base-level time and typically finds two or three new off-hooks each time it runs during peak traffic. In this walk-through, our call is one of these new off-hooks and our LEN has been loaded on the LSRH. When line scan completes, it returns to DTD.

11.5.3 Getting an OR and a POB

Including the D-priority interval above, main executes a sequence of base-level priority intervals DABACABA..., before executing the C-priority task dispenser (CTD). Among other functions, CTD unloads the LSRH and presents each entry, one at a time, as a task for a separate execution of DCL. In one of these separate executions, DCL processes the LEN of the call we are walking through the system.

DCL's first function is to obtain an idle OR, which will remain with our call until its completion. All the idle ORs are linked together on the originating register idle linked list (ORILL). If the ORILL is empty, DCL places the LEN back on the LSRH and returns to its task dispenser in a way that terminates LSRH unloading. There is no point in further unloading the LSRH if there are no more idle originating registers. If an OR is available, DCL continues. DCL uses our LEN and indexes into the line translation table to read our class of service. Among other things, DCL learns from our COS that we are a dial pulse single-party residential subscriber. The LEN and the COS are stored in the OR.

Then, DCL requests a POB for our call from the POB idle linked list (POBILL). If an idle POB is available, DCL continues in real time. But if no POB is available, DCL queues for an idle POB, and suspends. Queuing for a POB is implemented by a subroutine jump to a program we call POBNQ with the address of the OR as an argument. POBNQ places the return address in the OR and places the OR at the end of the POB queue linked list (POBQLL). POBNQ returns, not to the return address, but to the task dispenser, thus suspending this execution of DCL. Furthermore, POBNQ returns to CTD in a way that terminates LSRH unloading and, thus, postpones future executions of DCL. There is no point in further unloading the LSRH if there are no idle POBs.

A program we call POBDQ is a client of ATD that executes once for each OR on the POBQLL. If an idle POB is available, from the POBILL, then POBDQ jumps to the return address that had been stored in the OR and delivers the address of this idle POB. DCL continues processing our call as if there had never been a break in real time. If an idle POB had been available initially, the segment of DCL that follows would have continued to run under C priority with no break in real time. If DCL had suspended because of queuing for a POB, this same segment runs under A priority after the real-time break.

After obtaining an idle POB, DCL continues. The address of the OR is stored in the POB, and the address of the POB is stored in the OR. The LEN is also the address of the line's cut-off contact to the line signal distributors. DCL loads into the POB the peripheral order that opens our line's cut-off contact.

11.5.4 Calculating the Dial Connection

Next, DCL invokes a subroutine we call FCONN, which calculates the peripheral orders for fabric connections. DCL passes two parameters to FCONN: the OR address and a coded request for a connection to any idle dial pulse receiver. FCONN obtains an idle DPR by removing a circuit register (CR) from the dial pulse register idle linked list (DPRILL). By taking the CR from the top of the DPRILL, our call will use that DPR that has been idle longest; a method for distributing wear that is even better than using slipped multipling as in Step. Since the fabric address for any trunk or service circuit is a direct function of the address of its CR, FCONN calculates the trunk-side address of the desired connection. From the OR, FCONN reads the LEN, which it uses as the line-side address of the desired connection, and it locates the POB where it will place peripheral orders.

The fabric map is a table that has a bit for each ferreed switch in the entire fabric. The state of the bit matches the idle/busy status of the corresponding crosspoint. Using the line-side and trunk-side fabric addresses, FCONN calculates a first-choice path through the fabric. If all eight crosspoints for this path are not idle, as determined by the fabric map, then FCONN calculates a second-choice path. If no idle path is available, the CR is put back on the DPRILL, another DPR is selected, and FCONN tries to find a path between this one and our line appearance. After FCONN has made a certain number of attempts at a connection or if no idle DPR were available, FCONN aborts the call and returns to DCL with a code that indicates a failure. The DCL designer has a number of options at this *fail return* — probably the easiest action is to return the POB to its POBILL and then invoke POBNQ as if no POB had been available. The design of "failure legs" in call processing programs is typically more complicated than the "success legs."

If an idle DPR and a fabric path between it and our line are found, then FCONN loads the eight network controller peripheral orders into the POB, then a *delay* order, then the trunk signal distributor orders that close the dial tone relay and the cut-through relay in the DPR. In addition, the selected DPR and the fabric path are identified in the OR. If all went well in FCONN, it returns to DCL in a way that indicates success.

DCL places the starting address of its second segment in the OR and indicates whether DCL.2 runs under A, C, or E priority (C would be appropriate) and whether a POB is needed at the return (it is not). Then, this first segment of DCL loads the POB on the POB active linked list (POBALL) and returns to its task dispenser (CTD or ATD).

11.5.5 Connecting Dial Tone

The POB execution program (POBX) is a periodic ordered bits buffer client of the low-priority J-level task dispenser. POBX runs during every fifth J-level interrupt, every 25 milliseconds. Consider the next execution of POBX after the time that DCL loaded our POB on the POBALL. POBX sequences through all the POBs on the POBALL and also sequences through the peripheral orders on each such POB.

If an order is marked "executed," POBX continues to the next order in the POB. If an "unexecuted" order must be sent to a peripheral unit that has not received an order during this current execution of POBX, POBX outputs the order to this PU, marks the order "executed," and POBX continues to the next order. Each output order follows the sequence described in Section 11.3.3. If an unexecuted order must be sent to a PU that has already received an order during this current execution, the order is left unexecuted and POBX continues. If POBX reaches an unexecuted delay pseudo-order, and all prior orders in this POB have been executed, POBX marks the delay "executed" and proceeds to the next POB without processing subsequent orders in this POB. If POBX reaches an unexecuted delay pseudo-order or an end-of-POB pseudo-order, and some prior orders in this POB are still unexecuted, POBX simply proceeds directly to the next POB.

If POBX reaches an end-of-POB pseudo-order and all prior orders in the POB have been executed, POBX removes the POB from the POBALL, returns the POB to the POBILL if the OR indicates that a POB is not needed on the return, and places the address of the OR on the A-, C-, or E-priority POB execution results hopper (PERH) as determined by the priority of the return specified by the program that loaded the POB. POBX usually requires several executions to process all the orders in any one POB. It eventually would process our POB, would return it to the POBILL, and would place the address of our OR on the CPERH. When the dial tone relay in our DPR actually operates, we hear dial tone over the connection through the fabric. Talking battery is applied to our telephone through the windings of a ferrod, in one of the trunk scanners, associated with this DPR.

In its usual ABACABA... schedule, main eventually executes the C-priority task dispenser again. One of CTD's clients is the C-priority POB return program, which we call CPOBR. CTD executes CPOBR for each OR on the CPERH, one of which is ours. CPOBR reads the return address from our OR and performs a subroutine jump to DCL.2. Here, in DCL's second segment, the dial tone active bit in the CR is set to 1, the CR is placed on the trunk scan point active linked list (TSALL), the start address of DCL's third segment is loaded in the OR's return address word, integer 60 is placed in the OR's counter word, and the OR is placed on the 500-millisecond timing linked list for a 30-second timeout.

If DCL.3 is ever reached, the DPR is disconnected from the given line and the permanent signal tone is connected instead. DCL.2 returns to CPOBR and, eventually, CPOBR returns to CTD. ESS' response to our off-hook is finally completed and the system waits for us to dial our telephone.

11.5.6 Scanning Dial Pulses

Every other J-level interrupt, every 10 milliseconds, the pulse and digit scan program (P&DS) executes as a periodic client of the low-priority J-level ordered bits buffer task dispenser. P&DS executes for each circuit register on the TSALL. The CR for the DPR that is currently giving us dial tone is on this TSALL, and when P&DS executes on our behalf, it scans the trunk scanner that holds the ferrod that supervises our call. The new look at this ferrod is compared with the last look in the trunk status table, and this last look is updated.

If new look equals last look, then no further processing takes place. As we turn the dial on our telephone, the physical state of our line doesn't change. When we release the dial, the loop is broken and made every 100 milliseconds, nominally. The first make after a break is interpreted as the beginning of a new pulse and is further processed.

Three distinguished cases are the first pulse of the first digit, the first pulse of any subsequent digit, and pulses other than the first pulse in a digit. If the dial tone active bit in the OR is set when a dial pulse is detected, then this pulse is the first pulse in the first digit. In this case, the address of our OR is placed on the remove dial tone hopper (RDTH), and the dial tone active bit in the OR is cleared. If the pulse count in the CR is zero when a dial pulse is detected, then this pulse is the first pulse in some digit. In this case, the CR is removed from permanent signal or partial dial timing and is placed on the interdigit timing linked list (IDTLL). For all pulses, the pulse count in the CR is incremented and the interdigit timeout counter is set to 2 (explained below). When P&DS is done processing our call, it proceeds to the next CR on the TSALL. When P&DS has processed every CR on the TSALL, it returns to JTD, the J-priority task dispenser.

Following its ABACABA... schedule, main eventually executes BTD. One of BTD's clients is a short segment of the digit analysis program, DA.1, that BTD reexecutes with each successive OR on the RDTH. P&DS is still active on our call, at J level, while DA.2 processes our call, at base level. DA.1 acquires a POB from the POBILL, loads the order in this POB that causes the release of the dial tone relay in the dial pulse receiver to which we are connected, notes that no POB execution results return is needed, loads the POB on the POBALL, and returns to its task dispenser, BTD. Eventually, POBX, running at J level, executes the order in the POB and restores the POB to the POBILL. We no longer hear a dial tone.

A different periodic J-level client processes interpulse timing and determines when one digit ends and another begins. The interdigit timing (IDT) program executes during every 24th J-level interrupt, every 120 milliseconds. This timing list administrator executes in J level because, while the tolerance on interdigit timing is quite high, the latency tolerance for pulse timing is extremely small. ITD loops through the CRs on the IDTLL and decrements the counter word in each CR. When this counter word is 0, ITD reads the pulse count from the CR, stores it in the OR, and clears the CR's pulse count to 0. Furthermore, ITD removes this CR from the IDTLL and places it on the partial-dial timeout linked list (PDTLL), similar to permanent signal timeout, and places the address of our OR on the digit hopper. If a subsequent pulse is detected by P&DS, it is the first pulse of the next digit and P&DS removes the CR from the PDTLL and puts it back on the IDTLL.

Since P&DS resets this counter word to 2 every time a new pulse is detected, the only way the counter can count down to 0 is for IDT to visit the CR twice without a reset by P&DS. The actual interdigit timing interval is determined by examining the two worst cases. If IDT visited the CR immediately after P&DS reset it, then a second visit by IDT (without a subsequent reset by P&DS) would occur slightly more than 120 milliseconds after this first reset. However, if IDT visited the CR immediately before P&DS reset it, then slightly less than 240 milliseconds would have to elapse before IDT visited the CR twice after the reset. Thus, the interdigit interval T is $120 \text{ ms} < T < 240 \text{ ms}$; or, $T = 180 \pm 60$ milliseconds. This much time without a pulse is interpreted as the time between digits.

Following its ABACABA... sequence, main eventually executes CTD. One of CTD's clients is the principal segment of digit analysis, DA.2, which is executed by CTD for each OR on the digit hopper. The address of our OR is placed on the digit hopper, by IDT, one time for each digit that we dial from our telephone. DA.2 examines the digit counter in the OR, stores the new pulse count in the appropriate digit byte, increments the digit count, and clears the pulse count. We see that, while the individual pulses that constitute a digit are stored and counted in the DPR's CR at J level, the individual digits that constitute the called party's directory number are stored and counted in the call's OR at base level.

DA.2 examines the set of digits that have been dialed so far. Based on class of service, DA.2 predicts how many digits will be received. When DA.2 determines that a dialed number is completed, it loads the POB orders that disconnect the dialing connection and that connect the desired call. Changes in the national numbering plan are implemented as program changes in this segment of digit analysis. While we stop our call walk-through at this point, the rest of the processing of our call, while not obvious, is at least similar. Most readers are amazed at the number of different programs that have been involved in the supposedly simple tasks of detecting an off-hook, providing dial tone, and receiving dialed digits.

11.6 MAINTENANCE

ESS' high reliability results from (1) conservative hardware design and (2) maintenance software.

11.6.1 Design

ESS' reliability requirement allows a typical system to accumulate less than 1 hour of "downtime" in 40 years. Since each ESS system has so many parts, even using the most reliable (and expensive) components and the most conservative circuit design rules won't yield a system that nearly satisfies this strict requirement. Duplication of all vital subsystems, including the buses that interconnect them, provides sufficient statistical improvement as long as subsystem faults are quickly recognized, retried, isolated, diagnosed, repaired, and reinstated. Such rapid reinstatement requires program assistance for fault recognition and diagnosis. Because subsystem problems are handled automatically by programs, except for the actual repair, of course, circuits are designed to enable fault detection and program-controlled testing.

Besides affecting circuit design, this strict reliability requirement also affects the system's hardware architecture. The processor, the call store, the program store, and the memory bus are each separately duplicated. The least suspicious units from each subsystem pair are configured as the master processor system, and the remaining units from each subsystem pair are configured as the backup system. Similarly, the PU address bus, the CPD address bus, each CPD, each network controller (but not each crosspoint), each signal distributor controller (but not each relay), and each scanner controller (but not each ferrod) are duplicated. Every subsystem pair operates simultaneously in parallel, and check circuits verify that all key points have identical logical values.

While the ESS processor uses a 36-bit word, each memory unit accommodates a 43-bit word. Data is stored in memory using a single-error correcting, double-error detecting hamming code. If one of the 43 bits is read in error, the hamming circuitry not only corrects the error, but it also signals a request for fault diagnosis of the particular memory unit. If two of the 43 bits are read in error, the memory unit's mate is trusted to provide the correct data. Besides this special procedure in the memory units, the elaborate procedure for peripheral orders, described earlier, assists the identification of faulty peripheral units in the event of a failure in the peripheral subsystem.

11.6.2 Program Types

While this strict reliability requirement had a large impact on ESS' hardware circuits and architecture, it had an even greater impact on ESS' software architecture and program development effort. While the reader has hopefully been impressed with the size and complexity of call processing programs in ESS, maintenance programs make up the vast majority of the net ESS software size, and the typical maintenance program is no simpler than the typical call processing program. Maintenance programs are organized into five types, depending on the function and the deferability.

Emergency action. These programs respond immediately to their respective stimuli. They operate at interrupt levels A and B. EA.1 is a B-level program that is triggered by the timeout of the 1.4-second E-to-E sanity timer or if any C-, D-, or E-level interrupt lasts longer than 40 milliseconds. Any of these triggers indicates that the system is insane and a major alarm is sounded. EA.1 repeatedly attempts to configure a sane system by arranging different combinations of units, one from each mated pair, until some configuration is able to solve a complex puzzle. EA.0 is an A-level program that performs a total system-wide reset, including dropping all existing calls. It is initiated manually by the "red button" on the alarm control & display panel or by a timeout of EA.1.

Fault recognition. These programs operate in interrupt levels C through F, depending on the importance of the subsystem in which a fault is suspected. Fault recognition of the various peripheral units runs at F level, of the program store units at E level, of the call store units at D level, and of the processors at C level. Check-circuit mismatches, ASW failures, memory parity errors, and other hardware triggers initiate a corresponding interrupt. An *error* is a random event that is not expected to repeat, and a *fault* is physical problem that should be fixed. A typical fault recognition program tries to distinguish faults from errors, determines which unit from a mated pair is faulty, recovers processing capability by interchanging units between master and backup, removes the faulty unit from service (called quarantining), requests the diagnostic on this unit, and patches up the call processing that was interrupted.

Diagnostics. These programs are deferrable, but only for minutes, and not for hours. They operate as high-priority maintenance control (MAC) clients in E priority of base level. Like all base-level programs, these programs are partitioned into 10-millisecond segments. Like all MAC clients, only one segment runs each time main executes in E priority. After a unit has been completely diagnosed, the diagnostic program prints a numerical code on the ESS printer. The craftsman in the ESS office looks up this code in a fault dictionary, determines the circuit board on which the fault resides, and replaces this board. The circuit and program designers create this fault dictionary by systematically introducing single faults in the circuit and recording the program's response when it diagnoses the circuit with this single fault.

Exercise programs. These are extremely deferrable and fill any available processor execution time. They operate as low-priority MAC clients in E priority of base level. Some Exercise programs place test calls and verify that call processing programs perform as expected. In addition, the fault diagnostic programs run routinely and gradually diagnose every circuit in the system.

Audits. These programs are also extremely deferrable and also fill available processor execution time. They also operate as low-priority MAC clients in E priority of base level. Audits verify software links among ORs, CRs, and POBs. They verify that fabric connections indicated in ORs are consistent with the fabric map. They also ensure that every CR is accounted for — either active or on the CR's idle linked list. If some ILL should ever break, the trunks or service circuits, corresponding to the CRs lost off the end of the list, would be lost to the system forever. While such a break is unlikely, it is possible in 40 years of operation.

11.6.3 Procedure

When a fault occurs and is detected by the hardware, the corresponding fault recognition program runs immediately. This program quarantines the faulty unit and requests that the corresponding fault diagnostic program run. Beginning at this time, the working unit is not backed up and the system is vulnerable to a fault arising in this working unit. After the fault diagnostic program finally runs, it prints out a numerical result that the craftsman looks up in the fault dictionary to determine which circuit board to replace. After replacing the indicated circuit board, the craftsman requests the diagnostic again. If the unit passes, it is placed back in service. At this time, the working unit is backed up and the vulnerability interval is ended.

We have seen that emergency action programs run at the A and B levels of the interrupt hierarchy. Fault recognition programs run at levels C through F. Programs that perform i/o with the peripheral units run at H and J. Between F and H is interrupt level G, which is dedicated to programs that perform traffic monitoring.

11.7 DISCUSSION

11.7.1 Stored Program Control

The machine that executes the 1E's programs was very carefully named, especially in public. While I have called it a processor in this chapter, the Bell System was very careful to avoid using words like "computer" or "processor" when describing this component. They were concerned that, if they called it a processor or a computer, then some regulatory body might make them stop making it. So, it was formally called the stored program control (SPC) to avoid calling unwanted attention to it. Calling it a controller, and not a computer, put it in a category with markers as just another component of a telephone office.

The 1E would probably have been just as successful if a commercial computer had been used instead of the 1E's SPC. Special-purpose instructions, like the DZRMO, and the deliberate lack of hardware for multiplication made the SPC different from commercial computers. While AT&T also emphasized these differences, they really were minor. AT&T was determined that its SPC would be used in the 1E and that Western Electric would manufacture it.

11.7.2 New Features

Program control allowed the introduction of many novel non-POTS features, which had been contemplated for many years by Bell Labs' systems engineers. Many of these features couldn't be implemented at all in Step because of the progressive nature of its network, and while they might have been possible in Crossbar, they never could be implemented economically, partly because the subscription rate was expected to be low. Features implemented in software are more easily justified incrementally.

One such new feature is abbreviated dialing (AD). Each AD subscriber is allocated a table in the ESS' software. Then, the user specifies 10 AD telephone numbers and assigns each a one-digit code. These can be emergency numbers, frequently called numbers, or exceptionally lengthy numbers (like international numbers). When the AD subscriber dials #7 (or 7 after whatever code the local telco assigns for the AD prefix), the digit analysis program locates this user's AD table, extracts the seventh entry, and loads this number into the OR just as if the user had dialed it. The user loads his personal AD table by dialing an access code.

The real cost of providing abbreviated dialing is extremely low. Regulators determined the market price, which was set well above cost in yet another effort to subsidize POTS. Abbreviated dialing had some limited appeal, but not at the price that the regulators forced the telcos to charge. Now, abbreviated dialing is extremely common, but it is typically provided directly by the telephone. One can only wonder if such functions would have migrated to the CPE if the telcos had been able to price them appropriately when they were first introduced.

Another interesting 1E feature was the improved procedure for call tracing. It is very simple for some program to set a bit in a subscriber's class of service word and thereby trigger that the identity of every incoming call to this subscriber would be printed out. While this procedure doesn't have the dramatic appeal of trying to keep the "bad guy" on the line while the call is traced, calls haven't been traced the way you see it done in the movies for many decades.

Features beyond POTS are discussed in Chapters 15 and 18.

11.7.3 Modernizing Legacy Systems

It was clear from the Morris prototype (Section 11.1) that stored program control was the future of switching system control. But was an entirely new total system really necessary? The BOCs had a lot of capital invested in Step and, especially, in the newer Number 5 Crossbar. People in the BOCs wondered whether Step couldn't be modernized by intercepting the first selector link between the line finder and the first selector, and whether Number 5's markers couldn't just be replaced by computers, without completely replacing fabrics and line circuits. But, 1E developers in Bell Laboratories successfully resisted this and the 1E was developed as a total system — not as a modernization of existing legacy systems. In part this was appropriate because of new features and the reduced OAM&P costs from the 1E's maintenance strategy. Eventually, marker replacement occurred and step was computerized (more outside the United States than inside).

This disagreement illustrates the unusual power that Bell Labs had inside AT&T. Certainly, in today's environment, manufacturers make any product the carriers want — even if it's not in their long-term best interest. Was AT&T's technical conscience too strong, or even appropriate?

A final closing note: while I've never taken a formal poll, all LEC switchmen I have ever talked to have said that the 1E/1A was their favorite, the best switching system ever made — not just preferred over its electromechanical predecessors, but also preferred over its digital successors.

11.7.4 The Partition

The 1E has even greater separation of the system's controller from the rest of the system (called the process in this chapter). In Number 5 Crossbar, call control is split up among the two kinds of markers, originating registers, and other relatively intelligent hardware. Various control components are connected to the call during the time that their particular function is needed. The 1E shifts this partition in a different way because the system has virtually no intelligent hardware, except for the processor. All the control intelligence is in the processor's software and the dumb periphery is managed (really, micromanaged) by the processor. In hindsight, it seems bizarre to use an expensive main-frame computer for mundane functions like counting dial pulses.

From Step to Crossbar, we saw the monolithic system become partitioned into control and process. From Crossbar to the 1E, we see how this control became centralized. In subsequent systems, we'll see the control's programs become separated and specialized and how programs are scheduled by the processor's operating system in much the same that the Number 5 Crossbar scheduled the connections of various control units. These later systems will also show a little distribution of control intelligence among several mini-computers and even into per-card microprocessors. Not until

Chapter 18, where we discuss the Intelligent Network, do we see so much control distribution that the control subsystem actually returns to the Number 5 Crossbar structure of specialized components that are scheduled during the processing of the call.

EXERCISES

11.1 Interpolate Figure 11.10. (a) Calculate the percentages of execution time in J-priority, base-level call processing, and base-level maintenance, under a traffic load of $0.333T_{max}$. (b) Sketch the corresponding three-level oscilloscope trace (Figure 11.11) over 60 milliseconds (twelve 5-millisecond intervals). Assume that all 12 interrupts have the same duration. Begin your sketch with a J-level interrupt and let this be followed immediately by the start of an E-priority maintenance program that uses 10 milliseconds of execution time.

11.2 Interpolate Figure 11.10. (a) Calculate the percentages of execution time in J-priority, base-level call processing, and base-level maintenance, under a traffic load of $0.75T_{max}$. (b) Sketch the corresponding three-level oscilloscope trace (Figure 11.11) over 80 milliseconds (sixteen 5-millisecond intervals). Assume that all 16 interrupts have the same duration. Begin your sketch with a J-level interrupt and let this be followed immediately by the start of an E-priority maintenance program that uses 10 milliseconds of execution time.

11.3 Figure 11.11(b) shows an oscilloscope trace for J-priority, base-level call processing, and base-level maintenance, under a traffic load of $0.5T_{max}$. The figure shows that all eight indicated interrupts have the same duration — 2.5 milliseconds. Consider two modifications to the figure: (1) add a fourth line above the J line for H-priority interrupts, and (2) let the duration of the eight intervals be 2, 4, 6, 2.5, 1.5, 1, 2, and 1 milliseconds, respectively. Notice that the average duration is still 2.5 milliseconds. Let the fourth interrupt be partitioned into 1 millisecond of high-priority work and 1.5 milliseconds of low-priority work. Sketch the scope trace under these conditions. Hint: the transition from maintenance to call processing should shift to sometime during the sixth 5-millisecond interval.

11.4 Write the pseudo-code for the hopper unloading program. If the hopper is not empty, it reads from the unload address. Then, it must update this unload address the same way that the hopper-loading program updates the load address. Finally, if this program has just read the newest entry in the hopper, it must set the hopper-empty flag.

11.5 Write the pseudo-code for a program that deletes a given register from a one-way linked list. It must search the entire list, looking for the register that points to this given register. The first and last registers on the list require special handling. The program returns a fail flag if the list is empty or the prior register is not found.

11.6 Modify the pseudo-code for the programs that load and unload registers from a one-way linked list so they administer a two-way list. Write the pseudo-code for the simple program that deletes a given register from a two-way list.

11.7 Suppose six 20-word originating registers (ORs) reside physically in RAM, beginning at addresses 220, 240, ..., through 320, as shown in Table 11.2. The register that starts at 280 is busy, but all the others reside on a two-way idle linked list (ILL) whose head cell starts at 200. The 0 word in the head cell points to the first register on the list and the 1 word points to the last register on the list. The 0 word in each register is the forward pointer and the 1 word is the reverse pointer. The second column shows the current contents of the indicated memory locations. In the third column, show the changes to the original list after the register at 280 is added to the bottom of the list. In the fourth column, show the changes to the original list after the most idle register is deleted from the top of the list. In the fifth column, show the changes from the original list after the register at 260 is removed from the middle of the list.

TABLE 11.2. RAM for six ORs and ILL head cell

Address	Contents	Add 280	Delete top	Remove 260
200	240			
201	300			
220	300			
221	260			
240	320			
241	200			
260	220			
261	320			
280	xxx			
281	xxx			
300	300			
301	220			
320	260			
321	240			

11.8 Examine the space-time tradeoff between one-way and two-way linked lists. Consider the program that removes a register from a list. In a one-way or a two-way list, suppose the program needs 1 microsecond to remove a given register and patch the list. In a one-way list, suppose it uses an additional 1 microsecond to test if a given register is the one specified for removal and to move to the next one if it's not. (a) Consider a set of 10 registers. What is the difference in the number of required memory words if registers are placed on a two-way list instead of a one-way list? (b) If all registers are on the list, what is the average time to remove a given register from a two-way list? From a one-way list? Repeat the same two questions if there are 1000 registers on the list.

11.9 Suppose the J-level Task Dispenser must allocate execution time to nine different programs: P1 through P9. Table 11.3 shows the required frequency and expected duration of each program. Design an indexed ordered bits buffer for this task dispenser so that the expected total duration is uniform over all J-level interrupts.

TABLE 11.3. Frequency and duration of nine programs

Program	Runs every X ms	Last for Y ms
P1	5	.01
P2	10	.02
P3	10	.04
P4	20	.04
P5	20	.04
P6	40	.02
P7	40	.02
P8	40	.02
P9	40	.02

11.10 A small 1E office serves only 5000 lines. Assume that subscribers originate an average of one call every 10,000 seconds during the peak busy hour and that the average call duration, including time for dialing, is 150 seconds. Use $P_b \leq 0.015$ for all calculations. (a) How many originating registers are needed? (b) Assume 60% of the subscribers have dial pulse telephones and they average 10 seconds to dial a number and that Touch Tone subscribers average 4 seconds. If dial pulsing and Touch Toning use separate service circuits, how many DPRs and TTRs are needed? (c) Suppose DPRs and TTRs cost $300 apiece, but a combined DPR/TTR costs $400. Should this office use the specialized service circuits or the combined ones?

11.11 The office in the previous problem is so small that line scan would occur regularly every 100 milliseconds and would be seldom skipped. (a) In this small office, on the average, how many off-hooks would be detected in each execution of line scan during the peak busy hour? (b) How long should the line service request hopper be for $P_b \leq 0.015$? Since this calculation is "off the charts" in Section 5.2, you need to use the formula directly.

11.12 Consider a 36,000-line 1E office. Each of its 36 line scanners has 64 rows with 16 ferrods per row. (a) How many 16-line row scans are required to scan every line in the office? (b) Assume line scan's subprogram for processing one scan row has an average of eight instructions and that the processor's instruction cycle time is 0.5 microseconds. Ignoring loop overhead, what is the net accumulated base-level execution time for line scan? (c) What percentage of total real time is spent scanning lines? (d) If the system operates at 40% J level, how much elapsed time passes from the start of line scan to the end? This answer should show a range of values, depending on how often line scan is interrupted by J level.

11.13 Consider this same 36,000-line ESS office. Let each line go off-hook twice per hour during peak load. (a) What is the average number of off-hooks per second? (b) If line scan actually runs every 100 milliseconds, what is the average number of new off-hooks found per execution of line scan? (c) Using the answer to Exercise 11.12 (c), what is the value of the line scan overhead, the average percentage of real-time scanning lines per off-hook found? (d) If line scan is maximally deferred to run every 700 milliseconds, what is the revised average number of new off-hooks found in each execution? (e) What is the revised line scan overhead?

11.14 Consider a 1A office, which has a faster processor than a 1E office. At 2 A.M., when the processor handles practically 0 calls, assume the average J-level interrupt lasts 0.45 milliseconds. What percentage of execution time is spent in J level? In base level? Considering the base-level execution time only, assume that the entire ABACABADABACABABACABADABACAB sequence between an E priority and the next E priority requires 1.67 milliseconds while the various task dispensers look for work, and that the E-priority task dispenser executes one 10-millisecond program segment each time it gets a chance to run. At 2 A.M., what percentages of total real time are spent in J level? In A- through D-priority call processing? in E-priority maintenance?

11.15 Continuing the previous problem, assume that J-level work requires an additional 4% of total real time for every 500 Erlangs of additional system traffic load. And, assume that A- through D-priority call processing requires an additional 12% of total real time for every 500 Erlangs of additional system traffic load. Sketch the plot of percentage of real time versus traffic load for this hypothetical 1A. What is the maximum traffic load before the system crashes?

11.16 Still continuing this same problem, suppose the average line originates one call every 80 minutes during PBH. The average call duration is 5 minutes. How many lines can this 1A office handle?

11.17 Suppose the North American number plan is modified so that, after dialing a possible 0- or 1-prefix, the subscriber dials a digit indicating the number of digits that will follow. For example, an intraoffice call would be dialed as 4-ZZZZ and an unassisted long distance call would be dialed as 1-0-XXX-YYY-ZZZZ, where XXX is an area code, YYY is an office code, ZZZZ is an extension number, and 0 signifies that 10 digits will follow. (a) What are the dialed numbers for an unassisted local interoffice call? An operator-assisted toll call in the same area code? An operator-assisted toll call in another area code? (b) What would you recommend for direct-dialed international calls? (c) Describe, in detail, how a Step office would need to be changed so it could accommodate this new number plan. Describe the same for a Number 5 Crossbar office and a #1 ESS office. (d) A slight modification would be to dial the number of digits first, so it includes the 0- or 1- prefix, if needed. How does this change affect Step, Number 5 Crossbar, and the 1E?

SELECTED BIBLIOGRAPHY

Brooks, J., *Telephone — The First Hundred Years,* New York, NY: Harper & Row, 1976.

Carbaugh, D. H., et al., "No. 1 ESS Call Processing," *The Bell System Technical Journal,* Vol. XLIII, No. 5, Part 2, Sept. 1964, pp. 2483 – 2531.

Clarke, A. C., et al., *The Telephone's First Century — and Beyond,* New York, NY: Thomas Y. Crowell, 1977.

Downing, R. W., J. S. Nowak, and L. S. Tuomenoksa, "No. 1 ESS Maintenance Plan," *The Bell System Technical Journal,* Vol. XLIII, No. 5, Part 1, Sept. 1964, pp. 1961 – 2019.

Feiner, A., and W. S. Hayward, "No. 1 ESS Switching Network Plan," *The Bell System Technical Journal,* Vol. XLIII, No. 5, Part 2, Sept. 1964, pp. 2193 – 2220.

Freimanis, L., A. M. Guercio, and H. F. May, "No. 1 ESS Scanner, Signal Distributor, and Central Pulse Distributor," *The Bell System Technical Journal,* Vol. XLIII, No. 5, Part 2, Sept. 1964, pp. 2255 – 2282.

Fundamental Principles of Electronic Switching Circuits and Systems, New York, NY: Bell Telephone Laboratories, 1961.

Harr, J. A., E. S. Hoover, and R. B. Smith, "Organization of the No. 1 ESS Stored Program," *The Bell System Technical Journal,* Vol. XLIII, No. 5, Part 1, Sept. 1964, pp. 1923 – 1959.

Haugk, G., S. H. Tsiang, and L. Zimmerman, "System Testing of the No. 1 Electronic Switching System," *The Bell System Technical Journal,* Vol. XLIII, No. 5, Part 2, Sept. 1964, pp. 2575 – 2592.

Joel, A. E., Jr. (ed.), *Electronic Switching: Central Office Systems of the World,* New York, NY: IEEE Press, 1976.

Joel, A. E., Jr. (ed.), *Electronic Switching: Digital Central Office Systems of the World,* New York, NY: IEEE Press, 1982.

Keister, W., R. W. Ketchledge, and H. E. Vaughan, "No. 1 ESS: System Organization and Objectives," *The Bell System Technical Journal,* Vol. XLIII, No. 5, Part 1, Sept. 1964, pp. 1831 – 1844.

Noll, A. M., *Introduction to Telephones and Telephone Systems,* Norwood, MA: Artech House, 1986.

Nowak, J. S., "No. 1A — A New High Capacity Switching System," *Proceedings of the International Switching Symposium,* 1976, p. 131-1.

Rey, R. F. (ed.), *Engineering and Operations in the Bell System,* Second Edition, Murray Hill, NJ: AT&T Bell Laboratories, 1977.

Schindler, G. E., Jr. (ed.), *A History of Engineering and Science in the Bell System,* Murray Hill, NJ: Bell Telephone Laboratories, 1982.

Talley, D., *Basic Electronic Switching for Telephone Systems,* Rochelle Park, NJ: Hayden Book Company, 1975.

Talley, D., *Basic Telephone Switching Systems,* New York, NY: Hayden Book Company, 1969.

12

Private Networks

> *"Please, Mother, I'd rather do it myself."* — from a 1950s television commercial

While Chapters 3, 4, 9, and 10 emphasized how wire and carrier facilities are used in switched connections, this chapter emphasizes how they are used in nonswitched connections. A switching office's line side was first described in Chapter 3 and was continued in Chapter 10. A switching office's trunk side was first described in Chapter 4 and was continued in Chapter 9. This chapter's first three sections describe simple *private networks* — made from *leased lines, leased carrier,* and *leased trunks,* respectively. Section 12.3 includes a discussion of *foreign-exchange lines.* The pedagogy in these sections follows two example systems, which are progressively developed. The first section introduces these examples to illustrate simple systems. Then, the next two sections progressively embellish these two examples to illustrate more advanced principles and architectures.

Enterprise private networks, first mentioned in Chapter 10, are described fully in the fourth section. While most of today's long distance carriers own their own channels now, many typically got started as enterprise private networks. We'll see, in another bizarre twist in the story of switching, that most of AT&T's long distance competitors assembled their private networks by leasing lines from — of all places — AT&T. This chapter's last section covers the regulatory and technical issues of long distance competition. But before we discuss that, the fifth section describes the *provisioning* of digital carrier systems by *facility switching* with a digital access crossconnect system (DACS) and by Sonet *add-drop multiplexing.* Later, Chapter 17 compares facility switching, in general, to circuit and packet switching.

12.1 LEASED LINES

After discussing leased loop pairs, an example system is introduced and four different private network implementations are presented. Then, a second example system is introduced to illustrate the flexibility that leasing brings to the design of an enterprise private network. The fourth and fifth subsections discuss *pricing* and *node provisioning,* respectively.

12.1.1 Leased Loop Pair

As described in Section 3.4.4, the analog twisted pairs in the LEC's loop plant terminate on the main frame in the LEC's switching office. Because the LECs wisely overengineer their loop plant, many of these wire pairs don't terminate on line circuits or on the switching fabric. While some of these wire pairs are just idle, many are leased for uses other than switched telephone service.

From Section 2.1.2, recall that the first telephone office was built inside a company that provided security for banks. Burglar alarms were originally wired to this company so that personnel could provide remote security at night. During the day, the same wires were connected to a primitive switchboard. A hundred years later, instead of a security company letting its wires be used for telephony, the telephone company leases its wires for security — and many other applications.

Suppose A needs a permanent electrical connection to the CO, and 26-gauge copper twisted pair is sufficient for his application. Because the LEC's lines are practically ubiquitous, the telco could easily lease A the necessary pair. It would terminate on the main frame, but not on a line circuit. Suppose A needs a permanent electrical connection to B, and 26-gauge copper twisted pair is sufficient for their application. While the LEC's lines are practically ubiquitous, they form a star that homes on the LEC's CO. The telco could easily lease a pair to A, lease a pair to B, and connect the two pairs to each other at the main frame.

So, a pair that is used for telephone service, is crossconnected at the main frame from its wire pair (or equivalent channel) in the loop on one side to a line circuit and fabric appearance on the other side. Two pairs in the loop plant that have been leased for a permanent point-to-point connection are jumpered to each other at the main frame. The users at the other end of these leased lines never get dial tone.

12.1.2 A 1970s Data Entry Department

This is the first of two examples that will be continued and extended throughout the first three sections of this chapter. Both examples are from the data networking environment. We'll extend these examples to voice networking later in the chapter.

Consider a 1970s-vintage enterprise data network. Suppose some company's data entry department (DED) occupies two separate buildings: the data entry clerks reside in a comfortable downtown office building; and, in this instance of the example, the DED's computer center occupies a building directly across the street. Assume the data entry environment is old-fashioned — N dumb terminals connected to a time-shared host computer.

Let there be 96 dumb computer terminals in the office building and let their host computer be in the computer center building. The company's systems analysts determine that each terminal's channel to the host needs a 9.6-Kbps data rate, each way. The terminals are so busy that it is deemed inadvisable to use any type of intermediate data concentration like a LAN or a terminal concentrator in the office building. While this is slightly contrived, terminal concentrators and LANs don't always provide enough bandwidth reduction to justify themselves and parallel point-to-point connections may be optimal in many situations. How could the data entry department connect the 96 terminals to the host across the street? Some alternative implementations are presented.

1. DED could suspend a 96-pair drop cable over the street, if municipal ordinances allow it, or they could dig up the street and bury the cable. Unless the drop is very short and well shielded, they probably need a modem on each end of all 96 pairs.
2. DED could install a line-of-sight (microwave or infrared), transmission system across the tops of the two buildings. The transmission system must provide at least 96 separate channels so that each terminal's modem has a dedicated permanent channel to a port modem on the host across the street.
3. DED could buy a dial-up modem for each terminal and for each terminal port on the host computer and they could acquire a POTS line for each terminal and a 96-line hunt group for the host computer. This is a total of 96 POTS lines, a 96-line hunt group, and 192 modems. The terminal user would dial the host's telephone number, and the phone company's equipment would establish a voice-grade telephone connection to an idle line in the hunt group so the two modems can talk to each other.
4. They could lease 96 nonswitched private lines between the office building and the telephone company's main frame and another 96 between the computer center and the same main frame. The phone company would jumper the lines at the main frame so that each terminal modem had a dedicated permanent line, in two segments, to a port modem on the host across the street.

For a network whose nodes are this close together, the leased-line solution may not be the most economical, in terms of directly apparent cost. While the third implementation is probably the most expensive, the first two require significant maintenance and neither is very flexible if the department grows or shrinks, or one of the locations changes. This flexibility is demonstrated by the chapter's second example.

12.1.3 A 1980s Multibranch Bank

While this second example is also from data networking, it's a more modern problem — interconnecting LANs. Among their many useful functions, LANs provide concentration of data onto facilities. Often, a direct point-to-point connection, like in the DED example, is too inefficient and data concentration is cost-effective. While none of the implementations for the DED example was cost-effective, we will see cost-effective direct point-to-point solution as we extend that example.

One simple way to interconnect Ethernet LANs is through a pair-wise inter-LAN data path, called a bridge. The bridge has a gateway node on each of the two LANs that it bridges together, and each gateway knows the addresses of all the nodes on the other LAN. The two gateways are connected by a bridge link. Then, packets on one LAN that are destined for nodes on the other LAN are passed through the bridge, but intra-LAN packets are not.

Consider a bank that has two branch offices, both in the serving area of some LEC's CO. While each branch office has its own intraoffice LAN, the bank wishes to bridge the two LANs. So, a gateway is attached to each LAN, a private line is leased from each gateway to the LEC's main frame, and the two segments are crossconnected at the main frame. The two segments, one between the CO and each branch of the bank, constitute the bridge link. So, jumpering these two lines at the CO completes a primitive one-link metropolitan area network (MAN) that interconnects the two LANs.

Now, suppose the bank adds a third branch office, still in the serving area of the same CO. Since the bank wishes to interconnect the three LANs, a second gateway is attached to each existing LAN and the LAN in the new branch also needs two gateways. A private line is leased from each gateway to the LEC's main frame. The six segments are crossconnected, pair-wise, at the main frame into three bridge links, each with two segments. Jumpering these six leased lines pair-wise at the CO completes a three-link fully connected MAN that interconnects the three LANs.

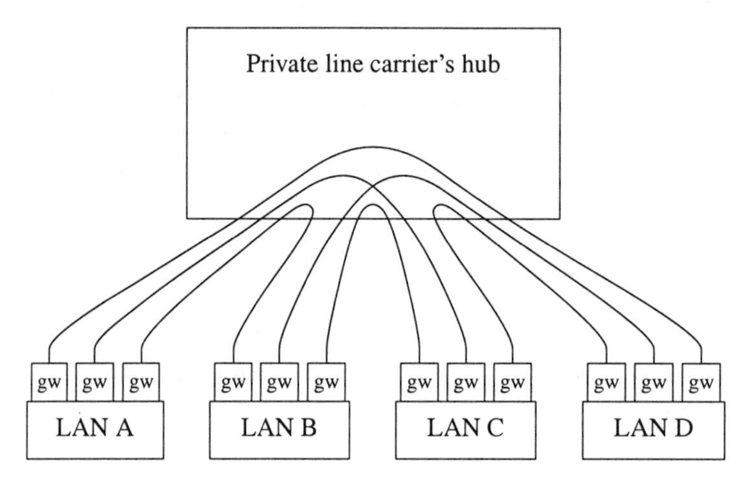

Figure 12.1. Bank with four branches.

Now, suppose the bank adds a fourth branch office, still in the serving area of the same CO. Figure 12.1 shows the configuration with four branches, 12 leased-line segments, and the interconnection at the CO. Jumpering these 12 lines pair-wise at the CO completes a six-link fully connected MAN that interconnects the four LANs.

While each LAN typically uses packet switching, there is no packet switching in the MAN; packets are transmitted over the MAN's channels but are not switched internally. While the private lines are leased from a carrier who typically also offers circuit switching service (such as the local telephone company), these particular channels are not circuit switched. These 12 channels are jumpered at the main frame; they are not connected to the hub's circuit switching equipment and they do not get dial tone service. These six jumpers in the hub constitute a primitive *facility switch.*

Now, suppose a new branch office with its own LAN opens up, or suppose the inter-LAN data traffic overloads these original 12 channels. In either case, the bank must lease more lines to these four branch offices and must arrange for more jumpering at the carrier's hub. While rearranging or augmenting the jumper wires is a primitive form of facility switching, its full significance is not clear from this example yet. This example shows that leased lines can be an important capability for any enterprise's telecom or computer networks.

12.1.4 Pricing

Many years ago the regulatory agencies established a tariff by which the telephone companies could lease out their lines for nonswitched service instead of for conventional switched service. Conventional switched lines came to be called dial tone circuits because, since a leased line does not have a fabric appearance, a leased line does not get dial tone. The tariff for leased lines was typically much higher than its actual cost because, as we have discussed repeatedly, the regulatory agencies have always worked with the telephone companies to find this kind of "cream" for the POTS subsidy.

In post-Carterfone America, leasing lines became another business where competitors saw an opportunity to sell some commodity at a price that would be greater than their cost, but that would be much lower than a market price that was deliberately set very high. We'll describe *loop bypass* when we discuss post-divestiture economics in Chapter 16. While this chapter continues under the assumption that lines are leased from the telephone company, the reader should be aware that the LEC is no longer the only carrier, especially in the enterprise market.

12.1.5 Node Provisioning

Suppose some enterprise needs a private line between two locations, A and B. If A and B are served by the same CO, then the private line has two segments — one from the main frame to A and the other from the main frame and B. The two segments are jumpered at the CO's main frame. Another potential price optimization could materialize someday.

Consider the case where the two locations are served not only by the same CO, but are served by the same distribution point (or even, pedestal) off this CO. Instead of jumpering the two segments at the main frame, they could be jumpered at this common distribution point. Neither segment of the private line has to even reach the main frame. While jumpering at the distribution point or pedestal means that the customer rents less copper, the administration of this jumper in the field has higher cost than administering the jumper inside the CO building.

Even if such localized jumpering does cost less overall, the regulators have consistently not allowed price to reflect individual cost in such local situations. Why should someone get a lower price for a private line because his two segments just happen to be served by the same distribution point, while someone else pays more because his two endpoints are not? But how can such things be regulated?

The tariff could be based on the distance between A and B. But, consider our data entry department, for example, where A and B are physically across the street from each other. If both locations are two miles from the CO, the total length of the two-segment leased line is 4 miles. But, two points A and B might be diametrically opposite each other in the CO serving area, with the CO in the center. For the same 4-mile length of copper and, presumably the same cost to the telco, the two locations could be 4 miles apart. Why would they pay a different price?

To complicate this issue, for years we allowed the unregulated "bypassers" (Section 16.5.4) to charge whatever the market would bear, also not based on cost. The price of private lines is one of the commodities most affected as we deregulate the telecom industry. We'll propose a solution to this whole issue of regulation in Section 22.4.

12.2 LEASED CARRIER

Since any telecommunications line might be implemented as a channel in a carrier system, a leased line might be implemented this way. The first subsection motivates this concept by showing how our multibranch bank's bridge links might be implemented with T1 by the telephone company. Having illustrated the cost savings and flexibility of such an implementation, the next subsection discusses the controversy of tariffing leased T1 and passing its benefits on to the customer. The last two subsections illustrate two more potential implementations for connecting our data entry department's dumb terminals to ports on the host computer: one using regular T1 and one using fractional T1.

12.2.1 T1 for the Multibranch Bank

Our example of the bank with N branches continues. Instead of assigning $N - 1$ separate analog twisted pairs between each branch office and the CO, the leased-line carrier (assume it's the telephone company) might implement this network over some kind of loop carrier system, analogous to T1, between each branch office and the CO. Suppose the cost of a leased T1 line is less than the cost of six individual private lines — a typical breakpoint, depending on the length of the line. While each line in Figure 12.1 represents a separate logical channel, the carrier might implement each group of $N - 1$ leased lines using a single physical T1 facility. For $N = 4$, as in Figure 12.1, T1 wouldn't be economical, at least because T1 typically uses three physical pairs (one for data flow in each direction and one spare). But, if the bank grew to have seven branches — or 12 or 20 — T1 would be a more economical implementation.

The carrier places one T1 muldem on the premises of each branch and assigns $N - 1$ channels on the DS0 side to local pairs that connect to the $N - 1$ bridge gateways in each branch. Also in each branch, the DS1 side of the T1 muldem connects to three physical pairs in the carrier's loop plant. The other end of these pairs terminates on the carrier's main frame. In the carrier's CO, the other ends of each triple-pair of leased lines is jumpered at the main frame to CO-internal pairs that connect to the mating T1 muldem in the CO. The assigned $N - 1$ channels on the DS0 side of the N muldems is wired back to the main frame, where they are crossconnected just as if they had been physical loop pairs.

For example, suppose $N = 7$, as shown in Figure 12.2. Each branch has six bridge gateways and a T1 muldem. Each triple of T1 lines between a branch and the CO carries six active channels (of the 24 available) instead of using six physical loop pairs. After demultiplexing at the CO, the 42 DS0 channels are crossconnected to form 21 bridge links, each link containing two segments.

While this implementation with T1 might cost the carrier less, the real advantage (to the carrier) occurs when the bank adds yet another branch. With $N = 7$ and using 42 leased physical lines, adding an eighth branch requires an eighth physical line in each of the original seven groups of lines. With $N = 7$ and using a T1 implementation, adding the eighth branch is accomplished by activating another logical channel in each T1 and by minor wiring changes in each branch and at the CO.

We know that carriers benefit from the economy of multiplexing, anyway. But also, as customers repeatedly change the number of lines they lease, the flexibility of the T1 implementation provides an additional significant cost savings for the carrier.

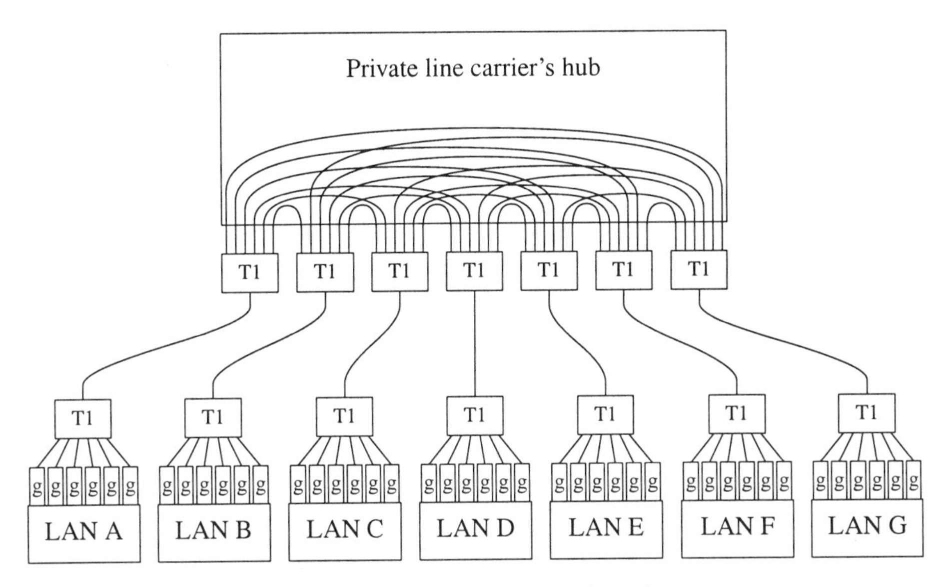

Figure 12.2. Bank with seven branches.

12.2.2 T1 Tariff Controversy

The technology of transmission carrier, particularly digital and most particularly T1, provided the pair-gain and economy of scale that made "bulk" digital channels much less costly than conventional voice-grade analog twisted pair. By reducing the cost of providing lines, but keeping price the same, leasing lines became an extremely important revenue source for the telephone companies.

For many years, the only commodity tariffed for leasing was the voice-grade line (or its digital equivalent). So, any cost savings were kept by the telephone company. Reiterating, regulators thought this was a good thing in that it provided revenue to subsidize POTS. But, enterprises, like the bank in our example, understood well the concept of "bulk rate" and they wanted the cost savings passed on to them.

Telecom managers and other industry users finally succeeded in forcing the regulators to tariff T1 directly and, by doing so, forcing the telcos to pass the economic benefit on to the enterprises that lease the lines. While the T1 tariff (the price) was still well above the cost of T1, it was still well below the tariff for some small number of individual lines (typically between 6 and 10). As we'll see later in this chapter, this T1 tariff enabled long distance competition.

Today, the United States has intense competition in this business. And, also today, an enterprise can lease a T3 or a Sonet OC3, or anything they want.

12.2.3 T1 for the Data Entry Department

T1 is also useful in the example of the data entry department, even though the leased facility links only two locations that are only across the street from each other. The simplicity of this example allows two different implementations (continued from Section 12.1.2). And, in each case, the implementation could be one where the telco uses its T1 to provide the channels or one where the enterprise purchases T1 muldems for use on telco leased lines.

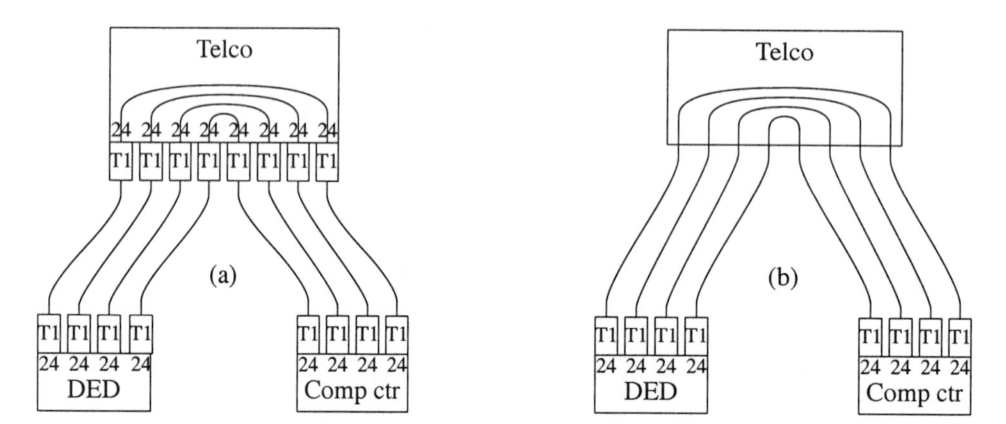

Figure 12.3. T1 configurations with: (a) jumpered DS0 (b) jumpered DS1.

5. For 96 separate channels, the DED could use four complete T1 systems (with 24 channels per T1) between the office building and the telephone company's main frame and another four between the computer center and the same main frame. In this implementation shown in Figure 12.3(a), with T1 terminals on both ends of each segment, the eight T1 systems are completely demultiplexed in the CO. At the main frame, the phone company would jumper the 96 channels from the office building to the 96 channels from the computer center.

6. With a simple point-to-point link like this, the T1s don't need to be demultiplexed at the CO. The DED could use four partial T1 systems between the office building and the telephone company's main frame and another four between the computer center and the same main frame. In this implementation shown in Figure 12.3(b), with T1 terminals only on the user end of each segment, the eight T1 systems are not demultiplexed in the CO, although repeaters may be required. At the main frame, the phone company would jumper the four T1 lines from the office building to the four T1 lines from the computer center.

In these fifth and sixth implementations above, modems convert the binary stream from the 96 host i/o ports on one side, or from the 96 terminals on the other side, into an analog signal whose bandwidth conforms to a voice-grade analog channel with 3.4 kHz bandpass. From the 96 modems on each side, these analog signals are transmitted to a T1 channel bank, where each is sampled and PCMed into a 64-Kbps stream. The 96 channels in each building are multiplexed up to four T1 streams, each at 1.544 Mbps. In the fifth implementation, these channels are all demultiplexed and converted back to analog (analog signals carrying binary data) at the CO's main frame. Because the sixth implementation simplifies the equipment and the jumpering required at the CO, intuitively, it seems like the best one.

Both implementations still seem to be inefficient, partly because of all the A-to-D conversion and partly because the T1 capacity doesn't seem to be used well. Data is converted from digital to analog (in the terminal's modem), back to digital (in the T1 channel bank), back to analog (in the T1 channel bank on the other end), and back to digital (in the computer's modem). While the net one-way digital capacity between each building is 4 × 1.544 Mbps = 6.176 Mbps, the net data rate required is only 96 × 9.6 Kbps = 921.6 Kbps, which is less than the capacity of one T1. A seventh implementation, one that is end-to-end digital, without the intermediate D-to-A-to-D-to-A-to-D conversion, might provide much more efficiency.

12.2.4 Fractional T1 for the DED

Conventional T1 has 192 data bits in a frame and formats each frame into twenty-four 8-bit words. Fractional T1 is implemented easily on leased T1 lines because these lines aren't switched. Fractional T1 can also be used on switched lines — but very carefully. The two requirements are: (1) the fractional frame partition must line up with the traditional partition of twenty-four 8-bit words (with four 2-bit fractional words per traditional 8-bit word, the format in our example does line up), and (2) the 8-bit digital channels must be switched digitally in every stage of switching they pass through. But, if the T1 stream is never D-to-A converted, as it wouldn't be in a leased line, the exact format of the 192-bit frame is irrelevant. In Fractional T1, the 192-bit frame can be formatted in almost any possible way (depending on the manufacturer).

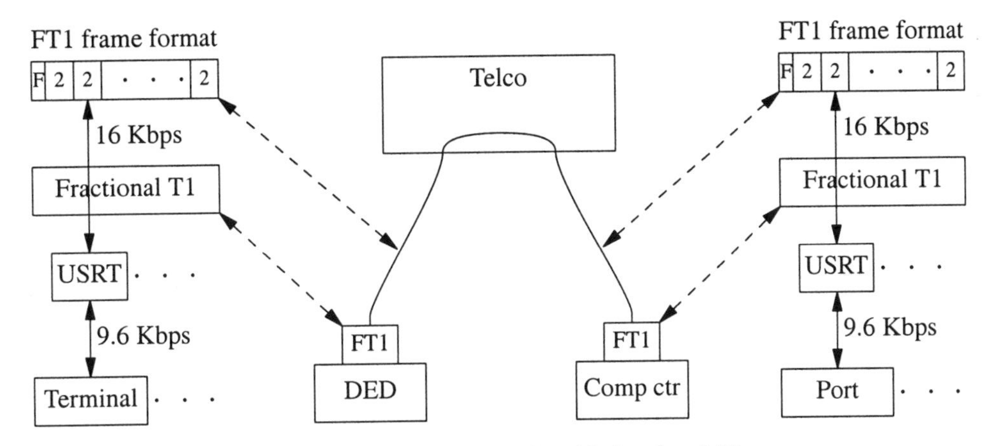

Figure 12.4. DED Example with fractional T1.

Using fractional T1 suggests a seventh implementation for our DED example, shown in Figure 12.4. The appropriate fractional T1 format for our DED example is to fracture the 192-bit frame into ninety-six 2-bit words. This gives 96 channels, and because each bit represents an 8-Kbps channel, each 2-bit channel supports 16 Kbps. All 96 channels, all the data between the 96 terminals and the host computer, fit onto one leased line with a fractional T1 muldem on each end.

Of course, the devil is in the details. Assume that a binary 9.6-Kbps stream can be transmitted directly over building wiring, in RS232C without a modem — between each of the 96 terminals and a fractional T1 in the office building, and between each of the 96 host ports and a fractional T1 in the computer center building. Near the respective fractional T1, each of the 192 local pairs is equipped with an intermediate device, called a universal synchronous receiver/transmitter (USRT). Each USRT communicates in RS232C at 9.6 Kbps on the terminal or host side and in RS232C at 16 Kbps on the fractional T1 side.

Since the end-to-end system guarantees that, even though a stream runs at 16 Kbps over the T1, the net data rate in each channel never exceeds 9.6 Kbps. So, the USRT's simple task is to receive an RS232C packet at one rate and retransmit it at the other rate. While it may seem that we've made the problem worse by shifting the 9.6-Kbps stream up to 16 Kbps, all the terminals in the office building, and all the terminal ports in the computer center, can easily communicate by leasing two segments of T1-grade lines, jumpering them at the main frame, and installing fractional T1 multiplexors in each building.

12.3 LEASED TRUNKS

This chapter's first section described how the telephone company could provision some of its lines, on a CO's line side, for a service other than POTS — leasing private lines. The example of the multibranch bank showed that significant networks can be assembled by leasing line-side channels from the telephone company or other carrier. But, by the definition of line side, all of this network's endpoints have to be within the serving area of the same switching office. Large networks and long point-to-point paths need to use channels, similarly dedicated for non-POTS use, but on the trunk side and even in the long distance network.

This section begins by describing how the telephone companies implement foreign exchange lines by provisioning some of their own lines and trunks. Then, in the last two subsections, the example of the data entry department is extended by showing how increasing the distance between the office and the computer center affects the provisioning of the leased channel between them.

12.3.1 Foreign Exchange Lines

Often, enterprises, and even individual subscribers, wish to have an office code, or even area code, that is different from the one the phone company would normally give them. Some of the more practical reasons are listed.

- An enterprise may wish to allow its customers to reach it by dialing a local number instead of making a toll call. Internet service providers (ISP), for example, offer their customers a local telephone number, even if the ISP's office doesn't reside in the local dialing area. This way, the ISP's customers avoid paying the LEC a connection charge for an Internet call that may have a long holding time by taking advantage of the LEC's flat-rate billing.
- The corporation may move into the serving area of another switching office, but their customers and the company stationery may still have the old telephone number.
- Many of an enterprise's outgoing calls might be concentrated into a small local region, but one that is outside the enterprise's own local region.

A less practical reason is that corporations and individuals may wish to improve their image by listing, in the phone book, a telephone number from a CO that serves a part of the region with more "snob appeal" than where the subscriber actually lives. All motivations arise whenever the telephone company is forced to split an NPA and re-assign an area code, for example, in New York City when the well-known 212 area code was changed from covering the entire city to only covering Manhattan. While people are always unhappy anyway, whenever the phone company changes their telephone number, in this case, many more people and corporations were especially unhappy because the number change revealed that their New York City address was not in Manhattan.

Obviously, however, everyone in America can't have a Beverly Hills, California, telephone number. So, to limit the interest and generate some revenue, the option to be assigned a "foreign" telephone number is tariffed. Even with local number portability (Section 18.4.4), this is nontrivial and has real cost. While computers, centralized procedures for number assignments, and database translations, all help simplify telephone number bookkeeping, it's still much simpler if subscribers are assigned office codes from their own home office. An even more difficult issue, routing incoming long distance calls, has three different implementations.

- Switching incoming calls as a permanent *call forward* requires that the subscriber have two telephone lines: the actual line, which might not be listed, and the listed number, which might not have a physical line. Calls to the listed number would be permanently forwarded to the actual line. With the second line in a different office, the call might be physically switched (tandemed) through the first office, where it ties up an additional fabric connection and a trunk in the trunk group between the two offices. Even if the network is smart enough to switch directly to the second office, because of appropriate interoffice signaling, every incoming call still must be processed by the call control software in both the target CO and the home CO.
- For most of the history of telephony, the only way to fake a telephone number was to lease what was called a foreign exchange line, commonly called an FX line. Using a private trunk, the subscriber is given a physical line appearance in the target office.
- With Intelligent Networking and Signaling System 7 (see Chapter 18), the translation of telephone number to line appearance can be moved further back in the network control. This is probably the ultimate solution for foreign exchange lines and many other types of local number portability. Not every office in the United States, or the world, is capable of this.

Figure 12.5 shows a configuration in which subscriber A is served physically by his home CO, but wishes to have a telephone number that makes it appear that he is served by some target CO. Before number portability, the only way A could have a telephone number from a different CO was by having a physical line appearance on the fabric of the target CO. In Figure 12.5, the home CO and the target CO are assumed to be geographically close enough to have direct trunking. The FX line is implemented by A leasing a private trunk in the trunk group between the two switching offices.

At the main frame in the home CO, if A is given the solid connection in Figure 12.5, then A's line appearance is in this office and A must be assigned a telephone number that is consistent with this office. However, if A is given instead the dashed connection in Figure 12.5, then A's line appearance is in the target office and A is easily assigned a telephone number that is consistent with this office. An idle trunk in the trunk group between the two COs is identified and permanently assigned to this permanent physical line. There is no trunk circuit on either end of this trunk and the trunk has no fabric appearance in either office. In the home CO, A's line is wired at the main frame to an in-

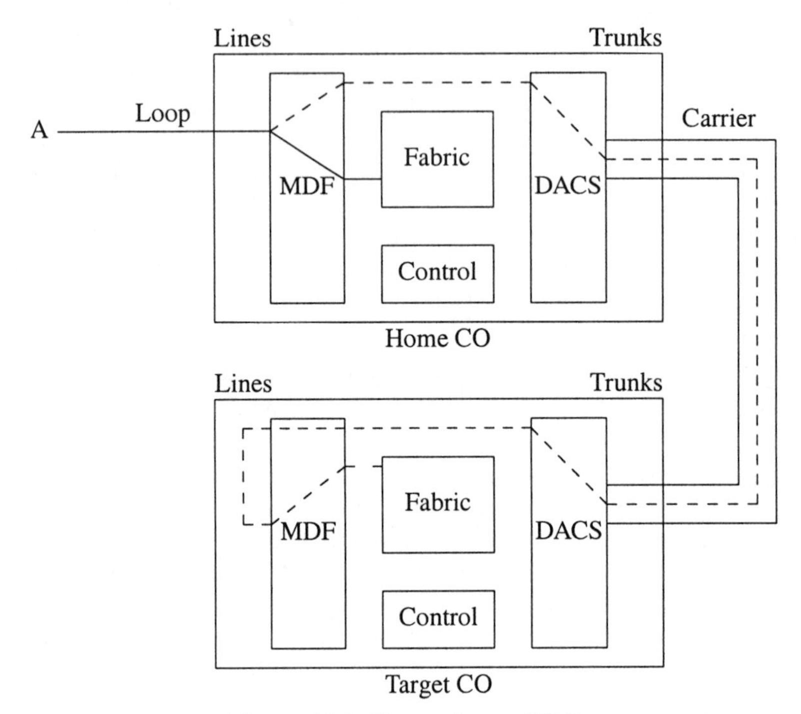

Figure 12.5. Illustrating an FX line.

house wire that bypasses the line circuits and fabric appearances. This bypass is wired to the leased trunk at the trunk side in a box labeled "DACS" in Figure 12.5. We'll discuss the details of this semi-permanent connection and the functions of a DACS in a later section — for now, just think of it as a kind of main frame for trunks.

In the target CO, the other end of this trunk is wired, at another DACS, to an internal wire that bypasses the trunk circuits and fabric appearances in this CO. The bypass is wired through the main frame from the office side to the loop side, jumpered on the loop side to another line termination, and wired back through the main frame to a line circuit (finally) and a line appearance on the target CO's switching fabric.

12.3.2 Relocating the DED's Computer Center

Return again to the example of the data entry department with two locations. The previous two sections listed seven different ways that the company could connect its 96 terminals to the host — assuming the company's office building and computer center are directly across the street from each other. Let's revise the example to something more realistic — let the office building be located in an industrial park or other professional district and let the computer center be located in a warehouse district. Consider the viability of all the previous implementations as a function of the separation of the company's two locations:

- Installing their own cable becomes less and less attractive as the two locations are farther and farther apart. While this option must always be considered in a campus environment, where the enterprise owns the property and the right-of-way, it is almost always impractical in a public environment.

- Any kind of line-of-sight transmission system must be seriously investigated for price, but only if the two locations are within line-of-sight. Cost of administration and maintenance must be weighed.
- Switched lines tie up line circuits and fabric crosspoints — in the PBX inside each of the enterprise's buildings and in the telephone company's switching offices on each end. So, their cost is relatively high compared to other solutions. However, if the enterprise can get the subsidized price for a POTS line, like we all can get in our residences, this could be a good choice. It's certainly the implementation we all choose at home. But, because enterprises have traditionally paid a premium price for POTS lines, this solution has been prohibitively expensive, especially if the two buildings are not within the same local flat-rate dialing area.
- The last four implementations — leased voice lines, leased demuxed T1 lines, leased connected T1 lines, and leased fractional T1 — required leasing private lines from some leased-line carrier like the telephone company or any of its competitors (discussed in Chapter 16). If the company's two locations are served by the same switching office, the resulting private line network is a simple star whose arms are joined at the CO's main frame, as in Figure 12.1. This network is more expensive and more complicated to manage if the enterprise's buildings are separated enough that they are served by different COs.

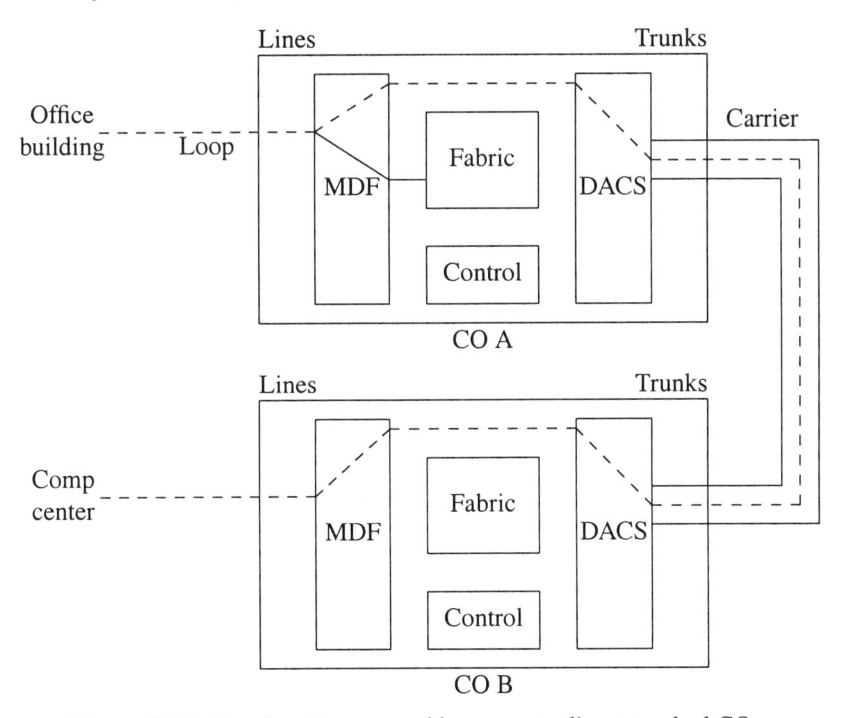

Figure 12.6. Two locations served by separate direct-trunked COs.

Figure 12.6 shows a simple configuration in which the two locations are not served by the same switching office. In this figure, however, while the two locations are served by separate COs, these two COs are assumed to be in the same local access transport area (LATA, Section 16.5.1) and to have direct trunking. The enterprise's two locations are each wired to a group of 96 private lines, or equivalent channels in a loop facility, or customized channels in some fractional T1 facility, which proceed to the main frame in their respective serving COs. Both serving COs are configured like the foreign exchange line's home CO in Figure 12.2. In both COs, these private lines bypass their respective fabrics and are wired at the DACS to trunks (or channels in trunk facilities) that proceed to the DACS in the CO that serves the other building. (In Section 12.4, I'll finally tell you what a DACS is.)

In the original example, with the two buildings across the street from each other, the private line has two segments — both on the loop side of the serving CO. In the example of the foreign exchange line, in the previous subsection, the private line also has two segments — one on the loop side of the serving CO and one in the trunk group directly between this CO and the target CO. In this example, illustrated in Figure 12.6, the private line has three segments — one on the loop side of each CO and one in the (direct) trunk group between the two COs. Depending on the specific implementation, the net private line, represented by the dashed line in Figure 12.6, could consist of:

- Ninety-six parallel twisted pairs, end-to-end;
- Four T1 lines in each of the three segments,
  - If all 12 T1 lines are completely demultiplexed at every node, then the bypass wires and jumpers in the main frames and DACSs are each 96 parallel twisted pairs
  - If the T1 lines are multiplexed only at the endpoints, then the bypass wires and jumpers in the main frames and DACSs are each eight parallel T1-grade twisted pairs (remember, each T1 requires two one-way twisted pairs);
- One T1 in each of the three segments, with a fractional T1 multiplexor on each end.

12.3.3 Further Separating the DED's Computer Center

In the previous subsection, as illustrated in Figure 12.6, our example enterprise's two buildings are served by separate COs, but these two COs have direct trunking. Now, let's make it just a little bit more complicated and assume that the two COs, while still in the same LATA, instead of having direct trunking, are interconnected through a common hub that we'll call a toll office.

Figure 12.7 shows this configuration, in which the two locations are served by separate switching offices that are toll or tandem connected. The company's two locations are each wired to a group of 96 private lines (or channels in a loop facility) that proceed to the main frame in their respective serving COs. Both serving COs are configured like foreign exchange line's home CO in Figure 12.2. In both COs, these private lines bypass their respective fabrics and are wired at the DACS to trunks (or channels in trunk facilities) that proceed to the DACS at the toll point. On the CO side of the toll office, tandem office, or other form of CO hub, the two sets of private lines are jumpered at the DACS. No switching is provided in any of the three offices.

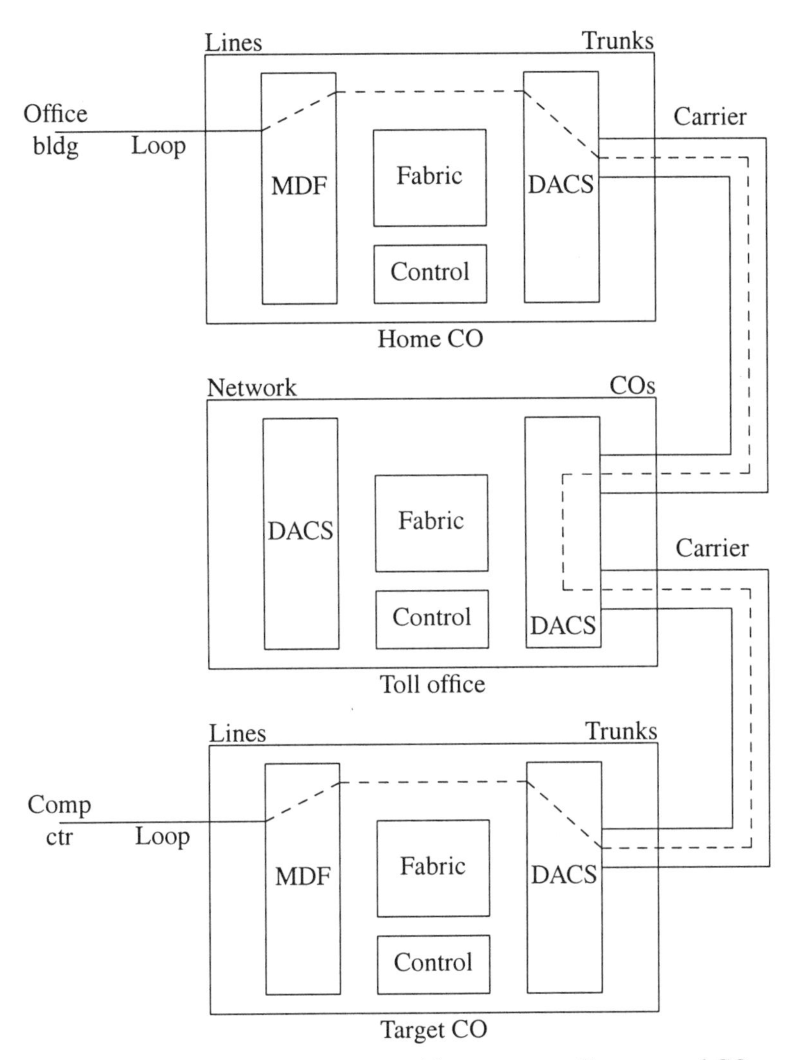

Figure 12.7. Two locations served by separate toll-connected COs.

Let the enterprise's two locations be so far apart that the COs that serve the two locations are not themselves served the same toll point or other hub and don't have direct trunking to any common tandem office or other CO. Someone in the enterprise's telecom department's staff would have to locate and price various private line segments among several COs and find the lowest-cost route. In a strictly hierarchical network architecture, the private-line architecture can be quite straightforward. However, in a more general network that is not strictly hierarchical, this takes the form of a *minimum spanning tree* problem. Further complicating matters, depending on the regulatory environment and the degree of competition, some of the private-line segments may be leased from different carriers.

Finally, suppose the two endpoints are far enough apart that they reside in different LATAs (described in Section 16.5.1) or different NPAs. In the current regulatory environment in the United States, the local exchange carrier may not provide long distance service and the interexchange carrier may not provide local service. The regulatory restrictions that were applied to switched service were also applied to leased service. So, an enterprise with two distant endpoints may have no choice but to lease segments from different carriers and to arrange for the segments to be semi-permanently provisioned.

12.4 ENTERPRISE NETWORKS

This section begins with a look at the requirements of an interlocation network for an enterprise with two locations — similar to our data entry department example, except for voice. Then, we expand the number of locations in this enterprise, similar to our growing multibranch bank. Physical and virtual enterprise voice networks are discussed. Finally, the example enterprise's data network is reconsidered and merged with the enterprise's voice network.

12.4.1 Tie Trunks

Consider an enterprise with two locations, each in the serving area of the same CO. Assume there are 96 telephones in each location. Look at several implementations:

- *CO-Centrex.* Each of the 96 telephones in each location is directly connected to Centrex service in the switching office. Any subscriber can easily call any other one and the numbering plan easily has three-digit dialing for extensions. The numbering plan could be location dependent, if desired, with extensions 100 – 199 in one location and 200 – 299 in the other. DID and AIOD are naturally included in Centrex.
- *192-line PBX in one location.* The 96 telephones that are in the same building as the PBX are directly connected over building wiring to line cards on the PBX. The 96 telephones in the other building are connected over 96 two-segment leased lines to line cards on the PBX. Dialing and numbering are the same as with Centrex. DID and AIOD would have to be included in the PBX. Outside calls are placed over a set of PBX trunks, which are accessed from the PBX by dialing 9. People in the location that doesn't have the PBX who make an outside call do this over a circuitous route: telephone to MDF to PBX to MDF.
- *96-line PBX in each building.* The 96 telephones in each building are directly connected over building wiring to line cards on the PBX in that building. There are no leased lines for extensions and no monthly Centrex bill. Each PBX has a set of dial-9 trunks for the extensions it serves. But how does employee A, in one building, call employee B, in the other building?
    - Since each PBX has CO trunks anyway, A dials 9 for an outside line and calls B in the PBX across the street. This procedure is especially annoying without DID. The eight-digit telephone call is no different from calling someone in another PBX at another enterprise. Furthermore, since the phone company charges for outgoing and incoming calls, this can get expensive, especially if the enterprise has a lot of internal calling.

- The employees' "sense of separation" can be reduced by using the 100 – 299 numbering plan described above, but having PBX software translate the three-digit number into the eight-digit number when necessary. This software service that makes separate PBXs feel like one PBX is called a virtual PBX. While the user interface is better than the case immediately above, the physical implementation and the cost are similar.
- Besides the CO trunks, the enterprise could provide a separate group of trunks that *tie* the two PBXs together. These tie trunks are implemented as private lines. In the typical PBX numbering plan, when A calls B in the other location, first A accesses a tie trunk by dialing 8, and then, with second dial tone, A dials B's three-digit extension.

While CO-Centrex almost always has a higher price than the equivalent PBX implementation, there is hidden economic advantage to being able to outsource the PBX and all its OAM&P. While the price of Centrex has traditionally been artificially higher than its true cost because of regulation for POTS subsidy, the Centrex tariff is gradually allowing the phone companies to be more competitive — in some states more than others. While the price of PBXs has traditionally been artificially lower than their true cost because U.S. tax law allows the depreciation of physical equipment (like a PBX) but not a service (like Centrex), look for the telcos to find ways to pass the depreciation of their CO switching equipment on to their Centrex customers. Furthermore, as the telephone companies move closer to a true user-programmed Intelligent Network (Chapter 18), expect the services available from Centrex in the future to be better than those available from PBXs (reminiscent of the days when only CO-Centrex could provide DID and AIOD). When the telecom industry is fully deregulated, corporate telecom managers across the United States will have to take another look at Centrex.

Because the two-location enterprise in this example would typically need only five or six tie trunks between these two particular PBXs, leasing tie trunks is much less expensive than leasing 96 private-line loop extensions. The savings from the line concentration is probably enough to offset the difference between operating two 96-line PBXs instead of one 192-line PBX. As long as the number of tie trunks in the trunk group is not overengineered and are well utilized, using tie trunks typically also costs less than the dial-9 implementation. Even blocking and lost economy of scale are not problems because software in many modern PBXs automatically converts an 8+ call into a 9+ call if all the tie trunks are occupied.

12.4.2 Enterprise Private Networks

Let's extend the example from the previous subsection. Assume that the enterprise above decided on the implementation that has a PBX in each location, with tie trunks. Suppose the enterprise grows and adds a third location. When the enterprise adds its third location, it installs a PBX in this new building and adds two new groups of tie trunks: a group between locations I and III and a group between II and III.

- Suppose the enterprise's three locations are all served by this same CO. The enterprise's network of tie trunks forms a logical ring — I to II to III and back to I. The physical architecture is a six-armed star — six arms because each link has two physical segments to the common CO. Like the three-branch bank's data network in the first section, little is gained by trying to optimize the architecture.

- Instead, suppose the three locations are separated enough so that each homes on a different CO. Then, the group of tie trunks is a set of multisegment leased channels. The logical architecture is still a three-sided ring. But because the physical architecture is also a ring, the enterprise's network of tie trunks might be optimally architected. If the three locations lie in a geographical line — for example, from Cleveland to Columbus to Cincinnati — it is probably wise to tandem the Cleveland-to-Cincinnati tie trunks through Columbus.

In large nationwide or global enterprises with many locations, telecom managers architect large enterprise private networks that may even have a hierarchical nature. The network links are groups of private lines. The two principal motivations are (1) an enterprise-wide numbering plan over the multiple locations and (2) a reduction of long distance charges in enterprises with heavy internal calling patterns. Because these leased-line links are very expensive, these telecom managers constantly monitor their utilization.

In large enterprises with many locations, the enterprise-wide numbering plan resembles the PSTN's plan. Each PBX gets a two- or three-digit office code, and A in New York calls B in Los Angeles by dialing 8-27-146, where 8 is the access code to the EPN, 27 is the office code for the PBX in LA, and 146 is B's extension number inside the LA PBX.

Section 18.2.2 examines *feature interaction* in great detail. An example of this is the interaction of two features: an enterprise's national private network built from leased lines and a PBX extension's ability to access the PSTN by dialing 9. Suppose an enterprise has locations in New York and Los Angeles. An employee in New York wishes to place a call to someone, not who works in the LA office, but who lives in NPA 312. This employee could dial into the PBX in LA, then dial 9, then dial a local call. Some enterprises don't want their employees to be able to do this. Other enterprises don't mind it. Still others make it a class of service perk, allowing it for some employees, but not for others. Still others are so lenient about this that they provide DID access for their tie trunks so that employees, from their homes, can call someone in LA over the enterprise private network. Such employee benefits may get abused if employees give these DID numbers to other people.

Enterprise private networks became quite popular and developed into an important segment of the telecom industry; they were especially targeted by the competitive carriers whose private line rates were not regulated (Section 12.6). The switched carriers, particularly AT&T's long distance carrier, responded by offering a *virtual private network* service. VPN is to EPN as Centrex is to PBX. An enterprise that subscribed to VPN didn't have its own private network, but it felt the same. The carrier offered the enterprise to use the same customized numbering plan they used with their EPN, but the carrier completed the enterprise's calls over the same shared trunks and switching systems that the public used.

Because the price was competitive with leased lines (remember, national leased lines are very pricey), many enterprises switched over to VPN. But many telecom managers remain concerned about sharing channels with the public (an EPN doesn't congest on Mother's Day) and with the potential for breach of security.

12.4.3 Integrated Voice/Data EPNs

In the first three sections of this chapter, the two running examples illustrate two enterprises' interlocation data networks. In the first two subsections of this section, the running example illustrates an enterprise's interlocation voice network. Most enterprises need interlocation private networks for voice and for data. Many enterprises administered completely separate networks, ignoring potential savings from economies of scale and scope, because voice networks were managed by the enterprise's telecom department, and data networks were managed by the enterprise's MIS department. It often took a reorganization before voice/data integration occurred. The first and most obvious step toward integration is to share the leased facilities between voice and data. The second, less obvious step is to use a common network.

Fractional T1 Tie Trunks. In the typical enterprise voice network, *N* PBXs are interconnected by groups of tie trunks. After the discussion in Section 12.2, any group of 6 – 24 tie trunks would be implemented over a leased T1 line. Most commercial PBXs have port cards that interface to internal voice channels on the fabric side of the card and directly to T1 carrier on the trunk side of the card.

If a particular group of tie trunks needs 14 trunks to handle the voice traffic, the other 10 channels in the T1 can be used for data. So, a fractional T1 on each end of this line would partition the T1 frame into 14 DS0 lines for the PBXs at each node, and the remaining 640 Kbps would be connected to the WAN gateway at each node.

Voice over Data WANs. WANs are no longer dumb networks in which LAN packets are transmitted over dumb point-to-point *bridge links* that connect LAN gateways to each other. Today, WANs are complex networks in which LAN packets are encapsulated in WAN packets of different types — notably IP, SMDS, frame relay, or ATM. The WAN's routers or switches packet-switch these WAN packets across the nation or even the world. Integration occurs by packetizing an enterprise's interlocation voice channels and sending the packetized voice over the same WAN. Voice over ATM and voice over IP are discussed in Chapter 19.

Similar to how enterprise voice networks are implemented as private networks or virtual private networks, a packet-switched WAN can be implemented as a private network, where the enterprise owns the routers and leases the lines, or as a shared WAN service provided by competing carriers. This kind of integration is motivated entirely by price, which, in this case, is almost completely independent of cost. When the market adjusts the price of leased channels, so that they are more consistent with their cost, this kind of integration may come to be seen historically as yet another fad (Section 19.5).

12.5 FACILITY SWITCHING

The thousands of channels that converge on a CO, or a node in a large private network, are assigned to various groups of trunks for switched service or groups of private lines. This procedure, called provisioning, is affected by changes in calling patterns, new housing construction, or reconfigurations of enterprises. *Facility switching* is an automated procedure for provisioning channels within transmission facilities and is such an important part of how the phone companies manage their resources that this management is collectively called OAM&P — for operation, administration, maintenance, and provisioning. The differences among facility switching, circuit switching, and packet switching will be discussed carefully in Chapter 17.

Next we discuss provisioning and two types of equipment for facility switching: the digital access crossconnect system (DACS) and the add-drop multiplexor (ADM). Section 12.5.4 presents an extended example. The last two subsections discuss *grooming,* a side effect and optimization of provisioning, and *consolidation,* an application of facility switching that is related to, but slightly different from, provisioning.

12.5.1 Provisioning

Section 3.4.3 describes the telephone company's procedure whenever a new subscriber's telephone is installed. Three different types of craftsmen are employed: an *installer* (when appropriate) connects a house-wiring pair at the telephone and at the protector block, a lineman jumpers the assigned loop segments at the pedestal and the distribution point, and a frameman jumpers the loop at the main distribution frame inside the CO. This procedure requires hundreds of personnel over a telco's region, and telephone companies have wanted to automate it for many years.

Consider replacing the main frame by some kind of automatic switching fabric. With perhaps 100,000 pairs on the MDF, this fabric would be extremely large and would have little concentration. Because many people keep the same telephone service for many years, the fabric's connections would have a very long average holding time. So, while the telephone company sees a lot of activity by framemen on its MDF, the average activity per line is quite low. The "automated main frame" has never proven economical.

The characteristics and economics of trunk-side channels are significantly different from the characteristics and economics of line-side channels. Because the number of channels on the trunk side is much smaller than on the line side, an automated provisioning fabric would have reasonable size. Because most of the trunk-side channels are multiplexed within carrier systems, a direct interface to the carrier would be an important feature and would further reduce the physical size. But, most relevant, trunk-side channels are arranged and rearranged much more frequently than line-side channels.

Trunk-side provisioning is changed frequently by the enterprises that lease trunks and by the telephone company itself.

- On the line side, leased lines are relatively short and inexpensive. So, underutilized lines are not a high priority for an enterprise's telecom manager and they tend to be leased almost semi-permanently. On the trunk side, however, leased trunks are longer and much more expensive. So, enterprise private networks receive considerable attention by professionals employed by the enterprise and most enterprises frequently change the number of private trunks under lease.
- The same concerns about over- or underutilization of private trunks in an EPN are felt, even more, by the telco itself about its own public switched network. Switched trunks also have high variation in traffic load because of different rates of traffic growth over different trunk groups, periodic fluctuations in calling behaviors, and increased susceptibility to hardware failures. Population shifts and gradual changes in calling patterns cause changes in the size of trunk groups. Time-of-day variations in calling patterns can be handled by shifting channels from one trunk group to another. It's even possible to automate the redeployment of channels based on measured changes in usage.

Enterprise network managers want this kind of fine control over the lines that they lease from the carriers. This high level of activity on the trunk side — frequently reassigning trunks from private to switched and reassigning trunk circuits and fabric appearances from a trunk in one group to a trunk in another — is all exacerbated because of "grooming" (Section 12.5.5).

12.5.2 Digital Access Crossconnect Systems

With the great penetration of digital carrier, especially on the trunk side, a digital implementation of a system that automates trunk-side crossconnect was destined to be very popular. Such a system is called a digital access crossconnect system (DACS). A DACS automates and simplifies the provisioning of both private and switched trunks. The following discussion assumes familiarity with the ADH and SDH hierarchies. These hierarchies are summarized in Table 12.1.

TABLE 12.1. The ADH and SDH hierarchies

Asynchronous digital hierarchy	Synchronous digital hierarchy
24 DS0s in a DS1 at 1.5 Mbps	1 DS3 in an OC1 at 52 Mbps
4 DS1s in a DS2 at 6 Mbps	3 OC1s in an OC3 at 155 Mbps
7 DS2s in a DS3 at 45 Mbps	4 OC3s in an OC12 at 620 Mbps
6 DS3s in a DS4 at 270 Mbps	4 OC12s in an OC48 at 2.5 Gbps

DACSs are named with two digits, from the DSx hierarchy. The first digit in the name represents the channel level to which the particular DACS can interface. The second digit in the name represents the level of channels that are internally crossconnected in the DACS. While you'll probably never see a real one, a four-port DACS 2/1 is illustrated in Figure 12.8. It provides a physical interface for four two-way DS2 channel-ends, or two two-way DS2 through-channels, shown on the four edges of the box in Figure 12.8. Because each DS2 stream carries four DS1 channels, the system provides a logical crossconnect for 16 two-way DS1 channel-ends, or eight two-way DS1 through-channels. This particular crossconnect pattern was established by typing, at the DACS's control terminal, a set of commands analogous to the column of eight commands tabulated at the right of Figure 12.8.

A DACS 3/3 provides an automated crossconnect for DS3 streams without demultiplexing any of the streams into subchannels. An extremely popular DACS is the DACS 3/1. An N-port DACS 3/1 provides a physical interface for N two-way DS3 channel-ends, or $N/2$ two-way DS3 through-channels. Since each DS3 stream carries 28 DS1 channels, the system provides a logical crossconnect for $N \times 28$ two-way DS1 channel-ends, or $N \times 14$ two-way DS1 through-channels. An N-port DACS 4/3 provides a physical interface for N two-way DS4 channel-ends, or $N/2$ two-way DS4 through-channels. Since each DS4 stream carries six DS3 channels, the system provides a logical crossconnect for $N \times 6$ two-way DS3 channel-ends, or $N \times 3$ two-way DS3 through-channels.

Is a DACS a switching fabric? A DACS switches carriers and subcarriers, but it doesn't switch by the call. A DACS moves channels from one trunk group to another, independently of what calls are going over those channels. Circuit switching is compared thoroughly against packet switching and facility switching in Chapter 17.

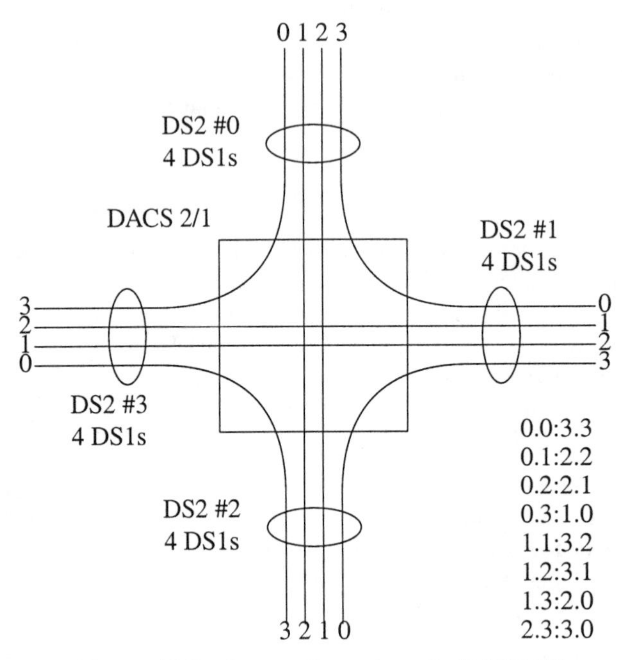

Figure 12.8. One configuration of a DACS 2/1.

Section 16.1 describes the structure of the old Bell System and the internal organization approximately common to Bell Laboratories and Western Electric. This internal organizational structure even affected architecture. The designers of switching systems were separated from the designers of transmission switching systems at a high level of Bell Labs' chain of command. The manufacturers of switching systems were separated from the manufacturers of transmission switching systems at a high level of WECo's chain of command. The DACS was designed, manufactured, marketed, and installed as a transmission product and not as a switching product. Circuit switching and facility switching had little in common, and people who understood one typically knew very little about the other.

A DACS has many different functions: trunk provisioning, private line assignment, add/drop, consolidation, and protection switching. These functions are described and illustrated in the following subsections.

12.5.3 Add/Drop Multiplexing

Transmission facilities typically have multiple hops. Consider an IXC's transmission facility between Cleveland and Cincinnati, which passes through a node in Columbus.

- If the Columbus node is used as a tandem switch, then a call from Cleveland to Cincinnati is switched at Cleveland to an outgoing Columbus trunk and multiplexed onto the facility. The call is demultiplexed off the facility in Columbus, switched there from the Cleveland incoming trunk to a Cincinnati outgoing trunk, and multiplexed back onto the facility again. The call is demultiplexed off the facility in Cincinnati and switched there to the appropriate local channel.

- If the facility carries direct channels between Cleveland and Cincinnati, they pass through Columbus, but wouldn't be demultiplexed or switched there. Trunk groups between Cleveland and Columbus, between Cleveland and Cincinnati, and between Columbus and Cincinnati would each use channels within this same facility. Some of the trunks out of Cleveland continue through to Cincinnati, but others *drop* off in Columbus. Some of the trunks into Cincinnati came all the way from Cleveland, but others were *added* in Columbus.

At every intermediate node along some multihop transmission facility, most of the facility's channels proceed through the node. But, some of the facility's channels drop off at the node and some of the node's channels must be added onto the facility.

This add/drop multiplexing can be accomplished by two separate pieces of equipment: a DACS and a mux. The node's DACS-X/Y aggregates all the DS-Y channels from the facility that should be added/dropped locally, into DS-X facilities that are provisioned locally. Each of these local DS-X facilities must then be muldemed to whatever channel sizes are needed locally — depending on whatever channel sizes are leased locally or can be interfaced directly to the node's switching fabric. If the facility carries a hierarchy of transmission levels, then drop/add can occur at any level of the transmission hierarchy without demultiplexing the entire facility down to its DS0 channels. An entire DS1 might be dropped out of a DS3 channel and another one added in its place, or an entire DS3 might be dropped out of an OC12 channel and another one added in its place.

For example, consider Figure 12.8. Suppose the DACS 2/1 is located in node #2 and DS2 #2 is a local facility. Then, the DACS has aggregated DS1s #0.1 and #0.2 in DS2 #0, DS1 #1.3 in DS2 #1, and DS1 #3.0 in DS2 #3 for local add/drop. If node #2 requires DS1 access to these channels, then DS2 #2 would have to be demultiplexed by an M12. If node #2 requires DS0 access to these channels, then these DS1s would have to be further demultiplexed by an M01. If node #2 requires DS0 access to 0.1 and 3.0 and DS1 access to 0.2 and 1.3, then only two of the four DS1s in DS2 #2 would have to be demultiplexed by M01s.

These two functions, DACS and mux, can be integrated into a single piece of equipment, called an add/drop multiplexor (ADM). Because of the relative simplicity of the way DS0 channels and DS1 "virtual tributaries" are packed into Sonet (SDH) frames, a Sonet ADM is considerably less expensive than the corresponding equipment for the ADH format. This simplicity, and the corresponding lower price of an ADM, probably had more to do with Sonet's success than any other factor.

After presenting an extended example in the next subsection, we will describe the subtle complexity of assigning channels to facilities that have intermediate add/drop multiplexors. Channel assignments can be optimized to reduce the complexity, and thereby the cost, of the intermediate DACSs and ADMs. This subtle design complexity almost vanishes if the facilities use the SDH.

12.5.4 Extended DACS/ADM Example

The usefulness of DACSs and ADMs is illustrated by an example. Consider an LEC's toll region, which has a toll hub, three local switching offices, and some IXC's point of presence. As shown in the inset at the upper center of Figure 12.9, logically, the five switching offices are assumed to be fully interconnected by direct trunking.

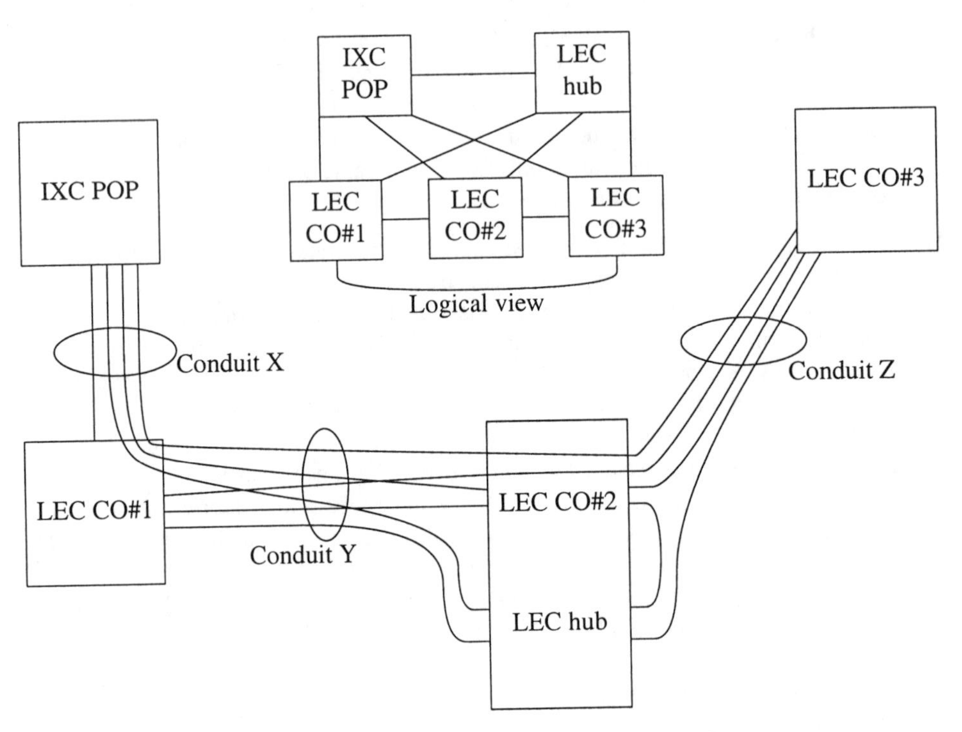

Figure 12.9. Trunks through a physical right-of-way.

The remainder of Figure 12.9 shows the physical connectivity. The figure shows the relative geographical location of four buildings that house the five offices. Assume the toll office resides in the same building as one of the local exchanges, but all the other offices are in separate buildings. Consistent with the real world, assume that the physical right-of-way conduits have the form of a minimum spanning tree. For this example, the conduit tree has three segments: conduit X between the POP and CO#1, conduit Y between CO#1 and the building that houses CO#2 and the toll point, and conduit Z between CO#3 and the building that houses CO#2 and the toll point.

Each line in Figure 12.9 represents a trunk group. The 10 trunk groups, shown logically in the inset at the upper center of Figure 12.9, are shown physically on the geographical layout of the buildings and their interconnecting conduits. Conduit X carries only four trunk groups — those that connect the POP to each of the other four offices in the region. Conduit Z also carries only four trunk groups — those that connect CO#3 to each of the other four offices in the region. Conduit Y carries six trunk groups — those that connect the POP with CO#2 and with the toll point and with CO#3 and those that connect CO#1 with CO#2 and with the toll point and with CO#3. Note that three of the 10 trunk groups pass through one intermediate building along their route, and the POP-to-CO#3 trunk group passes through two intermediate buildings.

For simplicity, assume each of the 10 trunk groups has the same size (240 trunks). Let 160 trunks in each group terminate on fabrics because they are used for switched calls and let 80 trunks in each group be leased. While 240 trunks can be carried by 10 T1 lines, the telco would probably not do this because there is no spare capacity and because they would probably use a higher level of multiplexing. If this is an active and high-growth region, the telco would probably use two DS3 systems in conduit X and in conduit Z, with a net capacity of $2 \times 28 \times 24 = 1344$ channels, to carry the $4 \times 240 = 960$ active channels in the four trunk groups in each conduit. They would probably use three DS3 systems in conduit Y, with a net capacity of $3 \times 28 \times 24 = 2016$ channels, to carry the $6 \times 240 = 1440$ active channels in the six trunk groups in this conduit.

While the telco only needs to equip 10 of the 14 DS2s to provide for the 960 channels in conduits X and Z and only 15 of the 21 DS2s to provide for the 1440 channels in conduit Y, such minimal provisioning has no immediate spare capacity. They might equip 12, 18, and 12 DS2s — in conduits X, Y, and Z, respectively — to allow 20% spare capacity and to simplify the grooming (defined in the following subsection). With 4, 6, and 4 trunk groups in the respective conduits, it is easy to assign 3 DS2s, with a net capacity of $3 \times 4 \times 24 = 288$ channels, to each trunk group (and 48 spare channels are equipped at the DS2 level in each group). Thus, each line in Figure 12.9 represents three DS2s, for each of the 10 groups. Assume that two of the DS2s in each group carry the 160 switched trunks and that the third DS2 in each group carries the 80 leased trunks.

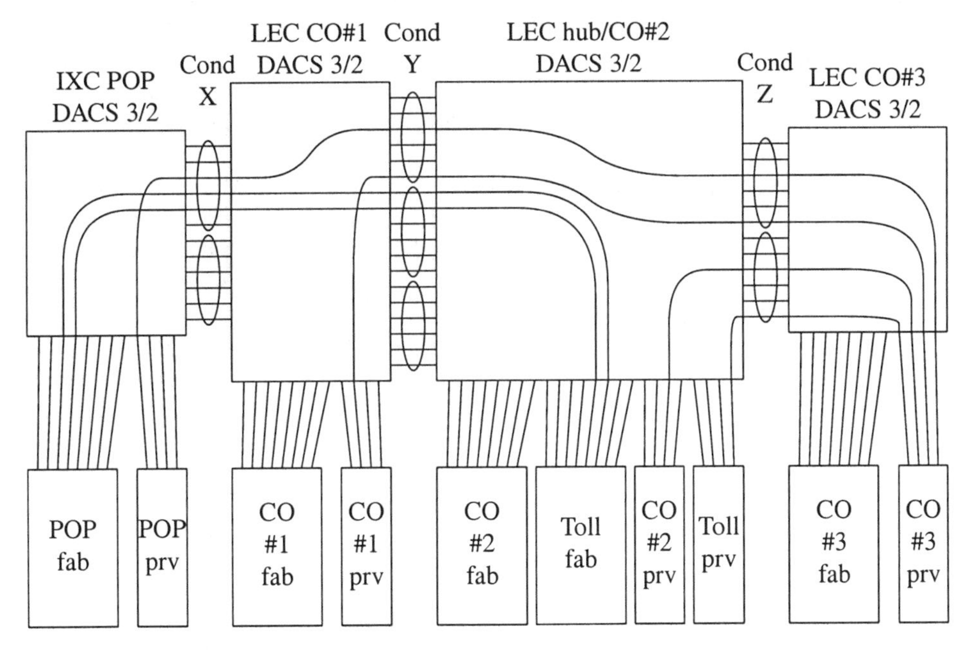

Figure 12.10. Partial assignment for DACSs in example.

Figure 12.10 illustrates this example logically, but in greater detail. Each of the four buildings has a DACS 3/2, shown across the top of the figure, to provide for the building's facility switching. In the conduits between these buildings, the ellipses represent the DS3s and the lines represent the DS2s (remember that only six of the DS2s in each DS3 are currently used). While each line in the previous figure represents three DS2s, in this figure each line represents one DS2. Each DACS terminates DS3 streams in the horizontal direction, internally crossconnects DS2 streams, and adds and drops DS2 streams out the bottom for the equipment in the local building.

The "prv" box, in each of the five offices in the four buildings, represents the four DS2 muldems for the $4 \times 80 = 320$ private trunks that terminate on this particular office. The "fab" box, in each of the five offices in the four buildings, represents the eight DS2 muldems for the $8 \times 80 = 640$ switched trunks that terminate on the fabric in this office. With 10 trunk groups in the region and 3 DS2s per trunk group, there are 30 DS2 routes in this region. To illustrate facility switching, the crossconnect assignments for six of these 30 DS2 routes are shown in Figure 12.10. The rest is left as an exercise (which is strongly recommended).

Consider the four groups of private trunks, at one DS2 per group of 80 trunks, that terminate on CO#3. The DS2 carrying private trunks between CO#3 and the toll point is the first (leftmost) line on "CO#3 prv" and the fourth (rightmost) line on "Toll prv." It crossconnects to the left side of the CO#3 DACS and to the right side of the Toll/CO#2 DACS at the first (bottom) DS2 within the first (bottom) DS3 in conduit Z. The DS2 carrying private trunks between CO#3 and CO#2 is the second line on "CO#3 prv" and the third line on "CO#2 prv." It crossconnects left in the CO#3 DACS and right in the toll/CO#2 DACS to the fourth DS2 within the first DS3 in conduit Z.

The DS2 carrying private trunks between CO#3 and CO#1 is shown on "CO#3 prv" as the third line and on "CO#1 prv" as the second line; it crossconnects left in the CO#3 DACS and right in the toll/CO#2 DACS to the first DS2 within the second DS3 in conduit Z and it crossconnects left in the Toll/CO#2 DACS and right in the CO#1 DACS to the first DS2 within the third DS3 in conduit Y. The DS2 carrying private trunks between CO#3 and the POP is shown on "CO#3 prv" as the fourth line and on "POP prv" as the first line — it crossconnects left in the CO#3 DACS and right in the toll/CO#2 DACS to the fourth DS2 within the second DS3 in conduit Z, it crossconnects left in the toll/CO#2 DACS and right in the CO#1 DACS to the fourth DS2 within the third DS3 in conduit Y, and it crossconnects left in the CO#1 DACS and right in the POP DACS to the fourth DS2 within the second DS3 in conduit X.

Now, consider the group of switched trunks between the toll point and the POP, with two DS2s for this group of 160 trunks. The two DS2s carrying these switched circuits are shown on "Toll fab" as the fifth and sixth lines and on "POP fab" as the third and fourth lines; they crossconnect to the left side of the toll/CO#2 DACS and to the right side of the CO#1 DACS at the fifth and sixth DS2s within the second DS3 in conduit Y and they crossconnect left in the CO#1 DACS and right in the POP DACS at the second and third DS2s within second DS3 in conduit X.

12.5.5 Discussion

The previous example surprised you, didn't it? Most readers have to read it several times. The general discussion in the first three subsections was so deceptively simple that the example caught you off guard. And that was not a hard example. In fact, it was a relatively simple example. Like so much of this science of switching, the devil's in the details. There are four important points.

1. Appreciate the role played by the DACS/ADM equipment. While the final result in the example above is complex, the procedure that leads to this final result is even more complex. Without automatic equipment, like a DACS/ADM, the crossconnects in Figure 12.10 would have to be implemented manually, using patch cords at a jack-panel similar to an operator's switchboard or using jumper

wires at a frame like the MDF. With a modern DACS, the crossconnects are implemented inside the DACS when the craftsman types a primitive command at a control console.

2. As complex as this previous example may seem, it is practically trivial compared to real-life problems. This example has only four nodes, it has a simple linear minimum spanning tree, and it has been extremely simplified by the assumption that all the trunk groups have the same size. Furthermore, it would be rare to allocate channels in DS2 "chunks" — the "finer" DS1 allocation is more typical.

3. Appreciate the hidden complexity that we won't discuss. Even with the automation that DACSs bring to the provisioning procedure, in the real world there is also considerable bookkeeping and coordination on all ends of all channels. Furthermore, in the real world, since all this detail is bound to have human error, the provisioning assignment has to be debugged. Centralizing the provisioning function, with remote access to the DACSs in all nodes, simplifies the provisioning procedure.

4. The example is even more deceptively simple because all the channels were neatly groomed. 240 channels per trunk group fit neatly into three DS2s and six DS2s fit neatly into one DS3. All we had to do in the previous subsection was analyze a problem whose channel provisioning assignment is artificially obvious. In the real world, someone has to synthesize the channel provisioning assignment, specify the transmission facilities, and specify the kind of facility switch in each node.

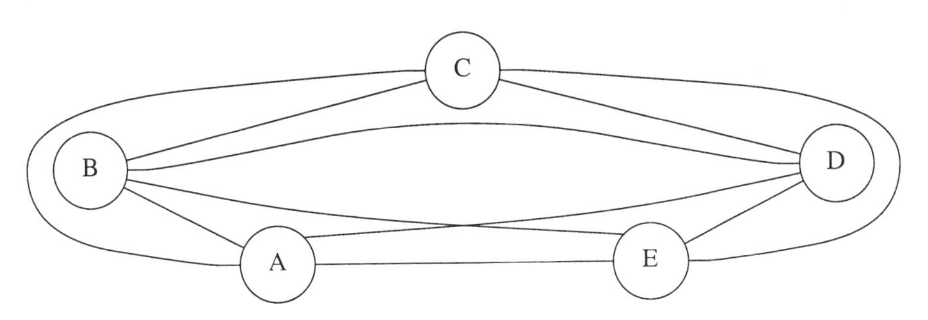

Figure 12.11. Example ring with five nodes.

12.5.6 Grooming

In the example from Section 12.5.4, each of the four DACSs performed two functions. They implemented the crossconnection of channel segments, among the various conduits, so that multisegment trunk groups could be connected together. Furthermore, they provided the implementation by which the channels that terminate in a given building are dropped out of the transmission facilities coming into the building and the channels that originate in a given building are added to the transmission facilities that leave the building. Since these two functions don't always optimize simultaneously, the overall optimization, called "grooming," can be elusive. The impact of good grooming is illustrated by another example.

Consider five switching nodes arranged in a ring, as shown in Figure 12.11. Let each office have two-way direct trunking to every other office. So, each node shares a trunk group with its nearest neighboring office counterclockwise, its nearest neighboring office clockwise, its next nearest neighboring office counterclockwise, and its next nearest neighboring office clockwise. Node C, for example, has direct trunks to B and to A at its counterclockwise port and direct trunks to D and to E at its clockwise port. Also, the trunks between B and D pass through C. So, the conduit between B and C carries segments of three trunk groups: trunks between B and C, trunks between A and C, and trunks between B and D.

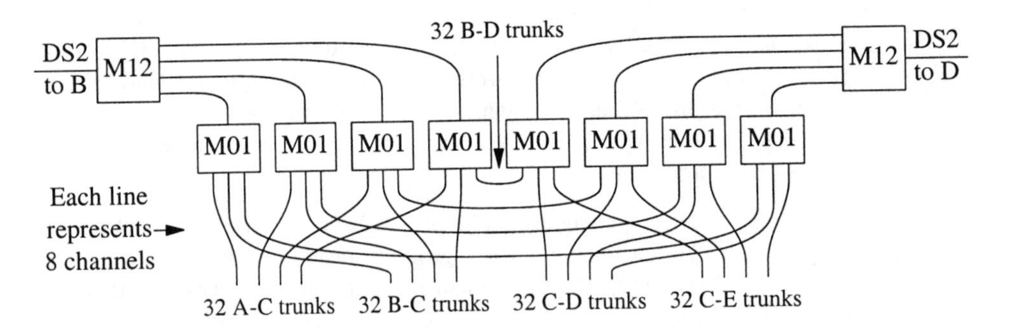

Figure 12.12. Illustrating poor grooming at node C.

Suppose each trunk group has 32 channels. Because each conduit carries three groups, each conduit needs 96 channels, organized as four T1s embedded in a T2. But how should the channels be groomed to the T1s? Examine two assignments.

Intuition might be obtained from *slipped multipling,* described in Section 6.4.4. Since there are four T1s between nodes C and B, node C might split the 32 channels in its B-to-C trunk group uniformly over the four T1s, eight channels in each — similarly for the 32 channels in the A-to-C trunk group and similarly for the 32 channels in the B-D trunk group. A similar uniform distribution could be used in the four T1s between C and D. Figure 12.12 illustrates how all the channels in node C would be multiplexed, demultiplexed, and crossconnected, using conventional T2 and T1 muldems.

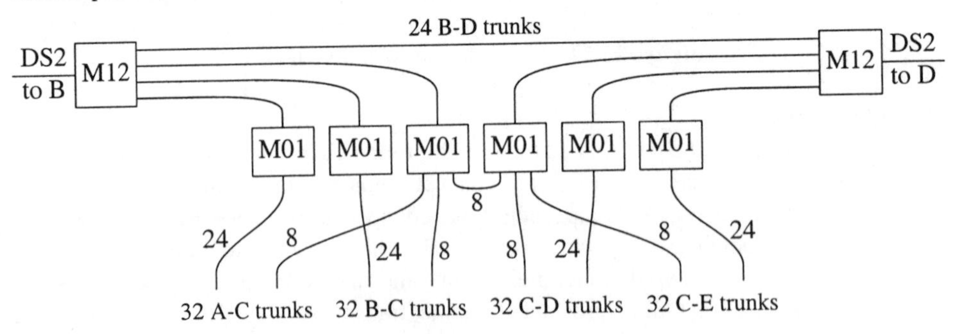

Figure 12.13. Illustrating good grooming at node C.

Instead of distributing channels uniformly, the other extreme might be to lump as many similar channels into the same T1 as possible. Figure 12.13 illustrates how all the channels in node C would be multiplexed, demultiplexed, and crossconnected, using conventional T2 and T1 muldems. The noticeable difference is that the T1 carrying 24 of the B-to-D channels doesn't have to be demultiplexed inside node C. For this grooming of the channels, node C has two fewer T1 muldems than the uniform grooming. This grooming saves 10 T1 muldems over the total network.

The telephone companies and many large enterprises have great need for people who are skilled at this kind of network design.

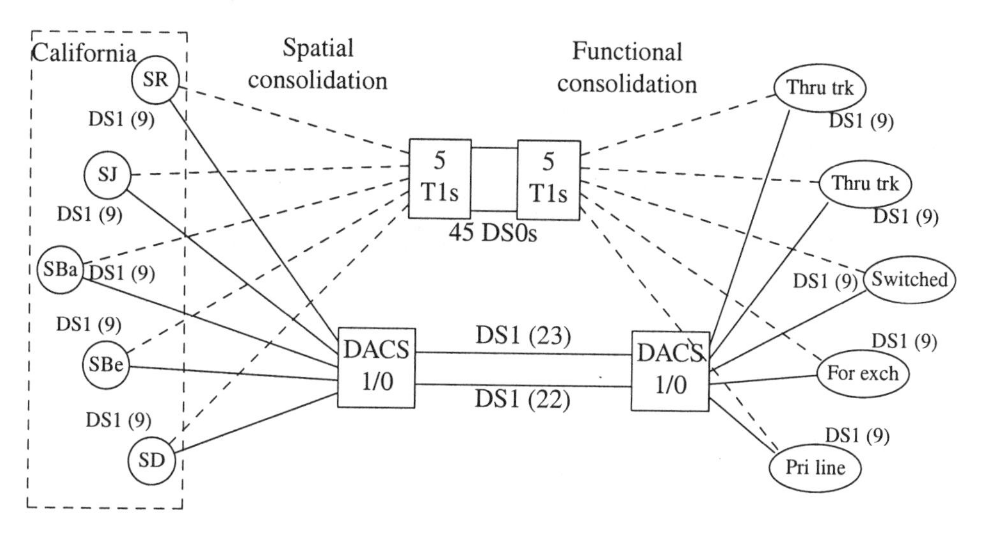

Figure 12.14. Spatial and functional consolidation.

12.5.7 Consolidation

The economic benefits of a DACS include conserving multiplexing equipment and optimizing the utilization of high-rate streams. These benefits are illustrated simply in Figure 12.14. The left side of Figure 12.14 shows an enterprise with locations in San Diego, Santa Barbara, San Bernardino, San Jose, and Santa Rosa. Each spatially separate building has nine channels that connect to equipment in Chicago, on the right side of the figure. If we assume that a DS1 is cheaper than nine DS0s (which would certainly be true between California and Illinois), there would be a T1 in each building, but, in each case, only nine of the 24 channels would be equipped. The right side of Figure 12.14 shows an enterprise network node in Chicago that terminates DS1 channels that provide five different functions. Assume that each separate function terminates nine channels, so 45 channels connect between various locations in California and various functional equipment in Chicago. Figure 12.14 shows two implementations for the cross-country route.

In the dashed implementation, across the top of Figure 12.14, the five T1 lines from California and the five T1 lines from Chicago meet somewhere in between, assume in Chicago. The five T1s from the west and the five from the east are demultiplexed, and the DS0s are crossconnected to provide the required connectivity. This obvious implementation requires five cross-country T1s. In the solid implementation, across the bottom of Figure 12.14, the five T1 lines from California converge on a DACS 1/0 in San Bernardino and the five T1 lines from Chicago converge on a DACS 1/0 in Chicago. At both DACSs, the five sparse T1s are crossconnected to two dense T1s, using 45 out of 48 possible channels. This less obvious implementation requires only two cross-country T1s. While the second implementation needs two DACS 1/0s, it saves 10 T1 muxes and a manual DS0 crossconnect panel. This process of compressing sparse local facilities into dense long-haul facilities is called consolidation. Two kinds of consolidation are apparent from Figure 12.14: spatial consolidation and functional consolidation. Often the economic benefit from using DACSs for consolidation can more than pay the price of the DACS. If it does, the other DACS functions are free.

12.5.8 Protection Switching

Protection switching is not performed by a DACS, but this is a good place to mention it. Transmission facilities typically have spare physical channels so that, if a wire breaks, the data stream can be moved to another wire. Muldems typically have spare transmitters and receivers so that, if a device breaks, another can be used in its place. Protection switching is the analog switching that automatically rearranges the assignments of physical channels to transmitters and receivers.

12.6 LONG DISTANCE COMPETITION

After revisiting the T1 tariff, this section describes the introduction of long distance competition. The second and third subsections discuss the emergence of MCI and Sprint. Then, the next two subsections describe the evolution to *equal access.* The last subsection describes the present system in the United States where every subscriber designates a *default carrier.*

12.6.1 The T1 Tariff Loophole

The T1 tariff, described earlier in this chapter, was extremely beneficial to every kind of enterprise. But it was extremely controversial and the Bell System opposed it vigorously. Their position was that only voice channels should be leased and tariffed. Ma Bell argued that bundling and economy of scale were the internal benefits of doing a mass business and that the bundles themselves should not go on the market. While Ma Bell, and the people who regulated her, seldom agreed about anything, they did agree on this one. By tariffing leased voice channels so that price was much higher than cost, the regulators found a lucrative way to force businesses to subsidize residential POTS. While this was almost a perfect scenario for the social engineers, businesses demanded the quantity discount for leased lines and the regulators were finally forced to accede.

After T1 was finally tariffed, the typical price for a leased T1 line was equivalent to the price for about six leased voice-grade channels. From the customers' perspective, for every set of 24 channels, about 18 of them (or three-fourths) were free. But, leased voice channels were so overpriced that even this quantity discount price was well above true cost. However, whether AT&T's attorneys never saw it, or they did see it but were prevented from fixing it, the T1 tariff had a fatal loophole (from AT&T's view) in it.

Suppose the phone company leases out a voice channel for $400 per month and leases out a T1 line over the same span for $2400 per month. You could start a small company, lease a T1 line from the telco for $2400 per month, and sublet its component voice channels for only $300 per month. You make a profit as long as you can sublet more than eight (one-third) of the component channels. You make a 200% profit if you can sublet all 24 channels. You have no up-front capital expenses because you simply lease lines from the phone company and they must lease to you. Your only voice-channel competitor is the phone company itself and they are prohibited by the tariff from lowering their price. Only in the United States could a tariff be written that allows someone to lease a quantity discount channel and then sublet its component channels in direct competition with the company that leased you the original discount channel.

12.6.2 MCI

MCI started as a small carrier that leased channels from its privately owned microwave carrier system. MCI founder, William McGowan, figured that MCI could expand beyond the region of its own carrier system by leasing T1 lines from AT&T and subletting the component voice channels. Because AT&T's tariffed rates were inflated and fixed by regulation, MCI could easily underprice Ma Bell for leased trunks. If you think Ma Bell was upset over the competition created by Carterfone, she was extremely upset by MCI's potential for cream skimming. AT&T appealed to the FCC, but the FCC's controversial Execunet decision ruled against AT&T's position. MCI took a significant share of the market for leased enterprise private networks.

Then, MCI invested in switching equipment and built a nationwide enterprise private network, using a lot of T1 lines leased from AT&T. MCI could now offer commercial switched long distance telephone service to public subscribers, enterprise and residential. The FCC decided that a little competition in the long distance market might be a good thing, they allowed MCI to continue, and they prevented AT&T from lowering its long distance rates — AT&T was regulated and MCI was not. Another advantage MCI had over AT&T was that MCI could limit its market to large cities and enterprises, where most of the nation's traffic is generated and where switching and transmission facilities are more efficiently utilized. AT&T, by mandate, had to serve the entire nation.

While long distance competition is probably a good thing in the long run for the United States, this author believes that this regulation was an unfair way to initiate it. MCI grew rapidly and quickly constructed its own transmission facilities anticipating that the regulators might change the rules. MCI now has its own physical network and a large market share.

12.6.3 Sprint

Even more bizarre than the MCI story, the creation of Sprint is reminiscent of Strowger's story. It begins with an application of time-domain reflectometry (TDR). Consider burying a long coaxial cable along a railroad track and transmitting an electronic pulse down the cable. Enough of the signal's evanescent wave exists outside the cable's shield and a train has so much mass of iron that some of the signal's intensity reflects from the position of the train back to the source. By measuring the time between the pulse transmission and receiving the reflected signal back, the location of the train can be computed. The Southern Pacific Railroad buried coax near all its tracks so they could monitor the exact location of trains for improved safety and scheduling. Since the Southern Pacific Railroad owned the right-of-way, they were able to lay a lot of coax at relatively low cost.

With many office locations all over the southwestern United States, the Southern Pacific Railroad had a large enterprise private network and a large monthly telephone bill to lease the lines. Noting that all their offices were located near their own railroad tracks, the company abandoned its leased-line network and multiplexed interoffice telephony with the TDR on their own coaxial cables. They built their own enterprise network from an infrastructure that they already owned and that was already in place — all they had to do was add the multiplexing equipment. When some of the other companies in the southwest United States found out about this network, they inquired about the price of leasing channels. When the Southern Pacific Railroad decided to get into the business of leasing channels on its coax, they formed a subsidiary. Using the initials of Southern Pacific Railroad for the first three letters, they called this subsidiary S-P-R-int.

Like MCI, Sprint expanded to nationwide service by leasing T1 lines from AT&T, and then Sprint also moved on to switched long distance service. The subsidiary was purchased from the parent railroad company and has been bought and sold several times. At the time that I'm writing this book, the 1996 Telecom Bill (Section 16.6.3) hasn't yet allowed the RBOCs into the long distance business. But before one of them can do this by buying Sprint, MCI (which was bought by WorldCom) is trying to buy Sprint.

12.6.4 Connecting to the Bell System

In the 1970s, the new long distance service providers, like MCI and Sprint, came to be called other common carriers (OCCs). With divestiture in 1984, AT&T's long distance business became just as "other" as MCI and Sprint to the new phone companies and the term "OCC" disappeared. All long distance service providers came to be called interexchange carriers (IXCs). But that's ahead of our story now; we'll cover this in Chapter 16. So, during the time before divestiture, all the OCCs had to interconnect to AT&T for three different reasons:

- Since OCCs did not provide local telephone service, their networks had to be accessible to Bell System customers;
- If an OCC didn't extend into a certain region, they might complete one of their calls into this region by dialing into AT&T's long distance network;
- If an OCC didn't own its own infrastructure, it leased T1 lines from AT&T.

One can imagine that MCI, Sprint, and the other OCCs didn't get much cooperation from Ma Bell. While an OCC would have preferred to connect into a CO on the trunk side (like they do today), AT&T wouldn't do anything that wasn't tariffed and the only tariffed service was on the line side. And AT&T was in no hurry to apply for a tariff. The only implementation available to any OCC was to look like a PBX. This meant a line-side connection to any CO, and it meant that the OCC had to be called by a seven-digit telephone number, and then the subscriber could call the desired party.

Dialing through a switching office into a PBX didn't work very well because the dialed digits weren't repeated (Chapters 3 and 4). While the telcos offered DID trunks (Section 10.5.2), they only repeated short extension numbers, not 10-digit long distance numbers. So, OCC access was practically limited to people who had Touch Tone telephones. Furthermore, as we'll see from the extended example in the next subsection, the subscriber's connection to an OCC tended to be quite circuitous and long. Without amplifiers on the line side, audio connections were attenuated.

The OCCs complained about having what they called equal access. There was little public support because the public didn't understand what the problem was. The OCCs were viewed as providing inferior quality and AT&T encouraged this perspective. The war was fought with TV ads.

AT&T ran an ad featuring a middle-aged couple trying to call their son Joey, who just went off to college. As new subscribers of an OCC, the ad showed them confused and annoyed about dialing two different telephone numbers, the first number to call the OCC through the telco and the second number to call Joey through the OCC. Then, MCI hired the very same actors to make an ad in which the same middle-aged couple, having switched back to AT&T, are aghast at receiving their first telephone bill. After this, it was rumored that AT&T would hire the very same actors back again and make a third ad. This time, having switched back to an OCC again, to save a few pennies, they would call Joey and have to shout into the phone, saying, "What did you say, Joey? Speak louder. We can't hear you." It was rumored that AT&T decided against this third ad because they anticipated a fourth ad, from MCI again, in which the attenuation problem is solved because MCI finally is granted equal access. Let's understand the architectural significance of equal access.

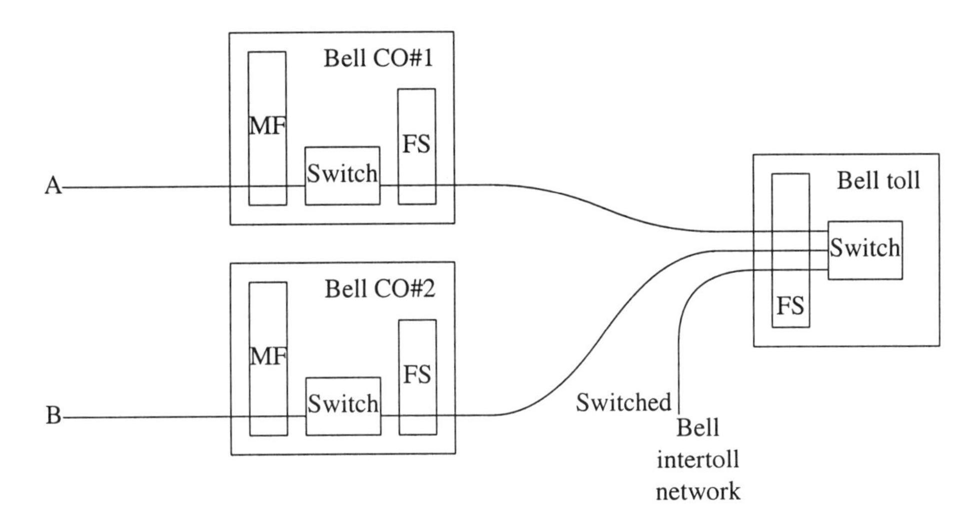

Figure 12.15. A and B make Bell System long distance calls.

12.6.5 Equal Access

Figure 12.15 shows two AT&T local subscribers, A and B, each served by separate Bell System switching offices. For simplicity, the two COs connect to a common toll office, also assumed to be part of the Bell System. A's line and B's line are wired through the main frames in their respective offices to appearances on the switch in each office. When A or B make a long distance call, they dial the called party's telephone number into their respective switch, the switch connects them to a toll trunk and outpulses the called party's telephone number to the toll office. At the toll office, the switch receives the called party's telephone number and continues placing the call over the Bell System's long distance network over a switched intertoll trunk.

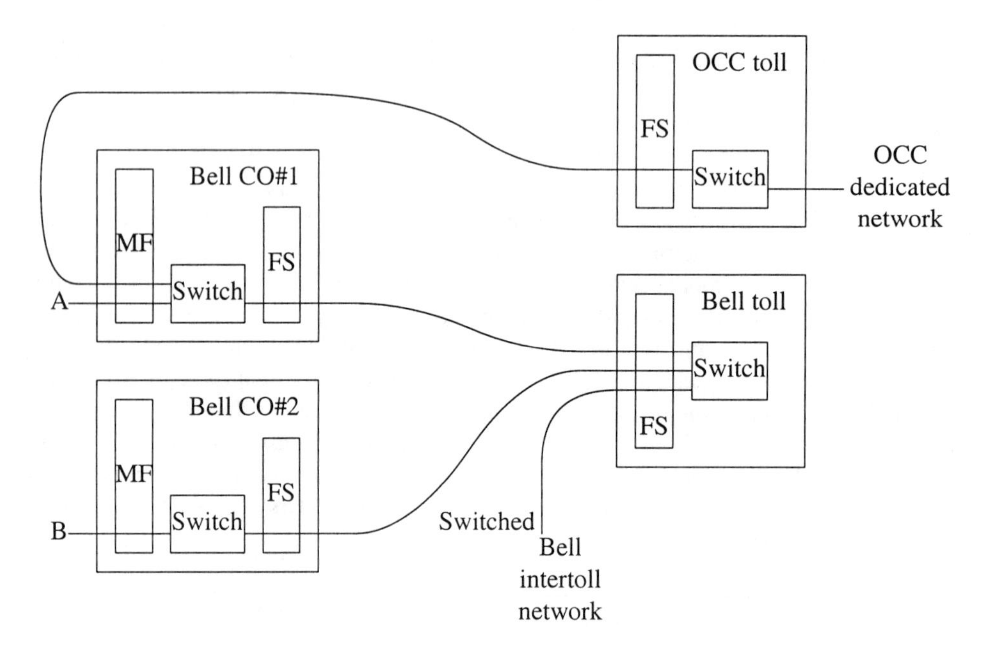

Figure 12.16. A makes an OCC long distance call.

In Figure 12.16, an OCC comes into the region, builds its own gateway office, and locates it in the serving area of this first CO. But, the CO-to-toll trunks to the OCC appear on the line side of the CO, while the CO-to-toll trunks to AT&T's network appear on the trunk side of the CO. If A makes the same call as before but uses the OCC instead, A first dials the seven-digit telephone number of the OCC. Since the OCC looks like a PBX, the first CO connects A to a hunt group on the line side of the CO.

The trunk seizure is detected by the OCC's gateway office and this office returns a second dial tone. Now A dials the called party's telephone number, through his home CO, into the OCC's gateway office. Here's where dial pulsing didn't always work and A was practically required to own a Touch Tone telephone. This gateway receives the called party's telephone number and places the call through the OCC's own long distance network over a switched OCC intertoll trunk.

Now, consider the case where B makes the same call, but uses the OCC, in the configuration of Figure 12.16. B first dials the seven-digit telephone number of the OCC into his home CO, the second CO in Figure 12.16. Without direct trunking between the two COs in Figure 12.16, CO#2 connects B to a toll trunk to the AT&T toll office, and then the toll point connects the call to a trunk that terminates on the CO#1. This first CO connects the incoming toll call to one of the PBX trunks in the OCC's hunt group and the call proceeds as above when A placed the call. While it does still work, the connection from B to the OCC gateway is much longer than A's connection (and more attenuated).

An even bigger problem occurs, from B's perspective, if the call from B to the hunt group is a toll call that is billed at AT&T's toll point. B would get a telephone bill from the OCC for the long distance call and another bill from Ma Bell for the access to the OCC. The OCC solves this problem by using a foreign exchange line.

In Figure 12.17, when B dials the seven-digit telephone number of the OCC, this

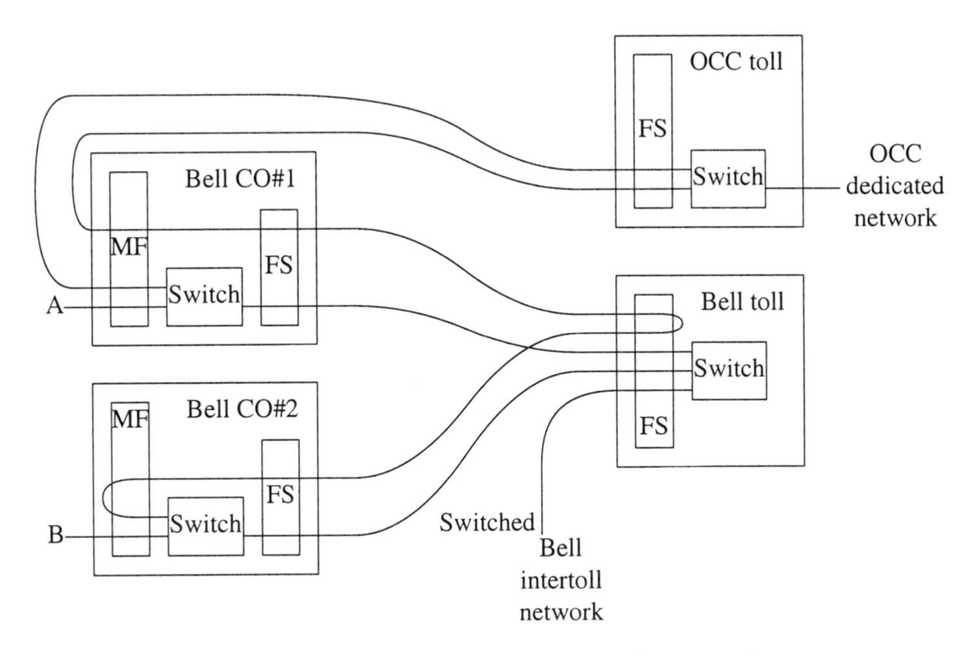

Figure 12.17. B makes an OCC long distance call.

second office connects B to an FX line in a local hunt group. This leased line has three segments: a leased trunk between the toll point and CO#2, a leased trunk between the toll point and CO#1, and a leased line between the OCC gateway and CO#1. The seizure is detected by the OCC's gateway and this office returns a second dial tone. Now B dials the called party's telephone number, through his home CO, into the OCC's gateway. This gateway receives the called party's telephone number and places the call through the OCC's own long distance network over a switched OCC intertoll trunk. But, as with A's OCC call above and B's OCC call without the foreign exchange line, the OCC can use this preferred route only if they have their own network in the called party's region.

Finally, Figure 12.18 shows how the OCC completes the call if they do not have their own network in the called party's region. At some point in the call, the OCC would have to complete the call over trunks leased from AT&T. For simplicity, we illustrate it right at this OCC gateway. Figure 12.18 illustrates the configuration that would always apply if the OCC didn't have any transmission infrastructure of its own.

12.6.6 Subscriber's Default IXC

The OCC's were finally given equal access. This had two far-reaching ramifications.

- The OCCs could connect to the trunk side of any class-5 central office, including AT&T's own class-5 offices. With their physical access on the trunk side, where the LECs normally controlled attenuation, the quality of an OCC's audio connection could now be at least equal to that of an AT&T long distance call.
- Since the OCCs no longer had to look like a PBX, OCC subscribers would no longer need to dial two complete telephone numbers when placing a long distance telephone call. But, the North American numbering plan had to be changed, again.

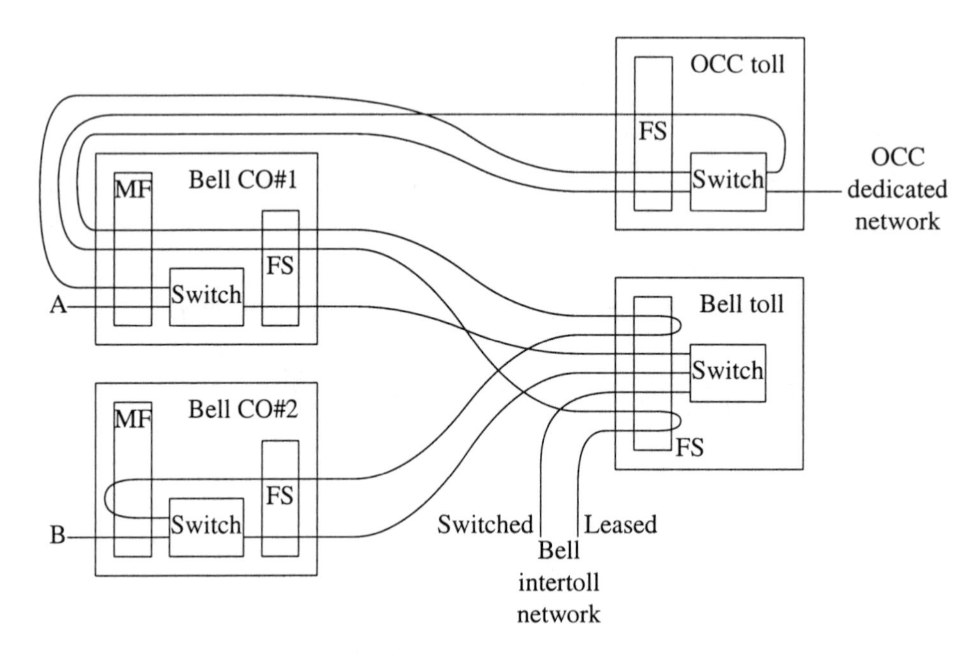

Figure 12.18. Completing A's or B's OCC call over a leased line.

Every IXC was assigned a dial prefix. This number wasn't a seven-digit telephone number, but rather a three-digit prefix that was inserted between the leading 1 and the called party's telephone number. Rather than increase the number of digits that would be required for every dialed long distance telephone call, every subscriber was allowed to specify one of the IXCs as his default IXC. If the subscriber wished to use his default carrier for a particular call, he didn't need to include any IXC selection prefix in the number he dialed. The only time a subscriber would need to include an IXC selection prefix in his dialed number was if he wished to use, for this particular call, another IXC — one that was not his default carrier.

But the North American numbering plan is not changed easily. Providing an additional field in every subscriber's class of service was not trivial, especially in electromechanical switching offices. And educating every subscriber in America on what a default carrier was, and on the procedure for specifying their default carrier, would require massive, and expensive, advertising.

Who paid for these changes? The respective manufacturers had to rewrite switching software for the 1E, the TSPS, the 4E, and similar offices. But they didn't just give away the new releases for free. The LECs paid for it, of course, but most of the LECs were BOCs. When they requested a slight rate increase from their respective PUCs to cover this cost, the PUCs typically refused to increase the price of POTS and offered instead to increase the tariff of some nonuniversal service like AT&T's long distance rates.

EXERCISES

12.1 Draw a block diagram for a case of the data entry department example, where the two locations are served by separate COs, but the two COs have direct trunking.

12.2 Draw a block diagram for a foreign exchange line between two serving COs that have a common toll point, but do not have direct trunking.

12.3 Draw a block diagram for a case of the data entry department example, where the two locations are served by two COs that are so separate that the lowest common point in the hierarchical network is a (class 3) regional center.

12.4 Consider a city whose streets lie in a rectangular north-south grid, like Manhattan. Let an enterprise have three buildings: A at the intersection of 34th Street and 9th Avenue, B at the intersection of 34th Street and 6th Avenue, and C at the intersection of 34th Street and 3rd Avenue. Let the common CO reside at 42nd Street and 6th Avenue. Assume 1 Erlang of two-way traffic intensity between each pair of locations and a 0.02 blocking probability. Assume that each private line between the CO and the buildings at 34th-and-3rd and at 34th-and-9th costs $100 per month and that each private line between the CO and the building at 34th-and-6th costs $50 per month. Calculate the monthly cost for the enterprise network under three different configurations: (a) three totally separate groups of point-to-point tie trunks; (b) point-to-point tie trunks from A-to-B and from C-to-B, but tie trunks from A-to-C are crossconnected through a DACS at B (not even connected to B's fabric); (c) point-to-point tie trunks from A-to-B and from C-to-B, but all tie trunks from A-to-C are tandem-switched at B (connected to B's fabric and circuit switched through it).

12.5 This problem illustrates the economy provided by tandem switching. Suppose the enterprise described in the previous exercise adds a fourth location — D, in Los Angeles. Assume that each private line from the serving CO in NYC to the building in LA costs $2000 per month and that all the price and load assumptions still apply from the previous exercise. Calculate the monthly cost for the Enterprise Network under two different configurations: (a) six totally separate groups of point-to-point tie trunks; (b) point-to-point tie trunks from A-to-B, A-to-C, B-to-C, and B-to-D, but all tie trunks from A-to-D and C-to-D are tandem-switched through B.

12.6 The Telecommunication Education and Research Network (TERN) was a nationwide private line network for use by university telecom programs in the United States. The initial paradigm was SMDS and the initial topology was a triangular backbone with nodes in Pittsburgh, Dallas, and Boulder. These three nodes were interconnected in a DS3 ring and each node served as a hub for a regional DS1 star. MCI provided all the national channels. The Pittsburgh node had DS3 connections to Dallas and to Boulder and DS1 connections to several universities in the northeastern United States. The Siemens SMDS switch at Pitt connected the 28 DS1s on one side to the two DS3s on the other side. This node was located at the University of Pittsburgh, in the Oakland section of Pittsburgh, several miles away from the MCI POP in downtown Pittsburgh. Bell Atlantic provided the local connection over an OC3 link that ran from the TERN node at Pitt to the Oakland CO to the downtown CO to the MCI POP. (a) Make a sketch of Pittsburgh; show these four nodes on Bell Atlantic's Sonet facility. Show DACSs inside each of the Bell Atlantic COs. Show one OC3 muldem inside Pitt's TERN node and another inside the MCI POP. Show one M13 inside Pitt's TERN node and another inside the MCI POP. Show the SMDS switch inside Pitt's TERN node. (b) Show TERN's two DS3 backbone links and 28 DS1 links, all terminating at the MCI POP. (c) Now show how these TERN links at the MCI POP are all provisioned to the SMDS switch at Pitt.

12.7 Complete the crossconnection assignments in the four DACSs of Table 12.1 for the extended example of Section 12.5.4.

12.8 In the extended example of Section 12.5.4, spare capacity was built into all three facilities, simplifying the DACS assignments and the add/drop. Assume for this exercise that the carrier provisions only enough facility capacity for the required channels. Let conduits X and Z carry 10 T2s (instead of two T3s each, as shown in Figure 12.7) and let conduit Y carry two T3s and an additional single T2 (instead of the three T3s shown in Figure 12.7). (a) Show that this facility capacity is sufficient. (b) Using whatever DACSs are necessary in the four buildings, allocate the same channels as in Section 12.5.4. If you groom the lone T2 in conduit Y wisely, you can save DACSs and muxes.

12.9 The right-of-way among four switching offices lies in a straight line. Sketch the geography and label the offices "W" through "Z" left to right. The four offices are fully connected by direct trunks. The W-to-Y trunk group passes through X, the W-to-Z trunk group passes through X and Y, and the X-to-Z trunk group passes through Y. Assume that each of the six trunk groups contains 816 trunks, in 34 DS1s. Each of the three physical spans (conduits) uses DS3 transmission. (a) How many DS1s are needed in the W-to-X conduit? In the X-to-Y conduit? In the Y-to-Z conduit? (b) Ignoring connectivity, what is the minimum number of DS3 systems in each conduit? (c) Because every office is directly trunked to each of the other three, how many DS1 muldems are needed in each office? (d) Ignoring grouping, what is the minimum number of M13s in each office? Achieving these minimum numbers from steps (b) and (d) simultaneously, if possible at all, typically requires very careful grooming. In this example, using a DACS 3/1 in offices X and Y, the channels can be groomed to achieve both minima. (e) On your sketch, inside the four offices show the three DS1 groups in each office, the four M13s in each office, and the DACS 3/1s in offices X and Y. (f) Assign 28 of the W-to-Z DS1s to DS3s that pass directly through X and Y without going through their DACSs. (g) Assign 28 of the W-to-X, X-to-Y, and Y-to-Z DS1s to DS3s that pass directly between these adjoining offices without going through their DACSs. (h) In the X-to-Y conduit, pack the remaining three DS3s so they carry 26, 26, and 28 DS1s respectively. (i) In office W, assign half of the W-to-Y DS1s and the remaining six W-to-X DS1s to one of the remaining DS3s and assign the remaining DS1s to the other remaining DS3. Following this same pattern, assign DS1s to DS3s in the remaining offices. (j) In office X, connect its DACS to the two remaining DS3s in the conduit to W, to the three remaining DS3s in the conduit to Y, and to the two remaining M13s in X. Following this same pattern, connect the DACS in Y to its remaining DS3s and M13s. (k) Consider the DS3 in the W-to-X conduit that carries six W-to-X DS1s and 17 W-to-Y DS1s. Inside the DACS in office X, assign these six W-to-X DS1s to the proper M13 in X, assign 14 of these W-to-Y DS1s to the DS3 in the X-to-Y conduit that carries 28 DS1s, and assign the remaining three W-to-Y DS1s to one of the partially filled DS3s in the X-to-Y conduit. Consider the corresponding DS3 in the Y-to-Z conduit and make the corresponding DS1 assignments inside the DACS in office Y. (l) Finish the DS1 assignments inside the two DACSs.

12.10 The right-of-way among four switching offices lies in a physical ring. The four offices are fully connected by direct trunks. Sketch the geography and label the four offices by the points of the compass. The N-to-E, E-to-S, and S-to-W conduits each hold three OC3s. The N-to-W conduit holds two OC3s and one OC1. Interoffice traffic requires four OC1 channels between every pair of offices; including the N-to-S and E-to-W trunk groups where there is no direct path. (a) How many OC1 muldems in each office? Add them to your sketch. (b) Also on your sketch, show four OC13 muldems in each office. One of the OC13 muldems in office N, and one in office W, is an ADM. Each of these ADMs terminates one OC3 from each of its office's conduits and terminates two OC1s within its local office. (c) Three of the N-to-S OC1s are muxed up to an OC3 that passes through office E without being terminated there. (d) One of the N-to-W OC1s is not muxed up to OC3, but uses the isolated OC1 in the N-to-W conduit. (e) One of the E-to-S OC1s goes around the ring the long way and, on this route, it passes through both ADMs. (f) One of the OC3s in the N-to-E conduit carries one N-to-E OC1, one E-to-W OC1, and one E-to-S OC1 (the "long" one). This OC3 terminates on the ADM in office N where the N-to-E OC1 is add/dropped and the other two are assigned to two of the three OC1s in one of the N-to-W OC3s. (g) Finish grooming the system.

12.11 Five switching nodes are connected in a physical ring. The physical link is a T3, which contains 28 embedded T1s. Each node has a direct logical channel to every other node, with two of its logical channels proceeding clockwise around the ring and two of its logical channels proceeding counterclockwise around the ring. Assume that each node contains a DACS 3/1 and that internode traffic is balanced. (a) Draw the network, showing all its logical channels. (b) What is the maximum number of T1s in each logical channel? (c) Show the general DACS assignment that would be used in each node.

12.12 Repeat the previous exercise except let the physical infrastructure be a Sonet ring. So, the link between each node is an OC1 and each node contains an OC1 ADM.

12.13 In the previous exercise, if one of the physical links has a cable cut, all the logical channels are redirected by quickly reconfiguring all five DACSs in the network. (a) Draw the network showing all its logical channels, ensuring that each node still has a direct logical channel to every other node. (b) What is the maximum number of T1s in each logical channel? (c) Show the general DACS assignment that would be used in each node.

SELECTED BIBLIOGRAPHY

Barth, P. A., "Customer Controlled Reconfiguration," *Telecommunications,* Vol. 20, No. 11, Nov. 1986.

Bear, E., "Designing an Embedded Voice over Packet Network Gateway," *Communication Systems Design,* Vol. 4, No. 10, Oct. 1998.

Bellamy, J., *Digital Telephony,* Second Edition, New York, NY: Wiley, 1991.

Cooley, K. D., et al., "Wideband Virtual Networks," *Telecommunications Magazine,* (issue on Switching Inside and Outside the Network), Vol. 21, No. 2, Feb. 1987.

Dziatkiewicz, M., "Universal Service — A Taxing Issue," *Wireless Business & Technology,* Vol. 5, No. 1, Jan. 1999.

Gould, E. P., and C. D. Pack, "Communications Network Planning in the Evolving Information Age," *IEEE Communications Magazine,* (issue on Network Planning), Vol. 25, No. 9, Sept. 1987.

Henck, F. W., and B. Strassburg, *A Slippery Slope — The Long Road to the Breakup of AT&T,* Westport, CT: Greenwood Press, 1988.

Hester, S., "Managing the Combined Virtual Network," *Data Communications,* Vol. 8, No. 12, Dec. 1979, pp. 41 – 53.

Holliman, G., and N. Cook, "Get Ready for Real Customer Net Management," *Data Communications Magazine,* (issue on Global Supernets), Sept. 21, 1995.

Lukens, M., "Subrate Services — The Vital Link in the T1 Chain," *Telecommunications Magazine,* (issue on Enhanced Networking), Vol. 20, No. 11, Nov. 1986.

Musselman, T., "T1 Network Management: A Strategic Perspective," *Telecommunications Magazine,* (issue on Network Management), Vol. 22, No. 2, Feb. 1988.

"One-Stop Telecomm Shopping," *Communications News,* Vol. 35, No. 2, Feb. 1998.

Phillips, B. W., "Making the Frame Relay Decision: User Criteria," *Telecommunications Magazine,* Vol. 29, No. 4, Apr. 1995.

Raack, G. A., E. G. Sable, and R. J. Stewart, "Customer Control of Network Services," *IEEE Communications Magazine,* Vol. 22, No. 10, Oct. 1984.

Rey, R. F. (ed.), *Engineering and Operations in the Bell System,* Second Edition, Murray Hill, NJ: AT&T Bell Laboratories, 1977.

Temin, P., with L. Galambos, *The Fall of the Bell System,* New York, NY: Cambridge University Press, 1987.

Weiss, M. B. H., and P. Bernt, *The Regulation and Deregulation of US Telecommunications,* Textbook under contract with Lawrence Erlbaum and Associates, Mahwah, NJ, C. Sterling (ed.).

13

Digital Circuit-Switching Concepts

> *"Nobody dast blame this man. You don't understand: Willy was a salesman. And for a salesman, there is no rock bottom to the life. He don't put a bolt to a nut, he don't tell you the law or give you medicine. He's a man way out there in the blue, riding on a smile and a shoeshine. And when they start not smiling back — that's an earthquake. And then you get yourself a couple of spots on your hat, and you're finished. Nobody dast blame this man. A salesman is got to dream, boy. It comes with the territory."* — from *Death of a Salesman*, by Arthur Miller

Digital switching is described in two chapters, this chapter and Chapter 14. Several different switching systems are described in Chapter 14, but first, some fundamental general concepts are presented in this chapter. The synergy between transmission and switching, and the marketing of that synergy, is discussed first. Sections 13.2 and 13.3 are a general discussion of time-division switching and time-multiplexed switching, respectively. The combination of these two provides an important network architecture that supports switching in both the space and time divisions and that is common to most modern digital switching systems. Integrated services digital network (ISDN) is discussed in Section 13.4 and the BORSCHT functions are described in the section after. Section 13.6 briefly describes the OAM&P functions and how they have been centralized and automated.

13.1 SYNERGY

Synergy is "the simultaneous action of separate agencies which, together, have greater total effect than the sum of their individual effects" [1]. The synergy between transmission and switching, and the marketing of that synergy, is discussed.

13.1.1 Synergy Between Switching and Transmission

During the 1970s and early 1980s in the United States, significant growth occurred in upscale suburban areas and in certain parts of the country. Shopping malls and industrial parks were constructed in the suburbs and many citizens migrated both out of the cities and into the "Sun Belt" and other parts of the United States. The telcos managed this growth and demographic shift through the use of pair-gain, using T1 and similar time-division multipexed (TDM) carrier systems. These systems appeared first in the interoffice trunk plant, where they provided additional channels economically, despite their relatively high price.

While the next generation toll office, described in the next chapter, obviously had to interface to analog metallic trunks, the decision to use digital technology in its internal fabric necessitated analog-to-digital (A-to-D) converters in every trunk interface. But what about the digital channels? It is obviously silly to convert digital transmission channels to analog in the toll-side terminal and then immediately convert back to digital in the internal fabric. So, not only does a digital toll switch provide for a direct interface to a digital channel, it also provides a direct interface to T1, without even demultiplexing.

Synergy occurs when two processes acting together have a greater effect than the sum of the effects of the two processes when they act alone. Digital carrier and digital switching provide an economic synergy because many A-to-D converters, multiplexors, and demultiplexors could be avoided if the digital carrier and the digital switch are installed simultaneously.

13.1.2 Marketing That Synergy

Suppose it's 1978 and you have just been promoted to Regional VP of Operations for Left Central Bell. Because the previous VP had been remiss about replacing old switching equipment, you are now in charge of several Step-by-Step offices that are increasingly unreliable, costly to maintain, and that cannot provide the new services that should be marketable in the more up-scale communities in your region. Suppose I am a salesman with a telephone equipment manufacturer and I try to sell you my company's new digital switching equipment to replace your old Steppers. The potential revenue from new services and the lower maintenance cost, however, can't quite justify replacing the old Step offices. I resort to describing a fanciful future network in which voice and data will be integrated onto a single ISDN infrastructure that Step won't be able to support. But you're skeptical about this.

Suppose, also, that the suburban communities in your region are growing, both in population and in telephone usage, so much that the trunk plant in your region, besides also being old and unreliable and costly to maintain, must be supplemented with additional trunks. Your transmission engineers have proposed handling the growth and upgrading the trunk equipment by installing T1 and T3 carrier systems. Furthermore, your people select my competitor's transmission equipment. You decide to start a large 4-year program to modernize and add channels to the trunk plant in your region.

I find out about this plan and ask to see you. Instead of asking you to buy my transmission equipment instead of my competitor's, I ask you to reconsider your decision to postpone buying my new digital switches. My competitor's digital switch is still in development. I argue that, when you install the new digital transmission equipment, you will need A-to-D and D-to-A converters on every trunk channel at the interface with your old Step switches. If you buy my switches now, before you begin your 4-year program of upgrading your trunk plant, you won't have to buy many of those A-to-D and D-to-A converters.

You recompute the financial numbers and you find that the money you saved in converters, when added to the money saved in decreased maintenance and the money gained through potential new services, is now enough to cost-justify buying and installing my switches. So, you and I do very well and I have hurt my competitor two ways. You modernize your switching equipment and also save money for your company, and I make a big commission. My competitor still sells you a lot of digital transmission equipment, but at a lower price because you don't buy his converters and my competitor loses a potential customer for his digital switches, whenever he's ready to start selling them.

This story illustrates the synergy between switching and transmission, how this synergy was marketed, the importance of demographics in the telephone business, and the emerging importance of timing when introducing a new product. Chapter 20 revisits the possibility of a similar synergy, one between optical fiber and photonic switches.

13.1.3 Digital Fever

The post-Carterfone PBX market became very competitive. Because AT&T had almost the entire market share in the late 1960s, it could only lose customers. The new manufacturers to the PBX market attracted many customers from new installations, the large base of CO-Centrex customers, and by replacing the old Step and Crossbar PBXs. Finally, they needed a strategy to win over the large embedded base of Dimension PBXs. In new installations and CO-Centrex converts, competing on price with AT&T in those days was like taking candy from a baby. In replacing old systems, new features and size reduction were pretty easy motivators for modernization. Dimension replacement was not going to be so simple. So, Northern Telecom, Rolm, and several others introduced PBXs in the mid-1970s that were aggressively advertised as being "digital."

Stoked by exquisite marketing hype, digital fever swept the PBX market. While the call features of the new digital PBXs weren't significantly different from those of the Dimension, digital PBXs replaced many Dimensions because of a market perception that digital PBXs were modern and Dimensions were old-fashioned, the economic advantage of purchasing over leasing (many Dimensions were leased — remember corporations subscribed to the Bell System's Centrex, independent of whether it was implemented as CO or CU), and the relatively unsophisticated level of corporate telecom managers.

Because AT&T had never really had to do any significant marketing, they underestimated the importance of marketing and it took a long time before they realized they were being out-marketed. Breaking from a long tradition, AT&T hired some executives from outside the company who had marketing expertise. One of them responded to this digital fever by claiming "Dimension is digital, too. Look, it has a digital computer in it." While those corporate telecom managers might have been relatively unsophisticated, they weren't totally naive, especially when helped by salesmen from Northern Telecom or Rolm. While AT&T's response was technically incorrect, AT&T assumed that its customers knew that it was the call features that were important, and not the fabric's technology, and that it was the digital nature of the control, and not the switching fabric, that provided these call features. But it was a disaster for AT&T. Market share plummeted and AT&T quickly started development of a line of digital PBXs (see the System 75 in Section 14.3).

After clobbering AT&T in the PBX market, by introducing digital PBXs with a successful marketing strategy, Northern Telecom extended this strategy into the CO switch market. But Northern Telecom's early success with their digital switch was based more on technology than on marketing. As the price of digital carrier systems decreased and the cost of installing wire increased, digital pair-gain began to prove in economically in the loop plant. A digital switch in the CO provides the same kind of economic synergy as a digital switch at the toll point.

Northern Telecom was very successful with the independent telcos, but even the Bell Operating Companies began to break ranks with their parent company and started buying Northern Telecom's DMS-10 instead of Western Electric's 1A. Bell Labs quickly started developing the 5E, a digital CO switch and introduced it in the early 1980s, barely in time to have a competitive switching product after divestiture. Despite Northern Telecom's head start in the digital CO market, the 5E gained immediate industry-wide acceptance after its introduction because a clever architecture provided the telcos an evolution strategy for new services (see Section 14.2.8).

What makes a switching system a digital switch? Is a digital switch any switching system, including Dimension and the 1A, that is controlled by a digital computer? Is a digital switch any switch that can be successfully marketed so the customer believes it is digital? Is a digital switch a switching system that has a digital fabric that switches digital signals without converting them to analog and that provides synergy with digital transmission equipment? While it is clear now that the last statement is the true definition of digital switching, it wasn't always so clear because the manufacturers that didn't have a real digital switch tried to define it differently. We will look at digital circuit-switching concepts in this chapter and several real digital switching systems in the next one.

13.2 TIME-DIVISION SWITCHING

Because channels are defined in digital carrier using the time division and because digital carrier is synergistic with digital switching systems, digital switching fabrics must be able to switch in space and time. Space/time switching is described in this section and the next. First, in this section, we describe switching in the time division only. Time-division switching is also called timeslot interchange. Then, in Section 13.3, we describe a transformation from the Clos fabric architecture to the time-space-time (TST) architecture that switches channels in both space and time and that is the underlying structure for most modern digital switching systems.

13.2.1 Timeslot Interchange

A "Lazy Susan," shown in Figure 13.1, is a circular turntable commonly found in Chinese restaurants. It is placed in the center of a larger stationary table. The patron's dinner plates are placed on the larger table, and the serving bowls containing the variety of ordered dishes are placed on the Lazy Susan. A patron may select a dish by rotating the Lazy Susan, instead of asking other patrons to pass it, and then may take a portion from the serving bowl and place it onto his dinner plate.

A one-stage TDM switching system with N timeslots is like a continually rotating Lazy Susan with N serving platters. If A calls B, then the system must locate two idle bowls on the Lazy Susan — t_{AB} for the channel from A's mouth to B's ear and t_{BA} for the channel from B's mouth to A's ear. With each revolution of the Lazy Susan, when t_{AB} is in front of A he must load a mouth sample on it, when t_{AB} is in front of B he must take an ear sample from it, when t_{BA} is in front of B he must load a mouth sample on it, and when t_{BA} is in front of A he must take an ear sample from it. If the Lazy Susan rotates at 8000 revolutions per second, in synchronism with the PCM sampling rate, then the bowls contain exactly one 8-bit PCM sample.

This analogy represents a two-stage switching system that performs simple selection

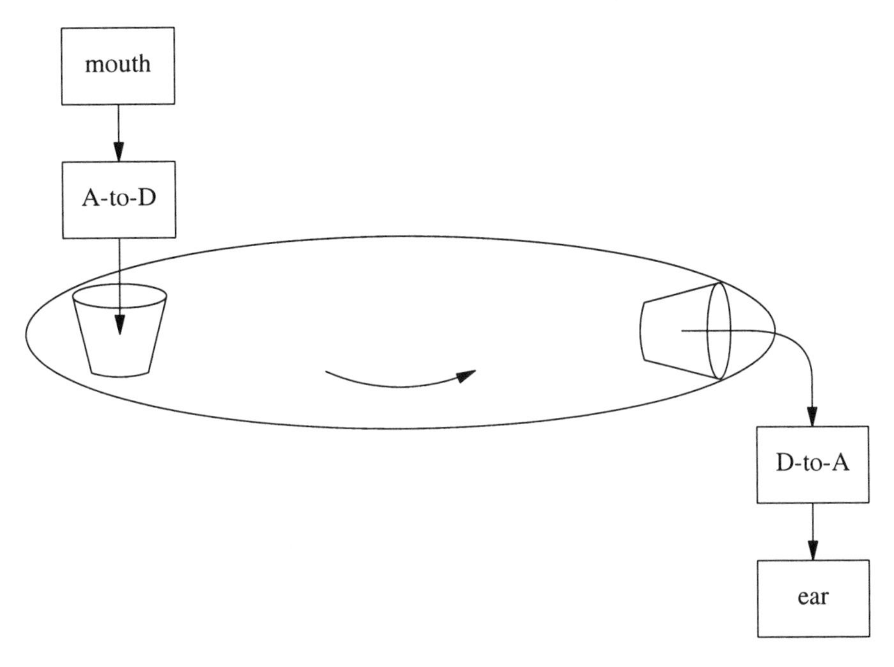

Figure 13.1. Lazy Susan analogy for timeslot interchange.

(Section 1.3), like tuning in a commercial radio or TV station (which uses FDM instead of TDM). The concentration of user's mouths to timeslots is a full-access first stage of switching, and the expansion of timeslots to user's ears is a full-access second stage. The system is nonblocking if it serves $N/2$ servers or fewer, and it may block if it serves more than $N/2$ users. In this context, the Lazy Susan performs a space-to-time conversion, or time-division multiplexing function, as part of its switching function.

In a time-only context, a single physical spatial input and a single physical spatial output are already time-multiplexed into N timeslots. If the assignment of channels to timeslots is different between the input and the output, the time switch in between must simultaneously move N chunks of data from their assigned timeslots in the input into their assigned timeslots in the output. If platters are loaded sequentially, they must be unloaded randomly (not really randomly, but according to the switching assignment); or they must be loaded randomly, if they are unloaded sequentially. In this context, without the corresponding TDM function, such a time switch is called a timeslot interchanger (TSI).

Figure 13.2 shows the obvious implementation, with a digital random access memory (RAM). The RAM has $3N$ words and an internal address counter. In this description, let the $3N$ words be labeled: $a_0 - a_{N-1}$, $e_0 - e_{N-1}$, and $o_0 - o_{N-1}$. The switching assignment, $i = A(j)$, is stored in words $a_0 - a_{N-1}$: if data from incoming timeslot i must be placed in outgoing timeslot j, then the binary code for j is stored in location a_i in RAM. If $N \leq 256$, then words $a_0 - a_{N-1}$ need only be 1 byte wide.

Words $e_0 - e_{N-1}$ and $o_0 - o_{N-1}$ have 8 bits each and are organized as a dual-access N-byte RAM; PCM samples in even-numbered input frames are written into bytes $e_0 - e_{N-1}$ and from odd-numbered input frames into bytes $o_0 - o_{N-1}$. Because such a

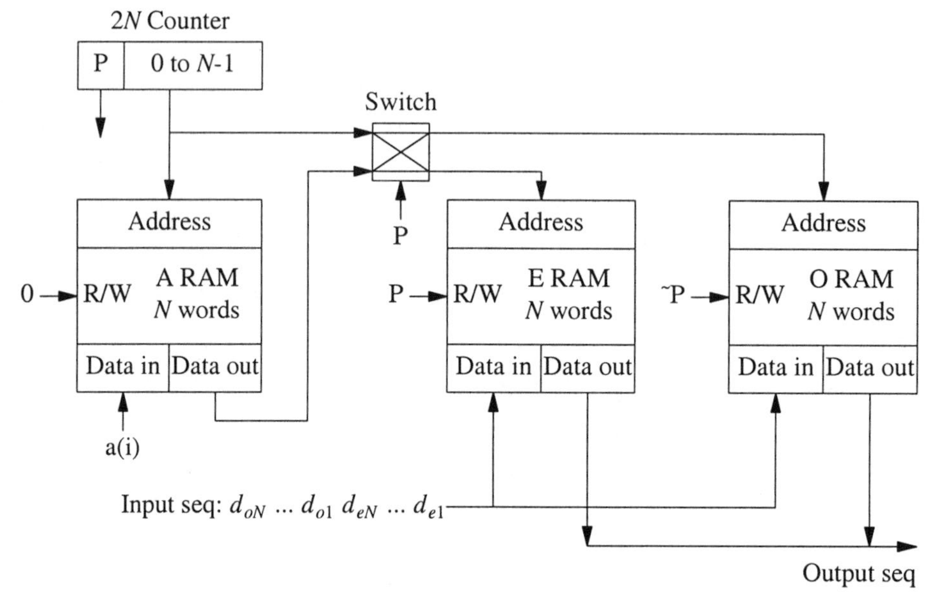

Figure 13.2. Timeslot interchanger.

TSI inserts one frame of delay from input to output, odd-numbered output frames are read from bytes $e_0 - e_{N-1}$ and even-numbered output frames from bytes $o_0 - o_{N-1}$. The first procedure, sequential write and random read, is described.

If f is an even number and $0 \leq i \leq N-1$, then data from incoming timeslot i of frame f is written into the byte at address e_i, and the data from incoming timeslot i of frame $(f+1)$ is written into the byte at address o_i. Since the data in incoming frame f will have been read and transmitted out during incoming frame $(f+1)$, data from incoming frame $(f+2)$ can use the same N bytes that were used by incoming frame f; and so forth. So, the write address is maintained by a simple binary counter that counts: $e_0, \cdots, e_{N-1}, o_0, \cdots, o_{N-1}$, increments at the start of every incoming timeslot, and resets to e_0 at the start of every even-numbered incoming frame.

The output side of the TSI is similar except that data is read from RAM instead of written into it, frames are delayed by one full frame, and counter-driven addressing is indirect through words $a_0 - a_{N-1}$ of the RAM. The read address is computed, using an address counter that counts from $a_0 - a_{N-1}$, increments at the start of every incoming timeslot, and resets to a_0 at the start of every incoming frame. At the start of every timeslot, the RAM at this address is read, identifying that incoming timeslot whose data should be switched and transmitted into this outgoing timeslot. This new address is read from block e during odd-numbered incoming frames and from the block o during even-numbered incoming frames. Data from this address is read and transmitted.

13.3 TIME-MULTIPLEXED SWITCHING

The fabrics described in Chapter 7 perform switching between physical channels on the left side of the fabric and physical channels on the right side. This is space-division switching. The TSI just described switches timeslots from the line on the left into the timeslots in the line on the right. This is time-division switching. But, space/time switching, switching in both the space and time divisions simultaneously, is more complicated than simply placing a space switch and a time switch together. This section begins with the classic Marcus transformation on the Clos architecture [2]. Then, time-multiplexed switching and space/time switching are described. Switching in multiple divisions is revisited in Chapter 21.

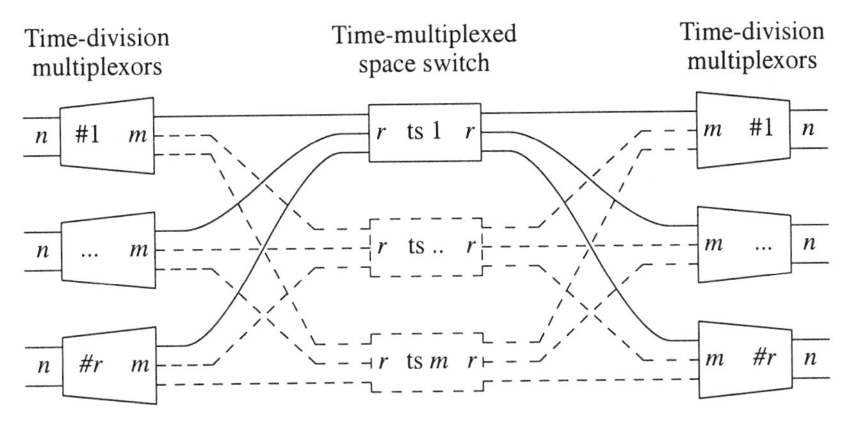

Figure 13.3. Marcus' transformation on the Clos architecture.

13.3.1 TST with TDM

Figure 13.3 shows a generalized $N \times N$ Clos fabric again, with $N = r \times n$ ports on its left and right edges. The left stage and the right stage each have r switches, each of which has n external ports on its outer edge and m ports for links on its inner edge. The center stage has m switches, each of which is $r \times r$. In Figure 13.3, all but one of the switches in the center stage is shown in dashed lines because they exist logically, but not physically. Similarly, the perfect-shuffle pattern between the outer stages and the center stage exists logically to all the switches in the center stage, but exists physically to only one of them.

Consider each of the $n \times m$ switches in the outer stages to be time-division multiplexors. Each has n physical channels on its outer edge and only one physical channel on its inner edge. Let real time on this single interior link be organized as a sequence of repeating frames and let these frames be organized into m sequential timeslots. A logical channel is defined by using one timeslot in each frame. Thus, this single physical channel is time multiplexed into m logical channels.

Data on each of the physical channels on the exterior ports is sampled at the sampling rate and frames are repeated on the interior link at the frame rate. If the frame rate equals the sampling rate, then the interior link provides m logical channels for the sampled data on the exterior ports. If an $n \times m$ space switch connects external port i to internal port g, $1 \leq i \leq n$ and $1 \leq g \leq m$, then a corresponding TDM unit samples the data from external port i and assigns the sampled data to timeslot g in the time-multiplexed link. Let the $2 \times r$ TDM units be synchronized together. Thus, the gth timeslot on each of the r a-links and on each of the r b-links are all simultaneous.

Figure 13.3 shows r such TDM units in the left stage. The interior edge of each unit consists of a single time-multiplexed a-link. These r a-links are wired to the left edge of a single $r \times r$ switch in the center stage, as shown by the solid lines in the center of Figure

13.3. Similarly, r time-multiplexed b-links are wired between the right edge of the $r \times r$ center switch and the left edge of each of the r TDM units in the right stage. At the a-links and at the b-links, the single time-multiplexed link at the interior edge of TDM unit j is wired to port j on the center switch, $1 \le j \le r$.

Let the single $r \times r$ switch in the center stage be time multiplexed. While a space-only switch has a single configuration by which its r left ports connect to its r right ports, a time-multiplexed switch (TMS) has m different configurations. The TMS assumes configuration 1 for a brief time, then configuration 2 for the same brief time, and so forth. After cycling through all m configurations, the pattern is repeated. This time-multiplexed operation is equivalent to m different switches in the center stage, each with its own configuration. Let this single $r \times r$ center stage TMS be synchronous with the timeslots in the a-links and b-links. The temporal duration of each configuration equals the duration of one timeslot and the duration of one complete cycle through all m configurations equals the duration of one frame.

Any space switch, like those described in Chapter 7, that is reconfigured in every successive timeslot is called a time-multiplexed (space) switch. This overall three-stage architecture in Figure 13.3, with a stage of TMS between two stages of TDM, is called the time-space-time (TST) architecture. TST is the basic architecture used in just about every one of today's commercial digital switching systems.

After describing path selection in a TST fabric, the third subsection describes another TST implementation, using TSI in the outer stages instead of TDM, and the fourth subsection describes Marcus' "dual" implementation, the space-time-space (STS) architecture.

13.3.2 Path Selection in TST Fabrics

Consider a regular space-only Clos fabric. Let external port $[i,j]$ be the ith exterior port on switch X_j in the left stage, and let external port $[o,p]$ be the oth exterior port on switch Z_p in the right stage; where $1 \le i \le n$, $1 \le j \le r$, $1 \le o \le n$, and $1 \le p \le r$. Connecting $[i,j]$ to $[o,p]$ over the gth path requires the operation of crosspoints $[i,g]$ in X_j, $[j,p]$ in Y_g, and $[g,o]$ in Z_p. Consider making the equivalent connection through an equivalent TST fabric that has TDM units in the outer stages and a single TMS in the center stage. In the left stage, TDM j assigns its exterior port i to timeslot g. In the center-stage, the TMS connects its left port j to its right port p during timeslot g. In the right stage, TDM p assigns timeslot g to its exterior port o.

The two theorems about Clos architectures apply equally well to TST fabrics. If $m \ge 2n - 1$ timeslots are provided in each link, then a TST fabric is wide-sense nonblocking. If $m \ge n$ timeslots are provided, then the TST fabric is rearrangeably nonblocking. Consider the path selection algorithm for a call between left port $[i,j]$ and right port $[o,p]$. There are m possible paths for this call, one path for each timeslot. Since the connection must use the specific a-link between the TMS and TDM j in the left stage and the specific b-link between the TMS and TDM p in the right stage, the selected timeslot must be idle in both these links. If $m \ge 2n - 1$, then the existence of such an idle timeslot is guaranteed. If $m \ge n$, then an idle timeslot can be found after possibly moving existing calls to alternate timeslots. If $m < n$, then there are three ways that the call might be blocked:

- If all m timeslots in the a-link are busy, then TDM j cannot assign a timeslot to exterior port i;
- If all m timeslots in the b-link are busy, then TDM p cannot assign a timeslot to exterior port o;
- Even if both exterior ports can be assigned timeslots, there may be no common idle timeslot nor a rearrangement of existing calls that can produce a common idle timeslot.

Let timeslot g be a common idle timeslot that is idle in the jth a-link and in the pth b-link. The path between them can be connected because the TMS is assumed to be a nonblocking fabric. If timeslot g is idle in both required links, then the jth left port and the pth right port are both idle in the gth configuration of the TMS. The path is connected by operating crosspoint $[j,p]$ during this gth configuration.

Controlling rearrangeably nonblocking fabrics, especially the problem of rearrangement order, was discussed in Section 7.3.3. If a rearrangeably nonblocking fabric is used as a TMS in a TST fabric, the problem with physical rearrangement order vanishes. Let an external port on TDM unit j on the left be connected to an external port on TDM unit p on the right. Then, in the TMS, left port j is connected to right port p during timeslot g. Suppose the addition of another call, between lines on different TDM units, uses this same timeslot g. Let the connection of this second call require rearranging the first call to an alternate path through the TMS. Because the calls are physically connected only during timeslot g, the virtually simultaneous rearrangement of the first call and connection of the second call is made in the memory of the fabric controller so that the TMS assumes its new configuration when timeslot g occurs in the next frame. If the rearrangement of the first call involves moving it to an alternate timeslot, then TDM units j and p must change their timeslot assignments in conjunction with the rearrangement in the TMS so that all three units move the call in the same frame.

Since the control algorithm for a fabric or subfabric is responsible for selecting a path for a new call given the configuration of existing connections, if some algorithm is controlling a TMS, then the bookkeeping is slightly complicated because each timeslot has its own configuration. The control algorithm for a TMS with m timeslots maintains m separate, and independent, network maps.

13.3.3 TST With TSI

The Clos architecture has two Marcus transformations, yielding the time-space-time (TST) fabric of Figure 13.3 and the space-time-space (STS) fabric that is described in the next subsection. But, the TST fabric has two implementations, depending on whether the modules in the first and third stages are TDMs or TSIs. The TDM implementation of the TST would be used in a class-5 switching office, a PBX, or any system that connects space channels to space channels through a fabric that uses time multiplexing. But, in any hierarchical network that uses digital time multiplexing in its transmission plant, channels in the higher network levels are not just space channels, they are defined in both space and time.

Using the TDM implementation of the TST in a class-4 office, for example, the external digital carrier systems would all have to be demultiplexed so that only space channels interfaced to the TST fabric. In Figure 13.4, each external input time-

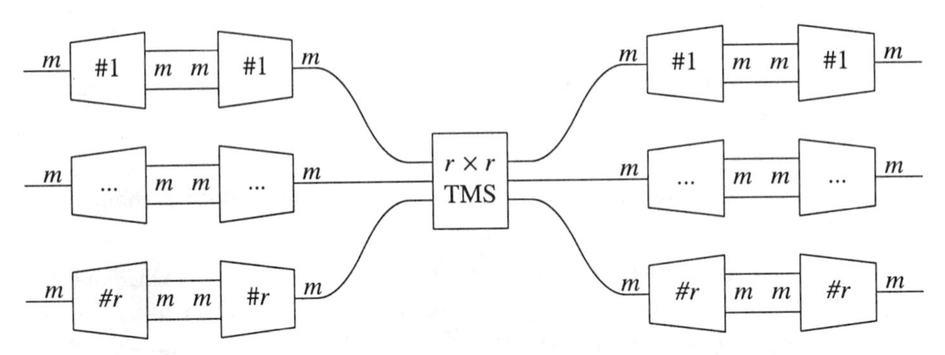

Figure 13.4. TST-TDM fabric interfacing space/time links.

multiplexed line on the left edge is demultiplexed to m space channels, and each external output time-multiplexed line on the right edge is remultiplexed from m space channels. If these external lines match the TST fabric's internal links, as in Figure 13.4 where all internal and external links have m timeslots, then each time multiplexor in the first stage of the TST fabric has a corresponding demultiplexor preceding it and each time demultiplexor in the third stage of the TST has a corresponding multiplexor following it. So, each space/time input line and output line interfaces to a first-stage module, comprised of a demux/mux tandem, which interfaces to a similar space/time fabric link. But a demux/mux tandem is one implementation of a TSI.

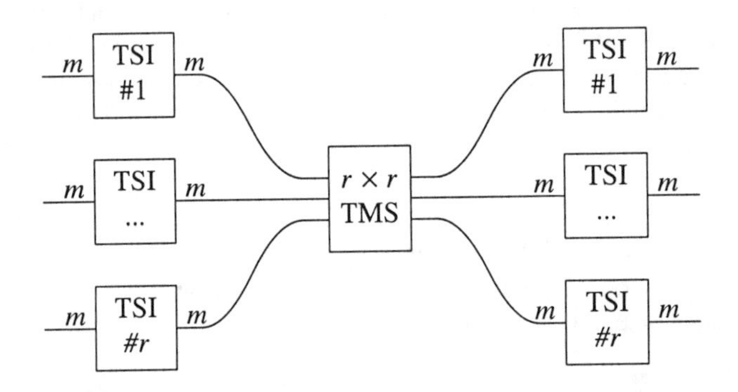

Figure 13.5. TST-TSI fabric interfacing space/time links.

Each demux/mux tandem in the Figure 13.4 is combined into a single TSI in Figure 13.5, resulting in a three-stage fabric that has a time-multiplexed (space) switch in the center stage and timeslot interchangers in the two outer stages, instead of time-division multiplexors. The external lines, on the left and right edges, are assumed to be time multiplexed. The fabric interfaces to channels that are defined in space and time and performs space switching and time switching on these channels in the same fabric.

Suppose data assigned to timeslot i in line j on the left edge must be switched into timeslot o in line p on the right edge; where $1 \le i \le m$, $1 \le j \le r$, $1 \le o \le m$, and $1 \le p \le r$. This connection has m paths through the total fabric, one path for each TMS timeslot through the center stage. Let path g be selected, where $1 \le g \le m$. In the left stage, TSI j assigns its left timeslot i to its right timeslot g. In the center stage, the TMS connects its left port j to its right port p during timeslot g. In the right stage, TSI p assigns its left timeslot g to its right timeslot o.

13.3.4 The STS Architecture

Time multiplexing and frequency multiplexing were presented in Section 1.2.3 — and wavelength multiplexing will be presented in Section 21.1.2 — as methods for sharing some physical resource. From this perspective of conserving a resource that is defined in space, the concept of space multiplexing seems odd. But, if we think of multiplexing simply as replication, then TDM and FDM and WDM are methods by which channels are replicated or multiplied in some stage of a system. So, if *X*DM is replication in the *X* division, then SDM is simply replication in the space division. A multipair cable, like house wiring or a feeder cable, is a space-multiplexed channel.

The TST/TSI architecture of Figure 13.5 establishes switched connections between a set of channels on the left that are defined in space and time and a set of channels on the right that are defined in space and time. Switching in these two divisions by the same fabric is called space/time switching. The TST structure in Figure 13.5 is only one of two implementations by which Marcus transforms the Clos for space/time switching. Figure 13.5 is generalized by observing that we have made two transformations to the three stages of the Clos architecture. The center-stage switch is a single switch that switches in one of the divisions, space, but is multiplexed in the other division, time. Each of the outer stages is a single switch that switches in one of the divisions, time, but is multiplexed in the other division, space. This generalization is even further generalized in Chapter 21, where photonic space/time/wavelength switching is discussed.

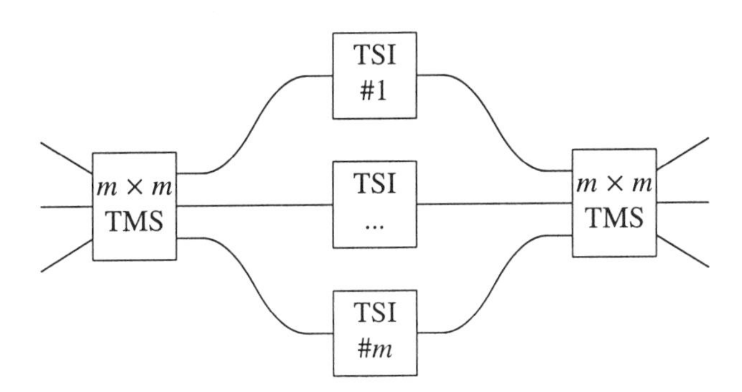

Figure 13.6. STS fabric interfacing space/time links.

Marcus' transformation still works if the order of multiplexing is reversed. Figure 13.6 shows a space-time-space (STS) fabric. The center-stage switches in the time division, but is multiplexed in the space division. Each of the outer stages is a single switch that switches in the space division, but is multiplexed in the time division.

Suppose data assigned to timeslot i in line j on the left edge must be switched into timeslot o in line p on the right edge; where $1 \le i \le r$, $1 \le j \le n$, $1 \le o \le r$, and $1 \le p \le n$. This connection has m paths through the total fabric, one path for each separate TSI in the center stage. Let path g be selected, where $1 \le g \le m$. In the left stage, the TMS connects its left port j to its right port g during timeslot i. In the center stage, TSI g assigns its left timeslot i to its right timeslot o. In the right stage, the TMS connects its left port g to its right port p during timeslot o.

13.4 ISDN

ISDN stands for Integrated Services Digital Network. ISDN's goal was to be a single *integrated* network that would provide circuit-switched connections for both voice and data. However, note that:

- ISDN isn't really a network. A user at an ISDN telephone and one at an analog POTS telephone are connected to the same digital network and can talk to each other. So, ISDN is more like a transport format, or the specification for a network interface, than it is a network. Because subscriber signaling is so unusual in ISDN, there are network-wide implications.
- While ISDN is digital, two of its most difficult implementation issues are analog: locating the BORSCHT functions and transmitting digital data over analog lines. These are problems because ISDN's principal medium (wire) and principal application (voice) are both inherently analog.
- While ISDN's ultimate purpose was to provide services, it took a long time after ISDN's introduction before any real user services were actually implemented. We'll postpone discussing the reasons until Chapter 15. While it's now outdated, people once joked that ISDN stood for "It Still Does Nothing."
- While ISDN can physically integrate voice and data onto one format, the principal switching paradigm (circuit switching) is optimal for voice. While ISDN includes a packet-switched paradigm, it is intended for signaling and low-rate user data, like key strokes. Most significant data traffic, like transferring large files, is better suited to a large packet-switched network, like the Internet, than to a circuit-switched network.

Reminiscent of the toll point in Chapter 9, when ISDN was first conceived, its many functions all made sense together. By the time ISDN was finally ready for the market, the market had changed so much that it wasn't clear that the combined functions were useful. Finally, in the mid-1990s, we may have gone full cycle. Consider the state of voice and data telecommunications in the early 1970s, when ISDN was born.

- Digital PBXs and digital COs were providing a dedicated codec to every voice line. However, there seemed to be a lot of potential synergy if this codec, and the other BORSCHT functions, could be migrated from the line card to the telephone or, at least, to the protector block.
- The very definition of "computer networking" then was that dumb terminals were connected to time-shared main-frame host machines. They were either directly wired or they were circuit-switched telephone connections where transport was byte oriented and used 300- or 1200-baud modems.

While voice was being digitized along the loop or in the switch and data was digital at the terminal and at the host, the universal network interface was the analog POTS line. The goals of the concept that eventually became ISDN were to provide end-to-end 64-Kbps clear digital connectivity, 2:1 pair-gain so that computer calls didn't busy the user's telephone, an out-of-band signaling channel to the user, a digital transport that eliminated the need for the modem, and a universal digital network interface that would be common for voice and for data. Rather than consolidate ISDN into one chapter, it makes more sense to discuss ISDN in its various contexts. Here, we'll discuss ISDN's implementation, and in Chapter 15 we'll talk about users and applications.

Most people's first exposure to ISDN is as a data format, $2B + D$, so that's the topic of the first subsection. Subsequent subsections will describe ISDN as an interface, then physically implementing this format in the loop, and finally ISDN as a switching control paradigm.

13.4.1 Data Format

The three ISDN rates and formats are called: basic rate, primary rate, and broadband. After lengthy international debate, Basic Rate (BR) ISDN was standardized worldwide at 144 Kbps. The format has 18 bits per frame and the usual 8000 frames per second. It holds two DS0 bearer (B) channels at 8 bits each per frame and another 2 bits per frame for a data (D) channel. The D-channel in this $2B + D$ format carries 16 Kbps and, for POTS calls, is used for signaling. While a user's voice call is typically assigned to one B-channel, a high-fidelity voice call could use both B-channels. At the minimum, an ISDN telephone is equivalent to a two-line key-set. Low-rate user data, like keystroke bytes or telemetry, can be packetized and can share the D-channel with the user's signaling. Circuit-switched data connections at 64 Kbps can be assigned to one or both B-channels.

Primary Rate (PR) ISDN was easily defined in North America at 1.544 Mbps. Recall that T1 has 24 DS0 bearer channels and performs signaling in one of three classical ways: inband using one-eighth of every channel in every frame, inband using one-eighth of every channel but only once per superframe, or out-of-band using the extended superframe message channel. While PR-ISDN is similar to T1, it has only 23 DS0 bearer channels, but they are 64 Kbps clear, and it uses the 24th DS0 channel for out-of-band signaling. While the PR-ISDN format is $23B + D$ and the BR-ISDN format is $2B + D$, the D-channels in each format have different bit rates: 16 Kbps in basic and 64 Kbps in primary. Broadband ISDN is about 150 Mbps and, at the time of the writing of the book, is synonymous with asynchronous transfer mode (ATM, Section 19.4).

Whether a DS0 channel is 56 Kbps clear, or 62.6 Kbps clear, or 64 Kbps clear, is not highly relevant when the channel carries digital voice. However, the ability to transmit one complete digital byte in each DS0 frame or timeslot is highly relevant in computer networking applications because of the simplification of the underlying hardware. Thus, because it supports 64-Kbps clear channels, the ISDN formats are well suited for transporting voice and data. The circuit/packet issue is discussed in Chapter 17.

13.4.2 Interfaces

The ISDN interface to customer premises equipment (CPE), like telephones, computers, and PBXs, is subtly complex. Figure 13.7 illustrates ISDN and non-ISDN CPE connected to one ISDN subscriber loop. Four different interfaces are present: called R, S, T, and U. Three different pieces of interface translation equipment are present: called TA, NT1, and NT2. While it seems overly complex at first glance, specifying four different interfaces and three separate pieces of equipment provides the flexibility needed for different implementations. The complexity is necessary and it's actually a clever design.

The S, T, and U interfaces are each just a different line coding of the BR-ISDN $2B + D$ format. But, the R interface is not necessarily even close to ISDN. R represents any non-ISDN interface, including analog POTS, RS-232C, or any proprietary internal

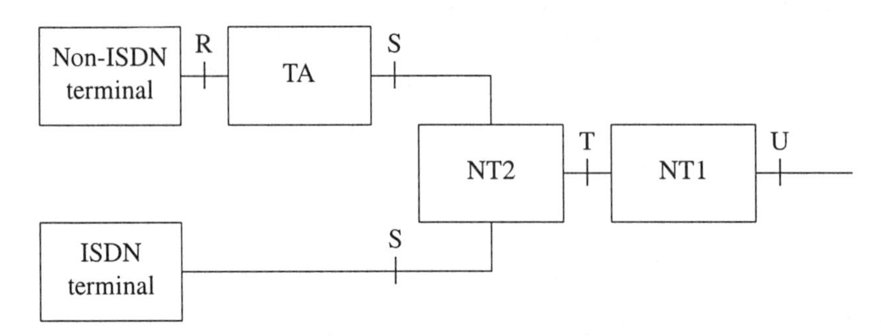

Figure 13.7. Illustrating the ISDN interfaces.

interface inside a switching system. TA represents the terminal adapter that would be required to make the non-ISDN CPE look like ISDN. For analog telephones, most of the BORSCHT functions are in the TA.

The U interface, a two-wire duplex implementation of BR-ISDN over the two-wire subscriber loop, is described in the next subsection. The S and T interfaces represent four-wire simplex implementations that would be found in the local wiring in a home or on one floor of an office building. NT1 is the necessary network terminal equipment, functionally equivalent to a modem, for translating between U and T.

The S and T interfaces are distinguished because of the complexity of connecting multiple CPE terminals to one ISDN line — the POTS equivalent of having extension telephones. The S interface is the four-wire simplex implementation of ISDN that carries data to and from each individual CPE terminal. The T interface is the four-wire simplex implementation of ISDN that carries combined data to and from the digital loop. NT2 is the network terminal equipment that directs incoming data on the T-side to appropriate lines on the S-side and merges outgoing data on the S-side to the common line on the T-side.

Consider the BORSCHT functions (described in Section 13.5) and the additional function of handling extensions. Battery is derived locally (with battery backup) and provided internally in an ISDN terminal or at the R-side of a TA for a non-ISDN terminal. Overvoltage protection must be present at the U-side of the NT1. Ringing is requested by a D-channel signaling message from the CO and provided internally by an ISDN terminal or generated by the R-side of a TA for a non-ISDN terminal. Supervision is signaled to the CO by a D-channel message produced internally by an ISDN terminal or by the S-side of a TA, by conventionally supervising the non-ISDN terminal on its R-side. Code conversion (A-to-D) is provided internally in an ISDN terminal or in the TA for a non-ISDN terminal. Hybridization (2W-to-4W conversion) is provided at the NT1 for the loop and in the TA if the non-ISDN terminal has a two-wire interface. Testing access is distributed in all equipment. Data distribution for extensions is performed by the NT2.

Consider the physical location of various components in the ISDN residence. ISDN to the office is left as an exercise. The NT1 is placed near the protector block. If the residence has only POTS telephones, or other CPE that have POTS interfaces, then the NT2 is unnecessary, the TA is near the NT1, and the house wiring is conventional analog R (which inherently handles extensions). If the residence has at least one ISDN terminal,

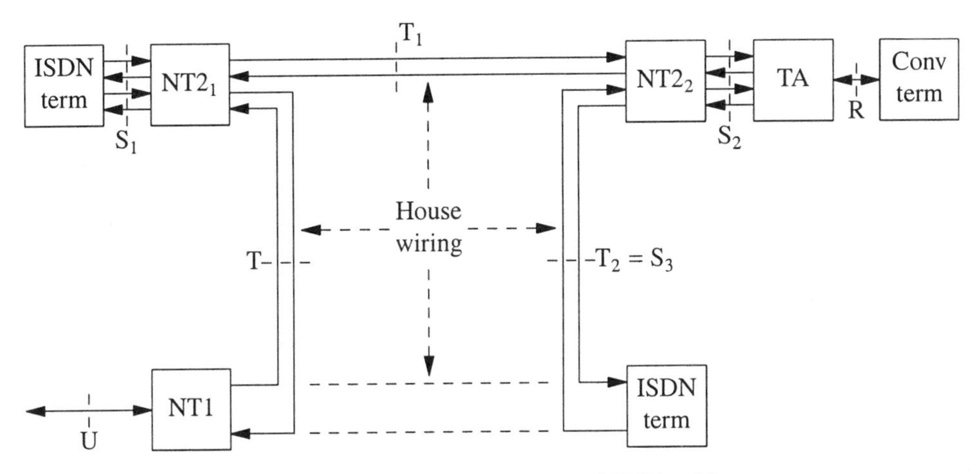

Figure 13.8. NT2 distribution in an ISDN residence.

then each non-ISDN terminal requires its own nearby TA and the house wiring is S or T. If the residence has only one terminal, then the NT2 is unnecessary and the house wiring is T. However, with two or more terminals in the residence, placing the NT2(s) is problematical.

Placing one NT2 near the NT1 at the protector block allows the house wiring to be S, but requires that the house be wired as a four-wire star. With conventional four-wire ring house wiring, the ring must be broken into a four-wire bus at the protector block, the NT2 function must be distributed into each TA and ISDN terminal along the bus (except the last one), and the bus must be physically interrupted by each NT2. Figure 13.8 shows an ISDN residence with two ISDN terminals and one non-ISDN terminal. Each line in the house wiring represents a simplex two-wire path. Figure 13.9 below shows how the equivalent NT2 function is split between the two physical NT2s in Figure 13.8.

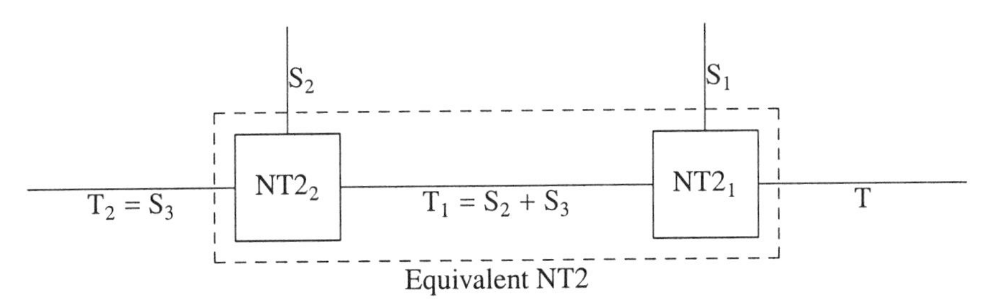

Figure 13.9. Distributed NT2 functionality.

13.4.3 BR-ISDN over an Analog Loop

The S and T interfaces are physically implemented directly on simplex four-wire paths because 144 Kbps is easily transmitted: (1) in one direction and (2) over short lengths of twisted pair, like over house wiring (Section 3.3) or a building's floor wiring (Section 10.2). But, if the twisted pair is inherently two-wire and it is long, like over a residential subscriber's loop (Section 3.4) or over campus wiring (Section 10.2), then the U interface's special line coding is needed. The NT1 equipment is the modem that translates between the simple and direct S and T interfaces used on local wires and the complex and indirect U interface used on long wire-pairs.

The bandwidth of metallic local loops is limited, especially long loops and loops with transmission (load coils, bridge taps, nonuniform gauge, etc.) anomalies that don't affect analog voice transmission, but greatly impair 144-Kbps digital transmission. The other big problem, beside limited bandwidth, that makes it difficult to transmit an ISDN signal over a metallic loop is that the local loop is full-duplex, but over a two-wire twisted pair. So, not only do the S and T interfaces use short lines, with relatively high bandwidth, the cables typically have enough pairs that the S and T interfaces can use four-wire transmission. The U interface, however, uses a long line, with relatively low bandwidth, in a loop plant that is wired for two-wire transmission.

Worldwide, the U-interface has two popular line-coding standards.

- One way to transmit full-duplex digital data over a two-wire twisted pair is to squander its bandwidth even more, a solution that works over short loops. Using "ping-pong" line coding, the line rate is slightly greater than two times the data rate and the channel never transmits data in both directions simultaneously. Eighteen bits are transmitted upstream in the first half of the frame time and 18 bits are transmitted downstream in the second half of the frame time. This solution is popular in regions or countries that have high population densities and, hence, short loops.

- Another way to transmit full-duplex digital data over a two-wire twisted pair is to increase the complexity (and, hence, the cost) of the line coder and decoder on both ends of the U interface's channel. In this standardized scheme: (1) two additional bits in each $2B + D$ frame, for framing and synchronization, increase the frame size to 20 bits and the line rate to 160 Kbps; (2) a four-level line code transmits two information bits with every line symbol and reduces the line rate to 80 KBaud; (3) an echo-canceler chip on each end of the line reduces near-end crosstalk enough to allow simultaneous digital transmission in both directions over the same pair. This solution is popular in regions or countries that have long loops and an economy that allows the endpoints to be more expensive.

Recently, new modulation techniques have effected significant advances in data modems, culminating in various implementations of so-called digital subscriber line (xDSL). Section 19.1 presents BR-ISDN's 144-Kbps line rate in the context of modems and xDSL.

13.4.4 Signaling

A direct metallic-pair BR-ISDN line terminates in a digital CO or PBX onto a specialized line card. The architecture is similar to the analog POTS case, but with obvious differences in the line interface, line card hardware, and subscriber signaling. The duplex U interface is translated to the simplex T interface by NT1 hardware on the line card. Still on the same line card, the two-way $2B + D$ T interface is muxed/demuxed into two mouth channels, two ear channels, and a two-way packet-switched signaling channel. The two two-way B-channels terminate on the system's fabric exactly like two digitized POTS lines.

The card's D-channel terminates on a message bus on the back-plane. On every card carrier for ISDN line cards, its D-channel message bus is connected to some kind of packet-switched network. The switch processor that serves this subscriber is also connected to this packet-switched network. Thus, by properly addressing the D-channel packets, the microprocessor in the subscriber's ISDN terminal or TA has a packet-switched data connection with this switch processor.

Suppose A, an ISDN subscriber, wishes to place a telephone call. From an analog telephone, conventional dialed digit signaling is still required. In POTS, a dial pulse receiver (DPR) or Touch Tone receiver (TTR) in the CO detects dialed digits. In ISDN, the digits could be detected in the TA on the user's premises and the TA would signal them to the switch processor over the D-channel. However, if this is too expensive, POTS signaling could be used instead. A lifts the handset, the R-side of the TA detects the supervision change, and the TA's microprocessor constructs a signaling packet. This off-hook packet, containing A's identification and class of service, is addressed to the switch processor that serves A's calls and is transmitted out the S-side of the TA down the D-channel to the switching office. This switch processor receives the packet, connects A's mouth samples to a TTR, and connects A's ear samples to DT.

From an ISDN telephone or integrated workstation, call origination does not use conventional digit detection. If A conventionally dials a telephone number, the terminal's internal microprocessor collects the digits. This microprocessor could also get the digits from an internal electronic directory or computer file that the subscriber accesses by a speed-dial telephone button or a point-and-click procedure on a display. However it gets the digits, the internal microprocessor constructs a signaling packet that contains A's identification and class of service and the telephone number of the party A wants to call. The packet is transmitted, using out-of-band signaling, to the same switch processor as above, except the packet requests final connection instead of connection for dial tone.

Unlike POTS, in ISDN, the signaling that identifies the called party does not have to come over the calling party's dedicated D-channel. The calling party is identified by data in the signaling packet and not by the physical location of the source of the data. If the electronic directory is located in a shared database, like a server, the user's telephone can communicate with this server over a typical LAN and the server can signal to the switch to initiate the call.

While an ISDN subscriber's signaling may still use the conventional inband signaling path for dialed digits, this path cannot be used for the alerting function (Section 1.5.6). Since the POTS ring signal is incompatible with ISDN's digital electronics, alerting must be performed by upstream signaling. Switching software transmits a packet over the D-channel: to the subscriber's ISDN telephone, where the internal microprocessor causes the telephone to ring by typically connecting an audio tone to a speaker; or to the subscriber's TA, where the microprocessor connects a ring generator located on the R-side of the TA to ring the subscriber's analog telephone. With upstream signaling over the D-channel from switching software to the subscriber's telephone, ISDN can easily implement new features, like calling-party identification, without a special modem.

Consider ISDN's multimedia capability.

- For voice communications, ISDN typically assigns the two B-channels for subscriber PCM, the equivalent of two voice lines to the same telephone, and uses the D-channel as an out-of-band signaling channel for both lines.

- For video communications, a full-motion video signal can be digitized and sufficiently compressed to fit on a 64-Kbps channel. While the picture quality and refresh rate are such that the image is barely acceptable, the cost and the ease of use more than compensate. Unfortunately, market price has not always been acceptable.
- For computer communications, ISDN presents an interesting choice. The ISDN subscriber can elect to use his D-channel, not only for telephony signaling between his CPE and switching software, but also as a data channel for low-rate computer communications. Or, the ISDN subscriber can elect to use one of his B-channels, not for PCM telephony, but as a 64-Kbps computer-communications channel.

13.4.5 ISDN's Network

If the BR-ISDN D-channel is used for computer communications, its effective bandwidth is less than 16 Kbps because the channel must be shared for signaling. While the D-channel has a higher bandwidth than many PC users have over a conventional modem, most residential users justify ISDN's typically high market price by using the higher-rate B-channel for computer communications. Frame relay was designed and offered as a computer communications protocol for use over the B-channel that was optimized for ISDN by taking advantage of telephony signaling. ISDN's designers and early proponents argued that BR-ISDN was an integrated network that would support voice communications, data communications, and integrated voice/data communications. In the mid- to late 1970s, the telecommunications industry seemed destined to ISDN.

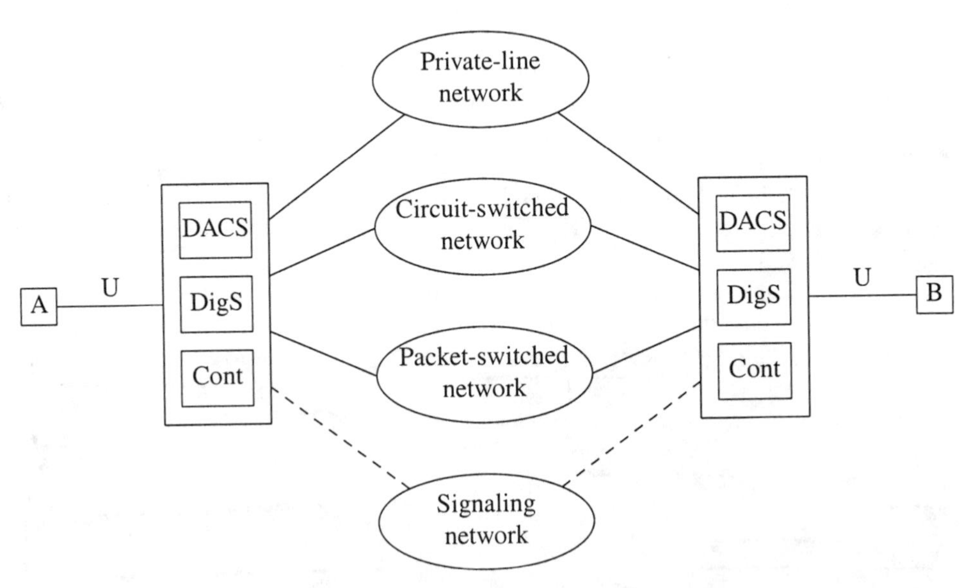

Figure 13.10. The national network from an ISDN perspective.

But ISDN never emerged as an end-to-end integrated network and the little success it has had is as an end-user interface. ISDN was never priced for success and ISDN terminals and instruments are still inexplicably more expensive than their counterparts. While frame relay has become extremely popular, instead of being circuit switched, it is usually used over permanent private lines or virtual private networks. Now, ISDN provides a common integrated interface for voice instruments and for computer communications instruments. Figure 13.10 shows that at the network end of this ISDN interface, voice and data and signaling are each typically separated at the PBX or CO and are each typically switched over separate logical networks. Why? Because computer communications simply isn't optimal over a circuit-switched network, it is more economical to optimize separate logical networks for each traffic type than to provide a single logical network that isn't optimal.

Recently, proponents of asynchronous transfer mode (it used to be called broadband ISDN) and proponents of the Internet protocol (IP) all argue that ATM or IP will "integrate" voice, data, video, and multimedia communications. Some skeptics wonder how this is different from ISDN's goal of more than 20 years ago and think history may repeat itself. But many people in the telecom industry believe that voice over packets is an easier problem than data over circuits. We'll discuss voice over ATM and voice over IP (VoIP), particularly the difference between VoIP and IP telephony, in Chapter 19.

13.5 BORSCHT

Of the many different functions associated with a POTS telephone call, seven of them have been especially identified because their implementation over an analog electrical line is much simpler than over a line that isn't completely so. A digital electronic line or an optical line, or a line that is mostly analog but has digital or optical segments, turns out to be an environment that greatly complicates the implementations of these seven functions. Because the first letters of these seven functions spell out the acronym BORSCHT (a soup made from beets, common to eastern Europe), they are called the BORSCHT functions.

13.5.1 Battery

The mouthpiece in the conventional analog telephone requires direct current for its operation. For many years, the switching office has provided talking battery over the subscriber's two-wire line, configured for DC as an electrical loop (Chapter 3). But because DC can't be transmitted digitally, lines that have analog breaks must provide talking battery from the subscriber-side terminal. The simplest procedure is to allocate several analog pairs between the office and the subscriber-side terminal (not just between the office and the office-side terminal) to bring DC around the digital segment. Talking battery is applied to the subscriber-side analog segment through a per-line circuit like a line circuit located in the subscriber-side terminal.

Telephones with LED lamps and displays and with digital components for storing telephone numbers can be powered by local electricity; but then these components fail when electric power fails. If they are powered by talking battery, either the total current drain (including line leakage) must be under 25 mA during on-hook or the components are not powered until the receiver is raised. Systems that are digital-to-the-home (or optical) will need to provide an analog pair for DC or will rely on local electricity with battery backup.

13.5.2 Overvoltage Protection

In a conventional analog system, lightening hits and other forms of overvoltage on the line must be isolated from the subscriber's telephones and house wires and from the switching equipment in the switching office. Overvoltage protectors are installed on the protector block in the subscriber's residence and also on the main frame in the switching office. In a line with analog breaks, each digital segment in the line must also be protected on both sides. Furthermore, digital circuits require greater protection than conventional analog equipment. Installing digital telephones in the residence may require changing the overvoltage protectors.

13.5.3 Ringing

The traditional signal for alerting the subscriber is a high-voltage AC signal that physically rings a bell in the telephone (Section 3.2.1). Acoustic power must be large enough for the signal to be audible and the acoustic power is traditionally provided directly by the electrical power of the ring signal. In Step, Panel, all versions of Crossbar, and even in the #1 ESS, telephones are rung by connecting the telephone's line to a ring generator through the system's fabric. The switching devices in the fabric must be capable of transmitting this high-voltage, and relatively high-power, signal. This procedure cannot work in a digital fabric and cannot work over a subscriber line that has a digital segment because power and high voltage cannot be transmitted directly using digital technology.

A digital telephone can be rung by transmitting a ring message over a message channel to the telephone. The telephone must have access to locally derived power in the case of a true bell or must produce an analog sound through a speaker (called a tone ringer). The switching office must know, by class of service, that this line must not be rung in the conventional way. Not only will it not work, but the high-voltage ring signal would probably destroy the digital electronics in the telephone.

The common situation with a conventional POTS telephone at the subscriber end of a loop that has a digital segment is even more complex. The digital transmission terminal on the subscriber end of the digital segment must be equipped with a ring generator and it must be able to apply ring signal without destroying the digital electronics in the terminal. The control unit in this subscriber-side terminal must receive a message when a telephone must be rung. This message comes, either directly from a processor in the office over some signaling channel, or from the control unit in the office-side terminal of the digital segment. In this common latter case, the office rings the line conventionally, this ring signal triggers a ring-detector circuit on the line's termination in the office-side terminal, the office-side controller sends a message over some digital channel to the controller in the subscriber-side terminal, and the analog segment is rung from the subscriber-side terminal.

13.5.4 Supervision

Supervision is the on-hook/off-hook status of the subscriber's telephone. In an analog metallic pair, supervision is signaled by the DC in the loop. You can't transmit DC, even at low power, directly over a digital segment. Supervision has been provided over digital spans in a variety of procedures. The problem is not so much how to do it, but that we have so many different ways to do it.

It would seem to be fairly simple to connect two digital carriers together in series; as long as the level of multiplexing is the same. But two T1s that use different supervision procedures are most simply interconnected by adding a short analog span between them. There is a lesson to be learned here; it represents a microcosm of the problems caused by having too many choices.

13.5.5 Coder and Decoder

The "codec" is the analog-to-digital converter. Since the telephone is inherently analog, any local loop with a digital segment must have a codec on the subscriber side of this segment. The problem is that there is no completely satisfactory place to put it.

In a local loop where the digital segment is installed solely for pair-gain, the codec resides in the digital terminal on the subscriber side of the segment and the loop is totally analog from this point to the telephone. However, as the subscriber end of the digital segment evolves to being closer to the subscriber, the idea of applying digital technology all the way to the telephone is tantalizing. If we put the codec in the telephone, we gain the potential advantage of digital time multiplexing at the telephone and in the house wiring. But we lose the simplicity of analog technology in implementing extension telephones on bus-wired house wiring and we require that the newly digitized subscriber discard all his analog telephones.

Putting the codec at the protector block may be a reasonable compromise in a house that has multi-pair house wiring. This way, the subscriber is digital-to-the-premises and can still use simple analog telephones in some rooms.

13.5.6 Hybrid

Recall that the hybrid is the transformer that performs the conversion from a full-duplex two-wire physical channel to a four-wire physical channel that is half-duplex on each pair. Most analog facilities are two-wire and most digital facilities are four-wire. Almost always, the drop-wire and the house wiring is two-wire. Since the telephone handset is four-wire, any analog telephone contains a hybrid (Section 3.1).

If a local loop with a digital segment contains an analog segment between the telephone and the subscriber side of the digital terminal, then at least two hybrids are necessary: one in the telephone and one at the terminal. If the digital loop could be digital to the telephone (including digital four-wire drop wire and house wiring), then no hybrid would be needed at all.

13.5.7 Testing

While busy/idle testing can be performed digitally, by examination of the supervision bit, many other forms of testing are inherently analog. If the subscriber's loop is interrupted by a digital segment, craftsmen cannot perform tests from the switching office that check for a permanently open circuit, permanently short circuit, or permanent short-to-ground in an analog segment on the subscriber's side of the digital segment, such as in the subscriber's drop-wire.

Other, more sophisticated tests that can also be performed over real physical wires from the switching office include testing for high impedance or leakage, testing bandpass, and checking for the number of telephone bells on the line (by measuring inductance). But, if a line has digital or optical segments, unless these segments are specially equipped, these tests must be performed from the subscriber's side of the digital or optical segment; from the vault, the pedestal, or the subscriber's premises.

13.6 OAM&P

The architectures of telephone switching systems have been optimized on the bases of technology, application, theory, and regulation. This book has covered these items in detail. But there is another important aspect — even automatic telephone systems must be manually operated, administered, maintained, and provisioned. Probably because the telecom industry has historically separated R&D and manufacturing from the carriers, even inside the Bell System, operation, administration, maintenance, and provisioning (OAM&P) historically have been considered after the system was designed. Because OAM&P has had little impact on architecture and because it is the purview of the carriers only, we will give it light coverage here. Do not interpret this to mean that OAM&P is unimportant, because it is an extremely important (and expensive) issue for the carriers.

- Operation includes all the day-to-day activities that are required to run the system.
- Administration includes all the routine manual and clerical activities required to keep up with customer changes, system changes, billing, and the like.
- Maintenance includes all the activities necessary to detect, locate, and repair faults and the routine steps taken to try to prevent them.
- Provisioning includes the routine activities required to change trunk groups and assign private lines (Chapter 12).

While some tasks (for example, reassigning line appearances on the main distribution frame) are easily categorized (MDF changes fall under administration), many tasks are not easily categorized and the distinctions are blurring more. The carriers treat these four activity categories collectively, using the term OAM&P as if it were a single name, similarly to how R&D usually specifies a single activity. Sometimes provisioning is not included with the other activities, leaving the others covered by the term OA&M.

OAM&P activities are typically very labor intensive. Many years ago, when the cost of labor was a smaller percentage of the carriers' total cost, the cost of OAM&P was not so significant. The automation and centralization of OAM&P became increasingly important. While OAM&P could have been automated and centralized before the digital age, it occurred at the same time as digital technology was introduced to telephony and the systems that automate and centralize are typically digital. The DACS is an example — while analog crossconnect could have been automated too, it wasn't. So, improved and cost-reduced OAM&P is another procedure that is synergistic with digital switching.

OAM&P has evolved through three generations. For the first hundred years, carriers performed ad-hoc OAM&P. Then, during the 1970s and 1980s, operation support systems (OSSs) were designed and installed in individual switching offices. Each of a large number of distinct OSSs, with an "alphabet soup" of acronyms, automated its specific OAM&P task. Then, after this second generation of OAM&P automation, third-generation OSSs were moved out of the switching offices and into large centers from which an entire region's OAM&P is performed.

Sections 5.4.1 and 9.3.2 showed how economy of scale centralized the telephone company's operators into large logical pools that serve a large region. The same principle has similarly centralized the telephone company's OAM&P personnel so that, now, most large telephone companies perform many of their automated OAM&P activities from regional centers. The design of these centers has an associated anecdote. The story goes that the design team at Bell Labs decided to name their system CHIMERA, an acronym for centralized hierarchical integrated maintenance equipment repair arrangement. AT&T's management liked this name, especially when they read that the Chimera was a fire-breathing monster from ancient Greek mythology that had a lion's head, a goat's body, and serpent's tail. The system was almost released to the BOCs with this name until someone, at the last minute, found that secondary definitions for Chimera include hodge-podge or an impossible or foolish fancy. But, despite this namer's opinion, centralized OAM&P has been beneficial for the LECs.

Let's look briefly at several OAM&P activities. The *administration of subscriber changes* includes loop wiring and MDF wiring (Section 3.4.4) and it includes the bookkeeping of assigning telephone numbers and class of service. While order-writing has become automated, these activities are still quite labor intensive. The subscriber's class of service has been administered quite differently as systems evolved through Step, Crossbar, and computer-controlled systems. After divestiture, it became even more complicated as telephone companies had to keep track of the subscriber's default interexchange carrier. Described in Chapter 18, migration toward a regional line information database (LIDB) will effect even further change in procedure.

Testing the analog and digital channels, on the line side and on the trunk side, is routine and also is done in response to direct complaints. In early systems, this testing was done physically at the line circuit or trunk circuit, by the craftsman who connected test equipment to the physical channel. As systems became more automated, the test equipment was located at the office's test center, such as the 1E's trunk and line test panel. There, the craftsman effects a connection through the fabric to the line or trunk under test. Now this testing is even more automated and it is performed from a regional center. Testing physical channels through a digital system is problematic, as discussed in the "T" of BORSCHT in the previous section.

Service observing was discussed briefly in Section 9.3.4. If the carrier uses this OAM&P practice, the service observing attendants interact directly with the craftsmen. While routine testing can detect some poor channels, many bad channels are located and repaired as a result of reports from customers or from service monitors. Besides locating channels with poor transmission, customer complaints and reports from service observing attendants can also be useful to locate underprovisioned trunk groups.

Many carriers have implemented complicated systems for *traffic monitoring* so that channel utilization is reported to a regional center. In Sonet, channel utilization is common-channel signaled over a data channel that is embedded in Sonet's line overhead. Traffic monitoring systems are evolving into regional subnetworks in much the same way that the system for call signaling evolved into Signaling System 7 (SS7, Section 18.1). Traffic monitoring information is useful in routing by the Advanced Intelligent Network and it is only a matter of time until this subnetwork fully merges into whatever SS7 evolves to.

The previous chapter described *provisioning* and how the DACS automated this activity. It won't be long before transmission facilities are automatically provisioned by software based on data collected by the carrier's traffic monitoring systems, Then, carriers may begin to allow their private-line customers the capability to log on to the carrier's DACSs and to provision their own facilities.

Call tracing is one of the most dramatic OAM&P tasks. We have all seen the kidnapping scene in a movie where the called party is asked by the police to keep the kidnapper on the phone long enough so that telephone company craftsmen can run around the office trying to locate the calling party's physical line. While this is how call tracing worked in Step and in Crossbar, it has been completely automated for many years. Even if story writers knew this, the old procedure makes for a much more dramatic story.

Faults in Step's switching fabric were typically located by subscriber complaints. Crossbar was more automated than Step and had an internal subsystem for testing its fabric (hence the sixth bar on the Crossbar switch). Whenever a fault was found, an alarm sounded and a small printer near the system's test panel dropped a trouble ticket. Beginning with the 1E, faults in the fabric and in every other part of the system are detected, located, and diagnosed by software. With many offices now unstaffed, trouble reports are delivered to the regional test center and personnel are dispatched, if necessary.

Now, partly because labor cost (particularly, the cost of benefits) has increased, and partly because automation has lowered other costs (like equipment), OAM&P has become an even more significant component of a carrier's cost. While the incumbent LECs have always been concerned about the reliability and survivability of their networks, increased competition in local telephony has spurred even more proactive trouble detection and repair. No LEC, incumbent nor competing, wants the stigma of not being reliable. The presentation in Chapter 19 of voice over ATM and VoIP will discuss how OAM&P is an extremely relevant factor in any carrier's or enterprise's decision about migrating toward a single integrated multimedia network based on either of these packet-switched paradigms. Since OAM&P is such a significant part of a network owner's cost, they will have to be careful to ensure that competition on price doesn't lead them to erode their network's reliability and survivability.

EXERCISES

13.1 An LEC considers replacing a Number 5 Crossbar office by a digital switch. Suppose the initial cost of the new switch, including installation, is $3M. Use a factor of 0.3 to relate first cost to annual cost. Suppose the digital switch is cheaper to operate by $200K per year. Suppose the LEC's long-range trunking plan calls for $10M worth of digital transmission equipment and that this equipment is 20% cheaper if A-to-D converters are not needed at the switch. (a) Should the Crossbar office be replaced? Suppose the LEC's long-range loop plan calls for $2M worth of digital transmission equipment and that this equipment is 25% cheaper if A-to-D converters are not needed at the switch. (b) Now should the Crossbar office be replaced?

13.2 An analog line concentrator resides logically between the line circuits and the line side of a switching fabric. One might be placed physically at the distribution point in the loop architecture or, even at the pedestal — in which case, the line circuits are also remoted. Or, one might be placed physically in the switching office, like the line finder stage in a Step fabric. Suppose a Crossbar switch, built-out to be 60×10, costs $1000. Suppose the A-to-D converters in a digital switch's line circuits cost $15 each. (a) Is it economical to split the line circuits in a digital switch into their analog and digital components and to use an in-office analog line concentrator between 60 analog line circuits and 10 ADCs? (b) Suppose the cost of remoting the 60 analog line circuits and the line concentrator is another $3000, but the loops between the remote point and the CO cost $70 apiece. Is remote line concentration economical?

13.3 Consider a TSI for a system with four timeslots per frame. Draw a block diagram — it should have four 2-bit words, eight 8-bit words, and a 3-bit binary counter. Walk the TSI through three consecutive frames of incoming data. Label the data in 12 consecutive incoming timeslots: A through L. Let the switching assignment be such that data from incoming timeslot 0 is output in timeslot 0, data from incoming timeslot 1 is output in timeslot 3, data from incoming timeslot 2 is output in timeslot 1, and data from incoming timeslot 3 is output in timeslot 2. For each timeslot, show on the block diagram the incoming data stream, the state of the address counter, the state of the read address, the contents of the RAM, and the outgoing data stream.

13.4 Consider a TSI for a system with only two timeslots per frame. Incoming timeslots are either interchanged, or not, from input to output through such a TSI. Consider a network with three such TSIs in series and a multiplexed format with four timeslots in a frame. Let the TSIs be structured so that the first and third TSIs interchange, or not, consecutive pairs of timeslots in each frame — first with second and third with fourth. The center TSI interchanges, or not, odd-numbered or even-numbered timeslots in each frame — first with third and second with fourth. This structure is the time equivalent of a 4×4 Benes network (Section 7.3.4). (a) Make two sketches of a 4×4 Benes network. (b) On each sketch, show a different fabric configuration for connecting the upper input to the lower output. (c) Corresponding to each configuration, show how the three-stage network of 2×2 TSIs has two configurations by which it can interchange the first and fourth timeslots in a frame.

13.5 Sketch a space-only Clos network with $n = 3$, $m = 5$, and $r = 3$. Sketch a TDM-type TST fabric that has $r = 3$ muxes on each edge, each mux serving $n = 3$ external ports. The 3×3 TMS and all the internal links have $m = 5$ timeslots. On each sketch, show a connection from the middle port on middle left switch to the middle port on the middle right switch, using the middle interior channel.

13.6 A TST network connects eight physical lines on each side, where each line has a DS1 format, but without the frame bit. (a) Draw a block diagram of the network. Show all modules and indicate whether their function is TDM, TSI, or TMS. Show all physical links. (b) Since the lines are lightly utilized, the network concentrates in its first stage and expands in its third stage. Assume that the external lines are only 25% occupied on the average, and that traffic distributes uniformly through the network. To ensure that the blocking probability = 0.01, how many timeslots are required in each of the network's internal links? How many timeslots are required in the TMS? (c) An incoming channel uses timeslot T in input line C. This channel's user requests a connection to an outgoing channel that uses timeslot V in output line X. If the switching network is totally idle, how many different paths could be used? (d) Let this connection be assigned to internal timeslot U. Specify which modules are used for this connection and, for each one, specify its internal switching assignment.

13.7 Sketch a TSI-type TST fabric that has a 3 × 3 TMS in the center and banks of TSIs on each edge. All ports, TSIs, links, and the TMS all have 24 timeslots. Timeslot 17 on left port 0 must be connected to timeslot 3 on right port 2 and internal timeslot 6 will be used. (a) Specify the timeslot interchanges performed in each TSI. (b) Specify the physical connections through the fabric during timeslot 6.

13.8 Repeat the previous problem for an STS fabric that has a 3 × 3 TMS on each edge and a bank of TSIs in the center.

13.9 An STS architecture is proposed for a new modular digital class-5 switching system. The system has a maximum of five SMs that are fully interconnected by inter-SM links in each direction. Center-stage TSI occurs in these inter-SM links. Each SM has a maximum of four TDM units, where each TDM unit serves a maximum of 2500 ports (lines, trunks, and service circuits) at 10:1 concentration. Each SM also has an internal 8 × 8 TMS with ports for links to the SM's internal TDMs and ports for the SM's inter-SM links. (a) Sketch the complete system, showing all SMs and their inter-SM links with TSIs. Inside each SM, show its TMS, all its TDM units, and all its internal links. (b) How many ports can be served by the largest system? (c) How many timeslots are required in the TM-TMS links? In each TMS? In each inter-SM link? In each TSI? (d) Where might this system be vulnerable to blocking?

13.10 Twenty-three BR-ISDN lines are multiplexed up to two PR-ISDN streams. What is the maximum average bit rate on the BR-ISDN D-channels so that the PR-ISDN D-channels can keep up with them?

13.11 Sketch a traditional Ethernet LAN configuration, a coax connected to a MAN gateway and to several PCs, each with a network interface card (NIC) and a tap. Sketch a residential BR-ISDN configuration: a local loop connected to several ISDN and analog appliances with intermediate NT1, NT2, and TA interfaces. Show the correlation between the components in each configuration.

13.12 Consider an ISDN office with an ISDN telephone, an analog POTS extension telephone, a UNIX work station, and a dock for a lap-top PC. (a) Show the R- and S-lines inside the office if the office has an internal TA and NT2 and one ISDN T-line from an NT1 in a wiring closet. (b) Repeat the problem for multiple S-lines into the office from a combined NT1/NT2 in the wiring closet.

13.13 Sketch block diagrams of four different configurations for ringing a telephone: (a) from the trunk side of the fabric, (b) from the line card, (c) from the user side of a digital loop carrier system, and (d) from the telephone itself. For each configuration, show the maximum digitization of the path between the telephone and the ring generator. For each configuration, describe the signaling path that might be required between the ring generator and the switching office's controller.

REFERENCES

[1] *Webster's New World Dictionary,* Third College Edition, New York, NY: Simon & Schuster, 1991.

[2] Marcus, M. J., "Space-Time Equivalents in Connecting Networks," *Proc. 1970 Int. Conference on Communications,* 1970, pp. 35.25 – 35.31.

SELECTED BIBLIOGRAPHY

"5ESS Switch — Global Technical Description" *AT&T Technical Journal,* No. 2, June 1990.

Aaron, M. R., "Digital Communications — The Silent (R)evolution?" *IEEE Communications Magazine,* Vol. 17, Jan. 1979, pp. 16 – 26.

Aidarous, S. E., P. A. Birkwood, and R. M. K. Tam, "An Architectural View for Integrated Network Operations," *IEEE Communications Magazine,* (issue on Network Planning), Vol. 25, No. 9, Sept. 1987.

Aldermeshian, H., "ISDN Standards Evolution," *AT&T Technical Journal,* (issue on ISDN), Vol. 65, No. 1, Jan./Feb. 1986, pp. 19 – 25.

Andrews, F. T., "Switching in a Competitive Market," *IEEE Communications Magazine,* (special issue on Switching in a Cooperative Environment), Vol. 29, No. 1, Jan. 1991.

Bastian, M., "Voice-Data Integration: An Architecture Perspective," *IEEE Communications Magazine,* (issue on Office Automation), Vol. 24, No. 7, Jul. 1986.

Bates, R., P. Abramson, and F. Noel, "Media for Voice-Data Integration," *IEEE Communications Magazine,* (issue on Office Automation), Vol. 24, No. 7, Jul. 1986.

Bellamy, J., *Digital Telephony,* Second Edition, New York, NY: Wiley, 1991.

Davis, H., P. Lashley, and C. Davis, "ISDN Adapter Design," *Communications Systems Design,* Vol. 2, No. 12, Dec. 1996.

Day, J. F., and A. Feiner, "Networking Voice and Data with a Digital PBX," *AT&T Technology Magazine,* (issue on Telecom 87), Vol. 2, No. 3, 1987.

Decina, M., and A. Roveri, *ISDN — Architectures and Protocols,* Milan, Italy: Italtel, 1986.

Frankel, D., "ISDN Reaches the Market," *IEEE Spectrum Magazine,* Vol. 32, No. 6, Jun. 1995.

Freuck, P., and J. Kutney, "An Approach to Introducing ISDN Basic-Access Terminal Adapters," *Telecommuncations,* Vol. 22, No. 2, Feb. 1998.

Geier, P., "ISDN: What's My Line?" *Communications News,* Vol. 35, No. 10, Oct. 1998.

Goeller, L. F., "The Great Integrated Digital Whoop-De-Doo," *Business Communications Review,* Nov./Dec. 1978.

Guinn, D. E., "ISDN — Is the Technology on Target?" *IEEE Communications Magazine,* (issue on ISDN: A Means Towards a Global Information Society), Vol. 25, No. 12, Dec. 1987.

Harrington, E. A., "Voice/Data Integration Using Circuit Switched Networks," *IEEE Transactions on Communications,* Vol. COM-28, Jun. 1980, pp. 781 – 793.

Hibner, D., "ISDN: Speeding Data Transfer," *Communications News,* Vol. 34, No. 1, Jan. 1997.

Holliman, G., and N. Cook, "Get Ready for Real Customer Net Management," *Data Communciations,* Vol. 24, No. 13, Sept. 21, 1995.

Hui, J. Y., *Switching and Traffic Theory for Integrated Broadband Networks,* Boston, MA: Kluwer, 1990.

Jacobsen, G. M., "OSSs — An Inevitable Evolution," *Telephony Magazine,* Vol. 222, No. 2, Jan. 13, 1992.

Joel, A. E., Jr., "Digital Switching — How It Has Developed," *IEEE Transactions on Communications,* Vol. 27, No. 7, Jul. 1979, pp. 948 – 959.

Joel, A. E., Jr., "Towards a Definition of Digital Switching," *Telephony,* Vol. 197, Oct. 1979, pp. 30 – 32.

Joel, A. E., Jr. (ed.), *Electronic Switching: Digital Central Office Systems of the World,* New York, NY: IEEE Press, 1982.

Jordan, B., "Network Management: How Integrated Is Your System?" *Telecommunications,* Vol. 22, No. 2, Feb. 1998.

Kajiwara, M., "Trends in Digital Switching System Architectures," *IEEE Communications Magazine,* (issue on Digital Switching), Vol. 21, No. 3, May 1983.

Lin, N., and C. Tzeng, "Full-Duplex Data Over Local Loops," *IEEE Communications Magazine,* (issue on Human Communication Systems), Vol. 26, No. 2, Feb. 1988.

Mack, J. E., and W. B. Smith, "Centralized Maintenance and Administration of Electronic Switching Systems," *Proceedings of the IEEE,* (issue on Telecommunications Circuit Switching), Vol. 65, No. 9, Sept. 1977.

Marcus, M. J., "The Theory of Connecting Networks and Their Complexity: A Review," *Proceedings of the IEEE,* (issue on Telecommunications Circuit Switching), Vol. 65, No. 9, Sept. 1977, pp. 1263 – 1271.

McDonald, J. C. (ed.), *Fundamentals of Digital Switching,* Second Edition, New York, NY: Plenum, 1990.

Nussbaum, E., and W. E. Noller, "Integrated Network Architectures — Alternatives and ISDN," *IEEE Communications Magazine,* (issue on ISDN), Vol. 24, No. 3, Mar. 1986.

Perucca, G., et al., "Advanced Switching Techniques for the Evolution to ISDN," *Telecommunications Magazine,* (issue on Switching Inside and Outside the Network), Vol. 21, No. 2, Feb. 1987.

Petterson, G., "ISDN: From Custom to Commodity Service," *IEEE Spectrum Magazine,* Vol. 32, No. 6, Jun. 1995.

Roca, R. T., "ISDN Architecture," *AT&T Technical Journal,* (issue on ISDN), Vol. 65, No. 1, Jan./Feb. 1986, pp. 4 – 17.

Ross, M. J., A. C. Tabbot, and J. A. Waite, "Design Approaches and Performance Criteria for Integrated Voice/Data Switching," *Proceedings of the IEEE,* Vol. 65, No. 9, Sept. 1997, pp. 1283 – 1295.

Rutkowski, A. M., "Emerging Network Switching Technology and Applications," *Telecommunications Magazine,* (issue on Switching Inside and Outside the Network), Vol. 21, No. 2, Feb. 1987.

Schindler, G. E., Jr. (ed.), *A History of Engineering and Science in the Bell System,* Murray Hill, NJ: Bell Telephone Laboratories, 1982.

Schneider, K., I. Frisch, and W. Hsieh, "Integrating Voice and Data on Circuit-Switched Networks," *EASCON,* Washington, DC, 1978, pp. 720 – 732.

Smith, L. D., "Operations Systems Support for Future Telephone Company Networks," *IEEE Communications Magazine,* (issue on Network Planning), Vol. 25, No. 9, Sept. 1987.

Stallings, W., *ISDN and Broadband ISDN with Frame Relay and ATM,* Fourth Edition, Upper Saddle River, NJ: Prentice Hall, 1999.

Stratnmeyer, C., "Voice/Data Integration — An Applications Perspective," *IEEE Communications Magazine,* (issue on ISDN: A Means Towards a Global Information Society), Vol. 25, No. 12, Dec. 1987.

Thompson, R. A., "An Experimental User-Resident Communications Controller Supporting Sub-Rate Circuit-Switched Service," *Proc. Int. Symposium on Subscriber Loops and Services,* Munich, Germany, Sept. 15 – 19, 1980, pp. 68 – 71.

Thompson, R. A., "Experimental Multiple-Channel Circuit-Switched Communications," *IEEE Transactions on Communications,* Vol. Com-30, No. 6, Jun. 1982, pp. 1399 – 1408.

Tomita, S., et al., "Some Aspects of Time-Division Data Switch Design," *Proceedings of the IEEE,* Vol. 65, No. 9, Sept. 1997, pp. 1295 – 1304.

Watkinson, B. G., and B. E. Voss, "The DMS-10 Digital Switching System," *IEEE Nat. Telecommunications Conference Rec.,* 1977, pp. 7.2.1 – 7.2.4.

14

Digital Switching Systems

"The Great Integrated Digital Whoop-De-Doo" — title of a 1978 essay by Leo Geller

This chapter describes three commercial digital switching systems: a digital toll switch in the first section, a class-5 digital local switch in the second section, and a digital PBX in the third section. As in Chapters 6, 8, and 11, details are taken from systems that were developed in and for the Bell System: the 4E, 5E, and System 75, respectively. Please don't interpret this as an advertisement for Lucent products. I wanted a representative system in each category and I selected these systems because their designers wrote thorough papers on how these systems work. In fact, I don't describe the latest releases of these systems because the technical details of the later systems are not as well published. The 4E's software isn't described because it's similar to that of the 1A. But, the chapter does present considerable detail of the software architectures of the original 5E and the System 75. Then, in the last section, a research system called XDS is described.

14.1 A DIGITAL TOLL SWITCHING SYSTEM

The toll point was discussed in Chapter 9 and the #1 Electronic Switching System in Chapter 11. ESS #4, now called the 4E, was specifically designed as a toll switch for a class-4 office [1 – 6]. The 4E's software is similar to the 1E's software, and the 4E even used the same processor that was used in the 1A. But the 4E's hardware was a significant departure from previous systems, using a digital electronic fabric with space and time switching.

14.1.1 Hardware Architecture

Figure 14.1 illustrates the hardware architecture of a digital toll switch. While the sidedness of a toll switch had local offices on one side and the long distance network on the other, all trunks are shown on the left of Figure 14.1, regardless of destination. Each individual (and phantom) metallic trunk terminates in a trunk circuit, called a voice frequency terminal (VFT) in the figure. Each analog carrier channel is demultiplexed (left-to-right), or multiplexed (right-to-left), and its demultiplexed analog subchannels also terminate on VFTs. Analog inband signaling is separated from voice at the VFTs

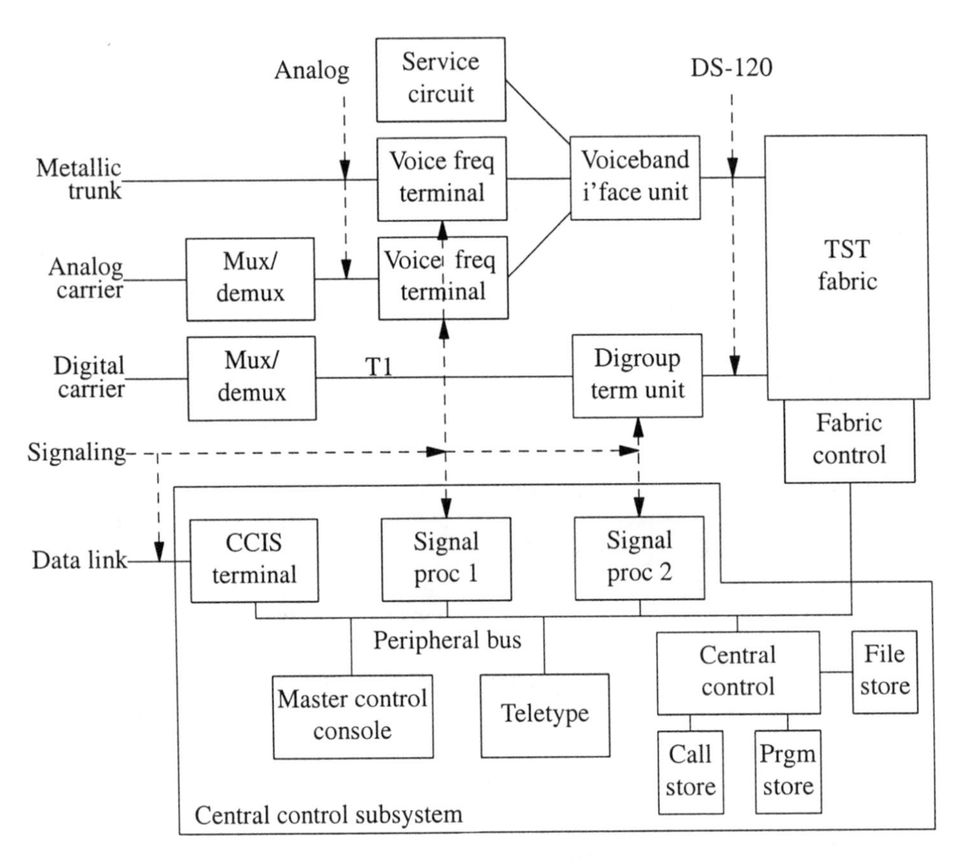

Figure 14.1. 4E block diagram.

and is handled by one of the signal processors (SP) in the processor subsystem. Each analog trunk channel and each service circuit, like a receiver or outpulser, terminates on a voice interface unit (VIU), where left-to-right analog signals are A-to-D converted and multiplexed and right-to-left digital signals are demultiplexed and D-to-A converted. This digital time-multiplexed channel uses a DS-120 format, which has 120 8-bit DS0 channels in each 125-μs frame, and terminates directly onto the switching fabric.

Each digital T1 carrier channel terminates directly on a digroup terminal unit (DTU). Each higher-order digital carrier channel (DS2, DS3, etc.) is demultiplexed (left-to-right), or multiplexed (right-to-left), to its component T1 subchannels, that also terminate on DTUs. Call signaling is separated from voice at the DTU and is handled by the other signal processor in the processor subsystem. Each DTU terminates five T1 channels and multiplexes (left-to-right), or demultiplexes (right-to-left), them to this same DS-120 format, which terminates directly onto the switching fabric.

The extension of the 1E/1A central control architecture to using a signal processor is further extended in the 4E to using two signal processors. One SP acts as an i/o front-end for signaling on analog trunks and the other for signaling on digital carrier. Out-of-band signaling, over data links to other switching offices, is handled by another special-purpose i/o front-end processor, called the CCIS terminal (common channel interoffice signaling is described in Section 18.1). Thus, in the 4E, the main processor spends most of its time in call processing. The 4E's central control is the same processor that was designed and built for the 1A, and the rest of the processor subsystem is similar to that of the 1A.

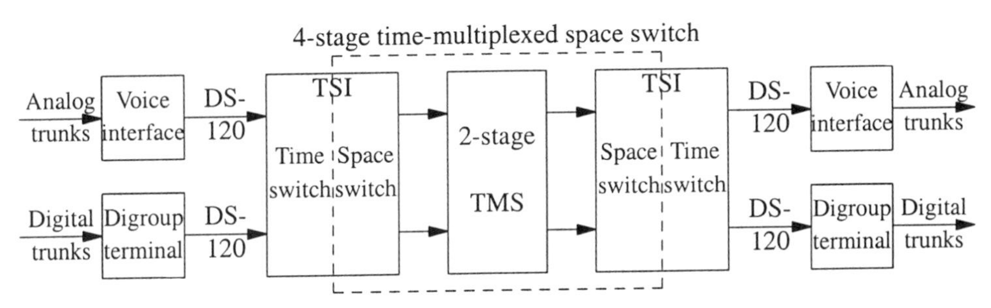

Figure 14.2. Time-space-time fabric.

14.1.2 Fabric

The 4E's time-space-time fabric is illustrated by the block diagram in Figure 14.2. The sidedness of the figure is mouth-to-ear and data moves from left-to-right across the figure. While two-way trunks are shown terminating once, on the left, in the preceding figure, they actually must terminate on both sides of the fabric, as shown in this figure.

The fabric has three physical stages of switching and three logical stages of switching, both shown horizontally across Figure 14.2. The multiple modules in each stage are not illustrated. The three physical stages consist of:

- A set of modules, called timeslot interchangers, in the leftmost stage;
- A similar set with similar names in the rightmost stage;
- And a set of modules, called time-multiplexed switches, in the center stage.

The number of physical modules in each stage depends on the size of the office, and the modules are interconnected by the traditional perfect shuffle (Section 7.2.7). Each timeslot interchanger module has two logical stages, called a time switch and a space switch. The three logical stages of switching consist of:

- The time switches from each of the timeslot interchangers in the physical stage at left form the logical multimodule TSI on the left;
- The time switches from each of the timeslot interchangers in the physical stage at right form the logical multimodule TSI on the right;
- The space switches from each of the timeslot interchangers in the physical stage at left, the two-stage TMS modules from the physical center stage, and the space switches from each of the timeslot interchangers physical stage at right all form the four-stage time-multiplexed space switch in the logical center stage.

The architecture is exactly that of the Clos network (Section 7.3.2) extended into a time-space-time structure (Section 13.3). The typical digital toll office terminates about 30,000 trunks, or their analog- or digital-multiplexed equivalents. The fabric has no concentration because trunk channels are already concentrated. If trunk utilization is so low that concentration might significantly raise the utilization of internal links in the fabric, then the cost of all those unused trunks is much greater than the cost of any extra fabric capacity. It makes more sense to eliminate trunks than to concentrate them. Thus, the throughput of a typical toll switch is about 30 K × 64 Kbps = 2 Gbps. While this throughput figure isn't so impressive by today's standards, it was an extremely large figure for a system in the mid-1970s.

14.1.3 Network Evolution

By the early 1970s, the Bell System's entire intertoll network had Number 4 Crossbar systems in all its switching nodes, from class 4 through class 1. Many were augmented by TSPS. When development of the #4 ESS was completed in 1976, a gradual installation program was begun for Number 4 Crossbar replacement. By 1981, over 50 Number 4 Crossbar systems were replaced by 4Es, about one-third of AT&T's total number of interexchange nodes. Number 4 Crossbar replacement was almost completed at the time of divestiture in 1984.

After the 5E (next section) was developed as a digital switch for class-5 offices, subsequent operator service position software (OSPS) was added to the 5E for controlling connections to toll operators. TSPS offices were gradually decommissioned and replaced by 5Es with OSPS. Later, another software package adapted local digital switches to toll switching applications. Now many intertoll, point-of-presence, and intra-LATA switching offices are modified digital local switches and the 4E no longer has a U.S. market. But, the North American interexchange network is still dominated by the #4 ESS.

While the emergence of other long distance companies (Section 10.6) was certainly not good for AT&T's long distance business, it was good for AT&T's manufacturing business (WECo) because they represented another market for the 4E (and, later, the 5E). After divestiture, the switching market became truly competitive, and Western lost some market share among the Baby Bells that had been captive markets when they were part of the Bell System. As many countries modernized their local and long distance networks, they leap-frogged over the 1E type of technology and replaced electromechanical switching offices by digital switches. A very large international market emerged suddenly and it became extremely competitive. There has been a shake-out of switch vendors over the past 20 years, not just within the United States, but globally.

14.2 A DIGITAL LOCAL SWITCHING SYSTEM

While the book is organized in a compromise between history and pedagogy, the sequence of these next two sections, covering a digital CO and a digital PBX, has been especially problematic. Since a digital PBX's size and functionality are similar to the size and functionality of a *switching module,* a component of a digital CO switch, hardware pedagogy suggests describing the digital PBX first. And because digital PBXs predated digital COs, in general, historical accuracy and continuity of their marketing suggest describing a digital PBX before describing a digital CO. In this chapter, this ordering is less important because the particular PBX and CO that are described in detail, the Bell System's System 75 and 5E, respectively, were relatively close contemporaries. However, because the software architecture of the selected PBX is a compromise between that of the 1E and the 5E, software pedagogy suggests describing the 5E's software before describing the compromise that followed. So, the digital CO is described first.

I'll describe the 5E here, not because it was the first digital CO switch, but because it has an interesting architecture and one that was widely publicized [7 – 15]. The hardware is described in the first five subsections and the software in Section 14.2.6. A call is walked through the system in Section 14.2.7 and the last subsection contains a general discussion.

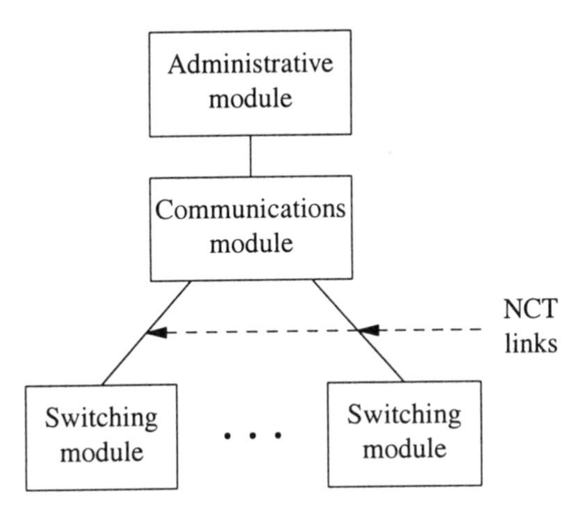

Figure 14.3. 5E block diagram.

14.2.1 Block Diagram

Figure 14.3 shows a high-level block diagram of the architecture of the #5 Electronic Switching System. This architecture is somewhere between being centralized, like the 1E, and being fully modular (Section 2.4.1). The 5E's core consists of an administrative module (AM), which contains a main-frame computer system, and a communications module (CM), which contains an intermodule switching stage and an interprocessor message switch. Because any digital switch, even with 0 lines and trunks, must have these two large subsystems, the typical architecture is only quasi-modular.

Lines and trunks are connected to switching modules (SM), which are partially self-contained. Each SM can terminate up to 4000 lines and trunks and one total 5E system can have almost 200 modules. While a 5E can physically terminate over half a million lines, processor real time constrains a typical 5E from ever being this large. While each switching module has its own processor, called a module control unit (MCU), the software is partitioned in a way that even an intramodule call requires a large amount of processing by programs that reside in the administrative module's processor (AP). Interprocessor communications between this main-frame AP in the AM and the module control unit in each SM uses channels through the CM — the same channels that can be used by voice calls.

The link between each SM and the central CM is called a network control and timing (NCT) link. Each NCT link consists of eight multimode optical fibers, each fiber carrying about 33 Mbps of digital signal. Since an SM typically resides in the switching office building with the AM and CM on which it homes, its NCT link is a short cable of optical fibers. Fiber's principal advantage over an equally short cable of wires is improved noise immunity. An SM, however, need not reside in the same office as its AM and CM, and, for such a remote switching module, its NCT link may be several miles long. The last subsection in this section discusses how this ability to "remote" a 5E's SM can bring some surprisingly large benefits to the telephone company.

Because the call processing for an intramodule call uses programs that reside in both the SM's MCU and the AM's processor, there is considerable interprocessor communication over the NCT link, especially during call setup. The call itself, however, if it's intramodule, can be connected completely within the SM, including a remote SM. Only intermodule calls require a talking connection through the CM.

Similar to other computer-controlled switching systems, like the 1E, and unlike electromechanical switches, like Step and Crossbar, digital switching systems have the physical appearance of a large computer center. Systems typically have many rows of attached cabinets, each the size and shape of a refrigerator. Instead of using raised floor for wiring, like a computer center, wires typically enter the cabinets through overhead troughs. The manufacturers' cabinets have characteristic colors: blue for Lucent, brown for Nortel, and so forth.

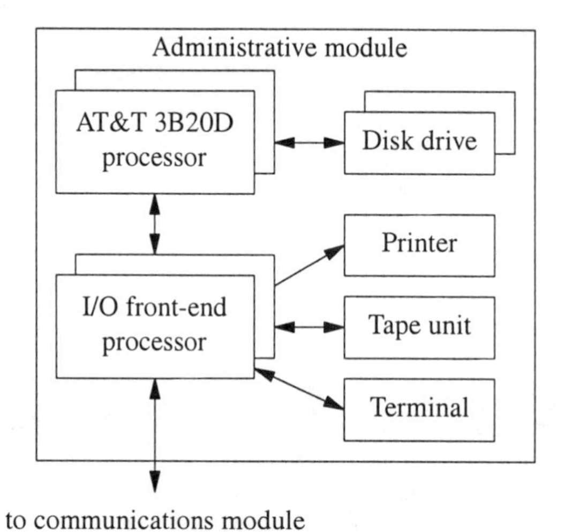

Figure 14.4. Administrative module.

14.2.2 Administrative and Communications Modules

Figure 14.4 shows the block diagram of the 5E's administrative module. The AM is a typical, but up-scale, main-frame computer center, with a main-frame processor connected to disk drives and to a front-end i/o processor. The main-frame processor was initially an AT&T 3B20D, a general-purpose computer running a variation of the UNIX operating system. Other computers have been used in newer models of the 5E. The disk drives hold (1) the system data, which is organized in a database and accessed through a database manager, and (2) low-use programs that need not reside permanently in the processor's random access memory.

The front-end i/o processor is a special-purpose machine. It acts as a peripheral to the main-frame processor for interoffice signaling data links, interprocessor signaling channels to the module control units via the communications module, and to the 5E's administrative control center. This control center contains printers, tape drives, and a computer terminal. To meet the reliability requirements of a switching office, the mainframe processor, i/o processor, and disk drives are duplicated. Components and subsystems, temporarily assigned as slaves, execute simultaneously with corresponding components and subsystems that are temporarily assigned as masters. *Matching circuits* verify that masters and slaves execute identically and initiate maintenance interrupts if they do not. Printers, tape units, and computer terminals are not duplicated.

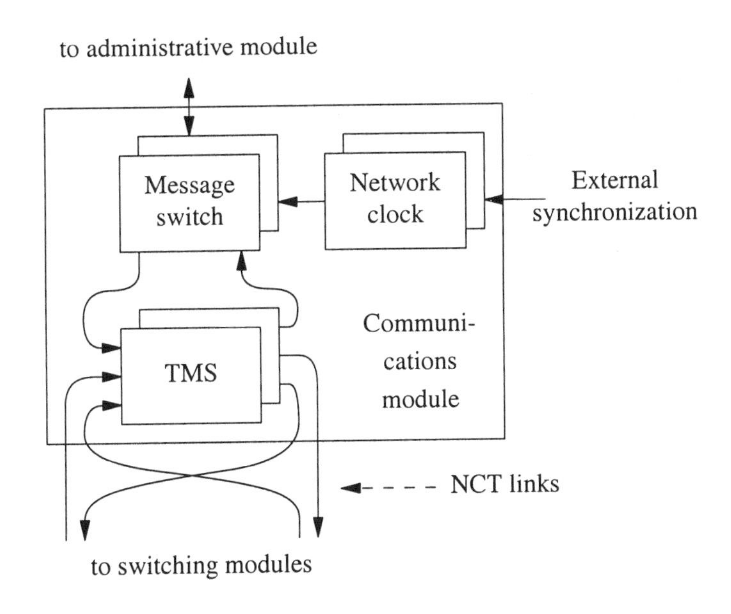

Figure 14.5. Communications module.

Figure 14.5 shows the block diagram of the 5E's communications module. The time-multiplexed switch is the center stage of the time-space-time switching fabric, typically found in any large digital switch. The TMS' sidedness is mouth-to-ear. Uplinks that carry mouth samples from the SMs, and processor output from the AM and MCUs, all terminate on the left side of the TMS. Downlinks that carry ear samples to the SMs, and processor input to the AM and MCUs, all terminate on the right side of the TMS. The AM and every SM each terminate two uplink fibers on the left side of the TMS and two downlink fibers on its right side. Since a fully equipped 5E has 191 SMs and one AM, the TMS is a 384 × 384 nonblocking distribution fabric. The TMS is time-multiplexed into 256 timeslots per frame, where each frame has the classical 125-μsecond duration.

The digital format on each fiber in each NCT link is time multiplexed — with 8000 frames per second, 256 timeslots per frame, and 16 bits per timeslot — for a total rate of 32.768 Mbps. While 8 bits per timeslot would be sufficient for conventional digital POTS, the 5E designers wisely overspecified each timeslot to allow for future expansion and non-POTS applications. With two uplink fibers and two downlink fibers (not counting duplication for reliability) between each SM and the TMS, and 256 timeslots in each fiber, each SM has access to 512 timeslots in each direction. A fully equipped TMS has a total throughput of 384 fibers × 32.768 Mbps per fiber = about 12 Gbps.

The TMS terminations for the AM connect to an interface unit, the *message switch* where the NCT link's format is converted to the data link format appropriate for the i/o processor front-end in the AM. The entire switching office is synchronized by a central *network clock,* which itself is synchronized by the North American T1 Stratum Clock. All three components are duplicated for reliability. Components and subsystems that are temporarily assigned as slaves execute simultaneously with corresponding components and subsystems that are temporarily assigned as masters. Matching circuits verify that masters and slaves execute identically and initiate maintenance interrupts if they do not.

The time-multiplexed switch is duplicated for reliability. The NCT links, containing two uplink fibers and two downlink fibers, are also duplicated, for a total of eight fibers in each NCT link. The A and B uplink (and downlink) fiber pairs can be terminated on the A and B copies of the TMS as A-A/B-B or A-B/B-A, under control of system configuration software (as in how the 1E configures dual processors, dual memory buses, and dual memory units).

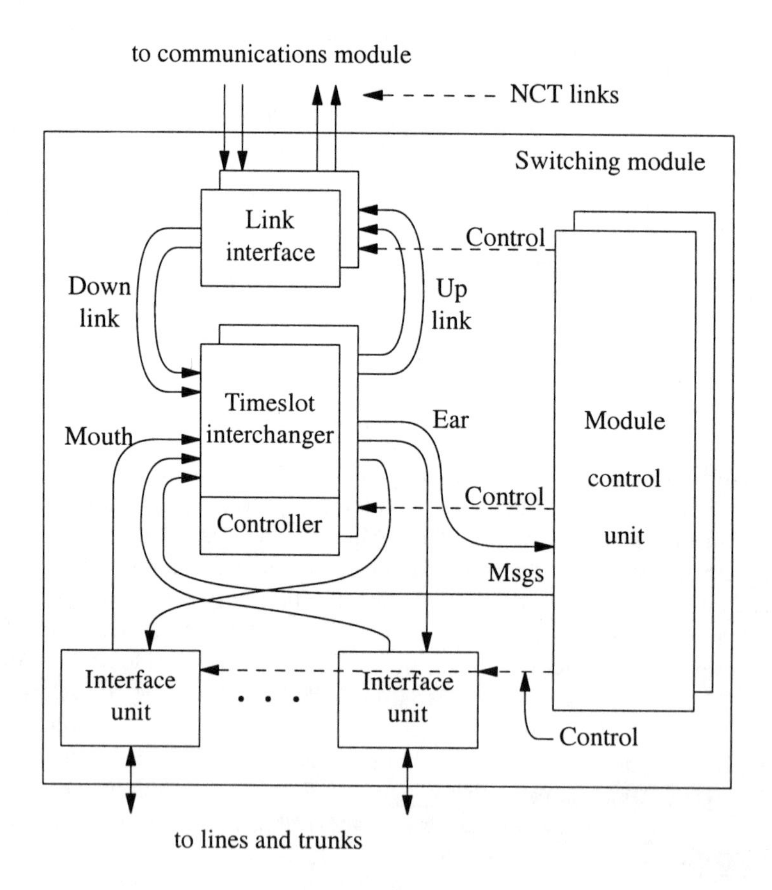

Figure 14.6. Switching module.

14.2.3 Switching Module

Figure 14.6 shows the block diagram of the 5E switching module. An SM's single stage of switching is implemented in its timeslot interchanger (TSI). One-way internal electronic channels that carry mouth samples from the terminals, and others that carry incoming data from the center stage, terminate on the input side (on the left) of the TSI. One-way internal electronic channels that carry ear samples to the terminals, and others that carry outgoing data to the center stage, terminate on the TSI's output side (on the right). While each set of channels carries 512 timeslots, the mouth channel has eight separate one-way subchannels that each carry 64 timeslots from a different interface unit (IU) to the TSI, the ear channel has eight separate one-way subchannels that each carry 64 timeslots to a different IU from the TSI, the incoming channel has two separate one-way subchannels that each carry 256 timeslots from the downlink interface, and the outgoing channel has two separate one-way subchannels that each carry 256 timeslots to the uplink interface.

With 256 timeslots per NCT link fiber, the number of fibers is doubled to provide 512 timeslots, doubled again to provide 512 timeslots in each direction, and doubled yet again for reliability (which is ignored in this discussion). The link interface (LI) circuit translates between the SM's internal data format and the NCT link's data format and converts between the SM's internal electronic signals and the NCT link's optical signals. Each one of the two LI uplink circuits reads from one of the two 16-bit parallel electronic outgoing channels and transmits this data over one of the two bit-serial NCT optical fiber uplinks. Each one of the two LI downlink circuits receives from one of the two bit-serial NCT optical fiber downlinks and writes this data onto one of the 16-bit parallel electronic incoming channels.

An IU is a subsystem whose printed wiring boards house the hardware interface between the office's internal channels and the lines and trunks that home on the office. Many different PWBs provide interfaces to the variety of subscriber lines and trunks, and other printed-wiring boards (PWBs) hold a variety of service circuits. All these PWBs interface on their interior side to a wiring back-plane that contains a mouth bus with 64 timeslots and an ear bus with 64 timeslots. Each IU interfaces externally to a maximum of 512 lines, trunks, and service circuits. So, an IU with only 64 ports would be nonblocking at the two most exterior edges of the switching fabric (there is no concentration nor expansion). An IU that is fully equipped with 512 ports provides 8:1 concentration and expansion between the PWBs and the mouth bus and ear bus, respectively. From Figure 5.6, 64 servers are seen to handle 47 Erlangs at a 0.01 blocking probability. Thus, an IU can be fully equipped if the average port in the IU carries less than $47 / 512 = 0.09$ Erlangs.

For an SM to meet a central office's severe reliability requirements, the NCT links, each link interface circuit, the timeslot interchanger, and the module control unit are all duplicated. Components and subsystems that are temporarily assigned as slaves execute simultaneously with corresponding components and subsystems that are temporarily assigned as masters. Matching circuits verify that masters and slaves execute identically and initiate maintenance interrupts if they do not. The IUs are not duplicated.

14.2.4 Switching Fabric

Let A and B speak to each other on an intramodule call. Suppose A homes on IU i and B homes on IU j in the same SM. Let the 5E's routing program assign the following network links to the half-duplex paths in each direction through the switching fabric: A's mouth to timeslot t on the mouth bus in IU i, B's ear to timeslot v on the ear bus in IU j, B's mouth to timeslot w on the mouth bus in IU j, and A's ear to timeslot y on the ear bus in IU i. Driver programs, in this SM's MCU, control the TSI, A's line card in IU i, and B's line card in IU j so that:

- The TSI moves A's digital mouth samples from timeslot t in mouth subchannel i to timeslot v in ear subchannel j and it moves B's digital mouth samples from timeslot w in mouth subchannel j to timeslot y in ear subchannel i;
- A's line card in IU i writes A's A-to-D converted mouth samples onto timeslot t in this IU's mouth bus and reads A's ear samples for D-to-A conversion from timeslot y in this IU's ear bus;
- B's line card in IU j writes B's A-to-D converted mouth samples onto timeslot w in this IU's mouth bus and reads B's ear samples for D-to-A conversion from timeslot v in this IU's ear bus.

An intermodule call is more complex than an intramodule call. Consider the intramodule call described above, except let A's line card reside in switching module M in IU <M.i> and let B's line card reside in switching module N in IU <N.j>. For simplicity, we'll only describe the "half-call," from A's mouth to B's ear. In addition to the network links that were assigned to the intramodule call above, let the 5E's routing program also assign: data from A's mouth to B's ear uses timeslot u in the TMS, and in NCT uplink fiber e (e = 0 or 1) from SM M to the TMS (call it uplink fiber <M.e>), and in NCT downlink fiber <N.f>. Driver programs, in the MCU of SM M, control the TSI in SM M. the LI in SM M, and A's line card in IU <M.i>. Driver programs, in the MCU of SM N, control the TSI in SM N, the LI in SM N, and B's line card in IU <N.j>. A driver program, in the main processor in the AM, controls the time-multiplexed switch in the communications module. These drivers cause the following actions on this half-call:

- A's line card in IU <M.i> writes A's A-to-D converted mouth samples into timeslot t on this IU's outgoing mouth bus;
- The TSI in SM M moves A's digital mouth samples from timeslot t of M's mouth subchannel <M.i> to timeslot u of M's outgoing subchannel <M.e>;
- During timeslot u, the TMS is configured so that port <M.e> on its input side is connected to port <N.f> on its output side;
- The TSI in SM N moves A's digital mouth samples from timeslot u of N's incoming subchannel <N.f> to timeslot v of N's ear subchannel <N.j>;
- B's line card in IU <N.j> reads B's ear samples for D-to-A conversion from timeslot v on this IU's incoming ear bus.

14.2.5 Distributed Control

The module control unit in the SM directly controls the TSI, the LI, and all the IUs. Within an SM, control is exerted over control channels that are separate from the channels that carry subscriber data. In contrast, signaling among the various MCUs and the AP is accomplished by messages that pass over message channels that are connected within the same channels that carry subscriber data.

If user A on SM M calls user B on SM N, call control software is distributed across three separate processors. Software that resides on the MCU in SM M controls A's line card and IU and the TSI in SM M and handles most of the call setup on the originating side. Software that resides on the MCU in SM N controls B's line card and IU and the TSI in SM N and handles most of the call termination. Software that resides on the AP controls the center-stage TMS, handles central functions like class of service determination and translating directory number to equipment location, and handles some of the functions not handled by the two MCUs. The three processors communicate with each other throughout the call over the interprocessor message channels embedded in the NCT links.

When installing or reprovisioning a modular switch, an interesting issue is whether similar types of ports should be *segregated* onto the same SM or *distributed* across all the SMs. The tradeoff is between increasing the number of intramodule calls versus the amount of software that must reside in the MCU. For example, ISDN lines and coin telephone lines typically would be segregated onto the same SM because an SM that has no such lines doesn't need the corresponding software in its MCU. However, service circuits would be distributed across all the SMs because we don't want to place an intermodule connection for subscriber dialing, unless all the digit receivers on the calling party's SM are busy.

14.2.6 Software

Figure 14.7 shows the high-level architecture of the 5E's software. Each of the bubbles represents a family of programs. Each program may execute as a single permanent copy on the AP, as a single permanent copy on every MCU, or as multiple copies (called processes) on the AP or any MCU.

- The peripheral control family of programs are drivers that control the switching system's hardware. Residing in any MCU as needed, some drivers control the lines, trunks, and service circuits in the IUs and others handle subscriber signaling to and from a telephone. A driver that controls time connections in a TSI resides in the MCU of the corresponding SM. A driver that controls space connections in the TMS resides in the AP.

- The routing and terminal administration (RTA) family is an intermediate layer of programs, higher than drivers that deal with the physical layer and lower than call processing that deals with features. One RTA program that establishes dial tone connections executes briefly as a process in the MCU for each new call. Another RTA program that establishes talking connections executes as a process in the AP for each new call, but for the duration of its call.

- The feature control (FC) family of programs handles call processing, at the feature level. A complete copy of one FC program executes as a separate process in each

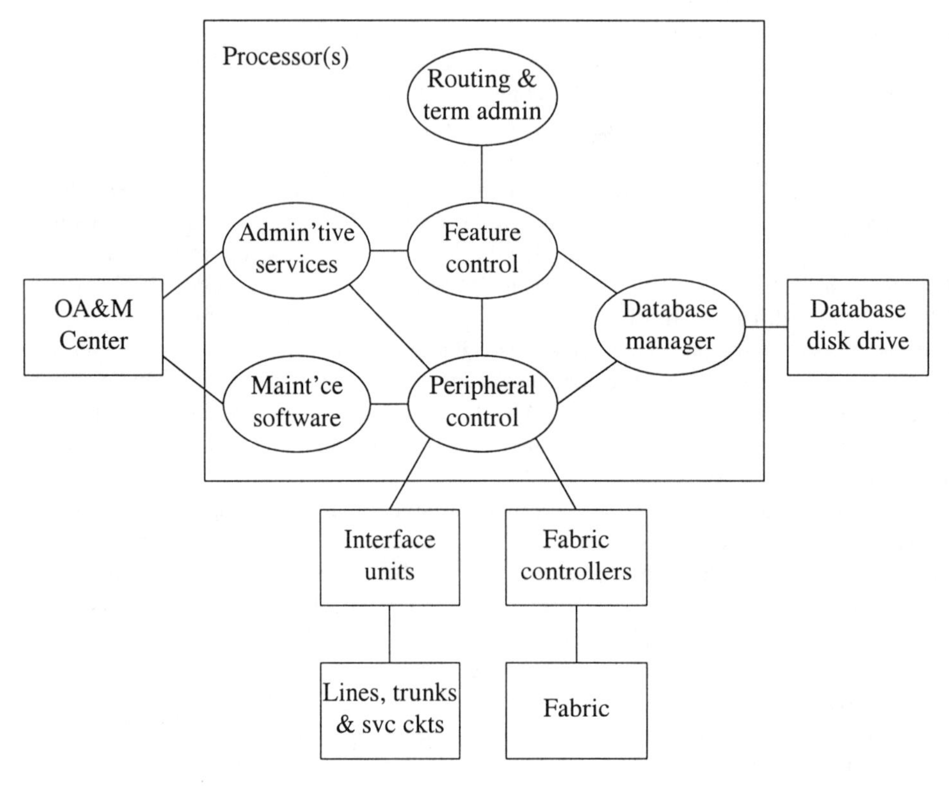

Figure 14.7. 5E software architecture.

user's MCU for each new call. While this controversial *process-per-call* architecture squanders memory and cycles in the MCUs, program design is simplified if the program only handles one call.

- The database manager executes as a single permanent process on the AP to provide a software interface for the storage and retrieval of data from the disk drive.
- The administrative services are programs that interact with the peripheral units in the AM control center and handle commands from administrative personnel and that come over data links from other offices or from the telephone company's centralized administrative center.
- Maintenance software is the family of programs that perform various testing and auditing functions. Each program executes as a process-per-test on the AP or MCU associated with the component under test.

The programs are written in C and crosscompiled and assembled into load modules for their respective processor. All the processes execute in a modern multiprocessing environment, under a proprietary version of the UNIX operating system. While programs still require multithreaded real-time breaks, as in the 1E, the 5E programmer designs simple UNIX commands into the C code, where the 1E programmer is responsible for establishing the details of real-time return. Furthermore, where the 1E programmer is responsible for the details of storing data for use by multiple programs, the 5E programmer invokes simple UNIX utilities that provide database access and interprocess communications (IPC). If two communicating processes reside on different processors, simple drivers extend the IPC into interprocessor communications.

While Figure 14.7 indicates a simple, but useful, block diagram, some of the programs execute on the AP and some on multiple MCUs, some as a single permanent process and others as multiple processes that are born and killed as needed. The assignment of a program to the AP or to the MCUs can be changed in successive releases (called generics) of the total 5E program. Programs may be reassigned to shift real-time load between the AP and the various MCUs or to reduce interprocessor communications through the CM's message switch. Since the processes that execute on an MCU depend on the SM's call activity, no two MCUs would ever simultaneously execute identical processes. But beyond this, programs that handle certain hardware or call features that are not equipped nor available on a given SM, would never execute on the corresponding MCU. Thus, segregating all the coin telephone line cards or all the ISDN lines in any modular office onto one SM has the distinct advantage that the programs that process coin telephone calls or ISDN calls don't have to execute, or even reside, on every MCU in the office.

14.2.7 Software Call Walk-Through

The operation of the 5E programs is illustrated by walking a call through the software. Let subscriber A, who homes on switch module M, call subscriber B, who homes on switch module N. The walk-through begins with A's off-hook to initiate the call and ends with B's off-hook to answer his ringing telephone. Figure 14.8 indicates the 17 sequential steps, described below, among many intercommunicating processes on three different processors. Since the public literature on the 5E is not very explicit, this walk-through may not be completely accurate. But, it's close enough that you'll get the general idea of how processes are born and die, how they communicate with each other, and how — despite the modern operating system environment — it still isn't as simple as you might have thought it would be.

1. Subscriber A lifts the receiver on his analog Touch Tone telephone. A current sensor on A's line card, which is housed in interface unit i in switch module M, detects this off-hook and signals the module control unit in this SM over the direct control channel between this MCU and this IU. This off-hook signal is received by a peripheral control program that executes as a permanent foreground process on every MCU.

2. This peripheral control foreground process instructs its MCU's operating system to create two additional processes. One is another peripheral control program process, that will act as a signaling interface (or driver) for this telephone only and only for the duration of this call. The second is a feature control program process, that will process the origination of this call for this subscriber only and only for the duration of this call.

3. A's new FC originating call process sends an interprocessor message to the permanent database manager process in the administrative processor. It sends A's line card location and requests A's class of service. The DBM finds A's class of

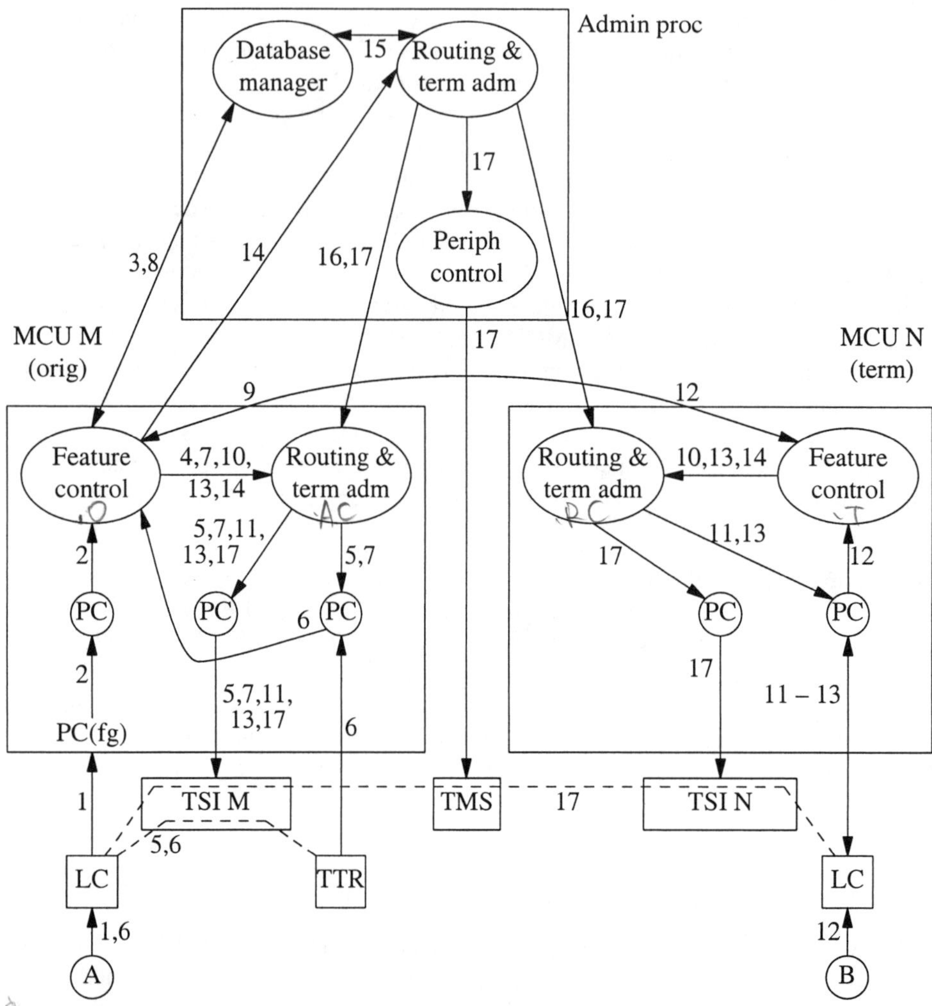

Figure 14.8. Call walk-through.

service in the database and returns it to A's FC call process by an AP-to-MCU message.

4. A's FC call process instructs its MCU's OS to create another process, a routing & terminal administration program process that will establish a single fabric connection, this one for dialing. This FC originating call process passes A's fabric address to this new RTA dial tone process and, based on A's COS, it also requests that A be connected to an idle Touch Tone receiver.

5. This RTA dial tone process locates an idle TTR and causes the birth of yet another peripheral control driver, this one for the TTR. Then, this RTA process finds a fabric connection between this TTR and A's line card. Figure 14.8 assumes that the RTA process has located an idle TTR in an IU on this SM and that a simple

intramodule connection will suffice. Otherwise, an intermodule connection would require the creation and execution of processes in the AP and the MCU in the TTR's SM (similar to the talking connection between A and B, shown below). This RTA dial tone process passes the connection request to the permanent peripheral control process that acts as a driver for this SM's timeslot interchanger. This PC TSI driver sends commands from the MCU to the TSI controller to establish the fabric connection, between A's line circuit and the selected Touch Tone receiver, shown in Figure 14.8 as the lower dashed line through this SM's TSI. A hears dial tone.

6. When A pushes a Touch Tone button on his telephone, analog Touch Tones are transmitted from his telephone, along his loop pair to his line circuit, where the analog signal is converted to PCM. The 64-Kbps PCM stream is transmitted from this line circuit into a timeslot to the TSI, through the TSI connection, and into a timeslot to the TTR, where each separate dialed digit is detected. The TTR signals each detected digit, over the IU-to-MCU link, to its unique PC TTR driver in the MCU. This driver signals each detected digit, now in the form of an interprocess message, to that unique feature control originating call process that stores and interprets A's dialed digits.
7. When this FC originating call process determines that dialing is complete, it signals the RTA dial tone process to disconnect the TSI connection between A's LC and the TTR. Then, this RTA dial tone process signals the PC TSI driver to disconnect the connection between A's line card and the TTR, kills the PC TTR driver, and then kills itself.
8. The FC originating call process requests the physical location of the called party by sending the directory number that A just dialed by an interprocessor message to the DBM in the AP. The DBM translates this directory number into B's fabric appearance by looking it up in the database. The DBM transmits B's fabric appearance back to A's FT call process by an interprocess message from the AP to the MCU in SM M.
9. A's FC originating call process determines that A has placed a call to B and that B is in this same switching office, but on a different switch module. This call process now proceeds to establish a ringing connection. It sends a message to the OS in the MCU in SM N, the SM where B is located, and requests the creation of an FC terminating call process to handle B's end of this call.
10. Simultaneously, each in its respective MCU, each FC call process spawns a new RTA process in its respective MCU, to handle each end of the ringing connection. A's FC originating call process spawns an RTA process in MCU M that will connect A to audible ring tone. B's FC terminating call process spawns an RTA process in MCU N that will connect B to a ring generator.
11. Simultaneously, each in its respective MCU, each RTA process establishes its end of the ringing connection. Because audible ring tone is not a BORSCHT function, the calling party hears audible ring tone via a fabric connection to a service circuit. The RTA process in MCU M, acting for A, locates an idle ART

service circuit and a connection through M's TSI between A's LC and this ART SC. This RTA process transmits instructions to the TSI driver in MCU M which passes them to the TSI controller. The connection is made and A hears audible ring tone. Meanwhile, the RTA process in MCU N, acting for B, spawns a peripheral control driver for B's telephone and transmits to this driver the command to ring B's telephone. Since ring is one of the BORSCHT functions, ringing cannot be transmitted through a digital fabric. So, the called party's telephone cannot be rung via a fabric connection to a service circuit (as in the 1E), but must be rung directly at the line circuit. This PC driver operates the relay in B's line card that places ring signal on B's loop pair and B's telephone rings.

12. B answers his telephone. The off-hook signal is detected in B's line card and then in the PC driver for this line card. This PC driver transmits a corresponding message to the FC call process that is handling the terminating side of this call. This FC call process transmits a corresponding message, by the interprocessor channel from MCU N to MCU M, to the call process in MCU M that is handling the originating side of this call.
13. Simultaneously, each in its respective MCU, each FC call process sends orders to its respective RTA process to disconnect its respective end of the ringing connection. The RTA process in MCU M, acting for A, transmits disconnect instructions to the TSI driver in MCU M which passes them to the TSI controller. The connection is broken and A no longer hears audible ring tone. Meanwhile, the RTA process in MCU N, acting for B, transmits disconnect instructions to the PC driver for B's LC. This PC driver releases the relay in B's line card that places ring signal on B's loop pair and B's telephone stops ringing. Now, both these RTA processes — the RTA process in MCU M that had connected A to ART and the RTA process in MCU N that had rung B's telephone — kill themselves.
14. Simultaneously, each in its respective MCU, each FC call process spawns a new RTA process in its respective MCU, to handle each end of the talking connection. A's FC originating call process spawns an RTA process in MCU M that will connect A to some as-yet-undetermined channel to the TMS. B's FC terminating call process spawns an RTA process in MCU N that will connect B to some as-yet-undetermined channel to the TMS. Since A's FC originating call process in MCU M knows this A-to-B call is intermodule, it also sends an interprocessor message to the AP and instructs its OS to create a new process for an RTA program that can establish an intermodule network connection through the TMS.
15. This new RTA process in the AP consults the network map in the database, via the DBM process in the AP, and locates an idle network path (two-way) between A on SM M and B on SM N.
16. This RTA process in the AP then uses interprocess messages to instruct A's RTA process in MCU M and B's RTA process in MCU N which pair of timeslots will be used through the TMS.
17. Simultaneously, each in its respective MCU or AP, each of these three co-operating RTA processes establishes its part of the talking connection between A

and B. The RTA process in MCU M, acting for A, locates a pair of connections through M's TSI: from A's mouth samples to the A-to-B timeslot in uplink M and from the B-to-A timeslot in downlink M to A's ear samples. This RTA process transmits instructions to the TSI driver in MCU M, which passes them to the TSI controller in SM M. Meanwhile, the RTA process in MCU N, acting for B, locates a pair of connections through N's TSI: from B's mouth samples to the B-to-A timeslot in uplink N and from the A-to-B timeslot in downlink N to B's ear samples. This RTA process transmits instructions to the TSI driver in MCU N which passes them to the TSI controller in SM N. Meanwhile, the RTA process in the AP orders connections in the TMS: from uplink M to downlink N during the A-to-B timeslot and from uplink N to downlink M during the B-to-A timeslot. This RTA process transmits instructions to the TMS driver in the AP which passes them to the TMS controller in the communications module. The connection is made; A and B can talk to each other.

14.2.8 Discussion

The 5E is an important switching system not only to the industry today, but also to the students of the discipline of switching. Many observations are made here that reflect the state of the industry and the state of the technology.

Common Control or Computer? While the common control units for the 1E and the 1A are computers, and state-of-the-art computers for their day, the regulatory environment of the 1960s and 1970s excluded AT&T from the computer business. Words like "processor" and, especially, "computer" were carefully and deliberately avoided in all the documentation and literature on these systems. The 4E was introduced during the changing regulatory environment of the late 1970s and early 1980s as AT&T began to think about, and talk about, entering the commercial computer business. Where the early literature and documentation on the 4E referred to its "common control," the later reports on this same system referred to the same unit as the "1A Processor." AT&T was feeling a little more daring and aggressive.

When AT&T finally did enter the computer business, one of its important marketing strategies was to emphasize to the computer marketplace that AT&T was not a newcomer to manufacturing computers. This was, of course, true, and, in fact, it could be legitimately argued that AT&T had been a pioneer in the computer industry for many years. The work on UNIX at Bell labs was a continuation of a long history of research and development in computers. When AT&T finally did get into the general-purpose computer business, the company decided to use its own UNIX operating system and its own top-of-the-line commercial computer in the 5E. This controller was boldly identified in the 5E's literature and documentation, not only as a "processor," but one with a commercial name.

Data LAN. Interprocessor communications requires a network especially designed for data applications. The 5E designers embedded their data network within the 5E's inter-module voice network. The logical alternative in the early 1980s would have been to use an Ethernet LAN.

- They would have avoided a special-purpose interface card in the computers because their computer family was already equipped with Ethernet interface cards.

Much more important, they would have avoided proprietary signaling formats and protocols; adopting evolving industry standards would have been much simpler in an industry standard LAN.

- But, Ethernet is a bus and suffers all the contention problems that are characteristic of bus architectures. Since the 5E's interprocessor data communication is point-to-point and star oriented, a physical bus, and even a logical bus, would have likely been a worse bottleneck than the LAN they have.

While the 5E designers wisely avoided using a bus-oriented data LAN for the interprocessor communication, using time-multiplexed channels in each NCT link, and switching these channels through the TMS in the CM, was probably not a good decision. Adding another couple of fibers to each NCT link, and terminating them directly on the message switch in the CM, would have provided a better growth strategy for the data LAN than reallocating timeslots from the intermodule voice network. The 5E's flexibility in allowing any process' execution site to shift between the AP and the various MCUs provides a means to optimize real time on these processors. Much of this flexibility is lost if these process execution sites are selected to reduce bottlenecks on the data LAN. Since processor real time is a more expensive resource than LAN real time, the LAN should have been overdesigned.

New Services Evolution. The ability to remote an SM gave the 5E a huge advantage. While remote switching modules (RSMs) have been deployed to implement line concentration, this was not their most significant application.

The telephone companies have historically had a public relations problem when introducing new services. Touch Tone was made available to subscribers served by Crossbar years before it was available to subscribers served by Step. Software-based services, like abbreviated dialing, were available initially only to subscribers served by the 1E, but not to subscribers served by electromechanical offices. Digital CO switches brought even more new services that could be made available to some subscribers but not to others, but the 5E's RSM provided a solution.

If a telco's toll region has at least one 5E CO, then RSMs can be installed in the other offices in the region. A subscriber served by a CO that doesn't offer a particular service can have his line appearance physically moved from the old switch to a 5E RSM in the same building. While he's now effectively served by a 5E, the 5E software doesn't require a change of telephone number. The gradual migration of subscribers to the RSM allows the telco to get more life out of its old switch. The use of 5E RSMs enables rapid deployment of ISDN (Section 13.4), calling party identification (Section 18.1.4), and other advanced subscriber services over an entire region instead of restricting them to the serving areas of selected COs.

Density of Line Cards. Each subscriber's line terminates on a telephone in the customer's side and on a line circuit in the office side. In Step, Panel, and Crossbar, these line circuits are mounted on the front surface of a panel that is mounted on an equipment frame. In this physical arrangement, each line circuit is accessible for wiring to the main frame and for repair if something breaks. In the 1E, line circuit components are physically distributed with the ferrod in a line scan unit and relays in other units. In digital systems, the line circuits are mounted on printed wiring cards that are plugged into slots in card carriers that are mounted on equipment frames. The straightforward implementation is for every subscriber to have a dedicated line circuit that resides on a dedicated line card. However, this may not be the most economic implementation.

The cost/price of a modern switch is discussed in Section 18.2.4. As discussed there, line cards typically represent more then half of the total cost/price of a switch's hardware (not counting software and not counting operating costs). So, the cost optimization of line circuits is highly motivated. One interesting optimization would be to place several line circuits onto one line card, as in a typical PBX (in the next section). However, this isn't so simple.

Consider the availability of a subscriber's line circuit. If a line circuit fails (presumably, a low-probability event), the corresponding subscriber is out of service until his line circuit is repaired or replaced. If his line circuit resides on a dedicated line card, the frameman simply pulls the card and plugs in another one. This subscriber's service isn't disrupted by the removal of his line card because he was out of service anyway. Consider if a line card holds 16 line circuits and one of them fails. If the frameman arbitrarily pulls the line card, any of the other 15 subscribers who may be busy would have an active call disrupted. If the frameman waits until some time when all 16 subscribers are simultaneously idle, the subscriber who is out of service will be out of service for a longer time. Since many state PUCs and many countries have very strict downtime requirements on the availability of central office POTS, the simplest way to meet these requirements is to place one line circuit per line card — despite its high price.

Some of the switch manufacturers have devised clever sparing arrangements for line cards so that they can hold multiple line circuits and so that availability requirements are still met. In a PBX, because no agency dictates availability, enterprises typically opt for low price over occasional service disruption, even if an active call is occasionally disrupted. So, a typical PBX line card holds at least eight line circuits. In many developing nations, where high service availability may be somewhat of a luxury, PBXs might provide CO switching as a cost optimization.

Another way to potentially economize on line circuits is to avoid providing one for every line. Especially when digital switches were first under development, the cost of the A-to-D converters was sufficiently high that thought was given to providing a stage of concentration between the analog lines and the BORSCHT functions. This stage of concentration would have to be analog switching (a line finder stage?). Some minimal analog line circuit would still be required for each line, for battery and supervision. While the A-to-D converter cost became cheap enough that analog concentration never proved in economically, some form of analog switching is still tempting; especially if it combines provisioning, automation of the main frame, line circuit sparing, and digital line circuit concentration.

The Partition. The 5E has two clear partitions: one between the AP and the MCUs and one between the MCUs and the hardware in the IUs. The low-level partition is similar to the partition in the 1E: put all the intelligence in the controller and micromanage the periphery. The 5E's high-level partition would be insightful if it were organized functionally. Instead the partition provides a means to shift real time among the processors and to reduce the data load on the interprocessor message network. We'll come back to this in Chapter 18.

In the previous subsection, an intraoffice call was walked through a digital switch; an interoffice call is easily extrapolated. If A had dialed the telephone number of someone in another exchange, his FC call process would have detected the foreign office code, area code, leading 1, or international 011 prefix. The call process would have requested a trunk group from the database manager instead of asking for a translation from directory number to fabric appearance. Then some RTA process would have located an idle trunk in this trunk group and the call would have completed. While this procedure seems natural, it bundles *route control* into the same program as *call control.* This has turned out to be problematic, as we'll see in Chapter 18.

14.3 A DIGITAL PBX

The introduction of digital PBXs into a highly competitive market was discussed in Sections 10.6 and 13.1. While the example digital PBX described here, the Bell System's System 75, was neither the first digital PBX nor even Western Electric's first digital PBX, its hardware and software architecture deserve special mention. While the System 75 is now marketed commercially under the name Definity G1, the principal references refer to this PBX by its old name, System 75, the name used throughout this chapter [16 – 20].

The first subsection describes this PBX's hardware architecture: in general, by a block diagram, by its physical design, and discusses its unusual characteristics. Section 14.3.2 describes the physical and blocking characteristics of its interconnection fabric, and Section 14.3.3 describes its three-tiered system of distributed processing. Port cards are discussed in the fourth subsection: by their generic block diagram, some interesting design decisions, and this system's unusual concept of universal port slots. Subscriber signaling and interprocessor signaling are covered in Sections 14.3.5 and 14.3.6, respectively. Then, Sections 14.3.7 and 14.3.8 describe the System 75's call model and its software architecture, respectively. A call is walked through the switch and the software in Section 14.3.9. Finally, Section 14.3.10 discusses architectural significance and philosophy.

14.3.1 Hardware Architecture

The System 75 PBX serves 100 – 1000 telephones in the relatively high-utilization environment of a typical enterprise. The size of its cabinet, the number of links in its fabric, its processor power, its signaling channel, and some aspects of its software architecture are all consistent with this maximum size and level of activity.

Figure 14.9 shows the System 75's block diagram. The six rectangular boxes in the upper left represent the printed-wiring boards (PWBs) in its *processor subsystem.* The four rectangular boxes in the lower left represent the *telephone subsystem* (the "process" in Figure 11.1). The system's line interfaces, trunk interfaces, and service circuits are packaged on PWBs, called port cards, that connect to a common TDM bus. The network interface card translates between the two buses.

These port cards and other PWBs have male contacts on one edge. When the cards are pushed into slots in a card carrier, the edge with the contacts plugs into a female connector on the inside of the back wall of the carrier. These edge connectors are all wired to a motherboard, a long PWB that runs along the entire back wall of the carrier. The intercard wiring, contained on the motherboard, is a simple 200-wire parallel bus.

The simplest and smallest possible system would reside on a single carrier, with an external tape unit and power supply. The cards in the processor subsystem plug into the

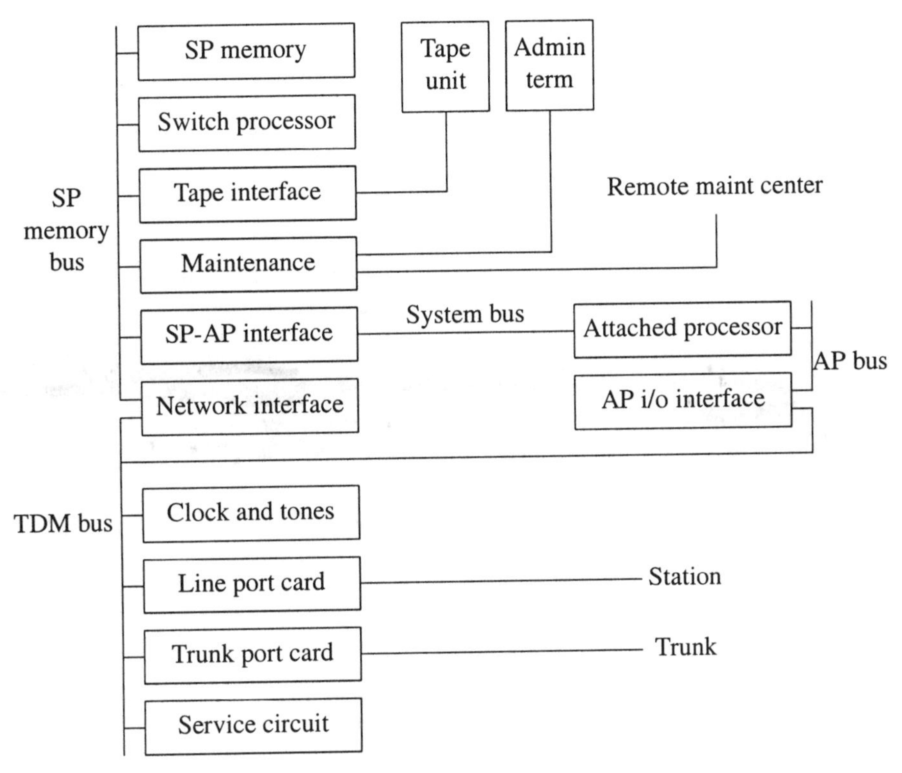

Figure 14.9. System 75 block diagram.

SP memory bus on the motherboard in the back of the left third of this carrier. All the port cards that house line interfaces, trunk interfaces, and service circuits plug into the TDM bus on the motherboard in the back of the right two-thirds of this same carrier. These two motherboards are separated and are bridged by the network interface card. This kind of carrier that holds the processor subsystem is called a control carrier, even though it also holds port cards. But, the number of lines that could be served by such a one-carrier system would be limited by the few carrier slots available for the port cards of the telephone subsystem.

The simplest and smallest commercially available System 75 comes in a small cabinet that holds two carriers, a tape unit, and a power supply. One carrier is a control carrier and the other is called a port carrier. Since the motherboard on the port carrier is completely TDM bus only, this carrier provides enough additional slots for port cards so the system can serve a couple of hundred lines. The TDM buses on each carrier are wired together by a ribbon cable.

The next larger cabinet, about the size and shape of a refrigerator, holds five carriers, a tape unit, and a power supply. One of the carriers must be a control carrier and the other two to four carriers are all port carriers. The TDM buses on all the equipped carriers are daisy-chained together by ribbon cables. While physically occupying as many as five carriers, the system has two logical buses: the SP memory bus for the processor subsystem and then TDM bus for the telephone subsystem. The cabinet also has a connector block, like a main frame, for interconnecting the lines and trunks to their respective line circuits and trunk circuits on port cards.

Figure 14.9 is different from the block diagrams of other systems.

- Except for the processor subsystem, the hardware has no logical intermediate structure. The system is a set of port cards attached to a bus with no switching modules or interface units or other subsystems.
- Unlike the block diagrams of other systems — in Chapters 6, 8, and 11 and Sections 14.1 and 14.2 — Figure 14.9 has no block labeled "Fabric." The fabric is part of the TDM bus and it's discussed in the next subsection.
- The processor subsystem is the middle layer of the three-tiered system of distributed processors that controls a System 75. The upper layer is the attached processor, shown at the right in Figure 14.9. The lower layer is a collection of microprocessors, one on each port card, not shown in Figure 14.9.

14.3.2 Interconnection Fabric

Most PBXs, like the Dimension (Section 10.4.1) or the System 75, are small enough that the switching fabric is just a simple physical resource that is time-shared. Dimension's fabric is a single time-multiplexed analog wire, bused throughout the system's back-plane so that all the system's PWBs are attached to it. The System 75 had the same designers and has a similar fabric, except it has a parallel-wire bus and it's digital. Each timeslot on this resource represents a junctor in the logical fabric. If A speaks to B over channel *m*, A's port card places A's mouth samples (a pulse amplitude modulated (PAM) pulse in Dimension or a pulse code modulated (PCM) byte in the System 75) onto this shared resource during timeslot *m*. B's port card reads this data from the resource during *m* and delivers it to B's ear. Such typical PBX *embedded fabrics* are:

- Full-access — each port card can access every timeslot;
- Unidirectional — B's mouth samples go to A's ear in a second timeslot;
- Two-stage — transmitters on N port cards concentrate down to M timeslots, $M \leq N$, which expand back out to receivers on the same N port cards.

Eighteen of the 200 wires in the motherboard are illustrated in Figure 14.10. The physical switching fabric resides on 16 of these wires, on two 8-bit parallel buses called the A bus and the B bus. Both buses are time multiplexed into 8000 frames per second and 256 timeslots per frame. Each *junctor channel* consists of eight parallel bits, placed on one of two buses during one of the 256 timeslots. Each junctor channel is identified by $X.m$, where X = A or B and $0 \leq m \leq 255$. The 512 channels are fully accessed by every port card. Over another wire in the motherboard, an 8-KHz frame clock provides frame synchronization to all the port cards. Over yet another wire, a 2.048-MHz timeslot clock provides bit synchronization for the two junctor channels in each timeslot.

The first five timeslots are used as a message channel between the switch processor and microprocessors on each port card. Some of the other junctor channels have dedicated telephony purposes, described later. About 480 channels remain to be used as one-way talking channels, for transmitting data from A's mouth to B's ear. Since the analog sampling rate is synchronized to the frame rate, and each junctor channel holds 8

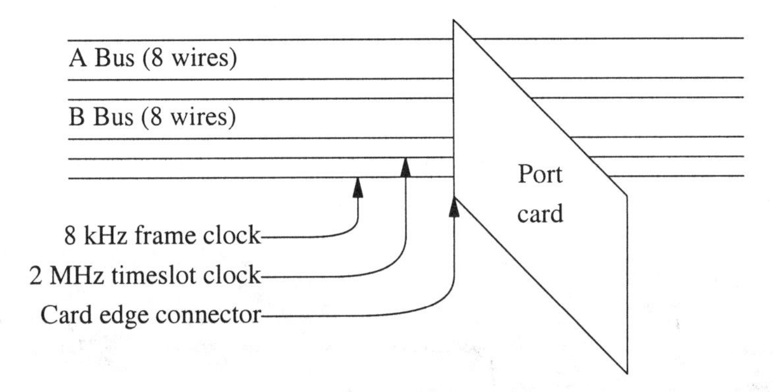

Figure 14.10. Universal motherboard.

bits, each two-way telephone conversation uses two junctor channels.

Two MHz is a nontrivial data rate over printed wires that extend over five carriers. The TDM bus is carefully terminated electrically to prevent reflections and the receivers and drivers on every port card are specially designed. The system is carefully (and rationally) synchronized so only one port card ever tries to transmit onto any bus in any timeslot. (Obviously, I don't think highly of networks that allow transmitters to transmit whenever they feel like it.)

A system that requires two junctor channels per conversation supports a maximum of 240 simultaneous two-way connections. Assuming relatively active telephones, 4-to-1 concentration suggests that a system supports a maximum of about 1000 lines. At eight lines per port card, a full system needs about 125 line cards. Adding a complement of port cards for the processor subsystem, service circuits, and trunks, the total number of port cards fills five carriers. The daisy-chained motherboards are electrically limited to five carriers and the switch processor's real-time capacity is reached by about 1000 busy telephones. So, the entire system is engineered and sized consistently.

Consider using a System 75 in a different market, like for residential telephony in a third-world country. To reduce cost, a PTT might use 16-to-1 concentration, even with party lines. There are two implementations:

- Remote analog concentrators might provide 4-to-1 line concentration between the analog drop wires and analog or digital loop channels. Then, these loop channels would terminate on line cards and be further concentrated, by 4-to-1, down to junctor channels.
- Or, instead of using analog line concentration, special port cards could be built to provide a line density of 32 lines per line card.

14.3.3 Distributed Processing

System 75 control uses three tiers of distributed processors. The optional attached processor resides at the highest layer, shown at the right of the block diagram in Figure 14.9. The optional AP might run application programs, including an on-line enterprise-wide telephony directory. This AP can be a special-purpose minicomputer, bought for this purpose, or it could be any existing general-purpose machine or server to which these applications are added. Since the AP has access to the software that controls the user's telephone, even if the user's PC or work station doesn't have a built-in telephone, complex integrated voice/data services can still be provided if the AP has a LAN-connection to the user's PC. This migration of advanced services to an external computer, available in enterprise switching for many years, is relatively new to residential customers through the Intelligent Network (Chapter 18). The more common features that should be expected in any enterprise switching system are not provided through the AP.

Conventional call control and most of the common features expected in enterprise switching are provided by software that resides in the switch processor, in the middle layer of the three-tiered architecture of distributed processors. Each PBX has a dedicated processor subsystem, on six PWBs that plug into the SP memory bus on the system's only control carrier. This processor subsystem includes the processor board (the original System 75 used an 8086 processor), the SP's RAM on a separate board, and various processor i/o devices that interface to a tape unit, an administrative terminal, a remote maintenance center, the optional AP, and a front-end to the system-wide signaling channel. This signaling channel, through the network interface, provides the IPC path between the switch processor and the microprocessors on every port card.

The third tier of distributed processing is a set of microprocessors, one on each port card, that control the hardware on their respective card. This microprocessor on each port card is called an angel and the switch processor is called the archangel.

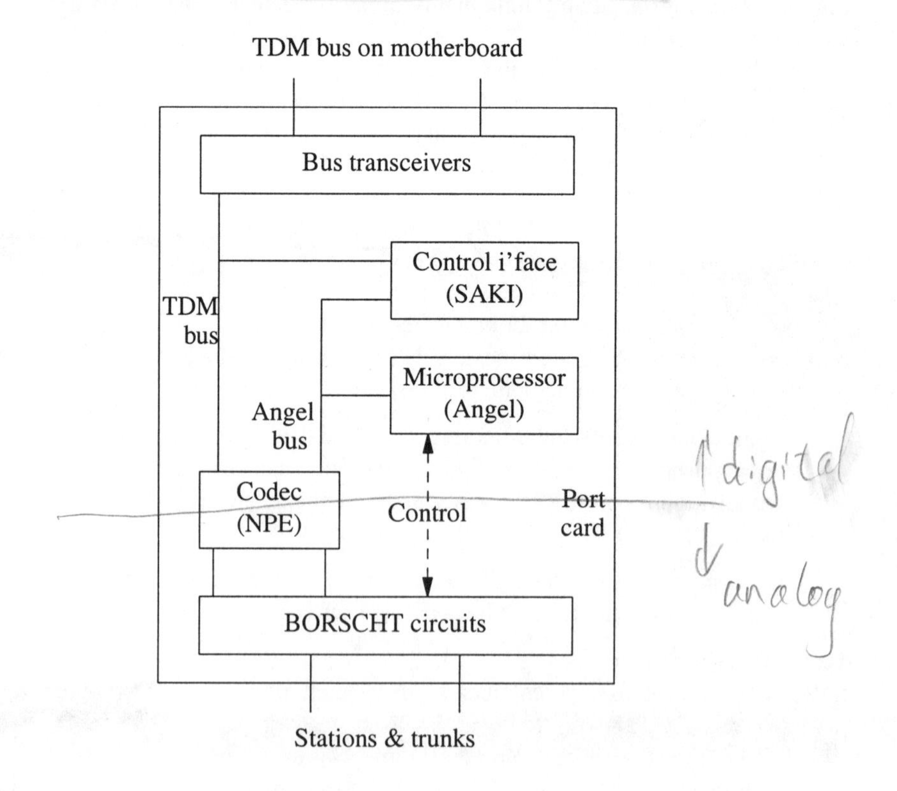

Figure 14.11. Generic port card.

14.3.4 Port Cards

Figure 14.11 illustrates a generic port card. Its architecture is a miniature version of the System 75's macro-architecture. Corresponding to the overall System 75's archangel, each port card has an angel. Corresponding to the overall System 75's two buses, the TDM bus and the SP memory bus, the printed wiring on each port card supports two parallel-wire buses: the extended TDM bus and a microprocessor i/o bus, called the angel bus. Corresponding to the overall System 75's network interface card, acting as a front-end message processor for the switch processor (archangel) and interfacing the system's TDM bus to its SP memory bus, each port card has a front-end message processor, called the SAKI (for sanity and control interface), for its angel that interfaces the card's angel bus to the system's extended TDM bus. On each port card, the extended TDM bus is read/written by transceivers, the SAKI, and the codec (if there is one). On each port card, the angel bus is read/written by the angel, the SAKI, and the codec (if there is one).

- The time-multiplexed A and B buses on the motherboard are extended onto every port card through its transceivers. Sixteen bus drivers and 16 receivers translate between the conventional TTL signaling on the port card and low-impedance electrical signaling on the back-plane.
- The angel is a one-chip microprocessor whose programs communicate with the archangel over the message channel and control the TDM assignments for the lines, trunks, or service circuits on its port card.
- The SAKI is an integrated circuit chip that acts as a buffer and format converter for signaling messages between the angel and the archangel. The SAKI reads from, and writes to, the TDM bus only during its first five timeslots.
- The codec, called the network processing element (NPE), performs analog-to-digital conversion between the TDM bus and all the analog lines and trunks that terminate on this port card or the service circuits that reside on it. The codec also provides a conferencing capability. The angel controls the codec's timing. Port cards that terminate digital telephones or T1 lines don't need a codec; they have an equivalent circuit that buffers and distributes PCM samples.
- A typical port card terminates more than one line or trunk. A typical analog line card terminates eight analog lines. Each analog line or analog trunk terminates on a line circuit or trunk circuit on the port card. These circuits collectively are called the BORSCHT circuits and they are directly controlled by the angel. Supervision is signaled directly from a BORSCHT circuit to the angel, and analog ringing is applied by a relay in the BORSCHT circuit under angel control.

Let user A's line terminate on the 11th card on carrier C and let user B's line terminate on the 12th card on carrier D. Suppose A's mouth is connected to B's ear through junctor channel B.100. Then, on card C.11, the NPE samples A's line and delivers his 8-bit PCM mouth sample to the B side of the transceivers at the beginning of timeslot 100. On card D.12, the transceivers read the TDM bus at the end of timeslot 100 and deliver the B side to the NPE which converts the data to the analog signal for B's ear.

Besides the port cards for analog lines, implied in Figure 14.11, other port cards house trunk and service circuits. Still other port cards provide interface for digital lines, like ISDN, or data lines, like RS-232C. Complicated voice/data terminals interface to two channels, one for the voice connection and one for the data connection, analogous to the two bearer channels in BR-ISDN. Is it a good idea to connect a client PC to a server on a host over a circuit-switched connection through a digital PBX? We'll discuss this in Chapter 17.

While the motherboard for the processor subsystem has specialized wiring for each different kind of PWB, the motherboard for the port cards is a 200-wire parallel bus. The back-plane wiring is identical in the back of every card slot. Thus, all port cards are plugged into one parallel-wire motherboard on the back-plane that supports the system's talking channels and the parallel-access message bus for signaling between the switch processor and the port cards. This feature, called universal port slots, allows port cards to be added easily without back-plane modification.

14.3.5 Subscriber Signaling

Dial pulse detection is performed in the BORSCHT circuit and each dial pulse is reported to the card's angel. The designers must have been tempted to place a Touch Tone detector on each analog line card. Or, because there is a digital signal processor (DSP) in the codec anyway (for conferencing), it must have been tempting to perform Touch Tone detection with this DSP. But they didn't do it, and the system performs Touch Tone detection conventionally through a network connection between a line card and a Touch Tone detector service circuit on its own port card. Touch Tone signals are A-to-D converted in the subscriber's NPE, PCM samples are written into their assigned timeslot, transmitted over the TDM bus on the back-plane, read by the Touch Tone detector port card, D-to-A converted in its codec, and delivered as analog signals to the analog filters on this port card.

Different from every other system we've seen so far, dial tone is not the responsibility of the Touch Tone detector. Only the dialing subscriber's mouth is connected to the TTD. Dial tone, audible ring tone, busy tone, reorder tone, plus all 16 Touch Tones and 15 MF tones are all produced by a tone port card. When the system is booted up, an initiation program in the SP assigns and connects each tone permanently to a TDM channel. So, the dialing subscriber hears dial tone when a call processing program in the SP causes his ear in his port card to be connected to the TDM channel that permanently carries dial tone. When the TTP reports digit reception to the SP, the dial-tone connection is disconnected.

14.3.6 Interprocessor Signaling Channel

Figure 14.12 illustrates interprocessor signaling between the archangel and all the angels. The network interface board in the processor subsystem buffers and formats signaling messages between the SP bus in the processor subsystem and the message channel within the TDM bus. On each port card, the SAKI chip (see Figure 14.11) performs this same function between the angel bus on the port card and this same message channel. The message channel architecture is a logical star, with the archangel at the logical hub, but on a physical bus. Angels don't talk to other angels.

Messages use the first five timeslots of every frame and this message channel is duplicated onto both buses. The byte in timeslot 0 identifies the message's destination address. Data, the body of the message, is limited to 3 bytes, which are placed into timeslots 1, 2, and 3. A check-sum byte is placed in timeslot 4. At three data bytes per

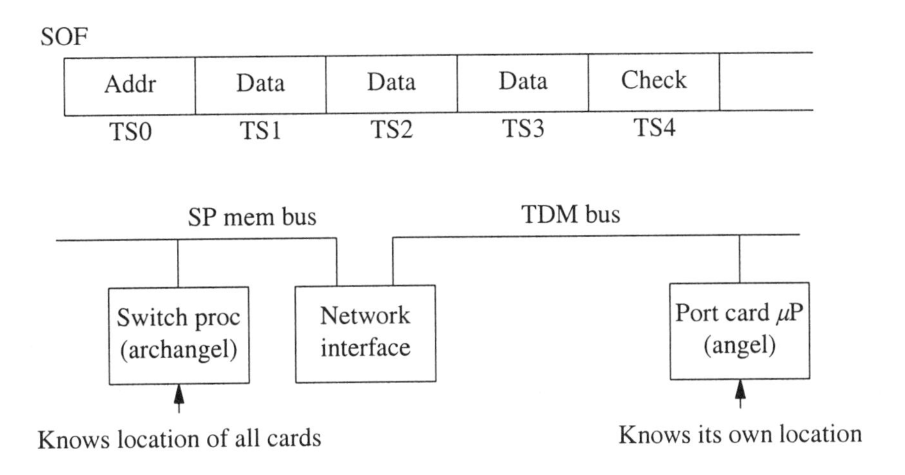

Figure 14.12. Control channel.

frame, the message channel's net throughput is:

$$8 \text{ bits/byte} \times 3 \text{ bytes/frame} \times 8000 \text{ frames/second} = 192 \text{ Kbps}$$

Each port slot has a unique 7-bit address that is hard-wired onto the motherboard. During the system's boot-up, the SAKI on each port card reads this address as its own. The maximum of 128 addresses exceeds the maximum number of port cards. The high-order bit in timeslot 0 holds an *address mode indicator*, and bits 1 – 7 of timeslot 0 hold the address of the angel that is receiving or transmitting the message.

- When the archangel sends a message downstream to some specific angel, the NPE places 0 in bit 0 of timeslot 0, this angel's address in bits 1 – 7 of timeslot 0, and 3 bytes of the message in timeslots 1 – 3. Long messages are sent in multiple frames.
- Periodically, the archangel polls the angels to determine if any have messages to send upstream. For this polling message, the NPE places 1 in bit 0 of timeslot 0, and a 4-bit *group address* in bits 3 – 6 of timeslot 0. The eight angels whose 4-bit subaddress matches these 4 bits, respond by placing a *request bit* in timeslot 2. Each angel places its request bit in a different bit position. Thus, the archangel can poll 128 angels, eight at a time, in just 16 frames.
- When an angel sends a message to the archangel, it waits to be polled first. Then, it waits until the archangel addresses it individually. After submitting a *request-to-send,* the angel cannot receive a message. The next time the angel sees its address in timeslot 0, the angel transmits its upstream message in timeslots 1 – 3. Bus contention is handled by *input queuing* in the SAKIs.

Downstream messages, archangel to angel, allow the program in the processor to control the LED and ringer on the telephones, control trunk circuits, set up and tear down network connections, and execute various maintenance tests. Upstream messages, angel to archangel, allow the port card to report supervision changes on trunks and telephones, button pushes on the telephones, and dial pulsing.

The designers provided for another 8-bit parallel bus, called the C bus, that was unused in early systems, but that allows the system to evolve to one with a separate channel devoted completely to signaling.

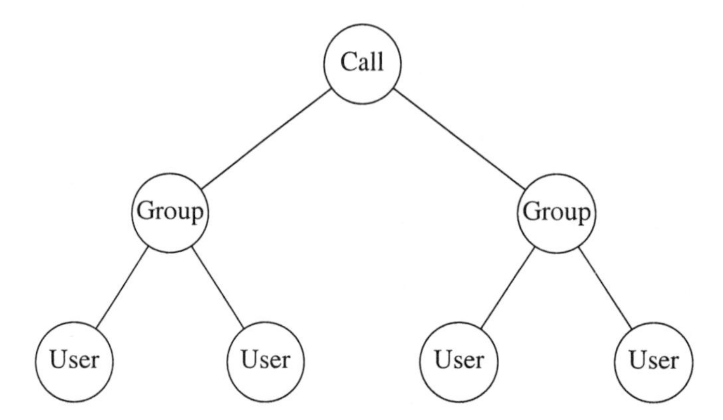

Figure 14.13. Call model.

14.3.7 Call Model

The System 75's call model is shown in Figure 14.13. It is best understood by considering a call through a tandem office. Consider that some subscriber, served by the office at the other end of trunk group A, wants to talk to another subscriber, served by the office at the other end of trunk group B. At the highest level of abstraction in Figure 14.13, a call in this tandem office is established by connecting some trunk from group A to some trunk from group B. In the intermediate level of abstraction in Figure 14.13, a particular trunk must be selected from within each group.

Besides trunk groups, users can also be grouped, particularly in an enterprise and most particularly in an enterprise that uses call pickup groups, like those offered by a key system. The members of a call pickup group are responsible for answering each others telephones whenever the called party is away. So, if engineer A calls accountant B, the software's call manager directs A's call to the accounting department's pickup group and instructs the software's group manager to connect this call to this group's member A. If A's telephone rings four times without answer, the group manager transfers the call to accountant C, who is next after B in the group's pickup chain. At the highest level, call control doesn't know about this and doesn't need to know.

The call model has become a very important software concept. We'll discuss it more in Section 18.4.3.

14.3.8 Software Architecture

System 75 software runs under a proprietary version of Unix, different from the 5E's version. While both versions are rich in interprocess communications utilities, the System 75's version of Unix supports a form of *object orientation.* "Object-oriented programming" is another of those terms, like "multiprocessing," that is not universally defined throughout the software discipline and industry.

Just as software architectures have different schemes for executing programs — batch, multitasking, and multiprocessing — software architectures also have different procedures for organizing data.

- Data can be placed in global tables, where all programs can read it and change it. We saw this scheme in Section 11.4.
- Data can be placed in centrally administered databases. Application programs request data from the database's manager and must have permission, especially to write new data. We saw this scheme in the Section 14.2.
- These databases can be effectively distributed so data is managed by the one application program that most often uses it. In this form of object-oriented programming, the program that "owns" the data can access it efficiently. Any other program can request the data from this owner, as if it were requesting the data from a database manager. We'll see this scheme in this section.

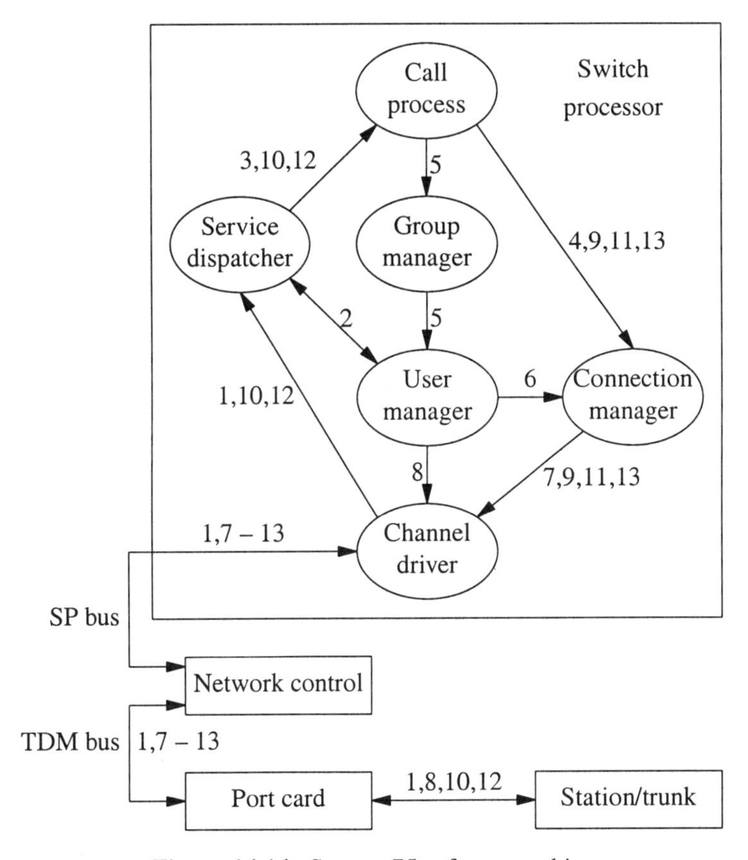

Figure 14.14. System 75 software architecture.

Figure 14.14 illustrates the System 75's software architecture. Call processes are programs that handle calls and maintain data about calls. The group manager is a program that handles pickup groups and maintains data about pickup groups. The user manager is a program that handles users and maintains data about users. The user manager is not the driver for the port cards and users' telephones; this software is located in the angels on each port card. The connection manager is a program that manages the system's fabric, maintains data about the state of the fabric, and acts as a driver for this fabric. The channel driver is a message processing driver between the switch processor's internal software IPC and the physical data channel over which the archangel talks with the angels.

We saw in Section 11.4 that every 1E program executes as a single process that performs its task for all clients. This multitasked architecture is very economical in its use of instruction space (program store) in the computer's memory. But it forces programmers to carefully store data, including loop indexes and other simple data, for different clients in different memory blocks. We saw in the previous section that every program executes as multiple processes; one is created for each new active client. While this multiprocessing architecture consumes a lot of instruction space in the computer's memory, it allows programmers to store data for different clients in simple "scratch" memory and registers. The System 75's software architects designed a compromise between these two extremes.

Like the multiprocessing environment, the System 75's call processing program runs as multiple processes. But like the multitasking environment, each process handles calls for multiple clients. Initially, this strategy might seem to have the worst characteristics of multiprocessing and multitasking. But the number of clients per process is not open-ended and potentially infinite; each process handles a maximum of 20 clients. This compromise architecture conserves instruction space compared to 100% multiprocessing while, at the same time, it allows programmers to store data in scratch memory (organized as indexed tables). Every time a programmer designs a memory word into his program, he must allocate an array of 20 words, one for each client. Each process of the call process program knows its ≤ 20 clients by an index: $0 \le i \le 19$.

In Figure 14.14, the service dispatcher is a task dispenser that assigns newly active clients and newly acquired data to its appropriate process of the call process program and passes the appropriate index with the data to this process. When the service dispatcher assigns a new call to a call manager process, the service dispatcher first tries to give the call to an active process that currently handles less than 20 calls. If all active call manager processes each handles exactly 20 calls, then the service dispatcher causes a new process to be created. Whenever a process is left with no calls to handle, the process is killed.

14.3.9 Call Walk-Through

A call is partially walked through the hardware and the software. The arrows in Figure 14.14 are labeled with the sequence numbers from the 13 steps in the description below.

1. The calling party, A, goes off-hook. In an analog telephone, the resultant DC status change is detected by a line sensor in the BORSCHT circuit in the port card. In a digital telephone, the off-hook is detected in the instrument and a digital signal is transmitted over a message channel from the telephone to the port card. In either case, the signal reaches the software in the port card's angel, which transmits a message to the service dispatcher program in the archangel (switch processor). This message is formatted into a packet by a driver program in the angel and output to the port card's SAKI chip, where it is stored for physical transmission in the first five timeslots over the TDM bus. The packet is physically received from the TDM bus by the network control card in the processor subsystem and input to the archangel over the switch processor's memory bus. In

the archangel, the packet is deformatted by the channel driver program and delivered by UNIX IPC to the service dispatcher program.

2. The service dispatcher program acquires A's class of service from the user manager program. In this object-oriented procedure, data about users belongs to the user manager. Because the identity of the port circuit and port card was included in the received off-hook message, the service dispatcher sends a UNIX IPC message to the user manager and requests class of service for the user at the identified physical address. The user manager receives this request and sends class of service back to the service dispatcher.

3. The service dispatcher program assigns the new call to a call process. If any existing call process is handling fewer than 20 calls, then this new call will be assigned to one of them. If all existing call processes are at full capacity, then the service dispatcher program spawns another call process and assigns the new call to it. A's physical address and class of service are sent to the responsible call process.

4. The call process reserves two timeslots on the TDM bus for eventual use by the two speaking directions: A's mouth to B's ear and B's mouth to A's ear. As the first step in processing any call, the call process sends a timeslot reservation command, including A's physical address, to the connection manager program. If no timeslots are available, the connection manager sends a no-acknowledgement (NAK) to the call process, which follows a program leg to connect A's ear to path-busy tone and await call abandonment. If timeslots are available, they are reserved against A's physical address and the connection manager returns an acknowledgement (ACK) to the call process.

5. In the second step of processing a call, the call process sends an origination command to the group manager, including A's physical address and class of service. While the call process could deal with the user manager directly, that would violate the call model's structure in which user $\subseteq$ group $\subseteq$ call. The group manager forwards the origination command to the user manager. The group manager also identifies all the other members of A's group and sends a status change command for each of these other users to the user manager.

6. The user manager orders the connection manager to initiate the procedure for receiving A's signaling. A's physical address and class of service are sent to the connection manager as part of the command. The connection manager uses class of service to determine whether signaling is via dial pulse, Touch Tone, or digital message channel. For Touch Tone, the connection manager identifies the physical address of an idle Touch Tone receiver card and uses A's physical address to identify the A's-mouth-to-B's-ear timeslot that had been previously reserved. If no Touch Tone receiver is idle, the connection manager would NAK the user manager and the call would be abandoned; otherwise, the call proceeds.

7. The connection manager sends network commands that connect A to the reserved Touch Tone receiver. One command instructs the angel processor on A's port card to place A's mouth samples in the A's-mouth-to-B's-ear timeslot on the TDM bus.

The other command instructs the angel processor on the Touch Tone receiver card to acquire its ear samples from this same timeslot. These two commands are sent by UNIX IPC to the channel driver, where they are packetized and output to the network controller over the switch processor's memory bus. The network controller transmits the packets to the angels on the two port cards over the first five timeslots on the TDM bus.

8. The user manager causes any status lamps to change on A's telephone and on every other telephone in A's group. Commands are sent to the angels on the port cards of every telephone in A's group. These commands also proceed, via the channel driver, the network controller, and the control channel on the TDM bus to the angels on the port cards of all users in the group. These port cards illuminate the line status lamps on these users' telephones.

9. The call process orders the connection manager to provide dial tone to A. In this third step of call processing, the call process sends a connect dial tone command, including A's physical address, to the connection manager. The connection manager sends a network command that instructs the angel on A's port card to acquire A's ear samples from the TDM bus timeslot that permanently carries dial tone. This command also proceeds, via the channel driver, the network controller, and the control channel on the TDM bus to the angel on A's port card.

10. A dials the Touch Tone digits for the desired terminating party, B (assumed intra-PBX). A's analog signals follow the same path and procedure that A's mouth samples will follow later, except that they are connected to the Touch Tone receiver card instead of B's port card. A's analog signal is A-to-D converted at his port card and the digital mouth samples are placed on the TDM bus in the A's-mouth-to-B's-ear timeslot. The Touch Tone receiver acquires its ear samples from this timeslot, and D-to-A converts A's signal back to analog Touch Tone signals. Tones are detected by analog processing and, for each tone, a message is transmitted from the TTR's angel to its SAKI to the TDM bus control channel to the network controller to the archangel's channel driver to the service dispatcher to the appropriate call process. After receiving the first digit, dial tone is disconnected when the call process instructs A's port card to cease acquiring ear samples (not shown on Figure 14.14). All A's dialed digits follow this same path to the call process, where they are translated.

11. When the call process determines that dialing is completed, the called party, B, is identified and rung (if idle). Translation from B's telephone number to his physical address is completed by a request to the user manager (not shown on Figure 14.14). B's busy/idle status is determined by a request to the connection manager (not shown on the figure). If B is idle, the call process sends out a series of commands: to the connection manager to disconnect the TTR from the A's-mouth-to-B's-ear timeslot and to make it available for another call, to the connection manager to connect A's ear to the TDM timeslot that permanently carries audible ring tone, and to the angel on B's port card to operate the ring relay in B's BORSCHT circuit. Notice that A hears audible ring tone through the fabric, but B is not rung through the fabric.

12. Similar to step 1 above, the called party B (not shown in Figure 14.14) answers his telephone by going off-hook. In an analog telephone, the resultant DC status change is detected by a line sensor in B's BORSCHT circuit and the ring relay is immediately released. A logical message is transmitted from B's angel to its SAKI to the TDM bus control channel to the network controller to the channel driver to the service dispenser to the call process assigned to this call.

13. This call process sends a series of commands to the connection manager: to disconnect A's ear from the audible ring tone timeslot and to connect it to the B's-mouth-to-A's-ear timeslot, to connect A's mouth to the A's-mouth-to-B's-ear timeslot, to connect B's ear to the A's-mouth-to-B's-ear timeslot, and to connect B's mouth to the B's-mouth-to-A's-ear timeslot. A and B are now talking to each other.

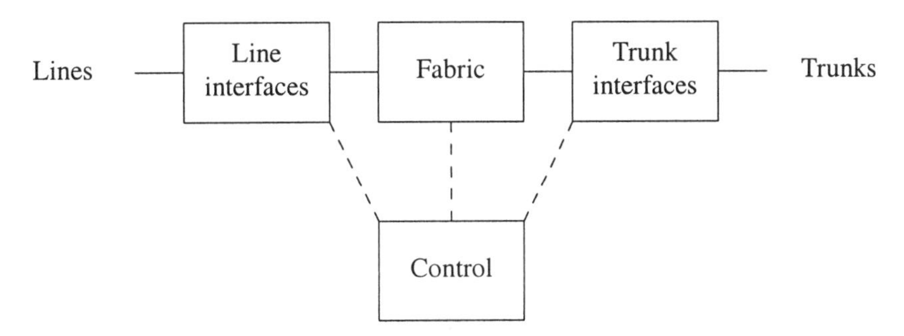

Figure 14.15. Generic block diagram of a switching office.

14.3.10 Discussion

We discuss the evolving separation of fabric and control and the System 75's partition.

Separation of Fabric and Control. The generic block diagram of a switching office, in Figure 14.15, shows a block for the fabric and a block for the control. But, historically, all systems haven't always conformed to this picture.

- Step-by-Step (Chapter 6) had no physically separate control subsystem. The logic for call control, switch control, and feature control were physically distributed throughout the architecture of the fabric.

- Panel (Section 2.5.2) and the various versions of Crossbar (Section 2.5.3, Chapter 8, and Section 9.2.5) match Figure 14.15 well if one redraws their block diagrams so the assorted registers and markers are gathered into a "control subsystem." Models 1 (Chapter 11) through 4 (Section 14.1) of the ESS, and TSPS (Sections 5.4.1 and 9.3.2), all match Figure 14.15 perfectly. These systems' processors were even called "central control."

- The modular architecture of today's digital switching systems, like the 5E (Section 14.2), does not match directly to Figure 14.15. While fabric and control are clearly separate in the overall system, they are also separate within the microcosm of each of the system's modules. Each module contains a piece of the fabric subsystem and a piece of the control subsystem. Figure 14.16 shows how a modular system's

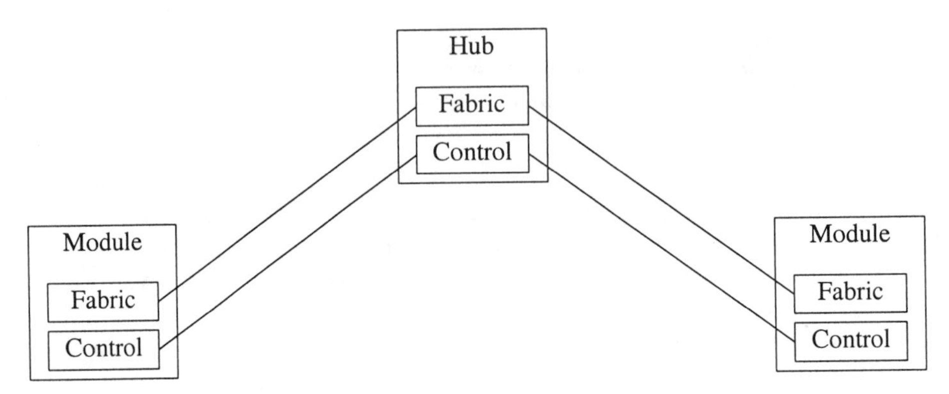

Figure 14.16. Fabric and control in a modular switching office.

physical architecture is congruent with the architecture of its distributed fabric and the architecture of its distributed control subsystem.

- While Step had no physically identifiable control subsystem, the System 75 has no physically identifiable fabric subsystem. Its logical fabric is physically distributed into the architecture of its control subsystem and interface cards.

The Partition. Placing software drivers in a microprocessor, which is physically on the port card with the hardware, has significant impact on the system's architecture. When component drivers are located in the switch processor, each type of component in the telephone process (Figure 11.1) requires a specialized wiring interface, called thick-wire control. When component drivers are located on the port card, because call processing software controls the telephone process over a common thin-wire message channel, the switch processor's entire periphery has an identical wiring interface.

- This common wiring interface enables the universal port slot. The system's OA&M is greatly simplified if port cards can be physically moved around and if a port card can be added without having previously reserved a prewired slot for it.
- This intermediate level of control offers a compromise between *flexibility* and *processor requirements.* Too much intelligence in the periphery (as in Crossbar) makes change difficult to implement. Too little (as in the 1E) places too much real-time burden on the main processor.

Even though the driver programs run on many duplicated processors, these processors are so inexpensive that the economy is sound. The tricky part, of course, is partitioning the programs so that just the right amount of control is placed in this firmware.

14.4 THE EXPERIMENTAL DIGITAL SWITCH

The experimental digital switch (XDS) was a research model of a digital switching fabric [21 – 22]. While developers of commercial systems actually did adopt several XDS concepts, the XDS architecture had far less impact than it deserved. Three significant XDS features are discussed in this section: the internal operation of the fabric; its ability to handle conferencing, broadcast, and subrate channels; and the intermodule network.

14.4.1 Module Architecture

A switching module consists of N line cards interconnected by a fabric. Those of us who worked on XDS debated about how many stages were in its fabric and about whether the fabric switched in the space or time division. It is not true that the main reason I've written this book is so that I can have the last word in this debate.

The logical design of a typical line card for a typical digital switching system includes an 8-bit memory that holds the user's mouth sample after digitization until it's ready to be placed into a timeslot on the motherboard, and another 8-bit memory that similarly buffers the user's ear sample in the other direction. In the physical design of a typical line card, these two 8-bit memories are on the card with the A-to-D converter, the BORSCHT circuits, and other hardware. In XDS, these two 8-bit memories associated with each line card, are physically separated and aggregated together.

Each XDS SM has two N-byte memory units, one that buffers N user mouth samples and one that buffers N user ear samples. Each memory unit is a RAM that is dual-ported and double-buffered. This kind of RAM allows each address to be read and written simultaneously. The line side of these memory units is time multiplexed into N timeslots. During timeslot n of an even-numbered frame, line card n writes its mouth sample into the A-half of word-n in the mouth-sample memory unit and reads its ear sample from the A-half of word-n in the ear-sample memory unit.

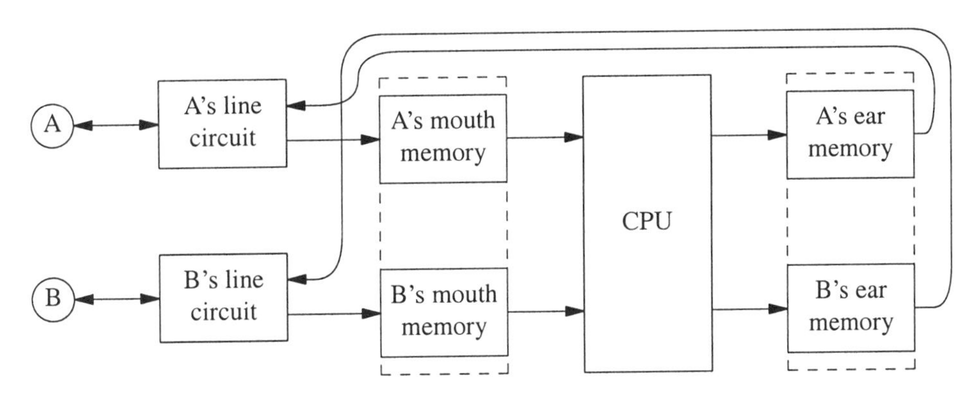

Figure 14.17. XDS switching module architecture.

Figure 14.17 indicates that the other port on each memory unit is connected to the memory bus of a high-speed 8-bit central processing unit (CPU). If this CPU can perform M 8-bit read/write operations in the time of one frame, then the CPU side of these memory units is time multiplexed into M timeslots. During timeslot m of an even-numbered frame, the CPU reads data from the B-half of word-m in the mouth-sample memory unit and writes it into the B-half of word-$c(m)$ in the ear-sample memory unit. Here, $c(m)$ is the switching assignment, the person to whom user m is connected. In odd-numbered frames, the line cards use the B-half of each double-buffered memory word and the CPU uses the A-half.

The CPU's ability to read from, and write to, random access memory is functionally equivalent to the function performed by a time multiplexor (read) and a time demultiplexor (write). The block labeled CPU in Figure 14.17 can be viewed to contain, logically, in series, left to right: a time multiplexor, one physical channel that is multiplexed M ways, and a time demux. Figure 14.17 is seen to closely resemble Figure 7.4(a), a two-stage fabric. But, instead of M space junctors interconnecting the two stages, the stages are interconnected by a single physical junctor that is time multiplexed into M timeslots. If $M = N$, then the fabric is nonblocking for conventional full-duplex intramodule calls.

Besides its multiplexing and demultiplexing functions, the CPU has an internal arithmetic/logic unit (ALU), and also special-purpose external hardware, capable of performing a variety of operations on its internal data between the time they are read from the left and written to the right. So, XDS's SM fabric architecture is a two-stage fabric, interconnected by a single physical intelligent junctor that is time multiplexed into M timeslots. New features and services can be implemented in XDS by programming the intelligent junctor. Among these added capabilities are broadcast, conferencing, and the ability to switch subrate channels.

14.4.2 Broadcast, Conferencing, and Subrate Channels

Every fabric we've discussed so far (and we're about two-thirds of the way through the book) is designed specifically for point-to-point connections and is very inefficient for point-to-multipoint connections. In XDS' fabric, broadcast connections are implemented easily and efficiently.

The bottleneck in the XDS fabric occurs at its junctors between the two stages and the significant shared resource there is the number of bus cycles the CPU can perform in one digital frame. Analogous to the operation of a timeslot interchanger, the CPU has two alternate junctor programs: (1) read incrementally from the mouth-sample RAM and write randomly into the ear-sample RAM according to the switching assignment; or (2) cycle incrementally through the ear-sample RAM and acquire the data for each successive ear-sample location by reading randomly from the mouth-sample RAM according to the inverse of the switching assignment. In either program, each half-duplex intelligent junctor requires three cycles. Using program 2, for each successive ear-sample address, B, (1) acquire the address of B's partner (A) in the connection (from another RAM that stores the inverse of the switching assignment), (2) read A's mouth sample from its RAM location (these two operations represent an indirect read), and (3) write this data into the RAM location for B's ear (the ear sample address is incremented during the write operation).

Using program 2 instead of program 1 enables broadcast connections at absolutely no additional cost in bus cycles or in the fabric's blocking characteristics. Every ear is connected to some mouth and each such connection requires three CPU bus cycles. That several ears might be connected to the same mouth is irrelevant. Although I can't think of a practical application (for voice), the fabric also supports a ring connection in which A talks to B and B talks to C and C talks to A (the ring is not limited to three nodes). Either program 1 or 2 supports a k-way ring equally well and the net cost is $3k$ CPU bus cycles. The fabric is inherently simplex and easily supports point-to-point, broadcast, ring, and other complex connections; as long as they are simplex. Because a point-to-point duplex call, in which A and B talk to each other, is just a pair of simplex calls, it is also easily implemented and the net cost is six CPU bus cycles (over two junctor timeslots).

Conferencing, even a simple three-way duplex call, is much more complex because each ear must hear the analog sum of the mouth samples of the other participants in the call. Conferencing is typically implemented in digital switching systems by connecting the participating parties through the fabric to a service circuit, called a conference bridge. A digital conference bridge can be implemented as: (1) k D-to-A converters interfaced to a k-way analog adder, (2) k μ-to-linear converter ICs interfaced to a k-way digital adder, or (3) as a k-input k-output digital signal processor (DSP) that is programmed to emulate the second implementation.

Implementing conferencing in a digital switch by installing m k-way conference bridges on the fabric requires that the number of conferees must be $\leq k$ in all conferences and that the number of simultaneous conference calls must be $\leq m$. The System 75 got around this by putting a DSP on every line card, a solution that is perhaps a bit of overkill. XDS demonstrated another solution in which the conferencing capability was built into each junctor at the heart of the fabric by adding the conference bridge DSP as a CPU peripheral.

Even without the DSP, the CPU has several useful internal operations that are useful for processing data as they traverse the junctor. By using mask, rotate, and OR operations, the user's 8-bit word could be arbitrarily partitioned into subrate channels, analogous to a fractional T1. The channel could also be partitioned temporally, as long as the user's equipment was synchronized to the frame clock that was examined by the junctor CPU. Using fractionalization to partition an 8-bit channel into eight 1-bit channels, one 64-Kbps channel could be partitioned into eight 8-Kbps subchannels. Each subchannel could be separately and independently switched in the XDS fabric and could be assigned to a different window in the user's PC. Switching subrate channels like this consumed many more CPU bus cycles than switching conventional full-rate channels. Determining the number of users that could be attached to one XDS SM with a certain probability of blocking became an interesting and difficult problem.

14.4.3 Intermodule Network

Constrained by the cycle-time capacity of the junctor processor, an XDS module can handle only a certain number of calls and, hence, only a certain number of telephones. So, a large system needs multiple XDS modules and a modular architecture. Like any digital switching system with a modular architecture, each XDS module has its own fabric, the junctor processor, and the modules' junctor processors are interconnected by a circuit-switched center stage that carries the users' PCM samples.

XDS' inter-SM center-stage network was modeled after Tandem Computers' interprocessor network. Tandem's architecture, designed for nonstop computing, consists of N parallel minicomputers that are interconnected by an extremely broadband bus network. Tandem's peripheral units, such as disk drives and terminal controllers, are all dual-ported and connect to any two of the N minis. Similarly, XDS' line card units were dual-ported and connected to any two of the N XDS modules. Tandem's interprocessor network, a simple bus with any extremely broad bandwidth, squanders bandwidth (appropriately, since the bus is a short LAN). Similarly, since it was viewed as a small part of the total system cost, XDS' inter-SM center-stage network was proposed to also have extremely broad bandwidth so that there would never be a bottleneck there even after a migration to broadband services.

Besides having its own fabric, the junctor processor, each XDS module also had its own switch processor, like any digital switching module would. Two separate and independent networks interconnected XDS's modules. The broadband center-stage network, described above, interconnected each modules' junctor processors. A packet-switched LAN, which carried interprocessor messages, interconnected the modules' switch processors. The XDS designers didn't think it was a good idea to embed IPC channels into the network used for switched data channels — as was done in the commercial systems described in this chapter.

14.5 DISCUSSION

It would be consistent with the Discussion section in prior chapters to discuss the partition and to summarize the software differences between the 1E generation of switches and the 5E/75 generation of switches. But, the 5E's and System 75's partitions were discussed in Sections 14.2 and 14.3, respectively, and the summary of software differences is postponed to Section 18.2. Instead, this concluding section examines network architectures for interprocessor communications, describes the characteristics that qualify a switch to be called "digital," and then discusses the market's failure to combine voice and data within some kind of integrated digital network.

14.5.1 Interprocessor Communications Networks

A modular architecture necessarily implies distributed processing. A system of distributed processors necessarily implies that interprocessor communications occurs over some network. This chapter has illustrated three separate ways to embed this interprocessor network into the switching system's architecture.

- In the 5E, the interprocessor network is deeply embedded within, and indistinguishable from, the voice fabric. Interprocessor messages traverse the same module timeslots, the same NCT links, and the same center-stage channels that users' mouth and ear samples traverse. As activity increases and interprocessor communication increases, timeslots are borrowed from the voice fabric. The problem, of course, is that the voice fabric is more heavily utilized when the system is extremely active.
- In the System 75, the interprocessor network is a logical partition of the system's voice fabric. Interprocessor messages use the first five timeslots of the system's TDM bus and users' voice samples use the last 251. This partitioning is permanent.
- In XDS, the interprocessor messages use a physically separate network from fabric used by the system's users' mouth and ear samples. If interprocessor communications habitually congests this network, the entire network could be upgraded as easily as any LAN is upgraded.

14.5.2 Integrating Voice and Data

While digital switches can support voice/data integration, it didn't happen nearly as much as was expected. Three observations are still significant today.

- The traffic characteristics of voice and data are sufficiently different that voice networks and data networks developed entirely different architectures and different types of switching equipment.

- Commercial user services that depend on real voice/data integration, like electronic Yellow Pages, have not been very successful in the marketplace anyway.
- Those real commercial voice/data user services that have been successful, like caller ID, have also been implemented on analog networks.

So, even though we have digital switching and digital computers, digital computers have rarely talked to each other through digital switches. Why not? Because: (1) traffic type matters at least as much as technology, and (2) networks must be architected for specific traffic types (Chapter 17). We still have to see whether these same observations apply to recent proposals for integrating voice, data, and video over packet-switched networks. We'll discuss this in Chapters 17 and 19.

EXERCISES

14.1 Compute the maximum throughput, in total bits per second, through the fabrics of a full-sized 4E, 5E, and System 75.

14.2 An IU in a 5E's SM is served by 64 timeslots in each direction and it interfaces to 512 ports, if fully equipped. (a) Requiring $P_B = 0.01$, for what percentage of the PBH can the average port be busy? (b) If ports are busy 15% of the time during PBH, how many IU interfaces can be equipped?

14.3 Analog concentration was proposed for the original 5E as a means to conserve A-to-D converters. Assume that a 5E line card costs $100. Assume that half of this cost is necessary per-line analog equipment and that half of this cost is digital equipment that is used only when a call is active. Let every line terminate on a per-line analog half-circuit and let active calls be analog concentrated to a bank of digital half-circuits. Assume that a fully equipped analog concentrator handles 256 lines at 0.08 Erlangs per line. (a) How much money is saved in digital half-circuits to cover the cost of the analog concentrator? (b) If the digital half-circuit costs $20 instead of $50, how much money is saved in digital half-circuits to cover the cost of the analog concentrator?

14.4 In a 5E, a calling party, wired to port 1 on IU 2 in SM 3, calls someone who is wired to port 8 on IU 7 in SM 6. Timeslots assigned for this call include: #50 in the mouth channel between IU 2 to the TSI in SM 3, #51 in the ear channel between IU 2 to the TSI in SM 3, #40 in the mouth channel between IU 7 to the TSI in SM 6, #41 in the ear channel between IU 7 to the TSI in SM 6, and #300 in the 5E's TMS. For each direction of data flow, describe the connections provided by the appropriate IUs in each SM, by the TSIs in each SM, and by the TMS.

14.5 Assume that a time-space-time switching fabric (like the 5E's) might be as large as $N = 128{,}000$ ports. Assume that the fabric's ports are 50% subscriber lines and 50% other circuits (service circuits, trunks, etc.). Eight different designs should be compared, based on four different proposed sizes of the center-stage TMS and two different port distribution strategies. The four proposed sizes are 16 × 16, 32 × 32, 64 × 64, and 128 × 128. The two proposed port distribution strategies are: (1) different types of ports uniformly distributed across all modules and (2) specialized modules in which all ports are of the same type. (a) In a full-size system, how many ports per TDM module correspond to each of the four TMS sizes? (b) Under which port distribution strategy are calls most likely to be intramodule? Under strategy (1), assume a 20:1 concentration between TDM ports and link timeslots. Under strategy (2), assume a 5:1 concentration between TDM ports and link timeslots. (c) For all eight cases, how many timeslots are required in each of the TDM-TMS links? Discuss the hardware and software implications of the two port distribution strategies. Which of the eight designs seems to be most practical?

14.6 Consider 5E SM which has its full complement of seven IUs. (a) If each IU's mouth and ear buses are 50% occupied during PBH, what is the average traffic intensity in each IU? (b) If the average holding time is 3 minutes per call per port, what is the call arrival rate per IU? (c) What is the call arrival rate for the entire SM? For each call, assume that call processing transmits an average of 10 interprocessor messages from the MCU to the AP. (d) How many messages per second are there between each MCU and the AP? (e) Assuming that the average message contains 10 bytes, what is the average data rate, in bits per second, in the message channel from each MCU to the AP? (f) How many DS0 channels are required at each SM? For each call, assume that call processing transmits an average of five interprocessor messages from the originating MCU to the terminating MCU. (g) How many messages per second leave and enter each MCU? (h) If there are 20 SMs, how many inter-SM message links are there? (i) Assuming that the average message contains 10 bytes, what is the average data rate, in bits per second, in the message channel from any MCU to any other MCU? (j) How many DS0 channels are required for each such channel?

14.7 For reasons that will be apparent in Chapter 18, suppose the 5E's feature control program is moved so it physically resides in, not just another processor, but a processor that is physically remote from the switching office. Discuss the impact on latency, the data network, the process-per-call implementation, and the ease of changing software releases.

14.8 Walk a BR-ISDN call through the 5E's software. Assume that "dialing" uses the D-channel.

14.9 Sketch a block diagram of what a System 75's BR-ISDN line card might look like. Assume eight lines to a card and that each line has an interface chip that demultiplexes its BR-ISDN frames. Show how the B-channels and D-channels reach the back-plane.

14.10 Continuing the previous problem, walk a BR-ISDN call through the System 75's software. Assume that "dialing" uses the D-channel.

14.11 Consider a heavily loaded System 75 and its message channel. (a) If 480 TDM channels are available for general use and they are 50% occupied during PBH, what is the average traffic intensity in the system? (b) If the average holding time is 3 minutes per call per port, what is the call arrival rate in the system? For each call, assume that an angel transmits an average of 10 interprocessor messages to the archangel and receives an equal number of interprocessor messages, including polling messages, from the archangel. (c) How many total messages per second is this System 75's angel-to-archangel message channel carrying in each direction? (d) How many total messages per second can the System 75's angel-to-archangel message channel possibly carry in each direction? (e) With eight ports on each port card, how many messages per second in each direction are processed by a SAKI?

14.12 Considering their software architectures, data ownership strategies, and operating systems, does the 5E or the System 75 seem like it has the better architecture for the following features: call coverage in a four-person office? Call forwarding? Call waiting?

14.13 Discuss modifying the System 75 and the 5E so that a System 75 is used as a 5E SM. Specifically address the following software issues. (a) How does the resulting AP-SP-angel three-level processor architecture affect the 5E's standard two-level architecture? (b) How would you settle the data ownership clash? Let each do its own thing? Emulate object managers on the 5E's database? How? Emulate database management by the System 75's object managers? How? (c) Are there any other significant problems?

14.14 The architecture of distributed processors in a modern digital switching system generally has three layers: (1) one large main-frame computer that serves all the system's modules, (2) a multiboard minicomputer that serves each module, and (3) a small microprocessor that serves each port card. But, the 5E's and the System 75's architectures of distributed processors only have two layers. (a) Which layers are missing in each of these systems? (b) How does this missing layer constrain the respective system?

14.15 The physical path used by interprocessor communications messages determines the physical structure of each system's interprocessor network. In the 5E and in the System 75, is this physical interprocessor network a bus, ring, or star?

14.16 If a system's flow of interprocessor messages is many-to-one, then the logical structure of the system's interprocessor network could be called a logical star. If a system's flow of interprocessor messages is many-to-many, then the logical structure of the system's interprocessor network could be called a logical bus. (a) Is the logical structure of the 5E's interprocessor network a logical bus or a logical star? (b) Is the logical structure of the System 75's interprocessor network a logical bus or a logical star?

14.17 Continue the three previous problems, applied to any general modular digital switch. (a) Could a layer of card-resident microprocessors connect to a module-resident switch processor by a physical star? (b) Would a layer of card-resident microprocessors ever need to be able to intercommunicate in a logical bus? (c) Could a layer of module-resident switch processors connect to a main-frame computer by a physical bus? (d) Does a layer of module-resident switch processors need to intercommunicate in a logical bus?

14.18 The four previous exercises have shown that, in both the 5E and the System 75, the physical architecture of their interprocessor network differs from its logical architecture. For each of these systems, does this difference arise (1) because the network between its layers of processors would naturally differ (previous problem) or (2) because of design choices?

14.19 Suppose the designers of the 5E and System 75 decide to change their respective systems so that interprocessor messaging is implemented over some conventional LAN architecture. Each system considers a bus, ring, or a star LAN. Explain which architecture seems best suited for each system.

14.20 When program 1 is used in XDS's junctor CPU, a 1-to-k broadcast call is implemented as k independent macros and the total call requires $3k$ CPU bus cycles. But when program 2 is used, the broadcast call is programmed as a loop. For each program, how many CPU bus cycles are required for a 1-to-k broadcast call? Don't forget that a cycle is needed to increment the index and another is needed to test for end-of-loop.

REFERENCES

[1] "No. 4ESS — Long Distance Switching for the Future," *Bell Telephone Labs Record,* Vol. 51, 1973, p. 226.

[2] *The Bell System Technical Journal,* (entire issue on No. 4 ESS), Vol. 56, No. 7, Sept. 1977, pp. 1015 – 1320.

[3] Abate, J. E., et al., "The Switched Digital Network Plan (No. 4 ESS)," *The Bell System Technical Journal,* Vol. 193, No. 17, Oct. 24, 1977, pp. 25 – 33.

[4] Bruce, R. A., P. K. Giloth, and E. H. Siegel, Jr., "No. 4 ESS — Evolution of a Digital Switching System," *IEEE Transactions on Communications,* Vol. Com-27, Jul. 1979, pp. 1001 – 1011.

[5] Jacobsen, C. R., and R. L. Simms, "Toll Switching in the Bell System," *Proc. Int. Switching Symposium,* 1976, pp. 132 – 134.

[6] Vaughan, H. E., A. E. Ritchie, and A. E. Spencer, "No. 4 ESS — A Full Fledged Toll Switching Node," *Int. Switching Symposium,* 1976.

[7] *5ESS Switch — Global Technical Description,* Iss. 2, American Telephone & Telegraph, Jun. 1990.

[8] Anderson, D. A., et al., "Call-Processing Software Structure" *AT&T Technical Journal,* (issue on 5ESS Switch Software), Jan. 1986, pp. 131 – 152.

[9] Andrews, F. T., Jr., and W. B. Smith, "No. 5 ESS — Overview," *ISS '81 Conference Pub.,* Vol. 3, 1981, Paper 31al.

[10] Brooks, J. M., and C. Blesch, "A Switch in Time," *AT&T Focus Magazine,* Jul./Aug. 1992.

[11] Carney, D. L., et al., "Architectural Overview," *AT&T Technical Journal,* (issue on The 5ESS Switching System) Vol. 64, No. 6, Part 2, Jul./Aug. 1985, pp. 1339 – 1356.

[12] Delatore, J. P., et al., "Operational Software," *AT&T Technical Journal,* (issue on The 5ESS Switching System,) Vol. 64, No. 6, Part 2, Jul./Aug. 1985, pp. 1357 – 1384.

[13] Hornbach, B. H., et al., "5ESS-2000 Switch: The Next Generation Switching System," *AT&T Technical Journal,* Vol. 72, No. 5, Sept./Oct. 1993, pp. 4 – 13.

[14] Kettler, H. W., et al., "The Application of Remote Switching Bell System Rural Modernization," *ISS '81 Conference Pub.,* Vol. 4, 1981, Paper 41c2.

[15] Martersteck, K. E., and A. E. Spencer, Jr., "Introduction," *AT&T Technical Journal,* (issue on The 5ESS Switching System) Vol. 64, No. 6, Part 2, Jul./Aug. 1985, pp. 1305 – 1314.

[16] "Computer Telephony: Say Goodbye to Key Phone Systems," *Communications News,* Vol. 35, No. 2, Feb. 1998.

[17] Baxter, L. A., et al., "System 75: Communications and Control Architecture," *AT&T Technical Journal,* Jan. 1985.

[18] Day, J. F., et al., "Looking over AT&T's Definity 75/85 Communications System," *AT&T Technology Magazine,* (issue on Electronic Messaging), Vol. 4, No. 2, 1989.

[19] Densmore, W., et al., "System 75: Switch Services Software," *AT&T Technical Journal,* Jan. 1985.

[20] Loverde, A. S., et al., "System 75: Physical Architecture and Design," *AT&T Technical Journal,* Jan. 1985.

[21] Lucky, R. W., "Flexible Experimental Digital Switching," *Proceedings of the International Seminar on Digital Communications,* Zurich, March 1978, pp. A4.1 – 4.

[22] Alles, H. G., "An Intelligent Network Processor for a Digital Central Office," *Proceedings of the International Seminar on Digital Communications,* Zurich, March 1978, pp. A5.1 – 6.

SELECTED BIBLIOGRAPHY

Belady, L. A., "Software Is the Glue in Large Systems," *IEEE Communications Magazine,* (issue on Telecom at 150), Vol. 27, No. 8, Aug. 1989.

Bellamy, J., *Digital Telephony,* Second Edition, New York, NY: Wiley, 1991.

Bergland, G. D., "A Guided Tour of Program Design Methodologies," *IEEE Computer Magazine,* Vol. 14, No. 10, Oct. 1981.

Bierman, E., "Changing Technology in Switching System Software," *Proceedings of the IEEE,* (issue on Telecommunications Circuit Switching), Vol. 65, No. 9, Sept. 1977.

Brand, J. E., and J. C. Warner, "Processor Call-Carrying Capacity Estimation for Stored Program Control Switching Systems," *Proceedings of the IEEE,* (issue on Telecommunications Circuit Switching), Vol. 65, No. 9, Sept. 1977.

Brooks, F. P., "No Silver Bullet: Essence and Accidents of Software Engineering," *IEEE Computer Magazine,* Vol. 20, No. 4, Apr. 1987.

Joel, A. E., Jr., "Digital Switching — How It Has Developed," *IEEE Transactions on Communications,* Vol. 27, No. 7, Jul. 1979, pp. 948 – 959.

Joel, A. E., Jr., "Towards a Definition of Digital Switching," *Telephony,* Vol. 197, Oct. 1979, pp. 30 – 32.

Joel, A. E., Jr. (ed.), *Electronic Switching: Central Office Systems of the World,* New York, NY: IEEE Press, 1976.

Joel, A. E., Jr. (ed.), *Electronic Switching: Digital Central Office Systems of the World,* New York, NY: IEEE Press, 1982.

Jovic, N., and G. W. Couturier, "Interprocessor Communication in Systems with Distributed Control," *Proceedings of the IEEE,* Vol. 65, No. 9, Sept. 1997, pp. 1323 – 1329.

Kajiwara, M., "Trends in Digital Switching System Architectures," *IEEE Communications Magazine,* (issue on Digital Switching), Vol. 21, No. 3, May 1983.

McDonald, J. C. (ed.), *Fundamentals of Digital Switching,* Second Edition, New York, NY: Plenum, 1990.

Rey, R. F. (ed.), *Engineering and Operations in the Bell System,* Second Edition, Murray Hill, NJ: AT&T Bell Laboratories, 1977.

Ritchie, D. M., and K. Thompson, "The UNIX Time-Sharing System," *The Bell System Technical Journal,* Vol. 57, No. 6, Part 2, Jul./Aug. 1978, pp. 1905 – 1929.

Schindler, G. E., Jr. (ed.), *A History of Engineering and Science in the Bell System,* Murray Hill, NJ: Bell Telephone Laboratories, 1982.

15

Human-Telecommunications Interaction

> *"I always wished my computer could be as easy to use as my telephone. I have gotten my wish. I no longer know how to use my telephone."* — attributed to Bjarne Stroustrup

After 14 long tedious chapters of studying the machinery of telecommunications, you have probably all forgotten the original purpose of communications stated in Chapter 1, "moving a concept from one brain to another brain." We've learned about the wires and other equipment that enable me to talk to you over the telephone. We read a lot about the "last mile" (local access) and the "last 100 meters" (the drop-wire), but precious little about the "last centimeter," how a new concept gets from a place outside your head to a place inside your brain. And even if you do know a little about human cognition and physiological psychology, you probably haven't seen this material presented from this perspective — the human as a component in a telecommunications system. This is a book about architectures and it's written at the systems level. This chapter describes the human user, himself a complex subsystem, the most important part of any switching (or computing) architecture.

This chapter discusses three topics: the human user of the system, the instruments by which this user interfaces to the system, and the services that this user receives from this system. The first five sections describe the human subsystem in the system. The first section shows several different block diagrams of systems where the human component is the real endpoint of the network, the source and sink of all the data. Sections 15.2 – 15.4 describe the human user as a system component with certain input/output characteristics. These sections examine respectively the human subsystem's three i/o ports — for audio, for video, and for data. These sections also describe the elementary appliance that provides the corresponding interface between the human user and the network. The audio appliance, the telephone, was described in Chapter 3, but the visual appliance, the television, and the primordial data appliance, the computer terminal, are described in Sections 15.3 and 15.4, respectively. Sections 15.2 – 15.4 describe the information format at its respective port and how the *cognitive bandwidth* of its respective port is modified by the network and by the instrument at the interface between the user and the network.

Generalizing from the notion of human-computer interaction, which describes how humans behave with their computers, Section 15.5 discusses cognition in human-telecommunications interaction. By researching side-tone (Chapter 3), the layout of the Touch Tone pad, the numbering plan (Chapter 1), and many other issues, the telecom industry has historically worked hard at its user interface. Section 15.5 describes two advanced human interfaces: the tree-menu-pointer interface and the GUI-WIMP. Section 15.6 describes how these cognitive concepts are implemented in the design of communications terminals. The final section describes "service" and the concept of being served. It presents the human user as a customer, someone who pays for the network and expects value for his money, the network's raison d'être, and (above all) someone who expects to be served by the network.

15.1 THE HUMAN COMPONENT

As defined in Chapter 1, "telecommunications" is a means of sharing remote information. But, more than moving bits from one machine to another, telecommunications is "moving concepts from one brain to another." The first subsection summarizes the three major telecommunications media and proposes a general model of the telecommunications environment. Two perspectives, ways of interpreting this model, are compared in the second and third subsections. The fourth subsection describes different ways that users are connected, through a network, to services.

15.1.1 The Telecommunications Environment

The three major telecommunications media are voice telephony, data, and broadcast video. Voice telephony is user-to-user, point-to-point, two-way communications. Data communications is also usually point-to-point and two-way, but it is usually user-to-service while telephony is usually user-to-user. Video communication is usually user-to-service also, but it is usually a broadcast one-way form of communications. Some typical characteristics of these three media are shown in Table 15.1.

TABLE 15.1. Tabulating some properties of the three major media

Medium	Parties	Fan-out	Direction	Regularity	Rate (bps)
Voice	User to user	Pt. to pt.	Balanced	Synchr.	10^5
Data	User to svc.	Pt. to pt.	Unbalan.	Bursty	$10^3 - 10^7$
Video	User to svc.	Broadcast	One way	Synchr.	10^8

While voice telephony and data communication are both usually two-way, telephony is usually balanced while data communication is usually unbalanced, with more data moving from service-to-user than from user-to-service. Telephony and video are typically synchronous, while data communication is typically bursty. The bandwidth required for telephony is specified by its digital equivalent of 64 Kbps. Full-motion video requires about 1000 times this bandwidth. Keyboard-to-computer data can be 100 times slower than voice telephony, while disk-to-processor data is typically 100 times faster. While radio is an alternate voice medium, with higher rate and fidelity than telephony, it more closely resembles the video medium. Telegram and electronic mail have some characteristics of data and some of voice telephony. Facsimile transmission has some characteristics of data and some of video. Videoconferencing has some characteristics of voice and some of video.

Figure 15.1 proposes a general model of the telecommunications environment. The

Figure 15.1. The telecommunications environment.

model in Figure 15.1 looks so simple that it would seem to be well understood and predictable. But, the telephone and the television were much more successful than their inventors could have ever imagined, each spawning huge industries. Both met market needs that weren't even known to exist at the time of their invention. Conversely, many examples of communications systems, apparently addressing obvious needs, have been commercial failures. This apparently simple telecommunications environment in Figure 15.1 is subtly complex and poorly understood.

Two different perspectives for Figure 15.1 are proposed: the window perspective and the interface perspective. These two perspectives are discussed in the following two subsections.

15.1.2 The Window Perspective

From this perspective, the purpose of the terminal and the network is to enable the user to perceive the service, or another user. If the user and the service were colocated, then the terminal and the network would not be needed. From this perspective, the terminal and the network are intermediaries and, as such, should be transparent. The terminal and the network represent a simple two-way window through which the user perceives the service.

For example, the Emergency Call System (ECS) is a specialized terminal that provides a specialized service. ECS is a special-purpose, originate-only, programmable telephone that adds its singular service to the residence. It is a small, modular system that connects to the network like a telephone. ECS is available in two commercial forms: Medical Alert and Smoke/Fire Alert. ECS links locally to a natural trigger mechanism: a manually activated trigger in the Medical Alert model and activation by the sound of the user's commercial smoke/fire detector in the Smoke/Fire Alert model. When triggered, the Emergency Call System places a telephone call to some party, typically the rescue squad or fire department, previously programmed by the user. When the called party answers, ECS recites an appropriate prerecorded message. The output and user feedback are by computer-synthesized voice, so there is no visual output display. User program input is from a console with fixed feature buttons.

If the user owned a private-line connection to the rescue squad or fire department, he could connect a trigger to the private-line at his end and a bell or light at the service's end. Using the public-switched telephone network instead of a private line requires additional equipment at the user's end, between the trigger and the telephone line. Illustrating this window perspective, ECS provides a natural and transparent way to make the switched connection resemble the private-line connection.

While the service is possible with a conventional telephone, which the user probably owns anyway, the general utility of the telephone makes it slower and less transparent, which justifies the ECS for many users. A general rule is observed: special-purpose terminals can be designed for improved interaction with specific services, but as a terminal is made more general-purpose, it becomes less efficient and less transparent.

15.1.3 The Interface Perspective

Two different perspectives for Figure 15.1 are the window perspective and the interface perspective. The former was discussed in the previous subsection. From the interface perspective, each box in Figure 15.1 represents a stage of transduction and each line represents an interface between successive transducers. From this perspective, a terminal is a transducer between the user's senses and the network's mechanisms. If the user and the service were colocated, certainly the network would not be needed. But, the user might be so incompatible with the service, or with another user, that a terminal must translate between them. For example, it is virtually impossible for a human and a computer, even in the same room, to communicate without a computer terminal.

All the components in Figure 15.1 are dynamic, including the user. One cause of the chaos in this field today, and a reason that standards are important for the future, is that each component in Figure 15.1 must be designed to interface with its adjacent components. But, they keep changing.

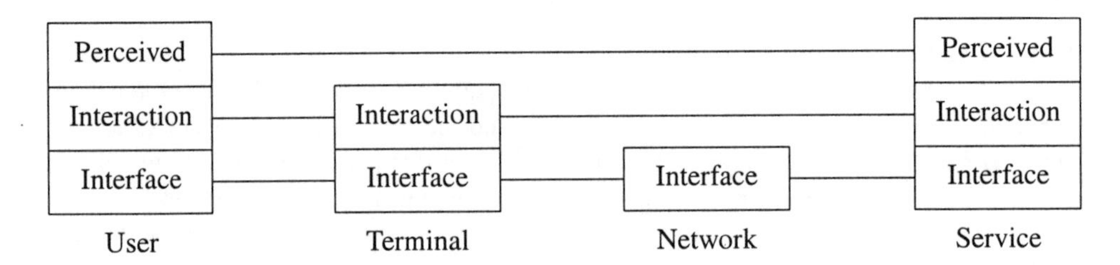

Figure 15.2. Layered processes in the communications environment.

Figure 15.2 is a schematic representation of a protocol stack with three layers. Call the layers:

- The *interface* layer;
- The *interaction* layer;
- The *perceived* layer.

If all four components from Figure 15.1 are present, then they must be physically and logically interconnected as indicated by the interface layer in each component in Figure 15.2. Once the network connection has been established, it is transparent. The user interacts with his terminal and his terminal interacts with the service, as indicated by the interaction layer in these three components in Figure 15.2. By the window perspective of the previous subsection, if the terminal is well designed, the user should barely perceive its presence. Instead, the user would perceive a direct communication with the service, as indicated by the perceived layer in these two components in Figure 15.2.

In circuit-switched networks, the user must first set up the call. This act has a different picture from Figure 15.2, which is relevant only in the steady state, during the call. Again, all four components from Figure 15.1 must be physically interconnected as indicated by the interface layer in each component in Figure 15.3. Because the network does not know which service the user will request, the connection is incomplete through the network during call setup. But, the user must be physically connected to the network through his terminal and every service must be connected to some port on the network.

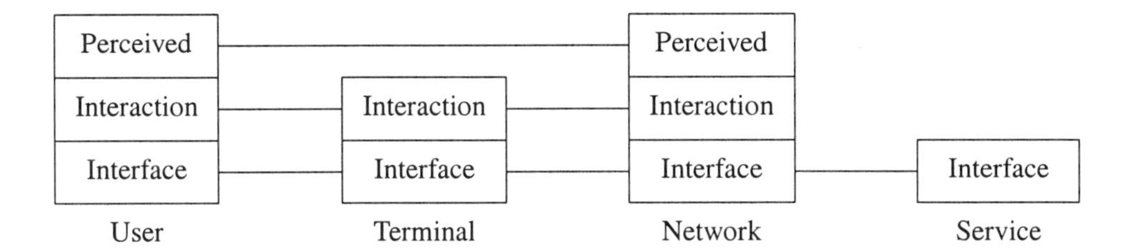

Figure 15.3. Call setup in the communications environment.

While the user still interacts with his terminal during call setup, the terminal interacts directly with the network, as indicated by the interaction layer in these three components in Figure 15.3. At a well-designed terminal, the user perceives a direct communication with some *connection request service* on some processor in the network, as indicated by the perceived layer in these two components in Figure 15.3.

When the user sets options in his terminal, an event even more primitive than call setup, he both interacts with, and directly perceives, his terminal. Thus, at different times, he is connected to his terminal at all three layers.

Both the window perspective and the interface perspective are correct. Any new terminal must be both a transparent window and a two-way transducer between the user and the network.

15.1.4 Connecting Users to Services

Figure 15.4 illustrates five different configurations in which users are connected to services. Figure 15.4(a) shows the simple configuration in which one user is connected to another user, like with POTS or any other point-to-point user-to-user connection. Instead of a user-to-user connection, the other four configurations illustrate user-to-service connections. Figure 15.4(b) shows the simplest configuration, in which a single user is connected to a single service, like a computer user who is logged in over some network or a "couch potato" watching television.

Figure 15.4(c) shows a more complex configuration in which many users are connected to the same service, like a telephone conference call or multiple simultaneous users processing electronic mail. In the conference call, the multiple users know about each other and are, in fact, interconnected by the service. In e-mail, the multiple users are all independent. While the figure accurately illustrates the physical connection, it also illustrates the logical connection if the e-mail program executes in a multitasking operating system. In a multiprocessing operating system, all users would have their own e-mail process and a more accurate logical illustration would show multiple services, one per user.

Figure 15.4(d) shows a single user again but, here, the terminal is a client-server workstation and the service is remote. For example, the user might be accessing the World Wide Web. The web browser executes on the client, the server provides software connectivity, and the particular web page under examination resides on some remote host. Because the client-to-server connection is probably over a data LAN and the server-to-host connection is probably over the Internet, the "network" is shown as two separate boxes. Figure 15.4(e) shows another configuration in which the user is connected to a

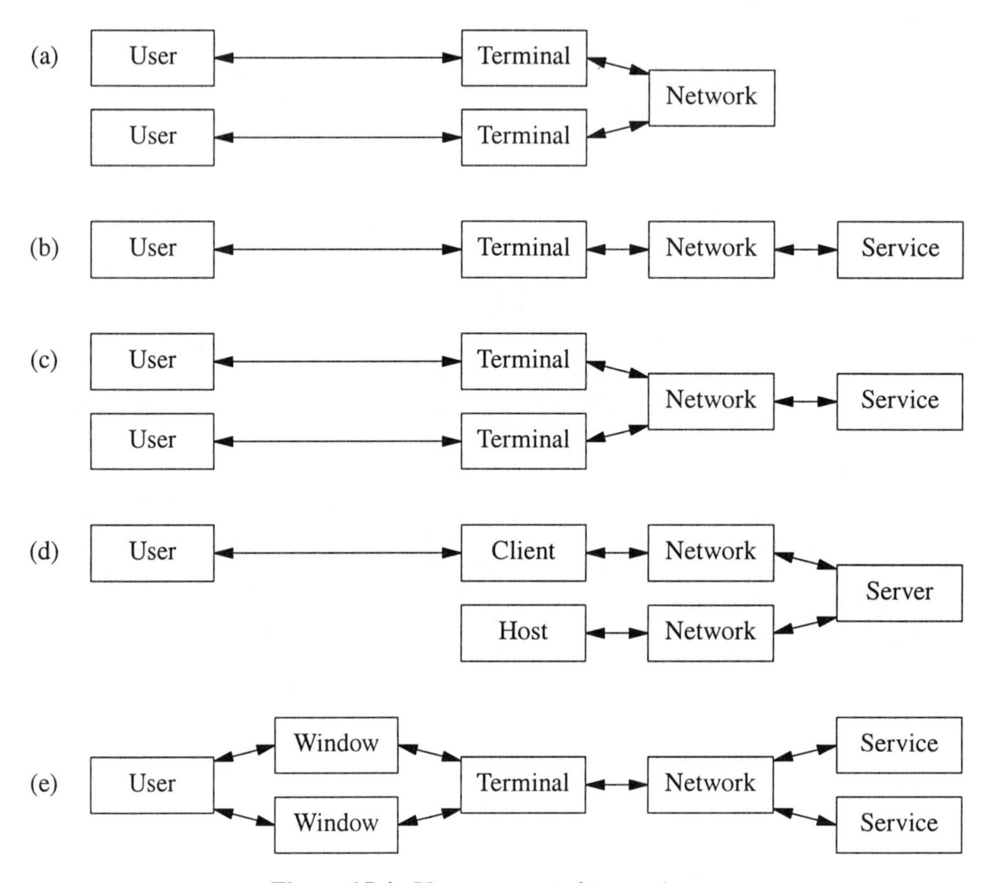

Figure 15.4. Users connected to services.

single host. But, here, the user executes many simultaneous service programs on the host and views each program through a separate window on the screen of his terminal.

While the discussion above is mostly at the physical layer, many of the telecommunications patterns illustrated in Figure 15.4 occur at multiple layers, analogous to the OSI Reference Model. The following sections of this chapter discuss human i/o, not only at the physical or physiological layers, but also at higher layers of communication. Just as the various protocol layers require their own software, the various layers of human communication require analogous processes, what I will call "brainware" throughout the rest of this chapter.

15.2 HUMAN AUDIO I/O

This section, and the two that follow, describe human sensory input and output (i/o). This section describes our audio i/o. Two subsequent sections describe our visual i/o and our data i/o, respectively. The human audio system has separate physiological ports — one for producing audio output and one for receiving audio input.

After a brief history of human audio i/o, the second and third subsections describe the human vocal apparatus, used for audio output, and its corresponding brainware. Then, Sections 15.2.4 and 15.2.5 describe the human ear, used for audio input, and its corresponding brainware. Language is discussed in Section 15.2.6 and the last subsection describes how the network and the terminal influence human audio i/o.

15.2.1 History

For most of the time that our species has been around, vision has been the most important sense for surviving — to find food and to avoid becoming some other creature's food. But, initially, audio was the more important medium for communication. Our species used word-of-mouth to communicate simple messages, to name things, and to name each other. Then, we developed complete spoken language and expressed ourselves and passed on our history and traditions from generation to generation.

When we began to use technology to record such information, it was visual, in drawings and in written language. This book considers written language to be data, a different form of physiological i/o. Technology for the audio medium is comparatively recent. The phonograph is only slightly more than 100 years old, radio is a little newer, and magnetic recording is only 50 years old. These were all analog technologies. Digital technology for audio is only 30 years old.

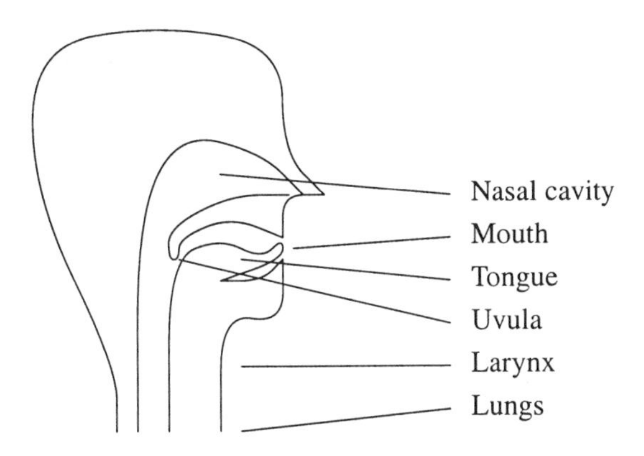

Figure 15.5. Cross-section of human head.

15.2.2 Audio Output Port

While we humans can produce audio output by stamping our feet, clapping, and making other noises, most of our audio output — and practically all of our audio information output — is vocal. In Figure 15.5, the human audio output system, our speaking mechanism, consists of our lungs and larynx, our mouth and tongue, our nasal cavity and uvula, our teeth and lips. This complex apparatus produces an acoustic wave from the mouth that has a complex time-varying recipe of harmonics. This acoustic signal requires a medium, like air, that can be locally compressed or rarified (evacuated). The acoustic signal produced by the mouth is a time-varying sequence of compressions and rarifications in the air. This signal travels away from the mouth through the air at the speed of sound (about 600 mph).

All sounds begin by exhaling from the lungs and modulating the air-stream by the rest of the vocal mechanism. The voiced sounds require exhaling through the larynx and tightening the vocal chords across the air stream, causing them to vibrate. The vibrating frequency determines the fundamental pitch, ranging normally between 100 and 300 Hz. People with thick vocal chords, like most males, have low pitch and their fundamental frequency is around 120 Hz. People with thin vocal chords, like children under 10, have high pitch and their fundamental frequency is around 300 Hz. Women typically have intermediate pitch and their fundamental frequency is around 220 Hz. We can control the tension of our vocal chords and, thereby, raise and lower our pitch — typically between half and double our normal pitch (plus and minus one full octave). Some languages, like Chinese, use pitch to distinguish the meanings of words that are otherwise identical. In English, we use pitch only for singing and for inflections, such as when we ask a question.

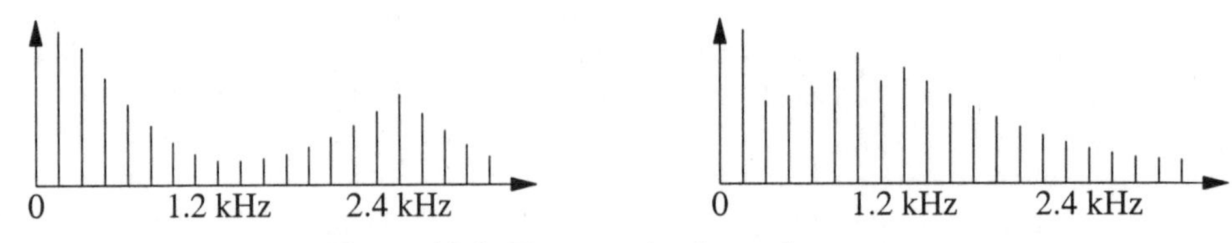

Figure 15.6. Two sustained vowel spectra.

Besides the fundamental frequency, the vibration waveform is very rich in harmonics. The interior of the mouth and nose acts as resonant cavities in which the various harmonics are attenuated differently. The human speaker can change the shape of this cavity and, thereby, control the sound that comes out of the mouth. Figure 15.6 shows two audio spectra, produced by the same fundamental frequency at 150 Hz, but by different mouth shapes — more detail on this below.

One simple control is the position of the uvula. All vowel sounds, plus "r" and "l," are formed in the mouth only because the uvula has closed off the nasal cavity. The sounds "m," "n," and "ng" are formed with the mouth and nose because the uvula is open. Try to make these sounds while holding your nose closed. In Figure 15.5, the head is speaking "ng" because the uvula is open and the tongue has sealed off the back of the mouth opening. If the lips seal off the mouth opening we say "m" and if the tongue seals off the front of the mouth, we say "n."

But, the real fine control mechanism that we have over cavity shape is to change the position of our lips and, especially, to change the position of our tongue inside our mouth. When we say "aah," we flatten our tongue to the bottom of our mouth (the doctor can see our throat better if our tongue is down flat). When we say "eee," we pull our tongue back toward our throat. When we say "ooo," we flatten our tongue but we purse our lips into a "kiss" position. Try to say "aah" with your lips pursed. If you say "eee" with your lips pursed, you get the German umlatted "u" sound.

The spectrum of the voiced sounds, comprised from harmonics of the vocal chord's fundamental frequency of vibration, typically extends out to 15 kHz. But, our controls have a remarkable effect on the shape of the spectrum of the acoustic waveform that is produced [1]. The spectrum of these sounds has hills and valleys. The spectrum of many sounds, including all the vowels, has three local maxima (hills) called formant frequencies. The two lowest ones are the most important. For a typical male voice, the first formant is 150 – 850 Hz and the second formant is 500 – 2500 Hz. The left graph of Figure 15.6 shows a typical audio spectrum for the "eee" sound, with the two lowest formant frequencies at 300 and 2400 Hz. The right graph of Figure 15.6 shows a typical audio spectrum for the "ooo" sound, with the two lowest formant frequencies at 900 and 1200 Hz. The exact value of these two formants and the magnitudes of the other harmonics will vary from speaker to speaker.

All of these sounds described so far are sustained — they can be continued or maintained for a long time. The "w" and "y" sounds are simply shortened versions of "ooo" and "eee," respectively. Any vowel can easily follow any other sound. Like vowels, the "r," "l," "m," "n," and "ng" sounds are also sustained, but they don't always easily follow every consonant — try saying "lrock" or "tnock."

TABLE 15.2. Classifying sounds from English

Type		Unvoiced	Voiced
Sustained	Mouth only		Vowels l, r
Sustained	Nasal		m, n, ng
Sustained	Fricative	h s sh f	wh z zh v
Dynamic	Fricative	ch k p t	j g b d

Another way we control the sound we make is to use turbulence. We can exhale harder than usual, making "hhh," or we can position our tongue or lips to create a small opening that we force air through. The resulting turbulence makes a "hiss" sound, which has a lot of random high frequencies. The class of sounds called the fricatives all use turbulence. The fricatives, especially "sss," have even higher frequencies than the voiced sounds, well beyond 20 kHz (where only dogs can hear it). They are distinguished by where the turbulence occurs inside the mouth and by whether the vocal chords are vibrating at the same time. Some of the fricatives are sustained, the sound can be held for a long time. Some of the fricatives are dynamic — the change of position makes the sound, and not the position itself. Try making a sustained "t" sound. To appreciate that "d" is just a voiced "t," compare the sounds "nnnnnt" and "nnnnnd." You'll notice that you have to stop humming the "nnnnn" to put the "t" after it, but you don't have to stop humming to put the "d" after it. Table 15.2 summarizes.

Finally, we also have volume control. If we exhale harder, our vocal chords vibrate with more intensity, the acoustic wave has greater compression and greater rarefaction, and we are louder and can be heard at a greater distance.

15.2.3 Speaking Brainware

While general *cognition* is discussed in more detail in Section 15.5, this subsection discusses the brainware responsible for human audio output. This mental processing is best appreciated by our experience.

- Observe infants as they learn to speak. Observe them learn to make sounds, then words, then sentences. Remember that we all learned to do these things without benefit of specific teaching.
- Consider having to write a computer program that can generate human speech from a given concept. Going further, imagine writing a computer program that could read the files that made this book and then could give a lecture on 1E software.

While speaking requires complex coordinated muscle control, through connecting nerve cables from brainware, we don't consciously think about it. In fact, most of you had to concentrate very hard on the exact position of your tongue as you read the previous subsection. We have the mental capability to relegate many repetitive tasks — like pumping our heart, breathing, walking, and speaking — to a subconscious layer in our brain. I like to think of this brainware as "drivers." While our heart and lung drivers were programmed in the womb (somehow), we program our walking and speaking drivers during our first year. We hear fellow humans make sounds and we know that we can make sounds. Then, we practice until we hear that the sounds we make are similar to the sounds we hear others make. Then, we "burn the code into ROM."

Now, as adults, when we want to say "Hello," we don't have to think about how to do it. We simply transmit some high-level command to the driver. With all the effort and technology that has been put into trying to make computer-generated speech sound natural, we humans seem to be always able to hear the difference anyway. So, it's all very remarkable and we don't understand it very well.

It seems that it's very important to program these drivers within some small window of time. If you didn't burn the code into ROM for making the sound of the German umlatted "o" at an early age, then you are forever doomed to have to think about making that sound. This explains why people who learn English later in life have foreign accents that are characteristic of their native language. For example, some people who grew up speaking Spanish or Italian put an extra vowel in front of English words that begin with "s" and some people who grew up speaking some Oriental languages struggle to distinguish "l" and "r." Similarly, when people who grew up speaking English learn another language later in life, they usually speak that language with an English accent.

15.2.4 Audio Input Port

Audio input is accomplished at the ears. We have two ears, partly for reliability and partly for discerning direction. On each side of our head, our hearing subsystem consists of the external horn and hole, the eardrum and several connecting bones and an internal organ.

Acoustic waves in the air are captured in the horn and focused into the earhole. Acoustic waves that do not travel through the air in our direction cannot be heard. Acoustic waves that are highly attenuated cannot be perceived. About an inch inside, the hole is sealed by a membrane, the eardrum. The successive compressions and rarefactions cause the eardrum to move in and out, respectively. Three tiny bones mechanically couple this vibration to another membrane attached to an internal organ, the cochlea.

The cochlea is shaped like a snail's shell, is filled with a fluid, and its interior wall is lined with small hairs. The acoustic signal is coupled into the cochlea and causes a pressure wave in the internal fluid to move the hairs. The position of the hairs is sensed by nerve endings in the cochlea and a parallel signal is transmitted over a cable of nerve fibers to brainware. There is evidence that built-in crosstalk in these nerve fibers allows for considerable signal processing along the nerve cable. It seems likely that the cochlea, audio nerve cable, and audio brainware perform a Fourier transform on the acoustic signal.

15.2.5 Hearing Brainware

We have all heard the philosophical question: "If a tree falls in the forest and nobody is present for miles around it, does it make a sound?" The answer, of course, is that it does not. When the tree falls and hits the ground, it produces an acoustic compression wave in the air. But, that is not a sound. A sound is a human perception, a signal inside our brain. It is not clear that we all perceive sound the same way. It is not clear that the signal from my audio nerve could be properly interpreted by your hearing brainware.

The hearing brainware receives a signal from both ears. If the signal from the left ear has greater intensity than the signal from the right ear, then we know that the sound come from our left. This direction-finding is automatic — we don't think about it.

Middle A on a piano produces a sound with a fundamental frequency of 440 Hz and a certain recipe of added harmonics. Middle A on a clarinet has the same fundamental frequency, but has a different recipe of harmonics. Our hearing brainware can tell us that the two notes are the same by comparing their fundamental frequencies, but we can also distinguish the piano from the clarinet by comparing the recipes of the sensed harmonics to a set of recipes that we have stored in memory. Even if we hear a clarinet play a note that we've never before heard a clarinet play, we can identify the sound as coming from a clarinet.

Similarly, we can determine that some human has said some word, like "Hello," by sensing the acoustic signal and comparing it to a stored version of the acoustic signal, even if the speaker is a total stranger whose voice we have never heard before. More remarkably, if the speaking human is known to us, we can identify who the speaker is by comparing the sensed acoustic signal to a set of recipes that we have stored in memory. And we can do this, even if this speaker says a word that we've never before heard this speaker say, or even any speaker say. In the next subsection, we'll show that it's even more remarkable. Not only do we do this so easily that we don't even think about it, but we learned how to do this by ourselves — nobody taught us. This is much more difficult than calculus — we can write computer programs to do calculus — and yet somebody has to teach us calculus; nobody teaches us to detect words and identify speakers.

We have the technology to record a clip of human speech. Suppose you and I each recorded: "halitosis" and "symbiosis." Now, imagine writing a computer program that analyzes these four recordings and tells us which pairs of recordings are the same word and which pairs of recordings were spoken by the same speakers. Similarly, consider having to write a computer program that converts acoustic waves in the air into computer data. Then, imagine that you could read this book to this program and it would be able to generalize the concepts to other domains. Or, the same computer program could hear instructions and learn math or geography or theology or how to drive a car or make a cake. Furthermore, this computer program doesn't start out with knowledge of the language — it learns it by observing patterns.

15.2.6 Language

Natural languages, their underlying grammars, and the *grammatical parsing* that is required to produce a sentence and to understand a sentence are so complex that *automata theory* is defined around the task. We humans are extremely good at processing natural language. We process language in real time, as we speak and as we listen. Language processing is such an easy task for our brain and such a difficult task for a computer, that it suggests that the inherent structure of the underlying processors is probably very different. The analogy between the human brain and the modern computer is probably extremely poor. Historically, technologists have tried to model the human body after the most complex machines of the day: the clock, the steam engine, and now the computer. They may all be equally poor models.

Perhaps the greatest insight into human language processing and the greatest indication of human propensity toward language comes from the amazing interaction between Ann Sullivan and Helen Keller [2]. When she was about one-and-a-half years old, Helen Keller suffered a severe fever that left her deaf and blind. While the child had certainly acquired basic oral language skills by this age, just as certainly, she couldn't have had any idea what written language was. Somehow, however, Ann Sullivan taught her to read Braille and to speak. The breakthrough occurred while little Helen had her hand in water and Ms. Sullivan wrote the letters "w-a-t-e-r" with her finger on the palm of Helen's other hand. Somehow, Helen knew that this sequence of tactile sensations was a name for water. Helen Keller's story should be required reading for all designers of systems that interact with humans.

Recall the difficulty you had when you tried to learn another language. Think about doing this without a teacher and without a book, like you did when you were an infant. Think about writing a computer program that you could read this book to and then, the program would work all the exercises at the end of each chapter.

15.2.7 Network Influence

The telephone is the human-telecom interface for audio, the transducer between the human and the network. Section 3.1 described how the telephone works: the mouthpiece or microphone is an acoustic-to-electrical transducer and the earpiece or speaker is an electrical-to-acoustic transducer. But, the transducer and the network deteriorate the signal by adding loss, echo, and noise. Amplifiers, echo-cancelers, and filters can compensate these impairments (within limits). But, the transducer and the network also limit the signal's bandwidth.

In music, an octave corresponds to a doubling of the frequency. Since middle A is 440 Hz then, an octave higher, high A is 880 Hz. This is the upper limit of most male singing voices. Many women can reach A above high A, at 1760 Hz. One octave above this, at 3520 Hz, is one of the highest notes on the piano keyboard. When this key is played on the piano, we hear this fundamental frequency and, in addition, we hear the recipe of harmonics that characterizes the sound of a piano. When this same note is played on the violin, we hear the same fundamental frequency but, in addition, we hear the recipe of harmonics that characterizes the sound of a violin. Most of our ears have sufficient audio bandwidth to detect the first five harmonics, when we hear the instruments directly.

However, if we listen to a recording on our stereos, or listen to a broadcast over FM radio, or over AM radio, the particular channel only passes a reduced number of the harmonics. The typical analog telephone connection, over a short wire-pair, has about 8 kHz of bandwidth. Over a long pair, such as a long distance wired connection, the bandwidth is reduced to under 4 kHz. Digital telephony uses this lowest common bandwidth, euphemistically called toll quality. So, if we listen to this high note played over a digital telephone channel, we only hear the fundamental and a highly attenuated second harmonic. We can still recognize the note, but we will have difficulty identifying whether the note was played on a piano, or on a violin, or on a tuning fork, because we don't hear all the harmonics, the complete acoustic information that our brainware needs.

We have a similar restriction listening to human voices over a telephone channel, especially children's and women's voices. Telephone bandwidth is sufficient for us to hear the first and second formant frequencies from all speakers. That's enough bandwidth that we can detect the words, especially in context. But, it's not enough bandwidth that we sound like ourselves. Everybody has a distinctive telephone voice. We can distinguish a voice at high fidelity from a voice over a telephone channel. Yet, we hear enough harmonics that we can identify a speaker, even one whom we have never before heard over the telephone.

While early speech compression techniques used to exacerbate this, more modern techniques (like those used in cell-phones and voice-over-IP) have a minor effect on fidelity. But, instead of 4:1 compressing a 4-kHz acoustic signal down so that it occupies the equivalent bandwidth of a 1-kHz channel, this author wishes the carriers would use compression technology to make the acoustic signal carried over a 4-kHz channel sound like it has 16 kHz of bandwidth. The telephone and the network restrict human audio capability in other ways besides limiting the bandwidth. One of the most important is that when we speak naturally, we can speak to many people at the same time. But, the network restricts our ability to broadcast.

15.3 HUMAN VISUAL I/O

Humans have a highly specialized seeing subsystem. Our eyes are physiological ports for visual input. But, while humans do produce a lot of visual output, in contrast to our visual input subsystem, we have no specific port for visual output, like fireflies or peacocks have. We have no being-seen subsystem. Our visual output port is distributed over our entire bodies. While we produce visual information by how we cross our arms or how we wiggle when we walk, most of our visual output is produced by hand gestures and by facial expressions. While all cultures do not agree on the assignment, we shake our heads up-and-down or back-and-forth to indicate positive and negative.

After a brief history of human visual i/o and a discussion of light, the third and fourth subsections describe the human eye, used for visual input, and its corresponding brainware. Television, the visual appliance, is discussed in Section 15.3.5 and the last subsection describes how the network and the terminal influence human visual i/o.

15.3.1 History

Technology for the visual medium is ancient. Cave drawings are many tens of thousands of years old and we've had written language for several thousand years. I consider written language to be data, which is discussed in the next section of this chapter. But, while human visual i/o handles full-motion video, these ancient media support still single images only.

Some significant modern technological advances in static single-image visual communications include the printing press about 500 years ago, the photographic camera about 150 years ago, the photocopier (xerography) about 40 years ago, and facsimile transfer (fax) about 30 years ago. While the moving picture was and is important, the single most significant modern technological advance in dynamic full-motion video communications is television, which has been commercial for only about 40 years. Digital visual media are extremely recent and still evolving.

15.3.2 Light

Electromagnetic radiation (EMR) is all around us. While much EMR is random, some of it carries information. While acoustic waves require a medium for transmission, and we once thought that electromagnetic (EM) waves did also (the mysterious fluid called ether), EM waves can travel through a vacuum — in fact, they travel best through a vacuum. Traveling EM waves are sinusoidal in distance and time. They have a frequency, measured in Hertz (cycles per second), and they have a wavelength, measured in meters per cycle. The relationship between frequency and wavelength is easy to see by performing algebra on their dimensions:

$$f \times w = \frac{\text{cycles}}{\text{second}} \times \frac{\text{meters}}{\text{cycle}} = \frac{\text{meters}}{\text{second}} = v$$

For electromagnetic waves in free space, the velocity v is the maximum speed of light, $c = 3 \times 10^8$ meters/second.

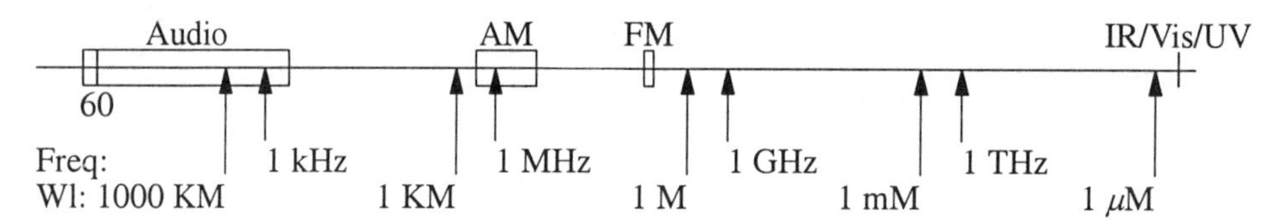

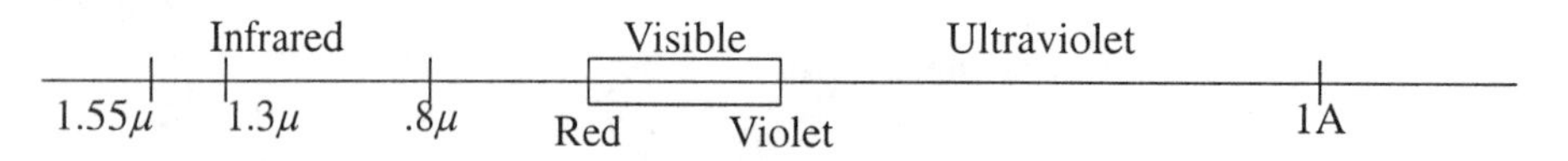

Figure 15.7. The electromagnetic spectrum.

EMR occurs over a huge spectrum, from the 60-Hz radiation of electrical power, to X rays and gamma rays. The upper line in Figure 15.7 shows a part of the electromagnetic spectrum. It is a log scale where frequency increases from left to right. Figure 15.7 also indicates some wavelengths at their corresponding frequencies. For some reason, the telecom community uses frequency when we speak about signals on the left side of this spectrum and we use wavelength when we speak about signals on the right side of this spectrum. On the far right of this broad spectrum at the top of Figure 15.7, is a small vertical line and the label, "IR/Vis/UV." This small region is expanded in the bottom of Figure 15.7.

The spectrum segment at the bottom of Figure 15.7 runs from 150 THz to 4750 THz in a log scale left to right. The frequencies are not shown — only some of the corresponding wavelengths. The short vertical line labeled "Vis" at the right end of the spectrum at the top of Figure 15.7 is expanded into the box labeled "Visible" on the spectrum segment at the bottom of Figure 15.7. This box, in the center of the lower line in Figure 15.7, shows the range of EMR that is perceptible to the human eye. We call this light, but it is just a narrow band in the EM spectrum. This visible spectrum is quite narrow, ranging from red's 0.66 microns to violet's 0.4 microns. EMR with wavelengths slightly larger than 0.66 microns is called infrared. EMR with wavelengths slightly smaller than 0.4 microns is called ultraviolet. Other relevant points near the visible spectrum are the ultraviolet signal at the far right whose wavelength equals 1 Angstrom Unit = 10^{-10}M and three infrared wavelengths at the far left that are relevant to optical fiber communications (Chapter 20).

15.3.3 Visual Input Port

Visual input is accomplished at the eyes. We have two eyes, partly for reliability and partly for discerning distance. The human visual input system consists of our eyes and the optic nerve, a cable that carries a signal to our visual brainware.

Our eye works like a camera. Light traveling toward our eye is focused by a variable lens onto the retina, at the back of the eye. We have muscle control to stretch this lens and change its thickness. This control allows us to correctly focus light from infinity or light from a close source. If the focal length of our lens is different from the front-to-back thickness of our eyeball, we correct the focus with eyeglasses. As we become older and the lens becomes less elastic, we have difficulty focusing on close objects and we need reading glasses or bifocals. The light that comes through the lens is projected, in focus, on the retina in exactly the same way that a transparency is projected onto a screen or the light through a camera lens is projected onto the photographic film inside the camera.

The retina, in the back of the eyeball, is covered with many closely spaced color-sensitive light sensors. Three different kinds of scattered sensors are sensitive to three different bands of wavelengths in the optical spectrum (lower line of Figure 15.7). Our visual brainware perceives the longest wavelengths, at the lower frequencies, as red and the shortest wavelengths, at the higher frequencies, as violet. Our visual brainware perceives signals, with wavelengths in between, as the other elementary (rainbow) colors. Our visual brainware perceives the absence of any visible signal as black and a uniform intensity over the entire visible spectrum as white.

The spacing of these sensors in the retina determines our visual *acuity,* the granularity of what we can see. In addition, some nerve endings are sensitive to static intensity and others are sensitive to changes of intensity. We can only see images that produce light or reflect light. We can't see an object whose light does not travel toward our eyes, or whose light is blocked by an opaque object, or whose light is extremely attenuated. If we could somehow increase our acuity and our sensitivity, both by an order of magnitude, the night sky would be very different from how we see it now.

Each light sensor is connected to a nerve ending and a parallel signal is transmitted to brainware along parallel nerve fibers in the optic nerve. As in the audio nerve, there is evidence that considerable processing occurs during this transmission. Brainware interprets the parallel signal: it processes the color and intensity for each picture element (pixel) and it assembles all the pixels into a scene. We see an image. And this all happens so quickly that we're ready to see another image in about 20 milliseconds.

15.3.4 Visual Brainware

When we're born, our eyes work immediately. Rapidly changing electrical signals travel along the many parallel optic nerves and reach our brain. But, there isn't any operational brainware there to interpret this. No wonder we cry the first time we open our eyes. As little babies, we are very busy in the first months of our lives, not only learning to interpret this vast signal, but burning ROM so that the interpretation process becomes yet another process that we don't have to consciously think about. Visual brainware smoothes the granularity of the parallel signal and blends the primary signals into a spectrum of perceived colors.

The signal processed by audio input brainware is a set of time-varying serial streams, one signal for each frequency in the cochlear filter. Audio brainware converts this into the perception of sound. The signal processed by visual input brainware is also a set of time-varying serial streams, one signal for each color-sensitive pixel in the retinal tessellation. Visual brainware converts this into the perception of a moving image.

Our visual process has limited time response. Some motions are so fast, they are just a blur. Try to follow the flight of a housefly or the cents column in the price as you pump gas — the machine displays every number from 0 to 9, but you can't see all of them. It appears that the temporal limitation is in the light sensors in the retina. They have persistence, spatial memory. While most of us can't see more than 40 images per second, some of us might get a headache in 50-Hz lighting but not in 60-Hz lighting.

Since we have two eyes, our brainware receives two slightly different representations of the same scene in front of us. We judge relative and absolute distance based on size and focus cues in each received image and from binocular differences in the two received monocular images. And we do all these things without thinking about them. We have the technology to scan a full-color image into a computer. But, imagine writing a computer program that can examine two slightly different 2-D views of the same 3-D scene and can tell us whether the red car is closer to, even with, or farther from the green bush. Computer programs find this difficult enough when the red car partially blocks the green bush, so it's really hard if the two objects are not overlapping. Imagine writing the program for a robot that plays center field or makes the three-point basketball shot or chips a golf ball onto the green.

As the next section discusses, when our eyes read text, we seem to connect the received visual signal into our language processor and not into our scene processor — or, maybe, they're the same processor. Deaf people can keep up with human audio speech using hand signing and lip reading. And some universal hand gestures "say" things, like to other motorists, that we wouldn't or can't say audibly. Finally, at the emotional level of cognition, don't underestimate the importance of body language and appearance and the amount of information that these things communicate (something of you that is put in my brain). One more way that the telephony is inferior to face-to-face communication is that, because we can't see the person we're talking to, we can't see each other's body language and we can't see how they're dressed.

15.3.5 Television

A television system has a video camera, which converts a full-motion image into an electronic signal, and a video monitor, which converts the electronic signal back into a full-motion image. We'll describe only the television monitor here [3 – 6]. The video camera works in a completely analogous way.

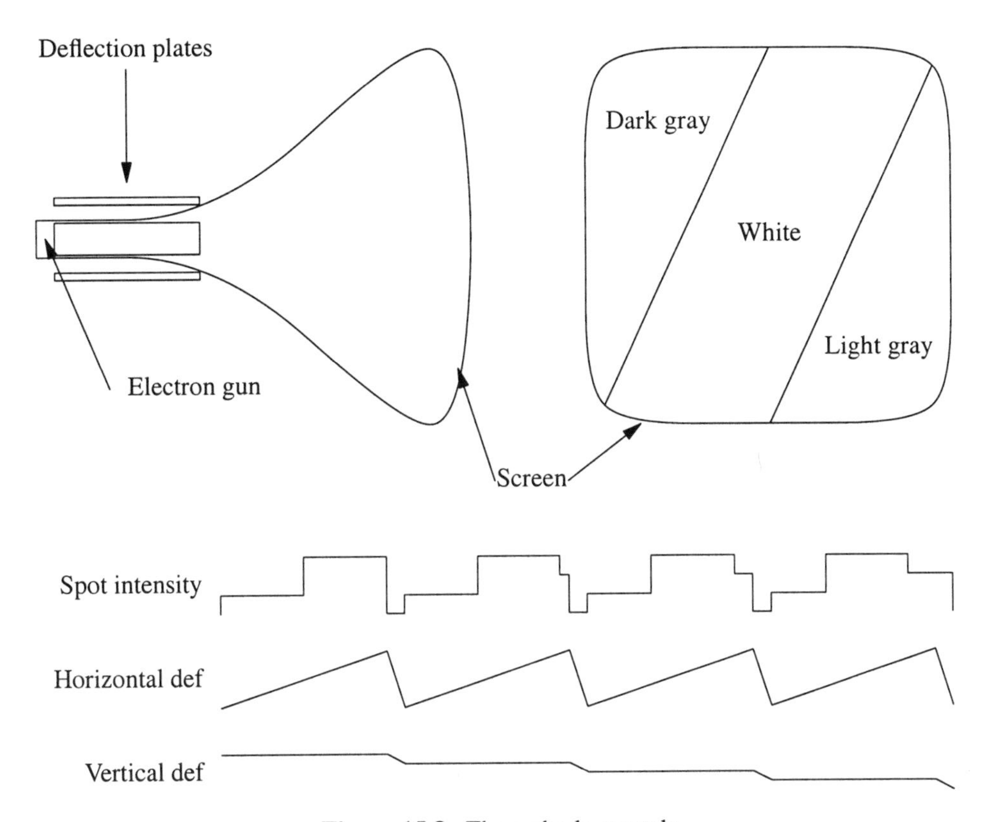

Figure 15.8. The cathode-ray tube.

The heart of the television receiver is the cathode-ray tube (CRT), or picture tube, shown in Figure 15.8 in cross-section at the upper left and, in the familiar front-view at the upper right. The interior surface of the screen is coated with a phosphorescent material that emits light when struck with electrons. In a black-and-white CRT, the emitted light is white. Electrons are emitted, in a narrow beam, from a "gun" at the back of the CRT's neck. The electron beam has three controls, illustrated by the three signal waveforms at the bottom of Figure 15.8. The beam's intensity is modulated by a signal to the electron gun. An intense beam produces a white spot where the beam hits the inside surface of the screen. A moderately intense beam produces a gray spot where the beam hits the inside surface of the screen. Shutting the beam off causes a black spot where the beam would have hit the inside surface of the screen.

The electron beam can also be steered, horizontally and vertically. Beam steering is typically accomplished inductively, not capacitively, with deflection coils, not deflection plates. Pedagogically, the capacitive process is easier to understand. So, Figure 15.8 shows two parallel vertical deflection (VD) plates, in cross-section, above and below the CRT's neck. Figure 15.8 shows one of two horizontal deflection (HD) plates in front of the CRT's neck. The second parallel HD plate is hidden behind the neck.

If the upper VD plate is more positive than the lower VD plate, the electric field deflects the electrons in the beam and bends the beam so that it strikes the screen above center. The more positive the upper VD plate is charged, the higher the beam strikes the screen. The more negative the upper VD plate is charged, the lower the beam strikes the screen. If the hidden HD plate is more positive than the visible HD plate, the electric field deflects the electrons in the beam and bends the beam so that it strikes the screen right of center. The more positive the hidden HD plate is charged, the more right of center the beam strikes the screen. The more negative the hidden HD plate is charged, the more left of center the beam strikes the screen.

The operation is seen by walking through the three signal waveforms at the bottom of Figure 15.8, from left to right. While the intensity is typically an analog signal, for simplicity, Figure 15.8 shows only four signal levels, corresponding to screen brightnesses of white, light gray, dark gray, and black. While the three signals endure as long as the monitor is on, the figure only shows four cycles. Assume that the initial value of the HD signal causes the electron beam to strike the screen at the left edge and that its most positive value causes the beam to strike the screen at the right edge. Assume that the initial value of the VD signal causes the electron beam to strike the screen at the top edge and that its most negative value causes the beam to strike the screen at the bottom edge. Now, since the three waveforms are synchronized, we can walk through all three signals, simultaneously, as time moves from left to right.

Initially, the beam strikes the screen in the upper left corner and the intensity of the spot is dark gray. As time advances, only the HD signal changes. As it becomes linearly more positive, the spot moves right linearly toward the center, still on the top edge and still dark gray. Almost halfway through the first cycle of all three waveforms, the intensity abruptly changes to white. Through the second half of this first cycle, as the HD signal becomes linearly still more positive, the spot moves linearly toward the upper right corner, still white. At this point in time, we have "painted" a narrow stripe across the extreme top of the screen. In the last moments of this first cycle the intensity abruptly changes to black, the HD signal changes linearly back to its initial value, and the VD signal changes linearly to a slightly less positive value. During this short horizontal retrace, the spot is repositioned on the left edge, slightly below the stripe that was just painted. Since the intensity is black, horizontal retrace is invisible.

Starting the second cycle, the beam's intensity changes abruptly to dark gray again when the beam strikes the screen slightly below the upper left corner. As time advances, only the HD signal changes. As it becomes linearly more positive, the spot moves right linearly toward the center, still slightly below the top edge and still dark gray. Almost halfway through this second cycle of all three waveforms, but a little earlier than in the previous cycle, the intensity abruptly changes to white. Through the second half of this second cycle, as the HD signal becomes linearly still more positive, the spot moves linearly to a position slightly below, and to the left of, the upper right corner, and still white. The intensity abruptly changes to light gray slightly before completely painting a second narrow stripe slightly below the top of the screen, just below the previous one. As in the first cycle, in the last moments of this second cycle, the spot is invisibly re-positioned to the left edge, again slightly below the stripe that was just painted.

The only differences between these first two cycles are that the second stripe is slightly below the first one and that the pattern is slightly different — the white band has shifted to the left and a light gray area has been introduced at the extreme right edge. Hopefuly, you can see the correlation between the first two cycles of all three signals and the extreme upper part of the pattern that appears on the face of the screen in Figure 15.8.

Hopefully, you can see the correlation to the remaining two cycles and, hopefully, you get the general idea.

The spot intensity signal is transmitted and the two deflection signals are synchronized internally. You can see how any noise that might be added to the spot intensity signal would be perceived as white flecks, or "snow," on the screen. You can see how an attenuated reflection would be perceived as a "ghost."

In the North American NTSC standard, a complete image has 524 stripes. After 524 cycles like the four in Figure 15.8, during its retrace, this time the VD signal returns to its full positive value instead of becoming slightly more negative. After this vertical retrace, a new image will be painted on the screen — or the same image will be refreshed or a slightly different image (to give the illusion of motion) will be painted.

But before the subsequent image can be painted, the previous image must be removed. Phosphorescent coatings can be manufactured to have a wide range of *persistence.* The actual persistence of each spot is very brief, only several milliseconds. Since the overall timing, described later, requires 127 μs to paint each line, only a few of the lines are actually illuminated at the same time. Our visual subsystem smoothes this "sliding window" into a single image and smoothes the repetition of single images into one continuously moving image. If it were not for this extremely short actual persistence, light pens wouldn't work (see next section).

For color television, the interior of the CRT is painted with three different phosphors that emit red (R), green (G), and blue (B) when struck by an electron. The CRT has three independent guns, one for the signal for each primary color. Colors are produced by simultaneously illuminating the three phosphor spots in each pixel with different intensities.

A video camera works just the opposite way. An image is scanned, one stripe at a time. This process is triplicated in a color camera, after the image is optically filtered through red, green, and blue for each signal. The 2-D image is serialized and and a cyclic intensity signal is produced — cyclic in frames-per-second and cyclic in lines-per-frame.

15.3.6 Network Influence

This subsection discusses physical bandwidth, techniques, inconsistency, and cognitive bandwidth.

Physical Bandwidth. The signal path from camera to monitor passes through several layers of processing, each of which adds noise and further restricts the electronic bandwidth. If a color camera's R, G, and B output signals are connected directly to some color monitor's R, G, and B electron guns, then the image seen by the camera appears on the monitor's CRT screen. The picture quality is excellent, even on a conventional television set. You see this kind of quality, the lack of snow and the sharp visual fidelity, on the CRT of a conventional PC.

At the television studio, a camera's R, G, and B signals are multiplexed together in a complicated format that is compatible with black-and-white sets. In the United States, the format details of the baseband signal and the bandwidth required of the television network were specified by the North American Television Standards Committee (NTSC). If this composite baseband signal is connected to the monitor's baseband signal input port, the signal is demultiplexed into its R, G, and B signals and the image seen by the camera appears on the monitor's CRT screen. This gives the picture quality of a monitor inside the studio, almost as good as the RGB picture and still much better than what we see on our home television set.

Also, at the studio, the composite baseband signal is modulated onto the carrier frequency that's been assigned for broadcast. If this modulated radio signal is connected directly to the monitor's receiver input, then the signal is demodulated by the monitor's tuner into composite baseband and the image seen by the camera appears on the monitor's CRT screen. Modulation, RF transmission, and demodulation all lower the picture's quality. Most of us are familiar with this picture quality because this is exactly what happens in a VCR, where the recorded signal is modulated onto the carrier frequency for channel 3 or channel 4.

Finally, if this modulated radio signal is broadcast from a transmitting antenna to a receiving antenna connected to the monitor, then the received signal is amplified by the monitor's receiver, then demodulated and demultiplexed, and the image seen by the camera appears on the monitor's CRT screen. Now the picture quality is what we're used to seeing with broadcast television — inferior to a VCR image because of the noise and attenuation introduced by broadcasting. Reception by antenna is usually worse than reception over cable because the CATV head end has a better receiving antenna than your TV and the coax is less susceptible to noise than free-space broadcast.

In telephony, the network characteristics are specified by the technology, primarily the bandwidth of a typical length of twisted pair. Fortunately, the technology didn't restrict the bandwidth too much. While we don't get high-fidelity audio, we get enough bandwidth that we can understand each other and we can identify speakers. In television, the baseband signal format and the required network bandwidth are specified by the visual limitations, in acuity and speed, of the human viewer. Fortunately, the human requirements were within the state-of-the-art of the technology at the time television was introduced. If the acuity and the temporal responsiveness of our visual input system were any better, the network requirements to provide satisfactory television might have been so great that television might have never succeeded.

Each horizontal stripe of the displayed picture is called a line and each set of 524 lines is called a frame (don't confuse a television frame with a T1 frame). Five hundred and twenty-four lines per frame is consistent with human visual acuity, but only when the screen size is limited to about 20 inches. Since NTSC television screens are 4×3 rectangles, screen size is measured diagonally and a 20-inch screen is actually 16×12 inches. With today's larger screens, 524 lines doesn't seem quite enough. New standards for high definition television (HDTV) have been developed and are beginning to show up in the market.

The required bandwidth depends directly on the required frame rate. For practical pricing, we must reduce the bandwidth required on the video equipment and in the networks that carry video signals. While 24 frames per second, like in old movies, is not enough, most of us can't see faster than 50 images per second. This limitation in our visual input system allows a compromise between high bandwidth and perceived flicker.

The classic NTSC format uses 30 frames per second, but interlaced, so it's actually 60 half-frames per second. In even-numbered frames, only the even-numbered lines are painted and, in odd-numbered frames, only the odd-numbered lines are painted. Using this clever scheme, our visual brainware doesn't see 30 frame-per-second flicker because the interlace blurs it. We perceive an acuity equivalent to 524 lines per frame even though only 262 lines are transmitted every 30th of a second. Interlace brings a four-fold decrease in required bandwidth. Two hundred and sixty-two lines every 30th of a second requires 127 μs per line.

Assuming, for now, the same density of picture elements (pixels) in each dimension, we must transmit $524 \times 4/3 = 699$ pixels per line; or 699 pixels per 127 μs; or 5.5 million pixels per second. Typical studio-quality video uses 5.5 MHz of bandwidth. Typical broadcast video channels have about 4.5 MHz of bandwidth. Typical television receivers have about 3.5 MHz of bandwidth. For the type of information that we typically receive over this medium, we have been satisfied with commercial television and it has become a huge industry of its own.

Different techniques for miscellaneous aspects of visual telecommunications are briefly summarized.

- *Image.* An image is a picture that doesn't move. The appearance of full motion is responsible for much of the bandwidth required for video transmission. When transmitting an image, the channel bandwidth, the picture's size and acuity, and transmission delay trade off against each other. We have several standards for facsimile transmission (fax).
- *Video and Audio Together.* At 12 kHz, the required capacity for even high-fidelity audio is incidental compared to the 4-MHz capacity required for video transmission. In commercial broadcast television, the analog audio signal is recorded and compressed, and then embedded into the baseband video signal during each line's retrace interval.
- *Light Pens and Touch Screens.* Since the phosphors are lit for a comparatively brief time during each video frame, a photodetector placed near the CRT screen would see light only when that particular part of the screen was being painted. Timing the detection of light with the geography of the image allows a processor to know the location of the detector. This technique underlies the light pen and some implementations of the touch-sensitive screen.
- *Point-to-Point Video.* Even though video transmission requires a broadband channel, commercial television can be radio broadcast because there are so few sources that radio spectrum can be assigned. But because only a limited number of point-to-point video connections can use radio spectrum, reaching a mass market will require a *tethered* access and switching.
- *Digital Video.* Digital video is just becoming available commercially as this book is going to print. The major benefits are improved picture quality and the ability to use digital techniques for compression. Without compression, an analog video signal, sampled at 8 MHz, requires about 48 Mbps with an acceptable 64-level gray scale, essentially a complete T3 channel. Digitizing the color composite baseband signal is subtly difficult and requires much more than 48 Mbps. Instead of A-to-D converting the classic composite baseband signal, digital television is much more effective if a codeword is transmitted for each picture element (pixel) in the frame. New standards have recently been developed.

- *Video Compression.* Video compression has been tried for years to enable transmission at reduced bandwidth. Most of us don't like video at 384 Kbps and find video over T1 to be barely acceptable. Compression hasn't been successful enough to create enough market that would lower the price of compression equipment. While there is great potential for even more compression, the ready availability of bandwidth for broadcast television has lowered the motivation. As with audio, compression can be used to squeeze conventional quality video into a smaller channel or to improve the quality of the video sent over a conventional channel.
- *Computer Imaging.* As computers have become faster, digital processing of pictures, first images and then even full-motion video, has migrated from hardware to software. Computer-generated images and, even, full-motion video have been extremely realistic since the *Star Wars* movies of the 1970s and early 1980s. Computer recognition of images, as in robotic vision, is still a long way off. We humans are amazingly good at recognizing faces by comparing the current image against something stored in memory.
- *Three-Dimensional Imaging.* An image can be perceived as three-dimensional if two slightly different views are presented to each eye. This can be accomplished with two screens or by time-sharing one screen and wearing blinking glasses. While holographic images are common, the transmission of full-motion holograms is still the stuff of science fiction.

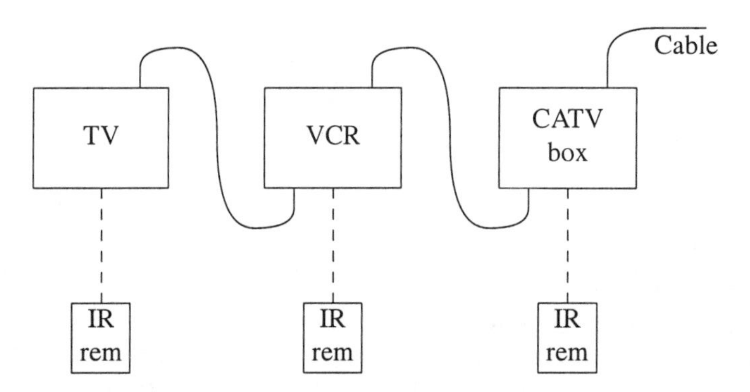

Figure 15.9. Television with VCR and cable.

Inconsistency. A configuration of commercial equipment, found in many homes, illustrates how such equipment can be difficult to manage and can even be inconsistent. Figure 15.9 shows a conventional television receiver, a video cassette recorder (VCR), and the set-top box for cable TV. Each piece of equipment comes with its own infrared remote controller. In the only wiring configuration that works, we connect the box's input to the cable and its video output to the antenna lead input on the VCR, and set the VCR to channel 3 or 4 permanently. We connect the VCR video output to the antenna lead input on the TV and set the TV to channel 3 or 4 permanently. We use the TV's IR remote to turn the TV on and off and adjust sound level, the VCR's IR remote to control tape or view from cable, and the box's IR remote to change channels; a nuisance, but acceptable.

Cable-ready television receivers work well with many CATV systems, eliminating the need for the box. But many people purchase them and bring them home, only to find that they don't work out. Caveat emptor. The configuration in Figure 15.9 requires a cable-ready VCR, not a cable-ready TV. Furthermore, if the CATV system scrambles its signals, the user must own a box anyway, for descrambling.

Many VCRs allow the user to record a program on one channel while viewing a program on a different channel. Many VCRs also may be preprogrammed to record, unattended, several programs on different channels at preset times. These features are possible, without reconfiguring, only when the VCR input receives all channels, as when connected to a home antenna. With cable, the user must also use rabbit ears or a rooftop antenna and must temporarily reconfigure the wiring. Switches are available, but the nuisance has increased. Furthermore, some security-conscious CATV systems require that the user's box upload a code and receive a downloaded descrambling key for each channel selection. In these systems, unattended recording from the VCR of cable-only programs is impossible, regardless of configuration.

The very important lessons to be learned are:

- Reliance on market evolution may lead to deadlocks;
- An architecture should have been planned (but, by whom?);
- Modular packaging doesn't always work;
- Privacy and security are difficult in broadcast systems;
- Interface specifications and standards are not enough;
- Services depend on the architecture of the network and the terminal equipment.

These lessons, especially the first two, generalize.

Cognitive Bandwidth. Finally, perhaps the biggest difference between natural vision and watching television is that, when we watch television, our field of view is constrained to a small screen. We perceive it as watching the world through a window or port hole. Natural vision has a much larger field of view. To understand this difference, imagine driving a car or playing basketball with your field of view confined to a small TV screen.

While large-screen television sets give us larger field of view, it's still a window, just a bigger one. But, because the large-screen television set and the conventional television set receive a signal from the same band-limited channel, we have an interesting tradeoff between acuity and field of view. Since the larger picture has the same number of pixels as the smaller one, the larger picture's quality is lower. It's as if the small window is open, but the large window has an insect screen across it. Section 22.6 describes an extreme video monitor and the extreme signal required to give it high quality.

15.4 HUMAN DATA I/O

Human data i/o is a little harder to appreciate because we don't have separate i/o ports specifically for data. This section describes human data processing, the primordial data appliance that we call a computer terminal, and how the network and this appliance influence our data i/o.

15.4.1 Human Data Processing

We humans communicate two kinds of data: telemetry and text. Since we didn't have text for our first million years, all our early data communication was telemetry. While primitive man whistled and yelled to his companions, more often, a more effective telemetry communication was hands-to-ears instead of mouth-to-ears. For manual output and audio input, we banged on something, like a drum, to make loud sounds. But, the most effective early telemetry communication was hands-to-eyes. Manual output and visual input — like smoke signals, semaphore, fire, and other forms of light — was the best way to communicate danger, food, or water — "one if by land and two if by sea."

Then, we invented text and written language. At the physiological layer, most of our text communication uses the same sensory ports that our telemetry communications uses. At the physiological layer, text is input with our eyes because we read with our eyes. The text output device is our fingers but, as with telemetry, we need an intermediate tool and medium. While the technologies that support data i/o are similar to the technologies that support audio i/o and visual i/o, they are different enough that this section is separated. It all started with the stylus and clay tablets, but it really took off with the use of papyrus and, then, paper. The printing press gave our text output a broadcast mode. Then, the typewriter allowed us to be neater. In modern times, we use an input appliance, called a computer terminal, which is a combination of a typewriter and a television.

While telemetry and text are input and output using the same physiological ports, we process telemetry and text differently in brainware. At the driver level, both kinds of data input are processed by our visual brainware. At the cognitive level, the two kinds of data — telemetry and text — are processed differently. Our text input port is our eyes, and while text is input visually at the physiological layer, at the cognitive layer text is much more like spoken language than like pictures. Human text processing probably uses much of the same brainware as speech processing. Thus, while text processing is visual at the physiological layer, text processing is more like audio at the cognitive layer. But, we don't parse it when somebody yells "duck," and we don't analyze the entire scene when we see a traffic light change from green to red.

Written language has had several forms in human history. In hieroglyphics and in Oriental picture writing (Kanji), each symbol represents a word from the spoken language. This causes a need for an enormous number of different characters and is particularly cumbersome when trying to build a keyboard. However, for common words it could be more efficient. For the number of times I typed "telecommunications" while writing this book, I should have defined a macro and let the word processor expand it.

Western languages, however, have a two-tiered structure. At the lower tier, the alphabet provides a written symbol for each *phoneme* in the language. At the upper tier, written words are constructed from symbols in the alphabet just like spoken words are constructed from phonemes. This technique allows much less memorization in mastering the written language and allows us to build a simple keyboard. There is considerable evidence that, as we become better readers, we recognize entire words at a time. So, on input, picture language and alphabet languages are probably equivalent. The major difference is in producing the text in each case.

Semaphore, like ASCII, builds on the elementary alphabet and would be impossible in picture languages. Deaf signing is a very efficient combination of the two types of written language. They have signs for most common words and less common words are spelled out. The deaf say "I love you" cleverly with one hand gesture, simultaneously configuring the fingers in the signs for "i," "l," and "y."

15.4.2 Computer Terminals — Hardware

A computer terminal is any appliance by which a user interfaces to a computer, and a communications terminal (described in Section 15.6) is any appliance by which a user interfaces to a network. In this book, computer terminals are a class of appliances that includes dumb terminals, but also are part of modern workstations and personal computers. Computer terminals have internal components, attached input and output devices, a display, and packaging. These features affect the computer terminal's capabilities, size, price, and quality. These characteristics, in turn, determine the usable service set and contribute to the user's perception of the computer terminal and the services. The rest of this subsection describes a computer terminal's hardware, including the internal components and the optional attached devices. The next two subsections describe the terminal's display and its packaging, respectively.

Components. Computer terminals are controlled by a variety of internal components. The most powerful types are general-purpose computers with i/o ports for humans. They have a processor, a large space of RAM and ROM, and fixed and optional special-purpose components as processor peripherals. Processor software typically runs in an operating system environment. More economical, but less powerful, architectures may be built around one-chip microcontrollers that typically do not have internal operating systems. Fixed special-purpose components typically include chip sets that interface to a keyboard, at least one data port, an analog or digital telephone line, and CRT electronics.

Commercially available one-chip keyboard controllers sense key operations on a coordinate grid pattern and transmit a serial code over the keyboard-to-terminal cable (or infrared link) to a UART that interfaces to the microprocessor. A typical chip set for an asynchronous data port includes a UART, RS-232C interface chip, and glue chips. Analog and digital telephone interface circuits are also commercially available.

Several manufacturers make chip sets that act as video controllers for ASCII characters and mosaic pixels. The explanation below uses typical numbers. The CRT screen is partitioned into a 24 × 80 area of picture elements (pixels). The corresponding pixel-RAM, with 24 × 80 = 1920 words, is written by the computer terminal's main microprocessor and is read by this chip set. Each pixel-word identifies which one of 256 patterns is displayed in the corresponding pixel, along with characteristics like intensity and blink rate. This set of 256 patterns includes the traditional alphanumeric characters, punctuation symbols, the two-letter symbols used to display ASCII control characters, a cursor symbol, and mosaic patterns for simple graphics. Each pattern is a 10 × 8 array of dots. Since a black-and-white display needs only 1 bit per dot, a pattern-ROM requires 256 × 80 / 8 = 2.56K bytes. The chip set delivers the entire displayed screen, consisting of 240 × 640 = 153,600 dots, in row-sequential bit-serial format to the CRT electronics, 30 times a second (about 4.6M dots per second). The chip set also provides for input from a light pen and translates to <x,y> position for output to the microprocessor.

Optional Devices. The computer terminal's external devices provide the network interface and provide for user input and output. Two categories of user input devices are data input devices and pointing devices.

- Data input devices include an alphanumeric keyboard or numeric keypad for text or numbers, a microphone for voice, and switches and knobs for control functions. Some computer terminals even have a camera for video input or a document scanner to read written or printed paper for fax transmission or optical character recognition.
- While keystrokes may be used to move a cursor and select from a menu, the mouse is a much better cursor moving device. Cursor-free pointing devices include light pens and touch-screens. These devices detect when the CRT's electron beam is under the pen or finger and the time of this event is translated into <x,y> coordinates.

User output devices include a speaker for voice, individual lamps or LEDs for status indications, a linear array of bit-mapped or seven-segment displays for short text messages, and a display screen for both text and image. While various flat-panel display technologies are used, especially on laptop PCs, the bulky CRT is still most common. A computer terminal may optionally include a printer, but the object frequently is to avoid or replace printed paper.

Physical design and shielding of internal circuits are important to reduce electromagnetic interference to the outside and among internal CRT, microprocessor, and telephone line. Standards are carefully upheld by the Federal Communications Commission and the Underwriter's Laboratory. Optional components include special-purpose hardware on cards, cartridges, and modules that are externally attached or internally installed. Optional software is conveniently sold on diskettes or in the familiar video-game packaging, ROM or battery-backed RAM in plug-in cartridges. Software download has become very practical over the Internet.

15.4.3 Display Format and Quality

The format and quality of graphical displays lie in a complex spectrum from static light-on-dark picture-element displays to full-motion color high-definition real-image video. The point on this spectrum selected for a computer terminal will strongly determine the services provided through the terminal, its cost, its internal architecture, and its network interface. Four overlapping neighborhoods on this spectrum are described.

Static light-on-dark displays of picture-elements, or dark-on-light, are used in early and inexpensive computer terminals. The screen is partitioned into a rectangular array, typically 24 × 80, of picture element (pixel) locations, each the space for a rectangular array of dots. The pixels are predefined, encoded, and stored internally, and the control is pixel-mapped, as described earlier. Pixel definition may vary: a pixel defined on a 7 × 5 array of large dots has much lower quality of font and granularity than one defined on a 12 × 9 array of smaller dots. But the CRT electronics of the latter requires more than triple the bandwidth of the former, and has higher cost. While the predefined set of pixels includes the usual alphanumeric characters, depending on the particular computer terminal and its intended application, this set can also include patterns for mosaic graphics. A more elite pixel display may expand the simple light-on-dark patterns to include gray scale and color. The communications format at the computer terminal's network interface is a simple message stream, such as the ASCII/RS-232C format. Such formats are extendable to cover expanded pixel sets, gray scale, or color.

Static displays of sketch-quality images, used in modern and expensive computer terminals and PCs, allow arbitrary shapes and lines, equivalent to fax. The control is directly bitmapped from an internal frame store whose capacity is 1 bit for every dot on the display. The bandwidth required of the CRT electronics is the same in pixel-mapping and bitmapping. But, while pixel-mapping requires only $24 \times 80 = 1920$ bytes = 15.36K bits of display memory, bitmapping requires at least 10 times this amount, depending on the dot density. Pixel-mapping is incorporated indirectly into a bitmapped display by writing pixel bit patterns into the frame store. More advanced, and more expensive, displays provide gray scale and color. Highly advanced PCs can store information about a three-dimensional image and sophisticated software can process hidden lines and surfaces, scaling, rotations, and perspective. While such computer terminals must be compatible with ASCII-like communications, they must also allow for entire bitmaps to be up or downloaded. While full-motion video (cartoons) can be generated internally, particularly for real-time rotations of three-dimensional images, downloading an entire bitmap 30 times per second requires a communications link of 6 Mbps, at least, with neither gray scale nor color, depending on the dot density.

Static displays of real images, equivalent to analog pictures like photographs, require a computer terminal similar to that above, but with higher resolution and finer granularity. The CRT is more expensive and the frame store is much larger. Most applications require the resolution of commercial television, at least. Since gray scale is necessary and color is common, increased quality (and cost) comes from raising the resolution. With gray scale and color, downloading even one static image can take an extremely long time over conventional voice/data communications systems. Alternatives are to use a high-rate LAN or to use conventional analog television networks and convert one frame to digital in the computer terminal with a device called a frame grabber.

Full-motion displays of real images require a television receiver and all its analog electronics. Color is expected and high-definition standards are under investigation.

Windows are logical displays on the same physical display. The common implementation is a rectangular area on a screen that directly corresponds to some communications link. These areas may overlap, and some may be entirely hidden by others. Windows are common on computer terminals and PCs, where each window communicates with a different active process on the host. Incorporation of sketch quality images in windows is now common on bitmapped computer terminals. Transmission of television with windows has been common for years, for example, in viewing baseball games. Windowing several channels on the same television receiver is also now commercially available. Coexistence of full-motion video windows and typical data windows on a computer terminal's screen is only restricted by the network interface.

15.4.4 Computer Terminals — Packaging

A terminal's packaging includes its size, modularity, styling, and even its name.

Size is an important characteristic. A computer terminal, or any kind of appliance, that overpowers an end table or occupies too much *footprint* on a countertop or desktop, will not be tolerated in the living room, kitchen, or office. An appliance that is so large that it must be considered to be a piece of furniture, like a large-screen television, must provide a very desirable service. Examples of the other size extreme for a terminal are pocket pagers and the Dick Tracy Wrist Radio.

Modular packaging is illustrated by many residential sound systems and the typical residential configuration of VCR and TV. The previous section warned of the risks of modular packaging in the discussion of the cable-ready TV. An even more modular architecture was proposed for Videotex.

Videotex is a prototype system that integrates voice, data, and static cartoon-quality images. Systems have been trialed in Great Britain, France, West Germany, Japan, and the United States. The prototypes have been modular; with a control unit and keyboard tied to the user's color television receiver. A Videotex color display features text and geometric and mosaic graphics. The human interface is by typed entry at the keyboard and uses tree traversal and menus for access and selection. While the concept of Videotex was usurped by the Internet, a discussion of the variety of communications interfaces in the different prototypes of Videotex is a worthwhile digression.

Potential communications interfaces were the user's video cable and the user's telephone line; one or both have been used. In alternate architectures:

- The remote centralized video database is dial-accessed, screen requests are transmitted upstream, and the requested image is transmitted downstream, all through the telephone network and without using video cable;
- Similar communication occurs over two-way video cable and the telephone network is not used;
- Screen requests are transmitted upstream over the telephone line and the requested image is transmitted downstream over one-way CATV cable;
- The entire database is repeatedly broadcast over one-way CATV cable and the controller grabs the desired image.

The style of most commercial computer terminals and PCs has been simple and linear, but not boxy. Besides being economical, molded plastic provides the desired look. While beige and light blue computer terminals have been commercially available, the preferred color seems to be off-white with charcoal trim. This typical styling and color have a high-tech, futuristic effect that may actually contribute to some users' resistance.

Commercial computer terminals and PCs have been given different kinds of names:

- Functional names like PC II, Silent 700, and DisplayPhone;
- High-tech names like Sceptre, Aerial, Presario 150, and Satellite Pro;
- License-plate names like ADM 2, VT 100, 510 BCT, and HP 2621.

These names are barriers for some people. The classic exception, and one of the reasons for their users' loyalty, is Apple.

15.4.5 Network Influence

Network influences include cognitive bandwidth, distributing intelligence, the evolving architecture, and the impact of photonics.

Cognitive Bandwidth. Besides not having its own dedicated sensory ports, text i/o is different from audio and visual i/o in another way. Humans frequently participate in direct human-to-human audio and visual communication. While we usually prefer to communicate with other humans directly like this, we can't always do it and sometimes we prefer an indirect method (last section of this chapter). Because of the difficulty or inconvenience caused by distance, we tolerate the relative inferiority of audio and visual telecommunication over the real thing. Humans do not, however, generally participate in direct human-to-human text communication. Text communication is almost always indirect, through some intervening machine. In classical text telecommunication, where I produce text on a laptop PC and you perceive text from a book, there is no interaction — it is one-way and one-time only. Modern text telecommunication over an intermediate network, electronic mail, is an improvement because it is interactive.

Only very recently has bandwidth become an issue in data communications. We become impatient when we send large files to each other over networks with low data rates. We're even starting to use data compression for speed-up as well as for encryption.

Distributing Intelligence. The network has had a large impact on the architecture of the generalized computer terminal, particularly the distribution of processing intelligence between the service and the teleterminals. The economically optimal architecture depends on the relative costs of computing and communicating and this changes.

Over the years the steadily declining price of digital electronics has dramatically reduced the cost of memory and computing. Although systems for digital transmission and digital switching also use this technology, the ratio of the cost of storing and processing bits to the cost of transmitting bits decreased steadily in the interval from 1955 to 1985. This trend was exacerbated because the cost of computing was competitive in an open market, while the cost of communicating had been largely regulated.

In a price-performance optimization, it follows that, over this interval, the distribution of intelligence between the terminal and the host (service) would reflect the state of this cost ratio. Gradually more intelligence was built directly into terminals and we saw the evolution of the computer terminal into the personal computer. Simply put, terminals became gradually more complex and self-contained because it became gradually more economical and convenient to provide local intelligence than to communicate with shared intelligence.

Besides a price-performance optimization, this architectural trend was also driven by the bandwidth demanded by users and services. The conventional 300/1200 Baud connection between terminal and host became gradually more inadequate. As local area networks came on the scene and provided greater bandwidth, at the burst rate, the architectural trend took a slight turn. With a personal computer, the user was responsible for administrative tasks that were conventionally performed by staff at centralized computing centers; notably, disk back-up. High-burst data rates over local area networks allowed the mass memory, previously in the personal computer, to migrate back to a centralized facility, called the file server, where it could be administered. A new kind of terminal, called a workstation, has evolved for this environment. A workstation is a terminal with an internal processor or a personal computer without a disk drive.

Evolving Architecture. A generalized terminal-to-host architecture is presented in Figure 15.10(a). Each of the four interfaces, labeled A – D in Figure 15.10(a), is a candidate for the boundary between the terminal on the user's premises and the host in some central location. Figure 15.10(b) – (e) represents architectures where the boundary between the terminal and the host has been selected at each of these candidate positions. The first three, in Figure 15.10(b) – (d), are classical configurations. The fourth, in Figure 15.10(e), is proposed for the future.

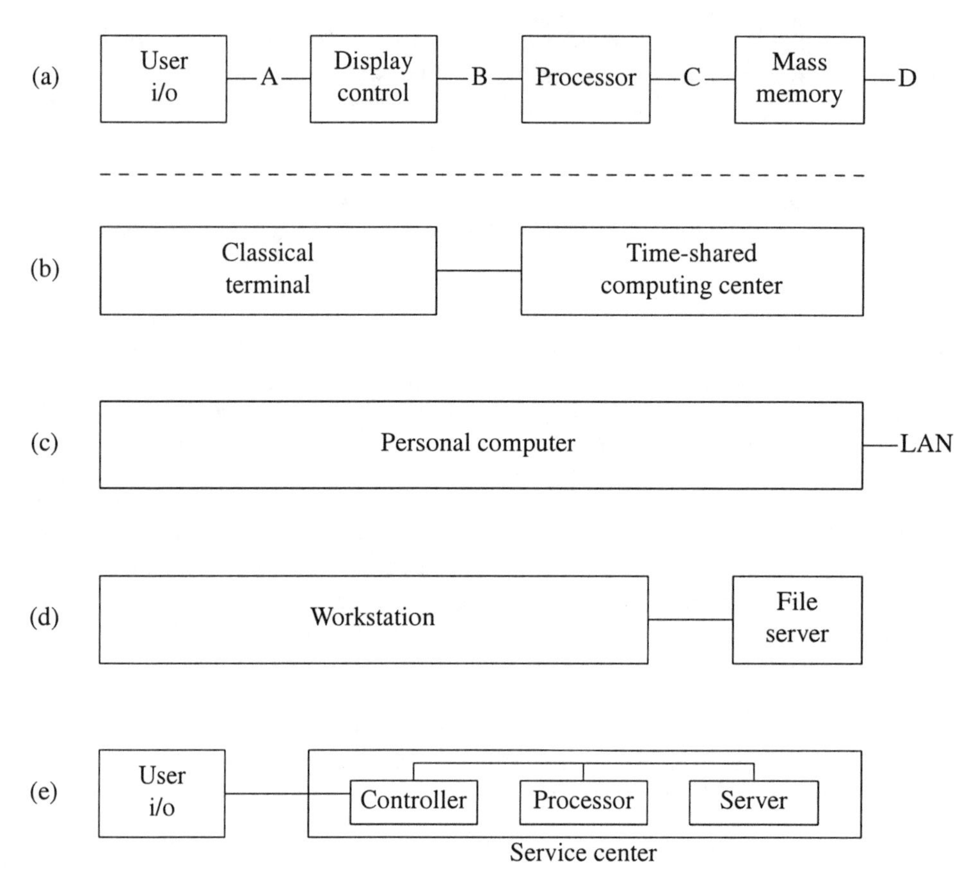

Figure 15.10. Four terminal-to-host interconnection alternatives.

Figure 15.10(b) illustrates the architecture for connecting a classical computer terminal to a classical computing center. The computer terminal on the user's premises contains a CRT and electronics, a keyboard, a byte-mapped memory, digital controller, and network interface. The computing center in a centralized location contains a general-purpose time-shared main-frame host computer with terminal interface peripherals and a cluster of disk drives. The network connection between the classical computer terminal and the central host system is typically circuit-switched and uses a low data rate. If a multiple-access local area network is used for this connection, the communication is still only point-to-point.

Figure 15.10(c) illustrates the architecture for interconnecting many personal computers over a local area network. The user has, on his premises, a complete stand-alone personal computer, including a floppy disk drive and a network interface. The multiple-access local area network need only support low-rate data transmission.

Figure 15.10(d) illustrates the architecture for connecting a modern workstation to a file server over a local area network. The workstation is equivalent to a computer terminal with an internal processor or to a personal computer without a disk drive. The file server in a centralized location contains the disk drives for all the workstations on the local area network. The local area network must support more data traffic and at a higher rate than the local area network in Figure 15.10(c). The connection between the user's workstation and the centralized file server could be circuit-switched and point-to-point. A multiple-access network is preferred if a workstation is served by more than one file server or if there is considerable terminal-to-terminal traffic.

Figure 15.10(e) illustrates a fourth alternative architecture. The equipment on the user's premises is even simpler than the classical computer terminal in Figure 15.10(b). The digital controller and possibly even the bitmapped memory are removed from the terminal and pooled in a centralized service center. This service center also includes the equipment found in the classical computing center in Figure 15.10(b). Communications requiring multiple access, for multiple file serving and for terminal-to-terminal data, uses the local area network internal to the service center. The network that interconnects the simple user's instrument with the terminal controller can be circuit-switched and point-to-point but it must support a very high data rate. Communications is squandered, but the terminal may be affordable and universal.

The Impact of Photonics. The advent of optical fiber and the accompanying photonics technology could dramatically reverse this architectural trend. While some people view this photonics technology as an evolutionary replacement for the physical point-to-point channels in existing architectures, others believe it will affect, in revolutionary ways, the architectures of communications networks and even computing systems.

> *Photonics technology will change telecommunications to the same degree that the transistor changed computing: improving throughput, cost, availability, and human capability by orders of magnitude.*

The impact of this technology is much more significant than just providing a good broadband physical layer. The virtually unlimited bandwidth of photonics technology can provide a channel between the terminal and the services that should:

- Dramatically affect the kinds of services that might be made available;
- Even redefine the boundary between terminal and network.

We've been on the leading edge of what could be a major revolution. But, by limiting photonics technology only to transport and diverting ourselves with near-term incremental technologies, we continue to postpone the revolution. Photonic switching is discussed in Chapters 20 and 21.

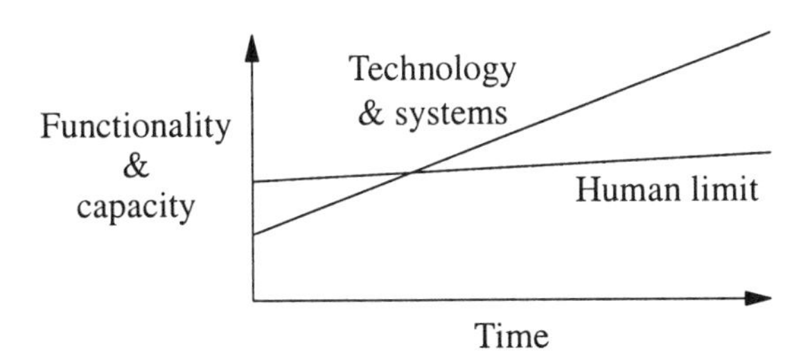

Figure 15.11. Challenging human capacity [7].

15.5 COGNITION

One of the fundamental problems facing the new technology is that its real impact may be limited by the cognitive capacity of humans. In Figure 15.11, the graph makes the point that the capability (however anybody chooses to define this) of our technology and the functionality (however anybody chooses to define this) of our systems has increased significantly over time. But, the cognitive capacity of humans (however anybody chooses to define that) has only risen slightly [7]. Most of us agree that the turn of the millennium is a time that is in the right side of the graph in Figure 15.11, where humans are making things that humans have trouble managing. We probably can't do much about improving human capability. But we can design our systems and services with the goal of operating within human cognitive limitations rather than overwhelming our human users.

We all agree that we would never design, build, sell, buy, accept, or use a system in which some critical manufactured component was overused and mismatched at every layer of design. But, any system in which the "human components" are overused and mismatched is similarly poorly designed. This cognitive issue has been around for a long time. Historically, and in contrast to the Computer Industry, the Telecom Industry has been very diligent about regarding its importance. They studied side-tone and they carefully laid out the Touch Tone pad. So, why wait until Chapter 15 in this book to discuss this? Because of our technology, the human interface has become an even more important issue than ever before.

> *Wherever our network is evolving, probably its most significant limitation is the cognitive overload of its human subsystems.*

This section deals with human cognition. The subsections are: a layered brainware architecture, the cognitive layer, memory, congruence, emotions, the user interface and methods for access and selection.

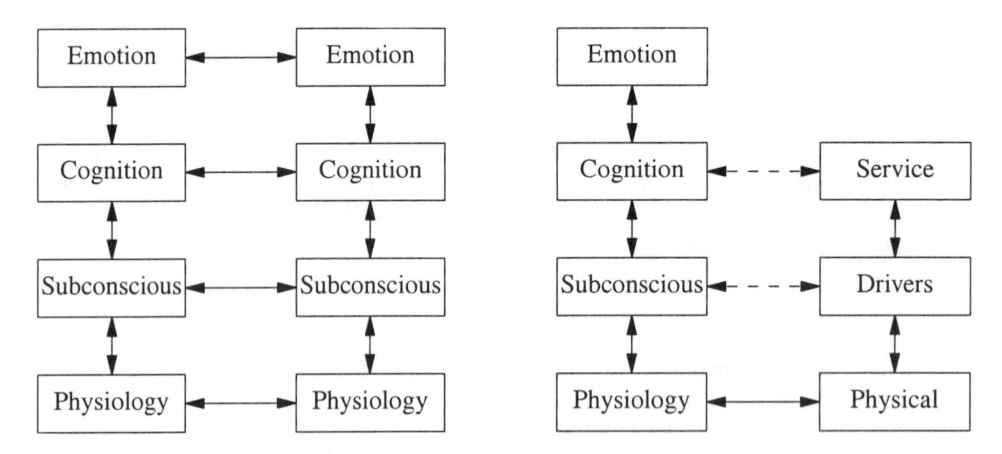

Figure 15.12. The human protocol stack.

15.5.1 A Layered Brainware Architecture

Figure 15.12 shows layered communication between two different pairs of communicating protocol stacks. On the left, two humans communicate, and on the right, a human communicates with a machine.

- The physiological layer resides at the lowest layer, corresponding to the physical layer in the OSI reference model. At this physiological layer, human beings communicate, with each other and with machines, by causing changes in their physical environment and by sensing the changes others have caused. We have i/o ports for sound, sight, smell, taste, and touch; the audio and visual ports were described in Sections 15.2 and 15.3.
- Human subconscious processing, corresponding to drivers in computer systems, takes place at the next-to-lowest layer. Through our input ports, our brains receive physical stimuli, and through our output ports, we produce physical actions that other humans can perceive. We described the brainware that drives the audio and visual ports in Sections 15.2 and 15.3. We have discussed physiology and subconscious brainware as much as we're going to. The rest of this section discusses the two upper layers.
- At the next-to-highest layer is human cognition, by which we process stimuli into understanding and process understanding into actions. As described in the next subsection, this cognitive layer has four sublayers.
- At the top of the stack are high-level human functions like knowledge, wisdom, and human emotion. We won't discuss knowledge or wisdom except to quote T. S. Eliot: "Where is the wisdom that is lost in knowledge and where is the knowledge that is lost in information?" Emotion is discussed in the third subsection.

There are physiological, subconscious, and cognitive differences between direct human-to-human communication (at the left in Figure 15.12), and communication between a human and a machine (at the right in Figure 15.12). As discussed in Section 15.1 and depicted in Figures 15.1 – 15.3, we humans perceive speaking to a machine either as a direct communication or as the human-telecommunication interface in an indirect human-to-human telecommunication over a network.

15.5.2 The Cognitive Layer

Telecommunications is more than moving bits from one machine to another and even more than moving sensory phenomena from one human to another — it is moving concepts from one human brain to another. True communication occurs at the cognitive layer in Figure 15.12. Here, information that is present in one human brain is made to be "in common" with another human brain. It is at this cognitive layer that we understand each other, not merely hear each other or see each other. This subsection describes four processing sublayers in this cognitive layer, then discusses human memory and the concept of *congruency.*

Processing Sublayers. Cognitive processing occurs at four internal sublayers: syntax, semantics, context, and background.

- *Syntax.* The left side of Figure 15.12 shows two humans communicating. As you read my book, we are communicating, but it's one-way — I'm communicating to you. Physiologically, I had to cause black marks to be put on white paper and you have to look at these black marks. Cognitively, for you to understand me, we have to first agree on an alphabet and on a language — the syntax.
- *Semantics.* At a deeper level of cognition, I have to do more than put down strings of random English words, even if they are in grammatically correct sentences. I have to convey a meaningful message — the semantics. "All positive electrons are green," is syntactically and semantically correct. It's even true because, by the rules of logic, a statement is true unless it can be proven false. While the statement violates something cognitive to all of us, the statement does convey a meaningful message.
- *Context.* While the statement, "The two wires are called the tip and ring," is syntactically and semantically correct and meaningful, it is much more sensible if it's read in the context of the telephone, the loop plant, or a switching fabric. It is out of place if found in the context of markers or 1E software scheduling. So, not only is structure required in the alphabet from which we make words and in the words from which we make sentences, and not only must every sentence convey meaning to the reader, but the sentences and paragraphs must be assembled within the framework of some higher-level contextual architecture.
- *Background.* The statement, "All positive electrons are green," violates what we know about electrons — they aren't positive and they don't have color. And, even if the syntactically and semantically correct statement, "The two wires are called the tip and ring," is placed in an appropriate context, it is still meaningless to anyone who doesn't know what a wire is, doesn't know how to count, or doesn't understand that we humans have a propensity to give names to things. As the one communicating to you, I make a large number of assumptions about you. I have had many life experiences and the memory of these experiences is available to me when I process any information that's new to me and as I output information for others. Not only have I assumed that your information processor works the same way that mine does, but I also assume that you and I have had many of the same life experiences. I have written this book for someone whose background is similar enough to mine that I didn't need to use 2000 pages.

When we humans communicate with each other, even if it's one-way like this book, we perform well at these four cognitive sublayers. We recognize mismatches in syntax, semantics, context, and background. We usually process ungrammatical statements anyway (error correcting), and at the other sublayers we ask questions of each other for clarification. This is why the classical classroom remains the ideal environment for teaching and learning. Even when the human-to-human communication is one-way, a good transmitter takes a lot of time to carefully outline and present material for the receiver — and this preparation is based on how we have learned that we like it presented to us. Some of this breaks down when humans communicate indirectly through a network, and much of this breaks down when humans communicate directly to machines (the right side of Figure 15.12).

Memory. This subsection briefly discusses human short-term memory, our cache, and the hypothesized structure of human relational memory, our database.

- *Short-term memory.* Apparently we have a short-term memory with a storage capacity for around 6 – 10 items. Most of us can remember a seven-digit telephone number we've never seen before, but only for a couple of minutes. Most of us can remember a 10-digit telephone number we've never seen before, but only if the area code is familiar, so it counts as one item. Most of us can't remember a 10-digit telephone number we've never seen before, not even for a couple of seconds. We get overloaded the first time we have to deal with a computer because there are too many items we might do and, since each item is new, we have to remember it while we're looking at the other items we might do. As we become familiar, the choices aren't in short-term memory any longer, especially if they're located similarly or are cognitively stimulating, like an icon would be.

- *Relational memory.* There is evidence that when we remember a concept, we link it to related concepts. We classify and sort and we organize things into a taxonomy, which seems to operate like a tree-structured relational database. A condor is a bird, which is an animal, which is a living thing, which is a. . . . When we learn about a new switching system, this new concept is closer to other switching systems we already know about than it is to birds that we know about. Texts on human cognition even define *learning* as "placing a new concept into our mind's relational tree."

Congruency. Communication works better if the transmitter and receiver are matched — the better they match, the better the communication. When dealing with a machine, a user is more comfortable if the interaction on the right side of Figure 15.12 feels more like the interaction on the left side of Figure 15.12. This subsection discusses cognitive match and service paradigms.

- *Cognitive match.* In the cognitive layer, where real communication occurs, we communicate best and easiest with a fellow human, face-to-face. At the lower sublevels of the cognitive layer, a fellow human respects our cognitive limitations because he has similar ones. Another human wouldn't expect us to remember a 13-digit number and knows we will struggle when we try to learn a new concept that's not related to anything we've ever seen before. At the higher sublevels of the cognitive layer, we communicate better if we have common experiences, common contexts, common backgrounds. If two communicating partners are more congruent, if their cognitive processes are more similar, they communicate better.

- *Service paradigms.* We also relate better to our machines if they perform services for us that are similar to familiar services. We like the new service to have a familiar name or a familiar icon or to behave in a familiar way. Common examples of service paradigms on computers are files, folders, file cabinets, deleting files by dragging their image to the wastebasket icon, and electronic mail. We see this extended, quite successfully, to paradigms in which new services are collected together in familiar ways. For example, many of our PCs show a collection of icons arranged in what is called a desktop and many collections of software applications are related to *office automation.*

Even if we have little in common, we can make assumptions about our partner's cognitive capacity and we can adjust these assumptions during the process based on questions, body language, and funny looks. No matter how dissimilar fellow humans may be from each other, we are all far better cognitive matches to any fellow human than to some machine. Besides having dissimilar syntax, semantics, and context from any machine, few machines are capable of having experiences and being able to draw upon experiences when processing new information. I know of no machine that adjusts its cognitive processing during a communication based on questions, body language, and funny looks. But, we will all communicate better with a machine that is as *congruent* as a machine can get, than with a machine that acts entirely like a machine. We will all communicate better with a machine that's been designed to be a cognitive match with our cognitive capabilities.

Consider congruency between two humans. You know, by your own experience, that you communicate better with fellow humans who have the same language, or have a similar age, profession, politics, religion, gender, culture, or education. This relationship is not strictly causal. Even if you and I have little in common, by this point in my book you have come to know me pretty well by now. You can probably accurately predict my life's experiences and my language, age, profession, gender, culture, and education — and probably even my politics and religion.

These paradigms are not always close enough to the real thing to be very successful cognitive matches. The desktop on my PC comes up by default looking more like a random collection of unrelated icons than the top of any desk I've ever seen. Furthermore, the icons on my PC's desktop have names and shapes that look more like advertising than like useful services. For example, I knew there had to be a calendar program on my PC's desktop, but it wasn't immediately clear by the icon names where it was.

Another problem with designing for congruency is that, while many users have common characteristics, all the potential users aren't all exactly the same. Because the vendors are so concerned about mass marketing equipment and services for the computer and telecom industries, the designers of the applications, those who even think about users, are forced to think that all users are the same.

15.5.3 Emotions

The upper layer in Figure 15.12 shows that when humans communicate, we acquire feelings. We might get emotional because of the information we have acquired or, more significant for this text, because of the process involved in acquiring the information. While you might not have expected to find the cognitive material in a book like this, you're probably very surprised to see a subsection (even a sentence) on emotion. But, how many times has somebody called you when you weren't home and did not leave a message on your voice-mail because (and they don't mind telling you this) "I hate talking to a machine."

One of the early experiments with *office automation,* performed in a bank in the mid-1970s, illustrated an interesting phenomenon because it was too far ahead of its time. A strong subliminal message was that bank executives at that time, and particularly if they were women, did not like to be seen performing any activity that others might think was typing. Unfamiliar PCs were confused by the user's peers with the typewriter, and their use was demeaning, and women were especially sensitive. Only 5 years after this experiment the PC not only became familiar, but people with PCs in their offices were seen by their peers and bosses as "rising stars" and "on the fast track." People placed computer terminals and personal computers on their desks, even if they didn't know how to use them, because of the image that it gave them.

Besides disliking machines and worrying about the image you project, human users can actually fear technology or see it as a job threat. If not properly taken into account, such emotions can cripple the interaction between user and machine. These emotional responses are extremely strong and are extremely important to consider when designing a system that deals directly with human users.

15.5.4 The User Interface

The user interface is the layer of hardware or software in any machine that interacts directly with the user at his lowest cognitive layers: physiology, subconscious brainware, syntax, and memory. The user interface typically has a great impact on the user's emotional response to a machine. While the user's cognitive and emotional perception of any kind of machine comes from the services that the machine provides, this perception also comes from the user's interaction with the machine and the service. This subsection describes the perceived friendliness that comes from general characteristics as superficial as packaging and color, and extends to shapes and names, and finally to congruency. The next subsection discusses the friendliness perceived by a machine's ease in providing access to services and in enabling the user to select the desired service, the procedures and command language involved when the user uses the machine and its services. This subsection provides examples of good and bad user interfaces, and then describes the user and his level of sophistication. The user interface is discussed in the role of being an agent or translator between the user (assumed to be naive) and the system (assumed to be difficult to use). The section closes by discussing the reasons behind the success of the familiar GUI-WIMP.

Good and Bad. An example of a good user-friendly design, showing an obvious cognitive match, is the power-seat control in many modern automobiles. It has two actuators, shaped like the physical cross-section of the seat bottom and the seat back, respectively. When you push the seat-bottom actuator toward the back of the car, the seat moves rearward; push it toward the front of the car and the seat moves forward. When you twist the seat-bottom actuator so the back moves down and the front moves up, the seat moves in a similar direction. When you rotate the seat-back actuator backwards, the seat back reclines; rotate it forward and the seat back moves toward the vertical.

Examples of poor designs abound, unfortunately: the control panel of most VCRs, the human interface to most library computer catalog services, and almost anything to do with a computer's operating system. I asked a class once, "Why are so many human interfaces so poorly designed?" One student, a woman, responded, "Because they're designed by men." It was a funny response and we all laughed, but is it true? Do machismo and "Spockian" logic, traditional male characteristics, make us think we should tough it out through pain, feel inadequate or stupid or feel like sissies if we complain, feel grateful to accept anything as long as we can see the logic in it? Wouldn't more sensitivity and estheticism, traditional feminine characteristics, improve our user interfaces? To the extent that these traditional characteristics of males and females really can be generalized, then her comment may have been painfully accurate.

The User. From either the window perspective or the interface perspective described in Section 15.1, Figure 15.1 implies that the users and services are the endpoints of some system of terminals and networks. We usually view such endpoints as boundary conditions or fixed parameters of a system's specification. Viewing services as fixed boundary conditions is a mistake because they are not necessarily fixed or given. Services are part of the total system and must be designed and optimized to conform to the networks, terminals, and users. Less obviously, viewing users as fixed boundary conditions is also a mistake [8].

While users cannot be designed and built, they can be selected and trained — within limits. A given system of services, terminals, and a network can be designed for a set of users who are assumed, or required, to have a certain minimum level of training, intelligence, and sophistication. While some, but certainly not all, of these qualities can be acquired, even training and selection cannot remove the great variation among users. While this is a large, frequently underestimated problem, there is even a larger problem.

Users are different from other system components, because they are human. Their performance varies daily, they get tired and bored, they have other things on their minds, and they have feelings. To the extent that some foibles are universal, the users represent a fixed boundary condition or, at least, a boundary condition that varies slowly in time. The only constant here is that this varies significantly.

The User's Agent. When dealing with anything complex, we have the choice of learning the complexity with which we must deal or of having some intermediary, or agent, act in our behalf. When you bring your automobile to a service station for maintenance, you are not expected to understand the complexity of the modern automobile, to diagnose the problem yourself, and to tell the mechanic exactly what you want him to do. Instead, you explain the symptoms to a service manager, who acts as your agent, and he writes the shop order. Even the most naive people can bring their automobiles to a service station for maintenance.

A similar situation arises when a new user sits at an unfamiliar terminal, which is connected to an unfamiliar computer, which runs an unfamiliar operating system, which includes an unfamiliar program for servicing electronic mail. This new user knows he has mail and he wants to read it. He is expected to deal with all the complexity himself and he is given volumes of poorly written documentation to help him. Typically he is given little human assistance and he is made to feel stupid when he requests this assistance. Most users are smart enough to learn it all, but mastering all this complexity is not their primary job. The user who takes the time to master it all has probably shirked on his primary job. Many users of computer-based services could use the assistance of an agent. This agent can assist the user in each direction of the transmission between the user and the service.

The transmission of data from the user includes typed commands that connect the terminal to the required host computer, log-in and password protocols to networks and hosts, setting transmission and terminal options, traversing a file directory, typed commands in the operating system's command language to invoke the desired service, character deletions and line deletions to recover from typing errors, and exit commands for graceful and nongraceful termination of a requested service. This list above does not include the commands used while dealing with the requested service itself — it is only all the surrounding complexity. Two ways that an agent can simplify these activities for the user are by:

- Allowing the user to select a service by pointing to an entry on a menu;
- Using icons instead of text to represent activities.

Some people believe that spoken input to computers is the panacea for these ills, but the user will still need to know all this complexity. And, having to speak computer jargon, instead of just typing it, may make it even worse.

The transmission of data to the user includes prompts for various kinds of user input, error messages, and the "garbage" or lack of response whenever terminal and host are mismatched in some way. The repertoire of error messages from most operating systems is classic for providing little assistance and for insulting the user. Since operating systems seem to be designed to be obnoxious, then an agent can, and should, be used to intervene.

This agent is some kind of program that interfaces between the user and the host operating system. It could reside on the host and be run as a kind of command interpreter and terminal driver. If it resides in the terminal, it can emulate any terminal for the host and it can also assist the user in connecting to the host.

The GUI-WIMP is familiar, by now, to most of us. GUI-WIMP stands for graphical user interface — windows, icons, menus, and pointers. The GUI-WIMP was developed by the remarkable group of researchers at Xerox PARC, was first commercialized by Apple, and is most familiar through Microsoft Windows. The GUI-WIMP has been successful because windows, icons, menus, and pointers are all nicely congruent to human users, and the action of point-and-click feels like a familiar paradigm.

- The basic window has several nicely designed icons for controlling the window's size, frames for controlling shape, a bar for dragging the window to another position on the screen, a title for whatever application is running inside the window, and labels for controlling files, editing, searching, and help. Each of these labels reveals a pull-down menu of commands, some of which are labels for another even deeper menu of commands. Then, whichever application runs inside the window typically adds its own collection of many more icons, labels, and possibly subwindows. It's a bit overwhelming for the initiate, but the consistency of icons and labels and their locations make the basic window easy to learn.
- The system of pull-down menus behind labels, even at multiple levels, is congruent to the human user's tree-structured memory. Some of the choices for labels could have been less ambiguous — for example, what's the difference between "search" and "find"? So, initiates often get confused in the detailed implementation, but most users experience *friendliness* in principle and learn quickly.
- The least congruent aspect of the GUI-WIMP is its pointer. The classic mouse uses critical desktop working area beyond that already required by a PC's monitor and keyboard, which can be problematic when the desktop's working area is already covered with papers and a coffee cup. But, the mouse-like pointing devices on laptops are so much less friendly that the classic mouse is the best pointer we have. The mouse would have been much friendlier for point-and-click if it had been left

simple. But, usually the mouse has two buttons and sometimes it has three. Sometimes the user should single-click and sometimes double-click. Again, we users learn to disambiguate, but we shouldn't have to. It should be more intuitive, more congruent.

Described in Section 15.6, the GetSet's user interface was slightly different from the GUI-WIMP. While GetSet's trees were very deep, there was little objection from users. GetSet's single window used a "relabelable button" as the pointing device. It restricted the number of available functions at any one time (a good thing), but restricted the geographical location of the labels (a bad thing). The pointing device with the best cognitive match is the touch-sensitive screen. But, we need a technology that tolerates greasy fingers and we probably need to mount the monitor into the desktop so the user won't be fatigued by reaching up all day long.

15.5.5 Access and Selection

This layer of user interface interacts with the user at a slightly higher cognitive level, at the layer of memory and semantics. The user's interaction with a service has four stages.

1. The user determines what services are available;
2. Then selects the desired service;
3. Then is served;
4. Then terminates this service and repeats.

The first two stages are an important part of the user interface to the service system and are discussed briefly. With computer services, the traditional approach is to provide the user a thick manual that describes all the commands and then the user selects a desired service by typing the command. Such manuals are backwards, in the same way that we are told to check the spelling of a word by looking it up in the dictionary. The user must know or suspect or guess the name of the service before he can select it.

The access of services is another one of those issues that is apparently obvious but deceptively complex. A service access mechanism using menu presentation is a good solution. If hundreds of services are available, the menu must be organized into a tree-like structure, so the user is not presented too many choices at the same time. The menu entries should be short unambiguous descriptions of the service, like "text editor," instead of a typical operating system command, like "vi." Tree-structured menus provide the user the ability to browse for services.

Selecting the desired service is easiest if the user is allowed to point to the desired service on the menu. Many pointing mechanisms are available. In the GetSet (next section), when a menu page is presented to the user, the entries are the labels for the buttons on the sides of the display screen. The user selects a service by traversing the tree-structured menu to the page on which this service is displayed and then pushing the corresponding button. Many modern work stations use a mouse-directed arrow to point to the desired service on a menu and then a click of a mouse button starts it. Touch-sensitive screens and light pens are also effective pointers.

The ease of access and selection is illustrated by the service: "call the person named X," as it might be implemented on a hypothetical user-friendly "integrated" terminal (next section). Let the user, a professor at the University of Pittsburgh, decide to call a colleague in the English department. One of the entries on the terminal's screen at the root of the tree-structured menu is "Directories." This user selects this item — by mouse point-and-click, or by touching it, or by pressing a nearby button — and a new menu page is displayed. One of the entries on this page is "Pitt Org" for the university's organization-chart directory. After selecting "Pitt Org," one of the entries on the subsequent page is "A&S," for the School of Arts and Sciences. After selecting "A&S," one of the entries on the subsequent page is "English." After selecting "English," the subsequent page displays the names of the faculty and staff in this department. This user pushes the button that is labeled by the name of the person to be called, lifts an associated telephone handset, and the call is placed. The user never sees the person's telephone number because it was never needed.

15.6 COMMUNICATIONS TERMINALS

A terminal is the "end of something." A bus terminal is a building at one end of a transportation route and a battery terminal is an electrical contact at one end of the battery. A transmission terminal is the multiplexing equipment at either end of a carrier link. A terminal illness is, well, you get the picture. A communications terminal is an instrument at one end of a communications channel [9 – 11].

Just as we users interact with a computer by using a computer terminal, then we would interact with a network by using a communications terminal. But, more than just interacting *with* a network, we would use a communications terminal to interact *through* the network, with another user or with some service. So, because a communications terminal is also an appliance by which we users interact with some remote service, the appliance can also be called a teleterminal. Because the most interesting, provocative, and challenging services the user receives are specifically multimedia, integrated, telecommunications services, the communications terminal is an evolution, generalization, integration, and synergism of the telephone, the computer terminal, and the television.

Different communications terminals are differentiated by the services provided to the user through the terminal, the physical equipment, the user interface, the network interface, and the distribution of intelligence. While many types of communications terminals have been hypothesized, proposed, and prototyped for 20 years (only a few were commercialized), the teleterminal remains a research concept. The most difficult issue is price and the complex market environment so that price could be reasonable.

The purpose of this section is to review, and try to bring order to, the state of the art in teleterminals and to predict future directions. After the first subsection discusses integration, the remaining subsections describe a particular research model, the GetSet.

15.6.1 Integration

Telecommunication occurs over physical and logical channels that reside in systems of channels, called networks. Independent networks have evolved, optimized for each of the three major telecommunications media: voice telephony, data, and broadcast video. Independent terminals have evolved similarly, also optimized for each of these media. These familiar special-purpose terminals are the telephone, the computer terminal, and the television.

The advantages of digital electronics technology have caused many different types of networks to be made from this same cloth, which suggests (to some people) the integration into one network. During the last 25 years, the study of integrated digital networks included the study of integrated terminals and integrated services; hence, the name Integrated Services Digital Network (Section 13.4). But the entire telecommunications industry continues to struggle with, and stumble over, just what it means to be integrated.

Four types of integration are suggested and we should always be careful to distinguish these four types.

- An *integrated network* carries all types of data for all types of media, not only over just a common physical infrastructure, but over a single common logical fabric. As was discussed in Section 13.4, ISDN was proposed for this years ago. As will be discussed in Chapter 19, ATM and IP are being proposed for this today.
- *Integrated access* provides a single common physical and logical interface and, even perhaps, a common local area network by which users can connect a variety of terminals to a variety of networks. But, the interior of the network is not necessarily integrated.
- An *integrated service* provides a logical combination of services, each of which has traditionally been offered over a single medium to a dedicated terminal. Services are discussed in the next section and "extreme" services are described in Chapter 22.
- An *integrated terminal* is the appliance by which, and through which, the user perceives an integrated service. The services don't have to be truly integrated, as long as the user perceives them to be integrated. A little later in this subsection, we will walk through four separate services: POTS, sending e-mail, reading e-mail, and a new feature for returning calls from an e-mail message. Even provided over separate networks, these four separate services are stitched together at a GetSet into what the user perceives to be a single integrated service.

These four types of integration could be viewed as being nested or they could be viewed as distinct implementations. This section deals with integration as it is perceived by the user at the teleterminal. We'll see that, if the user has an integrated communications terminal, then he doesn't need integrated services, access, or networks.

Integrated networks and access are distinguished and discussed at length in Chapters 19 and 22. But telephony's divestiture (next chapter) in the United States suggests an interesting hypothesis. Because many companies want to compete in the long distance business and competition in the local loop has been cool, perhaps integration is more important in the local portion of the PSTN than in the larger or higher-level layers of the network.

Figure 15.13 expands Figure 15.1 by showing the network box as having a single access to multiple networks. The multiple networks could be separately optimized: one for circuit-switched voice (POTS), one for packets carrying user's data (Internet), one for signaling (SS7), and others for other traffic types. With this kind of integration, the teleterminal has a simplified interface and the network has optimal performance for various traffic types.

The ultimate teleterminal will be some integration of the telephone, the computer

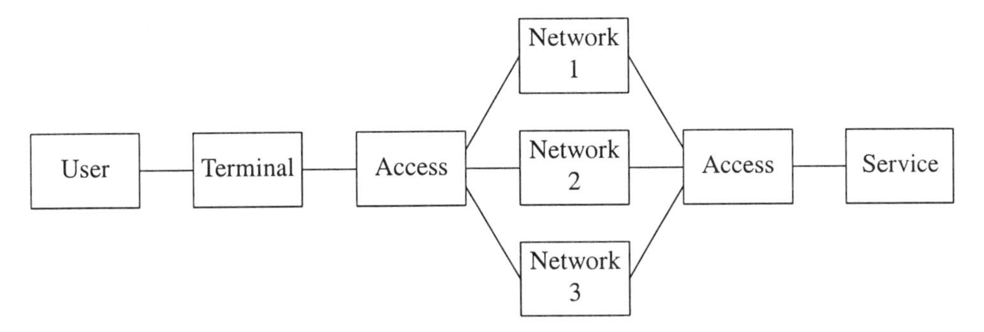

Figure 15.13. Integrated access.

terminal, and the television. It would provide users access to services that are an integration of telephony, data and computing, and video. We're in the early stages of seeing this three-way integration at the modern personal computer, but only when it is connected to a broadband network. The initial implementations of the teleterminal were just an integration of the telephone and the computer terminal and only provided access to integrated voice/data services.

DisplayPhone was such an integrated voice-data teleterminal, commercialized by Northern Telecom in the early 1980s. It had a telephone handset and a Touch Tone pad, a slide-out alphanumeric keyboard and a 9-inch CRT display, all integrated in an attractive package with a reasonably small footprint. For many years a DisplayPhone was cleverly showcased on J. R. Ewing's desk in the TV show *Dallas*. While DisplayPhone's user interface was better than that of most computer terminals of that time, it wasn't as creative as other contemporaries — GetSet (described next) and Apple's Macintosh. DisplayPhone was not a commercial success, mainly because it didn't provide quite enough functionality for its fairly high price.

15.6.2 GetSet

This subsection (and the next two) describes another integrated voice-data teleterminal, a research prototype called the GetSet [12 – 24]. The remainder of this section describes GetSet's hardware and its user interface. Then, we walk through an integrated voice-data service. The third subsection summarizes the prototype system architecture and the GetSet experiment.

GetSet had a regular telephone handset and an internal auto-dialer. It had a small alphanumeric keyboard that was adequate only for two-finger typing, and because it had no Touch Tone pad, the keyboard's number keys could be used for dialing. GetSet had a 5-inch CRT screen, which functioned as a small computer terminal display with room for only 16×32 ASCII characters. Twelve small buttons were mounted on GetSet's face, six on each of the CRT screen.

GetSet had three ports for wired connections: a telephone line, an RS-232C line, and a DC power line. In a typical installation, an external power supply and an external modem were hidden inside the home or office. GetSet had the footprint of a conventional office telephone. Its packaging and its name were described as "cute" — this was the desired effect. GetSet was showcased in the Bell System exhibit at the 1982 World's Fair in Knoxville, Tennessee.

GetSet was part of a research experiment, conducted at Bell Labs in the early 1980s, to determine the user's satisfaction with such an instrument and the services it could deliver. So, GetSet was implemented, not as a potential product, but as a prototype that could be easily deployed. The system architecture, dumb terminal connected to a host, was optimized for provision and changing of services.

GetSet's user interface is characterized.

- *Relabelable Buttons.* These are used as selecting devices, instead of using a mouse-and-cursor, as on conventional PCs. A touch-sensitive screen was tried on a later version and was thought to be the best method.
- *Menus.* Rather than expect the user to know a set of commands or, worse, to memorize a set of dial sequences, the set of functions that is currently available is displayed on a menu. Like ordering a meal at a French restaurant, the user selects the desired function by pointing to the item on the menu. While 12 function keys slightly exceed the user's short-term memory capacity, there was enough repetition and uniformity, that cognitive overload was never detected.
- *Trees.* Consistent with the simple model for learning, menus are organized in a tree-like structure, where each node in the tree is a menu of available related functions. On any menu at any node, some of the menu items are labels of commands and some of the menu items are labels of branches that lead to other nodes. Even though the preceding menus are not left on the screen, users seldom became lost in their tree.
- *Customization.* Each menu and the user's entire tree of menus are all customizable. To customize, the user had to be familiar with the menu tree's file structure, with the set of available program commands, and with the use of a text editor. Users built their own trees and labeled their own buttons, usually by editing an existing "starter tree."

GetSet was designed specifically to be small enough to occupy the same footprint that a telephone occupies on a desk or endtable. It was intended to replace the user's telephone, not his computer terminal nor personal computer. The set of services (Section 15.7.6) provided was consistent with this functionality.

15.6.3 Service Walk-Through

Consider a GetSet user named Alex. Let Figure 15.14 illustrate the base of Alex's customized menu tree and assume this screen is currently displayed on Alex's GetSet. Three buttons on the right side, labeled "Susan," "Dave," and "Home," originate telephone calls to these three people, respectively. While it's not clear from the label that the respective button's function is to make a telephone call, Alex knows what these buttons do because Alex specified their location and labeled them. Since Alex specifically put these three functions on the base of his menu tree, these three people are obviously important to Alex and are calls he makes frequently.

Suppose Dave is Alex's department head and Susan is the department secretary. The buttons are not labeled formally, by last name or by job function, as they would be found in a corporate directory or organization chart. They are labeled by the name that Alex uses when he addresses these two people. Alex's tree might have another "Susan"

Base	
☐ Personal Assistant	Special Dialing ☐
☐ General Directories	Top 10 ☐
☐ Personal Directories	Susan ☐
☐ New Services	Dave ☐
☐ Operating System	Home ☐
☐ -Explain-	-Lock- ☐

Figure 15.14. Base of the menu tree on Alex's GetSet © 1983 IEEE [14,23].

button, perhaps in a menu that is reached by selecting "Personal Directories" and then "Relatives" and then "Cousins." The location in the tree provides additional information to Alex about which Susan is called by which button. Even if his secretary is his cousin, it probably makes sense to Alex to have two different buttons in his tree, even if they call the same person.

In the first part of this walk-through, suppose Alex wants to call Susan. He lifts the handset off his GetSet and listens for dial tone. Alex pushes the "Susan" button, and because the GetSet has an internal autodialer, he hears the DTMF tones outpulsed over the line. In an ISDN implementation, Susan's telephone number would be signaled over the D-channel. A status message "Calling Susan" appears centered on the bottom of the screen. Alex hears an audible ring tone, but Susan doesn't answer. So, Alex decides to send Susan an electronic Call-Memo, requesting that she call him back.

Personal Assistant	
☐ Today's Appointments	Read Mail ☐
☐ Examine Appointments	Send Mail ☐
☐ Make an Appointment	Send General Call-Memo ☐
☐ Calendar This Month	Send Standard Call-Memo ☐
☐ Calendar Next Month	☐
☐ -Explain-	-To Base- ☐

Figure 15.15. Personal assistant node on Alex's menu tree © 1983 IEEE [14,23].

Alex pushes the "Personal Assistant" button, causing a traversal along a branch of his menu tree. Alex's GetSet buttons are relabeled to the screen shown in Figure 15.15. Most of the time, Alex sends the same Call-Memo by e-mail. This Call-Memo has his name and number on it and it requests that the recipient return the call. Alex has previously created this Call-Memo and he reuses it when he pushes the button for "Send Standard Call-Memo." The program responds by printing "Call-Memo sent to Susan" centered on the bottom of the screen.

In the second part of this walk-through, Susan returns the call to Alex. When Susan returns to her desk, the bottom line on her GetSet display reads, "You have new mail." While we assume here that the base of Susan's menu tree resembles Alex's (shown in Figure 15.14), it wouldn't have to. Susan's base certainly wouldn't have a "Susan" button and it probably wouldn't have an "Alex" button. Susan's base probably would have a button labeled "Department," which would transit to another node which certainly would have an "Alex" button.

Suppose Susan decides to read her new mail. Susan pushes the "Personal Assistant" button on the Base of her menu tree and then the "Read Mail" button on the subsequent Personal Assistant node, shown in Figure 15.15. Susan's GetSet displays the contents of her e-mail in-box, one new received message for each GetSet button, labeled by the name of the sender. Susan pushes the button labeled "Alex Bell," her screen clears, and Alex's standard Call-Memo scrolls onto her GetSet screen:

January 17, 1983. 9:45 A.M.
I tried to call you.
Please call me back on
Tel# MH6170.
—Alex Bell

Only the lower right button is labeled, as "Process Msg." After Susan presses this button, Alex's message is cleared away and her GetSet buttons are labeled with a set of options for processing his message. Instead of pushing "Delete" or "Save" or any of several others, suppose Susan decides to return the call right now. Susan lifts the handset off her GetSet, listens for dial tone, and she pushes the "Make Call" button. Alex's telephone number is automatically extracted from the Call-Memo, the proper dial prefixes are automatically adjusted, depending on the location of Susan's GetSet, and Susan hears the DTMF tones. Susan's GetSet displays Alex's Call-Memo again and a status message "Calling Alex Bell" appears centered on the bottom of the screen. Susan hears an audible ring tone and Alex answers. While they talk, Susan presses "Process Msg" again and then "Delete" on the subsequent screen.

While this walk-through may appear awkward in this description, it is really quite natural when performed from a real GetSet.

15.6.4 The GetSet Experiment

GetSet was the hardware instrument for the research of integrated voice/data services and the human-machine interface to them. This project was part of a 10-year experiment involving digital switching, teleterminals, and services [12]. Services included calling by name, organization chart access, access to a variety of public and personal directories, mail transactions including call memos, data access, calendar, reminders, games, and all the services typically found in a host computer/operating system environment.

While the GetSet appeared integrated to the user, it did not have an integrated network interface. Two separate communications ports are provided: an analog telephone line for voice communications and an RS-232C port for data communications. GetSet was configured as a dumb computer terminal and not as a stand-alone PC. While it had an internal microprocessor, its resident software only controlled the screen and keyboard, button labeling, the internal auto-dialer, and messaging with a host computer. Before using his GetSet, typically at the start of each day, the user had to connect to the host, log in from his GetSet, and execute a remote program that contained the user interface and services. Connection to the host computer was required throughout its operation. This implementation was deliberately selected to simplify the dissemination of new software to the more than 50 people who participated in the experiment and to make centralized directories more accessible. One probably would not use this architecture for a commercial product.

Several subsequent generations and implementations were tried.

- GetSet-80 had a 50% larger footprint, a low-profile 24 × 80 character display screen, a total of 25 function buttons on three edges of the screen, a touch-typing keyboard, and an interface that simulates a conventional six-button key telephone. GetSet-80 was more appropriate for office automation services than the original GetSet-32 and was used for several years in a field trial of voice/data services among executives and their secretaries.
- TouchTel was another variation. It had a 9-inch CRT screen and a touch-sensitive screen instead of relabelable buttons. TouchTel had more local intelligence and was cartridge programmable.

A great deal of effort went into building directories and providing links to corporate and local telephone directories. User behavior was carefully monitored and studied. A great deal was learned, product potential was high, but no product ever reached the market. While this experiment was performed in the early 1980s, it was at least two decades ahead of its time. Many of these functions are available from PCs now, but that's still just not a telephone. There's still a latent market for a good teleterminal if someone would make one at the right price.

While GetSet could have been implemented more easily in an integrated network, the services felt integrated. The only time users perceived a lack of integration was during the start-up procedure, with the login possibly after a modem call. The GetSet experiment revealed that, of the four types of integration described in the first subsection, integrated terminals were the most important for the user and integrated networks were the least important.

15.7 SERVICE

The first six sections of this chapter described the human user, the object, the one served. This section describes the concepts of service and of being served [25]. A telecommunications system that doesn't serve the human user is pointless and, if it doesn't serve him well, it is inferior. This section discusses generalized service, enhanced telecommunications services, and gives examples. It summarizes some lessons from the computer industry and describes implementations of an integrated voice/data service. Then it defines "service sets" and the "golden service" and discusses integration with image. We return to this subject in Chapter 22, where we describe some extreme services.

15.7.1 Being Served

It's been said:

> *"If you build a better mousetrap,*
> *the world will beat a path to your door."*

While this may be true, Ralph Waldo Emerson presents a vendor's orientation, and not a service orientation. Vendors ask, "Which of my products can I sell you?" Vendors try to make a better mousetrap, or a better network — some standard thing for a mass market. But users ask, "How might you serve me?" A user may need to get rid of some mice. If he goes to a hardware store, they will sell him a mousetrap, but buying a mousetrap may not be the best way to do it. If he goes to an exterminator, they will sell him an expensive treatment, but maybe a mousetrap was all he needed. The user wants to go to someone who will say, "Let me evaluate your problem and serve you in a way that optimizes price and performance."

Being served does not mean the user has received charity. I may pay somebody (a vendor) to cut my lawn. But I will pay more for somebody (lawn service) to permanently maintain my lawn without my having to call them — to cut it when necessary, not on their schedule. Users will pay to be served.

The dictionary has several related definitions for "service."

> *Performance of labor for the benefit of another.*
> *An act or means of supplying some general demand.*
> *Useful labor that does not produce a tangible commodity.*
> *A duty or ritual that is performed or required.*

Someone (or some vendor or carrier) who is totally self-centered is not likely to think about serving someone else. A person is likely to be truly served by a *servant,* someone who is *servile.* In their annual reports, few telecommunications corporations list "servility" as one of their outstanding characteristics. We use this word "service" in many different contexts, very few of which provide a real service.

- We say "at your service," but only as a formal greeting. We really don't mean it.
- Many people may be required by their government to perform military service or by their religion to perform a mission service. Many people don't feel served by either.
- Our utilities include water service, electric service, gas service, and plain old telephone service. To be fair, most of them come close to providing real service.
- Our dishes might be called a table service. But shouldn't that be a term for the waiter, and not the dishes?
- On the Sabbath, we might go to a worship service and when someone dies, we may go to a funeral service. But, it's not clear who's been served.
- Many stores have a desk for customer service and many automobile dealers have their service department. More modern examples and more to the point made here, most of the computer industry's hardware and software vendors have 800 numbers for customer service. Not all clients feel well served all the time.

- We buy gasoline at a service station, but we typically have to pump it ourselves. If we want to be served at a service station, we have to pay a lot more money.
- The U.S. government collects taxes from its citizens through a branch called the Internal Revenue Service. At the risk of being audited, I'll not comment on this name.
- The government's workforce is called the civil service. And politicians like to call themselves public servants. Unfortunately, some people in these two groups don't understand what service is.

15.7.2 Enhanced Telecommunications Services

The TS in POTS stands for Telephone Service. We humans do have a basic need to communicate, and the telecom industry does provide this service, but only to a small percentage of the world's population. Many of the 700 million of us who have telephones want more than POTS. The switch vendors and carriers have provided a very rich set of optional enhancements to POTS. This book calls these *features* and distinguishes them from *services.* An analogy in transportation is that an automobile provides a service, and power steering is a feature.

We users have shown a large, but limited, willingness to pay for many of the features that enhance POTS — if (1) the enhancement really serves us at (2) a fair price. Features like DTMF, one-touch dialing, and redial are handy, but the carrier's price is so high (to regulate a POTS subsidy) that the features' implementation has migrated into the electronics of the telephone. Features like call forwarding, which can only be implemented by the carriers, are priced far above their cost. Voice mail (which this book considers to be a service and not a feature) is priced so high by the carriers (by regulation, for POTS subsidy) that most subscribers purchase telephone answering machines instead.

Only very rarely does a totally novel service reach the market, like fire or the wheel. Most new services either replace a similar existing service or they replace or enhance a similar human activity. The automobile replaced the horse, but the horse replaced walking. So, whenever a new service reaches the market, the consumer compares it against the similar existing services, or the similar human activity, that the new service purports to replace or enhance. The consumer decides whether the new service is worth the price. Sometimes, vendors offer a service that consumers don't want, like PicturePhone. Often, vendors fail to offer a service that consumers do want — reread the Chauncey Depew letter in Section 2.1. Often, consumers don't know they want it when it's first offered, but the service meets some latent need that surprises the vendors and the consumers — like the telephone and the personal computer. When compared against existing services or the existing human activity, a new more expensive service must be (1) better or (2) different.

So, let's compare a generalized human activity, X, to the enhanced telecommunications service, tele-X, that purports to replace or enhance it.

- For any enhanced telecommunications service, tele-X, that corresponds to some existing service or human activity, X, tele-X always has the principal telecom advantage over X — participants can be physically separated. Users don't have to be colocated.

- While most enhanced telecommunications services are pricey, they are often cheaper than the corresponding human activity they replace. Calling your cousin in Missouri is much cheaper (if you both own a telephone) than going to Missouri for a visit. Even videoconferencing is cheaper than flying. Sending e-mail is cheaper (if you both own a PC and both have Internet access) than sending a letter. So, besides being cheaper than X, tele-X is also often more convenient than X.
- The service vendors, in trying to reach a large market, have sometimes made the services too general. One way to allow customization is to allow the user to build his service. For example, in enhanced 800 service, discussed in Section 18.3, users can specify call routing based on time-of-day and other factors. We'll discuss customization and market granularity in Section 22.5.
- So, why wouldn't we always select tele-X over X? Despite these significant advantages, these enhanced telecommunications services are sometimes disappointing. They sometimes compromise electronic bandwidth and, more significantly, the cognitive bandwidth described earlier in this chapter.

15.7.3 Examples

This subsection discusses several services. We compare their different implementations to the existing human activity they claim to replace or enhance. We also see the distinction between the vendor mentality and the service mentality.

Human Dialog. Consider having a dialog with a fellow human and compare three implementations: face-to-face, POTS, and e-mail. If you're proposing marriage, you probably need to do this face-to-face. If you need to talk to an employee because you have to fire him, you'd probably prefer to do it by e-mail, but it's probably not a good idea. If you're selling your car, you probably need to be, not only face-to-face, but co-located near your car. If you need an immediate answer, e-mail may be the worst choice, POTS may be a better choice, and paging him may be the best choice — but the optimal choice depends on the other party's typical behavior with his e-mail and his telephone. If you need to say something to someone who talks too much, but you don't have time for a bull session, you might send e-mail, even if you work in adjoining rooms. If a permanent record would be useful, send e-mail. The point is that, while these are similar implementations of the same service, they are also quite different. And, these different implementations of dialog are more or less appropriate for different applications of dialog.

Telephone Answering Systems. Consider the many implementations of the service that automatically answers your telephone.

- Another human might answer your phone. This person might be someone who lives in your home or he might be a coworker, part of a pick-up group (Section 14.3.7). This human could be a secretary or full-time operator, the attendant in a key telephone system (Section 10.3.2). This is the implementation that the calling party almost always prefers.
- Commercial telephone answering bureaus are available, typically for professionals like doctors. The doctor's receptionist initiates call forwarding at 5:00, so incoming calls will be forwarded to another telephone number, a number specific to

this particular client of this particular bureau. This second number's class of service is that of a PBX extension. The answering bureau is a computerized PBX with DID trunks. Since the switching office outpulses the forwarded call's extension number to the PBX, a form of calling party identification at the PBX allows an attendant there to know for whom the original call was intended. Most of the LECs offer a competing service.

- You can buy a telephone answering machine. While much cheaper than the two options above, the calling party doesn't talk to another human. Other drawbacks are: (1) a significant first cost, (2) must be replaced when it breaks, (3) doesn't work during a power failure, and (4) cannot answer your telephone for you when you're busy on the phone, only when you're idle. If the machine is software in your PC, then you must leave your PC powered up all the time.
- You can have voice mail. Most PBXs have this option and the LECs offer it for residential subscribers. For a monthly charge instead of a first cost, the service is similar to that of the machine, except it doesn't have drawbacks (2) – (4) described above. I have voice mail on my PBX extension and also on my home telephone.

Call Screening. Enabled by Signaling System 7 (Section 18.1), software in the switching office that serves some called party knows the calling party's telephone number. By subscribing to call screening, a subscriber provides a short list of telephone numbers and appears permanently busy to any incoming calls from any of these numbers. Then, this feature is generalized to calling party identification, at additional cost.

By providing a frequency-multiplexed signaling channel over the loop into the residence, the LEC could signal the calling party's telephone number to new equipment in the called party's home, before ringing his telephone. With calling party ID, the called party decides to answer the telephone, or not, based on knowledge of who's calling. If the LEC operates under the Intelligent Network (Chapter 18), the calling party's telephone number is even reverse-translated into the name listed in the telephone directory.

Now, at even more cost, the LECs offer a service to the people who make the annoying telephone calls, which allows their telephone numbers to be unlisted for identification. In this extreme example where a vendor mentality has taken over for a service mentality, do we expect soon that the LECs will offer (at still more cost) to sell the end-users some kind of unlisted ID override? One has the impression of dealing with an armament vendor who sells to both sides in a war — armor to one side, armor-piercing shells to the other side, then thicker armor to the first side, and so forth, for as long as the vendor can sustain it.

PicturePhone. The PicturePhone was prototyped in the 1960s, at great cost to the Bell System, and was even demonstrated at the World's Fair in New York in 1964. But PicturePhone had two marketing problems.

- Even though the PicturePhone instrument had an audio-only switch on it, the public perceived that the called party would have to be visible when answering every incoming telephone call. There was great concern over this, logical or not.

- PicturePhone was quite pricey, so pricey that market studies revealed that most consumers intended to wait to buy a PicturePhone until many of their friends and relatives owned one first. There seemed to be no way to get this cycle started.

PicturePhone is the archetype, the poster child, of the vendor mentality. Ma Bell expected her customers were going to buy this thing that provided a service very few people wanted.

ISDN. ISDN's "getting started" problem was even worse than PicturePhone's. To be served by ISDN, the user needs: (1) ISDN connectivity, (2) an ISDN instrument, and (3) an interesting and useful set of services that are not available with POTS. Most consumers were not going to switch to ISDN until the connectivity was priced reasonably, the instruments were priced reasonably, and reasonably priced services were actually working (not just promised). Since this required availability and pricing could only occur in a mass market, there seemed to be no way to get this cycle started either.

15.7.4 Lessons from the Computer Industry

In the early 1980s, user-oriented computer application programs — like early computer games, text editors, word processing programs, and spreadsheets — were constrained in two ways.

- Because the market was small, the marketplace was small. These early application programs had to be overly generalized so they could reach a mass market.
- While the technology of the first Apple computer and IBM's original PC seemed wonderful at the time, it was barely adequate to allow a realistic and rich user interface. So, these early application programs were barely adequate simulations of the existing human activity they tried to replace or augment.

Today, the computer industry provides: (1) high processor speed and large memory, (2) powerful and intuitive operating systems, and (3) an open and competitive marketplace (despite certain companies' efforts to constrain it) — all at (4) reasonable consumer-oriented prices. In this environment, application programs have become very creative. These application programs are less generalized — providing better service to smaller, more specific clientele — and they are more realistic simulations of the existing human activities they replace or augment. So, over the last 10 – 15 years, the industry for computer application programs has spiraled upward — driven by the mutual causalities of improved technology, more and better applications, an emerging competitive marketplace, and lower prices. If the telecom industry could only follow this example, its customers could be served as well, if not better, by telecommunications services. This discussion continues in Section 22.5.

15.7.5 An Integrated Voice/Data Service

The design of a terminal should be specified primarily by the set of services intended to be provided through it. Complicating these designs is the recent notion of integrated services. The integration of services is different from the union of sets of services. An example of a true integrated service is the coupling of directories with auto-dialing. The reader is, of course, familiar with the scenario in which a telephone user wishes to call someone whose number is not known. In the typical telephone environment, we think of this as using two distinct services.

For the first service, this user looks up the name of the desired party in some directory and retrieves the telephone number. If the user doesn't have the right directory, he must call some information operator. This operator has the right directory, looks up the name for the user, and gives the telephone number to the user, who typically writes it on a piece of scrap paper. Once the user has the called party's telephone number, for the second service, he places the call conventionally from his telephone. Wouldn't it be convenient if telephones were intelligent enough that the user could point to the entry in the directory or hold up the scrap of paper and the telephone would read the number and place the call? Most of us would pay a couple of dollars more for a telephone that could do this.

Many companies have recently computerized their corporate directory. The main advantage for the company is that those employees with computer terminals don't need a printed directory. The main advantage for the employee with a computer terminal is that this directory is current. Back to our scenario, in the first service the user types at his computer terminal the command for the program that accesses this directory, then he types the called party's name to that program, and then the program prints the telephone number on the computer terminal's screen. Then the user, acting as a simple transducer for the second service, reads the number from the computer terminal's screen and pushes the buttons on the telephone keypad. Wouldn't it be convenient if the telephone were intelligent enough to read the number directly off the computer terminal's screen and place the call? Most of us would pay a couple of dollars more for a telephone that could do this.

In this scenario, each of the two distinct services illustrates the window perspective. If the called party were colocated, the user wouldn't need the telephone or the network — at least for this call. If the directory were colocated with the user, he wouldn't need the computer terminal or the network connecting it to the computer, at least for finding this number. Another purpose for discussing this scenario is to illustrate an integrated service.

The suggested integrated service is "Ring the person named X (without bothering me about X's phone number)." The implementation suggested is some kind of technology by which a greatly modified telephone reads numbers off a CRT, but alternate implementations are available in more conventional technologies. In a second implementation, the computer on which the directory resides and the computer that controls the telephone center communicate about this user's intentions. In a third implementation, a slightly modified telephone acquires the telephone number because it has tapped the line between the computer terminal and the computer. As long as they cost the same, these three implementations would be virtually equivalent to the user. Providing this new service through existing familiar terminals and networks requires some modification to either or both. In a fourth implementation, these two separate terminals are merged into one integrated teleterminal, like a GetSet (previous section). It connects to the computer like a computer terminal and to the voice network like a telephone and it seems designed especially for this integrated service.

15.7.6 Service Sets

While a terminal may be specialized to one service, they are usually designed for a set of related services. Even the telephone is used for more and more services every year; and every year a new telephone comes out with another new button on it because the classic telephone was inadequate for some new service that somebody has added to the service set. A terminal is integrated if it interfaces to a union of historically independent service sets or to more than one telecommunications medium. An integrated terminal may be required because previously independent services have been integrated together to form a new service set.

For example, most people who work in an office environment expect their office telephone to provide a familiar set of telephone services. Sometimes the office telephone has six buttons on the bottom or an adjunct speaker to accommodate some services that are typically not available from their residence telephone. Similarly, these people expect a familiar set of text services to be available from their word processor or computer terminal. The union of these two historically independent service sets, the presentation of this combined service set through an integrated terminal, clever ways to integrate old services from each subset, and the invention of new integrated services has come to be called office automation (OA). An OA terminal resembles a computer terminal, but with an attached telephone handset. It's not a GetSet, because a GetSet is more like a telephone than like a computer terminal. In recent years, entire technical conferences and journals have been dedicated to OA. While OA has not been a large commercial success, it probably will be once the OA terminal becomes universal and the service set becomes fixed, understood, and familiar.

This relationship between the service set and the terminal has two parts. First, as already discussed, the terminal must have the necessary features so that the user may interface to every service in the service set. But, second, the set of services to which a user may interface from a given terminal must be complete. With OA, for example, if there is some telephone service or text service that some user wants but that the OA terminal cannot provide, then this user would need a telephone or word processor anyway. It is safe to say that no desk will have a telephone, a word processor, and an OA terminal on it; probably the OA terminal will go. For example, television receivers that answer the telephone can be criticized because the user needs a telephone in the room anyway; to originate calls. A more important issue is that any new service set, and corresponding integrated terminal, must meet a real need. But, the single most important issue is cost.

15.7.7 The Golden Service

The total commerce of so-called enhanced services is considered by many to be a potentially lucrative business; if it could ever get started. Worse than a chicken-and-egg problem, this total commerce is a three-way you-go-first scenario because the total commerce of providing new integrated services has three separate pieces. The position taken by each of the three pieces follows.

- The terminal manufacturers cannot produce a low-priced terminal until it can be mass produced. High sales volume requires a mass market demand, which will not happen until service software is available and networks are integrated, standardized, and real.

- The network providers cannot optimize their networks (to ISDN, perhaps) or offer such networks at low cost until many users own terminals and use existing services.
- The service vendors cannot afford to invest in the development of service software until integrated networks are real and many users own terminals.

A "golden" service is some wonderful service that consumers want so much that they will not only pay for this service but also buy the terminal and subscribe to an expensive network. If one such golden service exists, it is likely to be in the areas of entertainment, education, or security. But, so far, it has been elusive. The promise of many brass services has been insufficient to get the total business started.

With neither a ubiquitous terminal nor a golden service, the expected future sales of enhanced services must somehow subsidize the initial price of the terminal and network interface. In France, where the government controls the telephone network, terminals were made available at very low prices because of such subsidization and redirection of money saved by not printing directories. In the post-divestiture United States, price subsidization of terminals would require creative regulation — and that is unlikely, and inappropriate.

Further complicating an already complex market, some services that have received a little publicity may be resisted by vested interests. ISDN and digital out-of-band signaling provide the ability to display the telephone number, and even the identity, of the calling party on the screen of the called party's terminal during ringing. The user who subscribes to this service has the option of not answering undesired calls. When first proposed, this calling party identification service was tariffed in some states and banned in others. Field-trial users typically liked this service but enterprises that depend on mass calling and people with unlisted directory numbers were vocal opponents. A host of customized call screening services was imagined, depending on the legal and regulatory debate over calling party identification, with many countermanding and overriding service options. We discussed how the users on both sides of this issue might feel like they are dealing with an armaments vendor that alternately sells increasingly better armor to one side and increasingly better armor-piercing shells to the other, at increasing prices, of course.

15.7.8 Integration with Image and with Video

The integration of voice and data is only beginning to be understood, at least in terms of the service set and the integrated terminal. Complicating this progress, however, is the imminence of networks and terminals that provide image services and, even, full-motion video. Attempted integration of these poorly understood services and terminals with the only-slightly-better understood voice/data services and terminals has added to the confusion of the latter. But, the payoff is potentially huge.

The most primitive image terminal is the facsimile machine, which enables the remote copying of a piece of paper. The information includes text, not in ASCII format, but also sketches and script, like a signature. Photographs with gray scale didn't work in early versions of fax, but work reasonably well in later versions. The calling party inserts the original into the fax transmitter and dials the called party's fax machine, which produces a copy of that original. The information on the page is scanned, interpreted as a stream of black-and-white dots, converted into modem-like electronic signals, and carried over conventional telephone lines.

But, fax was conventionally a paper-in, paper-out process. One obvious extension of OA and integration of data with image is to deliver fax images to the screen of a terminal. Some recent PCs and word processors have optional software and external scanners that interface to fax, but the media are not integrated. It is relatively easy to convert an ASCII file into a formatted document and to transmit that document by fax. Receiving a text document by fax and storing it as an ASCII file, however, requires equipment for optical character recognition that is still not completely reliable.

Special-purpose terminals have been proposed and/or built for remote examination of X-ray photographs, drafting and reading blueprints, VLSI design, and a host of CAD/CAM applications. Some of these include gray scale and even color (Section 15.4). The next step in this integration progression is to extend this interaction to full-motion video.

Chapters 19 and 22 continue the discussion of integrated voice/data/video services.

EXERCISES

15.1 Consider other telecommunications media: broadcast radio voice, telegram, electronic mail, fax, and videoconferencing. Add these as rows to Table 15.1 and fill in the other columns for these new rows.

15.2 Describe how (a) the Windows operating system, (b) a conventional broadcast television set, and (c) an automobile's instrument panel present both the window and interface perspectives to its users.

15.3 Draw a diagram, similar to Figures 15.2 and 15.3, that represents the user setting the options on his PC.

15.4 We can double our voice's fundamental frequency by tightening our vocal cords. When we do this, do we double our formant frequencies? Explain.

15.5 A dipthong is a sound made by speaking two consecutive vowel sounds. For example, the English sound spelled "ow" (as in cow) is made by speaking "aaa" (as in cat), followed by "ooo" (as in boo). Identify other dipthongs and their constituent vowel sounds.

15.6 Devise a 5-bit binary code for the phonemes, the English sounds in Table 15.2. Identify 11 vowel sounds and add them to the table. Let the high-order bit distinguish the 16 phonemes at the bottom of the table from the 16 at the top. For the 16 phonemes at the bottom of the table, let two more bit positions separate voiced from unvoiced and separate sustained from dynamic, respectively.

15.7 What is the lowest frequency of the light we identify as red? What is the highest frequency of the light we identify as violet?

15.8 Assume that each of the seven colors of the rainbow has a uniform share of the wavelengths in the visible spectrum? Tabulate these seven colors and their corresponding range of wavelength.

15.9 A video image format is proposed that would have extremely high resolution, rate, and size. Suppose a video screen, that measures 4×3 meters, has 100 pixels per linear centimeter. Suppose each pixel requires 5-bit PCM for each of the three primary colors. If the entire screen is refreshed 50 times per second, what is the required bit rate?

15.10 An ASCII display with 24×80 characters uses a 5×7 dot matrix for each character. How large is the pattern ROM? How many dots on the entire screen? What is the bit-serial data rate to the CRT electronics?

15.11 Your professor tells you that you received an A+ on the midterm exam. Describe the cognitive processing at the levels of (a) syntax, (b) semantics, (c) context, (d) background, and (e) emotion.

15.12 Describe the processing tree for an N-function electronic wrist watch. Discuss the cognitive equivalence of the various branches.

15.13 Describe the congruency of Microsoft Outlook.

15.14 Speed dialing can be implemented by: (1) N separate buttons on the telephone instrument, (2) dialing short codes (*7), which are translated by a separate box (perhaps, your PC) in the home that intercepts the house wiring at the protector block, or (3) dialing short codes which are translated by software in the CO's switch processor. Discuss how the following issues affect each implementation: (a) use from an extension POTS telephone, (b) mnemonic techniques for remembering codes (if there are codes), (c) changing the assignment.

15.15 A third-party telephone answering service provider complains that DID extension numbers are misinterpreted by its equipment and sues the telephone company. The answer service claims that the telephone company is deliberately corrupting random DID numbers, in an attempt to win the third-party business' customers over to its own answer service. Suppose you are a consultant hired by the telephone company to write a formal legal deposition and to possibly testify in court as an expert witness. How would you defend the telephone company?

REFERENCES

[1] Lehiste, I. (ed.), *Readings in Acoustic Phonetics,* Cambridge, MA: MIT Press, 1967.

[2] Lash, J. P., *Helen and Teacher — The Story of Helen Keller and Anne Sullivan Macy,* New York, NY: Delacorte Press/Seymour Lawrence, 1980, pp. 54 – 55.

[3] Terman, F. E., *Elecronic and Radio Engineering,* New York, NY: McGraw Hill, 1955.

[4] Rzeszewski, T. S. (ed.), *Television Technology Today,* New York, NY: IEEE Computer Society Press, 1984.

[5] Netravali, A. N., and B. G. Haskell, *Digital Pictures — Representation and Compression,* New York, NY: Plenum Press, 1988.

[6] Goodenough, F., "95-V CB Process Builds RGB CRT-Driver Amp ICs," *Electronic Design,* Vol. 43, No. 14, Jul. 10, 1995.

[7] Card, S., "Human Interfaces," IEEE satellite broadcast, 1993.

[8] Smith, R. M., "A Model of Human Communication," *IEEE Communications Magazine,* (issue on Human Communication Systems), Vol. 26, No. 2, Feb. 1988.

[9] Thompson, R. A., "Emerging Terminals and the ISDN," complete session at the *IEEE Workshop on the Integrated Services Digital Network,* Oct. 1983; (summary paper presented at *IEEE Int. Conference on Communications*), Amsterdam, The Netherlands, May 14 – 17, 1984, Vol. 2, pp. 572 – 575.

[10] Thompson, R. A., "Teleterminals," *Yearbook of Science and Technology,* New York, NY: McGraw Hill, 1986.

[11] Thompson, R. A., "Communication Terminals," in *The Froehlich/Kent Encyclopedia of Telecommunications,* Vol. 3, Froehlich, F. E., and A. Kent (eds.), New York, NY: Marcel-Dekker, 1992, pp. 395 – 421.

[12] Bergland, G. D., "An Experimental Telecommunications Test Bed," *Proc. National Telecommunications Conference,* New Orleans, LA, Nov. 29 – Dec. 3, 1981, pp. F2.8.1 – F2.8.5; reprinted in *IEEE Journal on Selected Areas in Communications,* Feb. 1983.

[13] Hagelbarger, D. W., "Experiments with Teleterminals," *Proc. National Telecommunications Conference,* New Orleans, LA, Nov. 29 – Dec. 3, 1981, pp. F2.1.1 – F2.1.5; reprinted in *IEEE Journal on Selected Areas in Communications,* Feb. 1983.

[14] Thompson, R. A., "Accessing Experimental Telecommunications Services," *Proc. National Telecommunications Conference,* New Orleans, LA, Nov. 29 – Dec. 3, 1981, Vol. 3, pp. F2.2.1 – F2.2.5; reprinted in *IEEE Journal on Selected Areas in Communications,* Feb. 1983. © 1983 IEEE. Portions reprinted with permission.

[15] Allen, R. B., "Cognitive Factors in the Use of Menus and Trees: An Experiment," *Proc. National Telecommunications Conference,* New Orleans, LA, Nov. 29 – Dec. 3, 1981, pp. F2.5.1 – F2.5.5; reprinted in *IEEE Journal on Selected Areas in Communications,* Feb. 1983.

[16] Thompson, R. A., "User's Perceptions with Experimental Services and Terminals," *Proc. National Telecommunications Conference,* New Orleans, LA, Nov. 29 – Dec 3, 1981, Vol. 3, pp. F2.6.1 – F2.6.5; reprinted in *IEEE Journal on Selected Areas in Communications,* Feb. 1983.

[17] Smith, D. L., and R. D. Gordon, "An Access Tree Editor," *Proc. National Telecommunications Conference,* New Orleans, LA, Nov. 29 – Dec. 3, 1981, pp. F2.7.1 – F2.7.5; reprinted in *IEEE Journal on Selected Areas in Communications,* Feb. 1983.

[18] Klapman, R. N., "Enhanced Communications in an Executive Office," *Proc. National Telecommunications Conference,* New Orleans, LA, Nov. 29 – Dec. 3, 1981, pp. F2.4.1 – F2.4.5; reprinted in *IEEE Journal on Selected Areas in Communications,* Feb. 1983.

[19] Schell, W. M., "Control Software for an Experimental Teleterminal," *Proc. National Telecommunications Conference,* New Orleans, LA, Nov. 29 – Dec. 3, 1981, pp. F2.3.1 – F2.3.5; reprinted in *IEEE Journal on Selected Areas in Communications,* Feb. 1983.

[20] Bergland, G. D., "Experiments in Telecommunications Technology," *IEEE Communications Magazine,* Nov. 1982.

[21] Hagelbarger, D. W., R. V. Anderson, and P. S. Kubik, "Experimental Teleterminals — Hardware," *The Bell System Technical Journal,* Vol. 62, No. 1, Part 1, Jan. 1983, pp. 145 – 152.

[22] Bayer, D. L., and R. A. Thompson, "An Experimental Teleterminal — The Software Strategy," *The Bell System Technical Journal,* Vol. 62, No. 1, Part 1, Jan. 1983, pp. 121 – 144.

[23] Hagelbarger, D. W., and R. A. Thompson, "Experiments in Teleterminal Design," *IEEE Spectrum Magazine,* Vol. 20, No. 10, Oct. 1983, pp. 40 – 45. © 1983 IEEE. Portions reprinted with permission.

[24] Schwartz, T. A., "The AT&T Soft Touch-Sensitive Screen," *AT&T Technical Journal,* (issue on ISDN), Vol. 65, No. 1, Jan./Feb. 1986, pp. 62 – 67.

[25] Thompson, R. A., "Integration and User-Orientation of Broadband I. N. Services," *Proc. IEEE Intelligent Network Workshop '92.* Newark, NJ, Jun. 23 – 24, 1992, Session B3.1.

SELECTED BIBLIOGRAPHY

Baecker, R. M., et al., *Readings in Human-Computer Interaction: Toward the Year 2000,* Second Edition, San Francisco, CA: Morgan Kaufmann, 1995.

Burgess, P. N., and J. E. Stickel, "The Picturephone System: Central Office Switching," *The Bell System Technical Journal,* Vol. 50, 1971, p. 533.

Calhoun, C., "Technology's Global Village Fragments Community Life," *IEEE Spectrum Magazine,* (issue on Beyond 1984: Technology and the Individual), Vol. 21, No. 6, Jun. 1984.

Carey, J., "Consumer Adoption of New Communication Technologies," *IEEE Communications Magazine,* (issue on Telecom at 150), Vol. 27, No. 8, Aug. 1989.

"Designing the Human Interface," entire issue of *AT&T Technical Journal,* Vol. 68, No. 5, Sept./Oct. 1989.

Dix, A., et al., *Human-Computer Interaction,* New York, NY: Prentice Hall, 1993.

Eberts, R. E., *User Interface Design,* Englewood Cliffs, NJ: Prentice Hall, 1994.

Hasul, K., and S. Morita, "Man-Machine Interfaces in Office Communication Systems," *IEEE Communications Magazine,* (issue on Office Automation), Vol. 24, No. 7, Jul. 1986.

Hayward, W. S., Jr., and R. I. Wilkinson, "Human Factors in Telephone Systems and Their Influence on Traffic Theory, Especially with Regard to Future Facilities," *Sixth Int. Teletraffic Congress* (ITC), Paper 431, 1970.

"Human Factors and Behavioral Science," entire issue of *The Bell System Technical Journal,* Vol. 62, No. 6, Part 3, Jul./Aug. 1983.

Kitawaki, N., M. Honda, and K. Itoh, "Speech-Quality Assessment Methods for Speech-Coding Systems," *IEEE Communications Magazine,* Vol. 22, No. 10, Oct. 1984.

Larson, J. A., *Tutorial: End User Facilities in the 1980s,* New York, NY: IEEE Computer Society Press, 1982.

Lieberman, P., *Intonation, Perception, and Language,* Cambridge, MA: MIT Press, 1967.

Martin, J., *Design of Man-Computer Dialogues,* Englewood Cliffs, NJ: Prentice Hall, 1973.

McNinch, B., "ISDN: The Man Machine Interface," *IEEE Communications Magazine,* (issue on ISDN: A Means Towards a Global Information Society), Vol. 25, No. 12, Dec. 1987.

Murphy, N., "Safe Systems Through Better User Interfaces," *Embedded Systems Programming,* Vol. 11, No. 8, Aug. 1998.

Poole, D., A. Mackworth, and R. Goebel, *Computational Intelligence — A Logical Approach,* New York, NY: Oxford University Press, 1998.

Schindler, G. E., Jr. (ed.), *A History of Engineering and Science in the Bell System,* Murray Hill, NJ: Bell Telephone Laboratories, 1982.

Sproull, L., and S. Kiesler, *Connections — New Ways of Working in the Networked Organization,* Cambridge, MA: MIT Press, 1993.

Winner, L., "Myth-Information in the High-Tech Era," *IEEE Spectrum Magazine,* (issue on Beyond 1984: Technology and the Individual), Vol. 21, No. 6, Jun. 1984.

Yoshida, S., et al., "Interactive Multimedia Communications Systems for Next-Generation Education," *IEEE Communications Magazine,* Vol. 37, No. 3, Mar. 1999, pp. 102 – 106.

16

Breaking Up the Bell System

"They say that breaking u-up is ha-ard to do-oo." — from a 1950s rock & roll song

During the first half of the 1980s, the U.S. telecom industry was changed, dramatically and forever, by Computer Inquiry II (CI-2) and the Modified Final Judgement (MFJ) of the 1956 Consent Decree. Each of these regulations was initiated by a different branch of the U.S. federal government, two distinct branches that didn't seem to talk to each other and almost seemed to be competing with each other. No single person can tell you what really happened in the early 1980s. Much has been written by industry experts, who observed from the outside, or by insiders, who were perhaps a little too "inside." Like conflicting reports from witnesses of the same traffic accident, many different reports may be very inconsistent. This chapter is my view of what happened. I will present the view of only one person, a lowly employee inside AT&T's Bell Labs at the time. I admit that there's a little hindsight in this chapter, but not too much. While I certainly take the view that something had to happen as an inevitable result of the Carterfone case, I don't believe that what did happen was very rational, nor was it in the best interests of end-users.

This chapter begins with a study of the AT&T organization chart, which was very stable for the first 80 years of the 20th century. Sections 16.2 and 16.3, respectively, describe the events surrounding deregulation (specifically, CI-2) and divestiture (specifically, the MFJ). The impact of these events is discussed, including how the AT&T organization changed. Sections 16.4 and 16.5, respectively, describe how all these changes affected the resulting companies and even the architecture of the PSTN. Section 16.6 characterizes the 1990s telecom industry.

16.1 THE BELL SYSTEM

Before we can appreciate how the Bell System was broken up, we must understand the structure of the thing that was broken. With many billions of dollars in fixed assets and billions of dollars in annual revenue, AT&T's economy was bigger than that of most states in the United States and of most nations in the world. With more than one million employees, Ma Bell was more than just the telephone company — she was a vital part of America's economy. Figure 16.1 shows AT&T's organizational structure for most of the

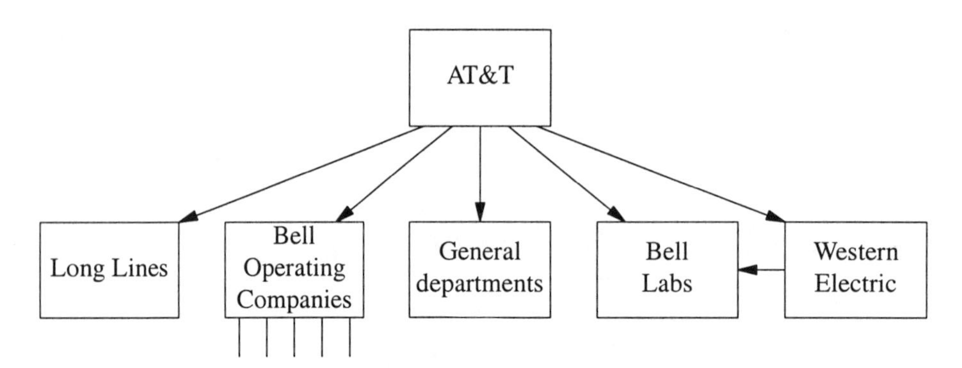

Figure 16.1. The Bell System: 1910 – 1980.

20th century. A small group of executives ran this enormous corporation as a collection of semi-autonomous corporations and departments.

The box in the lower mid-left represents all the separate corporations that provided local telephone service for their respective territories. Depending on whether you measure invested physical plant or count the numbers of employees, this one box represents between 75% and 85% of the total Bell System. Separate companies that were 100% owned included: New England Telephone, New York Telephone, New Jersey Bell, Bell of Pennsylvania, Diamond State Telephone (Delaware), Southern Bell, South Central Bell, Ohio Bell, Indiana Bell, Illinois Bell, Michigan Bell, Wisconsin Bell, Southwest Bell, Northwest Bell, Mountain Bell, Pacific Northwest Bell, and Pacific Telephone. Bell of Nevada was tied to Pacific Telephone. The four Chesapeake & Potomac companies — C&P of Maryland, C&P of Washington DC, C&P of West Virginia, and C&P of Virginia — were loosely related. Southern New England Telephone and Cincinnati Bell were only partially owned by AT&T.

All these corporations were regulated by the FCC and by the Public Utility Commissions of their respective states. These Bell Operating Companies collected revenues from their customers, including long distance calling revenue, taxes, and R&SE funds. From these revenues each BOC paid its internal expenses and bought new equipment (almost exclusively from Western Electric). Profits, federal excise taxes, R&SE revenue, payment for operational support, and long distance revenue were all delivered to AT&T.

While the Long Lines department was not a separate corporation, it ran AT&T's long distance business semiautonomously. This business was regulated by the FCC and the various state PUCs. Long Lines' revenues were collected by the Bell Operating Companies and were delivered via AT&T. These revenues paid for operational support, operating expenses, and allowed Long Lines to purchase new equipment (almost exclusively from Western Electric).

Western Electric (WECo) was a separate corporation that was 100% owned by AT&T. Western Electric was AT&T's manufacturing arm, with major factories in North Andover, Massachusetts; Kearny, New Jersey; Allentown and Reading, Pennsylvania; Columbus, Ohio; Indianapolis, Indiana; Raleigh, North Carolina; Atlanta, Georgia; Shreveport, Louisiana; Hawthorne, Illinois; Oklahoma City, Oklahoma; Denver, Colorado; and many other locations. Western's revenues came from the sale of equipment to Long Lines and the various BOCs. From these revenues, Western paid for the cost of manufacturing and provided revenue into Bell Labs for the R&D of new products.

AT&T's general departments were more of a supporting organization than a managing organization to the rest of the Bell System. It included the legal, finance, and marketing departments, and organizations that supported and centralized the Operating Companies' and Long Lines' operation of their telephone businesses and that coordinated long-range planning among the telephone businesses, Western Electric, and Bell Labs. Long Lines and the Operating Companies were internally taxed for this support.

Bell Telephone Laboratories Inc. was yet another internal corporation. Bell Labs was 50% owned by AT&T and 50% owned by Western Electric, an odd ownership arrangement that reflected the flow of resources into Bell Labs. Bell Labs employees responsible for pure research and for systems engineering were paid from the R&SE revenues that AT&T collected from its Operating Companies. Bell Labs employees responsible for product-oriented R&D were paid from Western Electric revenues. As strictly a cost center, Bell Labs didn't generate any revenue, except for royalties from patents dated before 1948. Principal locations included New York City, three in New Jersey — Murray Hill, Whippany, and Holmdel — and Naperville, Illinois. Smaller Bell Labs sites were located within many Western factories. Bell Labs was internally organized into seven technical areas: 1-research, 2-components, 3-customer premises equipment, 4-transmission, 5-switching, 6-military, and 7-internal staff. This compartmentalization affected architecture (Chapter 12) and the later reorganization after divestiture (Section 16.5).

Since the Bell System was much more technology driven than market driven, the general departments' marketing organization was small and had little corporate influence. And, while Bell Labs appears subservient in the organization chart, Bell Labs managers and researchers had a significant amount of influence over the entire Bell System. While tension and power struggles existed among Western, Bell Labs, and the general departments, it was comparatively gentle and healthy. After Carterfone (Section 10.6) there was significant growth in the size and political power of marketing and project management personnel in the PBX and consumer products areas of the general departments.

In Section 2.1 we described Bell System history as a microcosm of American history. We saw this theme in the 1960s (Section 10.6) and 1970s (Section 12.6) and we will see it again in the stories of deregulation and divestiture. This chapter describes how deregulation and divestiture broke up Ma Bell. We continue the theme that the telecommunications industry has been a microcosm of its time and we show continuity and inevitability from the regulatory events of the 1950s, 1960s, and 1970s.

16.2 DEREGULATION

Section 2.1 covered the Bell Systems' ancient history, up to the 1950s "Pax Romana." Later chapters covered subsequent regulatory history: Carterfone of 1968 in Section 10.6 and Execunet shortly afterwards in Section 12.6. These regulations allowed competition, in customer premises equipment and long distance service, respectively, without closing the separation between price and cost. The whole regulatory structure was made unstable and further change was inevitable. The Computer Inquiry II ruling in 1981 and 1984's Modified Final Judgment of the 1956 Consent Decree were orthogonal and contradictory regulatory changes from separate branches of the federal government. CI-2 is described in this section and the MFJ is described in the next. The FCC initiated CI-2, driven by:

- What the trade journals referred to as the "confluence" of the technologies of computers and communications;
- AT&T's desire to enter the computer business and to participate in this convolved market;
- AT&T's eagerness to offer so-called enhanced services (Section 18.3);
- The industry's concern with AT&T's potential to unfairly subsidize these ventures.

16.2.1 Background

The Carterfone and Execunet regulations had put AT&T in a difficult position. The prices that the Bell System were permitted to charge were set by regulatory bodies in such a way that luxury residential services and services predominantly used by businesses were artificially high and POTS was artificially low. While the competition that was encouraged in the 1970s really was "skimming the cream" that provided the POTS subsidy, by protesting this competition, AT&T created a public image that they were afraid of competition. Rather than win over public sympathy, they raised the public's suspicion that Ma Bell was a fat, lazy monopoly and that all telephone services were overpriced. So, while AT&T's regulatory position actually was unstable, their whining and lack of cooperation further eroded what little public support they had.

Then, in the 1970s, the trade press and the industry gurus began to discuss the confluence of communications and computing. They argued that digital electronic integrated circuits, as the common underlying technology in both industries, had the potential to provide wonderful new services. They argued, however, that the incompatibility of communications and computers was an architectural impediment to these new services and that this flawed architecture came about from the artificial separation of the communications and computer industries. So, they argued that these services could only be developed by companies that understood both communications and computing.

AT&T announced a new position of wanting to be freed from regulation so that they could provide these wonderful new services in a free market. I've always wondered whether AT&T orchestrated all this confluence talk behind the scenes. If they did, they did it masterfully and it never leaked that they did. If it all just happened serendipitously, they took advantage of it. For years, AT&T was very poor at marketing and public relations, but their PR here was superb: (1) AT&T changed its public image from being afraid of competition to wanting to be free to compete, (2) AT&T's arguments were made by technologists instead of by lawyers, (3) AT&T appealed to a future they wanted to be able to pursue instead of appealing to a past they wanted to preserve, and (4) AT&T was able to position its potential competitors as the ones who were afraid of competition.

AT&T's potential competitors in this future free market were now on the defensive: (1) they raised the concern that AT&T would use its telephone revenues as a cash cow to compete unfairly, (2) they reminded us of the reasons that Western Electric had been divested of all its nontelephone businesses in 1934, and (3) they worried aloud that AT&T might use its vast revenue stream to subsidize the R&D of the new services and to offer the new services at artificially low prices in order to drive out competitors. The federal government, helplessly watching while Japan was subsidizing its steel companies' competition with U.S. companies, was very sensitive to these arguments. AT&T had been so persuasive that the FCC granted AT&T some regulatory relief in 1982.

16.2.2 American Bell (Again)

In an FCC decision known as Computer Inquiry II, AT&T would be allowed to form a wholly owned subsidiary that would be completely free of any regulation and could compete in any businesses it chose. This corporation had to be held at arm's length — while AT&T could seed the new company with venture capital, AT&T could not put any more of its resources into the new company after it became profitable (it never did). This company was incorporated in 1982 and was named, historically (Section 2.1), American Bell International (ABI).

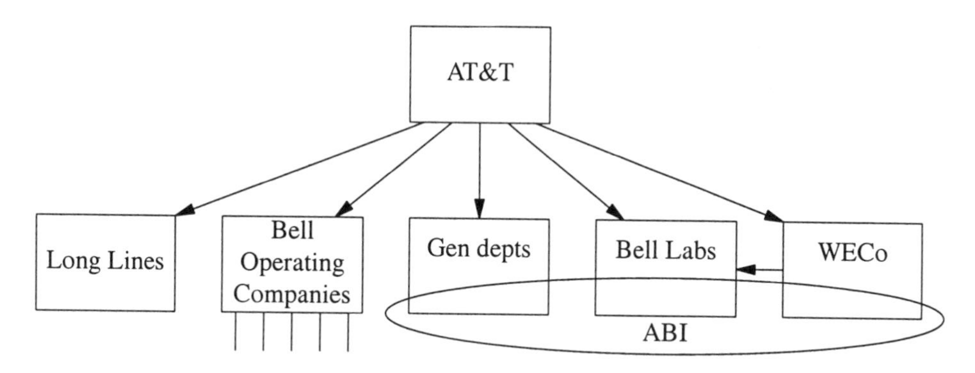

Figure 16.2. The secession of ABI.

Figure 16.2 shows how parts of the Bell System were separated to form ABI. Western Electric factories that manufactured telephones, modems, and PBXs were ceded to ABI, as were the consumer products area of Bell Labs and the part of AT&T's general departments that marketed AT&T's telephone and PBX businesses. ABI organized itself into separate lines of business (LoBs) for the PBX market, consumer electronics (telephones), the computer market, and a new nationwide enhanced computer network, called Net-1000. Marketing organizations from AT&T's general departments, factories from Western Electric, and product developers from Bell Labs were assigned to these various LoBs.

For all of its history to this time, AT&T had been technology driven, with much real power possessed by Bell Labs' executives and researchers. But ABI's new LoB managers were convinced, correctly, that ABI had to be market driven and not technology driven. Each LoB's marketing organization interpreted this to mean that its respective LoB would be marketing driven, not just market driven. Inside each LoB, the marketing organization evolved into product management and they were able to acquire funding control over their respective LoB's factories and development groups inside Bell Labs. This structure wasn't able to work in a Bell System culture. While Bell Labs was consistently able to hire America's best engineering school graduates, AT&T's marketing organization wasn't held in as high regard by America's best business school graduates. Inside AT&T, marketers were called marketeers, a derisive term sounding like "Mouseketeer" from a popular children's program.

Development engineers who were shifted from Bell Labs into ABI's LoBs felt little allegiance to their LoB and referred to themselves as Bell Labs employees, especially when recruiting new employees. Since some engineers were still using old Bell System pricing practices, some good ideas were killed early because costs were overestimated. Meanwhile, product management didn't win over the engineers' confidence when (1) marketers were hired from other companies with the idea that someone who could market soap powder could market PBXs, (2) many R&D projects were started and cancelled without explanation, (3) and product managers blamed Bell Labs' "country club" environment.

Besides the cost of CI-2 itself and the subsequent corporate reorganization, a vast amount of ABI's energy and corporate resources were wasted by all the power struggles, internal thrashing, and lost productivity. Finally, perceiving that ABI's internal power struggle between product management and R&D was hurting the company, the third partner, what had been Western Electric, took control of ABI — just in time for divestiture and the repeal of CI-2. While we'll never know ABI's total cost to AT&T, it must have been many billions of dollars. ABI's epitaph is this: some of the smartest people in the world failed to succeed at a business venture that almost everybody in the world thought was the world's inevitable future.

16.2.3 The Battle of the Titans

Recall that the gurus had argued that some great new services were potentially available in the technology, but that it would require companies that understood both communications and computing to fully exploit this potential. Only two U.S. corporations, one from each camp, had the resources to expand into the other camp — AT&T and IBM. The gurus predicted a future "Battle of the Titans," where the consumer would be the big winner. It turned out that IBM was extremely unsuccessful in the communications business and AT&T was extremely unsuccessful in the computer business. The Battle of the Titans was a stalemate and the consumer lost.

At the time, IBM held a very large share of the computer market and had just won (barely) its own antitrust battle with the Justice Department. Responding to AT&T's entry into the computer industry, IBM bought PBX manufacturer Rolm and became majority owner of Satellite Business Systems, Inc. After losing a lot of money in the communications business, IBM cut its losses and sold Rolm (to Siemens) and SBS. While other notable U.S. corporations, like RCA and GE, had failed to compete with IBM in the computer business, two more companies decided that this was the time to diversify and to enter the convolved computer/communications business. Exxon introduced a line of office products and quickly withdrew. Xerox, whose Palo Alto Research Center (PARC) had done the research that led to Ethernet and the GUI-WIMP user interface, also failed in their attempt to commercialize these ideas.

AT&T had been making computers and writing software for many years. But they were only for internal applications, as switching system controllers, and never for sale in the external commercial market. Bell Labs' researchers developed the UNIX operating system. UNIX ran on minicomputers, especially the popular PDP-11 made by the Digital Equipment Corporation (DEC). AT&T had released UNIX, free of charge, to so many universities that many computer science graduates had learned to appreciate UNIX's architecture, simplicity (relative to other operating systems in the 1970s), and features (like pipes and buffered i/o). So, AT&T had experience, a built-in market (for controllers of its own switching systems), name recognition, and a commercial market presence. Despite being so well positioned to succeed in the commercial computer market, ABI just could not do it. Perhaps if AT&T had bought DEC, as had been rumored, it might have been different. AT&T did eventually take over NCR, somehow believing that the merger of two struggling computer companies would produce a single successful one.

In its brief life, ABI was moderately successful in the PBX business, but lost a lot of AT&T's money in the computer business and in developing Net-1000. Furthermore, the cost of advertising the new company's name and rearranging personnel from AT&T to ABI was all wasted because CI-2 was made irrelevant by the divestiture of 1984.

16.3 DIVESTITURE

CI-2 and MFJ, initiated by separate branches of the federal government, were orthogonal and contradictory. The previous section described the Computer Inquiry II ruling of 1981. This section describes 1984's Modified Final Judgment (MFJ) of the 1956 Consent Decree and the background leading up to it. The MFJ was even more complex than CI-2 as it broke up the Bell System in the 1980s. MFJ gave us RBOCs, Bellcore, LATAs, access charges, bypassers, a reorganized AT&T, and most of the telecom industry's current structure. MFJ was the single most remarkable regulatory event of this century.

16.3.1 Background

By the early 1980s, the U.S. Supreme Court, the Federal Communications Commission, and the various state Public Utility Commissions had arrived at a regulatory structure that was inconsistent and unstable. Two other players in telecommunications regulation had not participated in this structure: the U.S. Congress and the Justice Department. While the U.S. Congress was conspicuously absent, and remained so until the mid-1990s, the U.S. Justice Department was still not satisfied.

The Justice Department had tried to divest Western Electric from AT&T ownership in 1911, in 1934, and again in 1948. Every time DoJ tried to do it, AT&T made a counteroffer. Why was DoJ so determined to remove AT&T's manufacturing branch? Why was AT&T so determined to keep it? DoJ had argued for 75 years that AT&T's ownership of Western Electric was "anticompetitive." DoJ argued that WECo overcharged for its equipment, causing AT&T to artificially inflate its stated investment in physical plant. AT&T argued that its stockholders deserved a competitive rate of return on capital investment. But, DoJ counterargued that AT&T's rate increases were squandered in poor management and nonproductive employees. If the Bell System had not had its own captive manufacturer, it would have to purchase on bids by competitive vendors and the result would be lowered telephone rates. But, the DoJ wouldn't be concerned about telephone rates, that was the FCC's domain.

Many people felt that the DoJ's motives were different but, for some reason, were never well explained to the public. U.S. tax law allows companies to depreciate capital equipment as a direct deduction before calculating taxes. If the original price of the equipment is artificially high, then the depreciation of that equipment is an artificially high tax deduction. Possibly, the DoJ's motive was suspicion of tax evasion, which was

DoJ's domain. There were claims that: R&SE revenue was finding its way into product development for WECo, large projects (like PicturePhone and Net-1000) wasted considerable money, and WECo overcharged the telcos for its products. There were claims that these things resulted in telephone rates that were too high and AT&T's taxes that were too low.

However, systems engineering and research at Bell Labs really were significant national treasures that gave the United States the world's best telephone system and many technological advances that boosted the country's economy. Were AT&T managers accountable to the American public? Some were not. Were phone rates higher than necessary? Probably a little bit. But, was the country well served by the Bell System? Undoubtedly. Ma Bell was a little "sick" and needed to be healed but, as the joke goes, "they cured the disease, but the patient died."

16.3.2 Business Set

Before we proceed, let's take a careful look, in hindsight, at the businesses that AT&T was in. In Chapter 9, we looked at the four colocated functions performed at the toll point and how it seemed so natural to perform all four of them at this place at a certain time. A similar thing is observed with AT&T's four lines of business. Sometimes the R&SE functions in Bell Labs seemed orthogonal to many of AT&T's real moneymaking businesses, but these four core businesses were also quite separable — or were they? These moneymaking businesses were:

- Local telephone service;
- Interexchange long distance telephone service;
- Design, manufacture, and sales of equipment for transmission and switching;
- Design, manufacture, and sales of customer premises equipment and consumer products.

Because AT&T had been in all four of these businesses for almost 100 years, these businesses could obviously coexist inside the same company. While we didn't know it at the time, that didn't mean that any arbitrary subset of these four businesses could coexist in the same company. Remember this point — we'll return to it.

16.3.3 Here We Go Again

In America in the mid- and late 1970s, the public and political mood was to deregulate industry and induce more competition, like with the airlines. So, under Assistant Attorney General William Baxter, the DoJ again set out to divest Western Electric from AT&T, claiming that it would stimulate competition. The divested WECo would have been in the last two of the four businesses in the list above, and what remained of AT&T would have been in the first two. Noting that the 1956 Consent Decree was meant to be a temporary solution, DoJ set out for a permanent solution. The case ended up in the federal court that was presided by Judge Harold Greene. Judge Greene is a remarkable individual, a gifted intellectual, a concerned and patriotic citizen, and is not the "villain" responsible for divestiture.

It became clear to AT&T management that something was going to have to be divested — there was nothing else to compromise. But, AT&T wanted to be free, eventually, of all regulation. They felt that the last three lines of business described above could eventually all be truly deregulated, but the first would always remain regulated. Furthermore, AT&T was losing money in certain BOCs and some executives believed that certain state PUCs had been so unreasonable for such a long time that it might make sense to give up. Internal rumors persisted that AT&T management was so upset with the California PUC that Pacific Tel was going to be unilaterally divested anyway.

So, AT&T offered a counterproposal to the DoJ and to Judge Greene. AT&T would rather give up the local telephone business than give up Western Electric. AT&T proposed splitting itself up so that the divested entity was in the first of the four lines of business in the list above, and what remained of AT&T would be in the last three. As a former law professor, Baxter had predicted this turn of events and the DoJ quickly agreed to this settlement. As long as the newly independent telephone companies would be free to purchase equipment in a competitive marketplace, the DoJ's agenda would be met.

Judge Greene had a different agenda, and he was the only one in the courtroom with the right agenda: would the U.S. public be well served by this arrangement? Judge Greene was surprised by it and wanted some time to consult industry experts and to think about it. He wasn't sure it was a good idea but, expressing considerable doubt, he did finally allow it, mainly because AT&T and the DoJ both agreed that it was a just and final solution. So the Modified Final Judgement of the 1956 Consent Decree was announced, to be effective in 1984.

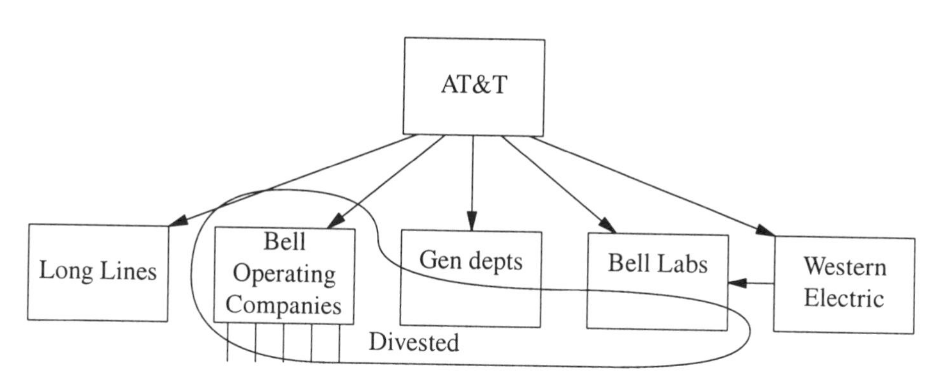

Figure 16.3. Parts divested from AT&T in 1984.

16.3.4 The Modified Final Judgement

Figure 16.3 shows the original AT&T organization chart and the parts that were divested by the MFJ in 1984. These included all of the Local Operating Companies, the parts of the old general departments that supported local telephony, and most of the pieces of Bell Labs that were supported by R&SE funds.

Rather than release the Bell Operating Companies as separate corporations, like the other independents, seven new holding companies, similar to GTE and Contel, were created ab ovo. These holding companies, called Regional Bell Operating Companies (RBOCs), were given jurisdiction over seven distinct regions of the United States and they were given ownership of the smaller BOCs. The support functions of AT&T's general departments were distributed among these seven RBOCs. They were created to be approximately equal in size, each one was just a little smaller than what remained of AT&T.

These seven holding companies went without official names for about a year. Clever people inside AT&T, noting that each region of the United States had a distinctive "bell" associated with it, gave an AT&T-internal name to each RBOC: Sleigh Bell, Liberty Bell, Southern Belle, Factory Bell, Cow Bell, Tinker Bell, and Taco Bell. It would be almost a year before marketing and PR people at these seven RBOCs officially named their respective companies. Divestiture took effect on January 1, 1984.

- New England Telephone and New York Telephone transferred from AT&T to Nynex.
- New Jersey Bell, Bell of Pennsylvania, Diamond State (Delaware) Telephone, and the four C&P telephone companies (Maryland, Virginia, Washington, DC, and West Virginia) transferred from AT&T to Bell Atlantic.
- Southern Bell and South Central Bell transferred from AT&T to Bell South.
- Ohio Bell, Indiana Bell, Illinois Bell, Michigan Bell, and Wisconsin Bell transferred from AT&T to Ameritech.
- Northwest Bell, Mountain Bell, and Pacific Northwest Bell transferred from AT&T to US West.
- Pacific Bell and Bell of Nevada transferred from AT&T to Pacific Telesis (PacTel).
- Southwest Bell, the largest of the BOCs, became an RBOC of its own, slightly changing its name to Southwest Bell Corporation.
- Because Southern New England Telephone and Cincinnati Bell had only been partially owned by AT&T, each became an independent telephone company.

Fortune Magazine lists the 500 largest corporations in the United States according to revenue — they are called the Fortune 500. Up until 1984, *Fortune Magazine* had never included AT&T on its list. If it had, AT&T would have easily been the *Fortune* #1, being bigger than several of America's other large corporations combined. But after divestiture, *Fortune Magazine* added AT&T and the seven new RBOCs to its list. The Bell System had been so large that these eight new listings were each among the 15 largest corporations in the United States. While the seven RBOCs were nicknamed the Baby Bells, these companies were not babies. AT&T was the biggest of the eight. Some large corporation that might have been *Fortune* #20 on December 31, 1983, would have found itself ranked as *Fortune* #28 on January 1, 1984.

16.4 EFFECTS OF DIVESTITURE

One major architectural and commercial effect of divestiture, the redefinition of long distance, is described in the next section. This section describes some other lasting effects of divestiture: corporate restrictions, the creation of Bellcore, and corporate re-organizations.

16.4.1 Restrictions

The rules of the MFJ restricted the eight pieces of the old Bell System from entering arbitrary businesses and limited the ways in which they could compete with each other.

- AT&T was not allowed to offer local telephone service, and the the RBOCs were not allowed to provide long distance service. The details of this restriction, including the redefinition of long distance and the architectural impact, are discussed in the next section.
- Essential to AT&T and to the DoJ, the RBOCs were precluded from any form of manufacturing. From the DoJ's perspective, the whole point of divestiture was that competition among manufacturers would drive down the price of switching equipment.
- Most telephone operators and attendant functions went to AT&T.
- Initially, the LECs would do all the billing. Even the subscriber's long distance bill, whether owed to AT&T or to any other long distance carrier, would be collected by the subscriber's LEC. The LEC would forward the revenue to each long distance carrier and the carriers paid the LECs to perform this service.
- While each RBOC could continue to publish its Yellow Pages directory in its own region, the RBOCs were precluded from competing in each other's regions for seven years. The RBOCs were also precluded for seven years from producing "Electronic Yellow Pages" (EYP), a business that many companies believed would be lucrative.

16.4.2 Bellcore

One major issue at divestiture was the disposition of R&SE revenue, described in Section 2.1.5. It was agreed that the R&SE function was important and that this "tax" on telephone users was an appropriate way to fund it. AT&T offered that its own R&D division, Bell Labs, could continue to perform this function. But the RBOCs argued that the R&SE function should be separated from a manufacturer's R&D division and should be performed by some organization that would have its allegiance to the nation's telephone users, who were paying the tax. So, a new R&D lab was created before divestiture actually occurred. This new R&D lab would report to the seven RBOCs, each of which would own a one-seventh share of the corporation. Bell Labs would be split up, with the WECo-funded people staying in AT&T and most of the R&SE-funded people going to the RBOC's new labs.

The seven RBOCs retained the rights to the Bell name and the blue bell logo. They were even nicknamed the Baby Bells. The new AT&T developed its own new logo, derisively nicknamed the Death Star, after the *Star Wars* movies of the era. So, the RBOCs expected that their new R&D lab would keep the revered Bell Labs name. But, AT&T fought hard to keep the Bell Labs name for its internal R&D organization. The RBOC's new lab was named Bell Communications Research, shortened to BellCoRe and, more simply, Bellcore. The AT&T subsidiary called Bell Labs and the RBOC-owned corporation called Bellcore were constantly confused by the public.

Entire organizations, particularly those responsible for systems engineering, were transferred from Bell Labs to Bellcore. Bell Labs' personnel who transferred to Bellcore would retain full seniority and benefits. Bellcore's new managers were also given the opportunity to recruit individuals directly from Bell Labs' pool of personnel. At a time when professional athletes were becoming free agents, engineers at Bell Labs found themselves in a similar position, as they negotiated raises and promotions for themselves.

My own personal experience was typical of the career decision that many people had to make. I had worked in the research area of Bell Labs and I was offered a lateral position in the research area of Bellcore. Without an enticing promotion either way, I had to decide in 1983 which company I thought would offer the better research opportunity in the long run. I predicted then that (1) Bell Labs' research area would probably become more "applied" (which I would have preferred) than Bellcore's and (2) as the RBOCs began to compete against each other, their cooperative ownership of Bellcore would be short-lived. So, I remained at Bell Labs, and while it wasn't a bad decision and I have no regrets, I was wrong about both predictions.

Most Bell Labs' R&SE personnel worked at Holmdel and Murray Hill in New Jersey. So, to minimize relocating people, Bellcore locations were built near each of these Bell Labs locations. But because divestiture arrived before the new buildings were completed, existing Bell Labs buildings were partitioned between these two companies. The large building in Holmdel was particularly interesting because it was already partitioned between Bell Labs and ABI, which was still required (under CI-2) to be at "arm's length." During this period, people who had once worked together closely, were now employed by three separated companies which lived, in a very odd arrangement, under the same roof. By the time the Bellcore personnel moved out of Holmdel, CI-2 was no longer enforced and the Holmdel building became an AT&T Bell Labs location exclusively. Now, of course, this building belongs to Lucent; but that's a later story.

While Bellcore retained its relationship with the RBOCs for a surprisingly long time, the time did finally come when the RBOC's mutual competition made it unfeasible to support common research. Bellcore was sold.

16.4.3 RBOC Reorganizations

Initially, the typical RBOC's organization chart resembled Figure 16.1. RBOCs had no long distance nor manufacturing divisions and their analog to an R&D division was their jointly owned Bellcore. Each RBOC contained a collection of BOCs, which initially remained essentially intact from their organization when inside the Bell System. Each RBOC also had a support organization taken from, and analogous to, AT&T's general departments. Each RBOC also had a set of arm's-length subsidiaries, analogous to ABI, for Yellow Pages, cellular, and other nonregulated businesses.

Gradually, the RBOCs integrated their BOCs together. Because SBC and PacTel each held one BOC, this integration was immediate. Bell South quickly eliminated the distinction between Southern Bell and South Central Bell and the other four RBOCs followed similarly. In Pennsylvania, now, while Bell of Pennsylvania, Inc., still exists for regulatory purposes, the name Bell of Pa has been almost completely replaced by the name Bell Atlantic on buildings, advertising, trucks, and the telephone bill.

After GTE acquired Contel, it was about the same size as each of the RBOCs. As this book is going to print, the RBOCs are consolidating. Southwest Bell, now called SBC, took over PacTel and SNETCo and is negotiating to take over Ameritech. Bell Atlantic has merged with Nynex and is negotiating to buy GTE. Expect more of this to occur.

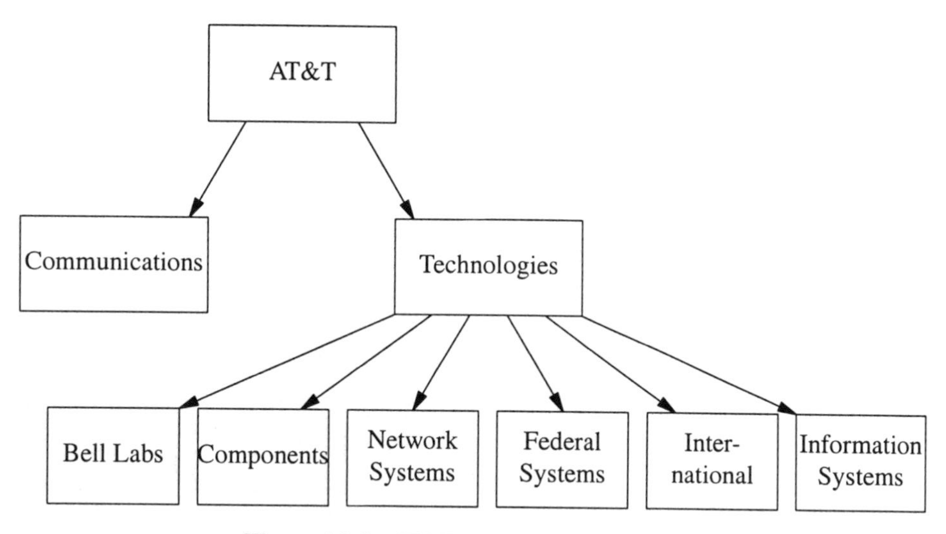

Figure 16.4. AT&T's organization: 1984.

16.4.4 AT&T's Reorganization

After divestiture and the subsequent repeal of CI-2, AT&T completely reorganized itself into two major divisions. Figure 16.4 shows AT&T's organization structure after divestiture. The Long Lines department remained essentially intact as one of the two major divisions of the new parent company. Changing its name to AT&T Communications, it continued to run AT&T's long distance business semiautonomously. This business was still regulated by the FCC and the various state PUCs. AT&T Communications' revenues were collected by the RBOCs and were delivered to AT&T. These revenues paid for operational support and operating expenses, and allowed Long Lines to purchase new equipment (still almost exclusively from the same part of Western Electric that they always dealt with — but now called AT&T Technologies' Network Systems Division).

After divestiture, AT&T's corporate management had to choose between running the company like ABI had been run, by people whose focus was marketing and product management, or like Western Electric had been run, by people whose focus was manufacturing. Western Electric management emerged as the leaders of the new parent company's other major division, given the name AT&T Technologies. While this "manufacturer's perspective" was probably not ideal, AT&T's marketing and product management organizations — especially inside ABI — had been so unsuccessful, that there really wasn't much choice. AT&T Technologies was subdivided into the six LoBs shown in Figure 16.4 — but they were not equal.

- Network Systems was the piece of the old WECo that focused on the technology of transmission and switching equipment. They had dedicated factories in Hawthorne, Illinois; Oklahoma City, Oklahoma; Columbus, Ohio; North Andover, Massachusetts; and Atlanta, Georgia. Network Systems was extremely profitable.

- Federal Systems was the piece of the old WECo that worked on government and military contracts. They had a dedicated factory in North Carolina and they were expected to be profitable.
- Information Systems was the new name for ABI. ATTIS included AT&T's PBX, consumer products, enhanced networking, and computer businesses. They had dedicated factories in Indianapolis, Shreveport, Little Rock, and Denver. ATTIS was expected to make a profit, but never did. Even though AT&T had divested its regulated component, ATTIS remained at arm's length for several years after divestiture because CI-2 was not repealed immediately by the FCC.
- International was a market-focused organization responsible for selling the products and services from the other divisions to non-U.S. markets. As international sales increased, particularly of the 5E, this organization became responsible for special products and specialized 5E software.
- Components was the piece of the old WECo that produced electronic components, mainly for internal products and systems. They had dedicated factories in Allentown and Reading, Pennsylvania. While the Components division was initially expected to make a profit, it didn't come easily — mainly because they had to produce so many proprietary components and replacement parts (like Strowger switches) for the other divisions. Soon, the Components division became a cost center.
- Bell Labs was always expected to be a cost center. They had no factories and were not expected to make a profit.

Inside Bell Labs, product development groups were funded directly by product managers within the appropriate LoB, and the research area was funded by an internal tax on all LoBs. Suppose you were a VP or AVP inside Bell Labs and your product manager told you to "do X," but your Bell Labs chain of command told you to "do Y." Which would you do? Silly question — you would follow the money and you'd do X. While AT&T had fought hard to keep the Bell Labs' name and the wholly owned company existed on paper, Bell Labs' influence on AT&T practically ceased. The president of Bell Labs once protested, "I am not a figurehead." But, in practice, he was.

Finally, the requirements of CI-2 were lifted and ATTIS was no longer held at arm's length. AT&T Technologies reorganized several times and some pieces even merged into AT&T Communications. For several years, the PBX business was merged with the long distance business. In a hostile takeover, AT&T acquired NCR and merged it with its own computer business, somehow believing that the sum of two struggling computer companies would produce one successful one. The parent company tried so many times to find an organizational structure that would work, that employees joked about the "org chart du jour." AT&T even got into the credit card business and employees joked that next they would enter the hotel and insurance businesses (like ITT did) and get out of the telecom business altogether.

16.5 LONG DISTANCE

Besides the other restrictions described in the previous section, the MFJ dramatically changed the structure and the business of long distance telephony in the United States. This section's three subsections describe LATAs, architectural changes in the long distance network, and the economics of the access charge.

16.5.1 LATAs

The MFJ carefully separated monopolistic local telephone service from competitive long distance telephone service. The new local telephone companies would not be permitted to complete long distance telephone calls. A new kind of carrier, the interexchange carrier (IXC), would not be permitted to provide local telephony. So, the MFJ had to carefully redefine what a long distance telephone call was.

It was in the RBOCs' best interest that they be allowed to complete any call between any pair of subscribers within their region and that long distance be defined as a telephone call between subscribers living in regions served by two different RBOCs. It was in the IXCs' best interest that RBOCs be allowed to complete intraoffice calls only and that long distance be defined as any call between subscribers served by different COs. A compromise was reached.

The MFJ defined 243 geographical regions, called local access transport areas (LATAs). A call between two subscribers who live in the same LATA can be completed entirely by the LEC. If the two subscribers in the same LATA are served by different phone companies (for example, one by a BOC and one by GTE), the two LECs can still complete the call without an IXC. A call between two subscribers who live in different LATAs must be completed by an interexchange carrier even if the two subscribers are served by the same LEC or even if the two subscribers live across the street from each other but the center of the street is the LATA boundary. If they reside in different LATAs, then the LEC cannot complete the call without invoking an IXC.

For an inter-LATA call, the calling party's LEC would connect the call from the switching office that serves the calling party to the nearest gateway office, called a point of presence (POP), of one of the IXCs. But which IXC should be selected by the originating party's LEC? Each of the IXCs was assigned a three-digit access code, which the calling party could dial before dialing the called party's telephone number. So, even though subscribers wouldn't have to dial 14 digits for every long distance telephone call, subscribers were asked to register with a default IXC. Then, if a subscriber dialed a long distance number without specifying a particular IXC, based on the subscriber's extended class of service, the LEC would ask the calling party's default IXC to complete this long distance call. The selected IXC would connect the call, through its long distance network, to its POP in the called party's LATA. The called party's LEC would complete the call, from the IXC's POP to the switching office that serves the called party.

LATAs are smaller than an RBOC region (there are seven of these) and typically smaller than a state (there are 50 of these). LATAs are much bigger than a CO serving area (there are 40,000 of these) and typically bigger than a class-4 toll region (there were 1500 of these). LATAs are roughly equivalent to number plan areas (NPAs), the regions with the same area code. Pennsylvania has six LATAs and a similar number of NPAs, roughy overlapping, but none exactly overlapping. New Jersey, with many area codes, is one complete LATA.

16.5.2 Rearchitecting the Long Distance Network

Section 4.1 described AT&T's classic hierarchical network. This classical structure was completely rearchitected after divestiture. First, as discussed at the end of Chapter 9, the toll point was disintegrated immediately after divestiture. Then, several years later, AT&T Communications (formerly Long Lines) flattened the network's hierarchy.

Chapter 9 described the toll point's evolution, including its disintegration at the time of divestiture. This subsection simply summarizes these events and shows how they conform to the MFJ's rearchitected network and separation of LECs from IXCs.

- Because the LECs retained responsibility for billing, the CAMA point (Section 9.2) — whether based in a class-4 office or a TSPS — logically went to the RBOCs.
- Because most of the operator functions went to AT&T, the attendants and supporting equipment — whether based in a class-4 office or a TSPS — logically went to AT&T.
- Because the LECs would be allowed to complete most intratoll calls, the class-4 office's hubbing function logically went to the RBOCs. The LEC's interoffice trunking is typically implemented physically by a Sonet ring and logically by a hub switch that is currently called a tandem switch.
- Because the LECs would not be allowed to complete most intertoll calls, the class-4 office's gateway function logically went to AT&T. The class-4 office, or its physical replacement, became AT&T's point of presence (POP) for the LATA.

After divestiture, AT&T lost the class-5 layer from its hierarchical network, and some of its class-4 layer. AT&T added new class-4 offices where they were needed and modified the class-4 offices that were retained. Each office in the modified class-4 layer became a gateway, called a point of presence, to the Long Lines network. For several years, AT&T retained this hierarchical structure with class-1 through class-4 offices.

Several years after divestiture, AT&T rearchitected its hierarchical network — flattening it and removing the distinction among classes 1 through 3. The new architecture implements what is called dynamic nonhierarchical routing (DNHR), in which any pair of POPs in the United States is interconnected by a maximum of three hops, through a maximum of two intermediate nodes. This DNHR architecture has two layers: a layer of POPs, which act as gateways on the edge of the network, and a layer of interior nodes, interconnected in a large mesh network.

As discussed in Sections 1.5 and 3.4.1, many years ago, the two wires of the local loop were named tip and ring after the parts of plug on an operator's patch-cord, shown in Figure 1.6. While the reason for the names has been long gone, the names have survived into the present age. Similarly, switching offices were named class 1 through class 5, after AT&T's traditional hierarchical interexchange network, described in Section 4.1. Many years after divestiture and several years after this classical interexchange network was flattened, people still refer to a local telephone switching office as a class-5 office.

16.5.3 Access Fee

One more important issue had to be resolved. If the Bell System would be broken up and if long distance calls and local calls would be the purviews of separate companies, then how could long distance telephony continue to provide a substantial subsidy for POTS?

One answer could have been that if the telephone business were going to be truly deregulated and competitive, then it should have *cost-based pricing,* there should be no more "cream skimming" and competitive loopholes, and POTS shouldn't be subsidized at all. But, not surprisingly, the regulators chose not to do this and, instead, they devised a very bizarre system of subsidies.

An alternate scenario was possible for subsidizing universal service. This is my favorite scenario and we'll come back to this in Section 22.4. Let the great majority of Americans, who can afford it, pay a price for POTS that allows their LEC monopoly to cover the cost of POTS and to also make a fair and reasonable profit. Then, if some layer of government thinks that some part of the population should get a discounted rate, let this branch of the government provide the money to these people that helps them pay for their POTS. Several generations ago, the U.S. federal government decided that certain qualified people deserved to have food at a discounted price. This decision could have been implemented by passing a law requiring that the price of certain foods (bread, milk, vegetables, ground beef) be sold below cost and that farmers and grocers could recover this loss by overcharging on other foods (steak, ice cream, potato chips). Such a system of price controls couldn't work in the competitive market that farmers and grocers operate in. So, instead, we use a system of food stamps, administered by the U.S. Department of Agriculture and paid for with tax revenue. Instead of providing an analogous regulatory structure for telecommunications, the regulators created yet another obtuse structure that was anticompetitive, inconsistent, unstable, and likely to fail.

Rather than really deregulate telephony, the long distance subsidization of POTS was exposed and formalized. Instead of relieving AT&T's long distance business from the POTS subsidy, the POTS subsidy was extended to MCI, Sprint, and the other common carriers. Each interexchange carrier is required to charge each of its subscribers a fixed monthly access fee. The subscriber pays this flat monthly charge for the right to be connected to the IXC's network, whether or not he uses this connection and regardless of how often he uses this connection. While there is some logic to such a charge, its cost isn't nearly as high as this access fee. Furthermore, why don't we have to pay for the right to be connected to the Internet? It's no different. For some reason, the regulators and politicians don't want to tax the Internet the same way that they want to tax long distance. This is just one of many inequities and another example of how we're not only still subsidizing POTS, but we're also subsidizing the Internet. More on this in Chapters 19 and 22.

But, the IXCs don't get to keep their access fees. For each of their subscribers, the IXC has to give the subscriber's access fee to whichever LEC each subscriber homes on. So, a LEC sends a monthly bill to a telephone subscriber that includes its own charges for local access, special services, and intra-LATA toll calls and also includes any IXCs' charges for their access fee and any inter-LATA toll calls. The LEC forwards the revenue for inter-LATA toll calls to the IXC, but the LEC keeps the IXC's access fee. This fee is a direct subsidy for POTS.

16.5.4 Bypass

This access fee regulation might have worked, but only in a closed market. The regulation not only allowed for, but it openly encouraged, competition in local access. This so-called loop competition was even further encouraged by the 1996 Telecommunications Act (discussed in the next section). Local access competition, as allowed in the MFJ and as clarified in the 1996 Telecom Act, was seen as a way to lower rates by forcing competition on price. In the residence market, several forms of competition were envisioned. These are discussed in greater detail in Chapter 19.

- Theoretically, CATV could provide POTS channels to the home by sub-multiplexing one or more 4-MHz video channels. Cable companies want to offer Ethernet connectivity in a similar way. A redesigned set-top box could have a port for Ethernet and another port that would interface to the analog wiring in the subscriber's home.
- While mobile telephony — especially, cellular and LEOS — all compete with the local loop, most subscribers use these technologies in addition to their local loop service at home. Few users actually desubscribe from their LEC. A new technology, called fixed wireless, could provide a simple cost-effective means to provide an alternate drop-wire for any company that wanted to seriously compete in the local loop.
- Gas and electric utilities in the United States are logically separating the service from the system that provides the service. Subscribers can buy gas or electricity from a different company than the company which owns the gas pipes or electric power lines. Similarly, subscribers can buy telephone service from a competitive local exchange carrier (CLEC), which would provide their telephone service over the LEC-owned local loop. In effect, the CLEC leases local loop facilities from the LEC and provides telephone service over this leased line.

The regulations that encourage loop competition do not require the cable company or the wireless provider or the CLEC to provide universal service. Because they don't need to subsidize POTS, they don't have to charge their customers the long distance access fee. A significant industry arose almost immediately, but its significant impact was in the enterprise market. The LECs refer to these companies as bypassers because, by using these companies for connection to their IXC, enterprises bypass the access charge that the IXCs are required to charge on behalf of the LECs. These companies refer to themselves as alternative access providers (AAPs), competitive access providers (CAPs), or competitive local exchange carriers (CLECs). The CAPs immediately installed facilities in commercial areas, typically installing optical fibers in the right-of-way conduits owned by gas and electric utilities. The CAPs concentrated on commercial regions because of the density of potential customers and their freedom from universal service. But, the main reason was that their principal market niche was to provide that direct connection between an IXC and its bypassed customers and their principal market advantage was access fee bypass. Since the IXCs weren't very interested in providing direct connections to residences, the CAPs principal business was to connect PBXs to IXCs.

While bypass violated the spirit of the access fee arrangement, AT&T Communications was the only IXC that cared about the LECs — not because AT&T was more altruistic than MCI and the other IXCs, but because the LECs were the best customers of AT&T Technologies (WECo). Realizing the dilemma that bypass caused AT&T, the smaller IXCs aggressively marketed LEC bypass to the enterprise market. This saved a lot of enterprises from paying the access fee, cost the LECs a lot of revenue for POTS subsidy, and caused AT&T to lose market share in the enterprise market.

16.6 THE POST-DIVESTITURE TELECOM INDUSTRY

Many of the RBOCs perceived AT&T simultaneously as their largest customer, their largest equipment vendor, and their largest competitor. This very unusual and complex relationship was exacerbated by bypass and was brought to a climax by the 1996 Telecom Act. This section describes AT&T's dilemma regarding bypass, AT&T's ironic decision to unilaterally and voluntarily divest itself of what used to be the Western Electric Company, and then discusses the 1996 Telecommunications Act.

16.6.1 AT&T's Dilemma

Before divestiture, AT&T was in the four separate businesses described in Section 16.3.2. These businesses made sense residing in a single corporation. After divesting the BOCs, AT&T retained three of these businesses, but they no longer made sense residing in the same company. With the RBOCs now as both a customer and a competitor, instead of as a partner, two of AT&T's remaining three businesses had a very complex inconsistency.

As discussed in the previous subsection, the regulators wanted to continue subsidizing POTS, but they also wanted competition in the local loop. To encourage loop competition, they freed the CLECs (or AAPs or CAPs or bypassers or whatever you call them) from universal service and from the long distance access fee that subsidized it. Only the existing LECs had to provide universal service and only the existing LECs had to put the access fee on their telephone bills. The new CLECs aggressively marketed enterprises, not residences. The CLECs principal market advantage was that their customers didn't have to pay the access fee, which an enterprise had to pay for each trunk. There was no direct advantage to an IXC whether an enterprise accessed long distance through a LEC or through a CLEC.

The LECs, of course, were extremely unhappy about this regulatory loophole, which they called bypass. The LECs couldn't stop the CLECs from bypassing them and the LECs had no argument or leverage to convince most of the IXCs to refuse to participate in bypass. But, the LECs did have leverage over AT&T. The implied threat from the LECs was that if AT&T encouraged bypass, then the LECs would buy more of their new switching and transmission equipment from Nortel, Siemens, Ericsson, and Alcatel. AT&T's Technologies division (what used to be Western Electric), particularly its network systems LoB, wanted to appease their largest customers, the LECs. MCI, Sprint, and the other IXCs, none of which had manufacturing divisions, saw an advantage in AT&T's reluctance to participate in bypass. They aggressively marketed bypass and AT&T lost more market share in the lucrative enterprise side of their long distance business. AT&T's Communications division also wanted to aggressively market bypass and AT&T's Technologies division wanted them not to. What is a CEO to do?

16.6.2 Final Irony

The Bell System of 1956 was self-consistent and stable. As discussed in Section 16.3.2, AT&T was in four separate businesses that made sense under the same roof: local telephony, long distance telephony, large telecommunications equipment manufacturing, and consumer products manufacturing. Breaking off the first business in 1984 left AT&T in the other three businesses, but they could not coexist under the same roof. So, AT&T had to break itself apart again. And this was a convenient time to stop losing money in the computer business.

After so many different concessions to the FCC and the DoJ over a period of about

70 years, just to keep control of Western Electric, in 1996 AT&T voluntarily and unilaterally let it go. After giving up so much over a period of about 15 years, to be allowed into the computer business, in 1996 AT&T voluntarily got out of it. In 1996, AT&T announced that it was splitting itself into three independent corporations.

- AT&T would retain its long distance business, the wireless business (from buying McCaw/Cellular One), and its new service ventures. Roughly, this was AT&T's original Long Lines department, before it was merged with some of AT&T's PBX and computer businesses.
- The piece of AT&T that had been AT&T Technologies, formerly the Western Electric Company, was left with all manufacturing except computers, the third and fourth businesses in the list above. But, instead of using the recognizable name Western Electric, the new company decided to call itself Lucent Technologies.
- The piece of AT&T that had been formed by merging ABI's own computer LoB with the acquired National Cash Register was divested from AT&T and given back the old name "NCR." Both AT&T and Lucent divested the computer business, which had lost billions of dollars since AT&T got into it.

The announcement came as a surprise to most people. The industry experts, who knew the historical context, thought this was an astounding turn of events. But people who had been inside AT&T during these 10 years, and who understood AT&T's bypass dilemma, knew this was just another inevitable part of the Carterfone chain reaction.

Bell Labs went with Lucent and AT&T started its own R&D subsidiary. So, now there are three easily confused but completely independent R&D companies: Bell Labs, Bellcore, and AT&T Labs. AT&T kept its "Death Star" logo, so Lucent came up with a new one, jokingly called "big red zero."

The conventional wisdom, and most industry experts, predicted that AT&T, finally freed from a manufacturer's inertia, would finally be able to live up to grand promise of convolution and provide those long-overdue integrated services. A small minority believed that Lucent, finally freed from AT&T's marketing, would finally be able to show its strength as one of the best manufacturers in the world. Which is right? The early returns show that Lucent is doing much better than AT&T, but we need more time before we can really answer.

16.6.3 The Telecommunications Act of 1996

For most of the 20th century, most of the American telecommunications industry's significant regulations have come from the FCC. Whenever the DoJ changed the rules, it was typically a major upheaval — particularly with the 1956 Consent Decree and its 1984 Modified Final Judgement. While the U.S. Congress usually avoided making rules for this industry, notable exceptions were the Telecommunications Acts of 1934 and of 1996.

As of the time that this book is being written, it is too early to evaluate the impact of the 1996 Telecommunications Act. While the act clarified some of the issues of the MFJ, it overturned some others — at least in principle. In particular, the 1996 Telecommunications Act changed the MFJ's rules that clearly precluded the LECs and the IXCs from providing telephone service in each other's domains.

- For 10 years, the RBOCs have been eager to offer long distance telephony, particularly within their respective regions where they already have transmission facilities in place anyway. Naturally, the IXCs opposed this. While new companies joining this market (Quest, Hyperion, etc.) usually have the large cost of installing fiber routes or leasing channels from existing carriers, the LECs wouldn't have to — they already have facilities (at least, inside their respective regions).
- For 10 years, the IXCs have been eager to expand their niche role and avoid competing on price by providing "bundled service" to the residence market. Naturally, the LECs opposed this. It's hard to blame the LECs for wanting to keep the significant sources of POTS subsidy when they are required to provide universal service and none of their competitors are.

The 1996 Telecommunications Act links these together. It specifies that an LEC qualifies to offer long distance service by demonstrating the existence of a sufficient level of local loop competition in its region. If this coupling works, the 1996 Telecommunications Act will go down in history as being a significant regulation. But most Americans, myself included, still don't have a CATV carrier or a CLEC offering service to their homes. LECs are now arguing that they are being as cooperative as possible, but that the IXCs, particularly AT&T, are deliberately dragging their feet about entering the CLEC market because it will keep the RBOCs out of the long distance business. AT&T argues, of course, that the RBOCs are still being anticompetitive. This kind of stalemate may make the Act irrelevant.

Now, many of the LECs are considering a reorganization that, from an historical perspective, looks like a microcosm of AT&T's divestiture in 1984. During the trial in Judge Greene's court, AT&T decided to divest itself of local telephony, believing that this part of their business was least desirable because it would always be regulated. Now, some of the LECs are looking at local access as only a part of their overall business and many of them see local access the same way that AT&T saw the LECs in the early 1980s. The local access business is less desirable because:

- It seems inevitable that it will always be entangled in some unworkable combination of regulation and competition;
- The large investment in the loop plant makes the owner appear unattractive to investors (and their computer programs) who look at simple figures, like corporate return on investment.

So, some of the LECs are considering ways to isolate themselves from the local access business, either by moving it into an arm's-length subsidiary or by completely divesting themselves of their loop plant. It seems that nobody wants to be stuck with universal service, at least not the way it's currently subsidized. Being required to provide the residential side of Internet access only exacerbates the problem. If the LECs do divest themselves of their loop plant, we would have yet another restructuring of the layers of telephone companies in the United States — probably with subsequent architectural impact.

Summarizing, between 1984 and 1996:

- The LECs provided universal service, local access, intra-LATA telephony, intra-LATA leased lines, and a variety of competitive services;
- The IXCs provided inter-LATA telephony and many of the same services.

After 1996, the IXCs can be CAPs, and once the level of loop competition is sufficient, the LECs can be IXCs. If the LECs move local access to local access carriers (LACs), then these LACs will provide universal service and local access, while the companies that are now LECs and IXCs could all be CAPs and they would be virtually indistinguishable. And, they will probably merge together even more.

16.7 DISCUSSION

So, has the U.S. public been well served by divestiture? While the conventional wisdom gives a resounding "yes" to this question, it's not a simple question to answer because (1) some features we enjoy now might have happened anyway (but might not have) and (2) the regulatory environment is not yet stable.

- Might the Bell System have offered consumer equipment, before Carterfone, that is as inexpensive as what we have today? Or are today's low prices an inevitable result of improved technology and prices even Western Electric might have provided if Ma Bell didn't have to subsidize universal service?
- Might the Bell System have provided long distance service, before Execunet, that is as inexpensive as what we have today? Remember that, if you include the access fee, you're paying more than seven cents per minute. Or are today's low prices an inevitable result of improved technology and prices Long Lines might have provided if Ma Bell didn't have to subsidize universal service? And, as IXCs merge and bundle service, will these low rates continue?
- Might the Bell System have provided wireless and cellular service that is as ubiquitous and inexpensive as what we have today? Or is this an inevitable technological phenomenon that Ma Bell might have provided with better interoperability, universality, and reliability?
- Might the Bell System have been more or less reliable and survivable, end-to-end, than today's network?
- Might the Bell System have provided universal 911? Some people in the United States still don't have it. Might Ma Bell have made it universal a long time ago?
- Might the Bell System have provided calling party ID, enhanced 800, and all the other new features and services? Might they have been as inexpensive as they are now? Might they have been implemented more optimally than they are now? Might the Bell System, before Carterfone, have provided SS7 access to PBXs and, even to telephones?
- Might the Bell System have impeded, encouraged, or even provided the Internet? Might it have been more or less ubiquitous? Might a pre-divestiture Internet still

have been protected from excise taxes? Might a pre-divestiture Internet still have had free local access?

- What about installations, repairs, directory information, and other customer-oriented services? Are these things better, cheaper, and easier now than they were under the Bell System? Are these things equally good or bad in the regulated and unregulated parts of the industry, and is this a fair indicator of the future? Was this change inevitable or did it result from regulation? Will these things be better or worse under CLECs, voice over cable, voice over Internet, and other variations?
- Might the economic growth in the telecommunications industry not have occurred without divestiture, or was it an inevitable result of new technology?
- Most of the these questions have to be asked separately for the enterprise market and the residence market.
    - While enterprises have clearly benefited from Carterfone, Execunet, deregulation, and divestiture, might many of the lower prices and improved services have occurred anyway if the Bell System's former regulators had adopted cost-based pricing?
    - While residential POTS service has become pricier since divestiture, the long distance access charge still holds it down. Might a pre-Carterfone monthly POTS bill have been lower, about the same, or greater than today's monthly POTS bill? Or tomorrow's?

Wherever you read "might" in the questions above, substitute "could," "would," or "should" and ask the question again. The answers to all the questions above should be considered in terms of whether the same answer will always continue, might change in the future, or will have to change in the future. None of these questions is easy to answer.

From 1956 to the present, the process had a lot of unnecessary steps. There was a lot of thrashing, and a lot of lawyers made a lot of money. But we didn't have a controlled experiment where we could test divestiture in the absense of many other variables. Many things have happened since divestiture, some bad and mostly good — but were they a direct result of divestiture or were they simply coincidences? Whichever, it's clear that Ma Bell's behavior in her early years, her uncompromising response to Carter, and her poor loser attitude in the 1970s all influenced her fate in the 1980s and beyond.

The end of Section 2.1 referred nostalgically to the 1950s. Correlate this discussion about telecommunications regulators with one of my favorite stories [1] — in the context of Willie Mays' famous catch in the 1954 Baseball World Series. After the Cleveland Indians got several men on base, the New York Giants brought in a left-handed relief-pitcher, Don Liddle, to pitch to Vic Wertz, an Indians' left-handed power hitter. The strategy called for a right-handed pitcher to replace Liddle and pitch to the right-handed batter who followed Wertz in the Indians' lineup. All baseball fans know that Wertz hit one of Liddle's pitches 500 feet to straightaway centerfield in the Polo Grounds and that Willie Mays caught it in full stride, over his shoulder, back to the plate. But, there is more to the story. Liddle was replaced, as he would have been no matter what Wertz did. Joking with his teammates, he strutted into the Giants' dugout and said proudly, "Well, I got my man."

EXERCISES

16.1 As this book is going to press, Bell Atlantic has already acquired Nynex and awaits FCC approval to acquire GTE. Furthermore, SBC has already acquired SNETCo and PacTel and awaits FCC approval to acquire Ameritech. If you could vote in the FCC, (a) would you approve Bell Atlantic's acquisition of GTE? (b) Would you approve SBC's acquisition of Ameritech? Explain why or why not.

16.2 If all these mergers are approved, the eight largest LECs in 1990 (seven RBOCs plus the merged GTE/Contel) will have been reduced to four by 2000. (a) Are the two smallest remaining LECs likely to merge with each other? (b) If not, which of these smallest two is most likely to be the next one acquired by one of the largest two? (c) If you could vote in the FCC, would you vote to approve this? Explain why or why not.

16.3 As this book is going to press, the FCC has not yet allowed the RBOCs to provide long distance service. If you could vote in the FCC, would you vote to approve? Explain why or why not.

16.4 If and when the RBOCs do enter the long distance business, which IXCs are most likely to be acquired quickly?

REFERENCE

[1] Will, G. F., *Men at Work — The Craft of Baseball,* New York, NY: Macmillan Publishing Company, 1990.

SELECTED BIBLIOGRAPHY

Ash, G. R., "Dynamic Network Evolution, with Examples from AT&T's Evolving Dynamic Network," *IEEE Communications Magazine,* Vol. 33, No. 7, Jul. 1995, pp. 26 – 39.

"AT&T 1993 Employee Annual Report," *AT&T Focus Magazine,* Jan. 1993.

Bell, T. E., "The Decision to Divest: Incredible or Inevitable?" *IEEE Spectrum Magazine,* (issue on The Future of Telecommunications), Vol. 22, No. 11, Nov. 1985.

Blalock, W. M., "North American Numbering Plan Administrator's Proposal on the Future of Numbering in World Zone 1," *Bellcore Letter,* Letter No. IL-92/01-013, Jan. 6, 1992.

Brands, H., and E. T. Leo, *The Law and Regulation of Telecommunications Carriers,* Norwood, MA: Artech House, 1999.

Brock, G. W., *Telecommunication Policy for the Information Age,* Cambridge, MA: Harvard University Press, 1994.

Brock, G. W. (ed.), *Toward a Competitive Telecommunication Industry,* Mahwah, NJ: Lawrence Erlbaum Associates, 1995.

Cortes-Comerer, N., "The Consumer Reels in the Aftermath," *IEEE Spectrum Magazine,* (issue on The Future of Telecommunications), Vol. 22, No. 11, Nov. 1985.

"Decade of Drama — Divided We Grew," *AT&T Focus Magazine,* Jan. 1992.

Dorros, I., "The New Future — Back to Technology," *IEEE Communications Magazine,* (issue on Telecommunications Regulation), Vol. 25, No. 1, Jan. 1987.

Dziatkiewicz, M., "Universal Service — A Taxing Issue," *Wireless Business & Technology,* Vol. 5, No. 1, Jan. 1999.

Fischetti, M. A., "Bypassing Is Big Business for Business," *IEEE Spectrum Magazine,* (issue of The Future of Telecommunications), Vol. 22, No. 11, Nov. 1985.

Geller, H., "Government Policy as to AT&T and the RBOCs: The Next 5 Years," *IEEE Communications Magazine,* (issue on Telecommunications Deregulation), Vol. 27, No. 1, Jan. 1989.

Gillett, S. E., and I. Vogelsang (eds.), *Competition, Regulation and Convergence,* Mahwah, NJ: Lawrence Erlbaum Associates, 1999.

Harrold, D. J., and R. D. Strock, "The Broadband Universal Telecommunications Network," *IEEE Communications Magazine,* (issue on Telecommunications Regulation), Vol. 25, No. 1, Jan. 1987.

Henck, F. W., and B. Strassburg, *A Slippery Slope — The Long Road to the Breakup of AT&T,* Westport, CT: Greenwood Press, 1988.

Lucas, J., "What Business Are the RBOCs In?" *TeleStrategies Insight Magazine,* Sept. 1992.

Marcus, M. J., "Technical Deregulation: A Trend in U.S. Telecommunications Policy," *IEEE Communications Magazine,* (issue on Telecommunications Regulation), Vol. 25, No. 1, Jan. 1987.

Marshall, C. T., "The Impact of Divestiture and Deregulation on Technology and World Markets," *IEEE Communications Magazine,* (issue on Telecommunications Deregulation), Vol. 27, No. 1, Jan. 1989.

McDonald, J. C., "Deregulation's Impact on Technology," *IEEE Communications Magazine,* (issue on Telecommunications Regulation), Vol. 25, No. 1, Jan. 1987.

Noll, A. M., *Introduction to Telephones and Telephone Systems,* Norwood, MA: Artech House, 1986.

Olson, J. E., "New Policies for a New Age," *IEEE Communications Magazine,* (issue on Telecommunications Regulation), Vol. 25, No. 1, Jan. 1987.

Perrin, S., "The CLEC Market: Prospects, Problems, and Opportunities," *Telecommunications,* Vol. 32, No. 9, Sept. 1998.

Shaw, J., *Telecommunications Deregulation,* Norwood, MA: Artech House, 1998.

Sikes, A. C., "Tommorrow's Communications Policies," *IEEE Communications Magazine,* (issue on Telecommunications Regulation), Vol. 25, No. 1, Jan. 1987.

Staley, D. C., "Domestic Roadblocks to a Global Information Highway," *IEEE Communications Magazine,* (issue on Telecommunications Deregulation), Vol. 27, No. 1, Jan. 1989.

Temin, P., with L. Galambos, *The Fall of the Bell System,* New York, NY: Cambridge University Press, 1987.

Thyfault, M. E., "AT&T Breaks Up Again — Split Decision," *Information Week Magazine,* Iss. 547, Oct. 2, 1995.

Tracey, L. V., "30 Years: A Brief History of the Communications Industry," *Telecommunications Magazine,* Vol. 31, No. 6, Jun. 1997.

Weiss, M. B. H., and P. Bernt, *The Regulation and Deregulation of US Telecommunications,* Textbook under contract with Lawrence Erlbaum and Associates, Mahwah, NJ.

Wolfson, J. R., "Computer III: The Beginning or the Beginning of the End for Enhanced Services Competition?" *IEEE Communications Magazine,* (issue on Telecommunications Deregulation), Vol. 27, No. 1, Jan. 1989.

Yoshizaki, H., "Breakup of AT&T and Liberalization of the Telecommunications Business," *IEEE Communications Magazine,* (issue on Telecommunications Deregulation), Vol. 27, No. 1, Jan. 1989.

17

Network and Switching Paradigms

> *"We hold these truths to be self-evident...."* — opening statement of the *U.S. Declaration of Independence,* probably written by Thomas Jefferson

This chapter addresses network topologies in the first two sections and switching paradigms in the last three. The first two sections cover topics in network paradigms. Section 17.1 compares the bus, ring, and star network topologies by their real capacity to deliver data and shows why the star topology is superior. Section 17.2 shows that LANs and WANs have different design optimizations, which means that big networks must be hierarchical — they can't be flat. The second two sections compare packet and circuit switching. Section 17.3 presents several different qualitative comparisons and Section 17.4 presents a quantitative comparison. Then, Section 17.5 adds facility switching to the set of compared paradigms and introduces what I call *capacity switching.*

This chapter's common thread is that the different topologies and the different paradigms really are different. Whether the differences among these topologies and paradigms are significant is discussed at the end of Chapter 19, where we discuss integration into one universal network under two different packet-switched paradigms.

17.1 CAPACITIES OF THE BUS, RING, AND STAR

This first section presents some obvious characteristics of bus, ring, and star architectures and plainly compares their real ability to deliver data. Most of the difficulty in specifying the analysis and comparing the performance of architectures is that we don't always measure the right thing. Is gigabit Ethernet better than 100-megabit ATM? How would we answer this? What do we mean by "better"?

The telephone industry has been using the star architecture for more than a hundred years and this section shows why: it is because the star is superior to the bus or the ring. The first subsection defines three different places in a network where data rate might be measured and it argues that, while none of these is really a good measure, one is better than the other two. The three subsequent subsections compare the bus, ring, and star with respect to this measure.

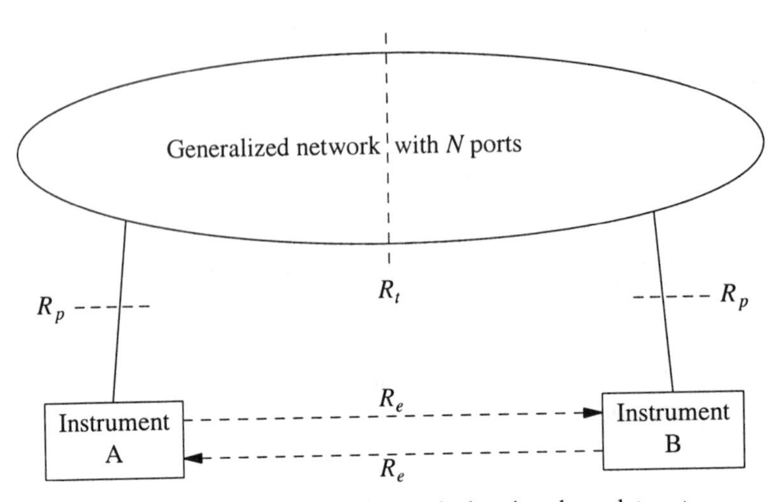

Figure 17.1. A generalized network showing three data rates.

17.1.1 A Network's Three Rates

Figure 17.1 shows two users' instruments connected to a generalized network. Data rate can be measured at three different places in the figure.

C_t is the network's gross *throughput capacity* and R_t is the network's actual gross *throughput data rate.* If the network has a single concentrated point for all traffic, like a bus or a star would, then C_t and R_t, respectively, are the total capacity and the total data rate through this point. If the network is a bus, then C_t is the bus' line rate. If the network has no single concentrated point, such as a mesh or a ring wouldn't, then C_t and R_t would be calculated by adding up the capacities and actual data rates, respectively, on all the network's independent internal links. Is C_t a good measure of a network's ability to carry data? R_t includes all the network's retransmitted data and C_t includes the required capacity for carrying it. While the capacity for high throughput and an actual observed high throughput data rate might indicate a good network, if this figure is high because the network frequently retransmits the same data, then this is not necessarily a good network.

C_p is the network's individual *port capacity* and R_p is the instrument's average real net *port data rate.* C_p is the *access rate,* the line rate of each user's instrument. R_p is the average rate of actual data transmitted in one direction between the user's instrument and the network. Is C_p a good measure of a network's ability to carry data? If the network switches by selecting (see Section 1.3), instead of using true switching, then C_p probably includes all the network's traffic, including traffic not necessarily destined for this instrument. C_p is more like a requirement for the instrument's transmitter and receiver than a benefit to the end user. While high port capacity and an actual observed high port rate might indicate a good network, if these figures are high because of unused data or overspecified need, then this is not necessarily a good network.

C_e is the network's effective *end-to-end capacity* and R_e is the network's average *end-to-end data rate.* C_e is the average capacity between any pair of ports and R_e is the average rate of data between any pair of ports. Is C_e a good measure of a network's ability to carry data? It probably is, but only if we assume that data is distributed uniformly throughout the network. While optimizing C_e isn't always the best strategy, it's a better measure than the capacities defined above. While high end-to-end capacity might indicate a good network, if $R_e \ll C_e$ for many pairs and $R_e \gg C_e$ for other pairs, then this is not necessarily a good network.

17.1.2 The Bus

Let the network consist of a single bus, connecting N instruments. Let C_t be the bus' specified line rate and let R_t be the bus' average gross rate of actual data. If the bus is 60% utilized, then $R_t = 0.6 \times C_t$. Since every instrument interfaces directly to the bus, $C_p = C_t$ and each instrument is required to have a line rate that is much greater than it needs for its own data. If there is no congestion on the bus (a ridiculous assumption), then the average available capacity per instrument is C_t / N and the average available end-to-end capacity for each pair of users is only $C_t / N(N-1)$.

TABLE 17.1. Measuring a bus with five sizes

	N=10	N=100	N=1K	N=10K	N=100K
C_t	100M	100M	100M	100M	100M
C_p	100M	100M	100M	100M	100M
R_p	10M	1M	100K	10K	1K
C_e	1.1M	10K	100	1	0.01

Consider the five examples in Table 17.1. Assume that the bus' line rate is 100 Mbps. See how the average end-to-end capacity becomes so small that at 100,000 instruments (a large class-5 serving area) the average end-to-end data rate is 1 bit every 100 seconds. When congestion is taken into account, the figures on the bottom line are much worse. The bus scales very poorly and it is a poor architecture, especially when many devices are connected to it.

The "Von Neumann bottleneck" arises because CPUs access their instructions and their data through the same address and data ports, each connected to memory by a bus topology. Computer architects cling to this bus architecture. When a computer architect designs a network (like Ethernet), it has a bus topology until it is ultimately redesigned. While there may be good reasons that CPUs can't be connected to RAM, ROM, and peripherals by star networks, these reasons don't apply to networks.

17.1.3 The Ring

Let the network consist of a one-way ring, connecting N nodes. Let C_p be the ring's specified link rate. One might argue that the ring's net throughput is $C_t = N \times C_p$. But, only a ring salesman would use this measure. We know that much of what is transmitted is the same data passing from node to node. If each packet goes halfway around the ring on the average, then it is transmitted $N / 2$ different times. R_p is the average rate that is available to an instrument that is wired to any node, or the useful link data rate. R_p is the link's line rate, C_p, divided by $N / 2$:

$$R_p = \frac{2 \times C_p}{N}$$

And, the effective net throughput is N times this value: $R_t = 2\ C_p$. As in the previous subsection, this analysis ignores congestion on the ring (again, a ridiculous assumption).

The average available end-to-end capacity for each pair of users is only:

$$C_e = \frac{R_t}{N(N-1)} = \frac{2 \times C_p}{N(N-1)}$$

The ring's useful pair-wise channel is twice as big as the equivalent channel on an equivalent bus.

TABLE 17.2. Measuring a ring with five sizes

	N=10	*N*=100	*N*=1K	*N*=10K	*N*=100K
C_t	1G	10G	100G	1T	10T
R_t	200M	200M	200M	200M	200M
C_p	100M	100M	100M	100M	100M
R_p	20M	2M	200K	20K	2K
C_e	2.2M	20K	200	2	0.02

Consider the five examples in Table 17.2. Assume that line rate on each of the ring's links is 100 Mbps. See how the average end-to-end rate becomes so small that at 100,000 instruments (a large class-5 serving area) the average end-to-end data capacity is 2 bits every 100 seconds. When congestion is taken into account, the figures on the bottom line are much worse. The ring scales a little better than the bus, but not much.

17.1.4 The (Active) Star

Let the network consist of an *N*-armed star, connecting *N* nodes to a single common hub. Let C_p be the specified link rate on each arm of the star. If we argue that the star's net throughput is $C_t = N \times C_p$, then, not only would we be correct, but this capacity is 100% useful: $R_t = C_t$. So, the average rate that is available to an instrument that is wired to any arm of the star is the link's full capacity. The useful link data rate is the link's line rate: $R_p = C_p$. As in the previous subsection, this analysis ignores congestion on the star but, in the star, this assumption isn't nearly as ridiculous. The average available end-to-end capacity for each pair of users is:

$$C_e = \frac{R_t}{N(N-1)} = \frac{R_p}{(N-1)}$$

The star's useful pair-wise channel is *N*-times bigger than the equivalent channel on an equivalent bus.

TABLE 17.3. Measuring a star with five sizes

	N=10	*N*=100	*N*=1K	*N*=10K	*N*=100K
C_t	1G	10G	100G	1T	10T
C_p	100M	100M	100M	100M	100M
R_p	100M	100M	100M	100M	100M
R_e	11M	1M	100K	10K	1K

Consider the five examples in Table 17.3. Assume that line rate on each of the star's arms is 100 Mbps. See how the average end-to-end rate becomes small but not nearly as small as in the bus or the ring. At 100,000 instruments (a large class-5 serving area) the average end-to-end capacity is 1000 bits every 1 second, 100,000 times faster than in the equivalent bus. The star scales much better than the bus or the ring.

17.2 HIERARCHICAL NETWORKS

This section deals with the proposal that a single flat network could interface to every kind of instrument in a single layer of networking. The first subsection observes a congruency among the large networks that carry telephone traffic, computer traffic, and video traffic. Then, after having a little fun with naming conventions for hierarchical networks, the last subsection reveals why big networks can't be flat — they must have at least two hierarchical levels.

17.2.1 Congruence

Congruence can be seen in three types of large networks, those that carry three distinctly different kinds of data traffic. At a general enough level of detail, big telephone networks, big computer networks, and big video networks all have a similar hierarchical architecture [1].

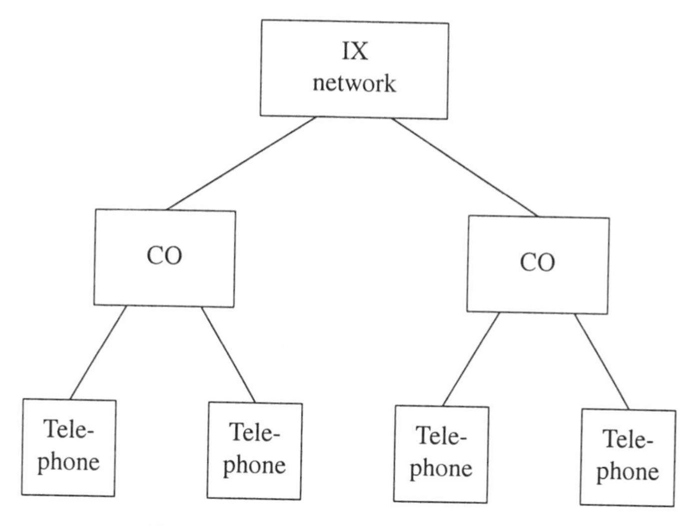

Figure 17.2. A big telephone network [1].

Figure 17.2 shows a very high level schematic diagram of a telephone network. While the PSTN in the United States has many more than only two layers in its hierarchical architecture, Figure 17.2 combines many of its real layers down to only two layers that are relevant to this section. Two significant points are observed.

- Telephones have access to switching offices by a local network, called the loop plant, that is owned and operated by a local exchange carrier (LEC). Partly because of the local loop's low bandwidth, its architecture allows the telephone's interface to the network to be inexpensive. An entire telephone costs $8.
- Networks for local and regional coverage are interconnected by LATA-wide transport networks, owned and operated by LECs, and by nationwide

transport/switching networks, owned and operated by interexchange carriers. Since these interexchange networks have high bandwidth and a highly multiplexed interface, the LEC's interface to the local or long distance toll network is very expensive.

One of the important functions of the layer labeled "CO" in Figure 17.2 is to connect local channels to interexchange channels. The cost of this gateway, which contains expensive multiplexors and broadband transmitters and receivers, is effectively shared by the LEC's users.

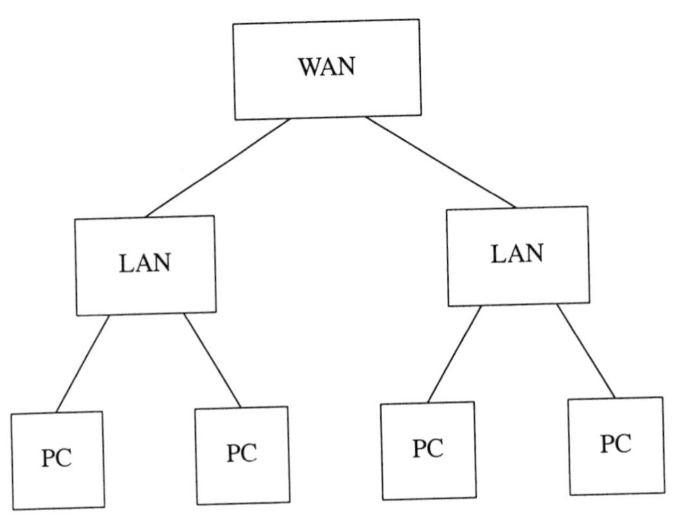

Figure 17.3. A big data network [1].

Figure 17.3 shows a very high level schematic diagram of a computer data network. While large computer networks like the Internet have many more than two layers in their hierarchical architecture, Figure 17.3 combines many of the real layers down to only two layers that are relevant to this section. Two significant points are observed.

- Personal computers have access to each other, workstations have access to servers, and dumb terminals have access to a host computer by a local network called a local area network (LAN). These LANs have architectures, bandwidth, and protocols that allow the PC's interface to the network to be very inexpensive.
- Local and metropolitan area networks are interconnected by nationwide transport networks, called wide area networks, like the Internet. Because these WANs have different protocols and channels than LANs and because they have very high bandwidth (typically, channels leased from the IXCs), the LAN/WAN interface is very expensive.

One of the important functions of the layer labeled "LAN" in Figure 17.3 is to convert local packets into nationwide packets. This requires a gateway, which contains expensive multiplexors, protocol converters, and broadband transmitters and receivers, whose cost is effectively shared by the LAN's users.

Figure 17.4 shows a very high level schematic diagram of a broadcast video network.

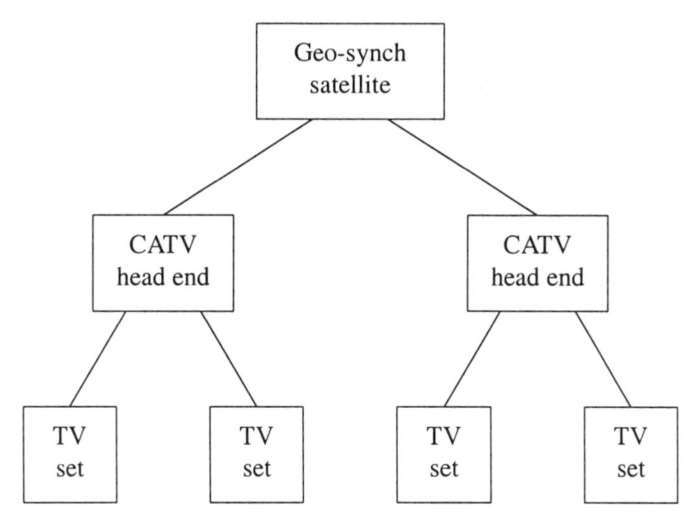

Figure 17.4. A big video network [1].

Video networks typically have layered hierarchical architectures, and the two layers shown in Figure 17.4 are realistic. Two significant points are observed.

- Television sets have access to CATV head ends by a local network called the "cable" that is owned and operated by a community antenna television provider. This cable's architecture, bandwidth, and carrier frequency selection allow the television set's interface to the network to be moderately inexpensive.
- Programming for local cable broadcast is transmitted over a one-way nationwide transport network, a set of geo-synchronous communications satellites. These satellite transmitters have high bandwidth and high carrier frequencies, which cause their highly multiplexed interface to be very expensive.

One of the important functions of the layer labeled "CATV head end" in Figure 17.4 is to connect local channels to nationwide channels. This requires a gateway, an earth station containing expensive multiplexors and broadband downlink receivers, whose cost is effectively shared by the cable's users.

17.2.2 A Hierarchical Naming Convention

Real telephone and computer networks have many more than only the two layers shown in the previous figures. AT&T's classic hierarchical long distance network, now obsolete, had five layers — six if you count a PBX, RSM, or CDO as a class-6 office. The IXCs have flattened their network architectures and AT&T's new dynamic nonhierarchical routing was described in Section 16.5.2. But, even in this flatter architecture, an inter-exchange call between parties served by distant PBXs is routed as follows: from the telephone to the PBX, to the CO, to the LEC's interexchange hub, to the IXC's POP, to the IXC's interior nodes, and back down again. By convention, residential telephone switching doesn't actively switch at the jack, protector block, pedestal, distribution point, or main frame, and enterprise switching doesn't actively switch at the jack, the wiring closet, the building wiring frame, or the PBX's main frame. But, if these wiring points are viewed as separating the loop network into hierarchical layers, the conventional telephone network is seen to have an eight-layer hierarchical structure.

Similarly, real computer networks have more than two layers. In the conventional enterprise, computers are interconnected by a local area network that typically covers at most one floor in a building. Typically, the multiple LANs in a building are interconnected by a building-wide network called a backbone. Typically, the multiple buildings on a campus or in an office park are interconnected by a metropolitan area network. Typically, many MANs are interconnected by a regional or national wide area network, which could be private or public. The Internet is a global public WAN.

This naming convention of LAN and MAN suggests generalizing the entire hierarchy by a uniform naming convention XAN, where X = I through P, instead of only L and M. While I admit I'm having a little fun with names here, there is a more relevant point to this. Eight layers of networking are described below, organized according to the physical size of the network. While this seems like a lot of layers, especially to the advocates of a flat network with only one layer, it isn't complete. Many of the layers described below contain sublayers.

Internal Area Networks. Devices, which are fabricated together onto an integrated circuit chip, are interconnected by simple networks, frequently buses fabricated from on-chip conducting paths. For example, a CPU architecture is a set of registers, a collection of gates and parallel paths (the on-chip network) by which data moves from one register to another, and a controller that decodes each op-code into a sequence of data movements. Short broadband networks, built from simple line-of-sight optical transmitters and receivers, may someday extend these on-chip networks to neighboring chips on a board. These on-chip networks are usually parallel and often are circuit switched, with each call connected for one clock cycle. The chips in my computers, telephones, and television set have such internal networks.

Junction Area Networks. CPU buses and other board-level parallel networks interconnect chips at their physical junction with a printed-wiring board. Motherboard buses interconnect the PWBs at their physical junction to a card carrier's back-plane inside a cabinet. These networks are usually buses, frequently circuit switched, but they could use a simple packet-switching paradigm. For example, the System 75's TDM bus, described in Section 14.3.2, circuit switches in each timeslot 6 – 255, but packet switches during timeslots 0 – 5. The workstation in my office contains such a standardized inter-board bus. (OK, I admit that "junction area network" isn't a good name, but I needed a "j-word." At least, I spelled it correctly; consider the next one.)

Kubicle Area Networks. While uncommon so far, many offices and cubicles almost have enough instruments that they might be interconnected within the cubicle, especially if the instruments have a common (ATM or ISDN?) interface. While current KANs are just simple wiring that provides many-to-one connections between instruments and jacks, future KANs might actually switch, probably as simple subnets of the LAN the office connects to. I recently installed an Ethernet hub in my office so that the single Ethernet cable that enters my office could be extended beyond my workstation to also connect to a separate port where I can dock my laptop computer, to a third port for my GetSet teleterminal (Section 15.6.2), and to two PCs in an adjoining graduate student office. While I frequently move data among my workstation, my laptop, and my GetSet, right now these data must transport to the floor-wide LAN outside my office to be switched. I don't have a television in my office and my telephone is still wired conventionally to a PBX, but these things could eventually change. My den at home is getting close to requiring the same connectivity as my office.

Local Area Networks. Networks in this layer cover a small building, like a residence, or a part of a large building, like a floor or a professional suite. In the typical residence environment, this layer covers the house wiring, typically two separate nonswitched networks — a cable of twisted pairs for telephone and a coax or twin-lead for television. These nonswitched infrastructures provide many-to-one passive paths between instruments inside the home and passive ports external to the home. In the typical enterprise environment, this layer covers the floorwide LAN and the building's backbone, typical of computer networks in the typical building of a typical enterprise. Unlike residential LANs, enterprise LANs do a lot of switching (usually, it's packet switching) because, unlike in the residence, data moves among the instruments attached to an enterprise LAN. Also falling within this layer are pick-up groups, hubbed off call directors or key systems, in the typical enterprise's voice network.

Metropolitan Area Networks. In telephony, this layer covers the loop plant in a community, and the corresponding wiring in a campus or large building, that connects residences to switching offices and PBX extensions to PBXs. For television, this layer is the network of coax and fiber that connects the CATV system's head end to the residence. In the typical enterprise's computer network, this layer describes the switched computer network (typically running token ring, FDDI, or ATM) that covers a campus, large building, or an industrial park.

National Area Networks. This layer covers the PSTN's interexchange networks, the LEC's intra-LATA networks, and the IXC's inter-LATA networks. It also covers an enterprise's various nationwide private and virtual-private networks for voice or for data and their integrated nationwide voice/data virtual private networks, typically running SMDS, frame relay, ATM, or IP. This layer also describes America's early quasi-public ARPA-Net, used exclusively for computer networking. This layer typically requires expensive transmission infrastructure, with all three forms of switching at many of its nodes. An exception, and a network that skips several of the layers above, is a one-way-only physical-star broadcast network implemented by a satellite in a geo-synchronous earth orbit (Section 19.3). The satellite's footprint can be as large as a nation and the network uses selection, but no switching. Television broadcast to the head end is to a MAN, television broadcast to a residential dish bypasses the MAN and goes directly to a LAN, and data broadcast to a pager bypasses KANs through MANs and goes directly to the instrument's JAN.

Orb Area Networks. Oceans and large deserts are large natural barriers to national networks. They have few internal sources or destinations of data. Like the open areas between buildings on a campus, the transmission systems that reside there only serve to interconnect switched networks — they don't do any switching themselves. Similarly, transoceanic links interconnect NANs that serve those countries on the ocean's rim by very few transmission links that are extremely expensive. This layer covers global networks, like the Internet, comprised of a mesh of NANs into an OAN. This layer also includes systems of satellites in low earth orbit, which may skip many of the above layers (Section 19.3).

(Inter)Planet Area Networks. While orbiting satellites and space stations are really

just part of the earth's OAN, we also have some limited communication to the Moon, Mars and other planets, and spaceships (*Viking* missions). Eventually, our earth's OAN will be to a PAN like a nation's NAN is to the OAN.

Well, that was fun. But, there was more purpose to this exercise. It's clear that we wouldn't necessarily use the same topology, switching paradigm, and protocols in a PAN that we might use in an IAN. But we probably would consider using the same topology, switching paradigm, and protocols in a JAN that we might use in an IAN; in a KAN that we might use in a LAN; or in a PAN that we might use in an OAN. So, where are the boundaries, the adjacent layers where we would have to consider not using the same topology, switching paradigm, and protocol on each side? The next subsection assures us that these boundaries exist.

17.2.3 Big Networks Can't Be Flat

At least two camps advocate evolving toward a broadband nationwide network that has a single-tiered nonhierarchical architecture, a flat network:

- The packet-switching community proposes to handle all types of communications by a common paradigm, either asynchronous transfer mode (ATM) or the Internet protocol (IP), as discussed in Chapter 19.
- The photonic-switching "dark fiber" community proposes to carry every type of communications over a separate wavelength-defined channel on an optical fiber (Section 21.1).

This section argues that telecommunications networks that are nationwide or global in scope cannot be fabricated economically in a single tier — they cannot be flat [1]. The discussion does not include the issue of paradigms and protocols — whether a single protocol can handle all layers of networking. The last two sections of Chapter 19 discuss such integrated networks. Here, we deal only with optimizing cost and bandwidth for users and carriers.

Except for the labels in the boxes, Figures 17.2 through 17.4 are seen to be identical. At this high level of abstraction, big telephone networks, big data networks, and big video networks have the same two-tiered hierarchical architecture. The upper tier is a single network that provides data transport over a large geographical area, possibly an entire nation or even the entire planet. The lower tier contains many separate geographically small networks, each of which provides access to the end equipment. This section refers to any network in the upper tier as a "nationwide transport network" and to any network in the lower tier as a "local access network." In between these two tiers is a network node that acts as a hub for the local access network and as a gateway for the nationwide transport network.

In all three cases, it is observed that channel utilization and interface cost are optimized differently in a local access network than in a nationwide transport network.

- It is observed that the *physical channel* in a local area network is naturally short and cheap and is a resource that may be squandered in order to optimize some more precious resource. In a nationwide transport network, the physical channel is naturally thousands of miles long, it is expensive, and it is the precious resource that must be optimized by the architecture.

- It is observed that a local access network's success depends heavily on its *cost of access,* particularly for the interface equipment directly associated with the appliance (telephone interface, PC LAN card, or CATV set-top box) that the user must buy or rent. The most successful local access networks operate at relatively low data rates in order to keep down the cost of the transmitters and receivers in the attached appliances. But in a nationwide transport network, the interface cost is typically much higher than in a local access network because they typically operate at much higher data rates. Interface cost is squandered in order to optimize channel utilization. But, since users interface indirectly to such networks, via local access networks, the expensive interface equipment is not required directly in each appliance. Its cost is shared by all the users on the local access network by placing the interface equipment in the hub/gateway.

TABLE 17.4. Channel characteristics to be optimized [1]

Component	Local access net.	National transport net.
Channel	Short, cheap	Long, expensive
Interface	Direct, cheap	Indirect, expensive
Optimization	Appliance interface	Physical channel

These observations are summarized in Table 17.4. Because local access networks must be optimized for interface cost and nationwide transport networks must be optimized for channel utilization, critical network parameters (data rate, for example) are inconsistent. This suggests that any network structure must have at least two tiers: a local access tier with inexpensive interface and a transport tier with efficient channels.

17.2.4 Packets and AONs

Using ATM or IP universally is discussed in more detail in Sections 19.4 and 19.5, and the dark-fiber all-optical network (AON) is described in Section 21.1. Both are introduced here as examples of technologies that purport to offer flat networks.

Packet switching is the most popular proposal for an end-to-end global multimedia network paradigm. ATM and IP are competing rivals: ATM is more likely to provide the necessary throughput, but IP is more universal. As discussed in Section 19.4, ATM's great advantage over other paradigms is its potential to provide reasonable network performance for voice, data, and video traffic. While not optimized for any of these data types, it appears that ATM may be usable for all of them. This property is especially important in the local access network, enabling a single physical and logical access for all data types and eventually offering a universal interface whose cost has the potential to be driven down by consumer marketing.

Some serious questions about ATM's ability to provide acceptable performance in a nationwide transport network remain unanswered. Perhaps the most serious questions about ATM, and any form of packet switching, are the related issues of scalability and a cost-effective growth plan. As discussed later in this chapter, packet switching in the nodes of a nationwide transport network can be compared to the procedure for collecting tolls on New Jersey's Garden State Parkway. It doesn't scale well because trying to increase the number of cars will cause even more congestion at the distributed toll gates. The physical medium of choice for nationwide transport, optical fiber, has a natural scalability and growth plan in its inherent bandwidth and the advances in the technology for exploiting this bandwidth. A photonic-switched nationwide transport network is proposed in Section 22.7.

All-optical networks have been proposed as an end-to-end wide area multimedia network paradigm by the research community in the last couple of years. The great advantage of AONs, and also guided-wave photonic-switched networks, over other paradigms is their ability to provide network transparency not only for data type and format but also (and most significantly) for data rate. This property is especially important in a nationwide transport network, where it enables a wide area switching paradigm that has the same built-in scalability and growth plan as the transmission plant. While recent research on optical fiber nonlinearities (Section 21.1.2) indicates that the use of wavelength-division multiplexing may be more restricted in transport networks than was anticipated by the AON architects, recent advances in optical time-division multiplexing (Section 21.1.1) suggest an optimistic future for all-optical transport networks that use some form of space-time-wavelength photonic facility switching.

The most serious concern about optics in the local access network is the cost of the receiver and, especially, the transmitter that must be placed in every office and in every home. This concern is exacerbated in a single-tier end-to-end AON, where the receivers and transmitters in the end nodes either must be wavelength-agile (in WDM systems) or must operate at the transport "rate du-jour" (in TDM systems). In either case, the users' cost seems unavoidably prohibitive.

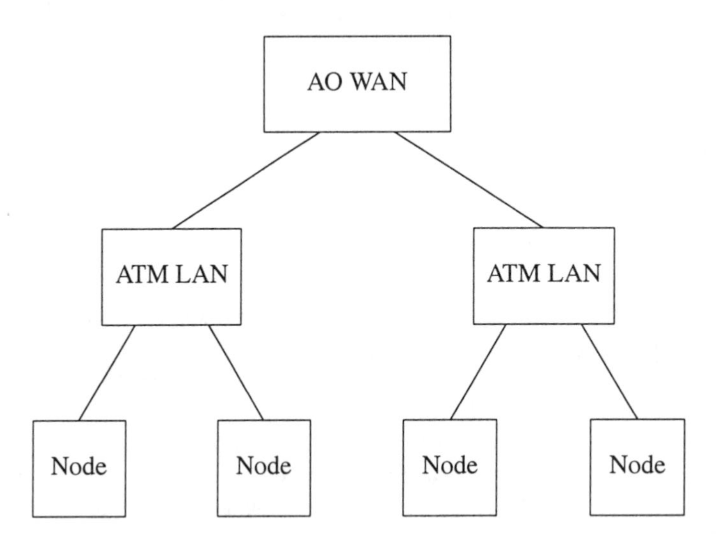

Figure 17.5. A big broadband network [1].

Summarizing then, it is argued here that:

- Packet switching (particularly ATM) makes sense in the local access network, but not necessarily in a nationwide transport network;
- Photonic switching (but probably not AONs) makes sense in a nationwide transport network, but not necessarily in a local access network.

These arguments do not diminish either paradigm because it was argued earlier that big networks must have at least two tiers anyway. Figure 17.5 suggests a two-tiered big network that uses ATM in the local access network and uses photonic switching in the nationwide transport network — and Chapter 22 develops this further. The interface between them is some kind of gateway that performs (digital) electronic-to-optical conversion, time and wavelength multiplexing, and an assignment of packets on the access side to channelized facilities on the transport side. The all-optical wide area network would be neither packet nor circuit switched, but would be some extension of facility switching, which I call capacity switching (last section of this chapter), in the space, time, and wavelength divisions (Chapter 22).

This section has argued qualitatively that telecommunications networks that are nationwide or global in scope cannot be fabricated economically in a single tier or from a single fabric; they cannot be flat.

17.3 PACKET AND CIRCUIT SWITCHING

This section defines and qualitatively compares packet switching and circuit switching. After summarizing several different kinds of switching, working definitions are given for packet switching and circuit switching. Then, after summarizing the conventional wisdom, the two switching paradigms are compared against automobile traffic and with regard to how they implement signaling.

17.3.1 Introduction

Packet, circuit, and facility switching are sometimes viewed as competing paradigms or as a hierarchy of paradigms. While some people view circuit switching as a subset of packet switching, most people view packet and circuit switching as competing paradigms, both supported by facility switching. Section 17.5 argues that packet, circuit, and facility switching can be viewed as a spectrum of paradigms. We begin discussing the relationship between circuit and packet switching by viewing them in a hierarchy, with bit switching added to the mix. We'll discuss how they all relate to facility switching later.

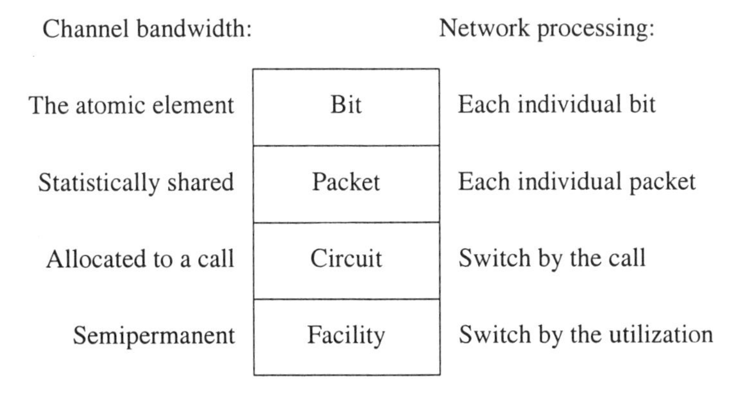

Figure 17.6. Comparing packet, circuit, and facility switching.

A *facility* is a set of channels, some of which are latent. As discussed in Section 12.5, the nonlatent channels represent a semipermanent super-channel with enough bandwidth to carry the required packet traffic, or they represent a semipermanent pool of channels that may be used temporarily as circuits. Latent channels are activated or active channels are deactivated, in a process known as facility switching, according to the utilization of the super-channel or pool. Facility switching is described in this context in Section 17.5.

A *circuit* is one channel within a facility, that is borrowed for the duration of a call or session. The circuit is a dedicated data path between two communicating parties. Data that either party transmits into the circuit is received only by the party at the other end. Two communicating parties are connected to the circuit, in a process known as circuit switching, during a setup interval at the beginning of a call. While a circuit can be designed to be optimal for a specific type of communication, whether the communicating parties underutilize the channel or fill it to capacity, they still pay the same for it.

A *packet* is a molecular chunk of stand-alone data. Every packet that ends up in some network must somehow make its destination known to this network. Because a circuit's destination is determined during setup, raw packets can be transmitted into circuits. In the absence of circuits, the destination must be attached to each packet, and then the network routes the packet by a process known as packet switching. While the user doesn't use any network resources beyond the duration of the packet, each individual packet must be separately processed. Packet switching and circuit switching are each optimal, but for different types of data communication.

A *bit* is an atomic piece of data. Showing this smallest element completes the hierarchy. In switching equipment and in transport equipment, we physically switch by the bit whenever a switching device changes state because of the value of a data bit. But, we don't logically switch by the bit in switching equipment, per se. While multiplexing equipment from the asynchronous digital hierarchy (DS2, DS3, etc.) interleaves data by the bit, one advantage of the synchronous digital hierarchy (SONET) is that it interleaves by the byte.

In a digital transmission paradigm, when A transmits data to B, this data has the form of a stream of bits. But, there is usually more structure to this stream, brought about by the nature of the channel, its user, and the data being transmitted. For example, in a simple point-to-point digital transmission from terminal A to host B, each time a key is tapped at A, the binary code (ASCII) for this key is placed inside a 9 – 12-bit digital word and this packet is transmitted at 1200 (or whatever) bps. If a dedicated point-to-point channel is expensive compared to other system costs, then permanently wiring A to B may be wasteful for three reasons:

- The channel carries no data during the short intervals between keystrokes;
- The channel is unused for long intervals if user A is not typing or not even working;
- The channel may become inadequate as A's connectivity and data rate requirements change.

These three points are addressed separately by packet switching, circuit switching, and facility switching, respectively.

17.3.2 Packet Switching

If some network of dedicated terminal-to-host point-to-point links is replaced by a bus-structured LAN, then this bus is statistically multiplexed, some protocol is employed to take care of collisions, and the packet format is expanded so that destination address is included with the data inside each packet. Packet switching is the generic process within the network by which each individual packet, which is transmitted into the network, is received by the correct destination.

Packet-switching hubs were implemented initially in software on general-purpose computers. But, network latency is enough of a problem without adding the huge delays characteristic of computer operating systems. Latency is improved by using a virtual circuit paradigm instead of the more classical connectionless datagram. Microprocessor-based hardware provides software control without operating system overhead, but the best speed-up comes with hardware-based fast-packet switches. But, as discussed back in Section 7.5, virtual-circuit packets switch faster when they are routed by special-purpose hardware. These packet fabrics frequently have the form of self-routing networks, which are reminiscent of the old Step-by-Step (Chapter 6) electromechanical telephone office equipment.

Regardless, if the packet's rate is uniform and size is constant, it may be more efficient and economical to define a channel based on the resulting uniform bit rate. If the network controller assigns one end of such a channel to a particular source X and the other end to a particular destination Y, then all the data that is transmitted into this channel by X would be automatically steered to Y, without requiring that the network examine each individual packet.

For example, in digital voice, an analog voice signal is sampled 8000 times per second and each analog sample is converted into one 8-bit digital word. If each 8-bit word is transmitted at the sample rate and each word is transmitted bit-serially, the result is a uniform 64-Kbps digital voice channel. In a T1 system, the 8-bit words from 24 different sources are accumulated 8000 times a second into a format called a frame, and one entire frame is transmitted at the sample rate, requiring 1.5 Mbps. Because the frame format has 24 timeslots of 8 bits each, each timeslot supports one 64-Kbps channel.

17.3.3 Circuit Switching

In a point-to-point network without packet switching, all the data transmitted by X into its assigned channel must still be delivered to Y. But, sometimes X wants to transmit to somebody other than Y and sometimes X is completely idle. So, the channel is not permanent — it is set up, usually at X's request, at the beginning of a call (also called a session) during which X transmits to Y, and only to Y. A circuit is such a temporary channel, which is connected for the duration of a call. Circuit switching is the process by which circuits are temporarily connected to two intercommunicating parties for their exclusive use.

Two precious telecommunications resources are (1) channel bandwidth and (2) network processing. Packet switching squanders network processing to optimize channel bandwidth. Packet switching is recommended whenever a source transmits packets with random start times, with random destinations, or that require a short packet-length latency. Packet-switching networks are sized according to statistics of packet size and arrival rate, and the indication of an overloaded packet network is that its packet queues overflow and packets are lost.

Circuit switching squanders channel bandwidth to optimize network processing. Circuit switching is recommended whenever a source transmits a uniform stream of many packets to the same destination. Circuit-switching networks are sized according to statistics of call duration and request rate, and the indication of an overloaded network is call blocking.

17.3.4 Conventional Wisdom

When packet switching and statistical multiplexing first appeared on the scene, you couldn't go to a conference without hearing some panel of experts debate the relative virtues of, and tradeoffs between, packet and circuit switching. The first question was, and continues to be, whether the cost and performance penalty from processing every packet is or is not compensated by the savings from the efficiency of statistically multiplexing bursty data. The second question was, and continues to be, whether the higher cost of the user interface (assuming packet interfaces cost more than circuit interfaces — and they may not) is or is not compensated by the lower cost of networking (assuming that a packet network costs less than a circuit network — and it may not).

TABLE 17.5. Properties of circuit and packet switching

Property	Circuit switching	Packet switching
Channel allocation	Dedicated	Shared
Overload's effect	Blocking	Congestion
Bandwidth required	Fixed	Dynamic
Overhead expended	During call setup	With every packet
Delay	Minimized	Large, random
Data arrive in order?	Always	Not necessarily
Error control	End-to-end	E-to-e and link layer
Flow control	No problem	Problematic

Some differences between packet and circuit switching are obvious. Table 17.5 summarizes the many conventional differences between circuit switching and packet switching. The relative costs of packet interfaces and circuit interfaces became moot because of VLSI technology. The components of both good and poor architectures can be implemented in VLSI and, thereby, can be made to have a low cost.

Neither the cost nor the performance penalties caused by processing every packet were ever appropriately measured. The savings caused by the greater bandwidth efficiency of statistically multiplexing bursty data depends on the data's burstiness, an elusive quantity. Nobody ever verified that packet switching costs less than circuit switching, even for moderately bursty data (whatever that means). The debate was never truly settled and the conferences stopped having them; it seemed like everybody lost interest. And voice and data networking each went its own separate way. The telephone industry continued doing circuit switching and the computer networking industry continued doing packet switching (with the notable exception of the dial-up modem).

Summarizing, the conventional wisdom is as follows:

Circuit switching is thought to be better for regular data streams.
Packet switching is thought to be better for bursty applications.

The remainder of this section continues this qualitative argument. The next section shows quantitatively that this conventional wisdom, or instinct, is correct. We'll add facility switching to the mix in the last section.

17.3.5 Lessons from Automotive Traffic

Telecommunications traffic is not perfectly analogous to automotive traffic, and the telecommunications infrastructure is not perfectly analogous to roads, which is the infrastructure for automotive traffic. But, the analogies are close enough that the telecommunications industry's systems engineers and architects might learn some lessons from our experiences while commuting to work and driving on long trips.

Transmission. Many different types of automotive traffic use our roads. But most joggers and most people who drive cars, trucks, buses, and bicycles wish that they had roads that were dedicated to their own particular type of traffic and that the others had to use other roads. Some traffic types go slower than other types. Some types of traffic make frequent stops, and some types of traffic want to go point-to-point without stopping. Some traffic types tolerate congestion because they spend so little time on the road or because they can weave between the other vehicles, while other traffic types cause congestion. For example, stopping all traffic in both directions when a school bus stops regards the unpredictability of the *pedestrian child* traffic type but disregards the impatience of the morning commuter.

We can't afford to give up the economies of scale and scope by providing separate roads for each traffic type, like we do when we build a railroad track or a subway. But, we do go partway. We provide bike paths, as much to provide safety for bicyclists as to remove the automobile congestion that bicycles cause. We provide busways, as much to remove the automobile congestion that buses cause as to provide efficient bus transportation that might attract more riders, further reducing automobile congestion. We provide bypasses and beltways, so through traffic can avoid the congestion of driving through a city or a business district. And, we provide limited-access superhighways, so that long distance drivers are not impeded by local drivers.

Another way that we provide separate paths for different traffic types is by allocating lanes (channels) in the same road (facility). The most obvious way we do this is by separating automotive traffic by its direction of travel. We realize how important this is to traffic flow whenever we have to take turns using a lane (ping pong) because one lane of a two-lane road is impassable. We prohibit trucks from using the left-hand lane of some highways because of the congestion they can cause, and we prohibit cars that carry only one person from using high-occupancy vehicle lanes because we want to try to change people's behavior.

What does this suggest about telecommunications traffic? While we may need to have a common infrastructure for all traffic types, it may be appropriate to separate different traffic types into different channels. Just as the network of roads that trucks can use is a logical subnetwork of the total automotive infrastructure, the network of channels that voice can use, or that packets can use, is a logical subnetwork of the total telecommunications infrastructure. Should we integrate them further? See Chapter 19.

Switching. After having compared the transmission of automotive traffic to telecommunications traffic, this subsection compares the switching of these two kinds of traffic. A street intersection is to automotive traffic as a switching office (or packet-switched router) is to telecommunications traffic.

Consider a four-lane street with two lanes in each direction. As you drive along, you approach an intersection with a traffic light; the light is red and vehicles are queued up in both lanes. If you want to go straight and the lanes are not marked, which lane do you get into? You might get behind the shortest line, but that won't necessarily be the quickest line through the intersection. You probably avoid the left lane (assuming we drive on the right) if you see cars in this lane with directional signals indicating a left turn, unless the light has a leading left-turn arrow. You probably avoid the right lane if you see a bus in this lane that has not yet reached a bus stop on the corner. If you see both, you may have to wait another cycle of the light.

This scenario illustrates the problem when we try to switch different types of traffic in the same fabric. We try to optimize too many things that compete with each other. On a one-way street, without oncoming traffic, left-turning traffic causes no more congestion than right-turning traffic. Even on a two-way street, left-turning traffic wouldn't impede any other traffic if we had a third lane; left-turning vehicles would congest only other left-turning vehicles. We place our bus stops on street corners in an effort to conserve parking spaces and to allow buses to load and unload when they have to be stopped at a light anyway. But buses wouldn't congest intersections if we moved our bus stops to the middle of a block instead. You can't simultaneously optimize bus traffic, through traffic, and parking spaces.

If intersections are analogous to switches, then city driving is analogous to packet switching. Each car is individually switched at each intersection. Wouldn't your morning commute be much more pleasant if you could reserve one timeslot on all the streets on your route to work before you ever left your home in the morning? Furthermore, because you get the timeslot for your car regardless of how many passengers you carry, would you feel like you wasted something if you drove alone? Such a reservation is analogous to circuit switching, switching by the call. I would like to be able to drive to work this way.

Collecting the Toll. New Jersey has two major toll highways: the New Jersey Turnpike and the Garden State Parkway. These two highways illustrate the two alternative techniques for collecting the toll.

- On each access ramp of the New Jersey Turnpike, before getting onto the highway, a driver gets a ticket that indicates the entry point. After getting off the turnpike, at a toll barrier on the access ramp, the driver submits the ticket and pays the particular toll for his route.
- On the Garden State Parkway, toll barriers are distributed along the highway. At each toll barrier, the driver slows and pays an identical toll.

Consider the advantages and disadvantages for each system. The parkway is more easily automated than the turnpike because every car pays the same toll, in coins, at each barrier. But the turnpike, because the toll booths are manual, can more easily differentiate the toll by traffic type. Because of the fixed locations of the parkway's toll barriers, vehicle A might drive 30 miles and pay one toll, while vehicle B might drive 15 miles and pay two tolls. So, the turnpike's method provides better quantization fairness. Still relevant to the analogy in this subsection, many toll roads (of both varieties) now offer new techniques so that subscribing drivers may pass quickly through the congestion caused by other drivers. But, most important, as anyone who has ever driven from northern New Jersey to the Jersey shore knows, the parkway's method congests badly under overload conditions and the turnpike's method provides a more scalable throughput.

Scalability. Scalability is an important consideration in designing tomorrow's telecommunications network. Consider two alternative ways to design an intersection and how each scales with increasing traffic load. Consider the intersection of Smallville Road, a lightly trafficked rural road, with State Route 999, a more heavily trafficked highway.

- In a perpendicular *signaled* intersection, traffic on 999 congests whenever a vehicle on 999 slows to turn onto Smallville Road. Worse, all traffic on 999 is periodically stopped by a traffic light, which allows vehicles on Smallville Road to get onto 999 or to cross it. Traffic on both roads is dramatically affected by this intersection.
- In a *bridged* interchange, vehicles exiting 999 can decelerate in an exit ramp and vehicles entering 999 from Smallville Road can merge with 999 traffic using an entrance ramp. Traffic on 999 that wishes to continue past the interchange simply zooms over the bridge that crosses Smallville Road, and traffic on Smallville Road that wishes to continue past the interchange simply goes under the 999 overpass. Traffic on both roads is minimally affected by this intersection.

If the traffic volume increases on both roads, we must increase the net vehicle capacity through the intersection. One way to scale the system is to raise the speed limit on both roads. While the bridged interchange scales nicely this way, the signaled interchange does not. We can't always scale a telecommunications network by increasing its data rate, but for different reasons, depending on the network's switching paradigm.

- An old-fashioned analog circuit-switched network with electromechanical crosspoints would scale this way as long as the increased data rate is within the analog channel's bandwidth. Raising the end-to-end data rate would be a graceful way to scale such a network.
- However, a conventional digital circuit-switched network does not scale this way because its time-multiplexed channels are predefined at a given data rate. Using this method to increase the digital PSTN's throughput by 10% requires raising the rate on each DS0 channel to 1.1×64 Kbps = 70.4 Kbps. Since such a change would be difficult to implement and expensive, raising the end-to-end data rate is not a graceful way to scale such a network.
- Packet-switched networks don't scale this way because each packet is individually processed in each router in the network. The routers would congest under this kind of scaling just as the signaled intersection above would congest. Because increasing the packet rate requires upgrading all the network's routers, raising the end-to-end data rate is not a graceful way to scale such a network.

The key difference among these three networks is that, in an analog network, its circuit-switched channels are transparent to data rate. Such a telecommunications network scales by raising end-to-end data rate just as nicely as the bridged interchange above scales to increased speed limit. We'll investigate this concept in Section 17.5 and see that it applies to certain kinds of photonic-switched networks in Chapters 20 and 21.

17.3.6 Headers and Signaling

In Section 1.1.2, switching was defined as "connecting channels for telecommunications." When multiplexing was described, we didn't formally define the term and we avoided the careful presentation of the difference between multiplexing and switching. Multiplexing is a kind of circuit switching, where many low-rate channels on one side of the multiplexor are assigned to subchannels on the other side that are part of one high-rate channel. But there's more to the distinction and a clear definition is elusive.

Some people would define these two terms, "switching" and "multiplexing," based on the OSI reference model. Within this framework,

- Switching is connecting channels in the network layer for the purpose of routing data;
- Multiplexing is connecting channels in the link layer for the purpose of selecting a path in the given route.

Using this distinction, in space-division circuit switching, selecting a trunk group for a call is switching, but selecting a particular trunk within the trunk group is multiplexing. I'm not particularly tied to the OSI stack, so I have defined switching more broadly. In this book, multiplexing is just a primitive form of switching that doesn't include a routing decision.

Similarly, earlier in this section we gave some working definitions of packet switching, circuit switching, and facility switching, but the distinctions are still fuzzy. While we recognize intuitively that these switching paradigms are different, it is very difficult to formally define what the differences are. This brief section suggests (but does not develop) the possibility that the significant difference between packet switching and circuit switching is in their signaling.

Signaling is the internal information (destination address, source address, class of service, etc.) that controls the user's information. Because signaling directs the switching nodes to establish the information's path, signaling occurs when the information's path is being established and signaling recurs for each temporal unit of the sequence of information. That is, signaling occurs at the beginning of the holding time. So,

- Signaling for facility-switched connections occurs at the beginning of each "lease";
- Signaling for circuit-switched connections occurs at the beginning of each call;
- Signaling for packet-switched connections occurs at the beginning of each packet.

Consider using *frequency of signaling* to characterize the respective switching paradigms. Virtual circuit switching is clearly within the paradigm of packet switching because, while some signaling occurs at the setup of the virtual call, the information in the virtual call would never reach its destination unless signaling also occurs with each individual packet. Time-assignment speech interpretation (TASI) is clearly within the paradigm of circuit switching because, while some signaling occurs during temporal breaks in the call's flow of information, this signaling does not occur with each byte. The call is broken into subcalls and TASI signaling occurs at the start of each subcall. Also, TASI signaling is confined to the TASI link and is not end-to-end. Because characterizing the switching paradigms by their frequency of signaling assigns these hard-to-categorize switching procedures to the paradigm that seems most intuitive, this characterization is very compelling.

It is also tempting to distinguish the paradigms, not by the frequency of their signaling, but by the implementations of their signaling. We observe that because packet-switched data units (packets) have headers, they can carry the signaling right in the packet, and because circuit-switched data units (bytes) do not have headers, they can't. But historically, circuits used to have a format that is completely analogous to a packet's format: the signaling occurred inside the circuit and at the beginning of the call. In circuit switching, this was called inband signaling. While packet switching has used this in-packet signaling implementation for its entire relatively brief history, signaling in circuit-switched networks has changed implementations several times. While many of these implementations have been described in several previous chapters, the next chapter fully describes the evolution of circuit-switched signaling from inband, to out-of-band, to common channel, to networked. So, the implementation of signaling is not as good a basis for defining the differences among facility, circuit, and packet switching as the frequency of signaling is.

This distinction, however, between the implementation of signaling and the frequency of signaling, begs a question. Perhaps out-of-band, common channel, and networked signaling could be modified for packet switching. Perhaps one or all of these signaling implementations might be appropriate in this alternate switching paradigm. The telephone industry learned many years ago that inband signaling is not a good way to do signaling in circuit-switched networks. Might not this lesson also apply to packet-switched networks? Does it make sense to transmit packet headers over one channel and the packet's body over a different one?

17.4 OPERATIONAL DOMAINS FOR CIRCUIT AND PACKET SWITCHING

While this section reopens the old circuit/packet debate, it tries to make the discussion scientific, instead of religious, by comparing the resources required when channels with a given size × utilization are circuit switched or packet switched [2,3].

The first subsection gives an initially fuzzy model for the required resources in circuit-switched and packet-switched networks and presents a two-dimensional size × utilization specification space. The second subsection shows how a boundary is found in this space such that circuit-switched networks need fewer resources than packet-switched networks on one side of the boundary and vice versa on the other side. The third subsection adds a simple model for packet overhead, finds a more accurate boundary, and compares it against the boundary drawn in the second subsection. The fourth subsection formalizes required resources as a measure that is a function of cost and several network performance parameters. The fifth subsection argues that the *specification space* must be much larger than only two dimensions, and many other candidate dimensions are proposed. Finally, the sixth subsection proposes evaluating other switching paradigms, like facility switching, and various categories of packet switching, like connectionless and connection-oriented, and ATM's VBR and CBR.

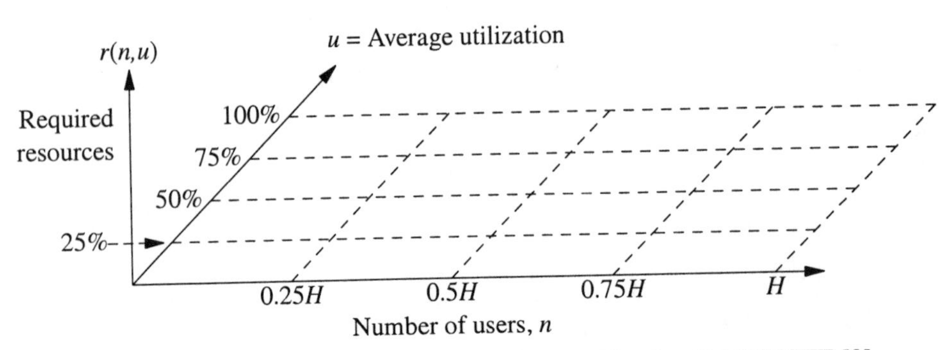

Figure 17.7. Required resources versus size and utilization © 1996 IEEE [2].

17.4.1 Network Operating Domain

Circuit-switched networks and packet-switched networks are compared objectively and analytically as surfaces in a three-dimensional evaluation space. The total resources required by some network's operation is plotted as a function of two network operational parameters: the number of users, n, and these users' average utilization, u. These parameters form a two-dimensional operational plane, $<n,u>$. Required resource is the third dimension, the dependent variable in this evaluation space. Twenty-five percent utilization means that the average capacity of all the channels in the network is only 25% used. Figure 17.7 shows lines in the u dimension representing this utilization, and also 50%, 75%, and 100% utilization. The network is assumed to behave poorly when some large number of users, H, try to use the network at 100% utilization. H and "poorly" are defined later. Figure 17.7 shows lines in the n dimension representing this number of users, and also 25%, 50%, and 75% of this number.

Surfaces will be developed that represent the resources required for operation of circuit-switched and packet-switched networks. The intersection of the surfaces for circuit switching and for packet switching defines two domains in the operation plane. The analysis will show that circuit switching uses fewer resources than packet switching in a domain that has many users, all at high average utilization, and that packet switching uses fewer resources than circuit switching in a domain with fewer users or with lower average utilization.

The resource required during a network's operation includes (1) the transmission bandwidth used in all the network's links and (2) the computer resource consumed by storing data in queues by processing packets and by setting up circuits. "Required resource" is an admittedly intuitive measure of the amount of a network's total capability that is actually used, consumed, allocated, or active during its operation.

In circuit-switched networks, the required resource covers the total channels that are allocated to calls, even if these channels are not fully utilized. In packet-switched networks, as the aggregate packet rate gets high, many packets will experience multiple transmissions or will not be optimally routed. Such effects cause the required resource to grow more than linearly.

The total resource required in a circuit-switched network is represented by the surface in Figure 17.8. Because circuit allocation counts, and not actual transmission, the required resource depends only on the number of calls, n, regardless of how efficiently these circuits are utilized. Thus, the required resource is independent of utilization, u,

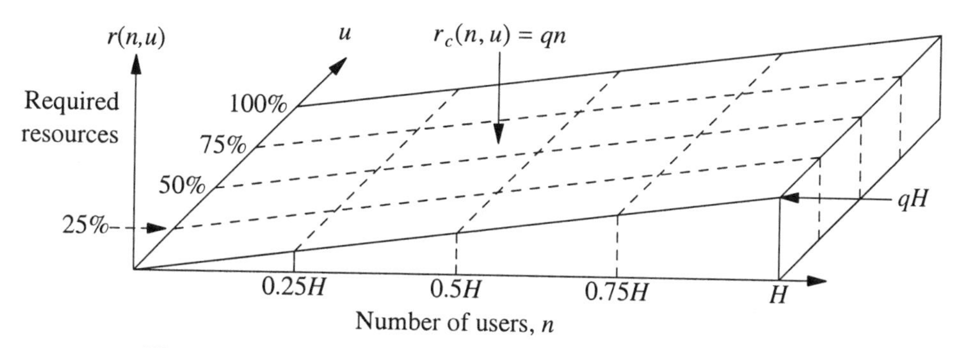

Figure 17.8. Resources for circuit switching © 1996 IEEE [2].

and increases linearly in n. This yields a wedge shape and the simple function:

$$r_c(n, u) = f(n \ only) = qn$$

The slope, q, in the n dimension represents the resource required for one of the network's standard channels. It is important to note that the cost of q has dropped significantly in the last several years.

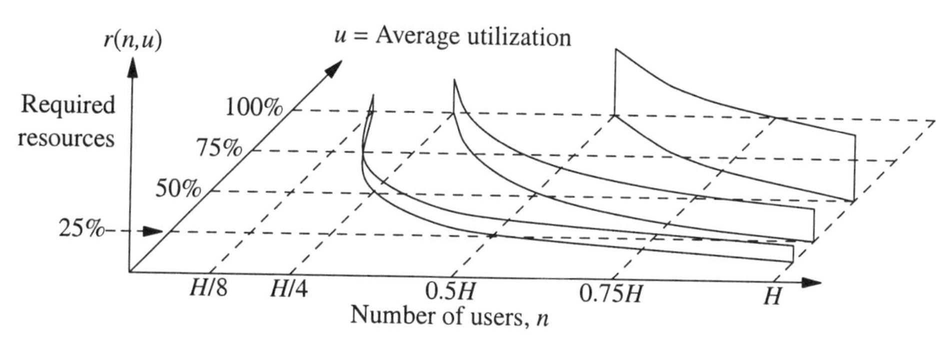

Figure 17.9. Loci of equal total load © 1996 IEEE [2].

The total load, t, offered to a packet-switched network is the sum of the packet rate delivered to, or provided by, each of the network's users. This total load, t, increases as the number of users increases and increases as these users' average utilization increases. Simply, $t = n \times u$. Figure 17.9 shows several curves of equal total load, each drawn as a "fence-line" in the <*n,u*> plane. The figure shows a "fence" constructed over each of these fence-lines. At any point on the <*n,u*> plane, the height of the fence that passes over <*n,u*> represents the resource, $r_p(n, u)$, required by a packet-switched network at the operating point <*n,u*>.

Each fence has a constant height because the total resource required by a packet-switched network would be a function only of t, the total load. As the value of equal total load increases, the required resource should increase and, so, the height of the corresponding fence increases. The fence heights increase more than linearly in t because a packet-switched network would experience greater retransmission, deflection, and queuing as t increases. If one could lay a smooth tentlike surface over these fence tops, each of the fence tops would be a contour of *equal required resource* on this surface.

This surface represents:

$$r_p(n, u) = r_p(n \times u) = r_p(t)$$

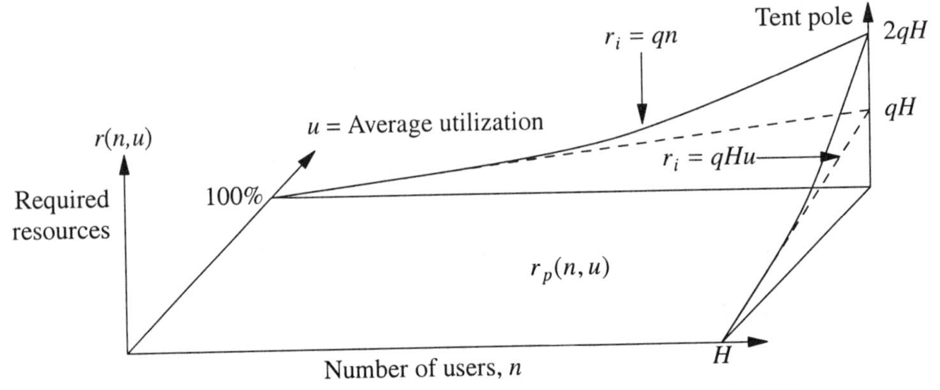

Figure 17.10. Resources for packet switching © 1996 IEEE [2].

A surface $r_p(n, u)$ is drawn in Figure 17.10 without the fences. The tentlike surface is "staked to ground" along the u axis and along the n axis. The tent is supported by a single tent pole at the point <H,1.0> in the upper right.

A different surface is indicated by the dashed lines in Figure 17.10. This surface also has zero height along both axes, but it increases linearly as n increases and increases linearly as u increases. The linear increase in n represents the resources required by an ideal circuit-switched network, in which the standard channels are defined by the users' utilization. At $u = 1.0$, this ideal circuit-switched network matches the circuit-switching wedge shown in Figure 17.8. The linear increase in u represents the resources required by an ideal packet-switched network that never needs to retransmit, deflect, nor queue a packet. The expression for this ideal network is:

$$r_i(n, u) = qnu$$

Notice that, in the ideal, circuit-switched and packet-switched networks perform identically. Practical circuit-switched networks are worse than ideal because standard channels may not be fully utilized. Practical packet-switched networks are worse than ideal because packets may be retransmitted, deflected, or queued. The surface for this ideal function reaches a maximum value of $q \times H$ above the point <H,1.0> in the operational plane. This value qH represents the resources required by a circuit-switched network that handles H users or for an ideal packet-switched network with H users, all at 100% utilization.

The resources required by a practical packet-switched network would always be greater than this ideal case. Required resources would be close to this ideal when n or u is small, because there would be few retransmissions, deflections, or data queues. Required resources would be much greater than this ideal when n and u are both large, because there would be many retransmissions, deflections, or data queues. So, the surface for $r_p(n, u)$ is greater than linear, as shown in Figure 17.10.

The surface $r_p(n, u)$ is tentlike, reaching a peak on the tent pole at <H,1.0>. Examination of Figure 17.10 shows that the peak of the ideal function, with value $q \times H$, is halfway up this tent pole. In this development, H is defined as the number of users, n, at which a practical packet-switched network requires $2qH$ resources when operating at 100% utilization. In other words, in a packet-switched network serving H users at 100%, 50% of the required resources are used for net throughput and 50% are consumed because of retransmissions, deflections, or data queues.

The development proceeds by assuming that the difference between the tent and the ideal surface is a quadratic in t. Not shown here, the coefficients of a general quadratic equation for $r_p(n, u)$ can be evaluated using the boundary conditions for the equation and its partial derivatives. The equation for the packet-switching tent is derived to be:

$$r_p(t) = \frac{q}{H}\, t^2 + qt, \text{ where } t = n \times u$$

The second term is the ideal function and the first term is the quadratic increment over this ideal.

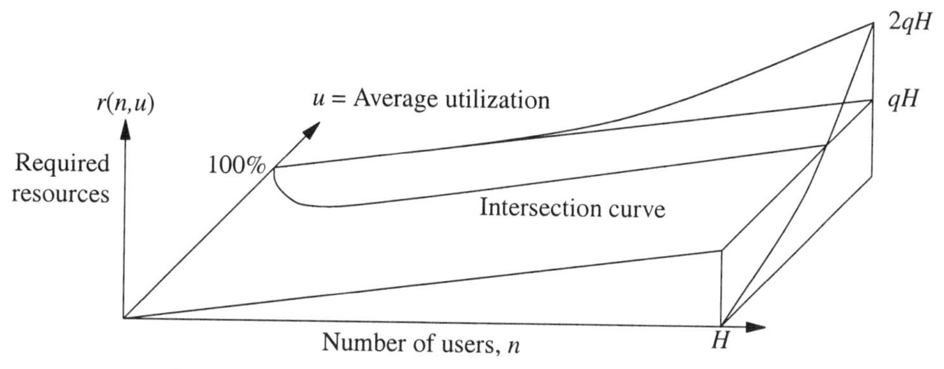

Figure 17.11. Comparing the surfaces © 1996 IEEE [2].

17.4.2 Comparison

The circuit-switching "wedge" from Figure 17.8 and the packet-switching tent from Figure 17.10 are drawn together in Figure 17.11. The tent and the wedge pass through each other to form a curve, labeled the "intersection curve" in Figure 17.11. The equation for this intersection curve is:

$$n = \frac{H \times (1 - u)}{u^2}$$

Figure 17.11 is redrawn with a different viewing angle in Figure 17.12, where the operational plane, <*n,u*>, lies in the plane of the paper. The r axis at <0.0,0.0> and the tent pole at <*H*,1.0> point straight up out of the paper. The intersection curve is shown, labeled "$r_p = r_c$," starting at <0.0,1.0>. The curve is concave up and, along the intersection curve, u decreases as n increases. Below this curve and to its left, the circuit-switching wedge rises linearly, out of the paper from left to right. In this domain, the packet-switching tent is below the wedge. Above this curve and to its right, the packet-switching tent rises out of the paper toward its peak at <*H*,1.0> in the upper right corner. In this domain, the circuit-switching wedge is below the tent.

This intersection curve partitions the operational plane, <*n,u*>, into two domains. Packet switching is more efficient than circuit switching in networks whose users operate

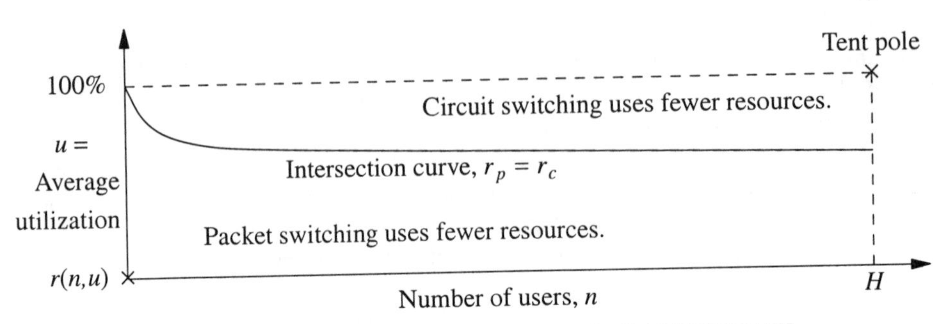

Figure 17.12. Changing the viewing angle © 1996 IEEE [2].

in the domain below and to the left of the intersection curve. Circuit switching is more efficient than packet switching in networks whose users operate in the domain above and to the right of the intersection curve.

Packet switching and circuit switching have their roles in the design, operation, and application of networks. This analysis attempts to quantify the boundary between the operational domains for each style of switching. While the notion of required resources is somewhat intuitive, the quadratic shape of the packet-switching surface may be a slight oversimplification, and the n axis is difficult to scale, the existence of these two domains is certainly true and the actual boundary between them is probably close to that derived here. Anyone considering a packet-switched network to support voice or video or any other high-utilization application must not ignore this.

One purpose of this analysis is to encourage additional research that might remove some of the intuitive nature of the analysis herein. An additional purpose is to encourage a similar analysis of a similar debate: maintaining several separate networks, each optimized for a specific type of data traffic (voice, video, file transfer, keystrokes, etc.) versus maintaining a single integrated network that is less than optimal for much of the traffic.

17.4.3 Accounting for Overhead

In circuit switching, each user receives a certain quantum of resource, independent of whether the user utilizes this resource continuously or in a bursty manner. In other words, the resource required for a call is independent of the utilization. It should be noted [3] that, in general, there is a difference between user utilization (v) and call utilization (u). The reason for this is that there is some inherent overhead in transmitting user information over the switching network. However, in the case of a circuit-switched network, once the circuit is set up, there is no extra overhead in transmitting the user's information (not counting the overhead, from higher layers of protocol, which is present in circuit- and packet-switched networks). Hence, $u = v$ in circuit switching. Assume that the minimum quantum of resources per user call is q units. The resource required for circuit-switching n calls, each requiring q units of resource and each with an average call utilization of u is: $r_c(n, v) = r_c(n, u) = qn$.

In the case of packet-switched networks, information is transferred in packets. Furthermore, unlike circuit-switched networks, each packet carries switching/routing information in it — not higher-layer protocol overhead, but the overhead needed to switch/route a packet through a network. If a packet has P total bits, with h bits of overhead, then the relationship between the user utilization (v) and call utilization (u) is:

$$u = \frac{P}{P - h} v = dv$$

Earlier in this section, the following relationship was developed for the resources required in a packet-switched network: $r_p(n, u) = an^2u^2 + qnu$, where the coefficient $a = q/H$ and H is defined as the number of users for which a practical packet-switched network requires $2qH$ resources to operate at 100% utilization. So, in terms of v:

$$r_p(n, v) = an^2d^2v^2 + qndv$$

Thus, the resources required for packet switching and for circuit switching are different. The boundary between the regions where packet switching is preferable over circuit switching and vice versa is calculated by setting $r_c(n, v)$ equal to $r_p(n, v)$:

$$an^2d^2v^2 + qndv = qn$$

From the above, the following relationship, that redefines the intersection boundary, is obtained:

$$n = \frac{H(1 - dv)}{d^2v^2}$$

As d becomes greater, the boundary curve in Figure 17.12 moves lower. So, as packets have a greater percentage of overhead, the region where circuit switching is better than packet switching increases.

17.4.4 A Better Measure?

The previous subsections compared the various paradigms using a measure called required resource. While this measure is intuitive, it is admittedly fuzzy. All we can truly say about it is that "big is bad." However, a truly scientific comparison requires a truly quantifiable measure. Two general quantities are: *cost* and *quality*. With cost, big is bad. With quality, big is good. While not a formal definition, an equation would have the general form:

$$\text{measure of badness} = \frac{\text{cost}}{\text{quality}}$$

Cost is the monetary value of a commodity as determined by net expense to produce it. *Price* is the monetary value of a commodity as determined by how much the market is willing to pay for it. Arguments can be made for using cost or price and the best measure is probably some combination or compromise. Furthermore, since a manufacturer's price is part of a carrier's cost, it is important to specify who's cost is important. Analysis from a manufacturer's or from a carrier's perspective would be different but, here, the relevant cost is the *net cost to the end user.* This is, after all, what ultimately determines the market demand.

Cost has many components, especially to the user at the bottom of the food chain, and it is important that all components be included in the analysis. Five components of cost are identified.

- *The Appliance.* The $8 telephone, the simple telephone network interface (between the house wiring and the drop wire), the ubiquitous RJ-11 jack, and the simplicity of adding extensions to analog-bus house wiring: all these factors will present a market barrier to the mass migration of voice to ATM or to IP. A significant component of the end user's cost is the price of internal wiring and jacks, various interface equipment, and especially the telecommunications instrument itself — the appliance. Delivery paradigms that are predicated on data compression (like LEOS telephony) offer lower cost for the channel, but require greater complexity in the appliance — the carriers might ignore this, but the end user won't. Because the appliance is the point of greatest replication, manufacturing economy of scale — particularly in electronics — can lower the appliance's price, but only in a mass market. The economics seem kind of circular.

- *Access.* Because the carrier has to bear the cost of bringing the network to the user, in any fair economy the user who subscribes to a network ought to pay to be connected to it, independent of how much he uses it. This is the logic behind the access fee that American telephone subscribers pay for interexchange service and this is, of course, the "missing price" in the current economics of Internet connection.

- *Call Processing.* In the United States, the regulations for the price of telephony permit carriers to charge only for the duration of completed calls. While the cost of ten 1-minute calls is much higher than the cost of one 10-minute call, the price is the same. The unrecovered cost of uncompleted calls (because the called party is busy) has created an artificial market for appliances with a repeat call feature. The signaling network's cost ought to cause a corresponding price to the end user and much of the SS7 activity occurs during call setup. Because circuits are switched at the beginning of a call, but packets are switched for every packet, ignoring this cost artificially favors the economics of packet switching.

- *Usage.* Besides the cost of call processing, the need for usage-sensitive pricing reflects another, better publicized, discrepancy between cost and price. Because there is a definite cost to the carrier for every bit transmitted over a packet-switched network and for holding time on circuit-switched calls, these also should be reflected into the cost of providing service.

- *Bit Rate.* While there is a cost for each bit transmitted, there is also a rate-based component to the cost. While 1000 circuit-switched 60-second calls at 64 Kbps each and 60,000 1-ms packets at 64 Mbps use the same facility for the same duration, it's not clear that their costs are the same.

Quality also has many components. While one's intuition might suggest that this would be complex, it may be easier to quantify than cost. Four components of quality are identified.

- *Realism.* The end user perceives a channel's realism differently, depending on whether the channel carries voice, video, or data. Voice realism is the acoustic fidelity of the channel and is perceived by how much the speaker sounds like his natural voice. Video realism is the naturalness of the time-varying image and requires high pixel density, high frame rate, and color. Data realism is the ability of a file to appear instantly. In all cases, higher bandwidth provides higher realism, but compression techniques can provide equivalent realism with channels with reduced bandwidth. Note also that increased realism requires a special appliance (like a high-definition television set) besides a special channel. Channels with high realism cost more.

- *Signal-to-Noise Ratio.* If noise power is high relative to signal power, voice crackles, video has snow, and data has bit errors. SNR can be improved by processing at the channel's endpoints — for example, with error correcting codes. SNR trades off against the channel's bandwidth. Channels with high SNR cost more.
- *Delay.* Latency between the time data enters a channel and when it exits is caused by path length and processing time. Channels with low delay cost more.
- *Reliability.* The user's access is almost always available and, when a channel is connected, it almost always remains connected. Users have historically cared about reliability, but many modern network designers are dismissing its importance. We'll return to this point in Chapters 19 and 22.

In all four components of channel quality, the scale is determined by human perception and it is not linear. For example, doubling the voice PCM sample rate and using a 128-Kbps channel would cause a noticeable improvement in voice fidelity, but using a 32-Kbps channel (either by using 4-bit PCM or 4000 samples per second) would be so poor that it would be unacceptable. Furthermore, the three components are not equally weighted. For example, delay is much more relevant to two-way voice communications than to video broadcast, and SNR is much more relevant to data communications than it is to voice or video. Similarly, regarding reliability, a network that is out of service twice a year may be perceived as extremely less reliable than one that is out of service once every 2 years.

17.4.5 The Specification Space

In earlier subsections, the specification space had two dimensions: size and utilization. While the measure may have been the fuzziest part of that analysis, limiting the specification space to only two dimensions may have been the greatest over-simplification. Many more dimensions are suggested. While they are not all equally important, and they are probably not mutually orthogonal, there are probably many more than those listed here.

- *Channel Utilization.* The measure probably depends on this parameter, the average density of bits in the channel, more than any other.
- *Size* is the number of users attached to some switch or LAN. When the number of users is small with respect to the network's capacity, statistical multiplexing performs well because congestion is minimal. Networks with a large number of users are more prone to uncertainty in paradigm tradeoffs.

- *Networking Layer.* The measure will be different depending on the position of the switch site in the network hierarchy:
    - In a data network's LAN, MAN, or WAN;
    - In the PSTN, at the level of a PBX, RSM, class-5 office, LEC inter-LATA hub, IXC POP, or IXC tandem switch.
- *Functional Layer.* Related to the physical layer above, perhaps the more significant distinction is the location of a network in a functional hierarchy, like the classical OSI stack. Different paradigms may measure differently at the physical layer, the link layer, and the network layer.
- *Location.* Related to position in the hierarchy, a different result might be expected, depending on whether the access point is inside a residence or an office building; or on whether the access point is physically near the network or is far enough to be served through a remote unit or concentrator.
- *Scope of the Calls.* Also related to the networking layer, a different measure might be expected — depending on the proportion of the calls through the network that are intraoffice or intra-LAN, interoffice or intra-MAN, or interexchange calls.
- *Application.* It may turn out that many of these differences simply stem from whether the switched channel is carrying: voice, single image, broadcast video, videoconference, keystrokes, file transfers, e-mail, web access, telemetry, or any other type of data. If so, this may suggest a separate network, optimized for each application with relevant architectural differences.
- *Participation.* The measure will also be a function of whether the interconnected parties participate in one-to-one communications, one-to-many (broadcast), one-to-some (narrowcast), or some-to-some (conference).
- *Bit rate* of the data streams. Networks carrying low data rates will probably have different measures than networks with high data rates. For example, packet switching at OC192 requires demultiplexing, while packet switching at OC3 may not.
- *Paradigm.* There certainly will be measurement differences across the analog/digital paradigms; and whether the data is is analog/digital and/or the channel is analog/digital. PCM is analog data over a digital channel, and modems carry digital data over an analog channel.
- *Physical Medium.* There could be differences depending on whether the network is fabricated from twisted pair, coax, fiber, or radio. For example, packet switching directly in optics is much less feasible than photonic circuit switching.
- *Multiplexing.* Related to the physical medium, the measure will be different depending on whether the physical medium multiplexes its channels in space, time, frequency, wavelength, or in the code division. Some forms of multiplexing are more or less amenable to packet switching.

- *Overflow Handling.* There may be measurement differences that depend on whether congestion is controlled by blocking, buffering, or deflection.
- *Compression.* Circuit-switched networks should probably measure better on uncompressed data, and packet-switched networks to measure better when data is compressed. There may be subtleties.
- *Holding Time.* Calls or sessions with long duration may measure differently from short calls. Certainly facility switching is optimized for extremely long holding times.
- *Services/Features.* The public-switched network is certainly optimized for POTS. The emerging Intelligent Network (Chapter 18) is predicated on circuit switching. As classic features from telephony are applied to video and data, packet switching and conventional broadcasting may have problems implementing them. Computer-telephony integration (Section 18.3.1) and IP telephony (Section 19.5) are opening this dimension. It may turn out that 800 service, call forwarding, and calling party identification are all easier in a packet-switching paradigm or that they may be very difficult to implement. We really don't know.
- *Age/Generation.* Electronic switches and Ethernets probably will measure differently from digital switches and ATM networks, not because of substantive differences, but because the newer systems integrate better or worse.
- *Signaling.* Significant measurement differences should be expected across networks that use differing procedures for signaling. Circuit-switched networks that still use inband signaling will undoubtedly measure poorer than those that use SS7. While packet switching doesn't have the same distinctions (inband, out-of-band, common channel, networked), there are some smaller differences (connectionless versus connection oriented) that should also measure differently.
- *Function.* Any networking paradigm will probably measure differently when it is performing different functions. Furthermore, different specialized parts of networks will probably measure differently. Examples of such functions include transmission, switching, multiplexing, selection, supervision, alerting, attending, and synchronizing — to name a few.
- *Topology.* Bus, ring, star, and mesh topologies will probably all measure differently.
- *Input Fan-out.* Another structural difference that may be relevant is how the network's inputs are fanned out inside the network. Fan-out can be accomplished by simple wiring (Section 7.1), splitting (as with coax), broadcast and select (as on a bus), and hand off (as in a ring).

17.4.6 Other Switching Paradigms

Initially, this section proposed a fuzzy measurement over a limited specification space. The two previous subsections have addressed its shortcomings. There is yet another shortcoming. The work compares only two paradigms: circuit and packet switching.

There are several categories of circuit switching and many categories of packet switching. While facility, protection, and capacity switching (defined in the next section) are similar to circuit switching, they are different enough that they need to be evaluated separately. Paradigms, like virtual circuits, are in between circuit and packet switching. Packet switching can be connection oriented or connectionless, or can be frame relay or cell relay. Even cell relay, like ATM, can have CBR or VBR. The byte- and bit-level interchange in multiplexors is a form of switching that should also be put to the same evaluation.

The industry is heading toward a layered structure that is fraught with overhead and bizarre interactions. We need analytical answers to questions like: does it make sense to have voice over IP over ATM over Sonet over optics? What is the real contribution of each of these layers?

17.5 FACILITY AND CAPACITY SWITCHING

This section adds facility switching to the previous comparisons and extends this concept to yet another switching paradigm, called capacity switching. The last subsection proposes that, rather than view the various switching paradigms in a hierarchy or as competing with each other, we might view them in a spectrum.

17.5.1 Facility Switching

Figure 17.6 showed four switching paradigms as a hierarchy. The two previous sections expanded on the middle two layers, packet and circuit switching. This section expands on the lowest layer from Figure 17.6.

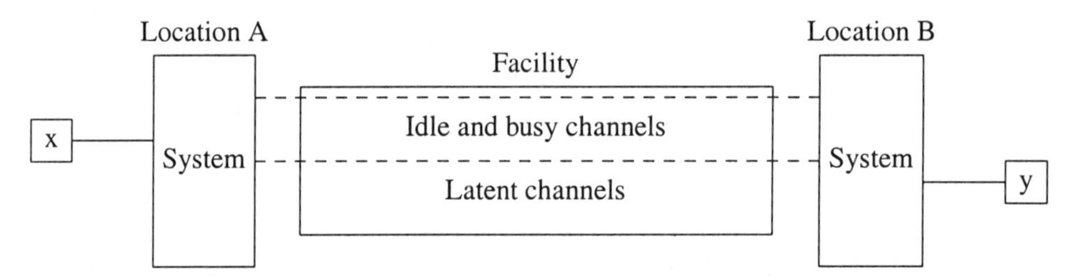

Figure 17.13. Telecommunications facilities.

A *facility* is a set of channels that can be aggregated into a single packet-bearing super-channel or that can be assigned to a pool of circuits. If a network becomes overloaded, either packet queues overflow too often or too many calls are blocked. Then, the capacity of some facility may need to be increased. For example, a circuit-switching network may have many channels (trunks) in some facility (trunk group) that interconnects two locations, A and B. While we usually think that a channel is either idle or busy, a third status is "latent," as shown in Figure 17.13. If x at location A calls y at location B, x gives y's address to the system at A and the network attempts to circuit switch x to y using some idle channel from the facility between A and B. If all the nonlatent channels are busy, the call is blocked. If too many calls between A and B are blocked, some of the latent channels in the facility would be added to the pool of usable channels, by the process of facility switching.

As discussed in Section 12.5, facility switching is a process by which channels are assigned semipermanently to augment the bandwidth of a packet link or to add channels to the pool of a circuit link. The network must monitor channel utilization and activate latent channels in overloaded links and deactivate active channels in underloaded routes. Other examples of latent channels, besides unused subchannels in multiplexed facilities, include unused (dark) fibers in deployed cables and unused optical bandwidth in fibers carrying signals at one wavelength.

17.5.2 Fast Provisioning

The qualitative and quantitative comparisons of packet switching and circuit switching tried to clarify the differences between these two paradigms. But, the net effect of this chapter has probably been to blur the distinction between these two switching paradigms and, probably, among all the paradigms. That's OK. It means there's still a lot of work to do.

While we saw some similarities between packet and circuit switching, we saw some subparadigms in the space between them. We're going to do this again. This section will show some similarities between facility and circuit switching and will introduce some subparadigms in the space between them. This subsection is a segue, taking us from facility switching toward circuit switching.

Facility switching with a DACS or ADM is used for provisioning (1) leased facilities or (2) trunks for interoffice trunk groups when interoffice traffic demands it. In either case, the holding time is in the order of months and the required switching speed is in the order of days. Section 12.5.7 mentioned protection switching as a case of facility switching that is used to reallocate channels to transmitters and receivers whenever a failure occurs. *Protection switching* occurs at the physical layer and, while its holding time is still quite long, its required switching speed is very fast. We see that, at least in terms of speed, protection switching is more like circuit switching than like facility switching. Consider provisioning on a faster schedule.

Channels, particularly long distance channels, might be reconfigured on a routine daily schedule depending on daily shifts of traffic. Since East Coast to the West Coast traffic is very light at 8 A.M. EST, people who use the telephone in the East Coast at this hour are most likely calling other people in the East Coast. As the morning progresses, East Coast callers tend to call more people gradually further west. So, while the telco (or the enterprise, if it's an EPN) certainly can't move the bandwidth around physically, they could consider reconfiguring the assignment of muldems to channels during the course of the day. Muldems might be moved so that north-south trunk groups on the East Coast might have more channels at 8 A.M.. And, as the day progresses, these muldems might be moved to east-west trunk groups. This procedure doesn't conserve bandwidth, but it might conserve muldems.

Obviously, such a procedure wouldn't be implemented by putting wheels on the muldems and physically rolling them around the switching office. A modified DACS could be used, that would be programmed using time of day. This kind of DACS would require more intelligence than just the simple command interpreter found in a conventional DACS.

What kind of switching is this? It reconfigures at the physical layer, like a protection switch, but it changes the capacity of trunk groups, like a facility switch. We could call it a fast facility switch. Now, let's take it one step closer to circuit switching.

17.5.3 Capacity Switching

Fast provisioning reallocates channels and bandwidth on a regular schedule, based on time of day. But, we could be even more clever than this. We could reallocate capacity as needed. I call this capacity switching.

- In a circuit-switching office, the switching software that monitors the utilization of trunk groups might be modified to request additional channels for some trunk group that is getting close to full capacity.
- In a packet-switching MAN or WAN, capacity switching is even easier. Because packets are queued up inside routers and gateways anyway, some program could monitor the queue depth associated with each route out of each node in the network. If a queue is getting close to overflow, this program would request an increase in the bandwidth of the packet channel that empties the queue.

Assume that (1) bandwidth is available in the facility that carries the trunk group or the packet stream and (2) unused muldems are available in the switching office or MAN/WAN node. Then, muldems could be reallocated in real time, dynamically changing the capacity of a CO's trunk groups or the data rate between two nodes in some WAN. Calls would never block and queues would never overflow unless (1) the facility ran out of bandwidth, (2) the office or node ran out of muldems, or (3) a sudden rush of calls or packets demanded capacity before the office or node had a chance to reallocate it. Capacity switching requires a new kind of fast DACS, some new programs inside existing controllers, and some increased signaling. We have to do more signaling because you can't change the bandwidth on one end of a facility without making sure that it's OK at the other end.

Tying this chapter together, Section 17.2 argued that a large network had to be hierarchical because different commodities had to be optimized at different layers. At the highest network layer in any national network, the bandwidth of its long expensive channels should be optimized. While multiplexing at the network's edges is important, capacity switching could further optimize this bandwidth inside the network. Furthermore, because capacity switching is useful for circuit switching and for packet switching, it would seem to be especially important in any integrated network that will support both paradigms. In fact, capacity switching may be the best reason to integrate circuit-switched and packet-switched networks — not under an integrated paradigm, but over an integrated network of facilities. Capacity switching could be implemented in today's PSTN or MANs and WANs. But, Section 22.7 will propose an implementation for a national capacity-switched network made from photonics technology.

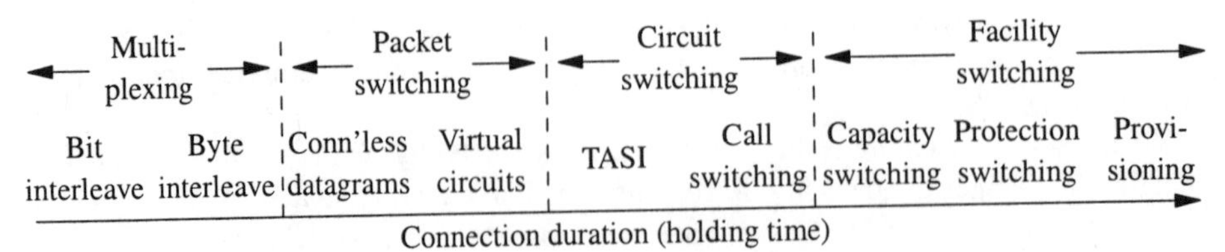

Figure 17.14. Switching paradigms as a spectrum.

17.5.4 Spectrum of Switching Paradigms

Figure 17.6 suggests that bit switching, packet switching, circuit switching, and facility switching reside in a hierarchy of switching paradigms. While this is partly true, it's not exactly accurate. The text of Section 17.3.1 suggests that packet switching and circuit switching are competing paradigms, both supported by facility switching. This is also partly true, but it isn't exactly accurate either.

Figure 17.14 suggests another relationship among these paradigms — which is also partly true but not exactly accurate. This relationship is a continuum, or spectrum, based on the duration of the connection. Obviously, we don't really have good definitions of these paradigms. So, naturally, it's hard to characterize their differences. One thing is clear — these paradigms are different. Each has its niche and its applications for which it is optimal. Remember this when we discuss integration at the end of Chapter 19.

EXERCISES

17.1 Consider a LAN that interconnects 30 instruments. (a) In a bus architecture, if the bus' line rate = 300 Mbps, compute C_t, C_p, R_p, and C_e. (b) In a ring architecture, if the ring's link rate = 300 Mbps, compute C_t, R_t, C_p, R_p, and C_e. (c) In a star architecture, if the star's link rate = 300 Mbps, compute C_t, C_p, R_p, and R_e.

17.2 Repeat the previous problem for a LAN that interconnects 300 instruments and whose line/link rates are 30 Mbps.

17.3 Make a sketch that shows how the CPU chip in your PC fits into the Internet. Sketch the IAN and JAN of your PC and identify the particular JAN. Sketch the KAN of your home or office. Sketch the LAN and MAN that connects your PC to the Internet and identify the particular products you use. Sketch the Internet as a NAN and OAN.

17.4 Consider some network that can support either VBR (packet switching) or CBR (circuit switching), but not both. Suppose that, when 1000 users send VBR packets at 100% of their utilization, this particular network loses 50% of its resources to queuing and congestion. For each of the following combinations of size and average utilization, should the network operate as VBR or as CBR?

- 100 users, each averaging 90% utilization;
- 200 users, each averaging 90% utilization;
- 400 users, each averaging 50% utilization;
- 1,000 users, each averaging 50% utilization;
- 4,000 users, each averaging 50% utilization;
- 4,000 users, each averaging 10% utilization;
- 40,000 users, each averaging 10% utilization.

17.5 Evaluate the circuit/packet boundary curve, $u = f(n)$, in Figure 17.12 for $n = H$, $n = H/2$, $n = H/4$, $n = H/8$, and $n = H/16$.

17.6 In one proposal for voice over IP (Section 19.5), an IP packet has 40 bytes of header and 10 bytes of payload. Sketch the adjusted circuit/packet boundary curve on Figure 17.12. Evaluate $u = f(n)$ for this adjusted curve for the same five values of n from the previous problem.

17.7 Develop the equation for the circuit/packet boundary in Figure 17.12 by assuming that $r_p(n, u)$ (Figure 17.10) is an exponential function of $n \times u$.

17.8 For any one of the additional dimensions proposed in Section 17.4.5, describe value along this dimension and discuss how the circuit/packet boundary surface varies from Figure 17.12 along the added dimension.

17.9 Propose some models for $r_x(n, u)$ for any one of the switching paradigms along the spectrum of Figure 17.14.

REFERENCES

[1] Thompson, R. A., "Big Networks Can't Be Flat," *Proc. 3rd Int. Conference on Telecommunications Systems,* Nashville, TN, Mar. 16 – 19, 1995, pp. 426 – 428.

[2] Thompson, R. A., "Operational Domains for Circuit- and Packet Switching," *IEEE Journal on Selected Areas in Communications,* Vol. 14, No. 2, Feb. 1996, pp. 293 – 297. © 1996 IEEE. Portions reprinted with permission.

[3] Thompson, R. A., and S. Banerjee, "Comparing the Throughputs of Circuit- and Packet-Switched Channels," (keynote address), *Proc. TeleTraffic '97,* Grahamstown, So. Africa, Sept. 8 – 11, 1997, pp. 9 – 15.

SELECTED BIBLIOGRAPHY

Adams, D., "Regulatory News — FCC Proposes Rules for Bell Company Long Distance," *Phone+,* Vol. 10, No. 12, Oct. 1996.

Alexander, S. B., et al., "A Precompetitive Consortium on Wideband All-Optical Networks," *IEEE Journal of Lightwave Technology,* May/Jun. 1993.

Boxer, A., "Where Buses Cannot Go," *IEEE Spectrum Magazine,* Vol. 32, No. 2, Feb. 1995.

Brackett, C. A., et al., "A Scalable Multiwavelength Multihop Optical Network: A Proposal for Research on All-Optical Networks," *IEEE Journal of Lightwave Technology,* May/Jun. 1993.

Chraplyvy, A. R., and R. W. Tkach, "Capacity Limits Due to Nonlinearties in Optical Networks," *IEEE LEOS Summer Topical Meeting,* 1994.

Christensen, B., "Building a Storage-Area Network," *Data Communications,* Vol. 27, No. 6, Apr. 21, 1998.

Comer, D. E., *Computer Networks and Internets,* Upper Saddle River, NJ: Prentice Hall, 1997.

"Connecting the Global Enterprise," *Telecommunications Magazine,* complete issue, Vol. 23, No. 11, Nov. 1989.

Decina, M., "Open Issues Regarding the Universal Application of ATM for Multiplexing and Switching in the B-ISDN," *Proc. IEEE Int. Communications Conference,* Denver, CO, Jun. 23 – 26, 1991, Vol. 3, pp. 1258 – 1264.

Gavish, B., "Low Earth Orbit Satellite Based Communication Systems — Research Issues," *Proc. 4th Int. Conference on Telecommunications Systems,* Nashville, TN, Mar. 21 – 24, 1996, pp. 412 – 419.

Gilhooly, D., "Unleashing the Baby Bells," *Telecommuncations,* Vol. 32, No. 2, Feb. 1998.

Green, P. E., et al., "All-Optical Packet-Switched Metropolitan-Area Network Proposal," *IEEE Journal of Lightwave Technology,* May/Jun. 1993.

Hawley, G. T., "Special Considerations for the Use of xDSL Technology for Data Access," *IEEE Communications Magazine,* Vol. 35, No. 3, Mar. 1997, pp. 56 – 60.

Jander, M., "Launching a Storage-Area Net," *Data Communications,* Vol. 27, No. 4, Mar. 21, 1998.

Jander, M., "Trading Up to Switches," *Data Communications,* Vol. 27, No. 3, Mar. 1998.

Joel, A. E., Jr., "What Is Telecommunications Circuit Switching?" *Proceedings of the IEEE,* (issue on Telecommunications Circuit Switching), Vol. 65, No. 9, Sept. 1977.

Kushner, C., "LEC Update — First Stage of FCC Regulation Complete, But Squabbles Continue," *Phone+,* Vol. 10, No. 12, Oct. 1996.

Lawyer, G., and D. Mendyk, "John McCain's Simple Agenda," *tele.com,* Vol. 3, No. 7, Jun. 1998.

Lawyer, G., and C. Weinschenk, "Telecom Act Progress Report," *tele.com,* Vol. 3, No. 2, Feb. 1998.

Livermore, E., et al., "Architecture and Control of an Adaptive High-Capacity Flat Network," *IEEE Communications,* Vol. 36, No. 5, May 1998.

McCain, J., "Telecom Reform: Avoiding the Terrible Two's," *Telecommunications,* Vol. 32, No. 1, Jan. 1998.

McDonald, J. C. (ed.), *Fundamentals of Digital Switching,* Second Edition, New York, NY: Plenum, 1990.

Mokhoff, N., "Transceiver Does Away with Central Hub in StarLAN," *Electronic Design Magazine,* Jun. 25, 1987.

Obuchowski, E. J., "Access Charge and Revenue Architecture," *AT&T Technical Journal,* Vol. 66, No. 3, May/Jun. 1987, pp. 73 – 81.

Rogers, C., "Telcos vs. Cable TV: The Global View," *Data Communications,* (issue on Global Supernets), Sept. 21, 1995.

Saruwatari, M., "Progress Toward Ultrahigh-Bit-Rate All-Optical TDM Transmission Systems," *IEEE LEOS Summer Topical Meeting,* 1994.

Sharony, J., T. E. Stern, and Y. Li, "The Universality of Multidimensional Switching Networks," *IEEE Transactions on Networking,* Dec. 1994.

Tanenbaum, A. S., *Computer Networks,* Third Edition, Upper Saddle River, NJ: Prentice Hall, 1996.

"Telecom Billing '92," *Conference on Developing Billing Systems for Tomorrow's Advanced Services,* Washington, DC, Jul. 23 – 24, 1992.

Thompson, R. A., "Standardization Didn't Help the Slide Rule," *Proc. IEEE Int. Communications Conference,* Chicago, IL, 1991.

Thyfault, M. E., "AT&T Breaks Up Again," *Information Week,* Oct. 2, 1995.

Thyfault, M. E., "Network Economy: Resurgence of Convergence," *Information Week,* Apr. 13, 1998.

Thyfault, M. E., and J. Gambon "The Telecom Bill: Why It Matters," *Information Week,* Jul. 17, 1995.

Weisman, D., "The Impact of Silicon Integration on the Communications Industry," *Telecommunications,* Vol. 32, No. 3, Mar. 1998.

"The Xerox Star — A Retrospective," *IEEE Computer Magazine,* complete issue, Vol. 22, No. 9, Sept. 1989.

18

Intelligent Network

> *"Ships that pass in the night, and speak [to] each other in passing; Only a signal shown and a distant voice in the darkness; So on the ocean of life we pass and speak [to] one another, Only a look and a voice; then darkness again and a silence."* — Henry Wadsworth Longfellow

During the last three decades of the previous millennium, America's PSTN has retained its conduit and most of its buildings, but they were gutted. With the significant exception of the loop plant, the PSTN's equipment and channels have been almost completely replaced during the last 30 years — some components more than once. During the 1970s, America's PSTN (predominantly, the Bell System) (1) replaced most electromechanical exchanges with electronic switching systems, which had programmed control but still switched analog signals; and (2) migrated many of its interoffice copper transmission facilities from analog transmission to digital transmission. During the 1980s, besides major changes in the industry's structure, the PSTN (1) replaced most remaining electromechanical exchanges with new digital switching systems (so that programmed control was practically ubiquitous); (2) almost completely evolved its interoffice copper facilities to digital transmission; and (3) started deploying optical fiber. During the 1990s, with even more changes in the industry, the PSTN (1) is now replacing many of the electronic switches with digital; and (2) continues to deploy optical fiber.

While physical changes in the PSTN's transmission and switching infrastructure have slowed during the 1990s, the PSTN's control structure has changed a great deal over a relatively short period of time. Control of the Public Switched Telephone Network has two separate, but interdependent, parts.

- *Switching control* is the human, wired, and/or programmed logic that processes telephone calls, locates and connects talking paths through switching nodes and networks, and provides various features and services. This process controls the switching functionality that provides call switching within the PSTN.
- *System control* is the human, wired, and/or programmed logic that supports the operation, administration, and maintenance of the network and that provisions its resources. This process controls the infrastructure and the distributed system that comprise the PSTN.

This book has little to say about system control or about OAM&P. This chapter describes the evolution of switching control, which has evolved synergistically with the evolutions of transmission and switching technology, the implementation of intelligence, the architectures of switching exchanges, signaling procedures and content, offered features and services, and even the definition of what a "call" is. While this chapter features the Intelligent Network, the IN is not the ultimate stage of this evolution. It is merely the current state of this ongoing evolution.

Because the evolutions of transmission and switching technology, the implementation of intelligence, and the architectures of switching exchanges have been covered thoroughly throughout this book, they don't need any more discussion. But because the coverage of signaling and offered features and services has been sporadic and cursory so far, these topics are discussed in this chapter in detail. The first section describes the evolution of signaling and the second section summarizes the evolution of switching software, including digressions on *feature interaction* and the software development process. Section 18.3 discusses *enhanced services* in general and the implementation of *800 calling* (America's term for what the rest of the world calls "green numbers") in particular. Section 18.4 describes America's current version of the IN, called the Advanced Intelligent Network, (AIN). I'll look into the crystal ball and describe my view of the future evolution of the IN in Section 18.5.

18.1 SIGNALING

This section's first subsection summarizes the evolution of signaling and its second subsection describes the SS7 network. The last two subsections illustrate the evolution of signaling by describing the evolving implementation of two different features: *busy tone* and *calling party identification.*

18.1.1 Evolution of Signaling

While signaling began as the transmission of the information that controls a call [1], distinct from the information carried by the call, signaling has come to include the transmission of information about the overall status of the network (system control). Over the past 30 years, switching control and signaling have evolved together through four generations, driving each other synergistically.

Inband signaling carries subscriber signaling, information from telephone to the switching system, and interoffice signaling, information between one switching system and another one. Inband signaling is used to transmit supervision, the on-hook/off-hook state of the subscriber's telephone. It is also used for number transmission and call progress indicators. Inband signaling is point-to-point and it uses the exact same channel as the content of the call. Inband signaling has two implementations: DC and AC.

- In DC inband supervision signaling, an off-hook telephone causes DC current to flow around the two wires of the subscriber's two-wire twisted pair telephone loop.
- In DC inband number transmission, dial pulsing causes very short interruptions of this DC current path for both subscriber signaling and interoffice signaling.
- AC inband number transmission has two different implementations. Subscriber signaling uses Dual-Tone Multi-Frequency (DTMF or "Touch Tone") to signal the

called party's telephone number from the subscriber's telephone to the switching system. Interoffice signaling uses Multi-Frequency (MF) pulsing to signal both the called and calling party's telephone number between offices.

- Progress indicators are signaled by audible AC signals that alert the user to the call's progress. These include: dial tone, ringing, audible ring tone, busy tone, reorder tone, and permanent signal "howl."

Out-of-band signaling carries the same signals and is also point-to-point. It uses a separate logical channel, which is distinct from the channel that carries the call's content. In E&M signaling, supervision is signaled by the presence/absence of an AC tone. While the signal is placed on the talk path's physical wires, it appears on both wires in parallel (with respect to ground), making it inaudible to the users and, hence, out of band.

The BORSCHT functions are seven telephone functions whose classical analog electrical implementations break down when a physical path has digital or optical segments. The S in BORSCHT represents "supervision." While the digital voice bearer channel is standardized now at

1 byte/frame = 8000 bytes/second = 64 Kbits/second

first-generation T1 systems allocated only 7 bits/frame for the pulse code modulated (PCM) voice and reserved 1 bit/frame to signal the state of supervision. This procedure has come to be called the 56-Kbps clear channel. Second-generation T1 systems use the superframe format, with six frames in a superframe. Since supervision is signaled once per superframe, instead of once per frame, the content channel can be assigned a 64-Kbps clear channel for the five frames in a superframe where supervision is not signaled. In these first two generations of T1, the supervision signaling is outside the logical talk path.

Another, more contemporary, example of out-of-band signaling occurs with calling party identification (CPI). If CPI is implemented without ISDN, a CPI display must be connected like an extension telephone to residence wiring. The called party's switching office signals the identity of the calling party to the called party's CPI display just before ringing the called party's telephone. The called party can read this display and know the identity of the person ringing his telephone before choosing to answer. The signaling occurs in a frequency-multiplexed channel on the subscriber's twisted pair loop. The signaling channel's bandpass is up around 12 kHz, within the bandpass of a typical twisted pair loop but outside the bandpass of the typical telephone.

Common channel signaling is also point-to-point, but it uses a distinct physical channel, a common data link, to carry all the signaling for all the calls that are assigned to the group of channels associated with the data link. Common channel signaling was first seen in automatic identified outward dialing (AIOD), described in Section 10.5.3, by which a PBX signals to the switching office on which it homes. It signals the identity of any PBX extension whenever it is switched to a dial-9 trunk. This way, the switching office can provide a detail bill to the extension if the user places a toll call. The common PBX-CO data link carries signaling that pertains to every trunk in the corresponding PBX-CO trunk group.

Third-generation T1 systems went to the extended superframe format, with 24 frames in an ESF. In ESF format, the sequence of T1 frame synchronization bits represents an 8000-bps channel with 24 bits per ESF. Because only half the bits are used for frame synch, the remaining 4000-bps channel can be used as a common supervision signaling channel for the 24 channels in the T1. All 24 voice channels are 64-Kbps clear channels. However, as the ESF signaling channel came to be used for more purposes than just supervision signaling, its 4-Kbps rate became inadequate.

In basic rate ISDN, the 144-Kbps physical channel is partitioned into 2B + D, where each bearer channel is 64-Kbps clear and the 16-Kbps data channel carries the signaling, including supervision changes for both B channels. Primary rate ISDN is really fourth-generation T1. Like ESF, it has 24 channels that are all 64-Kbps clear. However, one of these channels is allocated as the common D channel and the format is 23B + D. This common 64-Kbps signaling channel carries all the signaling that pertains to the other 23 channels in the facility. ISDN's D-channel signaling uses a standardized digital messaging format and a richer set of messages is available than in ESF.

Networked signaling is the logical next step in this evolution. The messaging format for point-to-point signaling is extended to include addresses, and the simple signaling messages are embedded inside connectionless packets. An entire packet-switched network, logically (but not physically) separated from the PSTN, is constructed with dedicated links and packet switching nodes. So far, this networked implementation is used only for interoffice signaling. Except for ISDN's common D channel, subscriber signaling is still implemented by inband or out-of-band techniques, even for calling party identification.

Since the first three forms of signaling use part of a channel or one of the channels in a trunk group, a switching office can "talk" only to those other offices to which it has *direct trunking*. When signaling was simple, that's all that was necessary. But now the software in one switching office must talk to the software in another switching office, to determine a called party's busy/idle status or to negotiate a route through the network or for a wide variety of other reasons. Modern signaling requires the ability for the software in any switching office to talk to the software in any other switching office in the country (indeed, in the world). So, every switching office is assigned an address, and office A signals to office B by sending a packet to B over a large packet-switched network.

The Bell System recognized the cost and performance benefits of networked signaling and was eager to implement it [2,3]. AT&T become impatient with the slow deliberations in the international standards bodies and went ahead and implemented its own proprietary protocol in the United States, called Common Channel Interoffice Signaling (CCIS). We see, by this name, that common channel signaling was not distinguished from networked signaling at the time. The Bell System installed some packet switching nodes around the United States and interconnected them by 56-Kbps clear channels, extracted from its own facilities, as an internal private packet-switched data network. When the international standard, called Signaling System 7, was finally completed, AT&T adopted it and adapted its CCIS network to conform. However, the legacy of 56-Kbps clear channels remains in the United States.

18.1.2 Signaling System 7

SS7 is an international standard that specifies a worldwide interoffice signaling network [4 – 7]. This section describes SS7's network topology and discusses the various SS7 protocol stacks. Some of the SS7 protocol will be apparent from the several call walk-throughs that will be presented in later sections.

Topology. The SS7 network is packet switched and message oriented. While the network is logically distinct from the circuit-switched and facility-switched PSTN, the SS7 network uses physical channels from the PSTN's transmission infrastructure. The SS7 network can be viewed as a logically separate network that has been provisioned within the PSTN. Since the SS7 network is "flat," having only one tier, SS7 messages are routed through many hops as they traverse the country.

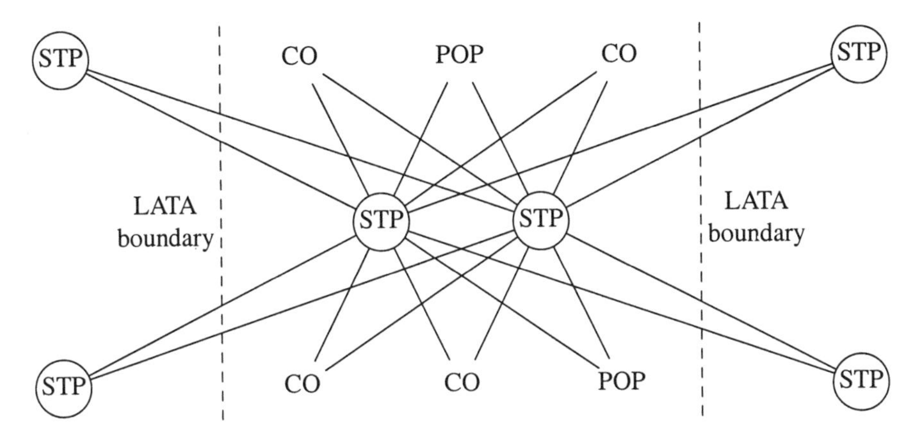

Figure 18.1. Illustrating the SS7 network.

A packet-switching node in the SS7 network is called a signal transfer point (STP). An STP is a general-purpose computer whose program performs packet-switched store-and-forward, addressing, and routing. At the time that this book is being written, the offered load on America's SS7 network requires about one STP in each LATA. For reliability and survivability, each LATA's STP is physically duplicated. A LATA's two STPs would never be housed in the same CO building, and while several SS7 links might use the same transmission facility, the links to the two STPs in one LATA are as physically diverse as possible.

Figure 18.1 illustrates. Within the LATA that an STP pair serve, each STP has a data link to every CO and every POP that is equipped for SS7 (not all COs in the United States are SS7 compatible). Each STP has data links to the STPs in adjoining LATAs. The minimum data link is specified at 56 Kbps in the United States and 64 Kbps in most other countries. If the SS7 packet traffic between a large CO and its STP is high enough to congest a 56-Kbps data link, then parallel links would be installed. Many of America's inter-STP links are such paralleled channels, called a link group.

Just like nobody owns the Internet, nobody owns the entire SS7 network. The regional BOC typically provides the STP pair in each LATA and provisions the 56-Kbps links to STPs in neighboring LATAs. These links typically use facilities leased from some IXC. For any of the other LECs in the LATA, or any of the IXCs that have POPs in the LATA, to connect to the SS7 network, they must arrange for SS7 addressing for their switching offices (following), provision channels between their switching offices and the nearest STP pair, and pay the STPs' owner for the right to use this STP pair. Some of the IXCs have their own STPs and, even their own private SS7 networks, while others lease time on LEC owned STPs and use LEC provided SS7 links (which may be leased from competing IXCs).

So, the LATA's STPs are gateways to the SS7 network in much the same way that an Internet service provider is a gateway for an end user's access to the Internet. And, the network of inter-STP links, the SS7's topology, has neither an overall plan nor a responsible architect.

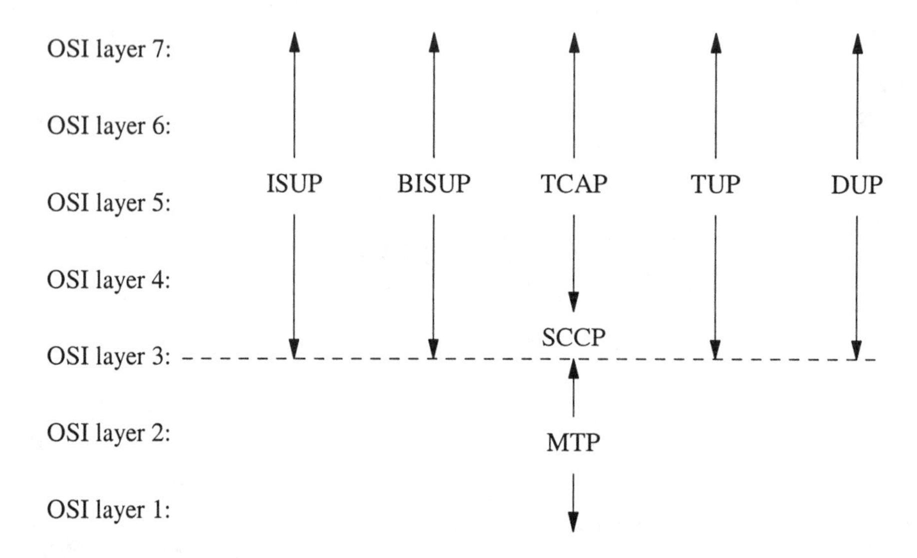

Figure 18.2. SS7 protocol stacks.

SS7 Standard Protocols. Signaling System 7 has five different standardized sets of interoffice signaling messages and their interoperating protocols. (With so many options, why are they called "standards"?) These SS7 protocols would be consistent with the seven layers of the OSI Reference Model (Section 18.3) except that (1) the functions normally found in OSI's layer 3, the network layer, are split up, and (2) the functions normally found in OSI layers 4 – 7 are merged.

The message transfer protocol (MTP) is standardized internationally for all SS7 standardized environments. MTP's three internal layers correspond to the three lowest layers of the OSI Reference Model. At the physical layer, MTP.1 is a DS0 (64 Kbps) or DS0A (56 Kbps) clear digital channel. At the link layer, MTP.2 is similar to the high-level data link control (HDLC) protocol, with flow control and CRC16 link error control. At the network layer, MTP.3 has three kinds of packets, called signaling units (SUs).

- Message SUs are variable length packets, 14 – 278 bytes long. MSUs route the higher-layer packets (from ISUP, TCAP, etc.) to the desired STP according to the STP's address, called a point code. While MSUs handle some other functions typically found in this layer, its addressing function is particularly lightweight.
- Link Status SUs have a 2-byte payload with a 5-byte header. LSSUs send information about link loading between STPs so that network management software can automatically provision/release DS0 channels for/from the link groups.

- Because SS7 links must be continuously occupied, an STP transmits a fill-in SU when it has no MSU or LSSU to transmit. FISUs have a 5-byte header with no payload.

The ISDN user part (ISUP) is the message set and its protocol for call setup signaling in those nations that typically adopt North American standards. ISUP packets are connectionless datagrams that are used for interoffice signaling during call setup, such as requesting the busy/idle status of a called party. A broadband version of ISUP, called BISUP, adds SUs to ISUP's to provide bandwidth allocation. Figure 18.2 also shows the European telephone user part (TUP) and data user part (DUP), neither of which is described here.

The transaction capability application part (TCAP) is the message set, and its protocol, for remote database access in those nations that typically adopt North American standards. TCAP packets are also connectionless datagrams, but these are used for special features, such as 800-number translation and most Intelligent Network signaling. "Old" TCAP is dividing itself into two protocol layers: "new" TCAP will reside only at the seventh, or application layer, and a new protocol will reside in layers 4 – 6, between "new" TCAP and SCCP. Yet another protocol, the *Intelligent Network Applications Protocol* (INAP) is described in Section 18.4.

The signaling connection and control part (SCCP) extends MTP.3 to end-to-end connectivity so that TCAP messages can reach a remote database and not just another STP. A corresponding function in ISUP enables interoffice signaling. A CO sends an ISUP packet to another CO by including the destination CO's STP point code and the CO's office code in the packet's address. The STPs have routing tables addressed by the office codes of the offices they serve.

These message sets have grown over the years to include the call triggers and many other signaling messages that are needed to implement the Intelligent Network. In the next couple of years, the SS7 message set will probably continue to increase, perhaps so much that it may need to be redefined. Furthermore, the signaling traffic load will continue to increase, especially as class of service migrates to a centralized line information database (next section) and local number portability (Section 18.4.4) is implemented.

It's likely that the data rate on many SS7 links will have to be greatly increased and that SS7's congestion control will have to be improved [8,9]. It's even possible that SS7 traffic might increase so much that more STPs might have to be added to the SS7 network. Proposals are being heard for converting the entire SS7 network over to a more robust physical and logical architecture. Asynchronous transfer mode (ATM), described in Section 19.4, in the next chapter, is probably an excellent candidate for carrying SS7 signaling messages.

18.1.3 Evolving Implementation of Busy Tone

The evolution of signaling is illustrated by the changing implementation of two features. Busy tone is described in this subsection and calling party identification is described in the next subsection. Consider four different cases in which subscriber A tries to call some other subscriber, who is busy on another call.

Two parties in the same office. Let A and B (as in Figure 4.1) home on the same switching office. After A dials B's telephone number, their common exchange knows that B is busy. This exchange establishes a talking connection, through its internal switching fabric, connecting A to a service circuit that provides the familiar audible busy tone to A's talk path. The holding time is very short, only a few seconds, long enough for A to hear busy tone and hang up.

The resources consumed are the CPU cycles for processing A's call by switching software in the processor inside their common office, the use of one busy tone circuit for the call duration, the use of one talking connection through this switching office's internal fabric for the call duration, and the CPU cycles for processing A's hangup by switching software in this same processor.

While the LEC provides the resources for this call, it earns no direct revenue because the call wasn't completed. Recall that, if A has flat-rate billing, the LEC wouldn't have earned any revenue on the call even if it had completed. Presumably, while these lost resources bring no direct revenue, the lost expense is recovered in the monthly fee A pays.

Direct-trunked offices with inband signaling. Let A and C (as in Figure 4.1) home on neighboring switching offices in the same LEC and let these two exchanges have direct trunking. After A dials C's telephone number, A's exchange knows that C homes on a different exchange, but doesn't know whether C is idle or busy — only C's exchange knows this. A's exchange connects A to its end of one of the trunks in the trunk group that goes to the exchange that serves C. C's exchange connects the other end of this trunk to a busy tone circuit in C's office. The holding time — on the busy tone circuit, the two fabric connections, and the interoffice trunk — is very short, long enough for A to hear busy tone and hang up.

The resources consumed for this call are the CPU cycles for processing this call by switching software in the processor inside A's exchange, the use of an MF outpulser in A's office and an MF receiver in C's office, the CPU cycles for processing this call by switching software in the processor inside C's exchange, the use of one trunk for the call duration, the use of a connection between A and the trunk through the internal fabric in A's switching office for the call duration, the use of one busy tone circuit for the call duration, the use of a connection between the trunk and the busy tone circuit through the internal fabric in C's office for the call duration, the CPU cycles for processing A's hang-up by switching software in the processor in A's office, and the CPU cycles for processing the disconnect by switching software in the processor in C's office.

While A's call to C requires more than twice the resources as A's call to B, the LEC still earns no direct revenue because the call wasn't completed. Again, if A has flat-rate billing and C is located inside A's flat-rate billing region, the LEC wouldn't have earned any revenue on the call even if it had completed. And again, while these lost resources bring no direct revenue, we assume the lost expense is recovered in the monthly fee A pays.

Separate LATAs with inband signaling. Let A and F home on switching exchanges that are very far apart (as in Figure 4.1). Let these two exchanges belong to different LECs and reside in different LATAs. As in the call from A to C, A's exchange knows that F homes on a different exchange, but doesn't know whether F is idle or busy — only F's exchange knows this. In this case, A's exchange knows that A's call to F is an inter-LATA call, requires an IXC for completion, and identifies A's default IXC. A's exchange connects A to its end of a trunk that goes to the POP in A's LATA for A's default IXC.

This IXC establishes an interexchange connection, possibly cross-country, between the POP in A's LATA and the POP in F's LATA. This POP in F's LATA connects the call to one end of a trunk that goes to F's exchange. F's exchange connects the other end of this trunk to a busy tone circuit in F's office. The holding time on this long multihop connection is very short, only a few seconds, long enough for A to hear busy tone and hang up.

This call requires the cooperation of three separate corporations: the LEC that serves A, the LEC that serves F, and A's default IXC. Consider the resources consumed by each telephone company.

- The resources consumed by A's LEC are the CPU cycles for processing this call by switching software in the processor inside A's exchange, the use of an MF outpulser in A's office, the use of one trunk between A's office and the IXC's POP for the duration of the call, the use of a connection between A and this trunk through the internal fabric in A's switching office for the call duration, and the CPU cycles for processing A's hang-up by switching software in the processor in A's office.
- The resources consumed by F's LEC are the CPU cycles for processing this call by switching software in the processor inside F's exchange, the use of an MF receiver in F's office, the use of one trunk between F's office and the IXC's POP for the duration of the call, the use of one busy tone circuit for the call duration, the use of a connection between the trunk and the busy tone circuit through the fabric in F's office for the call duration, and the CPU cycles for processing the disconnect by switching software in the processor in F's office.
- The resources consumed by A's default IXC are the CPU cycles for processing this call by switching software in the processors inside the POP in A's LATA and inside the POP in F's LATA and inside any intermediate tandem offices in the IXC's network through which this call is routed, the use of an MF outpulser and corresponding MF receiver in every office pair along the call's route, the use of one trunk in every hop for the call duration, the use of a connection from trunk to trunk through the fabrics in both POPS and in every intermediate tandem office for the duration, and the CPU cycles for processing the disconnect by switching software in the processors inside the POP in A's LATA and inside the POP in F's LATA and inside any intermediate tandem offices in the IXC's network through which this call is routed.

Even though the holding time is short, A's call to F consumes a lot of resources, especially if the call is cross-country. There are three telcos involved, and none of them earns any direct revenue, because the call wasn't completed. All three telcos would have earned revenue on this call if it had completed. While these lost resources bring no direct revenue, we can assume that A's LEC recovers the lost expense in the monthly fee A pays. We can also assume that F's LEC provides busy tone for A's LEC without charging that LEC for the service because A's LEC reciprocates and the favor is probably even. However, the IXC provides the most resources to this unbilled call and the IXC has no easy way to be reimbursed by A's LEC.

When the United States had only one long distance carrier, AT&T's monopoly, it didn't matter that this cost was never recovered. In those days, Ma Bell had many greater costs than this that were not recovered directly. All such costs were simply swept under the regulatory carpet. In the current system in the United States, the IXCs expect, rightly so, that the LECs will somehow reimburse them for the cost of such calls. The solution, of course, is to avoid completing such calls whenever the called party is busy.

Separate LATAs with networked signaling. Consider again the case where A and F home on switching exchanges that belong to different LECs and reside in different LATAs. This time, however, let the entire call use networked signaling. As in the previous case when A called F, A's exchange knows that F homes on a different exchange, but doesn't know whether F is idle or busy. In this case, however, A's exchange sends an SS7 packet to F's exchange and requests the busy/idle status of F's telephone. F's exchange returns an SS7 packet and indicates that F is busy. Then, as in the case when A called B, both homing on the same office, A's exchange establishes a talking connection, through its own internal switching fabric, connecting A to a service circuit that provides the familiar audible busy tone to A's talk path.

No trunks are involved in the call and no IXC is asked to do anything. Instead of completing a long distance call so A can listen to a busy tone generated in a distant office, A listens to a busy tone that is generated locally. The consumed resources are almost the same as in the first case above, where the called party homed on the same office as the calling party. The difference is a pair of cross-country SS7 packets. These two SS7 messages consume some CPU cycles in the switch processors in A's office and in F's office and in the STPs through which the packets are routed.

This section has presented an extended example but, hopefully, now the reader fully appreciates the operational and economical advantages of networked signaling.

18.1.4 Evolution of Calling Party Identification

The second feature examined is calling party identification. The evolution of this feature's implementation shows how the knowledge of who is making this call has penetrated deeper and deeper into the network as signaling evolved. Originally, the calling party's identity had to be signaled because toll calls were individually billed and the toll operator was located in an office remote from the calling party's central office.

In manual switching systems, the calling party's identity is unknown to the toll point after the subscriber is assigned to a toll trunk at the local office. In early automated switching systems, the calling party's identity was lost after the first stage of concentration (the line finders in Step, see Section 6.3). Before automated signaling, the only way the toll operator could know the calling party's identity was to ask for it, either from the local operator or directly from the subscriber. This process was the primordial signaling of the calling party's identity, definitely inband, from human to human.

The Bell System extended automation into its intertoll network in the 1950s and 1960s in a huge program that customers knew as Direct Distance Dialing (Section 9.2). One of DDD's significant and expensive steps was to provide for automatic number identification of the calling party for the new automatic message accounting (AMA) systems that computed the toll bill. Since the AMA equipment was typically located at the toll point, both called and calling telephone numbers had to be MF pulsed (inband signaling) from the calling party's local exchange to the newly automated toll point.

As the switching systems evolved to program control and the signaling evolved to networked signaling by which any CO can talk to any CO, it was simple to signal the calling party's identity all the way to the called party's local exchange. Once the called party's call processing program could be made aware of the identity of anyone trying to reach its client, a call screening feature could be offered to subscribers. In call screening, the subscriber creates a list of *persona non grata,* and anytime someone from this list tries to call this subscriber, call processing programs intercept the call and make the subscriber appear to be busy.

Still more recently, this feature has evolved into a popular feature called calling party identification. CPI requires an alphanumeric display on the subscriber's premises that connects to house wiring like an extension telephone. CPI also requires a downstream signaling channel that is frequency multiplexed onto the subscribers tip/ring pair. Whenever anyone tries to call a CPI subscriber, the local exchange signals the calling party's identity to the called party during the the 4-second interval between the first ring and the second ring of his telephone. In the simplest implementation, the CPI subscriber sees the calling party's telephone number on his display. Augmented by the Intelligent Network, as discussed later in this chapter, this telephone number can be reverse-translated back to the calling party's name, as listed in the telephone directory.

The obvious next step is that the LECs offer subscription to CPI anonymity for callers like those annoying people who call at dinnertime, trying to sell you something you don't need. One gets the impression of dealing with an armaments vendor who sells armor to one side in a war, armor piercing shells to the other side, then thicker armor to the first side, and so forth. Why not? Isn't this why we're deregulating?

18.2 SWITCHING SOFTWARE

This section reviews the different architectures of switching software (Chapters 11 and 14) and then describes two contemporary software issues: feature interaction and the software development process. The section closes by discussing the total cost of a telephone switching system.

18.2.1 Switching Software Architectures

Because switching systems were controlled by human logic and then by wired logic during the first 90 years of telephony, the evolution of switching software is a relatively recent phenomenon, occurring over the past 40 years [10 – 12]. Switching software has evolved in its *process structure* and its *data management.* Despite how improved software technology has simplified switching software, the functions that switching software must perform have become more complex. The Centrex dilemma is an example.

Process Structure. During the late 1950s and early 1960s, three large (even by today's standards) pioneering programming projects were developed concurrently in the United States by three different companies: IBM was developing the programs that would become the operating system for the 7090 family of computers; American Airlines (via MIT) was developing the initial version of the Sabre airline reservation program; and AT&T (via Bell Labs) was developing the first switching program for its #1 ESS. These were very exciting times. While the program developers tried to learn from each other, they had little precedence or experience. One common characteristic of all three projects was that completion and delivery schedules slipped by many years.

1E programs were written in a complex assembly language (Section 11.4.9). They were organized as a collection of re-entrant multitasked segments that executed in continuous real time, except for event-driven maintenance programs and the 5-ms timer-driven interrupt for i/o. While the programs were complex, the individual segments were short and well defined. The multiple segments of these reentrant multitasked programs had to be meticulously scheduled, and by the programmers. A single multitasked program performed its exclusive function for any client that needed this function and was properly scheduled. If a program took a real-time break at a segment boundary, the same program could repeat during the break, but on behalf of a different client.

The science and the art of designing programs have come a long way since those early pioneering times, partly because of increased technology and partly because computer memory has become a less precious resource. Typically, programmer salaries and development time are more precious to switching system manufacturers than computer memory. Multiprocessing operating systems are preferred over the multitasking software environment of the 1E mainly because programs are easier to design if reentrance need not be considered. Operating system environments are used so that program designers can concentrate on the direct function of the program they are designing and not worry about environmental issues like: real-time breaks, scheduling, interprocess messages, and many more. For similar reasons, programs are written in high-level languages. Since compiler inefficiency is not completely unimportant to manufacturers, intermediate-level languages, like C, are a good compromise.

Modern switching programs are written in a high-level language and the compiled load modules are organized as "processes," which are scheduled and controlled by a modern multiprocessing operating system. Without segmentation, the programs are long and complex. Processes are spawned and killed as needed. A common architecture is process-per-call, in which the operating system spawns a separate independent copy of a load module (process) of a call processing program for each new off-hook that is detected. While the same program might execute (pseudo) simultaneously with other instantiations of itself, each independent process only knows about one client.

This process-per-call software architecture means that a large program, like call processing, is replicated for every active call in a switching office. Replicating a multi-megabyte load module for several thousand independent calls requires a huge amount of program memory. It is especially inefficient because every process (copy of the load module) contains the program logic for every call feature, even if the process' client hasn't invoked the feature or doesn't even subscribe to the feature. As a memory optimization, a different software architecture, process-per-feature (per call), partitions the call processing load module into a simple POTS module and a collection of feature modules. The POTS process is spawned for every new call and a feature process is attached only when the client invokes the feature. While the memory savings is significant, feature interaction (next subsection) is exacerbated.

The program design process has even changed dramatically. The traditional flowchart design method has given way to modern structured design. While we usually associate the former with Fortran and the latter with Pascal, the design method is really independent of the language. Programmers are encouraged that the architecture of the software processes should be congruent with the architecture of the process being controlled. This last notion has led to considerable debate among switching software designers as to the exact nature of the call model. In fact, much of the disagreement between the advocates of the process-per-call architecture and the process-per-feature architecture is rooted in the more fundamental problem: the lack of a good model for the i/o between the user and the software.

Data Management. The nature of data management has also evolved since 1E software was developed. In the 1E, data is stored in globally accessible tables and formatted registers, that we would call data structures now. Because even internal data had to be stored carefully, a reentrant segment could not have internal scratchpad memory. Any program had permission to read or write any table at any time and every program designer needed to know the detailed structure of every table his program used. Programs allocated multiple blocks of memory and any of these memory blocks could be read and written by multiple programs. Every program kept track of its multiple clients and each client had to have a dedicated memory allocation, not only for global data (like dialed digits), but also for temporary internal data (like loop indexes).

In modern switching software, because each process only knows about a single client, programs can be designed using internal scratchpad memory. Furthermore, access to, and ownership of, global data is externally controlled and its detailed structure is hidden. One means of external control of data is through a structured database, where permission and read/write synchronization are maintained by the database manager. Programmers are freed from knowing the infinitesimal detail of the formats of the various data structures and all data inquiry is made through the database manager.

A logical evolutionary step would be to move the database to a special purpose back-end database machine. An even more centralized solution is a physically separated database processor connected by a network to serve a region. Among other things, the Intelligent Network (Section 18.4) is a move in this direction. A more distributed and less centralized means of external control of data is through *object orientation,* where different data is owned by different object managers. A process that requires some data that it doesn't own must request it by sending a message to the process that does own it.

One example of how data access has evolved is seen in the various implementations of class of service. In early electromechanical systems, COS included simple things like message/flat-rate billing, DP/TT subscriber signaling, or residence/coin/PBX. COS was implemented by wired logic in, or near, the subscriber's line circuit. In Step (Chapter 6), lines with similar COS are segregated into line finder groups. In Crossbar (Chapter 8), COS codes are wired into the line circuit and they are passed through the line link marker connector to the DT marker. In the 1E (Chapter 11), each subscriber has a word in a global COS table and the table is read by the reentrant dialing connection-lines program whenever its task dispenser gives it a new client. In the 5E (Section 14.2), each subscriber has a data structure that is maintained by a database program in the AP. Early in every new call, in the switching module's switch processor, the call's process requests the subscriber's COS by IPC. In the System 75 (Section 14.3), each subscriber has a data structure that is maintained by the user manager. The call process requests a subscriber's COS early in every new call.

With SS7 and IN, COS information is migrating to a specialized regional database, called the line information database (LIDB), so COS can be requested by programs outside the users' home switch. This will probably end up as a separate database machine, similar to the database machine that maintains ownership of 800 numbers — attached to, and addressable by, the SS7 network.

The procedure for passing data from one program to another has also evolved through many implementations: (1) placing real data or pointers to data in global tables that any program can read or write, (2) passing real data or pointers to data through subroutine calls, (3) data inheritance by "child" processes, (4) pipes and sockets, (5) OS-based interprocess communication, (6) interprocessor message communication among distributed processors in the same switching office, and (7) packet-switched messages over the SS7 network.

Complexity. The architecture of switching software has evolved from one that had monolithic reentrant multitasked segments to an architecture that had a process-per-call under a multiprocessing operating system. Access to global data evolved from random reading and writing of global tables to controlled access where data is carefully controlled and its structure is hidden. These two transitions simplified some of the program designer's tasks: keeping track of who the client is, keeping track of internal parameters, and accessing global data. But, even in modern systems, switching software has become practically too complex for human comprehension. There are a number of different reasons.

- Programs evolve, undergoing gradual changes, patches, and extensions — with an if clause added to one place in the program one year and a case added to a switch clause somewhere else in the program another year.
- Subsequent changes might be made to programs by an employee who is not the original designer of the program. Sometimes, the original designer may not even be available to answer questions.
- Sometimes, software development schedules cause programmers to compromise architectural purity.
- There may be too much functionality colocated in the same large tangle of software.

This last point is probably the most significant. In hindsight, switching software is reminiscent of how those four independent functions were colocated and entwined together in a 1960s toll point (Section 9.4). Consider the following illustrations.

- Just as the Y2K problem was built into MIS software, the numbering plan is built into switching software. For example, adding the 888 in-WATS prefix probably required adding another case to many switch clauses scattered throughout some program called "Digit Analysis" or "Feature Control", instead of simply changing a single entry in some database.
- The logic that controls a call, including managing features for a call, should have been separated from the logic that determines a route through the network.
- Even back in the early 1960s, 1E programmers had enough sense to isolate, into subroutines, the logic that knows the details of the 1E's hardware from the logic that performs more general functionality. The software architecture of more modern switches, like the 5E or the System 75, resembles the traditional Unix model: application programs that don't know about hardware working through *drivers* that do (see Figures 14.7 and 14.14). One manufacturer's dial tone

connection program, for example, should be so independent of hardware that this same program could be used in every switching system in the manufacturer's product line. That's almost universally not the case.

- Related to the previous issue, as corporations in the 1990s have partitioned themselves into finer and finer lines of business, different business units maintain different "grocery lists." So, a software change, like adding 888, may not be universal in simultaneous generics across a single manufacturer's PBXs, CO switches for its U.S. market, and CO switches for its international market.
- If a program like dial tone connection isn't written to be independent of the hardware of the switching system, it's even more unlikely that it would be independent of whether a call is POTS, coin, ISDN, cellular, or anything else.

Separating these functions, what I've called the partition, is discussed in Section 18.5.

The Centrex Dilemma. While many PBX manufacturers don't have the size or expertise to expand their business into the manufacture of large switches, Section 2.2.2 described that most manufacturers of large switches also make PBXs. Considering the technologies of these two kinds of products, their hardware and software are similar enough that it seems natural for a manufacturer of large switches to also manufacture PBXs. However, considering the market, not just the technology, because the market for switching systems is fragmented,

a PBX is not just a small switching office.

Market fragmentation is exacerbated by the complexity of switching software. Section 10.6 described how the Carterfone decision had the effect of creating a separate sub-market for enterprise switching. The two principal classes of vendors in this market are:

- The LECs, who offer Centrex service through their class-5 switching offices;
- The manufacturers, who sell PBXs for the customer unit (Section 10.5) implementation.

While Nortel, Siemens, Rolm, Lucent, NTT, Ericsson, Tai, and all the other PBX vendors make their own products for this enterprise market, the LECs don't (aren't allowed to) manufacture their own class-5 switching offices. In fact, until only relatively recently, the LECs didn't write the software that provides the Centrex service that allows them to compete in this enterprise market. To compete in this enterprise market, the LECs depend on the companies that manufacture their class-5 switching offices to provide them the necessary equipment and, until recently, the software that provides the features that the customers want. The problem is that most of these class-5 switch vendors also manufacture and market the very same PBXs the LECs are competing with. This created a level of distrust.

Studying the 5E's and 75's switching software in Chapter 14 should have revealed that PBX switching software is generally smaller and less complex than CO switching software. This smallness means that new features can be added to a PBX sooner than they can be added to a class-5 switching office. However, this undeniable truth is complicated by the distrust described in the previous paragraph.

- Would the LECs, which didn't appreciate the complexity of CO switching software and didn't trust the manufacturers anyway, be easily convinced that their manufacturers were being diligent at keeping their Centrex features competitive?
- Wouldn't a manufacturer, who has decided to aid its PBX LOB by deliberately delaying the release of Centrex feature software, use this complexity argument to justify delaying longer than necessary?

This Centrex dilemma continues in the "call diddling" scenario, described a little later.

18.2.2 Feature Interaction

A feature is any nonessential optional minor luxury that can be attached, at additional cost, to some basic service. When you buy a new car, the features include air conditioning, power windows and seats, a sports instrument panel, and many more. When you buy POTS, the features include Touch Tone, call forwarding, calling party identification, and many more.

This subsection presents some examples of how POTS features interact with each other, digresses into describing how new features are offered in legacy systems, and focuses on those feature interactions that are caused by inadequate signal commutation [13 – 17].

Examples. Unfortunately, telephone calling features are not independent. In fact, they depend on each other in complex and troublesome ways. Some of the causes and indications of feature interaction are described.

- Fuzzy boundaries around software modules make it difficult for a system integrator to fit a new feature program into a legacy system.
- Different features have different preconditions, assumptions, and environments. For example, many PBXs permit call transfer only on received calls that come from outside the PBX. While this constraint might make sense for the programmer, the users may not appreciate it. When a user attempts a call transfer on an internal received call or on a call he originated, he is not only thwarted, but the software may assume that he's trying to do something else. The user can be very surprised.
- Sequences of features may be identical or very subtly different. Consider A's, B's, and C's states in these two cases: (1) B calls A, A answers and they talk, C calls A and A has call waiting, A flashes for C's call; or (2) A calls B, B answers and they talk, A places B on hold, A calls C. In both cases, A is talking to C and B is waiting for A to finish. However, A originates in one case and terminates in the other; one case resulted from call waiting and the other from hold. One is tempted to say that A, B, and C are in equivalent states in both cases. But, are they? Does the equivalence hold up under every possible subsequent event that A, B, or C could transact? If you were a programmer, would you bet your job on it?
- The system's feature prioritization doesn't always match what the system's users expect, is seldom explained to the users, and might not even be known to the programmers. Feature priority may simply be a result of the programming and it may have never been planned.

- How can a user ignore an incoming feature? For example, suppose I'm in the process of adding parties to an important conference call, and I hear call waiting tone. Can I ignore the incoming call and continue adding parties? Recall that the usual procedure for adding new parties to a conference call is to flash for dial tone after each one is added, but the procedure for answering the incoming call is also to flash.
- Features often don't allow users to perform operations that seem perfectly logical to them. Continuing the example above, suppose I do answer the incoming call and decide that the caller should be added to the conference call. There is no simple way — I have to ask for his telephone number, ask him to hang up, and then I have to call him back.

New Features on Legacy Systems. Digressing from feature interaction for a moment, the problem of new features not being available on legacy systems is not new. When DDD first came out, it was gradually introduced so that subscribers that home on switch A might have DDD and subscribers that home on switch B don't have it. When Touch Tone first came out, it was not only gradually introduced, but it was introduced on different systems at different times. For many years, subscribers that homed on a Crossbar office could subscribe to Touch Tone, but subscribers that homed on Step offices could not. Then, program control enabled many new features like speed calling, call screening, and call forwarding. Again, subscribers that homed on a 1E office could subscribe for these new features, but subscribers that homed on electromechanical offices could not.

Today, this problem is resolved by using a remote switching module (RSM) off a 5E or other modular digital system. An RSM that homes on a remote digital switch can be placed in a building with a legacy system. Subscribers on the legacy system, who want a new feature, can be shifted from the legacy system to the RSM, without even having to change their telephone number.

Signal Commutation. If more than one subprogram or process is active on a given call when a signal is transmitted from the telephone, then some form of signal commutator must deliver this signal to the appropriate subprogram or process that controls the appropriate feature that may be relevant to a given call. Many feature interactions could be solved if a signal, such as flashing the hook-switch, could be delivered to that one subprogram or process that controls that specific feature which the user wishes to control. We don't have a good way to steer the signal to the appropriate feature module. Signal commutation problems are caused as much by the dumbness of the telephone as by inadequate software architecture. This is best seen by considering signal commutation in a different environment.

Consider a personal computer running a Windows-like operating system. Suppose the user has five separate windows open, displayed, and connected to active processes in the PC's software. If the user types Q on the PC's keyboard, each of the five active processes could potentially receive this Q, and if they did, each of the five active processes would probably take a different action in response to it. However, both intuitively and in the implementation, this Q not only is not delivered to all five processes, it is delivered to exactly one of them and it is exactly the one to which the user intends it to go. In the traditional GUI-WIMP user interface (Section 15.5.4), the destination of the

Q is selected by the user when he mouses the cursor into the window that corresponds to the process which should receive the Q.

Signal commutation is inadequate in telephony because we try to operate multiple feature processes without the logical and functional equivalents of the window and the mouse. The problem is exacerbated by the conventional software architecture in which all features are inside one common process-per-call. Here, the program designer prioritizes features for each signal as an N-squared problem, where N is the number of features. Software designers need a preference chain for telephone signals analogous to that used in Step-by-Step for line finder assignment (Section 6.3 and Exercise 6.3).

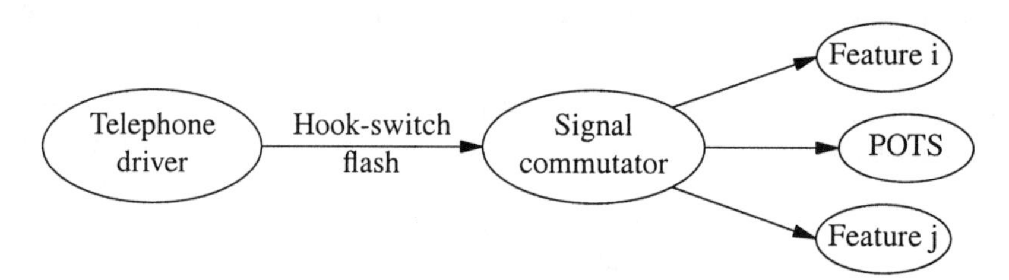

Figure 18.3. Signal commutation.

The telephone signal whose feature-process destination is most often ambiguous is the hook-switch flash. Figure 18.3 shows how feature programs could be organized around a process that acts like a signal commutator. The only two basic parts of call processing would be the commutator and the process that performs POTS. As the user invokes more features as part of his call, then additional feature processes would be appended as shown. When the user signals his call, such as by flashing the hook switch, the signal commutator's internal logic, with preestablished feature prioritization, steers this signal to the appropriate process. This architecture has a better chance of resolving feature interaction than the conventional process-per-call architecture, but this is by no means a solved problem.

18.2.3 Developing Switching Software

The evolution of switching software toward the Intelligent Network, and beyond it, depends as much on the process by which software is developed as it does on the software and its architecture [18 – 30]. Much of this subsection is anecdotal and is not implied to pertain to any one corporation. Two interrelated issues are (1) programmer productivity and (2) how this makes software cost a large part of the total cost of a switch.

Programmer Productivity. When 1E software was developed, the programming environment included assembly language, punched cards, infrequent system loads, and ad-hoc debugging. Beside these primitive tools, #1 ESS programmers had to tend to a lot of detail about data structures, concern themselves with scheduling and clients and re-entrance, and must understand the whole system. Programmers in that day produced only 1000 – 2000 lines of code per year.

The art of programming and the technology to support programmers has progressed significantly since then — with high-level languages and compilers, text editors and files, linking loaders, debugging aids, and formal development environments. Furthermore, with multiprocessing, operating systems, and databases, programmers now don't need to know detailed data structures, can write single client programs, and don't need to understand the whole system. Switching software has also become much more complicated since the early days of the #1 ESS. Today's typical programmer produces 1000 – 2000 lines of code per year. This productivity, which has been so constant over the years, amounts to about six lines per day [31] of designed, written, compiled, debugged code.

At this level of productivity, the typical software package found in a switch processor, with typically more than 10 million lines of source code, requires thousands of person-years of development effort. One reason for the low productivity is that many of these programs have been redesigned many times. And with the relatively high turnover common among programmers, programs are seldom redesigned by the same person. If you've ever tried to modify someone else's program, you understand.

Software Management. While switching software developers have a hard job, their supervisors have an even harder job [32,33]. Suppose you're the supervisor of a group of program developers and your group is assigned a program to develop. You estimate that the best programmer in your group could complete this job in 1 year, but upper management wants the job completed sooner than that. Possibly, two programmers could do the job in 6 months, but only if you had a second employee who was as good as your best programmer. But, 365 programmers can not finish the job in 1 day — in fact, this many people would cause it to take 1 year again. The diminishing return on increasingly paralleled effort is caused by two factors: (1) the programmers have different levels of achievement, and (2) they get in each other's way. Now, scale this problem up to a programming staff with, perhaps, 3000 programmers, and the crowd control problem is enormous. While everyone jokes that a baby can't be produced in 1 month by putting nine women on the job, management tries to do it anyway.

Further complicating this almost unmanageable process are the difficult logistics of scheduling software changes and of releasing, distributing, and installing new and modified programs into the field. Even with thousands of programmers, there are typically many more requests for program changes and development of new feature modules than can possibly be completed by a switch vendor's programming staff. So, requests are queued and prioritized. As programmers finish one task, they can be assigned the next task off the top of the list. Some changes or features that management thinks are urgent might receive almost immediate attention. Some changes or features that management thinks are relatively unimportant may remain on the list for years, and then, still may take 1 – 2 years to complete. Because it is not practical to release each change and each new feature module randomly, as they are completed, many are released together every 1 – 2 years as a new software generation, called a generic. This makes it all the more important for the programmers to meet schedules [34,35].

A Scenario. This complicated development process has been one of the drivers for the evolving Intelligent Network. This is all tied together by a hypothetical story. The following narrative is exaggerated so that all three players in this scenario are equally insulted.

> *A clever employee at Left Central Bell, a large LEC, invents a new calling feature that he calls "Call Diddling." LCB commissions a*

market survey and determines that many of its subscribers would pay handsomely for this feature. Now, many of LCB's switching systems are installations of the Super-Dooper Digital Switch, which is manufactured by Tortoise Telecom, Inc. So, LCB contacts its TT representative and requests that Call Diddling be added to the SDDS feature set in the next release. Since TT releases SDDS software generics every 2 years, and has just started a development cycle, the rep reports back to LCB that Call Diddling has been placed on the grocery list for the next cycle, which won't even begin for 2 years and won't complete for 4. After LCB presses him further, the rep adds that this list already has twice as many tasks as TT programmers can possibly complete. So, even after surrendering exclusive rights for Call Diddling, it probably won't make the cut for this next cycle anyway, which means that LCB would have to wait 6 years for it.

However, since PBX software is much easier to develop than CO software (at least, that's what TT keeps telling the LECs), TT adds Call Diddling to its PBX product line within 1 year. TT's PBXs compete directly with LCB's Centrex Service, which is now weaker without Call Diddling. Now, LCB knows very little about the software inside its own switching systems, mostly because TT doesn't release much detail, and LCB has little experience with large complex software and little appreciation that the SDDS software is as large and complex as it is. So, LCB thinks TT ought to be able to develop Call Diddling for the SDDS in only a month or two. LCB complains to the federal Department of Bullying and accuses TT of "irregularities." LCB requests that the DoB demand that all CO switch manufacturers must provide the capability for their customers, the carriers, to be able to design, write, debug, and install their own proprietary programs.

Of course, the DoB has its own agenda and LCB is playing right into it. The DoB wants to increase telephone competition, at any cost. So, the DoB uses this request from LCB to require that any CO switch must be programmable, not only by the manufacturer who built it, and not only by the carrier who owns it, but also by any third-party service provider, LEC competitor, software company, consultant, hacker, or any kid with a Nintendo.

The current policy in the United States proscribing who is allowed to write switching software is still unclear. While the scenario above may not be 100% accurate in depicting the actual sequence of events that led to the current policy, it's close. It's certainly clear the LECs have argued both sides of the issue at different times. When the LECs were arguing in favor of being allowed to write switching software, the manufacturers countered that naive personnel at the LECs didn't appreciate the complexity of switching software and might cause serious switch breakdowns. But, when the third-party vendors were arguing in favor of being allowed to write switching software, the LECs countered that naive personnel at such companies didn't appreciate the complexity of switching software and might cause serious switch breakdowns.

18.2.4 Cost Breakdown

The cost of a telephone switching system is proprietary to the manufacturer and the price is negotiated with each carrier. This cost must include the cost of hardware, manufacturing that hardware, designing it, and installing it. This cost also includes the cost of software development. Besides the initial cost of purchasing and installing the switch, the manufacturers incur the large recurring cost of redesigning software and producing new generics. The cost of a telephone switching system, to the carrier, includes the large initial price set by the manufacturer (which may have little resemblance to the true initial cost), the recurring price of new software releases and new hardware (which may have little resemblance to the true recurring cost), and the carrier's "people cost" of operating and maintaining the switch.

Computing total cost is further complicated by the need to combine initial cost and recurring cost appropriately. From Finance 101, an annuity (the annual payment) is related to the principal (the present value or initial cost) by

$$a = \frac{i\,P}{1 - v^n}\,;\ \text{where } v = \frac{1}{1+i}$$

and i is the compound interest rate, and n is the number of payments. While 25% might seem to be a high compound interest rate, a corporation borrows from stockholders and the annuity is actually the corporation's profit, which is heavily taxed. Assuming that a debt is paid off in 8 years at 25% interest, the equation shows that $a = 0.3004 \times P$. In other words, if $1000 is borrowed at 25% compound interest, paying off the loan in 8 years requires paying $300 per year. Also, Finance 101 teaches us that the time-varying value of money (because of compound interest, not inflation) makes it inappropriate to add costs directly that are incurred at different times. So, total cost is not just the sum of initial costs plus all the accrued annual costs. A system that costs $100,000 to purchase initially, and $10,000 per year for 10 years to operate, does not have a total cost of $200,000. The calculation is just not that simple.

It's difficult for a manufacturer to really know the total cost, or for a carrier to really know the total price or its internal cost. So, a breakdown of their cost or price is even more difficult. I made an educated guess at a high-level breakdown of the major cost components of a large CO switch, including initial and recurring cost, but not including operating cost. Representatives from several manufacturers and several carriers have seen the figures in Table 18.1 and nobody with any experience has found them unreasonable. So, let's hypothesize that the lifetime cost to the carrier of a modern digital switching system has four significant components:

TABLE 18.1. Estimated costs of a switching office

Price of line cards	20%
Price of all other HW	14%
Price of software	33%
Cost of operation	33%

The breakdown above is illustrated using the Finance 101 numbers discussed earlier. Assume that the price for a switching system is $500K for hardware and installation. Assume that the price of the initial software is $250K and that a new generic is released every year for $75K. Assume that the annual operating cost is $150K. The total cost can be calculated by casting all costs into equivalent annuities or into equivalent initial costs. The calculations assume a 25% interest rate and an 8-year lifetime, yielding an annuity-to-principal ratio (APR) = 0.3.

- Casting the two initial costs into their equivalent annual costs gives $150K for the hardware and $75K for the software. So, the sum of the annual software cost and the annual equivalent of the initial software cost is $150K. The total equivalent annual cost of the switch equals the total annual cost plus the equivalent annual cost = $450K per year.
- Casting the two annual costs into their equivalent initial costs gives $250K for the software and $500K for operation. So, the sum of the initial software cost and the initial equivalent of the annual software cost is $500K. The total of the two initial costs plus the two equivalent initial costs is $1.5M.

Whether comparing annual equivalents or initial equivalents, total cost is split evenly over hardware, software, and operation. Some surprising conclusions from these figures are:

- Sixty percent of the price of a switch's hardware is for line cards. This is typically overlooked when the cost of access is computed and by those who think CO switches can be replaced by the Internet. Every telephone — especially, an analog telephone — will always need a line card.
- Most of the remaining 40% of the price of a switch's hardware is for processors (APs and SPs). The switching fabric, the part most often optimized and the part that has been emphasized in this book, represents perhaps 15% of the price of the hardware and, hence, perhaps only 5% of the switch's total cost. This shows that the engineers are optimizing the minutia. This also shows that replacing only the fabric, perhaps by an ATM fabric still controlled by switching software, would be a minor change in the total cost.
- The total price of hardware is only about one-half of the total purchase price of the switch and is only about one-third of the lifetime cost of the switch.

18.3 ENHANCED SERVICES

This section introduces and motivates the evolution toward the Intelligent Network. The first subsection describes the evolution of 800 numbers in the United States and the third subsection presents another scenario that suggests taking 800 numbers a step further. In between these, the second subsection presents the 20-year story of *enhanced services.*

18.3.1 800-Calling

Wide area telephone service (WATS) was described in Section 9.1.3. But, to be more precise, WATS has two components, called out-WATS and in-WATS. While some telephone subscribers prefer message billing, in which they are billed for every minute of every call they originate, most subscribers have a billing arrangement in which they may place toll-free calls to parties within some local dialing region. The out-WATS subscriber pays a higher flat-rate monthly fee to enlarge this region into which he may place toll-free calls. The larger the toll-free calling region for outgoing calls, the larger is the flat-rate monthly fee.

While the calling party usually pays for a telephone call, in a collect call the called party agrees to pay the charges. A business, for example, as a convenience for its customers, might arrange for permanent collect calling, by which the called party always pays for all incoming calls. In many countries, such a permanent collect subscriber is said to have a "green number" or a "free phone." In the United States, these are called 800 numbers. In this arrangement, the called party automatically pays for any incoming calls that are placed by calling parties within some defined local dialing region. The in-WATS subscriber pays a higher flat-rate monthly fee to have an enlarged region from which permanent collect calls are toll-free. The larger the toll-free calling region for incoming calls, the larger is the flat-rate monthly fee. Many U.S. enterprises expand their toll-free in-WATS region to cover the entire country. In the United States, such green numbers with extended in-WATS were traditionally identified by assigning them 800 in the place of the area code.

The Bell System started offering 800 service back when AT&T's Long Lines division was the only long distance carrier. In this original 800 service, a customer called the listed 800 number and his local switch, on seeing the "0" in the second position, recognized the dialed number as a long distance call. The call was connected to a toll trunk and the called and calling numbers were signaled to the toll point. Call control software in AT&T's national network translated the 800 number into an actual telephone number and completed the call. As AT&T's interoffice signaling evolved to CCIS and SS7, the implementation of 800-number translation evolved to where a single (redundant) database system served the entire nation. In the United States, 800 service has evolved in two related ways: (1) its implementation and (2) the available optional features. This evolution is illustrated by following the progression of the 800 service offered to, and requested by, a hypothetical catalog retail company, whom we'll call Sellers.

1. Sellers, Inc. started its dial-up catalog retail business with a single call distribution center (CDC) in Chicago. A CDC functions like a PBX, with a listed directory number on a hunt group. Each incoming call to the hunt group is routed inside the CDC to whichever customer representative (rep) has been idle the longest. After these reps answer an incoming call, they ask for the calling customer's order and, in this era, they also had to ask for the customer's name and credit card number before they processed the order-entry manually.

2. When AT&T introduced 800 service, Sellers subscribed because customers appreciated that Sellers paid for their calls. Actually, AT&T marketed 800 service so well that Sellers' customers expected to not have to pay for their calls, Furthermore, while the telephone number 800-SELLERS cost more, the easily remembered number had a lot of market value.

In its initial implementation, 800 numbers were translated into directory numbers using a database that had to be installed into every class-4 switching office in the United States. In Figure 18.4(a), without networked signaling, the calling party dials the called party's 800 number into the local class-5 switching office. Because of the 800 prefix, this local office connects the call to a toll trunk and signals the called number (still an 800 number) to the toll point using some form of inband, out-of-band, or common channel signaling. The toll point translates the 800 number into its national directory number by dipping into its local copy of the national 800 number translation database. Based on the

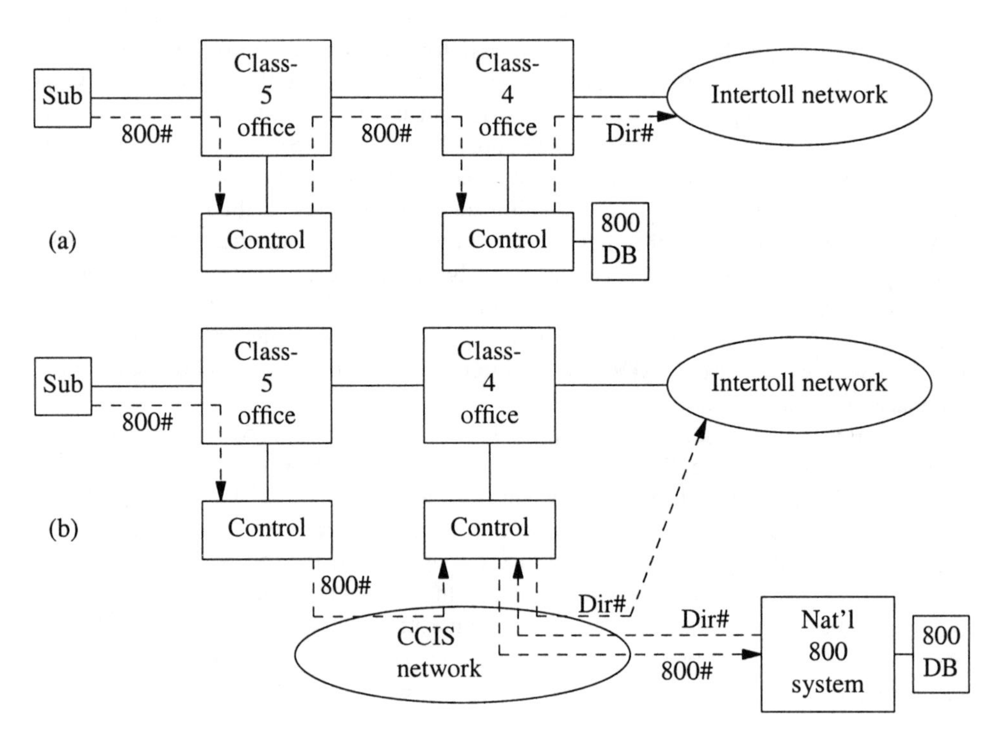

Figure 18.4. 800 calling (a) without CCIS and (b) with CCIS.

area code of the actual number, the call can now be routed as if the calling party had actually dialed this directory number. Since each toll point had its own copy of the common nationwide translation table, changes had to be distributed to every one of more than 1000 toll points (Table 4.1).

As 800 numbers became popular, maintaining over 1000 separate copies of this same database became tedious and expensive. With CCIS (Section 18.1.1), AT&T would be able to translate 800 numbers at one central database that served the entire United States. In Figure 18.4(b), with networked signaling, the calling party dials the called party's 800 number into the local class-5 switching office. Again because of the 800 prefix, this local office connects the call to a toll trunk and signals the called number (still an 800 number) to the toll point using networked signaling. The toll point translates the 800 number into its national directory number by a signaling exchange with this unique national 800 number translation database.

3. Meanwhile, Sellers automated its CDC by coordinating a data network. Each rep had a computer terminal that was connected by LAN to a host computer. The host provided a customer database and automated the order-entry. The rep simply asked for the calling customer's telephone number and typed this number at the terminal to access the customer's file of sales history and billing information.

4. As Sellers' business expanded to a national scope, the company's telecom manager faced an interesting tradeoff. Economy of scale on customer reps suggested that Sellers should continue to operate its nationwide business from that single CDC in Chicago.

But, in this era, long distance calls were still very expensive, especially from both coasts, where many of Sellers' customers lived. After careful analysis, Sellers decided to open six additional CDCs — so they had one in each of what is now an RBOC region.

Even though the initial 800 implementation, in Figure 18.4(a), had a separate physical copy of the 800-number translation database in each class-4 office, there was only one logical 800-number translation for the entire nation. So, each separate CDC had to be assigned a different 800 number. This number had to be carefully advertised on a regional basis, so customers called the CDC that made for the cheapest call.

5. Sellers complained to AT&T about needing seven different 800 numbers. Finally, AT&T introduced multilocation 800 service. With this option, two people in different parts of the United States might call the same 800 number, but AT&T could route each call to a different real telephone number, according to the calling party's area code.

In the initial implementation of 800 number translation, in Figure 18.4(a), with a database in each toll point, this enhancement required that more than 1000 databases had to have customized entries, a significant amount of expensive bookkeeping. By centralizing 800 number translation to a common national database, in Figure 18.4(b), each 800 number's multiple translations were stored in one place. By including the calling party's telephone number in the CCIS signaling to its national 800-number translation database system, the 800 number's translation could be different for different NPAs. The centralized implementation in Figure 18.4(b) was not affected by divestiture because, for several years after divestiture, AT&T was the only IXC that provided 800 service. For this post-divestiture period, the box labeled "Class-4 office" was AT&T's regional POP and the LEC had its tandem hub between the class-5 office and this POP.

Sellers subscribed immediately to multilocation 800 service because: (1) nationwide advertising was facilitated and (2) all seven CDCs were accessed from the same mnemonic 800-SELLERS number.

6. While Sellers was pleased with this, they had always had another complaint about 800 service. Sellers noticed that two kinds of calls were lost: (1) if all the reps at the assigned CDC were busy, then the calling customer got busy tone, even if other CDCs had available reps; and (2) if the assigned CDC were closed for the day — for example, at 5:01 P.M. on the East Coast — then the calling customer would not be served, even if CDCs in other time zones were still open. While routing a call to an alternate CDC is more expensive, Sellers would rather do that — and pay for it — than risk allowing this customer time to change his mind about the purchase or, even worse, call a competing catalog retailer. AT&T introduced enhanced 800 service by which 800 calls are routed in a preference chain, not only according to the calling party's area code, but also according to time-of-day and whether the preferred destination is busy. Sellers subscribed.

7. With the advent of long distance competition, the price of long distance calls went down. The price of 800 service didn't change much because, for many years, AT&T was the only IXC that provided it. The problem was that, in all the LEC's local switches, 800 numbers translated into long distance calls that the LEC routed to AT&T. If another IXC wanted to offer "Green Number Service," they couldn't use the 800 prefix — at least, not under the existing system that COs used for number translation. Using an alternate prefix had two big problems: (1) each IXC would have to have its own — and there were hundreds of IXCs — and (2) Sellers would never agree to change its number anyway.

But, finally, a solution emerged whereby other IXCs offered 800 service and customers could keep their 800 number and assign it to the IXC of their choice. The implementation requires an 800-number ownership database that is accessed by the LEC's switch before routing the 800 call to the appropriate IXC's POP.

8. CDC technology evolved to what is now called computer telephony integration (CTI) [36 – 40]. Several CTI vendors offered voice/data systems where a PBX is integrated with a client/server LAN and the PBX's SP is tightly coupled to the server. Three standards have been developed: CSTA in Europe, TSAPI by AT&T/Novell, and TAPI by Microsoft. In these systems, the company's databases for transaction processing and for customer sales history are tied to the CTI PBX. By subscribing to calling party ID on a CDC's hunt group, the CDC receives the telephone number of the calling party. Thus, at the same time that the CDC routes the incoming call to a rep, it also displays this customer's name, vital statistics, and sales history on the PC in front of this same rep. Sellers reequipped its seven sites with integrated voice/data CDCs and subscribed to calling party ID on all seven hunt groups.

9. While this is the current state of Sellers' 800 service — enhanced 800 with CTI — Sellers is still not satisfied. For example, if all reps are busy, an incoming call can be networked — voice and data together — by frame relay or some other WAN to another site. But, calls can't be automatically rerouted without the IN (Section 18.4). The scenario continues in the next subsection.

The 800 service is implemented from national databases and with a lot of SS7 signaling. In the United States, deregulation and competition have driven the price of 800 service so low that many residential subscribers have unlisted personal 800 numbers as a convenience for friends and family. The 800 service has become so popular that America's national numbering plan recently added 888 and 877 to triple the number of available toll-free numbers. When we continue this Sellers scenario later in this section, we'll see that 800 service implementation will change even more. But, first we digress and generalize. We must appreciate America's environment for telephone company provision of enhanced services.

18.3.2 Beyond POTS

Recall that a feature is any nonessential optional minor luxury that can be attached, at additional cost, to some basic service. An enhanced service is also a nonessential optional luxury. But, it isn't minor, and instead of being simply attached to some basic service, it replaces the basic service with a more elaborate one (Section 15.7.2). Centrex and 800 in-WATS are more than features — they are enhanced services.

Enhanced telephone services can be implemented:

- On the network by intelligent calling-party CPE;
- Through the network by intelligent called-party CPE;
- In the network by special intelligent network nodes;
- By the network by the network's own software.

For example, consider voice mail. I once owned a telephone answering machine. It added the voice mail enhanced service *onto* the network by providing it from intelligent CPE in my home that connected to the network like an extension telephone. It had an initial cost, but no recurring charge — at least, none that I recognized when I first bought it. It provided its service whenever I did not answer my telephone but, because of its location on the network's periphery, it could not provide its service whenever I was busy on the telephone. Like many devices in the consumer electronics market, telephone answering machines are typically designed to compete in this market on the basis of price. This can be problematic if the consumer happens to live in a neighborhood that is prone to lightning hits on the power line.

After lightning destroyed the second telephone answering machine in 1 year, I subscribed to my LEC's voice mail service. It gave me voice mail enhanced service *by* the network by providing it from hardware that is *in* the network and that the LEC maintains — not I. Furthermore, this service is hooked into switching software. It has a recurring monthly charge, but little initial cost. And, much to my delight, it provides its service, not only whenever I do not answer my telephone, but also whenever I am busy on the telephone — a feature of this service that I find very useful. The telephone answering service described in Section 15.7.3 is another implementation of voice mail, with even more features — for example, speaking to a real human. This implementation is *through* the network by intelligent called-party CPE.

The open network architecture [41 – 44] is a generalization of the controversy epitomized by the "call diddling" scenario. The carriers wanted the manufacturers to allow access for software written by the carriers. But, the manufacturers, concerned that the carriers' software might interfere with internal switching software and concerned about the finger pointing that might occur whenever a problem occurred, didn't want the carriers' software to be closely coupled to the switching software that the manufacturers provided with the switch. The third-party vendors wanted the carriers to allow access for enhanced services software, written by the third-party vendors. However, concerned that the third-party vendors' software might interfere with the operation of their switches and about the finger pointing that might occur whenever a problem occurred, the carriers didn't want third-party vendor's software to be closely coupled to the switching software inside their switches. The compromise was that the manufacturers would provide a standardized hardware/software interface between their own internal switching software and any external software that might be written by the carriers or by any third-party vendors. This general concept came to be called the open network architecture (ONA).

The ONA has the obvious potential for performance differences, depending on the nature of the interface and the degree of separation. An ONA interface could be designed by the same guidelines that were used in designing the post-Carterfone interconnect interface described in Section 10.6.3 — deliberately impeding the desired cooperation. Under the guise of reliability and security, competing software could be forced to interface through a heavyweight protocol and so much verification procedure and layers of hardware and software to incur noticeable delay. Such an ONA might meet the letter of the law, but not the intended spirit of the law.

Concerned about performance equality, the implementation of the ONA had to ensure that all the competing participants could offer their enhanced services under a similar performance. This requirement came to be called the comparably efficient interface (CEI). So, however ONA was implemented, CEI required that all competitors' software performed equally well, or equally poorly, as long as it was equal. This requirement for least common quality is reminiscent of using toll-grade voice as the standard for PCM. The Intelligent Network, discussed in the next section, evolved out of the CEI requirement on the ONA.

18.3.3 Customer Controlled Routing

The Sellers scenario continues because Sellers is still not completely satisfied with its 800 service. Sellers maintains an extensive database on its customers. Besides the obvious information (name, address, telephone number, credit card number, etc.), this database also includes a complete sales history and other personal information. Sellers' telecom manager describes the service he wants to a representative from his 800 service provider [45].

> *In New York City, Senora Rica, a long-standing and valued Sellers' customer, dials 800-SELLERS to purchase a new pair of matching lamp tables. Sellers wants its carrier to route her call to a Sellers' rep who: (1) is currently working (possibly at home), (2) is currently available and idle, (3) speaks Spanish (because Senora Rica prefers it), (4) has dealt with Senora Rica in the past, and (5) has a location that reduces Sellers' telephone bill. Assume that Senora Rica's call is routed to Pablo, who works from his home in El Paso. Pablo's telephone rings and Senora Rica's record appears immediately on the screen of Pablo's home PC.*
>
> *Pablo answers his telephone (in Spanish): "Hello. Welcome to Sellers, Senora Rica. This is Pablo again. How may I help you?" Senora Rica replies: "Hello, Pablo. Page 412 of the catalog shows some lamp tables. I want to buy two, made of oak." Pablo clicks on the catalog from his home PC, examines this page, and they discuss the particulars. Pablo examines Senora Rica's sales history and says (in Spanish): "Senora Rica, if I may make a suggestion, if you buy the lamp tables in cherry instead of oak, they will be a better match to the dining room set that you bought last month." Senora Rica replies: "It amazes me that you know this because, I remember that I bought that dining room set from Consuela. But, as usual, you have a good idea. I will buy them in cherry instead." Then, Pablo adds: "There is a nice cherry coffee table on page 415 that would go nicely with these cherry lamp tables." Senora Rica flips some pages, pauses, and says: "You are right, Pablo. That coffee table is nice. I will buy that also." They arrange for billing and shipping and the transaction ends.*

Consider what this scenario implies.

- This 800 call is routed, not only by the calling party's area code and by time-of-day and attendant availability, but now also by some customized information.
- This custom information is found in a private database, not necessarily in the carrier's database.

- Moving the CDC's functionality to the PSTN, not only enables reps to work at home, making them easier to hire, but also allows 800 clients to close down their physical CDCs and stop paying rent.
- Because no company wants to compete on price, 800 clients can compete on the basis of personal customer handling, like awareness of sales history and rep loyalty.
- More than just competitive advantage, such special handling can create more sales — like selling the coffee table to Senora Rica.

After describing this service scenario, Sellers' telecom manager discusses it with the telephone company's representative. [This service that Sellers wants and the subsequent discussion are not as fictional as it might appear in this subsection. While I have fictionalized the names and improved the dramatization, I actually witnessed such a discussion between a real telecom manager (TM) from a real company like Sellers and real representatives (CR) from real carriers.] Their discussion proceeds.

> *CR: You'd have to give us a little time to develop the software [notice, now, who's asking for time to develop software], but we should be able to route 800 calls this way.*
>
> *TM (interrupting): I want this service and I want it on my schedule, not on your schedule. Furthermore, I want it exclusively. This service will give us a competitive advantage and I don't want you offering it to my competitors.*
>
> *CR: You're being naive about this, but I suppose we can work something out. [Notice, now, who's calling someone else "naive."] But, you realize, of course, Sellers would have to let us access their database.*
>
> *TM: Would you let Sellers access your customer database?*
>
> *CR: Well, no. But, that's different. You might use our data for your own advantage. We have security concerns. We're the telephone company, after all.*
>
> *TM: I see. Let me see your credit card. [The Sellers telecom manager and the carrier's representative each open their wallets and each pulls out a credit card. The Sellers' TM holds up a Sellers credit card and the carrier's representative holds up his company's credit card.]*
>
> *CR: Uh-oh.*
>
> *TM: The telephone company, huh? When we don't compete, when your company gets out of every business that Sellers is in, then I'll think about letting you look into my database.*
>
> *CR: I can't promise that. But, in order to have this service, whoever routes this call must look at your data.*
>
> *TM: That's correct. And, since I'm not going to let you look at my data, then you have to let me control your network.*

CR: But, you can't develop switching software. It's much more complicated than you realize. You're being naive. [Notice, now, who's telling someone else that they don't appreciate the complexity of switching software.]

From this hypothetical dialog (actually, not so hypothetical), hopefully the reader sees consistency with the history of enhanced services, understands the modern-day lack of trust, and sees the motivation for allowing the users to create their own services and operate the network.

Recently, and probably motivated by competition in 800 service, some of the IXCs have provided some new services for their 800 clients. While still concerned about letting a client's software route their incoming calls, some IXCs provide databases in their networks where clients like Sellers can place selected information from their private customer database. This way calls can be routed by the carrier's software and the carrier doesn't need access to the client's entire customer database. Also, some carriers can route calls based on data in a client's database, but the client's usually put their more sensitive data in a separate database. Call routing based on customer-provided data enables services like calling an 800 number for a national pizza chain and being routed to the store that is closest to your home.

18.4 ADVANCED INTELLIGENT NETWORK

The evolution toward the Intelligent Network has been driven by:

- The need to simplify switching software through further modularization and partitioning;
- The optimization of centralizing the intelligence for, and control of, telephone services and features;
- The logical extension of call control features like busy handling, calling party ID, and the line information database (LIDB);
- The carriers' desire to program their own switches and offer strategic service in a competitive market;
- The implementation of ONA/CEI that enables the competitive deployment of 800 calling and other enhanced services, as independently from the network as practical;
- The logical extension of private networks to providing a platform so end users can implement their own private telephone services;
- Opening the enhanced service market to third parties and providing a friendly environment where they can create enhanced services.

18.4.1 IN Versions

While switching hardware hasn't changed much since the introduction of digital switching in the early 1980s, the physical architecture of switching software has evolved significantly during the 1990s. Beginning with a centralized approach to implement 800 calling and credit card calling, the PSTN's control structure has gradually evolved toward more centralized intelligence [46 – 53]. International standards bodies, technical workshops, several carriers, but mostly Bellcore, have contributed to an architecture of the Intelligent Network that has evolved through several architectural releases: IN 0, IN 0.1, IN 0.2, IN 1.0, and continuing.

One of the IN variations is called the Advanced Intelligent Network. While AIN is by no means universally deployed, AIN is described here as just one example of an IN implementation. AIN has had its own sequence of releases: AIN 0.0, AIN 0.1, and continuing. The carriers that use AIN are currently evolving to AIN 0.2. While AIN has been implemented by some carriers, and is used by many others, the opinion expressed here is that it is still in transition. While there may be more incremental changes to AIN 0.3, the industry is discussing the next major shift, to AIN 1.0. Because the partitioning of switching software will continue to evolve and the SS7 network may have to change soon, the AIN's physical architecture may have to evolve with it. This section's first two subsections present the AIN's physical architecture and then discuss AIN operation. The last two subsections describe local number portability and an extended example, respectively.

18.4.2 AIN Architecture

The AIN architecture is reminiscent of the architecture of ISDN interfaces described in Section 13.4.2 — it seems complex under a cursory examination, but it's quite elegant when you see how all the pieces fit together. However, the AIN architecture has to scale over greater utilization and will be subject to greater software partitioning than the architecture of ISDN interfaces. So, it may not evolve gracefully and may need further revision. We'll discuss the future of AIN in the next section, but the current state of its architecture right here. To do this, the AIN's complicated block diagram is presented and, then, each of its layers is described.

PSTN Control. Figure 18.5 shows a logical diagram of the AIN. Compare this figure to Figure 11.1, a generalized model for process control. Figure 11.1's upper box represents the process that is controlled, and its lower box represents the subsystem that controls this process. Figure 18.5 applies this process control model to America's PSTN, as equipped with AIN. A horizontal dashed line divides Figure 18.5 into upper and lower parts, corresponding to the two boxes in Figure 11.1. The upper part of Figure 18.5 shows the PSTN's switching infrastructure as the process that is controlled by the part underneath. The lower part of Figure 18.5 shows that the PSTN is controlled by a distributed control system, the AIN.

The traditional telephone network appears above the horizontal dashed line. This part of Figure 18.5 shows the CPE, implied to be telephones, for two subscribers, A and B. Each subscriber homes on a different local switching hub. These hubs are part of the generalized PSTN and they are interconnected by it. Voice subsystems, called intelligent peripherals, must be added to the PSTN infrastructure for AIN. They would be added as needed, but only one is shown in Figure 18.5. The switch processors in every hub of the PSTN are drawn below the line in Figure 18.5 to indicate that they are part of the distributed control system. Any switching hub whose switching software has been modified for AIN is renamed a service switching point (SSP).

AIN Layers. The architecture of the AIN, the distributed control system that controls the PSTN, appears below the horizontal dashed line. The AIN is a hierarchically distributed

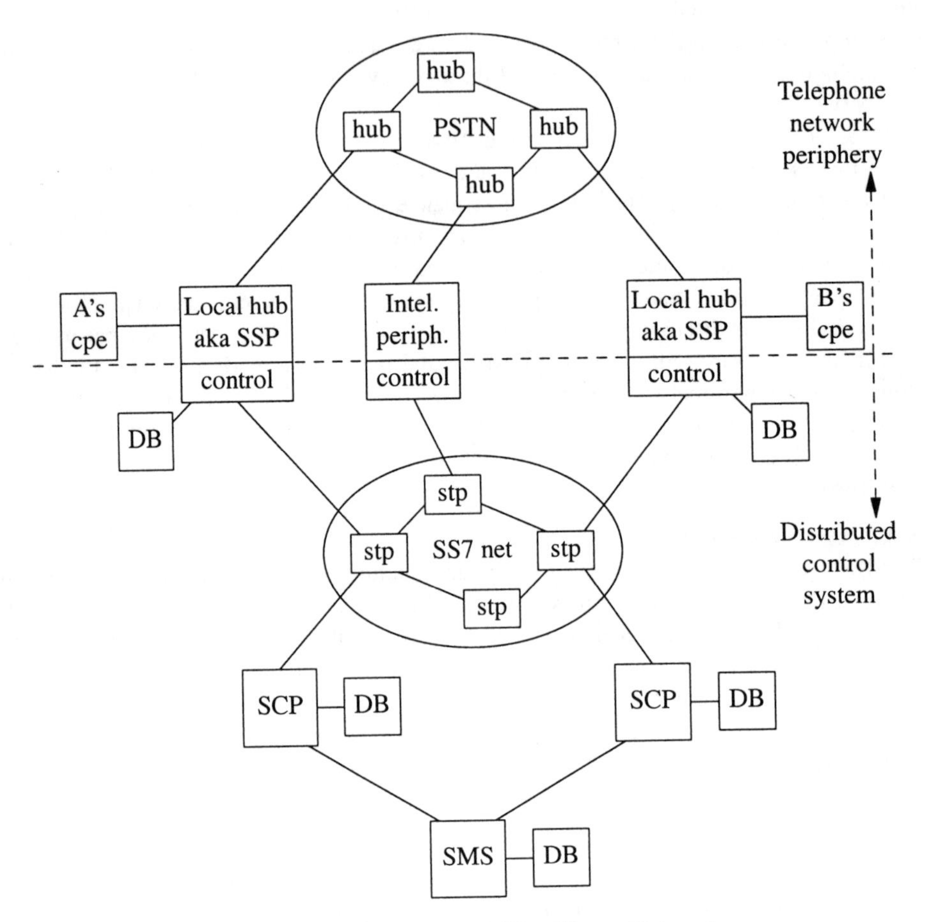

Figure 18.5. The Advanced Intelligent Network.

system of interconnected processors and databases. The system has four layers of special-purpose processors, each with its own database. The AIN's four layers are described, top-down.

Service switching point (SSP). An SSP is an AIN-compatible switching office. Every SSP is controlled by its own dedicated switch processor (SP). The 5E's SP is, itself, a distributed processor system. This SP physically resides with the SSP it controls, was manufactured as part of this SSP, and is programmed solely by the manufacturer of the SSP. In some AIN versions and some depictions of the America's AIN, the SSP is a separate processor that serves as an interface between the AIN and the switch processor inside the switching hub. While most of the 20,000 switching offices in the United States are SS7-compatible, many of these are not yet AIN-compatible SSPs.

While Figure 18.5 implies a single PSTN (like a Bell System), the United States has many competing parallel LECs and many competing parallel IXCs. So, to be more accurate, the CPE in Figure 18.5 could connect to a choice of local hubs and calls could be connected to a choice of IXCs. In Figure 18.5, the hubs shown in the PSTN circle might also be SSPs. Further muddying the picture: (1) some carriers, like independent

LECs, may choose to share some RBOC's AIN; (2) some carriers, like most IXCs, may choose to maintain their own AIN; and (3) some carriers may even choose to implement AIN functions from a proprietary architecture that doesn't resemble Figure 18.5. Furthermore, some day, PBX programs — and even programs in home PCs — might be allowed direct access to the SS7 network. So, in theory anyway, a PBX — or even a PC — could be an SSP.

Traditionally, the local database in a CO's switch processor holds class of service information for subscribers served by this CO, translates incoming called directory numbers to their physical line appearances, and translates outgoing office codes into the trunk group that has the physical channels for this call's route. Gradually, some of this information will migrate to a regional database, called the line information database (LIDB), like an SCP in the AIN architecture in Figure 18.5.

Intelligent peripheral. An intelligent periphal is a simple switching fabric populated by voice equipment like audio tone generators and equipment for recorded announcements and voice mail. The intelligent peripheral is the only IN component that is physically part of the PSTN.

Signal transfer point (STP). STPs were described earlier in Section 18.1. An STP is a packet-switched router, a node in the SS7 network. Even regions that don't use AIN have STPs for SS7 signaling. Considering reliability, the current SS7 traffic load, and the lengths of SSP-STP links, each LATA typically has two STPs. Because SS7 links use existing facilities, STPs are typically located in CO buildings that terminate a lot of channels, such as toll points, tandem switches, and large local switching offices. While SS7 traffic grows annually, STPs are so underutilized that the current set of STPs will be adequate for several more years.

Every SSP has at least one 56-Kbps SS7 link to both STPs in its LATA. Besides connecting to each SSP in its LATA, an STP must also connect to STPs in adjoining LATAs and to regional SCPs in the AIN. As SS7 traffic has increased, these low-rate links have congested and parallel 56-Kbps links have been added to the network as needed. As 800 service, cellular telephony, local number portability, and AIN services evolve and become more popular, SS7 traffic will increase even more, requiring even more paralleling of the 56-Kbps links. This poor scalability is a weakness in the SS7 architecture and we should expect to see a major change, maybe to T1, on these links. Some people advocate replacing the entire SS7 network with a more robust and scalable architecture, maybe ATM.

Service control point (SCP). An SCP is a special-purpose computer with a large associated database and many i/o ports for SS7 links. IN service programs reside in the SCP. As this book is going to press, those RBOCs, which have deployed AIN, typically have 10 or fewer SCPs in their entire region. Because few independent telcos or CLECs (CATV companies, for example) have deployed their own SCPs so far, they lease access and CPU cycles from the RBOC in their region. Many IXCs have evolved their 800 databases into SCPs and are deploying their own AIN SCPs on a national scale. Besides the SCPs dedicated to the 800-number database, other SCPs will become specialized and service specific. So, as class of service migrates into LIDBs, as cellular homing databases evolve into specific SCPs, and as independent telcos and competitive LECs deploy their own SCPs, the number of SCPs in a region will grow and so will SS7 packet traffic. But, when third-party providers deploy their own SCPs, when local number portability is deployed, and as AIN services get more popular, the number of SCPs will increase even more.

Just like STPs are paired for reliability, so are SCPs. Because each SCP has SS7 links to at least two STPs and has an SCCP address, any SSP can communicate with any SCP. For most applications, however, each SSP homes logically on one SCP. SCPs are typically located in large CO buildings, typically that also house an STP and a toll/tandem point. An SCP's operating system administers a set of application programs and a set of utility programs. One of these utilities handles TCAP messages, routing them between the SS7 link and the destination application program.

From this architecture, it is possible (even, likely) that intra-LATA calls would be completed using information obtained from an SCP that is located in a different LATA. So, one of the IXCs argued that, while the bearer channel may not cross the LATA boundary, the signaling does. This IXC argued that this is an inter-LATA call and, as such, must be completed by an IXC and not a LEC. While this IXC lost this argument, it does serve to indicate the level of nastiness and pettiness in America's telecom industry.

Service management system (SMS). An SMS provides the capability for administering customer information, like routing plans, by the telephone company or directly by the customer himself. Today, many enterprise telecom managers use the telco's SMS to continuously administer their enterprise private networks. In the AIN scheme, the SMS provides for:

- External access for service creation and debugging, typically over a modem-connected dial-up line through the PSTN;
- Access to the PSTN's various system control (OAM&P) functions for traffic statistics, billing, and other information that might be used in processing AIN services [54].

Currently, SMSs are not connected to the SS7 network, but this could change. Each SMS is dedicated to a unique set of SCPs and is connected to each of the SCPs it serves by a physical private line that runs X.25. Each SMS has a modem pool so that authorized external users, like third-party vendors, can log in.

Discussion. While the AIN can provide many new enhanced services, some present interesting challenges to implement. In Figure 18.5, consider a call in which A calls B. Let A have some originating IN call service whose program resides in the SCP at the left, and let B have some IN terminating IN call service whose program resides in the SCP at the right. These two physically separated service programs could interact with each other or, somehow, cancel each other. An example of interaction between two terminating services, that even causes trouble without AIN, is call forward deadlock, in which A forwards his calls to B and B forwards his calls to A.

Even in the simplified block diagram of Figure 18.5, which implies a single LEC, these service interactions are unresolved and still lurking out there. The problem is exacerbated when the interacting services are written by different AIN service providers, reside on SCPs manufactured by different hardware vendors, or are implemented on different architectural variations of the AIN.

18.4.3 AIN Operation

This AIN architecture is illustrated by describing *call triggers,* walking a call through the system, and showing how services are created. Extensions of AIN that will be described include: local number portability at the close of this subsection, a hypothetical example of a complex AIN service in the next subsection, and the next step after AIN in the next section.

Call Triggers. Just as the IN architecture has evolved through IN's various releases, the IN call model has also evolved. The IN call model is based on a finite-state machine, with a relatively small number of states and relatively sparse interconnectivity. The basic structure is characterized.

- A point in call (PIC) is the particular state of a call in whichever finite-state model is assumed or specified.
- A trigger detection point (TDP) is a PIC from which the call processing program may invoke some higher level of intelligence by sending a TCAP message to some SCP over the SS7 network.
- A trigger is a TDP that actually produces a query.

For example, at the dialing completed PIC, the TDP, "trigger on 800," doesn't cause a trigger unless the first four dialed digits are 1800. TDPs are potential triggers and the trigger occurs — or not — depending on the user's action, the user's class of service, whether the user has subscribed to a service, whether the service is currently deployed in this switching office, and many other issues. The set of all TCAP messages and the way they interact is called the Intelligent Network application protocol (INAP). In the OSI Reference Model, INAP is an application that resides above TCAP, which is in the application layer.

So, when a TDP actually triggers at some PIC, the appropriate TCAP or ISUP message is transmitted down the protocol stack. The program suspends and takes a real-time break. When the response packet is received, the program resumes at the PIC with the payload information available.

Call Walk-Through. AIN operation is illustrated by a call walk-through in Figure 18.6. Suppose users A and B in Figure 18.5 are the same Senora Rica and Pablo, respectively, from the service scenario presented by the Sellers' telecom manager. Suppose Senora Rica lives in New York City and Pablo works out of his home in El Paso, Texas.

The call begins when Senora Rica goes off-hook. Assume she homes on the Church Street CO in downtown Manhattan. If this CO is a 5E, then its Unix OS spawns, for Senora Rica, another process of the feature control program (Section 14.2.6) that handles dial tone connections and digit analysis. Senora Rica hears dial tone and she dials 1-800-SELLERS.

If this CO is an AIN-compatible 5E SSP, feature control has been modified to trigger on seeing an 800 number. On triggering, it consults the appropriate SCP for 800 ownership translation. A TCAP packet, containing the called number, is launched through the SS7's STP to the SCP in the lower left of Figure 18.6. This SCP is either the one that is assigned to serve this particular SSP (if it has the 800 ownership translation) or it is the special-purpose regional SCP that does have this translation. After receiving this

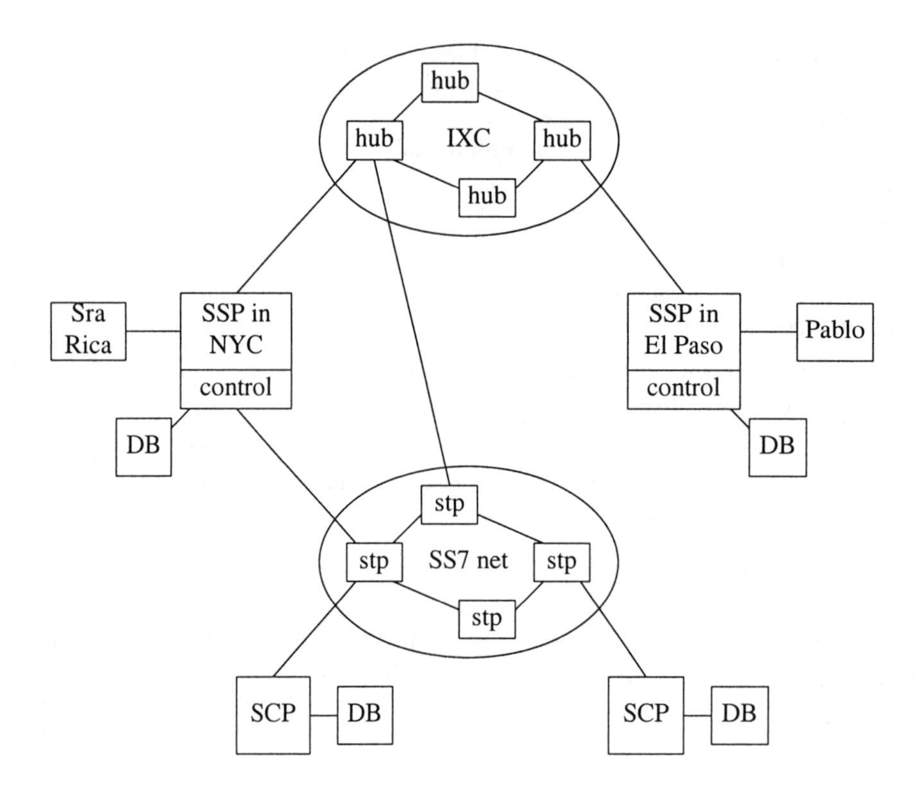

Figure 18.6. Walking a call through the AIN.

packet, an SCP program consults its local database and translates the 800 number into the identity of the IXC that owns the number.

The SCP transmits a TCAP packet back to the trigger point in Senora Rica's feature control process in the 5E processor in the Church Street SSP. This packet instructs the program to route the call to the IXC's regional POP, shown in the upper left inside the PSTN ellipse in Figure 18.6. Simultaneously, this SSP signals the call information to this POP, over the SS7 network shown in the lower center of Figure 18.6. If this POP is also an AIN-compatible 5E SSP, feature control has been modified to trigger on seeing an 800 number. On triggering, it consults its appropriate SCP for 800 number translation. A TCAP packet, containing the called number, is launched through the SS7's STP to the SCP in the lower right of Figure 18.6. This SCP is this IXC's special-purpose national SCP that translates the 800 numbers for its own clients (like Sellers). After receiving this packet, an SCP program consults its local database and attempts to translate the 800 number into its actual directory number.

However, for this call to 1-800-SELLERS, the SCP program triggers on this special number and an extraordinary procedure is invoked. This part of the walk-through depends on the implementation, on what Sellers' IXC allows its clients to do, and on how much Sellers trusts this IXC with confidential data.

- If Sellers trusts the IXC, Sellers would provide some customer information, which this IXC would install on its national 800 translation SCP. In this case, Senora Rica's 800 call would be routed to Pablo by the SCP in the lower right of Figure 18.6.
- If Sellers insists on keeping its customer information in its own database, then this SCP has to interrogate that database for any special instructions on how to route this particular calling party. But, SCPs can network directly only by SS7. So, the SCP would send a TCAP message to the IXC's SMS. There a program provided by the IXC, or by Sellers if this IXC allows it, sends an X.25 message to Seller's customer database. There, the 800 number is translated into Pablo's telephone number in Texas and this number is signaled back to the SMS and then to the SCP.

In either implementation, the program in the SCP sends a TCAP message to the trigger point in the IXC's POP in New York, the leftmost box in the ellipse at the top of Figure 18.6. The IXC would then route the call through its network to its POP in El Paso. There the call would be routed to Pablo through the LEC that serves his telephone.

Service Creation. The typical sequential procedure for creating a service has four steps [55,56].

1. The service's program is developed, debugged, and simulated off-line in the classical way that other programs are designed.
2. An executable module is downloaded to an SMS. Appropriate TDPs are installed in SSP programs, but they only trigger when the subscriber is one of a few beta users in the SSP. The TCAP packet distribution utility in the SCP is instructed to forward these packets to the SMS instead of to an application in this SCP. The program at the SMS executes when one of the beta users in the SSP triggers the service under test. The service program is beta tested and its developers monitor program activity by modem call to the SMS.
3. After documenting a successful beta test, the program developers request permission from the owner of SCP to run an alpha test. The set of test subscribers is increased and the program is installed to execute directly in the SCP instead of indirectly at the SMS. The program, now at the SCP, executes when one of the alpha users in the SSP triggers the service under test. The service program is alpha tested and its developers monitor program activity by examining messages that the SCP program sends to a file at the SMS.
4. Finally, the debugging triggers are removed from the service program at the SCP and the service is declared open for business.

Many software development environments have been created for IN software development. One such environment, called the service creation system (SCS) is extremely convenient and user-friendly for developing very simple services. SCS has a very simple interface, with a graphical language and point-and-click menus. The developed service executes directly on the SCS for debugging. More complex services need classical software development tools. But, who should be allowed to develop programs for the IN?

- Manufacturers?
- Carriers?
- Programmers contracted by the carriers?
- Programmers "certified" by the carriers?
- Competitive third-party vendors?
- Users building their own proprietary services (Sellers)?
- Free-ware benefactors in the Internet "underground"?
- Any kid with a Nintendo?
- Any "hacker" intent on bringing the system down?

How does the SCP's owner distinguish them?

18.4.4 Local Number Portability

LNP is a systematic approach by which a telephone subscriber can keep his phone number for his entire life, like he keeps his Social Security Number [56 – 60]. With LNP, a subscriber's logical number would be reconnected to a different physical line whenever he moves to another residence or changes carriers, analogous to the activity of a main frame or facility switching. LNP enables competition in the local loop because, without it, customers would have to change phone numbers every time they changed to another local carrier. LNP is a requirement of the 1996 Telecommunications Act.

The short-term solution for an LNP implementation uses call forwarding, but this consumes two telephone numbers — the listed number and the actual number. A ported logical number could never be someone else's physical number. The more optimal implementations, using the Intelligent Network and a significant amount of SS7 signaling, are problematic.

Depending on the implementation, LNP may require a regional database, similar to the 800-number database and the LIDB. The LNP database (LNPDB) translates the listed directory number, the number dialed to reach a given called party, into the actual physical number where the called party is logically connected to the PSTN. Using this approach does not consume two telephone numbers because physical and logical numbers are separated. The most significant problem is determining the optimal time/place during call routing to query the LNP database. Four proposals are under consideration.

- Let the originating carrier check for porting after receiving the called party's number from the originating subscriber. This procedure is the most logical but it requires a regional LNP database that can be queried by SS7 messages from any CO in the nation. While the LNPDB query might require interregional signaling, the query could be part of the busy/idle request that the originating carrier makes anyway. The problem is that all switching offices must cut over to LNP simultaneously. If A ports his home phone number, A cannot receive calls from B if B is served by a CO that doesn't have switching software, or that isn't SS7 compatible, or that hasn't installed the generic for LNP yet.

- Let the terminating carrier check for porting just before the arriving call (assuming inband signaling) is connected through the terminating switch. In this procedure, the LNP database only needs to be local, inside the CO, but tandeming the call's route could lead to circuitous routes, especially if the called party has ported his number to a distant CO, like across the country.
- If the next to last carrier does the LNPDB query, many of these circuitous routes might be avoided, especially if LNP is restricted to intra-LATA or intraregion. This procedure is a compromise between the first two.
- Query on release is similar to terminating carrier check except that the terminating CO signals to the originating CO to drop the switch train and completely reroute the call. Here, the LNPDB can be local and extra signaling occurs only on ported calls. However, because ported calls have significantly increased setup delay, the established LECs would have an advantage in loop competition.

As this book is going to press, this important issue is not yet resolved. LNP will have a significant impact on the SS7 network, especially if its implementation requires an LNP query to a remote database for every originating call to any number, ported or not. If remote LNP databases are required, perhaps one per region, then who's paying for it? But, LNP is required by the 1996 Telecom Act, so we will do it, somehow.

Once LNP is in place, however, some subscribers may want to reconnect their logical number to a different physical line every day, or even every hour. It might be convenient to reassign your logical telephone number to different nearby telephones during your daily activity. Subscribers may even expect to be able to port their logical number to a different line while they're using the telephone, analogous to the call hand-off in cellular telephony. This extension of LNP is called the personal communications number, which is discussed in the next chapter in the context of cellular communications.

18.4.5 Extended Example

The power of the AIN, and AIN's potential to really serve users, is illustrated by describing a sophisticated service. This example is purely hypothetical and is not an actual AIN service. In fact, it's not clear that the current state of the AIN is sufficiently advanced that this service could even be implemented — yet.

Imagine that the Educational Testing Service (ETS), the enterprise that administers SATs and GREs, wants to automate its procedure by which an approved admissions officer at some university acquires some applicant's official test result. Assume that an Internet implementation is deemed unacceptable because of concerns about security and the non-uniform level of automation among different admissions offices. Imagine that ETS and its community of users select an implementation based on telephone access.

So, ETS of Princeton, New Jersey offers dial-up access to a distributed database of SAT and GRE test results. The service is available to universities' authorized admissions officers. ETS equips seven sites, one in each RBOC region. Each site has a copy of the common database and a voice response unit (VRU). ETS keeps all sites up-to-date by broadcasting new data every night. The regional database is reached by dialing 1-800-CALLETS. An authorization database (ADB) is maintained by the Department of

Education in every state's capital city. Admissions officers are registered in the ADB by their telephone number and a personal identification number (PIN). The ADB also has a VRU.

First, a block diagram of the system's components is presented and then a call is walked through the service. The example shows an admissions officer at the University of Pittsburgh trying to reach an ETS site that happens to be located in Frederick, Maryland.

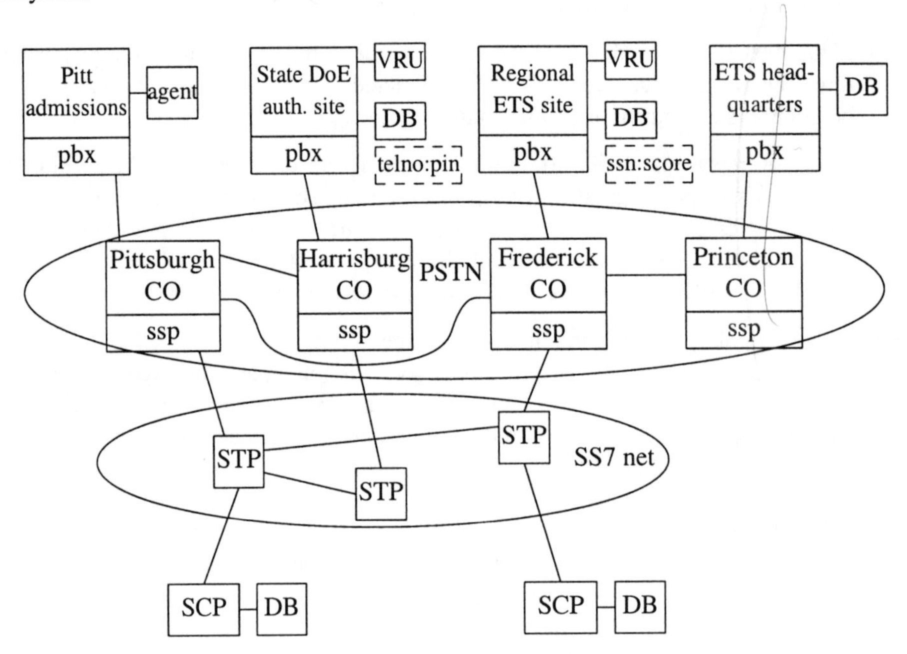

Figure 18.7. Hypothetical AIN service for ETS.

Figure 18.7 is a block diagram of the system components employed to implement this service for ETS. The top row shows: (1) the admissions office at the University of Pittsburgh (Pitt), with one admissions officer and Pitt's PBX; (2) the authorization site in Harrisburg, Pennsylvania, with its associated database and VRU; (3) the Bell Atlantic regional ETS site in Frederick, Maryland, with its associated database; and (4) ETS headquarters in Princeton, New Jersey, with its master database. The second row shows the switching offices associated with each of the locations in the top row, respectively. The four offices are compatible with SS7 and with AIN (they are SSPs). The third row shows part of the SS7 network, showing the STPs that are associated with the Pittsburgh, Harrisburg, and Frederick offices, respectively. The fourth row shows the SCPs, each with its associated database, that are associated with the Pittsburgh and Frederick offices, respectively. The SMSs aren't shown, but they would be in a fifth row, each with its associated database, connected to the SCPs.

While some clever designer might be able to implement this service now, AIN 0.1 may not have the necessary TDPs to provide this exact service. So, the service is proposed here in principle, but it might be possible on some future AIN release.

Ninette, an admissions officer at the University of Pittsburgh, needs to know the GRE score for Tim, a student who has applied to Pitt. She calls 1-800-CALLETS and we examine the details as we walk her call through the enhanced 800 service that some third-party vendor has implemented on the AIN exclusively for ETS. Such an 800 call has three separate parts and its detailed steps are grouped accordingly.

Calls to 1-800-CALLETS will be completed only if the caller is authorized. The originating CO, the SSP that serves the admissions officer's university, places a preliminary call to the appropriate authorization site. So, before Ninette can be connected to Frederick, she is connected first to the ADB in Harrisburg, Pennsylvania. This system prompts Ninette to enter her PIN at her Touch Tone telephone. The ADB verifies that this PIN matches the calling party ID, and if it does, the call is allowed to proceed.

1. Ninette dials 91-800-CALLETS into Pitt's PBX, which connects Ninette's telephone to an outside line and signals "18002255387 + Ninette's extension number" to the Pittsburgh CO. The Pittsburgh CO's conventional call processing program, seeing the leading 1 in the called number from a PBX trunk, verifies that this extension's class of service permits a long distance call.

2. Then, because the Pittsburgh CO is an IN-compatible SSP, call processing triggers on 800 and transmits an SS7 TCAP message containing "8002255387 + Ninette's extension number" to the SCP in the lower left of Figure 18.7. For simplicity, assume that this same SCP handles 800-number ownership and 800-number translation.

3. The SCP's TCAP utility routes the request to the 800 translator program. When this program translates the 800 number in its database, the call is recognized as an enhanced service, which must be processed by its corresponding program in the SCP (or SMS). This ETS program "dips" another database and translates the area code from Ninette's calling number into the telephone number of the authorization database in Harrisburg. This called number is returned to call processing in Pittsburgh.

4. Call processing in the Pittsburgh CO, which had been waiting for an 800-number translation, routes Ninette's call to the Harrisburg ADB, with Ninette's calling party ID.

5. The Harrisburg CO rings the ADB and transmits Ninette's CPI. The VRU answers the incoming call and Ninette is connected. After the VRU prompts Ninette for her PIN, she dials her PIN and the ADB verifies that the PIN matches the CPI in its database. The ADB signals the SCP that Ninette is authorized. Possible implementations for this step include: signaling over an ISDN line's D-channel to the Harrisburg CO, a data packet sent over some private network to the SCP, a dial-modem call to the SMS that serves this SCP, and probably others.

Ninette's call to the ADB Harrisburg is dropped and her call to ETS, now authorized, is connected to Frederick, Maryland. If the ETS site in Frederick were out of service or too busy, her call might be routed to Ameritech's or Nynex' ETS site instead.

6. The ETS program in the SCP transmits a disconnect/reconnect TCAP message to the Pittsburgh CO with the telephone number for ETS's database in Frederick. This message has to simulate a hang-up and a redial to call processing. Ninette's

call to the ADB is disconnected in Pittsburgh and the switch train to Harrisburg is dropped.

7. The ETS program transmits another SS7 message to the SCP that serves the Frederick CO, alerting an ETS program in this SCP that an incoming call with Ninette's CPI is authorized.

8. Call processing in the Pittsburgh CO routes Ninette's call to the ETS database in Frederick, with Ninette's calling party ID.

The CO in Frederick verifies that incoming calls to this ETS site are indeed authorized. Relying on the originating software to verify authorization leaves the ETS site vulnerable to unauthorized people who call the regular telephone number instead of the 800 number.

9. Call processing in the Frederick CO triggers on incoming call and sends a TCAP message, with Ninette's CPI, to its SCP requesting approval to proceed. This SCP verifies that the call is authorized.

10. The Frederick CO rings the ETS site and transmits Ninette's CPI. The VRU answers the incoming call and Ninette is connected.

The VRU at the ETS site prompts Ninette for the Social Security Number of the student whose test score she wants. After Ninette Touch Tones the SSN, the database retrieves the score, and the score is delivered to the VRU and spoken back to Ninette.

18.5 EVOLVING PARTITION

The switching partition is the boundary between a switching fabric and the system that controls it. Figure 11.1 shows the partition most clearly. The evolution of the partition has been followed in every chapter that described some telephone system — Chapters 6, 8, 11, and 14. Analogous to how the toll point disintegrated in Section 9.4, we've seen how switch control has moved further from the system being controlled and has even become a system of its own in the distributed processor architectures of most modern digital switches. This chapter has shown how control is evolving toward even further separation into distinct software and system components. This section proposes taking this even further by: (1) a local control architecture that uses physically distributed special-purpose processors, and (2) a software architecture whose logical functions are further disintegrated. If you've been following the partition subplot as you've been reading, you'll see that what's proposed here is simply a continuation of the momentum in the direction of the evolving partition.

18.5.1 The Physical Partition

Figure 18.8 shows a physical control architecture for a CO switch as a system of distributed processors on a LAN. While the figure shows a bus network, the bus is probably not the best architecture for this LAN. While the figure implies that co-located component processors are interconnected by a LAN, the component processors don't have to be in the same CO building and they could be interconnected by a MAN or WAN (NAN).

The manufacturer's processor contains the *drivers* for the hardware of the fabric and the PSTN. This processor is a simplified version of the switch processor found in every

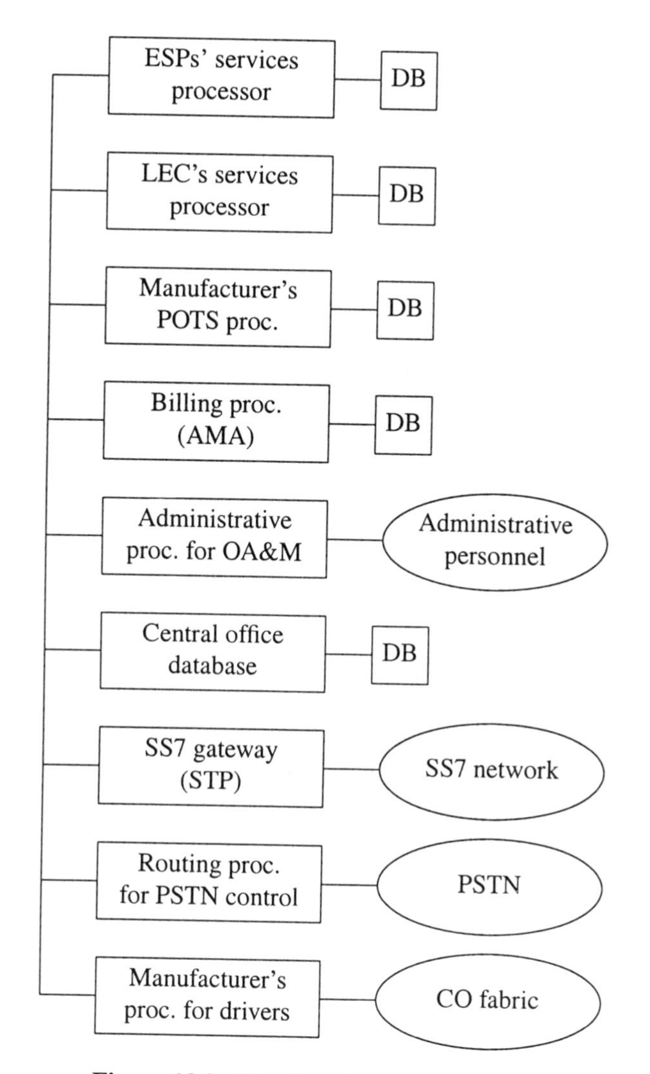

Figure 18.8. Distributed processor control.

5E switching module. Ideally, all switching fabrics and even all network architectures (including ATM and VoIP) would be controlled by the same set of primitive commands that would be issued as interprocess messages on the implied LAN.

Software for PSTN routing, POTS call processing, all features, and all enhanced services would be physically separated from this processor that physically controls the switch and the PSTN. Separate processors would house software provided by the manufacturer (POTS and most features), the LEC, and the third-party ESPs. Some enhanced services, like 800 calling, might reside on dedicated enhanced services processors and some might be pooled together on a common machine. The ESP processor in Figure 18.8 might be a single machine that is provided by the LEC and is shared by all ESPs or it might be a set of separate machines provided by each ESP. This set of processors is a generalization of the administrative processor found in the administrative module of every 5E. Recall that because Figure 18.8 is a logical block diagram, these services processors might be remote and shared by several COs — in which case they are more like an SCP than like an AP.

Some functions should be physically separated from the AP into special-purpose processors either because the function performs better on special-purpose hardware or because the function is asynchronous from call processing. A central office database machine would hold all translations and possibly class of service, analogous to the LIDB. A separate billing machine, analogous to the AMA, might be able to compute the telephone bill simply by monitoring interprocessor messages on the LAN. A separate administrative machine provides interface for LEC personnel for performing OAM&P and for remote access by the manufacturer.

Assuming the SS7 network evolves to two layers, the existing per-LATA STP-pair becomes a fault-tolerant tandem node, and also serves as a hub for each of its LATA's COs. A separate per-CO front-end machine acts as an SS7 gateway.

While Figure 18.8's level of physical fragmentation might seem extreme, there are many good reasons for it.

- One hundred 100-MIPS processors are cheaper than one 10K-MIPS processor. A high capacity CO LAN can make the physical separation unnoticeable.
- Several special-purpose processors, like database machines, can outperform one general-purpose processor.
- Because some software would be accessible to third-party vendors and some wouldn't be, security is easier if the sensitive software is physically separated.
- Placing all services software, from all vendors, in physically separated computers resolves the CEI issue.

18.5.2 The Logical Partition

In the Intelligent Network, we are witnessing the separation of feature control from the remainder of switching software. But this is only the beginning [61 – 67]. Some ATM advocates and some voice-over-IP (VoIP) advocates urge that route control and switch control should be separated from the remainder of switching software so that:

- The same software that finds circuit-switched routes through the PSTN can also be used to find virtual circuit routes for ATM packets and destination nodes for connectionless VoIP packets;
- If all switching software can be made independent of the hardware details of a specific switching office, then such software could interface to the driver for any type of office, presumably including an ATM switch or an IP router;
- Many new calling situations — such as connecting a packet-phone to a POTS phone or adding a POTS phone in audio-only conference to an existing multimedia call — will require even greater separation of switching software functions.

Figure 18.9 illustrates a finer partitioning of the functionality of telephony. It is not meant to be the final architecture and it does not address where the various software modules might reside, but it should serve as a point of discussion and should convince the

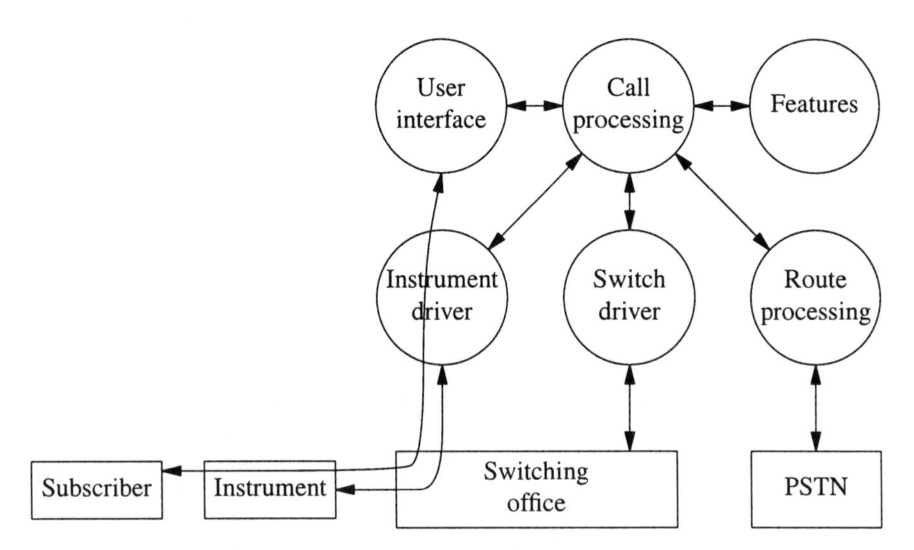

Figure 18.9. Partitioned call software.

reader that the AIN may not be the ultimate end and that even finer partitioning of switching software may follow.

Physical equipment. The hardware is shown across the bottom: from the user to the appliance to the switching office to the PSTN. Six different pieces of software are shown in the circles above. Here we discuss the functionality of these six pieces, the logical partition. The previous subsection described the computers where these programs might reside. The actual assignment of program modules to computers is left undone.

User interface. This process handles the dialog between the user and the system. It provides access to the set of features for which this user has subscribed, or has shown a propensity to use. It holds the user's custom numbering plan.

Instrument driver. This process handles the low-level i/o for the particular appliance that is used for the call. It is the driver for the telephone and defaults to assuming that the user uses an analog POTS phone. If the user's telephone identifies itself as being an ISDN phone, or as having a menu-driven display, or as being a multimedia workstation, or is otherwise different from — or more intelligent than — an analog POTS telephone, then this driver would respond in kind.

Call processing. This process controls the sequence of operations for originating or terminating a call. It knows about feature interaction and regulates which features are permitted to operate at any particular stage of any particular type of call.

Features. These modules control all the additional and special capabilities that enable the communicating parties to communicate in ways that are different from, or enhancements to, the conventional analog voice two-party point-to-point POTS call.

Switch driver. This process provides the translation between the generic commands used by the other programs and the actual details of operating the specific hardware in a given brand or model of switching equipment. Among other things, it provides for intraswitch connectivity: connecting A to a digit receiver, or A to B, or A to a trunk that goes to office X, for example.

Route processing. This process determines a circuit-switched trunk route through a multinode network.

EXERCISES

18.1 Discuss using ATM as a replacement for SS7's MTP.1, MPT.2, MTP.3, and SCCP. Would virtual circuits be more effective than connectionless packet switching for SS7 signaling? Is ATM's address space sufficient for SS7? Is the ATM cell large enough for most SS7 messages?

18.2 Consider how the implementation of call forwarding has changed because of modern signaling. Let user A be served by office X and let user B be served by office Y. Suppose B signals office Y that he wants his incoming calls forwarded to C, who happens to be served by office X. Now, suppose A calls B.

(a) With inband and out-of-band signaling, switching software in X has no way of knowing that B's incoming calls have been forwarded. Because only Y's software knows about B's call forwarding, office Y must complete the connection. Draw a block diagram showing A, B, C, X, and Y. Assume a trunk group between X and Y. Show A's call to B/C on this figure.

(b) With networked signaling, switching software in Y can tell the switching software in X about the true destination of A's call to B. Since X's software knows about B's call forwarding, office X can complete the connection. Draw a block diagram showing A, B, C, X, and Y. Assume a trunk group between X and Y. Show A's call to B/C on this figure.

18.3 Signal commutation could be disambiguated by using a voice response unit: "If you wish to hang up, please press 1; if you wish to add another conferee, please press 2; ...." Show a block diagram of the software architecture if provided by a switching system's manufacturer. Show the implementation by a third-party vendor on the IN. Do you think users would like this?

18.4 Describe an alternate implementation for disambiguating signal commutation that uses the calling party ID display unit.

18.5 Suppose you borrow \$1000 at 25% interest and pay off the loan at \$300.40 per year. Show that the loan is, indeed, paid off in 8 years by completing Table 18.2, continuing it for 8 years.

TABLE 18.2. Payment schedule

year	payment	interest owed	toward principal	amount still owed
0				1000.00
1	300.40	250.00	50.40	949.60
2	300.40	237.40	63.00	886.60
3	300.40			

In any year, part of the $300.40 annuity pays for interest and the remainder pays off some of the principal. The interest owed in year *n* is 0.25 times the amount of the principal remaining at the end of the previous year.

18.6 Consider several variations of the annuity example in Section 18.2.4. (a) Find the APR for 25% interest if the duration is 10 years instead of 8. (b) Find the interest rate for 0.3004 APR if the duration is 10 years instead of 8. (c) Find the APR for 8-years' duration if the interest rate is 20% instead of 25%. (d) Find the duration for 0.3004 A/P if the interest rate is 20% instead of 25%. (e) Find the duration for 25% interest if the APR is 0.25 instead of 0.3004. (f) Find the interest rate for 8-years duration if the APR is 0.25 instead of 0.3004.

18.7 Two switching systems are under consideration. It is decided to select on total cost. System A costs $600K for hardware and installation and $300K for initial software. System A's annual generics cost $50K and its annual operating cost is $100K. System B costs $400K for hardware and installation and $200K for initial software. System B's annual generics cost $100K and its annual operating cost is $200K. Which system should you buy?

18.8 A manufacturer considers modifying its switching system product by making a line card with an on-board microcontroller (Angel) and installing drivers on this processor. Suppose cost analysis reveals that the new line card will be 6% more expensive, but that software costs will decline by 4%. If the original product conforms to the cost breakdown given in Section 18.4.2, should the manufacturer make this change?

18.9 Currently, 800 numbers are decoded by double dipping. First, the calling party's CO dips the 800 ownership database to determine which IXC owns the number. Then, after signaling to this owning IXC, this IXC dips its own 800-translation database to determine the physical number. The LECs could eliminate the first dip by broadcasting the dialed number to all IXCs and requesting a message back from the IXC which owns the number. Even if broadcast were added to MTP, is this a good idea? Discuss.

18.10 Figure 18.10 shows a physical diagram of the buildings that house the six switching offices in Figure 18.1. Let the four COs and the two POPs be physically interconnected by a SONET ring, shown by the conduit connecting the six buildings. Place the LATA's two STPs in buildings with COs (not with POPs) so that: (1) each STP normally handles the load from three different exchanges, and (2) the lengths of the normally used data links is minimal. Show the exchange-STP links that are normally used. Then, add the links between each exchange and its backup STP. Each exchange must have SS7 links to two different STPs and, for reliability, the two links must use different facilities.

18.11 Continuing the previous exercise, assume: (1) one of the two buildings housing an STP also houses one of the SCPs that serves this LATA, and (2) the building with the other STP is the only building with direct facilities to this LATA's other SCP, in a neighboring LATA. Show the SS7 links between the two STPs in this LATA and the SCP that resides in this LATA. Show the SS7 links between the two STPs

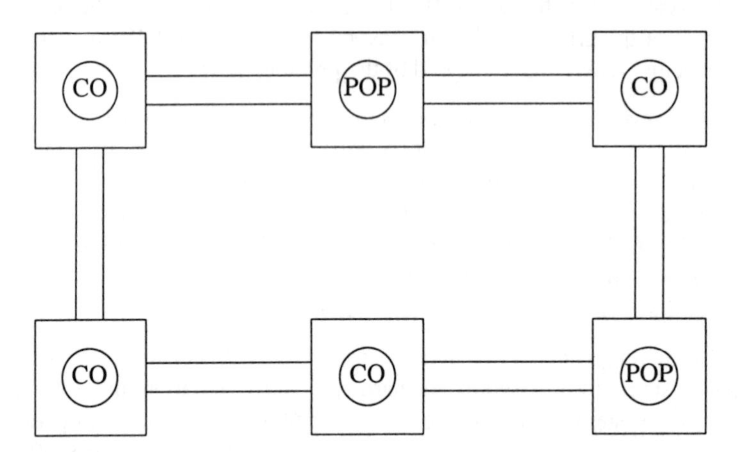

Figure 18.10. Six offices in a self-healing ring architecture.

in this LATA and their other SCP, in the neighboring LATA. For reliability, the two links from any STP to its respective SCPs cannot share common transmission facilities.

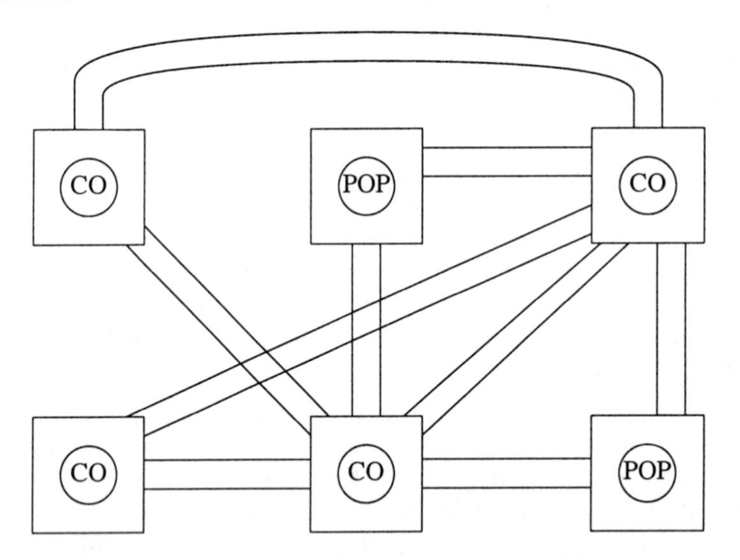

Figure 18.11. Six offices in a dual-hub star architecture.

18.12 Repeat the previous exercise except for a dual-hub star architecture of interoffice trunking. Assume that: (1) the CO building in the south-center contains a toll hub, (2) the CO building in the northeast also has direct facilities to all other exchange buildings in this LATA, (3) no other building pairs have direct facilities, and (4) all interoffice trunks are provisioned from transmission facilities that lie in the nine straight-line conduits shown in Figure 18.11. Place the two STPs in the obviously optimal buildings. Show the exchange-STP links that are normally used. Then, add the links between each exchange and its backup STP. Each exchange must have SS7 links to two different STPs and, for reliability, the two links must use different facilities.

18.13 Continuing the previous exercise, assume: (1) the building in the northeast also houses one of the SCPs that serves this LATA, and (2) the toll hub in the south-center is the only building with direct facilities to this LATA's other SCP, in a neighboring LATA. Show the SS7 links between the two STPs in this LATA and the SCP that resides in this LATA. Show the SS7 links between the two STPs in this LATA and their other SCP, in the neighboring LATA. For reliability, the two links from any STP to its respective SCPs cannot share common transmission facilities.

18.14 Consider the four previous exercises. In the LATA that uses a self-healing ring for interoffice trunking, count the total number of segments in all SSP-STP and STP-SCP links. In the LATA that uses a dual-hub star for interoffice trunking, count the total number of segments in all SSP-STP and STP-SCP links. Compare these two numbers relative to the number of conduits in each architecture.

18.15 Discuss the advantages and disadvantages of general-purpose SCPs versus special-purpose SCPs. Obviously, general-purpose SCPs are better if regional SCP TCAP traffic is light. But, as regional SCP TCAP traffic increases to the point of needing additional SCPs, the region could be split and the new ones could be general purpose or the region could be maintained and the new ones could serve special purposes like LIDB, cellular HLR/VLR, 800 ownership, etc. Consider point code addressing and the number of STP hops per TCAP message.

18.16 Walk an interoffice call through switching software, assuming the terminating party has call screening that is implemented on the IN.

18.17 Suppose an LEC offers calling party anonymity as a special service for subscribing originating parties. Modify the previous problem for an originating party, who has subscribed for this new service, whose call would have been screened otherwise.

18.18 Suppose an LEC offers calling party anonymity override as a special service for subscribing terminating parties. Modify the previous problem for a terminating party, who has subscribed for this new service, who receives an anonymous call that would have been screened otherwise.

18.19 Discuss the benefits and drawbacks to the following methods for breaking the call forwarding deadlock: (1) forwarded calls can't be forwarded, (2) any call forwarding program must receive the entire call forwarding chain list so it can avoid forwarding an incoming call to some telephone that has already forwarded this call, (3) any call forwarding program must backtrack through the call forwarding chain so it can avoid forwarding an incoming call to some telephone that has already forwarded this call, (4) call forwarding is implemented at a single SCP that serves America's entire PSTN, (5) any other approach you can think of.

REFERENCES

[1] Breen, C., and C. A. Dahlbom, "Signaling Systems for Control of Telephone Switching," *The Bell System Technical Journal,* Vol. 39, 1960, pp. 1381 – 1444.

[2] Dahlbom, C. A., "Signaling Systems and Technology," *Proceedings of the IEEE,* (issue on Telecommunications Circuit Switching), Vol. 65, No. 9, Sept. 1977, pp. 1349 – 1353.

[3] Miller, P. R., and R. E. Wallace, "Common Channel Interoffice Signaling (CCIS): Signaling Network," *The Bell System Technical Journal,* Vol. 57, No. 2, Feb. 1978, pp. 263 – 282.

[4] Hlawa, F., and A. Stoll, "Signaling for the Future (Explanation of Common Channel Signaling Based on CCITT System No. 7)," *Telephony,* Vol. 200, Feb. 9, 1981.

[5] Lawser, J. J., and P. L. Oxley, "Common Channel Signaling Network Evolution," *AT&T Technical Journal,* Vol. 66, No. 3, May/Jun. 1987, pp. 13 – 20.

[6] Modarressi, A. R., and R. A. Skoog, "Signaling System 7: A Tutorial," *IEEE Communications Magazine,* Jul. 1990, pp. 19 – 35.

[7] Russell, T., *Signaling System 7,* New York, NY: McGraw-Hill, 1995.

[8] Luetchford, J. C., N. Yoshikai, and T. H. Wu, "Network Common Channel Signaling Evolution," *Proc. XV Int. Switching Symposium, Vol. 2,* Berlin, Germany, Apr. 23 – 28, 1995, pp. 234 – 238.

[9] Sicker, D. C., "Future Signaling Architectures of Public Telecommunications Networks," unpublished Ph.D. Comprehensive Exam Paper — Telecommunications Program, School of Information Sciences, University of Pittsburgh, Nov. 6, 1997.

[10] Belady, L. A., "Software Is the Glue in Large Systems," *IEEE Communications Magazine,* (issue on Telecom at 150), Vol. 27, No. 8, Aug. 1989.

[11] Bernstein, L., "Software in the Large," *AT&T Technical Journal,* Vol. 75, No. 1, Jan./Feb. 1996, pp. 5 – 14.

[12] Bierman, E., "Changing Technology in Switching System Software," *Proceedings of the IEEE,* Vol. 65, No. 9, Sept. 1977, pp. 1329 – 1335.

[13] Aho, A. V., and N. D. Griffith, "Feature Interactions in the Global Information Infrastructure," *Proc. of the Conference on Foundations of Software Enineering,* 1995.

[14] Bouma, V., and H. Velthuijsen, *Introduction, Feature Interactions in Telecommunications Systems,* Amsterdam: IOS Press, May 1984.

[15] Cameron, E. J., and H. Velthuijsen, "Feature Interactions in Telecommunications Systems," *IEEE Communications Magazine,* Aug. 1993.

[16] Buyukdura, F. H., "Feature Interaction: Service and Functionality," *Proc. IEEE Intelligent Network Workshop '91,* Dallas, TX, Apr. 24 – 25, 1991, Session V.

[17] Zibman, I., et al., "An Architecture to Minimizing Feature Interactions in Telecommunications," *IEEE/ACM Transactions on Networking,* Vol. 4, No. 4, Aug. 1995, pp. 582 – 596.

[18] Belanger, D. G., S. G. Chappell, and M. Wish, "Evolution of Software Development Environments," *AT&T Technical Journal,* Vol. 69, No. 2, Mar./Apr. 1990, pp. 2 – 6.

[19] Bergland, G. D., "A Guided Tour of Program Design Methodologies," *IEEE Computer Magazine,* Vol. 14, No. 10, Oct. 1981.

[20] Bergland, G. D., and R. D. Gordon, *Tutorial: Software Design Strategies,* Second Edition, New York, NY: IEEE Computer Society, 1981.

[21] Bergland, G. D., et al., "Improving the Front End of the Software-Development Process for Large-Scale Systems," *AT&T Technical Journal,* Vol. 69, No. 2, Mar./Apr. 1990, pp. 7 – 21.

[22] Brooks, F., "No Silver Bullet, Essence and Accidents of Software Engineering," *IEEE Computer Magazine,* Vol. 20, No. 4, Apr. 1987, pp. 10 – 19.

[23] Dijkstra, E. W., *A Discipline of Programming,* Englewood Cliffs, NJ: Prentice Hall, 1976.

[24] Gries, D., *The Science of Programming,* New York, NY: Springer-Verlag, 1981.

[25] Jackson, M. A., *Principles of Program Design,* London, UK: Academic Press, 1975.

[26] Yeh, R. T. (ed.), *Current Trends in Programming Methodology, Volume I — Software Specification and Design,* Englewood Cliffs, NJ: Prentice Hall, 1977.

[27] Yeh, R. T., and K. M. Chandy (eds.), *Current Trends in Programming Methodology, Volume III — Software Modeling,* Englewood Cliffs, NJ: Prentice Hall, 1978.

[28] Yeh, R. T. (ed.), *Current Trends in Programming Methodology, Volume IV — Data Structuring,* Englewood Cliffs, NJ: Prentice Hall, 1978.

[29] Yourdon, E., *Techniques of Program Structure and Design,* Englewood Cliffs, NJ: Prentice Hall, 1975.

[30] Yourdon, E., and L. L. Constantine, *Structured Design — Fundamentals of a Discipline of Computer Program and Systems Design,* Second Edition, New York, NY: Yourdon Press, 1978.

[31] Lucky, R. W., "Collaboration," *IEEE Spectrum Magazine,* Reflections column, Nov. 1992.

[32] Beckett, J. T., et al., "Methods for Managing a Large Software Project," *AT&T Technical Journal,* (issue on 5ESS Switch Software), Jan. 1986, pp. 247 – 271.

[33] Metzger, P. W., *Managing a Programming Project,* Second Edition, Englewood Cliffs, NJ: Prentice Hall, 1981.

[34] Musa, J. D., "Software Reliability," *IEEE Spectrum Magazine,* (issue on Supercomputers: A Special Report), Vol. 26, No. 2, Feb. 1989.

[35] Kernighan, B. W., and D. M. Ritchie, *The C Programming Language,* Englewood Cliffs, NJ: Prentice Hall, 1978.

[36] Bell-Beam, D., "Computer Telephony Integration: More Than Call Centers — Really," *Phone+,* Vol. 11, No. 5, May 1997.

[37] Garrison, D., "The Future of CTI: Where Are We Headed?" *Telecommunications,* Vol. 32, No. 1, Jan. 1998.

[38] Harvey, D. E., S. M. Hogan, and J. Y. Payseur, "Call Center Solutions," *AT&T Technical Journal,* Vol. 70, No. 5, Sept./Oct. 1991, pp. 36 – 44.

[39] Reynolds, P., "The Science of Call Center Management," *Communications News,* Vol. 35, No. 10, Oct. 1998.

[40] "Telephony/CTI: Thomas Cook Global Services Travels Ahead," *Communications News,* Vol. 36, No. 1, Jan. 1999.

[41] Ianna, F., "Open Networking in a Competitive Environment," *Telecommunications Magazine,* (issue on Achieving Open Network Architectures), Vol. 21, No. 1, Jan. 1987.

[42] Rutkowski, A. M., "ONA Views and Plans: Bellcore," *Telecommunications Magazine,* (issue on Achieving Open Network Architectures), Vol. 21, No. 1, Jan. 1987.

[43] Rutkowski, A. M., "Open Network Architectures: An Introduction," *Telecommunications Magazine,* (issue on Achieving Open Network Architectures), Vol. 21, No. 1, Jan. 1987.

[44] Rutkowski, A. M., M. Gawdun, and N. Morley, "The RBOC's Views on ONA," *Telecommunications Magazine,* (issue on Achieving Open Network Architectures), Vol. 21, No. 1, Jan. 1987.

[45] Bonkowski, E., "Sears' Expectations of AIN," *Proc. IEEE Intelligent Network Workshop '91,* Dallas, TX, Apr. 24 – 25, 1991, Plenary Panel.

[46] Ambrosch, W. D., A. Maher, and B. Sasscer (eds.), *The Intelligent Network — A Joint Study by Bell Atlantic, IBM, and Siemens,* Berlin, Germany: Springer-Verlag, 1989.

[47] Berman, R. K., and J. H. Brewster, "Perspectives on the AIN Architecture," *IEEE Communications Magazine,* Vol. 30, No. 2, Feb. 1992, pp. 27 – 32.

[48] Bloom, P., and P. Miller, "Intelligent Network/2," *Telecommunications Magazine,* (issue on Achieving Open Network Architectures), Vol. 21, No. 1, Jan. 1987.

[49] Dreher, R., "A Network Architecture Supporting IN Feature Migration," *Proc. IEEE Intelligent Network Workshop '91,* Dallas, TX, Apr. 24 – 25, 1991, Session I.

[50] Duran, J. M., and J. Visser, "International Standards for Intelligent Networks," *IEEE Communications Magazine,* Vol. 30, No. 2, Feb. 1992, pp. 34 – 42.

[51] Faynberg, I., et al., *The Intelligent Network Standards,* New York, NY: McGraw-Hill, 1997.

[52] Russo, E. G., et al., "Intelligent Network Platforms in the U. S.," *AT&T Technical Journal,* Vol. 70, Nos. 3 – 4, Summer 1991, pp. 26 – 43.

[53] Thorner, J., *Intelligent Networks,* Norwood, MA: Artech House, 1994.

[54] Bennett, R. L., J. C. Chen, and T. B. Morawski, "Intelligent Network OAM&P Capabilities and Evolutions for Network Elements," *AT&T Technical Journal,* Vol. 70, Nos. 3 – 4, Summer 1991, pp. 85 – 98.

[55] Demestichas, P. P., et al., "Issues in Service Creation for Future Open Distributed Processing Environments," *Proc. 1999 IEEE Int. Conference on Communications,* (Session S8A-1: Topics in Communications Software), Vancouver, Canada, Jun. 6 – 10, 1999.

[56] Morgan, M. J., et al., "Service Creation Technologies for the Intelligent Network," *AT&T Technical Journal,* Vol. 70, Nos. 3 – 4, Summer 1991, pp. 58 – 71.

[57] Johnner, M., "How Local Number Portability Can Foster Competition," *Telecommunications,* Vol. 32, No. 2, Feb. 1998.

[58] Engelman, L., "Have Number, Will Travel," *Phone+,* Vol. 9, No. 10, Aug. 1995.

[59] Lin, Y-B., and H. C-H. Rao, "Number Portability for Telecommunication Networks," *IEEE Network,* Vol. 13, No. 1, Jan./Feb. 1999, pp. 56 – 62.

[60] Sicker, D., and M. B. H. Weiss, "Cost and Policy Implications of AIN-Based Local Number Portability Implementations," *Proc. 6th Annual Conference on Telecommunications Systems,* Nashville, TN, Mar. 1998.

[61] La Porta, T. F., and M. Veeraraghavan, "An Experimental Signaling Architecture and Modular Signaling Protocols," *Proc. XV Int. Switching Symposium, Vol. 2,* Berlin, Germany, Apr. 23 – 28, 1995, pp. 244 – 248.

[62] La Porta, T. F., et al., "Distributed Call Processing for Personal Communications Services," *IEEE Communications Magazine,* Vol. 33, No. 6, Jun. 1995, pp. 66 – 75.

[63] Lauer, G., J. Sterbenz, and I. Zibman, "Broadband Intelligent Network Architecture," IEEE Communications Society's *Proc. Intelligent Network Workshop,* 1995.

[64] Lu, C. C., "A Network-Independent Call Processing Architecture for Future Narrowband and Broadband Telecommunication Service," Ph.D. Dissertation, University of Pittsburgh, 1999.

[65] Minzer, S., et al., "Evolutionary Trends in Call Control," *Proc. XV Int. Switching Symposium, Vol. 2*, Berlin, Germany, Apr. 23 – 28, 1995, pp. 300 – 304.

[66] Sterbenz, J., and G. Lauer, "Issues in Session Control for Broadband Network Services," IEEE Communications Society's *Gigabit Networking Workshop*, 1996.

[67] Veeraraghavan, M., T. J. La Porta, and W. S. Lai, "An Alternative Approach to Call/Connection Control in Broadband Switching Systems," IEEE Communications Society's *Broadband Switching Symposium*, Apr. 1995, pp. 319 – 331.

SELECTED BIBLIOGRAPHY

Blalock, W. M., "North American Numbering Plan Administrator's Proposal on the Future of Numbering in World Zone 1," *Bellcore Letter*, Letter No. IL-92/01-013, Jan. 6, 1992.

Brand, J. E., and J. C. Warner, "Processor Call Carrying Capacity Estimation for Stored Program Control Switching Systems," *Proceedings of the IEEE*, (issue on Telecommunications Circuit Switching), Vol. 65, No. 9, Sept. 1977, pp. 1342 – 1349.

Brands, H., and E. T. Leo, *The Law and Regulation of Telecommunications Carriers*, Norwood, MA: Artech House, 1999.

Bretecher, Y., and B. Vilain, "The Intelligent Network in a Broadband Context," *Proc. Int. Switching Symposium*, Vol. 2, Apr. 1995, pp. 57 – 61.

Brock, G. W., *Telecommunication Policy for the Information Age*, Cambridge, MA: Harvard University Press, 1994.

Brock, G. W. (ed.), *Toward a Competitive Telecommunication Industry*, Mahwah, NJ: Lawrence Erlbaum Associates, 1995.

Claus, J., F. Lucas, and G. K. Helder, "New Network Architectural Functionality," *Telecommunications Journal*, Vol. 98, 1991, pp. 907 – 914.

Cronin, P., "An Introduction to TSAPI and Network Telephony," *IEEE Communications Magazine*, Vol. 34, No. 4, Apr. 1996, pp. 48 – 54.

Daoud, F., "Universal Broadband Mobility Target for Post-TINA Infrastructure: Toward the Fourth Generation," *IEEE Network*, Vol. 12, No. 4, Jul./Aug. 1998.

Deutschen, N. R., E. J. Bowers, and J. W. Lankford, "ASCC: The Impact of a Silver Bullet," *AT&T Technical Journal*, Vol. 75, No. 1, Jan./Feb. 1996, pp. 24 – 34.

Dorros, I., "The New Future — Back to Technology," *IEEE Communications Magazine*, (issue on Telecommunications Regulation), Vol. 25, No. 1, Jan. 1987.

Elixmann, M., et al., "Open Switch-Extending Control Architectures to Facilitate Applications," *Proc. Int. Switching Symposium*, Vol. 2, Apr. 1995, pp. 239 – 243.

Faggion, N., and T. Hua, "Personal Communications Services Through the Evolution of Fixed and Mobile Communications and the Intelligent Network Concept," *IEEE Network*, Vol. 12, No. 4, Jul./Aug. 1998.

Geller, H., "Government Policy as to AT&T and the RBOCs: The Next 5 Years," *IEEE Communications Magazine*, (issue on Telecommunications Deregulation), Vol. 27, No. 1, Jan. 1989.

Gillett, S. E., and I. Vogelsang (eds.), *Competition, Regulation and Convergence*, Mahwah, NJ: Lawrence Erlbaum Associates, 1999.

Harrold, D. J., and R. D. Strock, "The Broadband Universal Telecommunications Network," *IEEE Communications Magazine*, (issue on Telecommunications Regulation), Vol. 25, No. 1, Jan. 1987.

Heinmiller, W., R. Schwartz, and M. Stanke, "Solutions for Mediated Access to the Intelligent Network," *Proc. Int. Switching Symposium*, Vol. 2, Apr. 1995, pp. 217 – 221.

Henderson, K., "International Freephone Numbers Set to Ring This Month," *Phone+,* Vol. 11, No. 5, May 1997.

Homa, J., and S. Harris, "Intelligent Network Requirements for Personal Communications Services," *IEEE Communications Magazine,* Vol. 30, No. 2, Feb. 1992, pp. 70 – 76.

Jabbari, B., "Intelligent Network Concepts in Mobile Communications," *IEEE Communications Magazine,* Vol. 30, No. 2, Feb. 1992, pp. 64 – 69.

Jacobsen, G. M., "An Inevitable Evolution," *Telephony Magazine,* (special issue on OSSs: An Inevitable Evolution), Vol. 222, No. 2, Jan. 13, 1992.

Joel, A. E., Jr., and G. Spiro, "Bell System Features and Services," *Proc. Int. Switching Symposium,* Paris, France, 1979, pp. 1247 – 1255.

Kim, B. C., J. S. Choi, and C. K. Un, "A New Distributed Location Management Algorithm for Broadband Personal Communication Networks," *IEEE Transactions on Vehicular Technology,* Vol. 44, No. 3, Aug. 1995, pp. 516 – 524.

Laitenin, M., and J. Rantala, "Integration of Intelligent Network Service into Future GSM Networks," *IEEE Communications Magazine,* Jun. 1995, pp. 76 – 85.

Lucas, J., "What Business Are the RBOCs In?" *TeleStrategies Insight Magazine,* Sept. 1992.

Maastrick, V., and E. Schalk, "Call Modeling in a Broadband IN Architecture," *Proc. Int. Switching Symposium,* Vol. 2, Apr. 1995, pp. 340 – 344.

Marcus, M. J., "Technical Deregulation: A Trend in U.S. Telecommunications Policy," *IEEE Communications Magazine,* (issue on Telecommunications Regulation), Vol. 25, No. 1, Jan. 1987.

Marshall, C. T., "The Impact of Divestiture and Deregulation on Technology and World Markets," *IEEE Communications Magazine,* (issue on Telecommunications Deregulation), Vol. 27, No. 1, Jan. 1989.

Matlack, W. H., "Heaven... Almost Heaven," *Communications News,* Vol. 36, No. 6, Jun. 1999.

McDonald, J. C., "Deregulation's Impact on Technology," *IEEE Communications Magazine,* (issue on Telecommunications Regulation), Vol. 25, No. 1, Jan. 1987.

Messerschmitt, D. G., "The Future of Computer Telecommunications Integration," *IEEE Communications Magazine,* Vol. 34, No. 4, Apr. 1996, pp. 66 – 69.

Miller, J. A. and C. P. Cheng, "Designing Networking Solutions for the Nineties: A New Approach," *AT&T Technical Journal,* Vol. 70, No. 5, Sept./Oct. 1991, pp. 14 – 26.

Mishra, A., "Performance Scalability in Switching System Software," *Proc. 1999 IEEE Int. Conference on Communications,* (Session S8A-4: Topics in Communications Software), Vancouver, Canada, Jun. 6 – 10, 1999.

Musselman, T., "TI Network Management: A Strategic Perspective," *Telecommuncations,* Vol. 22, No. 2, Feb. 1998.

Olson, J. E., "New Policies for a New Age," *IEEE Communications Magazine,* (issue on Telecommunications Regulation), Vol. 25, No. 1, Jan. 1987.

O'Reilly Roche, M., "Call Party Handling Using the Connection View State Approach: A Foundation for Intelligent Control of Multiparty Calls," *IEEE Communications Magazine,* Vol. 36, No. 6, Jun. 1998, pp. 60 – 66.

Pandya, R., S. Tseng, and K. Basu, "Some Performance Benchmarks for the Design of Wireless Systems and Networks," *Proc. Int. Telecommunications Conference 15,* 1997, pp. 243 – 253.

Ramlakan, R. S., and S. H. Mneney, "A Study of the Transmission Buffer Within Common Channel Signaling System Number 7," *Proc. TeleTraffic '97,* Grahamstown, So. Africa, Sept. 8 – 11, 1997, pp. 295 – 302.

Rubinstein, C. B., G. J. Ryva, and P. S. Warwick, "Corporate Networking Solutions," *AT&T Technical Journal,* Vol. 70, No. 5, Sept./Oct. 1991, pp. 27 – 35.

Ryan, J. S., "CCITT Signalling System No. 6: A Story of International Cooperation," *Telecommunication Journal,* Vol. 41, No. 2, Feb. 1974, pp. 77 – 82.

Ryva, G. J., et al., "Corporate Networking: Evolution and Architecture," *AT&T Technical Journal,* Vol. 70, No. 5, Sept./Oct. 1991, pp. 2 – 13.

Sable, E. G., and H. W. Kettler, "Intelligent Network Directions," *AT&T Technical Journal,* Vol. 70, Nos. 3 – 4, Summer 1991, pp. 2 – 10.

Schopp, M., "User Modelling and Performance Evaluation of Distributed Location Management for Personal Communications Services," *IEEE Communications Magazine,* Jun. 1995.

Shaw, J., *Telecommunications Deregulation,* Norwood, MA: Artech House, 1998.

Sikes, A. C., "Tommorrow's Communications Policies," *IEEE Communications Magazine,* (issue on Telecommunications Regulation), Vol. 25, No. 1, Jan. 1987.

Staley, D. C., "Domestic Roadblocks to a Global Information Highway," *IEEE Communications Magazine,* (issue on Telecommunications Deregulation), Vol. 27, No. 1, Jan. 1989.

Stark, A. C., and P. M. Villiere, "Harnessing Intelligent Networks," *AT&T Technology Magazine,* (issue on Telecom '87), Vol. 2, No. 3, 1987.

Thompson, R. A., "The Structured Design of a Data Compression Program," *Proc. COMPSAC '78,* Chicago, IL, Nov. 13 – 16, 1978, pp. 430 – 434.

Thompson, R. A., and K. L. Montgomery, "The Synergy Between Photonic Switching and the Intelligent Network," *Proc. IEEE Intelligent Network Workshop '91.* Dallas, TX, Apr. 24 – 25, 1991, Session VI.

Thompson, R. A., "Integration and User-Orientation of Broadband I. N. Services," *Proc. IEEE Intelligent Network Workshop '92.* Newark, NJ, Jun. 23 – 24, 1992, Session B3.1.

van der Merwe, J. E., et al., "The Tempest — A Practical Framework for Network Programmability," *IEEE Network,* Vol. 12, No. 3, May/Jun. 1998.

Weiss, M. B. H., and P. Bernt, *The Regulation and Deregulation of US Telecommunications,* Textbook under contract with Lawrence Erlbaum and Associates, Mahwah, NJ.

Wells, G., "AIN R1.0 Trial Results," *Proc. IEEE Intelligent Network Workshop '92.* Newark, NJ, Jun. 23 – 24, 1992, Session P.1.

Willmann, G., and P. J. Kuhn, "Performance Modeling of Signaling System No. 7," *IEEE Communications Magazine,* Vol. 28, No. 7, Jul. 1990.

Wolfson, J. R., "Computer III: The Beginning or the Beginning of the End for Enhanced Services Competition?" *IEEE Communications Magazine,* (issue on Telecommunications Deregulation), Vol. 27, No. 1, Jan. 1989.

Wyatt, G. Y., et al., "The Evolution of Global Intelligent Network Architecture," *AT&T Technical Journal,* Vol. 70, Nos. 3 – 4, Summer 1991, pp. 11 – 25.

Yoshizaki, H., "Breakup of AT&T and Liberalization of the Telecommunications Business," *IEEE Communications Magazine,* (issue on Telecommunications Deregulation), Vol. 27, No. 1, Jan. 1989.

Zorpette, G., "The Telecommunications Bazaar," *IEEE Spectrum Magazine,* (issue on The Future of Telecommunications), Vol. 22, No. 11, Nov. 1985.

19

Evolving Infrastructure

> *"Curiouser and curiouser!"* — from *Alice in Wonderland,* by Lewis Carroll

This chapter describes some technological changes, some that are under way and some that are under test, to the PSTN infrastructure in the United States and worldwide. The first section presents several emerging "tethered" technologies on the line side — extensions of, and alternatives to, analog twisted pair. Sections 19.2 and 19.3 describe several wireless technologies — a hierarchy of terrestrial wireless telephony in Section 19.2 and telephony by satellite (especially in low earth orbit) in Section 19.3. Sections 19.4 and 19.5 discuss telephony over two packet paradigms — by asynchronous transfer mode (ATM) in Section 19.4 and by the Internet Protocol (IP) in Section 19.5.

19.1 TETHERED ACCESS

Chapter 16 described how the MFJ of 1984 and the Telecommunications Act of 1996 encouraged alternative access and loop competition. That chapter discussed how the economy of alternative access, while attractive in business districts, is quite unfavorable in residential, rural, and downscale communities. Alternative access is motivated by (1) IXCs trying to bring bundled service directly to end users, (2) LECs trying to offer broadband service as an alternative to CATV, and (3) CLECs trying to bypass the long distance access charge. This section and the next two describe some of the physical implementations for alternative access — this section describes copper implementations and the next two sections describe wireless implementations.

Alternative access is any technology for reaching end users that is different from conventional analog twisted pair. ISDN, described in Section 13.4, is such a technology. This section describes three other technologies that might provide alternative access over a "tether": asymmetric digital subscriber loop, hybrid fiber-coax, and fiber in the loop.

19.1.1 Asymmetric Digital Subscriber Loop

A low-rate channel can be frequency-multiplexed onto the conventional passband of the conventional analog twisted pair — a loop with no load coils. This channel, on a carrier frequency around 12 kHz, just above the voice band, carries the digitally modulated signal for calling party identification when CPI is implemented over analog twisted pair. The CPI display unit demultiplexes and demodulates this signal when it displays the calling party's telephone number.

Extending this concept, and that of BR-ISDN (Section 13.4), the LECs are investigating the earnings potential of (1) Internet access, when separated from universal POTS, and (2) alternative (to CATV) access for broadcast television. The high cost of loop replacement has motivated attempts to push very high data rates over twisted pair [1 – 8]. Many of the LECs have been experimenting with different techniques for transmitting a broadband data stream over this narrowband channel. They want to use their own loop plant to offer access at higher data rates than an analog modem's 56 Kbps or even ISDN's 128 Kbps.

The Shannon-Hartley law,

$$C = B \times \log_2(1 + S/N)$$

gives the theoretical limit for bit rate capacity C as a function of channel bandwidth B, signal power S, and noise power N. A good subscriber line has a signal-to-noise ratio around 45 dB (Section 3.4), corresponding to a power ratio (S/N) about 30,000. Then, the practical uppper limit of the quantity C/B is about 15 bits/second per Hertz, or 15 bits per cycle. Because of anti-aliasing filters, the actual analog bandwidth of the digital DS0 channel is about 3.4 kHz, with a very steep cutoff. So, a DS0 channel's theoretical bit rate capacity is around 15 bits/second per Hertz × 3.4 kHz = 51 Kbps. Since modern modems operate around this value, we can't expect to see much more improvement — over a DS0 channel. It is interesting to see that the bit rate capacity of the typical subscriber line is limited because of *digitization.*

While the bandwidth of the DS0 channel is around 3.4 kHz, the bandwidth of the *un-digitized* analog subscriber line, several thousand feet of 26-gauge copper twisted pair, is much higher. Figure 3.3 shows that, unlike the bandwidth of the DS0 channel, an *un-loaded* loop's analog bandpass falls off very gradually at high frequencies. If we can deal with the crosstalk, especially the near-end crosstalk, we should be able to effectively and practically transmit a 100-kHz analog signal over a long loop and a 5-MHz analog signal over a short loop. Allowing for higher noise power in the Shannon-Hartley formula, because of the increased bandwidth, we might be still able to attain a figure of about 8 – 10 bits per cycle. Attaining such high bit rates requires that the loop be devoid of load coils, bridge taps, and any kind of digital loop carrier.

So, different manufacturers have developed different equipment that uses different modulation techniques to transmit different rates of one-way and two-way digital data over different lengths of twisted pair, without needing amplifiers or repeaters. The two most popular technologies are: (1) carrierless amplitude and phase modulation, a form of quadrature amplitude modulation without carrier modulation; and (2) discrete multitone, a form of frequency-shift keying. The details of the modulation techniques are beyond the scope of this book. Protocols supported by different systems include ATM, IP, and frame relay.

If BB-ISDN can be transmitted at 150 Mbps over 100M of twisted pair, then ATM-to-the-desktop could use existing twisted pair on the floor of an office building. If a digital channel at 1 Mbps or more can be transmitted over 18 kft of twisted pair (Section 3.4), then broadband Internet access and compressed video could be delivered to most

American residences over existing twisted pair in the LEC loop plant. The service these systems provide is called digital subscriber loop.

A system whose upstream data rate equals its downstream data rate is said to be symmetric. But, asymmetric transmission rates help resolve the problem of near-end crosstalk. Because two-way digital transmission over a single twisted pair would probably use ping-pong (ping-pong can take advantage of asymmetry), and because most applications probably don't require symmetry, many of these prototype systems provide asymmetric digital subscriber loop (ADSL).

TABLE 19.1. XDSL variations

Name	Pair	Downstream rate (Mbps)	Upstream rate (Mbps)
IDSL	2w	0.128	0.128
CDSL	2w	1	0.128
SDSL	2w	0.768	0.768
HDSL	4w	1.5	1.5
ADSL min	4w	1.5	0.016
ADSL max	4w	8	0.640
VDSL min	4w	13	1.5
VDSL max	4w	52	2.3

Table 19.1 shows the variety of systems, at a variety of data rates, that have been prototyped as this book is going to press. Obviously these systems all have different costs and traverse different maximum loop lengths. For example, the maximum VDSL rate of 52 Mbps is limited to twisted-pair lines that are shorter than 300m, the typical line between an office and a wiring closet.

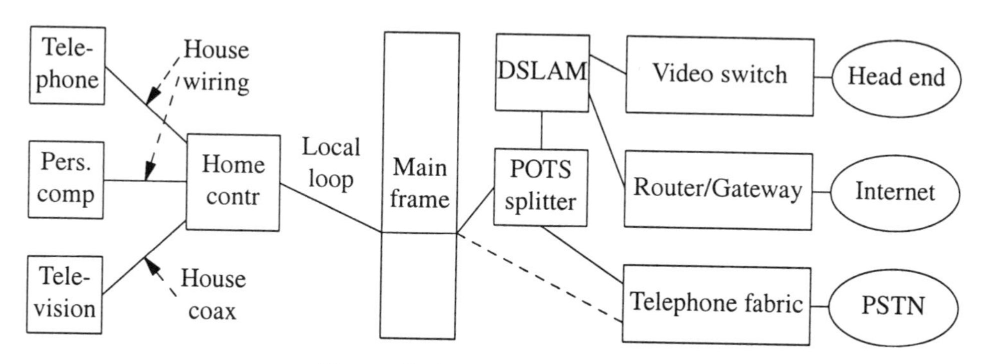

Figure 19.1. ADSL architecture.

Figure 19.1 illustrates only a generic block diagram for ADSL because different real systems have different variations on the detailed implementation. A telephone subscriber has three kinds of instruments in his residence: telephones are connected to pairs in the house telephone wiring, television sets are connected to the coax or twin-lead that runs throughout his residence, and PCs are connected to pairs in the house telephone wiring that are separate from the pairs used for telephones. Figure 19.1 represents all this by showing one of each type of instrument on the left of the figure. An ADSL home controller is placed near the protector block or telephone network interface (TNI) or replaces the CATV set-top box near one of the subscriber's TV sets. The residential coax or twin-lead and all telephone pairs in use are connected to the user side of this home controller. The office side of this home controller includes the network interface card (NIC) and it connects to the loop pair that's been assigned to this subscriber's ADSL service. This loop pair is conditioned by being cleared of any and all bridged taps and load coils and it must be analog all the way from the NIC to the main frame.

In the LEC's telephone switching building, this subscriber's previous wiring assignment, from main frame to switching fabric (the dashed line in the figure), is re-connected. The ADSL line is connected to a POTS splitter, which separates the ADSL signal from the conventional voice signal. The conventional voice signal from the subscriber's line is wired to a line card in the CO's telephone switch just like any non-ADSL line. The ADSL signal from the subscriber's line is wired to an ADSL multiplexor/demultiplexor, called the DSLAM. The subscriber's PC is connected, over an ADSL channel and through the DSLAM, to a LAN inside the CO to access a server for one of several competing Internet service providers, one of which is probably some arm's length subsidiary of the LEC. The subscriber views a television broadcast by signaling the request upstream for his ADSL channel to be switched to the requested program.

Various techniques, data rates, and vendors are competing for this market. There is unresolved controversy that some of the ADSL equipment, while not interfering with voice channels, may interfere with other ADSL channels in the same cable (see Exercise 19.1).

19.1.2 Cable Access

CATV carriers are the most likely competitors to the LECs because they already have alternate physical access to about 70% of all U.S. residences — certainly almost all of the up-scale residences. While they also face economic barriers right now, many are positioning themselves for head-to-head competition, at least in the enterprise market and with up-scale residential users. This preparation has a physical and a logical component [9,10].

Hybrid Fiber/Coax. Many CATV carriers are upgrading their physical plant so they will be ready to add voice and data to their video service. Section 1.4 and particularly Figure 1.3(a) described the typical CATV network as a star of buses of buses: the head end is the hub of a star of backbone links, each backbone link is a bus that is tapped by several local coaxial cables, each local cable is the bus that a residence taps into at the curb.

Just like the LEC's economic barrier at the drop wire, the CATV carriers can't easily justify replacing their coax along every street in every neighborhood. Many are, however, fortifying the backbone links that feed these neighborhood cables, by replacing the backbone coax with optical fiber. These systems, called hybrid fiber/coax (HFC), will have very broad bandwidth in their feeder plant.

The classical spectrum assignment on these coaxial cables was to place about 60 channels on carrier frequencies that have approximately 6-MHz spacing. The amplifiers in the network typically limit its bandwidth. Upgrading to fiber backbone and better amplifiers allows the network to carry 80 conventional channels between 50 and 550 MHz, upstream signals in the band below 50 MHz, digital high-definition signals in the range between 550 and 750 MHz, and future two-way services in the band between 750 MHz and 1 GHz. While 50 Mhz is sufficient upstream bandwidth for a neighborhood, it isn't enough for the region served by a backbone link. So, where the neighborhood coax meets the backbone, the simple tap must be replaced by active equipment capable of remultiplexing all the upstream data onto a separate fiber.

Services over coax. Some CATV carriers are beginning to offer logical services beyond video.

- Ethernet, designed to use coaxial cable as its physical medium anyway, can be offered over one of the cable's FDM channels. Access from a PC requires an Ethernet NIC, which can be connected by twisted pair, perhaps in the residential house wiring, to a simple modem in the set-top box.
- Telephone channels can be submultiplexed, in the time division (like a T1), onto another of the cable's FDM channels. Then, residential telephone wiring is also connected to the set-top box, which has an internal muldem and modem.

The video, data, and voice channels are separated from the feeder at the head end, where they are separately handled. But, HFC might also require voice and data switching in each distribution point (DP), where the fiber backbone meets the neighborhood cables. Voice channels may need to be concentrated at the DP, and by providing an Ethernet gateway at the distribution point, the Ethernet segment's length is confined to the neighborhood cable (which is still long enough to give considerable end-to-end delay).

19.1.3 Fiber in the Loop

It's important to distinguish fiber in the loop (FITL) from fiber to the home (FTTH). Many subscriber loops, especially in an enterprise like a university campus, have some optical fiber segments, particularly in the feeder portion of the loop architecture. This is FITL and it is typically motivated as an economical way to bring more POTS lines to the distribution point, the wiring closet, or the pedestal — not to bring more bandwidth to the end user. Fiber replacement of the drop wire and house wiring is very difficult to justify economically, without a golden service (Section 15.7.7). Such a golden service is assumed to be a sufficient economic driver for the extreme fiber loop network described in Section 22.6.

Just as fiber to the desk (FTTD) might be a wise investment, but only to certain desks in an enterprise, FTTH might be a wise investment, but only to up-scale homes where residents are more willing to pay for the services that would be available. Few LECs offer FTTH — partly because they're waiting for services to mature and for markets to develop and partly because they're concerned that some regulatory body may not allow selective placement. Some LECs are concerned that, if they offer FTTH, then some regulators might define it as a new universal service, and they could be required to provide it everywhere. They refer to this derisively as fiber to the cow (FTTC). While providing fiber to customers who could use it might be a good business opportunity for the LECs, many are going to wait until they are sure that the regulators will enforce its deployment across all access providers, not just the imbedded LECs.

19.2 WIRELESS TELEPHONY

Wireless systems are described by pedagogy similar to that used in Chapter 10, where we discussed the functional progression from a hunt group to a key system to a PBX to Centrex. This section discusses the increasing functionality of wireless systems from *cordless* to *fixed* to *mobile* to *cellular* to *personal communications numbers.* The final stage in this progression, *low earth orbit satellite systems,* is described in the next section. The first subsection discusses some common issues: power intensity, footprint, number of channels, channel reuse, and the biological effects of radiation.

In the following subsections, the issues that distinguish the various types of systems are the telephone's configuration, the analog or digital nature, the location in the architecture of the link that is transformed from wire to wireless, the nature of the base station that houses the antenna, the ability to roam/hand-off, and the degree of special handling in the PSTN.

19.2.1 Radio Power Intensity

This subsection discusses radio's attenuation, frequency reuse, and the effect of radiation on biological tissue.

Attenuation. Radio's forte as a communications medium is to support broadcast services. But, the common thread through all the wireless systems, described in this section and the next one, is that some point-to-point link, which had traditionally been a wire, is replaced by a point-to-point radio link. Using radio to support point-to-point communications requires a system architecture that is optimized to conserve radio spectrum. Footprint size and frequency reuse are managed by careful control of radio power levels.

First, consider an analogy. The sun is a continuously burning hydrogen bomb that produces relatively constant power. This solar power propagates away from the sun's surface into space, radially, in three dimensions. The earth, with a given size and a given distance from the sun, intercepts a small amount of this solar power. If the earth's orbit radius were doubled, we would capture one-fourth of the solar power we capture now — not because the solar power attenuates as it propagates but because, as this constant power propagates away from the sun, it dissipates to illuminate uniformly the surface area of a bigger concentric sphere. Any electromagnetic radiation in free space loses intensity inversely to the square of the distance between the transmitter and the receiver.

Receiver sensitivity, signal-to-noise ratio, and the amount of ambient noise all act together to determine the minimum signal intensity at the receiver so that the receiver can operate within its specification. Typically, in any radio system, the transmitter's signal power and the distance are the only two variables — all other parameters are usually fixed. If the transmitter's power is relatively small, then the transmitter-to-receiver distance must be small to compensate. This maximum distance determines the radius of a circle, centered at the transmitter, within which the receiver must reside. This circular region of acceptable operation is called the footprint. If you double the transmitter's power, you increase the footprint's radius by only 40%.

Frequency Reuse. Radio spectrum is a scarce and precious natural resource. A nation must conserve and manage its radio spectrum at least as wisely as it conserves and manages its energy, air quality, forests, and water. In the United States, the Federal

Communications Commission (FCC), a division of the federal government's legislative branch, manages radio spectrum. Appropriately, the FCC is more likely to allocate a large band of radio spectrum to an application that reaches thousands of people than to an application that only reaches several people. This fair policy makes it difficult to justify a lot of bandwidth for point-to-point radio links — unless the same band of frequencies can be shared by many people simultaneously.

If A talks to B in footprint 1 and C talks to D in footprint 2 and the two footprints don't overlap, then the two separate calls could be assigned to the same piece of the radio spectrum. In this way, the same frequency-defined channel is reused in many separate places. Generalizing the concept of TDM or FDM or WDM, this kind of channel sharing is space division multiplexing. There is an interesting tradeoff.

- If a footprint is very large, requiring high transmit power, the large population covered requires a lot of channels. But, because the footprint is large, few footprints are needed to cover the country. So, while few antennas and base stations are needed, the large amount of bandwidth required in each footprint is lightly reused.
- If a footprint is very small, requiring low transmit power, the small population covered requires only a few channels. But, because the footprint is small, many footprints are needed to cover the country. So, while many antennas and base stations are needed, the small amount of bandwidth required in each footprint is highly reused.

Frequency reuse is important to appreciate. We will generalize it into channel reuse in Section 22.8.

Biological Effect. Whether radio power intensity directly affects animal tissue is a very controversial and contentious question. There appears to be evidence that suggests a correlation between poor health and longtime exposure to high intensities of electromagnetic radiation [11]. We really don't know which exposure durations and power levels are safe. Not intending to alarm anyone or make a recommendation, I'm simply going to tell you how I behave.

While I, and the members of my family, get regular dental X-rays, we don't overdo it. And, I wear a hat in the sun. I own a cordless telephone and I use it as much as an hour a day, even though its antenna is very close to the soft tissue of my brain when I use the phone. I also own and use a cellular telephone, and even though its antenna is also close to my head when I use it, I limit my use to about an hour a month. I do not use my cellular telephone regularly from my car while commuting to work. If that were to become my practice, I would get an external antenna mounted outside my car.

19.2.2 Cordless Telephones

The instrument, the footprint, and the operation characterize cordless telephones.

Instrument. The traditional telephone has two parts: the handset contains the microphone and the speaker and it conforms to the human head; the base unit contains the hook switch, the dial, and most of the electronics. The two parts are traditionally connected by set of wires in a cable called the cord. In a cordless telephone, a point-to-point radio link replaces the cord.

The cordless telephone's base unit is placed in the residence like any telephone, connects to a wall jack by a four-wire cord with a modular plug, connects to local AC for power, and has an attached antenna. The antenna on the cordless handset typically telescopes into the body of the handset and is positioned directly above the user's ear when the handset is in use. Radio transmit power is very low and the technology has been analog for many years. Cordless operation typically fails when local power is out.

Footprint. At about 0.25 watts, the transmit power is so low that the typical usable footprint is only about 20 – 30m. Allocating radio spectrum for only 10 channels provides enough channels so that any subscriber can usually find an unused channel inside his footprint. This extremely narrow band of radio spectrum is well utilized because of the large amount of frequency reuse. Many people nationwide use the same channel simultaneously, each confined to his own footprint.

While channels are reused in different footprints, a cordless telephone's handset is keyed to its base unit. While a cordless subscriber might eavesdrop on another cordless subscriber's call in the same footprint, he cannot originate or terminate a call through another subscriber's base unit. And, he certainly cannot roam into another footprint. If an active cordless subscriber moves outside his own footprint, gets too far from his own base unit, this subscriber loses his connection just as if he had cut the handset cord on a conventional telephone.

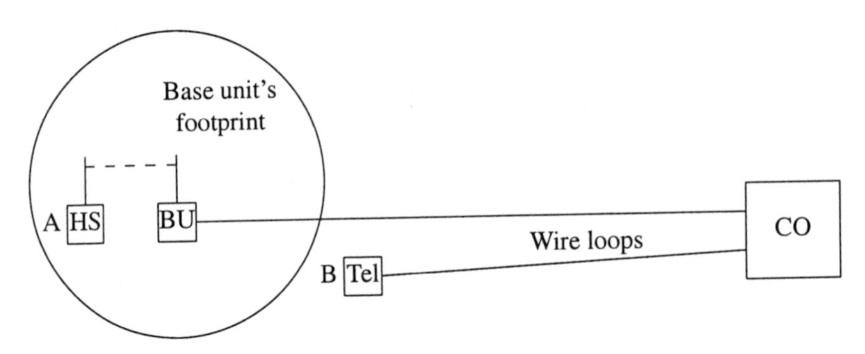

Figure 19.2. Cordless telephones in the PSTN.

Operation. Figure 19.2 illustrates cordless telephone operation in the context of the entire telephone system. A cordless telephone has a wireless cord between the handset (HS) and the base unit (BU). But, the base unit connects by wire to the CO, like any regular telephone. If A has a cordless telephone and A talks to B, whether A originates or B does, their call is connected at the CO in the usual way. The CO makes no hardware or software distinction, maintains no separate class of service, for cordless telephones.

19.2.3 Fixed Wireless

Wireless technology is used with cordless, mobile, cellular, and satellite so that the end user can move around, and is not tied to the end of a *tether.* Another reason for using wireless technology is to provision channels in environments that might otherwise be difficult because of terrain, right-of-way, or expense. Fixed wireless systems mount on a utility pole or antenna mast and provide wireless channels in such difficult environments. These channels are intended to be as short as a typical drop wire or as long as a typical

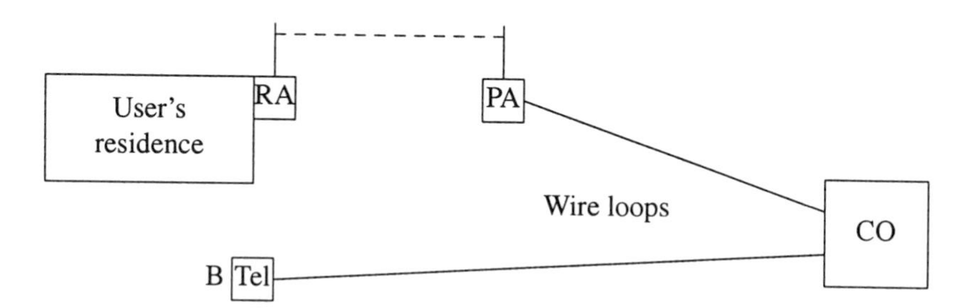

Figure 19.3. Fixed wireless in the PSTN.

loop pair. In Figure 19.3, one fixed wireless antenna is shown at a pedestal (PA) and the other is shown at the TNI on the user's residence (RA). This figure shows fixed wireless as replacing the tethered drop wire.

While the end users are not necessarily mobile, if the user's instrument is portable, the user could conceivably move around inside a residence, as with a cordless telephone. But, fixed wireless would not be deployed only for mobility — it would be deployed because it may be more economical than a drop wire in certain environments. The user's antenna (RA) connects directly to the user's house wiring and the user can use conventional telephones inside his residence.

Some of the LECs are looking at fixed wireless for provisioning universal service in rural areas. Some of the CLECs are looking at fixed wireless in competitive areas because they want to offer telephone service over their own physical channels, rather than lease loop pairs from the LEC. As this book is going to print, these fixed wireless systems are still in the early stages of development and deployment. Various vendors offer various technologies and bandwidths. Like a cordless telephone user, a subscriber with a fixed wireless last mile does not need a special class of service at the CO.

19.2.4 Mobile Telephony

Architecture and operation characterize mobile telephony.

Architecture. All the mobile telephones in a mobile telephone serving area are served by one antenna at the mobile base station (MBS). This serving area is the antenna's footprint and it is typically quite large — several miles in radius. The MBS's place in the PSTN architecture is analogous to a PBX or a community dial office (CDO) or an RSM. The MBS is an active switching point that has radio channels downstream to the users and wired channels upstream to the CO. Mobile is more like fixed wireless than like cordless because mobile's wireless link is equivalent to at least a drop wire, while cordless's wireless link is just the cord.

Operation. Figure 19.4 illustrates mobile telephone operation in the context of the entire telephone system. Subscriber A's mobile telephone (MT) is linked by radio to the MBS in the upper left of the figure as long as A remains within the MBS's footprint. This MBS acts like a PBX or RSM that homes on CO #1, even if A moves into the region served by a different CO (position D in the figure). If A originates or terminates a call with B, where A and B are served by the same CO, A's mobile channel is switched to a feeder channel at the MBS and this feeder channel is switched to B's line at CO #1.

If A originates or terminates a call with C, where A and C are served by different

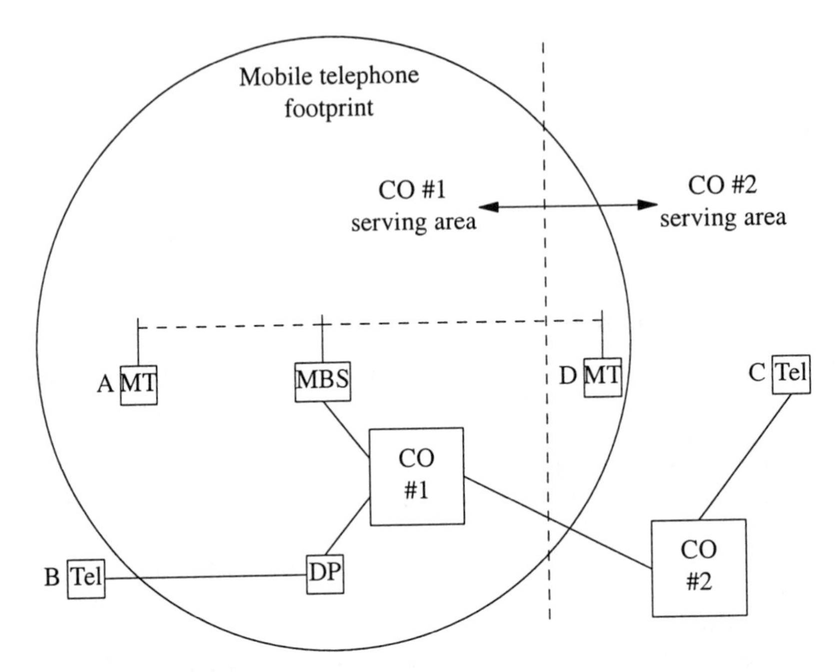

Figure 19.4. Mobile telephones in the PSTN.

COs, the call is switched through the PSTN as if B had called C. If A moves to position D, in the serving area of CO #2, A is still served by CO #1 because the MBS homes onto CO #1. The typical mobile telephone footprint is larger than a CO serving area. If A and D represent two different mobile subscribers inside the same footprint, the call may be connected at the MBS, if it has the intelligence to complete in intra-MBS call, or each user might be connected through to CO #1 and the call might have to be completed there.

A mobile telephone user may move outside the footprint. If so, he would have to be served by another MBS. If he is using his telephone when he moves across a footprint boundary, the call would be dropped. This is no different from taking a conventional telephone from your house, driving to another house, and plugging it in there. Mobile telephony is older technology, mainly analog. While most mobile phones have been replaced by cellular phones, mobile is still important in marine environments, called ship-to-shore. Mobile telephony's large footprint is necessary because antennas can't be built in open water to support cellular telephony's smaller footprints.

Mobile was designed for mobility within the footprint. The inconvenience of losing a call when crossing a boundary is minimized by making the footprint very large. But, this means that the radio transmit power is large. Typically, the mobile telephone's antenna is mounted on the body of the user's car or boat, safely removed from the user. With such a large footprint, there is less frequency reuse.

Mobile telephones need a class of service similar to that of a PBX extension. The CO must know to route incoming calls to a group of channels that connect to the MBS and to signal the called party's identity as in DID (Section 10.5.2).

19.2.5 Cellular Telephony

Cellular telephony is characterized by its local architecture, the system of *location registers,* the signaling that supports *roaming,* and the *hand-off* of an active call [12,13].

Local Architecture. Cellular telephones are like mobile telephones in that: (1) all the cell-phones inside the footprint, called a cell, are served by an antenna at the cell's base station (CBS); (2) this CBS has a place in the loop plant architecture analogous to a CDO, RSM, PBX, or distribution point — with radio channels on one side connecting to cell-phones and feeder channels on the other side connecting to some CO's main frame; (3) incoming and outgoing cellular calls are switched at the CBS between radio channels and feeder channels; and (4) a point-to-point wireless link replaces a traditional tether at the modular plug, house wiring, drop wire, and distribution wire — everything that connects a telephone to its distribution point.

While mobile telephone base stations typically home on a conventional CO, cellular telephone base stations home on a customized cellular switching office (CSO). A CSO is distinct from a conventional CO because (1) a cellular carrier in any region either competes with the region's LEC or it is an arm's-length subsidiary of the LEC, and (2) the control and signaling for cellular telephony is different from POTS. So, when a cell-phone user calls a conventional telephone, even if both parties are in the same house, it's an interexchange call.

A service region is "tiled" by adjoining cells (typically depicted as hexagons, but obviously circular) with a CBS at the center. The cell's footprint is smaller than a mobile footprint because radio power is lower in cellular telephony than in mobile telephony. Competing cellular providers have different tiling patterns. Cells have different sizes, depending on the number of users, the number of available frequencies, the amount of frequency reuse, the technology (analog, TDMA, CDMA, GSM), and regulation. Cell size is controlled by transmit power at the CBS and at the cell-phone. While all the cells in some serving area connect to a common cellular switching office, a carrier might have several CSOs in a metropolitan region.

A typical cellular telephone transmits at a power level between 0.6 and 3.0 watts. The power level is controlled by a signal from the CBS. If the CBS has trouble hearing the signal, it requests the cell-phone to increase its transmit power. If the CBS hears the signal well, it requests the cell-phone to decrease its transmit power in order to (1) decrease the cell-phone's signal power in adjoining cells and (2) to allow the cell-phone's battery to last longer before needing a charge.

While a cellular telephone's radio power is much higher than that of a cordless telephone, it is lower than that of a mobile telephone. A cell-phone's radio power is presumed to be low enough that the antenna is usually physically attached to the handset where it is near the user's head during use. A physically separated antenna is typically available as an option for the cautious high-usage subscriber. While the original cell-phones were analog, newer digital versions (TDMA, CDMA, and GSM) are now available. Conservation of battery charge is an additional constraint on the system architecture, signaling, and protocols.

One difference between mobile telephony and cellular telephony is that a cell is much smaller than a mobile telephony regional footprint. Where a mobile base station's footprint might cover the serving area of several COs, a single CO that provides cellular service typically has many cells. With a small footprint, there is a lot of frequency reuse, and hence, the FCC has allocated a lot of spectrum for cellular telephony. With enough radio spectrum, cellular telephone usage can be mass marketed and price competitive. Because the cell is small, however, the cellular subscriber is likely to move out of the footprint of one CBS and into the footprint of another. Where mobile telephony reduces the inconvenience of reconnecting by having a large footprint, cellular telephony eliminates this inconvenience by automatically reconnecting the call.

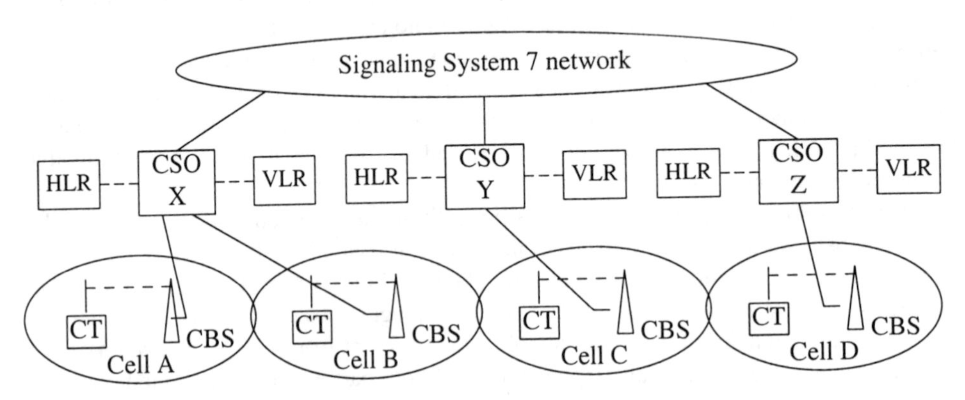

Figure 19.5. Signaling architecture to support roaming.

Location Registers. Providing outgoing and incoming telephone service to roaming subscribers requires more than just tiling a region with cells. While the system of CBSs and CSOs that support the cells could provide regional service, national service requires a third layer of architecture and a great deal of interoffice signaling. In Figure 19.5, cells A and B home on CSO X, cell C homes on CSO Y and cell D homes on CSO Z. Figure 19.5 shows two databases, an HLR and a VLR, logically associated with each CSO.

- The home location register (HLR) permanently stores the identities of all cell-phone subscribers whose physical address (home) is inside the region of cells served by the corresponding CSO. All of this carrier's subscribers who live inside cells A, B, and others served by CSO X, have entries in X's HLR. The carrier makes this entry when the party subscribes for service. One of the records in each home subscriber's entry is a pointer to the corresponding subscribers current location (by VLR address).

- The visitor location register (VLR) temporarily stores the identities of all cell-phone users who are currently located inside the region of those cells that are served by the corresponding carrier's CSO. All cell-phone subscribers, regardless of their home carrier, who are currently present inside cells A, B, and others served by CSO X, have entries in X's VLR. One of the records in each visiting subscriber's entry is a pointer to the corresponding subscribers home location (by HLR address). This entry is made by the system when the subscriber roams into one of this CSO's cells.

While Figure 19.5 indicates that each CSO has an associated HLR and VLR, the implied association is only logical. There are many different architectures of the physical association between each kind of database and all the CSOs in a region: one is that the HLR and VLR are physically inside the corresponding CSO; another is that one or both databases are physically located inside this carrier's nearest service control point (previous chapter) or on some other carrier's SCP. If some of a region's HLRs and VLRs are implemented at an SCP, even more SS7 signaling is required to support cellular telephony than is described below to support roaming.

Roaming. As long as it is powered on, a cellular telephone periodically broadcasts a location message, even if it's on-hook. All the CBSs that can receive this message negotiate to determine which CBS and CSO will be responsible for handling this cell-phone's outgoing and incoming calls. The decision is based on carrier, transmission type, and which CBS receives the signal at the highest power. As the subscriber moves from one cell to an adjoining cell, this responsibility shifts from one CBS to another and, occasionally, to another CSO or even another carrier. Sometimes, even the transmission technology will be changed, if the cell-phone is capable of more than one mode. A lot of radio spectrum is used, even when a cell-phone is on-hook, and a lot of interoffice signaling takes place.

Consider a Sprint digital cellular telephone subscriber, with a dual-mode cellular telephone, who lives in Pittsburgh in the 412 NPA. Let this subscriber's home cell correspond to cell D in Figure 19.5. Information about this subscriber is stored in the HLR associated with Sprint's CSO in Pittsburgh, corresponding to CSO Z in the figure.

1. Suppose this Sprint subscriber flies to Toledo and, after landing, he turns on his cell-phone. His cell-phone tries to communicate with a Sprint digital CBS. If it cannot find one, it tries to communicate with some other carrier's digital CBS with compatible technology. If there are none in Toledo, the cell-phone tries to communicate with some other carrier's analog CBS. Assume it finds an Ameritech CBS, corresponding to cell A in Figure 19.5. The cell-phone switches to analog mode and identifies its HLR address to this analog CBS.

2. Ameritech's CSO in Toledo, corresponding to CSO X in Figure 19.5, signals over the SS7 network to this subscriber's HLR, associated with Sprint's CSO in Pittsburgh. This subscriber's HLR in Pittsburgh, associated with CSO Z in the figure, is made to point to the VLR associated with CSO X in the figure. And, Ameritech's VLR, associated with CSO X, is made to point to Sprint's HLR, associated with CSO Z.

3. Suppose our subscriber roams further away from the CBS in cell A and moves toward the CBS that serves an adjoining cell, cell B in Figure 19.5. By periodically comparing the relative intensities of the signal received from our subscriber's cell-phone, the CBS in cell B is finally given responsibility for our subscriber. By definition of cell, he has roamed out of cell A and into cell B. The CSO that serves both cells knows which of its CBSs has responsibility.

4. If our subscriber originates a call while he is in cell B, CSO X would complete this call. Ameritech in Toledo would signal billing information to Sprint in Pittsburgh, using the pointer to CSO Z found in the subscriber's entry in CSO X's VLR.

5. If our subscriber terminates a call, however, it is more problematic. If someone (cellular or conventional) calls his cellular telephone number, the calling party's

telephone office signals with the Sprint CSO in Pittsburgh based on the called party's area code and office code. This CSO signals back to the calling party's office that the call should be completed to the Ameritech CSO in Toledo instead, based on the contents of the HLR — like a forwarded telephone call.

6. If our subscriber, now on-hook again, continues roaming, he may eventually move into cell C in Figure 19.5, which is served by a CBS that homes on CSO Y. Now, not only has our subscriber changed cells, but he has also changed CSO serving areas. This roam is handled one of two different ways, depending on whether Ameritech wants to optimize signaling or trunk channels.

    - CSO X might retain responsibility for incoming calls. X simply maintains a local pointer to CSO Y. An incoming call would still be routed to X because the HLR in Pittsburgh still points there, but X would tandem the call to Y and Y would complete the call.
    - CSO X might abdicate to Y. X would signal back to Pittsburgh that the HLR should point to Y. Our subscriber would be added to Y's VLR. If the VLRs for X and Y are located in the Toledo SCP, then this database change is merely logical.

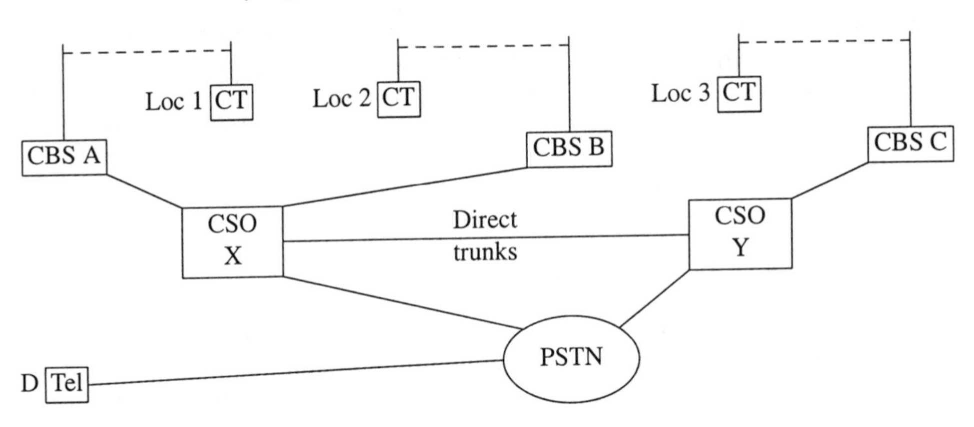

Figure 19.6. Cellular telephones in the PSTN.

Hand-Off. The really tricky part of cellular telephony occurs if the cellular subscriber is using his telephone (is off-hook) when he moves from one cell to another. The cellular telephone subscriber does not have to hang up and dial again, like the mobile telephone subscriber must. Instead, the PSTN automatically *hands off* the call, in real time, while the call is in progress. It's as if a residential subscriber instantaneously moved his telephone to the house across the street, while he was still in the middle of a conversation with someone.

Figure 19.6 illustrates a cellular subscriber as he roams through three different cells: the cells served by CBSs labeled A, B, and C. The two leftmost CBSs, A and B, home on CSO X, on the left in the figure, and CBS C homes on CSO Y, on the right.

1. In cell A in Figure 19.6, the subscriber's cellular telephone (CT) is within the cell (footprint) of the leftmost CBS. If the subscriber originates or terminates a call with D, his cellular link is switched to a feeder line at the left CBS, this feeder

line is switched at the leftmost CSO to a trunk that connects to the PSTN CO that serves subscriber D, and this trunk is switched at this PSTN CO to D's line.

2. As the cellular subscriber roams to cell B in Figure 19.6, the two leftmost CBSs compare their signal strengths from his cell-phone and they determine that he has roamed into the cell of the CBS in the center of Figure 19.6. The subscriber's new cellular link is switched to a feeder channel at this center CBS. Still in the leftmost CSO, the trunk that eventually leads to D is switched to this new feeder channel from the middle CBS and is disconnected from the previous feeder channel from the left CBS. The subscriber's previous connection in the left CBS is dropped.

3. Now, as the subscriber roams to cell C shown in Figure 19.6, the two rightmost CBSs compare their relative intensities of the signal received from this cell-phone and they determine that he has roamed into the cell of the CBS in the right of Figure 19.6. Note that this third CBS is served by another CSO. The subscriber's new cellular link is switched to a feeder channel at the right CBS. The call cannot be rerouted again to the new CSO through the PSTN because conventional switching offices are not capable of switching connections during a call. So, the call to D remains connected out of the left CSO in Figure 19.6. An idle trunk is allocated between the two CSOs. The rightmost CSO connects its end of this trunk to the cellular subscriber's new channel in the feeder cable to the rightmost CBS. The leftmost CSO takes down the previous connection between the trunk to D and the channel to the subscriber's previous CBS and, instead, connects the trunk to D to its end of the trunk to the other CSO.

19.2.6 Related Services

Three related wireless services are paging, cellular packet service, and so-called personal communications service.

Paging is wireless, but it isn't telephony and it isn't switched. It's a one-way broadcast system, hubbed from a satellite or from terrestrial antennas, depending on the system. The hub broadcasts a message, containing an address. If the receiver recognizes its address, then it beeps and/or displays a message, depending on the system. Two-way paging systems are under investigation.

Cellular data packet service is some standard packet-switched protocol, like frame relay, that is modified for use in cellular telephony. A laptop PC with a cellular packet modem has a data channel quite similar to that of a PC equipped with a conventional modem. This technology is becoming quickly popular for mobile data applications like express package delivery or police checks of license plate numbers.

Personal communications service (PCS) is different from the personal communications number (PCN) described next. Technically, PCS uses different spectrum than conventional cellular telephony, but the user doesn't perceive this difference. Functionally, PCS is a collection of telephone services that the carriers bundle with cellular telephony [14]. Depending on the vendor, additional services might include calling party identification, voice mail, call waiting, or paging.

19.2.7 Personal Communications Number

PCN is a proposed real-time version of local number portability (LNP), described in Section 18.4.4. If LNP is analogous to facility switching, then PCN would be analogous to circuit switching. PCN takes cellular telephony to the next step. With a personal communications number (PCN), telephone numbers are assigned to people, not telephones. After a PCN subscriber *logs on* from any nearby telephone, he will receive all incoming calls at this phone.

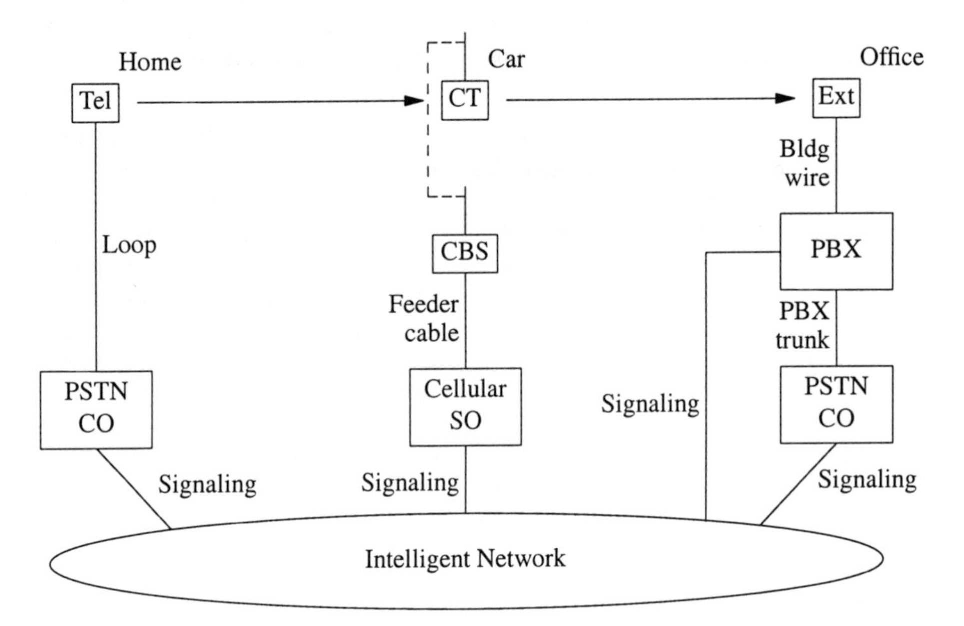

Figure 19.7. Illustrating a personal communications number.

Roaming. Figure 19.7 illustrates a PCN subscriber commuting to work in the morning.

1. When he came home the previous evening, he logged on from his conventional telephone in his home. From that time on, all incoming telephone calls to his personal communications number are directed by the PSTN to this home phone.

2. When he leaves for work in the morning, he logs on from his cellular car phone. From this time on, all incoming telephone calls to his personal number are directed by the PSTN to his cell-phone.

3. When he arrives at work in the morning, he logs on from the telephone in his office, an extension served by a PBX. From this time on, all incoming telephone calls to his personal number are directed by the PSTN to his DID PBX extension.

PCN might be implemented within the local exchange, and restricted to an LEC's local calling region or to a LATA, by simulating a call forward. Or, PCN might be implemented, similar to cellular telephony, using the Intelligent Network, described in the previous chapter. In the latter case, the PCN might be offered by a third-party vendor, which routes PCN calls through the LECs and IXCs.

Logging In. From a conventional telephone line, the log-on procedure consists of a caller-ID telephone call to a toll-free local number or to some PCN 800 number, where the user enters his personal communications number, followed by a password or PIN. From an ISDN line, the subscriber's telephone transmits a simple log-on signaling message over the D-channel upstream to the CO.

The requirement that the subscriber manually log on every time he temporarily relocates detracts from the service. Implementations that might facilitate logging on include:

- A simple device, which attaches as an extension to each telephone line like the caller ID display, that a single PCN subscriber can activate by pushing a button;
- A credit card reader, like those used in the supermarket or at a self-serve gasoline pump, that any subscriber could run a card through;
- A device, logically equivalent to the card reader, except it would be activated automatically by a small transmitter that the subscriber carries around in his pocket (it might double as a pen or pencil).

Automatic log-on procedures are fraught with implementation problems and would need very careful planning and easy provision for overriding. The subscriber carrying an automatic PCN log-in transmitter may not want to log on to every telephone he walks past or drives past. The etiquette of using the wall phone in a classroom or large conference room would have to be carefully thought out. A vibrating pager serves situations like this best.

19.3 LEOS SYSTEMS

While communications satellites have been used for voice telephony, they are more useful for data communications and are most useful for broadcast television. Their difficulty with voice comes from the inherent speed-of-light delay, caused by the typical satellite's geo-synchronous orbit. After providing some general background, this section describes telephony using a system of low earth orbit satellites (LEOS), where delay is not noticeable.

19.3.1 Communications Satellites

Man-made satellites were the stuff of science fiction for much of the first half of the 20th century. Years before it was thought to be realistic, forward-looking thinkers like Arthur C. Clarke and John Pierce hypothesized telecommunications by satellite. Then, in October 1957, the former Soviet Union launched Sputnik, and the space age began for real. The United States was the second human institution with enough economic power and technological know-how to put a satellite in orbit around the earth. The third such human institution with this much size and power was not a national government, but a U.S. corporation. The Bell System launched TelStar in 1962. In many ways, America's Bell System was more economically powerful and technologically advanced than most of the world's nations.

Government and other corporations were concerned about possible abuses by AT&T. So, the Communications Satellite Act formed a public/private corporation to manage communications satellites. ComSat was a government agency that was publicly owned, similar to the current form of the U.S. Postal Service. AT&T was one of the larger stockholders, with RCA, GTE, ITT, and NASA.

As international telephony increased in popularity, the undersea cables became congested. Satellite links were quite common for international telephone calls. But, users complained about the delay (explained below) and carriers laid undersea fiber. For a few years, the FCC had a policy that a random number of international calls had to go by satellite, but they eventually abandoned this and let the market determine the technology of the carrier. Now, most international calls use undersea fiber if it's available (like between North America and Europe), and satellite is used between parts of the world where terrestrial links are not available (like between North America and India).

19.3.2 Orbit Mechanics

When a natural or man-made satellite orbits the earth, the characteristics of the orbit are determined by the equilibrium between the gravitational force that pulls the satellite toward the earth and the centrifugal force that pushes the satellite away. The two most relevant characteristics of an orbit are (1) its *period* and (2) its *equivalent radius.* The period is the time required for a satellite to complete one revolution around the Earth. The equivalent radius is the orbit's average (real orbits are rarely perfect circles) distance from the center of the earth. A satellite's elevation is its average distance from the earth's surface, which equals the the orbit's equivalent radius minus the earth's radius. While a detailed discussion of orbit mechanics is outside the scope of this book, Kepler's third law states that an orbit's average radius and orbit period have a fixed relationship:

$$r^3 = k \times t^2$$

So, a satellite that completes one full orbit quickly must be close and a satellite with a large average radius must have a long period. About 250,000 miles away from the earth's center, the earth's moon has such a large average radius that a single orbital revolution requires about 28 days. Sputnik's orbit was only a couple of hundred miles above the earth's surface, so close that its orbit period was about 1.5 hours.

Between these two extremes, a satellite whose orbit has an equivalent radius about 30,000 miles (precisely 22,236 miles or 35,785 km above the earth's surface) makes one complete revolution around the earth in a single day. If such a satellite orbits the earth in the plane of the equator and in the same direction as the earth's rotation, it appears fixed in the sky when viewed from the earth's surface. This is the geo-synchronous orbit used by most communications satellites. The really significant advantage to such a large orbit is that a highly directional antenna, like the familiar parabolic dish, can be aimed from the earth at the "bird" once, and it doesn't need expensive electronics and motors for continuous tracking.

19.3.3 Geo-Synchronous Orbits

Geo-synchronous earth orbit satellites (GEOS) are an ideal technology for any form of one-way nonswitched broadcast communications, such as commercial broadcast television. Geo-synchronous communications satellites can also be used for point-to-point data communications, like paging, where the receiver selects (Section 1.3) by reading a packet header. They are less ideal for voice and data because these applications tend to be point-to-point and two-way.

The really significant disadvantage to geo-synchronous satellites is that they are so far away from the earth that speed-of-light delays make human dialog very awkward. The one-way speed-of-light delay between a GEOS and the earth's surface is 22,236 mi/186,000 mi/s = 0.12 seconds. Suppose A asks B a question. Let's ignore any internal processing delay in the satellite and also ignore any human response time. B doesn't hear the end of the question until 0.24 seconds after A completes it, and B responds. But, A doesn't hear the beginning of B's response until after another 0.24 seconds. So, A perceives a net delay between the end of his question and the beginning of B's response that is about 0.5 seconds greater than the normal delay in human processing. Including the processing time in the satellite and ground stations, the delay can be as great as one full second. This pause is noticeable and many users find it extremely annoying.

While GEOSs have been used for voice communications, and many zealots still advocate their use for voice, the system is unacceptable for regular human voice communication — not because it can't be done, but because people don't like it. People simply don't like talking over these channels, and with optical fiber, they don't have to.

19.3.4 Iridium Architecture

Communications by low earth orbit satellite (LEOS) have been proposed by several companies. This subsection describes the project called Iridium, originally proposed by Motorola, which is now a co-owner of the Iridium Company. While some of the features have changed recently, including the number of satellites in orbit, this section summarizes the project's original description [15].

While the original proposal called for 77 satellites, Iridium has 66 satellites in low earth orbits. The orbits are organized as six planes, with 11 satellites per plane. Each satellite is relatively large, weighing about 1100 pounds. The 11 satellites in each plane are in polar orbit, configured as a "string of pearls," each satellite equally spaced in its orbit by 360/11 = 32.7°. The six planes intersect both poles of the earth, and are all separated by approximately 180/6 = 30° — the spacing isn't uniform. Looking down at the North Pole, the six planes resemble 12 spokes. At an altitude of 785 km, the orbit period is about 2 hours.

The geometry is such that every point on the earth's surface always sees at least one satellite in the sky at greater than 10° above the horizon. When one satellite sets, another one rises — a new satellite enters the sky approximately every 10 minutes. Because the system's original configuration had seven planes and 77 satellites, the project got its name because it was thought to resemble the Iridium atom, which has 77 rotating electrons.

Each satellite casts a circular footprint that is slightly more than 2000 miles in diameter. These footprints barely tile the earth at the equator, with a great deal of overlap nearer the poles. Each footprint is partitioned into 37 cells, each slightly more than 400 miles in diameter. The pattern of cells in each footprint resembles a honey comb, with three concentric rings of cells surrounding the cell at the footprint's bull's eye. The rings contain 6, 12, and 18 cells, respectively, for a total of 37 cells in each footprint.

While each cell overlaps its neighbor cells, radio spectrum is assigned so that no neighboring cells use the same spectrum and spectrum is reused every seven cells. All cells are active while a satellite is directly over the earth's equator. But, as a satellite moves toward a pole, it shuts off some of the cells at the edge of its footprint to avoid overlapping the footprints of nearby satellites. While there are 77 × 37 = 2849 potential cells, only 1600 of them (about 56%) are on at any time, tiling the earth. Since Iridium needs about 40 cells to tile America's lower 48, it needs more than one satellite, especially since so many cells are off at the United States' latitude. The spectrum in any one cell would be reused 1600/7 = about 230 times worldwide, if the same spectrum were used worldwide (which, of course, it isn't).

Iridium uses so-called L Band, in the range 390 MHz to 1.55 GHz, for surface communications. PSK modulation is used over multiple carrier frequencies. Each carrier frequency is time multiplexed, providing about 150 channels per cell. At 2400 bps and some speech compression, each channel provides acceptable voice communications. Besides communicating with the earth's surface, either with telephones or with earth stations, each satellite can also communicate with other nearby satellites. Iridium uses so-called Ka Band, in the range 33 – 36 GHz, for intersatellite communications. Each satellite trunks to a maximum of two earth stations (but, there aren't 132 of them, yet) and to six neighboring satellites — the ones ahead and behind in the same plane and two in each adjacent plane.

19.3.5 Iridium Calls

Let A and B both be Iridium subscribers. If A calls B and both reside in the same footprint, the call is completed by the satellite that serves them both. If A and B reside in neighboring footprints, then the call is completed over the intersatellite link between the satellites that serve the respective parties. If A and B reside on opposite sides of the earth, then the call is completed over several intersatellite links tandem switched between intermediate satellites. If only A is an Iridium subscriber, but B is served by an LEC within the PSTN. Then, if A calls B, Iridium routes the call to that satellite which has a downlink to the earth station closest to B. The terrestrial part of the call would be routed through the PSTN; similarly, if B calls A.

The assignment of an Iridium telephone to a satellite is analogous to the assignment of a cell-phone to a cellular switching office, and the assignment of an Iridium telephone to an Iridium cell is analogous to the assignment of a cell-phone to a cellular base station. Because each satellite's footprint moves across the earth as the corresponding satellite moves through its orbit, Iridium telephones must be reassigned dynamically to different cells and different footprints. This reassignment is exactly like that of a roaming cell-phone, except the telephone doesn't move nearly as quickly as the equivalent CBSs and CSOs. Iridium will handle roaming and hand-offs. So, even though an Iridium cell is much larger than a cell-phone's cell, there would typically be more frequent hand-offs because the relative velocity between telephone and satellite is so much greater. Furthermore, there would be even more hand-offs as cells are shut off and turned on, or if the subscriber is in the seam between orbits, especially in the seam between the two counter-rotating planes.

At projected prices and guessing reasonably at costs, Iridium probably needs about half a million subscribers, worldwide, to break even. Tiling the United States with 40 cells, at 150 channels per cell and assuming a 2% utilization, Iridium could serve a maximum of 300,000 Americans, a penetration of only about 0.1%. The American subscriber base can only be about half of the needed worldwide base.

19.3.6 Other LEOS Systems

As this book is going to press, Iridium has gone into bankrupcy. Iridium was operational for less than 1 year. Was the Iridium concept a bad idea? Was the Iridium architecture wrong? Was the Iridium business poorly managed? We don't know yet.

Other competing systems have been proposed and are under development. Having observed Iridium's failure, maybe these sponsors will be cautious; maybe not. With different orbit patterns, other proposals require fewer satellites. But most other proposals do not provide intersatellite links. Functionally, these other LEOSs act like orbiting cellular base stations. They have a set of downlink channels to mobile telephones, another set of downlink channels to terrestrial switching offices. Because these satellites can only connect mobile instruments to the terrestrial PSTN, they can be viewed as large MANs or as CLECs offering loop replacement. But, while their service may be significantly less expensive, they will not avoid the regulators and the likelihood that access charges will be assessed. Iridium was a worldwide network that was completely independent of the PSTN except for calls to PSTN subscribers.

One particularly interesting and aggressive proposal is made by a company called Teledesic. They propose 840 satellites in low earth orbit and have the financial resources behind them to do it. The smaller footprint allows for a channel with much higher bandwidth (variable bit rate ATM between 16 Kbps and 2 Mbps on demand) than Iridium can possibly provide. Teledesic is targeting Internet access, not voice telephony.

19.4 ASYNCHRONOUS TRANSFER MODE

With no shortage of books on ATM, this section won't be very long. But rather than omit ATM completely, this brief section discusses ATM in the context of circuit switching and as a switching paradigm for voice [16]. This section describes ATM's roots, the basic principles of cell relay, and ATM's place in layered communications protocols and the OSI stack. Network integration is discussed in this section (for ATM) and continues in the next section (for IP) and in Section 22.2.

19.4.1 History and Development

Though ISDN's market penetration was disappointing, ISDN's advocates evolved it anyway and continued keeping the concept relevant. As discussed in Section 13.4, the original ISDN, what we now call basic rate, was designed with a circuit-oriented $2B + D$ format at 144 Kbps. Later, primary rate ISDN was designed for a data rate that was one order of magnitude higher than BR-ISDN. The designers chose a circuit-oriented $23B + D$ format at the T1 rate (in the United States) of 1.544 Mbps. Then, as vendors and carriers pondered broadband services, the ISDN community decided to design a third format, called broadband ISDN (BB-ISDN), for a data rate that would be two orders of magnitude higher than PR-ISDN — around 150 Mbps. The designers also chose to depart from the circuit-switched ISDN format.

While BR-ISDN's original goal was to integrate voice and data communications under a single paradigm and within a single network, it didn't happen. BB-ISDN's goal is to integrate voice, data, and video. As with the original BR-ISDN, BB-ISDN's goal was a single entire integrated network with one universal switching paradigm, not just integrated access to separate networks, each of which has a different switching paradigm. Research and exploratory prototypes of fast packet networks suggested that packet switching might be more universal than circuit switching as a paradigm for switching voice, data, and video through a single integrated network.

While BB-ISDN would target computer communications, it was designed by experts from the telecommunications industry, rather than the computer industry. Partly because of experience with, and understanding of, congestion in buses and rings (Section 17.1) and partly because of a natural telecom bias for the star architecture, the designers gave BB-ISDN a star architecture, with active switching at its hub. BB-ISDN's star topology is ironic because, at the same time, the more circuit-oriented Sonet architecture was being configured in a ring.

After selecting packet switching over circuit switching and selecting fixed-length packets over variable-length packets, BB-ISDN's designers still had to decide: (1) should the fixed-length packets be large and infrequent or small and frequent? and (2) should the packet's header be large enough to handle all overhead and signaling or should it be small, with its streamlined functionality augmented by call setup? While BB-ISDN's designers elected to use a small header, they compromised between big packets that would be efficient for carrying long files and small packets that would cause reduced delays with voice.

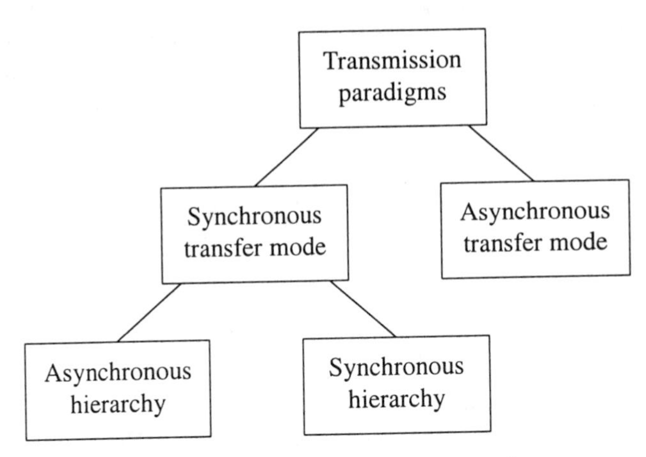

Figure 19.8. Related transmission paradigms.

Finally, many of the formats were renamed and Figure 19.8 shows the relationship among several formats. Circuit-oriented Sonet was renamed synchronous transfer mode (STM). Packet-switched broadband ISDN was renamed asynchronous transfer mode (ATM). A packet-switched data format that uses intermediate-sized packets came to be called cell relay. ATM and cell relay are special cases of packet switching, with all the characteristics of packet switching that were discussed in Chapter 17. In *Romeo and Juliet,* William Shakespeare writes:

> *"What's in a name? That which we call a rose, by any other name, would smell as sweet."*

19.4.2 Cell Relay

The ATM cell is a packet that contains 5 bytes of header and 48 bytes of payload. Cell relay is optimal for random communications of short files, applications like transaction processing. Long random files, which can be delivered efficiently over IP or frame relay, must be sliced up and delivered repetitively in many cells. Voice, traditionally circuit-switched at 8000 PCM-samples per second, is carried by packing a group of samples (not the full 48) into one cell and transmitting such cells less frequently.

While ATM's individual cells may not be optimal for long files or for voice, ATM's cell streams can be cleverly optimized for different types of traffic. Two of these stream types are called variable bit rate (VBR) and constant bit rate (CBR). During call setup before transmitting, the transmitting program first negotiates bandwidth and traffic type with the controlling software at the ATM hub. The ATM network provides a different guarantee of bandwidth, called quality of service (QoS), for each of these types of stream. Regular periodic traffic, like voice, is assigned to a CBR stream and the total bandwidth required for the stream of CBR cells is guaranteed. Random traffic, like file transfers, is sent with best effort over however much bandwidth is left over after what has been guaranteed for the CBR streams. It operates something like a fractional T1 (Section 12.2.4) except that ATM's data rates are much higher and ATM's QoS is capacity switched (Section 17.5), not facility switched.

While specific timeslots are not assigned to CBR streams, as in circuit switching, the equivalent bandwidth is still allocated. ATM advocates argue that this procedure is the best of both worlds (circuit switching and packet switching). ATM's detractors argue that because this procedure (1) has a call setup interval as in circuit switching and (2) allocates bandwidth as in circuit switching but (3) switches each individual cell by reading its header as in packet switching, then this procedure is the worst of both worlds. However you look at it, ATM, even in a CBR stream, is not circuit switching.

In the data LAN environment, ATM challenged Ethernet and won most of the early competitions. But, it was never clear how much of ATM's advantage could be attributed to cell relay. Certainly, much of ATM's advantage was attributed to its higher data rate and its star architecture. Responding to ATM's challenge, Ethernet has evolved first to a switched star architecture and, then, to a gigabit data rate. In the data WAN environment, ATM is challenging frame relay and IP, but is finding that both of these may also be too firmly entrenched. Responding to ATM's challenge and to the challenge from each other, frame relay and IP are also each evolving to their own respective second generation.

ATM can deliver repetitive streams and large files. But because it isn't optimized for either application, ATM will always face strong competition from switching paradigms that have been optimized for a specific application. ATM's advocates call it a jack of all trades, admitting that it may not be a master of any [17 – 21]. ATM's detractors call it a big duck — acknowledging that ducks can walk, fly, and swim (corresponding to voice, data, and video) but that ducks don't do any of these things very well.

19.4.3 ATM's Place

One wonders for which layer(s) ATM was designed and with which protocol(s) ATM must be accompanied? Some of ATM's detractors have called it a protocol without a layer, but that's not fair — these detractors are too wedded to the OSI stack. Perhaps ATM is a protocol that spans the first three OSI layers, but only partially spans each one.

For example, if some long ATM link needs underlying synchronization, does the link really need to be embedded in all of Sonet?

ATM has a set of optional lightweight attached protocols, which can be easily included as needed to provide for functions like synchronization and link-level forward-error correction. ATM can operate in native mode, directly on top of Sonet, or accompanied by an intermediate lightweight link-layer protocol. ATM has been proposed for use over wireless links, but the ATM cells might need to be embedded in some protocol with link-layer error correction because wireless links typically have a high bit-error rate. ATM has also been proposed to implement a kind of distributed capacity-switched network. Capacity switching was described at the very end of Chapter 17 and will be revisited in Section 22.7. While it's a clever idea, this application of ATM would run under Sonet and not over it.

ATM's ancestors, BR- and PR-ISDN, never achieved their goal of providing an integrated network for all types of traffic and all types of applications. Was it because the goal is fundamentally flawed? Perhaps we should let separate networks have their different switching paradigms, each optimized for their specific application. Integration is discussed further in the next section and in Section 22.2. One hears compelling arguments for voice-over-IP, other compelling arguments for IP over ATM, other compelling arguments for ATM over Sonet, and still other compelling arguments for Sonet over optics. Most people agree that it doesn't make sense to transmit voice-over-IP over ATM over Sonet over optics because such a protocol architecture probably uses more than half its physical bandwidth for overhead. Which layers are superfluous?

While some experts remain skeptical that ATM could (or should) become some kind of integrated network for all media, there are interesting possibilities for ATM for clever designers who can think outside the (OSI) box. Perhaps a better goal for ATM, like every other form of ISDN, is to be a single integrated interface that lets users have universal access to separate networks. I personally advocate some kind of integrated access network (Section 22.6) and believe that ATM is perfectly suited for such an application. In addition, I believe that the SS7 network will need to be replaced (Section 18.1) soon and I believe that ATM would be an excellent candidate for this application [22].

19.5 INTERNET TELEPHONY

While any switching paradigm or data protocol is capable of switching any kind of data, some are more optimal than others for different types of data. The previous section discussed asynchronous transfer mode and using cell relay as a switching paradigm for telephony. This section discusses another alternate architecture and switching paradigm that has been proposed for telephony: the Internet and the Internet protocol (IP). As a message-oriented packet-switched paradigm, the IP is optimal for file transfers and transaction processing. But, while IP isn't optimal for regular data streams like voice, it is capable of switching voice [23 – 25]. Currently, this application has some vocal advocates. Besides a general description of how, this section also discusses why. As in the previous section, this section presents IP, particularly Internet telephony, only as an overview and in the context of circuit switching.

The first two subsections of this section present the Internet's fundamental structure and the implementation of voice-over-IP, respectively. The third subsection describes three different configurations for Internet telephony, and the fourth subsection discusses the appliance interface in these configurations. After reminding us that transmitting voice is only a subset (the easiest part) of telephony, Section 19.5.5 describes the provision of telephone features, services, and enhanced services over an IP network. The sixth subsection discusses some regulatory issues. Sections 19.5.7 and 19.5.8 discuss the relative cost of IP telephony and some miscellaneous architectural issues, respectively. The last subsection summarizes, discusses, and editorializes.

19.5.1 The Internet

After describing the difference between an internet, an intranet, and the Internet, this subsection compares the Internet with the PSTN and then discusses Internet usage.

Internets, Intranets, and the Internet. Physically, an *internet* (with a lower-case "i") is a collection of routers that are interconnected by transmission channels. Logically, an internet is a message-oriented packet-switched network that has a common protocol, the Internet protocol (IP), and a common hierarchical addressing scheme. These routers act as gateways for some localized community of users and they serve to route other users' packets through the network.

In an enterprise's private *intranet,* the routers belong to the enterprise and the channels are part of the enterprise's private network. Section 12.4 described how some enterprises use private telephone and data networks to expedite their private business and how other enterprises use them to offer public service as a carrier — and how either type of enterprise might lease or own their private network. Similar dichotomies occur with intranets and internets. But, even an enterprise with a private internal intranet might provide gateways to other enterprise's intranets and to publicly accessible internets, including the Internet.

The *Internet* (upper-case "I") is a public global internet. A user or community of users joins the Internet by (1) placing an IP router/gateway on its LAN or MAN, (2) acquiring and assigning IP addresses for its router and other nodes on its LAN/MAN, and (3) acquiring the necessary channels to connect this new router to nearby routers that are already on the Internet. Some of these user communities are enterprises, like universities, whose users belong to the enterprise and whose participating computers are connected to the enterprise's LAN/MAN. Some of these user communities are commercial Internet service providers (ISPs), whose users subscribe for a fee. An individual, at a residential personal computer, typically connects to the Internet by modem through the PSTN. He calls the modem pool attached to the enterprise's or public ISP's LAN/MAN. In nations like the United States, which have regulated flat-rate billing, such a modem call is typically a local call for which the LEC receives no direct revenue.

Comparison with the PSTN. While the Internet and private internets and intranets are logically distinct from the PSTN, they are also intimately related to the PSTN physically, logically, and economically. Most of the physical channels that interconnect the Internet are leased from the PSTN's carriers, usually the IXCs. Local access to the Internet, particularly for residential users, is typically switched, not leased, through the LEC's network. This arrangement for local access is not optimal — far from it. Users use the PSTN for Internet access because it is convenient and because, in the United States and many other nations, telephone regulation is such that Internet users don't have to pay for this means of access.

While the PSTN and the Internet share the same physical facilities, which are typically owned and operated by the PSTN's carriers, the Internet differs radically from the PSTN in several ways.

- The PSTN switches its data through subsystems that the constituent carriers own and operate. The Internet switches its data by subsystems that its participating users own and operate.
- When the PSTN essentially was the Bell System, it had a well-defined architecture. After the Bell System disintegrated, the different carriers were free to architect their own pieces of the PSTN, but few departed from the common layered architecture. The Internet, on the other hand, has few layers and little architecture.
- When the PSTN was Ma Bell, she not only defined its architecture, but Bell Labs' systems engineers were the network's responsible architects. The Internet has no responsible architect. Nobody is responsible for end-to-end performance, reliability, or content. The Internet's advocates call this freedom and its detractors call it anarchy. Both are correct.

As a packet-switched WAN protocol, IP has no significant performance advantage over other modern layer-3 protocols — like ATM, SMDS, or frame relay. But, IP does have two very significant advantages over the others: (1) IP has a large addressing capability (and IPv6 makes it even bigger), and (2) the success of the Internet has caused IP to become so universal that IP addresses are almost as common as telephone numbers.

Usage and Growth. The Internet has been a worldwide phenomenon. Internet usage has increased enormously, rapidly, and steadily. While articles in the trade journals and expert keynote speakers give fantastic projections that this much growth will continue, their predictions may be unrealistic [26] — or they may not be. But, if Internet traffic does not taper off, it's not clear that the Internet's architecture will scale well. If it doesn't scale well, VoIP will be the application most affected because of its sensitivity to delay.

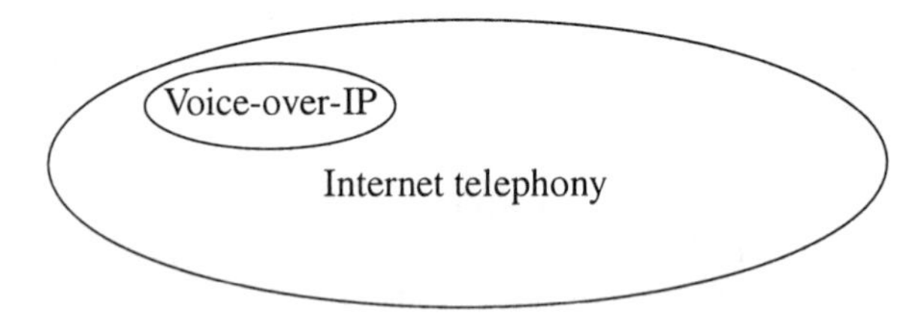

Figure 19.9. Voice-over-IP $\subseteq$ Internet telephony.

19.5.2 Voice-Over-IP

Internet telephony and voice-over-IP are not the same thing. Those who use these terms interchangeably are careless. This subsection on voice-over-IP is a subsection of this section on Internet telephony because voice-over-IP is a subset of Internet telephony, as shown in Figure 19.9. This subsection discusses only voice-over-IP, and only relatively briefly [27,28]. The first part describes one of several standards that have been proposed for VoIP's implementation. The second part discusses VoIP's principal technical issue —

delay. Later subsections describe the part of Internet telephony that is more than simply putting voice samples into IP packets.

Implementation. Several proposals exist for VoIP implementation. This discussion summarizes the International Telecommunications Union's (ITU) H.323 standard, using a G.729 Codec [29]. Analog voice is A-to-D converted into traditional PCM — 8000 samples per second at 8 bits per sample giving a 64-Kbps DS0 stream. While it might have been interesting for some carrier to use VoIP as a means for offering a higher-fidelity channel than circuit-switched voice offers, it clearly won't happen if VoIP is based on conventional PCM. Using silence suppression and modern source coding schemes, this traditional digital voice stream is compressed eight-fold, to about 8 Kbps. Approximately every 10 ms, 10 bytes (80 bits) from this compressed data stream are loaded into a packet's payload. The payload might be padded to 30 bytes.

Using the real-time transport protocol (RTP), sequence numbers and timestamps are placed inside the packet so the stream can be reassembled at the receive side. The user datagram protocol (UDP) transmits overhead for multiplexing, check-sum, data typing, and compression type to the receive side. The transmission control protocol (TCP) is used with IP to provide call control, call setup, and addressing. These three layers of processing (RTP, UDP, and TCP/IP) contribute 40 bytes to each packet's overhead. The VoIP packet is 50 – 80 bytes in length, where 40 of these bytes carry overhead. A significant part of the overhead carries the IP address.

There is not one standard for VoIP. So, one of the issues confronting its adoption is how two parties will be able to communicate if their PCs use differing standards or how one version of VoIP in a PC will interact with another version of VoIP in the network.

Delay. The Internet protocol provides connectionless datagram service. As a best-effort network, the Internet offers no QoS guarantees. These could be appended to IP, and might need to be if IP telephony is to work, or QoS guarantees could be provided by running IP over ATM. But, then why not just run voice directly over ATM? Packet delay and packet jitter are problematical.

The ITU has recommended a worst-case one-way delay of 150 ms for VoIP. This recommendation is consistent with user dissatisfaction over GEOS telephony (Section 19.3.3). This delay will be divided almost evenly three ways: transmit-side packet assembly, network delay, and receive-side packet disassembly. Assuming an average of five hops in an Internet route, the delay budget allows only 10 ms at each hop for transmission delay and propagation delay combined. An end-to-end TCP NAK will almost certainly cause the retransmitted packet to exceed the minimum delay. VoIP can be done inside the United States, with barely acceptable delay [30,31].

19.5.3 Configurations

This subsection describes three different configurations by which users might place a telephone call over an internet or the Internet [32 – 33]. Since many of the issues described in later subsections — like services, price, regulation, and reliability — depend on the particular configuration, we have to discuss these topologies before we discuss the other issues. The differences among the three topologies below are somewhat analogous to the ways a service is provided "on, through, in, or by" any network, as discussed in the previous chapter.

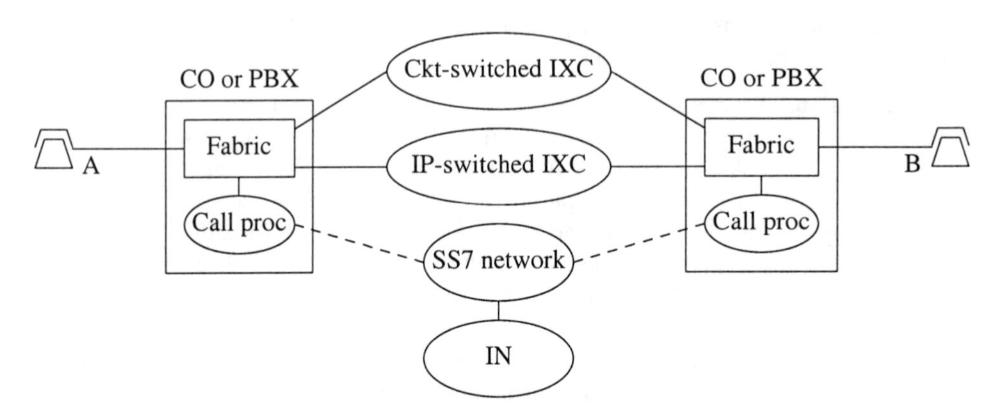

Figure 19.10. IP-switched interexchange carrier.

IP-Switched Interexchange Network. Figure 19.10 shows the conventional block diagram for long distance telephony. Suppose the calling party (A on the left), calls the called party (B on the right). Suppose both parties have conventional POTS or ISDN telephones. A signals B's telephone number into the switching office on the left. There, conventional LEC switching software, shown inside the "CO" box in the figure, processes A's call. This software module inside this CO accesses the SS7 network, as shown by the dashed lines on the figure, not only for busy-testing and other POTS functions, but especially if A places his call from a cellular telephone, dials an 800 number, or involves the Intelligent Network somehow. For a conventional long distance call, this call process in the LEC's CO connects A to his default interexchange carrier (IXC). For an 800 call, this call process connects A to the IXC that owns the 800 number.

A's call to B might use IP telephony if the IXC handling the interexchange part of the connection has a network that uses IP routing internally. Figure 19.10 shows two architecturally equivalent IXCs: the upper one uses conventional circuit switching and the lower one uses IP routing. This lower network is this IXC's private internet (with a lower case "i"). Some of the smaller IXCs in the United States use IP exclusively — but none of the large ones do, yet. If Figure 19.10 represents an LEC's intra-LATA network, then A's call to B is an interexchange call within the LATA. A's call to B over the upper network in Figure 19.10 represents an intra-LATA call over a conventional circuit-switched interexchange network. A's call to B over the lower network in Figure 19.10 represents the same call over an LEC's intra-LATA IP network. The LECs typically don't use IP for POTS calls.

If Figure 19.10 represents an enterprise with an enterprise private network (EPN), then A and B are members of the enterprise. A and B are behind PBXs, not COs, and A's call to B is an inter-PBX call over this EPN. If the enterprise's PBXs use circuit switching like a typical CO, then this figure is close to being representative. A's call to B over the upper network in Figure 19.10 represents an inter-PBX call over a conventional circuit-switched EPN. A's call to B over the lower network in Figure 19.10 represents the same call over an IP EPN. An enterprise might have two separate networks, as shown in Figure 19.10, or it might place voice and data over one common integrated IP network. If the enterprise's PBXs use IP switching, like some modern PBXs do, then the representation is closer to the ISP configuration described next. Figure 19.10 falls short of perfectly representing this PBX/EPN environment because PBXs typically don't access the SS7 network. Busy-testing and other POTS functions that are processed in the PSTN by ISUP messages must be handled by alternate methods or by ISUP messages inside the EPN.

This configuration in Figure 19.10 is the model perceived by the IXCs. They will migrate to IP switching if the cost of providing long distance telephony is lower using IP switching than using conventional circuit switching. This is not the model perceived by the ISPs, or by PC users.

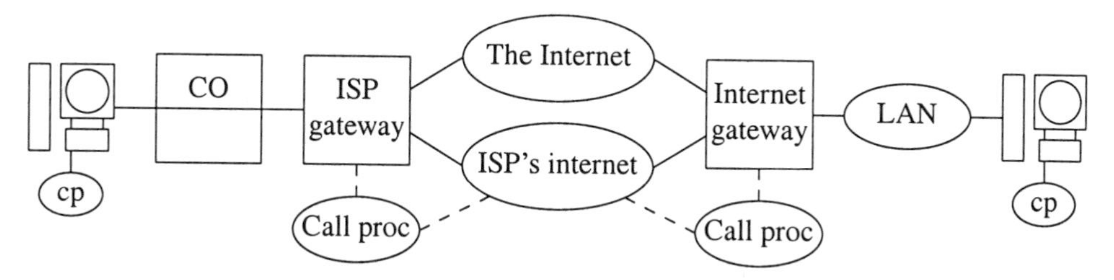

Figure 19.11. ISP-controlled internet.

VoIP through Middleware. Figure 19.10 showed a configuration for IP telephony between two users at telephones, where the call is processed by LEC switching software. The next two figures show configurations for IP telephony between two users at PCs, where the call is not processed by an LEC. The difference between these next two configurations is the location of the call-processing software — in the ISP's gateway/router or in the user's PC, respectively. IP telephony calls between two users, one at a telephone and one at a PC, will be discussed in a later subsection. While this figure and the next one show a CO on each end of the call, the only function for these COs is to provide local access between the PCs and their respective ISPs. To emphasize that the LEC's switching software processes only the local modem-calls between the end users and their ISPs, but not the call between A and B, the LEC software isn't shown on either figure.

Figure 19.11 shows the block diagram for a telephone call between A and B, each at a PC. In this configuration, middleware in the ISP's gateway/router processes calls. Suppose the calling party (A on the left) calls the called party (B on the right). From his PC, A originates a local modem-call to his ISP, through his LEC's CO. A indicates to his ISP his desire to make an IP telephone call. Like a typical e-mail or web connection via an ISP, a relatively dumb piece of software on the PC communicates with a relatively smart piece of middleware on the ISP's gateway/router. These processes are indicated on Figure 19.11. A signals B's IP address, not B's telephone number, into the ISP on the left. The ISP's call process determines B's IP gateway/router from his IP address and signals that IP device (which is assumed to be attached to this internet) to place a local modem-call to B's PC (which is assumed to be powered up, logged in, attended, and not busy on some other IP telephone call). All these assumptions will be addressed later. Middleware in B's IP gateway/router rings B's PC and sends A a corresponding message. B answers by signaling his ISP gateway/router, which sends a corresponding message to

A's IP address. A and B talk to each other by way of software that is similar to conventional FTP.

If the middleware inside A's IP gateway/router can access the SS7 network, then it might be able to use existing PSTN software for busy-testing and other POTS functions. Furthermore, SS7 access would greatly simplify middleware's call processing if A places his call from a cellular PC, or if A dials the IP equivalent of an 800 number, or if A involves the Intelligent Network somehow. If the middleware inside A's IP gateway/router cannot access the SS7 network, then busy-testing and other POTS functions that are processed in the PSTN by ISUP messages, must be handled by alternate methods or by ISUP messages inside the ISP's internet.

If Figure 19.11 represents an enterprise with an EPN, then A and B are members of the enterprise. A and B are behind PBXs, not COs, and A's call to B is an inter-PBX call over this EPN. Figure 19.11 does not represent the case where the enterprise's PBXs use conventional circuit switching like a typical CO — the previous figure is closer. However, if the enterprise's PBXs use IP switching, like some modern PBXs do, then the representation is closer to the ISP configuration of Figure 19.11. An IP-EPN call between two PCs inside some enterprise would look like Figure 19.11, where the enterprise's IP PBX takes the role of the ISP in the figure. PCs inside an enterprise wouldn't need LEC access to their IP PBX/gateway/router. An IP-EPN call between two telephones inside some enterprise would also look more like Figure 19.11 than like the previous figure because the call from A to B would use IP end-to-end from line card to line card. We'll see more on this later in this section.

This configuration in Figure 19.11 is the model perceived by the ISPs. They will offer long distance telephony if they can charge for it. They have some margin to raise their rates because this configuration allows the user to bypass the regulated long distance access charge. This is not the model perceived by the IXCs, or by PC users.

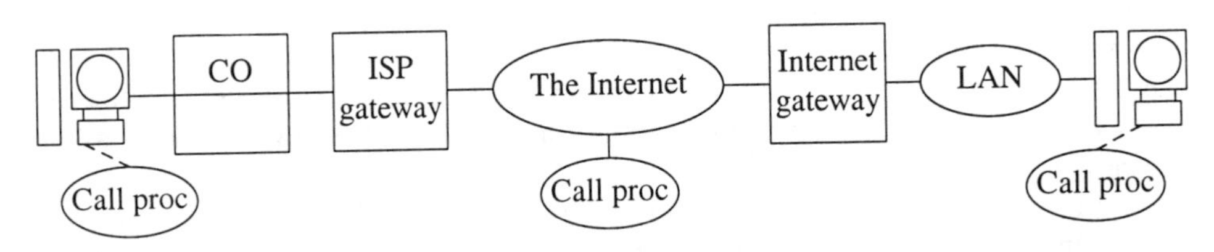

Figure 19.12. PC control over the Internet.

End-to-End Over the Internet. Figure 19.11 showed a configuration for IP telephony between two users, both at PCs, where call-processing software is located in the ISP's gateway/router and the IP network might be the ISP's private internet (with a lower-case "i"). In Figure 19.12, call-processing software is located only inside the user's PC, and the IP network is the Internet (with an upper-case "I"). To emphasize that call-processing software is local only, the "Call processing" ellipse, attached to each PC in Figure 19.12, is larger than the corresponding ellipse in Figure 19.11. There were no ellipses at all in Figure 19.10, because the user's devices were assumed to be dumb telephones. While Figure 19.12, and the previous one, show a CO on each end of the call, this CO's only function is to provide local access between the PCs and their respective ISPs. To emphasize that the LEC's switching software processes only the local modem-calls between the end users and their ISPs, LEC software isn't shown on the figure. To emphasize that the ISP's gateway/router doesn't know about telephony, no middleware is shown on Figure 19.12 either.

Figure 19.12 shows the block diagram for a telephone call between A and B, each at a PC. In this configuration, software inside the users' PCs processes the call. Suppose the calling party (A on the left), calls the called party (B on the right). From his PC, A originates a local modem-call to his ISP, through his LEC's CO. A establishes a two-way FTP connection to B's IP address in the usual way. A's ISP doesn't know, nor need not know, that the data being FTP-ed is VoIP. Like sending a fax from one PC to another, A's call process signals B's call process (B's PC is assumed to be powered up, logged in, attached to the Internet, attended, and not busy on some other IP telephone call). All these assumptions will be addressed later. B's call process rings B's PC and sends A a corresponding message. B answers by sending a message to A's IP address. A and B talk to each other over the same FTP connection that was used for signaling.

Since a PC-resident call-processing program would probably not be permitted direct access to the SS7 network, busy-testing and other POTS functions that are processed by ISUP messages must be handled by alternate methods or by ISUP messages over the Internet. Furthermore, this lack of SS7 access would greatly complicate local call processing if A places his call from a cellular PC, or if A dials the IP equivalent of an 800 number, or if A involves the Intelligent Network somehow.

This configuration in Figure 19.12 is the model perceived by PC users. PC-resident IP telephony software exists now that enables users at PCs to make voice calls anywhere in the world to other PCs over the Internet. Features and services are limited, but the PC user who is already paying for Internet access anyway perceives this worldwide long distance calling as free. He pays no long distance access fee, no usage charge, and the ISP gets no additional revenue. This is not the model perceived by the IXCs, or by the ISPs.

Summary. Much of the problem with discussing IP telephony is that the various participants have differing models and differing assumptions about what their own benefit will be for changing their telephone service to IP. Besides differing assumptions about cost, these three configurations have many other widely differing characteristics. There may be more than three configurations — I only thought of these three. The rest of this section addresses many other considerations and shows how they are inconsistent across the three configurations.

19.5.4 Interface

This subsection discusses one of the implementation issues that will differ across the three configurations just described. The users' appliances will interface differently in each of those configurations. Particular attention is to be paid to implementing the BORSCHT functions. This subsection asks many more questions than it attempts to answer.

Simple Telephone Interface. A telephone must interface to a line circuit that provides the BORSCHT functions. In the first configuration in the previous subsection, depicting an IP-switched IXC, the user's instrument is a POTS telephone and LEC provides the BORSCHT interface. This is the configuration we have today for an IP-switched IXC (like Quest, for example). It can continue as long as the LECs continue to provide (or, are required to continue to provide) network access over their loop, switched connections to the IXCs' gateways, and the per-line line circuit and all the BORSCHT functions. By emulating a simple dumb telephone, a PC can be used in the place of the telephone in Figure 19.10. Whether the user uses a telephone or a PC in Figure 19.10, the LEC's switching software routes the user's call to the IXC's gateway and the LEC charges its flat-rate monthly charge and the long distance access fee. The IP IXC charges a competitive long distance usage fee for its part in the call.

Simple Personal Computer Interface. A communicating residential personal computer must interface via a switched modem-to-modem call to an ISP. In Figures 19.11 and 19.12, the LEC provides network access over their loop, switched connections to the ISP, and some of the BORSCHT functions, and the LEC's switching software routes the user's modem-call to the ISP. Because the user still has to pay the LEC's monthly fee anyway, he will probably still use the LEC for local calls; the ISP has no motivation to change this.

The voice signal traverses some bizarre physical realizations. It begins as an analog electrical signal at the PC's microphone. Because IP telephony's A-to-D conversion (the "C" of BORSCHT) occurs in the PC, voice is transformed to a binary PCM signal, which might be a compressed form of the traditional 64-Kbps DS0 signal. This binary data is loaded into IP packets, still inside the PC. Then, an IP packet is transmitted by converting its bits into analog-modulated digital signals in the PC's modem. These signals are then A-to-D converted in the line circuit at the LEC into a DS0 signal (not the PCM of the voice signal, but the PCM of the modem signal). This DS0 channel is circuit switched at 64 Kbps to the ISP's line circuit, where the process is all reversed.

In the second configuration in the previous subsection, depicting an ISP that provides IP telephony through its own middleware, the ISP's switching software could route all of the user's telephone calls, local and long distance, over its own private internet or over the Internet. Besides its usual internet access fee, the ISP can also charge a competitive long distance usage fee. The ISP can charge a little more than they did without telephony because the user, who essentially has no IXC, does not pay the long distance access fee. The ISPs clearly prefer this configuration over the next one. Furthermore, because many of the implementation details are more easily resolved in this configuration over the next one, enterprises that elect to implement IP-switched EPNs will use this configuration. The PC interface, via a network interface card (NIC) to a LAN to the IP router, is analogous to Figure 19.11.

In the third configuration in the previous subsection, the PC's internal software routes long distance telephone calls over the Internet. The ISP charges only its usual internet access fee because, in this configuration, the ISP (like the LEC) does not distinguish IP telephony from any other form of IP data.

How does the IP telephony provider ring the called party? If the LEC rings the called party as a telephone, the called party's telephone line has to be idle. To the LEC, this means that the called party's modem must be idle. If the ISP rings the called party by sending a message to his PC, then the PC must be powered up and connected (by modem) to his ISP. Other telephony implementation details are similarly convoluted.

Crossed Appliances. Since a PC can emulate a telephone, a PC can be used in the IP-IXC configuration in Figure 19.10. But, the voice signal isn't packetized until it reaches the IXC, and the software inside the PC plays no role in improving the channel. A PC is a pretty expensive telephone, but if you have a PC anyway, then you don't have to buy a telephone unless you want to make calls during power outages or want to be able to roam around on a cordless or a cell-phone.

But, a telephone can't emulate a PC. You can't put a telephone into the place of the PC in Figures 19.11 and 19.12. True end-to-end IP telephony from a telephone requires an intermediate PC just like traditional telephony requires a TNI or protector block and CATV requires a set-top box. House wiring could connect to the residential PC, where IP conversion would take place — provided we can resolve at least as many BORSCHT problems as ISDN faces. And, if all the telephones in my home are wired to my PC, then doesn't my PC have to be permanently powered on and permanently connected to its ISP? Or, do I have to boot up and log in as part of placing a call or receiving an incoming call?

Finally, if telephones must be configured as in Figure 19.10 and PCs must be configured as in Figure 19.11 or 19.12, than how does a user at a telephone call a user at a PC?

19.5.5 Features and Services

Voice-over-IP is the procedure for putting voice samples into IP packets. Voice-over-IP is to Internet telephony as analog transmission and basic circuit switching are to analog POTS and as PCM is to digital switching. They correspond to the first three layers of the OSI Reference Model — and not even all of the third layer. But, just as telephony is much more than POTS, so Internet telephony is much more than voice-over-IP. The piece of IP telephony that we've only barely talked about so far is the implementation of those hundreds of features and services. This subsection simply lists only a few of the features and services and, as in the previous subsection, it raises more questions than it answers.

Basic POTS Features. The previous subsection discussed ringing and how it might be implemented in each of the three configurations of Section 19.5.3. But, ringing a PC, or a telephone that is connected to house wiring that interfaces to a PC, isn't the only basic POTS feature whose IP implementation is problematic. Most of the questions arise during call setup. Only a few of the many issues are raised here.

- *Signaling.* In Figure 19.10, obviously the IP-switched IXC is connected to the SS7 network just like any other IXC would be. If Figure 19.11 represents IP telephony, will any ISP be invited to connect to an STP? A kind of certification program might be developed, but it's probably not a good idea to open SS7 access to every mom-and-pop ISP in the world — the potential for malevolence is too great. The ISPs probably see an analogy to the "interface barrier" described in Section 10.6.3. If Figure 19.12 represents IP telephony, certainly every PC with an IP address will not be invited to connect to a nearby STP. If the ISPs or the Internet provide their own signaling network, will they try to be compatible or try to be proprietary? And, how does one cross between IP signaling and S7 signaling to complete a PC-to-telephone call?

- *Call progress tones.* In Figure 19.10, obviously the LEC is responsible for providing busy tone, audible ring tone, path busy tone, and permanent signal howl. But, in Figures 19.11 and 19.12, these tones, or the software-equivalent call progress indicator, has to be generated somewhere inside the IP network. In Figure

19.11, the ISP or the user's software could provide the progress indicators, but this ISP will expect additional revenue and the user's software might have difficulty knowing when an indicator must be provided. In both of these latter configurations, the issue becomes more problematic as the indicator has to be a real tone because the user is using a real telephone.

- *Busy.* Deeper than providing the audible tone, or other indication, that the called party is busy is the more fundamental question of just what does it mean for an IP telephone to be busy? The LEC thinks the line is busy if the PC's modem is connected to the ISP's modem. Certainly, this isn't how IP telephony wants to define "busy." If the called PC is disconnected from its ISP, should the called party's ISP try to make a modem-call to the called PC before giving a progress indicator? Do we all have to leave our PCs permanently powered on if we want to be able to receive telephone calls?

Optional POTS Features. Some optional traditional telephone features, like calling party ID, are actually easier to implement over an IP-switched network than over the PSTN — if the called party has a PC and it's powered on and modem-connected to the ISP. But, the implementations for some other features are not so obvious. Only one feature is discussed here to illustrate this issue — call forwarding.

Call forwarding is trivial for the IP-IXC in Figure 19.10 — the LEC forwards incoming calls, as it does for any IXC. In the configuration of Figure 19.11, the ISP's middleware can also forward incoming IP telephone calls for its clients if the ISP provides the service. The ISPs will provide this service for some added revenue from its users. In fact, this particular configuration offers the opportunity for a clever ISP to offer enhancements, like conditional call forwarding, in which the user could specify several lists of calling parties who he wishes to forward to different target destinations. This ability for an ISP to attract customers by offering better features works if customers are willing to pay for these features — the user can't think he has the third configuration. Furthermore, the ISP's feature can't interact with the corresponding LEC's feature. The end user with a single telephone line may need to ask his LEC to forward any POTS calls to his telephone number and may need to separately ask his ISP to forward any IP calls to his PC.

Call forwarding simply doesn't work without ISP middleware, unless the end user leaves his PC powered on and modem-connected. The reader who thought that Figures 19.11 and 19.12 were too similar to distinguish now sees the significant difference between these two configurations.

Conventional Telephone Services. Some traditional telephone services, like automatic answering (voice mail), may be easier to implement in an IP-switched network than on, in, by, or through the PSTN. The implementations for some other services that users will expect from any kind of carrier are not so obvious. Only one feature is discussed here to illustrate this issue — in-WATS.

In-WATS connections are trivial for the IP-IXC in Figure 19.10 — the LEC determines 800-number ownership, and if the IP-IXC owns the dialed number, the LEC connects the calling party to the IP-IXC, as it does for any IXC. In the configuration of Figure 19.11, the ISP's middleware can also provide in-WATS calling for its clients if the ISP provides the service. But, in this configuration, the calling party's ISP must determine 800-number ownership. This ISP must have access either to the PSTN's ownership database, by SS7 or by IP signaling to a new port on the database, or to some new database. This new IP 800-ownership database could be placed on the World Wide Web or the ISPs could cooperate in building and maintaining a proprietary database. Both implementations are problematic.

This particular service, in-WATS, is easily implemented on the Internet, without ISP middleware, if someone provides a web-based translation service. Each user could enter his own mnemonic 800 number and the corresponding IP address. In the public domain like this, will mnemonics be assigned first-come, first-served? The implementation is also fraught with security issues. Oversight requires an overseer and overseers expect to be paid.

Enhanced Services. Beyond conventional services, like in-WATS, a competitive IP-switched network must provide all the enhanced services that the PSTN can provide using the Intelligent Network [34]. Independent of configuration (I think), there seem to be two implementations:

- The Internet and all the ISPs and their internets have access to the SCPs in the existing IN — either by SS7 signaling or by some new port into an SCP. This implementation has security problems, issues with certification and permission, and vested interests that will expect payment.
- The IP telephone industry will develop their own competing platform for the provision of enhanced services.

In the latter scenario, the ISPs will have to cooperate with each other in order to succeed. They will have to duplicate all IN software and compete with the existing IN. They could compete by allowing end users access to user-controlled routing — that last service that Sellers requested in Section 18.3.3. This kind of participatory control would be relatively easy for an ISP because it's within the Internet's culture. It's been difficult for the PSTN's carriers to allow this, because it's countercultural for them.

19.5.6 Regulatory Issues

In the United States, an end user pays an access fee to his LEC for connectability to an interexchange carrier. We regulate telecommunications prices this way because (1) long distance telephony costs the LECs over and above local service and (2) we still wish to subsidize POTS. In the United States, an end user does not pay a similar access fee for connectability to the Internet. Somehow, America's regulators think these are different [35]. In a related issue, local and long distance telephony are also subject to state and federal excise taxes and, for some reason, the Internet is not.

The inequitable enforcement of the long distance access fee created a subindustry through which enterprises can bypass the access fee and still get IXC connectability. The residential user, too small to attract the bypassers' attention, has to pay the access fee. Because there is no Internet access fee that could be inequitably enforced and bypassed, the opposite thing has happened. Enterprises avoid ISPs altogether by installing their own routers onto the Internet. But, the residential user, too small to have his own router, has to subscribe to an ISP. The bypassers can operate within the price margin of the access fee and are seen by their customers as saving them money. Now, we're seeing the IXCs providing direct LEC bypass themselves. If such an IXC implements its network with IP switching, it can claim to be an ISP (not an IXC) and also bypass the long distance excise tax — giving itself even more price margin to work inside. The ISPs have no artificial margins to work within and are seen by their customers as another layer of cost — unless, by providing long distance service, they give residential customers a way to bypass the long distance excise tax.

IP telephony raises two regulatory issues: (1) extending the access fee to the Internet and (2) letting residential telephone users bypass the long distance access fee just like enterprises do. The first would impact Internet usage and the second would impact universal service by taking away another source of the POTS subsidy. Complicating the future regulatory environment, bureaucrats in the federal and state governments don't care now whether access fees are equitable, but they will care if their own revenues decline because of reduced income from long distance excise taxes. Would the regulators extend the telecommunications excise tax to the Internet without also extending the access fee to the Internet?

19.5.7 Cost

A global IP telephone call, over the Internet, has a lower price than a conventional global telephone call. Will it always be lower, or will it get even lower? And, is it low enough to offset the possibly reduced quality? The answer, of course, depends on the configuration. Many of my graduate students, those whose homes are overseas, use the Internet for voice calls to their home countries. They tell me that the quality of these calls is barely acceptable, but they use it anyway. For several reasons we must be careful about generalizing this behavior, and its potential users, too much.

- These students pay for Internet access as part of their university fees, and because they're in configuration 3 (end-to-end over the Internet), their Internet usage is free (to them). And, free is much cheaper than the price of a conventional global telephone call.
- These users are (1) extremely cost-conscious graduate students, (2) technophiles with a very high tolerance for the Internet's foibles, (3) homesick people who want to chat on the telephone for a long time, and (4) calling someone who also uses a PC with identical internal software.
- The VoIP services currently available on the Internet are not planned, engineered commercial services. Current VoIP demonstrations, prototypes, and beta trials have much better quality.

Most of the carriers aren't investigating VoIP or VoATM because they might be able to offer a higher quality of voice service than today's circuit switching allows. Some of them are investigating these alternative implementations for telephony because they might give the carrier some market advantage, like offering truly integrated services or finding price margins (bypassing fees and taxes) to attract customers. Most of the carriers are investigating these alternative implementations because they're hoping to lower their cost. Depending on their market niche or their configuration (Section 19.5.3), some carriers or ISPs might be better using VoIP, some might be better using VoATM, and some might be better using circuit switching.

Section 18.2.4 discussed the cost of owning a telephone switching system. Does replacing the circuit-switched fabric by an IP-switched fabric lower the overall cost?

Section 17.4 showed that packet switching a regular signal, like voice, should probably consume more resources than circuit switching it. But, Section 18.2.4 showed that the fabric is a small part of the overall cost, anyway.

Some cost analysis has shown that VoIP and VoATM use less transmission bandwidth than circuit-switched voice. This is counterintuitive because circuit-switched voice doesn't require transmitting packet overhead, which is especially large in VoIP. The difference may be explained because the payload in the IP packets and in the ATM cells carries highly compressed voice. Because most of the carriers have a lot of unused bandwidth anyway, the cost of transmission is a real cost, but one that's already been paid for. If a carrier is still contemplating packetizing its voice channels in order to conserve bandwidth, such a carrier might consider comparing this to the cost of compressing circuit-switched PCM or of using techniques like time-assignment speech interpolation [36].

Section 18.2.4 showed that one of telephony's significant costs is OAM&P. It's clear that operating, administering, managing, and provisioning a single network ought to cost less than performing all these tasks on several networks. One of the effects of integrating all traffic types onto one network is that the carrier can more easily integrate its OAM&P. However, because a carrier's voice network, ATM network, SMDS network, frame relay network, IP network, SS7 network, and any other networks all use a common underlying infrastructure, the OAM&P of the physical layer and the transmission systems is already integrated. But, if OAM&P integration is really a large cost savings for some carrier, then why integrate all the networks into IP, when it accounts for a small minority of the carrier's traffic and the OAM&P systems for circuit switching are already in place?

Another way a carrier can reduce its cost is to pass cost back to the end user. The user's end-to-end cost may increase, but the carrier may lower its cost. IP certainly provides this opportunity.

It should be less expensive for a newly created carrier to install IP routers and to implement an IP-only network than it is for an incumbent carrier to do this. Besides buying and installing its new IP routers, an incumbent carrier must also replace its existing circuit switches and must also surrender the lost investment. If IP telephony is really less costly than conventional circuit-switched telephony, a new IP-only carrier should be able to offer lower rates than the incumbents. At the time the book is going to press, this hasn't been the case.

19.5.8 Architecture

IP telephony's architectural issues include: topology, integration, and reliability.

Topology. Internet advocates refer to their network as a cloud. This is an accurate name because a cloud doesn't have a fixed structure and its components (subclouds?) aren't tied together by a regular topology. The cloud architecture, or lack of architecture, has a certain charm that is understandably appealing. While some forms of telecommunications might be optimal under such a flexible architecture, and we all appreciate the spontaneity of new services and the participatory involvement of Internet citizens, is this optimal for all our telecommunications needs?

Moving IP telephony off the Internet and onto enterprise IP-EPNs or ISP-owned private internets resolves some of the concerns, but removes a lot of the charm and the low cost — the things that appeal to most of the users. Another of IP telephony's architectural issues is that it's just not clear how an IP-telephone network is configured (Section 19.5.3) and how the Internet fits inside an overall telecommunications architecture — or does the overall telecommunications architecture fit inside the Internet?

Having a fixed architecture implies having an architect [37]. An architect exerts control over additions, changes, and possibly content, and is responsible for the network's quality and performance. For many years, the American telephone user financed an overall network architect through the R&SE tax on the telephone bill. Because the old Bell System would have never given us the World Wide Web, perhaps the job of Chief Network Architect should never have been assigned to the network's principal owner. But, that doesn't mean we didn't need any architect at all.

Integration. Section 15.6.1 discussed whether network, or access, services should be integrated. IP telephony is another instantiation of this old question. Now, however, with the success of the Internet, it's no longer a hypothetical or rhetorical question. But, there is no single answer. An enterprise may elect to integrate its EPN and may even elect to merge its LANs with its PBX. Residential owners of PCs may wish for integrated access: integrated physical access over (for example) XDSL or HFC, or integrated logical access over (for example) ATM or IP or circuit switching. A carrier may wish to present an integrated interface. Perhaps different carriers will do it differently and the issue will be settled in the marketplace.

Reliability. In the late 1980s, the U.S. National Academy of Science published a famous report in which a committee of telecommunications experts expressed a concern about the proliferation of buggy software releases and the consolidation of physical transmission resources into a relatively small number of physical nodes and channels. Their concern was that the PSTN might have increased vulnerability to outages because of software bugs, cut fibers, or failed nodes. This report caused considerable concern, especially when the United States experienced a large number of outages in the late 1980s and early 1990s.

We don't hear about network reliability any longer [38]. I recently attended a keynote address at a communications conference that was given by the CEO of a significant U.S. telecommunications company. He argued eloquently that the future network shouldn't be architected by old-fashioned guidelines, like optimizing for reliability. Is this what we really want? If some airline CEO said the same thing, would we all applaud?

If it were designed (and controlled) well, the cloud could have a lot of redundancy. A rich topology of interconnected routers has the potential to be more reliable than a lean topology of traditional switching systems — even if the routers are individually less reliable than the switches. Without an architect, we won't have that. Are we prepared as a nation to state unequivocally that it doesn't matter? Maybe some of the carriers will design their internal networks, either IP-switched internets or circuit-switched telephone networks, with reliability as a design goal and we'll let the marketplace decide.

19.5.9 Discussion

Chapter 17 developed a rationale that, for regular data traffic, packet switching generally uses more resources than circuit switching. Circuit switching is technically and architecturally a more effective switching paradigm than packet switching — for carrying regular traffic. Switching-by-the-call is technically and architecturally more effective than switching-by-the-packet or than switching-by-needed-capacity for carrying regular data between two parties who communicate their regular data for the duration of their session. A point-to-point network is technically and architecturally more effective than a broadcast network for carrying regular data among users who usually intercommunicate pair-wise. Certainly, then, a global point-to-point circuit-switched network that's been optimized for the standard DS0 channel and for switching-by-the-call is intuitively, technically, architecturally, and actually optimal for voice telephony. This PSTN may not be perfect and it may need to be modified, but its underlying architecture and switching paradigm are sound — for telephony.

The previous section presented some generalizations about asynchronous transfer mode, including a discussion of ATM's relative effectiveness at carrying voice, data, and video communications. While cell relay is a form of packet switching, the virtual-circuit nature of ATM and ATM's tiny packet overhead both suggest that voice-over-ATM — while still not as effective as circuit-switched voice — would probably be more effective than, for example, voice-over-Ethernet.

We are not saying here that Ethernet cannot be used for telephony. Voice-over-Ethernet can be made to have perfectly acceptable quality, as long as the network is a small LAN and it's lightly loaded. It's even better than a conventional telephone system for broadcast and conference calls. But, no commercially relevant carriers are using Ethernet for voice. Recently, interest has turned to voice over another message-oriented packet-switched protocol — the Internet protocol.

IP telephony's advocates don't debate the points of the three preceding paragraphs — they debate other points. But, perhaps telephony and architectural effectiveness are not the most important issues. Some relevant issues, like quality and cost, were discussed above. Should a carrier consider building its network on IP telephony? The following subsections offer some possible answers.

Minority Traffic Type. We transmitted and switched data through the circuit-switched PSTN and through enterprise's circuit-switched PBXs for many years. While circuit switching isn't optimal for most data communication applications, we did it this way anyway because data was the *minority traffic type* for many years. Inside most enterprises, data communications became significant enough, and the circuit switching of data was clumsy enough, that LANs were installed. But, for many years, there wasn't enough wide-area public data traffic to justify large public networks that could be optimized for data communications. The suboptimal use of circuit switching for wide-area data wasn't significant when data was a tiny percentage of the traffic on the PSTN.

As this book is going to press, wide-area data traffic is still small compared to voice traffic — in the United States and worldwide. But, wide-area data traffic is no longer such a tiny minority that it's suboptimality under circuit switching can still be ignored. Now that data has become a significant part of total traffic, the suboptimal use of circuit switching for data has become significant enough that we can cost-justify separate wide-area networks for data communications. But, will data communications traffic continue to grow until it not only passes voice traffic, not only reaches the point where voice traffic is the minority traffic type, but reaches the point where voice traffic is insignificant in comparison? If this happens, or if we can safely predict that it will happen soon, then IP telephony is a perfectly sensible thing for us to do.

Geographical Scope. Cost and perceived delay in an IP telephone call is determined directly by how many voice samples are placed in the payload of an IP packet. Placing more samples in a packet increases delay and decreases cost. One common proposal was optimized for a network that is physically as large as the United States and has the same number of nodes as the Internet in the United States.

- If Internet traffic continues to grow as rapidly as some advocates predict, then there will be many more nodes in the Internet. This will increase the perceived delay. If Internet traffic is going to taper off [26], then the standard will work. But, one of the arguments for changing voice traffic to IP is to take advantage of the increased growth of IP traffic.
- Using this choice of samples per packet will make IP telephony unacceptable for international telephone calls. Unless a new standard can be developed to provide acceptable quality for global VoIP, the carrier that elects to use the current standard is effectively positioning itself as a United States-only carrier. Any new standard with lower delay will need to be very clever to overcome the inherent delay caused by an IP packet's large overhead.

Summary. Voice-over-IP has been standardized and is available now, commercially and on the Internet. But, some implementation issues are still unresolved and general public acceptance of VoIP's delay is still a question, especially for international calling. The generalization to IP telephony, a complete telephone service based on the Internet protocol, is still in the prototype and beta-test stages. Many features and services have yet to be implemented and the interface with the PSTN, especially SS7 and the IN, remains open. Architectural issues like topology, scalability, and reliability are challenging. We're not sure how it will be regulated and taxed and even figuring its total cost is elusive. As carriers begin to offer complete telephone service over IP, the consumers will get their chance to be heard.

The relative ubiquity of IP software and of IP addresses is compelling, and so is integration. IP may be the way we finally achieve the 25-year-old dream of integration, one end-to-end network with a single switching paradigm for voice, data, and video.

EXERCISES

19.1 Consider each of the XDSL variations in Table 19.1, except for VDSL. Assume each signal is modulated using a line code such that the ratio of binary data rate to channel bandwidth is 100 bits per cycle. Assume that the maximum tolerable attenuation on a loop is 8 dB. Using Figure 3.4, what is the maximum distance for each of the XDSL variations? This exercise is contrived to fit Figure 3.4 and does not use realistic numbers.

19.2 Consider a two-pair cable and assume that all the noise on any one pair is caused by crosstalk from the other pair. Suppose an ADSL signal, after attenuation along one of the pairs, is received at -14 dBm. Suppose the net (crosstalk) noise power at the receiver is -30 dBm. (a) What is the signal-to-noise ratio (SNR) at the receiver? The simplest way to increase the SNR is to transmit at higher power. (b) By how many decibels should the ADSL transmitter be increased to achieve

SNR = 22 dB at the receiver? (c) How many times more wattage is required at the transmitter? Assume that an x-fold increase in the transmit power on one pair causes an x-fold increase in the crosstalk noise on the other pair. (d) What is the new net (crosstalk) noise power at the receiver on the other pair? (e) What is this other pair's new SNR if its transmitter remains at its original level? (f) If both transmitters operate at the increased power of part (b), what is the SNR at each receiver?

19.3 Repeat the previous exercise, except assume that each pair is susceptible to electromagnetic interference (EMI) noise in addition to the crosstalk noise. Suppose -25 dBm of EMI noise is detected at each receiver when both transmitters are quiet. This exercise, especially part (f), is much more difficult and is similar to Problem 4.11 from Bellamy's book.

19.4 Supppose an HFC distribution point is a hub for 15 neighborhood coaxial cables. Suppose the feeder cable to the head end contains two fibers: one carrying downstream signals and one carrying the aggregated upstream signals from each of the 15 neighborhood coaxial cables. Draw a block diagram showing the equipment in this distribution point.

19.5 A sunbather on the earth's equator exposes his back to the sun for 1 hour. Another sunbather at 45° latitude also exposes his back to the sun. How much longer must the second sunbather be in the sun to receive the same amount of energy? Consider only the differential distance from the sun and ignore the earth's axis tilt and the attenuation of solar energy in the earth's atmosphere. Because your answer will be inconsistent with our experience, obviously atmospheric attenuation is significant.

19.6 Repeat the previous exercise considering three sunbathers — on Venus, Earth, and Mars, respectively.

19.7 What percentage of the sun's total power output is captured by Earth?

19.8 What is the maximum number of carrier frequencies that would ever be needed so that any arbitrarily partitioned region would never have two adjoining cells that use the same frequency? (Think about the classic map coloring problem.)

19.9 Walk through the SS7 signaling for a call to an 800 number that is forwarded to a roaming cell-phone. Assume that all HLRs and VLRs are located in corresponding SCPs.

19.10 When a satellite's orbit is in equilibrium, centrifugal force equals gravitational force. Derive Kepler's third law from a free-body diagram.

19.11 A satellite is in a circular orbit with radius r_1 and makes one complete revolution in period t_1. If the satellite slows for some reason, it collapses to a smaller orbit. If the radius of its new orbit is one-fourth of its previous orbit, how does its new period compare to its old period?

19.12 Use the characteristics of the lunar orbit to evaluate k in the equation for Kepler's third law in Section 19.3.2. Using this value of k, compute the elevation of a geosynchronous orbit.

19.13 If an Iridium satellite orbits every 2 hours, how quickly does one cell's footprint move across the surface of the earth? What is the maximum time that an Iridium subscriber is served by a single cell?

19.14 If you are directly underneath an Iridium orbit, how often are you handed off to the next satellite? How often are you handed off to the next cell?

19.15 Assume that Iridium's six orbits are equally spaced, that satellites in adjoining orbits are exactly out of phase, and that satellites pass over the earth at 3° latitude every minute. Assume that all satellites over North America are orbitting South-to-North. (a) If one satellite is directly over Houston, Texas, over what location is the satellite to its north in the same orbit? (b) Over which longitudes are the adjoining orbits? (c) Over what locations are the nearest satellites that orbit over the U.S. Atlantic and Pacific coasts? (d) In this configuration, what is the route for an Iridium call from Boston to Los Angeles? (Assume the call tandems through the satellite over Houston.) (e) Ignoring the rotation of the earth under these orbits, what is the route for this same call at the beginning of each of the next 10 minutes? (f) By how many degrees longitude does the earth rotate during these 10 minutes? (g) Is the earth's rotation significant to this exercise?

19.16 Suppose an Ethernet LAN is connected to a 600-Mbps ATM backbone network. Suppose an Ethernet packet contains 1000 bytes, including its header. (a) How many ATM cells are required to transport this packet to another LAN? (b) What is the total number of bytes, including the headers of the ATM cells? (c) If the Ethernet packet is transmitted as a VBR stream, with minimum QoS = 10 Mbps and maximum QoS = 50 Mbps, what are the minimum and maximum transmission times for the entire packet? (d) What are the minimum and maximum transmission times for CBR transmission at 30 Mbps? (e) What are the minimum and maximum transmission times for best-effort transmission?

19.17 Consider packet efficiency in the two extreme cases of the VoIP packet payload. (a) For IP packets with a 40-byte header and a 10-byte payload, what percentage of the data transmission carries voice content? (b) What is the required transmission rate for each VoIP stream so the payload can carry the compressed 8-Kbps voice signal? (c) What percentage of the original DS0 stream is this VoIP stream? (d) For IP packets with a 40-byte header and a 30-byte payload, what percentage of the data transmission carries voice content? (e) What is the required transmission rate for each VoIP stream so the payload can carry the compressed 8-Kbps voice signal? (f) What percentage of the original DS0 stream is this VoIP stream?

19.18 Now, consider delay in the two extreme cases of the VoIP packet payload. (a) For IP packets with a 10-byte payload, how long does it take to load the payload of one packet from a compressed voice stream of 8000 bits per second? (b) If 50 ms is budgeted for delay at each endpoint, what percentage of this budget is used just assembling each packet's 10-byte payload? (c) For IP packets with a 30-byte payload, how long does it take to load the payload of one packet from a compressed voice stream of 8000 bits per second? (d) If 50 ms is budgeted for delay at each endpoint, what percentage of this budget is used just assembling each packet's 30-byte payload?

19.19 Assume an optical fiber's physical route, from New York to Los Angeles, is 3100 miles long. (a) Because light travels through fiber at two-thirds of light's ultimate speed, what is the propagation delay over this route? (b) If 50 ms is budgeted for the network delay in VoIP, what percentage of this budget is used just for propagation? (c) If the route passes through five nodes, how much processing delay is allowed in each node?

19.20 Consider a fourth configuration for Section 19.5.3 — that of an end-to-end provider with HFC access to the residence. (a) Discuss leveraging price with television service. (b) Describe the home wiring architecture if the CATV set-top box provides the IP/BORSCHT interface or if the home PC provides this interface. (c) Discuss the other issues raised in Section 19.5 as they relate to this configuration.

REFERENCES

[1] Aber, R., "xDSL Supercharges Copper," *Data Communications,* Mar. 1997.

[2] Cook, J. W., et al., "The Noise and Crosstalk Environment for ADSL and VDSL Systems," *IEEE Communications Magazine,* Vol. 37, No. 5, May 1999, pp. 73 – 78.

[3] Chen, W. Y., "The Development and Standardization of Asymmetrical Digital Subscriber Line," *IEEE Communications Magazine,* Vol. 37, No. 5, May 1999, pp. 68 – 72.

[4] Chen, W. Y. and D. L. Waring, "Applicability of ADSL to Support Video Dial Tone in the Copper Loop," *IEEE Communications Magazine,* Vol. 32, No. 5, May 1994, pp. 102 – 109.

[5] Flanagan, W. A., "Carrier Strategies for DSL," *Telecommunications Magazine,* Vol. 30, No. 12, Dec. 1996.

[6] Keyes, P. J., R. C. McConnell, and K. Sistanizadeh, "ADSL: A New Twisted-Pair Access to the Information Highway," *IEEE Communications Magazine,* (issue on Reshaping Communications through Technology Convergence), Vol. 33, No. 4, Apr. 1995.

[7] Maxwell, K., "Asymmetric Digital Subscriber Line," *IEEE Communications Magazine,* (issue on Broadband '96), Vol. 34, No. 10, Oct. 1996, pp. 100 – 106.

[8] Papir, Z., and A. Simmonds, "Competing for Throughput in the Local Loop," *IEEE Communications Magazine,* Vol. 37, No. 5, May 1999, pp. 61 – 66.

[9] Masud, S., "Cable Telephony: Say Hello to Your New Phone Company," *Telecommunications Magazine,* Vol. 32, No. 12, Dec. 1998.

[10] Paff, A., "Hybrid Fiber/Coax in the Public Telecommunications Infrastructure," *IEEE Communications Magazine,* (issue on Reshaping Communications through Technology Convergence), Vol. 33, No. 4, Apr. 1995.

[11] Bass, G., "Radar — Is Your Cell Phone Killing You?" *PC Computing,* Dec. 1999, pp. 62 – 64.

[12] "Advanced Mobile Phone Service," *The Bell System Technical Journal,* (entire issue), Vol. 58, No. 1, Jan. 1979.

[13] MacDonald, V. H., "The Cellular Concept," *The Bell System Technical Journal,* (special issue on Advanced Mobile Phone Service — AMPS), Vol. 58, No. 1, Jan. 1979, pp. 15 – 41.

[14] Kuruppillai, R., M. Dontamsetti, and F. J. Cosentino, *Wireless PCS,* New York, NY: McGraw-Hill, 1997.

[15] Read, B. B., "Iridium Launches Global Customer Service and Sales," *Call Center,* Vol. 11, No. 9, Sept. 1998.

[16] DePrycker, M., *Asynchronous Transfer Mode: Solution for Broadband ISDN,* Hemel Hempstead, UK: Ellis Horwood, 1991.

[17] Becker, D., "Which Services Will Work Better with ATM?" *Proc. IEEE Int. Communications Conference,* Denver, CO, Jun. 23 – 26, 1991, Vol. 3, pp. 1246 – 1250.

[18] Bonatti, M., F. Casali, and W. Maggi, "ATM Implementations for Large Scale Utilizations: What Is Needed Before '92 to Start in '95?" *Proc. IEEE Int. Communications Conference,* Denver, CO, Jun. 23 – 26, 1991, Vol. 3, pp. 1265 – 1270.

[19] Decina, M., "Open Issues Regarding the Universal Application of ATM for Multiplexing and Switching in the B-ISDN," *Proc. IEEE Int. Communications Conference,* Denver, CO, Jun. 23 – 26, 1991, Vol. 3, pp. 1258 – 1264.

[20] Long, J., and B. Turner, "Point/Counterpoint — ATM," *Computer Telephony,* Vol. 6, No. 9, Sept. 1998.

[21] Weick, G., H. Schmidt, and U. Ziegler, "Which Customer's Needs Have to Be Served by ATM?" *Proc. IEEE Int. Communications Conference,* Denver, CO, Jun. 23 – 26, 1991, Vol. 3, pp. 1238 – 1245.

[22] Franz, R., and K. Gradischnig, "ATM-Based SS7 for Narrowband Networks — A Step Towards Narrowband-Broadband Convergence," *Proc. Int. Switching Symposium '97,* 1997.

[23] "Internet Telephony," *IEEE Network,* (entire issue), Vol. 13, No. 3, May/Jun. 1999.

[24] Gbaguidi, C., et al., "Integration of Internet and Telecommunications: An Architecture for Hybrid Services," *IEEE Journal on Selected Areas in Communications,* Vol. 17, No. 9, Sept. 1999, pp. 1563 – 1579.

[25] Schoen, U., et al., "Convergence Between Public Switching and the Internet," *IEEE Communications Magazine,* Vol. 36, No. 1, Jan. 1998, pp. 50 – 65.

[26] Sobczak, J., "How Visionaries Lead Us Astray," *Business Communications Review,* Oct. 1998.

[27] Cray, A., "Voice Over IP: Hear's How," *Data Communications,* Vol. 27, No. 5, Apr. 1998.

[28] Gareis, R., "Voice Over IP Services: The Sound Decision," *Data Communications,* Vol. 27, No. 3, Mar. 1998.

[29] "Standardization and Characterization of G.729," *IEEE Communications Magazine,* (entire issue), Vol. 35, No. 9, Sept. 1997.

[30] Coviello, G. J., "Comparative Discussion of Circuit- vs. Packet-Switched Voice," *National Telecommunication Conference,* Alabama, Dec. 1978, pp. 12.1.1 – 12.1.7.

[31] Coviello, G. J., E. Lake, and G. Redinbo, "System Design Implications of Packetized Voice," *Int. Communications Conference,* Chicago, IL, Jun. 1976, pp. 38.3-49 – 38.3-53.

[32] Morgan, S., "The Internet and the Local Telephone Network: Conflicts and Opportunities," *IEEE Communications Magazine,* Vol. 36, No. 1, Jan. 1998, pp. 42 – 48.

[33] Bernier, P., "IP Telephony and the Enterprise — Opening the Gate," *Sounding Board,* Vol. 1, No. 2, Jul. 1998.

[34] Bowers, C. L., "Enhanced Services: Helping IP Telephony Grow," *Sounding Board,* Vol. 1, No. 2, Jul. 1998.

[35] Noll, A. M., "Tele-Perspectives," *Telecommunications Magazine,* Vol. 32, No. 11, Nov. 1998.

[36] Bullington, K., and J. Fraser, "Engineering Aspects of TASI," *The Bell System Technical Journal,* Vol. 38, Mar. 1959, pp. 353 – 364.

[37] Fontana, J., "As the 'Net Grows, Who's in Charge?" *Communications Week,* Aug. 11, 1997.

[38] "Telecommunications Quality," *IEEE Communications Magazine,* (entire issue), Vol. 26, No. 10, Oct. 1988.

SELECTED BIBLIOGRAPHY

Acampora, A. S., and J. H. Winters, "System Applications for Wireless Indoor Communications," *IEEE Communications Magazine,* (issue on Echo and Delay in Digital Mobile Radio), Vol. 25, No. 8, Aug. 1987.

Allen, D., "Opening Up Data Networks to a World of Voice Services," *Telecommunications Magazine,* Vol. 33, No. 3, Mar. 1999.

Anerousis, N., et al., "TOPS: An Architecture for Telephony over Packet Networks," *IEEE Journal on Selected Areas in Communications,* Vol. 17, No. 1, Jan. 1999, pp. 91 – 108.

Bainbridge, H., "Learning New Traditions," *Wireless Business & Technology,* Vol. 5, No. 1, Jan. 1999.

Baines, R., "Designing ADSL Devices," *Communication Systems Design,* Vol. 2, No. 11, Nov. 1996.

Bear, E., "Designing an Embedded Voice-over-Packet Network Gateway," *Communication Systems Design,* Vol. 4, No. 10, Oct. 1998.

Bell, T. E., "Telecommunications," *IEEE Spectrum Magazine,* (issue on Technology 1991), Vol. 28, No. 1, Jan. 1991.

Bellamy, J., *Digital Telephony,* Second Edition, New York, NY: Wiley, 1991.

Bernstein, L., and C. M. Yuhas, "Network Architectures for the 21st Century," *IEEE Communications Magazine,* Vol. 34, No. 1, Jan. 1996, pp. 24 – 28.

Brands, H., and E. T. Leo, *The Law and Regulation of Telecommunications Carriers,* Norwood, MA: Artech House, 1999.

Branson, K., "Cable Modems Charm Cable Companies Toward IP Telephony," *Sounding Board,* Vol. 1, No. 2, Jul. 1998.

Brock, G. W., *Telecommunication Policy for the Information Age,* Cambridge, MA: Harvard University Press, 1994.

Brock, G. W. (ed.), *Toward a Competitive Telecommunication Industry,* Mahwah, NJ: Lawrence Erlbaum Associates, 1995.

Brock, G. W., and G. L. Rosston (eds.), *The Internet and Telecommunications Policy,* Mahwah, NJ: Lawrence Erlbaum Associates, 1996.

Brodsky, I., "Wireless Data Networks and the Mobile Workforce," *Telecommunications Magazine,* Vol. 24, No. 12, Dec. 1990.

Cameron, W. H., C. LaCerte, and J. F. Noyes, "Integrated Network Operations Architecture and its Application to Network Maintenance," *IEEE Communications Magazine,* (issue on Echo and Delay in Digital Mobile Radio), Vol. 25, No. 8, Aug. 1987.

Carter, M., and G. Guthrie, "Pricing Internet: The New Zealand Experience," *Proc. 3rd Int. Conference on Telecommunications Systems,* Nashville, TN, Mar. 16 – 19, 1995, pp. 153 – 161.

Collet, R., "Can the Internet Continue to Function Under Its Current Structure?" *Telecommunications Magazine,* Vol. 32, No. 1, Jan. 1998.

Comer, D. E., *Computer Networks and Internets,* Upper Saddle River, NJ: Prentice Hall, 1999.

Comer, D. E., *Internetworking with TCP/IP,* Englewood Cliffs, NJ: Prentice Hall, 1988.

Comparetto, G., and R. Ramirez, "Trends in Mobile Satellite Technology," *IEEE Computer Magazine,* Feb. 1997.

Cooley, K. D., et al., "Wideband Virtual Networks," *Telecommunications Magazine,* (issue on Switching Inside and Outside the Network), Vol. 21, No. 2, Feb. 1987.

Cox, D. C., "Portable Digital Radio Communications — An Approach to Tetherless Access," *IEEE Communications Magazine,* (issue on Portable Communications), Vol. 27, No. 7, Jul. 1989.

Cox, D. C., "Wireless Personal Communications: What Is It?" *IEEE Personal Communications Magazine,* Vol. 2, No. 2, Apr. 1995.

Cray, A., and P. Heywood, "Building the Brave New World," *Data Communications,* Vol. 27, No. 14, Oct. 1998.

Daddis, G. E., and H. C. Torng, "A Taxonomy of Broadband Integrated Switching Architectures," *IEEE Communications Magazine,* Vol. 27, No. 5, May 1989.

Dawson, F., "Considering Cable Telephony — Despite VoIP Advances, Cablecos Have Their Skeptics," *Sounding Board,* Vol. 2, No. 2, Feb. 1999.

Dawson, F., "Incumbents, Ambitious Next-Gen Telcos Give Rise to Carrier-Class Gateways," *Sounding Board,* Vol. 1, No. 2, Jul. 1998.

Degan, J. J., Jr., G. W. R. Luderer, and A. K. Vaidya, "Fast Packet Technology for Future Switches," *AT&T Technical Journal,* Vol. 68, No. 2, Mar./Apr. 1989, pp. 36 – 50.

Diamond, S., "ATM: Merging Voice and Data," *Communications News,* Vol. 34, No. 1, Jan. 1997.

"DSL/MVL: High Speed Both Ways over POTS," *Communications News,* Vol. 35, No. 2, Feb. 1998.

Dwyre, D., "New Satellite Services to Bridge the Gap," *Telecommunications Magazine,* Vol. 32, No. 12, Dec. 1998.

Dziatkiewicz, M., "Universal Service — A Taxing Issue," *Wireless Business & Technology,* Vol. 5, No. 1, Jan. 1999.

Engleman, L., "Have Number, Will Travel," *Phone+ Magazine,* Vol. 9, No. 10, Aug. 1995.

Fischer, W., E. H. Goeldner, and N. Huang, "The Evolution from LAN/MAN to Broadband ISDN," *Proc. IEEE Int. Communications Conference,* Denver, CO, Jun. 23 – 26, 1991, Vol. 3, pp. 1251 – 1257.

Forgie, J., "Speech Transmission in Packet-Switched Store and Forward Networks," *AFIPS National Computer Conference,* Anaheim, CA, May 1975, pp. 137 – 142.

Freeman, R. L., *Telecommunication System Engineering,* Third Edition, New York, NY: Wiley, 1996.

Freeman, R. L., *Telecommunication Transmission Handbook,* Third Edition, New York, NY: Wiley, 1991.

Gillett, S. E., and I. Vogelsang (eds.), *Competition, Regulation and Convergence,* Mahwah, NJ: Lawrence Erlbaum Associates, 1999.

Gradischnig, K. D., "Trends of Signaling Protocol Evolution in ATM Networks," *Proc. XV Int. Switching Symposium, Vol. 2,* Berlin, Germany, Apr. 23 – 28, 1995, pp. 310 – 314.

Greene, T., "RBOCs Vow 1999 Will Be DSL's Year," *Network World,* Vol. 15, No. 33, Aug. 17, 1998.

Halls, D. A., and S. G. Rooney, "Controlling the Tempest: Adaptive Management in Advanced ATM Control Architecture," IEEE Communications Society's *Journal of Selected Areas in Communications,* Vol. 16, No. 3, Apr. 1998, pp. 414 – 423.

Harrington, E. A., "Voice/Data Integration Using Circuit Switched Networks," *IEEE Transactions on Communications,* Vol. COM-28, Jun. 1980, pp. 781 – 793.

Heywood, P., and T. C. Eng, "Global Supernets: Big Pipes, Big Promises... and One Big Problem," *Data Communciations,* (issue on Global Supernets), Vol. 24, No. 13, Sept. 21, 1995.

Homa, J., and S. Harris, "Intelligent Network Requirements for Personal Communications Services," *IEEE Communications Magazine,* Vol. 30, No. 2, Feb. 1992, pp. 70 – 76.

Honing, G. A., "A High-Performance Implementation of Signaling System No. 7," *Proc. XV Int. Switching Symposium, Vol. 2,* Berlin, Germany, Apr. 23 – 28, 1995, pp. 305 – 309.

Hotch, R., "ATM Now: The Way to Go," *Communications News,* Vol. 36, No. 1, Jan. 1999.

Hotch, R., "High-Speed Copper," *Communications News,* Vol. 36, No. 6, Jun. 1999.

Howald, R., and T. Funderburk, "Telephony Over HFC," *Communication Systems Design,* Vol. 3, No. 2, Feb. 1997.

Howard, P. K., *The Death of Common Sense — How Law Is Suffocating America,* New York, NY: Warner Books, 1994.

"Industry Mulls Future of PCS," *Telephony Magazine,* (90th anniversary issue), Vol. 220, No. 24, Jun. 17, 1991.

Issa, J., and R. Bieda, "The G.DMT and G.Lite Recommendations, Part 2," *Communication Systems Design,* Vol. 5, No. 6, Jun. 1999.

Jabbari, B., "Intelligent Network Concepts in Mobile Communications," *IEEE Communications Magazine,* Vol. 30, No. 2, Feb. 1992, pp. 64 – 69.

Jessup, T., "DSL: The Corporate Connection," *Data Communications,* Vol. 27, No. 2, Feb. 1998.

Kawamura, R., and I. Tokizawa, "Self-Healing Virtual Path Architecture in ATM Networks," *IEEE Communications Magazine,* (issue on Self-Healing Networks for SDH and ATM), Vol. 33, No. 9, Sept. 1995.

Kim, B. C., J. S. Choi, and C. K. Un, "A New Distributed Location Management Algorithm for Broadband Personal Communication Networks," *IEEE Transactions on Vehicular Technology,* Vol. 44, No. 3, Aug. 1995, pp. 516 – 524.

Kim, G., "IP Answers the Call (Center) — Reaching Casual Users," *Sounding Board,* Vol. 2, No. 2, Feb. 1999.

King, R., "ADSL: Mass Market or Bust," *tele.com,* Vol. 3, No. 11, Oct. 1998.

Korzeniowski, P., "Advance to Go," *tele.com,* Vol. 3, No. 12, Nov. 1998.

Krasner, N., "Homing In on Wireless Location," *Communication Systems Design,* Vol. 5, No. 6, Jun. 1999.

Laitenin, M., and J. Rantala, "Integration of Intelligent Network Service into Future GSM Networks," *IEEE Communications Magazine,* Jun. 1995, pp. 76 – 85.

La Porta, T. F., et al., "Distributed Call Processing for Personal Communications Services," *IEEE Communications Magazine,* (issue on Mobility in Intelligent Networks), Vol. 33, No. 6, Jun. 1995, pp. 66 – 75.

Laurent, T., "Viewpoint on Portability — Global Cellular Telecommunications: The Dawning of a New Era," *Portable Design,* Jun. 1996.

Littwin, A., "ADSL: Ready for Prime Time?" *Telecommunications Magazine,* Vol. 30, No. 12, Dec. 1996.

Llana, A., "Deploying DSL: DSL + ISP = CLEC," *Communications News,* Vol. 35, No. 11, Nov. 1998.

Loshin, P., "What's All the Noise About Switching Layers?" *Communications News,* Vol. 36, No. 1, Jan. 1999.

Lubin, M., and D. Lyon, "Technical Advances and Challenges in Wireless Communications," *Telecommunications Magazine,* Vol. 24, No. 12, Dec. 1990.

Lucas, J., "The Internet Predictions for 1997," *TeleStrategies Insight Magazine,* Oct. 1996.

Lucas, J., "What Business Are the RBOCs In?" *TeleStrategies Insight Magazine,* Sept. 1992.

Machlin, R., "The Internet: Redefining the WAN," *Telecommunications Magazine,* Vol. 32, No. 1, Jan. 1998.

Maier, G., et al., "Multistage WDM Passive Access Networks: Design and Cost Issues," *Proc. 1999 IEEE Int. Conference on Communications,* (Session S43-1: Optical System Design), Vancouver, Canada, Jun. 6 – 10, 1999.

Makris, J., "Danger: DSL Ahead," *Data Communications,* Vol. 27, No. 6, Apr. 1998.

Makris, J., "Gigabit Cabling: The Copper Stopper?" *Data Communications,* Vol. 27, No. 3, Mar. 1998.

Mandeville, R., and D. Newman, "ATM: Brains and Brawn," *Data Communications,* Vol. 27, No. 7, May 1998.

McCall, D., "Intranets: Last-Mile Connectivity," *Communications News,* Vol. 35, No. 11, Nov. 1998.

McQuillan, J., and V. Cerf, *A Practical View of Computer Communciations Protocols,* New York, NY: IEEE, 1978.

Mehrotra, A., *GSM System Engineering,* Norwood, MA: Artech House, 1997.

Miyahra, H., T Hasegawa, and Y. Teshigawara, "A Comparative Evaluation of Switching Methods in Computer Communication Networks," *Int. Communications Conference,* San Francisco, CA, Jun. 1975, pp. 6-6 – 6-10.

Moffett, R. H., "Echo and Delay Problems in Some Digital Communication Systems," *IEEE Communications Magazine,* (issue on Echo and Delay in Digital Mobile Radio), Vol. 25, No. 8, Aug. 1987.

Nederlof, L., et al., "End-to-End Survivable Broadband Networks," *IEEE Communications Magazine,* (issue on Self-Healing Networks for SDH and ATM), Vol. 33, No. 9, Sept. 1995.

Newman, D., M. Carter, and H. Holzbaur, "DSL: Worth Its Wait," *Data Communications,* Vol. 27, No. 8, Jun. 1998.

Obuchowski, E. J., "Access Charge and Revenue Architecture," *AT&T Technical Journal,* Vol. 66, No. 3, May/Jun. 1987, pp. 73 – 81.

Ochiogrosso, B., et al., "Performance Analysis of Integrated Switching Communication Systems," *National Telecommunications Conference,* Los Angeles, CA, Dec. 1977, pp. 12:4-1 – 12:4-13.

O'Keefe, S., "While RBOCs Drag Their Heels, CLECs, ISPs Mean Business," *Telecommunications Magazine,* Vol. 32, No. 12, Dec. 1998.

"One Network?" *Communications News,* Vol. 35, No. 3, Mar. 1998.

Pandya, R., "Emerging Mobile and Personal Communication Systems," *IEEE Communications Magazine,* (issue on Mobility in Intelligent Networks), Vol. 33, No. 6, Jun. 1995, pp. 44 – 52.

Patel, B. V., et al., "An Architecture and Implementation Toward Multiprotocol Mobility," *IEEE Personal Communications Magazine,* (issue on Global Wireless Communications), Vol. 2, No. 3, Jun. 1995.

Perucca, G., et al., "Advanced Switching Techniques for the Evolution to ISDN," *Telecommunications Magazine,* (issue on Switching Inside and Outside the Network), Vol. 21, No. 2, Feb. 1987.

Phillips, B. W., "Broadband in the Local Loop," *Telecommunications Magazine,* Nov. 1994.

Piercy, J., "ATM: Welcome to the Real World," *Communications News,* Vol. 35, No. 2, Feb. 1998.

Pollini, G. P., K. S. Meier-Hellstern, and D. J. Goodman, "Signaling Traffic Volume Generated by Mobile and Personal Communications," *IEEE Communications Magazine,* (issue on Mobility in Intelligent Networks), Vol. 33, No. 6, Jun. 1995, pp. 60 – 65.

Pollman, S., "Set-Top Terminal Design," *Communication Systems Design,* Vol. 3, No. 3, Mar. 1997.

Pottie, G. J., "System Design Choices in Personal Communications," *IEEE Personal Communications Magazine,* (issue on Global Wireless Communications), Vol. 2, No. 5, Oct. 1995, pp. 50 – 67.

Raymond, K. R., and E. A. Walvick, "Broadband, An Evolution," *Proc. IEEE Int. Communications Conference,* Denver, CO, Jun. 23 – 26, 1991, Vol. 3, pp. 1271 – 1276.

Rizzo, J. F., and N. Sollenberger, "Multitier Wireless Access," *IEEE Personal Communications Magazine,* (issue on Global Wireless Communications), Vol. 2, No. 3, Jun. 1995.

Roberts, L., "The Evolution of Packet Switching," *Proceedings of the IEEE,* Vol. 66, Nov. 1978, pp. 1307 – 1313.

Rogers, C., "Telcos vs. Cable TV: The Global View," *Data Communciations,* (issue on Global Supernets), Vol. 24, No. 13, Sept. 21, 1995.

Rubin, P., "The Real Space Race," *tele.com,* Jan. 1998.

Rudin, H., "Studies on the Integration of Circuit and Packet Switching," *Int. Conference on Communications,* Toronto, Canada, Jun. 1978, pp. 20.2.1 – 20.2.7.

Rutkowski, A. M., "Emerging Network Switching Technology and Applications," *Telecommunications Magazine,* (issue on Switching Inside and Outside the Network), Vol. 21, No. 2, Feb. 1987.

Scarcella, T., and R. V. Abbott, "Orbital Efficiency Through Satellite Digital Switching," *IEEE Communications Magazine,* (issue on Digital Switching), Vol. 21, No. 3, May 1983.

Schneider, K., I. Frisch, and W. Hsieh, "Integrating Voice and Data on Circuit Switched Networks," *EASCON,* Washington, DC, 1978, pp. 720 – 732.

Schopp, M., "User Modelling and Performance Evaluation of Distributed Location Management for Personal Communications Services," *IEEE Communications Magazine,* Jun. 1995.

Schwartz, M., "Network Management and Control Issues in Multimedia Wireless Networks," *IEEE Personal Communications Magazine,* (issue on Global Wireless Communications), Vol. 2, No. 3, Jun. 1995.

Shankar, B., "Boundless Europe: The Wireless Revolution," *Microwave Journal,* Sept. 1997.

Shaw, J., *Telecommunications Deregulation,* Norwood, MA: Artech House, 1998.

Shibutani, M., et al., "A Gigabit-to-the-Home (GTTH) System for Future Broadband Infrastructure," *Proc. Int. Conference on Communications '97,* (Session on High Speed Access over Wired Media), Montreal, Canada, Jun. 8 – 12, 1997.

Shumate, P. W., "Optical Lines to Homes," *IEEE Spectrum Magazine,* (issue on Supercomputers: A Special Report), Vol. 26, No. 2, Feb. 1989.

Skeimoto, T., and J. G. Puente, "A Satellite Time Division Multiple Access Experiment," *IEEE Transactions on Communication Technology,* No. 4, Aug. 1968, pp. 581 – 586.

Snow, A. P., *A Reliability Assessment of the Public Switched Telephone Network Infrastructure,* Ph.D. Dissertation, University of Pittsburgh, PA, 1997.

Snow, A. P., and M. B. H. Weiss, "Empirical Evidence of Reliability Growth in Large Scale Networks," *Journal of Network and System Management,* Vol. 5, No. 2, Jun. 1997, pp. 197 – 214.

Somers, D., "Making the Net More Intelligent," *Telecommunications Magazine,* Vol. 32, No. 1, Jan. 1998.

Spaniol, O., et al., "Impacts of Mobility on Telecommunication and Data Networks," *IEEE Personal Communications Magazine,* (issue on Global Wireless Communications), Vol. 2, No. 5, Oct. 1995.

Stallings, W., *High-Speed Networks — TCP/IP and ATM Design Principles,* Upper Saddle River, NJ: Prentice Hall, 1998.

Stallings, W., *ISDN and Broadband ISDN with Frame Relay and ATM,* Fourth Edition, Upper Saddle River, NJ: Prentice Hall, 1999.

Stallings, W., "IPv6: The New Internet Protocol," *IEEE Communications Magazine,* Vol. 34, No. 7, Jul. 1996, pp. 96 – 108.

Steele, R., "The Cellular Environment of Lightweight Handheld Portables," *IEEE Communications Magazine,* (issue on Portable Communications), Vol. 27, No. 7, Jul. 1989.

Stephens, D. C., J. C. R. Bennett, and H. Zhang, "Implementing Scheduling Algorithms in High-Speed Networks," *IEEE Journal on Selected Areas in Communications,* Vol. 17, No. 6, Jun. 1999, pp. 1145 – 1158.

Stephens, W. E., and T. C. Banwell, "155.52 Mb/s Data Transmission on Category 5 Cable Plant," *IEEE Communications Magazine,* (issue on Reshaping Communications through Technology Convergence), Vol. 33, No. 4, Apr. 1995.

Stewart, A., "Copper Cable: The End Is Not Near," *Communications News,* Vol. 35, No. 2, Feb. 1998.

Stewart, A., "Solving the Data-Access Riddle," *Communications News,* Vol. 36, No. 6, Jun. 1999.

Stiller, B., et al., "A Flexible Middleware for Multimedia Communication: Design, Implementation, and Experience," *IEEE Journal on Selected Areas in Communications,* Vol. 17, No. 9, Sept. 1999, pp. 1580 – 1598.

Sukow, R., "Wireless Local Loop Tradeoffs — Should Carriers Plan for End User Expectations in 2000... or 2010?" *Wireless Business & Technology,* Vol. 5, No. 1, Jan. 1999.

Tanenbaum, A. S., *Computer Networks,* Third Edition, Upper Saddle River, NJ: Prentice Hall, 1996.

Thyfault, M. E., "Merge Ahead," *Information Week,* Feb. 1, 1999.

Thyfault, M. E., "Network Economy: Resurgence of Convergence," *Information Week,* Apr. 13, 1998.

Titch, S., "It's Your Move," *tele.com,* Vol. 3, No. 12, Nov. 1998.

Tracey, L. V., "Voice Over IP: Turning Up the Volume," *Telecommunications Magazine,* Vol. 32, No. 3, Mar. 1998.

Transmission Systems for Communications, New York, NY: Bell Telephone Laboratories, 1964.

Valovic, T., "Wireless Communications: The Next Wave," *Telecommunications Magazine,* (issue on The Wireless Revolution), Vol. 24, No. 12, Dec. 1990.

Viren, M. A., and V. C. Rocca, "IADs: Leveling the Field," *Telecommunications Magazine,* Vol. 33, No. 3, Mar. 1999.

Wang, A., J.J. Werner, and S. Kallel, "Effect of Bridged Taps on Channel Capacity at VDSL Frequencies," *Proc. 1999 IEEE Int. Conference on Communications,* (Session S7-2: Last Mile Technologies), Vancouver, Canada, Jun. 6 – 10, 1999.

Warke, N., and M. Ali, "On the Next Generation of High-Speed Voiceband Modems," *Proc. 1999 IEEE Int. Conference on Communications,* (Session S7-5: Last Mile Technologies), Vancouver, Canada, Jun. 6 – 10, 1999.

Weiss, M. B. H., and J. Hwang, "Circuit Switched Telephony or Internet Telephony: Which Is Cheaper?" *Proc. 26th Annual Telecommunications Policy Research Conference,* Alexandria, VA., Sept. 1998.

Weiss, M. B. H., and P. Bernt, *The Regulation and Deregulation of US Telecommunications,* Textbook under contract with Lawrence Erlbaum and Associates, Mahwah, NJ, C. Sterling, (ed.).

"Where's the Internet Headed?" *Information Week,* Jul. 17, 1995.

Widjaja, I., and A. I. Elwalid, "Performance Issues in VC-Merge Capable Switches for IP over ATM Networks," *Proc. IEEE Infocom '98, Vol. 1,* (Session 3D: IP over ATM Networks), San Francisco, CA, Mar. 29 – Apr. 2, 1998, pp. 372 – 380.

Wilder, C., "Net Catalog Sales," *Information Week,* Feb. 1, 1999.

Willey, R., "IP QoS: A Top Performer in Its Field," *Communications News,* Vol. 36, No. 6, Jun. 1999.

Wirth, P. E., "Teletraffic Implications of Database Architectures in Mobile and Personal Communications," *IEEE Communications Magazine,* (issue on Mobility in Intelligent Networks), Vol. 33, No. 6, Jun. 1995, pp. 54 – 59.

Zimmerman, A. B., "The Evolution of the Internet," *Telecommunications Magazine,* Vol. 31, No. 6, Jun. 1997.

20

Photonic Switching in Space

"And God said, Let there be light: and there was light. And God saw the light, that it was good." — Genesis 1:4-5a, from the King James translation of the Bible

This chapter and the next one describe how data in optical form can be switched without ever converting the data out of optical form. This chapter describes photonic switching in the space division and the next chapter extends the discussion to include the time and wavelength divisions. Unlike optical computing, photonic switching (particularly in the space division) is not some technology that we can't see until the distant future. To emphasize the near-term possibility of photonic switching, this chapter deliberately describes some relatively old and proven devices and some sensible architectures that use these devices. Newer research devices are mentioned and one is discussed in a little detail in Section 21.5.1.

We'll see in these two chapters that photonic technology lends itself to call switching, capacity switching, protection switching, and provisioning — the right half of Figure 17.14. However, while photonic protection switches are commercially available, photonic circuit switches have not been commercially successful. Because photonic technology does not lend itself well to packet switching, almost all advances in circuit switching have been put on hold until we see how ubiquitous packet switching is going to become. This delay to reach commercialization should not be interpreted to mean that the technology and the architectures are not ready to be commercialized. Section 22.7 proposes using photonic technology to build a national capacity-switched all-optical network that would support circuits and packets.

This chapter describes circuit switching, in the space division, of optical signals using optical devices. The reader is assumed to be slightly familiar with the fundamental physics of optical fiber and of elementary devices like LEDs, lasers, and photodetectors. Section 20.1 is a long introduction — discussing background, motivation, synergy (Section 13.1), and definitions. A natural dichotomy of photonic switching devices is discussed in detail in Section 20.2. It develops some reasons for concentrating on certain kinds of devices and excluding others, and raises the importance of a specific data format. Section 20.3 describes several photonic switching devices — but only of one type. Section 20.4 is a lengthy description of many different architectures for fabrics that perform photonic switching in the space division only. Several photonic switching architectures are described, optimized for different photonic switching devices. Section 20.5 discusses photonic packet switching. Photonic switching in time and in wavelength and switching in multiple divisions are described in the next chapter.

20.1 INTRODUCTION

This lengthy introduction discusses some relevant background of optical communications and photonic switching — the characteristics and limitations, architectural impact, contemporary utilization, and the almost unbounded future. Very carefully, this section makes a strong case for circuit switching of optical signals in the space division, using optical devices. Photonics and optoelectronics are carefully distinguished.

20.1.1 Background

Corning Glass patented optical fiber in the 1970s. Its potential for telecommunications was immediately recognized and a worldwide effort followed that developed improved fiber, manufacturing techniques, necessary related electronics, connectors and splicing technology, practicality, applications, and education. During the 1980s, research laboratories (most notably NTT Labs in Japan and Bell Labs in the United States) competed vigorously to break one another's world record for the optical fiber transmission rate × distance product. Demand was so great and competition was so intense that the research state of the art was typically getting to the marketplace in only about 3 years.

It is remarkable that within 10 years of its invention, optical fiber was practical enough that it was being deployed by many telephone and CATV companies and many users. Within 20 years of its invention, optical fiber became the transmission medium of choice, worldwide. The real potential of optical fiber, and the justification for investing in its deployment, is its enormous and, even now, relatively unused bandwidth. While we're pushing the limits of the bandwidth of moving electrons along copper, it seems hard to imagine running out of information-carrying capacity over optical fiber.

Compare information moving along several different transmission media to water flowing through a pipe. Since a pipe with larger diameter allows more water to flow, pipe diameter is analogous to bandwidth in the transmission channel. If a conventional analog telephone channel is analogous to conventional household 3/8-inch copper tubing (0.25i id), then coaxial cable carrying 70 television signals is analogous to a 6-foot diameter sewer pipe. (This is not a comment about the content.) Admittedly, there's a bit of an apple/orange comparison, but the ratio of areas is equal to the ratio of the capacities of a 4-kHz telephone channel to 70 television channels at 4 MHz each. On this same scale, optical fiber's ultimate capacity is bigger than the Grand Canyon — not just the Colorado River at the bottom, but the entire canyon, filled to the brim with moving water. While we're not even close to being able to use all this capacity, the capacity we can use is higher than the competing media, we're progressing toward using more and more of it every year, and the end is not in sight (like it is with copper). Some claim that the optical fiber currently in the ground is future-proof.

20.1.2 Motivation

While optical fiber has several significant advantages, the most significant is its ability to carry data at very high rates. Fiber provides a relatively inexpensive medium for broadband channels. Fiber is also used for conventional digital voice channels, an application in which the channels are highly multiplexed. We see that transmission is migrating to optical fiber, that data rates are increasing and will continue to increase, and that the architecture of the PSTN is migrating toward having a reduced number of nodes that are interconnected by comparatively few physical channels, which are highly multiplexed. While this trend has an adverse effect on network reliability, as the United States witnessed in the early 1990s, this chapter describes architectures that support the trend.

Recall from Section 13.1 that the migration to digital switching was largely motivated by synergy, the conservation of A-to-D converters on the boundary between switching and transmission. Thus, one is tempted to justify the migration to photonic switching by a similar synergy, the conservation of E-to-O converters on this same boundary. Photonics technology, however, may not be advanced enough that we can realistically expect long-haul fiber to terminate directly on photonic equipment, without intermediate electronic equipment.

- *Repeaters.* While optical amplifiers, especially the Erbium-doped fiber amplifiers (EDFAs), are in practical usage, they are analog amplifiers: they amplify a digital signal; they can not regenerate it. While nonlinear photonic devices exist that might perform digital regeneration of an optical signal, they are far from practical in this application. So, at least some of an optical line's amplifiers might have to be electronic because they may have to do more than just amplify. Current thinking is that every third or fourth amplifier in a transmission line will have to be an electronic digital repeater. It seems logical to place such an electronic digital repeater at the ends of the optical links inside the network nodes.

- *Polarization alignment.* Many of the photonic technologies are sensitive to the polarization of the received optical signal. However, the polarization of a signal coming off an optical fiber is not necessarily 100% linear and its alignment is certainly not predictable. While polarization can be linearized and aligned mechanically, without electronics, it may be cheaper and more reliable to do the job with an optical repeater.

- *Wavelength demultiplexing.* We should expect, more and more, that an optical fiber will carry channels on more than one wavelength, multiplexed together. While wavelengths are demultiplexed optically, typical demultiplexors have so much optical loss that we need repeaters, or at least amplifiers, after demultiplexing.

- *Format conversion.* If line coding formats have to be converted (for example, from Manchester to RZ), this will require electronics. Converting frame formats (for example, North American to European DS1) also requires electronics. Facility switching and add-drop multiplexing with conventional ADH or SDH frame formats also require electronics.

- *Packet processing.* Computing cyclic redundancy code (CRC) checks, reading headers for routing, and other packet header computations all require electronics.
- *Synchronization.* We don't have an optical equivalent to the phase-locked loop and we certainly can't do frame synchronization and all the other types of synchronization that are required with complicated digital frame formats. If we need to synchronize a signal, we need to do it in electronics.

If any of these functions listed above, or any other functions that may have been omitted from this list, requires an electronic implementation, then they would probably all be implemented together in electronics. They would probably all be colocated in the node, on the transmission-switching boundary, and the synergy argument would fail.

So, if synergy is not the principal motivation for photonic switching, then,

> *Why would anyone consider switching photonically, using devices that still require more research, when one could switch electronically, with devices that are already proven and practical?*

Does this question sound familiar? Reread the second sentence of the third paragraph of the Chauncey Depew letter in Figure 2.1. Here are a few of the good reasons for switching optical signals directly.

- *Protection switching.* Sections 12.5.7 and 17.5 discussed the practice of installing physical analog switching between transmission facilities and their corresponding transmitters and receivers. This way if a facility is decommissioned or faulty, the expensive transmitters and receivers can be moved to another facility. This is called protection switching and it is even more common in optical facilities. So, if we're going to have some kind of photonic switching network anyway, as a protection switch, maybe its role should be expanded to include facility switching, capacity switching (Section 17.5), and even circuit switching.
- *Noise reduction.* Photons on fiber are inherently less prone to electromagnetic interference than electrons on copper. While most noise accumulates in the transmission facility, switching systems are also noisy environments. If noise and crosstalk become very serious problems, particularly as the bit error rate (BER) requirement drops to 10^{-12}, photonic switching may provide the answer.
- *Modularity.* Some firsthand personal experience has taught me an interesting and surprising lesson. The construction of the DiSCO prototype, described in the next chapter, entailed a great deal of complicated electronics and photonics equipment. In DiSCO's assembly and debugging, we spent far more time dealing with the electronics than with the photonics. The electronics needed to be wire-wrapped and we never did completely debug some problems in the electronics that I'm convinced were caused either by some bizarre heat effect or by some strange coupling through a power supply that might not have been perfectly regulated. The photonics components were all pigtailed and every time we connected a photonic component, it just worked the first time. As a newcomer to photonics and a veteran of electronics, I was expecting to have more trouble with the photonics than with the electronics. But, that is not what happened. While I'm not going so far as to

suggest the use of photonics over electronics because the implementation and system integration will be easier in photonics, I have no firsthand experience to refute this controversial point.

- *Immense connectivity.* The inherent three-dimensional nature of light suggests that a broad beam of light, or many parallel information-bearing rays, could be projected onto a two-dimensional plane of interconnected nano-processors, each with its own optoelectonics. One application is highly parallel picture processing for video generation, pattern recognition, and robotic vision. But, three-dimensional "free-space" systems have also been proposed to interconnect thousands of parallel optical signals. This potential benefit of optics is so relevant that we might even consider a photonic switching node, even if the transmission plant is all digital electronics and copper.
- *Immense data rate.* Because optical transmission avoids the detrimental reactive effects caused by parasitic inductors and capacitors in electronic transmission, optical channels have extremely high theoretical bandwidth. This very high channel capacity gives not only optical fiber, but also other photonic components, the potential to carry vast amounts of information at extremely high data rates. There is a related characteristic that may be even more important. Not only do optical fiber and other photonic components have extremely high channel capacity, but their construction is independent of the rate of the data that will go through them. Data rate is not part of the specification of such components, as it is with almost all electronic and copper components. Networks built with this technology have rate-scalable infrastructure.

These last two potential advantages that optics might have over electronics have the potential to be architectural breakthroughs. The ultimate goal, of course, is a system with both these characteristics — a physically small system that can provide rapid interconnection among thousands of optical fibers, each carrying 20 Gbps this year (and 50 Gbps 5 years later). Without a device breakthrough, this ultimate system will elude us because, so far, photonic devices come in a natural dichotomy that forces us to select one characteristic or the other. It appears that we can have immense connectivity or immense data rate, but we can't have both. The devices that have the potential for high connectivity can't support extremely high data rate, and the devices that have the potential for high data rate can't provide high connectivity.

20.1.3 Photonics and Optoelectronics

One of the drawbacks of electronic devices is that, by their nature, electrons move around inside them. Moving electrons are subject to resistance and reactance. We'll use the term "optoelectronics" to describe devices that operate externally on an optical signal but, when they do, a corresponding electronic signal moves around internally.

The term "photonics" suggests a family of technology based on the photon, analogous to how electronics technology is based on the electron. We'll use the term "photonics" to describe devices, which may be electronically controlled, but through which externally applied optical signals remain as optical signals internal to the device.

The goal of research in photonic switching is to use these new technologies to construct complete networks that would switch data in optical form, without conversion to electronics. While these same technologies might be used for optical computing, which is discussed briefly in this chapter, this chapter focuses on photonic switching. Because optical communications channels are defined in space, time, and wavelength, photonic switching in each of these three divisions is important. Photonic switching in the space division is described in this chapter. Photonic switching in the time division, the wavelength division, and in combinations of the three divisions, is described in the next chapter.

This chapter describes several exotic photonic switching devices that could significantly impact the future of telecommunications and computing. While interesting devices are described, the coverage is superficial and the chapter, like the book, is presented primarily at the architectural level. Because many of these photonic devices are immature, a mediocre device characteristic is seen, not as a "device problem," but as an "architectural challenge." Simply stated, if photonic devices don't have satisfactory characteristics, we will architect around their problems.

20.2 DEVICE DICHOTOMY

The devices that might be useful for photonic switching have a natural dichotomy. The optoelectronic and photonic devices that may be relevant to computing and communications come in two general categories. This section presents an interesting argument, based on the thermodynamics of data rate, that increases the relevance of one of these two types of devices, analog switching, and the block-multiplexed data format.

20.2.1 Two Categories of Devices

Two categories of devices are "free-space" and "guided-wave."

Free-Space Devices. Because these devices operate by some form of optoelectronic physical phenomenon, they regenerate any digital signals that pass through them but also impose internal reactance effects on them. Because, in this class of devices, each individual component is physically oriented normal to the surface of its substrate, they have a very high potential for integrability, but it is difficult to interconnect many neighboring devices into an integrated network. Because their input/output access is via the substrate's surface, many parallel devices can operate on many parallel signals, either as pixels in some common picture or as independent channels. Their two-dimensional parallel nature, surface access, and high potential integrability make them ideal for a free-space architecture that could provide immense connectivity. Regeneration makes them ideal for switching digital signals.

Guided-Wave Devices. Because these devices operate by some form of photonic physical phenomenon, they impose no reactance effects on signals that pass through them, but they cannot regenerate those signals. Because, in this class of devices, each individual component is physically oriented tangential to the surface of its substrate, they have a lower potential for integrability, but they are easily interconnected into complex integrated networks. Because their input/output access is via the substrate's edge, fewer parallel devices can operate on fewer parallel signals. The optical devices that will be described in the next section, optical fiber itself and both kinds of photonic amplifiers, can all support extremely high data rates. But, they are all analog devices and none of them integrates very well. Their sequential nature and reactance-free operation make them ideal for a network architecture that could switch a few signals each carrying an immense data rate and for networks where rate-independence is important. But, they perform analog switching, even on digital signals.

The most fundamental distinction between the optoelectronic devices and the guided-wave devices is illustrated next.

20.2.2 Thermodynamics of Data Rate

We present an interesting way to evaluate the long-term potential of a switching technology by looking at thermodynamic limitations imposed by data rate and by integration [1,2]. First, a chart of power versus rate is introduced and explained, and then, three limiting cases are explained. Then, integration is shown to exacerbate the thermodynamics (because many switching devices are all generating heat on the same chip) and analog switching is shown to have a significant advantage — when data is formatted using block multiplexing.

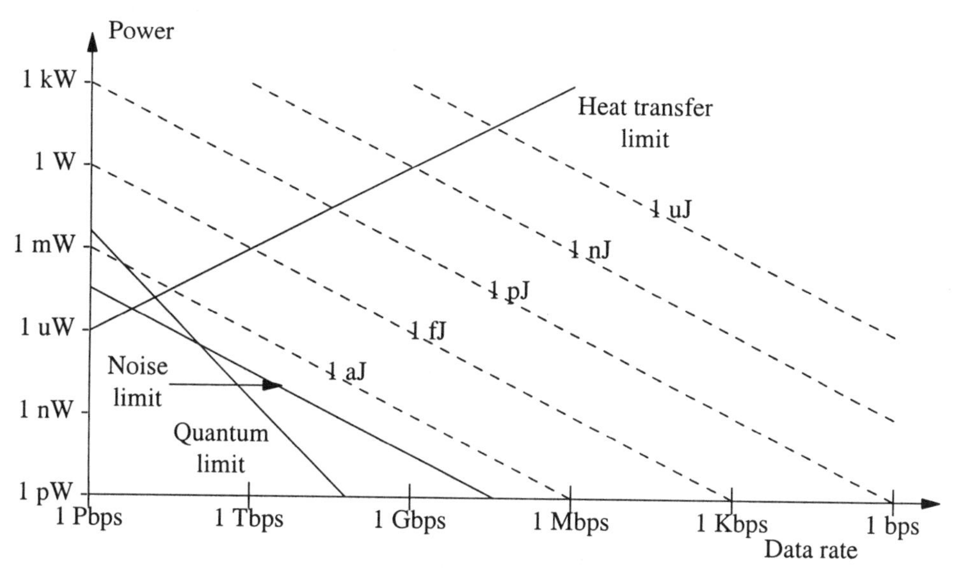

Figure 20.1. Power versus rate. © 1982 AT&T [1].

Figure 20.1 illustrates the thermodynamics of bit rate. The figure is also an exercise in the obscure regions of the metric system. Data rate is shown on the horizontal axis and total dissipated power is shown on the vertical axis. The dashed diagonal lines represent technology families, by the energy dissipated whenever a switch operates. For example, if data at 1 Gigabit per second is processed by a switch which dissipates 1 picojoule for each bit, then the total dissipated power is:

$$10^{-12} \text{ joules/bit} \times 10^{+9} \text{ bits/second} = 10^{-3} \text{ joules/second} = 1 \text{ milliwatt}$$

This algebraic result is obtained from Figure 20.1 by reading up from "1 Gbps" to the dashed line for "1 pJ" and then across to "1 mW." One milliwatt of dissipation is also obtained when a 1-nJ device operates at 1 Mbps, a 1-fJ device operates at 1 Tbps, or a 1-aJ device operates at 1 Pbps.

While we want to be able to process bits as fast as technology allows, a higher data rate implies more switch operations per second and, hence, more heat dissipated in the technology. While physical data transmission doesn't require switch operations, many switch operations are required in the transmitters, repeaters, and receivers along a transmission line and, especially, in the system equipment on the ends of the lines. Many switch operations are required whenever data is modulated and demodulated, encoded and decoded, compressed and decompressed, multiplexed and demultiplexed. These switch operations are unavoidable. Germane to this discussion, however, switch operations are also required anywhere data is switched. The number of switch operations per bit of data and hence the thermodynamics of the switching fabric, depends on the mode of switching, the data format, and the architecture of the fabric.

Figure 20.1 indicates that increasing the number of switch operations per second is accompanied by greater heat dissipation. Three solid lines on Figure 20.1 represent three different physical limits on bit processing.

- One of the lines that proceeds diagonally downward is labeled "Quantum Limit." This line represents the optical power that is carried by an optical signal at *f* bits per second that has only 10 photons in each bit and only one cycle of carrier in each bit.

$$P_q = 10\ e_p \text{ per second} = 10\ hf \times f = 10 \times 6.626 \times 10^{-34} \times f^2$$

  At 1 Pbps, the optical power in such a signal is 6.6 mW. Because Figure 20.1 is a log-log scale, a quadratic in f proceeds linearly to the right with slope = -2. Because we've assumed that the data rate equals the carrier frequency, signals in the high tera bits per second must be optical. While we can exceed one bit per cycle in today's electronic modems, it's hard to imagine doing it at high giga bits per second data rates. And, because no amount of energy can be lower than the energy of one photon, it's unlikely that any system can operate in the region to the lower left of this line.

- The other line that proceeds diagonally downward is labeled "Noise Limit." The power produced by random thermal noise is given by:

$$P_{TN} = \frac{e_{TN}}{\text{time}} = 4kT \times f$$

  where Boltzmann's Constant, $k = 1.38 \times 10^{-23}$J/°K. Because the components' ambient temperature is at least T = 300°K and since it would be difficult to detect a signal if its SNR is lower than 3 dB, this line represents $2\ P_{TN} = 32 \times 10^{-21} \times f$. No system can operate to the lower left of this line either.

- The line that proceeds diagonally upward is labeled "Heat Transfer Limit." This line represents the approximate limit for conducting heat away from existing semiconductor technology, about 1 watt per nanosecond. Any switching device that operates in the region to the upper left of this line, such as a 1-pJ device at 100 Gbps, accumulates heat faster than it can be conducted away — and overheats.

Given these three constraints, it appears from Figure 20.1 that the maximum possible data rate is about 30 Tbps and that this rate is achievable only if devices are developed that can switch at less than 1 aJ/bit. Existing devices are approaching 10 fJ/bit, which makes even 100 Gbps a problematic data rate.

In this context, consider two additional factors: integrated circuits and analog switching. The thermodynamics of these two factors significantly impacts the two principal characteristics that motivate photonic switching: connectivity and throughput.

20.2.3 Level of Integration

While the ability to integrate many devices onto a single chip is important for commercial practicality of any device, it is essential if we are to construct that projection plane required for a three-dimensional high-connectivity system. Many of the optical devices that show promise for a high level of integration are optoelectronic (not photonic) and digital.

The Self Electro-optic Effect Device (SEED) is such a device [3,4]. The SEED is an optical threshold device, typically implemented as an optical two-input NOR gate. As a true digital device, the SEED has most of the same advantages that electronic digital devices have, including that binary pulses are regenerated as they pass through. A SEED can perform a computer logic operation and it can serve as a simple switch in communications applications. Furthermore, integrated circuits have been built that have thousands of SEEDs on a chip.

Researchers have proposed (on paper) many optical computer architectures, and actual working prototypes have been built: one using SEEDs and another using switched directional couplers. While both prototypes are bit-serial computers, built with discrete components in a two-dimensional architecture, they are still important demonstrations of proof of concept. While optical computing is important research, there is still a lot to be done. It's clear that we will be executing programs on electronic computers for many years to come.

Three-dimensional free-space architectures for switching networks have also been prototyped, with their projection planes made from a variety of 1 × 1 optoelectronic devices: mechanical and liquid-crystal *shutters,* devices that operate like optical silicon controlled rectifiers (SCR), individual SEEDs, and integrated circuits of SEEDs. All these prototypes of switching fabrics are also important proofs of concept. They have illustrated the need for economical imaging optics if the three-dimensional architecture is ever to become practical, especially in multistage fabrics. But, they have illustrated another, more fundamental, problem.

The subtle disadvantage to being digital is that the switching device operates on every bit, and because it's also optoelectronic, electrons move around deep inside the device with every logic operation. So, each bit that passes through produces heat. If a digital optoelectronic device expends 10 pJ with each bit of throughput, then a discrete device operating at 10 Gbps dissipates 0.1 W of thermal energy. If 10,000 of them on a single chip could all operate at this data rate, then the chip would need to dissipate 1 kW. Because it can't, it would overheat. The reader is cautioned to be wary of the research reports on some of the switching technologies, including photonic technologies. Devices have been reported that can operate at 1 Gbps. It is also reported that these same devices have been integrated to more than 10,000 devices per chip. However, these two reports cannot be taken together to imply that 10,000 devices on one chip can all operate simultaneously at 1 Gbps. The digital optoelectronic devices that can be highly integrated, while they might be operated quickly as discrete components, have thermodynamic limitations when integrated together because, in the close proximity of an integrated circuit, all the devices on a single chip contribute to the chip's temperature.

20.2.4 Analog Switching for Throughput

In an analog device, like an electromechanical relay (Section 2.3.1), the signal that operates the device is different from the data that passes through the device.

> *If data is organized and formatted so that a "block" of consecutive bits is switched to the same switch port, then the device operates once per block, just before the beginning of each block, instead of having to operate on each individual bit.*

This block multiplexing format is discussed in more detail in Section 21.1.1. If such a device expends 10 pJ with each switching operation, then a discrete device operating on one timeslot every 100 ns, dissipates 0.1 mW of thermal energy, for any data rate. Even if 100 such devices on a single chip could all operate at this timeslot rate, then the entire chip would dissipate only 10 mW. We see that analog devices will excel at throughput and are consistent with serial time-multiplexed optical data streams that are confined to separate channels; that is, the existing real world.

We saw that, in the very act of regenerating, a digital device operates with each bit, and because it is also optoelectronic, electrons move around inside the device each time it operates. So, an analog device that switches a block with 1000 bits dissipates the same amount of heat as a digital device that switches 1 bit. A 1-pJ device will overheat if it operates at 1 Tbps but if it's analog and if it's switching 1000-bit blocks of data, it only has to operate for each block (not each bit). Because the device switches at a rate of 1 Gigablock per second, it operates quite comfortably. In fact, a chip with 10 or 100 such devices integrated together would still operate within the heat-transfer limit.

In the block multiplexed data format, data is packed into blocks at the line rate, but a *guard-band* interval is provided between consecutive blocks of data. No data is allowed in the guard band between the blocks. Two issues contribute to the required duration of the guard band: switching speed and variation in packet arrival time. We assume that data is organized into a block that is bunched into the temporal center of a timeslot. A time-multiplexed $N \times N$ switching network would have to be switched into its assigned configuration for a given timeslot at the beginning of that timeslot, before any of the data actually arrived at the switches. It would have to hold this configuration until all the data in every parallel block was throughput.

- The guard band must be long enough to allow all the switches time to operate and thus configure the state of the network.
- The data in each timeslot will not arrive at the switching network at exactly the same time. The different transmitters will not be perfectly synchronized, the fiber links will have different lengths, and the index of refraction in each link may be different. So, the guard band must also be long enough to allow for any variability in the arrival time of the data at the switching network.

20.2.5 Connectivity or Throughput?

We've seen that highly integrable digital devices, like SEEDs, are optimal for connectivity and that the slightly integrable analog devices are optimal for throughput. Figure 20.1 illustrates an interesting design space in which we can try to get a little of both. The bottom line is that, with existing technology, these four things — (1) three-dimensional connectivity, (2) digital regeneration, (3) integrated circuits, and (4) high data rate — do not scale together. Systems with only three of these characteristics, any three of the four, are not nearly as interesting as systems with all four. In terms of optimizing its application, we must choose between connectivity and throughput — we can't have both.

There just isn't a perfect photonic switching device. If there were, we could think about having high connectivity and high throughput in the same system. With today's devices, one must choose between connectivity and throughput. So, which is more like the real world?

- A photonic switching node that is built with highly integrated digital photonic devices can terminate thousands of fibers, but all the fibers must operate at very low data rates.
- A photonic switching node that is built with slightly integrated analog photonic devices can terminate fibers that operate at very high data rates, but it can only terminate tens of fibers.

Because the latter is more consistent with the evolving national network, the remainder of this chapter deals only with these systems. Despite how good it sounds to be digital and highly integrable, when it comes to photonic switching, scalability seems to favor the devices that allow high throughput, even if this means that they are analog devices and are only slightly integrable. This chapter and the next one emphasize this analog photonic switching.

20.3 GUIDED-WAVE SWITCHING DEVICES

Three guided-wave devices capable of photonic switching are described here: the semiconductor laser amplifier, the Mach-Zender interferometer, and the Lithium Niobate switched directional coupler.

20.3.1 Semiconductor Laser Amplifiers

A semiconductor laser amplifier (SLA) is just a semiconductor diode laser (SDL) with the following differences [5 – 7].

- The SDL is forward biased to produce a 1 and reverse biased to produce a 0. The SLA is forward biased to turn the amplifier on and reverse biased to turn it off.
- In a forward-biased SDL, photons are spawned by *spontaneous emission* in a confined physical channel that spans the junction, and photons (within the SDL's narrow gain bandwidth) are multiplied by *stimulated emission* each time they traverse the channel. While an SLA's forward-biased current is large enough to sustain multiplication by stimulated emission of those photons that enter the channel from the input fiber, this current is kept small enough to keep the diode's spontaneous emission (seen here as noise) at a reduced level. While an SLA is constructed like an SDL, it acts more like an LED than an SDL when there is no input signal.
- The SDL's channel is partially reflective (optically resonant) on both ends, so many of the photons in a signal bounce back and forth, being amplified for each pass down the channel and establishing the standing wave that gives the SDL its modes.

In the SLA, the chip's end-facets are given an antireflective coating so most photons make only one pass down the channel, from input fiber to output fiber, and the physical resonance that causes longitudinal modes is reduced.

- In an SDL, the photons that don't reflect are captured in the output fiber as an optical 1 pulse. But, SLAs are manufactured with two fiber "pigtails," one for the attenuated input signal and one for the amplified output signal. The input fiber couples an attenuated photonic signal into one end of the channel, and the amplified photonic signal couples from the other end of the channel into the output fiber. Some SDLs are also pigtailed with two fibers: one for the output signal and one that feeds the output signal back to the electronics to provide for signal stability and gain control.
- An SDL's output signal has longitudinal modes corresponding to the standing wavelengths that lie within the gain spectrum. An SLA's output signal is an amplified version of the input signal, corrupted by the SLA's spontaneous emission.

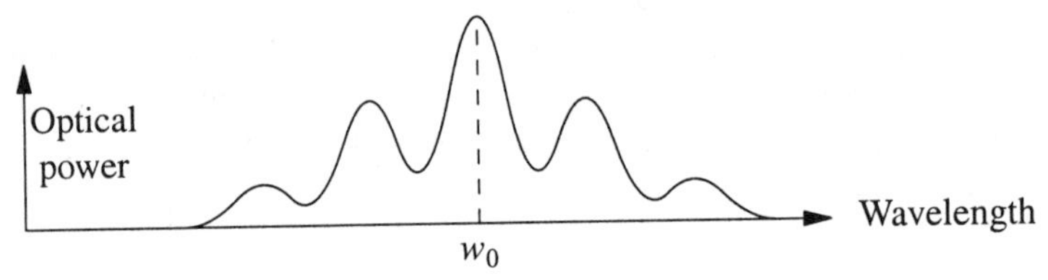

Figure 20.2. Longitudinal modes in a laser's spectrum.

The antireflective (AR) coating is similar to the radar-absorbing coating applied to spy planes. The boundary between two media is painted with a coating that is a quarter-wavelength thick and has a reflectivity such that the ideal net reflected signal is the sum of two signals of equal magnitude, 180° out of phase. An SLA with a poor AR coating, with high reflection at its edges, has a gain spectrum that resembles an SDL's power spectrum (Figure 20.2). As the AR coating improves, providing less reflection, the gain spectrum changes: (1) the center wavelength shifts, (2) the optical bandwidth increases, and (3) the shape of Figure 20.2 smooths such that the peaks decrease and the valleys increase. An SLA with a perfect AR coating, with no reflection at its edges, has a gain spectrum that resembles an LED's power spectrum.

While SLAs have been reported with gains around 20 dB, these are typically highly modal SLAs where the wavelength of the input signal is tuned to a high-gain mode — probably not a very practical arrangement because of the tuning. High-AR SLAs with smooth gain spectra have lower gain but have lower variation in gain over the range of wavelength. The optical bandwidth of an SLA's gain is typically about 50 nm, limited by the static nature of the antireflective coating.

Some of an SLA's gain is lost because of the coupling between the device's channel and the input and output fibers. A net gain of 6 – 10 dB is reasonable. The useful property of the SLAs is that the gain can be turned on and off rapidly by switching the current signal to the device. If an SLA is a switched amplifier, it is also an amplifying switch. Networks of SLA switches are discussed in the next section.

20.3.2 Mach-Zender Interferometer

While the ultimate speed of light is a true constant, the speed of light in a material — and many other optical properties of materials — are not true constants, but are complicated functions of the material's temperature, stress, optical intensity, and electric field. Any material in which these optical properties are especially dependent on electric field is said to be electro-optic. Crystalline Lithium Niobate, $LiNbO_3$, is a popular electro-optic material.

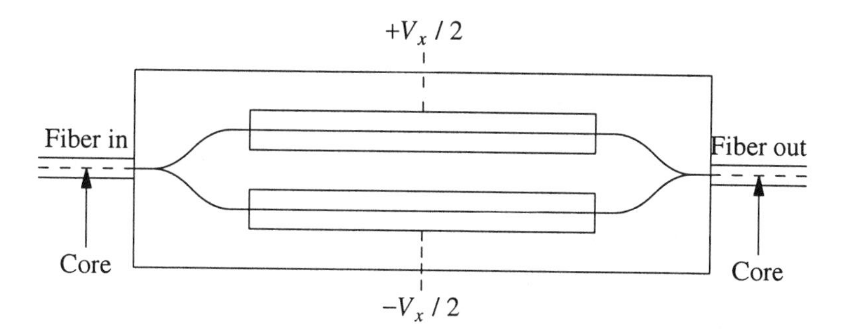

Figure 20.3. Mach-Zender interferometer.

A classical Mach-Zender interferometer (MZI) is fabricated using discrete components: two half-silvered mirrors for splitting and combining the optical signal and two separately controlled electro-optic wave-guides. The more useful fabrication is integrated on a chip of crystalline Lithium Niobate, as shown in Figure 20.3. By diffusing Titanium into the material, a pattern of parallel optical wave-guides, with passive splitters and combiners, is fabricated on the upper surface of a small crystalline chip of $LiNbO_3$. The MZI's input and output optical signals are carried over optical fibers whose ends are physically butt-coupled against opposite edges of the chip so that the fiber cores align with the ends of the diffused wave-guides.

The input signal splits so that equal halves of this signal traverse the two parallel optical wave-guides. Then, these two components of the signal are combined again to form the MZI's output optical signal. Recall that the light in each of these parallel wave-guides is a sinusoidal electromagnetic traveling wave. If the two wave-guides have equal length and equal index of refraction, then the two signal components add, in phase, at the output combiner, and the output signal exactly equals the input signal, except for any internal attenuation.

In the MZI's structure, however, each of these parallel optical wave-guides lies under the plate of a different capacitor. So, if a voltage is applied to the structure, each wave-guide resides in a different electric field. Because the material is electro-optic, this external voltage gives each wave-guide a slightly different index of refraction. This difference means that light travels a little faster in one wave-guide and a little slower in the other. Some exact voltage, V_x, changes the speed of light in the two wave-guides by just the right amount so that the two optical signals arrive at the output combiner exactly 180° out of phase. Because the two signals would exactly cancel, or interfere with, one another, the output signal would be completely extinguished. So, when the control voltage is 0, the input signal transmits through to the MZI's output, and when the control voltage = V_x, the input signal is blocked.

Since the electro-optic effect occurs extremely rapidly, the MZI switches very quickly, in even less than 1 ns, limited by the rate at which the capacitors can be charged. Obviously, V_x must have a different value for every different optical frequency (wavelength). So, an MZI works only on coherent light, with all the optical energy at the same frequency (wavelength). The MZI is a single-pole, single-throw switch in which an electronic control voltage switches an optical data signal. MZIs are commonly used as external modulators for lasers.

20.3.3 2 × 2 Lithium Niobate Switched Directional Coupler

This subsection discusses another analog photonic device, the switched directional coupler (SDC) made from Lithium Niobate [8 – 11]. The SDC's operation and fabrication are presented here and its considerable shortcomings are explained in detail. The SDC's story continues in the next section, which describes some novel switching networks that architect around the SDC's device problems.

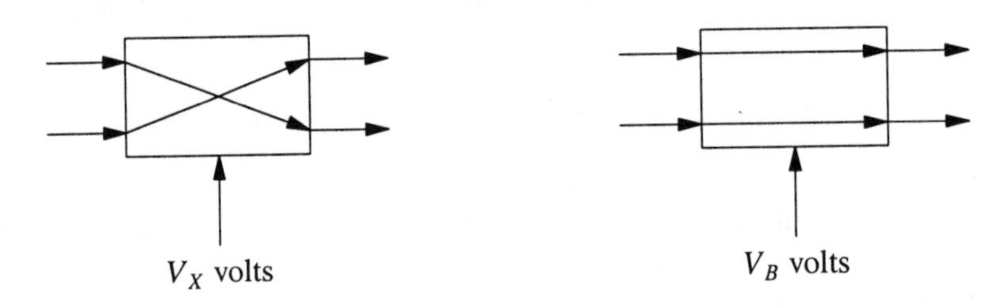

Figure 20.4. $LiNbO_3$ switched directional coupler.

The switched directional coupler, shown in Figure 20.4 is a little like the Mach-Zender interferometer described above and a little like a passive optical coupler (POC). While the MZI is a 1 × 1 device and the POC is a 1 × 2 device, the SDC is a 2 × 2 device, with two input fibers and two output fibers. Instead of being passive like the POC, the SDC is active like the MZI. Like the POC, any optical signal coming into the device on either input fiber can transfer, partially or fully, to the other wave-guide because of *evanescent wave coupling*.

SDCs are implemented in crystalline $LiNbO_3$ because Lithium Niobate happens to have a high electro-optic effect; its index of refraction is quite sensitive to any electric field. The parallel wave-guides lie between the parallel plates of a capacitor. The coupling of optical energy between wave-guides can be controlled electronically, as in an MZI.

SDCs have been made that switch using a variety of different physical processes over different electro-optic properties. While the more precise physics of SDCs is omitted here, all SDCs have a common effective operation. The connectivity between an SDC's two input fibers and its two output fibers can be controlled by an external voltage, as shown in Figure 20.4. With V_B volts applied, signals emerge from output ports on the same physical channel as their respective input ports. This is called the BAR state. With V_X volts applied, the input signals cross over to their opposite physical channels. This is called the CROSS state. The SDC is seen to be a single device that is an optical implementation of the beta element, discussed in Section 7.3.4.

The SDC has two very important properties.

- Because electrons don't move around whenever photons flow through an SDC, the flow of optical data produces no heat. But, electrons do have to move around when the device is switched between BAR and CROSS. So, if data is block multiplexed, then the only device switching occurs on the timeslot boundaries. This makes SDCs independent of data rate, so they don't have to be replaced whenever the network's data rate is increased.
- Like the MZI, the SDC switches very quickly between its two states. In block multiplexing, guard bands can be quite short. But, even if data is switched by the bit, the SDC can keep up with relatively high data rates.

An SDC's length, switching voltage, and switching speed are interrelated. Because SDCs can be fabricated to optimize any one parameter, if you read about new world's records in these parameters, you must be careful not to assume that one device can be fabricated that meets all these characteristics. In Lithium Niobate, a typical SDC is less than 1 cm long and less than 100 microns wide. A typical switching voltage is in the range of 5 – 100 volts. Obviously, a lower voltage is preferable, but that would require that the device be longer. While they can be operated in less than 100 ps, a more typical switching speed is a couple of nanoseconds.

Because the device works by changing an electric field and the wave-guides are embedded between the parallel plates of a capacitor, the applied switching voltage actually does charge a capacitor. This magnitude of this capacitor can also slow down the switching speed, but it is typically only a couple of picoFarads. SDCs work on single-mode light only, and the switching voltage for light that is polarized normal to the plane of the substrate is quite different from the switching voltage for light that is polarized tangential to the plane of the substrate. So, incoming light must have its polarization linearized and this polarization must be properly aligned before the light can enter the SDC's wave-guides.

As a true analog gate, an SDC not only doesn't regenerate its binary pulses, it introduces loss and crosstalk to them. Attenuation, crosstalk, and the sensitivity of the SDC's different parameters will be discussed later.

While several SDCs can be integrated onto one Lithium Niobate substrate, it would be difficult to get a hundred of them on a chip. Since wafers are cut from crystal cylinders, called boules, the chips are typically 3 – 4 inches long and about 1 cm wide. While the SDCs are narrow enough that many could fit top-to-bottom, the wave-guides must be separated at the chip's edge to accommodate the thickness of adjoining parallel optical fibers. SDCs cannot be densely packed along the length dimension because the interstage wiring (should we call it channeling?) uses up a significant amount of this dimension. Wave-guides diffused in the crystal that are allowed to bend at too great a radius of curvature have too much attenuation. So, architectures with many SDCs in the width dimension require more length to accommodate the increased curvature of the interstage wiring.

20.3.4 Lithium Niobate's Problems

Besides their analog nature, photonic SDCs were criticized for other inherent problems with Lithium Niobate technology, which made these devices appear to be problematic in practical implementations of photonic switching fabrics. Lithium Niobate's problems include high attenuation, high crosstalk, narrow optical bandwidth, dependence on polarization, low tolerance to small variations in the switching voltages, and variation of all parameters to temperature and even age. While this long list of problems makes Lithium Niobate seem impractical, it's not nearly as bad as it appears because there are really only two problems: attenuation and crosstalk. All the other problems are related to crosstalk and the crosstalk problem can be architecturally finessed.

This subsection carefully explains Lithium Niobate's problems. First, attenuation is discussed, and second, crosstalk is discussed. Then, Lithium Niobate's sensitivity problems are carefully explained. This subsection concludes with a philosophical discussion of the difference between a device solution and an architectural solution. Section 20.4 discusses how these problems have been solved — not by refining the devices, but by architecting around the problems — an architectural solution, called dilation.

Attenuation. Various sources of attenuation are given, followed by fabrication techniques and architectural techniques for dealing with the problem.

- *Coupling loss.* The largest source of attenuation in Lithium Niobate occurs at the chip boundary, where the fiber butts against the chip and the optical signal couples from the fiber's core into the wave-guide that is Titanium diffused into the Lithium Niobate. The pattern of the optical cross-section in a fiber's core differs from the corresponding pattern in one of Lithium Niobate's Titanium-diffused wave-guides. This modal mismatch causes a high attenuation where an optical signal couples between the fiber's core and the chip's wave-guide at the boundary where a fiber butts against the edge of a Lithium Niobate chip. When fabricating a chip, giving the wave-guide a more complex structure at the chip boundary can reduce the coupling loss. But, an architectural solution is even more effective. Coupling loss is worse when all the SDCs are discrete, each residing on its own chip of Lithium Niobate. Coupling loss can be greatly reduced from the discrete case if at least several adjacent stages can be integrated onto a single chip. Coupling loss is minimized if the entire fabric can be integrated onto a single chip. Even if it can't fit on one chip, coupling loss can be reduced if a network can be made to fit onto two butt-coupled chips — coupling from one wave-guide directly into another is less lossy than coupling from a fiber into a wave-guide.

- *Wave-guide loss.* Because the Titanium-diffused wave-guides are quite lossy, there is considerable loss along the wave-guides and internal to the SDCs. Manufacturing progress has lowered the loss in wave-guides and switches. Attenuation along the wave-guides and especially internal to the SDCs can be reduced by selecting a fabric architecture that has a reduced number of stages — like the Benes network (Section 7.3.4). Furthermore, because amplifiers can be simplified if we have approximately uniform loss in all paths, this suggests selecting a fabric architecture that has an equal number of stages in every path — again the Benes network.

- *Bend and crossover loss.* Loss increases wherever wave-guides bend. Loss in wave-guide bends can be reduced by making the bends with a relatively large radius of curvature. There is also a significant loss wherever wave-guides cross over each other. Wave-guides really pass through each other and photon's momentum keeps most of the desired optical power proceeding straight through the intersection. Crossover loss can be reduced by making the angle of intersection relatively large. Crossovers can be completely eliminated and wave-guide bends can be reduced by a careful choice of architecture, but the architectures that have no bends or crossovers are typically quite poor in other characteristics like connectivity, number of stages, and crosstalk. Increasing the bend radii and crossover angles requires increasing the spacing between adjacent columns of SDC switches. This effect is exacerbated in switch matrices with many rows of SDCs. But, fitting several stages on a single chip, to reduce coupling loss, requires decreasing the space between the SDCs. Unfortunately, we can't have it both ways. Fitting several stages on a single chip also requires decreasing the length of each SDC, which increases the switching voltage and the crosstalk. We see that fabrication is intricately tied to operating parameters.

Crosstalk. Crosstalk in a 2 × 2 switch is quantified by its extinction ratio (ER), a measurement that is similar to signal-to-noise ratio. Assume that two optical signals move left-to-right through an SDC, that all four paths through the switch have identical attenuation, and that the SDC is in the BAR state. Then, the signal detected at the upper right port has two components.

- The desired signal, detected at this upper right output port, is an attenuated version of the desired signal at the selected upper left input port. If P_s is the input power of the desired signal, then the signal detected at the desired output has power $r \times P_s$. The attenuation, r, is the fraction (hopefully close to 1) of the desired input signal that reaches the selected output and it's assumed to be identical in all four paths through the switch.

- Crosstalk is an undesired leakage of an attenuated version of the signal that is input to the lower left port and that is intended to be detected at the lower right port. The crosstalk has power $r \times x \times P_u$, where r is the expected attenuation in any path through the SDC, P_u is the input power of the undesired signal, and x is the fraction (hopefully close to 0) of the undesired input signal that reaches the unselected output, by traversing an unselected path through the SDC.

The SDC's extinction ratio is the ratio of these two components of the total signal detected at the upper right output port. If all selected paths have equal attenuations and $P_d = P_u$, then $ER = 1/x$, usually expressed in decibels.

Consider a signal that passes through K consecutive SDCs, each with identical loss, crosstalk, and extinction ratio. Accurately calculating the end-to-end extinction ratio is quite difficult because crosstalk in the stages near the output is more significant since it is not attenuated by as many subsequent stages of switches. If we ignore attenuation and only consider crosstalk, then crosstalk adds to the signal in each successive SDC and the end-to-end extinction ratio = K/x (be careful not to use decibels). It would actually be significantly better than this.

Sensitivity. After attenuation and crosstalk, Lithium Niobate's next most notorious problem is its high sensitivity to small variations in most of its operational parameters. While, initially, Lithium Niobate might seem discouraging and impractical, the discussion below shows that the parameter sensitivities arise because of the relatively high crosstalk requirement.

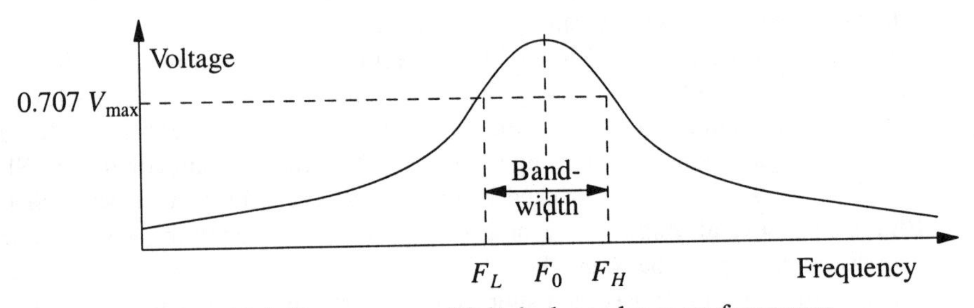

Figure 20.5. Power transmitivity's dependence on frequency.

Figure 20.5 shows a classical plot of transmitivity versus frequency, illustrating the definition of bandwidth. If the figure shows signal power transmitivity versus frequency for some channel, then this channel's bandwidth is the band of frequencies between F_L and F_H, the frequencies for which the power transmitivity is at least 50% of the peak mid-band transmitivity. If $F_H - F_L$ is very small, the channel is said to have high Q or to be narrowband — its transmitivity is extremely sensitive to frequency. Similar to how a signal's transmitted power depends on frequency and how this dependency specifies the channel's frequency sensitivity (quantified by the definition of bandwidth), Lithium Niobate's crosstalk depends on many operational parameters and this dependency specifies its sensitivity to these parameters.

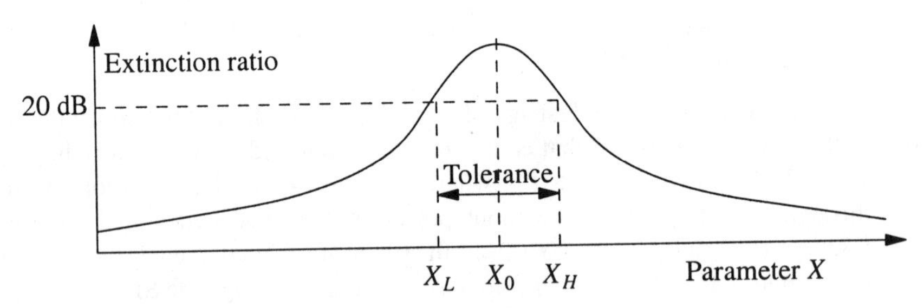

Figure 20.6. Crosstalk's dependence on parameters.

Figure 20.6 looks like the previous figure, only the definitions of the axes have been changed. Similar to how bandwidth is defined and specified in the previous figure, Figure 20.6 defines and specifies Lithium Niobate's tolerance (or intolerance) to any of several parameters. Where bandwidth is defined as the frequency range over which power is within some specification (greater than half the peak power), parameter tolerance is defined as the range in parameter value over which crosstalk is within some specification. The y axis represents the crosstalk in a Lithium Niobate SDC. Figure 20.6's x axis can represent any one of a number of some SDC's operational parameters.

- If x represents the switching voltage, then the figure shows that voltage X_0 is the ideal voltage to switch the SDC into its desired state, to maximize the ER. Some other voltage, far removed from X_0, is optimal for switching the SDC into its opposite state. For other voltages, the SDC is in some in-between state, neither BAR nor CROSS, where ER is unacceptably low at both output ports.
- Let x represent the signal's wavelength. Suppose the perfect optimum voltage is applied that places some SDC in its BAR (or CROSS) state. But, this optimum voltage depends on the wavelength, X_0, of the signal switched through the SDC. If the signal's wavelength changes, the SDC's extinction ratio worsens, unless the switching voltage changes correspondingly.
- Let x represent the signal's polarization. Suppose the perfect optimum voltage is applied that places some SDC in its BAR (or CROSS) state. But, this optimum voltage depends on the polarization, X_0, of the signal switched through the SDC. If the signal's polarization changes, the SDC's extinction ratio worsens, unless the switching voltage (or wavelength) changes correspondingly.
- Let x represent the switch's temperature. Suppose the perfect optimum voltage is applied that places some SDC in its BAR (or CROSS) state. But, this optimum voltage depends on the switch's temperature, X_0. If the switch's temperature changes, the SDC's extinction ratio worsens, unless the other parameters change correspondingly.
- Let x represent the length of time that a switch has been in its state. Suppose the perfect optimum voltage is applied that places some SDC in its BAR (or CROSS) state. But, this optimum voltage only pertains to the time, X_0, when the state change occurs. As the switch is in this state for a longer time, the SDC's extinction ratio worsens, unless the other parameters change correspondingly.

Lithium Niobate has a reputation for being extremely sensitive to relatively small changes in relatively many parameters. The problem lies in the y axis, not the x axis or the many x axes. While the scale in Figure 20.6 isn't shown, Lithium Niobate's real problem is that a reasonable ER of, say, 100:1 (20 dB) is high on the y axis. The two intercepts, corresponding to X_L and X_H, represent a narrow tolerance in the parameter represented on the x axis. Requiring that an SDC have an extinction ratio > 20 dB is what underlies Lithium Niobate's high sensitivity to all its operational parameters. If the SDC's required extinction ratio can be lower, then all the parameter dependencies can be desensitized. But wouldn't a lower extinction ratio requirement result in a poor bit error rate? Not necessarily. The previous subsection discussed that Lithium Niobate's attenuation problem has an architectural solution. So also does Lithium Niobate's crosstalk problem.

Device or Architectural Solution. Lithium Niobate's problems were exacerbated by an argument over who owned the problem — is it a device problem or is it a systems problem? Imagine a physicist and a systems engineer discussing this technology. The physicist tells the systems engineer, "Tell me the requirements, so I can build you a device that works within these parameters." The systems engineer tells the physicist, "Tell me the device's inherent characteristics, so I can build an architecture that works within these parameters."

Lithium Niobate's problems are not that this technology is overly sensitive to every one of these parameters. Its problem is that we've been requiring this technology to meet a crosstalk requirement that is unable to be met in a practical way. If we relax the crosstalk requirement, we remove the sensitivity to all the other parameters, and $LiNbO_3$ is OK. But, if we relax the crosstalk requirement, doesn't the fabric become impractical? That depends on the fabric's architecture, not on the devices. Repeat this: that depends on the fabric's architecture, not on the devices. We need an architecture for SDC switching networks in which the end-to-end crosstalk is better than the extinction ratio of each individual SDC in the network.

The breakthrough was not discovered in the physics lab. It was a casual observation by a systems researcher. We have crosstalk in these SDC fabrics, not because the devices are inadequate, but because we have two talking paths in each SDC. If the extinction ratio is too small in some switch, you can try to improve the switch or you can try to eliminate the source of the crosstalk. If there is only one switched connection set up through an SDC fabric, there will be no crosstalk and the end-to-end path through the switch will be extremely insensitive to operating parameters. If there are two switched connections set up through an SDC fabric, but the two connections have no SDCs in common (neither connection passes through the same SDC) there will be very little crosstalk and the end-to-end path through the switch will be extremely insensitive to operating parameters. It is only when some SDC inside the fabric supports two different switched connections that we have crosstalk. While this observation is obvious in hindsight, it had remarkable consequences.

20.4 PHOTONIC SPACE-DIVISION FABRICS

While many network architectures for photonic space switching have been proposed, only some of these have been prototyped. We have demonstrated the proof of concept that the technologies will work in a switching application. While much of the early work was based on classical space-switching architectures, significant new architectures have been proposed for time, space/time, and even space/time/wavelength switching. While optical packet-switching networks have been proposed, the technology is more easily applied to circuit switching.

This section describes several photonic switching fabrics that are constructed from guided-wave analog devices and that operate in the space division only. The *matrix architecture* is generalized four different ways and then, the dilated Benes network is shown to be a good architectural solution for Lithium Niobate's problems. The third subsection illustrates some switching architectures suitable for construction from SLAs.

20.4.1 Generalizing the Matrix Architecture for SDCs

Many different architectures have been proposed, and even prototyped, for switching fabrics built from SDCs. Some of them, even some that were prototyped, are really quite silly and came about only because the implementers were very naive about the theory of interconnection switching networks. Only two architectures are described here: the generalized matrix, in this subsection, and the dilated Benes, in the next one.

Figure 20.7 illustrates an $M \times N$ switching matrix using four different combinations of passive/active splitting/combining [12]. The matrix switches shown in the figure provide full connectivity for $M = 3$ input ports and $N = 3$ output ports. The discussion

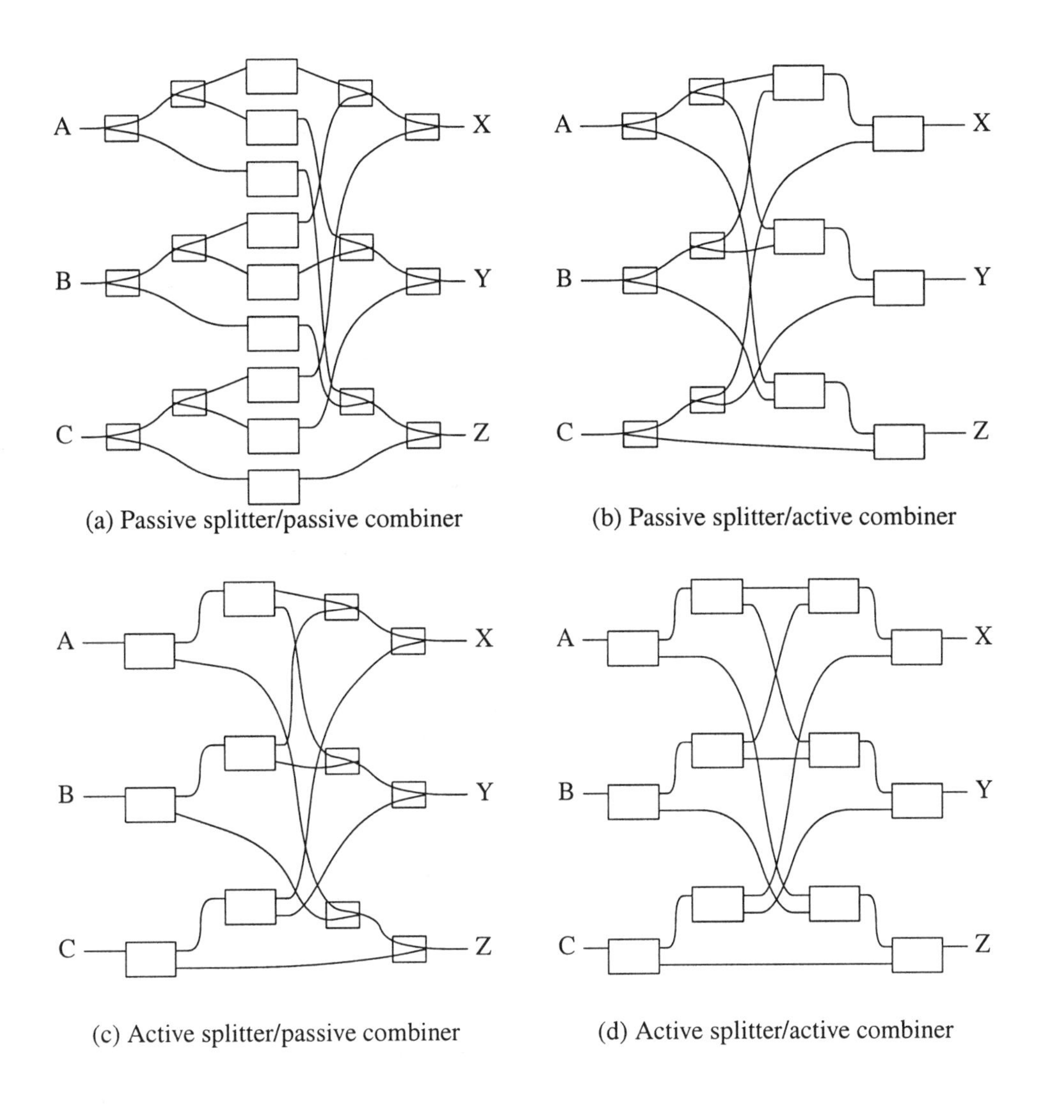

(a) Passive splitter/passive combiner

(b) Passive splitter/active combiner

(c) Active splitter/passive combiner

(d) Active splitter/active combiner

Figure 20.7. Four variations of a photonic switching matrix. © 1987 IEEE [12].

following assumes that $M = N$, but this isn't necessary. The four architectures are called: (a) the passive splitter/passive combiner (PS/PC), (b) the passive splitter/active combiner (PS/AC), (c) the active splitter/passive combiner (AS/PC), and (d) the active splitter/active combiner (AS/AC).

In Figure 20.7, a smaller box, with channels shown passing through it, represents a passive 1×2 splitter or combiner, while a larger box, with the channels shown interrupted, represents an active 1×2 switch, like an SDC. Each input or output port is connected to the root-side of a tree-like structure of passive or active 1×2 devices. Because these figures lack the rectilinear structure and "banjo wiring" of a traditional matrix switching architecture, they don't look like matrix switches. While the physical splitters and combiners are necessary because optical fiber doesn't "bus" well, they perform logically like a rectilinear bus. So, each structure in Figure 20.7 operates logically like a matrix switch.

Figure 20.7(a) is a 3 × 3 extension of Figure 7.1(c). It illustrates a simple 3 × 3 photonic switching matrix that uses an optical shutter or MZI as the switching device. An SDC with only one input and only one output wired can also be used as a 1 × 1 switch in this architecture. This 3 × 3 matrix requires nine switches, a generalized $M \times N$ matrix requires $M \times N$ SDCs, and a generalized $N \times N$ matrix requires N^2 SDCs. Because optical signals can't simply fan out, each of the $M = 3$ input signals must be passively split to the left side of $N = 3$ SDCs and each of the $N = 3$ output signals must be passively combined from the right side of $M = 3$ SDCs. This splitting and combining requires two PDCs for each port, for a total of 12 PDCs, and the structure is called the passive splitter/passive combiner.

Each of the SDCs in the center of Figure 20.7(a) can double as a 2 × 1 device by using two of its inputs instead of only one. So, one or both stages of passive splitting or combining can be absorbed into the switches in the center column, reducing not only the number of splitters and/or combiners, but also the total number of switches. These structures have been named by whether the input splitting and/or the output combining is passive or active.

Figure 20.7(b) is a passive splitter/active combiner. Each input signal is passively split to three junctors in the center of the network. From there, only one would be actively switched to the desired output. Figure 20.7(c) is an active splitter/passive combiner. Each input signal is actively switched to one of three junctors in the center of the network. From there it is passively combined with two other signals, presumably empty, to make the desired output signal. Figure 20.7(d) is an active splitter/active combiner. Each input signal is actively switched to one of three junctors in the center of the network. From there, it would be actively switched to the desired output.

Compare the four architectures in Figure 20.7 to each other and to the Benes (described next) [13,14].

- *Attenuation.* In any network architecture that uses PDCs, signals suffer 3 dB of attenuation in every PDC they pass through. So, consider generalizing the networks above to $N \times N$, let each PDC contribute 3 dB and each SDC contribute 1 dB, and ignore all losses except losses in PDCs and SDCs. Then, the AS/PC's and the PS/AC's losses are $4 \times \log_2 N$ dB, and the PS/PC's loss is $6 \times \log_2 N$ dB. The AS/AC's loss is $2 \times \log_2 N$ dB and the Benes' loss is even 1 dB less than the AS/AC's loss. So, for an 8 × 8 network, the PS/PC's loss is 18 dB, the AS/PC and PS/AC have loss of 12 dB, the AS/AC's loss is 6 dB, and the Benes' loss is 5 dB.

- *Switch count.* Generalizing to an N × N architecture, the PS/PC requires N^2 SDCs, or N SDCs per input port, and $2(N - 1)$ PDCs per port ($N - 1$ 1 × 2 switches in each binary tree). The AS/PC and the PS/AC require $N - 1$ SDCs per port and an equal number of PDCs. The AS/AC requires $2(N - 1)$ SDCs per port and the Benes requires $2 \log_2 N - 1$ SDCs per pair of ports. So, for an 8 × 8 network, the PS/PC has 64 SDCs and 112 PDCs, and the AS/PC and PS/AC have 56 of each. Needing no PDCs, the AS/AC has 112 SDCs, but the 8 × 8 Benes network has only 20 SDCs.

- *Crosstalk.* Using PDCs on the input side of a generalized matrix distributes input signals to many undesired places in the middle of the network. Using PDCs on the output side of a generalized matrix collects undesired signals from anywhere they might appear in the middle of the network. Consider an 8×8 network, ignore attenuation, and let each SDC's extinction ratio be 100 (20 dB). In the PS/PC and AS/PC, each output port collects 0.01P from each of two off-SDCs and their end-to-end SNR is 1/0.02 = 50 (not in dB). By its symmetry with the AS/PC, the PS/AC has the same value. But, in the AS/AC, since the highest crosstalk term passes through two off-SDCs, its end-to-end SNR is a little less than 1/0.0001 = 10K = 40 dB. Because a signal through the Benes accumulates 0.01P of crosstalk in each stage, the end-to-end crosstalk through a five-stage network is 0.05 and the end-to-end SNR is 1/0.05 = 20 = 13 dB. We will describe a modification of the Benes that brings its end-to-end SNR in line with that of the AS/AC.
- *Broadcast.* Because of all the PDCs, the PS/PC scores poorly in attenuation and switch count. While using PDCs on both sides of the network contributes to the PS/PC's low score on crosstalk, its redeeming quality is that it allows broadcast. Also having passive splitting on the input side, the PS/AC also allows broadcast.

20.4.2 Dilated Benes Architecture

The Benes architecture is illustrated in Figure 7.11 and was described in detail in Section 7.3.4. The Benes architecture brings several obvious benefits to photonic switching: it is constructed from a 2×2 switching device (the beta element), it has the lowest number of stages for any nonblocking network built from 2×2 switches, and all paths use the same number of switches. But, the Benes architecture is rich in wave-guide bends and wave-guide crossovers and it still has crosstalk. We solve Lithium Niobate's crosstalk by modifying the Benes architecture.

When we try to see in the dark, the pupils of our eyes become dilated. We automatically compensate for a lack of light by making the transparent opening larger, so more light can get into our eye. Similarly, we will architecturally dilate a switching network by providing a bigger optical channel for the light that goes through any switching network. Dilation is a transformation on any switching network architecture under which the network has more optical paths for the same amount of light as in the untransformed network. In this section, we will dilate the Benes architecture, but any known architecture can be dilated [15].

After describing dilation in general, the dilated Benes architecture is developed by an inductive construction: the elementary 2×2 network is developed, a recursive construction procedure is presented, and the induction is completed. Then, an 8×8 dilated Benes network (DBN) is illustrated and its prototype is described.

Dilation. Consider the 8×8 Benes network in Figure 7.11. Suppose all the ports on both sides are idle. Now, suppose the upper left port wishes to be connected to the upper right port. There are four different paths, one through each different 2×2 switch in the center column. Let the upper path be selected arbitrarily. Notice two things.

- While the path goes through one switch in each stage of the Benes, it uses only half of each 2×2 switch. Each switch in this path could also be used as part of one other path.

- There is absolutely zero crosstalk in this path, not because we picked a good path, but because there are no other calls through the network.

If a second connection is added to this first connection, there may (or may not be) any practical level of crosstalk, depending on whether the second path can be found through the fabric such that the paths use different switches in every stage; that is, no one switch is used for both paths. This may (or may not) be possible.

- Certainly if the second connection starts from the second port on the left side (or terminates on the second port on the right side), there will be crosstalk, because both paths must use half of the upper left switch (or upper right switch).
- If the second connection starts and terminates on the bottom half of both sides of the network, there will be crosstalk only if we select the path that uses the upper switch in the center column. If we select any of the other three paths for this connection, there will be practically zero crosstalk. There will still be some crosstalk, and the first connection will notice an increase in crosstalk, because some unwanted signal can leak through more than one switch.

Suppose an $N \times N$ Benes fabric is fully connected (has N connections through it). Then, because each column holds $N/2$ 2×2 switches, each switch in the entire fabric must support two paths. If the Benes fabric is implemented with photonic SDCs, each of the N complete paths through the fabric is exposed to crosstalk from some other path in each stage of the fabric.

Consider this observation. Do we have crosstalk because of the SDC's inadequacy? Or, do we have crosstalk because we're too greedy — we're trying to use both paths through every SDC?

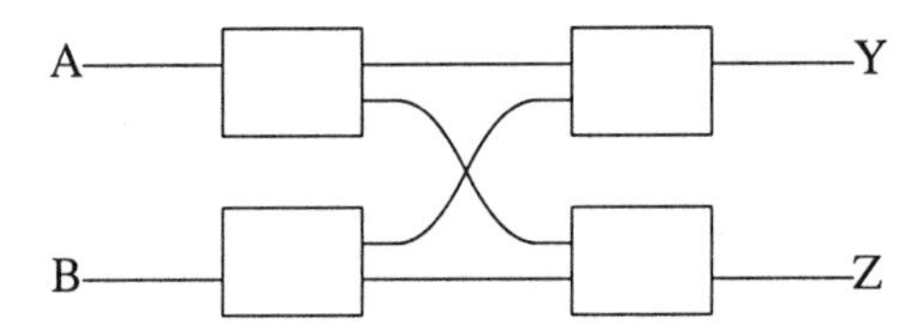

Figure 20.8. 2×2 dilated switching module.

2 × 2 Dilated Network Using SDCs. Dilation is illustrated on the most elementary network, a 2×2. Even though the network in Figure 20.8 has four switches, in two rows and two stages, it's functionally equivalent to a single SDC. The network has only two useful configurations, corresponding to an SDC's BAR and CROSS states. If all four switches are placed in their BAR state, A connects to Y and B connects to Z. If all four switches are placed in their CROSS state, A connects to Z and B connects to Y. Notice that, in both configurations, the two coexisting paths never share a common switch.

Consider the network of Figure 20.8 in its BAR configuration, with all four SDCs in their BAR states. Consider the four junctors in the center of the network. The upper junctor carries most of the signal that A transmits to Y. Some of this signal is attenuated in the upper left SDC, where some leaks to the junctor that proceeds to the lower right. The lower junctor carries most of the signal that B transmits to Z. Some of this signal is attenuated in the lower left SDC, where some leaks to the junctor that proceeds to the upper right. Because one input port on each first-stage SDC is unterminated, the unused path through each of these SDCs carries absolutely no photons. So, the signals in the two straight-across junctors are not corrupted by any crosstalk. But, the two diagonal junctors carry leakage signals to the second stage. Here, in this second stage, the actual signals are nominally switched to their respective output ports and the leakage signals are nominally switched to the SDCs' unterminated output ports. But, the leakage signals can leak onto the actual signals inside these second-stage SDCs. While there is some crosstalk in the output signals, the crosstalk is second order; it must leak through two SDCs.

Some observations are made.

- If each SDC's ER = X, then the network's end-to-end extinction ratio is X^2. So, for example, if system BER requires that end-to-end ER = 22 dB, then each SDC must operate with an individual switch ER = 11 dB. This figure is so easily attainable that SDC parameters can be greatly relaxed.
- The network of Figure 20.8 is *wide-sense* dilated, but only because the control algorithm is trivial (there is only one path for each pair of endpoints). While not clear from this example, dilated networks are generally *strict-sense* dilated because the network path must be carefully selected. In general, the number of paths through a dilated network is the same as the number of paths through its nondilated parent.
- Where the regular 2×2 Benes network has one single SDC, the 2×2 dilated Benes network has four SDCs. In general, a dilated network has more than twice as many switches as its nondilated parent.
- Where the regular 2×2 Benes network has one stage of switching, the 2×2 dilated Benes network has two stages. In general, a dilated network has one more stage (not double) than its nondilated parent.
- Compare Figure 20.8 to a 4×4 nondilated Benes network. Figure 20.8 results from doing two things to a 4×4 nondilated Benes network: (1) excise its center-stage, and (2) use only one port on each edge switch. This generalizes.

$N \times N$ Dilated Network by Recursion. In Benes' recursion, described in Section 7.3.4, an $N \times N$ Benes network is built by wiring together a column of 2×2 switches on the left, a column of 2×2 switches on the right, and a center stage consisting of two parallel identical $(N/2) \times (N/2)$ Benes networks. If the resulting $N \times N$ Benes network must provide a single connection, from one of the N left ports to one of the N right ports, it can choose from among $N/2$ different paths through the fabric. Any one of $N/2$ different paths can be selected.

In a similar construction, illustrated in Figure 20.9, an $N \times N$ dilated network of type X is built by wiring together a column of $1 \times K$ switches (commutators) on the left, a column of $K \times 1$ switches (decommutators) on the right, and a center stage consisting of K parallel identical $N \times N$ regular networks of type X. If each of the interior subnetworks is a regular $N \times N$ Benes fabric, then Figure 20.9 is the template to create an $N \times N$ dilated Benes network. If the $N \times N$ network in Figure 20.9 must provide a single connection, from one of the N left ports to one of the N right ports, its controller

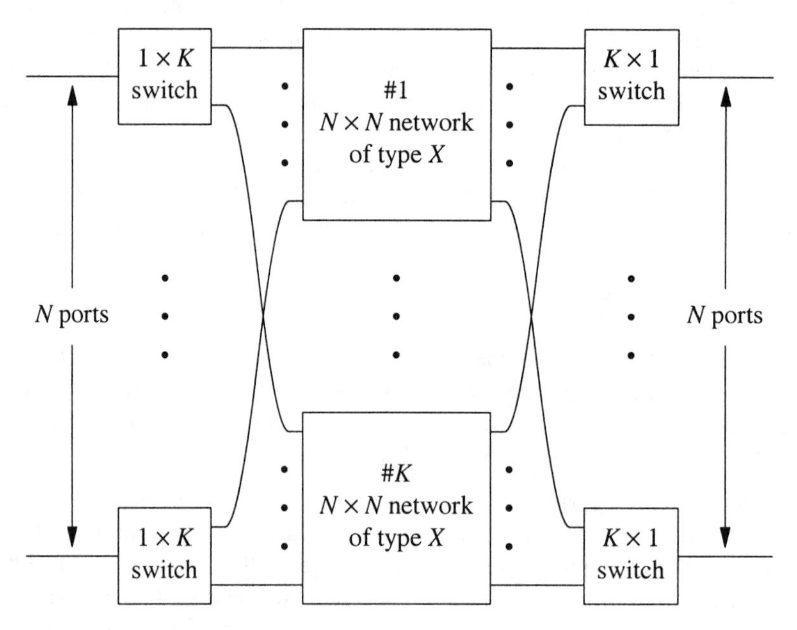

Figure 20.9. Template for dilation. © 1987 IEEE [15].

can choose from among $N/2$ different paths through each of the K interior subnetworks in the center stage. Any of the $K \times N/2$ selected paths uses half of one 2×2 switch in each stage. Obviously, if a network supports only one active path, this path cannot be affected by crosstalk from any other path through the network.

If a second connection is added to this first connection, there may (or may not) be crosstalk, depending on whether two paths can be found through the fabric such that the paths use different switches in every stage; that is, no one switch is used for both paths. If both paths use the same interior subnetwork, and both paths through this subnetwork have no switch in common, then the two paths will not mutually crosstalk. If switch sharing is unavoidable when two paths use the same interior subnetwork, crosstalk can still be avoided by routing the second connection through a different interior subnetwork; assuming that $K \geq 2$.

If the network in Figure 20.9 is fully connected (has N connections through it), we want to select K to be large enough so that none of the N paths through the fabric shares any switch with any of the other paths, under any switching configuration. Such fabrics, where every path through the fabric uses switches that are not shared by any other paths, are called dilated fabrics. If no switch in the network of Figure 20.9 ever carries more than one path, then there is no first-order crosstalk. While dilation is an optical term that implies that the switch is somehow open to more light than before, the concept could be applied to an electronic fabric if one were built from devices that have too much inherent electronic or electromagnetic crosstalk.

So, can this work for some finite value of K and if so, how large does K have to be? Obviously, K has to be larger than 1 — hopefully, not too much larger. With a maximum of N paths through an $N \times N$ network, replication by $K = N$ would have to completely eliminate all first-order crosstalk. But, replicating a nonblocking network $K = N$ times would produce a very inefficient and expensive fabric. While it's probably not obvious, we'll show that $K = 2$ is enough replication to implement dilation.

The Inductive Step. It's probably not intuitive that $K = 2$ would be sufficient for dilation. The convincing argument is a constructive *proof by induction.* Figure 20.8 illustrates the basis of the constructive induction.

- Figure 20.9 shows the architecture of a generalized dilated network. Figure 20.8 is seen to be a special case, where $N = 2$ and $K = 2$. In the leftmost column, each 2×2 SDC has an unused left port. Each represents a $1 \times K$ commutator, for $K = 2$. The rightmost column is similar and the stage containing K $N \times N$ switching modules in Figure 20.9 is empty in Figure 20.8.
- Figure 7.11 shows the architecture of a generalized Benes network. Figure 20.8 is seen to be a 4×4 Benes network, but with two modifications. (1) The center column of 2×2 switches is removed from the 4×4 Benes, and the columns adjacent to the center column are reconnected. (2) On the outer edge of the outer stages of the 4×4 Benes, only one of the two ports on each switch is used.

So, the switching module of Figure 20.8 operates as a 2×2 switch. The two inputs can be connected to the two outputs via a BAR or CROSS configuration of the module. In either configuration of the module, no two connections use the same switch. The module is dilated. When the module of Figure 20.8 is used in the inductive construction, all of its ports are used.

Now, use the template of Figure 20.9, with $K = 2$, to construct an $N \times N$ dilated Benes network. It has $K = 2$ interior subnetworks, each a regular $N \times N$ Benes network. Its leftmost and rightmost stages use 1×2 devices, which could be implemented as 2×2 SDCs with an unused port on their outer edge. So far the construction algorithm for an $N \times N$ dilated Benes network is the same as that for a $2N \times 2N$ regular Benes network except that in the dilated version, each of the 2×2 switches on the two edges is connected to only one external port. Obviously, a network can't be dilated if its edge switches are connected to two ports. It turns out that in the dilated Benes network the center column of switches is redundant and can be excised from the architecture without affecting its properties.

The architecture for any fabric constructed from 2×2 devices can be transformed into its equivalent dilated architecture. Furthermore, dilation can be accomplished by only requiring $K = 2$.

8 × 8 Dilated Benes Network. Figure 20.10 shows an 8×8 dilated Benes network [16,17]. The 2×2 SDCs in the leftmost and rightmost columns are connected to only one external port and have an unused port. The four interior columns of SDCs comprise two separate interior subnetworks, each a 4×4 dilated Benes network, but with connections to both ports on each edge switch. The network in Figure 20.10 would be a regular 16×16 Benes network if all 16 ports were connected on both edges and the center column of SDCs were replaced.

An $N \times N$ dilated Benes network has two more stages of switching, compared to a regular $N \times N$ Benes network, but one is compensated by the excised center column. While the number of stages is important in a photonic network because of the loss incurred in each stage, adding one stage of switching over the minimum number seems

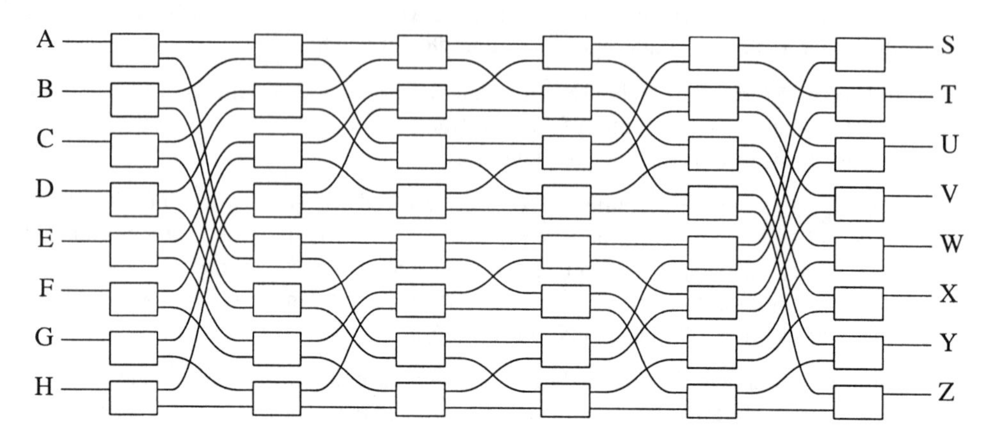

Figure 20.10. 8 × 8 dilated Benes network. © 1987 IEEE [15].

like a small penalty when it solves Lithium Niobate's crosstalk problem. An $N \times N$ dilated Benes network also has twice as many 2 × 2 switches per stage, compared to a regular $N \times N$ Benes network. However, recalling that SDCs are very thin, Lithium Niobate technology is easily squandered in this dimension. In practice, this has little effect on the level of integration in the fabrication of the photonic chip.

An 8 × 8 dilated Benes network was built, using Lithium Niobate SDCs, for the DiSCO project, described in the next chapter. The entire network is enclosed in a physical package, about 1 inch by 8 inches. The package has two ribbons of eight fibers each and 96 pins for the electrical control of the network's 48 SDCs. While the network's entire structure could not fit on one chip of Lithium Niobate, half of the structure could. Since Figure 20.10 is symmetric about a vertical line up the center, DiSCO's 8 × 8 DBN was built on two identical chips that were butt-coupled inside the package. Each input fiber required an individual polarization controller. The typical switching voltage was around 9V.

The four generalized matrix architectures and the regular Benes were all compared at the end of the previous subsection. The DBN's characteristics should be also compared. The DBN's attenuation would be comparable to that of the AS/AC, only slightly worse than the loss in the regular Benes network, which has the minimum number of stages for any fully connected network architecture built from beta elements. While the DBN's switch count is more than double that of the regular Benes, it's still better than any of the generalized matrix architectures. An 8 × 8 DBN has 48 SDCs. But, because of the shape of an SDC and its integrability, doubling the switch count over the regular Benes is not that expensive. Because, in a DBN, a signal accumulates second-order crosstalk in each SDC that it passes through, using the assumptions of this comparison, the DBN's end-to-end SNR is about 1/0.0006 = 1.667K = 33 dB, a little poorer than the AS/AC's end-to-end SNR, but much better than any of the others. The DBN can only provide point-to-point circuit-switched connections and cannot support broadcast.

Just as dilation is a useful architectural design to solve Lithium Niobate's crosstalk problem, it is similarly useful for solving Lithium Niobate's polarization problem. Optical signals can be separated into their orthogonal polarization components, each component can be switched through a separate network, and then the separate polarizations can be added together again. This doubling of the fabric completely eliminates the need for polarization adjusters and all the manual attention they require.

20.4.3 SLA Switching Networks

This subsection describes the semiconductor laser amplifier (SLA), not as an amplifier as we did in the previous section, but as a photonic switching device [18]. First, the SLA is shown to be a 1 × 1 switch with unusual properties. Then, a 2 × 2 switch is designed as a small fabric with four SLAs and its states and control are described. Then, many different matrix and modular photonic switching architectures are compared.

SLA as a 1 × 1 Optical Switch. While EDFAs have become the optical amplifier technology of choice, at 1.5 microns, the SLA still has an important role. Not only is the SLA the only feasible optical amplifier at 1.3 microns, but it has an important role in the future at both wavelengths because it can be used as an electronically controlled photonic switching device. While the Lithium Niobate SDC has many device drawbacks, the SLA is an excellent photonic switch.

- While the EDFA has an extremely slow switching time, the SLA can be turned on and off as quickly as an SDL, in the order of 1 ns.
- While the SDC has a very poor extinction ratio, requiring tight parameter tolerances or architectural cleverness, the SLA's ER is better than good. When an SLA is not forward biased, virtually none of the input photons (as low as -40 dB) reach the output fiber. When an SLA is forward biased, it not only allows a signal to pass, it amplifies any signals (by as much as +20 dB) that pass through it. Extinction ratios have been observed as high as +60 dB.

The SLA's most significant drawback is that, because it is only a 1 × 1 switch (single-pole single-throw), even simple network architectures need many devices. While conventional SLAs don't promise to integrate even as well as the SDCs, a highly integrable variation of the SLA, manufactured in Indium Phosphide, appears to be extremely promising.

Elementary 2 × 2 Network. A 2 × 2 switching module can be built easily and obviously with SLAs as the switching elements, shown in Figure 20.11. This figure is identical to Figure 7.1(c). Each optical input feeds a 1 × 2 splitter, and each optical output is fed from a 2 × 1 combiner. The module has four SLAs in a logical 2 × 2 array. The *i*th input is connected to the *j*th output by operating the <*i,j*>-th SLA.

If each SLA has enough gain to compensate for the net optical loss in the module, the output signal will have the same intensity as the input signal. For example, if fiber-to-device coupling losses and splitter/combiner losses are all -3 dB, then the SLAs in Figure 20.11 would each need +12 dB of optical gain. This is not unreasonable.

Switching States and Their Control. Besides having the potential for 0 net loss, this 2 × 2 optical switching module has another advantage over a 2 × 2 $LiNbO_3$ SDC. While the 2 × 2 SDC has only two states, BAR and CROSS, the 2 × 2 SLA module is more versatile.

- The 2 × 2 SLA module can be placed in a *broadcast* state. The *i*th input is connected to the 0 and 1 output by operating both the <*i*,0>th and the <*i*,1>th SLA. This broadcast state doesn't affect the net gain. The signal at the *i*th input appears

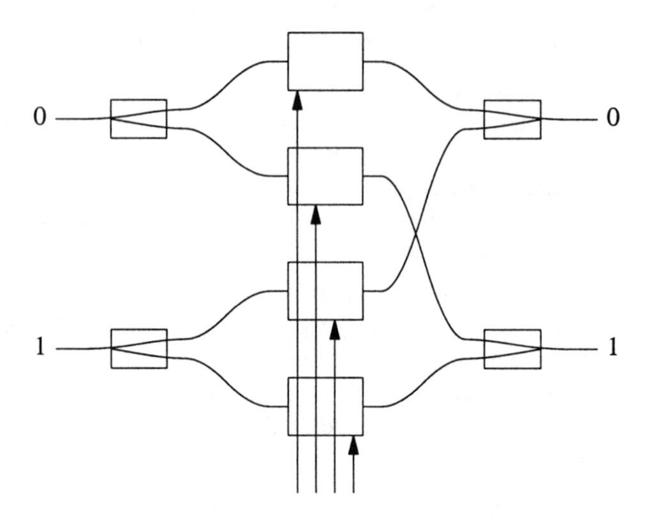

Figure 20.11. 2×2 SLA network. © 1988 IEEE [18].

on both outputs at the same power level. Of course, if the module is in one of its broadcast states, the other input signal is blocked.

- The SDC is typically in its BAR or its CROSS state, regardless of whether the inputs carry a true optical signal. While this simplifies the SDC's control, it means that any optical noise on either input to a switch always has a path through the switch to one of the outputs. The SLA module, on the other hand, can by placed in a "neither" state. This added flexibility ensures a quiet network.

The logical operation of a 2×2 network built from SLAs is summarized in Table 20.1.

TABLE 20.1. Control states for a 2×2 SLA network

0,0	0,1	1,0	1,1	Interpretation
0	0	0	0	Totally disconnected
1	0	0	0	Input 0 to output 0 only
0	1	0	0	Input 0 to output 1 only
1	1	0	0	Input 0 broadcast to both outputs
0	0	1	0	Input 1 to output 0 only
0	0	0	1	Input 1 to output 1 only
0	0	1	1	Input 1 broadcast to both outputs
1	0	0	1	BAR state
0	1	1	0	CROSS state

Matrix SLA Networks. The matrix construction of the 2×2 network can be generalized to construct an optical switching network of any size. Figure 20.7(a) shows the passive splitter/passive combiner architecture of a 3×3 photonic matrix network that uses SDCs as only 1×1 switches. A 3×3 SLA photonic matrix network has an identical architecture. Following the one-stage construction described in Section 7.2.1, an $N \times N$ SLA matrix network requires N^2 SLAs.

Generalizing from Figures 7.1(c) and 20.7, besides requiring N^2 SLAs, an $N \times N$ matrix network also requires $2N$ $1 \times N$ splitters, one for each input and one for each output. If a $1 \times N$ splitter is built from 1×2 splitters, then each such *splitter tree* requires $N - 1$ individual 1×2 splitters. So an $N \times N$ photonic matrix network needs $2(N - 1)$ 1 $\times$ 2 splitters. No matter what the implementation, any $1 \times N$ optical splitter (and combiner) has $3 \log_2 N$ dB of net loss.

Modular SLA Networks. The 2×2 SLA network can be generalized in another way. The 2×2 SLA network of Figure 20.11 can be encapsulated into a 2×2 photonic switching module. Then, this module can be used as a beta element in constructing a larger network, like an $N \times N$ Benes network.

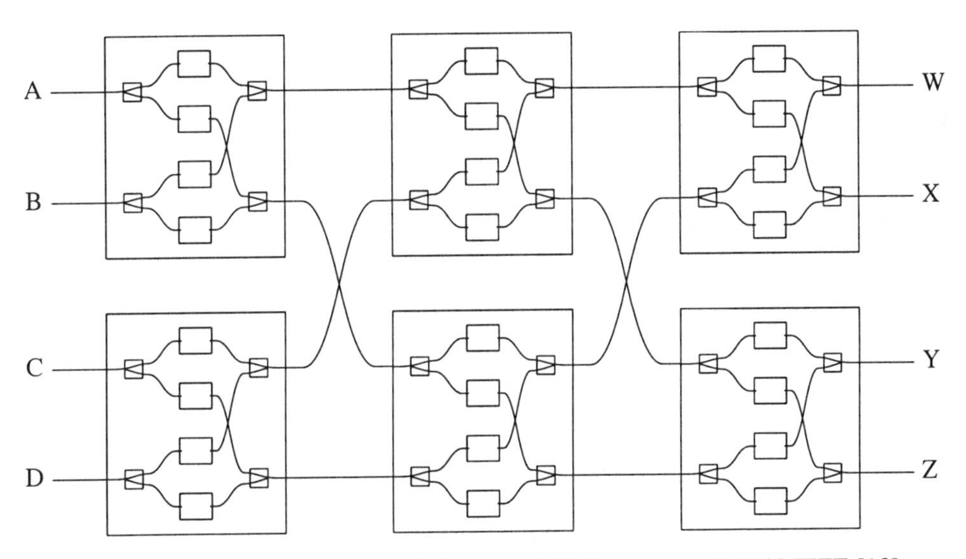

Figure 20.12. 4×4 Benes network from 2×2 SLA modules. © 1988 IEEE [18].

Figure 20.12 shows a 4×4 Benes network where each of the component modules is the 2×2 SLA matrix from Figure 20.11. The internal components inside each module are shown in the figure. Observe that each module's output comes from the 1-side of a 2 $\times$ 1 combiner and that each module's input goes into the 1-side of a 1×2 splitter. The module boundary disguises an architectural optimization: that both functions can be performed by a single 2×2 device.

This optimization is illustrated in Figure 20.13. Note that the interstage splitters and combiners are merged and the module boundaries are eliminated. While it doesn't look like one, Figure 20.13 is a 4×4 Benes network. Figure 20.12 is better than Figure 20.13 if it is desired to construct a stage as an integrated-circuit module. Then, the 4×4 Benes would be made from six identical modules. Figure 20.13 is better than Figure 20.12 if the network is built from discrete components or if the entire network is built as one integrated circuit.

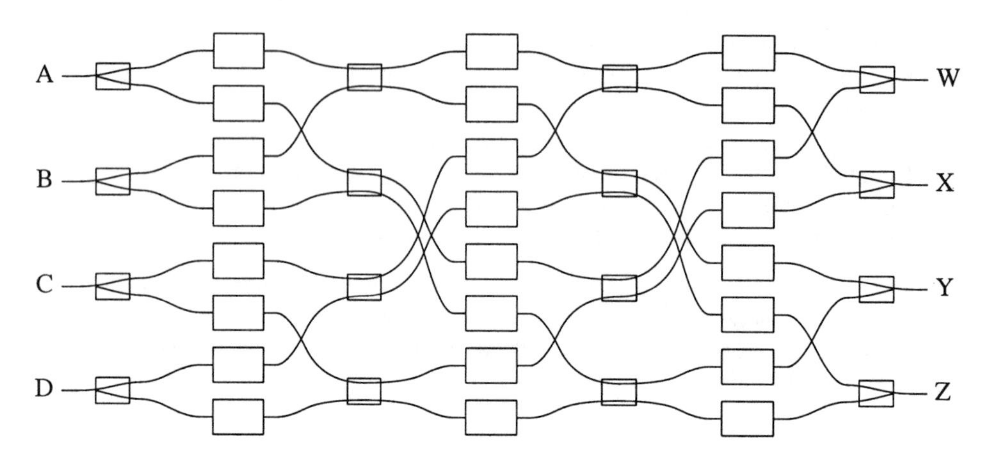

Figure 20.13. Nonmodular 4×4 Benes network from SLAs. © 1988 IEEE [18].

Comparing SLA Architectures. Comparing the relative characteristics of the different SLA network architectures is quite complicated. Table 20.2 summarizes only the numbers of switches and splitters that are needed in each respective architecture. Different fabric architectures impose different requirements on the reflectivity, gain, and other physical requirements of the SLA devices that are used to construct each fabric. The details are beyond the book's intended scope.

When an SLA is operated as a switch, it is an analog switch, similar to an SDC. The SLA's big advantage over the SDC is that the SLA amplifies the signals that flow through it. While the SLA is a 1×1 switch and the SDC is a 2×2, 2×2 structures can be fabricated using SLAs and passive optical couplers.

In a relatively new technology, network architectures are diffused onto an integrated circuit, typically fabricated in Indium Phosphide (InP). Available components include wave-guides, splitters, combiners, and fiber couplers; the wave-guides themselves are the 1×1 switches. This InP technology appears (at least, to this author) to hold the future of photonic switching.

TABLE 20.2. Architectural and device specifications for SLA networks

Size	Network	# SLAs	# Splitters
2-by-2	Matrix/Benes	4	4
4-by-4	1-Stage Matrix	16	24
4-by-4	Modular Benes	24	24
4-by-4	Non-Mod Benes	24	16
8-by-8	1-Stage Matrix	64	112
8-by-8	Modular Benes	80	80
8-by-8	Non-Mod Benes	80	48
16-by-16	1-Stage Matrix	256	480
16-by-16	Modular Benes	224	224
16-by-16	Non-Mod Benes	224	128

20.5 PHOTONIC PACKET SWITCHING

While the previous section and the next chapter are based on circuit switching in photonics, this section describes packet switching in optical technology. Packet switching further exacerbates the thermodynamic problem discussed in Section 20.2. Not only must every bit, at least in the packet's header, be individually switched, the entire header typically must also be stored and operated on with software. While a format for packet switching, optimized for low thermodynamics, is described in this section, packet switching is seen to have an inherent disadvantage at extremely high data rates.

Besides requiring that every packet be individually processed, packet switching's other major drawback is the lack of a really good way to handle congestion and collisions. Packet congestion can be handled by software or by hardware. If handled by hardware, packets can either be (1) queued up in memories or (2) deflection-routed into the network. Because there is no photonic software, photonic networks have to use one of the hardware-based solutions. Since the serial delay-line is the only practical form of optical memory, packet queuing in optical packet-switched networks must use optical fiber as memory. While optical fiber can be dedicated for storing packets, a photonic network has fiber between its nodes anyway that could be used for memory and transmission. So, most proposed packet-switched photonic networks use deflection routing.

This section's two subsections describe two different styles of architecture for photonic packet switching, each reminiscent of previously described systems. The first subsection describes a progressive packet-switching network, reminiscent of Step-by-Step (Chapter 6), for use with conventional packet formats. The second subsection describes a technique based on a nonconventional header multiplexing.

20.5.1 Self-Routed Packet-Switched Networks

Assume that packets have the conventional format, with the header at the beginning. Further assume that a binary-coded destination address is placed at the very beginning of the header. Even more restrictive, assume that this address is not an absolute end-marked name for the node but, instead, is the complete relative routing direction that steers a packet from its source through some network to its destination.

Assume the source node got this information during the call setup of the virtual-circuit connection and installed it into the packet's header before launching the packet into the network. Let this address contain a field for each node the packet passes through, in the order of the hops. Let the entry in each node's field be the address of this node's exit port that routes the packet to the next node in the path. Examine one node in this network and assume that each arriving packet's header begins with the binary code for the proper exit port from this node. Let the node strip this information away so that, after exiting the node, the packet's header begins with the code for the next node's exit port.

Inside the network node, incoming packets find their way to the assigned output port by routing themselves through the node's internal switching fabric. The fabric's basic structure has the Banyan architecture discussed in Section 7.5. When a packet arrives at a 2×2 switch in the Banyan, the first bit determines which output the remainder of the packet should use. The steering bit is stripped off and the packet proceeds on to the next stage of the network. This logical operation can't be implemented in photonics (at least, not in the foreseeable future). So, the implementation must be in optoelectronics, perhaps SEEDs, or in electronics with O-to-E conversion on the fabric edges.

Because each switch in the Banyan has two inputs, two packets could arrive simultaneously. Both these packets could be destined to the same switch output. While packet headers could be examined ahead of time so that collisions could be avoided by buffering at the node's input, this seems extraordinarily complicated, especially at high data rates. If the Banyan is electronic, packet-sized buffers could be implemented as part of each 2 × 2 switch, but with considerable performance and cost penalty. So, in photonics, deflection routing seems like the best, if not the only, solution.

Consider one arbitrary node in a large network. Let this node have an internal Banyan, a fabric of 2 × 2 switches. If two packets arrive at the same 2 × 2 switch and both packets are destined for this switch's same output port, then the switch delivers one of the packets to the wrong port. Here are three (are there more?) alternative strategies.

- *External network-wide deflection.* The packet that goes to the wrong port on this switch will progress on to the wrong output port on the internal Banyan. From here, it is transmitted to the wrong next node. If the packet's header also contains the destination address, and not just the virtual circuit's path, then this wrong node can correct the address and redirect the packet. This procedure doesn't seem amenable to photonics.

- Instead of letting the packet go to a wrong node, the node can make another attempt to steer the packet to the correct node. If a packet is steered to some switch's wrong port, then the switch might have the internal logic to change the rest of the bits in the address and, thereby, steer the packet to an output port on the Banyan that does not lead on to another node. There are two internal strategies.

    - *Feedback around the Banyan.* Some of the Banyan's output ports can feed back around to its input ports. This is especially effective when the Banyan is combined in tandem with a Batcher sorting network, as in two prototype electronic packet switching networks, called Starlite and Sunshine.

    - *Iterate the Banyans.* Banyans can be cascaded serially so that the correct ports all lead to the corresponding next node and the wrong ports lead to a subsequent Banyan still in this same node.

All these schemes are more amenable to electronic technology than to photonic technology. However, somebody should try to build one using SEEDs.

20.5.2 Header Multiplexing

Recall that block multiplexing is a modification of the classical digital multiplexing format that makes circuit switching more amenable to photonic technology. Similarly, header multiplexing is a modification of the conventional statistical multiplexing format that makes packet switching more amenable to optoelectronic technology. The conventional packet format is ordered so that the packet's header precedes the user's data, carried behind in the packet's body. In electronic packet switching, the header is processed after the entire packet is stored in the receiving node. In photonic packet switching, while the header must still be processed electronically, we want the body to remain in optical form and we want to avoid storing it in the receiving node.

With header multiplexing, the header for packet $n + 1$ coincides with the previous, or nth, packet. This format allows the receiving node's processing unit ample time to determine a packet's next destination before the data actually arrives. Header multiplexing can be implemented in space, time, or wavelength. The conventional packet format is easily converted to the header multiplexed format in the time domain. The header is stripped off an arriving packet and is delivered to the receiving node's processing unit, but the packet's body is delayed by one timeslot in a fiber delay line. With high data rates, even this conventional format may need to be modified by the addition of a temporal guard band between the header and the body.

In space or wavelength, packet headers and packet bodies use separate time-synchronized channels. In timeslot t, the *header channel* carries the header for the packet that will arrive on the *body channel* in timeslot $t + 1$. The header channel and the body channel can be two separate optical fibers or two separate wavelengths on the same fiber. It is important to see that, while a packet's header is much smaller than its body, they occupy the same amount of time on their respective channels. So, the header channel can not only have a lower line rate than the body channel, but the header channel's line rate can remain constant even if the packet's body size is doubled (and also doubling the optical line rate on the body channel).

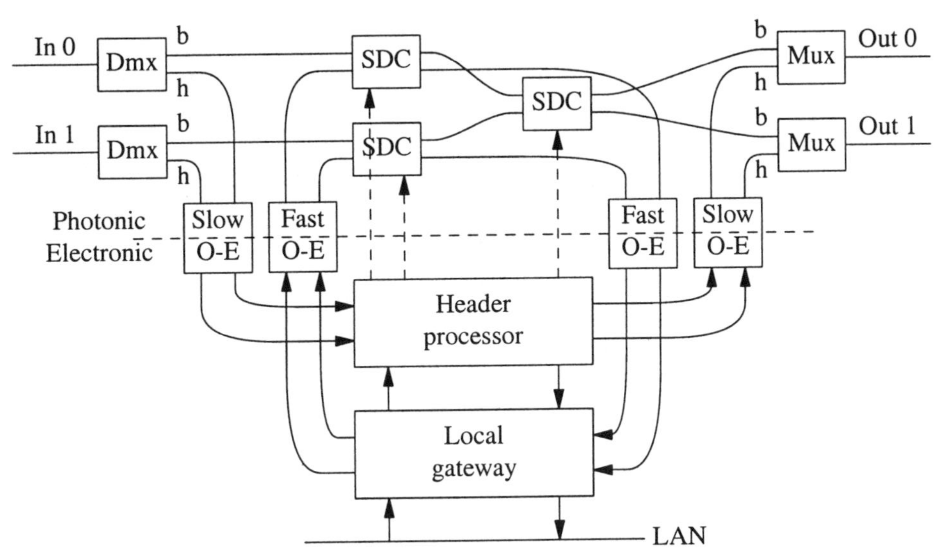

Figure 20.14. 2×2 Add-Drop packet-switching element. © 1994 IEEE [19].

Figure 20.14 shows a 2×2 switching element that performs packet switching in a photonic environment with header multiplexing [19]. Two mutually synchronized input packets enter the element over two separate physical channels at the left of Figure 20.14. Both packets' headers are O-to-E converted and are delivered to the electronic header processing unit (HPU). If the headers are wavelength multiplexed, as assumed in the figure, then they are simply demultiplexed, as shown. If the header channels are physically separated, then the node has no demultiplexors and the input header channels would connect directly to the O-to-E converters. If the packet format is conventional, but with a header-to-body guard band, then the headers are switched, instead of being demultiplexed. Figure 20.14 assumes that each packet's header arrives one timeslot ahead of the body of the packet for which it provides the signaling. If each packet's header and body arrive in the same timeslot, then the body of the packet would be delayed, by inserting the appropriate length of fiber between the demultiplexors and the two leftmost SDCs.

The HPU has an entire timeslot in which to process both packet headers. If either (or both) packet's final destination is the LAN for which this node is the gateway, then the HPU puts the corresponding leftmost SDC(s) in its CROSS state and the packet's body would be O-to-E converted and dropped to the local gateway processor. At the same time, if the local gateway processor has an appropriately addressed local packet in its internal queue, then this packet would be E-to-O converted and added for output transmission in the place of the retained packet. If neither packet is local, then the HPU puts both leftmost SDCs in their BAR states and both incoming packets proceed to the rightmost SDC.

The HPU routes the two incoming packets, or any local packets that are added in their place, by putting the rightmost SDC in its BAR or CROSS state, appropriately. If both packets are destined for the same output link, then one would be deflected. If deflection is not tolerated, one of the packets could be stored locally in the gateway and transmitted in a later timeslot.

Figure 20.14 shows a photonic packet-switching node as if it is a node in a large WAN. A similar architecture, without the local gateway, could be used to implement a 2 × 2 switching element for use in a Banyan or other self-directed packet-switched fabric. There are many possibilities, some of which might be practical.

EXERCISES

20.1 Compare the bandwidths of a 4-kHz analog telephone channel, a coaxial cable carrying 70 4-MHz television signals, and two different ways for measuring optical fiber's ultimate bandwidth. Make the comparisons by using analogies to pipe diameters. (a) Let 3/8-inch copper tubing represent the 4-kHz analog telephone channel. Using its inside diameter of 1/4 inches, compute the ratio of bandwidth to square feet. Use this ratio for the other three analogies. (b) Compute the corresponding pipe diameter for a coax carrying 280 MHz. (c) Designating the narrow bandwidth of optical fiber as a 60-nm window around 1.3 microns and a 90-nm window around 1.5 microns, find the optical frequencies for each of the four limiting wavelengths and the frequency bandwidth of each window. Then, compute the corresponding pipe diameter for the sum of both windows. (d) Designating the broad bandwidth of optical fiber as the complete wavelength window between 0.6 microns and 1.6 nm, find the optical frequencies for each of these limiting wavelengths and the frequency bandwidth of this window. Then, compute the corresponding pipe diameter for this window.

20.2 Consider a semiconductor device that dissipates 1 nJ each time it operates. (a) How much is the maximum data rate, limited by thermal transfer? (b) If some integrated circuit contains 10,000 such devices, operating concurrently, how much is the maximum data rate, limited by thermal transfer? (c) If a single discrete device operates in analog mode on 1000-bit blocks of data, how much is the maximum data rate, limited by thermal transfer? (d) If some integrated circuit contains 100 devices, operating as in part (c), how much is the maximum data rate, limited by thermal transfer?

20.3 What are the extinction ratios, in decibels, for a 2×2 switch that separates its input signals by 80/20, 95/5, 99/1, 99.9/0.1?

20.4 Appproximate the curve in Figure 20.6 by the Gaussian probability function. Let X_0 on Figure 20.6 correspond to the Gaussian mean and let X_H and X_L correspond to a Gaussian tolerance of plus/minus 0.1 standard deviation. Scale Figure 20.6 by assuming that some parameter X is specified at a value of 100 and that the 20-dB extinction ratio requires a numerical tolerance of $98.5 < X < 101.5$. (a) What is the corresponding standard deviation? (This is approximately the Gaussian scale for IQ.) Suppose dilation opens the equivalent Gaussian tolerance to plus/minus 1.0 standard deviations. (b) What is the corresponding numerical tolerance?

20.5 Consider a 2×2 matrix switch built with SDCs and passive splitters and combiners. Corresponding to Figure 20.7, sketch the four implementations: PS/PC, PS/AC, AS/PC, and AS/AC. In the PS/PC, use the SDCs as 1×1 switches. In the other three, use the SDCs as 1×2 and/or 2×1 switches.

20.6 Assume that both input signals to a 2×2 optical switching fabric have power = P, that passive splitters and combiners are perfect at 50/50, and that the SDCs perform an 80/20 separation of input signals. Compare the signal-to-crosstalk ratio at either output from five different architectures for a 2×2 fabric built using SDCs: a single SDC used as a 2×2 switch and the four implementations of the matrix switch from the previous exercise.

20.7 Sketch a 4×4 dilated Benes network. Input 0 can be connected to output 0 over two different paths. Arbitrarily select the upper path and show it on your sketch. While input 1 could be connected to output 1 over two different paths in an idle network, in the presence of the previous 0-to-0 connection, only one of these paths is usable (and it's dilated). (This illustrates that dilation is *strict-sense* — see Section 7.3.1. Arbitrarily selected paths are not necessarily dilated; an intelligent control algorithm has to be used.) Add the dilated 1-to-1 connection to your sketch. Input 3 can be connected to output 3 over two different connections and, even in the presence of the previous 0-to-0 and 1-to-1 connections, both paths are dilated. Arbitrarily select the lower path and show it on your sketch. Now assume that the 1-to-1 connection is released. Redraw your sketch, showing only the 0-to-0 and 3-to-3 connections. Now, suppose input 1 must be connected to output 2. While there are two connections, how many are dilated? (This illustrates that dilation in the dilated Benes is only *rearrangeable* — see Section 7.3.3.) Show a rearranged network pattern that supports 0-to-0, 1-to-2, and 3-to-3, all over dilated paths.

20.8 Control of a 2×2 network built from semiconductor laser amplifiers (SLA) is illustrated in Table 20.1. Design a digital control unit for this network. This is a difficult digital design problem and shouldn't be attempted by a beginner. Because there are nine states in the table, the network must be controlled by a

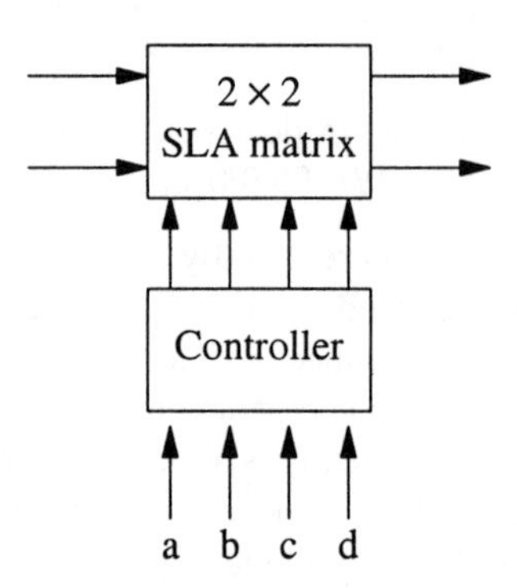

Figure 20.15. Controller for an SLA 2 × 2 switch.

4-bit word, abcd, the inputs to the controller. Since there are four SLAs to control, the controller has four output signals, one to each SLA. Assign the four input variables as: a = broadcast or not, b = input(s) are connected to corresponding or opposite output(s), c = enable input 0 or not, d = enable input 1 or not. Show the 4-bit code that specifies each of the nine states in Table 20.1. Taking advantage of *don't-care* code words, design the controller. [I found a solution requiring only five two-input logic gates. Can you find it? Can you beat it? There is a lot of opportunity with how don't-cares are assigned.]

20.9 Consider optical packets at the node shown in Figure 20.14. Suppose three consecutive packets arriving over In 0 are destined for Out 0, the local node, and Out 0, respectively. Synchronous with these three packets, suppose three consecutive packets arriving over In 1 are destined for the local node, Out 1, and Out 0, respectively. Suppose several packets are queued up in the local gateway that are destined for both output ports. Assume that, while incoming packets may have to be deflected to the wrong output, local packets wouldn't be transmitted unless the correct output is available. Walk the incoming and outgoing packets through the node. Describe the state of all SDCs in each packet timeslot and identify any deflected packets.

REFERENCES

[1] Smith, P. W., "On the Physical Limits of Digital Optical Switching and Logic Elements," *The Bell System Technical Journal,* Vol. 61, No. 8, Oct. 1982, pp. 1975 – 1993. © 1993 AT&T.

[2] Smith, P. W., "On the Role of Photonic Switching in Future Communications Systems," *IEEE Circuits and Devices Magazine,* May 1987.

[3] Cloonan, T. J., et al., "A 3D Crossover Switching Network Based on S-SEED Arrays," in *Photonic Switching II,* pp. 196 – 199, K. Tada and H. S. Hinton (eds.), Berlin, Germany: Springer-Verlag, 1990.

[4] Hinton, H. S., et al., "Photonic Switching Fabrics Based on S-SEED Arrays," in *Photonic Switching II,* pp. 30 – 39, K. Tada and H. S. Hinton (eds.), Berlin, Germany: Springer-Verlag, 1990.

[5] Jopson, R. M., and G. Eisenstein, "Optical Amplifiers for Photonic Switches," in *Photonic Switching*, pp. 84 – 87, T. K. Gustafson and P. W. Smith (eds.), Berlin, Germany: Springer-Verlag, 1988.

[6] Kobayashi, S., and T. Kimura, "Semiconductor Optical Amplifiers," *IEEE Spectrum*, May 1984.

[7] Thylen, L., P. Granestrand, and A. Djupsjobacka, "Optical Amplification in Switching Networks," in *Photonic Switching*, pp. 212 – 215, T. K. Gustafson and P. W. Smith (eds.), Berlin, Germany: Springer-Verlag, 1988.

[8] Alferness, R. C., "Guided-Wave Devices for Optical Communication," *IEEE Journal of Quantum Electronics*, Vol. QE-17, No. 6, Jun. 1981, pp. 946 – 955.

[9] Hinton, H. S., "Photonic Switching Using Directional Couplers," *IEEE Communications Magazine*, Vol. 25, No. 5, May 1987, pp. 16 – 26.

[10] Izutsu, M., et al., "Lithium Niobate Devices and Their Role in Photonic Switching," in *Photonic Switching II*, pp. 318 – 328, K. Tada and H. S. Hinton (eds.), Berlin, Germany: Springer-Verlag, 1990.

[11] Schmidt, R. V., and R. C. Alferness, "Directional Coupler Switches, Modulators, and Filters Using Alternating $\Delta\beta$ Techniques," *IEEE Transactions on Circuits and Systems*, Vol. CAS-26, Dec. 1979, pp. 1099 – 1108.

[12] Spanke, R. A., "Architectures for Guided-Wave Optical Space Switching System," *IEEE Communications Magazine*, Vol. 25, No. 5, May 1987. © 1987 IEEE. Portions reprinted with permission.

[13] Lu, C. C., and R. A. Thompson, "The Double-Layer Network Architecture for Photonic Switching," *IEEE Journal of Lightwave Technology*, Vol. 12, No. 8, Aug. 1994, pp. 1482 – 1489.

[14] Payne, W. A., and H. S. Hinton, "System Considerations for the Lithium Niobate Photonic Switching Technology," in *Photonic Switching*, pp. 196 – 199, T. K. Gustafson and P. W. Smith (eds.), Berlin, Germany: Springer-Verlag, 1988.

[15] Padmanabhan, K., and A. N. Netravali, "Dialed Networks for Photonic Switching," *IEEE Transactions on Communications*, Vol. COM-35, No. 12, Dec. 1987, pp. 1357 – 1365. Also in *Photonic Switching*, pp. 142 – 145, T. K. Gustafson and P. W. Smith (eds.), Berlin, Germany: Springer-Verlag, 1988. © 1987 IEEE. Portions reprinted with permission.

[16] Thompson, R. A., et al., "An Experimental Modular Switching System with a Time-Multiplexed Photonic Center-Stage," *Proc. Photonic Switching Topical Meeting, OSA*, Salt Lake City, UT, Mar. 1 – 3, 1989, pp. 92 – 93.

[17] Thompson, R. A., J. J. Horenkamp, and G. D. Bergland, "Photonic Switching of Universal Timeslots," *Proc. XIII Int. Switching Symposium*, Stockholm, Sweden, May 27 – Jun. 1, 1990, Vol. 1, pp. 165 – 170.

[18] Evankow, J. D., Jr., and R. A. Thompson, "Photonic Switching Modules Designed with Laser Diode Amplifiers," *IEEE Journal on Selected Areas in Communications*, Vol. 6, No. 7, Aug. 1988, pp. 1087 – 1095. © 1988 IEEE. Portions reprinted with permission.

[19] Blumenthal, D. J., R. J. Feuerstein, and J. R. Sauer, "Multihop 2x2 All-Optical Photonic Packet Switch," *IEEE Conference on Communications*, 1994, pp. 1364 – 1368. © 1994 IEEE. Portions reprinted with permission.

SELECTED BIBLIOGRAPHY

Any good physics textbook should provide further background in optics. For much greater detail, there are any number of good textbooks that specialize in optical communications. A very readable textbook on optical communications is *Fiber Optic Communications*, by Joseph C. Palais. Greater depth can be found in *Optical Fiber*

Telecommunications, edited by Stewart E. Miller and Alan G. Chynoweth, or *Fiber Optic Networks,* by Paul E. Green, Jr.

Many of the devices described in this chapter and the next are not recent enough, or of enough general interest, to have found their way into textbooks. Yet, most of this work on photonic switching devices represents group efforts and the life's work of many very talented individuals. Besides citing some individual papers below, it is more appropriate to cite the researchers themselves. Anyone seeking further background on these photonic devices should browse through the collected works of the following researchers, whom I acknowledge for their contributions, both to this book and my own research: R. C. Alferness, E. DeSurvire, R. M. Jopson, D. A. B. Miller, M. J. O'Mahony, P. Smith, and J. Veselka.

The reader interested in photonic switching should read Scott Hinton's book, *An Introduction to Photonic Switching Fabrics,* which covers many of these topics in greater detail. As in the previous paragraph, besides citing specific references, it seems more useful to cite people: A. Hill, H. S. Hinton, A. Huang, D. K. Hunter, H. Jordan, I. Kaminow, S. Knauer, C. T. Lea, C. C. Lu, G. Luderer, C. M. Qiao, S. Ramanan, G. Richards, J. Sauer, R. A. Thompson, and L. Thylen.

Ailawadi, N. K., et al., "Broadband Photonic Switching Using Guided-Wave Fabrics," *IEEE LTS,* May 1991, pp. 38 – 43.

Alferness, R. C., "Waveguide Electrooptic Switch Arrays," *IEEE Journal on Selected Areas in Communications,* Vol. 6, No. 7, Aug. 1988, pp. 1117 – 1130.

Bell, T. E., "Fiber Optics," *IEEE Spectrum Magazine,* (25th Anniversary Issue), Vol. 25, No. 11, 1988.

Bergmann, E. E., A. M. Odlzko, and S. H. Sangani, "Half Weight Block Codes for Optical Communications," *AT&T Technical Journal,* Vol. 65, No. 3, May/Jun. 1986.

Bianchini, R. P., Jr., and H. S. Kim, "Design of a Nonblocking Shared-Memory Copy Network for ATM," *IEEE Infocom '92,* 1992, pp. 876 – 885.

Boncek, R. K., et al., "1.24416 Gbits/s Demonstration of a Transparent Optical ATM Packet Switch Node," *Electronics Letters,* Vol. 30, No. 7, Mar. 31, 1994, pp. 579 – 580.

Chaffee, C. D., *The Rewiring of America — The Fiber Optics Revolution,* Boston, MA: Academic Press, 1988.

Cisneros, A., and C. A. Brackett, "A Large ATM Switch Based on Memory Switches and Optical Star Couplers," *ICC '91,* 1991, pp. 721 – 728.

DeBosio, A., C. DeBernardi, and F. Milindo, "Deterministic and Statistic Circuit Assignment Architectures for Optical Switching Systems," *Topical Meeting on Photonic Switching, Technical Digest Series,* Vol. 13, OSA, 1987, pp. 35 – 37.

DeBosio, A., et al., "ATM Photonic Switching Node Architecture Based on Frequency Switching Techniques," in *Photonic Switching II,* pp. 300 – 303, K. Tada and H. S. Hinton (eds.), Berlin, Germany: Springer-Verlag, 1990.

Elion, H. A., and V. N. Morozov, *Optoelectronic Switching Systems in Telecommunications and Computers,* New York, NY: Marcel Dekker, 1984.

Eiselt, M., et al., "Experimental Optical ATM 2 x 2 Switching Node," *European Conference on Optical Communications,* 1992, pp. 373 – 376.

Eng, K. Y., "A Photonic Knockout Switch for High-Speed Packet Networks," *IEEE Journal on Selected Areas in Communications,* Vol. 6, No. 7, Aug. 1988, pp. 1107 – 1116.

Evankow, J. D., and R. M. Jopson, "Nondestructive Measurement of Length Dependence of Gain Characteristics in Fiber Amplifiers," *IEEE Transactions Photonics Technology Letters,* Vol. 3, No. 11, Nov. 1991.

Fortenberry, R., et al., "Photonic Packet Switch Using Semiconductor Optical Amplifier Gates," *Electronics Letters,* Vol. 27, No. 14, Jul. 1991, pp. 1305 – 1307.

Fye, D. M., "Practical Limitations on Optical Amplifier Performance," *Journal of Lightwave Technology,* Vol. LT-2, No. 4, Aug. 1984.

Giacopelli, J. N., et al., "Sunshine: A High-Performance Self-Routing Broadband Packet Switch Architecture," *IEEE Journal on Selected Areas in Communications,* Vol. 9, No. 8, Oct. 1991, pp. 1289 – 1298.

Green, P. E., Jr., *Fiber Optic Networks,* Englewood Cliffs, NJ: Prentice Hall, 1993.

Ha, W. L., et al., "Photonic Fast Packet Switching at 700 Mbps," *OFC '91,* 1991.

Haas, Z., "The Staggering Switch: An Electronically Controlled Optical Packet Switch," *IEEE Journal of Lightwave Technology,* Vol. 11, No. 5/6, May/Jun. 1993, pp. 925 – 937.

Hardy, S., "Ascend Enters Optical Networking Fray," *Lightwave,* Aug. 1998.

Hinton, H. S., "Photonic Switching Technology Applications," *AT&T Technical Journal,* Vol. 66, No. 3, May/Jun. 1987, pp. 41 – 53.

Hinton, H. S., "Architectural Considerations for Photonic Switching Networks," *IEEE Journal on Selected Areas in Communications,* Vol. 6, No. 7, Aug. 1988, pp. 1209 – 1226.

Hinton, H. S., *An Introduction to Photonic Switching Fabrics,* New York, NY: Plenum Press, 1993.

Huang, A., and S. Knauer, "Starlite: A Wideband Digital Switch," *IEEE Globecom,* 1984, pp. 121 – 125.

Ikeda, M., "Tandem Switching Characteristics for Laser Diode Optical Switches," *Electronics Letters,* Vol. 21, No. 6, Mar. 14, 1985.

Jacobs, I., "Basics of Lightwave," *Telephone Engineer & Management,* Feb. 1985, pp. 74 – 80.

Jajszczyk, A., and H. T. Mouftah, "Photonic Fast Packet Switching," *IEEE Communications Magazine,* Vol. 31, No. 2, Feb. 1993, pp. 58 – 65.

Karol, M. J., M. G. Hluchj, and S. P. Morgan, "Input Versus Output Queueing on a Space-Division Packet Switch," *IEEE Transactions on Communications,* Vol. COM-35, No. 12, Dec. 1987, pp. 1347 – 1356.

Kasahara, K., "In Pursuit of Perfection," *Photonics Spectra,* Vol. 32, No. 2, Feb. 1998.

Kobayashi, H., B. L. Mark, and Y. Osaki, "Call Blocking Probability of All-Optical Networks," *Proc. 1st IEEE Int. Workshop on Broadband Switching Systems,* Poznan, Poland, Apr. 19 – 21, 1995, pp. 186 – 200.

Kuroyanagi, S., T. Shimoe, and K. Murakami, "Photonic ATM Switching Network," in *Photonic Switching II,* pp. 296 – 299, K. Tada and H. S. Hinton (eds.), Berlin, Germany: Springer-Verlag, 1990.

Lambert, P., and F. Barbetta, "The Shape of Light," *tele.com,* Vol. 3, No. 9, Aug. 1998.

Laude, J. P., K. Liddane, and S. Slutter, "Diffraction Gratings Stretch Fiber's Capacity," *Photonics Spectra,* Vol. 30, No. 2, Feb. 1996.

"Lightwave Technology Market Forecast," *Lightwave,* Dec. 1993, pp. 30 – 47.

Marrackchi, A., *Photonic Switching and Interconnects,* New York, NY: Marcel Dekker, 1994.

Marx, B. R., "Opto-Electronics Research Thrives in Europe," *Laser Focus World,* Vol. 32, No. 12, Dec. 1996.

Miller, S. E., and A. G. Chynoweth (eds.), *Optical Fiber Telecommunications,* Orlando, FL: Academic Press, 1979.

Nakagawa, K., "Role of Optical Amplifiers in Realizing All-Optical Communication Networks," in *Photonic Switching II,* pp. 14 – 29, K. Tada and H. S. Hinton (eds.), Berlin, Germany: Springer-Verlag, 1990.

Palais, J. C., *Fiber Optic Communications,* Fourth Edition, Englewood Cliffs, NJ: Prentice Hall, 1998.

"Photonics in the Information Age," special issue of *Photonics Spectra,* Feb. 1995.

Renaud, M., et al., "Network and System Concepts for Optical Packet Switching," *IEEE Communications Magazine,* Vol. 35, No. 4, Apr. 1997, pp. 96 – 102.

Sato, Y., K. Aida, and K. Nakagawa, "An Erbium-Doped Fiber Active Switch," in *Photonic Switching II*, pp. 122 – 125, K. Tada and H. S. Hinton (eds.), Berlin, Germany: Springer-Verlag, 1990.

Schwartz, M. I., "Optical Fiber Transmission — From Conception to Prominence in 20 Years," *IEEE Communications Magazine,* (issue on 100 Years of Communications Progress), Vol. 22, No. 5, May 1984.

Shibata, J., and T. Kajiwara, "Optoelectronic ICs," *IEEE Spectrum Magazine,* (issue on Supercomputers: A Special Report), Vol. 26, No. 2, Feb. 1989.

Shimazu, Y. and M. Tsukada, "Ultrafast Photonic ATM Switch with Optical Output Buffers," *IEEE Journal of Lightwave Technology,* Vol. 10, No. 2, Feb. 1992, pp. 265 – 272.

Simon, J. C., "Semiconductor Laser Amplifier for Single Mode Optical Fiber Communications," *Journal of Optical Communications,* Vol. 4, No. 2, 1983.

Suematsu, Y., and K. I. Iga, *Introduction to Optical Fiber Communications,* New York, NY: Wiley, 1982.

Thompson, R. A., "Optical Fiber Communications," in *Encyclopedia of Computer Science and Technology, Vol. 41,* pp. 235 – 266, Kent, A., and J. G. Williams (eds.), New York, NY: Marcel Dekker, 1999.

Thompson, R. A., "Traffic Capabilities of Two Rearrangeably Nonblocking Photonic Switching Modules," *AT&T Technical Journal,* Vol. 64, No. 10, Dec. 1985, pp. 2331 – 2373.

Tsuda, H., Y. Sakai, and T. Kurokawa, "Optical Flip-Flops Using Two Nonlinear Etalons," in *Photonic Switching II,* pp. 118 – 121, K. Tada and H. S. Hinton (eds.), Berlin, Germany: Springer-Verlag, 1990.

Wakao, K., et al., "InGaAsP/InP Optical Switches Embedded with Semi-Insulating InP Current Blocking Layers," *IEEE Journal on Selected Areas in Communications,* Vol. 6, No. 7, Aug. 1988, pp. 1199 – 1204.

Yasui, T., and K. Kikuchi, "Photonic Switching System/Network Architectural Possibilities," in *Photonic Switching*, pp. 24 – 35, T. K. Gustafson and P. W. Smith (eds.), Berlin, Germany: Springer-Verlag, 1988.

Yeh, Y., M. G. Hluchyj, and A. S. Acampora, "The Knockout Switch: A Simple, Modular Architecture for High-Performance Packet Switching," *IEEE Journal on Selected Areas in Communications,* Vol. SAC-5, No. 8, Oct. 1987, pp. 1274 – 1283.

Zhong, W. D., Y. Onozato and J. Kaniyil, "Copy Network with Shared Buffers for Large-Scale Multicast ATM Switching," *IEEE/ACM Transactions on Networking,* Vol. 1, No. 2, Apr. 1993, pp. 157 – 165.

21

Photonic Switching in Time and Wavelength

"It is a capital mistake to theorize before one has data." — spoken by Sherlock Holmes in *Scandal in Bohemia,* by Arthur Conan Doyle

Chapter 20 described photonic switching devices and photonic switching in the space division. After discussing the practical limitations of optical multiplexing in Section 21.1, the rest of this chapter describes techniques and fabric architectures for photonic switching in the time division, in the wavelength division, and in combination with the space division. While photonic switching in the space division may be ready for commercialization, photonic switching in time and wavelength is not. The architectures described in this chapter are paper designs — realistic protoypes have not yet been built.

Optical timeslot interchange is presented in Section 21.2. Section 21.3 describes switching fabric architectures for photonic switching in the space and time divisions and two research examples are presented. Section 21.4 discusses optical switching of channels that are defined by their wavelength, and Section 21.5 discusses switching fabric architectures for photonic switching in space and wavelength. Section 21.6 describes elementary switching fabrics for photonic switching in all three divisions — space, time, and wavelength. Some of this material on photonic switching carries over to the next chapter, where two photonic-switched networks are proposed: a metropolitan photonic-switched access network and a wide area photonic-switched infrastructure network.

21.1 OPTICAL MULTIPLEXING

Conventional transmission practices are described including digital data formats, time-multiplexing and Sonet, and wavelength multiplexing (including the limitations imposed by fiber's nonlinearities). While frequency modulation (FM) is a viable multiplexing technique in optics, the two more practical truly optical multiplexing techniques are time multiplexing and wavelength multiplexing. This section tries to show that the vast useable bandwidth of optical fiber is only slightly accessible by using time-division multiplexing alone, or by using wavelength-division multiplexing alone. Only the combination of these two multiplexing techniques allows optimal utilization of this vast resource with reasonable data rates, reasonable channel separation, and consistency with the bizarre characteristics of optical fiber.

21.1.1 Optical Time-Division Multiplexing

Optical TDM formats, and their limitations, are described.

Optical TDM Formats. Even while the carriers were only first installing single-mode optical fiber, they knew that they were utilizing only a very small portion of the fibers' bandwidth. The promise that there would be so much room for growth was one of the reasons for selecting fiber over more conventional media and even has motivated replacement by fiber. The first attempt to increase the utilization of this bandwidth was by deploying time-division multiplexing.

If N tributary channels, each at R bps, are TDMed together, then the multiplexed stream must run at $N \times R$ bps. Typically, the main stream rate is even higher because it also carries control and synchronization information. Different time multiplexing granularities are possible.

The most obvious approach is to *bit interleave* the tributary data onto the main stream, as is done in the DS2 layer, and above, of the so-called asynchronous digital hierarchy (ADH). One problem with this technique is that the control processor in the multiplexor and in the demultiplexor must operate at least as quickly as the main stream's data rate. Thus, bit interleave is problematic at rates above 200 Mbps and extremely difficult at rates above 2 Gbps — at least, if conventional techniques are used. All-optical bit-interleave has been demonstrated in the laboratory by simply merging two bit streams, each carrying narrow pulses that are mutually out of phase. While even 50-Gbps data rates have been demonstrated, this technique will not be practical until it provides enough intersymbol accuracy to support simple bit-synchronization [1].

In *byte interleave,* each tributary contributes 8 consecutive bits to the main stream, instead of only 1. While the electronics close to the main stream's transmitter and receiver must operate at the main stream's data rate, the control processor operates on 8-bit parallel data at one-eighth of the main stream's data rate. This scheme, which was used in DS1, the first digital time multiplexing system, was readopted for the OC-1 through OC-48 layers of Sonet, also called the synchronous digital hierarchy (SDH). Since the Sonet integer represents multiples of 51.84 Mbps, an OC-48 stream runs at 2.5 Gbps and its control processor must operate at 300 MHz, approaching the limit of processor speed, without going to a super-computer.

While optical fiber will allow much higher data rates, the transmitter and receiver electronics, whose architecture is specified by the data format, is seen to be the bottleneck. Because the solution appears to be to use bigger chunks of data in the time multiplexing format, the next obvious step is to use *cell interleave,* in which each tributary contributes 53 consecutive bytes to the main stream, instead of only 1. Typically in such systems, the cell's destination address is included as part of the cell and packet switching is employed. Packet switching is problematic if done directly off the mainstream data because the processors that read the address and steer the cell must operate so quickly. At the end of the previous chapter, we saw that photonics lends itself only incrementally to packet switching. While cell interleave could still be used in a circuit-switched system. it may be simpler to allocate chunks as a fixed duration of time instead of as a fixed number of bits.

In *block multiplexing,* the main stream's repetition interval, called a frame, is divided into a fixed number of timeslots at every layer of multiplexing, regardless of the number of bits in a frame. For example, in a block multiplexing format with 8000 frames/second and 250 timeslots/frame, each timeslot has duration 488 ns. Then, if each timeslot carries one cell, the net data rate is:

$$8 \text{ bits/byte} \times 53 \text{ bytes/cell} \times 250 \text{ cells/frame} \times 8000 \text{ frames/second} = 848 \text{ Mbps}$$

If three such streams are multiplexed together, instead of having 750 timeslots in the resulting frame, the multiplexed frame still has 250 timeslots, but each timeslot carries three cells instead of only one. If the line's actual data rate is a little faster than the nominal required rate, the data in each timeslot can be bunched a little tighter inside each timeslot and time gaps can be inserted between successive timeslots. This guard band can be used for delay compensation and to allow time for switching operations. While block multiplexing hasn't been used yet, it may be the format that allows time-multiplexed data rates to approach 10 Gbps and perhaps go a little beyond this.

Optical TDM Limitations. After discussing how modal and chromatic dispersion affect intersymbol interference, the practical limits of time-division multiplexing are discussed.

Early optical fibers had core diameters of 40 – 70 microns and had an attenuation minimum in the 800 – 850-nm range of the optical spectrum. Early optical transmitters used light-emitting diodes, whose optical signal has a broad physical beam and a broad optical spectrum, typically in the 800 – 850-nm range. The spectral match between fibers and LEDs means that optical signals are minimally attenuated and the physical match between the fiber's large core size and the LED's large beam size means efficient coupling of optical power. However, the large capture angle and thick core mean that optical signals traverse the fiber in a multiplicity of spatial EM modes (and the fiber is called multimode fiber, MMF). But because different modes traverse a length of fiber at different velocities, a pulse containing many modes will "widen" as it moves along a fiber. Because this modal dispersion limits how close together consecutive pulses may be, it limits the data rate. But, the problem is solved by using single-mode fiber (SMF).

Today's optical fibers have core diameters around 8 microns and additional attenuation minima around 1300 nm and around 1500 nm. These newer minima are much better than the old minimum at 800 nm. While optical transmitters at 1300 nm and, especially, at 1500 nm, have more complex chemistry and are costlier than those at 800 nm, the spectral match between new fibers and transmitters means that optical signals are even less attenuated than in the old fibers. While the small core size means that only a small percentage of an LED's broad optical beam actually couples into the SMF, this small core supports only the fundamental spatial EM mode. With only one mode, there is no modal dispersion. However, the relatively large optical bandwidth of an LED means that signals traverse the fiber in a relatively broad optical spectrum. But because different wavelengths traverse a length of fiber at different velocities, a pulse containing many wavelengths will widen as it moves along a fiber. Because this chromatic dispersion limits how close together consecutive pulses may be, it limits the data rate. This problem is solved by (1) operating at the dispersion minimum wavelength of 1300 nm, (2) doping the fiber to shift its dispersion minimum wavelength, or (3) using a transmitter that has a very narrow optical spectrum.

This better transmitter is the semiconductor diode laser (SDL). Besides having a narrower optical spectrum than an LED, the SDL also has a narrower physical beam than an LED. The SDL is also more expensive than an LED. While the optical signal from an SDL has a much narrower optical spectrum than that from an LED, it is still broad enough that data rate is still limited by chromatic dispersion. Because an SDL is *coherent,* it produces a signal whose optical spectrum has a structure with 4 – 8 maxima separated by 2 – 5 nm. Each of these maxima is called a longitudinal mode and an SDL with such a spectrum is called a multimode laser (MML). Confusing the laity, an SDL's longitudinal modes bear no relationship to an MMF's spatial modes. SDLs can be constructed, at considerable additional expense, so that only one of these longitudinal modes is produced; these are called single-mode lasers (SML). Today, extremely high data rates of 2.4 Gbps are obtained by using an SML with SMF in the 1300-nm region. But because any longitudinal mode has spectral length, about 1 – 2 nm of optical bandwidth, chromatic dispersion is still not entirely irrelevant, even in the 1300-nm region.

Summarizing, we have gradually chipped away at the physical characteristics of optical transmission that limit data rate. First, modal dispersion was completely solved and now chromatic dispersion has been attacked. Today's data rates are as much limited by the technology of optical transmitters and receivers as by the physical characteristics of optical fiber. Two important observations are:

- Every step of the way (from MMF to SMF, from LEDs to SDLs, from MMLs to SMLs) has been accomplished by dramatically increasing the cost of the endpoint of the optical transmission system.
- Today's *direct detection,* in which a 30-Gbps channel may require 100-GHz bandwidth, squanders optical bandwidth to simplify the transmitters and receivers. But, even if we conserved optical bandwidth by using a more complicated line coding, TDM alone would still not come close to utilizing fiber's full capacity.

While currently practical and still having more potential, time multiplexing is limited by fiber characteristics that cause modal and chromatic dispersion and by practical data rate limitations in the transmitter and receiver. While optical TDM in excess of 50 Gbps has been reported in research experiments, it is clear that TDM will not be the multiplexing format that leads to optimal bandwidth utilization.

21.1.2 Wavelength-Division Multiplexing

The research community's attention has recently turned to network architectures that use wavelength multiplexing, instead of TDM, to try to increase the utilization of optical fiber's bandwidth. Over the past 5 years several network architects have reported research designs for so-called all-optical networks (AONs). These AONs are LANs, MANs, and WANs in which data would be carried end-to-end over a passive network of optical fiber (so-called "dark fiber") by using hundreds, or thousands, or even tens of thousands, of different optical wavelengths. But, as interesting and provocative as these architectures are, other researchers argue that the underlying assumptions may be inconsistent with the physics of optical fiber [2 – 5]. Like TDM, WDM is also limited by fiber's characteristics — four different nonlinearities in optical fiber that cause *cross-channel interference* among optical wavelengths. These nonlinearities constrain the number of wavelengths that can be transmitted simultaneously over an optical fiber, especially a long fiber.

- *Stimulated Raman scattering.* The natural vibration of the fiber's molecules interacts with the optical signals in the fiber to cause a Doppler shift. Optical energy at short wavelengths shifts to a longer wavelength in the presence of any other optical signals at the longer wavelength. While this effect is enhanced by Erbium-doping to provide a physical mechanism for optical amplification (a good thing), it also causes cross-channel interference between two WDM signals in the same fiber (a bad thing). Even at extremely low optical power per channel, this effect limits the practical number of WDM channels to the high-100s — maybe as many as 1000. At higher channel power, which is more practical, the allowable number of channels drops even lower, but the other effects become even more constraining.

- *Carrier-induced cross-phase modulation.* Because of CICPM, an optical signal travels faster in a "lit" fiber than it does in a "dark" fiber. While this phenomenon is essential for sustaining optical solitons (a good thing), it also causes an odd coupling among multiple WDM channels and cross-channel interference (a bad thing). While the theory is not well developed, the empirical rule of thumb is that the product of number of channels times power per channel (in milliwatts) must be less than 21. Thus, at 0.1 mW per channel, there can only be 210 channels, and at 1 mW per channel, there can only be 21 channels.

- *Four-wave mixing.* This nonlinearity causes two or more optical signals to intermodulate, producing new optical signals at a variety of unusual sum and difference frequencies. While there is an interesting research opportunity to derive an optimal nonuniform spacing of the WDM channels, the practical implications limit per-channel power to less than 1.5 mW when the number of channels is in the range of 5 – 20.

- *Stimulated Brillioun scattering.* Naturally occurring acoustic waves in the optical fiber interact with the optical signals in the fiber to cause reflections and wavelength shifting (to higher wavelengths). This nonlinearity limits the power per channel to 2 or 3 mW, independent of the number of WDM channels. This nonlinearity is the practical limiting case when the number of channels is in the range of 1 – 5.

Each of these four nonlinearities limits a different region of the relationship between power per channel and number of channels. Because, in a WAN, power per channel determines a link's repeater spacing and the link's length, the constraint imposed by these four nonlinearities is transformed into a constraint on the relationship between the number of WDM channels and the lengths of the links in the WAN. The net result of these four nonlinearities is a physical constraint on the number of WDM channels in a fiber and the allowable length of a fiber link (with amplification). For example, with a state-of-the-art repeater and a 60-mile repeater spacing, it turns out that

$$N \times L \leq 12K$$

where N is the number of WDM channels and L is the link length in miles. For example, a fiber carrying 100 WDM channels can only be 120 miles long, and a fiber that is 1200 miles long can only support 10 WDM channels. But network architects are still proposing AONs with a physical size and number of wavelengths that exceed these limits. It is clear that the full impact of these nonlinearities is not well known nor well understood by many of these architects.

As this book is going to press, however, laboratory experiments and even transmission products have been announced in which this number × length constraint has been exceeded [6]. While details are being kept secret, there are three possibilities: (1) the analysis leading to the constraint is incorrect, (2) the claims made by these new products are false, or (3) someone has figured out a way to compensate for one or more of these nonlinearities. The reader is encouraged to watch this carefully. In particular, if some kind of *nonlinearity compensation* has been invented, is the technique so expensive that it can only be used in large transmission systems, or is it inexpensive enough that it can be more universally applied — thus making dark-fiber LANs and WANs practical?

21.1.3 Optical Solitons

Optical solitons are pulses that last longer than conventional physics would suggest. They were first observed as water waves in a canal and have been observed and predicted in many media, including optics. The existence of solitons in optical fiber was predicted many years ago and described mathematically. Then they were observed in the laboratory and, then, demonstrated in reproducible experiments. Initially a physical phenomenon with only academic interest, optical solitons now appear to be not only potentially practical, but they may be the physics by which we may come to utilize more of fiber's capacity.

The second of those four nonlinearities described previously was carrier-induced cross-phase modulation. The effect of this nonlinearity on optical transmission, that it causes an optical signal to travel faster in a lit fiber than it does in a dark fiber, was presented above as detrimental to wavelength multiplexing and a hindrance to bandwidth utilization. But this same nonlinearity can be a benefit in the time domain. It is possible to tune the wavelength, optical spectrum, optical power intensity, and temporal width of an optical pulse so that a fiber's CICPM nonlinearity exactly cancels the fiber's chromatic dispersion. Chromatic dispersion causes the slower colors in a pulse to lag behind the faster colors and the pulse spreads in time. However, CICPM causes these slower colors to speed up because the faster colors have illuminated the fiber ahead of them. Such pulses, called optical solitons, are the most likely way to approach 10-Gbps data rates and even higher.

21.1.4 Architectural Impact

While many architectures were proposed in many publications and considerable corporate and government research funding was spent on this endeavor, most of it has been for naught. It is clear that WDM, also, will not be the multiplexing format that leads to optimal bandwidth utilization. The primary purpose of this section has been to clarify these limitations in TDM and in WDM, caused by the physical characteristics of optical fiber, and to show that the combination of TDM and WDM is essential to optimally utilize fiber's bandwidth [7]. The secondary purpose of this section is to encourage those who perform research on network architecture, and those who support it, to adopt the causality that architecture does not come first. Architecture is a design phase that should be dependent on the market, applications, technology, and the laws of physics. It is naive and professionally lazy to propose an architecture and just assume that the physicists will be able to make it work.

Consider three network sizes:

- These nonlinearities do not constrain an intraoffice LAN, where the links are short. But, in this environment, the cost of wavelength-agile lasers and/or receivers probably makes WDM AONs too expensive compared to other optical and electronic LAN architectures that ought to provide acceptable performance in such a small network.
- A typical fiber-to-the-home MAN is more interesting. With maximum loop lengths around 4 miles, a WDM AON architecture should support 3000 WDM channels. This may be enough for a typical central office serving area. As with the intraoffice LAN, the links are short enough that full bandwidth utilization is not required economically. Use of WDM AONs in an FTTH MAN is an economic issue that, again, comes down to the cost of wavelength-agile endpoints.
- But, the WDM AON architecture just will not work in a U.S.-size WAN. For link lengths to be long enough to build a practical network, the nonlinearities constrain the number of WDM channels. With long links, where the economics requires optimizing the bandwidth utilization, WDM is no better than TDM at filling the available spectrum.

It has been shown that neither TDM nor WDM alone will be successful at even barely approaching optical fiber's enormous channel capacity. However, the combination of TDM and WDM is quite promising. Suppose we assume a basic OC-3 (155 Mbps) channel and wish to achieve 150 Gbps (only 5% of optical fiber's low-attenuation capacity) on a single fiber in the next 5 – 10 years.

- TDM requires a single transmitter and receiver capable of 150 Gbps.
- WDM requires 1000 channels (for a 12-mile maximum link length).

However, if the basic OC-3 channels are multiplexed up to OC-96, or to some other structure at about 5 Gbps, then only 30 of these channels are required to achieve 150 Gbps, and the WAN's links can be 400 miles long.

Hopefully, the reader can see the futility of the TDM-only and WDM-only approaches in attempting to utilize fiber's capacity. Hopefully, the reader can see the necessity to use both types of multiplexing to achieve even 5% of fiber's low-attenuation capacity, let alone 100% of it, let alone to begin to achieve more than 10% of its higher-attenuation capacity.

21.2 PHOTONIC TIME-DIVISION SWITCHING

The goal of photonic switching is to switch signals while allowing them to remain in their optical format. The previous chapter illustrated many techniques for photonic switching in the space division. But, in today's real world, and probably into the foreseeable future, signals on optical fibers are, and will be, time multiplexed. So, to ever attain synergy and *rate independence,* we have to perform photonic switching in the time division also. The basic building block is a timeslot interchanger (TSI) that operates on signals that are still in an optical format.

Because we really can't switch in the time division, in optics or electronics, channels are converted from time to space, switched in the space division, and then converted back from space to time. In an electronic TSI, described in Section 13.2, random-access memory addressing performs the time-to-space conversion, and storage in digital memory provides the delay necessary so that data can be moved from one timeslot to another. In optics, *commutation* performs the time-to-space conversion and simple segments of fiber provide the necessary delay.

Note that data cannot be switched forward in time because there are no physical delays with negative value. While this seems obvious and may sound silly, we can perform the equivalent of switching forward in the space and wavelength divisions. So, the time division is different from the other two; it has this constraint the others don't have.

The following assumptions and constraints are assumed.

- While the figures in this section show one physical input fiber and one physical output fiber, the switching fabrics are really $M \times M$ because the input fiber and the output fiber are each assumed to be time multiplexed into M timeslots per frame.
- Assume that timeslots use the block-multiplexed frame format, described in the previous section.
- Assume *frame integrity*. Data can be interchanged among timeslots within a frame, but once specific data is multiplexed together into some frame, all of it must remain within this same frame. Frame integrity is subtly important in a rearrangeable network. Without it, data can be lost during reconfigurations. This will be illustrated later in this chapter. For now, assume that, while entire frames may be delayed, once individual data is placed in a frame, it can never cross the frame's boundary.

The first and second subsections show two architectures that could work. When we examine them in detail, however, we find that first one has too much fiber to be realistic. And, when we "cost them out," we have to wonder if either one could be practical. Because the first architecture is simpler, it might be favored. But, the third and fourth subsections show optimizations on the second architecture that may actually be practical. No full-scale prototype photonic TSI has ever been built.

21.2.1 With Passive Delays

Figure 21.1 shows a schematic of one architecture for a photonic TSI. The figure provides photonic timeslot interchange for a time-multiplexed transmission line that has $M = 3$ timeslots per frame. In Figure 21.1, let optical data move from left to right, entering the photonic TSI from the input fiber at the left and exiting the photonic TSI onto the output fiber at the right. The first stage, at the left in Figure 21.1, is a $1 \times D$ space-division commutator ($D = 5$ in the figure) that switches data, during its timeslot, from the input fiber to any of D ports at this commutator's output. The commutator is a $1 \times D$ switch that could be built as a binary tree with $D - 1$ switching devices that are each 1×2, such as 2×2 SDCs each with an unused input port. Each of the

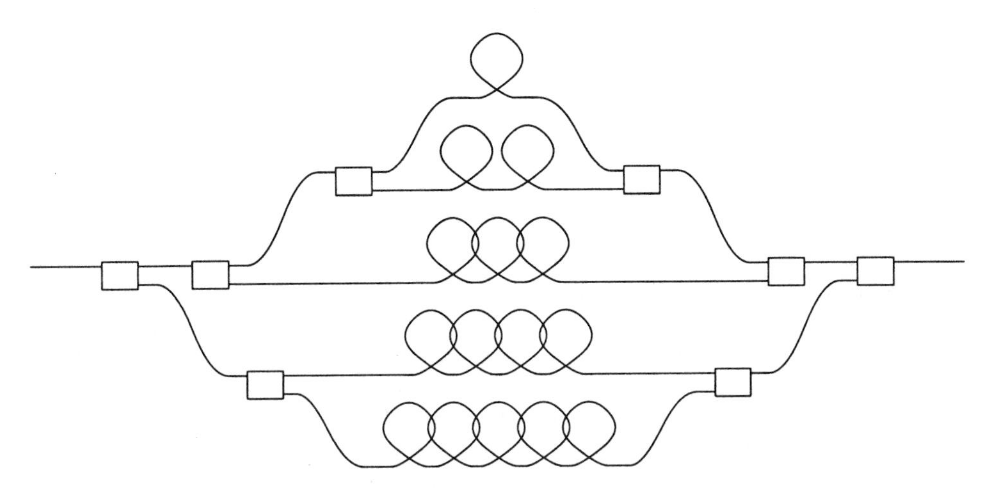

Figure 21.1. Photonic TSI using delays with different lengths. © 1987 IEEE [8].

commutator's D ports is connected to a different fiber segment, each with a different length, corresponding to a time delay equal to an integer number of timeslot periods. If T is the duration of one timeslot, then the fiber segments have delays of T, $2T$, $3T$, ..., DT.

The last stage, at the right in Figure 21.1, delivers the data, emerging from the D fiber delay lines, to the single output fiber. Note that no two fiber delay lines will produce data simultaneously. This last stage could be a $D \times 1$ passive optical coupler or, as shown in the figure, a $D \times 1$ space-division commutator, just like the input commutator, except reversed. The switched implementation is preferred over the passive implementation because it has lower loss.

While it might seem that the number of different delay lengths, D, would equal the number of timeslots in a frame, M, it's subtly more complicated than this. Allowing for the possibility that the last timeslot in the input frame might need to be placed into the first timeslot in the output frame, and in order to maintain frame integrity, the system must always store an entire input frame before any parts of this frame can be transmitted out. So,

- Data that remains in its original timeslot must be delayed by M timeslots, or MT seconds;
- Data that switches from timeslot M to timeslot 1 (one frame later) must be delayed by one timeslot, or T seconds;
- Data that switches from timeslot 1 to timeslot M (one frame later) must be delayed by $2M - 1$ timeslots, or by $(2M - 1)T$ seconds. So,

$$D = 2M - 1.$$

Consider an example. If frames in the i/o fibers have 256 timeslots, then $D = 511$. If the frame rate is the conventional 8000 frames per second, then the frame's duration is 125 μseconds. Each timeslot's duration is:

$$\frac{125\ \mu\text{sec/frame}}{256\ \text{timeslots/frame}} = 488 \text{ ns per timeslot}$$

The segment length for a delay of one timeslot is:

$$L = \frac{3 \times 10^8 \text{ m/sec}}{1.45} \times 488 \text{ ns/timeslot} = 101 \text{ meters per timeslot}$$

Is this practical? The longest fiber segment has length $L = 511 \times 101$ m = 51.6 km, which is long enough that the delay line may need a repeater or, at least, an amplifier. However, the entire set of 511 fiber segments has, if you think about it, 256 times as much total length as this one longest fiber. So, the total amount of fiber in all 511 delay lines is 256×51.6 km = 13,200 km, enough to reach about one-third of the way around the earth or from Seattle to Singapore. This architecture isn't realistic for this example and, generally, is practical only for systems with fewer timeslots and/or smaller frames.

Besides being unrealistic for this example, this architecture is also relatively inefficient and costly. This architecture has D fiber delay lines, but only M of them (about half) contain data at one time and the remaining $M - 1$ are empty. Even the delay lines that are used, are not filled to capacity. For this example with 256 timeslots, only about 2% of the slots actually hold any data (see Exercise 21.11). But, if we assume fiber is cheap, these inefficiencies are unimportant.

With D ports, a commutator contains $D - 1 = (2M - 1) - 1$ SDCs. With two commutators, the system has $2 \times (2M - 2) = 4M - 4$, approximately four SDCs per timeslot. And, the SDCs are not cheap. The next architecture, by using approximately three SDCs per timeslot, is slightly cheaper, but it will be seen to be considerably more complicated to operate than the passive photonic TSI. Subsequent subsections will solve these difficulties and further reduce the cost.

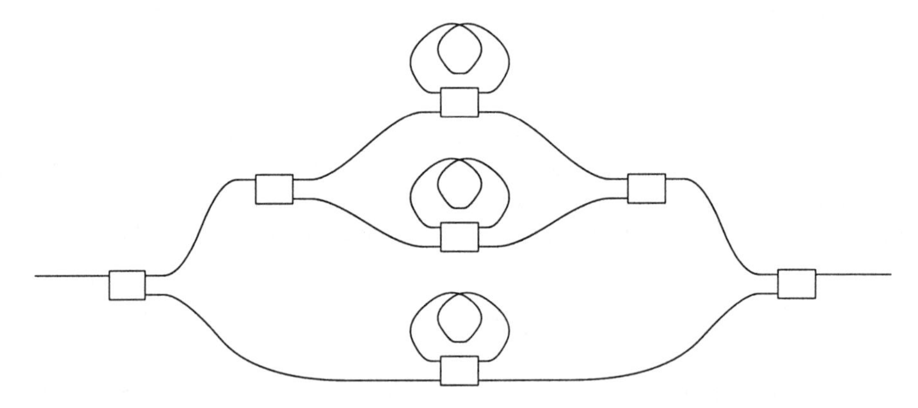

Figure 21.2. Photonic TSI using identical active delays. © 1987 IEEE [8].

21.2.2 With Active Delays

The architecture in Figure 21.2 is similar to the previous architecture except:

- It has M delay elements instead of $D = 2M - 1$ delay elements;
- All the delay lines in Figure 21.2 have the same length = T;
- Each delay element in Figure 21.2 is much more complicated than the simple fiber delay lines in the previous system.

Each delay element consists of a segment of optical fiber, with a length providing $1T$ of delay, that is fed back on itself through an SDC [8]. If this controlling SDC is in the BAR state, any optical data in the loop recirculates once per timeslot. So, this delay element can store all the data that is contained within a single timeslot.

Assume that each fiber loop contains data from some previous input timeslot. Consider that particular delay element which contains the data that must be placed into the current timeslot in the output fiber. At the beginning of this timeslot, this element's SDC is switched to the CROSS state and the data emerging from the fiber loop proceeds to the output commutator and onto the output fiber. Simultaneously, data in this same timeslot, but coming off the input fiber, is steered through the input commutator to this same delay element. Because the controlling SDC is in the CROSS state, this new input data proceeds right into the same fiber delay line that the output data is exiting. By the time the first optical bit of this new data reaches the end of the fiber delay line, the old data has completely exited onto the output fiber and the last bit in this input timeslot is just entering the loop. If the controlling SDC is switched to the BAR state during the guard band, all the data from this input timeslot reenters the loop and recirculates once for each timeslot of required subsequent delay.

This architecture uses its M delay elements more efficiently than the previous architecture uses its $2M - 1$ delay lines because each of the M delay elements always has data in it. And because each delay line is one timeslot in duration, all the slots are filled and the total length of fiber is 100% used. No optical TSI with frame integrity could require less fiber. Furthermore, with only $M - 1$ SDCs in each commutator and one SDC in each delay element, this architecture requires $2 \times (M - 1) + M = 3M - 2$, approximately three SDCs per timeslot, or about 25% fewer than the architecture using passive delays.

Returning to the example with 125-ms frames and 256 timeslots in a frame, recall that each timeslot has 488-ns duration and the segment length for a delay of one timeslot is 101m. While the previous architecture had too much fiber to be practical, perhaps this architecture might be practical? The total amount of fiber in all 256 delay lines is $256 \times 101\text{m} = 26$ km, about 16 miles. If all 26 km of fiber are coiled into a circle with 0.4-m diameter, the resulting fiber bundle has

$$\frac{26 \text{ km}}{0.4\pi} = 20,526 \text{ loops}$$

If we assume that an optical fiber's cross-section is a square, 125 microns on each edge, then the resulting fiber bundle has a square cross-section whose sides are

$$\sqrt{20,526} \times 125 \times 10^{-6} = 1.8 \text{ cm}$$

or, about 0.7 inch. So, the total amount of fiber fits into a torus that has a 0.7-inch square cross-section and a 0.4-m diameter. It fits easily inside a pizza box (my pet measure for defining a practical amount of fiber).

While this second architecture uses 25% fewer SDCs than the first one and it has a more realistic amount of fiber than the first one, it still has three significant problems.

- The control algorithm is quite complicated (Exercise 21.8).

- While three switches per timeslot is 25% lower than the cost of the previous architecture, it is still quite expensive and doesn't scale well to a frame format with thousands of timeslots.

- Data in each reentrant delay is vulnerable to attenuation and crosstalk in the 2×2 switch that controls the delay element. A device solution would be to implement these switches with SLAs, using the architecture in Figure 20.11. However, if the controlling 2×2 switches are implemented with a single SDC, then an architectural solution is required. Careful study of a prototype has provided a good understanding of the attenuation and crosstalk.

    - For a delay of m timeslots, an optical signal passes through the delay element's controlling SDC $m + 1$ times and suffers the corresponding loss $m + 1$ times. (And, the variation in loss is almost as problematic as the large magnitude of the highest loss.)

    - Because the signal emerging from the delay element passes through the controlling SDC at the same time that the new signal enters into the delay element, the output (which is attenuated) is vulnerable to crosstalk from the input (which is not), especially if implemented in $LiNbO_3$.

The next two subsections show the solution to two of the problems. Section 21.2.3 solves the SNR problem by making the delay elements even more complicated and increasing the number of SDCs per timeslot. Section 21.2.4 shows how to reduce the SDC count to well below its original value.

21.2.3 Reducing the Crosstalk

Two techniques could reduce the SNR due to crosstalk from the input signal interfering with the output signal.

- The stored signal should pass through fewer SDCs, or the same SDC fewer times, while it traverses through its delay element.

- The input and output signals should not pass through the same SDC simultaneously. There are a couple of ways to do this:

    - Another delay element could be added so that the input signal doesn't have to go onto the same element that produces the output signal;

    - Each delay element could be made more complex so the input and output signals don't pass through the same SDC when the element does its simultaneous input/output.

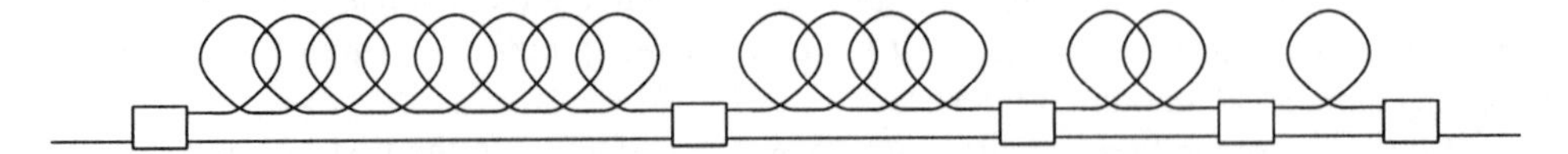

Figure 21.3. Feed-forward multistage delay element [9].

Figure 21.3 shows a more complex delay element [9]. (Note: This is just one single delay element and a complete photonic TSI requires many of these.) Its architecture reduces the number of switch crossings and provides separate SDCs for the input and output. If the maximum delay required of a delay element is $D = 2M - 1$, then let K be the smallest integer such that $2^K > 2M - 1$. Then, let the delay lines in the multistage element have lengths corresponding to delays $= 2^k \times T$ for all $0 \le k \le K - 1$. In other words, the delay lines have delays equal to the unit delay T times all the powers of 2: $k = T, 2T, 4T, 8T, \ldots, 2^{K-1}T$. Note that there are K different delays and they all feed forward, not backward. $K = 4$ in Figure 21.3 and the delays have length T, $2T$, $4T$, and $8T$. The delay element of Figure 21.3 would be used in a system with $5 \ge M \ge 8$ timeslots. Consider the following observations.

- A signal that passes through all K delay lines is delayed by $2^K - 1$, which is $\ge 2M - 1$. The maximum delay in the delay element of Figure 21.3 is $15T$.
- For an arbitrary delay of m timeslots, the binary code for m indicates which delays should be included in the signal's path and which should not. The control signal to the corresponding SDCs is computed from this code. For a delay of $6T$ in Figure 21.3, the binary code for 6 (0110), indicates that the required delays are those with lengths $4T$ and $2T$ only. The five SDCs must be assigned to BAR (0) and CROSS (1) according to the control code 01010, left to right across Figure 21.3.
- Every signal passes through each SDC exactly one time, passing through exactly $K + 1$ SDCs. The number of switch crossings is logarithmic in M and the loss is the same small value for any delay. For any delay between 1 and 15 in Figure 21.3, a signal passes through five SDCs.
- A new incoming signal can safely enter such a delay element at the same time that its internal signal exits, because the two signals would never pass through the same SDC at the same time.

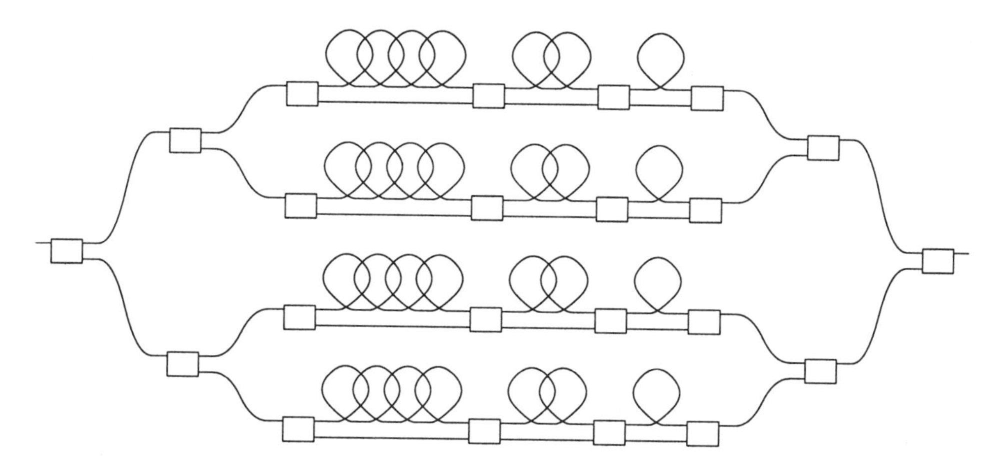

Figure 21.4. Active feed-forward photonic TSI for $M = 4$. © 1988 IEEE [10].

Figure 21.4 shows a complete active photonic TSI built using these multistage feed-forward delay elements. The transmission line is assumed to have only four timeslots per frame. With $M = 4$ timeslots, the TSI needs $M = 4$ delay elements and each delay element's maximum delay is $2M - 1 = 7$ timeslots. So, $K = 3$ and each delay element has $K + 1 = 4$ SDCs and $K = 3$ delay lines, with lengths T, $2T$, and $4T$.

Consider a switching assignment in which the four incoming timeslots emerge in reverse order. The first timeslot is delayed by $7T$, the second is delayed by $5T$, the third by $3T$, and the fourth by $1T$. The first timeslot enters the upper element, whose switches are set to BBBB. The second timeslot enters the second element, which is set to BXXB. The third timeslot enters the third element, which is set to XXBB. The fourth timeslot enters the fourth element, which is set to XBXB. In the next frame, as the data emerges from the fourth element, the first incoming timeslot enters this fourth element, which is set to BBBB. As the second outgoing timeslot is filled from the third element, the second incoming timeslot enters this element, which is set to XXBB. And so forth.

Consider a more substantial example. To interchange 100 timeslots per frame, an active feed-forward photonic TSI for 100 timeslots needs 100 delay elements, each capable of providing the maximum delay = $199T$. So, $K = 8$ and each delay element has $K + 1 = 9$ SDCs and $K = 8$ delay lines, and each delay element's maximum delay = $255T$. Moving data from the first timeslot to the last timeslot requires the maximum delay of $199 = 128 + 64 + 4 + 2 + 1$, with control code = 11000111. Keeping data in its same timeslot requires a delay of $100 = 64 + 32 + 4$, with control code = 01100100. With 99 SDCs in each commutator and 9 SDCs in each delay element, this TSI has a total of $2 \times 99 + 100 \times 9 = 1098$ SDCs, or about 11 per timeslot (consistent with the general formula of $K + 3$ SDCs per timeslot).

For M timeslots, a photonic TSI that is built with M such feed-forward active delay elements should have a greatly reduced crosstalk over one that is built with M single-loop active delay elements. Photonic TSIs that handle large numbers of timeslots could be built using SDCs that have easily attainable extinction ratios. But, as with most architectural solutions, solving one problem can often exacerbate some other problem. In this case, solving the crosstalk problem exacerbates the switch-count. An M-timeslot photonic TSI, which is built with multistage feed-forward delay elements, requires $K + 1$ SDCs in each of M delay elements and $M - 1$ SDCs in each of two commutators, for $M \times (K + 1) + 2 \times (M - 1) = M\ (K + 3) - 2$ SDCs, or approximately $4 + \log_2 M$ switches per timeslot. The SDC cost is much higher than even the first architecture.

21.2.4 Reducing the Switch-Count

Section 7.2 demonstrated that, in an $N \times N$ space-division fabric, the total count of switches can be reduced by increasing the number of stages in the fabric. So, might this concept apply also to a time-division fabric?

Figure 21.5 shows a three-stage space-division network with the Clos architecture. The figure identifies a particular pair of input/output ports and illustrates a particular path that connects these two ports spatially through the network. The connection's input port is #g on module #h in the first stage. The connection's output port is #k on module #j in the third stage. Of the m possible paths through this fabric, the selected path passes through module #i in the center stage. Making space-division connections in three different modules, one in each stage, connects the path.

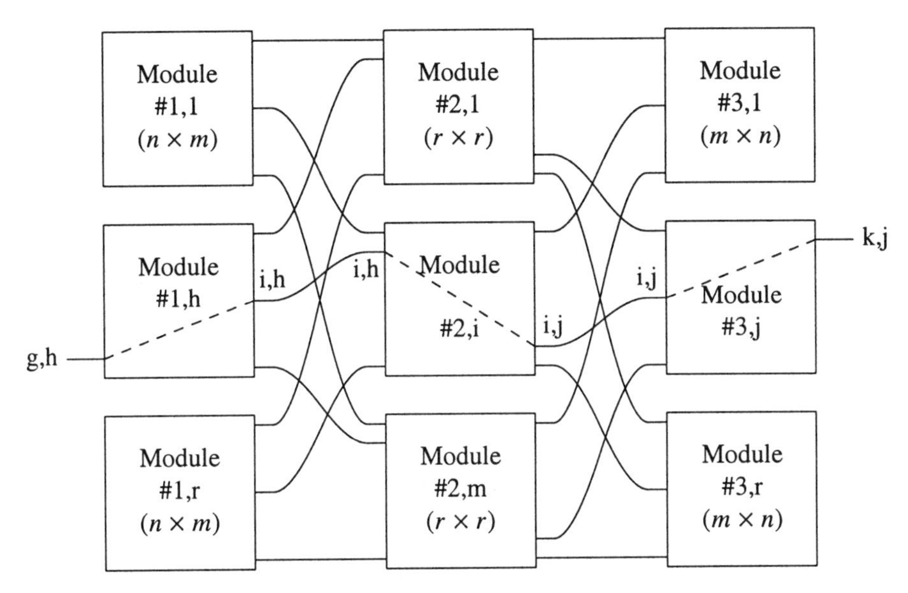

Figure 21.5. Three-stage Clos network. © 1988 IEEE [10].

- In module #h in the first stage, left port #g is connected to right port #i. This action connects input <g,h> to a-link <i,h>.
- In module #i in the center stage, left port #h is connected to right port #j. This action connects a-link <i,h> to b-link <i,j>.
- In module #j in the third stage, left port #i is connected to right port #k. This action connects b-link <i,j> to output <k,j>.

Figure 21.6 shows a proposed time-division network with three stages [10]. The signal's frame structure has an additional layer, with timeslots belonging to subframes and subframes belonging to frames. The figure identifies the pair of input/output timeslots that corresponds to the pair of ports in the Clos network in the previous figure. The figure also illustrates the time path, corresponding to the space path through the Clos network, that connects these two ports temporally through the network. The connection's input timeslot is #g in subframe #h entering the first TSI. The connection's output timeslot is #k on subframe #j exiting the third TSI. Of the m possible paths through this fabric, the selected path passes through timeslot #i in both subframes. Making time connections in the three separate timeslot interchangers connects the path.

- The first TSI interchanges timeslots within a subframe. Within subframe #h, the contents of timeslot #g are moved into timeslot #i. This action moves input data from timeslot <g,h> to timeslot <i,h> in an intermediate frame that corresponds to an a-link.
- The second TSI interchanges a given timeslot from one subframe to another. The contents of timeslot #i in subframe #h are moved into timeslot #i in subframe #j. This action moves the data from timeslot <i,h> in the a-link frame to timeslot <i,j> in an intermediate frame that corresponds to a b-link.

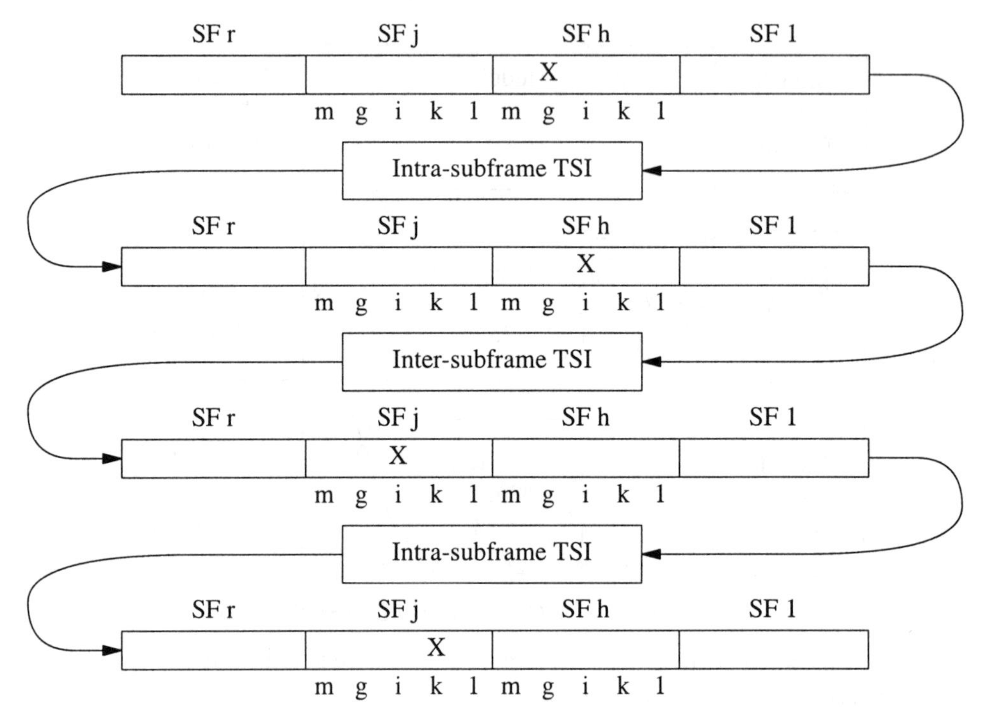

Figure 21.6. Corresponding three-stage timeslot interchanger. © 1988 IEEE [10].

- The third TSI, like the first one, interchanges timeslots within a subframe. Within subframe #j, the contents of timeslot #i are moved into timeslot #k. This action moves the data from timeslot <i,j> in the b-link frame to timeslot <k,j> in the output frame.

Each timeslot in a frame of the input line of Figure 21.6 corresponds to an input port in the Clos network of the previous figure. Each timeslot in a frame of the output line of Figure 21.6 corresponds to an output port in the Clos network of the previous figure. Selecting a *time path* in Figure 21.6 that uses intermediate timeslot #i corresponds to selecting a space path in the previous figure that uses module #i in the center stage of the Clos network.

Because the two figures above represent completely analogous fabrics, the Clos inequalities, described in Section 7.3.2, pertain also to the three-stage time fabric in Figure 21.6. Specifically, the time fabric of Figure 21.6 is rearrangeably nonblocking as long as $M = N$. That is, the number of subframes equals the number of timeslots in a subframe. Thus, a line with 256 timeslots should have 16 subframes, with 16 timeslots per subframe, and a line with 10,000 timeslots should have 100 subframes, with 100 timeslots per subframe.

Now, consider a truly significant example, one with 10,000 timeslots in a frame. The frame is partitioned into 100 subframes, each with 100 timeslots. The previous subsection concluded with an example of a TSI that can handle 100 timeslots. It had 1098 SDCs, about 11 per timeslot. In this example, the first and third TSI stages are exactly this same TSI. The center stage TSI is also the same except that every delay line in every delay element is 100 times longer than the corresponding delay line in the first or third-stage TSI. In all three stages of TSI, the total number of SDCs = 3 × 1098 = 3294, or about 0.3 SDCs per timeslot.

21.3 PHOTONIC SWITCHING IN SPACE AND TIME

We have already described architectures for photonic switching in the space division and in the time division. This section describes photonic switching architectures that operate on signals in both the space and time divisions simultaneously. Observing that the Marcus Transformation (Section 13.3) applies to photonics as well as to electronics, a research prototype of a photonic time-space-time architecture is described. Then, in the second subsection, instead of providing separate network stages for time and space switching, the Hunter Transformation is shown to "diffuse" space and time switching throughout a multistage space/time fabric.

21.3.1 Using Marcus's Transformation

In the first subsection, Marcus's Transformation is illustrated by a photonic TDM-TMS-TDM system that was actually prototyped. The second subsection discusses the generalization of the architecture to a true TST or STS system.

DiSCO. When the System 75 PBX was described in Section 14.3, it was shown to resemble a single switching module of a digital switching system. In a research prototype called DiSCO, for distributed switching with centralized optics, a maximum of eight PBXs can be interconnected through a center stage, in the same way that switching modules are interconnected in a digital switching system [11,12]. DiSCO's center stage is the 8 × 8 dilated Benes network, described in Section 20.4.2 and illustrated in Figure 20.10, operated as a time-multiplexed switch.

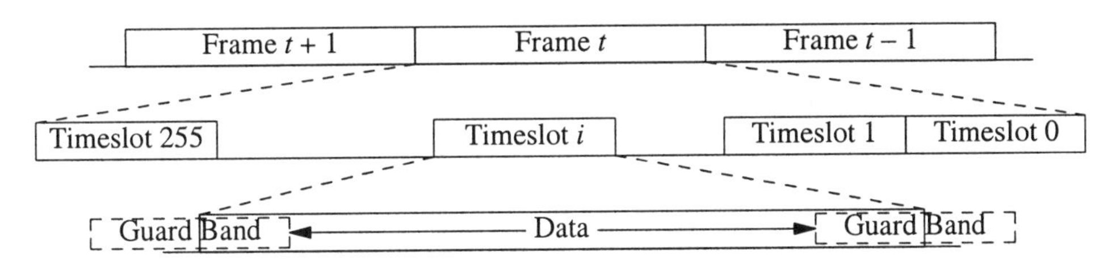

Figure 21.7. DiSCO's time hierarchy [11].

The PBXs and the center stage are all synchronized to a common 8-kHz frame clock and to the System 75's conventional 256 timeslots per frame, a hierarchical time structure illustrated in the top two rows of Figure 21.7. Because it operates as a TMS, the center stage reconfigures at the beginning of each timeslot, every 125 ms / 256 = 488 ns. A guard band of about 10 ns was provided around each timeslot boundary to allow switching time for the Lithium Niobate SDCs in the 8 × 8 dilated Benes prototype (about 2 – 3 ns) and to allow for variation in the center stage arrival time for the data that would be nominally centered in each timeslot.

Figure 21.8 shows a block diagram of the DiSCO system. Each of the switching modules is a System 75 PBX, conventional except for the addition of a link interface

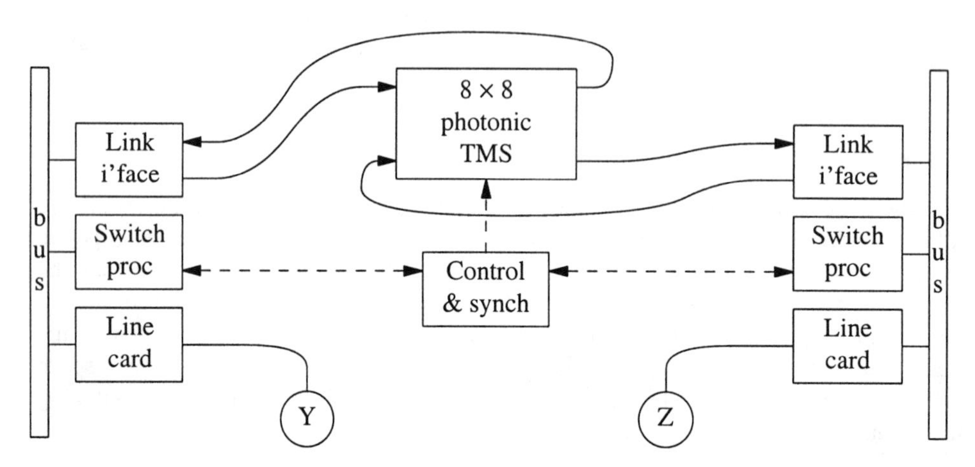

Figure 21.8. DiSCO block diagram [11].

card, capable of communicating with the photonic center stage. This link interface card interfaces with the PBX's TDM bus and has parallel-to-serial shift registers, a laser diode transmitter, and a photodetector. One side reads from its TDM bus and transmits, serially and optically, over an uplink to the input side of the photonic center stage TMS. The other side receives, serially and optically, over a downlink from the output side of this photonic center stage and writes onto its TDM bus. The links are cut to a length that corresponds to a delay of 488 ns, exactly one PBX timeslot.

While never implemented in the System 75's software, a hypothetical call walk-through is presented. User Y, on switching module (SM)1, goes off-hook and calls user Z, on SM2. SM1's connection manager and SM2's connection manager signal back and forth over a signaling path through the center stage. Timeslots 249 – 255 (at the center stage) are dedicated to IPC among the eight SMs.

The two connection managers negotiate a pair of idle *channel triplets,* one triplet for each way of the talk path. To simplify the explanation that follows, let's ignore the A and B buses of the System 75 and assume a single 8-bit TDM bus. In each triplet, if M is the selected center-stage timeslot, then timeslot $M - 2$ is needed in the speaker's SM and timeslot $M + 2$ is needed in the listener's SM. Let timeslot 100 be the center stage timeslot selected for Y's communication to Z, and let timeslot 200 be the center stage timeslot selected for Z's communication with Y. Because the center stage, as a Benes network, is (rearrangeably) nonblocking, end-to-end paths exist from SM1 to SM2 in timeslot 100 and from SM2 to SM1 in timeslot 200. Let timeslot 98 in SM1 and timeslot 102 in SM2 be assigned to the samples from Y's mouth to Z's ear, and let timeslot 198 in SM2 and timeslot 202 in SM1 be assigned to the samples from Z's mouth to Y's ear. After Z answers his telephone, three pairs of connections are established.

- On SM1, user Y's line card is instructed to connect Y's mouth samples to timeslot 98 and Y's ear samples to timeslot 202.
- On SM2, user Z's line card is instructed to connect Z's mouth samples to timeslot 198 and Z's ear samples to timeslot 102.

- The center stage controller is instructed to connect its input 1 to output 2 during timeslot 100 and to connect its input 2 to output 1 during timeslot 200.

Then Y's line card places Y's mouth samples onto SM1's TDM bus at the beginning of timeslot 98 so they can be read by SM1's link interface card at the end of 98. SM1's LI card transmits Y's mouth samples to the center stage during the next timeslot and they arrive during timeslot 100. They switch through the center stage and reach SM2's LI card during timeslot 101. They are placed onto SM2's TDM bus at the beginning of timeslot 102 so they can be read by B's line card at the end of 102. The call walk-through is similar for Z's communication to Y during timeslots 198 – 202.

Because Lithium Niobate requires single-mode signals, the up and downlinks must be single-mode fiber and the transmitters in the link interface cards must be lasers. In constructing the prototype, the slowest commercially available laser transmitters had a bit rate of only 180 Mbps. Using the block-multiplexed format and transmitting the contents of the A- and B-buses (after a start bit) requires only 17 bits / 180 Mbps = 94 ns; the rest of the 488-ns timeslot is guard band. Note that, because the channels on the A- and B-buses in the same timeslot are *gang-switched* together, if the channel on SM1's A-bus in timeslot M is used for a call to SM2, then the channel on SM1's B-bus also goes to SM2. The data transmitted during this window is limited to 17 bits because the PBX's back-plane only has 16 bits of data in each timeslot. If data could bypass this back-plane bottleneck, then the window could be filled with as much data as can be transmitted in 488 ns.

In a second experiment, a video interface card was built. The front-end ADC converts an analog video signal and stores the data in an elastic (FIFO) store. The back-end transmits the data optically at about 1.6 Gbps over the uplink. PBX software would instruct the video card which timeslots to use. Sampling the video signal at 8 MHz, produces a 64-Mbps stream, or 8000 bits per PBX frame. Transmitting 616 data bits per timeslot requires 13 timeslots per frame. Transmitting at 1.57 Gbps, 617 bits (including a start bit) requires 617 bits / 1.57 Gbps = 394 ns out of the 488 ns in each timeslot. So, the guard band has duration 488 – 394 = 94 ns, more than enough time to allow jitter in the arrival time of data at the center stage and to allow for the switches to operate. Uncompressed digitized video was successfully switched through the same center stage.

The idea of transmitting at different rates (and even different wavelengths) generalizes; such timeslots are called universal timeslots. As long as transmitter and receiver agree on the contents, the switch and the software don't even need to know. For example, suppose it is desired to switch one entire T1 frame in each DiSCO timeslot. If the contents of each T1 frame are transmitted at 500 Mbps, the time required is 386 ns, which still allows for a 102-ns guard band around each timeslot. If the guard band can be as small as 40 ns, then each T1 frame can be expanded to fill 448 ns and the line rate can be reduced to 193 bits / 448 ns = 438 Mbps.

The concept of universal timeslot (System 75 in Section 14.3.4) extends beyond allowing a variable data rate to include also the possibility of having several wavelengths, as long as they are within the SDC's spectrum. Separate wavelengths cannot be switched to different destinations, however; they must all be gang-switched together. We'll look at separately switching wavelength-defined channels in the next section of this chapter.

True TST and STS. As just discussed, the DiSCO center stage performs time-multiplexed switching on eight fibers, each with 256 timeslots. The three-stage time-space-time fabric is fully connected and nonblocking except for the concentration in its eight modules. The only time switching occurs is when conversations are assigned to timeslots in these concentrating time multiplexors. There is no true timeslot interchange, as in the 5E's switching module. If the DiSCO system terminated optical fibers that were already time multiplexed, instead of electronic narrowband telephone lines, the corresponding photonic time-space-time network would require a bank of photonic TSIs in the first and the third stages. Or, an equivalent space-time-space photonic network would require a TMS in the first and the third stages and a bank of photonic TSIs in the center stage. In either case, each bank of photonic TSIs needs eight separate TSIs, each one capable of interchanging 256 optical timeslots.

Consider the three-stage photonic TSI, described in Section 21.2.4. Instead of seeing the frame as 256 timeslots, the frame would be seen as having 16 subframes, each with 16 timeslots. Then, each stage of photonic TSI must interchange among 16 time channels. Consider making each photonic TSI using the architecture with the feed-forward delay elements. A feed-forward delay element, capable of providing the necessary delay for a subframe with 16 timeslots or for a frame with 16 subframes, requires five delays and six SDCs. A single photonic TSI, capable of interchanging 16 time channels, requires 15 SDCs in its input and its output commutators and 6 SDCs in each of its 16 delay elements, for a total of 126 SDCs, or approximately $4 + \log_2 16 = 8$ switches per timeslot. A three-stage TSI, capable of interchanging among 256 timeslots, requires $3 \times 126 = 378$ SDCs and a bank of eight such TSIs requires $8 \times 378 = 3024$ SDCs. Recalling that DiSCO's 8×8 photonic dilated Benes TMS has 48 SDCs, a complete three-stage photonic TST network has $3024 + 48 + 3024 = 6096$ SDCs and the STS version has $48 + 3024 + 48 = 3120$ SDCs.

In each photonic TSI, each feed-forward delay element requires five fiber loops (with delays of $1T$, $2T$, $4T$, $8T$, and $16T$) or enough fiber to provide a total delay of $31T$ seconds. Each complete photonic TSI, capable of interchanging among 16 time channels, requires enough fiber to provide $16 \times 31T = 496T$ seconds of delay. In the three-stage photonic TSI, because the first and third stages interchange timeslots within a subframe, T is the duration of one timeslot. However, because the middle photonic TSI interchanges timeslots across subframes, T is 16 times the duration of one timeslot. A three-stage TSI, capable of interchanging among 256 timeslots, requires $2 \times 496T + 16 \times 496T = 8928T$ seconds, where T is the duration of one timeslot. A bank of eight such TSIs requires $8 \times 8928 = 71{,}424T$ seconds of delay, where T is the duration of one timeslot. A complete three-stage photonic TST network needs enough fiber to provide $2 \times 71{,}424\ T = 142{,}848T$ seconds of delay and the STS version needs enough fiber to provide $71{,}424T$ seconds of delay. The STS version uses half the total fiber and almost half the SDCs as the TST version of the Marcus network for this system.

The total fiber delay in one complete three-stage photonic TSI was shown above to be $8928T$ seconds, where T is the duration of one timeslot. At 256 timeslots per frame, this total delay is

$$8928 \text{ ts} \times 488 \text{ ns/ts} = 4.36 \text{ ms or } \frac{8928 \text{ ts}}{256 \text{ ts/frame}} = 34.875 \text{ frames}$$

In the previous section, we showed that one frame (256 timeslots) of delay required a torus of coiled fiber with 0.4-m diameter and square cross-section of about 1.8 cm. For

35 frames of delay, the square cross-section must be increased to $\sqrt{35} \times 0.018\text{m} = 0.106\text{m}$, about 4.2 inches; about three times too big to fit inside a pizza box. One complete TSI stage, containing eight of these photonic TSIs, requires about 20 times too much fiber to be practical. But, we can make it all fit in a pizza box if we can make the frames 1/20th the size, by increasing the frame rate to 160,000 frames per second.

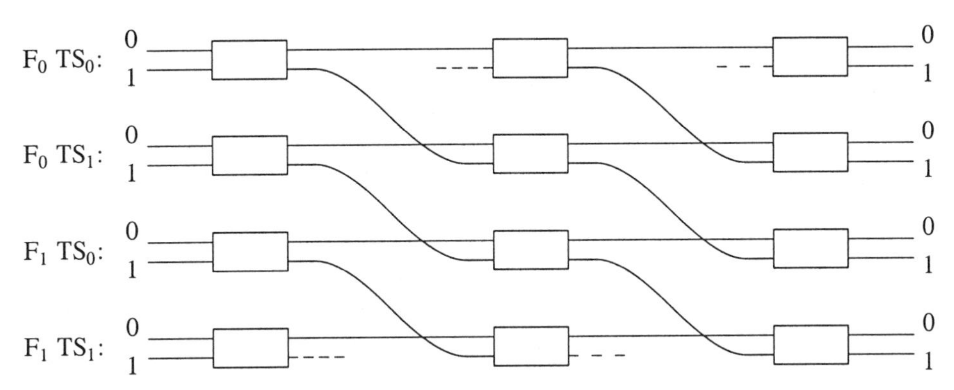

Figure 21.9. Space/time diagram for the most elementary case. © 1993 IEEE [13].

21.3.2 Using Hunter's Transformation

Combining space and time switching in a single fabric is illustrated by considering the most elementary case: two distinct optical fibers, each carrying only two time-multiplexed channels per frame [13,14]. Hunter's transformation is effected through a series of three diagrams. Figure 21.9 is the first diagram, illustrated using this simple example. The figure represents a space/time network that is finite in its space dimension, but is infinite in its time dimension. Each row in the diagram represents the entire space dimension. For this simple example, each row contains two input fibers, two output fibers, and three stages of 2×2 switches. But, instead of time multiplexing the same physical network, each row represents the time dimension during a different timeslot.

Figure 21.9's vertical axis represents the infinite time dimension. The figure ignores the framing of timeslots and simply assumes that timeslots occur in an infinite series. So, the diagram has one row for each timeslot and the timeslot that corresponds to each row is space multiplexed over two physical channels. If one of the two channels in the ith timeslot must be switched to some channel in the jth timeslot, then, in the ith row, this channel must be space switched to some junctor that is wired to this jth row — later in time. None of these interrow junctors goes "uphill," because data cannot be switched earlier in time. In this example, because channels are time switched forward by only one timeslot per stage, the interrow junctors in Figure 21.9 drop down by only one row. But, in general, interrow junctors could span more than one row on the space/time diagram.

For Figure 21.9 to represent a real network, the network would have to be infinite; otherwise it could only be used for a fixed number of timeslots and then would have to be restarted. None of the physical components is time multiplexed. Each row is used once, during its assigned timeslot, and is never used again. Besides ignoring frames, we're also ignoring frame integrity, but we'll come back to this later in this section.

A practical space/time switching system would not have to be represented by a diagram that is infinite in the time dimension because timeslots are framed. The time

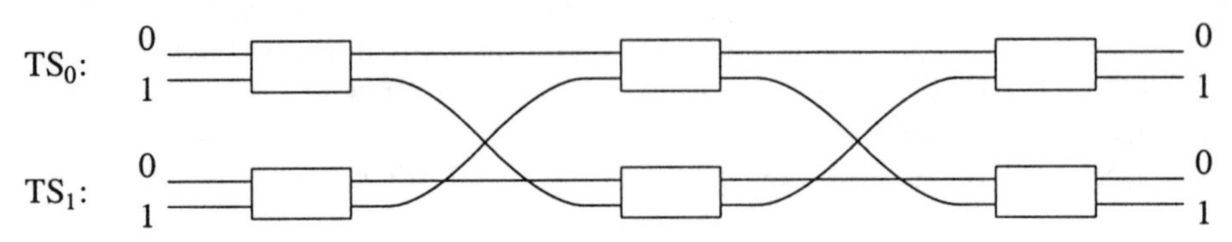

Figure 21.10. Framing time on the space/time diagram. © 1993 IEEE [13].

switching is not arbitrary, but is cyclic, repeating with every new frame. Furthermore, no timeslot would ever be switched so as to skip a frame. So, if the frame has N timeslots, then all the time switching can be represented by only N contiguous rows of Hunter's space/time diagram. The simplification caused by framing is illustrated in Figure 21.10.

In the previous space/time diagram, interrow junctors that proceed to some timeslot in a subsequent frame are wired uphill in Figure 21.10 to this same timeslot, but in the current frame. Where time passes vertically down the previous space/time diagram, time passes cyclically in Figure 21.10 from the top row to the bottom row and then back to the top row again. This diagram is more practical than the previous space/time diagram because it is finite. But, it is unrealistic because it allows forward time shifting, and while it time multiplexes the physical network in every frame, it requires a separate physical network (row) for each timeslot in a frame. So, we still need to constrain time to flow backward only and we still need to time multiplex the physical network for every timeslot.

Because this most elementary example has only two timeslots per frame, only two rows of Hunter's space/time diagram are required to illustrate the complete switching capability. The architecture of Figure 21.10 resembles a 4×4 Benes architecture, where the rows of the Benes architecture correspond to timeslots in the space/time system and the junctors in a Benes network that move from row i to row j correspond to a time shift of $|j - i|$ timeslots. In general, for a system with N fibers and M timeslots, its framed space/time diagram would have N fibers in each row and M rows. While this example is based on the Benes architecture, the time/space diagram can be based on any architecture. So, the number of stages and the interswitch and interrow wiring would depend on the network architecture that was selected as the basis.

Hunter's space/time diagram is transformed into a physical system by two observations.

- The physical time-multiplexed switching network corresponds to one single row of the diagram.
- The interrow junctors correspond to delays with duration equal to the number of rows skipped in the diagram.

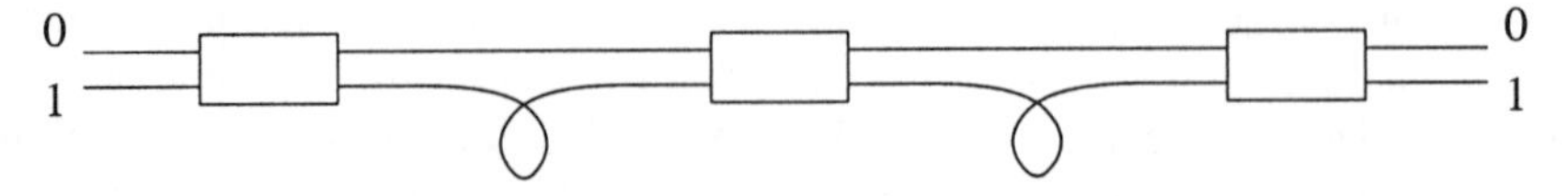

Figure 21.11. A $< 2, 2 > \times < 2, 2 >$ space/time switching fabric. © 1993 IEEE [13].

For our elementary example, the framed space/time diagram in the previous figure corresponds to a physical implementation that has two delays, each with duration of one timeslot. Figure 21.11 illustrates the space/time switching fabric that corresponds to the framed space/time diagram in the previous figure. This fabric provides space/time switching for a system that has two space channels, with two timeslots apiece.

Examining the three figures in this subsection, consider connecting input channel <1,0>, the first timeslot on the lower input fiber, to output channel <0,1>, the second timeslot on the upper output fiber. This connection requires a delay of one timeslot and it has two paths: one where the delay occurs in the a-links and one where the delay occurs in the b-links. Every connection has two paths through Figure 21.11 for exactly the same reason that every connection has two paths through a 4 × 4 Benes network. Consider the fabric of Figure 21.11.

- For the first path, all three switches are in the CROSS state. During timeslot 0, the path proceeds from the lower input fiber to the upper a-link and then to the lower b-link, where it's delayed by one timeslot. Then, during timeslot 1, the path proceeds from the lower b-link to the upper output fiber.
- For the second path, the three switches are in the BAR, CROSS, and BAR states, respectively, from left to right. During timeslot 0, the path proceeds from the lower input fiber to the lower a-link, where it's delayed by one timeslot. Then, during timeslot 1, the path proceeds from the lower a-link to the upper b-link and then to the upper output fiber.

Frame integrity can be illustrated, even in this elementary example. In the example above, the data from channel <1,0> never moved to another frame when it switched to channel <0,1>, on either path. Rearranging paths during the transmission stream wouldn't be problematic except that the switching assignment is backward in time. Consider, instead, switching data from channel <0,0> to channel <0,0>. Again, this connection has two paths, one path through each of the two center stage switches in the corresponding Benes network, the framed space/time diagram in Figure 21.10. But, now the difference between the two paths is subtly different. If the three switches are each in their BAR state, the data from input channel <0,0> moves through the upper a- and b-links and arrives at output channel <0,0> during the same frame. If the three switches are in the CROSS, BAR, and CROSS states, respectively, the data from input channel <0,0> moves through the lower a- and b-links, is delayed by one timeslot in each, and arrives at output channel <0,0>, but one frame later. The two paths are not exactly identical because any rearrangements would lose or repeat a timeslot. Hunter's transformation can be performed with frame integrity, resulting in a slightly more complicated physical network; but we'll skip that here.

21.4 PHOTONIC WAVELENGTH-DIVISION SWITCHING

In recent years, research in photonic switching has focused on wavelength-multiplexed networks. This section first describes wavelength-only networks and then a proposed network architecture, called ShuffleNet, that could be implemented as a space/wavelength network. The third and fourth subsections discuss the need for, and implementation of, wavelength interchange — especially as the space division is added to the architecture.

21.4.1 Wavelength-Only Networks

Most of the current research proposals are variations on the concept of providing a large passive optical network (PON), also known as dark fiber, that would be wavelength multiplexed. Transmitters and receivers would have to be wavelength agile. To initiate a connection, the calling party arranges for an idle wavelength and his transmitter is tuned to transmit at this wavelength. All receiving parties would have to be alerted to the channel's wavelength and their receivers would have to be tuned to this wavelength. The simplicity of operation is extremely appealing, especially if the network is implemented as a single large bus, but there are problems.

- *Scope.* While the concept of using a single network is appealing because of its simplicity, each channel does have an associated cost. It seems inefficient to consume one entire nationwide channel just to call someone who lives across the street.

- *Bandwidth.* A U.S. nationwide network, with 10:1 concentration, requires about 30 million channels. The frequency spectrum in optical fiber for low attenuation is about 15 THz around 1.3 microns and about 15 THz around 1.5 microns. Spreading these 30 million channels over this 30 THz leaves about 1 MHz per channel. Even a 50% guard band leaves 500 kHz per channel, which, while not enough bandwidth for uncompressed video, is still plenty of bandwidth. So, bandwidth, per se, is not a problem.

- *Accuracy.* Consider an optical channel that is centered exactly on wavelength equal to 1300 nm, has frequency bandwidth equal to 0.5 MHz, and has a frequency guard band of 0.5 MHz. In the frequency domain, the channel's center frequency is $f_c = c / 1.3 \times 10^{-6} = 230$ THz. If we divide the guard band evenly between the transmitter and the receiver, the laser transmitter's center frequency and the receiver filter's center frequency must both lie in the range: $f_c \pm 0.25$ MHz, or about $\pm 0.0000001\%$. It's not clear that this much stability is even possible, and even if it were, it's not clear it would be practical or affordable.

- *Fiber nonlinearities.* The first section of this chapter described how fiber's nonlinearities impose a limit on the number of different wavelength-defined channels that can be transmitted simultaneously on an optical fiber. There is still debate on whether these limitations are merely theoretical, applying only to channels that carry very high data rates over very long distances. If these nonlinearities are actually relevant to practical systems, then they will limit the number of channels, possibly even more than accuracy will limit the number.

It appears to be more realistic if a PON is implemented as a layered network or a multihop network. A regional PON could serve few enough subscribers that the number of wavelengths might be reasonable and channels could be reused for local calls. In such an architecture, an interoffice or inter-LAN call would be switched to a higher layer in a hierarchical network architecture. A long distance call might be switched through many hops and layers across the country.

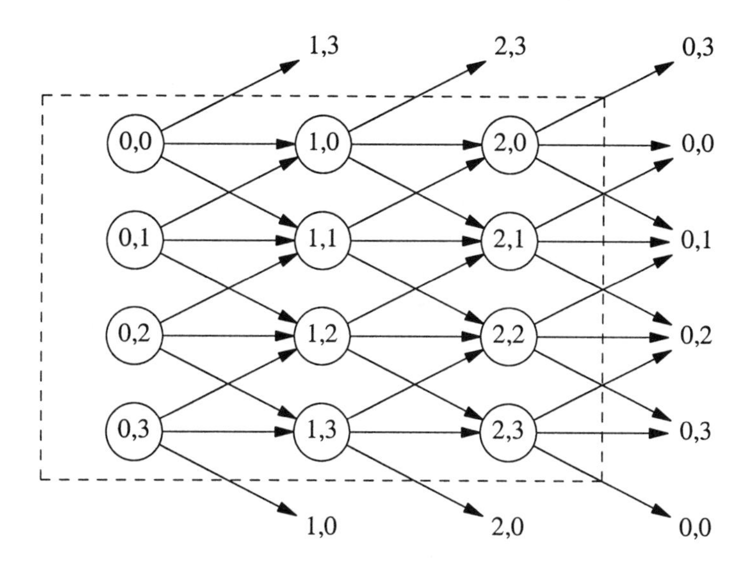

Figure 21.12. ShuffleNet.

21.4.2 ShuffleNet

ShuffleNet is a multihop architecture depicted logically, not physically, in Figure 21.12. While ShuffleNet is presented here as a space/wavelength network with an optical implementation, it generalizes to any division(s) and to electronic implementations. ShuffleNet is a two-dimensional toroidal network — a ring of super-nodes (horizontally, in Figure 21.12), where each super-node is a stack of nodes (vertically, in Figure 21.12). In the Manhattan street network (MSN), each column of nodes is also a ring, but in ShuffleNet the nodes in each super-node aren't connected to each other at all. There might be a separate LAN inside each super-node to handle local calls and the ShuffleNet could be viewed as MAN or WAN. Instead of a single gateway between LANs, in ShuffleNet some nodes in one super-node are connected to some nodes in the next super-node around the ring. The general ShuffleNet structure is uniform, with three parameters.

- One parameter is the number of super-nodes, n, and the ShuffleNet in Figure 21.12 has $n = 3$ super-nodes.
- Another parameter is the number of nodes in each super-node, k, and the ShuffleNet in Figure 21.12 has $k = 4$ nodes per super-node. The 0th super-node and the 0th and 3rd nodes in each super-node are shown twice in this figure to simplify drawing the channel connectivity between adjacent nodes.
- The degree, d, of each node is the number of channels that enter the node or that exit the node. Any node's incoming channels all come from nodes to its left and its outgoing channels all go to nodes to its right. In Figure 21.12, $d = 3$.

A great deal of research has gone into determining the maximum number of hops, and other performance measurements, as functions of n, k, and d. But, we'll skip that here. Instead, let's focus strictly on some various implementations.

- In the obvious implementation, let Figure 21.12 represent a physical diagram. Each node has d incoming links (physical wires, fibers, or radio channels) and d outgoing links. The incoming links come from d different nodes in the column to the left and the outgoing links go to d different nodes in the column to the right. This implementation allows the nodes in each super-node to be spatially separated.
- The implementation implied by this section of this chapter is one in which adjacent super-nodes (columns) are interconnected by a single physical channel, like one optical fiber. The set of $k \times d$ (= 12 in the example) channels between each super-node (column) are wavelength multiplexed onto this one physical channel. So, each super-node has a wavelength demultiplexor on its input side and a wavelength multiplexor on its output side. Each node is assumed to be close to its super-node's demultiplexor and multiplexor so that it can connect there to its internode input and output channels.
- The implementation above, with wavelength-defined channels on an optical fiber, can be generalized to the frequency domain on copper or radio, to the time domain on glass, copper, or radio, or to combinations of these.

ShuffleNet is a provocative architecture and it relies on photonic technology to provide its most interesting implementation: using wavelength multiplexing. But, ShuffleNet does not necessarily switch signals in their photonic format and is only incremental with respect to photonic technology.

21.4.3 Wavelength Interchange — Why?

For several years, the research community has debated whether such a large optical network can really be passive or whether active switching is needed for practicality. A particularly important, controversial, and contentious question is whether such a network requires wavelength interchange. This subsection discusses why (or why not) and the next subsection discusses how.

Rather than give a definitive answer, two contrived examples provide convincing arguments for each answer.

Example where WLI is superfluous. Consider a large nonhierarchical PON. Assume there are thousands of users and each user has direct access to the PON as if it were a single optical bus. Assume each user's appliance has a tunable laser and a receiver with a tunable optical filter. For A to transmit to B, an idle wavelength is selected and then A's laser and B's receiver filter are set to this wavelength.

This architecture is exactly analogous to the two-stage full-access network of Figure 7.4(a). The set of available wavelengths corresponds to the K junctors, where tuning the laser corresponds to the first stage of switching and tuning the receiver filter corresponds to the second stage of switching. Wavelength interchange corresponds to a $K \times K$ switching fabric, installed as a center stage in Figure 7.4(a). A center stage serves no purpose if the switching stages are full access (Section 7.2.2), corresponding to every laser and filter being able to tune to any wavelength. People who see a large full-access PON in their crystal ball insist that wavelength interchange is unnecessary, and they are correct.

Example where WLI makes a big difference. Consider a hierarchical network with local PONs at the lowest layer. An LPON might serve a campus or an office building or a residential area similar to that served by a CATV coax or by a telephone switching office. The advantage that an architecture with many LPONs has over an architecture comprised of a single large PON is that a local call in LPON A can use the same wavelength as a local call in LPON B. This wavelength reuse is analogous to the frequency reuse discussed in Section 19.2.1.

Suppose these LPONs are interconnected in a star architecture in which one two-way optical fiber connects a gateway on each LPON to a photonic switch at a central hub. Assume that all communication is two-way on the same wavelength. One two-way fiber from each of N LPONs converges on the hub. Now, consider a simple system that has three LPONs, each with three users, and let each fiber link carry three wavelengths. Since each LPON has three users and three channels to the hub, it seems intuitive that there would be no blocking.

Let a user on LPON A and a user on LPON C be connected. Arbitrarily, they are assigned wavelength 1. Then, let a user on LPON B and a second user on LPON C be connected. They can't use wavelength 1 because it's already busy on LPON C so they are arbitrarily assigned to wavelength 2. Third, let a second user on LPON A and a second user on LPON B be connected. They can't use wavelength 1 because it's already busy on LPON A and they can't use wavelength 2 because it's already busy on LPON B, so they must be assigned to wavelength 3. The current status is that each LPON has one remaining idle user and one remaining idle wavelength. The problem, of course, is that the remaining idle wavelengths are a different wavelength on each LPON. None of the third users on each LPON can be connected — without being able to convert wavelengths at the hub.

Discussion. So, does a large network with wavelength-defined channels require wavelength interchange? It depends on the architecture and the assumptions. In Section 14.2 and again in the discussion of DiSCO, when a user on SM1 of a modular switching system and a user on SM2 of the same system are connected together, the network control software locates a common idle timeslot. If the modules have timeslot interchangers (like in the 5E), then the connecting timeslot must only be idle on 1's up-link and on 2's down-link. However, if the modules do not have timeslot interchangers (as in DiSCO), then the connecting timeslot must be idle on the internal buses in both modules. There would be little constraint if the center stage had timeslot interchangers.

Now, consider connecting user 1 in Boston to user 2 in San Diego over a nationwide network that has many hops, either because it is hierarchical or it has many physical nodes. Assume there are four intermediate nodes for this connection — in Pittsburgh, Chicago, Denver, and Phoenix. Without wavelength interchange, network control software must locate a common idle channel, a single common wavelength that is unused in all five links. With wavelength interchange at the nodes, network control software must locate any idle wavelength in each link. I can't prove it, but this chapter proceeds under the assumption that

just as true space/time switching requires timeslot interchange,
true space/wavelength switching requires wavelength interchange.

21.4.4 Wavelength Interchange — How?

While the analogies among space, time, and wavelength are not perfect, all three divisions share the common characteristic that a complete interchanger is a system of converters.

- In the *space division,* a junctor shifts one single channel from row i inside some switching fabric to row j. A single junctor is a *space position converter.* A complete switching fabric has many junctors and it shifts many channels simultaneously from their row on the fabric's input to an assigned row on the fabric's output. A complete fabric is a *space position interchanger.*
- In the *time division,* an optical delay line, or a delay element, shifts one single channel from timeslot s in some frame to timeslot t. A single delay element is a *timeslot converter.* Section 21.2 developed architectures for complete photonic timeslot interchangers, made with many delay elements, that shift many channels simultaneously from their timeslot in the TSI's input frame to their assigned timeslots in the TSI's output frame.
- Similarly, the *wavelength division* has an elementary converter that operates on a single channel and a complete interchanger that operates on many channels simultaneously. In the wavelength division, a wavelength converter shifts one single channel from wavelength v on some fiber to wavelength w. A complete wavelength interchanger (WLI), made with many wavelength converters, shifts many channels simultaneously from their wavelength in the WLI's input fiber to their assigned wavelengths in the WLI's output fiber.

Two general approaches to wavelength interchange are optoelectronic and photonic. This subsection concentrates on optoelectronic WLI because photonic WLI is still very much a research area. So, this subsection begins with an optoelectronic wavelength converter.

If an input signal has only one wavelength-defined channel, then such a repeater's input side requires a photodetector and a bit-discriminating amplifier. If the output signal also has only one wavelength-defined channel, then the internal electronic signal is transmitted with a tunable laser. But, it is subtly different from its analogies in space or time because, unlike space or time junctors, this wavelength junctor does not shift its input signal to some other channel that is a relative distance away. Instead, it redirects its input signal to a specific channel. We'll come back to this.

Now, instead, suppose the input signal has L different wavelength-defined channels, only one of which is to be converted. While we can fully extract one space or time-defined channel, we don't have a good way to fully extract one wavelength-defined channel. So, all L channels have to be demultiplexed; that is, converted from being wavelength separated to being space separated. Then, the selected channel is converted as above, while the other $L - 1$ channels are merely amplified (either electronically or optically). Then, all L channels are remultiplexed.

Now, suppose all L of the wavelength-defined input channels might potentially or generally require wavelength conversion. Such a true wavelength interchanger is functionally analogous to a timeslot interchanger. In this case, and still using an optoelectronic implementation, after demultiplexing, photodetecting, and amplifying, each separated internal electronic signal is transmitted out on a different wavelength.

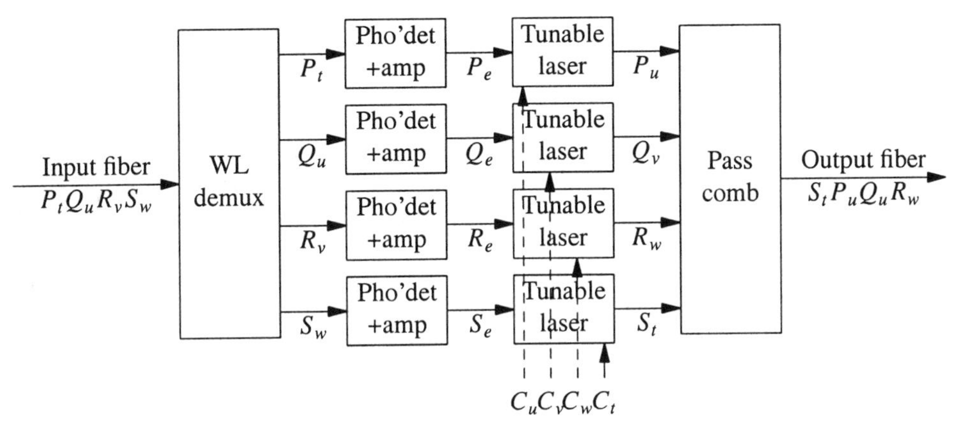

Figure 21.13. A 4 × 4 (absolute) wavelength interchanger. © 1996 IEEE [15].

Figure 21.13 shows a full 4 × 4 wavelength interchanger. Signals *P*, *Q*, *R*, and *S* enter the WLI on wavelengths *t*, *u*, *v*, and *w*, respectively. In the switching assignment illustrated, *w* is switched to *t* and each of the other three channels is switched up by one wavelength. So, signals *P*, *Q*, *R*, and *S* exit the WLI on wavelengths *u*, *v*, *w*, and *t*, respectively.

Two different internal implementations can be imagined.

- Each of the *L* laser transmitters can be permanently assigned to a specific wavelength. After demultiplexing the input signal, each of the *L* wavelength-defined channels is individually photodetected and amplified. Then, each of these *L* electronic signals must be routed to its assigned laser by an internal $L \times L$ space-division switching network. The optical signals produced by the *L* lasers can then be combined in a wavelength multiplexor. A passive combiner could be used instead of the multiplexor, but for any significant value of *L*, the passive combiner will have more loss.

- Each of the *L* laser transmitters can be tunable. After demultiplexing the input signal, again each of the *L* wavelength-defined channels is individually photo-detected and amplified. But here, each of these *L* electronic signals is routed permanently to the same tunable laser for any switching assignment. Switching is accomplished by controlling the tunable lasers so that each one transmits at the assigned wavelength. But, now, the four converted optical signals cannot be wavelength multiplexed because, in an optical multiplexor, each wavelength must always be assigned to its corresponding spatial input port. Combining at the output would require an $L \times L$ space-division photonic switch between the lasers and the wavelength multiplexor or an $L \times 1$ passive combiner must be used instead of a wavelength multiplexor.

All these optoelectronic implementations share the drawback that, not only are the wavelengths fixed, but the data rate is as well. Digital electronic designers always ask for the data rate as one of the design parameters and then design the electronic component for that specified data rate. You can't make a microprocessor run faster than its rated speed and the same thing applies to the electronics inside an optoelectronic component.

To allow for data rate agility in the network, a rate-independent implementation is preferred.

Recalling optical fiber's nonlinearities, discussed in Section 21.1, optical signals at wavelength w_1 are sometimes (annoyingly) converted, or partially converted, to another wavelength w_2. Some exotic photonic devices can accentuate these nonlinearities and convert a signal, without intervening electronics, from w_1 to w_2. Some can even convert a set of signals, simultaneously but not arbitrarily.

For example, let a set of signals on wavelengths $w_1, w_2, \ldots, w_L$ be input to one such device, which has two externally controlled states. In state 1, the signals emerge unchanged and in state 2, they emerge reversed to $w_L, w_{L-1}, \ldots, w_1$, respectively. Other fixed patterns of interchange are possible, but the device cannot arbitrarily interchange wavelength-defined channels the way a TSI can arbitrarily interchange time-defined channels. Completely flexible channel assignment corresponds to *full access,* and completely flexible channel interchange corresponds to *full connectivity* in the corresponding logical stage of switching. While interesting network architectures are probably possible with such *limited connectivity* switching stages, these photonic devices are a long way from being practical.

21.4.5 Differences Among Space, Time, and Wavelength

While space switching, timeslot interchange, and wavelength interchange are analogous operations [16,17], there are many practical differences among them. While it's tempting to simply map architectures from one domain into another one, this is quite naive. We saw this difficulty in the previous subsection, when we tried to map the Benes space switch into the space/time domain. Because signals in time-defined channels cannot physically be moved backward in time, the corresponding Benes network has no junctors that move upward in the interstage wiring. The wavelength domain has similar constraints.

Another research photonic device that performs wavelength mapping has the constraint that signals can only be shifted to higher wavelengths (analogous to the physics of an Erbium-doped fiber amplifier). While this constraint has been observed to be analogous to being constrained from shifting time-defined channels to earlier timeslots, this observed analogy is also naive. Since timeslots are framed, time-defined channels are cyclic. Even though timeslot 23 comes after timeslot 0, data from timeslot 23 can be shifted to timeslot 0 by moving it into a later frame or by inserting one frame of delay. Because wavelength-defined channels are not cyclic, we don't have the analogous architectural trick and one-way-only wavelength shifting is a serious architectural constraint.

But, there is yet another subtlety that distinguishes wavelength interchange from space switching and timeslot interchange.

- If the *i*th switching module on a modular digital switching system is physically moved from its location to another building, as a remote switching module, it is still the *i*th module. While there may be some implementation problems with synchronization and increased delay in the junctors, physically moving the module doesn't effect the architecture of the center stage TMS. In its CROSS state, a 2×2 space switch places the data on its input ports into the opposite corresponding output ports and this identical device is used all through a Benes network. So, space switching is *relative* and doesn't depend on absolute location.

- A T1 line has 24 time-defined channels. The framing bit synchronizes the T1 timeslot interchanger, but not by the absolute value of Greenwich Mean Time. And while two T1s may be out of phase with each other, a simple delay aligns them temporally. In its CROSS state, a 2×2 timeslot interchanger places the data in any incoming timeslots into the corresponding opposite outgoing timeslots, and a network of $N \times N$ timeslot interchange could be constructed from an iteration of identical 2×2 components. So, time-defined channels are also ordinal and time switching is also relative.
- Wavelength-defined channels, however, are not ordinal — there is no *i*th wavelength. In its CROSS state, a 2×2 wavelength interchanger would be expected to place the data on its incoming wavelengths onto the corresponding opposite output wavelengths. So, if wavelength w_1 carried signal s_1 and wavelength w_2 carried signal s_2, and both wavelengths were multiplexed on a fiber that was input to a 2×2 WLI in its CROSS state, the output fiber should have wavelength w_1 carrying signal s_2 and wavelength w_2 carrying signal s_1. But, how do the tunable lasers know the opposite signal's wavelength? A given 2×2 WLI can be adjusted for a specific pair of wavelengths, but a network of $N \times N$ wavelength interchangers cannot be built from an iteration of identical 2×2 components. Wavelength-defined channels are absolute and wavelength switching is not relative. This is a very significant difference.

21.5 PHOTONIC SWITCHING IN SPACE AND WAVELENGTH

We've assumed that true space/wavelength switching requires WLI just as true space/time switching requires TSI. But, true space/time switching also requires a time-multiplexed switch, so doesn't it seem that true space/wavelength switching would require a wavelength multiplexed switch? A TMS is a space switch that allows a different switching configuration for each timeslot. So, if a WMS exists, it would be a space switch that allows a different switching configuration for each wavelength.

After describing a technology that promises to provide a wavelength-multiplexed switch, the second and third subsections generalize Marcus's and Hunter's Transformations, respectively, from space/time to space/wavelength.

21.5.1 Wavelength-Multiplexed Switch

While a TMS supports different switching configurations, they don't coexist simultaneously but rather during different timeslots. So, a TMS' only significant requirement is that it be able to reconfigure rapidly between its timeslots. However, a WMS supports different switching configurations that must coexist simultaneously in the same physical switch. While a WMS is much more difficult to imagine than a TMS, the technology exists to do it. It's just not quite practical; at least, not yet.

An acousto-optic tunable filter (AOTF) is a 2×2 photonic device, a variation on the switched directional coupler [18 – 20]. In the AOTF and in the Lithium Niobate SDC, coupling between the channels is controlled by changing the index of refraction, n, in the material. However, while n is changed in Lithium Niobate by controlling the electric

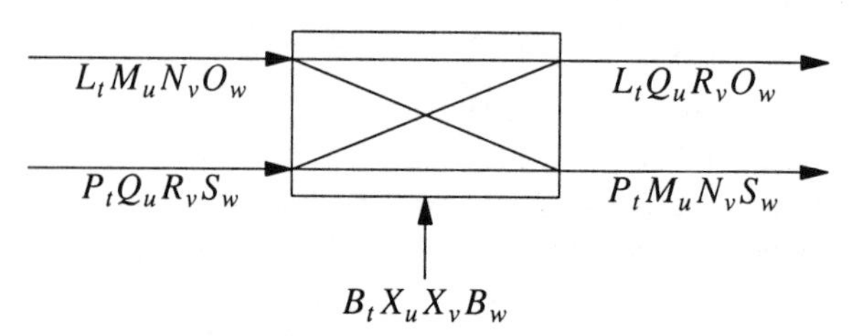

Figure 21.14. Acousto-optic tunable filter.

field in the material, n is changed in the AOTF by controlling the mechanical stress in the material. One easy way to control mechanical stress is by physical vibration, at acoustic frequencies. With no acoustic signal, the AOTF is in the BAR state. When an acoustic signal is resonant on the AOTF's substrate, the AOTF is placed into the CROSS state, but only for a specific optical wavelength — which depends on the frequency of the acoustic wave. If several acoustic waves are induced simultaneously, then the AOTF is in its CROSS state for several optical wavelengths simultaneously — and in its BAR state for any other wavelengths.

The AOTF's utility as a WMS is illustrated in Figure 21.14. The figure shows an AOTF, interfacing to two input optical fibers on the left, to two output optical fibers on the right, and to a single frequency-multiplexed electrical control on the bottom. The upper input fiber carries four wavelength-multiplexed signals (L, M, N, and O) on wavelengths t, u, v, and w, respectively. The lower input fiber carries four wavelength-multiplexed signals (P, Q, R, and S) on the same four wavelengths, respectively. So, four wavelength-defined channels enter the AOTF on each of the two input fibers and four wavelength-defined channels exit the AOTF on each of the two output fibers.

But, in the example illustrated in Figure 21.14, let the AOTF switch the channels so that the signals on wavelengths t and w pass straight through the AOTF and the signals on wavelengths u and v cross over. AC control voltages induce the appropriate acoustic waves in the AOTF to place it in its CROSS state for wavelengths u and v. But, the AOTF is in its BAR state for wavelengths t and w. So, signals on wavelengths t and w are switched through the AOTF so that L_t and O_w exit on the upper output fiber and P_t and S_w exit on the lower output fiber. Signals on wavelengths u and v are switched through the AOTF so that M_u and N_v exit on the lower output fiber and Q_u and R_v exit on the upper output fiber.

The four wavelengths on each fiber must be identical because, remember, wavelength-defined channels are absolute. Within this restriction, research models of an AOTF have operated as a WMS successfully for around five wavelengths. As a 2×2 switch, AOTFs can be assembled into Benes architectures.

21.5.2 Marcus's Transformation in Space and Wavelength

Section 13.3 showed how Marcus transformed a Clos architecture into a time-space-time or a space-time-space implementation. The necessary ingredients are a TSI and the TMS. Marcus's Transformation extends easily from the time division to the wavelength division, transforming any Clos architecture into a wavelength-space-wavelength or a space-wavelength-space implementation. The necessary ingredients would be a WLI and the WMS.

Just as the TST implementation can avoid TSIs if every outer-stage module doubles as a time multiplexor, then the WSW implementation can avoid WLIs if every outer-stage module doubles as a wavelength multiplexor. The outer wavelength stages of such a WSW implementation would be modules, in which external signals are assigned to wavelengths (it's a multiplexor and not a true switch). The wavelength-defined channels are wavelength multiplexed onto up-links that go to a center stage WMS and are demultiplexed off down-links that come from a center stage WMS. At the center stage, a Benes network of AOTFs terminates the wavelength-multiplexed up-links from each module on its left side and the wavelength-multiplexed down-links to each module on its right side. While such a wavelength analogy to DiSCO has never been prototyped, a prototype with four modules and four wavelengths should be feasible. The center stage would be a 4×4 Benes network, using AOTFs as the six 2×2 beta elements. If all six AOTFs operate on the same four wavelengths, the center stage would be a 4×4 WMS, multiplexed over these same four wavelengths.

However, if the WSW network is used as a complete switching node in a network where the links are already multiplexed, then true WLI is required in the wavelength stages of the WSW or SWS implementations. While neither a WLI nor a WMS is easy or inexpensive, WLI seems to be the more expensive, the one requiring more manual attention, and the one presenting more architectural constraints (especially if implemented in photonics instead of optoelectronics). The SWS implementation is probably better than the WSW because SWS only has one stage of WLIs and WSW has two.

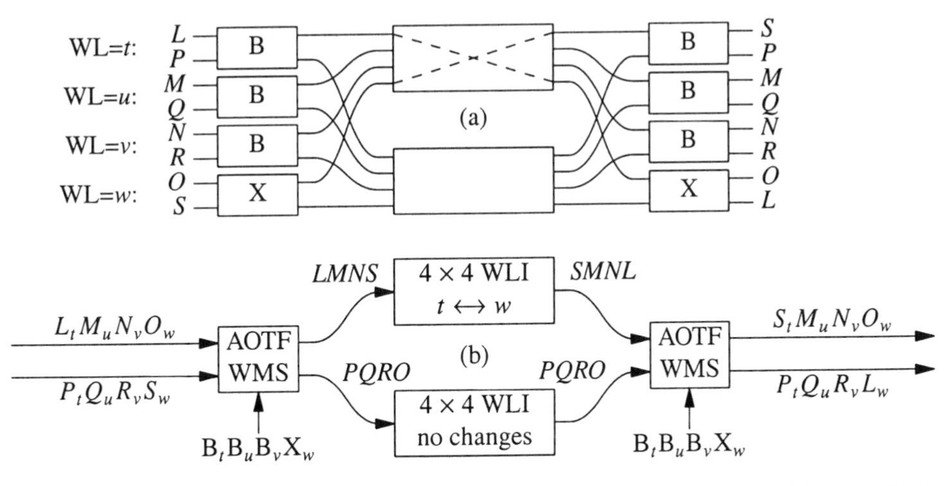

Figure 21.15. $< 2,4 > \times < 2,4 >$ space-wavelength-space network. © 1996 IEEE [15].

A complete 8×8 space/wavelength switching network is shown in Figure 21.15(b), and its corresponding space-only Clos network is shown in Figure 21.15(a). Figure 21.15(b) has two fibers on each end, each carrying four wavelength-defined channels. Each edge-switch in the corresponding Clos network also has two fibers on each end, but the Clos has four spatially distinct switches in each outer stage, one corresponding to each of the four separate wavelengths. The eight incoming channels are labeled L through S. Channels L through O are assigned to upper fibers in parts (a) and (b) and channels *P* through *S* to lower fibers on both figures. In Figure 21.15(b), *L* and *P* are both assigned to wavelength *t*, *M* and *Q* to *u*, *N* and *R* to *v*, and *O* and *S* to *w*. Since the first and third stages of the Clos network in part (a) are wavelength multiplexed in this extension of Marcus's Transformation, the corresponding stages in part (b) require a single 2×2 AOTF WMS. Because the center stage of the Clos network in part (a) has two 4×4 modules, the center stage of part (b) requires two WLIs, each capable of fully interchanging signals among four wavelengths.

In the switching assignment shown in Figure 21.15(a) and (b), signals *M* through *R* remain in their respective space/wavelength channels, but signals *L* and *S* are interchanged in both the space and wavelength of their respective channels. One of this assignment's many network configurations is illustrated in Figure 21.15(a) and (b). In the Clos network of part (a), the 2×2 switches in the first and third stages are placed in BAR, BAR, BAR, and CROSS, respectively, down each column. So, in Figure 21.15(b), the input and output AOTFs are placed into their BAR state for wavelengths *t, u,* and *v* and into their CROSS state for wavelength *w*. In the Clos network of part (a), the upper 4×4 center stage switch interchanges its upper and lower ports and switches its two middle ports straight across (not shown), and the lower 4×4 center stage switch switches all its ports straight across (not shown). So, in Figure 21.15(b), the upper WLI is configured so that incoming signals on wavelengths *t* and *w* exit the WLI on the other signal's wavelength but incoming signals on wavelengths *u* and *v* remain on these wavelengths, and the lower WLI is configured so that incoming signals on all four wavelengths remain on their same wavelengths.

The channels are easily traced through the Clos network of Figure 21.15(a), but the SWS implementation in Figure 21.15 (b) shows a little more detail.

- Because the left-stage WMS is in its BAR state for wavelengths *t, u,* and *v* then channels *L, M, N, P, Q,* and *R* remain on their respective levels at the b-links. But because this WMS is in its CROSS state for wavelength *w*, channels *O* and *S* exit this AOTF on their opposite levels at the b-links. Channels *L, M, N,* and *S* are assigned to wavelengths *t, u, v,* and *w*, respectively, as they enter the upper WLI, and channels *P, Q, R,* and *O* are similarly assigned as they enter the lower one.

- Because the upper WLI interchanges wavelengths *t* and *w* but doesn't affect channels on wavelengths *u* and *v*, then channels *S, M, N,* and *L* are assigned to wavelengths *t, u, v,* and *w*, respectively, as they exit this upper WLI. Because the lower WLI doesn't interchange any wavelengths, the exiting channel assignment is the same as how it entered.

- Because the right-stage WMS is in its BAR state for wavelengths *t, u,* and *v*, then channels *S, M, N, P, Q,* and *R* remain on their respective levels at the outputs. But because this WMS is in its CROSS state for wavelength *w*, channels *L* and *O* exit this AOTF on their opposite levels at the outputs. Channels *S, M, N,* and *O* are assigned to wavelengths *t, u, v,* and *w*, respectively, as they enter the upper output fiber, and channels *P, Q, R,* and *L* are similarly assigned as they enter the lower one.

We see that signals that were input on channels *L* and *S* emerge on each other's channel and that no other signals are switched.

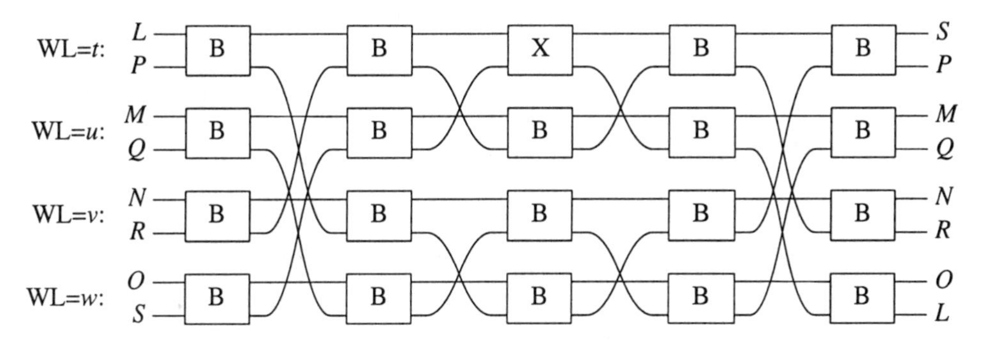

Figure 21.16. Modified 8 × 8 Benes architecture. © 1996 IEEE [15].

21.5.3 Hunter's Transformation in Space and Wavelength

Hunter's Transformation also extends to space/wavelength switching. Here the absolute nature of WLI allows an optimization of the architecture. The 8 × 8 Benes architecture in Figure 7.11 is redrawn in Figure 21.16, but with two minor changes.

- In the first and last columns, interchange the 2 × 2 switches in the second and third rows. This structural reconfiguration of the switches doesn't affect the logical operation of the network and this particular transformation, when generalized, converts a Banyan network into a form called an Omega network (Section 7.5).
- In every switch in the network, connect its ports so that its upper port connects to a junctor that is wired to the previous or subsequent switch in the same row and its lower port connects to a junctor that is wired to the previous or subsequent switch in some other row. This structural reconfiguration of the junctor wiring doesn't affect the logical operation of the network and this particular transformation, when generalized, converts any network into what we call its slipped counterpart.

Then, let this slipped and Omega-ed Benes network, shown in Figure 21.16, represent the logical architecture of a $<2,4> \times <2,4>$ space/wavelength network, where each row corresponds to a different wavelength. Junctors in the Benes architecture that shift a signal among rows represent wavelength interchangers, for these specific wavelengths. Because the wiring between the first and second columns (a-links) and between the fourth and fifth columns (d-links) all switch a signal to its same row or shift it so $t \longleftrightarrow v$ and $u \longleftrightarrow w$, the WLI that corresponds to these places need not be general, but only has to be capable of these specific wavelength interchanges. Because the wiring between the second and third columns (b-links) and between the third and fourth columns (c-links) all switch a signal to its same row or shift it so $t \longleftrightarrow u$ and $v \longleftrightarrow w$, the WLI that corresponds to these places also need not be general, but only has to be capable of these specific wavelength interchanges.

The logical modified 8 × 8 Benes architecture from the previous figure maps into the physical $<2,4> \times <2,4>$ architecture shown in Figure 21.17. Each of the five physical 2 × 2 switches in the figure is a wavelength-multiplexed switch, implemented as an acousto-optic tunable filter. The four wavelength interchangers, wired between these AOTFs, are simplified so they only perform those wavelength interchanges indicated on Figure 21.17.

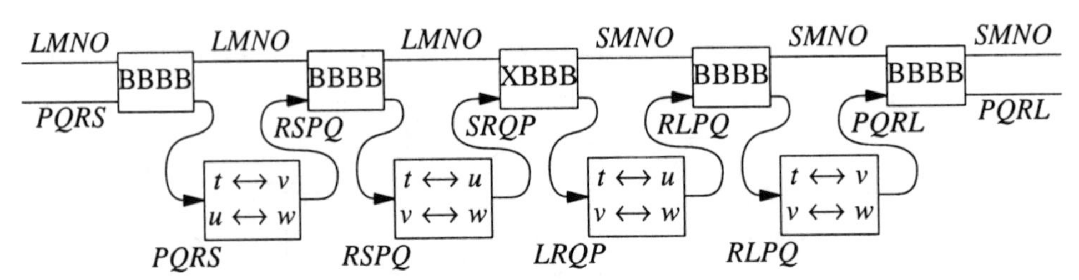

Figure 21.17. $\langle 2,4 \rangle \times \langle 2,4 \rangle$ Hunter architecture. © 1996 IEEE [15].

The switching assignment used to illustrate the Marcus-Transformed space/wavelength network in the previous subsection is also used here to illustrate this Hunter-Transformed space/wavelength network. Recall that input signals in space/wavelength channels L and S interchange into each other's channel and that channels M through R are not changed. One possible configuration of this switching assignment is illustrated on the modified Benes network in the previous figure. Nineteen beta elements are placed into their BAR states and only one, the first SDC in the third stage, is placed in its CROSS state. The reader should trace and verify that the eight connections are established.

Each of the AOTF-based WSMs in Figure 21.17 must be assigned to its BAR or CROSS state for each of the four wavelengths. The assignment for the example configuration is shown in the figure. The BAR/CROSS assignment for wavelength i in AOTF j in Figure 21.17 is identical to the assignment for switch i in column j from the previous figure. Consider channels M, Q, and L:

- In Figure 21.16 channel M is switched straight across the modified Benes network. So, in Figure 21.17, M remains on the upper rail straight across the five AOTFs, never changing from its original wavelength, u.
- In Figure 21.16 channel Q cannot be switched straight across the modified Benes network because M has that path. Instead, Q uses a path that starts at the second row and bounces around the third and fourth rows before returning to the second row. Analogously, in Figure 21.17, Q must take the lower rail, which causes it to pass through all the WLIs. So, Q starts in wavelength u and bounces around in wavelengths v and w before returning to wavelength u.
- In Figure 21.16 channel L is switched halfway straight across the modified Benes network, then drops to the second row, and then to the fourth row. Analogously, L remains on the upper rail halfway across Figure 21.17 and, there, switches to the lower rail, causing it to pass through the last two WLIs. So, L remains on wavelength t until the last two WLIs switch it to wavelengths u and w, respectively.

21.5.4 Comparison

We compare the Marcus-Transformed and Hunter-Transformed implementations. The Hunter implementation above requires five AOTFs. While the Marcus SWS implementation of Figure 21.15(b) only requires two, the Marcus WSW implementation would only require one. But, AOTFs are not the most critical components.

The Marcus SWS implementation in Figure 21.15(b) needs two 4×4 nonblocking fully connected wavelength interchangers, and the Marcus WSW implementation requires four of these. While the Hunter implementation also requires four wavelength interchangers, they are extremely specialized. The first and fourth WLIs are permanently wired internally so that $t \rightarrow v$, $u \rightarrow w$, $v \rightarrow t$, and $w \rightarrow u$. This WLI needs no control signal, has no internal switching, and its outputs are permanently remultiplexed. The second and third WLIs are similarly permanent, assigned so that $t \rightarrow u$, $u \rightarrow t$, $v \rightarrow w$, and $w \rightarrow v$.

While AOTFs are too "researchy" to price them out, the Hunter implementation appears to be the clear winner.

21.6 ELEMENTARY PHOTONIC SPACE/TIME/WAVELENGTH SWITCHING

This chapter and the previous one have described photonic switching in single switching divisions: space only, time only, and wavelength only. This chapter has also described photonic switching in two of the three possible pairs of switching divisions: space and time together and space and wavelength together. While we don't discuss time and wavelength together, this section describes photonic switching in all three divisions together: space, time, and wavelength. This chapter has presented photonic circuit switching right at the state of the art and has even described some system possibilities that have never been prototyped. Because this section is even more forward-looking, it can't get into it in too much depth; the depth just isn't there [15,21].

This section examines only the most elementary case of space/time/wavelength photonic switching network architectures: an 8×8 fabric. Each side of the fabric has two (space) fibers and each fiber carries four multiplexed channels: defined over two wavelengths and over two timeslots. The timeslots are synchronized, so each timeslot holds four channels: over two fibers and over two wavelengths. The two wavelengths are identical in each fiber (analogous to time's synchronization), so each wavelength carries four channels: over two fibers and over two timeslots.

The first two subsections examine the Marcus and Hunter transformations one more time (for the last time in this book). If I have tried to make one impression on the reader throughout this book, it is that architectures cannot be designed in a vacuum. These two subsections are written in deliberate contradiction to this advice. These subsections arrive at some interesting results and describe some interesting architectures — but, they can't be built. The last subsection shows one possible implementation of one of this section's architectures.

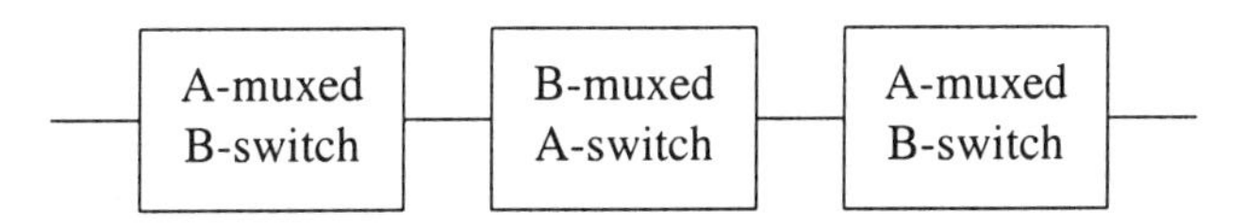

Figure 21.18. Generalized two-division Clos/Marcus network. © 1996 IEEE [15].

21.6.1 By Marcus's Transformation

Figure 21.18 illustrates a generalization of Marcus's Transformation of the Clos network architecture. Consider two arbitrary switching divisions, A and B. The Clos network in Figure 21.18 switches channels, defined in the A and B divisions, between the input carrier and the output carrier. Each of the two exterior stages of switching is an A-multiplexed B-switch, capable of switching channels within the B division independently over each different value of A. The center stage of switching is a B-multiplexed A-switch, capable of switching channels within the A division independently over each different value of B. To connect an incoming signal in channel <*a,b*> to an outgoing signal in channel <*c,d*>, an idle intermediate B-channel *e* is used. While A = *a*, the first stage switches *b* to *e*. While B = *e*, the center stage switches *a* to *c*. And, while A = *c*, the third stage switches *e* to *d*.

For example, let A = space and B = time. Then, each of the two exterior stages is a space-multiplexed time-switch. Because space multiplexing is just physical replication, each of these stages is a bank of TSIs. And, the center stage is a time-multiplexed space-switch, a TMS. For this example, Figure 21.18 is the classical TST architecture. Reversing A and B to time and space, respectively, converts Figure 21.18 to the STS architecture. We see a general structure from which we could even build a time/wavelength network.

The Benes network architecture was derived from the Clos architecture by nesting successively smaller networks within the Clos' center stage. Dilation was illustrated by a similar nesting. After illustrating a three-stage timeslot interchanger, also based on the Clos, we showed that TSI structure could be extended by nesting. Here, we nest the Clos again by adding a third division, C.

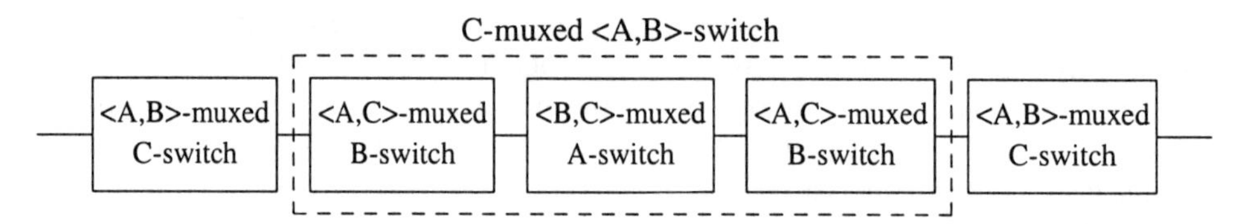

Figure 21.19. Clos/Marcus network in three divisions. © 1996 IEEE [15].

Figure 21.19 illustrates an extension of the previous figure to a third division. Consider three arbitrary switching divisions: A, B, and C. The Clos network in Figure 21.19 switches channels, defined in the A, B, and C divisions, between the input carrier and the output carrier. Each of the two exterior stages of switching is an <A,B>-multiplexed C-switch, capable of switching channels within the C division independently over each different value of A and B. The center stage (in the dashed box) of switching is a C-multiplexed <A,B>-switch, capable of switching channels within the <A,B> division independently over each different value of C. But, this dashed box is the network of the previous figure, except multiplexed over C. So, the stages of the previous figure, besides being multiplexed as shown, must also be multiplexed over C.

How does this work in the three photonics divisions? The photonic divisions of space, time, and wavelength can be assigned to the generalized divisions A, B, and C in any order. But, after examining all six permutations, we find that one or two of the five stages must include replicated (space-multiplexed) wavelength-multiplexed TSIs and space-switches that are both time and wavelength multiplexed. We don't have the technology to do these things — at least, not directly.

21.6.2 By Hunter's Transformation

Figure 21.16 shows how an 8×8 Benes architecture, modified by being Omega-ed and slipped, corresponds to a $< 2, 4 > \times < 2, 4 >$ Hunter network in space and wavelength. It can obviously also be made to correspond to a $< 2, 4 > \times < 2, 4 >$ Hunter network in space and time. In this section, this same architecture is made to correspond to the most elementary three-division network: one with two space channels, each with two timeslots per frame and each carrying two wavelengths.

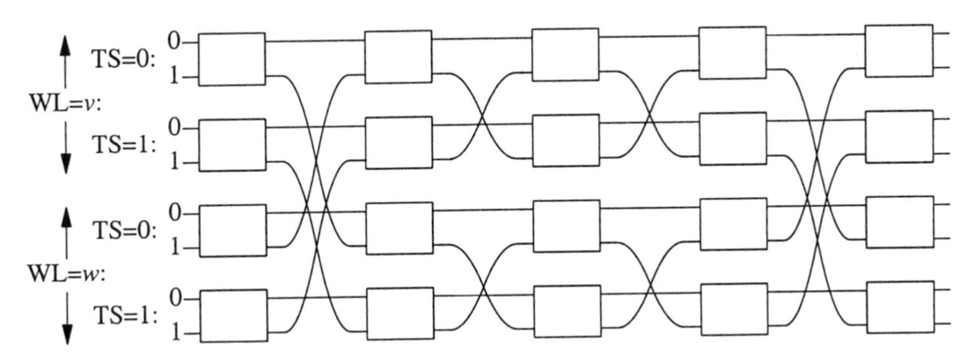

Figure 21.20. Modified Benes $< 2, 2, 2 > \times < 2, 2, 2 >$ network. © 1996 IEEE [15].

Figure 21.16 is repeated here in Figure 21.20 except, instead of labeling the four rows so they correspond to four wavelengths, in Figure 21.20 they are labeled so they correspond to two timeslots in each of two wavelengths. While the two space channels, 0 and 1, must be assigned directly to the ports of the 2×2 switches in each row, the assignment of the time and wavelength channels is arbitrary. Instead of the assignment shown in Figure 21.20, alternating rows could have been assigned to wavelengths v and w (instead of to timeslots) and the upper and lower pairs of rows could have been assigned to timeslots 0 and 1 respectively (instead of to wavelengths). Using the assignment shown in the figure, observe the interstage wiring.

- Between the first and second stages (the a-links) and between the fourth and fifth stages (the d-links) all signals in these junctors either: (1) remain on the exact same space/time/wavelength channel or they (2) switch to the opposite wavelength (staying in the same timeslot). Thus, the corresponding three-dimensional network's a-links and d-links will have only WLIs and no TSIs.
- Between the second and third stages (the b-links) and between the third and fourth stages (the c-links) all signals in these junctors either: (1) remain on the exact same space/time/wavelength channel or they (2) switch to the opposite timeslot (staying on the same wavelength). Thus, the corresponding three dimensional network's b-links and c-links will have only TSIs and no WLIs.

The modified Benes network in the previous figure converts into the Thompson/Hunter network shown in Figure 21.21. While the WLIs are simple, they aren't nearly as simple as the delay elements. If time and wavelength are interchanged in labeling of the modified Benes, then the WLIs and the delay elements would switch places in Figure 21.21.

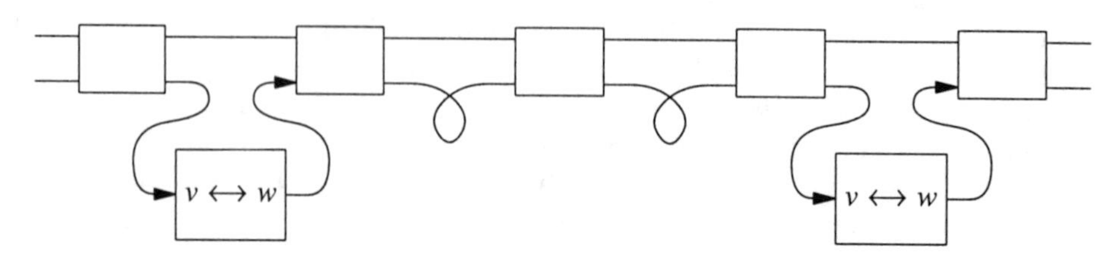

Figure 21.21. A $<2,2,2> \times <2,2,2>$ Thompson/Hunter network. © 1996 IEEE [15].

Consider a signal on wavelength v, in timeslot 0, and on fiber 0 (the upper input fiber) in Figure 21.21. Consider connecting this signal to its exact corresponding channel on the output side of Figure 21.21. Because each connection has four paths through the corresponding Benes network, this connection should also have four paths through the Thompson/Hunter network of Figure 21.21. One connection passes straight across the upper rail in Figure 21.21, the second passes through both delays, the second passes through both WLIs, and the fourth connection passes through both delays and both WLIs.

21.6.3 Reality Check

While Figure 21.21 looks quite elegant and might make for a nice paper, anyone who tried to prototype it would find that it can't be built. One subtlety is that both WLIs must be able to be time multiplexed and both the delays must be able to be wavelength multiplexed. Both technologies allow this. The more subtle problem is that each of the five 2×2 switches in Figure 21.21 must be both time multiplexed and wavelength multiplexed. No existing technology can do this. Clearly, a Lithium Niobate SDC can't be wavelength multiplexed. While it's not so obvious that an AOTF can't be time multiplexed, it can't. With a switching speed of several milliseconds, time multiplexing with AOTFs would require a guard band of several milliseconds duration on the boundary of every timeslot. This just wouldn't be feasible.

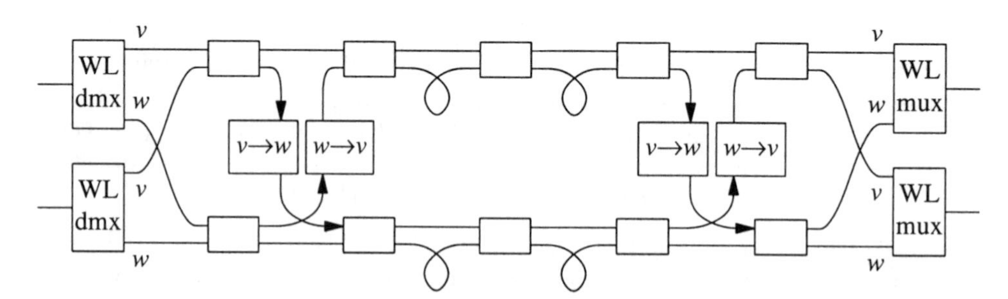

Figure 21.22. <2,2,2> STW Thompson/Hunter network using SDCs. © 1996 IEEE [15].

The Thompson/Hunter network from Figure 21.21 is redrawn in Figure 21.22. The revised architecture has two parallel tracks, each a copy of the original network. Wavelength v is demultiplexed off both incoming fibers and all four channels use the upper track and wavelength w uses the lower track. Wavelength interchangers are wired so that signals also change tracks. Then, both wavelengths are remultiplexed at the output.

While the block diagram in Figure 21.22 is more than twice as complicated as that in the previous figure, the advantage here is that none of the 2×2 switches in Figure 21.22 has to be wavelength multiplexed. So, the switches can be implemented using Lithium Niobate switched directional couplers. In general, such an implementation would require a track for each wavelength.

A similar diagram can be drawn with a time multiplexor and a time demultiplexor at its respective edges, instead of the wavelength counterparts. Wavelength interchangers would remain *in track,* but the delay elements would be wired so that signals that switched to other timeslots would also change tracks. Here, because none of the 2×2 switches would have to be time multiplexed, they could be implemented using acousto-optic tunable filters. This implementation would require a track for each timeslot.

Because there are likely to be many more timeslots than wavelengths and because SDCs are more practical than AOTFs, the implementation of Figure 21.22 appears to be more practical. But, there may be architectural tricks that reverse this. The Marcus-Transformed network from this section's first subsection may be subject to similar implementations. This work remains undone.

EXERCISES

21.1 Consider a block-multiplexed optical transmission format with 8000 frames per second and 256 timeslots per frame. Each channel carries a complete T1 signal and the line rate is 500 Mbps. How large is the guard band in each timeslot?

21.2 Consider modifying a Sonet OC-12 system for block multiplexing by eliminating a tributary to make room for guard bands. Maintaining the same transmission bit rate, guard bands can be accommodated by multiplexing 11 OC-1 streams, instead of the conventional 12. Let block-multiplexed timeslots be defined as the time allotted to three columns in an OC-1 frame — equivalent to one DS1 virtual tributary. Because these timeslots are fixed at all levels of multiplexing, an OC-11 stream carries 33 OC-1 columns in each of its timeslots. (a) How many timeslots are in such a block-multiplexed frame? (b) Including its guard band, how long is each timeslot? (c) How many bits are in each timeslot? (d) Compute the data duration in each timeslot. (e) How big is each timeslot's guard band?

21.3 Convert Sonet to block multiplexing using wavelength multiplexing instead of eliminating a tributary. This method also simplifies the removal of overhead signaling. Sonet OC-48 could be wavelength multiplexed on two wavelengths: wavelength w_o carries the three columns of section overhead (SOH) and line overhead (LOH) in every Sonet frame and wavelength w_d carries the 87 columns of data (and path overhead, POH). (a) What is the minimum data rate on w_o? (b) If w_d still runs at the OC-48 rate of 2.48832 Gbps, and if all 87 data columns in each OC-48 frame are transmitted as a single timeslot, how long is the block of data in each timeslot? (c) How long is the guard band in each frame?

21.4 Modifying the previous exercise, suppose each of the 87 data columns in each OC-48 frame is transmitted as a separate timeslot. (a) As a fraction of a frame, including its guard band, how long is each timeslot? (b) How long is the block of data inside each timeslot? (c) How long is the guard band in each timeslot?

21.5 Modifying the previous two exercises, assume that a block-multiplexed timeslot is defined by the duration of one OC-1 data byte. (a) With $87 \times 9 = 783$ data bytes in each OC-1 frame, how many timeslots in such a block-multiplexed frame? (b) As a fraction of a frame, including its guard band, how long is each timeslot? Let these timeslots be fixed at all levels of multiplexing except that such a block-multiplexed OC-N stream has N bytes in each of its timeslots. (c) Transmitting an OC-48 stream at its conventional 2.48832 Gbps, how long is the block of data inside each timeslot? (d) How long is the guard band in each timeslot?

21.6 A fiber transmission line is 720 miles long. It has a state-of-the-art optical repeater every 60 miles. Thus, the line has 12 segments, each 60 miles long, with a repeater at the end of each segment. Some repeaters can be modified (at considerable expense) to demultiplex, separately amplify each channel, and re-multiplex. Call these multiplexing repeaters and call the others simple repeaters. How many wavelength-defined channels are supported over the line under each of these conditions: (a) the only multiplexing occurs at the transmitter, the only demultiplexing occurs at the receiver, and the 11 intermediate repeaters are all simple; (b) there is one multiplexing repeater in the middle of the line with five simple repeaters on each side; (c) two of the 11 repeaters are multiplexing repeaters; (d) three of the 11 are multiplexing repeaters; (e) every other repeater is a multiplexing repeater; and (f) every repeater is a multiplexing repeater?

21.7 Consider a time-multiplexed photonic link with four timeslots per frame. (a) Sketch a passive-delay photonic TSI (similar to Figure 21.1) for this link. Suppose the switching assignment is $0 \to 3$, $1 \to 1$, $2 \to 0$, and $3 \to 2$. (b) Walk the four timeslots in one frame through your architecture.

TABLE 21.1. Beginning the active-delay photonic TSI control

Input	Delay 0	Delay 1	Delay 2	Delay 3	Output
a_0	[.]				
b_0	a_0	[.]			
c_0	a_0	b_0	[.]		
d_0	a_0	b_0	c_0	[.]	
a_1	a_0	b_0	$[c_0]$	d_0	c_0
b_1	a_0	$[b_0]$	a_1	d_0	b_0
...					

21.8 Consider the same time-multiplexed link as in the previous exercise. (a) Sketch a photonic TSI that uses active delay elements (similar to Figure 21.2) in which a single loop with length $1T$ is fed backward. (b) For the same switching assignment as in the previous exercise, show the operation of this TSI over 12 consecutive timeslots. Table 21.1 has a row for each new timeslot coming off the input link. It also shows which timeslots are coming off each of the active delay lines and which timeslot is entering the output link. In each row of this table, the bracketed entry occurs in the column that represents the delay element used for this row's i/o operation. The quantity inside the brackets is the data that is delivered to the output from this delay element at the same time that the input data is placed into this delay element. In the first frame the four delay lines are arbitrarily loaded in order. Then, when timeslot a from frame 1 is coming off the input link, timeslot c from frame 0 is loaded onto the output link (into the output link's timeslot 0). So, timeslot a from frame 1 takes the place of timeslot c of frame 0 on delay 2. Complete the next six rows of this table.

21.9 Consider, for a third time, the same time-multiplexed link as in the two previous exercises. (a) Sketch a photonic TSI that uses active delay elements in which multistage loops with length $2^k \times T$ are fed forward (similar to Figure 21.4). Assume that the ith timeslot is always steered to the ith delay element. (b) For the same switching assignment as in the two previous exercises, what is the binary control word for each delay element?

21.10 Consider, for a fourth time, the same time-multiplexed link as in the three previous exercises. In this case, assume that the four timeslots are organized as two timeslots in each of two subframes. (a) Sketch a three-stage photonic TSI (similar to Figure 21.6) where each TSI uses active multistage delay elements (similar to those in the previous exercise, except for $M = 2$). (b) Sketch a spatial Clos network with $n = m = r = 2$ (it's a 4×4 Benes) and find one configuration of this network for this same switching assignment that has been used in the three previous exercises. (c) Using the equivalent configuration for the three-stage photonic TSI, walk the four timeslots in one frame through the three stages.

21.11 Derive formulas for the occupancy of the delay lines in the various photonic TSI architectures. During the operation of the various photonic TSIs described in Section 21.2, what fraction of the timeslots is occupied in all the fiber delays: (a) for the passive-delay photonic TSI in Figure 21.1? (b) For the active single-loop photonic TSI in Figure 21.2? (c) For the active feed-forward photonic TSI in Figure 21.4? (d) For the three-stage photonic TSI in Figure 21.6?

21.12 Draw a graph of effective channel rate versus link transmission rate for a DiSCO-like block-multiplexed system. The y axis is the average bit rate of each TDM-ed channel, from 0 to 150 Mbps. The x axis is bit rate R that the transmitter uses in every timeslot, from 200 Mbps to 2.5 Gbps. Assume the same frame format used in DiSCO, but with an 88-ns guard band. (a) Draw lines corresponding to N timeslots per channel, for $N = 1, 2, 4, 8$, and 16. (b) Identify practical configurations of $< R, N >$ for channel rates equivalent to DS1, CEPT1, DS2, DS3, SONET OC-1,and Sonet OC-3.

21.13 Consider two fibers, each with two timeslots per frame. Let data a, b, c, and d be assigned as follows: a to timeslot 0 on fiber 0, b to timeslot 1 on fiber 0, c to timeslot 0 on fiber 1, and d to timeslot 1 on fiber 1. Using the same switching assignment from the exercises above: a comes out in timeslot 1 on fiber 1, b comes out in timeslot 1 on fiber 0, c comes out in timeslot 0 on fiber 0, and d comes out in timeslot 0 on fiber 1. Show how to implement this switching assignment on a Benes/Hunter space/time network. Hint: make c emerge without delay. There will not be frame integrity. Specify the three-bit control code that is applied to the three SDCs, in each of the two timeslots. Walk one frame through the network.

21.14 Consider a wavelength-multiplexed photonic link with four wavelengths. Suppose the switching assignment is $0 \to 3$, $1 \to 1$, $2 \to 0$, and $3 \to 2$. Specify the laser control in the WLI shown in Figure 21.13.

21.15 Consider two fibers, each with two wavelengths. (a) Draw a $<2,2> \times <2,2>$ SWS network similar to Figure 21.15 (except with two wavelengths per fiber, instead of four). Let data a, b, c, and d be assigned as follows: a to wavelength 0 on fiber 0, b to wavelength 1 on fiber 0, c to wavelength 0 on fiber 1, and d to wavelength 1 on fiber 1. Using the same switching assignment from the exercises above: a comes out on wavelength 1 on fiber 1, b comes out on wavelength 1 on fiber 0, c comes out on wavelength 0 on fiber 0, and d comes out on wavelength 0 on fiber 1. Show how to implement this switching assignment on your sketch. (b) Specify the two-bit control code that is applied to each of the AOTFs. (c) Walk one frame through the network.

21.16 Repeat the previous exercise using the Benes/Hunter architecture. (a) Draw a $<2,2> \times <2,2>$ Benes/Hunter network similar to Figure 21.17 (except with two wavelengths per fiber, instead of four). Show how to implement this same switching assignment on your sketch. (b) Specify the two-bit control code that is applied to each of the AOTFs. (c) Walk one frame through the network.

21.17 There are nine ways that space, time, and wavelength can be assigned to A, B, and C in Figure 21.19. Draw a three-dimensional five-stage $<2,2,2> \times <2,2,2>$ Marcus/Clos network for each of these nine configurations. Are any of these nine impossible, impractical, or difficult to implement? Does any configuration appear more practical than the others?

21.18 Consider an elementary $<2,2,2> \times <2,2,2>$ STW Thompson/Hunter network implemented from SDCs that are fabricated in Lithium Niobate. The network performs WDM at its inputs and outputs and has two parallel tracks, one track for each wavelength. The input channel on <fiber 1, timeslot 0, wavelength 1> must be connected to the output channel on <fiber 0, timeslot 1, wavelength 0>. Draw N separate block diagrams of this network, one diagram indicating each of the N different paths that could be used to complete the connection.

21.19 Repeat the previous exercise for an STW Thompson/Hunter network, implemented from AOTFs and with TDM at its inputs and outputs.

21.20 Consider extending the elementary $<2,2,2> \times <2,2,2>$ STW Thompson/Hunter network by extending only one of the divisions to four channels. There are three cases, corresponding to <4,2,2>, <2,4,2>, and <2,2,4>. Draw a network corresponding to Figure 21.22 for each of these three cases. Rank these three implementations by their practicality.

21.21 Repeat the previous exercise, but for an STW Thompson/Hunter network, implemented from AOTFs and with TDM at its inputs and outputs.

REFERENCES

[1] Saruwatari, M., "Progress Toward Ultrahigh-Bit-Rate All-Optical TDM Transmission Systems," *IEEE LEOS Summer Topical Meeting,* 1994.

[2] Chraplyvy, A. R., "Limitations on Lightwave Communications Imposed by Optical Fiber Nonlinearities," *IEEE Journal of Lightwave Technology,* Oct. 1990.

[3] Chraplyvy, A. R., and R. W. Tkach, "What Is the Actual Capacity of Single-Mode Fiber in Amplified Lightwave Systems?" *IEEE LEOS Summer Topical Meeting,* Jul. 1993.

[4] Chraplyvy, A. R., and R. W. Tkach, "Capacity Limits Due to Nonlinearities in Optical Networks," *IEEE LEOS Summer Topical Meeting,* 1994.

[5] Marcuse, D., A. R. Chraplyvy, and R. W. Tkach, "Dependence of Cross-Phase Modulation on Channel Number in Fiber WDM Systems," *IEEE Journal of Lightwave Technology,* May 1994.

[6] Simmons, J. M., and A. A. M. Saleh, "The Value of Optical Bypass in Reducing Router Size in Gigabit Networks," *Proc. 1999 IEEE Int. Conference on Communications,* (Session S16A-4: Information Transport Using Optical Architecture), Vancouver, Canada, Jun. 6 – 10, 1999.

[7] Alexander, S. B. et al., "A Precompetitive Consortium on Wideband All-Optical Networks," *IEEE Journal of Lightwave Technology,* Vol. 11, No. 5/6, May/Jun. 1993, pp. 714 – 735.

[8] Thompson, R. A., and P. P. Giordano, "An Experimental Photonic Time-Slot Interchanger Using Optical Fibers as Reentrant Delay-Line Memories," *IEEE Journal of Lightwave Technology,* Vol. LT-5, No. 1, Jan. 1987, pp. 154 – 162. © 1987 IEEE. Portions reprinted with permission.

[9] Thompson, R. A., "Optimizing Photonic Variable-Integer-Delay Circuits," in *Photonic Switching,* pp. 158 – 166, T. K. Gustafson and P. W. Smith (eds.), Berlin, Germany: Springer-Verlag, 1988.

[10] Thompson, R. A., "Architectures with Improved Signal-to-Noise Ratio in Photonic Systems with Fiber-Loop Delay Lines," *IEEE Journal on Selected Areas in Communications,* Vol. 6, No. 7, Aug. 1988, pp. 1096 – 1106. © 1988 IEEE. Portions reprinted with permission.

[11] Thompson, R. A., et al., "An Experimental Modular Switching System with a Time-Multiplexed Photonic Center-Stage," *Proc. Photonic Switching Topical Meeting, OSA,* Salt Lake City, UT, Mar. 1 – 3, 1989, pp. 92 – 93.

[12] Thompson, R. A., J. J. Horenkamp, and G. D. Bergland, "Photonic Switching of Universal Timeslots," *Proc. XIII Int. Switching Symposium,* Stockholm, Sweden, May 27 – Jun. 1, 1990, Vol. 1, pp. 165 – 170.

[13] Hunter, D. K., and D. G. Smith, "New Architectures for Optical TDM Switching," *IEEE Journal of Lightwave Technology,* Vol. 11, No. 3, Mar. 1993, pp. 495 – 511. © 1993 IEEE. Portions reprinted with permission.

[14] Hunter, D. K., P. J. Legg, and I. Andonovic, "Architecture for Large Dilated Optical TDM Switching Networks," *IEE Proc. J,* Vol. 140, No. 5, Oct. 1993, pp. 337 – 343.

[15] Thompson, R. A., and D. K. Hunter, "Elementary Photonic Switching Modules in Three Divisions," *IEEE Journal on Selected Areas in Communications,* Vol. 14, No. 2, Feb. 1996, pp. 362 – 373. © 1996 IEEE. Portions reprinted with permission.

[16] Kobayashi, H., and I. P. Kaminow, "Dualities Among Space, Time and Wavelength in All-Optical Networks," *Proc. Photonics Switching Meeting,* Salt Lake City, UT, Mar. 1995.

[17] Sharony, J., T. E. Stern, and Y. Li, "The Universality of Multidimensional Switching Networks," *IEEE/ACM Transactions on Networking,* Vol. 2, No. 6, Dec. 1994.

[18] Cheung, K. W., "Acousto-optic Tunable Filters in Narrowband WDM Networks: System Issues and Network Applications," *IEEE Journal on Selected Areas in Communications,* Vol. 8, No. 6, Aug. 1990, pp. 1015 – 1025.

[19] Cheung, K. W., et al., "Wavelength-Selective Circuit and Packet Switching Using Acousto-Optic Tunable Filters," *Proc. Globecom '90,* San Diego, CA, Dec. 2 – 5, 1990, pp. 1541 – 1547.

[20] Kanayama, Y., N. Goto, and Y. Miyazaki, "Integrated Optical Switches for Wavelength-Division-Multiplex Communication Using Collinear Acousto-Optic Interactions," in *Photonic Switching II,* pp. 245 – 248, K. Tada and H. S. Hinton (eds.), Berlin, Germany: Springer-Verlag, 1990.

[21] Thompson, R. A., "An Elementary Three-Division Photonic Switching Network Architecture," *Proc. 1st IEEE Int. Workshop on Broadband Switching Systems,* Poznan, Poland, Apr. 19 – 21, 1995, pp. 201 – 209.

SELECTED BIBLIOGRAPHY

Bouillet, E., and K. Bala, "The Benefits of Wavelength Interchange in WDM Rings," *Proc. Int. Conference on Communications '97,* (Session on WDM Optical Networks), Montreal, Canada, Jun. 8 – 12, 1997.

Brackett, C. A., "Dense Wavelength Divison Multiplexing Networks: Principles and Applications," *IEEE Journal on Selected Areas in Communications,* Vol. 8, No. 6, Aug. 1990, pp. 948 – 864.

Brackett, C. A., et al., "A Scalable Multi-Wavelength Multi-Hop Optical Network: A Proposal for Research on All-Optical Networks," *IEEE Journal of Lightwave Technology,* May/Jun. 1993.

Chidgey, P. H., "Multi-Wavelength Transport Networks," *IEEE Communications Magazine,* Vol. 32, No. 12, Dec. 1994, pp. 28 – 35.

Djupsjobacka, A., "Time Division Multiplexing Using Optical Switches," *IEEE Journal on Selected Areas in Communications,* Vol. 6, No. 7, Aug. 1988, pp. 1227 – 1231.

Fabianek, B., et al., "Optical Network Research and Development in European Community Programs: From RACE to ACTS," *IEEE Communications Magazine,* Vol. 35, No. 4, Apr. 1997, pp. 50 – 56.

Gerla, M., et al., "Minimum Distance Routing in the Bidirectional Shufflenet," *Proc. IEEE Infocom '98, Vol. 1,* (Session 1D: Switching I), San Francisco, CA, Mar. 29 – Apr. 2, 1998, pp. 102 – 109.

Gerstel, O., R. Ramaswami, and G. Sasaki, "Cost Effective Traffic Grooming in WDM Rings," *Proc. IEEE Infocom '98, Vol. 1,* (Session 1C: WDM Networks), San Francisco, CA, Mar. 29 – Apr. 2, 1998, pp. 69 – 77.

Goodman, M. S., and H. Kobrinski, "Dynamic Wavelength Tuning for Broadband Optical Packet Switching," in *Photonic Switching II,* pp. 258 – 265, K. Tada and H. S. Hinton (eds.), Berlin, Germany: Springer-Verlag, 1990.

Green, P. E., et al., "All-Optical Packet-Switched Metropolitan-Area Network Proposal," *IEEE Journal of Lightwave Technology,* May/Jun. 1993.

Harai, H., M Murata, and H. Miyahara, "Performance of All-Optical Networks with Limited-Range Wavelength Conversion," *Proc. Int. Conference on Communications '97,* (Session on WDM Optical Networks), Montreal, Canada, Jun. 8 – 12, 1997.

Krishnaswamy, R. M., and K. N. Sivarajan, "Design of Logical Topologies: A Linear Formulation for Wavelength Routed Optical Networks with No Wavelength Changers," *Proc. IEEE Infocom '98, Vol. 2,* (Session 7D: Optical Networks), San Francisco, CA, Mar. 29 – Apr. 2, 1998, pp. 919 – 930.

Liu, G., K. Y. Lee, and H. F. Jordan, "Time Division Multiplexed de Bruijn Network and ShuffleNet for Optical Communications," *Proc. IEEE Infocom '94, Vol. 3,* Toronto, Canada, Jun. 12 – 16, 1994, pp. 1244 – 1251.

Lutkowitz, M., "The Advent of WDM and the All-Optical Network: A Reality Check," *Telecommunications,* Vol. 32, No. 7, Jul. 1998.

Nakazawa, M., "Telecommunications Rides a New Wave," *Photonics Spectra,* Vol. 30, No. 2, Feb. 1996.

Nesset, D., T. Kelly, and D. Marcenac, "All-Optical Wavelength Conversion Using SOA Nonlinearities," *IEEE Communications Magazine,* Vol. 36, No. 12, Dec. 1998, pp. 56 – 61.

Nosu, K., and H. Toba, "Optical FDM Technologies for Future Lightwave Networks," in *Photonic Switching II,* pp. 250 – 253, K. Tada and H. S. Hinton (eds.), Berlin, Germany: Springer-Verlag, 1990.

O'Mahony, M. J., "Optical Multiplexing in Fiber Networks: Progress in WDM and OTDM," *IEEE Communications Magazine,* Vol. 33, No. 12, Dec. 1995, pp. 82 – 88.

"Optical Multiplexed Networks," special issue of *IEEE Communications Magazine,* Vol. 32, No. 12, Dec. 1994.

Parsons, N. J., N. F. Whitehead, and M. Erman, "Multidimensional Photonic Switches," in *Photonic Switching II,* pp. 364 – 369, K. Tada and H. S. Hinton (eds.), Berlin, Germany: Springer-Verlag, 1990.

Pease, R., "NTT Brings Fiber-to-the-Curb via Passive Optical Networking," *Lightwave,* Aug. 1998.

Roberts, E., "On a New Wavelength," *Data Communications,* Vol. 26, No. 15, Nov. 1997.

Ramaswami, R., "Multiwavelength Lightwave Networks for Computer Communication," *IEEE Communications Magazine,* Feb. 1993, pp. 78 – 88.

Roberts, E., "On a New Wavelength," *Data Communications,* Vol. 26, No. 15, Nov. 1997.

Ryan, J. P., "WDM: North American Deployment Trends," *IEEE Communications Magazine,* Vol. 36, No. 2, Feb. 1998, pp. 40 – 44.

Senior, J. M., M. R. Handley, and M. S. Leeson, "Developments in Wavelength-Division Multiple-Access Networking," *IEEE Communications Magazine,* Vol. 36, No. 12, Dec. 1998, pp. 28 – 36.

Sharma, V., and E. A. Varvarigos, "Limited Wavelength Translation in All-Optical WDM Mesh Networks," *Proc. IEEE Infocom '98, Vol. 2,* (Session 7D: Optical Networks), San Francisco, CA, Mar. 29 – Apr. 2, 1998, pp. 893 – 901.

Spirit, D. M., A. D. Ellis, and P. E. Barnsley, "Optical Time Division Multiplexing: Systems and Networks," *IEEE Communications Magazine,* Vol. 32, No. 12, Dec. 1994, pp. 56 – 62.

Subramaniam, S., M. Azizoglu, and A. K. Somani, "On the Optimal Placement of Wavelength Converters in Wavelength-Routed Networks," *Proc. IEEE Infocom '98, Vol. 2,* (Session 7D: Optical Networks), San Francisco, CA, Mar. 29 – Apr. 2, 1998, pp. 902 – 909.

Tang, K. W., "CayleyNet: A Multihop WDM-Based Lightwave Network," *Proc. IEEE Infocom '94, Vol. 3,* Toronto, Canada, Jun. 12 – 16, 1994, pp. 1260 – 1267.

Thompson, R. A., "Architectures for Photonic Space- and Time-Switching Systems in Lithium Niobate," *Proc. 1990 Int. Topical Meeting on Photonic Switching,* Kobe, Japan, Apr. 12 – 14, 1990, pp. 208 – 211.

Thompson, R. A., "The Dilated Slipped Banyan Switching Network Architecture for Use in an All-Optical Local-Area Network," *IEEE Journal of Lightwave Technology,* Vol. 9, No. 12, Dec. 1991, pp. 1780 – 1787.

Thompson, R. A., "The Impact of Fiber Nonlinearities on the Architecture of Optical Data Networks," *Proc. 4th Int. Conference on Telecommunications Systems,* Nashville, TN, Mar. 21 – 24, 1996, pp. 500 – 502.

Trischitta, P. R., and W. C. Marra, "Applying WDM Technology to Undersea Cable Networks," *IEEE Communications Magazine,* Vol. 36, No. 2, Feb. 1998, pp. 62 – 66.

Tucker, R. S., et al., "4-Gb/s Optical Time-Division Multiplexed System Experiments Using Ti:$LiNbO_3$ Switch/Modulators," in *Photonic Switching,* pp. 208 – 211, T. K. Gustafson and P. W. Smith (eds.), Berlin, Germany: Springer-Verlag, 1988.

Turley, S., "Technology Close-Up — Exploring the New Communications Landscape," *Photonics Spectra,* Vol. 32, No. 3, Mar. 1998.

Van Belle, M., "Photonics in Telecommunications — Just Over the Horizon Waits the All-Optical Network," *Photonics Spectra,* Vol. 32, No. 8, Aug. 1998.

Yao, J. G., "Transmission Characteristics of a Semiconductor Laser Amplifier Based Optical Delay Line Memory," *Proc. Int. Conference on Communications '97,* (Session on Advanced Optical Transmission Technologies), Montreal, Canada, Jun. 8 – 12, 1997.

Yates, J., J. Lacey, and D. Everitt, "Blocking in Multiwavelength TDM Networks," *Proc. 4th Int. Conference on Telecommunications Systems,* Nashville, TN, Mar. 21 – 24, 1996, pp. 535 – 541.

22

Network of the Future

"There is a tide in the affairs of men which, taken at the flood, leads on to fortune; omitted, all the voyage of their life is bound in shallows and in miseries. On such a full sea are we now afloat, and we must take the current when serves or lose our ventures." — from *Julius Caesar*, by William Shakespeare

I am often asked questions like these:

"When will we finally all have ISDN?"
"Describe computing and communications 10 years from now."
"When will we have fiber to the home?"

I am always careful to preface any answer to such questions with the caveat that I am willing to predict the technology and services "that are possible" in some given time frame. But, I will never predict which technologies and services "will be available" in some given time frame. Technology is predictable. The telecom industry is not. Consistent with this caveat, this chapter describes what could happen, and not what will happen.

The book closes, finally, by looking into the future. This chapter's first five sections are extremely philosophical, but bear with it. Section 22.1 frames the discussion of the future in the context of the past, and the second section discusses integrated networks. Section 22.3 describes the evolving user, including a redefinition of universal service, and the fourth section reviews some of the regulatory choices we must make before any future planning can even be considered. Section 22.5 is an essay on how synergies among bandwidth, intelligence, and openness enable new services. This chapter's last three sections describe three extreme networks. Sections 22.6 and 22.7 propose two future architectures, for the local loop and for the national network, respectively. The final section specifies a global telecommunications network and a "Vail-like" commitment to achieve it.

22.1 HISTORICAL CONTEXT

This section summarizes the evolving PSTN and the lessons we should learn.

22.1.1 The Evolving PSTN

Some of us Americans remember the first-generation Public Switched Telephone Network (PSTN) — characterized by analog electrical transport, manual control and signaling, and patch-cord switching. Initially, it was even more primitive than its contemporary, the Public Telegraph Network, in its physical implementation and its operation. While the telegraph network was well established before the telephone network was even started, the telephone network evolved with new technology and grew rapidly. Meanwhile, the telegraph network remained stagnant and it declined.

Beginning about 1890 and ending about 1960, the North American and European PSTN infrastructures gradually evolved to an electromechanical generation — characterized by dial telephones, inband signaling, relay-logic control, and automatic switching. Beginning in the mid-1960s, these PSTNs began to evolve to a digital electronic generation: first to digital transmission, digital computer control, and out-of-band signaling; then to digital switching equipment; and then to networked signaling.

In the United States, digital transmission and software-controlled switching offices are almost universal now, most telephone companies in the United States have replaced most of their analog switches (including the #1 and #1A ESS) by digital switches, and the evolution to networked signaling is almost completed. Very few U.S. subscribers are still served by central offices that do not use SS7 signaling. Worldwide, the evolution is further behind, but is moving along quite quickly. Many developing nations have modern telephone technology but lack the revenue to invest in the loop plant that gives their citizens access to it. So, wireless telephones are much more popular outside the United States than inside.

By the 1940s, the FCC allocated radio spectrum for a few regional channels for broadcast voice and video services. While radio spectrum is not manufactured, it is a precious resource and should be viewed as part of the nation's telecommunications infrastructure. An alternate network, using satellites for wide area communications and coaxial cable for local distribution, has evolved and many users have migrated to CATV, and away from this radio infrastructure, for their broadcast services.

Data networking started with simple wiring for connecting computer terminals to time-sharing main frames. As intelligence migrated from the host to the terminal, the data networks evolved to LANs, like Ethernet, with statistical multiplexing and high-burst data rates. Wide area data networking, allowing LANs to interconnect, followed in both the private and public domains. Data networks have evolved rapidly in recent years.

As the world's analog-to-digital migration is completing, the PSTN currently finds itself in the early stages of, not one but, four new important evolutions. These simultaneous evolutions are not only mutually dependent, but they are strangely coupled — sometimes supporting each other, and sometimes competing with one another.

- Beginning in the early 1980s, the infrastructure began to evolve to a photonic generation, characterized by physical transport based on optical fiber. As just discussed, photonics could penetrate much deeper.
- Beginning in the mid-1980s, wireless access to the PSTN evolved from cordless telephones, mobile radio, and pagers to cellular systems with car phones and complex systems that handle call hand-off automatically. Low earth orbit satellite systems may serve to completely bypass the conventional land-based infrastructure.

- Beginning in the late 1980s, the infrastructure began to evolve to an intelligent network — characterized by distributed processing of the network's stored-program control and open-network access to third-party applications providers.
- Beginning in the mid-1990s, voice traffic began to migrate to packet-switched networks like frame relay, IP (the Internet), and asynchronous transfer mode (ATM).

22.1.2 Lessons from History

The history of the evolutions of these networks yields many interesting and relevant observations.

The *manual-to-mechanical* and *mechanical-to-digital* migrations were both very significant to U.S. commerce. Both contributed powerful upgrades to our infrastructure and global products from the U.S. telecommunications industry. Both of these migrations were driven predominantly by technology push, with very little influence by market pull. The *electronic-to-photonic* migration seems to be sharing these characteristics.

Very little of this has occurred without government involvement. Federal, state, and local governments have supported and regulated the monopolies for telephone and CATV. The FCC controls usage of the radio infrastructure. Federal funding built the Internet, and federal regulation continues to force the LECs to subsidize it. Future regulatory issues are probably more complex and more important than future technological issues. While total deregulation is an ideal, today's piecemeal deregulation is very awkward. But, some regulation is probably necessary to guarantee the provision of universal service.

Successive generations have been directly compatible with more media than their predecessors. But, little true compatibility actually occurred either because regulation prevented a merger of infrastructures or there was not enough economic justification. While the electromechanical generation could have easily integrated the telegraph medium, regulation prevented it. While the digital generation is more compatible with the data medium than the mechanical generation was, the migration was not pulled by this increased compatibility. Both the data medium and the digital generation of the PSTN were born from the same underlying technology — digital electronic silicon integrated circuits. Compatibility seems to be merely a temporal coincidence.

The various uni-media networks have been made indirectly compatible with other media by interface equipment. Modems, fax machines, and video compression equipment allow the PSTN to be used for data, image, and video, respectively. While a network, which currently requires a modem to interface data, could evolve to a truly integrated voice/data network via the ISDN paradigm, it has been very slow to do so. Users seem willing to use modems to interface with the PSTN or to use a data-only uni-media network. Packetized voice interfaces and digital imaging equipment allow data networks to be used for voice and image, respectively. Because of their broadcast natures, the radio and CATV infrastructures have not been used much for voice or data. Now, however, it is clear that compatibility with the video medium is an important factor that is pulling the electronic-to-photonic migration.

While the technology for transport gradually evolved from analog to digital, the physical layer (26-gauge copper twisted pair) was largely unchanged for 100 years. In the evolution to photonics, transport's physical layer is changing (the technology is still digital) from copper wire to optical fiber.

The migration of the switching function has consistently lagged the migration of the corresponding transport function. Switching is more complex than transport. When transport migrates, it changes its technology (for example, from analog to digital) and maybe its physical implementation (for example, from copper to fiber). However, when switching migrates, it changes not only its technology and physical implementation, but it may even change its paradigm (for example, from circuits to packets). Whether the electronic-to-photonic migration changes the switching function with its transport function, like the previous migrations did, depends on whether the enabled applications can pull it.

Not only has the rate of successive migrations accelerated, but the time to complete a generation changeover has shortened. The manual-to-mechanical migration took twice as long to complete as the mechanical-to-digital migration. While the electronic-to-photonic migration of the transport function has moved rapidly in the long distance plant, the total migration of transport to photonics may take a very long time because of the need to raise the capital to fiber the loop. In contrast, migration to wireless interfaces and to the Intelligent Network could be quick — because they're being pulled by the market. Perhaps, we're migrating quicker, not because technology is accelerating, but because the market wants it.

Negroponte has observed a remarkable transition. Years ago, narrowband data (voice) was carried by tether (wire) and broadband data (video) by wireless (radio). In the near future, narrowband data might be wireless and broadband data might be carried by tether (fiber).

The Modified Final Judgment of the 1956 Consent Decree, agreed upon in Judge Green's court for 1984, made a permanent and controversial change to the infrastructure. While some people would like to go back to the "good old days" of Ma Bell, it is still too early to fairly evaluate the technological and economic impact of divestiture. While the causality of divestiture and competitiveness could be argued, they are correlated. Under the old rules, we probably would not be experiencing four new concurrent network evolutions.

It will be interesting to observe the current migrations and monitor their places in history. But, it is more relevant and more interesting to try to decide which outcomes are in the nation's best interest, and in humanity's best interest, and to encourage and influence the network's evolution toward these outcomes.

22.2 INTEGRATION

This section addresses this 20-year-old *integration question*. We won't answer it here because answers are hard to reach. But, we'll try to shed light on the question in hopes that it will be addressed logically and in the consumers' best interest. After reviewing the historical context and motivation in the first two subsections, the third subsection describes different categories of integration. The last two subsections discuss the relevant issues: cost, in the fourth subsection, and other issues, possibly even more important than cost, in the fifth.

22.2.1 Historical Context

For many years, the voice and data segments of the telecommunications industry each operated its own paradigm over the same physical infrastructure. Telecommunications engineers have debated the relative merits of circuit switching versus packet switching for decades. Never reaching any conclusion, this debate quieted in the late 1980s and the two camps simply agreed to disagree. Voice communications continued to use circuit switching and computer communications continued to use packet switching.

The voice industry has used circuit switching for more than 100 years. Even though it has changed and continues to change, such as converting to digital and now to optics, the PSTN has always been optimized for voice traffic. Even today, in the United States, about 80% of the traffic on the PSTN is voice. While the various telephone companies want to provide data service because data shows more growth than voice, they are slow to physically change their infrastructures. A whole generation of switching experts has come to believe that data should either fit into a voice paradigm or find a network of its own. This attitude is changing but not necessarily for the right reasons.

The computer networking industry has been using packet switching for about 25 years. Packet-switched networks are optimized for bursty data traffic. During the last 20 years, data traffic has grown much more rapidly than voice and as an ever-increasing percentage of the total traffic in the PSTN. While many gurus predict this rate will continue, there are also reasons to believe that it will flatten [1]. A whole generation of computer scientists has been taught that circuit switching is an anachronism that a bunch of dinosaurs insist on perpetuating for voice service. This attitude has never changed.

So, today's telecommunications networks are *uni-media.*

- *Voice.* The PSTN, while constantly evolving and now composed of multiple parallel competing networks, has always been optimized for the voice medium. While broadcast voice has been carried by radio for many years, recent advances in cellular technology have provided useful and popular wireless access to the PSTN for point-to-point voice service. Separate networks for point-to-point voice are possible over CATV and over low earth orbit satellite systems. Modems and fax machines are interfaces that allow data transfer and image transfer over the ubiquitous PSTN.
- *Video.* A simple system consisting of several one-way 5-MHz FDM channels, allocated from the nation's radio spectrum, has been used effectively for the conventional broadcast video medium. Currently, users are gradually migrating to a newer system that uses satellites and regional CATV networks. Furthermore, high-definition television (HDTV) is easier to implement on the CATV system. While existing video networks offer broadcast service, a limited amount of video jukebox and videoconferencing traffic will require a kind of point-to-point service, called narrowcast.
- *Data.* Data networks have evolved from terminal-to-host networks, optimized for keystrokes and data entry, to LAN-based file-server networks, optimized for electronic mail (e-mail) and file transfer. These LANs are now interconnected by wide area and global area long distance networks, like the Internet, to extend LAN applications and to support important new applications, like electronic funds

transfer (EFT) and electronic data interchange (EDI). Image communications, especially high-quality medical images, is a new source of file transfer traffic.

Should these uni-media networks be integrated? A telecommunications paradigm, called the Integrated Services Digital Network (ISDN), emerged in the mid-1970s out of a perceived need to support voice and data applications over a common networking paradigm. For a variety of reasons, ISDN has been slow to emerge as a product. Today, emerging applications — teleconferencing, telecommuting, groupware, and other forms of so-called multimedia communications — suggest the need for one integrated network paradigm that can support voice, computer data, and video. A new networking paradigm, asynchronous transfer mode (ATM), has emerged from this perceived need. It remains to be seen if ATM meets a real need or will be as slow to evolve as ISDN has been.

22.2.2 Motivation

New and changing applications and the predicted increased demand for more bandwidth suggest that a new telecommunications infrastructure will be needed for the data and video media; and possibly even for voice, especially voice applications that are integrated with data and/or video. Whether this new infrastructure supports a single networking paradigm, like ATM, or supports various virtual uni-media networks, the cost of constructing a nationwide broadband telecommunications infrastructure suggests that there won't be too many of them (maybe just one) and that we'd better plan it well.

There is nothing inherently wrong with having separate uni-media networks. Since they are multiplexed within the same physical network, this infrastructure has economy of scope and scale. But, the separate uni-media networks are optimized for their applications. In the early 1990s, ATM advocates proposed to integrate all applications under not just a common infrastructure, but also a common switching paradigm. But, a lot has changed, recently.

- ATM's promise in the WAN may be difficult to implement.
- Ethernet and IP are just not fading away quietly.
- Transmission bandwidth is much cheaper than it used to be.
- Data compression is more expensive than was predicted.
- The new services promised by the emerging Intelligent Network (Chapter 18) are predicated on call switching.
- The embedded base of switching systems is too new and too expensive to simply replace.
- Carriers are reluctant to deoptimize their networks away from voice traffic.

Thus, it is even more important, now, to understand the tradeoffs between circuit and packet switching, to find out if integration might cause some loss of net throughput, and, if so, by how much.

22.2.3 Categories of Integration

Network integration is an attempt to bring all the media to one common network. But, what are we trying to integrate? Three different categories of integration are described.

Interfacing. While a medium can be transmitted over a network that is not optimal for it, some kind of interface may be required. Because of ubiquity and ease of access, the PSTN has been used for many media besides voice. Data has been transmitted over the PSTN, even a digital PSTN, for many years by the use of a modem interface in order to translate the data spectrum into the PSTN's passband. Fax equipment is really just an interface between the image medium and an audio network. Video has also used the PSTN for many years, classically interfacing to 45-Mbps DS3 channels through A-to-D conversion equipment. More recently, reduced-quality video or the use of video compression equipment on the user's premises allows the use of a smaller channel. For most television transmission and MAN and WAN data applications, the PSTN has simply been a vendor of leased lines for private networks. Most user-oriented data and fax applications have simply used circuit switching.

Now, packet-switched voice is an important consideration. ATM is probably better suited for voice, but IP is more common. This section will make the point several times that packet-switched voice is only one component (the easiest one) of packet-switched telephony. Besides requiring an interface, it also typically requires assigning high priority to voice packets within the network. While packet-switched voice is more feasible when speech compression is used, the cost of user equipment is increased, and significantly if quality is not compromised. People are even using voice digitizers to create large files and then shipping these files by Internet. While this is inconvenient for real-time two-way voice conversation, it is a cheaper way to implement a one-way call overseas. This economics is artificial because users must pay for overseas use of the PSTN directly, but they don't pay anything directly for use of the Internet. While packet-switched fax is not common, it seems to be a more rational integration than packet-switched voice.

Super-Medium. A second form of integration is to merge the media together into one super-medium. Multimedia is a new single medium whose applications are each a combination of conventional voice, data, image, and video services into one integrated service. But, there is a subtle and important difference between *multimedia* and *multiple media.*

The characteristics of multimedia traffic are not an average of voice, data, image, and video traffic characteristics. Multimedia will have its own characteristics and it is too new for us to understand its behavior. A multimedia network is one that would be appropriate for multimedia traffic, when we finally come to understand it. A multiple-media network is one that is appropriate for each separate medium and, hopefully, with some efficiency.

Super-Network. The most obvious form of integration is to build some new network that is suitable for multiple media. The ISDN was, and still is, an attempt to integrate voice and data onto a common network. We must ask why it has not yet been commercially successful, and if it ever does become successful, we must ask why it took so long. ISDN suffers from a lack of market pull and the typical cost of $17 a month more than regular telephone service has not helped pull ISDN into the home.

But, beyond cost and other nontechnical issues, a difficult technical question remains. If one is going to construct a single voice/data infrastructure, should its architecture and switching paradigm be:

- Optimized for the majority of the traffic (voice)?
- Optimized for the high-growth traffic (data)?
- Optimized for the traffic one hopes to attract (video)?
- Some compromise that's not optimal for any of them?

The overwhelming concern among many experts is that such a super-network may end up like any jack of all trades — he is often the master of none. We must take care that our attempt to architect an infrastructure that is optimal for all data types doesn't give us, instead, an infrastructure that is optimal for no data types.

	Public LEC/LAN	Public IXC/WAN	Private network
Equipment	?	?	?
Bandwidth	?	?	?
Access	?	?	?
OAM&P	?	?	?
Change	?	?	?

Figure 22.1. Cost framework for evaluating integration.

22.2.4 Cost

Cost is an extremely complicated issue, as complicated as the technology itself. Figure 22.1 is a blank spreadsheet for evaluating different components of cost in different types of networks. The figure implies that the cost space is two-dimensional, with 5 components $\times$ 3 types = 15 entries. While there are probably more than five relevant components, and there may be more than three relevant types of networks, these additions only expand the spreadsheet in their respective dimensions.

Cost is even more complicated than this because it has at least one additional dimension. The cost space in Figure 22.1 really has 30 entries, because the entire spreadsheet has to be evaluated under two different kinds of integration: (1) physical only and (2) physical/logical. In *physical only* integration, different types of networks share the same physical infrastructure — the kind of integrating we do today. In *physical/logical* integration, there is only one type of network — the kind of integrating advocated for ISDN for more than 20 years and now advocated for ATM and for IP. Let's discuss just a few of the entries.

- *Bandwidth in a public IXC/WAN.* Often the bandwidth comparisons are like comparing apples to oranges. If packet switching's statistical multiplexing really does save relevant bandwidth, then bring back TASI. If the speech compression typically found in integrated networks really does save relevant bandwidth, then compress the circuit-switched data.

- *Access in a public LEC/LAN.* The only fair cost basis is the end user's net cost. Cost of access can't ignore any new equipment the subscriber has to buy. Access can be integrated without integrating the entire network. Their costs need to be determined separately. Many of the experts who think an overall integrated network is silly are big advocates of integrating the access.
- *OAM&P in any public network.* The carrier's OAM&P cost is surprisingly high. The argument that one network is cheaper to manage than *N* networks is true at the logical level of the network. The physical infrastructure has to be managed regardless of how many logical networks are embedded in it. And, if managing one network really is simpler and cheaper than managing several, then revert to one network, the one that is already easiest to manage — the PSTN.
- *Enterprise private networks.* EPNs probably make the most sense to integrate. The enterprise's cost is the carrier's price. It makes sense for an enterprise to integrate all traffic onto whatever logical network the carriers are discounting. Integrating voice onto frame relay is simply responding to price.
- *Changing any public network.* This includes migrating and scalability. A large up-front investment is required to create an integrated network and to abandon an established logical network. Then, annual growth of traffic load requires that bandwidth and new equipment be added to the network. If growth is predicted to be small, then integration seems unmotivated. If growth is predicted to be large, then a scalable architecture is important. Scalability is discussed in the next subsection.

Most of the pro-integration arguments we read in the trade press are based on cost. But, I've never seen real cost figures, have you? You're not going to see them here either. We're not going to put any dollar entries into the spreadsheet of Figure 22.1 — it's far too difficult a task. Besides, nobody reveals their real costs anyway — at least, not to me. The point made here is that those who claim that integration saves cost, or doesn't, have probably oversimplified.

Finally, we have to ask, is cost even relevant? In the United States, the IXCs are charging a flat rate to call anywhere. If channels really aren't cheap enough to squander this way, the IXCs have overbuilt and/or the market has forced them into this pricing structure. If channels, even in the WAN, really are a commodity, then cost may be irrelevant.

22.2.5 Other Issues

Other issues include telephony, scalability, performance, reliability, quality, and format.

Telephony. Data-over-ISDN was a small subset of, and the easy part of, computer networking over ISDN. Developing technology that puts a large file into a D-channel is the easy part. Providing an economical end-to-end service for a significant number of communicating computer users is the hard part. Similarly, (Section 19.5) voice-over-IP is a small subset of Internet telephony and voice-over-ATM is a small subset of ATM telephony. Developing technology that puts PCM samples into a packet or cell is the easy part. Telephony also requires addressing, signaling, universal service, hundreds of complicated features, and some extremely complex services — like call forwarding, 800 numbers, calling party ID, cellular, and local number portability.

Telephony over any form of integrated network requires either less than total integration, compatibility with SS7 and the IN, or a whole lot of hard work. Telephony over any form of end-to-end integrated network requires some way of serving the huge number of people who don't have a PC at home and some way of interfacing to the $8 telephone.

Scalability. We read everywhere about how big the Internet is. But, Internet traffic is a relatively small part of all computer communications traffic in the United States and computer communications traffic is small compared to voice traffic. The migration of all forms of computer communications, or all forms of total communications, to one transmission/switching paradigm requires that the "winning" paradigm be scalable. If it doesn't scale well, not only will it fail to support the applications it has won from other networks, it will disrupt its own applications.

The sum of all computer communications traffic is still very small compared to all voice traffic. Even in the United States, where we have a high density of computers, voice still represents over 80% of all the data in the infrastructure. The migration of telephone traffic to one packet-switched integrated network will really disrupt this one network unless its architecture is very carefully designed to be scalable.

Neither of the contenders for this integrated network show the potential to scale well under an order-of-magnitude increase in offered load. Only the manufacturers of IP routers or of ATM switches are looking forward to this magnitude of increase. Reread Section 17.3.5 where telecommunications traffic is compared to automotive traffic.

Performance Tradeoffs. In specifying a network, bit error rate trades off against delay. This tradeoff is seen at the link layer, if error correction is deemed necessary. If we try to decrease a network's BER from 10^{-6} to 10^{-12} to improve its performance for large file transfers, the corresponding increased delay may make this same network less acceptable for voice. This tradeoff is also seen at the physical layer, where frame buffers are enlarged to lower the probability of dropping a frame. In the extreme, even physical integration may be problematic. If we demand high performance, we could conceivably need, not just different switching paradigms for voice and data, but even different physical infrastructures.

Reliability. For over 100 years, telephone switching systems have been designed against a very strict reliability requirement. The specification for the 1E was "1 hour of down-time in 40 years." This requirement was accepted throughout the Bell System as part of its internal motivation to provide the best service possible. While telecommunications under the old Bell System had its flaws (for example, Ma Bell would have never given us the Internet), the Bell System did regulate its own reliability and did become extremely embarrassed whenever an outage occurred. That oversight is gone now. Nobody's governing the network's reliability any longer and this move to an integrated network should be a concern to anyone who thinks the PSTN should be reliable.

Quality. We digitized audio recording because it reduced noise and gave us much more fidelity — it improved quality. We're digitizing television, even at greatly increased bandwidth, because it will give us better picture quality. When we digitized telephony to the standard B-channel, noise was greatly reduced, but so was fidelity. We gave the local call, which used to have about 8 kHz of bandwidth, the same low fidelity as the toll call. But, we did this in the 1960s because digital bandwidth was expensive.

Now, more than a generation later, digital bandwidth can be squandered and we're proposing to further degrade the quality of the standard telephone channel. We should be willing to sacrifice even more audio quality if we get something in return — like the mobility we get from a cell-phone. But, why should we be willing to sacrifice the quality of the POTS channel? Shouldn't the PSTN be evolving to a network that provides a standard telephone channel whose fidelity is equivalent to 12 kHz of analog bandwidth instead of to 3.6 kHz of analog bandwidth?

IP or ATM. IP is the most universal of the integrated network candidates and has a well-established addressing scheme. ATM is better adapted to support the variety of requirements, but ATM is complex and is not as universally deployed.

Summary. The case for integration is still evolving and will ultimately depend on the business justification.

22.3 EVOLVING USER

Chapter 15 described the general users of traditional telecommunications systems. This section distinguishes two specific kinds of users: the maximum user is identified by marketeers as the potential purchaser of new telecommunications products, and the minimum user is identified by regulatory bodies to receive service that is (1) basic, (2) universal, and (3) subsidized. This section's first subsection describes today's different networks and our maximum users' access to them. The second subsection describes how this maximum user goes about getting served (anyway) by a telecom industry that may not be trying to meet his needs. The third subsection discusses the minimum user and a proposed redefinition of universal service.

22.3.1 Network Access

Figure 22.2 shows a four-tiered picture of today's network architecture. The figure has four tiers because the user is part of the network. The four layers are (1) the user, (2) access, (3) local switching, and (4) the WAN. The figure shows five different kinds of access channels: narrowband wireless, data LAN, POTS loop, cable, and broadband wireless. The figure shows seven different gateways between the LAN and the WAN: a low earth orbit satellite, a cellular switching office, a data WAN gateway, a conventional telco switching office, a CATV head end, a terrestrial broadcast antenna, and a geo-synchronous satellite. From the user's perspective, these are seven different hubs, representing seven different companies (or sets of companies), that are reached from the five different access networks.

- Currently, we use a narrowband wireless channel for direct access to a LEOS, a cellular (voice) CO, or a data WAN gateway (via modem or CDPD).
- Currently, we use data LANs for direct access to data WANs, but as LAN speeds evolve upward and LAN protocols become more appropriate for voice and video, we're beginning to see people using LANs, not only to access the traditional voice and video gateways, but also to obtain their voice and video service over a data

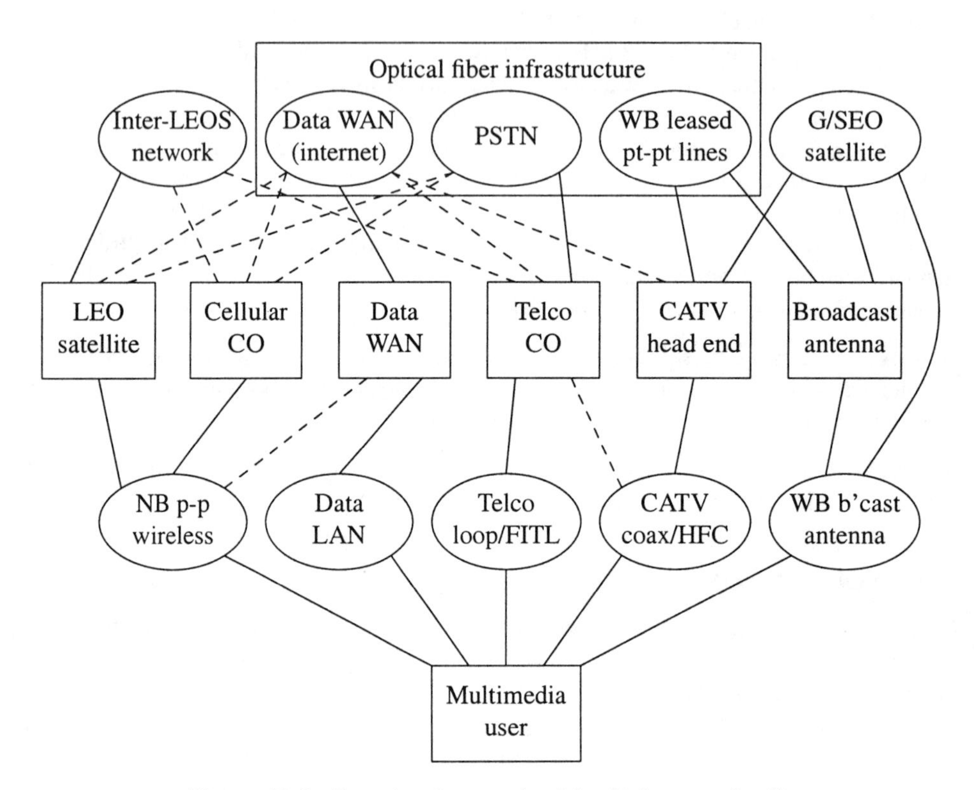

Figure 22.2. Two-tiered network with a little more detail.

WAN. While quality is poor, people are willing to tolerate it because of the low price. When and if the Internet price ceases to be artificially low, the practice of using data networks for voice and video should stop.

- Currently, we use telephone lines for access to the (voice) PSTN, but also, via modems, for access to data MANs and WANs. We also use a different kind of modem for access to FAX image.

The legendary (but, hopefully, neither imaginary nor lonely) multimedia user is shown at the bottom of Figure 22.2. Again, I use myself as the typical user. While I own three multiple-media instruments, I don't do any true multimedia communicating — not yet. I don't have all the boards, the connectivity, or the video camera; I don't have a good enough reason to make the investment; and I don't know anybody to whom I could place a multimedia call. But, like all of us, I'm evolving toward it. This is a very significant point: the migration to true multimedia calling will be very gradual.

The lines radiating out from this user (me) in Figure 22.2 represent access networks, on which most of us have ports today. While these access networks are generally separate (physically, functionally, and corporately), they are all slowly evolving toward one another's capabilities.

Today, I have narrowband point-to-point wireless access for the cellular telephone that I usually keep in my car. Today, this wireless link connects me to a cellular CO.

And, while I might subscribe to a LEOS system when it's offered, I probably won't because my current small roaming requirements probably won't justify LEOS' higher price, especially if cellular roaming fees are dropped in favor of one-rate pricing. But, many people probably will subscribe to LEOS and LEOS service could be very popular.

While the workstation in my office can be expanded for multimedia, today it is equipped and connected only for data and incoming image. This workstation sits on a building-wide LAN that has evolved in only a few years from (1) a one-tiered 10-BASE-5 Ethernet, to (2) a two-tiered LAN in which the 10-BASE-5 LAN is a backbone for several floor-wide 10-BASE-2 Ethernets, to (3) an ATM backbone for the floor-wide 10-BASE-2 LANs, and, eventually, to (4) a one-tiered all-ATM LAN. This kind of evolution is typical of many enterprises. This evolving building-wide LAN has a gateway onto the University of Pittsburgh's (Pitt) campus-wide MAN, which has itself evolved in only a few years from Ethernet to FDDI and probably to ATM. Other MAN ports include a modem pool and a gateway to a TCP/IP WAN, the Internet.

Besides being equipped for data and incoming image, my laptop and my PC at home are both equipped for telephone, fax, and modem. My PC is physically connected to a telephone line. My laptop has two ports: I use a port for a telephone line when I have my laptop at home and a port for an Ethernet connection when I have my laptop in my office. My Internet access over a telephone connection comes by a modem call through the PSTN to the modem pool at Pitt. If I didn't have an account on a computer with Internet access, I would place my modem call to one of the Internet access providers. The point is that my computers are connected by Ethernet when I'm at work and by telephone line when I'm at home. My data network access at work is better than the telephone line access at home. I've inquired about ISDN to my home, but it's not available yet. Even if it were, I'd have to think carefully about the monthly ISDN bill and about the price of a different board in my PC. I will probably be an early subscriber to ADSL or cable modem (Section 19.1) when it becomes available where I live.

Since I live in a suburban residential area, there is no alternate access provider and the local cable operator does not offer telephone or data service — not yet. So, I have no choice but to use the telephone company for conventional telephone service (I probably would anyway). My telephones are wired to a Bell Atlantic switching office, equipped with a Lucent 5E, via an RSM. The number of separate lines in my home has evolved up and down (see Section 3.3) over the years as my family and needs have evolved. Today I have two telephone lines into my home. Initially, the functional dichotomy was to use one for voice and the other for modem calls from the computer. But, the economics of telephone pricing motivated a different functional dichotomy in which we use one line for originating calls and the other for terminating calls.

- Today, my listed number has the minimal free calling area because I seldom use it to originate calls. Since this is the line over which most people call me, I've subscribed for the telco's voice mail (see Section 15.7.3) and for call forwarding on this line only.
- Today, my second line, which has an unlisted number, has no special terminating features because it seldom terminates a call. But, since I usually originate on this line, it has an extra large free calling area — large enough that a call to Pitt is free, especially a modem call. (Yes, I am guilty of unfairly using telco resources for

modem calls that have extraordinarily long holding times and I will continue to do this until the rate structure changes.) Both lines have the COS for Touch Tone. The kitchen wall phone, the busiest phone in the house, has programmable dialing and two-line selection. By using two-line wall jacks, the other phones in the house, including the cordless, can use either line (not as conveniently).

- But, the trickiest thing we did was to specify a different default IXC on each line. Now we can take advantage of whatever gimmick each IXC employs as a loss-leader without paying the corresponding penalty. If you're going to have two lines anyway, subscribing to two different IXCs will help defray the cost.

Pittsburgh's hills cause me to have poor reception of terrestrial broadcast television signals, even with a rooftop antenna. So, I need cable or direct satellite just to receive basic TV. Since I don't subscribe to any of these enhanced TV channels and I can't get local channels over direct satellite, I subscribe to cable. Since it only costs a little more than minimum to get the sports and educational channels, we get them also. I have one television set in my house and its adjuncts include a set-top box, a VCR, and an A-B switch. Since my cable provider does not remodulate the local channels, my TV and VCR can receive these channels without the set-top box. So, I split the cable signal to both the set-top box and VCR and then use the A-B switch to select the TV input from either the set-top box or the VCR. This way I can record anything on channels 2 – 13 at the same time that I view something on any channel.

22.3.2 Getting Served

So, why have I described this typical maximum user in so much detail?

- To illustrate that this user has a complex set of instruments, access channels, and services. This guy is way beyond POTS and he's not that unusual.
- To illustrate that this user's set of instruments, access channels, and services have evolved very slowly. He did not throw out all his existing appliances, channels, and services (and buy all new ones) on the same day.
- To illustrate that this gradual evolution toward this current set of instruments, channels, and services has been motivated by exactly two things — *offered service* and *price.* This user doesn't buy anything unless it provides some service that he thinks is worth the price.
- To illustrate that even this especially astute user doesn't care about the underlying technology. This user will not pay more for ATM if he can get the services he cares about over Step at a lower price. He has no brand loyalty and he not only takes advantage of pricing gimmicks, he gets a perverse pleasure in beating them at their game.

It is important that the telecommunications industry serve its customers well and that these users perceive it that way. For this to happen, we need open communication between the users and the providers. If the competitive market isn't satisfying users, other advocacy may be needed — perhaps in the form of strong and active user groups or, as a last resort, government intervention. Future regulatory issues are discussed in the next section.

22.3.3 Extending Universal Service

Telephony was a luxury during its first 40 years, but has become a virtual necessity during the past 80 years. Vail's concept of POTS is the ability to initiate and receive local calls using a black dial telephone. POTS has been sufficient if you need to:

- Get help in an emergency;
- Communicate with friends and relatives in our mobile society;
- Get important information;
- Be socially connected.

These things are tied closely enough to "life, liberty, and the pursuit of happiness" that government should see to it that they are available for all citizens. These things are as important as national defense, public education, protection from crime, and other mandates of government. This is universal service. But, if some citizen wants to do these things over a very large area, or wants to do them more conveniently, or wants to extend the scope to nonessential information (like entertainment), then this is not universal service and it is outside the scope of things that government should ensure. This citizen should not expect that what he wants should be available everywhere, for every citizen, and he should expect to pay more for it.

In the United States, Vail's concept of universal service in every home has been essentially achieved for many years. But, in recent years, it is no longer clear that POTS is sufficient service to ensure life, liberty, and the pursuit of happiness in the United States. It is argued that the information age is now, that information is power, and that the gap between haves and have-nots is widening because the haves have better access to better information than the have-nots. Most people believe that the widening gap is not a good thing and that government has a role in helping the have-nots keep up. One way government can do this is to redefine the minimum telecommunications service to include equity in information access and, then, to cause this new universal service to be implemented.

There are two issues: (1) How exactly is the new universal service defined? and (2) How does government enable it? History teaches us some valuable lessons.

Redefining Universal Service. Vail's original universal service was not defined as one 26-gauge twisted pair into every residence in America. It was defined as the capacity to provide analog voice telephony over a small dialing area from one black dial telephone. Universal service was not defined then by technology or bandwidth and probably shouldn't be defined this way now. Universal service was defined then by the capacity to provide some minimum service and probably should be defined this same way now.

Enabling the New Universal Service. While Vail's original universal service was mandated by government, it was not enabled directly by some government bureaucracy. The United States got the best telephone service in the world through free enterprise and capitalism. Government's role was to encourage private industry and to exercise control over the benevolent monopoly. The United States succeeded at universal service by subsidizing it, but within a closed economic system.

Because we have broken up this closed economic system, this method of subsidization cannot work. Because we can't expect the industry to subsidize universal service, old or new, we must at least assume that the free market will cause the service to be priced fairly. Then, if government thinks some people need help in affording this minimum service at its fair price, government has to help them pay this fair price. For example, expecting long distance customers to subsidize POTS through their access fee was poorly conceived.

However it is redefined, implemented, and enabled, the new definition of universal service must include adequate access to the information on the Internet.

22.4 REGULATORY ISSUES

This section of the book discusses some of the alternative choices that we must make as we plan the network of our future. Should the future network be integrated or should it be optimized for special traffic types? Should the future network be centrally administered or should it be more like the Internet? Should the future network be operated by some kind of enterprise that exists to serve the network's users or should it be operated by multiple enterprises in a free market or should it be operated somewhere in between? Should future network regulation continue to subsidize some kind of universal service or should the network be openly competitive? Should content be controlled or should any kind of data be allowed? These kinds of questions must be discussed before we plan our future or make a deliberate decision that we don't want to plan our future.

22.4.1 Integrated or Optimized?

Should all traffic types use one single common network, whose architecture and switching paradigm are compromises for the different traffic types? Or, should many networks be specialized, each of whose architecture and switching paradigm is optimized for some particular traffic type? There is no simple answer, but the answer may be different if it's being asked about a physical network or a logical network, or if it's being asked about a network that provides direct access to users or interconnectivity among regions.

Access networks may need to be physically integrated in order to serve users who typically are too frugal or too rural to attract competitive LECs (or any LECs). In commercial serving areas or residential areas that serve wealthy users or dense populations, LECs will naturally compete with physically distinct networks. Access networks that are logically integrated would probably have a competitive advantage over those that are not because they would have a single common interface, without needing translation, at the user's appliance and at the serving area's hub switch. This would seem to be the logical place for ATM in the network.

Interoffice networks also need to be physically integrated because of the cost of the long channels. But, voice, telemetry, various types of video, and computer communications for file transfer, transaction processing, and other applications all have very different switching requirements. So, these interoffice networks need to be logically optimized for the types of data they carry. Since this arena will undoubtedly be competitive, several competing interoffice networks, which might start out physically and logically integrated, would probably migrate toward different logical optimizations depending on the network's mix of traffic. But, even so, these logical networks would necessarily need a common infrastructure.

Very large interregional networks need to be physically integrated even more than interoffice networks. But, these networks are so big, and their channels are so long and expensive, that they would need also to be logically integrated. However, their integration would be optimized, not for switching, but to support scalably broadband optical transmission.

If and when we reach a good decision on these questions, should the decision be mandated? And, if so, by whom? Or, should the network just evolve in a free market?

22.4.2 Centrally Managed or Self-Managed?

Should the network be administered by some central governing body or is this unnecessary, or even counter-productive? Is asking about "the network" like asking about "the airline"? In the United States, the Internet is an example of a self-managed network. It was originally designed to provide computer communications among universities and other research-oriented enterprises. The participating enterprises were expected, then, to be altruistic good citizens of the network. Now that we have opened the Internet to market forces — to enterprises who owe allegiance to stockholders, to advertisers, and to all manner of content providers — can we still make these same assumptions?

If you believe that "the network" needs to be be managed, then, by whom? And, how much power should the network's manager have? Should somebody police its use and force the road hogs and abusers to pay more? Should somebody be responsible for ensuring adequate bandwidth? For ensuring reliability? For providing security and privacy? For keeping its use in the public's best interest? For controlling the content? For protecting the network from terrorists and hackers? For an overall architecture? For long-range planning? For a numbering plan? Or, don't we need these things? Or, will the market automatically provide them? These are important questions which are all difficult to answer.

Wouldn't a self-managed network be less expensive than a centrally managed network, like the PSTN, especially if it's partially subsidized by the PSTN? If left to make free-market decisions, typically based on price, would the users continue to support a managed and reliable network? Is it in the users' long-term best interest if market forces select self-managed networks until none of the competing networks are centrally managed, including that network which may be partially supporting and subsidizing some of the other networks?

Having implied that managed networks are good, are PTTs good network managers? Was Ma Bell a good network manager? Enterprises that have managed networks for our species have hisorically been kindly and maternalistic to their well-behaved users, who were happy to receive whatever service they got. But, historically, these managers have not been open to change or willing to share responsibility. Ma Bell would never have given us the Internet as we know it — partly because Ma Bell would have been required to overcharge for it as a means to subsidize POTS, but mostly because Ma Bell just didn't have the mindset to allow the network's users the kind of shared responsibility and governance that the Internet has allowed. Ma Bell would have never allowed some third party to put the World Wide Web on her network.

Can we have a benevolent dictatorship? Can we have the good side of Ma Bell to manage our network for us without having Ma Bell's dark side keeping us from touching it? And, please, can't we do this within the private sector?

22.4.3 Subsidized or Competitive?

Does universal service "promote the general welfare" in the same sense as intended by the writers of America's Constitution? If we agree with Theodore Vail that it does, then can universal service happen naturally in a free market or a self-managed network? Or, should universal service be supported by, or required by, the government or the network's manager (if there is one)? Consider some alternative subsidization policies that have been used, or could be used, in the telecommunications industry.

- As a commercial enterprise, one carrier provides all telecommunications service in a closed economic system. Prices are regulated, not based on cost, but to ensure that everyone has minimal service at low price. But, when this was done in the United States, the carrier, the Bell System, was slow to respond to market need and America's commercial enterprises (the users who were typically overcharged) resented the high prices. This won't work in a global economy and only works in a closed market. America would never go back to it, even if it could work.
- Assign one carrier in each region to be responsible for providing universal service. Require that this common carrier subsidize universal service by selling other services at a premium. But, then, allow (and even encourage) other carriers, who are not responsible for providing universal service, to compete with this common carrier in these artificially priced businesses. Users might be forced to pay a premium to access nonminimal services so the common carriers can continue to subsidize universal service, but there would be loopholes to bypass this also. This describes the post-Carterfone Bell System and today's system of RBOCs — it's not very fair and it won't work in the long run.
- Some government agency, which is not expected to make a profit, provides universal service. Free enterprise provides nonminimal services. But, no country with PTT-telephony ever had better telephone service than the United States had under the Bell System, and most countries with PTT-telephony are privatizing their networks.
- We could divide the telecommunications business into "hot" and "cold" zones, based on geography and type of service. We could have a completely free market system except that any company, seeking a charter to serve a hot zone, would be legally required to also serve some cold zone. We would have to police this, but it might work.
- We could completely decouple "welfare" from the telecom industry and let the entire telecom industry be a completely free market. If the government thinks some people should have discounted service, let the government help these users pay for it, at market price. Some government agency could administer the distribution of telecom stamps, analogous to food stamps, which could be redeemed when the user pays his telephone bill.

This last method is my favorite approach to provide universal service. Expecting the telecom industry to provide universal service is analogous to expecting grocery stores to give food to the needy. We don't provide food for the needy this way — instead, government assists the needy to pay the fair market price.

22.4.4 Competing on Price

In any free-market economy, the consumers of the goods and services naturally prefer direct competition on price. But, the suppliers of these goods and services naturally try to avoid competing directly on price alone. The enterprises in the telecommunications industry are no different. When the only feature that distinguishes your product from a competitor's product is the product's price, the result is lowered prices (good for the consumer) and lowered profits (bad for the supplier). Competing by offering unique features, better customer service, or more convenience can benefit both the consumer and the supplier. Instead, a supplier can still avoid competing on price, by trying to compete on the basis of real or perceived value or on real or perceived quality, by bundling products together, or by offering a "loss leader" to the market.

As an example of bundling with a loss leader, suppose supermarket A offers meat on sale but raises the prices of its produce, and supermarket B offers produce on sale but raises the prices of its meat. While some consumers will buy meat at A and produce at B, many consumers have neither the inclination nor the time to do this and will do all their shopping at the one supermarket that is perceived to be the best supplier. Now, in a deregulated free market, the telecommmunications carriers are engaging in many of these classical marketing strategies. Earlier, I described how I specify a different default IXC on each of the telephone lines in my home so that I can take advantage of whatever loss-leader one of them might offer, without having to pay the corresponding penalty. Many consumers won't do this and prefer one-stop shopping. We're also seeing bundling extend to multiple media, where the consumer could obtain local and long distance service, Internet access, and CATV all from a single carrier. Some consumers will prefer this and some will not.

22.5 SYNERGIES

Three interrelated changes in the PSTN will affect and effect the services that the evolving network provides [2].

- *Broadband.* Enabled by lightwave technology (Chapters 20 and 21) and by improved multiplexing and line-coding techniques, bandwidth is increasingly available — and at decreasing price. As the access to this vast pool of bandwidth becomes broader, some users will be able to purchase not only lower-priced DS0 channels, but reasonably priced DS1 and, even, OC1 channels.
- *Intelligent.* Enabled by the gradual migration of call control, the evolving Intelligent Network (Chapter 18) has enabled optional services like enhanced 800, cell-phone roaming, and caller ID. Many more new services, like local number portability, should follow.
- *Open.* Enabled by divestiture and other forms of deregulation, a slowly evolving group of third-party vendors will accelerate the provisioning of (1) even more enhancements to POTS and (2) new services that take advantage of broadband channels. Current examples include the Internet and its service providers, the independent cellular carriers, and the vendors of voice-mail systems and similar POTS adjuncts.

The evolution of any one of these three characteristics depends on the growth of the other two. If the total system is properly managed and regulated, we will have an exciting future ahead of us.

> *The evolution and expansion of the bandwidth, the control, and the openness of the PSTN are interdependent, coupled together by the market of evolving users.*

After presenting the postulates that form the basis of this essay, the second and third subsections show how evolving bandwidth and evolving control support each other. The fourth subsection describes how openness supports, and is supported by, both these two evolutions. But, beyond just supporting each other, the relationship is shown to be synergistic. More than just a broadband intelligent open network, the new network is a flexible platform for providing many interesting new services that are not available, or even possible, with today's network. This section concludes by discussing integrated services that are broadband, intelligently controlled, and provided in an open market. Five extreme examples are presented.

22.5.1 Postulates

Nine statements are postulated, without discussion. Each supports this section's hypothesis.

- Logical equivalence, in a physical distributed architecture, requires nearly equivalent performance.
- The network should be carefully layered, physically and logically.
- Four principal network costs are per-line equipment, long distance channels, OAM&P, and software development.
- Software has *hardened* and hardware has *softened.* Changing the modem on my PC was much easier than upgrading to Windows 98. The silver bullets that repel the software complexity monster are (1) hardware and (2) modularity, not more software.
- Since 100 10-MIPS processors cost much less than one 1000-MIPS processor, processing and control should be distributed.
- Twenty years of research on *distributed databases* has still not resulted in universally acceptable ways to do this.
- Photonics is an enabling technology that will affect humanity the way the transistor has affected humanity.
- While we must ensure system reliability and universal service, a user-oriented telecommunications market should be allowed to take its course.
- There is a lucrative market for new services that really serve users, and there is a lucrative market for connecting users to these services. These two markets are naturally decoupled, just like the grocery industry is decoupled from the trucking industry. Users will be optimally served if bundling and cross-subsidization are discouraged — maybe even illegal.

22.5.2 Photonics Supports the IN

Recent advances in photonic switching promise to undergird the implementation of the Intelligent Network. Inexpensive broad bandwidth and short network latency reduce the network's impact on performance and availability, potentially realizing the desired *comparable efficiency* between service software resident on switch processors and service software resident on processors very remote in the network. Chapter 18 described the PSTN's intelligence and signaling and discussed the potential evolution of the architecture of the software, the databases, and the processes that control the network. We saw that location and allocation is evolving — driven by its complexity, competition, and new services. While network intelligence remains logically at the switch nodes, the SS7 network allows it to be physically distributed. It is extremely important that we correctly architect:

the physical and logical location and allocation of the intelligence that controls the network and provides the new services.

Switch processes seem destined to be simple drivers. Service software is destined to be independent of hardware. Processing seems destined to be distributed, with (1) a distributed cluster of peer processors inside each switching node with special-purpose processors (for DB, OA&M, billing, etc.) and (2) a separate network-wide cluster of processors on which multiple vendors can implement competing services. Data seems destined to be centralized per region or even per nation.

A future processor architecture for each switching node was proposed at the end of Chapter 18. A LAN might interconnect several small special-purpose processors: the manufacturer's processor that drives the switching hardware, the carrier-owned processor that controls the internode network, a DB processor, an OA&M processor, an AMA processor, a carrier-owned processor that provides the LEC's services, an STP gateway to the "SS8" network, and a gateway to the external processors that provide enhanced services. It seems better to bring the call to the software than to bring software to the call.

But, the road to this new IN has roadblocks.

- The I of CEI, which has been called the ONA, needs further hardware and software definition and standardization.
- The CE of CEI requires that we remove the remoteness penalty so that services on remote processors perform as well as services on local processors.

For both these roadblocks, we need a signaling network that provides interprocessor communications that is broadband, has small latency, and is inexpensive. The SS7 network, at least in its current implementation, isn't it. So, we need to migrate the SS7 network to broader bandwidth and smaller latency. But, with all that unused bandwidth in all those fibers out there, there is no excuse for allowing the signaling network to be a constraint. So, we see that the emerging photonic network supports the emerging Intelligent Network.

22.5.3 IN Supports Photonics

In the reciprocal relationship, the Intelligent Network potentially provides a distributed network control engine that could make photonic networks really perform. Section 20.5 described how optical packet switching is problematic and that rate independence is probably an important design consideration. We saw that block multiplexing (in space, time, and wavelength) is a good match for analog-switched photonic devices. Capacity switching seems a lot like circuit switching — which is thought to be old-fashioned, inflexible, and less than optimal by some. But, an evolved IN and a better signaling network could provide the flexibile call setup and end-to-end control that would be required to make photonic capacity switching the paradigm of the future.

Much of the PSTN's long distance infrastructure has already evolved to photonics. During the late 1980s and early 1990s, the transmission rate in this optical network accelerated like the bit density on integrated-circuit memory chips — doubling every 3 years. In both cases, the acceleration is driven by technology, competition, and demand. The optical rate increase slowed during the mid-1990s while more carriers built parallel national networks. Recently, the practical rate abruptly quadrupled, from 2.5 Gbps to 10 Gbps, as wavelength multiplexing became practical. Physical data rate could continue accelerating for 10 more years, as long as the market demands it.

As discussed in Section 21.1, the interexchange part of the PSTN is beginning to define its channels, not only in space and time, but also in wavelength. As described in the previous two chapters, photonic switching (also in space, time, and wavelength) enables switching to follow the technological evolution of transmission. As data rates increase and additional wavelengths are added, a truly scalable network can evolve by simply replacing transmitters and receivers on the network's edge. But, as discussed in Chapter 20, an optical network that uses packets, bit interleaving, or digital devices may require significant changes (internal to the network) as fiber's physical rate gradually increases. It won't scale well.

A scalable architecture is proposed later in this chapter. It is intended for the highest layer of a large network — like a national network (NAN) or a global network (OAN). The proposed network has two layers: a layer of regional gateways and a layer of high-throughput tandem nodes. While the details are described later, this network architecture will require a significant amount of overall end-to-end control. Whether routing a packet, connecting a call, or adjusting the node-to-node capacity, the network's distributed controllers will have to signal to each other. While even a national network that routes IP or ATM packets at its highest layer will also require a significant amount of node-to-node and end-to-end signaling, a photonic-switched network probably requires more of it. This greater amount of signaling and control would be a serious drawback to a photonic-switched network if the network of distributed controllers had to be constructed *ab ovo*. It's not a drawback because we already have a signaling network and a distributed collection of processors, the Intelligent Network. We see that the emerging Intelligent Network supports the emerging photonic network.

The migration of switching software and network control from switch processors to the AIN has been called a brain transplant on the PSTN. But, the PSTN's infrastructure is also evolving and, whether photonic switching or packet switching is adopted, we're even advocating an evolution to an entirely new switching paradigm at the highest layer. If the IN can control this new network as well as it controls the old one, even during the transition, then should we call this a body transplant (keeping the brain) on the PSTN? Even though today's network control architecture and network infrastructure are both evolving, they are evolving toward a mutually supportive (even synergistic) total system.

22.5.4 Openness Supports Both

This evolution of network control to *open intelligence* and of the network fabric to broad bandwidth provides a vast potential for new services. Corresponding to these two evolutionary trends are two kinds or revolutionary services: (1) custom call control (way past 800) by the network and (2) broadband delivery (video and high-speed data) through the network. But, who provides these services?

We asked a similar question in Section 19.5 when we asked who provides the network control? For all of the benefits the Bell System gave America, the Bell System would have never given us the Internet. If we leave service provision to the carriers, they will surely bundle them up with a lot of other stuff. History has shown that this potential for broadband intelligent services won't develop unless this arena becomes an open marketplace.

One of my favorite stories is about the marketing consultant who was hired many years ago to help the Cunard Lines improve its business. He posed a simple question to the company's executives and Board of Directors: "What business are you in?" Annoyed with this silly question from such an expensive consultant, they replied: "We're in the *transportation business,* of course." The consultant told them: "That answer is why your business is suffering. If people want transportation between Europe and the United States, would they select a slow ship over a fast airplane? No, unless it's cheaper. Since you have no advantage over an airplane, you have to make your fare lower than that of an airplane." The consultant continued: "You have optimized your company for the wrong business. Don't think of a ship as a means of transportation. Optimize your company as a *resort business* and think of your ships as luxury hotels. If people want to spend a few days in a luxury hotel, would they select one of your ships over a conventional resort? Yes, because your luxury hotels float and move. Customers can check in to one of your luxury hotels in the United States, and check out in Europe. Since you have a big advantage over a conventional resort, you can make your fare even higher than that of a conventional resort." Cunard Lines followed this wise advice. They made their ships more luxurious, not less. They made their fares higher, not lower. This was an expensive consultant who earned his pay and Cunard Lines improved its business because:

they knew what business they were in and
they optimized their company for this business.

A telecommunications carrier, especially an LEC in the United States, is in the business of *selling channels.* Since the regulators make it hard to be in any other business, anyway, the business should be optimized to sell channels. While POTS sells a lot of channels, services sell even more channels. How is a company that is optimized for selling channels more likely to succeed?

- By building services themselves and selling them directly?
- By creating a rich marketplace in which other companies, third-party vendors that are optimized for building services, can operate successfully:

- By paying just enough to cover the carrier's cost;
- By doing what they are good at — building services;
- By placing their services in this marketplace;
- By selling their services (and the carrier's channels) to end users?

This potentially evolving network has synergies among (1) broad bandwidth, (2) intelligent control, and (3) open access to third-party service providers.

22.5.5 Better Services

This subsection continues the discussion of services from Section 15.7. In terms of providing real services to real users, the telecom industry is about 15 years behind the computer industry. Today's telecom services are constrained in the same two ways that user-oriented computer application programs were in the 1980s: (1) they are overly generalized to reach a mass market, and (2) they are barely adequate simulations of the existing human activity they try to replace or augment [3].

Making the future network broadband, intelligent, and open should have the same effect on telecommunications services that (1) high processor speed and large memory, (2) powerful and intuitive operating systems, and (3) an open and competitive marketplace had on computer application programs since the early 1980s. In this future broadband intelligent open network, telecommunications services could become (1) less generalized and (2) more realistic.

Better Customization. The broadband intelligent open network will serve a larger clientele and will give them less expensive access to a better network. So, future service providers could provide services that are more customized to specific applications and for specific users. While a single implementation of a service may be too general to meet anyone's needs, services don't have to be customized for individuals — that would be too expensive. Service constituencies might be grouped by:

- Age (children, teenagers, the elderly, etc.);
- Class (up-scale, poor, blue-collar, middle income, etc.);
- Occupation (accountants, lawyers, librarians, nurses, etc.);
- Interests (sports, collectors, games, crafts, etc.);
- Handicap (blind, deaf, retarded, people with dyslexia, etc.);
- Affliction (alcoholics, diabetics, people with Alzheimer's disease, etc.);
- Support groups (caregivers, parents, charitable organizations, etc.);
- Religion (various organized creeds and denominations);
- Travelers (route planning, reservations, advice, etc.);
- Many other categories.

For example, the World Wide Web is a database for almost all constituencies except the computer illiterate. But, customized excellent services would be much more expensive to implement than simply putting data on the web and they might require much more bandwidth than the Internet provides. TV programmers understand about reaching smaller constituencies. Instead of trying to design television programs with mass appeal, they design television programs specifically for: children, sports enthusiasts, people who like action, people who like drama, and for many other kinds of special-interest groups. The variety show, which appealed to all viewers in the 1950s, appealed to so few viewers in the 1980s that it's gone. Attempts to generalize are not very successful. For example, attempts to attract more women to watch the Super Bowl or the Olympics by broadening the coverage are resented by some men who typically want to see more action and less up close and personal.

Two related issues are (1) diffusion and (2) granularity. If the constituency is geographically clustered, then specialized services can be located in each cluster and the network isn't as important. For example, services in Spanish would be more popular in the southwestern United States than in Minnesota. If the constituency is geographically dispersed, then specialized services can't be optimally located and the network becomes very important. For example, services for diabetics would have to be highly distributed or accessible through the network.

More Realism. As discussed in Section 15.7, today's telecommunications services are typically barely adequate simulations of existing human activities. If you don't think so, poll the public and ask how many people like using voice mail. The current market attempts to convince users to replace one of their existing activities, like traveling, by some equivalent telecom service, like videoconferencing. Unfortunately, these equivalent telecom services are far inferior to the real activity they hope to replace.

Consider third-party services, as we see them now and as they could be in the future. Today, services like bank-by-phone, voice response units, and answer service are extremely simple, dependent on the telephone's Touch Tone pad for their control, and use audio bandwidth *through* the network. Future equivalent services, and many more that are more interesting and more useful, could be much more complex, could assume that the user has a better controller, and could run over broader bandwidth. While they would still operate through the network, parts of the service might be embedded *in* the network and might be partially controlled *by* the network. Or, instead of the network controlling the service, the service might control the network. Such services require a vast army of vendors, competing in an open market. While manufacturers and carriers might participate in the competition, the third-party vendors would be more likely to meet customer need, especially if the market were open to global providers and imported software. This is an extremely interesting market for what could become a booming cottage industry. But, more likely than any of these scenarios, the customer himself is most likely to meet his own need if he can put his own service in place.

Compensation. When using a service that replaces some human activity, the users are connected to a machine that poorly simulates the real activity. While these artificial services provide the stated telecom advantages, they have too many disadvantages, because the machine is narrowband, unintelligent, and closed.

- While migration to broadband removes the bandwidth compromise, the artificial services will probably still feel impersonal, sterile, and, well, artificial. So, if we can't completely solve the problem, maybe we can add other features that give it offsetting advantages.

- Migration to an intelligent network gives the machine the potential for advantages that even the real service can't provide.
- Migration to an open network lets the market compete on the best service, and if none of the offered services is satisfactory, lets you build your own.

If the future network is broadband, intelligent, and open, future service providers can make services seem less sterile than services that are provided over a narrowband, unintelligent, closed network. But, these services will still be inferior to the corresponding human activity, unless the new services provide some additional capability that isn't available from the existing human activity. If we can't overcome the features that are lost through telecommunications, then let's add new features that might compensate for those that are lost. Then, telecommunications services in the broadband intelligent open network could be not only more like the existing human activity, but they could even be better than the existing human activity. Each of the extreme examples in the next subsection suggest advantages that could make a future telecom service better than its corresponding real human activity.

22.5.6 Five PANSs

In contrast to POTS, we propose PANS, for personally accessible new service or pretty awesome new service. Five PANSs are described [2], some of which are beginning to emerge on the Internet. After reading how these proposed PANSs are so much more imaginative than the equivalent services that have been typically available, I encourage you to add your imagination to what I've started. The reader is urged to lobby for the right network and the right to install services into it — and then to get busy making competitive services that are even better than the five described here.

Tele-Bridge. This is a four-way voice/data call through intermediating service software, like an outsourced multimedia conference call. This is one example where services can be found on the Internet that are very close to what's described here. If you feel like playing bridge, you don't have to be colocated with the rest of the foursome. You simply call an 800 number or find the web site and wait to be assigned to a game. Since Tele-Bridge knows your capability, the service assigns you to a game with three other people, from anywhere in the world, who have been waiting to play with someone at a similar level of skill. If you're a frequent user, you might play against these same people again. While the game could be played without associated voice connectivity, bidding and other aspects of the game might seem sterile. At the players' option, the game could be strictly serious or it could have an associated audio chat channel.

How might Tele-Bridge be so much better than real bridge that the most avid bridge fans would occasionally choose Tele-Bridge over real bridge, even with an existing bridge group who all live in the same neighborhood?

- Tele-Bridge automatically matches players with similar ability.
- Tele-Bridge removes any possibility of cheating, secret codes, or table talk.
- Tele-Bridge automatically shuffles cards and deals them, and grease from potato chips won't get on the cards.

- The host doesn't have to clean the house before the game, prepare any food, or clean up after the game.
- You can play any time of day — even in the middle of the night, a group would always be waiting somewhere in the world to form a foursome.
- Nobody has to get dressed up — you can wear your pajamas.
- While participants could identify themselves, it might be more fun to have a *nom de guerre,* like "The Cleveland Clubber" or "Heart Wrencher," and remain semi-anonymous.
- Tele-Bridge automatically announces regular tournaments, seeds the tournaments, and automatically negotiates the schedules for the players.
- If the game is poker instead of bridge, Tele-Poker would automatically handle all the betting and money transfers.

Specific details depend on the implementation of the terminal and the channel. While the game could be played in a data-only medium, a text-only screen format would be extremely uninteresting. If local software displayed graphics for the cards played and cards in the player's own hand, then only simple data would have to be transmitted as cards are played. However, exemplifying the tradeoff between channel requirements and terminal complexity, if the user has a dumb terminal instead of a PC, the service would have to transmit an entire updated screen after each play. Using a mouse or touch-screen with an image of the cards would be superior to using a keyboard to select the card to be played from your hand. Except for the ability to read faces, especially when playing poker, four-way video probably doesn't add much to this service and may even detract from the mystery of not knowing who's playing.

Tele-Indy. This is a multiplayer game in which participants drive around a virtual track and race against each other. Excellent versions of this game are currently available in arcades and implementations with lower quality are available on the Internet. Tele-Indy is also an N-way data call through intermediating service software. If you feel like racing, you don't have to be colocated with the rest of the field or go to an arcade as with the equivalent computer game. You simply call an 800 number or go to a web page and wait for the next race to start. Each player selects a color and number for his car, which is how he is perceived by the other drivers in the race. The central software maintains the overview of the race, but each player sees his own view of the race as a full-motion video cartoon. The player's CRT simulates the front windshield, the driver's console, and the rear-view mirrors. Other variations include a bumper-car simulation, tank warfare, or a three-dimensional N-way airplane dogfight.

Except for being safer, Tele-Indy could never really be better than actually driving a race car in the Indianapolis 500 — except that most people never get the opportunity do that. But, Tele-Indy would be much better than the equivalent computer game, even one found in a video-game arcade.

- The number of entrants would be much larger than most computer games allow.

- Most multiplayer video games don't give each driver an individual view of the race.
- Acoustic separation enables customized sound also, as each player would hear a car pass him.

Specific details depend on the implementation of the terminal and the channel. However, exemplifying the tradeoff between channel requirements and terminal complexity, if the user has a dumb terminal or television, instead of a PC, the service would have to have a switched video connection. At a PC, local software would simulate the view of the race for its driver. All participants would communicate steering wheel and gas pedal information to the central software, which would communicate the position of all cars to the local software in each driver's PC. If the channel has enough bandwidth and each driver has a camera aimed at himself, then the drivers could not only see the other cars in the race, but would see each car's driver through the windshield. Then, facial expressions and hand gestures could liven the game even more.

Tele-Diagnosis. This is a semipermanent many-to-one telemetry-data connection that reverts to a broadband connection or a virtual 911 call whenever the situation warrants. Many medically oriented web sites exist on the Internet and several commercial products have come close to what's described here. While Tele-Diagnosis couldn't possibly be better than competent medical care or full-time nursing, it could at least be affordable and adequate. The impersonal nature of the service would be compensated by the privacy it offers and the ability for the patient to remain home.

Specific details depend on the implementation of the attached devices and the channel. While the goal of Tele-Diagnosis is to provide remote diagnosis of medical conditions and remote health monitoring, the coverage would depend on the channel's bandwidth. The health monitoring extreme would be a remote equivalent to the monitoring provided in a hospital's intensive care unit, possibly including a video link. Furthermore, as long as the long distance data connection is reliable, the patient could shop around for expertise or price. This service might extend to receiving a remote physical examination, getting a remote second opinion for a medical condition, or emergency diagnosis and triage on a patient in an ambulance. The coverage and diagnosis would be provided by appropriate medical personnel, who could also be remote, but could be augmented by an on-line expert system.

Tele-Meeting. This is an N-way voice/data/video call through intermediating service software. While videoconferencing has been available for many years, its deployment has been disappointing — because a "conference" is too general. Tele-Meeting is specifically designed to replace the meeting — not some generalized conference.

How might a Tele-Meeting be so much better than real meeting that it would occasionally be preferred over a real meeting, even by a group who all work together in the same building?

- Attendees who choose not to physically attend a meeting wouldn't have to use saving airfare as an excuse, because it would be socially acceptable.
- The virtual meeting could take place in a window on the attendee's PC — the same PC the attendee wishes to do work at, during the meeting.

- The attendee can pay variable attention to the meeting. When someone is speaking whose opinion you don't really care to hear, you can work on your book or browse the web or play Tele-Bridge. This kind of behavior is considered rude in a physical meeting when, in fact, it is people who talk too much who are really the ones who are rude.
- Tele-Meetings would be arranged by an automatic scheduler, which negotiates the time of the meeting by consulting the computer-based calendars of all attendees.
- A silent message channel is available from each attendee directly to the meeting's moderator. This allows attendees to request closure or a high priority in the speaker order.
- More useful forms of voting would be available. Weighted voting allows some attendees to have a more powerful vote than others (for example, at a stockholder's meeting). Analog voting allows attendees to vote 60/40 instead of yes/no.
- The minutes of the previous meeting, the agenda, and other handouts would be available on other windows of all attendees' PCs. Timers could be displayed to discourage time wasters.

Because the Tele-Meeting probably takes place in a professional setting, attendees could be expected to have a PC or workstation. While voice and data connectivity are important, video connectivity might also be useful.

Tele-Teaching. This is another more specific variation on the overly general videoconference. It is also an *N*-way voice/data/video call through intermediating service software. Unlike the Tele-Meeting, here the video link is very important. Tele-Teaching goes way beyond distance education, which tends to be a monologue from a talking head augmented by a remote blackboard.

How might Tele-Teaching be made so good that, besides being used as a remote virtual classroom, it might even be used as a teaching aid inside the real classroom?

- Multiple small rooms, equipped with Tele-Teaching and staffed with a teaching assistant, might be a friendlier learning environment than a large lecture hall.
- Lectures and classroom events would be recorded and repeatable.
- Lectures might be perceived to be given by cartoon characters who could be selected for role modeling or for acquiring improved attention.
- Seeing the lecture outline, images, and the blackboard from a PC would facilitate the students' note-taking. The truly skilled teacher could become very effective at presenting a lecture through multiple windows.
- Students could use a silent channel to the teacher as a means to control pace and to ask questions. The teacher could ask questions for class-wide voting to determine what the class already knows or whether more time is needed on some subject.
- Teachers could quickly issue a display of clapping frogs to the PCs of students who answer questions correctly in class. The whole educational process could be more fun.

- Reading assignments, homework, examinations, and grading could all be automated.

Since Tele-Teaching probably takes place in an educational setting, attendees could be expected to have a PC or workstation.

22.6 RESIDENTIAL FIBER STAR INFRASTRUCTURE

This section proposes an architecture for an extreme network infrastructure that could serve a residential area similar to that served by a traditional telephone switching office or CATV head end. While the application may be futuristic and even unrealistic, the development here is an example of good design methodology. Networks should be architected by carefully understanding users, services, and technology first, and then creating an optimal system that matches the three at low cost. This section walks us slowly and pedantically through a design process that considers users, applications, and evolution before architecture.

22.6.1 Assumptions and Services

The design principle illustrated here is: first, specify the service, and second, architect a system to provide it.

The following is a list of technology assumptions.

- Active switching (with a remote concentrator, community dial office, or remote switch unit) can be distributed into the loop plant.
- The maximum practical data rate on a single fiber is 10 Gbps.
- The maximum number of fibers that can terminate on one photonic integrated-circuit component is 32.
- Fiber links are half-duplex.
- Residential telephone wiring can support digital transmission at several megabits per second on each twisted pair.

The following is a list of user, service, and regulatory assumptions.

- A serving area consists of many thousands of subscribers — more than one 10,000-line office, but fewer than 10 offices.
- Ultrahigh-definition digital video (UDV) is proposed. The viewing screen is thin enough and light enough to hang on a wall with a molly and a hook. This screen measures 4 feet by 6 feet, with a liquid-crystal panel, like that on a laptop PC. With 30 color pixels per linear inch, this screen has $(4 \times 12 \times 30) \times (6 \times 12 \times 30) =$ 3.11 million pixels. Using 16 bits to encode the color and intensity of each pixel allows for $2^6 = 64$ levels of intensity and $2^{10} = 1024$ colors. Refreshing the entire screen 40 times per second, the date rate required to support a video image on such a screen is $3.11 \text{ M} \times 16 \times 40 = 1.99$ Gbps.

- UDV is the golden service for fiber-to-the-home. The up-scale market pull for UDV service will be sufficient to economically enable the installation of another loop infrastructure.
- UDV programming will be so expensive to produce that there will only be 10 UDV program providers nationally.
- Conventional high-definition digital television (DTV) signals can be economically digitized without compression. Using 12 bits to encode the color and intensity of each pixel allows for 32 levels of intensity and 128 colors. At 4 MHz of equivalent analog bandwidth, the required data rate is $2 \times 4M \times 12 = 96$ Mbps. This picture will be much nicer than A-to-D converted baseband. Subscribers can select from 100 downstream channels. Subscribers can view/record four channels simultaneously, either by owning four different instruments or by allocating four windows on a single DTV screen. Subscribers can transmit one DTV channel upstream at this rate.
- While 8-bit companded PCM is satisfactory for audio, 12 bits would be even better and it could be linear, without companding. However, 4 kHz of bandwidth has never been satisfactory. Upgrading telephony to 16 kHz of bandwidth, the data rate required to support high-fidelity telephony is $2 \times 16K \times 12 = 384$ Kbps. Most people would be willing to pay a little bit more for high-fidelity telephony. This rate defines the elementary broadband channel (EBC), which supports this new up-scale POTS and which is the building block for *N*-channel data connections. Fiber loop subscribers would have the capacity for many EBCs into their homes and offices.
- Subscribers can allocate four of these new EBCs for downstream Internet access at 1.5 Mbps. One channel is probably satisfactory for upstream access. People would be willing to pay a bit more for 1.5-Mbps downstream Internet access. There is capacity for even much higher bandwidth if the subscriber wishes to pay for it.
- This service is way beyond POTS and it is not universal. It will be legal to offer an extremely high grade of service to a serving area that is densely populated by people who can afford it, without being required to offer the identical service to people who can't afford it or in regions that are sparsely populated.

22.6.2 Residence's Service, Equipment, and Wiring

If the user owns a UDV monitor and subscribes to UDV service, the net downstream signal into his home has capacity for one UDV channel at 2.0 Gbps. From the 10 downstream UDV channels available, the user selects the particular channel he wishes to view by signaling upstream. Since the subscribers can't be expected to afford a UDV camera, they will not be transmitting a UDV signal.

Besides this UDV signal, the net downstream signal into every user's home needs the capacity for four conventional DTV channels at 100 Mbps each. Any of these four channels can carry switched or broadcast DTV downstream. From the 100 broadcast DTV channels available and any switched DTV channels that the user has set up, he selects the particular four channels he wishes to view by signaling upstream. To allow the subscriber to transmit switched DTV upstream, the upstream signal has capacity for one DTV channel.

Besides the UDV signal and the four downstream DTV channels, the net downstream signal into every user's home has capacity for far more (250) EBCs than could possibly be used. Besides the upstream DTV channel, the net upstream signal has capacity for far more (125) EBCs than could possibly be used.

So, the net downstream data capacity is 4×100 Mbps + 250×0.4 Mbps = 500 Mbps for the residential user who does not subscribe to UDV and is 500 Mbps + 2 Gbps = 2.5 Gbps for the user who does subscribe to UDV. The net upstream capacity is 100 Mbps + 125×0.4 Mbps = 150 Mbps. While the upstream signal wouldn't necessarily use the Sonet format (more likely, it would use native ATM), by using a standard transmit rate, like OC-3, optical transmitters would be more affordable for mass deployment.

In the residence, equipment called the residence controller (RC) provides a physical and logical interface between the upstream and downstream fibers on the network side and the specialized house wiring on the residence side.

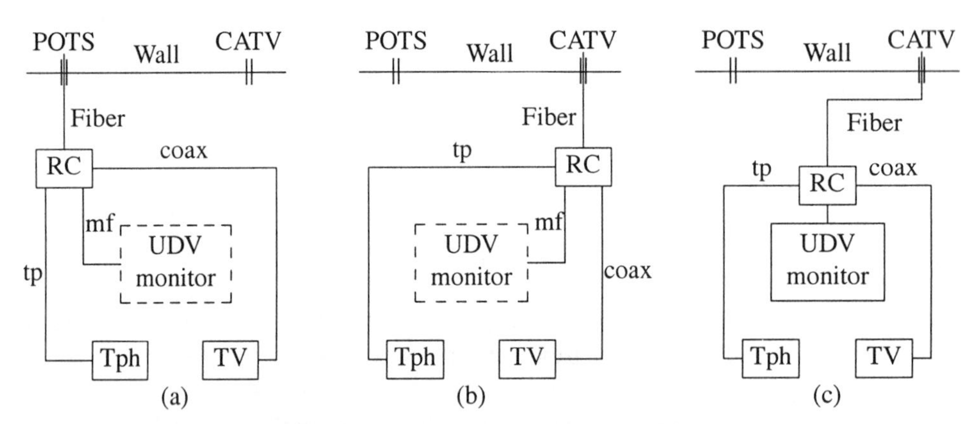

Figure 22.3. Distribution inside the residence.

Figure 22.3 illustrates how the residential controller connects to the network and to the residence's different specialized house wiring networks.

If the user does not own a UDV monitor, he would not subscribe to UDV service and his downstream signal would operate at only 500 Mbps. The carrier would provision an inexpensive circuit for transmitting data into his home and the user's RC would have an inexpensive receiver. This user's relatively simple residential controller would be located physically near where the two fibers penetrate the house, either (a) at the protector block or (b) at the CATV cable entry. When this user finally purchases a UDV monitor, he must request upgraded service from the carrier and he must upgrade his residential controller.

For the UDV subscriber, his more complicated residential controller might still also be located at either of these places indicated by Figure 22.3(a) or (b), in which case some kind of physical link, probably multimode fiber, carries the UDV signal from the RC to the UDV monitor. Alternatively, this more complicated RC might be physically included (c) in the UDV monitor, in which case the two fibers, and the specialized internal house wiring, would all terminate at the UDV monitor.

Conventional residential multipair telephone wiring also terminates on the residential controller. The RC is capable of time-space switching multiple digitized POTS channels or multiple 400-Kbps EBCs, in each direction, between the fibers on the network side and the various twisted pairs. The upstream channels are multiplexed onto the upstream fiber to the pedestal and the downstream channels are demultiplexed from the downstream fiber from the pedestal. All the residence's analog POTS telephones would be connected to those twisted pairs that terminate on BORSCHT circuits in the RC. All the residence's EBC instruments would be connected to those twisted pairs that terminate on EBC circuits in the RC.

Since most residences are already wired with conventional coaxial cable, this cable would be used to carry digital and analog TV channels between the RC and the various video devices in the home. Since all video signals in the network are digital (DTV), any analog signals in the residence's coax would have to be D-to-A converted in the RC. The residential coax carries five internal 100-Mbps DTV channels, four of which are frequency-division multiplexed at the controller. These four internal DTV channels carry downstream DTV channels, which are demultiplexed off the downstream fiber from the pedestal. Typically, all four of these downstream channels carry broadcast DTV channels, but one of these four channels might carry downstream switched DTV instead, when the subscriber is using this service. If the residence still has analog TV sets, they are connected to this same coax and incoming digital channels are converted to analog in the RC and transmitted over FDM channels that are separate from the five digital channels in the residence's coax.

The subscriber's high-definition digital television sets can be tuned to any of the residence's five internal DTV channels by selection at the set. If the set is capable of displaying more than one program, in various windows on the screen, then these displays would be controlled at each individual set. If a user in the residence wishes to view a broadcast program that is not one of the four currently transmitted into the home, then this user must signal upstream to arrange for the desired program to be transmitted downstream and must specify which of the four downstream channels should be used. Viewing a program from an analog TV set is similar except the viewer must also arrange for the desired program to be D-to-A converted at the RC and simulcast in analog format on some other FDM channel over the residence's coax.

Two of the five internal DTV channels are used whenever the subscriber makes a switched DTV call. One of the internal DTV channels carries the downstream signal, taken from one of the four downstream channels from the network. The subscriber's digital camera, or A-to-D converted analog camera, can be plugged into any coax jack. When activated, it produces a frequency-multiplexed signal onto the fifth internal DTV channel on the residential coax, from which it is converted and switched onto the lone upstream 100-Mbps DTV channel to the network.

Personal computers are connected to one or more EBCs in the network. For example, if the user wants data access at a T1 rate, he must arrange to gang together six EBCs. However, residence wiring may prove to be a bottleneck. A PC with a conventional modem is connected to an analog twisted pair in the residence's telephone wiring and is limited to the modem's rate, say 56 Kbps. Or, new EBC modems that might be made available for use with two pair in the residence's telephone wiring could provide 384 Kbps each way. A PC with a conventional CATV/Ethernet modem is connected to an analog FDM channel in the residence's coaxial cable. This analog FDM channel is converted at the RC and switched to an appropriate number of EBCs in the fiber to the network.

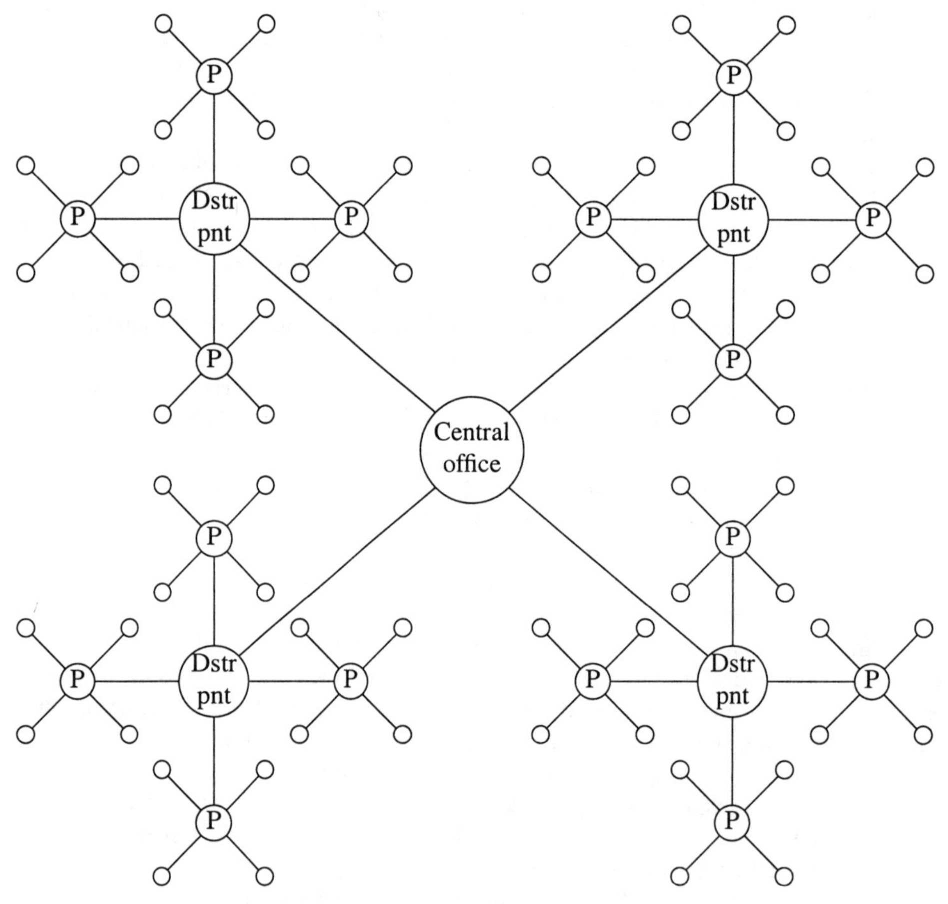

Figure 22.4. The star^3 topology — top view.

22.6.3 Circuit-Switched Fiber Star^3

A three-tiered switched fiber star is proposed as an architecture for fiber to the home, where each home doesn't merely tap a bus. The bandwidth required and the revenue generated are both assumed to be sufficient to justify that each home terminates a fiber end. Figure 22.4 illustrates the top view of the topology. Based on the preceding assumptions, the topology probably cannot flatten to two tiers, but it would be worth examining. And, while the topology might require four tiers to meet all the assumptions, it seems that it will work with only three tiers. The degree of the various hubs may not be optimal, but Figure 22.4 shows a first draft.

- The primary hub, the central office serves a maximum of eight secondary hubs (only four are shown in Figure 22.4).

- Each of these eight secondary hubs, the distribution points, serves a maximum of 32 tertiary hubs (only four tertiary hubs are shown for each "Dstr pnt" in Figure 22.4).
- Each of these $8 \times 32 = 256$ tertiary hubs, the pedestals, serves a maximum of 128 residences (only four residences are shown for each "P" in Figure 22.4).

Hubs that must deal with higher bandwidths have lower degree. Each of the eight distribution points serves a maximum of $32 \times 128 = 4K$ residences. A central office serves a maximum of $8 \times 32 = 256$ pedestals and $8 \times 32 \times 128 = 32K$ residences. Intuitively, the pedestal appears to be the most critical point, where the economics will be most difficult to justify.

The following subsections assume that this LAN is circuit switched. Channels are separated by category and circuit switched in nodes at all four layers of switching. These three subsections will show that this paradigm probably isn't very good and concludes that a cell-switched paradigm would probably be much better. The detailed architecture is worth studying, but don't extrapolate this conclusion to the entire network. It simply reinforces the proposal that *access should be integrated.*

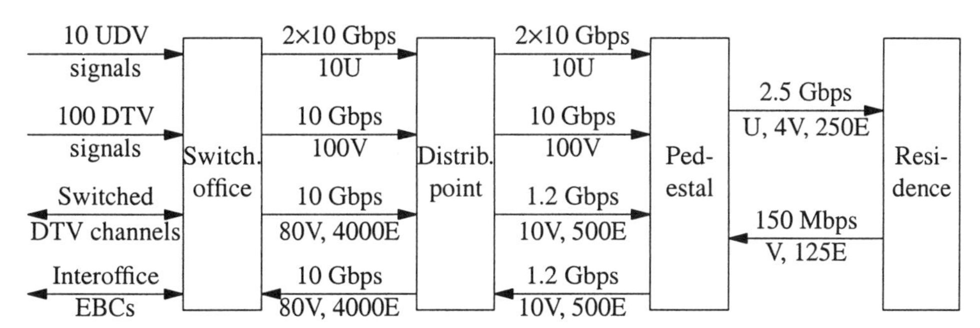

Figure 22.5. The star^3 topology — side view.

22.6.4 Data Flow

Figure 22.5 shows a side view of the architecture. This subsection and the next two examine the data flow between successive nodes, the links between these nodes, and the switching inside the nodes. First, this subsection examines the flow of each type of data.

The 10 broadcast UDV channels are delivered to every central office separately, via satellite or via some national broadband fiber infrastructure. Each incoming channel operates at 2.0 Gbps. The CO multiplexes these 10 channels onto two 10-Gbps channels and broadcasts this signal over 16 fibers, two to each of the 8 distribution points (DP) it serves. Each DP receives this multiplexed UDV signal over two fibers from the CO and it simply rebroadcasts the signal to the 32 pedestals it serves. Each pedestal receives this multiplexed UDV signal over two fibers from its DP and it demultiplexes the signal back down to its constituent 10 UDV channels at 2.0 Gbps each. Any user who wishes to view some UDV channel signals his pedestal over his upstream fiber, and his pedestal then switches the desired channel to his downstream fiber. The user's controller demultiplexes the UDV channel off this downstream fiber and directs it to the user's UDV monitor.

The 100 conventional high-definition broadcast DTV channels are also delivered to the CO via satellite or via some national broadband fiber infrastructure. Each incoming channel operates at 100 Mbps. The CO multiplexes these 100 channels onto one 10-Gbps channel and broadcasts this signal over eight fibers, one to each of the eight DPs it serves. Each DP receives this multiplexed DTV signal over one fiber from the CO and it simply rebroadcasts the signal to the 32 pedestals it serves. Each pedestal receives this multiplexed DTV signal over one fiber from its DP and it demultiplexes this signal back down to its constituent 100 broadcast DTV channels at 100 Mbps each. Any user who wishes to view some conventional high-definition broadcast DTV channel signals his pedestal over his upstream fiber, and his pedestal then switches the requested signal to one of the four DTV channels in his downstream fiber. Then, the user's controller places this signal onto one of five FDM channels on the user's in-house coax. The user selects the desired channel from the DTV monitor.

While subscribers also have capacity for one switched DTV channel in each direction, their utilization would probably be small. Perhaps, initially only a small percentage of users would use these channels at any one time. But, as more subscribers buy cameras, the utilization would increase. So, while each pedestal terminates one potential switched DTV channel in each direction to/from each of the 128 subscribers it serves, it can concentrate this capacity. Since this service probably wouldn't ever be more than 10% utilized in peak busy hour, the infrastructure must be designed for this ultimate use. Initially, since all the channels wouldn't have to be provisioned, perhaps as few as 10 V-channels, a total of 1 Gbps, would be sufficient each way between the DP and each pedestal it serves. Even with this 10:1 concentration at the pedestal, another 4:1 concentration would probably be acceptable at the DP — requiring $32 \times 10 / 4 = 80$ V-channels, a total of 8 Gbps, between the DP and the CO.

While subscribers have the capacity for hundreds of 400-Kbps elementary broadband channels in both directions, the average busy-hour utilization by each subscriber might be one upstream channel and two downstream channels. So, while each pedestal has hundreds of incoming and outgoing EBCs from each of the 128 subscribers it serves, it can concentrate this capacity down to: 125 channels (50 Mbps) of total upstream capacity and 250 channels (100 Mbps) of total downstream capacity. Since concentrating the E-channels doesn't really save much, Figure 22.5 shows the full complement of 4000 E-channels between the CO and each of its DPs and overprovisions by providing 500 E-channels between the DP and each of its pedestals.

22.6.5 Network Links

Second, consider the links between each node in detail. Links on the network side of the CO include the 10 downstream broadcast UDV channels, the 100 downstream broadcast DTV channels, interoffice trunks for switched DTV channels in each direction, and interoffice trunks for EBCs in each direction.

The cable between the central office and each of its eight distribution points has five fibers, all running at 10 Gbps: two for the 10 downstream UDV channels, one for the 100 downstream broadcast DTV channels, one for the downstream switched DTV channels and EBCs, and one for the upstream switched DTV channels and EBCs.

The cable between any distribution point and each of its 32 pedestals also has five fibers: two at 10 Gbps for the 10 downstream UDV channels, one at 10 Gbps for the 100 downstream broadcast DTV channels, one at 1.2 Gbps for the downstream switched DTV channels and EBCs, and one at 1.2 Gbps for the upstream switched DTV channels and EBCs.

The cable between the pedestal and each of its 128 residences has two fibers, one for each direction. The downstream fiber at 2.5 Gbps carries one UDV channel at 2.0 Gbps, four DTV channels at 100 Mbps each, and additional capacity for 250 of the 400-Kbps EBCs. The upstream fiber at 150 Mbps carries one DTV channel at 100 Mbps and additional capacity for 125 of the 400-Kbps EBCs.

22.6.6 Circuit-Switched Paradigm

Now, consider the switching requirements in each type of node. This subsection examines a configuration based entirely on circuit switching — possibly photonic, in space, time, and wavelength. We'll see that this is probably not a good implementation.

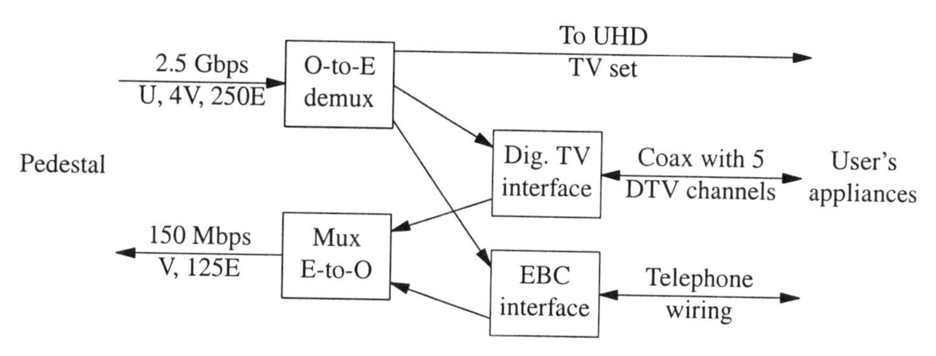

Figure 22.6. Switching at the residential controller.

Consider the switching requirements for the residential controller shown in Figure 22.6.

- If the user subscribes to UDV service, his lone 2.0-Gbps UDV channel is demultiplexed from the downstream fiber and is facility switched to the subscriber's UDV display.
- The four downstream DTV channels are demultiplexed from the downstream fiber and are FDM'ed onto the subscriber's coax.
- If active on a switched DTV call, one of the four downstream channels carries the downstream half of the call. The upstream half of the call is demultiplexed off the house coax and is multiplexed onto the upstream fiber to the pedestal.
- Conventional twisted pair in the residence can carry conventional two-way analog telephone signals, one-way ISDN signals, or one-way EBCs. These analog or digital electronic signals interface to the controller. Downstream channels are demultiplexed from the downstream fiber and upstream channels are multiplexed onto the upstream fiber.

Consider the switching requirements for the pedestal shown in Figure 22.7.

- The 10 UDV channels arrive downstream over two fibers from the DP. The 10 channels are demultiplexed. Then, in a broadband switch with broadcast capability, any of the 10 UDV channels can be switched to any and all of the 128 residential downstream interface units.

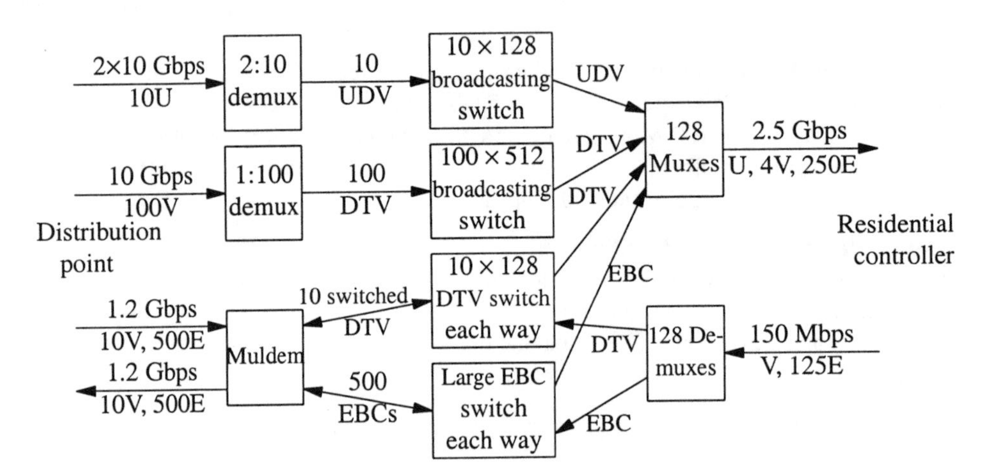

Figure 22.7. Switching at the pedestal.

- The 100 broadcast DTV channels arrive downstream over one fiber from the DP. The 100 broadcast channels are demultiplexed. Then, in a middleband switch with broadcast capability, any of the 100 DTV channels can be switched to any and all of the 128 residential downstream interface units.

- The downstream switched DTV and EB channels are demultiplexed off the downstream fiber from the DP. The upstream switched DTV and EB channels are demultiplexed off the upstream fiber from the RC. The downstream switched DTV channels, downstream EBCs, upstream switched DTV channels, and upstream EBCs are switched in separate point-to-point fabrics. The upstream switched DTV and EB channels are multiplexed onto the upstream fiber to the DP. The downstream switched DTV and EB channels are delivered to the 128 residential downstream interface units.

- Each of the downstream residential interface units multiplexes the user's UDV channel, four downstream DTV channels, and EBCs onto the residence's downstream fiber. Each of the upstream residential interface units demultiplexes its user's upstream DTV and B channels from its upstream fiber.

Consider the switching requirements for the distribution point shown in Figure 22.8.

- The three fibers carrying downstream broadcast channels are simply repeatered and broadcast to each of the pedestals that are served by this DP.

- Channels in the fiber carrying switched DTV and EB channels from the CO are switched to channels carrying DTV and EB channels in the fibers to each of the DP's pedestals. Channels in the fiber carrying switched DTV and EB channels from the 128 pedestals are switched to channels carrying DTV and EB channels in the corresponding fiber to the CO.

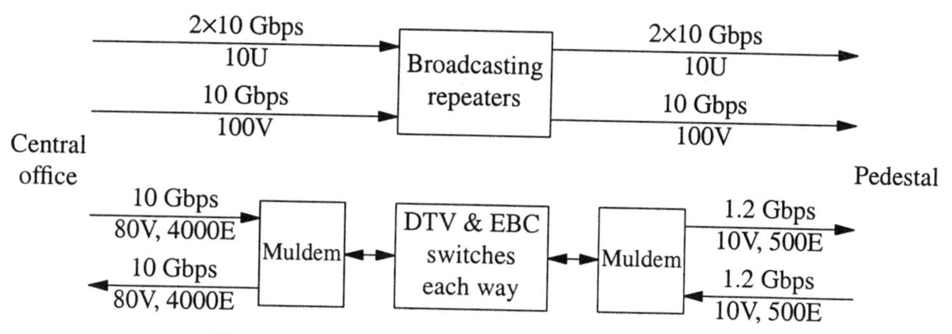

Figure 22.8. Switching at the distribution point.

Consider the switching requirements for the central office shown in Figure 22.9.

- Ten incoming external UDV broadcast signals, at 2.0 Gbps each, are multiplexed in space, time, and wavelength onto two space channels at 10 Gbps each. Each signal is broadcast onto eight downstream fibers, each wired to a different DP.
- One hundred incoming external DTV broadcast signals, at 100 Mbps each, are multiplexed in time and wavelength up to 10 Gbps. This signal is broadcast onto eight downstream fibers, each wired to a different DP.
- Switched DTV channels, at 100 Mbps each, are switched in space, time, and wavelength in each direction. The line side terminates $8 \times 80 = 640$ V-channels in each direction, probably time/wavelength multiplexed on eight internal fibers, one to/from each DP. The details are left as an exercise.
- Switched EBCs, at 400 Kbps each, are switched in space, time, and wavelength in each direction. The line side terminates $8 \times 4000 = 32{,}000$ E-channels in each direction, probably time/wavelength multiplexed on eight internal fibers, one to/from each DP. The details are left as an exercise.

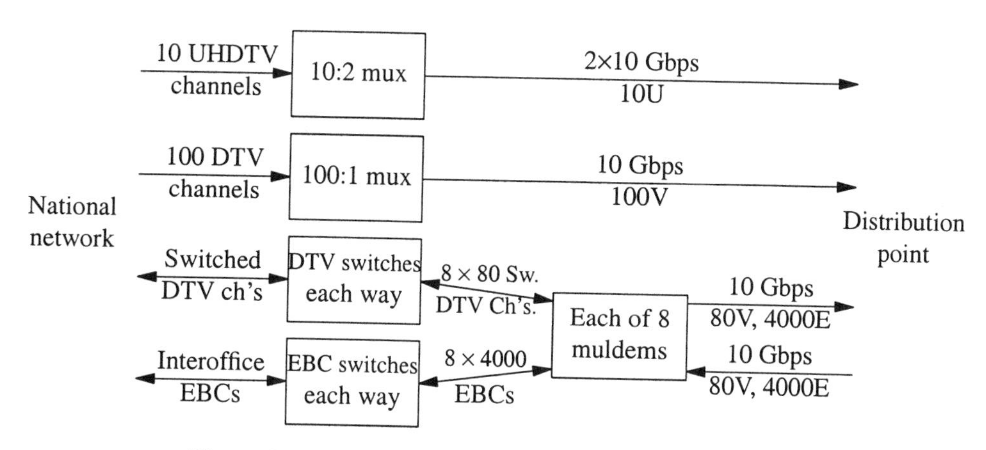

Figure 22.9. Switching at the central office.

- Many of these E-channels carry six regular B-channels. So, the EBC switch in Figure 22.9 would act as an ADM for these channels, which would be switched in the CO's conventional telephone system. Again, the details are left as an exercise.

Intuitively, the infrastructure appears to be sound, but the switching paradigm looks like it would be too expensive.

22.6.7 Cell-Switched Paradigm

Observe the following.

- Because several different specialized switching networks are required, *economy of scope* is lost at every node in the network.
- Broadcast capability, which is difficult in circuit switching, is required in the UDV and DTV fabrics inside the pedestal.
- Intuitively, the pedestal is the most economically critical hub, and in this circuit-switched paradigm, the pedestal requires a significant amount of expensive equipment.

Because this application must deliver multiple types of data, circuit switching is probably not an ideal paradigm. Intuitively, this network seems to be the ideal application in which to use some switching paradigm that provides economy of scope over all data types. ATM appears to be the ideal switching paradigm for this environment. But, just because this conclusion seems to apply in this local network, doesn't mean that it applies in the national network. By this point in this book, you're smarter than this.

22.7 NATIONAL PHOTONIC CAPACITY-SWITCHED NETWORK

An extreme network is proposed here that provides nationwide connectivity, at very high data rates, for all data types. While the photonic network described in the previous subsection is admittedly a bit of a stretch and may be unrealistic, the photonic network described here is seriously proposed to be completely logical and practical.

The most likely application of photonic switching is in a national, or even global, infrastructure network that transmits and switches its data in optical form. A physical infrastructure network is easily partitioned into many different logical special-purpose networks that would be optimized for different traffic types or different paradigms. Such a photonic-switched network must be flexible in, or even transparent to, the data format and the data rate. This section describes the architecture for such a network.

The first subsection describes a simple version that was proposed 15 years ago and the second subsection extends this initial proposal as the upper layer of a two-tiered national network. Sections 22.7.3 and 22.7.4 discuss the capacities of the network's links and the network's capacity-switched (Section 17.5) nature. The last subsection describes the network's lower layer as a set of *overlay networks.*

22.7.1 AmosNet

Figure 22.10 illustrates a prototype of a photonic-switched network that was proposed by Amos Joel in the mid-1980s [4]. Amos proposed placing a photonic time-multiplexed switch (TMS) in Dallas. This network hub would be connected, by fibers in each direction, to nodes in Los Angeles, St. Louis, Atlanta, and Houston. This TMS would

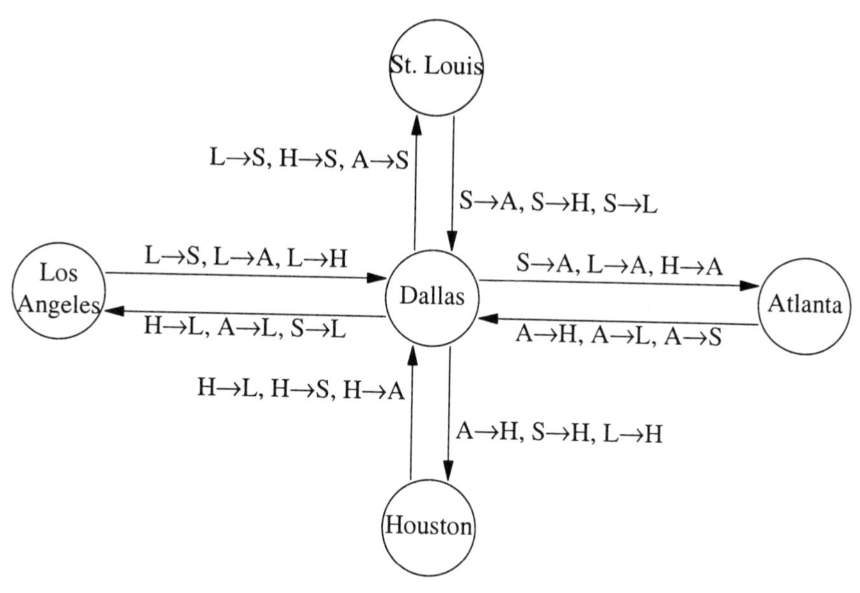

Figure 22.10. Amos Joel's proposed prototype [4].

have three timeslots in each frame and the switch would be reconfigured during a temporal guard band at the beginning of each timeslot. Since each frame holds 2.4 Gbps / 8K fps = 300K bits/frame, each timeslot holds about 100,000 bits, the equivalent of 12,500 DS0 channels or more than 18 DS3 channels.

- During the first timeslot, incoming data from each of the four nodes is switched at Dallas to its next adjacent clockwise node: Los Angeles to St. Louis, St. Louis to Atlanta, Atlanta to Houston, and Houston to Los Angeles.
- During the second timeslot, incoming data from each of the four nodes is switched at Dallas to its node directly across the switch: Los Angeles to Atlanta, St. Louis to Houston, Atlanta to Los Angeles, and Houston to St. Louis.
- During the third timeslot, incoming data from each of the four nodes is switched at Dallas to its next adjacent counterclockwise node: Los Angeles to Houston, St. Louis to Los Angeles, Atlanta to St. Louis, and Houston to Atlanta.

In Figure 22.10, the frames indicated on the network links illustrate how each node would transmit and receive its data in each frame.

So, if X in St. Louis wants to talk to Y in Houston, the switching equipment in St. Louis would have to place X's mouth samples into the second timeslot with other data that is bound from St. Louis to Houston. Switching equipment in Houston would have to know where to find the data for Y in the corresponding arriving timeslot in each frame. Y's address could be transmitted inside a packet with X's mouth samples or a sub-channel could be assigned for the duration of X's call to Y. While data in AmosNet is certainly not packet switched, it really isn't circuit switched either because the photonic TMS in Dallas doesn't switch by the call. The switch in Dallas is a capacity switch, as defined in Section 17.5. Each of the four edge-nodes is a regional hub for circuit-switched voice and packet-switched data traffic; for signaling, telemetry, video, and multimedia traffic; for DS-X, Sonet, IP, frame relay, SMDS, X.25, ATM, and any other format. At least one other layer of switching resides in each hub's region.

Since the corresponding frames in each inbound fiber would have to arrive in Dallas during the timeslot's temporal guard band, some kind of control synchronizes and adjusts the data launch times at each of the four nodes. Furthermore, the timeslot boundaries are variable and some kind of control adjusts the boundaries as traffic varies. AmosNet is seen to be a simple photonic implementation of a network that illustrates three concepts from Chapter 17: (1) big networks can't be flat, (2) the New Jersey Turnpike has more scalable throughput than the Garden State Parkway, and (3) fast-facility capacity switching. While a great deal would have been learned by implementing this trial, unfortunately AmosNet was never prototyped.

The photonic-switched national infrastructure proposed in this section is based on AmosNet, but is extended to (1) multiple hubs, (2) multiple gateways per hub, (3) more than three frame partitions, (4) frame partitions based on time and wavelength, and (5) a more detailed lower layer.

22.7.2 Two-Tiered National Photonic Infrastructure

The total area served is divided into $M \times N$ geographical districts with approximately equal long distance traffic load. In this example, the area served is the continental United States, America's 48 contiguous states. With no better algorithm for selecting M and N, let's simply take an educated guess at two numbers and see if they work out. If they don't work out, then pick different numbers. Let $M = 8$ and $N = 6$. (These numbers will work out — but that still doesn't mean they're the best numbers.) So, the continental United States is divided into $M \times N = 48$ geographical districts with approximately equal total long distance traffic load. While we shouldn't assume that traffic load is uniform over population, if we did this, then each region has about 5 million people. Place one gateway node into each of the $M \times N = 48$ districts.

Then, assign each of the $M \times N = 48$ districts to a region. Each of the N regions contains M contiguous districts. For this example, $N = 6$ regions each contain $M = 8$ districts. Place one tandem node into each of the $N = 6$ regions. Now, connect the $M = 8$ gateway nodes in each region to its tandem node in a star architecture (of course). Connect the $N = 6$ tandem nodes to each other in a mesh architecture.

Figure 22.11 illustrates the proposed architecture. The network's $N = 6$ tandem nodes are located in Philadelphia, Atlanta, Chicago, Dallas, Phoenix, and Boise. Of the $M = 8$ gateway nodes that home on the tandem node in Philadelphia, Figure 22.11 shows only five of them — in Boston, New York City, Washington, Syracuse, and Pittsburgh. Figure 22.11 only shows four of the gateway nodes that home on each of the other tandem nodes. Backup links, connecting each gateway node to a tandem node outside its region, are not shown in Figure 22.11.

Each of the tandem nodes in the "corners" (Philadelphia, Atlanta, Phoenix, and Boise) is connected to its three closest neighboring tandem nodes. For example, Phoenix has tandem links to Boise, Dallas, and Chicago. Each of the tandem nodes in the "center" (Chicago and Dallas) is connected to all the other tandem nodes. This particular tandem-node connection pattern is only suggested — others may be better.

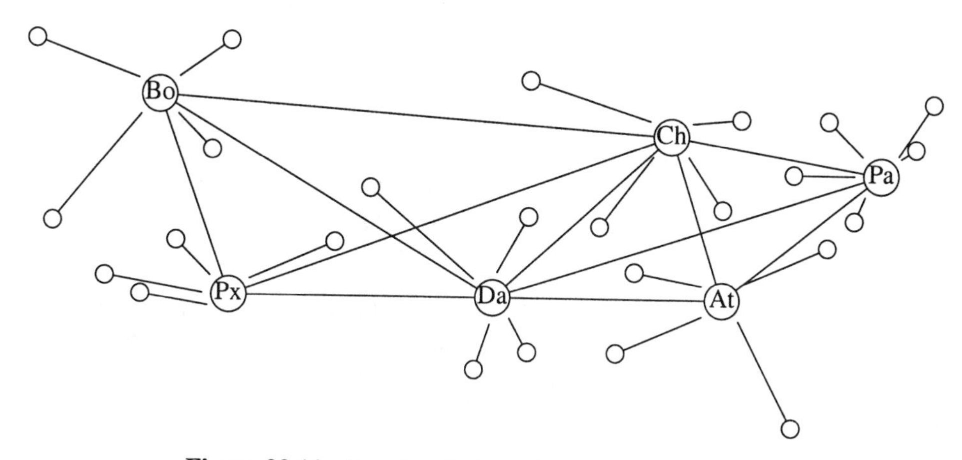

Figure 22.11. Two-tiered photonic national infrastructure.

Examine the function of each class of node in more detail.

- Each tandem node is (analog) photonic switched and capacity switched. Highly multiplexed optical fibers connect each of these tandem nodes to neighboring tandem nodes and to the set of gateway nodes inside its geographical region, and, for reliability, to some gateway nodes in neighboring regions. Besides routing data through the node for other nodes, each tandem node also drops and adds data for the gateway nodes that home on it. All the fiber facilities would be multiplexed to the fullest extent that the technology, competition, and demand require. Changes in multiplexing are assumed to occur frequently and such changes should be transparent at these tandem nodes.
- Each gateway node is a gateway to this high-rate nationwide infrastructure network, providing interdistrict connectivity for all the networks in the gateway's geographical district. Each gateway node interfaces to every type of network (circuit-switched POTS, video, X.25, SMDS, frame relay, IP, ATM, SS7, etc.) in its district. Each gateway node connects to every switching office, PBX, ISP, MAN, and CATV head end in its district and, for reliability, to some of these in neighboring districts. The gateway nodes could also serve as a hub switch for intradistrict connectivity.

For example, the operation is illustrated by walking through a POTS call. Let residential telephone subscriber A in North Andover, Massachusetts, call residential telephone subscriber B in San Clemente, California. A's mouth samples are routed from his telephone, to his serving switching office in North Andover, and over a trunk to the gateway node in Boston. There, A's mouth samples are switched to a multiplexed super-channel with other traffic of all types that are bound for the San Diego gateway. This super-channel might switch through tandem nodes in Philadelphia, Dallas, and Phoenix. In Phoenix, this entire super-channel from Boston is switched to the gateway node in San Diego, where it is demultiplexed. As circuit-switched POTS, A's mouth samples are trunked to a telephone switching office in San Clemente, and on to B's telephone. A residence-to-residence call is routed through a maximum of seven switching nodes and an enterprise-to-enterprise call through a maximum of nine.

22.7.3 Link Capacities

Within each district, every class-5 local switching office, every private network, every MAN and WAN, every Internet service provider, and possibly even every CATV head end provisions its interdistrict channels to its nearest gateway node and, for fault tolerance, to one other gateway node in some neighboring district. These channels carry every type of traffic between this district and any other district, analogous to how an LEC directs inter-LATA traffic to the IXC POP within a LATA.

At each gateway node, every type of outbound traffic from the district's sources is sorted according to the district of its destination, but not according to its type. The outbound data is switched and multiplexed into 47 optical super-channels that are multiplexed by space, time, and wavelength. These "chunks" of the facilities are called super-channels in the subsection because they are comprised of finer, more elementary channels, as described in the next subsection. Furthermore, the analysis in this subsection is simplified by ignoring the backup connections or by assuming they are implemented as idle spares.

So, these 47 super-channels are transmitted to the tandem node that serves this gateway node. At any tandem node, all the incoming multiplexed super-channels are switched (in space, time, and wavelength) to multiplexed super-channels that proceed to other gateway nodes in this region or to multiplexed super-channels that connect to other tandem nodes.

Assume for the moment that each of the 48 gateway nodes in Figure 22.11 handles the same amount of traffic in each direction. Since intradistrict traffic of every type would be handled by the district's carrier for this type of traffic, any traffic that reaches a gateway node is interdistrict traffic. Assume that the typical user has interdistrict busy-hour traffic of 0.06 Erlangs per typical telephone user and 1 Kbps per typical data user. Then, a district's 5 million users produce an average of

$$(0.06 \times 5\text{M} \times 64\ \text{Kbps}) + (5\text{M} \times 1\ \text{Kbps}) = 24\ \text{Gbps}$$

in each direction. If we assume that interdistrict traffic is distributed uniformly over the other 47 districts, then each gateway node transmits 24 / 47 = 0.5 Gbps to every other district.

Since each gateway node homes on one (principal) tandem node, 32 Gbps in each of these links should provide enough capacity, while still accounting for traffic fluctuations, occupancy, queuing, and short-term growth. So, each tandem node has uplinks from eight gateway nodes and downlinks to eight gateway nodes, with maximum capacity of 32 Gbps in each link. At each tandem node, 7 / 47 = 15% of all traffic is intraregion to/from its gateways and flows from/to other gateways served by this same node. So, 85% of all traffic, $0.85 \times 8 \times 24$ Gbps = 163 Gbps, is interregion to/from other tandem nodes.

Examine the highest level of topography in Figure 22.11.

- Each of the four corner nodes has a north-south trunk link to the corner node on the same coast and has east-west trunk links to both center nodes. Since no corner node would normally carry tandem traffic, the total traffic leaving (and entering) each of these corner nodes is only the 163 Gbps that comes from (goes to) the gateways it serves.

- Each corner node's north-south link normally carries one-fifth of the traffic in/out of these nodes, or 163 Gbps / 5 = 32 Gbps. The center node's north-south link would also normally carry 32 Gbps. Forty Gbps should be sufficient capacity in these links while still accounting for traffic fluctuations, occupancy, queuing, and short-term growth.
- Each corner node's two east-west links normally carries half of the remaining four-fifths of the traffic in/out of these nodes, or 163 Gbps × 0.4 = 64 Gbps. Eighty Gbps should be sufficient capacity in these links while still accounting for traffic fluctuations, occupancy, queuing, and short-term growth.
- Each of the two center nodes (Chicago and Dallas) connects to all of the other five nodes. The 64 Gbps of traffic over each link to the corner nodes contains 32 Gbps of local traffic and 32 Gbps of tandem traffic.
- For fault tolerance, all calculated capacity requirements for link transmission and node throughput would have to be doubled.

22.7.4 Capacity Switching

According to the discussion above, the links between the tandem nodes each carry 40 super-channels, one super-channel for each district outside the region served by the particular tandem node. A link between a tandem node and any gateway node it serves carries 47 super-channels, one super-channel for each of the other districts. In both types of links, these super-channels are comprised by aggregating a set of *elementary channels* together — and this is why they are called super-channels.

Suppose each elementary channel operates at the OC1 rate of 50 Mbps. Then, each east-west intertandem link has the capacity for 80 Gbps / 50 Mbps = 1600 of these, but only uses 64 Gbps / 50 = 1280 of them, on the average. Each north-south intertandem link has the capacity for 40 Gbps / 50 Mbps = 800 of these, but only uses 32 Gbps / 50 = 640 of them, on the average. And, each tandem-to-gateway link has the capacity for 32 Gbps / 50 Mbps = 640 of these, but only uses 24 Gbps / 50 = 480 of them, on the average.

If all the traffic loads equal the average load, then these elementary channels would be capacity switched through the two tiers of the network so that every gateway node has a 10-channel super-channel to each of the other 47 gateway nodes. These small super-channels provide 500 Mbps between every pair of gateway nodes, which equals the average gateway-to-gateway rate calculated previously. Such an assignment leaves a pool of unused elementary channels in each link.

If data with a particular source and destination begins to block, queue, or otherwise congest in any gateway, then this gateway arranges with the other nodes to acquire additional elementary channels, from the pool of spare capacity in each intermediate link and node. These additional elementary channels increase the size of the super-channel between the given source gateway and given destination gateway, and lower the congestion. A similar mechanism would serve to reduce the capacity of an oversized super-channel by releasing some of its elementary channels back to the idle pool in the super-channel's constituent links and nodes.

While the control processes and the necessary signaling are not trivial, none of this seems impractical. The most interesting issue, which makes or breaks the proposed architecture, is whether the network is able to respond quickly enough to changes in traffic loads. It seems intuitive that the SS7 network would be inadequate. It seems intuitive that fairly simple look-ahead schemes could be implemented to make this work. A lot needs to be done.

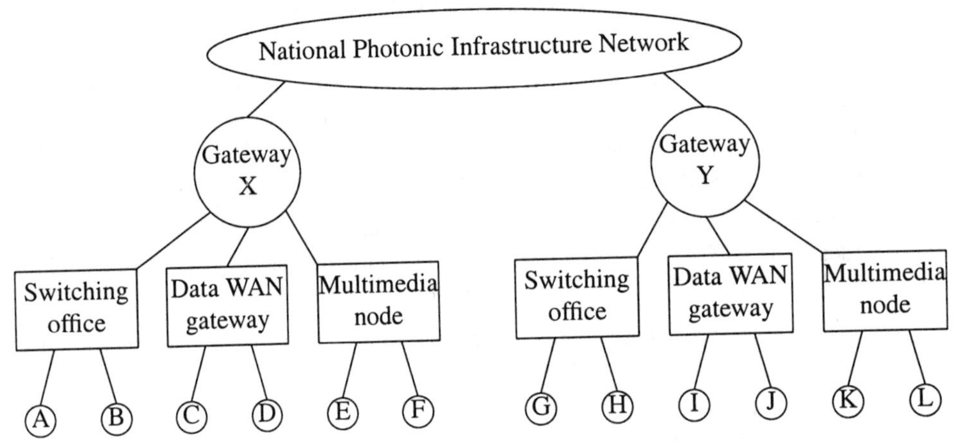

Figure 22.12. Specialized overlay networks.

22.7.5 Overlay Networks

Each of the 48 gateway nodes offers a capacity-switched direct super-channel to each of the other 47 gateway nodes. If the super-channel's bandwidth is insufficient enough that congestion or blocking begin to occur, the network automatically makes the super-channel bigger. What this architecture loses in economy of scale, it gains in economy of scope.

Figure 22.12 illustrates the operation of a capacity-switched national network.

- Let telephone user A in district X place a POTS telephone call to telephone user H in district Y. A's LEC connects A to gateway node X, where his POTS channel is multiplexed onto a 64-Kbps circuit within the national X-to-Y super-channel. All data in this super-channel is delivered to the gateway node in district Y, where A's channel is demultiplexed and connected through H's LEC to H.
- Let computer user I in district Y download a file to computer user D in district X. Let I's computer and D's computer each be connected to respective district-wide networks that are price and performance optimized for file transfer — connectionless packet-switched networks like Ethernet LANs connected to MANs using frame relay or IP. I's LAN and MAN transmit I's connectionless packets to gateway node Y, where his packets are multiplexed up onto a statistically multiplexed channel within the national Y-to-X super-channel. All data in this super-channel is delivered to the gateway node in district X, where I's packets are demultiplexed and transmitted through D's MAN and LAN to D.

- Multimedia traffic is not the simple union of all traffic types — it is a type of traffic that is distinct from the other traffic types and has characteristics that are intermediate to voice and data. Let multimedia user F in district X place an integrated voice/data/video call to multimedia user L in district Y. F's district-wide ATM carrier connects F's cells to gateway node X, where his ATM cells are multiplexed up onto a broadband ISDN CBR channel within the national X-to-Y super-channel. All data in this super-channel is delivered to the gateway node in district Y, where F's channel is demultiplexed and connected through L's district-wide ATM carrier to L.

Besides voice, file transfer, and multimedia, other traffic types might generate enough traffic to justify implementing district-wide networks that have switching paradigms that are optimized for this particular traffic type. These additional traffic types might include transaction processing, high-definition digital video, and telemetry. While such specialized networks appear to make sense in a given district, separate nationwide networks for each traffic type may be difficult to justify. The underlying premise here is that very large networks, like a U.S.-sized NAN, have to operate at high data rates without a lot of intermediate demultiplexing and switching — and especially not packet switching.

When voice was the predominant traffic type, it made sense to have a nationwide network that was optimized for voice and to carry the other traffic types over this network at less than optimal performance. Now that neither voice nor file transfer is predominant and both types of traffic are so different, it may not make sense for one traffic type to carry the other one. Instead, as proposed here, yet another traffic type is proposed, one that is optimized for a national network at very high data rates. Then, all traffic types, while separated inside their respective districts, are all carried over this nationwide transport network. After bashing integrated networks in Chapter 17 and earlier in this chapter, isn't this an integrated network? Yes, it is — but this one scales well.

22.8 GLOBAL TELECOMMUNICATIONS NETWORK

This section describes the planning for a third extreme network. Section 22.6 proposed a network that is extreme in its bandwidth and the previous section proposed a network that is extreme in its connectivity and scope. This section proposes a network that is extreme in its size — a network that might provide a new universal service to every human on earth.

22.8.1 A Global Vail-Like Commitment

As discussed in Section 2.1.4, AT&T President Theodore Vail stated the Bell System's goal for universal service about 100 years ago. By setting the goal of bringing affordable POTS to every family in the United States, he sized the network's ultimate architecture. While Vail didn't say it this way, he specified that Bell's national network had to be able to grow to sufficient capacity that it could bring the analog equivalent of 64 Kbps per line to 100 million (10^8) people, at an average penetration of five people per line (considering several people per residence and party lines), at an average usage occupancy of about 0.8%. While the Bell System's initial network didn't have the capacity for 64K × 100M × 0.2 × 0.008 = 10^{10} = 10 Gbps of throughput, it had an architecture that not only was capable of growing to this capacity with the technology of that day, but did grow to this capacity and well beyond it. In the United States, we even outgrew Vail's vision because of increases in population (more than doubled), penetration (the United States has more telephones than people), and occupancy (including holding times for Internet connections). Furthermore, the United States has several competing networks, each at least as big as Vail's original specification, and we're looking seriously at raising the per-line bandwidth.

Now it is time for, not the United States or any other single nation, but the entire human species, to make a similar commitment. We need a new network, a redefined universal service, and a global Vail-like commitment to achieve them. Let's be as far-reaching today as Vail was 90 years ago. We must raise the network's user base by almost two orders of magnitude, to 6 billion (6×10^9) people.

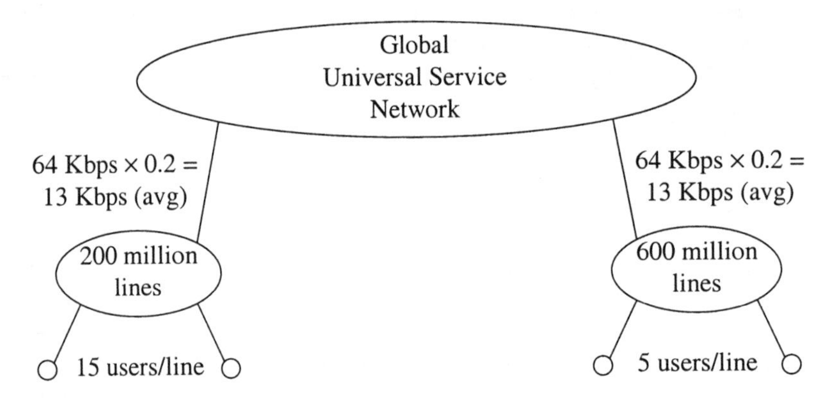

Figure 22.13. The ultimate global network.

As shown in Figure 22.13, if we bring POTS to that 50% of the world's population who have never made a single telephone call, even at 15 people per line, the network needs 200 million telephone lines, worldwide. Assuming five people per line for the other half of the population, the network needs another 600 million lines. Now, the global network serves the 50% of us who are "haves" with 700 million lines.

But, data connectivity should increase the average holding time to much more than Vail's network had — perhaps to a 2% occupancy. Taking all this together, the specified throughput capacity of the network is $64K \times 0.8G \times 0.2 = 10^{13}$ = 10 Tbps, worldwide. This new network requires three orders of magnitude more capacity than Vail's network and it must cover two orders of magnitude more territory than the network Vail proposed 90 years ago. While this is not the size of the network we must build (yet), it is the size to which the network's architecture must grow — gracefully.

While the "service" in "universal service" is specified by the per-capita bandwidth and the occupancy, the "universal" in "universal service" is specified by the number of ports on the network's edge and by its affordability. In the United States, manufacturers and carriers salivate over projections that up-scale users might be willing to pay \$1000 – \$10,000 for customer premises equipment and as much as \$100 per month for connectivity and usage. But, these are unrealistic figures for many Americans. And, if these figures price America's have-nots out of the market, how could they work for a global universal service?

Let's provide global universal telephone/Internet service at the minimal, most basic level, for $10 per month to a voice/data appliance priced at $200. For voice-only service, the monthly rate could be even lower (because the holding time is much lower) and the appliance could be even more affordable. While $8 telephones are commercially available in the United States, such a telephone assumes a typical local infrastructure with RJ-11 jacks, house wiring, and a drop wire to a pedestal. But, this local infrastructure is the principal economic barrier in developing countries. So, our universal telephone has to be wireless, and perhaps $25 is a reasonable goal for its price.

If we can agree that this is our goal, then anyone who proposes a network architecture that can't grow to this capacity or that can't offer service within an order of magnitude of these prices has, by definition, proposed a niche network. Niche networks have their place and a lot of money can be made in the niches. But, niche networks can't provide universal service.

22.8.2 Alternative Technologies

Figure 22.13 doesn't describe this network in enough detail that we can make significant decisions about technology and architecture. But, Figure 22.13 does describe this network in enough detail, surprisingly, that we can ask some interesting questions — which have implied answers.

- Can the network in Figure 22.13 be a single dark-fiber all-optical network? If the 10^{13}bps = 10 Tbps is optical, that is more bandwidth than we'll ever get practically from one optical fiber. Besides, we could never even get close to having that many wavelengths on one optical fiber that is long enough to cover the entire planet. And, only a few people could afford the instrument because of the price of its interface to the network.

- Can the network in Figure 22.13 even be flat? If Section 17.2 didn't settle this already, obviously a public-switched global network can't be one big Ethernet. But, could it be a worldwide terrestrial star network with a single switch (probably ATM) at its single hub? Obviously, we have to take advantage of calling patterns. We have to have short channels and long channels. It seems silly to think that, economically, we could allocate a global channel to party A who wants to call party B across the street.

- Can the network of Figure 22.13 be the Internet, a large mesh of IP routers? Can the network of Figure 22.13 be a large mesh of ATM switches? Is packet switching of any kind a practical implementation for this network? One of these could be a contender, but probably not at the network's highest layer.

- Can the network in Figure 22.13 be satellite based? Satellites give everyone equal access to the network. But, if the 10 Tbps is wireless of any type, it will require a lot of footprints and a lot of frequency reuse. Since voice will never be acceptable over GEOS, we must look at LEOS. A large collection of satellites in low earth orbit may be a reasonable implementation, if it handles the global interconnections.

- Can the network of Figure 22.13 be built like the capacity-switched photonic network that was proposed in the previous section? This is probably the most scalable implementation at the highest layer, but it requires an access network.

> How can we afford to run telephone lines to the half of the world's population who have never placed a telephone call?

What's implied? How about using low earth orbit satellites, probably running some version of ATM that's been modified for wireless, for access only? How about using a capacity-switched photonic network as the upper layer?

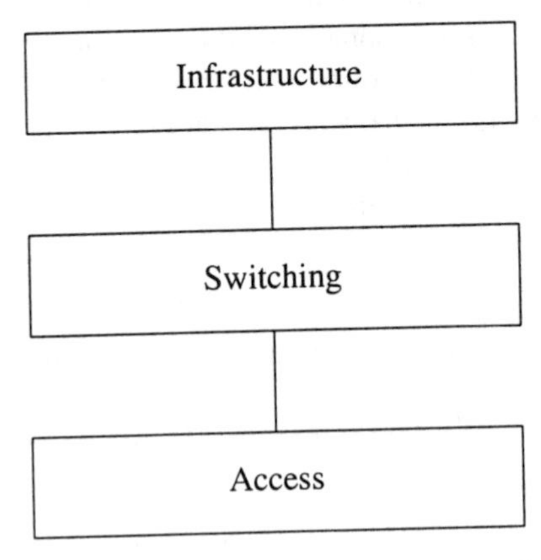

Figure 22.14. Layers of a big network.

22.8.3 Network Layers

A big network, one that spans many thousands of miles and several billion users, needs at least three separate layers (shown in Figure 22.14) for access, for switching, and for infrastructure.

Access. Assume that the network's lowest layer tiles the earth with physical and logical serving areas that hold about 100,000 lines each. This size is within an order of magnitude of the area served by switching office building, the area served by a CATV head end, the area served by a cellular switching office, and the cell footprint of a low earth orbit satellite. This is also roughly the number of lines that can be served by a practical computer system before it runs out of real time. This seems, somehow, to be a natural quantum, a similar level of granularity across several different technologies.

The network's interface to the CPE on its edge should be optimized to offer the end user low overall end-to-end price. Network access should, first and most importantly, enable minimal total price for its minimal universal service. Second, but within the constraints of the first requirement, it should integrate the user's CPE for other services to a common low-priced interface.

This access layer has both a physical and a logical component. Physically, we have to get as much use as we can from our embedded base of twisted pairs and coaxial cable, but not if it means further separating the information haves from the information have-nots. And, we have to be careful not to invest so much in upgrading our embedded base, that we embed it deeper and further postpone the inevitable migration to fiber access.

Physically, we need an inexpensive telecommunications interface device, a merger between a set-top box and a telephone network interface, that allows house wiring and house coax on one side to interface to drop wires, CATV coax, and anything else on the other side. Functioning as a dumb gateway/PBX, it has to be battery backed, with optional PWBs. Logically, ATM seems like the best choice of protocol, but at a low enough data rate that it's mass-marketable. And, it needs a workable strategy for gradual migration from Ethernet in the office and from POTS in the residence.

But, is access the biggest impediment? While, economically, it may be, there are several good ways to provide 64-Kbps access to 100,000 people, all of whom live relatively close together. It may have a significant price, but most of us can conceive of several different technologies that have the capability. Most of us see several ways for subscriber I in serving area R to call subscriber J who is also in serving area R. It's not so clear how subscriber I in serving area R calls subscriber K who resides in a serving area on the other side of the earth.

A global network that serves 1 billion lines needs upper layers that are capable of interconnecting 10,000 of these serving areas.

Upper Layers. One big problem is addressing. While we have a global telephone numbering plan and a global scheme for IP addresses, neither seems appropriate for this large network. While we're not going to solve this problem here, we are going to look at the network's topology. Section 17.2 asserted that "big networks can't be flat." If a big network needs at least two layers, then how many layers does it need? What is the function of each layer? Which layers are physical and which are logical?

The previous section described a capacity-switched photonic network. Since its districts contain about 5 million lines each, each district served by a gateway node covers 50 of the serving areas described in this section. Obviously, some kind of regional network interconnects all the serving areas in some region and connects them all to the region's gateway node. But, is there still another layer? Does the architecture of Figure 22.11 scale to the entire planet? My intuition tells me that it doesn't and that yet another layer is required above this capacity-switched photonic network. This really interesting and provocative problem is left open.

This section opened with the plea that we make a Vail-like commitment to build a global network. But, who is the "we" who should make the commitment?

22.9 EPILOGUE

Well, you're finally done. I hope you liked my book and I hope you learned something. I know it was long but, I thought the WHOLE story had to be told. I worked hard on this project, not just on the facts and the presentation, but also on the organization, the style, and the interwoven themes. I hope you think it was worth it.

If you don't agree with my conclusions, I hope it isn't because I didn't present them clearly and I hope it isn't because of hype and spin from those who are trying to sell you different conclusions. If you don't agree with my conclusions because of sound economic and architectural reasoning, then that's OK; maybe I'm wrong. And, if I am wrong, then dismiss the conclusions but, please, don't dismiss the entire book or the methodologies contained inside it.

If you learned something between these covers, then I'm pleased. If I made you think outside the box (even just for today) then I'm glad I put in the effort. But, I hope I've changed you fundamentally — so you can never look at something again without: (1) questioning how and why it got that way, (2) doubting fact-less hype and those who use it, and (3) reminding yourself that "it works" is not the same as "it's good."

I'm sure there are still altruists, scholars, and genuine seekers after wisdom and truth out there. If you are one and if you have something constructive to say, then I'd like to hear from you. Please send me e-mail at *rat@tele.pitt.edu.*

EXERCISES

22.1 Describe the following people in the context of Section 22.3 and Figure 22.2: (a) yourself, (b) the "connected" person you would like to be, and (c) the most "connected" person you know.

22.2 Propose a "minimal" Internet connection and a "minimal" Internet that might be included with voice in universal service.

22.3 Consider the human activity of reading a book and propose a corresponding new service, Tele-Reading. Discuss which aspects of real reading would probably be compromised in Tele-Reading. Discuss what new features, not available with real reading, could be available in Tele-Reading. Describe how Tele-Reading might be customized for reading books like this one instead of just books in general or might be customized for some of the groups mentioned in Section 22.5.5.

22.4 Think of some other human activity, X, and propose its corresponding Tele-X. Discuss which aspects of X would probably be compromised in Tele-X. Discuss what new features, not readily available with X, could be made available in Tele-X. Describe how Tele-X might be customized.

22.5 For the photonic circuit-switched Star^3 shown in Figure 22.5, specify how the various links should be multiplexed in space, time, and wavelength and which blocks should be electronically or photonically switched.

22.6 Provide further details for what might be inside some of the blocks in Figure 22.6 through Figure 22.9.

22.7 DS0 channels (ISDN's B-channels) can be time multiplexed inside one of the E-channels described in Section 22.6. In which nodes of Figure 22.5 should this multiplexing and demultiplexing take place?

22.8 Section 22.6 concludes that ATM is probably a better paradigm than circuit switching for accessing the service described in Section 22.6.1. (a) Discuss whether this conclusion implies that ATM should also be used for the signals and channels at the extreme left side of Figure 22.5. (b) Since UDV might be problematic over ATM, redesign the residential Star^3 architecture of Section 22.6 so that it provides UDV, for those customers who subscribe to it, over a photonic circuit-switched subnetwork and the rest of the services over an ATM subnetwork.

22.9 Redesign the AmosNet architecture of Figure 22.10, adding a fifth edge node in Pittsburgh. Specify the modified frame format on the links, the four interconnection configurations in Dallas, and their schedule.

22.10 Redesign the AmosNet architecture of Figure 22.10, adding an active spare hub in Oklahoma City for reliability. Specify the frame format on the added links and the timeslot schedules at each of the four edge nodes.

22.11 Using a current census and trying to conform to state boundaries, specify the approximate boundaries of the 48 districts, all having approximately equal population, described in Section 22.7.2. Group these districts into the six regions described in Section 22.7.2.

22.12 Add an international gateway in Baltimore as a tandem node to the architecture of Figure 22.11. Redraw the network's links and respecify the size of the average capacity-switched channels and super-channels.

22.13 Specify the assignment of backup links between each gateway node and a secondary tandem node so that no tandem node in Figure 22.11 connects to more than 10 secondary gateway nodes.

22.14 Suppose the 48 gateway nodes are assigned backup links to secondary tandem nodes as in Exercise 22.13. If the tandem node in Chicago were to fail, specify the modified average capacity-switched channels and super-channels for the other links in the network of Figure 22.11.

22.15 Redesign the photonic infrastructure network of Figure 22.11 by specifying the locations of the tandem nodes and the average capacity-switched channels and super-channels for: (a) four regions of 12 districts each, (b) eight regions of six districts each, or (c) 12 regions of four districts each.

22.16 Redesign the photonic infrastructure network of Figure 22.11 for larger districts: 24 districts with about 10 million people each. Specify the locations of the tandem nodes and the average capacity-switched channels and super-channels for: (a) four regions of six districts each, or (b) six regions of four districts each.

22.17 Redesign the photonic infrastructure network of Figure 22.11 for a three-tiered topology. Let the lower 48 states be partitioned into four fully connected super-regions. Let each super-region contain a super-tandem node and three regions. Let each region contain a subtandem node and be partitioned into four districts. Let each district contain a gateway node. Specify the locations of the super-tandem nodes and subtandem nodes then specify average capacity-switched channels, super-channels, and super-super-channels.

22.18 Suppose a company has offices in Hartford, Baltimore, Pittsburgh, and Chicago. Draw its frame relay EPN as an overlay network on Figure 22.11.

22.19 Resize the global network of Figure 22.13 by assuming that 1 billion people have one person per line, 1 billion people have five people per line, and 4 billion people have 15 people per line.

22.20 Resize the global network of Figure 22.13 by assuming that each line is busy 1% of the time on 64-Kbps voice and 1% of the time on 10-Kbps computer data.

22.21 Apply the photonic network described in Section 22.7 to the global network described in Section 22.8. (a) Define 48 global districts of approximately equal traffic load as 24 developed-nation districts each with about 62.5 million people, and 24 third-world districts each with about 187.5 million people. (b) Organize these 48 districts into a topology analogous to Figure 22.11, except with 12 global regions of four districts each. (c) Assuming 1% interregion traffic, specify the average capacity-switched channels and super-channels for this global network.

REFERENCES

[1] Sobczak, J., "How Visionaries Lead Us Astray," *Business Communications Review,* Oct. 1998.

[2] Thompson, R. A., and K. L. Montgomery, "The Synergy Between Photonic Switching and the Intelligent Network," *Proc. IEEE Intelligent Network Workshop '91.* Dallas, TX, Apr. 24 – 25, 1991, Session VI.

[3] Thompson, R. A., "Integration and User-Orientation of Broadband I. N. Services," *Proc. IEEE Intelligent Network Workshop '92.* Newark, NJ, Jun. 23 – 24, 1992, Session B3.1.

[4] Joel, A. E., Jr., *Photonic Switching,* United States Patent #4,736,462, Apr. 5, 1988.

SELECTED BIBLIOGRAPHY

Acampora, A. S., "The Scalable Lightwave Network," *IEEE Communications Magazine,* Vol. 32, No. 12, Dec. 1994, pp. 36 – 42.

"Adapting State Regulation of Telephone Companies to Industry Change," *ICA White Paper,* Boston, MA: ICA, 1989.

Anderson, B., "Telecommunications Regulation in the States," *IEEE Communications Magazine,* Vol. 33, No. 12, Dec. 1995, pp. 78 – 80.

Bainbridge, H., "Learning New Traditions," *Wireless Business & Technology,* Vol. 5, No. 1, Jan. 1999.

Bernt, P., and M. Weiss, *International Telecommunications,* Carmel, IN: Sams Publishing, 1993.

Brands, H., and E. T. Leo, *The Law and Regulation of Telecommunications Carriers,* Norwood, MA: Artech House, 1999.

Brock, G. W., *Telecommunication Policy for the Information Age,* Cambridge, MA: Harvard University Press, 1994.

Brock, G. W. (ed.), *Toward a Competitive Telecommunication Industry,* Mahwah, NJ: Lawrence Erlbaum Associates, 1995.

Canavan, J. E., "Universal Service Policy in the United States: Where Do We Go from Here?" *Telecommunications,* Vol. 31, No. 11, Nov. 1997.

Chen, T. M., and D. G. Messerschmitt, "Integrated Voice/Data Switching," *IEEE Communications Magazine,* (issue on Integrated Voice and Data Communications), Vol. 26, No. 6, Jun. 1988.

Clarke, A. C., "Extra-Terrestrial Relays: Can Rocket Stations Give Worldwide Radio Coverage?" *Wireless World,* Oct. 1945.

Cray, A., and P. Heywood, "Building the Brave New World," *Data Communications,* Vol. 27, No. 14, Oct. 1998.

Culver, D., "Poverty Lines," *tele.com,* Vol. 3, No. 9, Aug. 1998.

Davies, K. P., "HDTV Evolves for the Digital Era," *IEEE Communications Magazine,* Vol. 34, No. 6, Jun. 1996, pp. 110 – 112.

Dorros, I., "The New Future — Back to Technology," *IEEE Communications Magazine,* (issue on Telecommunications Regulation), Vol. 25, No. 1, Jan. 1987.

Dziatkiewicz, M., "Universal Service — A Taxing Issue," *Wireless Business & Technology,* Vol. 5, No. 1, Jan. 1999.

Farrel, J., and M. L. Katz, "Public Policy and Private Investment in Advanced Telecommunications Infrastructure," *IEEE Communications Magazine,* Vol. 36, No. 7, Jul. 1998, pp. 87 – 92.

Gareis, R., "Brave New World Betrayed," *Data Communications,* Vol. 27, No. 14, Oct. 1998.

Gareis, R., and P. Heywood, "Tomorrow's Networks Today," *Data Communications Magazine,* (issue on Global Supernets), Vol. 24, No. 13, Sept. 21, 1995.

Geller, H., "Government Policy as to AT&T and the RBOCs: The Next 5 Years," *IEEE Communications Magazine,* (issue on Telecommunications Deregulation), Vol. 27, No. 1, Jan. 1989.

Geller, H., "U.S. Domestic Telecommunications Policy in the Next Five Years," *IEEE Communications Magazine,* (issue on Telecom at 150), Vol. 27, No. 8, Aug. 1989.

Gillett, S. E., and I. Vogelsang (eds.), *Competition, Regulation and Convergence,* Mahwah, NJ: Lawrence Erlbaum Associates, 1999.

Green, P. E., Jr., "Toward Customer-Useable All-Optical Networks," *IEEE Communications Magazine,* Vol. 32, No. 12, Dec. 1994, pp. 44 – 49.

Harrold, D. J., and R. D. Strock, "The Broadband Universal Telecommunications Network," *IEEE Communications Magazine,* (issue on Telecommunications Regulation), Vol. 25, No. 1, Jan. 1987.

Hayashi, K., "NTT's Open Network Declaration from a Historical Perspective," *IEEE Communications Magazine,* Vol. 34, No. 7, Jul. 1996, pp. 54 – 60.

Heywood, P., and T. C. Eng, "Global Supernets: Big Pipes, Big Promises... and One Big Problem," *Data Communications Magazine,* (issue on Global Supernets), Vol. 24, No. 13, Sept. 21, 1995.

Hiltz, S. R., and M. Turoff, *The Network Nation — Human Communication via Computer,* Reading, MA: Addison-Wesley, 1978.

Howard, P. K., *The Death of Common Sense — How Law Is Suffocating America,* New York, NY: Warner Books, 1994.

Hundt, R. E., and G. L. Rosston, "Alternative Paths to Broadband Deployment," *IEEE Communications,* Vol. 36, No. 7, Jul. 1998.

King, R., "Last on Line — To Have and Have Not," *tele.com,* Vol. 3, No. 9, Aug. 1998.

King, R., "The Universal Service Crisis — Agrarian Myths," *tele.com,* Vol. 3, No. 7, Jun. 1998.

Langford, G., "Planning the Multi-National Network," *Telecommunications,* Vol. 23, No. 11, Nov. 1989.

Lawyer, G., "The Universal Service Crisis — Rural Retreat," *tele.com,* Vol. 3, No. 7, Jun. 1998.

Lin, L. Y., E. Karasan, and R. W. Tkach, "Layered Switch Architectures for High-Capacity Optical Transport Networks," *IEEE Journal on Selected Areas in Communications,* Vol. 16, No. 7, Sep. 1998, pp. 1074 – 1080.

Lucas, J., "What Business Are the RBOCs In?" *TeleStrategies Insight Magazine,* Sept. 1992.

Malek, M., "Integrated Voice and Data Communications Overview," *IEEE Communications Magazine,* (issue on Integrated Voice and Data Communications), Vol. 26, No. 6, Jun. 1988.

Marcus, M. J., "Technical Deregulation: A Trend in U.S. Telecommunications Policy," *IEEE Communications Magazine,* (issue on Telecommunications Regulation), Vol. 25, No. 1, Jan. 1987.

Marshall, C. T., "The Impact of Divestiture and Deregulation on Technology and World Markets," *IEEE Communications Magazine,* (issue on Telecommunications Deregulation), Vol. 27, No. 1, Jan. 1989.

McCain, J., "Telecom Reform: Avoiding the Terrible Two's," *Telecommunications,* Vol. 32, No. 1, Jan. 1998.

McDonald, J. C., "Deregulation's Impact on Technology," *IEEE Communications Magazine,* (issue on Telecommunications Regulation), Vol. 25, No. 1, Jan. 1987.

McKenna, R. B., and D. L. Poole, "Data Communications: Where Regulators Clash with Reality," *IEEE Communications Magazine,* Vol. 36, No. 7, Jul. 1998, pp. 96 – 99.

Miki, T., "The Potential of Photonic Networks," *IEEE Communications Magazine,* Vol. 32, No. 12, Dec. 1994, pp. 23 – 27.

Nederlof, L., et al., "End-to-End Survivable Broadband Networks," *IEEE Communications Magazine,* (issue on Self-Healing Networks for SDH and ATM), Vol. 33, No. 9, Sept. 1995.

Noll, A. M., "Tele-Perspectives," *Telecommunications,* Vol. 32, No. 11, Nov. 1998.

Olson, J. E., "New Policies for a New Age," *IEEE Communications Magazine,* (issue on Telecommunications Regulation), Vol. 25, No. 1, Jan. 1987.

Perez, F., "The Case for a Deregulated Free Market Telecommunications Industry," *IEEE Communications Magazine,* Vol. 32, No. 12, Dec. 1994, pp. 63 – 70.

Perrin, S., "The CLEC Market: Prospects, Problems, and Opportunities," *Telecommunications,* Vol. 32, No. 9, Sept. 1998.

Personick, S. D., "Towards Global Information Networking," *Proceedings of the IEEE,* Vol. 81, No. 11, Nov. 1993, pp. 1549 – 1557.

Pollman, S., "Set-Top Terminal Design," *Communication Systems Design,* Vol. 3, No. 3, Mar. 1997.

Rogers, C., "Telcos vs. Cable TV: The Global View," *Data Communications Magazine,* (issue on Global Supernets), Vol. 24, No. 13, Sept. 21, 1995.

Russo, K., "Tariff Watch," *Telecommunications,* Vol. 32, No. 12, Dec. 1998.

Shaw, J., *Telecommunications Deregulation,* Norwood, MA: Artech House, 1998.

Sikes, A. C., "Tommorrow's Communications Policies," *IEEE Communications Magazine,* (issue on Telecommunications Regulation), Vol. 25, No. 1, Jan. 1987.

Snow, A. P., *A Reliability Assessment of the Public Switched Telephone Network Infrastructure,* Ph.D. Dissertation, University of Pittsburgh, PA, 1997.

Snow, A. P., and M. B. H. Weiss, "Empirical Evidence of Reliability Growth in Large Scale Networks," *Journal of Network and System Management,* Vol. 5, No. 2, Jun. 1997, pp. 197 – 214.

Staley, D. C., "Domestic Roadblocks to a Global Information Highway," *IEEE Communications Magazine,* (issue on Telecommunications Deregulation), Vol. 27, No. 1, Jan. 1989.

Tanzillo, K., "FCC to Teach Old Tricks to New Dogs," (an interview with Mark Reed, FCC Chairman) *Communications News,* Vol. 33, No. 7, Jul. 1996, pp. 12 – 13.

"Telecommunications Quality," *IEEE Communications Magazine,* entire issue, Vol. 26, No. 10, Oct. 1988.

Thompson, H. B., "The Supercarrier Syndrome: Good for the Industry?" *Telecommunications,* Vol. 32, No. 1, Jan. 1998.

Valovic, T., "Public and Private Networks: Who Will Manage and Contol Them?" *Telecommuncations,* Vol. 22, No. 2, Feb. 1998.

Weiss, M. B. H., and P. Bernt, *The Regulation and Deregulation of US Telecommunications,* Textbook under contract with Lawrence Erlbaum and Associates, Mahwah, NJ, C. Sterling (ed.).

Wilson, M. R., "The Quantitative Impact of Survivable Network Architectures on Service Availability," *IEEE Communications,* Vol. 36, No. 5, May 1998.

Wolfson, J. R., "Computer III: The Beginning or the Beginning of the End for Enhanced Services Competition?" *IEEE Communications Magazine,* (issue on Telecommunications Deregulation), Vol. 27, No. 1, Jan. 1989.

Yoshizaki, H., "Breakup of AT&T and Liberalization of the Telecommunications Business," *IEEE Communications Magazine,* (issue on Telecommunications Deregulation), Vol. 27, No. 1, Jan. 1989.

About the Author

Richard A. Thompson is a professor in, and codirector of, the graduate program in telecommunications at the University of Pittsburgh. He has expertise in analog, digital, packet, and photonic switching; in digital and optical transmission; and in human-telecommunications interaction. Before coming to Pitt in 1989, Dr. Thompson worked for more than 20 years for Bell Laboratories, mostly in the Research Area at Murray Hill, New Jersey. At Bell Labs, he had a principal role in the AIOD, TSPS, XDS, GetSet, optical TSI, and DiSCO projects, which are all described in this book. Dr. Thompson also was a professor of electrical engineering at Virginia Tech, worked for the Royal Division of Litton Industries, and has had a part-time appointment at the National Institute of Standards and Technology.

Dr. Thompson's fascination with telephone switching led him to collect information about it for most of his life. This book is his "opus magnum." Dr. Thompson has taught telephone switching for new Bell Labs employees, for operating engineers in training at Bell Labs, as a college course at Virginia Tech, as a special course for the North Jersey section of the IEEE, as a college course at the University of Pittsburgh, and in many special and customized courses. He has published more than 100 papers in numerous research journals and conferences. Rich has been a member of the IEEE since 1963 and has served it in many roles. He is currently an officer in the Communications Society's Technical Committee on Switching and Routing.

Dr. Thompson is a 1960 graduate of Scotch Plains-Fanwood High School in New Jersey and he earned his B.S. degree in electrical engineering from Lafayette College in 1964. He earned his M.S., also in electrical engineering, from Columbia University in 1966 and his Ph.D. in computer science from the University of Connecticut in 1971.

Rich and Sandy have been married for 35 years. They have two married children, Pam and Jeff, and six grandchildren, Tim, Michael, Mary, Peter, Justin, and Theresa.

Index

Recent Titles in the Artech House Telecommunications Library

Vinton G. Cerf, Senior Series Editor

Access Networks: Technology and V5 Interfacing, Alex Gillespie

Achieving Global Information Networking, Eve L. Varma, Thierry Stephant, et al.

Advanced High-Frequency Radio Communications, Eric E. Johnson, Robert I. Desourdis, Jr., et al.

Asynchronous Transfer Mode Networks: Performance Issues, Second Edition, Raif O. Onvural

ATM Switches, Edwin R. Coover

ATM Switching Systems, Thomas M. Chen and Stephen S. Liu

Broadband Network Analysis and Design, Daniel Minoli

Broadband Networking: ATM, SDH, and SONET, Mike Sexton and Andy Reid

Broadband Telecommunications Technology, Second Edition, Byeong Lee, Minho Kang, and Jonghee Lee

The Business Case for Web-Based Training, Tammy Whalen and David Wright

Communication and Computing for Distributed Multimedia Systems, Guojun Lu

Communications Technology Guide for Business, Richard Downey, Seán Boland, and Phillip Walsh

Community Networks: Lessons from Blacksburg, Virginia, Second Edition, Andrew M. Cohill and Andrea Kavanaugh, editors

Component-Based Network System Engineering, Mark Norris, Rob Davis, and Alan Pengelly

Computer Telephony Integration, Second Edition, Rob Walters

Desktop Encyclopedia of the Internet, Nathan J. Muller

Digital Modulation Techniques, Fuqin Xiong

E-Commerce Systems Architecture and Applications, Wasim E. Rajput

Enterprise Networking: Fractional T1 to SONET, Frame Relay to BISDN, Daniel Minoli

Error-Control Block Codes for Communications Engineers, L. H. Charles Lee

FAX: Facsimile Technology and Systems, Third Edition, Kenneth R. McConnell, Dennis Bodson, and Stephen Urban

Guide to ATM Systems and Technology, Mohammad A. Rahman

A Guide to the TCP/IP Protocol Suite, Floyd Wilder

Information Superhighways Revisited: The Economics of Multimedia, Bruce Egan

Internet E-mail: Protocols, Standards, and Implementation, Lawrence Hughes

Introduction to Telecommunications Network Engineering, Tarmo Anttalainen

Introduction to Telephones and Telephone Systems, Third Edition, A. Michael Noll

IP Convergence: The Next Revolution in Telecommunications, Nathan J. Muller

The Law and Regulation of Telecommunications Carriers, Henk Brands and Evan T. Leo

Marketing Telecommunications Services: New Approaches for a Changing Environment, Karen G. Strouse

Multimedia Communications Networks: Technologies and Services, Mallikarjun Tatipamula and Bhumip Khashnabish, editors

Packet Video: Modeling and Signal Processing, Naohisa Ohta

Performance Evaluation of Communication Networks, Gary N. Higginbottom

Practical Guide for Implementing Secure Intranets and Extranets, Kaustubh M. Phaltankar

Practical Multiservice LANs: ATM and RF Broadband, Ernest O. Tunmann

Protocol Management in Computer Networking, Philippe Byrnes

Pulse Code Modulation Systems Design, William N. Waggener

Service Level Management for Enterprise Networks, Lundy Lewis

SNMP-Based ATM Network Management, Heng Pan

Strategic Management in Telecommunications, James K. Shaw

Successful Business Strategies Using Telecommunications Services, Martin F. Bartholomew

Telecommunications Department Management, Robert A. Gable

Telecommunications Deregulation, James Shaw

Telephone Switching Systems, Richard A. Thompson

Understanding Modern Telecommunications and the Information Superhighway, John G. Nellist and Elliott M. Gilbert

Understanding Networking Technology: Concepts, Terms, and Trends, Second Edition, Mark Norris

Videoconferencing and Videotelephony: Technology and Standards, Second Edition, Richard Schaphorst

Visual Telephony, Edward A. Daly and Kathleen J. Hansell

Wide-Area Data Network Performance Engineering, Robert G. Cole and Ravi Ramaswamy

Winning Telco Customers Using Marketing Databases, Rob Mattison

World-Class Telecommunications Service Development, Ellen P. Ward

For further information on these and other Artech House titles, including previously considered out-of-print books now available through our In-Print-Forever® (IPF®) program, contact:

Artech House
685 Canton Street
Norwood, MA 02062
Phone: 781-769-9750
Fax: 781-769-6334
e-mail: artech@artechhouse.com

Artech House
46 Gillingham Street
London SW1V 1AH UK
Phone: +44 (0)20 7596-8750
Fax: +44 (0)20 7630-0166
e-mail: artech-uk@artechhouse.com

Find us on the World Wide Web at:
www.artechhouse.com